Permutations If $P(n, r)$ (where $r \leq n$) is the number of permutations of n elements taken r at a time, then

$$P(n, r) = \frac{n!}{(n - r)!}.$$

Combinations If $\binom{n}{r}$ denotes the number of combinations of n elements taken r at a time, where $r \leq n$, then

$$\binom{n}{r} = \frac{n!}{(n - r)!r!}.$$

Probability in a Binomial Experiment If p is the probability of success in a single trial of a binomial experiment, the probability of x successes and $n - x$ failures in n independent repeated trials of the experiment is

$$\binom{n}{x} p^x (1 - p)^{n-x}.$$

Mean The mean of the n numbers, $x_1, x_2, x_3, \ldots, x_n$, is

$$\bar{x} = \frac{\sum (x)}{n}.$$

Standard Deviation The standard deviation of the n numbers, $x_1, x_2, x_3, \ldots, x_n$, with mean $\bar{x}$, is

$$s = \sqrt{\frac{\sum (x - \bar{x})^2}{n - 1}}.$$

Binomial Distribution Suppose an experiment is a series of n independent repeated trials, where the probability of a success in a single trial is always p. Let x be the number of successes in the n trials. Then the probability that exactly x successes will occur in n trials is given by

$$\binom{n}{x} p^x (1 - p)^{n-x}.$$

The mean is

$$\mu = np,$$

and the standard deviation is

$$\sigma = \sqrt{np(1 - p)}.$$

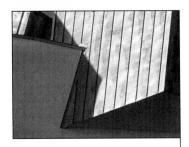

6th EDITION

Mathematics with Applications

IN THE MANAGEMENT, NATURAL, AND SOCIAL SCIENCES

Margaret L. Lial
American River College

Thomas W. Hungerford
Cleveland State University

Charles D. Miller

HarperCollinsCollegePublishers

Sponsoring Editor: George Duda
Development Editor: Sandi Goldstein
Project Editor: Carol Zombo
Art Administrator: Jess Schaal
Text and Cover Design: Lucy Lesiak/Lesiak-Crampton Design
Cover Photo:© Richard Bryant/Esto/Arcaid
Photo Researcher: Sandy Schneider
Production Administrator: Randee Wire
Compositor: Interactive Composition Corporation
Printer and Binder: R.R. Donnelley & Sons
Cover Printer: The Lehigh Press, Inc.

Mathematics with Applications in the Management, Natural, and Social Sciences, Sixth Edition, Copyright © 1995 by HarperCollins College Publishers.

Library of Congress Cataloging-in-Publication Data

Lial, Margaret L.
    Mathematics with applications in the management, natural, and
social sciences / Margaret L. Lial, Thomas W. Hungerford, Charles D.
Miller—6th ed.
        p.   cm.
    Includes index.
    ISBN 0-673-46943-3
    1. Mathematics.   I. Hungerford, Thomas W.   II. Miller, Charles David,   III. Title.
    QA37.2.L5   1995
    510—dc20   94-13576
                                                                                    CIP

    6 7 8 9 10-DOC-01 00 99 98

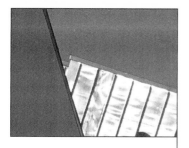

Preface

Mathematics with Applications, Sixth Edition, is designed to provide mathematical topics needed by students in the fields of business management, social science, and natural science. The book is written at a level that makes it accessible to such students. Topics are presented by proceeding from what is already known to new material, from concrete examples to general rules and formulas. Almost every section includes pertinent applications. The only prerequisite we assume is a course in algebra. Chapter 1 provides a thorough review of algebra, and a pretest is included in the *Instructor's Guide* that accompanies the text to help determine how much review is needed for a particular student or class.

We have retained popular features from earlier editions: extensive examples; exercises keyed to the text; realistic and timely applications; end-of-chapter case studies; margin problems; highlighted rules, definitions, and summaries; and pedagogical use of color.

Use of Technology

In this new edition, we have introduced boxes with references to graphers (graphing calculators or computers), and we discuss how they apply to the topic at hand. These are designed so an instructor may or may not choose to integrate them into the course. Where appropriate, exercise sets include clearly labeled exercises especially for graphers. Each such exercise is identified with ▦ . The opening page of each applicable chapter includes a list of available technology resources that are appropriate for the chapter.

New Exercises

More than one-third of the exercises are new to this edition, including drill and practice problems as well as a variety of interesting applications. Many of the applications are based on current real data and come from a variety of areas of interest to students. Several new case studies that explore contemporary issues and themes are included as well. This edition includes new conceptual and writing exercises that require students to demonstrate an understanding of the concepts that goes beyond computational skill. Several exercise sets also include connection exercises, identified by ◀▷, that integrate concepts and skills developed earlier with those just introduced or that connect the concepts presented within a chapter. Four answer checkers helped ensure that the highest level of accuracy was maintained.

NEW CONTENT HIGHLIGHTS

- The material in Chapter 2 was rearranged to flow from topics that are familiar to those that are not. The chapter now begins with linear equations and slope, then discusses functions and their applications. The introduction to functions has been enhanced.
- The general discussion of graphing polynomials by hand has been expanded to help students better understand the shape of these graphs.
- The section on logarithmic functions has been rewritten to emphasize base 10 and base e logarithms.
- In Chapter 6, we rewrote the introduction to systems of equations to reflect current terminology and to include examples of "nonsquare" systems.
- The presentation of linear programming in Section 7.4 has been revised so the mechanics are discussed first, then the reasons given. The applications are presented separately in the next section. We now introduce duality and minimization problems in Section 7.6, before nonstandard problems are covered in Section 7.7.
- The topics within some of the sections in Chapter 8 were rearranged to improve the flow and keep related topics together.
- Limits at infinity and horizontal asymptotes are now discussed in Chapter 12 with curve sketching.

NEW FEATURES

Several new features, designed to assist students in the learning process, have been integrated into this edition. These features are illustrated on the following two pages. In addition, changes in format enhance the book's pedagogical features and increase its accessibility.

- *For Graphers* These notes permit the integration of graphing calculators and computers into the courses using this text.
- *Conceptual and Writing Exercises* To complement the drill and application exercises, several exercises that require a deeper understanding of the concepts introduced in a section are included in almost every exercise set. Nearly half of these require the student to respond by writing a few sentences.
- *Connection Exercises* Exercises that require students to use skills or techniques covered in previous sections or chapters are included in some section exercises and review exercise sets. These exercises are identified with ◀▷ .
- *Graphing Calculator/Computer Exercises* Exercises identified with ▦ are included in appropriate exercise sets for those using graphers. These exercises demonstrate how graphers can be used to clarify and illustrate concepts.

---

32   CHAPTER 1 FUNDAMENTALS OF ALGEBRA

 Solve each inequality. Graph each solution.

(a) $|y - 2| > 5$

(b) $|3k - 1| \geq 2$

(c) $|2 + 5r| - 4 \geq 1$

Answers:

(a) All numbers in $(-\infty, -3)$ or $(7, \infty)$

(b) All numbers in $\left(-\infty, -\frac{1}{3}\right)$ or $[1, \infty)$

(c) All numbers in $\left(-\infty, -\frac{7}{5}\right)$ or $\left[\frac{3}{5}, \infty\right)$

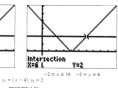

FOR GRAPHERS

One method for solving absolute value equations with a grapher is to graph each side of the equation simultaneously. Some graphers have a key labeled ABS that gives the absolute value of a quantity. This function is accessed through a special menu on other graphers. For instance, in Example 2, graph $y_1 = |x - 4|$ and $y_2 = 2$. Remember to key in ABS($x - 4$), not ABS $x - 4$. Then locate the x-values of the points where these two graphs intersect (are equal). Figure 1.20 shows the x-values of the intersection points are 2 and 6. To solve the inequality $|x - 4| < 2$, from the graph in Figure 1.20 decide on which intervals the graph of $y_1 = |x - 4|$ is below (and therefore less than) the graph of $y_2 = 2$. As the figure shows, this happens for $2 < x < 6$. You could also solve this inequality by graphing $y = |x - 4| - 2$ and locating the intersection points with the x-axis.

$-2 \leq x \leq 10, \ -2 \leq y \leq 6$ $-2 \leq x \leq 10, \ -2 \leq y \leq 6$
$y_1 = |x - 4|, y_2 = 2$
FIGURE 1.20

▶ **EXAMPLE 6** Solve $|2 - 7m| - 1 > 4$.

First add 1 on both sides.

$$|2 - 7m| > 5$$

Now use property (4) from above to solve $|2 - 7m| > 5$ by solving the inequality

$$2 - 7m < -5 \quad \text{or} \quad 2 - 7m > 5.$$

Solve each part separately.

$$-7m < -7 \quad \text{or} \quad -7m > 3$$
$$m > 1 \quad \text{or} \quad m < -\frac{3}{7}$$

The solution, all numbers in $\left(-\infty, -\frac{3}{7}\right)$ or $(1, \infty)$, is graphed in Figure 1.21. ◀

FIGURE 1.21

9.1 PERMUTATIONS AND C

29. How many area codes would be possible if all restrictions on the second digit were removed?

30. A problem with the plan in Exercise 29 is that the second digit in the area code now tells the phone company equipment that a long-distance call is being made. To avoid changing all equipment, an alternative plan proposes a 4-digit area code and restricting the first and second digits as before. How many area codes would this plan provide?

31. Still another alternative solution is to increase the local dialing sequence to 8 digits instead of 7. How many additional numbers would this plan create? (Assume the same restrictions.)

32. Define permutation in your own words.

Use permutations to solve each of the following problems. (See Examples 5–7.)

33. A baseball team has 20 players. How many 9-player batting orders are possible?

34. In a game of musical chairs, 12 children will sit in 11 chairs arranged in a row (one will be left out). In how many ways can the 11 children find seats?

35. From a carton of 12 cans of a soft drink, 2 are to be selected for testing. In how many ways can this be done?

36. In an election with 3 candidate for one office and 5 candidates for another office, how many different ballots may be printed?

37. From a pool of 7 secretaries, 3 are selected to be assigned to 3 managers. In how many ways can they be selected?

38. A chapter of union Local 715 has 35 members. In how many different ways can the chapter select a president, a vice-president, a treasurer, and a secretary?

39. The television schedule for a certain evening shows 8 choices from 8 to 9 P.M., 5 choices from 9 to 10 P.M., and 6 choices from 10 to 11 P.M.. In how many different ways could a person schedule that evening of television viewing from 8 to 11 P.M.? (Assume each program that is selected is watched for an entire hour.)

40. In a club with 15 members, how many ways can a slate of 3 officers consisting of president, vice-president, and secretary/treasurer be chosen?

Use combinations to solve each of t
(See Examples 8–9 and 12–13.)

41. Management Five items are t from the first 50 items on an asse the defect rate. How many differ can be chosen?

42. Social Science A group of 3 s from a group of 12 students to ta biology.
(a) In how many ways can this
(b) In how many ways can the gr part be chosen?

43. Natural Science From a group of 16 smokers and 20 nonsmokers, a researcher wants to randomly select 8 smokers and 8 nonsmokers for a study. In how many ways can the study group be selected?

44. Five cards are drawn from an ordinary deck. In how many ways is it possible to draw
(a) all queens;
(b) all face cards (face cards are the Jack, Queen, and King);
(c) no face card;
(d) exactly 2 face cards;
(e) 1 heart, 2 diamonds, and 2 clubs.

Exercises 45–64 are mixed problems that may require permutations, combinations, or the multiplication principle. (See Examples 10, 11, and 14.)

45. Use a tree diagram to find the number of ways 2 letters can be chosen from the set {L, M, N} if order is important and
(a) if repetition is allowed;
(b) if no repeats are allowed.
(c) Find the number of combinations of 3 elements taken 2 at a time. Does this answer differ from (a) or (b)?

46. Repeat Exercise 45 using the set {L, M, N, P}.

47. Explain the difference between a permutation and a combination.

48. Padlocks with digit dials are often referred to as "combination locks." According to the mathematical definition of combination, is this an accurate description? Explain.

2.4 EXERCISES   **129**

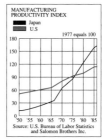

MANUFACTURING
PRODUCTIVITY INDEX

■ Japan
▨ U.S

1977 equals 100

Source: U.S. Bureau of Labor Statistics
and Salomon Brothers Inc.

34. Management   The graph gives U.S. imports and exports in billions of dollars over a five-year period. Estimate the break-even point.

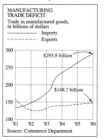

MANUFACTURING
TRADE DEFICIT

Trade in manufactured goods,
in billions of dollars

——— Imports
- - - - - Exports

$293.8 billion

$148.7 billion

Source: Commerce Department

◆ **35. Management**   Canadian and Japanese investment in the United States in billions of dollars in 1980 and 1990 are shown in the chart.*

* Data reprinted with permission from *The World Almanac and Book of Facts,* 1992. Copyright © 1991. All rights reserved. The World Almanac is an imprint of Funk & Wagnalls Corporation.

	1980	1990
Canada	12.1	27.7
Japan	4.7	108.1

(a) Assuming the change in investment in each case is linear, write an equation in function form giving the investment in year x for each country. Let x be the number of years since 1980.
(b) Graph the functions from part (a) on the same coordinate axes.
(c) Find the intersection point where the graphs in part (b) cross and interpret your answer.

◆ **36.** Social Science   The median family income (in thousands of dollars) for the white population in the United States is given by $f(x) = (4/3)x + 8$, where x is the number of years since 1973. The median family income (in thousands of dollars) for the black population in the United States is given by $f(x) = (2/3)x + 6$.
(a) Graph both functions on the same coordinate axes.
(b) Do the graphs intersect? If so, in what year was median family income the same for white and black populations?
(c) What can you infer from the two graphs in part (a)?

Management   *Suppose that you are the manager of a firm. You are considering the manufacture of a new product, so you ask the accounting department to produce cost estimates and the sales department to produce sales estimates. After you receive the data, you must decide whether to go ahead with production of the new product. Analyze the following data (find a break-even point) and then decide what you would do. (See Example 7.)*

37. $C(x) = 80x + 7000;$ $R(x) = 95x;$ no more than 400 units can be sold.
38. $C(x) = 65x + 9500;$ $R(x) = 80x;$ no more than 600 units can be sold.
39. $C(x) = 140x + 3000;$ $R(x) = 125x$ (Hint: what does a negative value of x mean?)
40. $C(x) = 1750x + 95,000;$ $R(x) = 1750x$
41. Management   The revenue in millions of dollars from sales of a home supplies outlet is given by $R(x) = .3x$. The profit in millions of dollars from sales of x units is given by $P(x) = .2x - .5$.
(a) Find the cost function.
(b) Find $C(7)$.
(c) What is the marginal cost?

MATHEMATICS OF FINANCE

eposits $10,000 at the beginning of ears in an account paying 5% com-. He then puts the total amount on account paying 6% compounded other 9 years. Find the final amount e entire 21-year period.

ds $10,000 in 8 years.
should he deposit at the end of each ompounded quarterly so that he will 00?
quarterly deposit if the money is % compounded quarterly.

55. Harv's Meats knows that it must buy a new deboner machine in 4 years. The machine costs $12,000. In order to accumulate enough money to pay for the machine, Harv decides to deposit a sum of money at the end of each 6 months in an account paying 6% compounded semiannually. How much should each payment be?

56. Karin Sandberg wants to buy an $18,000 car in 6 years. How much money must she deposit at the end of each quarter in an account paying 12% compounded quarterly so that she will have enough to pay for her car?

◆ **57.** In a recent state lottery, the jackpot was $27 million. An Australian investment firm tried to buy all possible combinations of numbers, which would have cost $7 million. (In fact, the firm ran out of time and was unable to buy all combinations.) Suppose the investment firm had accomplished its goal and held the only winning ticket in the lottery. Assume that the firm would receive the jackpot in payments of $1.35 million paid at the end of each year for 20 years.
(a) Suppose each jackpot payment received by the firm is invested at 8% interest compounded annually.

How many years would it take until the value of this investment would be greater than the amount the firm would have if it had simply invested its original $7 million at the same rate (instead of buying lottery tickets)? (*Hint:* Experiment with different values of n, the number of years.)
(b) How many years would it take in part (a) at an interest rate of 12%?

58. Diane Gray sells some land in Nevada. She will be paid a lump sum of $60,000 in 7 yr. Until then, the buyer pays 8% simple interest quarterly.
(a) Find the amount of each quarterly interest payment.
(b) The buyer sets up a sinking fund so that enough money will be present to pay off the $60,000. The buyer wants to make semiannual payments into the sinking fund; the account pays 6% compounded semiannually. Find the amount of each payment into the fund.
(c) Prepare a table showing the amount in the sinking fund after each deposit.

59. Joe Seniw bought a rare stamp for his collection. He agreed to pay a lump sum of $4000 after 5 yr. Until then, he pays 6% simple interest semiannually.
(a) Find the amount of each semiannual interest payment.
(b) Seniw sets up a sinking fund so that enough money will be present to pay off the $4000. He wants to make annual payments into the fund. The account pays 8% compounded annually. Find the amount of each payment.
(c) Prepare a table showing the amount in the sinking fund after each deposit.

5.4 PRESENT VALUE OF AN ANNUITY; AMORTIZATION

Suppose that at the end of each year, for the next 10 years, $500 is deposited in a savings account paying 7% interest compounded annually. This is an example of an ordinary annuity. The **present value** of this annuity is the amount that would have to be deposited in one lump sum today (at the same compound interest rate) in order to produce exactly the same balance at the end of 10 years. We can find a formula for the present value of an annuity as follows.

COURSE FLEXIBILITY

The book can be used for a variety of courses, including the following:

Finite Mathematics and Calculus (one year or less) Use the entire book; cover topics from Chapters 1–4 as needed before proceeding to further topics.

Finite Mathematics (one semester or two quarters) Use as much of Chapters 1–4 as needed, and then go into the topics of Chapters 5–10 as time permits and local needs require.

Calculus (one semester or quarter) Use Chapters 1–4 as necessary and then use Chapters 11–14.

College Algebra with Applications (one semester or quarter) Use Chapters 1–8 with the topics of Chapters 7 and 8 being optional.

Chapter interdependence is as follows.

Chapter	Prerequisite
1 Fundamentals of Algebra	None
2 Functions and Graphs	None
3 Polynomial and Rational Functions	Chapter 2
4 Exponential and Logarithmic Functions	Chapter 2
5 Mathematics of Finance	Chapter 4
6 Systems of Linear Equations and Matrices	None
7 Linear Programming	Chapters 2, 6
8 Sets and Probability	None
9 Further Topics in Probability	Chapter 8
10 Introduction to Statistics	Chapter 8
11 Differential Calculus	Chapters 2–4
12 Applications of the Derivative	Chapter 11
13 Integral Calculus	Chapters 11–12
14 Multivariate Calculus	Chapters 11–13

SUPPLEMENTS

FOR THE INSTRUCTOR

Our extensive supplemental package includes an instructor's guide, answer manual, and software. A number of other related materials are available, and are also listed below.

Instructor's Guide with Tests and Solutions   The *Instructor's Guide with Tests and Solutions* includes a lengthy set of test questions for each chapter, organized by section, plus answers to all of the test questions. It also includes one pretest, in a short-answer format. This manual also contains solutions to the even-numbered section exercises, which, with the *Student's Solution Manual* together provide detailed, worked-out solutions to each exercise in the book. Finally, this guide includes a list of all conceptual, writing, challenging, connection, and grapher exercises.

Instructor's Answer Manual   This manual includes answers to every exercise in the book. This manual also includes a list of all conceptual, writing, challenging, connection, and grapher exercises.

Printed Test Forms   This manual includes three different but equivalent tests for each chapter generated using the HarperCollins Test Generator/Editor for Mathematics (described below).

HarperCollins Test Generator/Editor for Mathematics with QuizMaster   Available in IBM and Macintosh versions, the test generator is fully networkable. The test generator enables instructors to select questions by objective, section, or chapter, or to use a ready-made test for each chapter. The editor enables instructors to edit any preexisting data or to create their own questions. The software is algorithm driven, allowing the instructor to regenerate constants while maintaining problem type, providing a nearly unlimited number of available test or quiz items in multiple-choice and/or open-response formats for one or more test forms. The system features printed graphics and accurate mathematics symbols. **QuizMaster** enables instructors to create tests and quizzes using the Test Generator/Editor and save them to disk so students can take the test or quiz on a stand-alone computer or network. **QuizMaster** then grades the test or quiz and allows the instructor to create reports on individual students or entire classes.

Overhead Transparencies   A set of two-color transparencies is available to help enhance lectures.

FOR THE STUDENT

Student's Solution Manual   This book provides solutions to the odd-numbered section exercises, odd-numbered chapter review exercises, and all case study exercises. (ISBN 0-673-46944-1)

Topics in Finite Mathematics: An Introduction to the Electronic Spreadsheet, by Sam Spero, Cuyahoga Community College (ISBN 0-065-00300-4), is a user-friendly guide designed to introduce students to the various ways one can approach problem solving with spreadsheets. Knowledge of spreadsheets is not assumed, and the approach is adaptable to all spreadsheet programs.

The Electronic Spreadsheet and Elementary Calculus, by Sam Spero, Cuyahoga Community College (ISBN 0-673-46595-0), is a companion to *Topics in Finite Mathematics: An Introduction to the Electronic Spreadsheet.* This guide helps students get started with graphing and problem solving by means of the spreadsheet. As with the companion volume, knowledge of spreadsheets is not assumed, and the approach is adaptable to all spreadsheet programs.

Interactive Mathematics Tutorial Software with Management System   This innovative package is available in IBM (both DOS and Windows formats) and Macintosh versions and is fully networkable. As with the Test Generator/Editor, this software is algorithm driven, which automatically regenerates constants so a student will not see the numbers repeat in a problem type if he or she revisits any particular section. The tutorial is objective-based, self-paced, and provides unlimited opportunities to review lessons and to practice problem solving. If students give a wrong answer, they can request to see the problem worked out and get a textbook page reference. The program is menu-driven for ease of use, and on-screen help can be obtained at any time with a single keystroke. Students' scores are automatically recorded and can be printed for a permanent record. The optional **Management System** lets instructors record student scores on disk and print diagnostic reports for individual students or classes. This software may also be purchased by students for home use. (Macintosh version ISBN 0-673-55815-0; IBM version 0-673-55814-2)

GraphExplorer provides students and instructors with a comprehensive graphing utility, and is available in IBM and Macintosh formats.

StatExplorer (IBM and Macintosh), helps students enhance their understanding of statistics by exploring a wide range of statistical representations including graphs, centers and spreads, and transformations.

Explorations in Finite Mathematics (IBM format only), by David Schneider, University of Maryland (ISBN 0-673-46932-8), contains on one disk a wider selection of routines than in any similar software supplement. Included are utilities for Gaussian elimination, matrix operations, graphical and simplex methods for linear programming problems, probability, binomial distribution, simple and compound interest, loan and annuity analysis, finance table, difference equations, and more. Refined monitor display for fractions, color capabilities, choice of exact or approximate calculations with matrices, and refined printing capabilities further set this apart from other programs.

Visual Calculus (IBM format only), by David Schneider, University of Maryland, (ISBN 0-673-99015-X), is intended to enhance the visualization of the fundamental concepts of calculus. Although it has most of the computing power of standard calculus packages, it focuses on the teaching of concepts rather than on the calculations. The program can be run with a CGA, Hercules, MCGA, EGA, or VGA monitor, and can be used with or without a mouse.

Matrix with Linear Programming (IBM format only), by MaylinDittmore (ISBN 0-06-501266-6), is designed to assist the student in any course of study that involves the use of matrices. MATRIX was created not only to help the student with tedious calculations associated with matrices, but also to help them gain an understanding of and appreciation for real-world problems that can be analyzed and solved using matrices.

Developmental Mathematics: Graphing Calculator Investigations, Dennis C. Ebersole, Northampton County Area Community College (ISBN 0-06-501439-1)

College Algebra and Trigonometry: Graphing Calculator Investigations, Dennis C. Ebersole, Northampton County Area Community College (ISBN 0-06-500888-X)

These are intended to supplement a standard text by providing investigations that help students visualize key concepts, look for patterns, generalize and apply concepts.

RELATED BOOKS

Mathematics with Applications, Sixth Edition, is one text within the complete line of Lial/Miller mathematics for management offerings: *Finite Mathematics*, Fifth Edition, *Calculus with Applications*, Fifth Edition, *Calculus with Applications, Brief Version*, Fifth Edition, and *Finite Mathematics and Calculus with Applications*, Fourth Edition.

ACKNOWLEDGMENTS

We wish to thank the following instructors who reviewed the manuscript and made many helpful suggestions for improvement.

Michael J. Bradley, *Merrimack College*
James F. Brown, *Midland College*
James E. Carpenter, *Iona College*
Faith Y. Chao, *Golden Gate University*
Jan S. Collins, *Embry-Riddle University*
Gordon Feathers, *Iona College*
Richard E. Goodrick, *University of Washington*
Kay Gura, *Ramapo College of New Jersey*
Joseph A. Guthrie, *University of Texas at El Paso*
Arthur M. Hobbs, *Texas A & M University*
Miles Hubbard, *St. Cloud State University*
June Jones, *Macon College*
Akihiro Kanamori, *Boston University*
Robert A. Moreland, *Texas Tech University*
Elizabeth Polenzani, *Pasadena City College*
Norman Rittgers, *Pasadena City College*
Gordon Shilling, *University of Texas at Arlington*
Joan M. Spetich, *Baldwin-Wallace College*
William D. Stark, *Navarro College*
Giovanni Viglino, *Ramapo College of New Jersey*
Bhushan Wadhwa, *Cleveland State University*

We also wish to thank those who did an excellent job checking all the answers for us: Michael Bradley, Merrimack College, James Carpenter, Iona College, Dennis Kern, Sul Ross State University, and Gordon Shilling, University of Texas at Arlington.

Special thanks go to Jim Eckerman, American River College, who wrote the *Appendix on Graphing Calculators*; to Paul Eldersveld, College of DuPage, who did an outstanding job coordinating all the print ancillaries; to Paul Van Erden, American River College, who created an accurate and complete index for us; and to James Walker, American River College, who carefully compiled the index of applications. We also thank the fine, professional staff at HarperCollins for their assistance and contributions to this book: George Duda, Sandi Goldstein, Carol Zombo, Linda Youngman, Kevin Connors, and Ed Moura.

Margaret L. Lial
Thomas W. Hungerford

Contents

Index of Applications

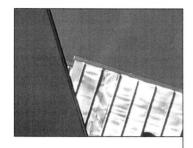

To The Student

There are many pedagogical features in this text which will enhance your understanding of the concepts.

   Side problems are one of these features that will help you to learn new concepts and reinforce your understanding. They are referred to in the text by numbers within colored squares: **1**. When you see that symbol, you should work the indicated problem at the side before going on.

   One of the main reasons for learning mathematics is to be able to use it to solve practical problems. As its name implies, the purpose of this book is to show how to use mathematics to solve applied problems. However, for many students, learning how to use mathematical skills in real-world applications is the most difficult task they face. A common difficulty students have with applied problems is trying to do everything at once. It is usually best to attack the problem in stages as outlined below:

Solving Applied Problems

1. Decide on the unknown. Name it with some variable that you *write down*. Many students try to skip this step. They are eager to get on with the writing of the equation. But this is an important step. If you don't know what the variable represents, how can you write a meaningful equation or interpret a result?
2. Draw a sketch or make a chart, if appropriate, showing the information given in the problem.

3. Decide on a variable expression to represent any other unknowns in the problem. For example, if x represents the width of a rectangle, and you know that the length is one more than twice the width, then *write down* that the length is $1 + 2x$.

4. Using the results of Steps 1–3, write an equation that expresses a condition that must be satisfied

5. Solve the equation.

6. Check the solution in the words of the *original problem*, not just in the equation you have written.

CHAPTER 1

Fundamentals of Algebra

TECHNOLOGY RESOURCES

Developmental Mathematics: Graphing Calculator Investigations, Ebersole

College Algebra and Trigonometry: Graphing Calculator Investigations, Ebersole

This book deals with the application of mathematics to business, social science, and biology. Almost all of these applications begin with some real-world problem that is solved by first writing appropriate equations or other mathematical relationships that describe the problem. The properties of algebra are then used to simplify the relationships and solve the equations, producing a solution to the real-world problem.

Because algebra is so vital to the study of applications of mathematics, we begin with a review of some of the fundamental ideas of algebra. If you have not used your algebraic skills for some time, it is important for you to study the review material in this chapter carefully; your success in covering the material that follows will depend upon your algebraic skills.

1.1 THE REAL NUMBERS

Only real numbers will be used in this book.* The names of the most common types of real numbers are as follows.

The Real Numbers

Natural (counting) numbers	$1, 2, 3, 4, \ldots$
Whole numbers	$0, 1, 2, 3, 4, \ldots$
Integers	$\ldots, -3, -2, -1, 0, 1, 2, 3, \ldots$
Rational numbers	All numbers of the form p/q, where p and q are integers, with $q \neq 0$
Irrational numbers	Real numbers that are not rational

*Not all numbers are real numbers. An example of a number that is not a real number is $\sqrt{-1}$.

1

1 Draw a number line and graph the numbers -4, -1, 0, 1, 2.5, and $13/4$ on it.

Answer:

The three dots used in this box show that the numbers continue indefinitely in the same way. The relationships among these types of numbers are shown in Figure 1.1. Notice, for example, that the integers are also rational numbers and real numbers, but the integers are not irrational numbers.

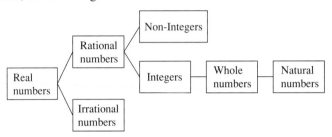

FIGURE 1.1

One example of an irrational number is π, the ratio of the circumference of a circle to its diameter. The number π can be approximated by writing $\pi \approx 3.14159$ ($\approx$ means "is approximately equal to"), but there is no rational number that is exactly equal to π. Another irrational number can be found by constructing a triangle having a 90° angle, with the two shortest sides each 1 unit long, as shown in Figure 1.2. The third side can be shown to have a length that is irrational ($\sqrt{2}$ units). Many numbers have square roots that are irrational numbers; in fact, if a number is not the square of a rational number, then its square root is irrational.

FIGURE 1.2

The various types of numbers used in this book can be illustrated with a diagram called a **number line.** Each real number corresponds to exactly one point on the line and vice-versa. A number line with several sample numbers located (or **graphed**) on it is shown in Figure 1.3.*   **1**

FIGURE 1.3

---

* The use of margin problems is explained in the section "To the Student" preceding this chapter.

2 Name all the types of numbers that apply to the following.

(a) -2

(b) $-5/8$

(c) π

Answers:

(a) Integer, rational, real

(b) Rational, real

(c) Irrational, real

3 Use a calculator to approximate the following numbers to three significant digits.

(a) $\sqrt{850}$

(b) $-4/7$

(c) $\pi/3$

Answers:

(a) 29.2

(b) $-.571$

(c) 1.05

▶ **EXAMPLE 1**   What kind of number is each of the following?

(a) 6

The number 6 is a natural number, whole number, integer, rational number, and real number.

(b) $3/4$

This number is rational and real.

(c) $\sqrt{8}$

Because 8 is not the square of a rational number, $\sqrt{8}$ is irrational and real. ◀ **2**

All real numbers can be written in decimal form. Rational numbers in decimal form either terminate or repeat. For example,

$$.5, \quad .125, \quad 1.333\ldots, \quad 2.412412\ldots$$

are rational numbers. Irrational numbers in decimal form neither terminate nor repeat. There is no way to write an irrational number exactly as a decimal.

We can use a calculator to approximate an irrational number. An important question is, "How many digits should be given in an approximation?" In some cases, we may choose to give an approximation with an arbitrary number of significant digits. In calculations, we use the following rule:

Round the answer to a calculation to the *least* number of significant digits in any of the given numbers.

A significant digit includes any digit after the decimal point, but not zeros at the end of a whole number. Thus 27.0 has three significant digits, but 270 has two significant digits. ◀ **3**

Some basic properties of the real numbers follow.

Properties of the Real Numbers

For all real numbers, a, b, and c, the following properties hold true.

Commutative properties $a + b = b + a$ $ab = ba$

Associative properties $(a + b) + c = a + (b + c)$ $(ab)c = a(bc)$

Identity properties   There exists a unique real number 0, called the **additive identity,** such that

$$a + 0 = a \quad \text{and} \quad 0 + a = a.$$

There exists a unique real number 1, called the **multiplicative identity,** such that

$$a \cdot 1 = a \quad \text{and} \quad 1 \cdot a = a.$$

4 Name the property illustrated in each of the following examples.

(a) $(2 + 3) + 9$
$= (3 + 2) + 9$

(b) $(2 + 3) + 9$
$= 2 + (3 + 9)$

(c) $(2 + 3) + 9$
$= 9 + (2 + 3)$

(d) $(4 \cdot 6)p = (6 \cdot 4)p$

(e) $4(6p) = (4 \cdot 6)p$

Answers:

(a) Commutative property

(b) Associative property

(c) Commutative property

(d) Commutative property

(e) Associative property

Inverse properties

For each real number a, there exists a unique real number $-a$, called the **additive inverse of a,** such that

$$a + (-a) = 0 \quad \text{and} \quad (-a) + a = 0.$$

If $a \neq 0$, there exists a unique real number $1/a$, called the **multiplicative inverse,** such that

$$a \cdot \frac{1}{a} = 1 \quad \text{and} \quad \frac{1}{a} \cdot a = 1.$$

Distributive property $a(b + c) = ab + ac$

▶ **EXAMPLE 2**   The following statements are examples of the commutative properties. Notice that the order of the numbers changes from one side of the equals sign to the other, so that the order in which two numbers are added or multiplied is unimportant.
(a) $(6 + x) + 9 = (x + 6) + 9$
(b) $(6 + x) + 9 = 9 + (6 + x)$
(c) $5 \cdot (9 \cdot 8) = (9 \cdot 8) \cdot 5$
(d) $5 \cdot (9 \cdot 8) = 5 \cdot (8 \cdot 9).$ ◀

▶ **EXAMPLE 3**   The following statements are examples of the associative properties. Here the order of the numbers does not change, but the placement of the parentheses does change. This means that when three numbers are to be added, the sum of any two can be found first, then that result is added to the remaining number.
(a) $4 + (9 + 8) = (4 + 9) + 8$
(b) $3(9x) = (3 \cdot 9)x$
(c) $(\sqrt{3} + \sqrt{7}) + 2\sqrt{6} = \sqrt{3} + (\sqrt{7} + 2\sqrt{6}).$ ◀ **4**

The identity properties give some special properties of the numbers 0 and 1. Since 0 preserves the identity of a real number under addition, 0 is the identity element for addition. In the same way, 1 preserves the identity of a real number under multiplication and is the identity element for multiplication.

▶ **EXAMPLE 4**   By the identity properties,
(a) $-8 + 0 = -8,$
(b) $(-9)1 = -9.$ ◀

▶ **EXAMPLE 5**   By the inverse properties, the statements in parts (a) through (d) are true.
(a) $9 + (-9) = 0$ **(b)** $-15 + 15 = 0$

5 Name the property illustrated in each of the following examples.

(a) $2 + 0 = 2$

(b) $-\dfrac{1}{4} \cdot (-4) = 1$

(c) $-\dfrac{1}{4} + \dfrac{1}{4} = 0$

(d) $1 \cdot \dfrac{2}{3} = \dfrac{2}{3}$

Answers:

(a) Identity property

(b) Inverse property

(c) Inverse property

(d) Identity property

6 Use the distributive property to complete each of the following.

(a) $4(-2 + 5)$

(b) $2(a + b)$

(c) $-3(p + 1)$

(d) $(8 - k)m$

(e) $5x + 3x$

Answers:

(a) $4(-2) + 4(5) = 12$

(b) $2a + 2b$

(c) $-3p - 3$

(d) $8m - km$

(e) $(5 + 3)x = 8x$

(c) $-8 \cdot \left(\dfrac{1}{-8}\right) = 1$

(d) $\dfrac{1}{\sqrt{5}} \cdot \sqrt{5} = 1$ ◀

Note   There is no real number x such that $0 \cdot x = 1$, so 0 has no inverse for multiplication. **5**

One of the most important properties of the real numbers, and the only one that involves both addition and multiplication, is the distributive property. The next example shows how this property is applied.

▶**EXAMPLE 6**   By the distributive property,
(a) $9(6 + 4) = 9 \cdot 6 + 9 \cdot 4$
(b) $3(x + y) = 3x + 3y$
(c) $-8(m + 2) = (-8)(m) + (-8)(2) = -8m - 16$
(d) $(5 + x)y = 5y + xy.$ ◀

Note   As shown in Example 6(d), by the commutative property, the distributive property can also be written as $(a + b)c = ac + bc.$ **6**

Comparing two real numbers requires symbols that indicate their order on the number line. The following symbols are used to indicate that one number is greater than or less than another number.

| $<$ means *is less than* | $\leq$ means *is less than or equal to* |
| $>$ means *is greater than* | $\geq$ means *is greater than or equal to* |

The following definitions show how the number line is used to decide which of two given numbers is the greater.

For real numbers a and b,
if a is to the left of b on a number line, then $a < b$;
if a is to the right of b on a number line, then $a > b$.

▶**EXAMPLE 7**   Write *true* or *false* for each of the following.
(a) $8 < 12$
This statement says that 8 is less than 12, which is true.
(b) $-6 > -3$
The graph of Figure 1.4 shows that -6 is to the *left* of -3. Thus, $-6 < -3$, and the given statement is false.

FIGURE 1.4

7 Write *true* or *false* for the following.

(a) $-9 \leq -2$

(b) $8 > -3$

(c) $-14 \leq -20$

Answers:

(a) True

(b) True

(c) False

8 Graph all integers x such that

(a) $-3 < x < 5$

(b) $1 \leq x \leq 5$.

Answers:

(a)
-3 -2 -1 0 1 2 3 4 5

(b)
0 1 2 3 4 5 6

9 Graph all real numbers x such that

(a) $-5 < x < 1$

(b) $4 < x < 7$.

Answers:

(a)
-5   1

(b)
4   7

(c) $-2 \leq -2$
Because $-2 = -2$, this statement is true. ◀ **7**

A number line can be used to draw the graph of a set of numbers, as shown in the next few examples.

▶ **EXAMPLE 8** Graph all integers x such that $1 < x < 5$.
The only integers between 1 and 5 are 2, 3, and 4. These integers are graphed on the number line in Figure 1.5. ◀ **8**

FIGURE 1.5

▶ **EXAMPLE 9** Graph all real numbers x such that $1 < x < 5$.
The graph includes all the real numbers between 1 and 5 and not just the integers. Graph these numbers by drawing a heavy line from 1 to 5 on the number line, as in Figure 1.6. Open circles at 1 and 5 show that neither of these points belongs to the graph. ◀ **9**

FIGURE 1.6

A set that consists of all the real numbers between two points, such as $1 < x < 5$ in Example 9, is called an **interval**. A special notation called **interval notation** is used to indicate an interval on the number line. For example, the interval including all numbers x, where $-2 < x < 3$, is written as $(-2, 3)$. The parentheses indicate that the numbers -2 and 3 are *not* included. If -2 and 3 are to be included in the interval, square brackets are used, as in $[-2, 3]$. The chart below shows several typical intervals, where $a < b$.

Inequality	*Interval Notation*	*Explanation*
$a \leq x \leq b$	$[a, b]$	Both a and b are included.
$a \leq x < b$	$[a, b)$	a is included, b is not.
$a < x \leq b$	$(a, b]$	b is included, a is not.
$a < x < b$	(a, b)	Neither a nor b is included.

Interval notation is also used to describe sets such as the set of all numbers x, with $x \geq -2$. This interval is written $[-2, \infty)$.

 Graph all real numbers x such that

(a) $[4, \infty)$;

(b) $[-2, 1]$.

Answers:

(a)

(b)

▶ **EXAMPLE 10** Graph the interval $[-2, \infty)$.

Start at -2 and draw a heavy line to the right, as in Figure 1.7. Use a solid circle at -2 to show that -2 itself is part of the graph. The symbol, ∞, read "infinity," *does not* represent a number. It simply indicates that *all* numbers greater than -2 are in the interval. Similarly, the notation $(-\infty, 2)$ indicates the set of all numbers x with $x < 2$. ◀

FIGURE 1.7

ORDER OF OPERATIONS   We avoid possible ambiguity when working problems with real numbers by using the following *order of operations,* which has been agreed on as the most useful. This order of operations is used by computers and many calculators.*

Order of Operations

If parentheses or square brackets are present:

1. Work separately above and below any fraction bar.
2. Use the rules below within each set of parentheses or square brackets. Start with the innermost and work outward.

If no parentheses are present:

1. Find all powers and roots, working from left to right.
2. Do any multiplications or divisions in the order in which they occur, working from left to right.
3. Do any additions or subtractions in the order in which they occur, working from left to right.

▶ **EXAMPLE 11**   Use the order of operations to simplify the following.

(a) $\dfrac{-9(-3) + (-5)}{2(-2)^3 - 5(3)}$

Work separately above and below the fraction bar. Recall that $(-2)^3$ means $(-2)(-2)(-2) = -8$. Follow the order of operations.

$$\frac{-9(-3) + (-5)}{2(-2)^3 - 5(3)} = \frac{-9(-3) + (-5)}{2(-8) - 5(3)} = \frac{27 + (-5)}{-16 - 15} = \frac{22}{-31} = -\frac{22}{31}$$

---

* Some calculators may not follow this convention. To see if yours does, use it to work Examples 11–12 and compare the results. If your calculator works differently, consult the instruction manual (or your instructor) for directions.

11 Simplify the following.

(a) $4^2 \div 8 + 3^2 \div 3$

(b) $[-7 + (-9)](-4) - 8(3)$

(c) $\dfrac{-11 - (-12) - 4 \cdot 5}{-2^3 - (-3)(-5)}$

(d) $\dfrac{36 \div 4 \cdot 3 \div 9 + 1}{9 \div (-6) \cdot 8 - 4}$

Answers:

(a) 5

(b) 40

(c) $\dfrac{19}{23}$

(d) $-\dfrac{1}{4}$

12 Evaluate the following if $m = -5$ and $n = 8$.

(a) $-2mn - 2m^2$

(b) $\dfrac{4(n - 5)^2 - m}{m + n}$

Answers:

(a) 30

(b) $\dfrac{41}{3}$

(b) $-(3 - 5) - [2 - (3^2 - 13)]$

Start with the innermost parentheses. Evaluate the power first: $3^2 = 3 \cdot 3 = 9$. Now follow the order of operations.

$$-(3 - 5) - [2 - (3^2 - 13)] = -(3 - 5) - [2 - (9 - 13)]$$
$$= -(-2) - [2 - (-4)] = 2 - [6] = -4$$

◀ **11**

▶ **EXAMPLE 12** Use the order of operations to evaluate each expression if $x = -2$, $y = 5$, and $z = -3$.

(a) $-4x^2 - 7y + 4z$

Use parentheses when replacing letters with numbers.

$$-4x^2 - 7y + 4z = -4(-2)^2 - 7(5) + 4(-3)$$
$$= -4(4) - 7(5) + 4(-3) = -16 - 35 - 12 = -63$$

(b) $\dfrac{2(x - 5)^2 + 4y}{z + 4} = \dfrac{2(-2 - 5)^2 + 4(5)}{-3 + 4}$

$$= \dfrac{2(-7)^2 + 20}{1} = 2(49) + 20 = 118 \quad ◀ \ \textbf{12}$$

ABSOLUTE VALUE Distance is always given as a nonnegative number. For example, the distance from 0 to -2 on a number line is 2, the same as the distance from 0 to 2. The **absolute value** of a number a gives the distance on the number line from a to 0. Thus, the absolute value of both 2 and -2 is 2. We write the absolute value of the real number a as $|a|$. For example, the distance on the number line from 9 to 0 is 9, as is the distance from -9 to 0. (See Figure 1.8.) By definition, $|9| = 9$ and $|-9| = 9$.

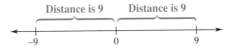

Distance is 9   Distance is 9

−9   0   9

FIGURE 1.8

The facts that $|9| = 9$ and that $|-9| = 9 = -(-9)$ suggest the following algebraic definition of absolute value.

Absolute Value

For any real number a,

$$|a| = a \qquad \text{if } a \geq 0$$
$$|a| = -a \qquad \text{if } a < 0.$$

For every real number, $|a| \geq 0$.

13 Find the following.

(a) $|-6|$

(b) $-|7|$

(c) $-|-2|$

(d) $|-3 - 4|$

(e) $|2 - 7|$

Answers:

(a) 6

(b) -7

(c) -2

(d) 7

(e) 5

Note the second part of the definition: for a negative number, say -5, the negative of -5 is the positive number $-(-5) = 5$. Similarly, if a is any negative number, then $-a$ is a *positive* number. Thus, for every real number a, $|a|$ is nonnegative.

▶**EXAMPLE 13**   Find each of the following.

(a) $|5|$

Since $5 > 0$, then $|5| = 5$.

(b) $|-5| = 5$

(c) $-|-5| = -(5) = -5$

Evaluate within the absolute value bars first, as we do with parentheses.

(d) $|0| = 0$

(e) $|8 - 9|$

First simplify the expression inside the absolute value bars.

$$|8 - 9| = |-1| = 1$$

(f) $-|-4 - 7| = -|-11| = -11$ ◀ **13**

1.1 EXERCISES

Name all the types of numbers that apply to each of the following. (See Example 1.) Approximate each irrational number using three significant digits.

1. 8   **2.** -5 **3.** -1 **4.** 0   **5.** 2/3

6. $-4/5$ **7.** $-\sqrt{3}$ **8.** $\sqrt{15}$ **9.** 3π **10.** $2/\pi$

Label each of the following as true or false.

11. Every integer is a rational number.

12. Every integer is a whole number.

13. Every whole number is an integer.

14. No whole numbers are rational.

Identify the properties that are illustrated in each of the following. Some may require more than one property. Assume all variables represent real numbers. (See Examples 2–6.)

15. $4(-6) = -6(4)$

16. $3 + (-3) = 0$

17. $-7 + 0 = -7$

18. $3 + (-3) = (-3) + 3$

19. $0 + (-7) = -7 + 0$

20. $8 + (12 + 6) = (8 + 12) + 6$

21. $[5(-8)](-3) = 5[(-8)(-3)]$

22. $8(m + 4) = 8m + 8 \cdot 4$

23. $x(y + 2) = xy + 2x$

24. $8(4 + 2) = (2 + 4)8$

25. How is the additive inverse property related to the additive identity property? the multiplicative inverse property to the multiplicative identity property?

26. Explain the distinction between the commutative and associative properties.

Graph each of the following on a number line. (See Examples 8 and 9.)

27. All integers x such that $-5 < x < 5$

28. All integers x such that $-4 < x < 2$

29. All whole numbers x such that $x \leq 3$

30. All whole numbers x such that $1 \leq x \leq 8$

31. All natural numbers x such that $-1 < x < 5$

32. All natural numbers x such that $x \leq 2$

Graph the following intervals. (See Example 10.)

33. $(3, \infty)$ **34.** $(-\infty, 5)$ **35.** $(-\infty, -2]$ **36.** $[-4, \infty)$ **37.** $(-8, -1)$

38. $(2, 6)$ **39.** $[-1, 10]$ **40.** $[-5, 5]$ **41.** $[-2, 2)$ **42.** $(3, 7]$

Evaluate each expression using the order of operations given in the text. (See Example 11.)

43. $9 - 4^2 - (-12)$

44. $8^2 - (-4) + 11$

45. $-2(9 - 8) + (-7)(2)^3$

46. $6(-5) - (-3)(2)^4$

47. $(4 - 2^3)(-2 + \sqrt{25})$

48. $[3^2 - (-2)][\sqrt{16} - 2^3]$

49. $6^2 - 3\sqrt{45}$

50. $(1 + \sqrt{3})^2$

Evaluate each of the following if $p = -2$, $q = 4$, and $r = -5$. (See Example 12.)

51. $-3(p + 5q)$ **52.** $2(q - r)$ **53.** $\dfrac{q + r}{q + p}$ **54.** $\dfrac{3q}{3p - 2r}$ **55.** $\dfrac{\dfrac{q}{4} - \dfrac{r}{5}}{\dfrac{p}{2} + \dfrac{q}{2}}$ **56.** $\dfrac{\dfrac{3r}{10} - \dfrac{5p}{2}}{q + \dfrac{2r}{5}}$

Evaluate each of the following. (See Example 13.)

57. $-|-4|$ **58.** $-|-2|$ **59.** $|6 - 4|$ **60.** $|3 - 17|$

61. $-|12 + (-8)|$ **62.** $|-6 + (-15)|$ **63.** $|8 - (-9)|$ **64.** $|-3 - (-2)|$

65. $|8| - |-4|$ **66.** $|-9| - |-12|$ **67.** $-|-4| - |-1 - 14|$ **68.** $-|6| - |-12 - 4|$

In each of the following problems, fill in the blank with either $=$, $<$, or $>$, so that the resulting statement is true.

69. $|5|$ ____ $|-5|$ **70.** $|3|$ ____ $|-3|$ **71.** $-|7|$ ____ $|7|$

72. $-|-4|$ ____ $|4|$ **73.** $|10 - 3|$ ____ $|3 - 10|$ **74.** $|6 - (-4)|$ ____ $|-4 - 6|$

75. $|-2 + 8|$ ____ $|2 - 8|$ **76.** $|3 + 1|$ ____ $|-3 - 1|$ **77.** $|3| \cdot |-5|$ ____ $|3(-5)|$

78. $|3| \cdot |2|$ ____ $|3(2)|$ **79.** $|3 - 2|$ ____ $|3| - |2|$ **80.** $|5 - 1|$ ____ $|5| - |1|$

81. In general, if a and b are any real numbers having the same sign (both negative or both positive), is it always true that $|a + b| = |a| + |b|$? Explain your answer.

82. If a and b are any real numbers, is it always true that $|a + b| = |a| + |b|$? Explain your answer.

83. If a and b are any two real numbers, is it always true that $|a - b| = |b - a|$? Explain your answer.

84. For which real numbers b does $|2 - b| = |2 + b|$? Explain your answer.

Social Science   *Use inequality symbols to rewrite each of the following statements, which are based on a recent article in the **Sacramento Bee** newspaper.* Using x as the variable, describe what x represents in each exercise and then write an inequality. Example: At least 4000 foreign students attend the University of Southern California (USC). Let x represent the number of foreign students attending USC. Then x $\geq$ 4000.*

* "State colleges drawing foreigners despite cuts" by Lisa Lapin from *The Sacramento Bee,* December 2, 1992. Copyright, The Sacramento Bee, 1994. Reprinted by permission.

85. Foreign students contribute more than $1 billion annually to the California economy.

86. More than 60% of the international students come from Asian countries.

87. Less than 7.5% of foreign students in the United States now come from Middle Eastern countries.

88. No more than 10% of the foreign students in the United States originate in Japan.

89. California has more than 13% of all foreign students in the United States.

90. Foreign students must prove they have at least $22,000 in cash to spend here each year.

Social Science *Sociologists measure the status of an individual within a society by evaluating for that individual the number x, which gives the percentage of the population with less income than the given person, and the number y, the percentage of the population with less educa-*

tion. The average status is defined as (x + y)/2, while the individual's status incongruity is defined by |(x − y)/2|. People with high status incongruities would include unemployed Ph.D.'s (low x, high y) and millionaires who didn't make it past the second grade (high x, low y).

91. What is the highest possible average status for an individual? The lowest?

92. What is the hightest possible status incongruity for an individual? The lowest?

93. Jolene Rizzo makes more money than 56% of the population and has more education than 78%. Find her average status and status incongruity.

94. A popular movie star makes more money than 97% of the population and is better educated than 12%. Find the average status and status incongruity for this individual.

---

1 Is -4 a solution of the following equations?

(a) $3x + 5 = -7$

(b) $2x - 3 = 5$

(c) Is there more than one solution of the equation in (a)?

Answers:

(a) Yes

(b) No

(c) No

1.2 LINEAR EQUATIONS AND APPLICATIONS

One of the main uses of algebra is to solve equations. An **equation** states that two mathematical expressions are equal. Examples of equations include $x + 6 = 9$, $4y + 8 = 12$, and $9z = -36$. The letter in each equation, the unknown, is called the **variable.** Sometimes an equation must be solved that has more than one variable. As a general rule, the first few letters of the alphabet, a, b, c, and so on, are used to represent constants, while letters such as x, y, and z are used for variables.

A **solution** of an equation is a number that can be substituted for the variable in the equation to produce a true statement. For example, substituting the number 9 for x in the equation $2x + 1 = 19$ gives

$$2x + 1 = 19$$
$$2(9) + 1 = 19 \quad \text{Let } x = 9$$
$$18 + 1 = 19. \quad \text{True}$$

This true statement indicates that 9 is a solution of $2x + 1 = 19$. **1**

Equations that can be written in the form $ax + b = c$, where a, b, and c are real numbers, with $a \neq 0$, are called **linear equations.** Examples of linear equations include $5y + 9 = 16$, $8x = 4$, and $-3p + 5 = -8$. Examples of equations that are *not* linear include $|x| = 4$, $2x^2 = 5x + 6$, and $\sqrt{x + 2} = 4$.

2 Solve the following.

(a) $3p - 5 = 19$

(b) $4y + 3 = -5$

(c) $-2k + 6 = 2$

Answers:

(a) 8

(b) −2

(c) 2

The following properties are used to solve equations.

Properties of Equality

For any real numbers a, b, and c.

(a) if $a = b$, then $a + c = b + c$.   **Addition property of**
(The same number may be   **equality**
added to both sides of an equation.)

(b) if $a = b$ and $c \neq 0$, then $ac = bc$.   **Multiplication property of**
(The same nonzero number may be   **equality**
multiplied on both sides of an
equation.)

Recall that subtraction is defined in terms of addition: for all real numbers a and b,

$$a - b = a + (-b);$$

and division is defined in terms of multiplication: for all real numbers a and all nonzero real numbers b,

$$\frac{a}{b} = a \cdot \frac{1}{b}.$$

For this reason, special properties for subtraction and division are not needed, as the following examples show.

▶ **EXAMPLE 1** Solve the linear equation $5x - 3 = 12$.

Using the addition property of equality, add 3 to both sides. This isolates the term containing the variable on one side of the equals sign.

$$5x - 3 = 12$$
$$5x - 3 + 3 = 12 + 3 \quad \text{Add 3 to both sides}$$
$$5x = 15$$

To get $1x$ instead of $5x$ on the left, use the fact that $(1/5) \cdot 5 = 1$, and multiply both sides of the equation by $1/5$.

$$5x = 15.$$

$$\frac{1}{5}(5x) = \frac{1}{5}(15) \quad \text{Multiply both sides by } \frac{1}{5}$$

$$1x = 3$$
$$x = 3$$

The solution of the original equation, $5x - 3 = 12$, is 3. Check the solution by substituting 3 for x in the original equation. ◀ **2**

3 Solve the following.

(a) $3(m - 6) + 2(m + 4)$
$= 4m - 2$

(b) $-2(y + 3) + 4y$
$= 3(y + 1) - 6$

Answers:

(a) 8

(b) -3

▶**EXAMPLE 2**　Solve $2k + 3(k - 4) = 2(k - 3)$.

First simplify this equation using the distributive property. By this property, $3(k - 4)$ is $3k - 3 \cdot 4$, or $3k - 12$. Also, $2(k - 3)$ is $2k - 2 \cdot 3$, or $2k - 6$. The equation can now be written as

$$2k + 3(k - 4) = 2(k - 3)$$
$$2k + 3k - 12 = 2k - 6.$$

On the left, $2k + 3k = (2 + 3)k = 5k$, again by the distributive property, which gives

$$5k - 12 = 2k - 6.$$

One way to proceed is to add $-2k$ to both sides.

$$5k - 12 + (-2k) = 2k - 6 + (-2k) \qquad \text{**Add** $-2k$ **to both sides**}$$
$$3k - 12 = -6$$
$$3k - 12 + 12 = -6 + 12 \qquad \text{**Add 12 to both sides**}$$
$$3k = 6$$
$$\frac{1}{3}(3k) = \frac{1}{3}(6) \qquad \text{**Multiply both sides by** $\frac{1}{3}$}$$
$$k = 2$$

The solution is 2. Check this result by substituting 2 for k in the original equation.　◀　**3**

FOR GRAPHERS

A grapher* can be used to solve equations in two ways. One way is to write the equation with one side equal to 0. Then let y equal the expression on the other side. For the equation in Example 2 (using x in place of k), let $y = 2x + 3(x - 4) - 2(x - 3)$. To solve $y = 0$, graph $y = 2x + 3(x - 4) - 2(x - 3)$ and look for any value of x where the graph intersects the x-axis. As Figure 1.9(a) shows, $y = 0$ at $x = 2$. The ⎡range⎤ or ⎡window⎤ key is used to determine the viewing rectangle, the portion of the graph that will show on the screen. To get a reasonable graph, it is important to choose an appropriate viewing rectangle. This may require some trial and error.

Graphers allow us to input more than one equation and see the graphs on the same grid. Thus another method of solving the equation in Example 2 is to graph the expressions on each side of the equation simultaneously, letting $y_1 = 2x + 3(x - 4)$ and $y_2 = 2(x - 3)$. See Figure 1.9(b). Then use ⎡trace⎤ to find the x-value where the two graphs intersect. ⎡zoom⎤ may also be used, if necessary, to get a more accurate answer. In some cases an exact answer

---

　* We use the term grapher here to mean a graphing calculator or computer with graphing software.

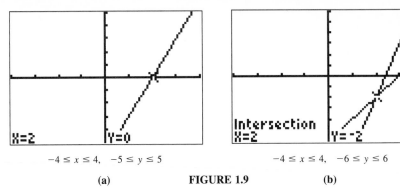

$$-4 \le x \le 4, \quad -5 \le y \le 5$$

(a)   FIGURE 1.9   **(b)**

$$-4 \le x \le 4, \quad -6 \le y \le 6$$

cannot be obtained. In such cases ⌐zoom⌐ may be used to get an accurate approximation. Look for the answer to remain unchanged to the desired number of decimal places after repeated zooming.*

The next three examples show how to simplify the solution of linear equations involving fractions. We solve these equations by multiplying both sides of the equation by a **common denominator,** a number that can be divided (with remainder 0) by each denominator in the equation. This step will eliminate the fractions.

Caution   Because this is an application of the multiplication property of *equality,* it can be done *only in an equation.*

▶**EXAMPLE 3**   Solve $\dfrac{r}{10} - \dfrac{2}{15} = \dfrac{3r}{20} - \dfrac{1}{5}$.

Here the denominators are 10, 15, 20, and 5. Each of these numbers can be divided into 60; therefore, 60 is a common denominator. Multiply both sides of the equation by 60.

$$60\left(\frac{r}{10} - \frac{2}{15}\right) = 60\left(\frac{3r}{20} - \frac{1}{5}\right)$$

Use the distributive property to eliminate the denominators.

$$60\left(\frac{r}{10}\right) - 60\left(\frac{2}{15}\right) = 60\left(\frac{3r}{20}\right) - 60\left(\frac{1}{5}\right)$$

$$6r - 8 = 9r - 12$$

Add $-6r$ and 12 to both sides.

$$6r - 8 + (-6r) + 12 = 9r - 12 + (-6r) + 12$$

$$4 = 3r$$

---

* Some graphers will find the intersection automatically. Check your manual for details.

4 Solve the following:

(a) $\dfrac{x}{2} - \dfrac{x}{4} = 6$

(b) $\dfrac{2x}{3} + \dfrac{1}{2} = \dfrac{x}{4} - \dfrac{9}{2}$

Answers:

(a) 24

(b) -12

5 Solve the equation

$$\dfrac{5p + 1}{3(p + 1)} = \dfrac{3p - 3}{3(p + 1)} + \dfrac{9p - 3}{3(p + 1)}.$$

Answer:
1

Multiply both sides by 1/3 to get the solution.

$$r = \dfrac{4}{3}$$

Check this solution in the original equation. ◀ **4**

▶ **EXAMPLE 4** Solve $\dfrac{4}{3(k + 2)} - \dfrac{k}{3(k + 2)} = \dfrac{5}{3}.$

Multiply both sides of the equation by the common denominator $3(k + 2)$. Here $k \neq -2$, since $k = -2$ would give a 0 denominator, making the fraction undefined.

$$3(k + 2) \cdot \dfrac{4}{3(k + 2)} - 3(k + 2) \cdot \dfrac{k}{3(k + 2)} = 3(k + 2) \cdot \dfrac{5}{3}$$

Simplify each side and solve for k.

$$4 - k = 5(k + 2)$$
$$4 - k = 5k + 10 \qquad \textbf{Distributive property}$$
$$4 - k + k = 5k + 10 + k \qquad \textbf{Add } k \textbf{ to both sides}$$
$$4 = 6k + 10$$
$$4 + (-10) = 6k + 10 + (-10) \qquad \textbf{Add } -10 \textbf{ to both sides}$$
$$-6 = 6k$$
$$-1 = k \qquad \textbf{Multiply by } \dfrac{1}{6}$$

The solution is -1. Substitute -1 for k as a check.

$$\dfrac{4}{3(-1 + 2)} - \dfrac{-1}{3(-1 + 2)} \overset{?}{=} \dfrac{5}{3}$$
$$\dfrac{4}{3} - \dfrac{-1}{3} \overset{?}{=} \dfrac{5}{3}$$
$$\dfrac{5}{3} = \dfrac{5}{3}$$

The check shows that -1 is the solution. ◀ **5**

Caution Because the equation in Example 4 has a restriction on k, it is *essential* to check the solution.

▶ **EXAMPLE 5** Solve $\dfrac{x}{x - 2} = \dfrac{2}{x - 2} + 2.$

Multiply both sides of the equation by $x - 2$, assuming that $x - 2 \neq 0$. This gives

$$x = 2 + 2(x - 2)$$
$$x = 2 + 2x - 4$$
$$x = 2.$$

6 Solve each equation.

(a) $\dfrac{3p}{p+1} = 1 - \dfrac{3}{p+1}$

(b) $\dfrac{8y}{y-4} = \dfrac{32}{y-4} - 3$

Answer:
Neither equation has a solution.

7 Solve for x.

(a) $2x - 7y = 3xk$

(b) $8(4 - x) + 6p$
$= -5k - 11yx$

Answers:

(a) $x = \dfrac{7y}{2 - 3k}$

(b) $x = \dfrac{5k + 32 + 6p}{8 - 11y}$

Recall the assumption that $x - 2 \neq 0$. Because $x = 2$, we have $x - 2 = 0$, and the multiplication property of equality does not apply. To see this, substitute 2 for x in the original equation—this substitution produces a 0 denominator. Since division by zero is not defined, there is no solution for the given equation. ◀ **6**

Sometimes an equation with several variables must be solved for one of the variables. This process is called **solving for a specified variable.**

▶ **EXAMPLE 6**   Solve for x: $3(ax - 5a) + 4b = 4x - 2$.
Use the distributive property to get

$$3ax - 15a + 4b = 4x - 2.$$

Treat x as the variable, the other letters as constants. Get all terms with x on one side of the equals sign, and all terms without x on the other side.

$$3ax - 4x = 15a - 4b - 2 \qquad \textbf{Isolate terms with } x \textbf{ on the left}$$
$$(3a - 4)x = 15a - 4b - 2 \qquad \textbf{Distributive property}$$
$$x = \frac{15a - 4b - 2}{3a - 4} \qquad \textbf{Multiply by } \tfrac{1}{2}$$

The final equation is solved for x, as required. ◀ **7**

▶ **EXAMPLE 7**   The formula

$$A = \frac{24f}{b(p + 1)}$$

gives the approximate annual interest rate for a consumer loan paid off with monthly payments.* Here f is the finance charge on the loan, p is the total number of payments, and b is the original balance of the loan. Solve the formula for p.

Treat p as the variable and the other letters as constants. The goal is to isolate p on one side of the equals sign. Begin by multiplying both sides of the formula by $p + 1$. (Because p is the number of payments, p is greater than 0 here, so $p \neq -1$.)

$$A = \frac{24f}{b(p + 1)}$$

$$(p + 1)A = (p + 1)\frac{24f}{b(p + 1)}$$

$$(p + 1)A = \frac{24f}{b}$$

---

* This formula is not accurate enough for the requirements of federal law.

8 Solve $J\left(\dfrac{m}{k} + a\right) = m$ for k.

Answer:

$k = \dfrac{Jm}{m - Ja}$

Now think of undoing the operations that have been performed on p. Multiply both sides of the equation by $1/A$ to undo the multiplication by A.

$$\frac{1}{A}(p + 1)A = \frac{1}{A} \cdot \frac{24f}{b}$$

$$p + 1 = \frac{24f}{Ab}$$

(We must assume $A \neq 0$. Why is this a very safe assumption here?) Finally, undo the addition of 1 to p by adding -1 on each side to get

$$p = \frac{24f}{Ab} - 1,$$

a result solved for p. ◄ **8**

APPLIED PROBLEMS   One of the main reasons for learning mathematics is to be able to use it to solve practical problems. However, for many students learning how to use mathematical skills in real-world applications is the most difficult task they face. Hints that may help with applications are given in the rest of this section.

A common difficulty with applied problems is trying to do everything at once. It is usually best to attack the problem in stages as follows.

Solving Applied Problems

1. Decide on the unknown. Name it with some variable that you *write down*. Many students try to skip this step. They are eager to get on with the writing of the equation. But this is an important step. If you don't know what the variable represents, how can you write a meaningful equation or interpret a result?
2. Draw a sketch or make a chart, if appropriate, showing the information given in the problem.
3. Decide on a variable expression to represent any other unknowns in the problem. For example, if x represents the width of a rectangle, and you know that the length is one more than twice the width, then *write down* that the length is $1 + 2x$.
4. Using the results of Steps 1–3, write an equation that expresses a condition that must be statisfied.
5. Solve the equation.
6. Check the solution in the words of the *original problem*, not just in the equation you have written.

The following examples illustrate this approach.

▶**EXAMPLE 8** If the length of a side of a square is increased by 3 centimeters, the new perimeter is 40 centimeters more than twice the length of the side of the original square. Find the length of a side of the original square.

Step 1 What should the variable represent? To find the length of a side of the original square, let

$$x = \text{length of a side of the original square.}$$

Step 2 Draw a sketch, as in Figure 1.10.

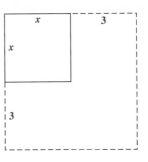

FIGURE 1.10

Step 3 The length of a side of the new square is 3 centimeters more than the length of a side of the old square, so

$$x + 3 = \text{length of a side of the new square.}$$

Now write a variable expression for the new perimeter. Since the perimeter of a square is four times the length of a side,

$$4(x + 3) = \text{the perimeter of the new square.}$$

Step 4 Write an equation by looking again at the information in the problem. The new perimeter is 40 more than twice the length of a side of the original square, so the equation is

$$\begin{pmatrix} \text{the new} \\ \text{perimeter} \end{pmatrix} \text{ is } 40 \begin{pmatrix} \text{more} \\ \text{than} \end{pmatrix} \begin{pmatrix} \text{twice the side of} \\ \text{the original square} \end{pmatrix}$$

$$4(x + 3) \quad = \quad 40 \quad + \quad 2x.$$

Step 5 Solve the equation.

$$4(x + 3) = 40 + 2x$$
$$4x + 12 = 40 + 2x$$
$$2x = 28$$
$$x = 14$$

9 **(a)** A triangle has a perimeter of 45 centimeters. Two of the sides of the triangle are equal in length, with the third side 9 centimeters longer than either of the two equal sides. Find the lengths of the sides of the triangle.

(b) A rectangle has a perimeter which is five times its width. The length is 4 more than the width. Find the length and width of the rectangle.

Answers:

(a) 12 centimeters, 12 centimeters, 21 centimeters

(b) Length is 12, width is 8

Step 6   Check the solution using the wording of the original problem. The length of a side of the new square would be $14 + 3 = 17$ centimeters; its perimeter would be $4(17) = 68$ centimeters. Twice the length of the side of the original square is $2(14) = 28$ centimeters. Because $40 + 28 = 68$ centimeters, the solution checks with the words of the original problem. ◀ **9**

▶**EXAMPLE 9**   Chuck travels 80 kilometers in the same time that Mary travels 180 kilometers. Mary travels 50 kilometers per hour faster than Chuck. Find the speed of each person.

Use the steps given earlier.

Step 1   Use x to represent Chuck's speed and $x + 50$ to represent Mary's speed, which is 50 kilometers per hour faster than Chuck's.

Step 2 and 3   Constant velocity problems of this kind require the distance formula

$$d = rt,$$

where d is the distance traveled in t hours at a constant rate of speed r. The distance traveled by each person is given, along with the fact that the time traveled by each person is the same. Solve the formula $d = rt$ for t.

$$d = rt$$

$$\frac{1}{r} \cdot d = \frac{1}{r} \cdot rt$$

$$\frac{d}{r} = t$$

For Chuck, $d = 80$ and $r = x$, giving $t = 80/x$. For Mary, $d = 180$, $r = x + 50$, and $t = 180/(x + 50)$. Use these facts to complete a chart, which organizes the information given in the problem.

	d	r	t
Chuck	80	x	$\dfrac{80}{x}$
Mary	180	$x + 50$	$\dfrac{180}{x + 50}$

Step 4   Because both people traveled for the *same time,* the equation is

$$\frac{80}{x} = \frac{180}{x + 50}.$$

10 **(a)** Tom and Dick are in a run for charity. Tom runs at 7 mph and Dick runs at 5 mph. If they start at the same time, how long will it be until they are 1/2 mile apart?

(b) In part (a), suppose the run has a staggered start. If Dick starts first, and Tom starts 10 minutes later, how long will it be until they are neck and neck?

Answers:

(a) 15 minutes (1/4 hour)

(b) Tom runs 25 minutes

11 An investor owns two pieces of property. One, worth twice as much as the other, returns 6% in annual interest, while the other returns 4%. Find the value of each piece of property if the total annual interest earned is $8000.

Answer:
6% return: $100,000; 4% return: $50,000

Step 5   Multiply both sides of the equation by $x(x + 50)$.

$$x(x + 50)\frac{80}{x} = x(x + 50)\frac{180}{x + 50}$$

$$80(x + 50) = 180x$$

$$80x + 4000 = 180x$$

$$4000 = 100x$$

$$40 = x$$

Step 6   Since x represents Chuck's speed, Chuck went 40 kilometers per hour. Mary's speed is $x + 50$, or $40 + 50 = 90$ kilometers per hour. Check these results in the words of the original problem.   ◀ **10**

▶ **EXAMPLE 10**   A financial manager has $14,000 to invest for her company. She plans to invest part of the money in tax-free bonds at 6% interest and the remainder in taxable bonds at 9%. She wants to earn $1005 per year in interest from the investments. Find the amount she should invest at each rate.

Let x represent the amount to be invested at 6%, so that $14,000 - x$ is the amount to be invested at 9%. Interest is given by the product of principal, rate, and time in years ($i = prt$). Summarize this information in a chart.

Investment	Amount Invested	Interest Rate	Interest Earned in 1 Year
Tax-free Bonds	x	6% = .06	.06x
Taxable Bonds	$14,000 - x$	9% = .09	.09$(14,000 - x)$
Totals	14,000		1005

Because the total interest is to be $1005,

$$.06x + .09(14,000 - x) = 1005.$$

Solve this equation.

$$.06x + 1260 - .09x = 1005$$

$$-.03x = -255$$

$$x = 8500$$

The manager should invest $8500 at 6%, and $14,000 - $8500 = $5500 at 9%.   ◀ **11**

1.2 EXERCISES

Solve each equation. (See Examples 1–5.)

1. $3x + 5 = 20$

2. $4 - 5y = 9$

3. $.6k - .3 = .5k + .4$

4. $2.5 + 5.04m = 8.5 - .06m$

5. $\dfrac{2}{5}r + \dfrac{1}{4} - 3r = \dfrac{6}{5}$

6. $\dfrac{2}{3} - \dfrac{1}{4}p = \dfrac{3}{2} + \dfrac{1}{3}p$

7. $2a - 1 = 3(a + 1) + 7a + 5$

8. $3(k - 2) - 6 = 4k - (3k - 1)$

9. $2[x - (3 + 2x) + 9] = 2x + 4$

10. $-2[4(k + 2) - 3(k + 1)] = 14 + 2k$

11. $\dfrac{3x}{5} - \dfrac{4}{5}(x + 1) = 2 - \dfrac{3}{10}(3x - 4)$

12. $\dfrac{4}{3}(x - 2) - \dfrac{1}{2} = 2\left(\dfrac{3}{4}x - 1\right)$

13. $\dfrac{5y}{6} - 8 = 5 - \dfrac{2y}{3}$

14. $\dfrac{x}{2} - 3 = \dfrac{3x}{5} + 1$

15. $\dfrac{m}{2} - \dfrac{1}{m} = \dfrac{6m + 5}{12}$

16. $-\dfrac{3k}{2} + \dfrac{9k - 5}{6} = \dfrac{11k + 8}{k}$

17. $\dfrac{4}{x - 3} - \dfrac{8}{2x + 5} + \dfrac{3}{x - 3} = 0$

18. $\dfrac{5}{2p + 3} - \dfrac{3}{p - 2} = \dfrac{4}{2p + 3}$

19. $\dfrac{3}{2m + 4} = \dfrac{1}{m + 2} - 2$

20. $\dfrac{8}{3k - 9} - \dfrac{5}{k - 3} = 4$

Solve each equation for x. (See Example 6. In Exercises 25 and 26, recall that $a^2 = a \cdot a$.)

21. $4(a - x) = b - a + 2x$

22. $(3a + b) - bx = a(x - 2)$

23. $5(b - x) = 2b + ax$

24. $bx - 2b = 2a - ax$

25. $x = a^2x + ax - 3a + 3$

26. $a^2x - 2a^2 = 3x$

Solve each equation for the specified variable. Assume all denominators are nonzero. (See Example 7.)

27. $PV = k$ for V

28. $i = prt$ for p

29. $V = V_0 + gt$ for g

30. $S = S_0 + gt^2 + k$ for g

31. $A = \dfrac{1}{2}(B + b)h$ for B

32. $C = \dfrac{5}{9}(F - 32)$ for F

33. $\dfrac{1}{R} = \dfrac{1}{r_1} + \dfrac{1}{r_2}$ for R

34. $m = \dfrac{Ft}{v_1 - v_2}$ for v_2

Use a calculator to solve each of the following equations. Round to the nearest hundredth.

35. $9.06x + 3.59(8x - 5) = 12.07x + .5612$

36. $-5.74(3.1 - 2.7p) = 1.09p + 5.2588$

37. $\dfrac{2.63r - 8.99}{1.25} - \dfrac{3.90r - 1.77}{2.45} = r$

38. $\dfrac{8.19m + 2.55}{4.34} - \dfrac{8.17m - 9.94}{1.04} = 4m$

Natural Science   *In the metric system of weights and measures, temperature is measured in degrees Celsius (°C) instead of degrees Fahrenheit (°F). To convert back and forth between the two systems, use*

$$C = \dfrac{5(F - 32)}{9} \quad and \quad F = \dfrac{9}{5}C + 32.$$

In each of the following exercises, convert to the other system. Round answers to the nearest tenth of a degree if necessary.

39. 20°C **40.** 100°C **41.** 59°F **42.** 86°F **43.** 100°F **44.** 40°C

45. Make a complete list of the steps needed to solve any linear equation. (Some equations will not require every step.)

46. When solving an equation that includes a variable denominator, what condition must the solution satisfy?

47. Refer to Exercise 25. Suppose someone tells you that there is no reason to solve for x, because the left side of the equation is already equal to x. Is this correct? Explain.

48. A computer that usually sells for x dollars has been discounted 30%. Which of the following does *not* represent its sale price? Explain why.

(a) $x - .30x$ (b) $.70x$ (c) $\dfrac{7x}{10}$

(d) $x = .30$

Management *Example 7 introduced the formula for the approximate annual interest rate of a loan paid off with monthly payments:*

$$A = \frac{24f}{b(p + 1)}.$$

Use this formula to find the value of the variables not given in each of the following. Round A to the nearest percent and round other variables to the nearest whole numbers. (This formula is not accurate enough for the requirements of federal law.)

49. $f = \$800$, $b = \$4000$, $p = 36$; find A

50. $f = \$60$, $b = \$740$, $p = 12$; find A

51. $A = 8\%$, $b = \$2000$, $p = 36$; find f

52. $A = 5\%$, $b = \$1500$, $p = 24$; find f

53. $A = 6\%$, $f = \$370$, $p = 36$; find b

54. $A = 10\%$, $f = \$490$, $p = 48$; find b

Management *When a loan is paid off early, a portion of the finance charge must be returned to the borrower. By one method of calculating finance charge (called the rule of 78), the amount of unearned interest (finance charge to be returned) is given by*

$$u = f \cdot \frac{n(n + 1)}{q(q + 1)}$$

where u represents unearned interest, f is the original finance charge, n is the number of payments remaining when the loan is paid off, and q is the original number of payments. Find the amount of the unearned interest in each of the following.

55. Original finance charge = $800, loan scheduled to run 36 months, paid off with 18 payments remaining

56. Original finance charge = $1400, loan scheduled to run 48 months, paid off with 12 payments remaining

Solve each applied problem. (See Examples 8–10.)

57. A closed recycling bin is in the shape of a rectangular box. Find the height of the bin if its length is 18 feet, its width is 8 feet, and its surface area is 496 square feet.
(a) Choose a variable and write down what it represents.
(b) Write an equation relating the height, length, and width of a box to its surface area.
(c) Solve the equation and check the solution in the wording of the original problem.

58. The length of a rectangular label is 3 centimeters less than twice the width. The perimeter is 54 centimeters. Find the width. Follow the steps outlined in Exercise 57.

59. A puzzle piece in the shape of a triangle has a perimeter of 30 centimeters. Two sides of the triangle are each twice as long as the shortest side. Find the length of the shortest side.

60. A triangle has a perimeter of 27 centimeters. One side is twice as long as the shortest side. The third side is seven centimeters longer than the shortest side. Find the length of the shortest side.

61. The distance between New York and London is (about) 3500 miles. If an airplane has a cruising speed of 490 mph and a tail wind of 20 mph up to the point of no return, which then increases to 50 mph, how many miles out will it reach the point of no return? (A tail wind blows in the same direction as the plane. The point of no return is the point on the flight where it will take the same amount of time to fly on to the destination as to fly back to the starting point.)

62. On vacation, Le Hong averaged 50 mph traveling from Denver to Minneapolis. Returning by a different route that covered the same number of miles, he averaged 55 mph. What is the distance between the two cities, if his total traveling time was 32 hours?

63. Russ and Janet are running in the Apple Hill Fun Run. Russ runs at 7 mph, Janet at 5 mph. If they start at the same time, how long will it be before they are 1/2 mile apart?

64. If the run in Exercise 63 has a staggered start and Janet starts first, with Russ starting 10 minutes later, how long will it be before he catches up with her?

65. Joe Gonzalvez received $52,000 profit from the sale of some land. He invested part at 5% interest, and the rest at 4% interest. He earned a total of $2290 interest per year. How much did he invest at 5%?

66. Weijen Luan invests $20,000 received from an insurance settlement in two ways: some at 6%, and some at 4%. Altogether, she makes $1040 per year interest. How much is invested at 4%?

67. Maria Martinelli bought two plots of land for a total of $120,000. On the first plot, she made a profit of 15%. On the second, she lost 10%. Her total profit was $5500. How much did she pay for each piece of land?

68. Suppose $20,000 is invested at 5%. How much additional money must be invested at 4% to produce a yield of 4.8% on the entire amount invested?

69. Cathy Wacaser earns take-home pay of $592 a week. If her deductions for taxes, retirement, union dues, and medical plan amount to 26% of her wages, what is her weekly pay before deductions?

70. Barbara Dalton gives 10% of her net income to the church. This amounts to $80 a month. In addition, her paycheck deductions are 24% of her gross monthly income. What is her gross monthly income?

Natural Science   *Exercises 71 and 72 depend on the idea of the octane rating of gasoline, a measure of its antiknock qualities. Actual gasoline blends are compared to standard fuels. In one measure of octane, a standard fuel is made with only two ingredients: heptane and isooctane. For this fuel, the octane rating is the percent of isooctane; i.e., a gasoline with an octane rating of 98 has the same antiknock properties as a standard fuel that is 98% isooctane.*

71. How many liters of 94 octane gasoline should be mixed with 200 liters of 99 octane gasoline to get a mixture that is 97 octane?

72. A service station has 92 octane and 98 octane gasoline. How many liters of each gasoline should be mixed to provide 12 liters of 96 octane gasoline for a chemistry experiment?

 Use a grapher to verify the solutions in each of the following exercises.

73. Exercise 9   **74.** Exercise 10   **75.** Exercise 19   **76.** Exercise 18

1.3 LINEAR INEQUALITIES

An **inequality** is a statement that two mathematical expressions are *not* equal. Usually an inequality indicates that one expression is less than (or greater than) another. Inequalities are very important in applications. For example, a company wants revenue to be *greater than* costs and must use *no more than* the total amount of capital or labor available.

A **linear inequality,** such as $2m + 1 < 7$, is solved by simplifying it to the form $m < k$, for some number k. The following properties are used to simplify an inequality.

1 **(a)** First multiply both sides of $-6 < -1$ by 4, and then multiply both sides of $-6 < -1$ by -7.

(b) Multiply both sides of $9 \geq -4$ first by 2, and then by -5.

(c) First add 4, and then add -6 to both sides of $-3 < -1$.

Answers:

(a) $-24 < -4$; $42 > 7$

(b) $18 \geq -8$; $-45 \leq 20$

(c) $1 < 3$; $-9 < -7$

Properties of Inequality

For real numbers a, b, and c,

(a) if $a < b$, then $a + c < b + c$
(b) if $a < b$, and if $c > 0$, then $ac < bc$
(c) if $a < b$, and if $c < 0$, then $ac > bc$.

Throughout this section, definitions are given only for $<$; but they are equally valid for $>$, $\leq$, or $\geq$.

Caution   Pay careful attention to part (c): if both sides of an inequality are multiplied by a negative number, the direction of the inequality symbol must be reversed. For example, starting with the true statement $-3 < 5$ and multiplying both sides by the positive number 2 gives

$$-3 \cdot 2 < 5 \cdot 2,$$

or

$$-6 < 10,$$

still a true statement. On the other hand, starting with $-3 < 5$ and multiplying both sides by the negative number -2 gives a true result only if the direction of the inequality symbol is reversed:

$$-3(-2) > 5(-2)$$
$$6 > -10. \quad \blacksquare$$

▶ **EXAMPLE 1**   Solve $3x + 5 > 11$. Graph the solution.
First, add -5 to both sides.

$$3x + 5 + (-5) > 11 + (-5)$$
$$3x > 6$$

Now multiply both sides by 1/3.

$$\frac{1}{3}(3x) > \frac{1}{3}(6)$$
$$x > 2$$

(Why was the direction of the inequality symbol not changed?) As a check, note that 0, which is not part of the solution, makes the inequality false, while 3, which is part of the solution, makes it true.

$$3(0) + 5 > 11 \qquad\qquad 3(3) + 5 > 11$$
$$5 > 11 \quad \text{False} \qquad\qquad 14 > 11 \quad \text{True}$$

2 Solve the following. Graph each solution.

(a) $5z - 11 < 14$

(b) $-3k \leq -12$

(c) $-8y \geq 32$

Answers:

(a) $z < 5$

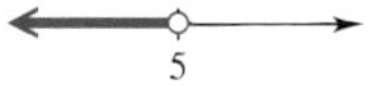

(b) $k \geq 4$

(c) $y \leq -4$

3 Solve the following. Graph each solution.

(a) $8 - 6t \geq 2t + 24$

(b) $-4r + 3(r + 1) < 2r$

Answers:

(a) $t \leq -2$

(b) $r > 1$

The solution is the inverval $(2, \infty)$, which is graphed in Figure 1.11. An open circle at 2 shows that 2 is not included. ◀ **2**

FIGURE 1.11

▶ **EXAMPLE 2** Solve $4 - 3x \leq 7 + 2x$.

$$4 - 3x \leq 7 + 2x$$
$$4 - 3x + (-4) \leq 7 + 2x + (-4)$$
$$-3x \leq 3 + 2x$$

Add $-2x$ to both sides. (Remember that *adding* to both sides never changes the direction of the inequality symbol.)

$$-3x + (-2x) \leq 3 + 2x + (-2x)$$
$$-5x \leq 3$$

Multiply both sides by $-1/5$. Since $-1/5$ is negative, change the direction of the inequality symbol.

$$-\frac{1}{5}(-5x) \geq -\frac{1}{5}(3)$$
$$x \geq -\frac{3}{5}$$

Figure 1.12 shows a graph of the solution $[-3/5, \infty)$. The solid circle in Figure 1.12 shows that $-3/5$ is included in the solution. ◀ **3**

FIGURE 1.12

FOR GRAPHERS

A grapher can be used to solve the inequality in Example 2 as follows. Write the inequality with 0 on one side: $4 - 3x - 7 - 2x \leq 0$. Graph $y = 4 - 3x - 7 - 2x$. (See Figure 1.13.) Because $y = 4 - 3x - 7 - 2x$, the y-values of the points on the graph show where $4 - 3x - 7 - 2x$ is positive

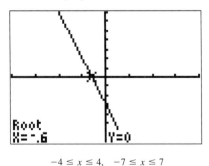

$-4 \leq x \leq 4, \quad -7 \leq x \leq 7$

FIGURE 1.13

 Solve each of the following. Graph each solution.

(a) $9 < k + 5 < 13$

(b) $-6 \leq 2z + 4 \leq 12$

Answers:

(a) $4 < k < 8$

(b) $-5 \leq z \leq 4$

or negative. The part of the graph below the x-axis shows where $y < 0$; the part above the x-axis shows where $y > 0$. The x-value where $y = 0$ is the dividing point. The graph shows that $y = 4 - 3x - 7 - 2x \leq 0$ when $x \geq -3/5$. This is the same solution found algebraically. Note that the graph does not tell you whether the endpoint is included in the solution. You must decide whether the endpoint satisfies the given inequally—that is, whether the inequality involves $<$ or $\leq$. The algebraic method is simpler for linear inequalities, but the graphical method is quite useful for inequalities of higher degree.

▶ **EXAMPLE 3** Solve $-2 < 5 + 3m < 20$. Graph the solution.

The inequality $-2 < 5 + 3m < 20$ says that $5 + 3m$ is *between* -2 and 20. We can solve this inequality with an extension of the properties given above. Work as follows, first adding -5 to each part.

$$-2 + (-5) < 5 + 3m + (-5) < 20 + (-5)$$
$$-7 < 3m < 15$$

Now multiply each part by $1/3$.

$$-\frac{7}{3} < m < 5$$

A graph of the solution, $(-7/3, 5)$, is given in Figure 1.14. ◀ **4**

FIGURE 1.14

FOR GRAPHERS

An inequality such as $-2 < 5 + 3x < 20$ in Example 3 can be solved with a grapher by slightly altering the grapher method given above. Here, it is not possible to get 0 on one side because there are three "sides." Instead, graph three expressions, -2, $5 + 3x$, and 20, simultaneously. Use ⎢trace⎥ to approximate the x-values where the graph of $5 + 3x$ intersects the graphs for -2 and 20. Use ⎢zoom⎥ to get more accurate values. Then decide on what interval the graph of $5 + 3x$ lies between the graphs of -2 and 20. See Figure 1.15. (Don't forget to decide about the endpoints separately.)

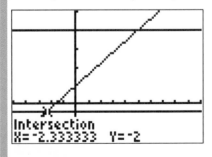

$-5 \leq x \leq 10, \quad -10 \leq y \leq 25$

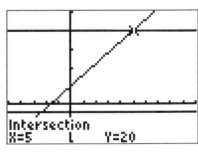

$-5 \leq x \leq 10, \quad -10 \leq y \leq 25$

FIGURE 1.15

5 Solve each inequality. Graph each solution.

(a) $5p - 10 < -20$ or $2p + 6 > 8$

(b) $1 - 4y \geq 5$ or $2 - 3y \leq -7$

Answers:

(a) All numbers in $(-\infty, -2)$ or $(1, \infty)$

(b) All numbers in $(-\infty, -1]$ or $[3, \infty)$

6 In Example 5, what Celsius temperatures correspond to 5°F to 95°F?

Answer:
−15°C to 35°C

The solutions of some inequalities result in graphs with separate parts, as shown in the next example.

▶ **EXAMPLE 4**   Solve $3x - 5 < 7$ or $2x - 1 > 13$. Graph the solution.

Begin by solving each inequality separately, keeping the "or" between the two solutions.

$$
\begin{array}{ccc}
3x - 5 < 7 & \text{or} & 2x - 1 > 13 \\
3x < 12 & \text{or} & 2x > 14 \\
x < 4 & \text{or} & x > 7
\end{array}
$$

The final inequality says that x may be any number less than 4 *or* any number greater than 7. The solution includes the numbers graphed in Figure 1.16. There is no shortcut way to write the solution $x < 4$ or $x > 7$. In interval notation the solution is written "all numbers in $(-\infty, 4)$ or $(7, \infty)$," which shows that numbers that satisfy the given inequality may be in *either* interval. ◀ **5**

FIGURE 1.16

▶ **EXAMPLE 5**   The formula for converting from Celsius to Fahrenheit temperature is

$$F = \frac{9}{5}C + 32.$$

What Celsius temperature range corresponds to the range from 32°F to 77°F?

The Fahrenheit temperature range is $32 < F < 77$. Since $F = (9/5)C + 32$,

$$32 < \frac{9}{5}C + 32 < 77.$$

Solve the inequality for C.

$$32 < \frac{9}{5}C + 32 < 77$$

$$0 < \frac{9}{5}C < 45$$

$$0 < C < \frac{5}{9} \cdot 45$$

$$0 < C < 25$$

The corresponding Celsius temperature range is 0°C to 25°C. ◀ **6**

A product will break even, or produce a profit, only if the revenue R from selling the product at least equals the cost C of producing it, that is if $R \geq C$.

▶**EXAMPLE 6**   A company analyst has determined that the cost to produce and sell x units of a certain product is $C = 20x + 1000$. The revenue for that product is $R = 70x$. Find the values of x for which the company will break even or make a profit on the product.

Solve the inequality $R \geq C$.

$$R \geq C$$
$$70x \geq 20x + 1000 \qquad \text{Let } R = 70x, C = 20x + 1000$$
$$50x \geq 1000$$
$$x \geq 20$$

The company must produce and sell 20 items to break even and more than 20 to make a profit.   ◀

1.3   EXERCISES

1. Explain how to determine whether an open circle or a solid circle is used when graphing the solution of a linear inequality.

2. The three-part inequality $p < x < q$ means "p is less than x and x is less than q." Which one of the following inequalities is not satisfied by any real number x? Explain why.

(a) $-3 < x < 5$ (b) $0 < x < 4$

(c) $-7 < x < -10$ (d) $-3 < x < -2$

Solve each inequality. Graph each solution in Exercises 3–38. (See Examples 1–4.)

3. $-8k \leq 32$ **4.** $-6a \leq 36$ **5.** $-2b > 0$ **6.** $6 - 6z < 0$

7. $3x + 4 \leq 12$ **8.** $2y - 5 < 9$ **9.** $-4 - p \geq 3$ **10.** $5 - 3r \leq -4$

11. $7m - 5 \leq 2m + 10$ **12.** $6x - 2 > 4x - 8$

13. $m - (4 + 2m) + 3 < 2m + 2$ **14.** $2p - (3 - p) \leq -7p - 2$ **15.** $-2(3y - 8) \geq 5(4y - 2)$

16. $5r - (r + 2) \geq 3(r - 1) + 5$ **17.** $3p - 1 < 6p + 2(p - 1)$ **18.** $x + 5(x + 1) > 4(2 - x) + x$

19. $-7 < y - 2 < 4$ **20.** $-3 < m + 6 < 2$ **21.** $8 \leq 3r + 1 \leq 13$

22. $-6 < 2p - 3 \leq 5$ **23.** $-4 \leq \dfrac{2k - 1}{3} \leq 2$ **24.** $-1 \leq \dfrac{5y + 2}{3} \leq 4$

25. $z + 1 \leq 2$ or $z - 5 \geq 1$ **26.** $2y + 2 \leq 5$ or $3y + 4 \geq 22$

27. $6m + 4 \geq 4 + m$ or $2m + 6 < -2 - 2m$ **28.** $5 - 2t > 3$ or $8 - 4t < 2 - 2t$

29. $\dfrac{3}{2}b - 2 < 4$ or $\dfrac{3}{4}b + \dfrac{1}{3} > \dfrac{19}{3}$ **30.** $\dfrac{x}{4} + 3 < \dfrac{9}{4}$ or $\dfrac{x}{2} - 4 > -\dfrac{7}{2}$

31. $\frac{3}{5}(2p + 3) \geq \frac{1}{10}(5p + 1)$

32. $\frac{8}{3}(z - 4) \leq \frac{2}{9}(3z + 2)$

33. $42.75x > 7.460$

34. $15.79y < 6.054$

35. $8.04z - 9.72 < 1.72z - .25$

36. $3.25 + 5.08k > 0.76k + 6.28$

37. $-(1.42m + 7.63) + 3(3.7m - 1.12) \leq 4.81m - 8.55$

38. $3(8.14a - 6.32) - (4.31a - 4.84) > .34a + 9.49$

Solve each of the following applied problems. (See Example 5.)

39. Natural Science Federal guidelines require drinking water to have fewer than .050 milligrams per liter of lead. A test using 21 samples of water in a midwestern city found that the average amount of lead in the samples was .040 milligrams per liter. All samples had lead content within 5% of the average. Did all the samples meet the federal requirement?
  (a) Select a variable and write down what it represents.
  (b) Write a three-part inequality to express the sample results.
  (c) Answer the question.

40. Management The federal income tax for an income of $18,551 to $44,900 is 28% times (net income − 18,550) + $2782.50.
  (a) State what the variables represent.
  (b) Write this income bracket as an inequality.
  (c) Write the tax range in dollars for this income bracket as an inequality.

41. Social Science A standard intelligence test has an average score of 100. According to statistical theory, about $\frac{1}{3}$ of the scores are above 112 or below 88. About 5% of the scores are below 76 or above 124. Write each of these ranges of scores as an inequality using "or."

42. Management In 1993 Pacific Bell charged $.40 for the first minute plus $.31 for each additional minute (or part of a minute) for a dial-direct call from Sacramento to North Tahoe, California.* How many minutes could a person talk for no more than $2.00?

43. A student has a total of 970 points so far in her algebra class. At the end of the course she must have 81% of the 1300 points possible in order to get a B. What is the lowest score she can earn on the 100-point final to get a B in the class?

44. A nearby business college charges a tuition of $6440 annually. Tom makes no more than $1610 per year in his summer job. What is the least number of summers that he must work in order to make enough for one year's tuition?

Management *In Exercises 45–50, find all values of x where the following products will at least break even. (See Example 6.)*

45. The cost to produce x units of wire is $C = 50x + 5000$, while the revenue is $R = 60x$.

46. The cost to produce x units of squash is $C = 100x + 6000$, while the revenue is $R = 500x$.

47. $C = 85x + 900$; $R = 105x$

48. $C = 70x + 500$; $R = 60x$

49. $C = 1000x + 5000$; $R = 900x$

50. $C = 2500x + 10,000$; $R = 102,500x$

51. Bill Bradkin went to a conference in Montreal, Canada for a week. He decided to rent a car and checked with two rental firms. Avery wanted $56 per day, with no mileage fee. Hart wanted $216 per week and $.28 per mile (or part of a mile). How many miles must Bradkin drive before the Avery car is the better deal?

52. After considering the situation in Exercise 51, Bradkin contacted Lowcost Rental Cars and was offered a car for $198 per week plus $.30 a mile (or part of a mile). With this offer, can he travel more miles for the same price as Avery charges? How many more or fewer miles?

---

*Pacific Bell White Pages, January 1992–January 1993

Use a grapher to solve these inequalities.

53. $4(3x - 2) \geq .6(2x - 1)$

54. $\dfrac{7}{3}(z + 1) \leq \dfrac{2}{9}(2z - 5)$

55. $2x - \dfrac{3x - 2}{4} > .8x + 7$

56. $\dfrac{7x + 4}{3} - 2 < .3x + \dfrac{2x - 1}{5}$

---

1 Solve each equation.

(a) $|y| = 9$

(b) $|r + 3| = 1$

(c) $|2k - 3| = 7$

Answers:

(a) $9, -9$

(b) $-2, -4$

(c) $5, -2$

2 Solve each equation.

(a) $|r + 6| = |2r + 1|$

(b) $|5k - 7| = |10k - 2|$

Answers:

(a) $5, -7/3$

(b) $-1, 3/5$

1.4 ABSOLUTE VALUE EQUATIONS AND INEQUALITIES

Recall from Section 1.1 that the absolute value of the number a, written $|a|$, gives the distance on a number line from a to 0. For example, $|4| = 4$, and $|-7| = 7$. In this section equations and inequalities involving absolute value are discussed.

▶**EXAMPLE 1**   Solve the equation $|x| = 3$.
   There are two numbers whose absolute value is 3, namely 3 and -3. The solutions of the given equation are 3 and -3.   ◀

▶**EXAMPLE 2**   Solve $|p - 4| = 2$.
   This equation will be satisfied if the expression inside the absolute value bars, $p - 4$, equals either 2 or -2:

$$p - 4 = 2 \quad \text{or} \quad p - 4 = -2.$$

Solving these two equations produces

$$p = 6 \quad \text{or} \quad p = 2,$$

so that 6 and 2 are solutions for the original equation. As before, check by substituting in the original equation.   ◀   **1**

▶**EXAMPLE 3**   Solve $|4m - 3| = |m + 6|$.
   The quantities in absolute value bars must either be equal or be negatives of one another to satisfy the equation. That is,

$$4m - 3 = m + 6 \quad \text{or} \quad 4m - 3 = -(m + 6)$$
$$3m = 9 \qquad\qquad 4m - 3 = -m - 6$$
$$m = 3 \qquad\qquad 5m = -3$$
$$m = -\dfrac{3}{5}.$$

Check that the solutions for the original equation are 3 and $-3/5$.   ◀   **2**

Solve each inequality. Graph each solution.

(a) $|x| \leq 1$

(b) $|y| \geq 3$

Answers:

(a) $[-1, 1]$

(b) All numbers in $(-\infty, -3]$ or $[3, \infty)$

4 Solve each inequality. Graph each solution.

(a) $|p + 3| < 4$

(b) $|2k - 1| \leq 7$

Answers:

(a) $(-7, 1)$

(b) $[-3, 4]$

The next examples show how to solve inequalities with absolute value.

▶ **EXAMPLE 4**   Solve each inequality.

(a) $|x| < 5$
Because absolute value gives the distance from a number to 0, the inequality $|x| < 5$ is true for all real numbers whose distance from 0 is less than 5. This includes all numbers from -5 to 5, or numbers in the interval $(-5, 5)$. A graph of the solution is shown in Figure 1.17.

FIGURE 1.17

(b) $|x| > 5$
In a similar way, the solution of $|x| > 5$ is given by all those numbers whose distance from 0 is *greater* than 5. This includes the numbers satisfying $x < -5$ or $x > 5$. A graph of the solution, all numbers in

$$(-\infty, -5) \quad \text{or} \quad (5, \infty),$$

is shown in Figure 1.18.   ◀ **3**

FIGURE 1.18

The examples above suggest the following generalizations.

Assume a and b are real numbers with b positive.

1. Solve $|a| = b$ by solving $a = b$ or $a = -b$.
2. Solve $|a| = |b|$ by solving $a = b$ or $a = -b$.
3. Solve $|a| < b$ by solving $-b < a < b$.
4. Solve $|a| > b$ by solving $a < -b$ or $a > b$.

▶ **EXAMPLE 5**   Solve $|x - 2| < 5$.
Replace a with $x - 2$ and b with 5 in property (3) above. Now solve $|x - 2| < 5$ by solving the inequality

$$-5 < x - 2 < 5.$$

Add 2 to each part, getting the solution

$$-3 < x < 7,$$

which is graphed in Figure 1.19.   ◀ **4**

FIGURE 1.19

5 Solve each inequality. Graph each solution.

(a) $|y - 2| > 5$

(b) $|3k - 1| \geq 2$

(c) $|2 + 5r| - 4 \geq 1$

Answers:

(a) All numbers in $(-\infty, -3)$ or $(7, \infty)$

(b) All numbers in $\left(-\infty, -\dfrac{1}{3}\right]$ or $[1, \infty)$

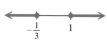

(c) All numbers in $\left(-\infty, -\dfrac{7}{5}\right]$ or $\left[\dfrac{3}{5}, \infty\right)$

FOR GRAPHERS

One method for solving absolute value equations with a grapher is to graph each side of the equation simultaneously. Some graphers have a key labeled ABS that gives the absolute value of a quantity. This function is accessed through a special menu on other graphers. For instance, in Example 2, graph $y_1 = |x - 4|$ and $y_2 = 2$. Remember to key in ABS$(x - 4)$, not ABS $x - 4$. Then locate the x-values of the points where these two graphs intersect (are equal). Figure 1.20 shows the x-values of the intersection points are 2 and 6. To solve the inequality $|x - 4| < 2$, from the graph in Figure 1.20 decide on which intervals the graph of $y_1 = |x - 4|$ is below (and therefore less than) the graph of $y_2 = 2$. As the figure shows, this happens for $2 < x < 6$. You could also solve this inequality by graphing $y = |x - 4| - 2$ and locating the intersection points with the x-axis.

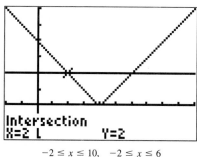

$-2 \leq x \leq 10, \quad -2 \leq x \leq 6$

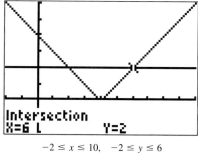

$-2 \leq x \leq 10, \quad -2 \leq y \leq 6$

$y_1 = |x - 4|, y_2 = 2$

FIGURE 1.20

▶ **EXAMPLE 6**   Solve $|2 - 7m| - 1 > 4$.

First add 1 on both sides.

$$|2 - 7m| > 5$$

Now use property (4) from above to solve $|2 - 7m| > 5$ by solving the inequality

$$2 - 7m < -5 \quad \text{or} \quad 2 - 7m > 5.$$

Solve each part separately.

$$-7m < -7 \quad \text{or} \quad -7m > 3$$

$$m > 1 \quad \text{or} \quad m < -\frac{3}{7}$$

The solution, all numbers in $\left(-\infty, -\dfrac{3}{7}\right)$ or $(1, \infty)$, is graphed in Figure 1.21.

◀ **5**

FIGURE 1.21

6 Solve each inequality.

(a) $|5m - 2| > -1$

(b) $|2 + 3a| < -3$

(c) $|6 + r| > 0$

Answers:

(a) All real numbers

(b) No solution

(c) All real numbers except -6

7 Write each statement using absolute value.

(a) m is at least 3 units from 5.

(b) t is within .01 of 4.

Answers:

(a) $|m - 5| \geq 3$

(b) $|t - 4| \leq .01$

▶**EXAMPLE 7**   Solve $|2 - 5x| \geq -4$.

The absolute value of a number is always nonnegative. Therefore, $|2 - 5x| \geq -4$ is always true, so that the solution is the set of all real numbers. Note that the inequality $|2 - 5x| < -4$ has no solution, because the absolute value of a quantity can never be less than a negative number.   ◀ **6**

Absolute value inequalities can be used to indicate how far a number may be from a given number. The next example, illustrates this use of absolute value.

▶**EXAMPLE 8**   Write each statement using absolute value.

(a) k is at least 4 units from 1.

If k is at least 4 units from 1, then the distance from k to 1 is greater than or equal to 4. See Figure 1.22(a). Since k may be on either side of 1 on the number line, k may be less than -3 or greater than 5. Write this statement using absolute value as follows:

$$|k - 1| \geq 4.$$

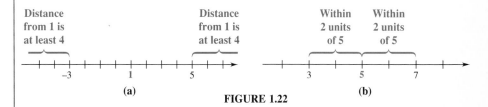

FIGURE 1.22

(b) p is within 2 units of 5.

This statement means that the difference between p and 5 must be less than or equal to 2. See Figure 1.22(b). Using absolute value notation, the statement is written as

$$|p - 5| \leq 2. \quad ◀ \;\; \textbf{7}$$

1.4 EXERCISES

Solve each equation. (See Examples 1–3.)

1. $|2h + 1| = 5$

2. $|4m - 3| = 12$

3. $|6 - 2p| = 10$

4. $|-5x + 7| = 15$

5. $\left|\dfrac{5}{r - 3}\right| = 10$

6. $\left|\dfrac{3}{2h - 1}\right| = 4$

7. $\left|\dfrac{6y + 1}{y - 1}\right| = 3$

8. $\left|\dfrac{3a - 4}{2a + 3}\right| = 1$

9. $|3y - 2| = |4y + 5|$

10. $|1 - 3z| = |z + 2|$

11. $|2 - 5m| = |9 + 3m|$

12. $|3 - 2x| = |4 - 5x|$

Solve each inequality. Graph each solution. (See Examples 4–7.)

13. $|x| \geq 5$

14. $|p| > 7$

15. $|m| < 1$

16. $|r| \leq 4$

17. $|a| < -2$

18. $|b| > -5$

19. $|x| - 3 \leq 7$

20. $|r| + 3 \leq 10$

21. $|2x + 5| < 3$

22. $\left|x - \frac{1}{2}\right| < 2$

23. $|3m - 2| > 4$

24. $|4x - 6| > 10$

25. $|3z + 1| \geq 7$

26. $|8b + 5| \geq 7$

27. $\left|5x + \frac{1}{2}\right| - 2 < 5$

28. $\left|x + \frac{2}{3}\right| + 1 < 4$

---

29. Explain why it is incorrect to write the absolute value inequality $|x| > 6$ as any of these: $-6 > x > 6$, $-6 > x < 6$, or $-6 < x > 6$.

30. Without actually going through the solution process, we can say that the equation $|5x - 6| = 6x$ cannot have a negative solution. Explain why this is true.

31. Physical Science The temperatures on the surface of Mars in degrees Celsius approximately satisfy the inequality $|C + 84| \leq 56$. What range of temperatures corresponds to this inequality?

32. Natural Science Dr. Tydings has found that, over the years, 95% of the babies he has delivered have weighed y pounds, where $|y - 8.0| \leq 1.5$. What range of weights corresponds to this inequality?

33. Physical Science The industrial process that is used to convert methanol to gasoline is carried out at a temperature range of 680° F to 780°F. Using F as the variable, write an absolute value inequality that corresponds to this range.

34. Physical Science When a model kite was flown across the wind in tests to determine its limits of power extraction, it attained speeds of 98 to 148 feet per second in winds of 16 to 26 feet per second. Using x as the variable in each case, write absolute value inequalities that correspond to these ranges.

Write each of the following statements using absolute value. (See Example 8.)

35. x is within 4 units of 2.

36. m is no more than 8 units from 9.

37. z is no less than 2 units from 12.

38. p is at least 5 units from 9.

39. If x is within .0004 units of 2, then y is within .00001 units of 7.

40. y is within .001 units of 10 whenever x is within .02 units of 5.

In Exercises 41 and 42, write each inequality using absolute value.

41. Natural Science The number of milligrams of a certain substance per liter in drinking water samples all tested within .05 of 40 milligrams per liter.

42. Social Science When administering a standard intelligence test, we expect about $1/3$ of the scores to be more than 12 units above 100 or more than 12 units below 100.

Use a grapher to solve the following inequalities.

43. $|3x - 5| > 4$

44. $|1 - x| > x + 1$

45. $|2x + 3| < x - 1$

46. $|3x + 5| < |x - 1|$

47. $2 + 3|x + 3| < 5|x - 1|$

48. $2|x - 4| > 3|x + 2| - 1$

1 Evaluate the following.

(a) 6^3

(b) 5^4

(c) 1^7

(d) $\left(\dfrac{2}{5}\right)^3$

Answers:

(a) 216

(b) 625

(c) 1

(d) 8/125

2 Evaluate the following.

(a) $3 \cdot 6^2$

(b) $5 \cdot 4^3$

(c) -3^6

(d) $(-3)^6$

(e) $-2 \cdot (-3)^5$

Answers:

(a) 108

(b) 320

(c) -729

(d) 729

(e) 486

1.5 POLYNOMIALS

Before studying polynomials, it is helpful to review the definitions and rules of exponents.

EXPONENTS   As mentioned in Section 1.1,

$$3^2 = 3 \cdot 3 \quad \text{and} \quad 2^3 = 2 \cdot 2 \cdot 2.$$

This section gives a more general meaning to the symbol a^n. First, recall that n is the **exponent** (or **power**) in a^n, and a is the **base.**

If n is a natural number and a is any real number, then

$$a^n = a \cdot a \cdot a \cdots a,$$

where a appears as a factor n times.

▶ **EXAMPLE 1**   Evaluate the following.

(a) $8^3 = 8 \cdot 8 \cdot 8 = 512$
Read 8^3 as "8 cubed."
(b) $5^2 = 5 \cdot 5 = 25$
Read 5^2 as "5 squared."
(c) $4^5 = 4 \cdot 4 \cdot 4 \cdot 4 \cdot 4 = 1024$
Read 4^5 as "4 to the fifth power."
(d) $\left(\dfrac{3}{4}\right)^2 = \dfrac{3}{4} \cdot \dfrac{3}{4} = \dfrac{9}{16}$ ◀ **1**

Caution   A common error in using exponents occurs with expressions such as $4 \cdot 3^2$. The exponent of 2 applies only to the base 3, so that

$$4 \cdot 3^2 = 4 \cdot 3 \cdot 3 = 36.$$

On the other hand,

$$(4 \cdot 3)^2 = (4 \cdot 3)(4 \cdot 3) = 12 \cdot 12 = 144,$$

and so

$$4 \cdot 3^2 \neq (4 \cdot 3)^2.$$

Be careful, too, to distinguish between expressions like -2^4 and $(-2)^4$.

$$-2^4 = -(2^4) = -(2 \cdot 2 \cdot 2 \cdot 2) = -16$$
$$(-2)^4 = (-2)(-2)(-2)(-2) = 16$$

and so

$$-2^4 \neq (-2)^4. \quad \textbf{2}$$

3 Simplify the following.

(a) $5^3 \cdot 5^6$

(b) $(-3)^4 \cdot (-3)^{10}$

(c) $(5p)^2 \cdot (5p)^8$

Answers:

(a) 5^9

(b) $(-3)^{14}$

(c) $(5p)^{10}$

By the definition of an exponent,

$$3^4 \cdot 3^2 = (3 \cdot 3 \cdot 3 \cdot 3)(3 \cdot 3) = 3^6.$$

This suggests the following property for the product of two powers of a number.

If m and n are natural numbers and a is a real number, then
$$a^m \cdot a^n = a^{m+n}.$$

▶**EXAMPLE 2**   Simplify the following.

(a) $7^4 \cdot 7^6 = 7^{4+6} = 7^{10}$

(b) $(-2)^3 \cdot (-2)^5 = (-2)^{3+5} = (-2)^8$

(c) $(3k)^2 \cdot (3k)^3 = (3k)^5$

(d) $(m + n)^2 \cdot (m + n)^5 = (m + n)^7$ ◀ **3**

POLYNOMIALS   A polynomial is an algebraic expression like

$$5x^4 + 2x^3 + 6x, \qquad 8m^3 + 9m^2 - 6m + 3, \qquad 10p, \quad \text{or} \quad -9.$$

More formally, a **polynomial** in one variable is an expression of the form

$$a_n x^n + a_{n-1} x^{n-1} + \cdots + a_1 x + a_0,$$

where $a_0, a_1, a_2, \ldots,$ and a_n are real numbers, n is a whole number, and $a_n \neq 0$. However, 0 is considered a polynomial, called the **zero polynomial.** For example, the polynomial

$$8x^3 + 9x^2 - 6x + 3$$

is of the form

$$a_n x^n + a_{n-1} x^{n-1} + \cdots + a_1 x + a_0,$$

with $n = 3$, $a_n = a_3 = 8$, $a_{n-1} = a_2 = 9$, $a_1 = -6$, and $a_0 = 3$. Expressions that are *not* polynomials include

$$8x^3 + \frac{6}{x}, \qquad \frac{9 + x}{2 - x}, \quad \text{and} \quad \frac{-p^2 + 5p + 3}{2p - 1}.$$

A nonzero expression involving only multiplication (or division), such as $9p^4$, is called a **term.** The number 9 is called the **coefficient,** p is the **variable,** and 4 is the **exponent.** (Exponents are discussed in more detail later in this chapter.) The **degree of a nonzero term** with only one variable is the exponent on the variable. For example, the term $9p^4$ has degree 4. The **degree of a polynomial** is the highest degree of any of its nonzero terms. Thus, the degree of $-p^2 + 5p + 3$ is 2. No degree is assigned to the zero polynomial because it has no nonzero terms.

4 Add or subtract.

(a) $(-2x^2 + 7x + 9)$
$+ (3x^2 + 2x - 7)$

(b) $(4x + 6) - (13x - 9)$

(c) $(9x^3 - 8x^2 + 2x)$
$-(9x^3 - 2x^2 - 10)$

Answers:

(a) $x^2 + 9x + 2$

(b) $-9x + 15$

(c) $-6x^2 + 2x + 10$

ADDITION AND SUBTRACTION   Two terms having the same variable and the same exponent are called **like terms;** other terms are called **unlike terms.** Polynomials can be added or subtracted by using the distributive property to combine like terms. Only like terms can be combined. For example,

$$12y^4 + 6y^4 = (12 + 6)y^4 = 18y^4$$

and

$$-2m^2 + 8m^2 = (-2 + 8)m^2 = 6m^2.$$

The polynomial $8y^4 + 2y^5$ has unlike terms, so it cannot be further simplified. Polynomials are subtracted using the fact that $a - b = a + (-b)$. The next example shows how to add and subtract polynomials by combining terms.

▶ **EXAMPLE 3**   Add or subtract as indicated.

(a) $(8x^3 - 4x^2 + 6x) + (3x^3 + 5x^2 - 9x + 8)$
　　Combine like terms.

$(8x^3 - 4x^2 + 6x) + (3x^3 + 5x^2 - 9x + 8)$ 　　　　　Commutative
$= (8x^3 + 3x^3) + (-4x^2 + 5x^2) + (6x - 9x) + 8$　　property and
　　　　　　　　　　　　　　　　　　　　　　　　　associative
　　　　　　　　　　　　　　　　　　　　　　　　　property

$= 11x^3 + x^2 - 3x + 8$ 　　　　　　　　　　Distributive
　　　　　　　　　　　　　　　　　　　　property

(b) $(-4x^4 + 6x^3 - 9x^2 - 12) + (-3x^3 + 8x^2 - 11x + 7)$
$= -4x^4 + 3x^3 - x^2 - 11x - 5$

(c) $(2x^2 - 11x + 8) - (7x^2 - 6x + 2)$
　　Use the definition of subtraction: $a - b = a + (-b)$. Here a and b are polynomials, and $-b$ is

$$-(7x^2 - 6x + 2) = -7x^2 + 6x - 2.$$

Now perform the subtraction.

$$(2x^2 - 11x + 8) - (7x^2 - 6x + 2)$$
$$= (2x^2 - 11x + 8) + (-7x^2 + 6x - 2)$$
$$= -5x^2 - 5x + 6 \quad ◀ \ \boxed{4}$$

MULTIPLICATION   The distributive property is also used to multiply polynomials. For example, the product of $8x$ and $6x - 4$ is found as follows.

$$8x(6x - 4) = 8x(6x) - 8x(4) \quad \text{**Distributive property**}$$
$$= 48x^2 - 32x \quad\quad x \cdot x = x^2$$

5 Find the following products.

(a) $-6r(2r - 5)$

(b) $(8m + 3)(m^4 - 2m^2 + 6m)$

(c) $(5k - 1)(2k + 3)$

(d) $(7z - 3)(2z + 5)$

Answers:

(a) $-12r^2 + 30r$

(b) $8m^5 + 3m^4 - 16m^3$
     $+ 42m^2 + 18m$

(c) $10k^2 + 13k - 3$

(d) $14z^2 + 29z - 15$

6 Write an expression for profit if revenue is $7x^2 - 3x + 8$ and cost is $3x^2 + 5x - 2$.

Answer:
$4x^2 - 8x + 10$

The product of $3p - 2$ and $5p + 1$ can be found by using the distributive property twice.

$$\begin{aligned}
(3p - 2)(5p + 1) &= (3p - 2)(5p) + (3p - 2)(1) \\
&= 3p(5p) - 2(5p) + 3p(1) - 2(1) \\
&= 15p^2 - 10p + 3p - 2 \\
&= 15p^2 - 7p - 2
\end{aligned}$$

A shortcut for finding the product of two binomials like $(3p - 2)(5p + 1)$ is called **FOIL** for First, Outer, Inner, Last. For example,

$$\begin{aligned}
& \quad\quad\;\; \mathbf{F} \quad\quad\;\; \mathbf{O} \quad\quad\;\; \mathbf{I} \quad\quad\;\; \mathbf{L} \\
(5m + 6)(m - 4) &= 5m(m) + 5m(-4) + 6(m) + 6(-4) \\
&= 5m^2 - 20m + 6m - 24 \\
&= 5m^2 - 14m - 24.
\end{aligned}$$

With practice you should be able to add the outer and inner products mentally and write the three terms of the product without the intermediate steps.

▶ **EXAMPLE 4**   Find each product.

(a) $\begin{aligned}[t] 2p^3(3p^2 - 2p + 5) &= 2p^3(3p^2) + 2p^3(-2p) + 2p^3(5) \\ &= 6p^5 - 4p^4 + 10p^3 \end{aligned}$

(b) $\begin{aligned}[t] (3k - 2)(k^2 + 5k - 4) &= 3k(k^2 + 5k - 4) - 2(k^2 + 5k - 4) \\ &= 3k^3 + 15k^2 - 12k - 2k^2 - 10k + 8 \\ &= 3k^3 + 13k^2 - 22k + 8 \end{aligned}$

(c) $(6t + 5)(5t - 1) = 30t^2 + 19t - 5$

Use FOIL. Mentally combine the outer and inner products to get

$$-6t + 25t = 19t. \quad ◀ \quad \boxed{5}$$

The profit from the sales of an item is the difference between the revenue R received from the sales and the cost C to sell the item. Thus, profit P is found from the equation $P = R - C$.

▶ **EXAMPLE 5**   Suppose the cost to sell x compact discs is $2x^2 - 2x + 10$ and the revenue from the sale of x discs is $5x^2 + 12x - 1$. Write an expression for the profit.

Because profit equals revenue less cost, or $P = R - C$, the profit is

$$\begin{aligned}
P &= (5x^2 + 12x - 1) - (2x^2 - 2x + 10) \\
&= 3x^2 + 14x - 11. \quad ◀ \quad \boxed{6}
\end{aligned}$$

1.5 EXERCISES

Evaluate each of the following. (See Example 1.)

1. 3^4

2. 5^3

3. $\left(\dfrac{5}{3}\right)^3$

4. $\left(\dfrac{3}{4}\right)^4$

5. -6^2

6. -3^4

7. $(-3)^4$

8. $(-6)^3$

9. $7 \cdot 4^3$

10. $8 \cdot 3^4$

11. $-2 \cdot 4^5$

12. $-5 \cdot 2^5$

13. Explain how the value of -3^2 differs from $(-3)^2$. Do -3^3 and $(-3)^3$ differ in the same way? Why or why not?

14. Describe the steps used to multiply 4^3 and 4^5. Is the product of 4^3 and 3^4 found in the same way? Explain.

Simplify each of the following. Leave answers with exponents. (See Example 2.)

15. $2^4 \cdot 2^3$

16. $3^8 \cdot 3^3$

17. $(-5)^2 \cdot (-5)^5$

18. $(-4)^4 \cdot (-4)^6$

19. $(-3)^5 \cdot 3^4$

20. $8^2 \cdot (-8)^3$

21. $(2z)^5 \cdot (2z)^6$

22. $(6y)^3 \cdot (6y)^5$

Add or subtract as indicated. (See Example 3.)

23. $(3x^3 + 2x^2 - 5x) + (-4x^3 - x^2 + 8x)$

24. $(-2p^3 - 5p + 7) + (-4p^2 + 8p + 2)$

25. $(-4y^2 - 3y + 8) - (2y^2 - 6y - 2)$

26. $(7b^2 + 2b - 5) - (3b^2 + 2b - 6)$

27. $(2x^3 - 2x^2 + 4x - 3) - (2x^3 + 8x^2 - 1)$

28. $(3y^3 + 9y^2 - 11y + 8) - (-4y^2 + 10y - 6)$

Find each of the following products. (See Example 4.)

29. $-9m(2m^2 + 3m - 1)$

30. $2a(4a^2 - 6a + 3)$

31. $(3z + 5)(4z^2 - 2z + 1)$

32. $(2k + 3)(4k^3 - 3k^2 + k)$

33. $(6k - 1)(2k - 3)$

34. $(8r + 3)(r - 1)$

35. $(3y + 5)(2y - 1)$

36. $(5r - 3s)(5r + 4s)$

37. $(9k + q)(2k - q)$

38. $(.012x - .17)(.3x + .54)$

39. $(6.2m - 3.4)(.7m + 1.3)$

40. $2p - 3[4p - (3p + 1)]$

41. $5k - [k + (-3 + 5k)]$

42. $(3x - 1)(x + 2) - (2x + 5)^2$

43. $(4x + 3)(2x - 1) - (x + 3)^2$

44. Suppose one polynomial has degree 3 and another also has degree 3. Find all possible values for the degree of their (a) sum (b) difference (c) product.

45. If one polynomial has degree 3 and another has degree 4, find all possible values for the degree of their (a) sum (b) difference (c) product.

46. Generalize the results of Exercise 45: Suppose one polynomial has degree m and another has degree n, where m and n are natural numbers with $n < m$. Find all possible values for the degree of their (a) sum (b) difference (c) product.

47. Find $(3x + 2y)(3x - 2y)$. Then state in words a formula for the result of the product $(a + b)(a - b)$.

48. Find $(2x - 5)^2$. Then state in words a formula for finding $(a - b)^2$.

Write an expression for profit given the following expressions for revenue and cost. (See Example 5.)

49. Revenue: $5x^3 - 3x + 1$; cost: $4x^2 + 5x$

50. Revenue: $3x^3 + 2x^2$; cost: $x^3 - x^2 + x + 10$

51. Revenue: $2x^2 - 4x + 50$; cost: $x^2 + 3x + 10$

52. Revenue: $10x^2 + 8x + 12$; cost: $2x^2 - 3x + 20$

1 Factor out the greatest common factor.

(a) $12r + 9k$

(b) $75m^2 + 100n^2$

(c) $6m^4 - 9m^3 + 12m^2$

(d) $3(2k + 1)^3 + 4(2k + 1)^4$

Answers:

(a) $3(4r + 3k)$

(b) $25(3m^2 + 4n^2)$

(c) $3m^2(2m^2 - 3m + 4)$

(d) $(2k + 1)^3(7 + 8k)$

1.6 FACTORING

The reverse of polynomial multiplication, where a polynomial is written as a product of other polynomials, is called **factoring.** For example, one way to factor the number 18 is to write it as the product $9 \cdot 2$. Since $18 = 9 \cdot 2$, both 9 and 2 are called **factors** of 18. The integer factors of 18 are 2, 9, -2, -9, 6, 3, -6, -3, 18, 1, -18, -1.

The distributive property is used to factor polynomials. It is customary to restrict factoring to *integer* coefficients in order to avoid an infinite number of possible factors. We follow that convention in this book.

GREATEST COMMON FACTOR The algebraic expression $15m + 45$ is made up of two terms, $15m$ and 45. Each of these terms can be divided by 15. In fact, $15m = 15 \cdot m$ and $45 = 15 \cdot 3$. By the distributive property,

$$15m + 45 = 15 \cdot m + 15 \cdot 3 = 15(m + 3).$$

Both 15 and $m + 3$ are factors of $15m + 45$. Since 15 divides into all terms of $15m + 45$ and is the largest number that will do so, it is called the **greatest common factor** for the polynomial $15m + 45$. The process of writing $15m + 45$ as $15(m + 3)$ is called **factoring out** the greatest common factor.

▶**EXAMPLE 1** Factor out the greatest common factor.

(a) $12p - 18q$

Both $12p$ and $18q$ are divisible by 6, and

$$12p - 18q = 6 \cdot 2p - 6 \cdot 3q = 6(2p - 3q).$$

(b) $8x^3 - 9x^2 + 15x$

Each of these terms is divisible by x.

$$8x^3 - 9x^2 + 15x = (8x^2) \cdot x - (9x) \cdot x + 15 \cdot x$$
$$= x(8x^2 - 9x + 15)$$

(c) $5(4x - 3)^3 + 2(4x - 3)^2$

The quantity $(4x - 3)^2$ is a common factor. Factoring it out gives

$$5(4x - 3)^3 + 2(4x - 3)^2 = (4x - 3)^2[5(4x - 3) + 2]$$
$$= (4x - 3)^2(20x - 15 + 2)$$
$$= (4x - 3)^2(20x - 13). \quad ◀ \quad \boxed{1}$$

2 Factor the following.

(a) $r^2 - 5r - 14$

(b) $3m^2 + 5m - 2$

(c) $6p^2 + 13pq - 5q^2$

Answers:

(a) $(r - 7)(r + 2)$

(b) $(3m - 1)(m + 2)$

(c) $(2p + 5q)(3p - q)$

FACTORING TRINOMIALS   Factoring is the opposite of multiplication. Because the product of two binomials is usually a trinomial, we can expect factorable trinomials (that have terms with no common factor) to have two binomial factors. Thus, factoring trinomials requires using FOIL backwards.

▶**EXAMPLE 2**   Factor each trinomial.

(a) $4y^2 - 11y + 6$

To factor this trinomial, we must find integers a, b, c, and d such that

$$4y^2 - 11y + 6 = (ay + b)(cy + d).$$

By using FOIL, we see that $ac = 4$ and $bd = 6$. The positive factors of 4 are 4 and 1 or 2 and 2. Since the middle term is negative, we consider only negative factors of 6. The possibilities are -2 and -3 or -1 and -6. Now we try various arrangements of these factors until we find one that gives the correct coefficient of y.

$$(2y - 1)(2y - 6) = 4y^2 - 14y + 6 \quad \text{Incorrect}$$
$$(2y - 2)(2y - 3) = 4y^2 - 10y + 6 \quad \text{Incorrect}$$
$$(y - 2)(4y - 3) = 4y^2 - 11y + 6 \quad \text{Correct}$$

The last trial gives the correct factorization.

(b) $6p^2 - 7pq - 5q^2$

Again, we try various possibilities. The positive factors of 6 could be 2 and 3 or 1 and 6. As factors of -5 we have only -1 and 5 or -5 and 1. Try different combinations of these factors until the correct one is found.

$$(2p - 5q)(3p + q) = 6p^2 - 13pq - 5q^2 \quad \text{Incorrect}$$
$$(3p - 5q)(2p + q) = 6p^2 - 7pq - 5q^2 \quad \text{Correct}$$

Finally, $6p^2 - 7pq - 5q^2$ factors as $(3p - 5q)(2p + q)$. ◀ **2**

Note   In Example 2, we chose positive factors of the positive first term. Of course, we could have used two negative factors, but the work is easier if positive factors are used.

The method shown above can be used to factor a **perfect square trinomial,** one that is the square of a binomial. The binomial can be predicted by observing the following patterns that always apply to a perfect square trinomial.

$$\left.\begin{array}{l} x^2 + 2xy + y^2 = (x + y)^2 \\ x^2 - 2xy + y^2 = (x - y)^2 \end{array}\right\} \quad \textbf{Perfect square trinomials}$$

3 Factor each trinomial.

(a) $4m^2 + 4m + 1$

(b) $25z^2 - 80zt + 64t^2$

Answers:

(a) $(2m + 1)^2$

(b) $(5z - 8t)^2$

▶ **EXAMPLE 3** Factor each trinomial.

(a) $16p^2 - 40pq + 25q^2$

Because $16p^2 = (4p)^2$ and $25q^2 = (5q)^2$, use the second pattern shown above with $4p$ replacing x and $5q$ replacing y to get

$$16p^2 - 40pq + 25q^2 = (4p)^2 - 2(4p)(5q) + (5q)^2$$
$$= (4p - 5q)^2.$$

Make sure that the middle term of the trinomial being factored, $-40pq$ here, is twice the product of the two terms in the binomial $4p - 5q$.

$$-40pq = 2(4p)(-5q)$$

(b) $169x^2 + 104xy^2 + 16y^4 = (13x + 4y^2)^2$, since $2(13x)(4y^2) = 104xy^2$.

◀ **3**

FACTORING BINOMIALS Three special factoring patterns are listed below. Each can be verified by multiplying on the right side of the equation. These formulas should be memorized.

$x^2 - y^2 = (x + y)(x - y)$	**Difference of two squares**
$x^3 - y^3 = (x - y)(x^2 + xy + y^2)$	**Difference of two cubes**
$x^3 + y^3 = (x + y)(x^2 - xy + y^2)$	**Sum of two cubes**

▶ **EXAMPLE 4** Factor each of the following.

(a) $4m^2 - 9$

Notice that $4m^2 - 9$ is the difference of two squares, since $4m^2 = (2m)^2$ and $9 = 3^2$. Use the pattern for the difference of two squares, letting $2m$ replace x and 3 replace y. Then the pattern $x^2 - y^2 = (x + y)(x - y)$ becomes

$$4m^2 - 9 = (2m)^2 - 3^2$$
$$= (2m + 3)(2m - 3).$$

(b) $128p^2 - 98q^2$

First factor out the common factor of 2.

$$128p^2 - 98q^2 = 2(64p^2 - 49q^2)$$
$$= 2[(8p)^2 - (7q)^2]$$
$$= 2(8p + 7q)(8p - 7q)$$

(c) $x^2 + 36$

The *sum* of two squares usually cannot be factored. To see this, check some possibilities.

4 Factor the following.

(a) $9p^2 - 49$

(b) $y^2 + 100$

(c) $9r^2 + 12r + 4 - t^2$

(d) $81x^4 - 16y^4$

Answers:

(a) $(3p + 7)(3p - 7)$

(b) Cannot be factored

(c) $(3r + 2 + t)(3r + 2 - t)$

(d) $(9x^2 + 4y^2)(3x + 2y) \cdot$
$(3x - 2y)$

5 Factor the following.

(a) $a^3 + 1000$

(b) $z^3 - 64$

(c) $100m^3 - 27z^3$

Answers:

(a) $(a + 10)(a^2 - 10a + 100)$

(b) $(z - 4)(z^2 + 4z + 16)$

(c) $(10m - 3z)(100m^2 + 30mz + 9z^2)$

$$(x + 6)(x + 6) = (x + 6)^2 = x^2 + 12x + 36$$
$$(x + 4)(x + 9) = x^2 + 13x + 36$$

Any product of two binomials will always have a middle term unless it is the *difference* of two squares.

(d) $4z^2 + 12z + 9 - w^2$

Notice that the first three terms can be factored as a perfect square.

$$4z^2 + 12z + 9 - w^2 = (2z + 3)^2 - w^2$$

Written in this form, the expression is the difference of squares, which can be factored as

$$(2z + 3)^2 - w^2 = [(2z + 3) + w][(2z + 3) - w]$$
$$= (2z + 3 + w)(2z + 3 - w).$$

(e) $256k^4 - 625m^4$

Use the difference of two squares pattern twice, as follows:

$$256k^4 - 625m^4 = (16k^2)^2 - (25m^2)^2$$
$$= (16k^2 + 25m^2)(16k^2 - 25m^2)$$
$$= (16k^2 + 25m^2)(4k + 5m)(4k - 5m). \quad \blacktriangleleft \quad \boxed{4}$$

▶**EXAMPLE 5**   Factor each of the following.

(a) $k^3 - 8$

Use the pattern for the difference of two cubes, since $k^3 = (k)^3$ and $8 = (2)^3$, to get

$$k^3 - 8 = k^3 - 2^3 = (k - 2)(k^2 + 2k + 4).$$

(b) $m^3 + 125 = m^3 + 5^3 = (m + 5)(m^2 - 5m + 25)$

(c) $8k^3 - 27z^3 = (2k)^3 - (3z)^3 = (2k - 3z)(4k^2 + 6kz + 9z^2) \quad \blacktriangleleft \quad \boxed{5}$

▶**EXAMPLE 6**   Factor each of the following.

(a) $12x^2 - 26x - 10$

Look first for a common factor. Here there is a common factor of 2: $12x^2 - 26x - 10 = 2(6x^2 - 13x - 5)$. Now try to factor $6x^2 - 13x - 5$. Possible factors of 6 are 3 and 2 or 6 and 1. The only factors of -5 are -5 and 1 or 5 and -1. Try various combinations. You should find the trinomial factors as $(3x + 1)(2x - 5)$. Thus,

$$12x^2 - 26x - 10 = 2(3x + 1)(2x - 5).$$

(b) $16a^2 - 100 - 48ac + 36c^2$

Factor out the common factor of 4 first.

6 Factor.

(a) $6x^2 - 27x - 15$

(b) $18 - 8xy - 2y^2 - 8x^2$

Answers:

(a) $3(2x + 1)(x - 5)$

(b) $2(3 - 2x - y) \cdot$
$(3 + 2x + y)$

$$16a^2 - 100 - 48ac + 36c^2 = 4[4a^2 - 25 - 12ac + 9c^2]$$
$$= 4[(4a^2 - 12ac + 9c^2) - 25] \quad \text{Rearrange terms and group.}$$
$$= 4[(2a - 3c)^2 - 25] \quad \text{Factor the trinomial.}$$
$$= 4(2a - 3c + 5)(2a - 3c - 5) \quad \text{Factor the difference of squares.}$$

◀

Caution Remember always to look first for a common factor. **6**

FACTORING BY GROUPING When a polynomial has more than three terms, it can sometimes be factored by a method called **factoring by grouping.** For example, to factor

$$ax + ay + 6x + 6y,$$

collect the terms into two groups so that each group has a common factor.

$$ax + ay + 6x + 6y = (ax + ay) + (6x + 6y)$$

Factor each group, getting

$$ax + ay + 6x + 6y = a(x + y) + 6(x + y).$$

The quantity $(x + y)$ is now a common factor, which can be factored out, producing

$$ax + ay + 6x + 6y = (x + y)(a + 6).$$

It is not always obvious which terms should be grouped. Experience and repeated trials are the most reliable tools for factoring by grouping.

▶**EXAMPLE 7** Factor by grouping.

(a) $mp^2 + 7m + 3p^2 + 21$

Group the terms as follows.

$$mp^2 + 7m + 3p^2 + 21 = (mp^2 + 7m) + (3p^2 + 21)$$

Factor out the greatest common factor from each group.

$$(mp^2 + 7m) + (3p^2 + 21) = m(p^2 + 7) + 3(p^2 + 7)$$
$$= (p^2 + 7)(m + 3) \quad p^2 + 7 \text{ is a common factor.}$$

(b) $2y^2 - 2z - ay^2 + az$

Grouping terms as above gives

$$2y^2 - 2z - ay^2 + az = (2y^2 - 2z) + (-ay^2 + az)$$
$$= 2(y^2 - z) + a(-y^2 + z).$$

7 Factor by grouping.

(a) $a^3 - a^2b^2 + ab - b^3$

(b) $6x - 8xy - 3y + 4y^2$

Answers:

(a) $(a - b^2)(a^2 + b)$

(b) $(3 - 4y)(2x - y)$

We need a common factor in the parentheses. Notice that the expression $-y^2 + z$ is the negative of $y^2 - z$, so the terms should be grouped as follows.

$$2y^2 - 2z - ay^2 + az = (2y^2 - 2z) - (ay^2 - az)$$
$$= 2(y^2 - z) - a(y^2 - z) \qquad \text{Factor each group}$$
$$= (y^2 - z)(2 - a). \qquad \text{Factor out } y^2 - z.$$

◀ **7**

1.6 EXERCISES

Factor out the greatest common factor in each of the following. (See Example 1.)

1. $12x^2 - 24x$

2. $5y - 25xy$

3. $r^3 - 5r^2 + r$

4. $t^3 + 3t^2 + 8t$

5. $2m - 5n + p$

6. $4k + 6h - 5c$

7. $6z^3 - 12z^2 + 18z$

8. $5x^3 + 35x^2 + 10x$

9. $25p^4 - 20p^3q + 100p^2q^2$

10. $60m^4 - 120m^3n + 50m^2n^2$

11. $3(2y - 1)^2 + 5(2y - 1)^3$

12. $(3x + 7)^5 - 2(3x + 7)^3$

13. $3(x + 5)^4 + (x + 5)^6$

14. $3(x + 6)^2 + 2(x + 6)^4$

Factor each of the following completely. Factor out the greatest common factor as necessary. (See Examples 2–4 and 6–7.)

15. $2a^2 + 3a - 5$

16. $6a^2 - 48a - 120$

17. $x^2 - 64$

18. $x^2 + 17xy + 72y^2$

19. $9p^2 - 24p + 16$

20. $3r^2 - r - 2$

21. $r^2 - 3rt - 10t^2$

22. $ax - ay + bx - by$

23. $m^2 - 6mn + 9n^2$

24. $8k^2 - 16k - 10$

25. $4p^2 - 9$

26. $8r^2 + r + 6$

27. $3x^2 - 24xz + 48z^2$

28. $9m^2 - 25$

29. $a^2 + 4ab + 5b^2$

30. $6y^2 - 11y - 7$

31. $3x - 3y - ax + ay$

32. $4y^2 + y - 3$

33. $3a^2 - 13a - 30$

34. $3k^2 + 2k - 8$

35. $21m^2 + 13mn + 2n^2$

36. $81y^2 - 100$

37. $20y^2 + 39yx - 11x^2$

38. $12s^2 + 11st - 5t^2$

39. $64z^2 + 25$

40. $p^2q^2 - 10 - 2q^2 + 5p^2$

41. $y^2 - 4yz - 21z^2$

42. $49a^2 + 9$

43. $3n^2 - 4m^2 + m^2n^2 - 12$

44. $y^2 + 20yx + 100x^2$

45. $121x^2 - 64$

46. $4z^2 + 56zy + 196y^2$

47. $24a^4 + 10a^3b - 4a^2b^2$

48. $10x^2 + 34x + 12$

49. $18x^5 + 15x^4z - 75x^3z^2$

50. $16m^2 + 40m + 25$

51. $4z^2 + 4z + 1 - t^2$

52. $y^2 + 10y + 25 - z^2$

53. $m^2 - n^2 + 2n - 1$

54. $a^2 - 4b^2 - 4b - 1$

55. $5m^3(m^3 - 1)^2 - 3m^5(m^3 - 1)^3$

56. $9(x - 4)^5 - (x - 4)^3$

57. When asked to factor $6x^4 - 3x^2 - 3$ completely, a student gave the following result: $6x^4 - 3x^2 - 3 = (2x^2 + 1)(3x^2 - 3)$. Is this answer correct? Explain why.

58. When can the sum of two squares be factored? Give examples.

Factor each of the following. (See Example 5.)

59. $a^3 - 216$ **60.** $b^3 + 125$ **61.** $8r^3 - 27s^3$ **62.** $1000p^3 + 27q^3$

63. $64m^3 + 125$ **64.** $216y^3 - 343$ **65.** $1000y^3 - z^3$ **66.** $125p^3 + 8q^3$

67. Explain why $(x + 2)^3$ is not the correct factorization of $x^3 + 8$ and give the correct factorization.

68. Describe how factoring and multiplication are related. Give examples.

---

1 What values of the variable make each denominator equal 0?

(a) $\dfrac{5}{x - 3}$

(b) $\dfrac{2x - 3}{4x - 1}$

(c) $\dfrac{x + 2}{x}$

(d) Why do we need to determine these values?

Answers:

(a) 3

(b) 1/4

(c) 0

(d) Because division by 0 is undefined.

1.7 RATIONAL EXPRESSIONS

Later chapters of the book contain work with algebraic fractions. Examples of these fractions, called **rational expressions,** include

$$\frac{8}{x - 1}, \quad \frac{3x^2 + 4x}{5x - 6}, \quad \text{and} \quad \frac{2 + \dfrac{1}{y}}{y}.$$

Because rational expressions involve quotients, it is important to keep in mind values of the variables that make denominators 0. For example, 1 cannot be used as a replacement for x in the first rational expression above, and 6/5 cannot be used in the second one, since these values make the respective denominators equal 0. **1**

OPERATIONS WITH RATIONAL EXPRESSIONS The rules for operations with rational expressions are the usual rules for fractions.

Operations with Rational Expressions

For all mathematical expressions P, $Q \neq 0$, R, and $S \neq 0$.

(a) $\dfrac{P}{Q} = \dfrac{PS}{QS}$ **Fundamental property**

(b) $\dfrac{P}{Q} \cdot \dfrac{R}{S} = \dfrac{PR}{QS}$ **Multiplication**

(c) $\dfrac{P}{Q} + \dfrac{R}{Q} = \dfrac{P + R}{Q}$ **Addition**

(d) $\dfrac{P}{Q} - \dfrac{R}{Q} = \dfrac{P - R}{Q}$ **Subtraction**

(e) $\dfrac{P}{Q} \div \dfrac{R}{S} = \dfrac{P}{Q} \cdot \dfrac{S}{R}, R \neq 0.$ **Division**

2 Write each of the following in lowest terms.

(a) $\dfrac{12k + 36}{18}$

(b) $\dfrac{15m + 30m^2}{5m}$

(c) $\dfrac{2p^2 + 3p + 1}{p^2 + 3p + 2}$

Answers:

(a) $\dfrac{2(k + 3)}{3}$ or $\dfrac{2k + 6}{3}$

(b) $3(1 + 2m)$ or $3 + 6m$

(c) $\dfrac{2p + 1}{p + 2}$

The following examples illustrate these operations.

▶**EXAMPLE 1** Write each of the following rational expressions in lowest terms (so that the numerator and denominator have no common factor with integer coefficients except 1 or -1).

(a) $\dfrac{12m}{-18}$

Both $12m$ and -18 are divisible by 6. By operation (a) above,

$$\frac{12m}{-18} = \frac{2m \cdot 6}{-3 \cdot 6}$$

$$= \frac{2m}{-3}$$

$$= -\frac{2m}{3}.$$

(b) $\dfrac{8x + 16}{4} = \dfrac{8(x + 2)}{4} = \dfrac{4 \cdot 2(x + 2)}{4} = \dfrac{2(x + 2)}{1} = 2(x + 2).$

The numerator, $8x + 16$, was factored so that the common factor could be identified. The answer could also be written as $2x + 4$, if desired.

(c) $\dfrac{k^2 + 7k + 12}{k^2 + 2k - 3} = \dfrac{(k + 4)(k + 3)}{(k - 1)(k + 3)} = \dfrac{k + 4}{k - 1}$ ◀ **2**

The values of k in Example 1(c) are restricted to $k \neq 1$ and $k \neq -3$. From now on, such restrictions will be assumed when working with rational expressions.

▶**EXAMPLE 2**

(a) Multiply $\dfrac{2}{3} \cdot \dfrac{y}{5}$.

Use operation (b) above; multiply the numerators and then the denominators.

$$\frac{2}{3} \cdot \frac{y}{5} = \frac{2 \cdot y}{3 \cdot 5} = \frac{2y}{15}$$

The result, $2y/15$, is in lowest terms.

(b) $\dfrac{3y + 9}{6} \cdot \dfrac{18}{5y + 15}$

Factor where possible.

$$\frac{3y + 9}{6} \cdot \frac{18}{5y + 15} = \frac{3(y + 3)}{6} \cdot \frac{18}{5(y + 3)}$$

$$= \frac{3 \cdot 18(y + 3)}{6 \cdot 5(y + 3)} \qquad \textbf{Multiply numerators and denominators}$$

3 Multiply.

(a) $\dfrac{3r^2}{5} \cdot \dfrac{20}{9r}$

(b) $\dfrac{y-4}{y^2-2y-8} \cdot \dfrac{y^2-4}{3y}$

Answers:

(a) $\dfrac{4r}{3}$

(b) $\dfrac{y-2}{3y}$

4 Divide.

(a) $\dfrac{5m}{16} \div \dfrac{m^2}{10}$

(b) $\dfrac{2y-8}{6} \div \dfrac{5y-20}{3}$

(c) $\dfrac{m^2-2m-3}{m(m+1)} \div \dfrac{m+4}{5m}$

Answers:

(a) $\dfrac{25}{8m}$

(b) $\dfrac{1}{5}$

(c) $\dfrac{5(m-3)}{m+4}$

$$= \frac{3 \cdot 6 \cdot 3(y+3)}{6 \cdot 5(y+3)} \qquad 18 = 6 \cdot 3$$

$$= \frac{3 \cdot 3}{5} \qquad \text{Write in lowest terms}$$

$$= \frac{9}{5}$$

(c) $\dfrac{m^2+5m+6}{m+3} \cdot \dfrac{m^2+m-6}{m^2+3m+2}$

$$= \frac{(m+2)(m+3)}{m+3} \cdot \frac{(m-2)(m+3)}{(m+2)(m+1)} \qquad \text{Factor}$$

$$= \frac{(m+2)(m+3)(m-2)(m+3)}{(m+3)(m+2)(m+1)} \qquad \text{Multiply}$$

$$= \frac{(m-2)(m+3)}{m+1} \qquad \text{Lowest terms}$$

$$= \frac{m^2+m-6}{m+1} \quad \blacktriangleleft \; \boxed{3}$$

▶ EXAMPLE 3

(a) Divide $\dfrac{8x}{5} \div \dfrac{11x^2}{20}$.

As shown in operation (e) above, invert the second expression and multiply.

$$\frac{8x}{5} \div \frac{11x^2}{20} = \frac{8x}{5} \cdot \frac{20}{11x^2} \qquad \text{Invert and multiply}$$

$$= \frac{8x \cdot 20}{5 \cdot 11x^2} \qquad \text{Multiply}$$

$$= \frac{32}{11x} \qquad \text{Lowest terms}$$

(b) $\dfrac{9p-36}{12} \div \dfrac{5(p-4)}{18}$

$$= \frac{9p-36}{12} \cdot \frac{18}{5(p-4)} \qquad \text{Invert and multiply}$$

$$= \frac{9(p-4)}{12} \cdot \frac{18}{5(p-4)} \qquad \text{Factor}$$

$$= \frac{27}{10} \qquad \begin{array}{l}\text{Multiply and write} \\ \text{in lowest terms}\end{array} \quad \blacktriangleleft \; \boxed{4}$$

▶ EXAMPLE 4 Add or subtract as indicated.

(a) $\dfrac{4}{5k} - \dfrac{11}{5k}$

5 Add or subtract.

(a) $\dfrac{3}{4r} + \dfrac{8}{3r}$

(b) $\dfrac{1}{m - 2} - \dfrac{3}{2(m - 2)}$

(c) $\dfrac{p + 1}{p^2 - p} - \dfrac{p^2 - 1}{p^2 + p - 2}$

Answers:

(a) $\dfrac{41}{12r}$

(b) $\dfrac{-1}{2(m - 2)}$

(c) $\dfrac{-p^3 + p^2 + 4p + 2}{p(p - 1)(p + 2)}$

As operation (d) above shows, when two rational expressions have the same denominators, subtract by subtracting the numerators and keeping the common denominator.

$$\frac{4}{5k} - \frac{11}{5k} = \frac{4 - 11}{5k} = -\frac{7}{5k}$$

(b) $\dfrac{7}{p} + \dfrac{9}{2p} + \dfrac{1}{3p}$

These three denominators are different; operation (c) for addition requires the same denominators. Find a common denominator, one which can be divided by p, $2p$, and $3p$. A common denominator here is $6p$. Rewrite each rational expression, using operation (a), with a denominator of $6p$. Then, using operation (c), add the numerators and keep the common denominator.

$$\frac{7}{p} + \frac{9}{2p} + \frac{1}{3p} = \frac{6 \cdot 7}{6 \cdot p} + \frac{3 \cdot 9}{3 \cdot 2p} + \frac{2 \cdot 1}{2 \cdot 3p} \qquad \text{Operation (a)}$$

$$= \frac{42}{6p} + \frac{27}{6p} + \frac{2}{6p}$$

$$= \frac{42 + 27 + 2}{6p} \qquad \text{Operation (c)}$$

$$= \frac{71}{6p}$$

(c) $\dfrac{k^2}{k^2 - 1} - \dfrac{2k^2 - k - 3}{k^2 + 3k + 2}$

Factor the denominators to find a common denominator.

$$\frac{k^2}{k^2 - 1} - \frac{2k^2 - k - 3}{k^2 + 3k + 2} = \frac{k^2}{(k + 1)(k - 1)} - \frac{2k^2 - k - 3}{(k + 1)(k + 2)}$$

The common denominator is $(k + 1)(k - 1)(k + 2)$. Write each fraction with the common denominator.

$$\frac{k^2}{(k + 1)(k - 1)} - \frac{2k^2 - k - 3}{(k + 1)(k + 2)}$$

$$= \frac{k^2(k + 2)}{(k + 1)(k - 1)(k + 2)} - \frac{(2k^2 - k - 3)(k - 1)}{(k + 1)(k - 1)(k + 2)}$$

$$= \frac{k^3 + 2k^2 - (2k^2 - k - 3)(k - 1)}{(k + 1)(k - 1)(k + 2)} \qquad \text{Subtract fractions}$$

$$= \frac{k^3 + 2k^2 - (2k^3 - 3k^2 - 2k + 3)}{(k + 1)(k - 1)(k + 2)} \qquad \text{Multiply } (2k^2 - k - 3)(k - 1)$$

$$= \frac{k^3 + 2k^2 - 2k^3 + 3k^2 + 2k - 3}{(k + 1)(k - 1)(k + 2)} \qquad \text{Polynomial subtraction}$$

$$= \frac{-k^3 + 5k^2 + 2k - 3}{(k + 1)(k - 1)(k + 2)} \qquad \text{Combine terms} \quad ◀ \ \boxed{5}$$

6 Simplify each complex fraction.

(a) $\dfrac{t - \dfrac{1}{t}}{2t + \dfrac{3}{t}}$

(b) $\dfrac{\dfrac{m}{m+2} + \dfrac{1}{m}}{\dfrac{1}{m} - \dfrac{1}{m+2}}$

Answers:

(a) $\dfrac{t^2 - 1}{2t^2 + 3}$

(b) $\dfrac{m^2 + m + 2}{2}$

COMPLEX FRACTIONS   Any quotient of two rational expressions is called a **complex fraction.** Complex fractions can be simplified by the methods shown in the following examples.

▶**EXAMPLE 5**   Simplify each complex fraction.

(a) $\dfrac{6 - \dfrac{5}{k}}{1 + \dfrac{5}{k}}$

Multiply both numerator and denominator by the common denominator k.

$$\dfrac{6 - \dfrac{5}{k}}{1 + \dfrac{5}{k}} = \dfrac{k\left(6 - \dfrac{5}{k}\right)}{k\left(1 + \dfrac{5}{k}\right)} \qquad \text{Multiply by } \dfrac{k}{k}$$

$$\dfrac{k\left(6 - \dfrac{5}{k}\right)}{k\left(1 + \dfrac{5}{k}\right)} = \dfrac{6k - k\left(\dfrac{5}{k}\right)}{k + k\left(\dfrac{5}{k}\right)} \qquad \text{Distributive property}$$

$$= \dfrac{6k - 5}{k + 5} \qquad \text{Simplify}$$

(b) $\dfrac{\dfrac{a}{a+1} + \dfrac{1}{a}}{\dfrac{1}{a} + \dfrac{1}{a+1}}$

Multiply both numerator and denominator by the common denominator of all the fractions, in this case $a(a + 1)$. Doing so gives

$$\dfrac{\dfrac{a}{a+1} + \dfrac{1}{a}}{\dfrac{1}{a} + \dfrac{1}{a+1}} = \dfrac{\left(\dfrac{a}{a+1} + \dfrac{1}{a}\right)a(a+1)}{\left(\dfrac{1}{a} + \dfrac{1}{a+1}\right)a(a+1)} = \dfrac{a^2 + (a+1)}{(a+1) + a} = \dfrac{a^2 + a + 1}{2a + 1}.$$

As an alternative method of solution, first perform the indicated additions in the numerator and denominator, and then divide.

$$\dfrac{\dfrac{a}{a+1} + \dfrac{1}{a}}{\dfrac{1}{a} + \dfrac{1}{a+1}} = \dfrac{\dfrac{a^2 + 1(a+1)}{a(a+1)}}{\dfrac{1(a+1) + 1(a)}{a(a+1)}} = \dfrac{\dfrac{a^2 + a + 1}{a(a+1)}}{\dfrac{2a + 1}{a(a+1)}}$$

$$= \dfrac{a^2 + a + 1}{a(a+1)} \cdot \dfrac{a(a+1)}{2a + 1} = \dfrac{a^2 + a + 1}{2a + 1} \qquad ◀ \boxed{6}$$

1.7 EXERCISES

Write each of the following in lowest terms. Factor as necessary. (See Example 1.)

1. $\dfrac{8x^2}{40x}$

2. $\dfrac{27m}{81m^3}$

3. $\dfrac{20p^2}{35p^3}$

4. $\dfrac{18y^4}{27y^2}$

5. $\dfrac{5m+15}{4m+12}$

6. $\dfrac{10z+5}{20z+10}$

7. $\dfrac{4(w-3)}{(w-3)(w+3)}$

8. $\dfrac{-6(x+2)}{(x-4)(x+2)}$

9. $\dfrac{3y^2-12y}{9y^3}$

10. $\dfrac{15k^2+45k}{9k^2}$

11. $\dfrac{8x^2+16x}{4x^2}$

12. $\dfrac{36y^2+72y}{9y}$

13. $\dfrac{m^2-4m+4}{m^2+m-6}$

14. $\dfrac{r^2-r-6}{r^2+r-12}$

15. $\dfrac{x^2+3x-4}{x^2-1}$

16. $\dfrac{z^2-5z+6}{z^2-4}$

Multiply or divide as indicated in each of the following. Write all answers in lowest terms. (See Examples 2 and 3.)

17. $\dfrac{4p^3}{49} \cdot \dfrac{7}{2p^2}$

18. $\dfrac{24n^4}{6n^2} \cdot \dfrac{18n^2}{9n}$

19. $\dfrac{21a^5}{14a^3} \div \dfrac{8a}{12a^2}$

20. $\dfrac{2x^3}{6x^2} \div \dfrac{10x^2}{15x}$

21. $\dfrac{2a+b}{2c} \cdot \dfrac{15}{4(2a+b)}$

22. $\dfrac{4(x+2)}{w} \cdot \dfrac{3w}{8(x+2)}$

23. $\dfrac{15p-3}{6} \div \dfrac{10p-2}{3}$

24. $\dfrac{6m-18}{18} \cdot \dfrac{20}{4m-12}$

25. $\dfrac{2k+8}{6} \div \dfrac{3k+12}{2}$

26. $\dfrac{5m+25}{10} \cdot \dfrac{12}{6m+30}$

27. $\dfrac{9y-18}{6y+12} \cdot \dfrac{3y+6}{15y-30}$

28. $\dfrac{12r+24}{36r-36} \div \dfrac{6r+12}{8r-8}$

29. $\dfrac{4a+12}{2a-10} \div \dfrac{a^2-9}{a^2-a-20}$

30. $\dfrac{6r-18}{9r^2+6r-24} \cdot \dfrac{12r-16}{4r-12}$

31. $\dfrac{k^2-k-6}{k^2+k-12} \cdot \dfrac{k^2+3k-4}{k^2+2k-3}$

32. $\dfrac{n^2-n-6}{n^2-2n-8} \div \dfrac{n^2-9}{n^2+7n+12}$

33. In your own words, explain how to find the least common denominator for two fractions.

34. Describe the steps required to add three rational expressions. You may use an example to illustrate.

Add or subtract as indicated in each of the following. Write all answers in lowest terms. (See Example 4.)

35. $\dfrac{3}{5z} - \dfrac{2}{3z}$

36. $\dfrac{7}{4z} - \dfrac{5}{3z}$

37. $\dfrac{r+2}{3} - \dfrac{r-2}{3}$

38. $\dfrac{3y-1}{8} - \dfrac{3y+1}{8}$

39. $\dfrac{4}{x} + \dfrac{1}{3}$

40. $\dfrac{6}{r} - \dfrac{3}{4}$

41. $\dfrac{2}{y} - \dfrac{1}{4}$

42. $\dfrac{6}{11} + \dfrac{3}{a}$

43. $\dfrac{1}{6m} + \dfrac{2}{5m} + \dfrac{4}{m}$

44. $\dfrac{8}{3p} + \dfrac{5}{4p} + \dfrac{9}{2p}$

45. $\dfrac{1}{m-1} + \dfrac{2}{m}$

46. $\dfrac{8}{y+2} - \dfrac{3}{y}$

47. $\dfrac{8}{3(a-1)} + \dfrac{2}{a-1}$

48. $\dfrac{5}{2(k+3)} + \dfrac{2}{k+3}$

49. $\dfrac{2}{5(k-2)} + \dfrac{3}{4(k-2)}$

50. $\dfrac{11}{3(p+4)} - \dfrac{5}{6(p+4)}$

51. $\dfrac{2}{x^2-2x-3} + \dfrac{5}{x^2-x-6}$

52. $\dfrac{3}{m^2-3m-10} + \dfrac{5}{m^2-m-20}$

53. $\dfrac{2y}{y^2+7y+12} - \dfrac{y}{y^2+5y+6}$

54. $\dfrac{-r}{r^2-10r+16} - \dfrac{3r}{r^2+2r-8}$

55. $\dfrac{3k}{2k^2+3k-2} - \dfrac{2k}{2k^2-7k+3}$

56. $\dfrac{4m}{3m^2+7m-6} - \dfrac{m}{3m^2-14m+8}$

In each of the following exercises, simplify the complex fraction. (See Example 5.)

57. $\dfrac{1 + \dfrac{1}{x}}{1 - \dfrac{1}{x}}$

58. $\dfrac{2 - \dfrac{2}{y}}{2 + \dfrac{2}{y}}$

59. $\dfrac{\dfrac{1}{x+h} - \dfrac{1}{x}}{h}$

60. $\dfrac{\dfrac{1}{(x+h)^2} - \dfrac{1}{x^2}}{h}$

61. $\dfrac{1 + \dfrac{1}{1-b}}{1 - \dfrac{1}{1+b}}$

62. $\dfrac{m - \dfrac{1}{m^2-4}}{m+2}$

1.8 INTEGER EXPONENTS

In Section 1.5 we showed that

$$a^m \cdot a^n = a^{m+n}.$$

In this section, we will develop further definitions and properties of exponents. We begin with a rule for quotients. By definition,

$$\frac{6^5}{6^2} = \frac{6 \cdot 6 \cdot 6 \cdot 6 \cdot 6}{6 \cdot 6} = 6 \cdot 6 \cdot 6 = 6^3.$$

Because there are 5 factors of 6 in the numerator and 2 factors of 6 in the denominator, the quotient has $5 - 2 = 3$ factors of 6. In general, if $m > n$, then

$$\frac{a^m}{a^n} = a^{m-n}.$$

Using this rule,

$$\frac{3^5}{3^5} = 3^{5-5} = 3^0.$$

On the other hand, $3^5/3^5 = 1$. Therefore, it is reasonable to define $3^0 = 1$.

1 Evaluate the following.

(a) 17^0

(b) 30^0

(c) $(-10)^0$

(d) $-(12)^0$

Answers:

(a) 1

(b) 1

(c) 1

(d) -1

Zero Exponent

If a is any nonzero real number, then

$$a^0 = 1.$$

The symbol 0^0 is not defined.

▶ **EXAMPLE 1**   Evaluate the following.

(a) $6^0 = 1$

(b) $(-9)^0 = 1$

(c) $-(4^0) = -(1) = -1$ ◀ **1**

If m is less than n in the quotient rule given above, the result has a negative exponent. For example,

$$\frac{3^2}{3^4} = 3^{2-4} = 3^{-2}.$$

However,

$$\frac{3^2}{3^4} = \frac{3 \cdot 3}{3 \cdot 3 \cdot 3 \cdot 3} = \frac{1}{3^2},$$

which suggests that 3^{-2} should be defined to be $1/3^2$. Thus, we have the following definition of a negative exponent.

Negative Exponent

If n is a natural number, and if $a \neq 0$, then

$$a^{-n} = \frac{1}{a^n}.$$

▶ **EXAMPLE 2**   Evaluate the following.

(a) $3^{-2} = \frac{1}{3^2} = \frac{1}{9}$

(b) $5^{-4} = \frac{1}{5^4} = \frac{1}{625}$

(c) $9^{-1} = \frac{1}{9^1} = \frac{1}{9}$

(d) $-4^{-2} = -\frac{1}{4^2} = -\frac{1}{16}$

2 Evaluate the following.

(a) 6^{-2}

(b) -6^{-3}

(c) -3^{-4}

(d) $\left(\dfrac{5}{8}\right)^{-1}$

(e) $\left(\dfrac{1}{2}\right)^{-4}$

Answers:

(a) $1/36$

(b) $-1/216$

(c) $-1/81$

(d) $8/5$

(e) 16

3 Evaluate the following.

(a) $\left(\dfrac{7}{3}\right)^{-2}$

(b) $\left(\dfrac{5}{9}\right)^{-3}$

Answers:

(a) $9/49$

(b) $729/125$

(e) $\left(\dfrac{3}{4}\right)^{-1} = \dfrac{1}{\left(\dfrac{3}{4}\right)^{1}} = \dfrac{1}{\dfrac{3}{4}} = \dfrac{4}{3}$

(f) $\left(\dfrac{2}{3}\right)^{-3} = \dfrac{1}{\left(\dfrac{2}{3}\right)^{3}} = \dfrac{1}{\left(\dfrac{2^{3}}{3^{3}}\right)} = 1 \cdot \dfrac{3^{3}}{2^{3}} = \dfrac{3^{3}}{2^{3}} = \dfrac{27}{8}$ ◀ **2**

Parts (e) and (f) of Example 2 involve work with fractions that can lead to error. For a useful shortcut with such fractions, use the properties of division of rational numbers and the definition of a negative exponent to get

$$\left(\frac{a}{b}\right)^{-n} = \frac{1}{\left(\dfrac{a}{b}\right)^{n}} = \frac{1}{\left(\dfrac{a^{n}}{b^{n}}\right)} = 1 \cdot \frac{b^{n}}{a^{n}} = \frac{b^{n}}{a^{n}} = \left(\frac{b}{a}\right)^{n}.$$

For all positive integers n and nonzero real numbers a and b,

$$\left(\frac{a}{b}\right)^{-n} = \left(\frac{b}{a}\right)^{n}.$$

▶**EXAMPLE 3** Evaluate each of the following.

(a) $\left(\dfrac{1}{3}\right)^{-2} = \left(\dfrac{3}{1}\right)^{2} = 9$

(b) $\left(\dfrac{3}{4}\right)^{-3} = \left(\dfrac{4}{3}\right)^{3} = \dfrac{64}{27}$ ◀ **3**

The definitions and properties discussed above are summarized below, together with three power rules that follow from the definition of an exponent. These rules and definitions should be memorized.

Definitions and Properties of Exponents

For any integers m and n, and any real numbers a and b for which the following exist,

(a) $a^{m} \cdot a^{n} = a^{m+n}$ **Product property**

(b) $\dfrac{a^{m}}{a^{n}} = a^{m-n}$ **Quotient property**

(c) $(a^{m})^{n} = a^{mn}$

(d) $(ab)^{m} = a^{m} \cdot b^{m}$ $\Big\}$ **Power properties**

(e) $\left(\dfrac{a}{b}\right)^{m} = \dfrac{a^{m}}{b^{m}}$

4 Simplify the following.

(a) $9^6 \cdot 9^{-4}$ **(b)** $\dfrac{8^7}{8^{-3}}$

(c) $\dfrac{14^9}{14^{12}}$ **(d)** $(13^4)^{-3}$

(e) $(6y)^4$ **(f)** $\left(\dfrac{3}{4}\right)^{-2}$

Answers:

(a) 9^2 **(b)** 8^{10}

(c) $1/14^3$ **(d)** $1/13^{12}$

(e) 6^4y^4 **(f)** $4^2/3^2$

5 Simplify the following. Give answers with only positive exponents. Assume all variables represent nonzero real numbers.

(a) $\dfrac{(t^{-1})^2}{t^{-5}}$ **(b)** $\dfrac{(3z)^{-1}z^4}{z^2}$

(c) $(p^{-5}q)^{-1}$

(d) $\dfrac{(3a)^{-2}(b^2c^3)^4}{(ab)^{-3}c^4}$

Answers:

(a) t^3 **(b)** $z/3$

(c) p^5/q **(d)** $\dfrac{ab^{11}c^8}{3^2}$

6 Simplify. Give answers with positive exponents.

(a) $\dfrac{3^{-5} \cdot 3^{-2}}{3^{-6} \cdot 3^3}$

(b) $4^{-1} + 2^{-2}$

(c) $\dfrac{3^{-1} + 2^{-2}}{2^{-2}}$

Answers:

(a) $1/3^4$ **(b)** $1/2$

(c) $7/3$

(f) $a^0 = 1$

(g) $a^{-n} = \dfrac{1}{a^n}$

(h) $\left(\dfrac{a}{b}\right)^{-n} = \left(\dfrac{b}{a}\right)^n.$

▶ **EXAMPLE 4** Use the properties of exponents to simplify each of the following. Write answers with positive exponents.

(a) $7^{-4} \cdot 7^6 = 7^2$ **Property (a)**

(b) $\dfrac{9^{14}}{9^{-6}} = 9^{20}$ **Property (b)**

(c) $(2m^3)^4 = 2^4 \cdot (m^3)^4 = 2^4m^{12}$ **Properties (c) and (d)**

(d) $(3x)^4 = 3^4x^4$ **Property (d)**

(e) $\left(\dfrac{9}{7}\right)^{-6} = \left(\dfrac{7}{9}\right)^6 = \dfrac{7^6}{9^6}$ **Properties (h) and (e)** ◀ **4**

▶ **EXAMPLE 5** Use the definition of negative exponents and the properties of exponents to simplify each of the following expressions. Give answers with only positive exponents. Assume all variables represent nonzero real numbers.

(a) $\dfrac{(m^3)^{-2}}{m^4} = \dfrac{m^{-6}}{m^4} = m^{-6-4} = m^{-10} = \dfrac{1}{m^{10}}$

(b) $\dfrac{(2y)^3y^{-1}}{y^2} = \dfrac{2^3y^3y^{-1}}{y^2} = \dfrac{2^3y^2}{y^2} = 2^3$

(c) $(x^2y^{-3})^4 = (x^2)^4(y^{-3})^4 = x^8y^{-12} = \dfrac{x^8}{y^{12}}$

(d) $\dfrac{(2y)^{-3}(y^2z)^3}{(yz)^{-2}z^3} = \dfrac{2^{-3}y^{-3}y^6z^3}{y^{-2}z^{-2}z^3} = \dfrac{2^{-3}y^3z^3}{y^{-2}z} = 2^{-3}y^5z^2 = \dfrac{y^5z^2}{2^3}$ ◀ **5**

▶ **EXAMPLE 6** Simplify. Write the result with positive exponents.

(a) $\dfrac{2^{-3} \cdot 2^{-1}}{2^4 \cdot 2^{-7}} = \dfrac{2^{-4}}{2^{-3}} = 2^{-4-(-3)} = 2^{-1} = \dfrac{1}{2}$

(b) $2^{-1} + 3^{-1} = \dfrac{1}{2} + \dfrac{1}{3} = \dfrac{5}{6}$

(c) $\dfrac{2^{-1} - 3^{-2}}{2^{-1}} = \dfrac{\dfrac{1}{2} - \dfrac{1}{3^2}}{\dfrac{1}{2}} = \dfrac{\dfrac{1}{2} - \dfrac{1}{9}}{\dfrac{1}{2}} = \dfrac{\dfrac{9}{18} - \dfrac{2}{18}}{\dfrac{1}{2}} = \dfrac{7}{18} \cdot \dfrac{2}{1} = \dfrac{7}{9}$

◀ **6**

1.8 EXERCISES

Evaluate each of the following. Write all answers without exponents. (See Examples 1–3.)

1. 5^0 **2.** 8^0 **3.** 6^{-1} **4.** 10^{-3} **5.** 2^{-5} **6.** 5^{-2} **7.** -4^{-3}

8. -7^{-4} **9.** $(7.94)^{-3}$ **10.** $(12.5)^{-2}$ **11.** $\left(\frac{1}{3}\right)^{-2}$ **12.** $\left(\frac{1}{6}\right)^{-3}$ **13.** $\left(\frac{2}{5}\right)^{-4}$ **14.** $\left(\frac{4}{3}\right)^{-2}$

15. Explain why $-2^{-4} = -1/16$, but $(-2)^{-4} = 1/16$.

16. Explain the reason a negative exponent is defined as a reciprocal: $a^{-n} = 1/a^n$.

Simplify each of the following. Write all answers using only positive exponents. (See Examples 4 and 5.)

17. $\dfrac{4^{-2}}{4^3}$ **18.** $\dfrac{6^{-3}}{6}$ **19.** $\dfrac{9^{-4}}{9^{-3}}$ **20.** $\dfrac{7^{-5}}{7^{-8}}$ **21.** $4^{-3} \cdot 4^6$ **22.** $5^{-9} \cdot 5^{10}$ **23.** $7^{-5} \cdot 7^{-2}$

24. $9^{-1} \cdot 9^{-3}$ **25.** $\dfrac{8^9 \cdot 8^{-7}}{8^{-3}}$ **26.** $\dfrac{5^{-4} \cdot 5^6}{5^{-1}}$ **27.** $\dfrac{10^8 \cdot 10^{-10}}{10^4 \cdot 10^2}$ **28.** $\dfrac{2^{-4} \cdot 2^{-3}}{2^6 \cdot 2^{-5}}$ **29.** $\left(\dfrac{5^{-6} \cdot 5^3}{5^{-2}}\right)^{-1}$ **30.** $\left(\dfrac{8^{-3} \cdot 8^4}{8^{-2}}\right)^{-2}$

Simplify each of the following. Assume all variables represent positive real numbers. Write answers with only positive exponents. (See Examples 4 and 5.)

31. $\dfrac{z^5 \cdot z^2}{z^4}$ **32.** $\dfrac{k^6 \cdot k^9}{k^{12}}$ **33.** $\dfrac{2^{-1}(p^{-1})^3}{2p^{-4}}$ **34.** $\dfrac{(5x^3)^{-2}}{x^4}$

35. $(q^{-5}r^2)^{-1}$ **36.** $(2y^2z^{-2})^{-3}$ **37.** $(2p^{-1})^3 \cdot (5p^2)^{-2}$ **38.** $(5^{-1}m^2)^{-3} \cdot (3m^{-2})^4$

39. $\dfrac{7^{-1} \cdot 7r^{-3}}{7^2 \cdot (r^{-2})^2}$ **40.** $\dfrac{12^3 \cdot 12^5 \cdot y^{-2}}{12^{-1} \cdot (y^{-3})^{-2}}$ **41.** $\dfrac{6k^{-4} \cdot (3k^{-1})^{-2}}{2^3 \cdot k^2}$ **42.** $\dfrac{8p^{-3} \cdot (4p^2)^{-2}}{p^{-5}}$

43. $\dfrac{(2x)^{-2}(x^{-1}y)^{-3}}{(xy)^{-2}y^2}$ **44.** $\dfrac{(3z^{-1})^2(z^2w)^{-2}}{(2z^{-1}w^{-2})^3}$ **45.** $(m^4p^{-2})^2 \cdot (m^{-1}p^2)$ **46.** $(z^3x^5)^{-1} \cdot (z^{-4}x^{-2})$

Evaluate each expression, letting $a = 2$, $b = -3$, and $c = 0$. (See Example 6.)

47. $a^3 + b$ **48.** $b^3 + a^2$ **49.** $-b^2 + 3(c + 5)$ **50.** $2a^2 - b^4$

51. $a^7b^9c^5$ **52.** $12a^{15}b^6c^8$ **53.** $a^b + b^a$ **54.** $b^c + a^c$

55. $a^{-1} + b^{-1}$ **56.** $b^{-2} - a^{-1}$ **57.** $a^{-b} + b^{-c}$ **58.** $a^{2b} + b^b$

Perform all indicated operations and write answers with positive exponents. Assume all variables represent positive real numbers. (See Example 6.)

59. $2^{-1} - 3^{-1}$ **60.** $4^{-1} + 5^{-1}$ **61.** $(6^{-1} + 2^{-1})^{-1}$ **62.** $(5^{-1} - 10^{-1})^{-1}$

63. $\dfrac{3^{-2} - 4^{-1}}{4^{-1}}$ **64.** $\dfrac{6^{-1} + 5^{-2}}{6^{-1}}$ **65.** $\dfrac{a^{-1} + b^{-1}}{(ab)^{-1}}$ **66.** $\dfrac{p^{-1} - q^{-1}}{(pq)^{-2}}$

1 Evaluate the following.

(a) $16^{1/2}$

(b) $16^{1/4}$

(c) $-256^{1/2}$

(d) $(-256)^{1/2}$

(e) $-8^{1/3}$

(f) $243^{1/5}$

Answers:

(a) 4

(b) 2

(c) -16

(d) Not a real number

(e) -2

(f) 3

1.9 RATIONAL EXPONENTS AND RADICALS

In this section the definition of a^n will be extended to include rational values of n, such as $1/2$ and $5/4$, and not just integer values. In order to do this, we need some terminology.

There are two numbers whose square is 16, 4 and -4. The positive one, 4, is called the **square root** (or second root) of 16.* Similarly, there are two numbers whose fourth power is 16, 2 and -2. We call 2 the **fourth root** of 16. This suggests the following generalization.

If n is even, the **nth root of a** is the positive real number whose nth power is a.

All nonnegative numbers have nth roots for every natural number n, but *no negative number has a real, even nth root.* For example there is no real number whose square is -16, so -16 has no square root.

We say that the **cube root** (or third root) of 8 is 2 because $2^3 = 8$. Similarly, since $(-2)^3 = -8$, we say -2 is the cube root of -8. Again, we can generalize.

If n is odd, the **nth root of a** is the real number whose nth power is a.

Every real number has an nth root for every *odd* natural number n.

RATIONAL EXPONENTS   We can now define rational exponents. If they are to have the same properties as integer exponents, we want $a^{1/2}$ to be a number such that

$$(a^{1/2})^2 = a^{1/2} \cdot a^{1/2} = a^{1/2+1/2} = a^1 = a.$$

Thus, $a^{1/2}$ should be a number whose square is a and it is reasonable to *define* $a^{1/2}$ to be the square root of a (if it exists). Similarly, $a^{1/3}$ is defined to be the cube root of a, and $a^{1/n}$ is defined to be the nth root of a (if it exists).

▶**EXAMPLE 1**   Evaluate the following roots.

(a) $36^{1/2} = 6$ because $6^2 = 36$

(b) $-100^{1/2} = -10$

(c) $-(225^{1/2}) = -15$

(d) $625^{1/4} = 5$ because $5^4 = 625$

(e) $(-1296)^{1/4}$ is not a real number, but $-1296^{1/4} = -6$ because $6^4 = 1296$

(f) $(-27)^{1/3} = -3$

(g) $-32^{1/5} = -2$ ◀ **1**

---

*Sometimes the positive square root is called the *principal* square root.

2 Evaluate the following.

(a) $16^{3/4}$

(b) $25^{5/2}$

(c) $32^{7/5}$

(d) $100^{3/2}$

Answers:

(a) 8

(b) 3125

(c) 128

(d) 1000

For more general rational exponents, the symbol $a^{m/n}$ should be defined so that the properties for exponents still hold. For the first power property to hold,

$$(a^{1/n})^m \text{ must equal } a^{m/n}.$$

This suggests the following definition.

For all integers m and all positive integers n, and for all real numbers a for which $a^{1/n}$ is a real number,

$$a^{m/n} = (a^{1/n})^m.$$

▶ **EXAMPLE 2**   Evaluate the following.

(a) $27^{2/3} = (27^{1/3})^2$

$= 3^2 = 9$

(b) $32^{2/5} = (32^{1/5})^2$

$= 2^2 = 4$

(c) $64^{4/3} = (64^{1/3})^4$

$= 4^4 = 256$

(d) $25^{3/2} = (25^{1/2})^3$

$= 5^3 = 125$ ◀ **2**

It can be shown that all the properties given earlier for integer exponents also apply to rational exponents.

▶ **EXAMPLE 3**   Simplify the following. Assume all variables represent positive real numbers.

(a) $\dfrac{27^{1/3} \cdot 27^{5/3}}{27^3} = \dfrac{27^{1/3+5/3}}{27^3}$ **Product property**

$= \dfrac{27^2}{27^3} = 27^{2-3}$ **Quotient property**

$= 27^{-1} = \dfrac{1}{27}$ **Definition of negative exponent**

(b) $81^{5/4} \cdot 4^{-3/2} = (81^{1/4})^5 \cdot (4^{1/2})^{-3}$ **Definition of $a^{m/n}$**

$= 3^5 \cdot 2^{-3}$ **Definition of $a^{1/n}$**

$= \dfrac{3^5}{2^3}$ or $\dfrac{243}{8}$ **Definition of negative exponent**

(c) $6y^{2/3} \cdot 2y^{1/2} = 12y^{2/3+1/2} = 12y^{7/6}$

(d) $\left(\dfrac{3m^{5/6}}{y^{3/4}}\right)^2 \cdot \left(\dfrac{8y^{-3}}{m^6}\right)^{2/3} = \dfrac{3^2 m^{5/3}}{y^{3/2}} \cdot \dfrac{8^{2/3}y^{-2}}{m^4}$ **Power properties**

$= \dfrac{9m^{5/3}}{y^{3/2}} \cdot \dfrac{4}{m^4 y^2}$ **Definition of a^{-n}**

$= \dfrac{36m^{(5/3)-4}}{y^{(3/2)+2}}$ **Product and quotient properties**

3 Simplify. Assume all variables represent positive real numbers.

(a) $6^{2/5} \cdot 6^{3/5}$

(b) $\dfrac{8^{2/3} \cdot 8^{-4/3}}{8^2}$

(c) $3x^{1/4} \cdot 5x^{5/4}$

(d) $\left(\dfrac{2k^{1/3}}{p^{5/4}}\right)^2 \cdot \left(\dfrac{4k^{-2}}{p^5}\right)^{3/2}$

(e) $a^{5/8}(2a^{3/8} + a^{-1/8})$

Answers:

(a) 6

(b) $1/8^{8/3}$ or $1/2^8$

(c) $15x^{3/2}$

(d) $32/(p^{10}k^{7/3})$

(e) $2a + a^{1/2}$

4 Simplify.

(a) $\sqrt[3]{27}$

(b) $\sqrt[4]{625}$

(c) $\sqrt[6]{64}$

(d) $\sqrt[3]{\dfrac{64}{125}}$

Answers:

(a) 3

(b) 5

(c) 2

(d) 4/5

$$= \frac{36m^{-7/3}}{y^{7/2}} = \frac{36}{m^{7/3}y^{7/2}} \qquad \textbf{Definition of } a^{-n}$$

(e) $m^{2/3}(m^{7/3} + 2m^{1/3}) = (m^{2/3+7/3} + 2m^{2/3+1/3}) = m^3 + 2m$ ◄ **3**

Most roots of integers are irrational numbers that can be approximated with a calculator. A calculator with a square root key, $\boxed{\sqrt{x}}$, will give not only square roots, but by pressing it twice, fourth roots; by pressing it a third time, eighth roots; and so on (because $(x^{1/2})^{1/2} = x^{1/4}$, the fourth root of x, and $[(x^{1/2})^{1/2}]^{1/2} = x^{1/8}$, the eighth root of x).

Most calculators have a key labeled $\boxed{y^x}$ (or $\boxed{x^y}$ or $\boxed{\wedge}$). This key allows you to find any (appropriate) number to any power. For example, to evaluate $81^{5/4}$, enter 81, press y^x, and enter 1.25 (for 5/4). You should get 243. However, the calculator may not always give the value of an expression in the most useful form. In Example 3(b) we found that $81^{5/4} \cdot 4^{-3/2} = 3^5/2^3$ or 243/8. The calculator gives it as 30.375, which may not be as useful. Some calculators give an error message if you try to calculate a negative number to an odd power (such as $(-8)^{1/3}$). In that case, use a positive base and change the sign of the answer.

RADICALS   The nth root of a was denoted above as $a^{1/n}$. An alternative notation for nth roots uses **radicals.**

If n is an even natural number and $a \geq 0$, or n is an odd natural number,

$$a^{1/n} = \sqrt[n]{a}.$$

The symbol $\sqrt{}$ is a **radical sign,** the number a is the **radicand,** and n is the **index** of the radical. The familiar symbol $\sqrt{a}$ is used instead of $\sqrt[2]{a}$.

▶**EXAMPLE 4**   Simplify the following.

(a) $\sqrt[4]{16} = 16^{1/4} = 2$

(b) $\sqrt[5]{-32} = -2$

(c) $\sqrt[3]{1000} = 10$

(d) $\sqrt[6]{\dfrac{64}{729}} = \dfrac{2}{3}$ ◄ **4**

The symbol $a^{m/n}$ also can be written in an alternative notation using radicals.

For all rational numbers m/n and all real numbers a for which $\sqrt[n]{a}$ exists,

$$a^{m/n} = \left(\sqrt[n]{a}\right)^m \quad \text{or} \quad a^{m/n} = \sqrt[n]{a^m}.$$

5 Simplify.

(a) $\sqrt{3} \cdot \sqrt{27}$

(b) $\sqrt{\dfrac{3}{49}}$

(c) $\sqrt[5]{\sqrt[4]{3}}$

Answers:

(a) 9

(b) $\dfrac{\sqrt{3}}{7}$

(c) $\sqrt[20]{3}$

Notice that $\sqrt[n]{x^n}$ cannot be written simply as x when n is even. For example, if $x = -5$,

$$\sqrt{x^2} = \sqrt{(-5)^2} = \sqrt{25} = 5 \neq x.$$

However, $|-5| = 5$, so that $\sqrt{x^2} = |x|$ when x is -5. This is true in general.

For any real number a and any natural number n,

$$\sqrt[n]{a^n} = |a| \text{ if } n \text{ is even}$$

and

$$\sqrt[n]{a^n} = a \text{ if } n \text{ is odd.}$$

To avoid this difficulty that $\sqrt[n]{a^n}$ is not necessarily equal to a, we shall assume that all variables in radicands represent only nonnegative numbers, as they usually do in applications.

The properties of exponents can be written with radicals as shown below.

For all real numbers a and b and positive integers m and n for which all indicated roots exist,

(a) $\sqrt[n]{a} \cdot \sqrt[n]{b} = \sqrt[n]{ab}$

(b) $\dfrac{\sqrt[n]{a}}{\sqrt[n]{b}} = \sqrt[n]{\dfrac{a}{b}} \quad (b \neq 0)$

(c) $\sqrt[m]{\sqrt[n]{a}} = \sqrt[mn]{a}.$

To see why property (c) is true, write the expression using exponents.

$$\sqrt[m]{\sqrt[n]{a}} = (a^{1/n})^{1/m} = a^{(1/n)(1/m)} = a^{1/(mn)} = \sqrt[mn]{a}$$

▶ **EXAMPLE 5** Simplify the following.

(a) $\sqrt{6} \cdot \sqrt{54} = \sqrt{6 \cdot 54} = \sqrt{324} = 18$

Alternatively, simplify $\sqrt{54}$ first.

$$\sqrt{6} \cdot \sqrt{54} = \sqrt{6} \cdot \sqrt{9 \cdot 6}$$
$$= \sqrt{6} \cdot 3\sqrt{6} = 3 \cdot 6 = 18$$

(b) $\sqrt{\dfrac{7}{64}} = \dfrac{\sqrt{7}}{\sqrt{64}} = \dfrac{\sqrt{7}}{8}$

(c) $\sqrt[7]{\sqrt[3]{2}} = \sqrt[21]{2}$ ◀ **5**

SIMPLIFYING RADICALS When working with numbers, it is customary to write them in their simplest form. For example, $10/2$ is written as 5, $-9/6$ is

6 Simplify.

(a) $\sqrt{80}$

(b) $\sqrt[3]{32m^4p^6q^2}$

Answers:

(a) $4\sqrt{5}$

(b) $2mp^2\sqrt[3]{4mq^2}$

7 Simplify. Assume all variables represent positive real numbers.

(a) $5\sqrt[3]{4x^2z} - 3\sqrt[3]{4x^2z}$

(b) $\sqrt{27m^3n} + 2m\sqrt{12mn}$

Answers:

(a) $2\sqrt[3]{4x^2z}$

(b) $7m\sqrt{3mn}$

written as $-3/2$, and $4/20$ is written as $1/5$. Expressions with radicals also should be written in their simplest form. The *simplified form* of a radical is defined as follows.

Simplified Radicals

An expression with radicals is said to be simplified when all the following conditions are satisfied.

1. The radicand has no factors with powers greater than or equal to the index of the radical.
2. The index of the radical and the power of the radicand have no common factor except 1.
3. All denominators are rational numbers.
4. All indicated operations have been performed (if possible).

▶ **EXAMPLE 6**   Simplify each of the following. Assume all variables represent positive real numbers.

(a) $\sqrt{175} = \sqrt{25 \cdot 7} = \sqrt{25} \cdot \sqrt{7} = 5\sqrt{7}$

(b) $\sqrt[3]{81x^5y^7z^6} = \sqrt[3]{27 \cdot 3 \cdot x^3 \cdot x^2 \cdot y^3 \cdot y^3 \cdot y \cdot z^3 \cdot z^3}$

$\qquad = \sqrt[3]{(3^3x^3y^3y^3z^3z^3)(3x^2y)}$

$\qquad = 3xyyzz\sqrt[3]{3x^2y} = 3xy^2z^2\sqrt[3]{3x^2y}$ ◀ **6**

Radicals with the same radicand and the same index, such as $3\sqrt[4]{11pq}$ and $-7\sqrt[4]{11pq}$, are called **like radicals.** Only like radicals can be added or subtracted. As shown in part (b) of the next example, it is sometimes necessary to simplify radicals before adding or subtracting. As before, assume that all variables represent only positive numbers.

▶ **EXAMPLE 7**   Simplify the following.

(a) $3\sqrt[4]{11pq} - 7\sqrt[4]{11pq} = (3 - 7)\sqrt[4]{11pq} = -4\sqrt[4]{11pq}$

(b) $\sqrt{98x^3y} + 3x\sqrt{32xy}$

First remove all perfect square factors from under the radical. Then use the distributive property, as follows.

$\sqrt{98x^3y} + 3x\sqrt{32xy} = \sqrt{49 \cdot 2 \cdot x^2 \cdot x \cdot y} + 3x\sqrt{16 \cdot 2 \cdot x \cdot y}$

$\qquad = 7x\sqrt{2xy} + (3x)(4)\sqrt{2xy}$

$\qquad = 7x\sqrt{2xy} + 12x\sqrt{2xy} = 19x\sqrt{2xy}$ ◀ **7**

Multiplying radical expressions is much like multiplying polynomials.

▶ **EXAMPLE 8**   Multiply the following.

(a) $(\sqrt{2} + 3)(\sqrt{8} - 5) = \sqrt{2}(\sqrt{8}) - \sqrt{2}(5) + 3\sqrt{8} - 3(5)$ **FOIL**

$\qquad = \sqrt{16} - 5\sqrt{2} + 3(2\sqrt{2}) - 15$

8 Multiply

(a) $(\sqrt{5} - \sqrt{2})(3 + \sqrt{2})$

(b) $(\sqrt{3} + \sqrt{7})(\sqrt{3} - \sqrt{7})$

Answers:

(a) $3\sqrt{5} + \sqrt{10} - 3\sqrt{2} - 2$

(b) -4

9 Rationalize the denominator.

(a) $\dfrac{2}{\sqrt{5}}$

(b) $\sqrt{\dfrac{3}{2}}$

Answers:

(a) $\dfrac{2\sqrt{5}}{5}$

(b) $\dfrac{\sqrt{6}}{2}$

$$= 4 - 5\sqrt{2} + 6\sqrt{2} - 15$$
$$= -11 + \sqrt{2}$$

(b) $(\sqrt{7} - \sqrt{10})(\sqrt{7} + \sqrt{10}) = (\sqrt{7})^2 - (\sqrt{10})^2$
$$= 7 - 10 = -3 \quad \blacktriangleleft \quad \boxed{8}$$

RATIONALIZING THE DENOMINATOR Before calculators were easily available, it was useful to **rationalize denominators** (write denominators with no radicals) because it was easy to calculate results like $\sqrt{2}/2 = 1.414/2 = .707$ mentally, but more difficult to compute $1/\sqrt{2} = 1/1.414$ without pencil and paper. However, there are other good reasons to rationalize denominators (and sometimes numerators). Some examples will occur in Chapter 11 of this text. The process of rationalizing the denominator is explained in the next examples.

▶ **EXAMPLE 9** Rationalize each denominator.

(a) $\dfrac{4}{\sqrt{3}}$

To rationalize the denominator, multiply by 1 in the form $\sqrt{3}/\sqrt{3}$ so that the denominator is $\sqrt{3} \cdot \sqrt{3} = 3$, a rational number.

$$\frac{4}{\sqrt{3}} \cdot \frac{\sqrt{3}}{\sqrt{3}} = \frac{4\sqrt{3}}{3}$$

(b) $\sqrt{\dfrac{3}{5}}$

$$\sqrt{\frac{3}{5}} = \frac{\sqrt{3}}{\sqrt{5}} \cdot \frac{\sqrt{5}}{\sqrt{5}} = \frac{\sqrt{15}}{5} \quad \blacktriangleleft \quad \boxed{9}$$

▶ **EXAMPLE 10** Rationalize each denominator.

(a) $\dfrac{1}{1 - \sqrt{2}} \cdot$

A useful approach here is to multiply both the numerator and denominator by the **conjugate*** of the denominator, in this case $1 + \sqrt{2}$. As suggested by Example 8(b), the product $(1 - \sqrt{2})(1 + \sqrt{2})$ is rational.

$$\frac{1}{1 - \sqrt{2}} = \frac{1(1 + \sqrt{2})}{(1 - \sqrt{2})(1 + \sqrt{2})} = \frac{1 + \sqrt{2}}{1 - 2}$$
$$= \frac{1 + \sqrt{2}}{-1} = -1 - \sqrt{2}$$

---

* The conjugate of $a\sqrt{m} + b\sqrt{n}$ is $a\sqrt{m} - b\sqrt{n}$.

10 Rationalize the denominator.

(a) $\dfrac{4}{1 + \sqrt{3}}$

(b) $\dfrac{\sqrt{5}}{\sqrt{x} - \sqrt{5}}$, $x \neq 5$

Answers:

(a) $-2 + 2\sqrt{3}$

(b) $\dfrac{\sqrt{5x} + 5}{x - 5}$

(b) $\dfrac{\sqrt{2}}{\sqrt{2} + \sqrt{y}}$, $y \neq 2$

$$\frac{\sqrt{2}}{\sqrt{2} + \sqrt{y}} = \frac{\sqrt{2}}{\sqrt{2} + \sqrt{y}} \cdot \frac{\sqrt{2} - \sqrt{y}}{\sqrt{2} - \sqrt{y}}$$

$$= \frac{2 - \sqrt{2y}}{2 - y} \quad \blacktriangleleft \;\; \boxed{10}$$

1.9 EXERCISES

Evaluate each of the following. Write all answers without exponents. Write decimal answers to the nearest tenth. (See Examples 1–2.)

1. $49^{1/2}$ **2.** $8^{1/3}$ **3.** $(7.51)^{1/4}$ **4.** $(68.93)^{1/5}$ **5.** $27^{2/3}$ **6.** $24^{3/2}$

7. $(947)^{2/5}$ **8.** $(58.1)^{3/4}$ **9.** $-64^{2/3}$ **10.** $-64^{3/2}$ **11.** $(9/25)^{1/2}$ **12.** $(2401/16)^{1/4}$

13. $81^{-3/4}$ **14.** $16^{-5/4}$ **15.** $(16/9)^{-3/2}$ **16.** $(8/27)^{-4/3}$ **17.** $\left(\dfrac{27}{64}\right)^{-1/3}$ **18.** $\left(\dfrac{121}{100}\right)^{-3/2}$

Simplify each of the following. Write all answers using only positive exponents. Assume all variables represent positive real numbers. (See Example 3.)

19. $5^{1/3} \cdot 5^{5/3}$ **20.** $18^{4/7} \cdot 18^{3/7}$ **21.** $8^{2/3} \cdot 8^{-1/3}$ **22.** $12^{-3/4} \cdot 12^{1/4}$

23. $\dfrac{16^{1/5} \cdot 16^{-2}}{16^{-2/5} \cdot 16^{4/5}}$ **24.** $\dfrac{2^{-4/3} \cdot 2}{2^{5/3} \cdot 2^{-2/3}}$ **25.** $\dfrac{9^{-5/3}}{9^{2/3} \cdot 9^{-1/5}}$ **26.** $\dfrac{3^{5/3} \cdot 3^{-3/4}}{3^{-1/4}}$

27. $(2p)^{1/2} \cdot (2p^3)^{1/3}$ **28.** $(5k^2)^{3/2} \cdot (5k^{1/3})^{3/4}$ **29.** $\dfrac{(mn)^{1/5}(m^2n)^{-2/5}}{m^{1/3}n}$ **30.** $\dfrac{(r^2s)^{3/2}(rs^2)^{1/4}}{r^3s}$

31. $p^{2/3}(2p^{1/3} + 5p)$ **32.** $2z^{1/2}(3z^{-1/2} + z^{1/2})$ **33.** $(m + 2)^{1/2}[4(m + 2)^{-1/2} - 3(m + 2)^{3/2}]$

34. $(k - 1)^{-1/2}[2(k - 1)^{3/2} + (k - 1)^{1/2}]$

35. Explain how to rationalize the denominator of $\sqrt[5]{3/2}$.   **36.** Describe the steps required to multiply $\sqrt[5]{2}$ and $\sqrt{2}$.

Simplify each of the following. Assume that all variables represent positive real numbers.
(See Examples 4–6 and 9.)

37. $\sqrt[3]{64}$ **38.** $\sqrt[4]{625}$ **39.** $\sqrt[5]{-243}$ **40.** $\sqrt[7]{-128}$ **41.** $\sqrt{48}$ **42.** $\sqrt{72}$

43. $-\sqrt[4]{162}$ **44.** $-\sqrt[4]{1296}$ **45.** $-\sqrt{\dfrac{16}{3}}$ **46.** $-\sqrt{\dfrac{5}{27}}$ **47.** $-\sqrt[3]{\dfrac{3}{2}}$ **48.** $-\sqrt[3]{\dfrac{4}{5}}$

49. $\sqrt{8z^5x^8}$ **50.** $\sqrt{24m^6n^5}$ **51.** $\sqrt{m^2n^7p^8}$ **52.** $\sqrt{81p^5x^2y^6}$ **53.** $\sqrt{\dfrac{2}{3x}}$ **54.** $\sqrt{\dfrac{5}{3p}}$

55. $\sqrt{\dfrac{x^5y^3}{3^2}}$ **56.** $\sqrt{\dfrac{g^3h^5}{r^3}}$ **57.** $\sqrt[3]{\sqrt{4}}$ **58.** $\sqrt[4]{\sqrt[3]{2}}$ **59.** $\sqrt[6]{\sqrt[3]{x}}$ **60.** $\sqrt[8]{\sqrt[4]{x}}$

Simplify each of the following. Assume that all variables represent positive real numbers.
(See Examples 7 and 8.)

61. $5\sqrt{2} - 3\sqrt{8} + \sqrt{72}$

62. $6\sqrt{3} - 2\sqrt{27} - 3\sqrt{45}$

63. $-\sqrt{24} + 7\sqrt{54} - 3\sqrt{96}$

64. $3\sqrt{28p} - 4\sqrt{63p} + \sqrt{112p}$

65. $9\sqrt{8k} + 3\sqrt{18k} - \sqrt{32k}$

66. $3\sqrt[3]{16} - 4\sqrt[3]{2}$

67. $\sqrt[3]{2} - \sqrt[3]{16} + 2\sqrt[3]{54}$

68. $2\sqrt[3]{3} + 4\sqrt[3]{24} - \sqrt[3]{81}$

69. $\sqrt[3]{32} - 5\sqrt[3]{4} + \sqrt[3]{108}$

70. $(\sqrt{2} + 3)(\sqrt{2} - 3)$

71. $(\sqrt{5} + \sqrt{2})(\sqrt{5} - \sqrt{2})$

72. $(\sqrt[3]{11} - 1)(\sqrt[3]{11^2} + \sqrt[3]{11} + 1)$

73. $(\sqrt[3]{7} + 3)(\sqrt[3]{7^2} - 3\sqrt[3]{7} + 9)$

74. $(3\sqrt{2} + \sqrt{3})(2\sqrt{3} - \sqrt{2})$

75. $(4\sqrt{5} - 1)(3\sqrt{5} + 2)$

Rationalize the denominator of each of the following. Assume that all variables represent non-negative numbers and that no denominators are 0. (See Example 10.)

76. $\dfrac{3}{1 - \sqrt{2}}$ **77.** $\dfrac{2}{1 + \sqrt{5}}$ **78.** $\dfrac{4 - \sqrt{2}}{2 - \sqrt{2}}$ **79.** $\dfrac{\sqrt{3} - 1}{\sqrt{3} - 2}$ **80.** $\dfrac{p}{\sqrt{p} + 2}$ **81.** $\dfrac{\sqrt{r}}{3 - \sqrt{r}}$

82. Management   The decline in sales in the absence of promotional activities can often be expressed as

$$S = k(2.72)^{-.01t},$$

where S represents sales at time t, measured in years, and k represents sales at $t = 0$. Find S for the following values of t and k.
(a) $k = 1000$, $t = 1$
(b) $k = 1000$, $t = 5$
(c) $k = 50,000$, $t = 2$

83. Management   The theory of economic lot size shows that, under certain conditions, the number of units to order to minimize total cost is

$$x = \sqrt{\dfrac{kM}{f}}.$$

Here k is the cost to store one unit for one year, f is the (constant) setup cost to manufacture the product, and M is the total number of units produced annually. (See Section 12.3.) Find x for the following values of f, k, and M.
(a) $k = \$1$, $f = \$500$, $M = 100,000$
(b) $k = \$3$, $f = \$7$, $M = 16,700$
(c) $k = \$1$, $f = \$5$, $M = 16,800$
(d) $k = \$3$, $f = \$750$, $M = 30,000$

84. Social Science   The threshold weight T for a person is the weight above which the risk of death increases greatly. One researcher found that the threshold weight in pounds for men aged 40–49 is related to height in inches by the equation $h = 12.3T^{1/3}$. What height corresponds to a threshold of 216 pounds for a man in this age group?

85. Social Science   The length L of an animal is related to its surface area S by the equation

$$L = \left(\dfrac{S}{a}\right)^{1/2},$$

where a is a constant that depends on the type of animal. Find the length of an animal with a surface area of 1000 square centimeters if $a = 2/5$.

1 Solve the following equations.

(a) $(y - 6)(y + 2) = 0$

(b) $(5k - 3)(k + 5) = 0$

(c) $(2r - 9)(3r + 5)(r + 3) = 0$

Answers:

(a) 6, −2

(b) 3/5, −5

(c) 9/2, −5/3, −3

1.10 QUADRATIC EQUATIONS

An equation with 2 as the highest exponent on the variable is called **quadratic.** Some examples of quadratic equations are

$$2x^2 + 5x - 1 = 0, \quad 3m^2 = 6 - m, \quad y^2 = 4, \quad \text{and} \quad 2m^2 - m = 0.$$

If a, b, and c are real numbers, with $a \neq 0$, then

$$ax^2 + bx + c = 0,$$

is a **quadratic equation.**

(If $a = 0$, the equation becomes $bx + c = 0$, which is linear.) A quadratic equation written in the form $ax^2 + bx + c = 0$ is said to be in **standard form.**

The simplest method of solving a quadratic equation, but one that is not always easily applied, is by factoring. This method depends on the *zero-factor property.*

Zero-Factor Property

If a and b are real numbers, with $ab = 0$, then $a = 0$ or $b = 0$ or both.

▶**EXAMPLE 1**   Solve the equation $(x - 4)(3x + 7) = 0$.

By the zero-factor property, the product $(x - 4)(3x + 7)$ can equal 0 only if at least one of the factors equals 0. That is, the product equals zero only if $x - 4 = 0$ or $3x + 7 = 0$. Solving each of these equations separately will give the solutions of the original equation.

$$x - 4 = 0 \quad \text{or} \quad 3x + 7 = 0$$
$$x = 4 \quad \text{or} \quad 3x = -7$$
$$x = -\frac{7}{3}$$

The solutions of the equation $(x - 4)(3x + 7) = 0$ are 4 and −7/3. Check these solutions by substitution in the original equation. ◀ **1**

▶**EXAMPLE 2**   Solve $6r^2 + 7r = 3$.

Write the equation in standard form as

$$6r^2 + 7r - 3 = 0.$$

Now factor $6r^2 + 7r - 3$ to get

$$(3r - 1)(2r + 3) = 0.$$

2 Solve the following.

(a) $y^2 + 3y = 10$

(b) $2r^2 + 9r = 5$

(c) $4k^2 = 9k$

Answers:

(a) $2, -5$

(b) $1/2, -5$

(c) $3/2, 0$

3 Solve each equation.

(a) $p^2 = 21$

(b) $(m + 7)^2 = 15$

(c) $(2k - 3)^2 = 5$

Answers:

(a) $\pm\sqrt{21}$

(b) $-7 \pm \sqrt{15}$

(c) $(3 \pm \sqrt{5})/2$

By the zero-factor property, the product $(3r - 1)(2r + 3)$ can equal 0 only if
$$3r - 1 = 0 \quad \text{or} \quad 2r + 3 = 0.$$
Solve each of these equations separately to find that the solutions of the original equation are $1/3$ and $-3/2$. Check these solutions by substitution in the original equation. ◀ **2**

An equation such as $x^2 = 5$ has two solutions, $\sqrt{5}$ and $-\sqrt{5}$. The same idea is true in general.

Square-Root Property

If $b > 0$, then the solutions of $x^2 = b$ are $\sqrt{b}$ and $-\sqrt{b}$.

The two solutions are sometimes abbreviated $\pm\sqrt{b}$.

▶ **EXAMPLE 3** Solve each equation.

(a) $m^2 = 17$

By the square-root property, the solutions are $\sqrt{17}$ and $-\sqrt{17}$, abbreviated $\pm\sqrt{17}$.

(b) $(y - 4)^2 = 11$

Use a generalization of the square-root property, working as follows:
$$(y - 4)^2 = 11$$
$$y - 4 = \sqrt{11} \qquad \text{or} \quad y - 4 = -\sqrt{11}$$
$$y = 4 + \sqrt{11} \qquad\qquad y = 4 - \sqrt{11}.$$
Abbreviate the solutions as $4 \pm \sqrt{11}$. ◀ **3**

As suggested by Example 3(b), any quadratic equation can be solved using the square-root property if the equation can be written in the form $(x + n)^2 = k$. For instance, to write $4x^2 - 24x + 19 = 0$ in this form, we need to write the left side as a perfect square, $(x + n)^2$. To get a perfect square trinomial, first add -19 to both sides of the equation, then multiply both sides by $1/4$, so that x^2 has a coefficient of 1.
$$4x^2 - 24x + 19 = 0$$
$$4x^2 - 24x = -19$$
$$x^2 - 6x = -\frac{19}{4} \qquad (*)$$

Adding 9 to both sides will make the left side a perfect square, so the equation can be written in the desired form.

$$x^2 - 6x + 9 = -\frac{19}{4} + 9$$

$$(x - 3)^2 = \frac{17}{4}$$

Use the square-root property to complete the solution.

$$x - 3 = \pm\sqrt{\frac{17}{4}} = \pm\frac{\sqrt{17}}{2}$$

$$x = 3 \pm \frac{\sqrt{17}}{2} = \frac{6 \pm \sqrt{17}}{2}$$

The two solutions are $\dfrac{6 + \sqrt{17}}{2}$ and $\dfrac{6 - \sqrt{17}}{2}$.

Notice that the number 9 added to both sides of equation (*) is found by taking $(1/2)(6) = 3$ and squaring it: $3^2 = 9$. This always works because

$$\left(x + \frac{b}{2}\right)^2 = x^2 + 2\left(\frac{b}{2}\right)x + \left(\frac{b}{2}\right)^2 = x^2 + bx + \left(\frac{b}{2}\right)^2.$$

The process of changing $4x^2 - 24x + 19 = 0$ to $(x - 3)^2 = 17/4$ is called **completing the square.*

▶ **EXAMPLE 4**   Solve $9z^2 - 12z - 1 = 0$.

Get a coefficient of 1 in the given equation by multiplying both sides by $1/9$.

$$z^2 - \frac{4}{3}z - \frac{1}{9} = 0$$

$$z^2 - \frac{4}{3}z = \frac{1}{9} \qquad \textbf{Add 1/9 on both sides.}$$

Find half of $-4/3$ and square it.

$$\frac{1}{2}\left(-\frac{4}{3}\right) = -\frac{2}{3} \quad \text{and} \quad \left(-\frac{2}{3}\right)^2 = \frac{4}{9}$$

$$z^2 - \frac{4}{3}z + \frac{4}{9} = \frac{1}{9} + \frac{4}{9} \qquad \textbf{Add 4/9 to both sides.}$$

$$\left(z - \frac{2}{3}\right)^2 = \frac{5}{9} \qquad \textbf{Factor on the left; add on the right.}$$

---

*Completing the square is discussed further in Section 3.1.

4 Solve by completing the square.

(a) $m^2 + 2m = 3$

(b) $3r^2 - r = 1$

Answers:

(a) $1, -3$

(b) $\dfrac{1 \pm \sqrt{13}}{6}$

$$z - \frac{2}{3} = \sqrt{\frac{5}{9}} \quad \text{or} \quad z - \frac{2}{3} = -\sqrt{\frac{5}{9}} \qquad \textbf{Square-root property}$$

$$z - \frac{2}{3} = \frac{\sqrt{5}}{3} \quad \text{or} \quad z - \frac{2}{3} = -\frac{\sqrt{5}}{3} \qquad \textbf{Simplify.}$$

$$z = \frac{2}{3} + \frac{\sqrt{5}}{3} \quad \text{or} \quad z = \frac{2}{3} - \frac{\sqrt{5}}{3}.$$

These two solutions may be written as

$$z = \frac{2 + \sqrt{5}}{3} \quad \text{or} \quad z = \frac{2 - \sqrt{5}}{3},$$

abbreviated as $(2 \pm \sqrt{5})/3$. ◀ **4**

The method of completing the square can be used on the general quadratic equation,

$$ax^2 + bx + c = 0 \quad (a \neq 0),$$

to convert it to one whose solution can be found by the square-root property. This will give a general formula for solving any quadratic equation. Going through the necessary algebra produces the following important result.

Quadratic Formula

The solutions of the quadratic equation $ax^2 + bx + c = 0$, where $a \neq 0$, are given by

$$x = \frac{-b \pm \sqrt{b^2 - 4ac}}{2a}.$$

FOR GRAPHERS

An important feature of a grapher is that it can be programmed to evaluate expressions such as the one in the quadratic formula. Because this is a formula that is used frequently throughout mathematics and its applications, it is worth programming. Refer to your manual for details.

When a solution is an irrational number, as in the examples above, the grapher gives an approximation that can be accurate to as many as 8–10 decimal places (within the limits of the machine or the software). For practical purposes, this is more useful than an exact answer, such as $2 + \sqrt{3}$.

A grapher can be used to solve a quadratic equation by using the method described in Section 1.2. For instance, to solve $9x^2 - 12x - 1 = 0$, the equation in Example 4, graph $y = 9x^2 - 12x - 1$. Look for the x-values where the graph crosses the x-axis.

5 Use the quadratic formula to solve each of the following equations.

(a) $3x^2 + 11x - 4 = 0$

(b) $2z^2 - 7z - 4 = 0$

Answers:

(a) $1/3, -4$

(b) $-1/2, 4$

▶**EXAMPLE 5** Solve $x^2 - 4x - 5 = 0$ by the quadratic formula.

The equation is already in standard form (it has 0 alone on one side of the equals sign) so that the letters a, b, and c of the quadratic formula can be identified. The coefficient of the squared term gives the value of a; here $a = 1$. Also, $b = -4$ and $c = -5$. (Be careful to get the correct signs.) Substitute these values into the quadratic formula.

$$x = \frac{-(-4) \pm \sqrt{(-4)^2 - 4(1)(-5)}}{2(1)}$$ Let $a = 1, b = -4, c = -5$

$$= \frac{4 \pm \sqrt{16 + 20}}{2}$$ $(-4)^2 = (-4)(-4) = 16$

$$x = \frac{4 \pm 6}{2}$$ $\sqrt{16 + 20} = \sqrt{36} = 6$

The $\pm$ sign represents the two solutions of the equation. First use $+$ and then use $-$ to find each of the solutions.

$$x = \frac{4 + 6}{2} = \frac{10}{2} = 5 \quad \text{or} \quad x = \frac{4 - 6}{2} = \frac{-2}{2} = -1$$

The two solutions are 5 and -1. Check by substituting each value in the original equation. ◀ **5**

Caution When using the quadratic formula, remember the equation must be in the form $ax^2 + bx + c = 0$. Also, notice that the fraction in the quadratic formula extends under *both* terms in the numerator. Be sure to add $-b$ to $\pm\sqrt{b^2 - 4ac}$ *before* dividing by $2a$.

▶**EXAMPLE 6** Solve $x^2 + 1 = 4x$.

First add $-4x$ to both sides, to get 0 alone on the right side.

$$x^2 - 4x + 1 = 0$$

Now identify the letters a, b, and c. Here $a = 1, b = -4$, and $c = 1$. Substitute these numbers into the quadratic formula

$$x = \frac{-(-4) \pm \sqrt{(-4)^2 - 4(1)(1)}}{2(1)}$$

$$= \frac{4 \pm \sqrt{16 - 4}}{2}$$

$$= \frac{4 \pm \sqrt{12}}{2}$$

$$= \frac{4 \pm 2\sqrt{3}}{2}$$ $\sqrt{12} = \sqrt{4 \cdot 3} = \sqrt{4} \cdot \sqrt{3} = 2\sqrt{3}$

6 Find exact and approximate solutions for the following.

(a) $y^2 - 2y = 2$

(b) $x^2 - 6x + 4 = 0$

Answers:

(a) Exact: $1 + \sqrt{3}, 1 - \sqrt{3}$; approximate: $2.732, -.732$

(b) Exact: $3 + \sqrt{5}, 3 - \sqrt{5}$; approximate: $5.236, .764$

$$= \frac{2(2 \pm \sqrt{3})}{2} \qquad \textbf{Factor } 4 \pm 2\sqrt{3}$$

$$x = 2 \pm \sqrt{3}.$$

The two solutions are $2 + \sqrt{3}$ and $2 - \sqrt{3}$.

The exact values of the solutions are $2 + \sqrt{3}$ and $2 - \sqrt{3}$. In many cases a decimal approximation of these solutions is needed. Use a calculator to find that $\sqrt{3} \approx 1.732$, so that (to the nearest thousandth) the solutions are

$$2 + \sqrt{3} \approx 2 + 1.732 = 3.732$$

or

$$2 - \sqrt{3} \approx 2 - 1.732 = .268. \quad \blacktriangleleft$$

Caution   In Example 6, be careful to factor $4 + 2\sqrt{3}$ *before* dividing by the denominator 2.

$$\frac{4 + 2\sqrt{3}}{2} \ne \frac{2 + 2\sqrt{3}}{1} \quad \boxed{6}$$

The above quadratic equations all had two different solutions. This is not always the case, as the following example shows.

▶ **EXAMPLE 7**   Solve the following.

(a) $9x^2 - 30x + 25 = 0$

Here $a = 9$, $b = -30$, and $c = 25$. By the quadratic formula,

$$x = \frac{-(-30) \pm \sqrt{(-30)^2 - 4(9)(25)}}{2(9)}$$

$$= \frac{30 \pm \sqrt{900 - 900}}{18}$$

$$= \frac{30 \pm 0}{18}$$

$$= \frac{30}{18} = \frac{5}{3}.$$

The given equation has only one solution. (The rational solution indicates that this equation could have been solved by factoring.)

(b) $x^2 - 6x + 10 = 0$

Because $a = 1$, $b = -6$, and $c = 10$,

7 Solve the following equations.

(a) $9k^2 - 6k + 1 = 0$

(b) $4m^2 + 28m + 49 = 0$

(c) $2x^2 - 5x + 5 = 0$

Answers:

(a) $1/3$

(b) $-7/2$

(c) No real number solutions

$$x = \frac{-(-6) \pm \sqrt{(-6)^2 - 4(1)(10)}}{2(1)}$$

$$= \frac{6 \pm \sqrt{36 - 40}}{2}$$

$$= \frac{6 \pm \sqrt{-4}}{2}.$$

Recall from Section 1.9 that $\sqrt{-4}$ is not a real number, so there are no *real number* solutions to this equation. ◀ **7**

▶**EXAMPLE 8**   A landscape architect wants to make an exposed gravel border of uniform width around a small shed behind a company plant. The shed is 10 feet by 6 feet. He has enough gravel to cover 36 square feet. How wide should the border be?

Follow the steps given in Section 1.2 for solving applied problems. A sketch of the shed with border is given in Figure 1.23. Let x represent the width of the

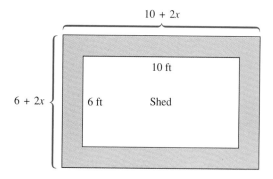

FIGURE 1.23

border. Then the width of the large rectangle is $6 + 2x$ and its length is $10 + 2x$. We must write an equation relating the given areas and dimensions. The area of the large rectangle is $(6 + 2x)(10 + 2x)$. The area occupied by the shed is $6 \cdot 10 = 60$. The area of the border is found by subtracting the area of the shed from the area of the large rectangle. This difference should be 36 square feet, giving the equation

$$(6 + 2x)(10 + 2x) - 60 = 36.$$

Solve this equation with the following sequence of steps.

$$60 + 32x + 4x^2 - 60 = 36$$
$$4x^2 + 32x - 36 = 0$$
$$x^2 + 8x - 9 = 0$$
$$(x + 9)(x - 1) = 0$$

8 The length of a picture is 2 inches more than the width. It is mounted on a mat that extends 2 inches beyond the picture on all sides. What are the dimensions of the picture if the area of the mat is 99 square inches?

Answer:

5 inches by 7 inches

9 Solve each equation.

(a) $9x^4 - 23x^2 + 10 = 0$

(b) $4x^4 = 7x^2 - 3$

Answers:

(a) $-\sqrt{5}/3, \sqrt{5}/3, -\sqrt{2}, \sqrt{2}$

(b) $-\sqrt{3}/2, \sqrt{3}/2, -1, 1$

The solutions are -9 and 1. The number -9 cannot be the width of the border, so the solution is to make the border 1 foot wide.   ◀ **8**

Some equations that are not quadratic can be solved as quadratic equations by making a suitable substitution. Such equations are called **quadratic in form.**

▶**EXAMPLE 9**   Solve $4m^4 - 9m^2 + 2 = 0$.
   Use the substitutions

$$x = m^2 \quad \text{and} \quad x^2 = m^4$$

to rewrite the equation as

$$4x^2 - 9x + 2 = 0.$$

This quadratic equation can be solved by factoring.

$$4x^2 - 9x + 2 = 0$$
$$(x - 2)(4x - 1) = 0$$
$$x - 2 = 0 \quad \text{or} \quad 4x - 1 = 0$$
$$x = 2 \quad \text{or} \quad 4x = 1$$
$$x = \frac{1}{4}$$

Because $x = m^2$,

$$m^2 = 2 \qquad \text{or} \quad m^2 = \frac{1}{4}$$
$$m = \pm\sqrt{2} \quad \text{or} \quad m = \pm\frac{1}{2}.$$

There are four solutions: $-\sqrt{2}, \sqrt{2}, -1/2,$ and $1/2$.   ◀ **9**

In Example 9, we saw that a fourth degree equation had 4 solutions. Every equation of degree n (for natural numbers n) has up to n real number solutions.

▶**EXAMPLE 10**   Solve $4 + \dfrac{1}{z - 2} = \dfrac{3}{2(z - 2)^2}$.
   First, substitute u for $z - 2$ to get

$$4 + \frac{1}{z - 2} = \frac{3}{2(z - 2)^2}$$
$$4 + \frac{1}{u} = \frac{3}{2u^2}.$$

10 Solve each equation.

(a) $\dfrac{13}{3(p+1)} + \dfrac{5}{3(p+1)^2} = 2$

(b) $1 + \dfrac{1}{a} = \dfrac{5}{a^2}$

Answers:

(a) $-4/3, 3/2$

(b) $(-1 + \sqrt{21})/2$, $(-1 - \sqrt{21})/2$

11 Solve each of the following equations for the indicated variable. Assume no variable equals 0.

(a) $k = mp^2 - bp$ for p

(b) $r = \dfrac{APk^2}{3}$ for k

Answers:

(a) $\dfrac{b \pm \sqrt{b^2 + 4mk}}{2m}$

(b) $\pm\sqrt{\dfrac{3r}{AP}}$ or $\dfrac{\pm\sqrt{3rAP}}{AP}$

Now multiply both sides of the equation by the common denominator $2u^2$; then solve the resulting quadratic equation.

$$2u^2\left(4 + \frac{1}{u}\right) = 2u^2\left(\frac{3}{2u^2}\right)$$

$$8u^2 + 2u = 3$$

$$8u^2 + 2u - 3 = 0$$

$$(4u + 3)(2u - 1) = 0$$

$$4u + 3 = 0 \quad \text{or} \quad 2u - 1 = 0$$

$$u = -\frac{3}{4} \quad \text{or} \quad u = \frac{1}{2}$$

$$z - 2 = -\frac{3}{4} \quad \text{or} \quad z - 2 = \frac{1}{2} \qquad \text{Replace } u \text{ with } z - 2.$$

$$z = -\frac{3}{4} + 2 \qquad\qquad z = \frac{1}{2} + 2$$

$$z = \frac{5}{4} \quad \text{or} \quad z = \frac{5}{2}.$$

The solutions are $5/4$ and $5/2$. Check both solutions in the original equation.   ◀ **10**

The next example shows how to solve an equation for a specified variable when the equation is quadratic in that variable.

▶ **EXAMPLE 11**   Solve $v = mx^2 + x$ for x. (Assume $m \neq 0$.)

The equation is quadratic in x because of the x^2 term. Use the quadratic formula, first writing the equation in standard form.

$$v = mx^2 + x$$

$$0 = mx^2 + x - v$$

Let $a = m$, $b = 1$, and $c = -v$. Then the quadratic formula gives

$$x = \frac{-1 \pm \sqrt{1^2 - 4(m)(-v)}}{2m}$$

$$x = \frac{-1 \pm \sqrt{1 + 4mv}}{2m} \qquad ◀ \; \textbf{11}$$

1.10 EXERCISES

Solve each equation. (See Examples 1 and 2.)

1. $(x + 3)(x - 12) = 0$ **2.** $(p - 16)(p - 5) = 0$ **3.** $y^2 + 15y + 56 = 0$ **4.** $k^2 - 4k - 5 = 0$

5. $2x^2 = 5x - 3$ **6.** $2 = 12z^2 + 5z$ **7.** $6r^2 + r = 1$ **8.** $3y^2 = 16y - 5$

9. $2m^2 + 20 = 13m$ **10.** $10a^2 + 17a + 3 = 0$ **11.** $m(m - 7) = -10$ **12.** $z(2z + 7) = 4$

13. $9x^2 - 16 = 0$ **14.** $25y^2 - 64 = 0$ **15.** $16x^2 - 16x = 0$ **16.** $12y^2 - 48y = 0$

Solve the following equations by the square-root property or by completing the square. (See Examples 3 and 4.)

17. $(r - 2)^2 = 7$ **18.** $(b + 5)^2 = 8$ **19.** $(4x - 1)^2 = 20$ **20.** $(3t + 5)^2 = 11$

21. $q^2 + q = 12$ **22.** $y^2 + 24 = 11y$ **23.** $4z^2 - 4z = 1$ **24.** $11p^2 - 11p + 1 = 0$

Solve the following equations. If the solutions involve square roots, give both the exact and approximate solutions.(See Examples 1–7.)

25. $2x^2 + 5x + 1 = 0$ **26.** $3x^2 - x - 7 = 0$ **27.** $4k^2 + 2k = 1$ **28.** $r^2 = 3r + 5$

29. $5y^2 + 6y = 2$ **30.** $2z^2 + 3 = 8z$ **31.** $6x^2 + 6x + 5 = 0$ **32.** $3a^2 - 2a + 2 = 0$

33. $2r^2 - 7r + 5 = 0$ **34.** $8x^2 = 8x - 3$ **35.** $6k^2 - 11k + 4 = 0$ **36.** $8m^2 - 10m + 3 = 0$

37. $2x^2 - 7x + 30 = 0$ **38.** $3k^2 + k = 6$ **39.** $5m^2 + 5m = 0$ **40.** $8r^2 + 16r = 0$

Give all real number solutions of the following equations. (See Examples 9 and 10. Hint: In Exercise 45, let $u = p - 3$.)

41. $z^4 - 2z^2 = 15$ **42.** $6p^4 = p^2 + 2$ **43.** $2q^4 + 3q^2 - 9 = 0$ **44.** $4a^4 = 2 - 7a^2$

45. $6(p - 3)^2 + 5(p - 3) - 6 = 0$ **46.** $12(q + 4)^2 - 13(q + 4) - 4 = 0$ **47.** $2(p + 5)^2 - 3(p + 5) = 20$

48. $4(2y - 3)^2 - (2y - 3) - 3 = 0$ **49.** $1 + \dfrac{7}{2a} = \dfrac{15}{2a^2}$ **50.** $5 - \dfrac{4}{k} - \dfrac{1}{k^2} = 0$

51. $-\dfrac{2}{3z^2} + \dfrac{1}{3} + \dfrac{8}{3z} = 0$ **52.** $2 + \dfrac{5}{x} + \dfrac{1}{x^2} = 0$

Students often confuse expressions with equations. In (a) add or subtract the expressions as indicated. In (b) solve the equation. Compare the results.

53. (a) $\dfrac{6}{r} - \dfrac{5}{r - 2} - 1$ **(b)** $\dfrac{6}{r} = \dfrac{5}{r - 2} - 1$ **54. (a)** $\dfrac{8}{z - 1} - \dfrac{5}{z} - \dfrac{2z}{z - 1}$ **(b)** $\dfrac{8}{z - 1} = \dfrac{5}{z} + \dfrac{2z}{z - 1}$

Solve the following problems. (See Example 8.)

55. An ecology center wants to set up an experimental garden. It has 300 meters of fencing to enclose a rectangular area of 5000 square meters. Find the length and width of the rectangle.
  (a) Let x = the length and write an expression for the width.
  (b) Write an equation relating the length, width, and area, using the result of part (a).
  (c) Solve the problem.

56. A shopping center has a rectangular area of 40,000 square yards enclosed on three sides for a parking lot. The length is 200 yards more than twice the width. Find the length and width of the lot. Let x equal the width and follow steps similar to those in Exercise 55.

57. A landscape architect has included a rectangular flower bed measuring 9 ft by 5 ft in her plans for a new building. She wants to use two colors of flowers in the bed, one in the center and the other for a border of the same width on all four sides. If she can get just enough plants to cover 24 sq ft for the border, how wide can the border be?

58. Joan wants to buy a rug for a room that is 12 feet by 15 feet. She wants to leave a uniform strip of floor around the rug. She can afford 108 square feet of carpeting. What dimensions should the rug have?

59. In 1991 Rick Mears won the (500 mi) Indianapolis 500 race. His speed (rate) was 100 mph (to the nearest mph) faster than that of the 1911 winner, Ray Harroun. Mears completed the race in 3.74 hours less time than Harroun. Find Mears's rate to the nearest whole number.

60. Chris and Josh have received walkie-talkies for Christmas. If they leave from the same point at the same time, Chris walking north at 2.5 mph and Josh walking east at 3 mph, how long will they be able to talk to each other if the range of the walkie-talkies is 4 mi? Round your answer to the nearest minute.

61. Management The manager of a bicycle shop knows that the cost of selling x bicycles is $C = 20x + 60$ and the revenue from selling x bicycles is $R = x^2 - 8x$. Find the break-even value of x.

62. Management A company that produces breakfast cereal has found that its operating cost in dollars is $C = 40x + 150$ and its revenue in dollars is $R = 65x - x^2$. For what value(s) of x will the company break even?

63. Physical Science If a ball is thrown upward with an initial velocity of 64 feet per second, then its height after t seconds is $h = 64t - 16t^2$. In how many seconds will the ball reach
(a) 64 ft?   **(b)** 28 ft?
(c) Why are two answers possible?

64. Physical Science A particle moves horizontally with its distance from a starting point in centimeters at t seconds given by $d = 11t^2 - 10t$.
(a) How long will it take before the particle returns to the starting point?
(b) When will the particle be 100 centimeters from the starting point?

Solve each of the following equations for the indicated variable. Assume all denominators are nonzero, and that all variables represent positive real numbers. (See Example 11.)

65. $S = \dfrac{1}{2}gt^2$ for t

66. $a = \pi r^2$ for r

67. $L = \dfrac{d^4 k}{h^2}$ for h

68. $F = \dfrac{kMv^2}{r}$ for v

69. $P = \dfrac{E^2 R}{(r + R)^2}$ for R

70. $S = 2\pi rh + 2\pi r^2$ for r

For each of the following equations, use a grapher to solve the equation. Then solve the equation algebraically and compare the results.

71. $2x^2 = x - 1$

72. $3x^2 + 2 = x$

---

73. What limitation on solving equations with a grapher is suggested by Exercises 71 and 72?

1 Solve the following. Graph the solution.

(a) $y^2 + 2y - 3 < 0$

(b) $2p^2 + 3p - 2 < 0$

Answers:

(a) $(-3, 1)$

(b) $(-2, 1/2)$

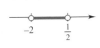

1.11 POLYNOMIAL AND RATIONAL INEQUALITIES

We begin our discussion of polynomial inequalities with **quadratic inequalities**; inequalities of the form $ax^2 + bx + c > 0$ (or $<$, or $\leq$, or $\geq$). The highest exponent on the variable is always 2. Examples of quadratic inequalities include

$$x^2 - x - 12 < 0, \quad 3y^2 + 2y \geq 0, \quad \text{and} \quad m^2 \leq 4.$$

A method of solving quadratic inequalities is shown in the next few examples.

▶**EXAMPLE 1** Solve the quadratic inequality $x^2 - x - 12 < 0$.

The solution of this inequality will include all numbers that make $x^2 - x - 12$ *negative* (< 0). Thus, we are concerned only with the *sign* of the expression. The sign of the expression will change from negative to positive or positive to negative only on either side of those x values that make the expression 0. To find these x values; we solve the *equation* $x^2 - x - 12 = 0$.

$$x^2 - x - 12 = 0$$
$$(x + 3)(x - 4) = 0$$
$$x + 3 = 0 \quad \text{or} \quad x - 4 = 0$$
$$x = -3 \quad \text{or} \quad x = 4$$

These numbers, -3 and 4, separate the number line into three regions, as shown in Figure 1.24.

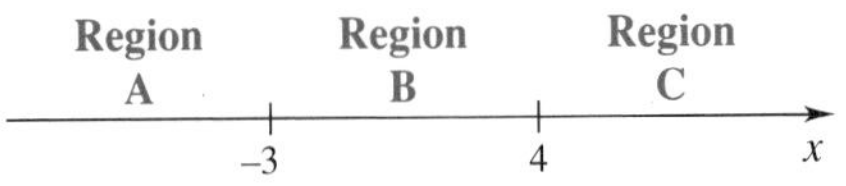

Region A	Region B	Region C

$$\overset{\text{}}{\underset{-3 \qquad\qquad 4}{\rule{4cm}{0.4pt}}} \quad x$$

FIGURE 1.24

Because the polynomial will change sign only at a point where its value is 0, the sign of the polynomial will be the same (either positive or negative) throughout a given region. To find the sign of the polynomial in each region, substitute any value from a region for x in the polynomial. For example,

in region A, let $x = -5$: $(-5)^2 - (-5) - 12 = 18 > 0$;

in region B, let $x = 0$: $(0)^2 - (0) - 12 = -12 < 0$;

in region C, let $x = 5$: $(5)^2 - (5) - 12 = 8 > 0$.

The only region where $x^2 - x - 12$ is negative (<0) is region B, so the solution is $(-3, 4)$, as shown in Figure 1.25. ◀ **1**

FIGURE 1.25

2 Solve each inequality. Graph each solution.

(a) $k^2 + 2k - 15 \geq 0$

(b) $3m^2 + 7m \geq 6$

Answers:

(a) All numbers in $(-\infty, -5]$ or $[3, \infty)$

(b) All numbers in $(-\infty, -3]$ or $[2/3, \infty)$

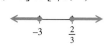

▶ **EXAMPLE 2**   Solve the quadratic inequality $r^2 + 3r \geq 4$.

First rewrite the inequality so that one side is 0.

$$r^2 + 3r \geq 4$$
$$r^2 + 3r - 4 \geq 0 \qquad \text{Add } -4 \text{ to both sides}$$

Now solve the corresponding equation.

$$r^2 + 3r - 4 = 0$$
$$(r - 1)(r + 4) = 0$$
$$r = 1 \quad \text{or} \quad r = -4$$

Test a number from each region shown in Figure 1.26.

Let $x = -5$ from region A: $(-5)^2 + 3(-5) - 4 = 6 > 0$.
Let $x = 0$ from region B: $(0)^2 + 3(0) - 4 = -4 < 0$.
Let $x = 2$ from region C: $(2)^2 + 3(2) - 4 = 6 > 0$

We want the inequality to be ≥ 0, that is, positive or 0. The solution includes numbers in region A and in region C, as well as -4 and 1, the endpoints. The solution, which includes all numbers in the intervals $(-\infty, -4]$ or $[1, \infty)$, is graphed in Figure 1.26. ◀ **2**

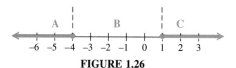

FIGURE 1.26

Although the inequality in the next example is not a quadratic inequality, it can be solved in the same way. Any **polynomial inequality** in the form $P(x) > 0$ or $P(x) < 0$ that can be factored can be solved in this way.

▶ **EXAMPLE 3**   Solve $q^3 - 4q > 0$.

Solve the corresponding equation by factoring.

$$q^3 - 4q = 0$$
$$q(q^2 - 4) = 0$$
$$q(q + 2)(q - 2) = 0$$
$$q = 0 \quad \text{or} \quad q + 2 = 0 \quad \text{or} \quad q - 2 = 0$$
$$q = 0 \qquad\qquad q = -2 \qquad\qquad q = 2$$

These three numbers separate the number line into the four regions shown in Figure 1.27.

FIGURE 1.27

3 Solve each inequality. Graph each solution.

(a) $m^2 - 9m > 0$

(b) $2k^3 - 50k \leq 0$

Answers:

(a) All numbers in $(-3, 0)$ or $(3, \infty)$

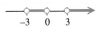

(b) All numbers in $(-\infty, -5]$ or $[0, 5]$

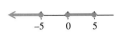

Test a number from each region.

$$\text{If } q = -3, (-3)^3 - 4(-3) = -15 < 0.$$
$$\text{If } q = -1, (-1)^3 - 4(-1) = 3 > 0.$$
$$\text{If } q = 1, (1)^3 - 4(1) = -3 < 0.$$
$$\text{If } q = 3, (3)^3 - 4(3) = 15 > 0.$$

The numbers that make the polynomial > 0, or positive, are in the intervals

$$(-2, 0) \quad \text{or} \quad (2, \infty),$$

as graphed in Figure 1.27. ◀ **3**

 FOR GRAPHERS

The inequalities discussed in this section can be solved with a grapher in the same way as the linear inequalities in Section 1.4. The graphical method has more of an advantage over the algebraic method when solving these inequalities than with linear inequalities, since even the algebraic method presented here refers to the number line. To solve $x^3 - 4x > 0$, for example, graph $y = x^3 - 4x$ and look for the interval where the graph is above the x-axis. (That is, where $y > 0$.) As Figure 1.28 indicates, this occurs in the intervals $(-2, 0)$ and $(2, \infty)$.

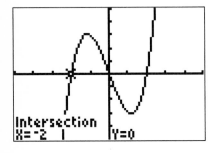

$-5 \leq x \leq 5, \quad -5 \leq y \leq 5$

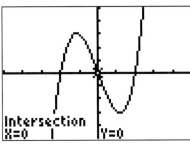

$-5 \leq x \leq 5, \quad -5 \leq y \leq 5$

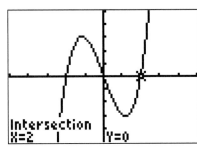

$-5 \leq x \leq 5, \quad -5 \leq y \leq 5$

FIGURE 1.28

RATIONAL INEQUALITIES Inequalities with quotients of algebraic expressions are called **rational inequalities.** These inequalities can be solved in much the same way as polynomial inequalities.

▶**EXAMPLE 4** Solve the rational inequality $\dfrac{5}{x + 4} \geq 1$.

Write the corresponding equation and get one side equal to 0.

$$\frac{5}{x + 4} = 1$$

$$\frac{5}{x + 4} - 1 = 0$$

4 Solve each inequality.

(a) $\dfrac{3}{x - 2} \geq 4$

(b) $\dfrac{p}{1 - p} < 3$

(c) Why is 2 excluded from the solution in part (a)?

Answers:

(a) $(2, 11/4]$

(b) All numbers in $(-\infty, 3/4)$ or $(1, \infty)$

(c) When $x = 2$, the fraction is undefined.

Write the left side as a single fraction.

$$\frac{5}{x + 4} - \frac{x + 4}{x + 4} = 0 \qquad \text{Get a common denominator}$$

$$\frac{5 - (x + 4)}{x + 4} = 0 \qquad \text{Subtract fractions}$$

$$\frac{5 - x - 4}{x + 4} = 0 \qquad \text{Distributive property}$$

$$\frac{1 - x}{x + 4} = 0$$

The quotient can change sign only when the denominator is 0 or when the numerator is 0. This happens when

$$1 - x = 0 \quad \text{or} \quad x + 4 = 0$$
$$x = 1 \quad \text{or} \quad x = -4.$$

As in the earlier examples, test a number from each of the regions determined by 1 and -4 in the original inequality.

$$\text{Let } x = -5: \ \frac{5}{-5 + 4} = \frac{5}{-1} = -5 \ngeq 1.$$

$$\text{Let } x = 0: \ \frac{5}{0 + 4} = \frac{5}{4} \geq 1.$$

$$\text{Let } x = 2: \ \frac{5}{2 + 4} = \frac{5}{6} \ngeq 1.$$

The test shows that numbers in $(-4, 1)$ satisfy the inequality. With a quotient, the endpoints must be considered individually to make sure that no denominator is 0. In this inequality, -4 makes the denominator 0, while 1 satisfies the given inequality. Write the solution as $(-4, 1]$. ◀

Caution As suggested by Example 4, be very careful with the endpoints of the intervals in the solution of rational inequalities. **4**

▶ **EXAMPLE 5** Solve $\dfrac{2x - 1}{3x + 4} < 5$.

Write the corresponding equation, then get 0 on one side. Begin by subtracting 5 on both sides and combining the terms on the left into a single fraction.

$$\frac{2x - 1}{3x + 4} = 5$$

$$\frac{2x - 1}{3x + 4} - 5 = 0 \qquad \text{Get 0 on one side}$$

$$\frac{2x - 1 - 5(3x + 4)}{3x + 4} = 0 \qquad \text{Subtract}$$

$$\frac{-13x - 21}{3x + 4} = 0 \qquad \text{Combine terms}$$

5 Solve each rational inequality.

(a) $\dfrac{3y - 2}{2y + 5} < 1$

(b) $\dfrac{3c - 4}{2 - c} \geq -5$

Answers:

(a) $(-5/2, 7)$

(b) All numbers in $(-\infty, 2)$ or $[3, \infty)$

Set the numerator and denominator each equal to 0 and solve the two equations.

$$-13x - 21 = 0 \quad \text{or} \quad 3x + 4 = 0$$

$$x = -\frac{21}{13} \quad \text{or} \quad x = -\frac{4}{3}$$

Use the values $-21/13$ and $-4/3$ to divide the number line into three intervals. Test a number from each interval in the original inequality. The quotient is negative for numbers in $(-\infty, -21/13)$ or $(-4/3, \infty)$. Neither endpoint satisfies the given inequality. ◀

Caution In problems like Examples 4 and 5, we cannot begin by simply multiplying both sides by the denominator to simplify the inequality, because we do not know whether the variable denominator is positive or negative. **5**

1.11 EXERCISES

Solve each of the following quadratic inequalities. Graph each solution. (See Examples 1 and 2.)

1. $(x + 5)(2x - 3) \leq 0$
2. $(5y - 1)(y + 4) > 0$
3. $r^2 + 4r > -3$
4. $z^2 + 6z < -8$

5. $4m^2 + 7m - 2 \leq 0$
6. $6p^2 - 11p + 3 \geq 0$
7. $4x^2 + 3x - 1 > 0$
8. $3x^2 - 5x > 2$

9. $6m^2 + m > 1$
10. $10r^2 + r \leq 2$
11. $2y^2 + 5y \leq 3$
12. $3a^2 + a > 10$

13. $x^2 \leq 25$
14. $y^2 \geq 4$
15. $p^2 - 16p > 0$
16. $r^2 - 9r < 0$

Solve the following inequalities. (See Example 3.)

17. $x^3 - 9x \geq 0$
18. $p^3 - 25p \leq 0$
19. $6k^3 - 5k^2 < 4k$
20. $2m^3 + 7m^2 > 4m$

21. A student solved the inequality $p^2 < 16$ by taking the square root of both sides to get $p < 4$. She wrote the solution as $(-\infty, 4)$. Is her solution correct?

Solve the following rational inequalities. (See Examples 4 and 5.)

22. $\dfrac{y + 2}{y - 4} \leq 0$

23. $\dfrac{r - 3}{r - 1} \geq 0$

24. $\dfrac{z + 6}{z + 3} > 1$

25. $\dfrac{a - 2}{a - 5} < -1$

26. $\dfrac{1}{3k - 5} < \dfrac{1}{3}$

27. $\dfrac{1}{p - 2} < \dfrac{1}{3}$

28. $\dfrac{2y + 3}{y - 5} \leq 1$

29. $\dfrac{a + 2}{3 + 2a} \leq 5$

30. $\dfrac{7}{k + 2} \geq \dfrac{1}{k + 2}$

31. $\dfrac{5}{p + 1} > \dfrac{12}{p + 1}$

In each of the following inequalities, the intervals on the number line are $(-\infty, 2)$, $(2, 5)$, and $(5, \infty)$. Without actually solving the inequality, state whether the numbers 2 and 5 will be included in or excluded from the solution of the inequality.

32. $(x - 2)(x - 5) \geq 0$

33. $(x - 5)(x - 2) < 0$

34. $\dfrac{x - 2}{x - 5} > 0$

35. $\dfrac{x - 5}{x - 2} \geq 0$

Solve the following problems.

Management *A product will break even or produce a profit only if the revenue from selling the product at least equals the cost of producing it. Find all x-intervals in Exercises 36 and 37 for which the product will at least break even.*

36. The cost to produce x underwater cameras is $C = 80x + 10,000$; the revenue is $R = 150x$.

37. The cost to produce x chocolate bars is $C = 125x + 5000$; the revenue is $R = 150x$.

38. An analyst has found that his company's profits, in hundreds of thousands of dollars, are given by $P = 3x^2 - 35x + 50$, where x is the amount, in hundreds of dollars, spent on advertising. For what values of x does the company make a profit?

39. The commodities market is very unstable; money can be made or lost quickly on investments in soybeans, wheat, and so on. Suppose that an investor kept track of her total profit, P, at time t, in months, after she begain investing, and found that $P = 4t^2 - 29t + 30$. Find the time intervals where she has been ahead.

40. The manager of a large apartment complex has found that the profit is given by $P = -x^2 + 250x - 15,000$, where x is the number of apartments rented. For what values of x does the complex produce a profit?

41. **Physical Science** A physicist has found that the velocity of a moving particle is given by $2t^2 - 5t - 12$, where t is time in seconds since he began his observations. (Here t can be positive or negative; think of t seconds before his observations began.) Find the time intervals in which the velocity has been negative.

42. **Physical Science** A projectile is fired from ground level. After t seconds its height above the ground is $220t - 16t^2$ feet. For what time period is the projectile at least 624 feet above the ground?

 Use a grapher to solve the following inequalities.

43. $3.1x^2 - 7.4x + 3.2 > 0$

44. $.5x^2 - 1.2x < .1$

45. $\dfrac{2x - 1}{x - 5} > 2x - 1$

46. $\dfrac{x + 3}{x - 5} < x + 3$

CHAPTER 1 SUMMARY

KEY TERMS AND SYMBOLS

1.1
$\approx$ is approximately equal to
π pi
$|a|$ absolute value of a
number line
graph
real number
natural (counting) number
whole number
integer
rational number
irrational number
properties of the real numbers
additive inverse
multiplicative inverse
interval
interval notation
order of operations
absolute value

1.2 variable
linear equation
properties of equality
solving for a specified variable

1.3 $<$ is less than
$\leq$ is less than or equal to
$>$ is greater than
$\geq$ is greater than or equal to
linear inequality
properties of inequality

1.5 a^n a to the power n
exponent or power
base
polynomial

zero polynomial
term
coefficient
degree of a nonzero term
degree of a polynomial
like terms
FOIL

1.6 factoring
factor
greatest common factor
difference of two squares
difference of two cubes
sum of two cubes
factoring by grouping
perfect square trinomial

1.7 rational expression
operations with rational
  expressions
complex fraction

1.8 $a^{-n} = \dfrac{1}{a^n}$ negative exponent

properties of exponents

1.9 $a^{1/n}$ nth root of a
$\sqrt{a}$ square root of a
$\sqrt[n]{a}$ nth root of a
radical
radical sign
radicand
index
simplified radical
like radicals
rationalizing the denominator
conjugate

1.10   quadratic equation   1.11   quadratic inequality
        zero-factor property   polynomial inequality
        square-root property   rational inequality
        completing the square
        quadratic formula

KEY CONCEPTS

Absolute Value

Assume a and b are real numbers with $b > 0$.

The solutions of $|a| = b$ or $|a| = |b|$ are $a = b$ or $a = -b$.

The solutions of $|a| < b$ are $-b < a < b$.

The solutions of $|a| > b$ are $a < -b$ or $a > b$.

Factoring

$$x^2 + 2xy + y^2 = (x + y)^2 \qquad x^3 - y^3 = (x - y)(x^2 + xy + y^2)$$

$$x^2 - 2xy + y^2 = (x - y)^2 \qquad x^3 + y^3 = (x + y)(x^2 - xy + y^2)$$

$$x^2 - y^2 = (x + y)(x - y)$$

Rules for Radicals

Let a and b be real numbers, n and k be positive integers, and m be any integer for which the following exist.

$$a^{m/n} = \sqrt[n]{a^m} = (\sqrt[n]{a})^m \qquad \sqrt[n]{a^n} = |a| \text{ if } n \text{ is even} \qquad \sqrt[n]{a^n} = a \text{ if } n \text{ is odd}$$

$$\sqrt[n]{a} \cdot \sqrt[n]{b} = \sqrt[n]{ab} \qquad \frac{\sqrt[n]{a}}{\sqrt[n]{b}} = \sqrt[n]{\frac{a}{b}} \ (b \neq 0) \qquad \sqrt[k]{\sqrt[n]{a}} = \sqrt[kn]{a}$$

Rules for Exponents

Let a, b, r, and s be any real numbers for which the following exist.

$$a^{-r} = \frac{1}{a^r} \qquad\qquad a^0 = 1 \qquad\qquad \left(\frac{a}{b}\right)^r = \frac{a^r}{b^r}$$

$$a^r \cdot a^s = a^{r+s} \qquad\qquad (a^r)^s = a^{rs} \qquad\qquad a^{1/r} = \sqrt[r]{a}$$

$$\frac{a^r}{a^s} = a^{r-s} \qquad\qquad (ab)^r = a^r b^r \qquad\qquad \left(\frac{a}{b}\right)^{-r} = \left(\frac{b}{a}\right)^r$$

Solving Quadratic Equations

Let a, b, and c be real numbers.

Factoring: If $ab = 0$, then $a = 0$ or $b = 0$ or both.

Square-Root Property: If $b > 0$, then the solutions of $x^2 = b$ are $\sqrt{b}$ and $-\sqrt{b}$.

Quadratic Formula: The solutions of $ax^2 + bx + c \ (a \neq 0)$ are

$$x = \frac{-b \pm \sqrt{b^2 - 4ac}}{2a}.$$

CHAPTER 1   REVIEW EXERCISES

Name the numbers from the list $-12, -6, -9/10, -\sqrt{7}, -\sqrt{4}, 0, 1/8, \pi/4, 6, \sqrt{11}$ *that are*

1. whole numbers;

2. integers;

3. rational numbers;

4. irrational numbers.

For Exercises 5-10, choose all words from the following list that apply: natural numbers, whole numbers, integers, rational numbers, irrational numbers, real numbers, undefined.

5. 0   **6.** $\sqrt{36}$ **7.** -1 **8.** $\dfrac{5}{8}$ **9.** $\sqrt{15}$ **10.** $\dfrac{3\pi}{4}$

Write the following numbers in numerical order from smallest to largest.

11. $-7, -3, 8, \pi, -2, 0$

12. $\dfrac{5}{6}, \dfrac{1}{2}, -\dfrac{2}{3}, -\dfrac{5}{4}, -\dfrac{3}{8}$

13. $|6 - 4|, -|-2|, |8 + 1|, -|3 - (-2)|$

14. $\sqrt{7}, -\sqrt{8}, -|\sqrt{16}|, |-\sqrt{12}|$

Write without absolute value bars.

15. $-|-6| + |3|$ **16.** $|-5| + |-9|$ **17.** $7 - |-8|$ **18.** $|-2| - |-7 + 3|$

Graph each of the following on a number line.

19. $x \geq -3$ **20.** $-4 < x \leq 6$ **21.** $x < -2$ **22.** $x \leq 1$

Use the order of operations to simplify.

23. $(-6 + 2 \cdot 5)(-2)$ **24.** $-4(-7 - 9 \div 3)$ **25.** $\dfrac{-8 + (-6)(-3) \div 9}{6 - (-2)}$ **26.** $\dfrac{20 \div 4 \cdot 2 \div 5 - 1}{-9 - (-3) - 12 \div 3}$

Solve each equation.

27. $2x - 5(x - 4) = 3x + 2$

28. $5y + 6 = -2(1 - 3y) + 4$

29. $\dfrac{2z}{5} - \dfrac{4z - 3}{10} = \dfrac{-z + 1}{10}$

30. $\dfrac{p}{p + 2} - \dfrac{3}{4} = \dfrac{2}{p + 2}$

31. $\dfrac{2m}{m - 3} = \dfrac{6}{m - 3} + 4$

32. $\dfrac{15}{k + 5} = 4 - \dfrac{3k}{k + 5}$

Solve for x.

33. $5ax - 1 = x$ **34.** $6x - 3y = 4bx$ **35.** $\dfrac{2x}{3 - c} = ax + 1$ **36.** $b^2x - 2x = 4b^2$

Management   *Solve each of the following problems.*

37. A computer printer is on sale for 15% off. The sale price is $425. What was the original price?

38. To make a special mix for Valentine's Day, the owner of a candy store wants to combine chocolate hearts which sell for $5 per pound with candy kisses which sell for $3.50 per pound. How many pounds of each kind should be used to get 30 pounds of a mix that can be sold for $4.50 per pound?

39. A real estate firm invests the $100,000 proceeds from a sale in two ways. The first portion in invested in a shopping center that provides an annual return of 8%. The rest is invested in a small apartment building with an annual return of 5%. The firm wants an annual income of $6800 from these investments. How much should be put into each investment?

40. For a particular line of tools, a hardware store has average monthly costs in dollars of $C = 55x + 180$ per unit and corresponding revenue of $R = 100x$ per unit, where x is the number of units sold. How many units must be sold for the store to break even on those tools for the month?

Solve each of the following inequalities.

41. $-6x + 3 < 2x$

42. $12z \geq 5z - 7$

43. $2(3 - 2m) \geq 8m + 3$

44. $6p - 5 > -(2p + 3)$

45. $-3 \leq 4x - 1 \leq 7$

46. $0 \leq 3 - 2a \leq 15$

Solve each equation.

47. $|m - 3| = 9$

48. $|6 - x| = 12$

49. $\left|\dfrac{2 - y}{5}\right| = 8$

50. $\left|\dfrac{3m + 5}{7}\right| = 15$

51. $|4k + 1| = |6k - 3|$

52. $|2x - 5| = |3x + 5|$

Solve each inequality.

53. $|b| \leq 8$

54. $|a| > 7$

55. $|x| \leq -4$

56. $|y| > -1$

57. $|2x - 7| \geq 3$

58. $|4m + 9| \leq 16$

59. $|5k + 2| - 3 \leq 4$

60. $|3z - 5| + 2 \geq 10$

61. Describe the steps you would use to solve

$$\dfrac{ab - c}{d} = 4 \text{ for } a.$$

62. Explain how the steps used to solve $3 - 5x = 18$, $3 - 5x < 18$, $|3 - 5x| = 18$, and $|3 - 5x| < 18$ differ. Compare the solutions.

Perform each of the indicated operations.

63. $(3x^4 - x^2 + 5x) - (-x^4 + 3x^2 - 8x)$

64. $(-8y^3 + 8y^2 - 3y) - (2y^3 - 4y^2 - 10)$

65. $-2(q^4 - 3q^3 + 4q^2) + 4(q^4 + 2q^3 + q^2)$

66. $5(3y^4 - 4y^5 + y^6) - 3(2y^4 + y^5 - 3y^6)$

67. $(5z + 2)(3z - 2)$

68. $(8p - 4)(5p + 3)$

69. $(4k - 3h)(4k + 3h)$

70. $(2r - 5y)(2r + 5y)$

71. $(6x + 3y)^2$

72. $(2a - 5b)^2$

73. $(4x - 3)^3$

74. $(2x + 3)^4$

75. $(2k - 1)(3k^2 - 4k + 1)$

76. $(3k + 4)(-k^4 + k^3 - 3)$

Factor as completely as possible.

77. $2kh^2 - 4kh + 5k$

78. $2m^2n^2 + 6mn^2 + 16n^2$

79. $3a^4 + 13a^3 + 4a^2$

80. $24x^3 + 4x^2 - 4x$

81. $10y^2 - 11y + 3$

82. $8q^2 + 3m + 4qm + 6q$

83. $4a^2 - 20a + 25$

84. $36p^2 + 12p + 1$

85. $144p^2 - 169q^2$

86. $81z^2 - 25x^2$

87. $8y^3 - 1$

88. $125a^3 + 216$

Perform each operation.

89. $\dfrac{4x}{5} \cdot \dfrac{35x}{12}$

90. $\dfrac{5k^2}{24} - \dfrac{75k}{36}$

91. $\dfrac{c^2 - 3c + 2}{2c(c - 1)} \div \dfrac{c - 2}{8c}$

92. $\dfrac{p^3 - 2p^2 - 8p}{3p(p^2 - 16)} \div \dfrac{p^2 + 4p + 4}{9p^2}$

93. $\dfrac{2y - 10}{5y} \cdot \dfrac{20y - 25}{12}$

94. $\dfrac{m^2 - 2m}{15m^3} \cdot \dfrac{5}{m^2 - 4}$

95. $\dfrac{2m^2 - 4m + 2}{m^2 - 1} \div \dfrac{6m + 18}{m^2 + 2m - 3}$

96. $\dfrac{x^2 + 6x + 5}{4(x^2 + 1)} \cdot \dfrac{2x(x + 1)}{x^2 - 25}$

97. $\dfrac{6}{15z} + \dfrac{2}{3z} - \dfrac{9}{10z}$

98. $\dfrac{5}{y - 2} - \dfrac{4}{y}$

99. $\dfrac{2}{5q} + \dfrac{10}{7q}$

100. $\dfrac{\dfrac{1}{x} - \dfrac{2}{y}}{3 - \dfrac{1}{xy}}$

---

101. Describe the steps needed to find the sum of the following expression:

$$\frac{2a + b}{4a^2 - b^2} + \frac{5a}{2a - b}.$$

102. Give some examples of corresponding rules for exponents and radicals, and explain how they are related.

103. Give two ways to evaluate $125^{2/3}$ and then compare them. Which do you prefer? Why?

Simplify each of the following. In Exercises 104–125, write all answers without negative exponents. Assume all variables represent positive real numbers.

104. 5^{-3}

105. 10^{-2}

106. -7^0

107. -3^{-1}

108. $\left(-\dfrac{6}{5}\right)^{-2}$

109. $\left(\dfrac{2}{3}\right)^{-3}$

110. $4^6 \cdot 4^{-3}$

111. $7^{-5} \cdot 7^{-1}$

112. $\dfrac{9^{-4}}{9^{-3}}$

113. $\dfrac{6^{-2}}{6^3}$

114. $\dfrac{9^4 \cdot 9^{-5}}{(9^{-2})^2}$

115. $\dfrac{k^4 \cdot k^{-3}}{(k^{-2})^{-3}}$

116. $4^{-1} + 2^{-1}$

117. $3^{-2} + 3^{-1}$

118. $125^{2/3}$

119. $128^{3/7}$

120. $9^{-5/2}$

121. $\left(\dfrac{144}{49}\right)^{-1/2}$

122. $\dfrac{5^{1/3} \cdot 5^{1/2}}{5^{3/2}}$

123. $\dfrac{2^{3/4} \cdot 2^{-1/2}}{2^{1/4}}$

124. $(3a^2)^{1/2} \cdot (3^2a)^{3/2}$

125. $(4p)^{2/3} \cdot (2p^3)^{3/2}$

126. $\sqrt[3]{27}$

127. $\sqrt[4]{625}$

128. $\sqrt[5]{-32}$

129. $\sqrt[6]{-64}$

130. $\sqrt{24}$

131. $\sqrt{63}$

132. $\sqrt[3]{54p^3q^5}$

133. $\sqrt[4]{64a^5b^3}$

134. $\sqrt{\dfrac{5n^2}{6m}}$

135. $\sqrt{\dfrac{3x^3}{2z}}$

136. $\sqrt[3]{\sqrt{5}}$

137. $\sqrt[4]{\sqrt[3]{10}}$

138. $2\sqrt{3} - 5\sqrt{12}$

139. $8\sqrt{7} + 2\sqrt{28}$

140. $5\sqrt[3]{16r^2s^3} - s\sqrt[3]{54r^2}$ **141.** $2\sqrt[4]{16mn^2} - \sqrt[4]{81mn^2}$ **142.** $(\sqrt{5} - 1)(\sqrt{5} + 1)$ **143.** $(\sqrt{7} - \sqrt{3})(\sqrt{7} + \sqrt{3})$

144. $(2\sqrt{5} - \sqrt{3})(\sqrt{5} + 2\sqrt{3})$ **145.** $(4\sqrt{7} + \sqrt{2})(3\sqrt{7} - \sqrt{2})$ **146.** $\dfrac{\sqrt{2}}{1 + \sqrt{3}}$ **147.** $\dfrac{4 + \sqrt{2}}{4 - \sqrt{5}}$

Solve each equation. Give only real number solutions.

148. $(b + 7)^2 = 5$ **149.** $(2p + 1)^2 = 7$ **150.** $2p^2 + 3p = 2$ **151.** $2y^2 = 15 + y$

152. $x^2 - 2x = 2$ **153.** $r^2 + 4r = 1$ **154.** $2m^2 - 12m = 11$ **155.** $9k^2 + 6k = 2$

156. $2a^2 + a - 15 = 0$ **157.** $12x^2 = 8x - 1$ **158.** $2q^2 - 11q = 21$ **159.** $3x^2 + 2x = 16$

160. $6k^4 + k^2 = 1$ **161.** $21p^4 = 2 + p^2$ **162.** $2x^4 = 7x^2 + 15$ **163.** $3m^4 + 20m^2 = 7$

164. $3 = \dfrac{13}{z} + \dfrac{10}{z^2}$ **165.** $1 + \dfrac{13}{p} + \dfrac{40}{p^2} = 0$

166. $\dfrac{15}{x - 1} + \dfrac{18}{(x - 1)^2} = -2$ **167.** $\dfrac{5}{(2t + 1)^2} = 2 - \dfrac{9}{2t + 1}$

Solve each equation for the specified variable.

168. $p = \dfrac{E^2R}{(r + R)^2}$ for r **169.** $p = \dfrac{E^2R}{(r + R)^2}$ for E **170.** $K = s(s - a)$ for s **171.** $kz^2 - hz - t = 0$ for z

Solve each inequality.

172. $r^2 + r - 6 < 0$ **173.** $y^2 + 4y - 5 \geq 0$ **174.** $2z^2 + 7z \geq 15$ **175.** $3k^2 \leq k + 14$

176. $\dfrac{m + 2}{m} \leq 0$ **177.** $\dfrac{q - 4}{q + 3} > 0$ **178.** $\dfrac{5}{p + 1} > 2$ **179.** $\dfrac{6}{a - 2} \leq -3$

180. $\dfrac{2}{r + 5} \leq \dfrac{3}{r - 2}$ **181.** $\dfrac{1}{z - 1} > \dfrac{2}{z + 1}$

182. If $a < b$, on what x-interval is $(x - a)(x - b)$ positive? Negative? Where is the product zero? Explain.

183. On what x-interval is $(x - a)^2$ positive? Negative? Where is it zero? Explain.

Work each applied problem.

184. A recreation director wants to fence off a rectangular playground beside an apartment building. The building forms one boundary, so she needs to fence only the other three sides. The area of the playground is to be 11,250 square meters. She has enough material to build 325 meters of fence. Find the length and width of the playground.

185. Two cars leave an intersection at the same time. One travels north and the other heads west traveling 10 mph faster. After 1 hour they are 50 miles apart. What were their speeds?

Imagine two refrigerators in the appliance section of a department store. One sells for $700 and uses $85 worth of electricity a year. The other is $100 more expensive but costs only $25 a year to run. Given that either refrigerator should last at least 10 years without repair, consumers would overwhelmingly buy the second model, right?

Well, not exactly. Many studies by economists have shown that in a wide range of decisions about money—from paying taxes to buying major appliances—consumers consistently make decisions that defy common sense.

In some cases—as in the refrigerator example—this means that people are generally unwilling to pay a little more money up front to save a lot of money in the long run. At times, psychological studies have shown, consumers appear to assign entirely whimsical values to money, values that change depending on time and circumstances.

In recent years, these apparently irrational patterns of human behavior have become a subject of intense interest among economists and psychologists, both for what they say about the way the human mind works and because of their implications for public policy.

How, for example, can the United States move toward a more efficient use of electricity if so many consumers refuse to buy energy-efficient appliances even when such a move is in their own best interest?

At the heart of research into the economic behavior of consumers is a concept known as the discount rate. It is a measure of how consumers compare the value of a dollar received today with one received tomorrow.

Consider, for example, if you won $1,000 in a lottery. How much more money would officials have to give you before you would agree to postpone cashing the check for a year?

Some people might insist on at least another $100, or 10 percent, since that is roughly how much it would take to make up for the combined effects of a year's worth of inflation and lost interest.

But the studies show that someone who wants immediate gratification might not be willing to postpone receiving the $1,000 for 20 percent or 30 percent or even 40 percent more money.

In the language of economists, this type of person has a high discount rate: He or she discounts the value of $1,000 so much over a year that it would take hundreds of extra dollars to make waiting as attractive as getting the money immediately.

Of the two alternatives, waiting a year for more money is clearly more rational than taking the check now. Why would people turn down $1,400 dollars next year in favor of $1,000 today? Even if they needed the $1,000 immediately, they would be better off borrowing it from a bank, even at 20 percent or even 30 percent interest. Then, a year later, they could pay off the loan—including the interest—with the $1,400 and pocket the difference.

The fact is, however, that economists find numerous examples of such high discount rates implicit in consumer behavior.

While consumers were very much aware of savings to be made at the point of purchase, they so heavily discounted the value of monthly electrical costs that they would pay over the lifetime of their dryer or freezer that they were oblivious of the potential for greater savings.

Gas water heaters, for example, were found to carry an implicit discount rate of 100 percent. This means that in deciding which model was cheapest over the long run, consumers acted as if they valued a $100 gas bill for the first year as if it were really $50. Then, in the second year, they would value the next $100 gas bill as if it were really worth $25, and so on through the life of the appliance.

Few consumers actually make this formal calculation, of course. But there are clearly bizarre behavioral patterns in evidence.

Some experiments, for example, have shown that the way in which consumers make decisions about money depends a great deal on how much is at stake. Few people are willing to give up $10 now for $15 next year. But they are if the choice is between $100 now and $150 next year, a fact that would explain why consumers appear to care less about many small electricity bills—even if they add up to a lot—than one big initial outlay.

EXERCISES

1. Suppose a refrigerator that sells for $700 costs $85 a year for electricity. Write an expression for the cost to buy and run the refrigerator for x years.

2. Suppose another refrigerator costs $1000 and $25 a year for electricity. Write an expression for the total cost for this refrigerator over x years.

3. Over 10 years which refrigerator costs the most? By how much?

4. In how many years will the total costs for the two refrigerators be equal?

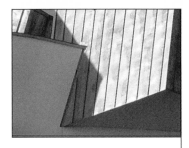

CHAPTER 2

Linear Equations, Functions, and Graphs

TECHNOLOGY RESOURCES

GraphExplorer

College Algebra and Trigonometry: Graphing Calculator Explorations, Ebersole

Developmental Mathematics: Graphing Calculator Explorations, Ebersole

Functions are an extremely useful way of describing many real-world situations in which the value of one quantity varies with, depends on, or determines the value of another. For example, the graph in Figure 2.1 shows the temperature (in degrees Fahrenheit) at each time during a single day (with time measured in hours after midnight).

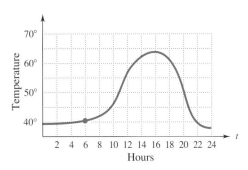

FIGURE 2.1

The graph relates the two variables, time and temperature. With each time t for $0 \le t \le 24$, the graph associates a temperature. For instance, the time $t = 6$ is associated with the temperature $40°$ since the point $(6, 40)$ is on the graph. This process of associating with each value of one variable (time) a specific value of a second variable (temperature) is an example of a *function*.

In the first part of this chapter, we discuss equations of lines, which define the simplest type of function. Then we introduce the basic ideas of functions, functional notation, applications, and graphs of functions. Nonlinear functions will be considered in Chapters 3 and 4.

1 Locate $(-1, 6)$, $(-3, -5)$, $(4, -3)$, $(0, 2)$, and $(-5, 0)$ on a coordinate system.

Answer:

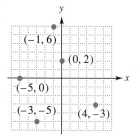

2.1 LINEAR EQUATIONS

Because of their simplicity, linear equations are used in many applications to describe relationships between two variables. We shall see some of these applications in this chapter. A **linear equation** is any equation that can be put in the form $ax + by = c$, where either a or b (but not both) may be 0. For example,

$$2x + 5y = 21, \quad x = 7, \quad \text{and } y = -3$$

are linear equations.

We can draw a picture of a linear equation, called its *graph*, by plotting pairs of values of x and y that satisfy the equation. The pairs of numbers are written as *ordered pairs*. An **ordered pair** of numbers is written between parentheses with the x-value first, as (x, y). Ordered pairs are graphed on a **Cartesian coordinate system**, as shown in Figure 2.2. The horizontal number line is called the **x-axis,** and the vertical number line is the **y-axis.** The point where the number lines cross is the zero point on both lines and is called the **origin.**

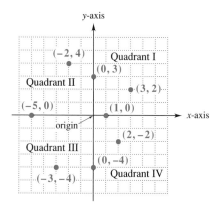

FIGURE 2.2

Each point in a Cartesian coordinate system corresponds to an ordered pair of real numbers. Several points and their corresponding ordered pairs are shown in Figure 2.2. In the point $(-2, 4)$, for example, -2 is the **x-coordinate** and 4 is the **y-coordinate.** From now on, instead of referring to "the point corresponding to the ordered pair $(-2, 4)$," we will say "the point $(-2, 4)$."

The x-axis and y-axis divide the plane into four parts, or **quadrants.** For example, Quadrant I includes points whose x- and y-coordinates are both positive. The quadrants are numbered as shown in Figure 2.2. The points of the axes themselves belong to no quadrant. **1**

The **graph** of an equation in two variables is the set of all ordered pairs that satisfy the equation. There are infinitely many ordered pairs that satisfy a linear equation; hence, its graph has infinitely many points. It can be shown that the graph of a linear equation is a straight line. A straight line is determined by two points, so we need to plot only two points to determine the graph of a linear equation; however, we recommend plotting three points, using the third point as a check.

Some points that are useful in plotting a graph of any equation are the points where the graph crosses the axes. An *x*-**intercept** of a graph is the *x*-coordinate of a point where the graph crosses the *x*-axis (the *y*-coordinate of this point is 0 since it is on the axis). Similarly, a **y-intercept** is the *y*-coordinate of a point where the graph crosses the *y*-axis (the *x*-coordinate of this point is 0).* For example, the intercepts are marked on the graph in Figure 2.3.

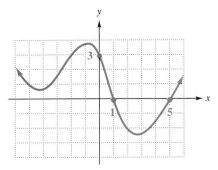

x-intercepts 1 and 5
y-intercept 3
FIGURE 2.3

▶**EXAMPLE 1**   Graph $y = -2x + 5$.
   Find the *y*-intercept by letting $x = 0$ in the equation $y = -2x + 5$ and solving for *y*.

$$y = -2x + 5$$
$$y = -2(0) + 5 \qquad \text{Let } x = 0$$
$$y = 5$$

The *y*-intercept is 5, leading to the ordered pair $(0, 5)$. In the same way, the *x*-intercept is found by letting $y = 0$ and solving for *x*.

$$0 = -2x + 5 \qquad \text{Let } y = 0$$
$$2x = 5 \qquad \text{Add } 2x \text{ to both sides}$$
$$x = \frac{5}{2} = 2\frac{1}{2} \qquad \text{Multiply both sides by } \frac{1}{2}$$

---

*Some instructors prefer to define the intercepts to be the *points* where the graph crosses the axes. Either definition provides essentially the same information.

2 Find the intercepts of each of the following. Graph the lines.

(a) $3x + 4y = 12$

(b) $5x - 2y = 8$

Answers:

(a) x-intercept 4
y-intercept 3

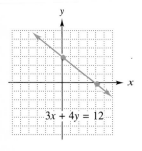

(b) x-intercept 8/5
y-intercept -4

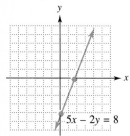

The x-intercept is 5/2, or $2\frac{1}{2}$, so the graph goes through (5/2, 0). These two intercepts lead to the graph of Figure 2.4.

As a check, a third point can be found by choosing another value of x (or y) and finding the corresponding value of the other variable. Check that (1, 3), (2, 1), (3, -1), and (4, -3) among other points, satisfy the equation $y = -2x + 5$ and lie on the line of Figure 2.4. ◀ **2**

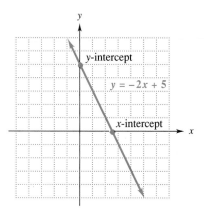

FIGURE 2.4

FOR GRAPHERS

To use a grapher to graph the equation of a line, we must first solve the equation for y, because graphing calculators and most computer programs are designed to accept equations solved for y. The reason for this will become apparent in Section 2.3.

▶**EXAMPLE 2**   Graph each equation.

(a) $y = -3$.

The equation $y = -3$, or equivalently, $0x + y = -3$, always gives the same y value, -3, for any value of x. Therefore, no value of x will make $y = 0$, so the graph has no x-intercept. Since $y = -3$ is a linear equation with a straight line graph, and since the graph cannot cross the x-axis, the line must be parallel to the x-axis. For any value of x, the value of y is -3, and so the graph is the horizontal line parallel to the x-axis, with y-intercept -3, as shown in Figure 2.5.

3 Graph each equation.

(a) $y = -5$

(b) $x - 4 = 0$

Answers:

(a)

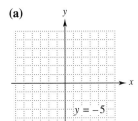

(b)

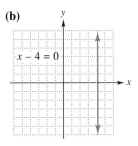

(b) $x = -1$.

Here x is always -1 and y can be any number. Find some ordered pairs: $(-1, 0)$, $(-1, 2)$, and $(-1, 4)$ all satisfy the equation. Plot these ordered pairs and draw a line through them to get the graph in Figure 2.6.   ◄ **3**

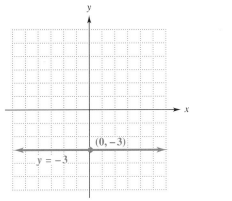

FIGURE 2.5

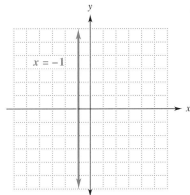

FIGURE 2.6

Example 2 suggests the following.

The graph of $y = k$, where k is a real number, is the horizontal line having y-intercept k.

The graph of $x = k$, where k is a real number, is the vertical line having x-intercept k.

▶ **EXAMPLE 3**   Graph $y = -3x$.

Begin by looking for the x-intercept.

$$0 = -3x \qquad \text{Let } y = 0$$
$$0 = x, \qquad \text{Multiply both sides by } -\frac{1}{3}$$

The corresponding ordered pair is $(0, 0)$. Letting $x = 0$ leads to exactly the same ordered pair, $(0, 0)$. Two different points are needed to determine a straight line, and the intercepts have led to only one point. Get a second point by choosing some other value of x (or y). For example, if $x = 2$,

$$y = -3x = -3(2) = -6,$$

Graph

(a) $x = 5y$;

(b) $5x = y$.

Answers:

(a)

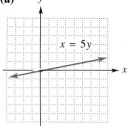

(b)

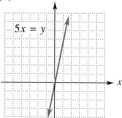

giving the ordered pair (2, −6). These two ordered pairs, (0, 0) and (2, −6), were used to get the graph in Figure 2.7. ◀

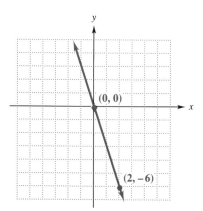

FIGURE 2.7

Linear equations can be very helpful in setting up a mathematical model for a real-life situation. In almost every case, linear (or any other reasonably simple) equations provide only *approximations* to the real-world situations. However, these approximations are aften remarkably useful.

SUPPLY AND DEMAND The supply of and demand for an item are usually related to its price. Producers will supply large numbers of the item at a high price, but consumer demand will be low. As the price of the item decreases, consumer demand increases, but producers are less willing to supply large numbers of the item. The curves that show the quantity that will be supplied at a given price and the quantity that will be demanded at a given price are called **supply and demand curves.** Straight lines often appear as supply and demand curves, as in the next example. In applications, we often use as variables letters that suggest what they represent rather than the letters x and y. In supply and demand problems, we use p for price and q for quantity. We will discuss the economic concepts of supply and demand in more detail in later chapters.

▶**EXAMPLE 4** Greg Odjakjian, an economist, has studied the supply and demand for aluminum siding and has determined that price per unit,* p, and demand, q, are related by the linear equation

$$p = 60 - \frac{3}{4}q.$$

---

*An appropriate unit here might be, for example, one thousand square feet of siding.

5 Suppose price and demand are related by $p = 100 - 4q$.

(a) Find the price if the demand is 10 units.

(b) Find the demand if the price is $80.

(c) Write the corresponding ordered pairs.

Answers:

(a) $60

(b) 5 units

(c) (10, 60); (5, 80)

(a) Find the demand at a price of $40 per unit.

Let $p = 40$.

$$p = 60 - \frac{3}{4}q$$

$$40 = 60 - \frac{3}{4}q \qquad \text{Let } p = 40$$

$$-20 = -\frac{3}{4}q \qquad \text{Add } -60 \text{ on both sides}$$

$$\frac{80}{3} = q \qquad \text{Multiply both sides by } -\frac{4}{3}$$

At a price of $40 per unit, 80/3 (or $26\frac{2}{3}$) units will be demanded; this gives the ordered pair (80/3, 40). (It is customary to write the ordered pairs so that price comes second.)

(b) Find the price if the demand is 32 units.

Let $q = 32$.

$$p = 60 - \frac{3}{4}q$$

$$p = 60 - \frac{3}{4}(32) \qquad \text{Let } q = 32$$

$$p = 60 - 24$$

$$p = 36$$

With a demand of 32 units, the price is $36. This gives the ordered pair (32, 36).

(c) Graph $p = 60 - \frac{3}{4}q$.

Use the ordered pairs (80/3, 40) and (32, 36) to get the demand graph shown in Figure 2.8. Only the portion of the graph in Quadrant I is shown, because the supply and demand functions are meaningful only for positive values of p and q. **5**

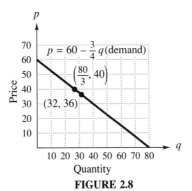

FIGURE 2.8

(d) From Figure 2.8, at a price of $30, what quantity is demanded?

Price is located on the vertical axis. Look for 30 on the p-axis and read across to where the line $p = 30$ crosses the demand graph. As the graph shows, this occurs where the demand is 40.

(e) At what price will 60 units be demanded?

Quantity is located on the horizontal axis. Find 60 on the q-axis and read up to where the vertical line $q = 60$ crosses the demand graph. This occurs where the price is $15 per unit.

(f) What quantity is demanded at a price of $60 per unit?

The point $(0, 60)$ on the demand graph shows that the demand is 0 at a price of $60 (that is, there is no demand at such a high price). ◀

▶**EXAMPLE 5**   Suppose that the economist of Example 4 concludes that the price and supply of siding are related by

$$p = \frac{3}{4}q.$$

(a) Find the supply if the price is $60 per unit.

$$60 = \frac{3}{4}q \qquad \text{Let } p = 60$$

$$80 = x$$

If the price is $60 per unit, then 80 units will be supplied to the marketplace. This gives the ordered pair $(80, 60)$.

(b) Find the price per unit if the supply is 16 units.

$$p = \frac{3}{4}(16) = 12 \qquad \text{Let } q = 16$$

If the supply is 16 units, then the price is $12 per unit. This gives the ordered pair $(16, 12)$.

(c) Graph $p = \frac{3}{4}q$.

Use the ordered pairs $(80, 60)$ and $(16, 12)$ to get the supply graph shown in Figure 2.9.

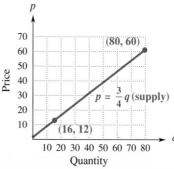

FIGURE 2.9

(d) Use the graph in Figure 2.9 to find the price at which 60 units will be supplied.

Quantity is located on the horizontal axis. The vertical line $q = 60$ crosses the supply graph when the price is $45 per unit. ◀

Graphing both the supply and demand curves on the same axes makes it easy to answer certain questions.

▶**EXAMPLE 6** The supply and demand curves of Examples 4 and 5 are shown in Figure 2.10, with the supply curve in color.

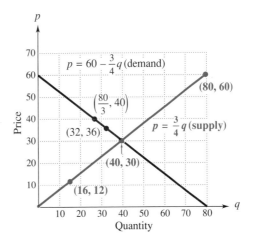

FIGURE 2.10

(a) Use Figure 2.10 to determine whether there is a surplus or shortage of supply at a price of $40 per unit.

Find 40 on the vertical-axis. Read across to where the line $p = 40$ crosses the supply graph. Since the supply graph lies to the right of the demand graph at that point, supply is greater than demand and there is a surplus of supply.

(b) For what quantity does demand exceed supply?

The demand graph lies above the supply graph so demand is greater than supply when the quantity supplied or demanded is in the interval $[0, 40)$. ◀

As shown by the graphs of Figure 2.10, both the supply and the demand functions pass through the point $(40, 30)$. If the price of the siding is more than $30, the supply will exceed the demand. At a price less than $30, the demand

6 The demand for a certain commodity is related to the price by $p = 80 - (2/3)q$. The supply is related to the price by $p = (4/3)q$. Find

(a) the equilibrium demand;

(b) the equilibrium price.

Answers:

(a) 40

(b) 160/3

will exceed the supply. Only at a price of $30 will demand and supply be equal. For this reason, $30 is called the *equilibrium price*. When the price is $30, demand and supply both equal 40 units, the *equilibrium supply* or *equilibrium demand*. By definition, the **equilibrium price** of a commodity is the price at the point where the supply and demand graphs for that commodity cross. The **equilibrium demand** is the demand at that same point; the **equilibrium supply** is the supply at that point. By definition, the equilibrium supply and the equilibrium demand are equal.

▶**EXAMPLE 7**   Use algebra to find the equilibrium supply, demand, and price for the aluminum siding. (See Examples 4 and 5.)

The equilibrium supply is found when the prices from both supply and demand are equal. From Example 4, $p = 60 - (3/4)q$; in Example 5, $p = (3/4)q$. Set these two expressions for p equal to get the linear equation

$$60 - \frac{3}{4}q = \frac{3}{4}q.$$

$$240 - 3q = 3q \qquad \text{Multiply both sides by 4}$$
$$240 = 6q \qquad \text{Add } 3q \text{ to both sides}$$
$$40 = q.$$

The equilibrium supply is 40 units, the same answer found above. This is also the equilibrium demand. To find the equilibrium price, substitute 40 for q in either the supply or demand function and solve for p. As we saw above, the equilibrium price is $30.   ◀   **6**

FOR GRAPHERS

The equilibrium price and supply/demand can also be found with a grapher. Use your grapher to graph the supply and demand equations in Examples 4 and 5, written as $y = (3/4)x$ and $y = 60 - (3/4)x$. Then use trace and zoom to determine the x- and y-values of the point of intersection of the two graphs (40, 30).

2.1   EXERCISES

Graph each of the following linear equations. (See Examples 1–3.)

1. $2x - y = 7$ **2.** $3x - y = 4$ **3.** $x + y = 4$ **4.** $x - y = 5$

5. $2x + 5y = 10$ **6.** $6x + 3y = 18$ **7.** $4x + y = 8$ **8.** $x + 3y = 9$

9. $x - 5 = 0$ **10.** $y + 2 = 0$ **11.** $y = 3x$ **12.** $y = -x$

13. $x + 2y = 0$ **14.** $x - 3y = 0$ **15.** $5y - 3x = 12$ **16.** $2x + 7y = 14$

17. $8x + 3y = 10$ **18.** $9y - 4x = 12$ **19.** $x = 2y + 3$

20. What is the equation of the *x*-axis? the *y*-axis?

21. What is meant by the graph of an equation?

22. Describe the *x*-intercept and the *y*-intercept of a graph.

Management   *Use the supply and demand curves graphed below to answer Exercises 23–26. (See Examples 4–6.)*

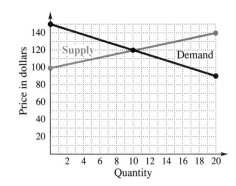

23. At what price are 20 items supplied?

24. At what price are 20 items demanded?

25. Find the equilibrium supply and the equilibrium demand.

26. Find the equilibrium price.

Management   *Work the following exercises. (See Examples 4–7.)*

27. Suppose that the demand and price for a certain brand of shampoo are related by

$$p = 16 - \frac{5}{4}q.$$

where *p* is price, in dollars, and *q* is demand. Find the price for a demand of
(a) 0 units;   **(b)** 4 units;   **(c)** 8 units.
Find the demand for the shampoo at a price of
(d) $6;   **(e)** $11;   **(f)** $16.
(g) Graph $p = 16 - (5/4)q$. Suppose the price and supply of the shampoo are related by

$$p = \frac{3}{4}q.$$

where *q* represents the supply, and *p* the price. Find the supply when the price is
(h) $0;   **(i)** $10;   **(j)** $20.

(k) Graph $p = (3/4)q$ on the same axes used for part (g).
(l) Find the equilibrium supply.
(m) Find the equilibrium price.

28. Let the supply and demand for radial tires in dollars be given by

$$\text{supply: } p = \frac{3}{2}q \quad \text{and} \quad \text{demand: } p = 81 - \frac{3}{4}q.$$

(a) Graph these on the same axes.
(b) Find the equilibrium demand.
(c) Find the equilibrium price.

29. Let the supply and demand for bananas in cents per pound be given by

$$\text{supply: } p = \frac{2}{5}q \quad \text{and} \quad \text{demand: } p = 100 - \frac{2}{5}q.$$

(a) Graph these on the same axes.
(b) Find the equilibrium demand.
(c) Find the equilibrium price.
(d) On what interval does demand exceed supply?

30. Let the supply and demand for sugar be given by
$$\text{supply: } p = 1.4q - .6$$
and
$$\text{demand: } p = -2q + 3.2,$$
where *p* is in dollars.
(a) Graph these on the same axes.
(b) Find the equilibrium demand.
(c) Find the equilibrium price.
(d) On what interval does supply exceed demand?

31. In a recent issue of *Business Week,* the president of Insta-Tune, a chain of franchised automobile tune-up shops, says that people who buy a franchise and open a shop pay a weekly fee (in dollars) of

$$y = .07x + \$135$$

to company headquarters. Here *y* is the fee and *x* is the total amount of money taken in during the week by the tune-up center. Find the weekly fee if *x* is
(a) $0;   **(b)** $1000;   **(c)** $2000;
(d) $3000.   **(e)** Graph *y*.

32. In a recent issue of *The Wall Street Journal,* we are told that the relationship between the amount of money that an average family spends on food eaten at home, *x*, and

the amount of money it spends on eating out, y, is approximated by the model $y = .36x$. Find y if x is
(a) $40; (b) $80; (c) $120; (d) Graph y.

Management *When revenue and cost equations are given, the break-even point is the x-value where revenue equals cost.*

33. For x thousand policies, an insurance company claims that their monthly revenue in dollars is given by $R = 125x$ and their monthly cost in dollars is given by $C = 100x + 5000$.
    (a) Find the break-even point.
    (b) Graph the revenue and cost equations on the same axes.
    (c) From the graph, estimate the revenue and cost when $x = 100$ (100 thousand policies).

34. The owners of a parking lot have determined that their weekly revenue and cost in dollars are given by $R = 80x$ and $C = 50x + 2400$, where x is the number of long-term parkers.
    (a) Find the break-even point.
    (b) Graph R and C on the same axes.

(c) From the graph, estimate the revenue and cost when there are 60 long-term parkers.

35. **Management** Ral Corp. has an incentive compensation plan under which a branch manager receives 10% of the branch's income after deduction of the bonus but before deduction of income tax.* Branch income for 1988 before the bonus and income tax was $165,000. The tax rate was 30%. The 1988 bonus amounted to
    (a) $12,600 (b) $15,000
    (c) $16,500 (d) $18,000

Management *The lost value of equipment over a period of time is called* **depreciation.** *The simplest method for calculating depreciation is* **straight-line depreciation.** *The annual straight-line depreciation on an item that cost x dollars with a useful life of n years is* $D = (1/n)x$. *Find the depreciation for items with the following characteristics.*

36. Cost: $12,482; life 10 yr

37. Cost: 39,700; life 12 yr

38. Cost: $145,000; life 28 yr

_____
* Uniform CPA Examination, May, 1989, American Institute of Certified Public Accountants

2.2 SLOPE AND THE EQUATIONS OF A LINE

As mentioned in the previous section, the graph of a straight line is completely determined by two different points on the line. The graph of a straight line also can be drawn knowing only *one* point on the line *if* the "steepness" of the line is known, too. The number that represents the "steepness" of a line is called the *slope* of that line.

To see how slope is defined, start with Figure 2.11, which shows a line passing through the two different points $(x_1, y_1) = (-3, 5)$ and $(x_2, y_2) = (2, -4)$. The difference in the two x values,

$$x_2 - x_1 = 2 - (-3) = 5$$

in this example, is called the **change in x.** The Greek letter Δ (delta) is used to denote change. The symbol Δx (read "delta x") represents the change in x. In the same way, Δy represents the **change in y.** In this example,

$$\Delta y = y_2 - y_1 = -4 - 5 = -9.$$

1 Find the slope of the line through

(a) $(6, 11)$, $(-4, -3)$;

(b) $(-3, 5)$, $(-2, 8)$.

Answers:

(a) $7/5$

(b) 3

The **slope** of the line through the two points (x_1, y_1) and (x_2, y_2), where $x_1 \neq x_2$, is defined as the quotient of the change in y and the change in x, or

$$\text{slope} = \frac{\text{change in } y}{\text{change in } x} = \frac{\Delta y}{\Delta x} = \frac{y_2 - y_1}{x_2 - x_1}.$$

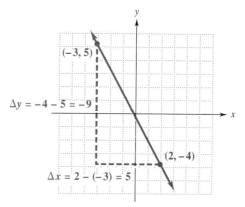

FIGURE 2.11

The slope of the line in Figure 2.11 is

$$\text{slope} = \frac{\Delta y}{\Delta x} = \frac{-4 - 5}{2 - (-3)} = -\frac{9}{5}.$$

Using similar triangles from geometry, it can be shown that the slope is independent of the choice of points on the line. That is, the same value of the slope will be obtained for *any* choice of two different points on the line.

▶ **EXAMPLE 1** Find the slope of the line through the points $(-7, 6)$ and $(4, 5)$. Let $(x_1, y_1) = (-7, 6)$. Then $(x_2, y_2) = (4, 5)$. Use the definition of slope.

$$\text{slope} = \frac{\Delta y}{\Delta x} = \frac{5 - 6}{4 - (-7)} = -\frac{1}{11}$$

The slope also could have been found by letting $(x_1, y_1) = (4, 5)$ and $(x_2, y_2) = (-7, 6)$. In that case,

$$\text{slope} = \frac{6 - 5}{-7 - 4} = \frac{1}{-11} = -\frac{1}{11},$$

the same answer. ◀ **1**

Caution When finding the slope of a line, be careful to subtract the *x*-values and *y*-values in the same order. For instance, consider the slope of the line through the points (2, 3) and (4, 9).

Correct	Incorrect
2nd point − 1st point	2nd point − 1st point
$\dfrac{9 - 3}{4 - 2} = 3$	$\dfrac{9 - 3}{2 - 4} = -3$
2nd point − 1st point	1st point − 2nd point
1st point − 2nd point	1st point − 2nd point
$\dfrac{3 - 9}{2 - 4} = 3$	$\dfrac{3 - 9}{4 - 2} = -3.$
1st point − 2nd point	2nd point − 1st point

The slope of a line is a measure of the steepness of the line. Figure 2.12 shows examples of lines with different slopes. Lines with positive slopes go up as *x* goes from left to right along the *x*-axis, while lines with negative slopes go down as *x* goes from left to right.

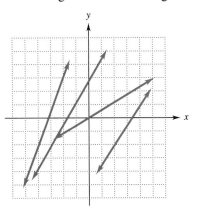

Lines with positive slope

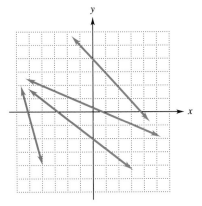

Lines with negative slope

FIGURE 2.12

▶ **EXAMPLE 2** Find the slope of the line $3x - 4y = 12$.

The slope is found using two different points on the line. The intercepts can be used for the two points. First let $x = 0$ and then let $y = 0$ to find the intercepts.

$$
\begin{array}{ll}
\text{If } x = 0, & \text{If } y = 0 \\
3x - 4y = 12 & 3x - 4y = 12 \\
3(0) - 4y = 12 \quad \text{Let } x = 0 & 3x - 4(0) = 12 \quad \text{Let } y = 0 \\
-4y = 12 & 3x = 12 \\
y = -3 & x = 4
\end{array}
$$

2 Find the slope of

(a) $8x + 5y = 9$;

(b) $8y = x$.

Answers:

(a) $-8/5$

(b) $1/8$

3 Use $x = 0$ and $y = 0$ to get two ordered pairs and then find the slope of each line.

(a) $2x + 3y = 5$

(b) $4x + 5 = -6y$

Answers:

(a) $(0, 5/3)$ and $(5/2, 0)$; $-2/3$

(b) $(0, -5/6)$ and $(-5/4, 0)$; $-2/3$

4 Tell if the lines in each of the following pairs are *parallel, perpendicular,* or *neither.*

(a) $x - 2y = 6$ and
$2x + y = 5$

(b) $3x + 4y = 8$ and
$x + 3y = 2$

(c) $2x - y = 7$ and
$2y = 4x - 5$

Answers:

(a) Perpendicular

(b) Neither

(c) Parallel

These results give the ordered pairs $(0, -3)$ and $(4, 0)$. Now the slope can be found from the definition.

$$\text{slope} = \frac{0 - (-3)}{4 - 0} = \frac{3}{4} \quad \blacktriangleleft \quad \boxed{2}$$

We shall assume the following facts without proof.

Two nonvertical lines are parallel whenever they have the same slope.

Two lines, where neither is vertical, are perpendicular whenever the product of their slopes is -1.

▶ **EXAMPLE 3** Tell whether each of the following pairs of lines are *parallel, perpendicular,* or *neither.*
(a) $2x + 3y = 5$ and $4x + 5 = -6y$
Find the slope of each line by first finding two points on each line. The intercepts are a good choice. **3**
From Problem 3 at the side, both slopes are $-2/3$, so the lines are parallel.
(b) $3x = y + 7$ and $x + 3y = 4$
Verify that the slope of $3x = y + 7$ is 3 and the slope of $x + 3y = 4$ is $-1/3$. Since $3(-1/3) = -1$, these lines are perpendicular.
(c) $x + y = 4$ and $x - 2y = 3$
The slope of the first line is -1 and of the second line is $1/2$. Because the slopes are not equal and their product is $-1/2$, not -1, the lines are neither parallel nor perpendicular. $\blacktriangleleft$ **4**

SLOPE-INTERCEPT FORM A generalization of the method of Example 2 can be used to find the equation of a line, given its y-intercept and slope. Assume that a line has y-intercept b, so that its goes through $(0, b)$. (See Figure 2.13.)

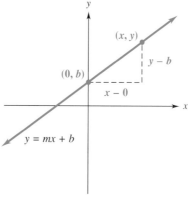

FIGURE 2.13

5 Find an equation for the line with

(a) y-intercept -3 and slope $2/3$;

(b) y-intercept $1/4$ and slope $-3/2$.

(c) Are these lines parallel, perpendicular, or neither?

Answers:

(a) $y = \dfrac{2}{3}x - 3$

(b) $y = -\dfrac{3}{2}x + \dfrac{1}{4}$

(c) Perpendicular

Let the slope of the line be m. If (x, y) is any point on the line *other* than $(0, b)$, using the definition of slope with the points $(0, b)$ and (x, y) gives

$$m = \frac{y - b}{x - 0}$$

$$m = \frac{y - b}{x}$$

$$mx = y - b \qquad \text{Multiply by } x.$$

from which

$$y = mx + b. \qquad \text{Add } b. \text{ Reverse the equation.}$$

This discussion is summarized as follows.

Slope-Intercept Form

If a line has slope m and y-intercept b, then

$$y = mx + b$$

is the **slope-intercept form** of the equation of the line.

▶**EXAMPLE 4** Find an equation for the line with y-intercept $7/2$ and slope $-5/2$.

Use the slope-intercept form with $b = 7/2$ and $m = -5/2$.

$$y = mx + b$$

$$y = -\frac{5}{2}x + \frac{7}{2} \quad ◀ \quad \boxed{5}$$

The slope-intercept form of the equation of a line shows that the slope can be found from the equation by solving for y. In this form, the coefficient of x is the slope and the constant term is the y-intercept.

For instance, we found in Example 2 that the slope of the line $3x - 4y = 12$ is $3/4$. This slope can also be found by solving for y.

$$3x - 4y = 12$$

$$-4y = -3x + 12 \qquad \text{Add } -3x$$

$$y = \frac{3}{4}x - 3 \qquad \text{Divide by } -4$$

The coefficient of x gives the slope, $3/4$.

6 Find the slope and
y-intercept for

(a) $x + 4y = 6$;

(b) $3x - 2y = 1$.

Answers:

(a) Slope $-1/4$; y-intercept $3/2$

(b) Slope $3/2$; y-intercept $-1/2$

7 Use the slope and
y-intercept to graph

(a) $2x + 5y = 10$;

(b) $-x + y = 4$.

Answers:

(a)

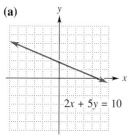

$2x + 5y = 10$

(b)

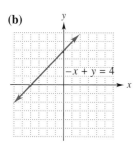

$-x + y = 4$

▶ **EXAMPLE 5** Find the slope and y-intercept for each of the following lines.

(a) $5x - 3y = 1$

Solve for y.

$$5x - 3y = 1$$
$$-3y = -5x + 1$$
$$y = \frac{5}{3}x - \frac{1}{3}$$

The slope is $5/3$ and the y-intercept is $-1/3$.

(b) $-9x + 6y = 2$

Solve for y.

$$-9x + 6y = 2$$
$$6y = 9x + 2$$
$$y = \frac{3}{2}x + \frac{1}{3}$$

The slope is $3/2$ and the y-intercept is $1/3$. ◀ **6**

The slope and y-intercept of a line can be used to draw the graph of the line, as shown in the next example.

▶ **EXAMPLE 6** Use the slope and y-intercept to graph $3x - 2y = 2$.

Solve for y.

$$3x - 2y = 2$$
$$-2y = -3x + 2$$
$$y = \frac{3}{2}x - 1$$

The slope is $3/2$ and the y-intercept is -1.

To draw the graph, first locate the y-intercept -1, as shown in Figure 2.14. Use the slope to find a second point on the graph. If m represents the slope, then

$$m = \frac{\Delta y}{\Delta x} = \frac{3}{2}.$$

If x changes by positive 2 units ($\Delta x = 2$), then y will change by positive 3 units ($\Delta y = 3$). Find the second point by starting at the y-intercept graphed in Figure 2.14 on the next page and moving 2 units to the right and 3 units up. After this second point has been located, a line can be drawn through it and the y-intercept. Because $3/2 = -3/-2$, a second point also can be located by moving 2 units to the left (-2) and 3 units down (-3). This is a good way to check the line found by using $m = 3/2$. ◀ **7**

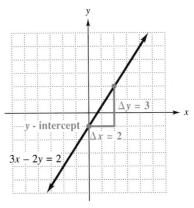

FIGURE 2.14

POINT-SLOPE FORM The slope-intercept form of the equation of a line involves the slope and the y-intercept. Sometimes, however, the slope of a line is known, together with one point (perhaps *not* the y-intercept) that the line goes through. The *point-slope form* of the equation of a line is used to find an equation in this case. Let (x_1, y_1) be any fixed point on the line and let (x, y) represent any other point on the line. If m is the slope of the line, then, by the definition of slope,

$$\frac{y - y_1}{x - x_1} = m,$$

or

$$y - y_1 = m(x - x_1).$$

Point-Slope Form

If a line has slope m and passes through the point (x_1, y_1), then

$$\boldsymbol{y - y_1 = m(x - x_1)}$$

is the **point-slope form** of the equation of the line.

▶**EXAMPLE 7** Find an equation of the line with the given slope and going through the given point.

(a) $(-4, 1)$, $m = -3$

Use the point-slope form because a point the line goes through, together with the slope of the line, is known. Substitute the values $x_1 = -4$, $y_1 = 1$, and $m = -3$ into the point-slope form.

8 Find an equation of the line having the given slope and going through the given point.

(a) $m = -3/5$, $(5, -2)$

(b) $m = 1/3$, $(6, 8)$

Answers:

(a) $5y = -3x + 5$

(b) $3y = x + 18$

9 Find an equation of the line through

(a) $(2, 3)$ and $(-4, 6)$;

(b) $(-8, 2)$ and $(3, -6)$.

Answers:

(a) $2y = -x + 8$

(b) $11y = -8x - 42$

$$y - y_1 = m(x - x_1)$$
$$y - 1 = -3[x - (-4)] \qquad \text{Let } y_1 = 1, m = -3, x_1 = -4$$
$$y - 1 = -3(x + 4)$$
$$y - 1 = -3x - 12 \qquad \text{Distributive property}$$
$$y = -3x - 11$$

(b) $(3, -7)$, $m = 5/4$

$$y - y_1 = m(x - x_1)$$
$$y - (-7) = \frac{5}{4}(x - 3) \qquad \text{Let } y_1 = -7, m = \frac{5}{4}, x_1 = 3$$
$$y + 7 = \frac{5}{4}(x - 3)$$
$$4(y + 7) = 5(x - 3) \qquad \text{Multiply both sides by 4}$$
$$4y + 28 = 5x - 15 \qquad \text{Distributive property}$$
$$4y = 5x - 43 \qquad \blacktriangleleft \ \boxed{8}$$

The point-slope form can also be used to find an equation of a line given two different points that the line goes through. The procedure for doing this is shown in the next example.

▶ **EXAMPLE 8** Find an equation of the line through $(5, 4)$ and $(-10, -2)$.
Begin by using the definition of slope to find the slope of the line that passes through the two points.

$$\text{slope} = m = \frac{-2 - 4}{-10 - 5} = \frac{-6}{-15} = \frac{2}{5}$$

Use $m = 2/5$ and either of the given points in the point-slope form. If $(x_1, y_1) = (5, 4)$, then

$$y - y_1 = m(x - x_1)$$
$$y - 4 = \frac{2}{5}(x - 5) \qquad \text{Let } y_1 = 4, m = \frac{2}{5}, x_1 = 5$$
$$5(y - 4) = 2(x - 5) \qquad \text{Multiply both sides by 5}$$
$$5y - 20 = 2x - 10 \qquad \text{Distributive property}$$
$$5y = 2x + 10.$$

Check that the result is the same when $(x_1, y_1) = (-10, -2)$. $\qquad \blacktriangleleft \ \boxed{9}$

▶ **EXAMPLE 9** Find an equation of the line through $(8, -4)$ and $(-2, -4)$.
First find the slope.

$$m = \frac{-4 - (-4)}{-2 - 8} = \frac{0}{-10} = 0$$

10 Find an equation of the line through $(-2, 5)$ and $(7, 5)$. Describe the graph of this line.

Answer:
$y = 5$ A horizontal line with all y-values 5.

11 Find an equation of the line through $(-5, 1)$ and $(-5, 7)$. Describe the graph of this line.

Answer:
$x = -5$ A vertical line with all x-values -5.

Choose, say, $(8, -4)$ as (x_1, y_1), and use the point-slope form.

$$y - y_1 = m(x - x_1)$$

$$y - (-4) = 0(x - 8) \qquad \text{Let } y_1 = -4, m = 0, x_1 = 8$$

$$y + 4 = 0 \qquad\qquad 0(x - 8) = 0$$

$$y = -4 \quad \blacktriangleleft$$

As shown in the previous section, $y = -4$ represents a horizontal line with y-intercept -4.   **10**

▶ **EXAMPLE 10**   Find an equation of the line through $(4, 3)$ and $(4, -6)$.
Begin by finding the slope.

$$m = \frac{-6 - 3}{4 - 4} = \frac{-9}{0}$$

Division by 0 is not defined, so the slope is undefined. Graphing the given ordered pairs $(4, 3)$ and $(4, -6)$ and drawing a line through them gives a vertical line. From the last section, vertical lines have equations of the form $x = k$, where k can be any real number. Since the x-coordinate of the two ordered pairs given above is 4, the desired equation is $x = 4$.   ◀ **11**

Examples 9 and 10 suggest the following.

The slope of every horizontal line is 0.

The slope of every vertical line is undefined.

A summary of the equations of lines discussed in this section follows.

Equation	Description
$ax + by = c$	If $a \neq 0$ and $b \neq 0$, the line has x-intercept c/a and y-intercept c/b.
$x = k$	**Vertical line**, x-intercept k, no y-intercept, undefined slope
$y = k$	**Horizontal line**, y-intercept k, no x-intercept, slope 0
$y = mx + b$	**Slope-intercept form**, slope m, y-intercept b
$y - y_1 = m(x - x_1)$	**Point-slope form**, slope m, the line passes through (x_1, y_1)

2.2 EXERCISES

In each of the following exercises, find the slope, if it is defined, of the line. (See Examples 1, 2, and 5.)

1. through $(-5, 4)$ and $(3, 7)$

2. through $(-1, -3)$ and $(4, 8)$

3. through $(2, 5)$ and $(0, 6)$

4. through $(9, 0)$ and $(12, 15)$

5. through the origin and $(-4, 6)$

6. through the origin and $(8, -2)$

7. through $(-1, 4)$ and $(-1, 8)$

8. through $(-3, 5)$ and $(2, 5)$

9. $y = 2x$

10. $y = 3x - 2$

11. $5x - 9y = 11$

12. $4x + 7y = 1$

13. parallel to $2y - 4x = 7$

14. perpendicular to $6x = y - 3$

15. through $(-1.978, 4.806)$ and $(3.759, 8.125)$

16. through $(11.72, 9.811)$ and $(-12.67, -5.009)$

17. What is true of the slopes of perpendicular lines?

Tell whether each pair of lines is parallel, perpendicular, or neither. (See Example 3.)

18. $4x - 3y = 6$ and $3x + 4y = 8$

19. $2x - 5y = 7$ and $15y - 5 = 6x$

20. $3x + 2y = 8$ and $6y = 5 - 9x$

21. $x - 3y = 4$ and $y = 1 - 3x$

22. $4x = 2y + 3$ and $2y = 2x + 3$

23. $2x - y = 6$ and $x - 2y = 4$

Find the slope and y-intercept of each of the following lines. (See Examples 2 and 5.)

24. $y = -3x + 4$

25. $2y = x + 5$

26. $4x + 3y = 12$

27. $6x - 9y = 14$

28. $4x + y = 0$

29. $2x - 3y = 0$

30. $y + 3 = 0$

31. $x = 2$

32. Match each equation with the line that most closely resembles its graph.
(*Hint:* Consider the signs of m and b in the slope-intercept form.)
(a) $y = 3x + 2$
(b) $y = -3x + 2$
(c) $y = 3x - 2$
(d) $y = -3x - 2$

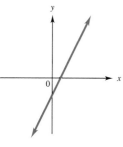

A

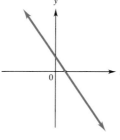

B

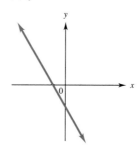

C

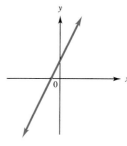

D

Graph the line through the given point with the given slope. (See Example 6.)

33. $(-1, 2)$, $m = -2/3$ **34.** $(-4, -3)$, $m = 5/4$ **35.** $(-2, -2)$, $m = 2$ **36.** $(-2, 3)$, $m = -1/2$

37. $(8, 2)$, $m = 0$ **38.** $(2, -4)$, $m = 0$ **39.** $(6, -5)$, undefined slope   **40.** $(-8, 9)$, undefined slope

Find an equation in the form by $= ax + c$ for the line with the given y-intercept and slope. (See Example 4.)

41. 4, $m = -\dfrac{3}{4}$

42. -3, $m = \dfrac{2}{3}$

43. -2, $m = -\dfrac{1}{2}$

44. $\dfrac{3}{2}$, $m = \dfrac{1}{4}$

45. $\dfrac{5}{4}$, $m = \dfrac{3}{2}$

46. $-\dfrac{3}{8}$, $m = \dfrac{3}{4}$

Find an equation in the form by $= ax + c$ for each of the following lines. (See Examples 4 and 7–10.)

47. Through $(5, 1)$, $m = -1$ **48.** Through $(0, 3)$, $m = -3$ **49.** Through $(-2, 3)$, $m = 3/2$

50. Through $(3, 2)$, $m = 1/4$ **51.** Through $(0, 1)$, $m = -2/3$ **52.** Through $(-1, 1)$, and $(2, 5)$

53. Through $(-8, 4)$, and $(-8, 6)$ **54.** Through $(2, -5)$, and $(4, -5)$ **55.** Through $(-1, 3)$, and $(0, 3)$

In each of the following problems, assume that the data can be approximated closely by a straight line. Find the equation of the line. Then answer the question.

56. Management   The sales of a small company were $27,000 in its second year of operation and $63,000 in its fifth year. Let y represent sales in year x. What were the sales in the fourth year?

57. Management   Consumer prices in the United States at the beginning of each year, measured as a percent of the 1967 average, have produced a graph that is approximately linear, In 1984 the consumer price index was 300% and in 1987 it was 333%. Let y represent the consumer price index in year x, where $x = 0$ corresponds to 1980. In what year will the consumer price index reach 350%?

58. Physical Science   Temperatures of 32° and 212° Fahrenheit correspond to temperatures of 0° and 100° Centigrade. Let y be the Centigrade temperature corresponding to Fahrenheit temperature x. An increase in temperature of 1° Fahrenheit corresponds to what change in degrees Centigrade?

59. Physical Science   Suppose a baseball is thrown at 85 miles per hour. The ball will travel 320 feet when hit by a bat swung at 50 miles per hour and will travel 440 feet when hit by a bat swung at 80 miles per hour. Let y be the number of feet traveled by the ball when hit by a bat swung at x miles per hour. (*Note:* This is valid for $50 \leq x \leq 90$, where the bat is 35 inches long, weights 32 ounces, and strikes a waist-high pitch so the place of the swing lies at 10° from the diagonal.*) How much farther will a ball travel for each mile per hour increase in the speed of the bat?

60. Social Science   The number of farms in the United States declined from 6 million in 1920 to 2 million in 1980.** Let y be the number of farms (in millions) x years after 1900. How many farms were there in 1960?

61. Social Science   The average size of farms in the United States increased from 100 acres in 1920 to 700 acres in 1980. Let y be the average size x years after 1900. In what year was the average size 400 acres?

62. Social Science   The worldwide consumption of cigarettes increased from 2.5 trillion in 1960 to 4 trillion in 1980. Let y be the consumption of cigarettes (in trillions) x years after 1940. In what year will the consumption reach 5.5 trillion?

---

*Adair, Robert K; The Physics of Baseball; Harper & Row; 1990.

** The information for Exercises 60–63 was taken from *Mathematics and Global Survival* by Richard H. Schwartz, 2nd Edition; Ginn Press, 1991.

63. **Natural Science**   The amount of tropical rain forests in Central America decreased from 130,000 square miles to about 80,000 square miles from 1969 to 1985. Let y be the amount (in ten thousands of square miles) x years after 1965. How large will the rain forests be in the year 1997?

64. **Natural Science**   In 1990, the Intergovernmental Panel on Climate Change predicted that the average temperature on the earth would rise. .3°C per decade in the absence of international controls on greenhouse emissions.* The average global temperature was 15°C in 1970. Let y be the average global temperature t years after 1970. Scientists have estimated that the sea level will rise by 65 centimeters, if the average global temperature rises to 19°C. From your equation, when will this occur?

*See *Science News*, June 23, 1990, p. 391.

2.3 FUNCTIONS

To understand the origin of the concept of function, we consider some "real life" situations in which one numerical quantity depends on, corresponds to, or determines another.

▶**EXAMPLE 1**   The amount of income tax you pay depends on the amount of your income. The way in which the income determines the tax is given by the tax law.   ◀

▶**EXAMPLE 2**   The weather bureau records the temperature over a 24-hour period in the form of a graph (see Figure 2.15). The graph shows the temperature that corresponds to each given time.   ◀

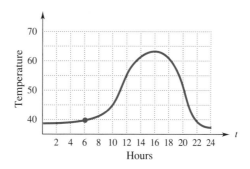

FIGURE 2.15

▶**EXAMPLE 3**   Suppose a rock is dropped straight down from a high point. From physics we know that the distance traveled by the rock in t seconds in $16t^2$ feet. So the distance depends on the time.   ◀

The first common feature shared by these examples is that each involves two sets of numbers, which we can think of as a set of inputs and a set of outputs.

	Set of Inputs	*Set of Outputs*
Example 1	All incomes	All tax amounts
Example 2	Hours since midnight	Temperatures during the day
Example 3	Seconds elapsed after dropping the rock	Distances rock travels

The second common feature is that in each example there is a definite *rule* by which each input determines an output. In Example 1 the rule is given by the tax law, which specifies how each income (input) determines a tax amount (output). Similarly, the rule is given by the time/temperature graph in Example 2 and by the formula (distance $= 16t^2$) in Example 3.

Each of these examples could be represented by an idealized calculator that has a single operation key and can receive or display any real number. When a number is entered (*input*), and the "rule key" is pressed, an answer is displayed (*output*). (See Figure 2.16.) The formal definition of function has these same common features (input/rule/output), with a slight change of terminology.

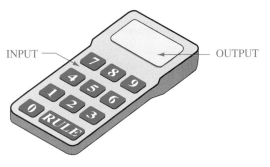

FIGURE 2.16

A **function** consists of a set of input numbers called the **domain,** a set of output numbers called the **range,** and a rule by which each input (number in the domain) determines exactly one output (number in the range).

The domain in Example 1 consists of all possible income amounts; the rule is given by the tax law, and the range consists of all possible tax amounts. In Example 2 the domain is the set of hours in the day (that is, all real numbers from 0 to 24); the rule is given by the time/temperature graph, which shows the temperature at each time. The graph also shows that the range (the temperatures that actually occur during the day) includes all the numbers from 38 to 63.

1 Do the following define functions?

(a) The correspondence defined by the rule $y = x^2 + 5$.

(b) The $\boxed{1/x}$ key on a calculator

(c) The correspondence between a computer, x, and several users of the computer, y

Answers:

(a) Yes

(b) Yes

(c) No

In Example 2, for each time of day (number in the domain) there is one and only one temperature (number in the range). Notice, however, that it is possible to have the same temperature (number in the range) corresponding to two different times (numbers in the domain).

> By the rule of a function, each number in the domain determines *one and only one* number in the range. But several different numbers in the domain may determine the same number in the range.

In other words, exactly one output is produced for each input, but different inputs may produce the same output.

▶**EXAMPLE 4** Which of the following rules describe functions?
(a) Use the optical reader at the checkout counter of the supermarket to convert codes to prices.
  For each code, the reader produces exactly one price, so this is a function.
(b) Enter a number in a calculator and press the $\boxed{x^2}$ key.
  This is a function because the calculator produces just one number x^2 for each number x that is entered.
(c) Assign to each number x the number y given by this table.

x	1	1	2	2	3	3
y	3	-3	5	-5	8	-8

  Since at least one x-value corresponds to more than one y-value, this table does not define a function.
(d) Assign to each number x the number y given by this equation: $y = 3x - 5$.
  Because the equation determines a unique value of y for each value of x, it defines a function. ◀ **1**

The preceding examples show that there are many ways to define a function. Almost all of the functions used in this book will be defined by equations, as in part (d) of Example 4, and *x will be assumed to represent the input variable.*
  The domain and range may or may not be the same set. For instance, in the function given by the x^2 key on a calculator, the domain consists of all numbers (positive, negative, or 0) that can be entered in the calculator, but the range consists only of *nonnegative* numbers (since $x^2 \geq 0$ for every x). In the function given by the equation $y = 3x - 5$, both the domain and range are the set of all real numbers because of the following *agreement on domains.*

Unless otherwise stated, assume that the domain of any function defined by an equation is the largest set of real numbers that are meaningful replacements for the input variable.

2 Do the following define y as a function of x?

(a) $y = -6x + 1$

(b) $y = x^2$

(c) $x = y^2 - 1$

(d) $y < x + 2$

Answers:

(a) Yes

(b) Yes

(c) No

(d) No

3 Give the domain and range.

(a) $y = 3x + 1$

(b) $y = x^2$

Answers:

(a) Both are the set of all real numbers

(b) Domain: set of all real numbers; range $[0, \infty)$

For example, suppose

$$y = \frac{-4x}{2x - 3}.$$

Any real number can be used for x except $x = 3/2$, which makes the denominator equal 0. By the agreement on domains, the domain of this function is the set of all real numbers except $3/2$.

▶ **EXAMPLE 5** Decide whether each of the following equations defines y as a function of x. Give the domain and range of any functions.

(a) $y = -4x + 11$

For a given value of x, calculating $-4x + 11$ produces exactly one value of y. (For example, if $x = -7$, then $y = -4(-7) + 11 = 39$.) Because one value of the input variable leads to exactly one value of the output variable, $y = -4x + 11$ defines a function. Both x and y may take on any real-number values, so both the domain and range are the set of all real numbers, written $(-\infty, \infty)$.

(b) $y^2 = x$

Suppose $x = 36$. Then $y^2 = x$ becomes $y^2 = 36$, from which $y = 6$ or $y = -6$. Since one value of x can lead to two values of y, $y^2 = x$ does not define y as a function. ◀ **2**

▶ **EXAMPLE 6** Find the domain and range for each of the functions defined as follows.

(a) $y = x^4$

Any number may be raised to the fourth power so the domain is $(-\infty \, \infty)$. Because $x^4 \geq 0$ for every value of x, the range is the interval $[0, \infty)$.

(b) $y = \sqrt{6 - x}$.

For y to be a real number, $6 - x$ must be nonnegative. This happens only when $6 - x \geq 0$, or $6 \geq x$, making the domain the interval $(-\infty, 6]$. The range is $[0, \infty)$ because $\sqrt{6 - x}$ is always nonnegative.

(c) $y = \frac{1}{x + 3}$

Because the denominator cannot be 0, $x \neq -3$ and the domain consists of all numbers in the intervals,

$$(-\infty, -3) \quad \text{or} \quad (-3, \infty).$$

Because the numerator can never be 0, $y \neq 0$. There are no other restrictions on y, so the range consists of the numbers in $(-\infty, 0)$ or $(0, \infty)$. ◀ **3**

Caution Notice the difference between the rule $y = \sqrt{6 - x}$ in Example 6(b) and the rule $y^2 = 6 - x$. For a particular x-value, say 2, the radical expression in the first rule represents a single positive number. The second rule, however, when $x = 2$, produces *two* numbers, $y = 2$ or $y = -2$.

FUNCTIONAL NOTATION   In actual practice, functions are seldom presented in the style of domain, rule, range, as they have been here. Functions are usually denoted by a letter (f is frequently used). If x is an input (number in the domain), then $f(x)$ denotes the output number that the function f produces from the input x. The symbol $f(x)$ is read "f of x". The rule is usually given by a formula, such as $f(x) = \sqrt{x^2 + 1}$. This formula can be thought of as a set of directions.

Name of function   Input number
$$f(x) = \sqrt{x^2 + 1}$$
Output number   Directions that tell you what to do with input x in order to produce the corresponding output $f(x)$; namely, "square it, add 1, and take the square root of the result."

For example, to find $f(3)$, (the output number produced by the input 3), simply replace x by 3 in the formula:

$$f(3) = \sqrt{3^2 + 1} = \sqrt{10}.$$

Similarly, replacing x by -5 and 0 shows that

$$f(-5) = \sqrt{(-5)^2 + 1} = \sqrt{26} \quad \text{and} \quad f(0) = \sqrt{0^2 + 1} = 1.$$

These directions can be applied to any quantities, such as $a + b$ or c^4 (where a, b, c are real numbers). Thus, to compute $f(a + b)$, the output corresponding to input $a + b$, we square the input [obtaining $(a + b)^2$], add 1 [obtaining $(a + b)^2 + 1$] and take the square root of the result:

$$f(a + b) = \sqrt{(a + b)^2 + 1} = \sqrt{a^2 + 2ab + b^2 + 1}.$$

Similarly, the output $f(c^4)$ corresponding to the input c^4 is computed by squaring the input $[(c^4)^2]$, adding 1 $[(c^4)^2 + 1]$, and taking the square root of the result:

$$f(c^4) = \sqrt{(c^4)^2 + 1} = \sqrt{c^8 + 1}.$$

▶ **EXAMPLE 7**   Let $g(x) = -x^2 + 4x - 5$. Find each of the following.

(a) $g(-2)$
Replace x with -2.

$$g(-2) = -(-2)^2 + 4(-2) - 5$$
$$= -4 - 8 - 5$$
$$= -17$$

(b) $g(x + h)$
Replace x by the quantity $x + h$ in the rule of g.

$$g(x + h) = -(x + h)^2 + 4(x + h) - 5$$
$$= -(x^2 + 2xh + h^2) + (4x + 4h) - 5$$
$$= -x^2 - 2xh - h^2 + 4x + 4h - 5$$

4 Let $f(x) = 5x^2 - 2x + 1$.
Find the following.

(a) $f(1)$

(b) $f(3)$

(c) $f(1 + 3)$

(d) $f(1) + f(3)$

(e) $f(m)$

(f) $f(x + h) - f(x)$

Answers:

(a) 4

(b) 40

(c) 73

(d) 44

(e) $5m^2 - 2m + 1$

(f) $10xh + 5h^2 - 2h$

5 A developer estimates that the total cost of building x large apartment complexes in a year is approximated by

$$A(x) = x^2 + 80x + 60,$$

where $A(x)$ represents the cost in hundred thousands of dollars. Find the cost of building

(a) 4 complexes;

(b) 10 complexes.

Answers:

(a) $39,600,000

(b) $96,000,000

(c) $g(x + h) - g(x)$
Use the result from part (b) and the rule for $g(x)$.

$$g(x + h) - g(x) = (-x^2 - 2xh - h^2 + 4x + 4h - 5) - (-x^2 + 4x - 5)$$
$$= -2xh - h^2 + 4h$$

(d) $\dfrac{g(x + h) - g(x)}{h}$

The numerator was found in part (c). Divide it by h as follows.

$$\frac{g(x + h) - g(x)}{h} = \frac{-2xh - h^2 + 4h}{h}$$
$$= \frac{h(-2x - h + 4)}{h}$$
$$= -2x - h + 4 \quad \blacktriangleleft$$

The quotient found in Example 7(d)

$$\frac{g(x + h) - g(x)}{h}$$

is important in calculus, and we will see it again in Chapter 11. **4**

Caution Functional notation is *not* the same as ordinary algebraic notation. You cannot simplify an expression such as $f(x + h)$ by writing $f(x) + f(h)$. To see why, consider the answers to Problems 4(c) and (d) at the side, that show that

$$f(1 + 3) \neq f(1) + f(3).$$

▶ **EXAMPLE 8** Suppose the sales of a small company have been estimated to be

$$S(x) = 125 + 80x,$$

where $S(x)$ represents the total sales in thousands of dollars in year x, with $x = 0$ representing 1990. Estimate the sales in each of the following years.

(a) 1990

Since $x = 0$ corresponds to 1990, the sales for 1990 are given by $S(0)$. Substituting 0 for x gives

$$S(0) = 125 + 80(0) \qquad \textbf{Let } x = 0$$
$$= 125.$$

Since $S(x)$ represents sales in thousands of dollars, the sales would be estimated as 125×1000, or $125,000, in 1990.

(b) 1994

To estimate sales in 1994, let $x = 4$.

$$S(4) = 125 + 80(4) = 125 + 320 = 445,$$

so that sales should be about $445,000 in 1994.   ◀   **5**

2.3 EXERCISES

Which of the following rules define y as a function of x? (See Examples 1–5.)

1.

x	3	2	1	0	−1	−2	−3
y	9	4	1	0	1	4	9

2.

x	9	4	1	0	1	4	9
y	3	2	1	0	−1	−2	−3

3.

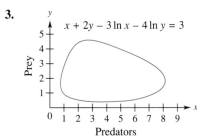

$x + 2y - 3 \ln x - 4 \ln y = 3$

4.*

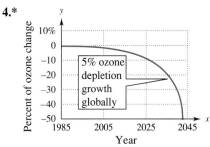

5. $y = x^3$

6. $y = \sqrt{x - 1}$

7. $x = |y + 2|$

8. $x = y^2 + 3$

9. $y = \dfrac{-1}{x - 1}$

10. $y = \dfrac{4}{2x + 3}$

11. In your own words, give a definition of *function, domain,* and *range.* Give examples.

12. List three ways to represent a function and give a "real-life" example of each.

Give the domain and range of each function defined as follows. (See Examples 5 and 6.)

13. $f(x) = 4x - 1$

14. $f(x) = 2x + 7$

15. $f(x) = x^4 - 1$

16. $f(x) = (2x + 5)^2$

17. $f(x) = \sqrt{-x} + 3$

18. $f(x) = \sqrt{4 - x}$

19. $f(x) = \dfrac{1}{x - 1}$

20. $f(x) = \dfrac{1}{x - 3}$

21. $f(x) = |5 - 4x|$

22. $f(x) = |-x - 6|$

For each of the following functions, find (a) f(4) (b) f(−3) (c) f(0) (d) f(a). (See Example 5.)

23. $f(x) = 6$

24. $f(x) = 0$

25. $f(x) = 2x^2 + 4x$

26. $f(x) = x^2 - 2x$

27. $f(x) = \sqrt{x + 3}$

28. $f(x) = \sqrt{5 - x}$

* Graph regarding ozone depletion from *The Sacramento Bee*, August 23, 1988. Copyright, The Sacramento Bee, 1988. Reprinted by permission.

For each of the following functions, find (a) $f(p)$, (b) $f(-r)$, and (c) $f(m + 3)$. (See Example 7.)

29. $f(x) = 5 - x$

30. $f(x) = 3x + 7$

31. $f(x) = \sqrt{4 - x}$

32. $f(x) = \sqrt{-2x}$

33. $f(x) = x^3 + 1$

34. $f(x) = 2 - x^3$

35. $f(x) = \dfrac{3}{x - 1}$

36. $f(x) = \dfrac{-1}{2 + x}$

For each of the following find

$$\frac{f(x + h) - f(x)}{h}.$$

37. $f(x) = 2x - 4$

38. $f(x) = 2 - 3x$

Work the following exercises.

39. **Management** China's economy has been booming in recent years, which has caused its Asian neighbors to begin to worry about the increasing competition. The graph* shows two functions representing China's exports and imports (in billions of dollars) over the years 1986 to 1991.

GROWING TRADE SURPLUS

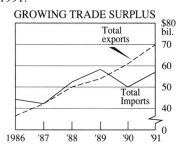

(a) Estimate the amount of China's imports in 1986; in 1991.
(b) Estimate the amount of China's exports in 1986; in 1991.
(c) What was the difference between imports and exports in 1991?
(d) What is the significance of the points where the graphs of the two functions cross?

---

*Chart from "China's Booming Economy," *from U.S. News and World Report,* October 19, 1992. Copyright © 1992 by U.S. News & World Report. Reprinted by permission

** Graph "Energy Use Down, Savings Up" from "The Greenhouse Effect" by the Union of Concerned Scientists. Reprinted by permission.

40. **Management** The graph** shows two functions of time. The first, which we will designate $I(t)$, shows the energy intensity of the United States, which is defined as the thousands of BTUs of energy consumed per dollar of Gross National Product (GNP) in year t (in 1982 dollars). The second, which we will designate $S(t)$, represents the national savings per year (also in 1982 dollars) due to a decreased use of energy.

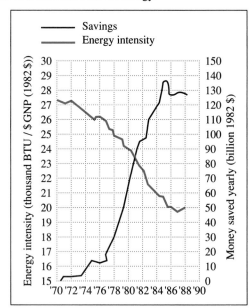

(a) Find $I(1976)$ (b) Find $S(1984)$.
(c) Find $I(t)$ and $S(t)$ at the time when the two graphs cross. In what year does this occur?

(d) What significance, if any, is there to the point at which the two graphs cross?

41. Management   A chain-saw rental firm charges $7 per day or fraction of a day to rent a saw, plus a fixed fee of $4 for resharpening the blade. Let $S(x)$ represent the cost of renting a saw for x days. Find each of the following.

(a) $S\left(\dfrac{1}{2}\right)$ **(b)** $S(1)$ **(c)** $S\left(1\dfrac{1}{4}\right)$ **(d)** $S\left(3\dfrac{1}{2}\right)$

(e) What does it cost to rent a saw for $4\dfrac{9}{10}$ days?

(f) A portion of the graph of $y = S(x)$ is shown here. Explain how the graph could be continued.

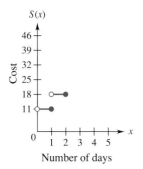

(g) What is the domain variable?

(h) What is the range variable?

(i) Write a sentence or two explaining what part (c) and its answer represent.

(j) We have left $x = 0$ out of the graph. Discuss why it should or shouldn't be included. If it were included, how would you define $S(0)$?

42. Natural Science   The table gives estimates of the percent of ozone change from 1985 for several years in the future, if chlorofluorocarbon production is reduced 80% globally.

Year	Percent
1985	0
2005	1.5
2025	3
2065	4
2085	5

(a) Plot these ordered pairs on a grid.

(b) The points from part (a) should lie approximately on a straight line. Use the pairs (2005, 1.5) and (2085, 5) to write an equation of the line.

(c) Letting $f(x)$ represent the percent of ozone change and x represent the year, write your equation from part (b) as a rule that defines a function.

(d) Find $f(2065)$. Does it agree fairly closely with the number in the table that corresponds to 2065? Do you think the expression from part (c) describes this function adequately?

(e) Describe the domain and range of this function (not just the part shown in the table) in words.

2.4 APPLICATIONS OF LINEAR FUNCTIONS

When mathematics is used to solve a real-world problem, the first step is to write one or more mathematical relationships (equations or inequalities) that describe the situation. In this phase of problem solving, if the result is to be useful, it is important to include all the pertinent information. Setting up these relationships requires a solid understanding of the situation to be described, as well as a knowledge of the most useful mathematics available.

Earlier in this chapter we studied linear equations and were introduced to the concept of a function. In the examples and exercises of Section 2.3, we found that linear equations usually define functions. (Only linear equations of the form $x = k$ do not define functions.)

A **linear function** is a function whose rule can be written in the form

$$f(x) = ax + b$$

for some constants a and b.

In this section we show how linear functions can be applied to a variety of real-world situations.

TEMPERATURE   One of the most common linear relationships found in everyday situations deals with temperature. Recall that water freezes at 32° Fahrenheit and 0° Celsius, while it boils at 212° Fahrenheit and 100° Celsius. The ordered pairs (0, 32) and (100, 212) are graphed in Figure 2.17 on axes showing Fahrenheit (F) as a function of Celcius (C). The line joining them is the graph of the function.

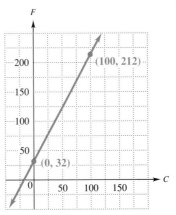

FIGURE 2.17

▶ **EXAMPLE 1**   Derive an equation relating F and C.

To derive the required linear equation, first find the slope using the given ordered pairs, (0, 32) and (100, 212).

$$m = \frac{212 - 32}{100 - 0} = \frac{9}{5}$$

The F-intercept of the graph is 32, so by the slope-intercept form the equation of the line is

$$F = \frac{9}{5}C + 32.$$

1 Find a linear equation describing the sales for store B from Example 2.

Answer:
$y = 30{,}000x + 50{,}000$

This equation can be written to give C in terms of F.

$$C = \frac{5}{9}(F - 32) \quad \blacktriangleleft$$

COMPARING RATES OF CHANGE   One way to compare the change in some quantity in two or more situations is to compare the rates at which the quantity changes in the two cases. If the increase or decrease in a quantity can be approximated (or modeled) by a linear function, we can use the work in Section 2.2 to find the rate of change of the quantity with respect to time.

▶**EXAMPLE 2**   The chart below shows sales in two different years for two stores of a major chain of discount stores:

Store	Sales in 1992	Sales in 1995
A	$100,000	$160,000
B	50,000	140,000

A study of company records suggests that the sales of both stores have increased linearly (that is, the sales can be closely approximated by a linear function). Find a linear equation describing the sales for Store A.

To find a linear equation describing the sales, let $x = 0$ represent 1992 so that 1995 corresponds to $x = 3$. Then, by the chart above, the line representing the sales for Store A passes through the points $(0, 100000)$ and $(3, 160000)$. The slope of the line through these points is

$$\frac{160{,}000 - 100{,}000}{3 - 0} = 20{,}000.$$

Using the point-slope form of the equation of a line gives

$$y - 100{,}000 = 20{,}000(x - 0)$$
$$y = 20{,}000x + 100{,}000$$

as an equation describing the sales of Store A.   ◀ **1**

AVERAGE RATE OF CHANGE   Notice that the sales for Store A in Example 2 increased from $100,000 to $160,000 over the period from 1992 to 1995 representing a total increase of $60,000 in 3 years.

$$\text{Average rate of change in sales} = \frac{\$60{,}000}{3} = \$20{,}000 \text{ per year}$$

This is the same as the slope found in Example 2. Verify that the average annual rate of change in sales for store B over the 3-year period also agrees with the slope of the equation found in Problem 1 at the side. Management needs to watch

2 A certain new anticholesterol drug is related to the blood cholesterol level by the linear model

$$y = 280 - 3x,$$

where x is the dosage of the drug (in grams) and y is the blood cholesterol level.

(a) Find the blood cholesterol level if 12 grams of the drug are administered.

(b) In general, an increase of 1 gram in the dose causes what change in the blood cholesterol level?

Answers:

(a) 244

(b) A decrease of 3 units

the rate of change in sales closely in order to be aware of any unfavorable trends. If the rate of change is decreasing, then sales growth is slowing down, and this trend may require some response.

Example 2 illustrates a useful fact about linear functions. If $f(x) = mx + b$ is a *linear* function, then the **average rate of change** in y with respect to x (the change in y divided by the corresponding change in x) is the slope of the line $y = mx + b$. In particular, the average rate of change of a linear function is constant. (See Exercises 14 and 15 in this section for examples of functions where the average rate of change is not constant.)

▶**EXAMPLE 3**   Average family health care costs in dollars y over the 10-year period of the 90s are expected to increase each year as given by

$$y = 510x + 4300,$$

where x is the number of years since 1990. By this estimate, how much will health care cost the average family in 1996? What is the average rate of increase?

Let $x = (1996 - 1990) = 6$. The health care costs will be

$$y = 510(6) + 4300 = 7360 \quad \text{or} \quad \$7360.$$

The average rate of increase in cost is given by the slope of the line. The slope is 510, so this model indicates that each year costs will increase \$510.   ◀ **2**

COST ANALYSIS   The cost of manufacturing an item commonly consists of two parts. The first is a **fixed cost** for designing the product, setting up a factory, training workers, and so on. Within broad limits, the fixed cost is constant for a particular product and does not change as more items are made. The second part is a *cost per item* for labor, materials, packing, shipping, and so on. The total value of this second cost *does* depend on the number of items made.

▶**EXAMPLE 4**   Suppose that the cost of producing clock-radios can be approximated by the linear model

$$C(x) = 12x + 100.$$

where $C(x)$ is the cost in dollars to produce x radios. The cost to produce 0 radios is

$$C(0) = 12(0) + 100 = 100,$$

or \$100. This amount, \$100, is the fixed cost.

Once the company has invested the fixed cost into the clock-radio project, what is the additional cost per radio? To find out, first fnd the cost of a total of 5 radios:

$$C(5) = 12(5) + 100 = 160,$$

3 The cost in dollars to produce x kilograms of chocolate candy is given by $C(x)$, where in dollars

$$C(x) = 3.5x + 800.$$

Find each of the following.

(a) The fixed cost

(b) The total cost for 12 kilograms

(c) The marginal cost per kilogram

(d) The marginal cost of the 40th kilogram

Answers:

(a) $800

(b) $842

(c) $3.50

(d) $3.50

or $160. The cost of 6 radios is

$$C(6) = 12(6) + 100 = 172,$$

or $172. The sixth radio costs $172 - $160 = $12 to produce. In the same way, the 81st radio costs $C(81) - C(80) = \$1072 - \$1060 = \$12$ to produce. In fact, the $(n + 1)$st radio costs

$$C(n + 1) - C(n) = [12(n + 1) + 100] - [12n + 100] = 12,$$

or $12 to produce. Because each additional radio costs $12 to produce, $12 is the variable cost per radio. The number 12 is also the slope of the cost function, $C(x) = 12x + 100$.   ◄

Example 4 can easily be generalized. Suppose the total cost to make x items is given by the linear cost function $C(x) = mx + b$. Then the fixed cost (the cost which occurs even if no items are produced) is found by letting $x = 0$.

$$C(0) = m \cdot 0 + b = b$$

Thus, the fixed cost is the y-intercept of the cost function.

In economics, **marginal cost** is the rate of change of cost. Marginal cost is important to management in making decisions in areas such as cost control, pricing, and production planning. If the cost function is $C(x) = mx + b$, then its graph is a straight line with slope m. Since the slope represents the average rate of change, the marginal cost is the number m.   **3**

In Example 4, the marginal cost (slope) 12 was also the cost of producing one more radio. We now show that the same thing is true for any linear cost function $C(x) = mx + b$. If n items have been produced, the cost of the $(n + 1)$st item is the difference between the cost of producing $n + 1$ items and the cost of producing n items, namely, $C(n + 1) - C(n)$. Because $C(x) = mx + b$,

$$
\begin{aligned}
\text{Cost of one more item} &= C(n + 1) - C(n) \\
&= [m(n + 1) + b] - [mn + b] \\
&= mn + m + b - mn - b = m \\
&= \text{marginal cost.}
\end{aligned}
$$

This discussion is summarized as follows.

In a **linear cost function** $C(x) = mx + b$, m represents the marginal cost and b the fixed cost. The marginal cost is the cost of producing one more item.

Conversely, if the fixed cost is b and the marginal cost is always the same constant m, then the cost function for producing x items is $C(x) = mx + b$.

4 The total cost of producing 10 units of a business calculator is $100. The marginal cost per calculator is $4. Find the cost function, $C(x)$, if it is linear.

Answer:
$C(x) = 4x + 60$

▶**EXAMPLE 5**   The marginal cost to produce an anticlot drug is $10 per unit, while the cost to produce 100 units is $1500. Find the cost function $C(x)$, given that it is linear.

Since the cost function is linear, it can be written in the form $C(x) = mx + b$. The marginal cost is $10 per unit, which gives the value for m, leading to $C(x) = 10x + b$. To find b, use the fact that the cost of producing 100 units of the drug is $1500, or $C(100) = 1500$. Substituting $x = 100$ and $C(x) = 1500$ into $C(x) = 10x + b$ gives

$$C(x) = 10x + b$$
$$1500 = 10(100) + b$$
$$1500 = 1000 + b$$
$$500 = b.$$

The cost function is $C(x) = 10x + 500$, where the fixed cost is $500. ◀ **4**

If $C(x)$ is the total cost to manufacture x items, then the **average cost** per item is given by

$$\overline{C}(x) = \frac{C(x)}{x}.$$

In Example 4, the average cost per clock-radio is

$$\overline{C}(x) = \frac{C(x)}{x} = \frac{12x + 100}{x} = 12 + \frac{100}{x}.$$

Note what happens to the term $100/x$ as x gets larger.

$$\overline{C}(100) = 12 + \frac{100}{100} = 12 + 1 = \$13$$

$$\overline{C}(500) = 12 + \frac{100}{500} = 12 + \frac{1}{5} = \$12.20$$

$$\overline{C}(1000) = 12 + \frac{100}{1000} = 12 + \frac{1}{10} = \$12.10$$

As more and more items are produced, $100/x$ gets smaller and the average cost per item decreases.

▶**EXAMPLE 6**   Find the average cost per unit to produce 50 units and 500 units of the anticlot drug in Example 5.

The cost function from Example 5 is $10x + 500$, so the average cost per unit is

$$\overline{C}(x) = \frac{C(x)}{x} = \frac{10x + 500}{x} = 10 + \frac{500}{x}.$$

5 In Problem 4, at the side, find the average cost per calculator to produce 100 calculators.

Answer:
$46

If 50 units of the drug are produced, the average cost is

$$\overline{C}(50) = 10 + \frac{500}{50} = 20,$$

or $20 per unit. Producing 500 units of the drug will lead to an average cost of

$$\overline{C}(500) = 10 + \frac{500}{500} = 11,$$

or $11 per unit. ◀ **5**

BREAK-EVEN ANALYSIS   The **revenue** $R(x)$ from selling x units of a product is the product of the price per unit p and the number of units sold (demand) x so that

$$R(x) = px.$$

The corresponding **profit** $P(x)$ is the difference between revenue $R(x)$ and cost $C(x)$. That is,

$$P(x) = R(x) - C(x).$$

A company can make a profit only if the revenue received from its customers exceeds the cost of producing its goods and services. The number of units at which revenue just equals cost is the **break-even point.**

▶**EXAMPLE 7**   A firm producing poultry feed finds that the total cost $C(x)$ of producing x units is given by

$$C(x) = 20x + 100.$$

Management plans to charge $24 per unit for the feed.
(a) How many units must be sold for the firm to break even?
    The firm will break even (no profit and no loss) as long as revenue just equals cost, or $R(x) = C(x)$. From the given information, since $R(x) = px$ and $p = \$24$,

$$R(x) = 24x.$$

Substituting for $R(x)$ and $C(x)$ in the equation $R(x) = C(x)$ gives

$$24x = 20x + 100,$$

from which $x = 25$. The firm breaks even by selling 25 units. The graphs of $C(x) = 20x + 100$ and $R(x) = 24x$ are shown in Figure 2.18. The break-even point (where $x = 25$) is shown on the graph. If the company produces more than 25 units (if $x > 25$), it makes a profit. If it produces less than 25 units, it loses money.

6 For a certain magazine, $C(x) = .70x + 1200$, where x is the number of magazines sold. The magazine sells for $1 per copy. Find the break-even point.

Answer:
4000 magazines

(b) What is the profit if 100 units of feed are sold? Use the formula for profit $P(x)$.

$$P(x) = R(x) - C(x)$$
$$= 24x - (20x + 100)$$
$$= 4x - 100$$

Then $P(100) = 4(100) - 100 = 300$. The firm will make a profit of $300 from the sales of 100 units of feed. ◀ 6

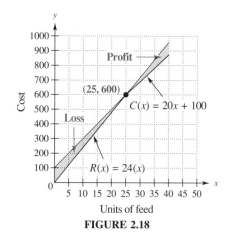

FIGURE 2.18

2.4 EXERCISES

Physical Science *Use the formula derived in Example 1 for conversion between Fahrenheit and Celsius in Exercises 1–4.*

1. Convert each temperature.
 (a) 58°F to Celsius   **(b)** 50°C to Fahrenheit

2. Convert each temperature.
 (a) 98.6°F to Celsius   **(b)** 20°C to Fahrenheit

3. Find the temperature at which the Celsius and Fahrenheit temperatures are numerically equal.

4. Derive the formula giving Celsius temperature in terms of Fahrenheit temperature.

5. Management Assume that the sales of a certain appliance dealer are approximated by a linear function. Suppose that sales were $850,000 in 1987 and $1,262,500 in 1992. Let $x = 0$ represent 1987.
 (a) Find an equation giving the dealer's yearly sales.
 (b) Estimate the sales in 1999.

 (c) The dealer estimates that a new store will be necessary once sales exceed $2,170,000. When is this expected to occur?

6. Management Assume that the sales of a certain automobile parts company are approximated by a linear function. Suppose that sales were $200,000 in 1985 and $1,000,000 in 1992. Let $x = 0$ represent 1985 and $x = 7$ represent 1992.
 (a) Find the equation giving the company's yearly sales.
 (b) Estimate the sales in 1996.
 (c) The company wants to negotiate a new contract with the automobile company once sales reach $2,000,000. When is this expected to occur?

7. Social Science The number of children in the U.S. from 5 to 13 years old decreased from 31.2 million in 1980 to 30.3 million in 1986.
 (a) Write a linear function describing this population y in terms of year x for the given period.

(b) What was the average rate of change in this population over the period from 1980 to 1986?

8. Social Science   Over a recent 3-year period, medical expenses in the U.S. rose in a linear pattern from 10.3% of the gross national product in year 0 to 11.2% in year 3.

(a) Assuming that this change in medical expenses continues to be linear, write a linear function describing the percent of the gross national product devoted to medical expenses, y, in terms of the year, x.

(b) Give the average rate of change of medical expenses from year 0 to 3. Compare it with the slope of the line in part (a). What do you find?

9. What is meant by the *average rate of change?* How is it found?

10. Management   In deciding whether to set up a new manufacturing plant, company analysts have established that a reasonable function for the total cost to produce x items is

$$C(x) = 500,000 + 4.75x.$$

(a) Find the total cost to produce 100,000 items.

(b) Find the rate of change of the cost of the items to be produced in this plant.

11. Management   Suppose the sales of a particular brand of electric guitar satisfy the relationship

$$S(x) = 300x + 2000,$$

where $S(x)$ represents the number of guitars sold in year x, with $x = 0$ corresponding to 1987.
Find the sales in each of the following years.

(a) 1987   **(b)** 1990   **(c)** 1991

(d) The manufacturer needs sales to reach 4000 guitars by 1996 to pay off a loan. Will sales reach that goal?

(e) Find the annual rate of change of sales.

12. Social Science   In psychology, the just-noticeable-difference (JND) for some stimulus is defined as the amount by which the stimulus must be increased so that a person will perceive it as having just barely increased. For example, suppose a research study indicates that a line 40 centimeters in length must be increased to 42 centimeters before a subject thinks that it is longer. In this case, the JND would be $42 - 40 = 2$ centimeters. In a particular experiment, the JND (y) is given by

$$y = .03x,$$

where x represents the original length of the line. Find the JND for lines having the following lengths.

(a) 10 centimeters   **(b)** 20 centimeters

(c) 50 centimeters   **(d)** 100 centimeters

(e) Find the rate of change in the JND with respect to the original length of the line.

13. Social Science   Most people are not very good at estimating the passage of time. Some people's estimations are too fast, and others' are too slow. One psychologist has constructed a mathematical model for actual time as a function of estimated time: if y represents actual time and x estimated time, then

$$y = mx + b,$$

where m and b are constants that must be determined experimentally for each person.

Suppose that for a particular person, $m = 1.25$ and $b = -5$. Find y if x is

(a) 30 minutes;   **(b)** 120 minutes;

(c) Find the rate of change of actual time with respect to estimated time for this person. Is this person's estimate of time too fast or too slow?

14. Management   The graph below shows the total sales, y, in thousands of dollars from the distribution of x thousand catalogs. Find the average rate of change of sales with respect to the number of catalogs distributed for the following changes in x.

(a) 10 to 20   **(b)** 10 to 40

(c) 20 to 30   **(d)** 30 to 40

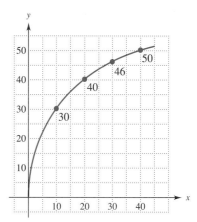

15. Management The graph below shows annual sales (in units) of a typical product. Sales increase slowly at first to some peak, hold steady for a while, and then decline as the product goes out of style. Find the average annual rate of change in sales for the following changes in years.

(a) 1 to 3   (b) 2 to 4   (c) 3 to 6
(d) 5 to 7   (e) 7 to 9   (f) 8 to 11
(g) 9 to 10   (h) 10 to 12

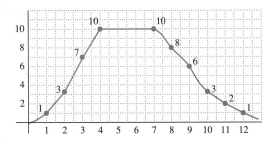

16. What is the meant by the marginal cost of a product? the fixed cost?

Management *Write a cost function for each of the following. Identify all variables used. (See Example 5.)*

17. A chain-saw rental firm charges $12 plus $1 per hour.

18. A trailer-hauling service charges $45 plus $2 per mile.

19. A parking garage charges 35¢ plus 30¢ per half hour.

20. For a 1-day rental, a car rental firm charges $14 plus 6¢ per mile.

Management *Assume that each of the following can be expressed as a linear cost function. Find the appropriate cost function in each case. (See Example 5.)*

21. Fixed cost, $100; 50 items cost $1600 to produce.

22. Fixed cost, $1000; 40 items cost $2000 to produce.

23. Marginal cost, $120; 100 items cost $15,800 to produce.

24. Marginal cost, $90; 150 items cost $16,000 to produce.

25. Management The total cost (in dollars) to produce x algebra books is $C(x) = 5.25x + 40,000$.
(a) What is the marginal cost per book?
(b) What is the total cost to produce 5000 books?

26. Management In Exercise 10, we were given the following function for the total cost in dollars to produce x items.

$$C(x) = 500,000 + 4.75x$$

(a) Find the marginal cost per item of the items to be produced in this plant.
(b) Find the average cost per item.

27. What is the break-even point for a product? How is it found?

28. Social Science College costs at private four-year colleges (excluding room and board) rose in a linear pattern from $7000 in 1987–88 (year 0) to $11,000 in 1993–94 (year 6).
(a) Assuming that this change in college costs continues to be linear, write a linear function describing the cost, y, in terms of the year, x.
(b) Give the average rate of change of college costs from year 0 to 6. Compare it with the slope of the line in part (a). What do you find?

Management *Assume that each of the following can be expressed as a linear cost function. Find (a) the cost function; (b) the revenue function; (c) the break-even point. (See Example 7.)*

	Fixed Cost	Marginal Cost per Item	Item Sells for
29.	$500	$10	$35
30.	$180	$11	$20
31.	$250	$18	$28
32.	$1500	$30	$80

33. Management The graph shows the productivity of U.S. and Japanese workers in appropriate units over a 35-year period. Estimate the break-even point (the point at which workers in the two countries produced the same amounts).*

---

*The figure for Exercises 33 and 34 appeared in *The Sacramento Bee*, December 21, 1987. Reprinted by permission of The Associated Press.

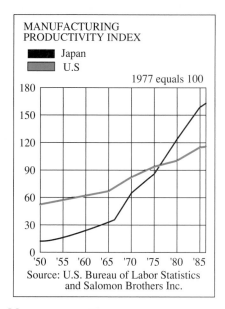

MANUFACTURING
PRODUCTIVITY INDEX

■ Japan
■ U.S

1977 equals 100

Source: U.S. Bureau of Labor Statistics
and Salomon Brothers Inc.

34. Management   The graph gives U.S. imports and exports in billions of dollars over a five-year period. Estimate the break-even point.

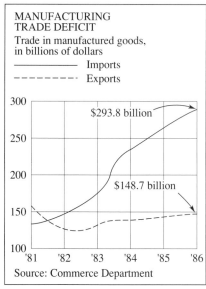

MANUFACTURING
TRADE DEFICIT

Trade in manufactured goods,
in billions of dollars

———— Imports
-------- Exports

$293.8 billion

$148.7 billion

Source: Commerce Department

35. Management   Canadian and Japanese investment in the United States in billions of dollars in 1980 and 1990 are shown in the chart.*

	1980	*1990*
Canada	12.1	27.7
Japan	4.7	108.1

(a) Assuming the change in investment in each case is linear, write an equation in function form giving the investment in year x for each country. Let x be the number of years since 1980.

(b) Graph the functions from part (a) on the same coordinate axes.

(c) Find the intersection point where the graphs in part (b) cross and interpret your answer.

36. Social Science   The median family income (in thousands of dollars) for the white population in the United States is given by $f(x) = (4/3)x + 8$, where x is the number of years since 1973. The median family income (in thousands of dollars) for the black population in the United States is given by $f(x) = (2/3)x + 6$.

(a) Graph both functions on the same coordinate axes.

(b) Do the graphs intersect? If so, in what year was median family income the same for white and black populations?

(c) What can you infer from the two graphs in part (a)?

Management   *Suppose that you are the manager of a firm. You are considering the manufacture of a new product, so you ask the accounting department to produce cost estimates and the sales department to produce sales estimates. After you receive the data, you must decide whether to go ahead with production of the new product. Analyze the following data (find a break-even point) and then decide what you would do. (See Example 7.)*

37. $C(x) = 80x + 7000$; $R(x) = 95x$;   no more than 400 units can be sold.

38. $C(x) = 65x + 9500$; $R(x) = 80x$;   no more than 600 units can be sold.

39. $C(x) = 140x + 3000$; $R(x) = 125x$ (Hint: what does a negative value of x mean?)

40. $C(x) = 1750x + 95,000$; $R(x) = 1750x$

41. Management   The revenue in millions of dollars from sales of x units at a home supplies outlet is given by $R(x) = .3x$. The profit in millions of dollars from sales of x units is given by $P(x) = .2x - .5$.

(a) Find the cost function.

(b) Find $C(7)$.

(c) What is the marginal cost?

42. Management The profit function in millions of dollars for x million units of the new blue jello is $P(x) = .7x - 25.5$. The cost is given by $C(x) = .9x + 25.5$.
(a) Find the revenue function.
(b) Find $R(10)$.
(c) Find the break-even point.

43. Management The sales of a certain furniture company in thousands of dollars are shown in the chart below.

x (year)	y (sales)
0	48
1	59
2	66
3	75
4	80
5	90

(a) Graph this data, plotting years on the x-axis and sales on the y-axis. (Note that the data points can be closely approximated by a straight line.)
(b) Draw a line through the points $(2, 66)$ and $(5, 90)$. The other four points should be close to this line. (These two points were selected as "best" representing the line that could be drawn through the data points.)
(c) Use the two points of (b) to find an equation for the line that approximates the data.
(d) Complete the following chart.

Year	Sales (actual)	Sales (predicted from equation of (c))	Difference, Actual Minus Predicted
0			
1			
2			
3			
4			
5			

(e) Use the result of (c) to predict sales in year 7.
(f) Do the same for year 9.

44. Social Science Some scientists believe there is a limit to how long humans can live.* One supporting argument is that during the last century, life expectancy from age 65 has increased more slowly than life expectancy from birth, so eventually these two will be equal, at which point, according to these scientists, life expectancy should increase no further. In 1900, life expectancy at birth was 46 years, and life expectancy at age 65 was 76. In 1975, these figures had risen to 75 and 80, respectively. In both cases, the increase in life expectancy has been linear. Using these assumptions and the data given, find the maximum life expectancy for humans.

45. Management In the profit-volume chart below, EF and GH represent the profit-volume graphs of a single-product company for 1989 and 1990, respectively.**

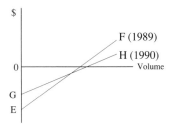

If 1989 and 1990 unit sales prices are identical, how did total fixed costs and unit variable costs of 1990 change compared to 1989?

	1990 total fixed costs	1990 unit variable costs
a.	Decreased	Increased
b.	Decreased	Decreased
c.	Increased	Increased
d.	Increased	Decreased

*See *Science,* November 15, 1991, pp 936–38.
** Uniform CPA Examination, May, 1991, American Institute of Certified Public Accountants

2.5 OTHER USEFUL FUNCTIONS AND THEIR GRAPHS

For each input number in the domain of a function, we can form the ordered pair consisting of this input and the corresponding output number. The set of all such ordered pairs (considered as points in the coordinate plane) is the **graph** of the function. In other words, the graph of a function f consists of all possible ordered pairs $(x, f(x))$, where x is any number in the domain of f. The graph of a function provides a geometric picture of the function, which is often useful.

As we saw in Section 2.2, the graph of a linear function is a straight line. In this section we consider functions whose graphs consist of straight line segments or other curves. These functions are used as models in a variety of applications.

Recall from Chapter 1 that the absolute value of every number is nonnegative. The function defined by $y = |x|$ is called an **absolute value function.**

▶**EXAMPLE 1**   Graph each function.

(a) $f(x) = |x|$

From the definition of $|x|$,

$$f(x) = |x| = \begin{cases} x & \text{if } x \geq 0 \\ -x & \text{if } x < 0. \end{cases}$$

The graph will consist of portions of the two lines with equations $y = x$ and $y = -x$. For $x \geq 0$, we graph $y = x$, and for $x < 0$, we graph $y = -x$. The two partial lines meet at $(0, 0)$, which is an endpoint for one of them. The graph is shown in Figure 2.19. Note that the domain is $(-\infty, \infty)$ and the range is $[0, \infty)$.

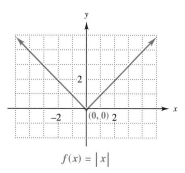

$$f(x) = |x|$$

FIGURE 2.19

(b) $f(x) = |3x + 4|$

Using the definition of absolute value gives

$$f(x) = \begin{cases} 3x + 4 & \text{if } 3x + 4 \geq 0 \\ -(3x + 4) & \text{if } 3x + 4 < 0. \end{cases}$$

1 Graph each function.

(a) $f(x) = 2 - |x|$

(b) $f(x) = |5x - 7|$

Answers:

(a)

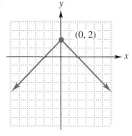

(b)

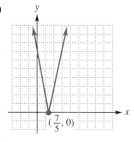

The inequality $3x + 4 \geq 0$ is satisfied whenever $x \geq -4/3$, and $3x + 4 < 0$ is satisfied whenever $x < -4/3$. If $x = -4/3$, $y = 0$, so the graph will consist of two lines that meet at $(-4/3, 0)$. The line with equation $y = 3x + 4$ has slope 3; the line with equation $y = -(3x + 4) = -3x - 4$ has slope -3. Use the point $(-4/3, 0)$ and the two slopes to graph the two partial lines as shown in Figure 2.20. This function also has domain $(-\infty, \infty)$ and range $[0, \infty)$.

◀ **1**

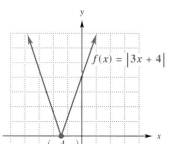

FIGURE 2.20

The graphs of the absolute value functions in Example 1 are made up of portions of two different lines. Such functions are called **piecewise functions.** Other piecewise functions are defined with different equations for different parts of the domain.

▶ **EXAMPLE 2** Graph each function.

(a) $f(x) = \begin{cases} x + 1 & \text{if } x \leq 2 \\ -2x + 7 & \text{if } x > 2 \end{cases}$

Again, the graph consists of parts of two lines. For $x \leq 2$, use the equation $y = x + 1$ to find the ordered pairs in the first table below, including the endpoint value $x = 2$ for one of the pairs. For $x > 2$, use the equation $y = -2x + 7$ to find the ordered pairs in the second table.

$x \leq 2$				$x > 2$			
x	-2	0	2	x	2	3	4
$y = x + 1$	-1	1	3	$y = -2x + 7$	3	1	-1

Note that even though 2 is not in the interval $x > 2$, we found the ordered pair for that endpoint because the graph will extend right up to that point. Because this endpoint $(2, 3)$ agrees with the endpoint for the interval $x \leq 2$, the two parts of the graph are joined at that point as shown in Figure 2.21.

2 Graph $f(x)$, where

$$f(x) = \begin{cases} -2x + 5 & \text{if } x < 2 \\ x - 4 & \text{if } x \geq 2 \end{cases}.$$

Answer:

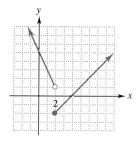

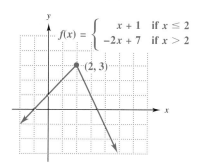

FIGURE 2.21

(b) $f(x) = \begin{cases} x & \text{if } x \leq 0 \\ x + 1 & \text{if } x > 0 \end{cases}$

The ordered pairs $(-2, -2)$, $(-1, -1)$, and $(0, 0)$ satisfy $y = x$. The endpoint is $(0, 0)$. For $y = x + 1$, some ordered pairs are $(1, 2)$ and $(2, 3)$. The ordered pair $(0, 1)$ is an endpoint, but is *not* part of the graph, as indicated by the open circle in Figure 2.22. ◄ **2**

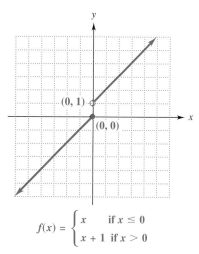

$$f(x) = \begin{cases} x & \text{if } x \leq 0 \\ x + 1 & \text{if } x > 0 \end{cases}$$

FIGURE 2.22

Caution   In Example 2(a), notice that we did not graph the entire lines but only those portions with domain as given. Graphs of these functions should *not* be two intersecting lines.

The **greatest integer function,** written $y = [x]$, is defined by saying that $[x]$ is the greatest integer less than or equal to x. For example, $[8] = 8$, $[-5] = -5$, $[\pi] = 3$, $[12\frac{1}{9}] = 12$, $[-2.001] = -3$, and so on.

3 Graph $y = [\frac{1}{2}x + 1]$.

Answer:

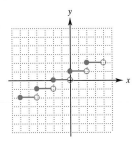

4 Assume that the post office charges 30¢ per ounce, or fraction of an ounce, to mail a letter. Graph the ordered pairs (ounces, cost).

Answer:

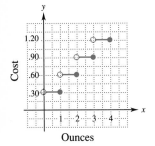

▶**EXAMPLE 3**   Graph $y = [x]$.

For x in the interval $0 \leq x < 1$, the value of $[x] = 0$. For values of x where $1 \leq x < 2$, $[x] = 1$, and so on. Thus, the graph, as shown in Figure 2.23, consists of a series of line segments. In each case, the left endpoint of the segment is included, and the right endpoint is excluded. The domain of the function is the set of all real numbers, and the range is the set of integers.   ◀ **3**

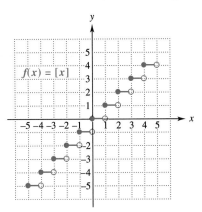

FIGURE 2.23

The greatest integer function, graphed in Figure 2.23, is an example of a **step function.**

▶**EXAMPLE 4**   An overnight delivery service charges $25 for a package weighing up to 2 pounds. For each additional pound or fraction of a pound there is an additional charge of $3. Let $D(x)$ represent the cost to send a package weighing x pounds. Graph $D(x)$ for x in the interval $(0, 6]$.

For x in the interval $(0, 2]$, $y = 25$. For x in $(2, 3]$, $y = 25 + 3 = 28$. For x in $(3, 4]$, $y = 28 + 3 = 31$, and so on. The graph, which is that of a step function, is shown in Figure 2.24.   ◀ **4**

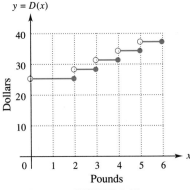

FIGURE 2.24

FOR GRAPHERS

The absolute value function is often a built-in function in a graphing utility. It is designated by $|x|$ or abs(x). If neither of these forms is available, the fact that $|x| = \sqrt{x^2}$ can be used.

Piecewise functions can be graphed with some graphers by inputing each function separately with its domain and graphing them simultaneously. Check the manual for your grapher. The graph will not distinguish between open and closed endpoints, so it is still important for the user to understand what the graph should look like.

Some graphers have a built-in function for the greatest integer function.

▶ **EXAMPLE 5**   Business graphs are often made up of portions of straight lines. For example, the graph in Figure 2.25 shows the per capita personal income in the greater Sacramento area from 1980 to 1992. To get this graph, points are plotted for different dates and connected with straight line segments. The graph suggests that, after a steady rise in income from 1980 to 1990, incomes leveled off and increased much more slowly from 1990 to 1992.   ◀

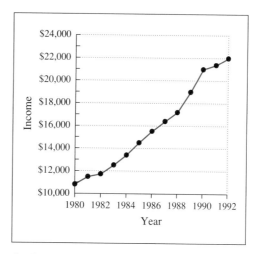

Graph showing per capita personal income in the greater Sacramento area, *Economic Profile Greater Sacramento Area* from Sacramento Area Commerce and Trade Organization, Summer, 1992. Reprinted by permission.

FIGURE 2.25

5 Graph $f(x) = \sqrt{4 - x}$.

Answer:

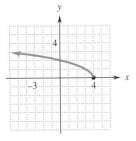

The graphs of many functions do not consist of straight line segments. In such cases it may be necessary to plot additional points to determine a pattern. When you are uncertain about the graph of a function, follow this procedure to graph it by hand.

Graphing a Function by Point Plotting

1. Determine the domain of the function.
2. Select a few numbers in the domain of f (include both negative and positive ones when possible) and compute the corresponding values of $f(x)$.
3. Plot the points $(x, f(x))$ computed in step two. Use these points and any other information you may have about the function to make an "educated guess" about the shape of the entire graph.
4. Unless you have information to the contrary, assume that the graph is continuous (unbroken) wherever it is defined.

▶ **EXAMPLE 6**   Graph $g(x) = \sqrt{x + 1}$.

Because the rule of the function is defined only when $x + 1 \geq 0$ (that is, when $x \geq -1$), the domain of g is the interval $[-1, \infty)$. Use a calculator to get a table of ordered pairs, such as the one in Figure 2.26. Plot the points and connect them in order (as x increases) to get the graph in Figure 2.26. ◀ **5**

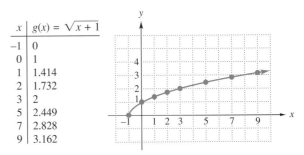

x	$g(x) = \sqrt{x + 1}$
−1	0
0	1
1	1.414
2	1.732
3	2
5	2.449
7	2.828
9	3.162

FIGURE 2.26

If a graph is to represent a function, each value of x from the domain must lead to exactly one value of y. In the graph in Figure 2.27, the domain value x_1 leads to *two* y-values, y_1 and y_2. Since the given x-value corresponds to two different y-values, this is not the graph of a function. This example suggests the **vertical line test** for the graph of a function.

6 Does this graph represent a function? How can you tell?

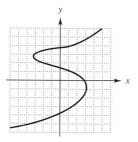

Answer:
No; a vertical line crosses the graph at more than one point.

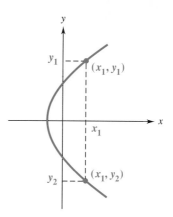

FIGURE 2.27

Vertical Line Test

If a vertical line intersects a graph at more than one point, the graph is not the graph of a function.

▶**EXAMPLE 7**   Use the vertical line test to decide which of the graphs in Figure 2.28 are graphs of functions.

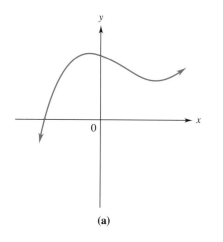

(a)

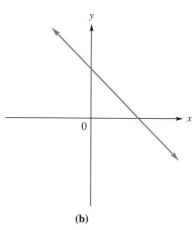

(b)

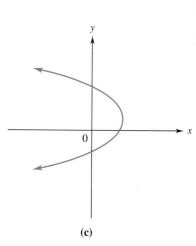

(c)

FIGURE 2.28

(a) Every vertical line intersects this graph in at most one point, so this is the graph of a function.
(b) Again, each vertical line intersects the graph in at most one point, showing that this is the graph of a function.
(c) It is possible for a vertical line to intersect the graph in part (c) twice. This is not the graph of a function. ◀ **6**

2.5 EXERCISES

Which of the following are graphs of functions? (See Example 7.)

1.

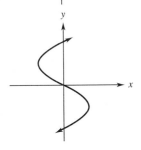

2.

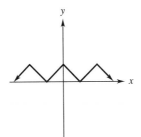

3.

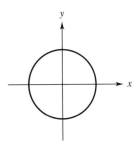

4.

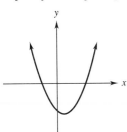

5.

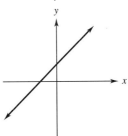

6.

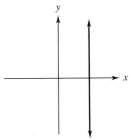

Graph each of the following. (See Example 1.)

7. $y = |x - 4|$

8. $y = |x + 4|$

9. $y = |-4 - x|$

10. $y = |4 - x|$

11. $y = |2x + 5|$

12. $y = |3 - 4x|$

13. $y = -|x|$

14. $y = -|x - 1|$

15. $y = |x| - 2$

16. $y = |x| + 3$

---

17. Compare the graphs of Exercises 7, 13, and 15 to the graph of $f(x) = |x|$.

Graph each of the following functions. (See Example 2.)

18. $y = \begin{cases} x + 2 & \text{if } x \le 1 \\ 3 & \text{if } x > 1 \end{cases}$

19. $y = \begin{cases} 2x - 1 & \text{if } x < 0 \\ -1 & \text{if } x \ge 0 \end{cases}$

20. $y = \begin{cases} 3 - x & \text{if } x \le 0 \\ 2x + 3 & \text{if } x > 0 \end{cases}$

21. $y = \begin{cases} x + 5 & \text{if } x \le 1 \\ 2 - 3x & \text{if } x > 1 \end{cases}$

22. $y = \begin{cases} |x| & \text{if } x \le 2 \\ -x & \text{if } x > 2 \end{cases}$

23. $y = \begin{cases} -|x| & \text{if } x \le 1 \\ 2x & \text{if } x > 1 \end{cases}$

Graph each of the following functions. (See Example 3.)

24. $y = [x - 3]$

25. $y = [x + 2]$

26. $y = [-x]$

27. $y = -[x]$

28. $y = [2x]$

29. $y = [3x]$

30. Describe how the *y*-values of the greatest integer function are determined for negative *x*-values.

Graph each of the following functions. (See Example 6.)

31. $y = \sqrt{x}$

32. $y = \sqrt{-x}$

33. $y = \sqrt{2x - 1}$

34. $y = \sqrt{2x + 1}$

35. $y = \sqrt{x} - 1$

36. $y = \sqrt{x} + 1$

Solve the following problems. (See Examples 4 and 5.)

37. The charge to rent a Haul-It-Yourself Trailer is $25 plus $2 per hour or portion of an hour. Find the cost to rent a trailer for
  (a) 2 hours; **(b)** 1.5 hours; **(c)** 4 hours;
  (d) 3.7 hours.
  (e) Graph the ordered pairs (hours, cost).

38. A delivery company charges $3 plus 50¢ per mile or part of a mile. Find the cost for a trip of
  (a) 3 miles; **(b)** 3.4 miles; **(c)** 3.9 miles;
  (d) 5 miles.
  (e) Graph the ordered pairs (miles, cost).
  (f) Is this a function?

39. A college typing service charges $3 plus $7 per hour or fraction of an hour. Graph the ordered pairs (hours, cost).

40. A parking garage charges $1 plus 50¢ per hour or fraction of an hour. Graph the ordered pairs (hours, cost).

41. A car rental costs $37 for 1 day, which includes 50 free miles. Each additional 25 miles, or portion, costs $10. Graph the ordered pairs (miles, cost).

42. For a lift truck rental of no more than 3 days, the charge is $300. An additional charge of $75 is made for each day or portion of a day after 3. Graph the ordered pairs (days, cost).

43. Social Science United States health care costs as a percent of the gross national product from 1960 to 1992 are given by

$$y = \begin{cases} .22x + 5.5 & \text{for 1960 to 1985} \\ .29x + 3.75 & \text{for 1985 to 1990.} \end{cases}$$

Let $x = 0$ represent 1960 and graph this function. What does the graph suggest about health care costs?

44. Social Science Personal taxes in the United States from 1960 to 1990 are approximated by

$$f(x) = \begin{cases} 7.9x + 50.4 & \text{from 1960 to 1975} \\ 35.4x - 361.6 & \text{from 1975 to 1990.} \end{cases}$$

Let $x = 0$ represent 1960 and graph the function. What happened to personal taxes in 1975?

45. Natural Science The snow depth in Michigan's Isle Royale National Park varies throughout the winter. In a typical winter, the snow depth in inches is approximated by the following function.

$$f(x) = \begin{cases} 6.5x & \text{if } 0 \le x \le 4 \\ -5.5x + 48 & \text{if } 4 < x \le 6 \\ -30x + 195 & \text{if } 6 < x \le 6.5 \end{cases}$$

Here, *x* represents the time in months with $x = 0$ representing the beginning of October, $x = 1$ representing the beginning of November, and so on.
  (a) Graph $f(x)$.
  (b) In what month is the snow deepest? What is the deepest snow depth?
  (c) In what months does the snow begin and end?

46. Natural Science A factory begins emitting particulate matter into the atmosphere at 8 A.M. each workday, with the emissions continuing until 4 P.M. The level of pollutants, $P(t)$, measured by a monitoring station 1/2 mile away is approximated as follows, where *t* represents the number of hours since 8 A.M.

$$P(t) = \begin{cases} 75t + 100 & \text{if } 0 \le t \le 4 \\ 400 & \text{if } 4 < t < 8 \\ -100t + 1200 & \text{if } 8 \le t \le 10 \\ -\dfrac{50}{7}t + \dfrac{1900}{7} & \text{if } 10 < t < 24 \end{cases}$$

Find the level of pollution at
(a) 9 A.M.   (b) 11 A.M.   (c) 5 P.M.
(d) 7 P.M.   (e) Midnight.
(f) Graph $y = P(t)$.
(g) From the graph in part (f), at what time(s) is the pollution level highest? lowest?

47. Management The **elasticity of demand** is the percent by which the demand for a product changes as price changes. For example, an elasticity of -1 means that demand changes at the same rate as price, so that a 10% price increase will cause a 10% drop in demand. An elasticity of $-.4$ means that a 10% price increase will cause a .4 of 10% = 4% drop in demand. Recently, there has been much controversy over the elasticity of demand for gasoline. It is now agreed that the short-term elasticity is small (you still have to get to work tomorrow), but the longer term elasticity is much higher (your next car will be much more fuel efficient). One recent projection of gasoline elasticity is as follows, where $e(t)$ represents elasticity at time t measured in years from some base year.

$$e(t) = \begin{cases} -.10 & \text{for} & 0 \le t \le 2 \\ -.25 & \text{for} & 2 < t \le 5 \\ -.50 & \text{for} & 5 < t \le 8 \\ -.75 & \text{for} & 8 < t \le 12 \\ -1.10 & \text{for} & 12 < t \le 16 \end{cases}$$

For example, a 10% price increase now will cause a .75 of 10% = 7.5% drop in demand in years 9 through 12.
(a) Graph $y = e(t)$.
(b) Give the domain and range for e.

48. Management Normally, an increase in price will produce an increase in the supply of an item. However, there are a few items where an increase in price produces a *decrease* in supply. An example is labor—in developing countries with few consumer goods available, an increase in wage rates can actually lead to workers putting in fewer hours. Such situations produce *backward-bending* supply curves. Some economists now feel that oil may be on a backward-bending supply curve—as the price increases, some exporting nations can get sufficient revenue for their needs with the sale of fewer barrels of oil. To see how these curves work, let x be the number of millions of barrels of oil produced in

some fixed time period, and let p be the price per barrel. Based on certain published data, the supply curve must be graphed as follows.
(a) For $10 \le x \le 12$, graph $p = (5/2)x$.
(b) For $12 < x \le 14$, graph $p = x + 18$.
(c) For $32 \le p \le 38$, graph $x = 14$.
(d) For $11 \le x \le 14$, graph $p = -2x + 66$.

49. The graph below, based on data given in *Business Week* magazine, shows the number of aircraft with airfones since 1984.*
(a) Is this graph that of a function?
(b) What does the domain represent?
(c) Estimate the range.

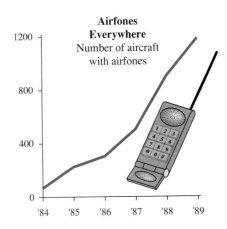

Data: GTE Airphone, Inc.

50. Graph

$$y = x + 4 \text{ and } y = (x^2 + 3x - 4)/(x - 1).$$

Do you see a difference between the graphs? Should there be a difference? If so, describe the difference?

51. Graph $y = |x|$. Then graph $y = |x - 2|$, $y = |x| + 2$, and $y = 2|x|$. How are these four graphs related? How do they differ?

---

*"Airfones Everywhere" reprinted from February 15, 1990 issue of *Business Week* by special permission, copyright © 1990 by McGraw-Hill, Inc.

KEY TERMS AND SYMBOLS

2.1 linear equation
graph
ordered pair
Cartesian coordinate system
x-coordinate
y-coordinate
quadrant
x-intercept
y-intercept
supply and demand curves
equilibrium price
equilibrium supply/demand

2.2 Δx change in x
Δy change in y
m slope
slope-intercept form
point-slope form

2.3 function
domain
range
$f(x)$ "f of x"

2.4 linear function
average rate of change
fixed cost
marginal cost
linear cost function
average cost
revenue
profit
break-even point

2.5 absolute value function
piecewise function
greatest integer function
step function
vertical line test

KEY CONCEPTS

If $p = f(q)$ gives the price per unit when x units can be supplied, and $p = g(q)$ gives the price per unit when q units are demanded, then the **equilibrium price, supply,** and **demand** occur at the q-value such that $f(q) = g(q)$.

The **slope** of the line through the points (x_1, y_1) and (x_2, y_2), where $x_1 \neq x_2$ is $m = \dfrac{y_2 - y_1}{x_2 - x_1}$. Nonvertical **parallel lines** have the same slope and **perpendicular lines,** if neither is vertical, have slopes with a product of -1.

The line with equation $ax + by = c$ has x-intercept c/a and y-intercept c/b.
The line with equation $x = k$ is vertical with x-intercept k, no y-intercept, and undefined slope.
The line with equation $y = k$ is horizontal, with y-intercept k, no x-intercept, and slope 0.

The line with equation $y = mx + b$ has slope m and y-intercept b.
The line with equation $y - y_1 = m(x - x_1)$ has slope m and goes through (x_1, y_1).

A **function** consists of a set of input numbers called the **domain,** a set of output numbers called the **range,** and a rule by which each number in the domain determines exactly one number in the range.

A **linear cost function** has equation $C(x) = mx + b$ where m is the **marginal cost** (the cost of producing one more item) and b is the **fixed cost.** If $R(x)$ is the revenue function, the **break-even point** is the x-value such that $C(x) = R(x)$.

If a vertical line intersects the graph of a function in more than one point, the graph is not that of a function.

CHAPTER 2   REVIEW EXERCISES

Graph each of the following linear equations.

1. $5x - 3y = 15$ **2.** $2x + 7y - 21 = 0$ **3.** $y + 3 = 0$ **4.** $y - 2x = 0$

5. In your own words, define the slope of a line.

In Exercises 6–15, find the slope of the line.

6. through $(-1, 4)$ and $(2, 3)$ **7.** through $(5, -3)$ and $(-1, 2)$

8. through $(7, -2)$ and the origin   **9.** through $(8, 5)$ and $(0, 3)$

10. $2x + 3y = 30$ **11.** $4x - y = 7$

12. $x + 5 = 0$ **13.** $y = 3$

14. parallel to $3x + 8y = 0$ **15.** perpendicular to $x = 3y$

16. Graph the line through $(0, 5)$ wih $m = -2/3$.   **17.** Graph the line through $(-4, 1)$ with $m = 3$.

18. What information is needed to determine the equation of a line?

Find an equation for each of the following lines.

19. Through $(5, -1)$, slope $2/3$ **20.** Through $(8, 0)$, slope $-1/4$

21. Through $(5, -2)$ and $(1, 3)$ **22.** Through $(2, -3)$ and $(-3, 4)$

23. Undefined slope, through $(-1, 4)$ **24.** Slope 0, through $(-2, 5)$

25. x-intercept -3, y-intercept 5   **26.** x-intercept $-2/3$, y-intercept $1/2$

27. Management   The supply and demand for a certain commodity are related by

     supply: $p = 6q + 3$ demand: $p = 19 - 2q,$

where p represents the price at a supply or demand of q units. Find the supply and the demand when the price is as follows.

(a) $10   **(b)** $15   **(c)** $18

(d) Find the equilibrium price.

(e) Find the equilibrium quantity (supply/demand).

28. Management   For a particular product, 72 units will be supplied at a price of $34, while 16 units will be supplied at a price of $6.

(a) Write a linear supply equation for this product.

(b) The demand for this product is given by the equation $p = 12 - .5q$. Find the equilibrium price and quantity.

29. What is a function? A linear function?

30. What ordered pair corresponds to the notation $f(2) = 5$? Which number is the input and which is the output?

31. Write the notation $f(-7) = 10$ as an ordered pair. Which number is the input and which is the output?

Which of the following rules defines a function?

32.

x	3	2	1	0	1	2
y	8	5	2	0	-2	-5

33.

x	2	1	0	-1	-2
y	5	3	1	-1	-3

34. $y = \sqrt{x}$

35. $x = |y|$

36. $x = y^2 + 1$

37. $y = 5x - 2$

For each function defined as follows, find (a) f(6), (b) f(−2), (c) f(p), (d) f(r + 1).

38. $f(x) = 4x - 1$

39. $f(x) = 3 - 4x$

40. $f(x) = -x^2 + 2x - 4$

41. $f(x) = 8 - x - x^2$

---

42. Let $f(x) = 5x - 3$ and $g(x) = -x^2 + 4x$. Find each of the following.

(a) $f(-2)$ **(b)** $g(3)$ **(c)** $g(-k)$

(d) $g(3m)$ **(e)** $g(k - 5)$ **(f)** $f(3 - p)$

43. Social Science   The percent of children living with a never-married parent was 4.2 in 1960 and had risen to 30.6 in 1990. Write a linear equation in function form to express this relationship giving the percent as a function of the year. Let 1960 correspond to $x = 0$. According to this function, what percent of children were living with a never-married parent in 1995? Does this seem reasonable?

44. Social Science   In 1960 the percent of children living with two parents was 87.7. In 1990 72.5% were living with two parents. The decrease has been approximately linear. Write an equation giving the percent y as a function of the year x, with 1960 corresponding to $x = 0$. Use the equation to estimate the number of children living with two parents in 1995.

45. What is the marginal cost to manufacture a product? the fixed cost?

Management   *In Exercises 46–49, find the following.*

(a) *the linear cost function*

(b) *the marginal cost*

(c) *the average cost per unit to produce 100 units*

46. Eight units cost $300; fixed cost is $60.

47. Fixed cost is $2000; 36 units cost $8480.

48. Twelve units cost $445; 50 units cost $1585.

49. Thirty units cost $1500; 120 units cost $5640.

50. Management   The graph shows the percent of car owners who own foreign cars since 1975. Estimate the average rate of change in the percent over the following intervals.

(a) 1975 to 1983   **(b)** 1983 to 1987

(c) 1975 to 1991   **(d)** 1987 to 1991

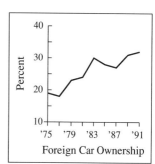

Foreign Car Ownership

51. **Management** The cost of producing x units of a product is $C(x)$, where $C(x) = 20x + 100$. The product sells for \$40 per unit.
(a) Find the break-even point.
(b) What revenue will the company receive if it sells just the number of units from part (a)?
(c) Find the profit function.

52. **Management** A product can be sold for \$25 per unit. The cost of producing x units in a certain plant is $C(x) = 24x + 5000$.
(a) Can this item make a profit?
(b) Suppose the profit function is $P(x) = 8x - 5000$. What is the revenue function?
(c) What is the break-even point using the profit function from part (b)?

Graph each of the following functions.

53. $f(x) = |x| - 3$

54. $f(x) = -|x| - 2$

55. $f(x) = -|x + 1| + 3$

56. $f(x) = 2|x - 3| - 4$

57. $f(x) = [x - 3]$

58. $f(x) = \left[\frac{1}{2}x - 2\right]$

59. $f(x) = \begin{cases} -4x + 2 & \text{if } x \le 1 \\ 3x - 5 & \text{if } x > 1 \end{cases}$

60. $f(x) = \begin{cases} 3x + 1 & \text{if } x < 2 \\ -x + 4 & \text{if } x \ge 2 \end{cases}$

61. $f(x) = \begin{cases} |x| & \text{if } x < 3 \\ 6 - x & \text{if } x \ge 3 \end{cases}$

62. Let f be a function that gives the cost to rent a floor polisher for x hours. The cost is a flat \$3 for cleaning the polisher plus \$4 per day or fraction of a day for using the polisher.
(a) Graph f.
(b) Give the domain and range of f.
(c) David Fleming wants to rent a polisher, but he can spend no more than \$15. At most how many days can he use it?

63. A trailer hauling service charges \$45, plus \$2 per mile or part of a mile.
(a) Is \$90 enough for a 20-mile haul?
(b) Graph the ordered pairs (miles, cost).
(c) Give the domain and range.

64. **Social Sciences** The birth rate per thousand in developing countries, for 1775–1977, can be approximated by

$$f(x) = \begin{cases} 42 & \text{from 1775 to 1925} \\ 67.5 - .17x & \text{from 1925 to 1977.} \end{cases}$$

Graph the function. Let $x = 0$ correspond to 1775. What does the graph suggest about the birth rate in developing countries?

A challenging problem facing the United States is the effect of rising health-related costs on the nation's long-term economic prosperity. In 1940 health care absorbed $4 billion, a mere 4 percent of our gross national product (GNP). In 1990 these costs were $666 billion, 12.2 percent of the GNP.

We can find the average percent rate of increase per year from these figures as follows. Let y represent the percent of GNP spent on health-related costs in year x.

$$\frac{\Delta y}{\Delta x} = \frac{12.2 - 4}{90 - 40} = \frac{8.2}{50} = .164$$

Thus, on the average, each year since 1940 the percent of GNP spent on health costs has increased by an average of .164 or 16.4%.

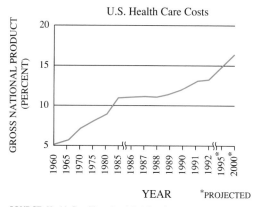

U.S. Health Care Costs

SOURCE: Health Care Financing Administration.

FIGURE 1

Figure 1 shows U.S. health care costs as a percent of the GNP from 1960 to 1992 and projections for the years up to 2000. By reading figures for the percent of GNP from the graph for selected years, we can find a function that will give

the approximate percent of GNP expended in any year. The graph can be approximated reasonably well for prediction purposes by a straight line. To write the equation of a linear function, we need to know two points on the line. Suppose we choose $x = 60$ (for 1960), with $y = 5\%$ (approximately) and $x = 92$ (for 1992), with $y = 13\%$ (approximately). This gives the ordered pairs (60, 5) and (92, 13). The slope is

$$m = \frac{13 - 5}{92 - 60} = \frac{8}{32} = \frac{1}{4} = .25$$

Now we will use the point-slope form to get the equation of the line that will define the function.

$$y - 5 = .25(x - 60)$$
$$y - 5 = .25x - 15$$
$$y = .25x - 10$$

To verify that this equation will give values of y that are reasonable, let us find y for $x = 90$ and compare with the figures given at the beginning of this application.

$$\text{If } x = 90, \quad y = .25(90) - 10 = 12.5.$$

These results are reasonably close to the given value.

Using the function to predict the percent of the GNP that will be spent on health costs in 1997, let $x = 97$ in the equation.

$$\text{if } x = 97, \quad y = .25(97) - 10 = 14.25$$

This is slightly lower than the value from the graph in Figure 1, which is closer to 15. Notice that the graph is increasing more sharply after 1992. Because of this the slope (or rate of change) in the equation is probably too small for accurate predictions much beyond 1992.

---

*From *Scientific American,* "Health Care Reform," Rashi Fein, November 1992, pages 46 and 49.

EXERCISES

1. Select the ordered pairs corresponding to 1985 and 1992 and write an equation for the line that approximates the graph in Figure 1. Does this equation produce a better approximation for 1997 than the prediction in the text.

2. Give some reasons that might explain the discrepancy in the average rate of increase in the percent of GNP of .164 found first and the slope of .25 found next.

3. Just by looking at the graph in Figure 1, describe the rate of increase in percent of GNP over the years from 1960 to 1992 and the expected change from 1992 to 2000.

Marginal Cost—Booz, Allen and Hamilton*

Booz, Allen and Hamilton is a large management consulting firm. One of the services it provides to client companies is profitability studies, which show ways in which the client can increase profit levels. The client company requesting the analysis presented in this case is a large producer of a staple food. The company buys from farmers, and then processes the food in its mills, resulting in a finished product. The company sells both at retail and under its own brands, and in bulk to other companies who use the product in the manufacture of convenience foods.

The client company has been reasonably profitable in recent years, but the management retained Booz, Allen and Hamilton to see whether its consultants could suggest ways of increasing company profits. The management of the company had long operated with the philosophy of trying to process and sell as much of its product as possible, since they felt this would lower the average processing cost per unit sold. However, the consultants found that the client's fixed mill costs were quite low, and that, in fact, processing extra units made the cost per unit start to increase. (There are several reasons for this: the company must run three shifts, machines break down more often, and so on.)

In this case, we shall discuss the marginal cost of two of the company's products. The marginal cost (cost of producing an extra unit) of production for product A was found by the consultants to be approximated by the linear function

$$y = .133x + 10.09,$$

where x is the number of units produced (in millions) and y is the marginal cost.

For example, at a level of production of 3.1 million units, an additional unit of product A would cost about

$$y = .133(3.1) + 10.09$$
$$\approx \$10.50.^{\dagger}$$

---

* Case study, "Marginal Cost—Booz, Allen and Hamilton" supplied by John R. Dowdle of Booz, Allen & Hamilton, Inc. Reprinted by permission.

† The symbol "≈" means *is approximately equal to.*

At a level of production of 5.7 million units, an extra unit costs $10.85. Figure 1 shows a graph of the marginal cost function from $x = 3.1$ to $x = 5.7$, the domain to which the function above was found to apply.

The selling price for product A is $10.73 per unit, so that, as shown on the graph of Figure 1, the company was losing money on many units of the product that it sold. Since the selling price could not be raised if the company was to remain competitive, the consultants recommended that production of product A be cut.

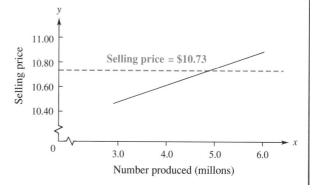

FIGURE 1

For product B, the Booz, Allen and Hamilton consultants found a marginal cost function given by

$$y = .0667x + 10.29,$$

with x and y as defined above. Verify that at a production level of 3.1 million units, the marginal cost is $10.50, while at a production level of 5.7 million units, the marginal cost is $10.67. Since the selling price of this product is $9.65, the consultants again recommended a cutback in production.

The consultants ran similar cost analyses of other products made by the company, and then issued their recommendations: the company should reduce total production by 2.1 million units. The analysts predicted that this would raise profits for the products under discussion from $8.3 million annually to $9.6 million, which is very close to what actually happened when the client took this advice.

EXERCISES

1. At what level of production, x, was the marginal cost of a unit of product A equal to the selling price?

2. Graph the marginal cost function for product B from $x = 3.1$ million units to $x = 5.7$ million units.

3. Find the number of units for which marginal cost equals the selling price for product B.

4. For product C, the marginal cost of production is

$$y = .133x + 9.46.$$

   (a) Find the marginal cost at a level of production of 3.1 million units; of 5.7 million units.
   (b) Graph the marginal cost function.
   (c) For a selling price of $9.57, find the level of production for which the cost equals the selling price.

CHAPTER 3

Polynomial and Rational Functions

TECHNOLOGY RESOURCES
GraphExplorer

College Algebra and Trigonometry: Graphing Calculator Investigations, Ebersole

Not all real world applications can be adequately modeled by the linear and piecewise functions used in Chapter 2. In many cases, a function whose graph is a curve rather than a straight line is needed. In this chapter we consider some of the most useful such functions: *polynomial functions,* which are defined by polynomial expressions, and *rational functions,* which are defined by quotients of polynomials.

3.1 QUADRATIC FUNCTIONS

Our study of polynomial functions begins with *quadratic functions,* whose rules are given by quadratic polynomials.

A function f is a **quadratic function** if
$$f(x) = ax^2 + bx + c,$$
where a, b, and c are real numbers with $a \neq 0$.

Quadratic functions are especially good models for many situations in which a quantity takes a maximum or a minimum value. They may also be used to describe supply and demand curves, cost, revenue, and profit, as well as other quantities.

The function defined by $f(x) = x^2$ is the simplest quadratic function. To see that it actually is quadratic, let $a = 1$, $b = 0$, and $c = 0$ in the definition $f(x) = ax^2 + bx + c$.

149

Graph each of the following parabolas.

(a) $f(x) = x^2 - 4$

(b) $f(x) = x^2 + 5$

Answers:

(a)

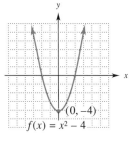

$f(x) = x^2 - 4$

(b)

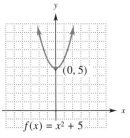

$f(x) = x^2 + 5$

 EXAMPLE 1 Graph the quadratic function defined by

$$f(x) = x^2.$$

Graph the equation $y = x^2$ by choosing several negative, zero, and positive values for x, and then finding the corresponding values for y, as in the table with Figure 3.1. The resulting points can then be plotted, with a smooth curve drawn through them, as in Figure 3.1. The domain of the function defined by $f(x) = x^2$ is the set of all real numbers, while the range is the set of all nonnegative real numbers. ◀

x	y
2	4
1	1
0	0
−1	1
−2	4

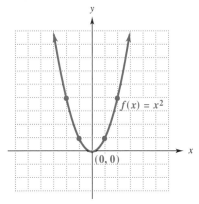

FIGURE 3.1

The curve in Figure 3.1 is called a **parabola.** The lowest point on this parabola, (0, 0), is called the **vertex.** It can be shown that every quadratic function has a graph that is a parabola. If the graph of Figure 3.1 were folded in half along the y-axis, the two halves of the parabola would match exactly. This means that the graph of a quadratic function is *symmetric* about a vertical line through the vertex: this line is the **axis of the parabola.** The axis of the parabola in Figure 3.1 is $x = 0$, the y-axis.

Parabolas have many useful properties. Cross sections of radar dishes and spotlights form parabolas. Discs often visible on the sidelines of televised football games are microphones having reflectors with parabolic cross sections. These microphones are used by the television networks to pick up the shouted signals of the quarterbacks.

EXAMPLE 2 Graph the quadratic function defined by $g(x) = x^2 + 2$.

For any value of x that might be chosen, the corresponding value of y will be 2 more than for the parabola $f(x) = x^2$ graphed in Figure 3.1. This means that the graph of the parabola $g(x) = x^2 + 2$, as shown in Figure 3.2, is shifted 2 units upward compared to the graph of $f(x) = x^2$. For example, the point (0, 2) is the vertex of the parabola $g(x) = x^2 + 2$, while (0, 0) is the vertex of the parabola $f(x) = x^2$. The axis is still the line $x = 0$. ◀ ◀ **1**

2 Graph $f(x) = -(x - 3)^2$.

Answer:

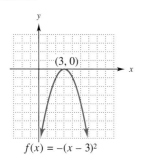

$f(x) = -(x - 3)^2$

x	y
2	6
1	3
0	2
−1	3
−2	6

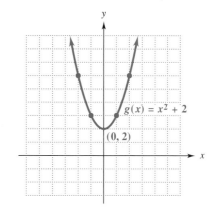

FIGURE 3.2

▶**EXAMPLE 3**   Graph the function defined by $h(x) = -(x + 2)^2$.

The table in Figure 3.3 shows several ordered pairs that belong to the function. These ordered pairs show that the graph of $h(x) = -(x + 2)^2$ is "upside down" in comparison to $f(x) = x^2$ (because of the negative sign) and also shifted 2 units to the left. Here the vertex, $(-2, 0)$, is the *highest* point on the graph. The axis of this parabola is the line $x = -2$. ◀ **2**

x	y
0	−4
−1	−1
−2	0
−3	−1
−4	−4

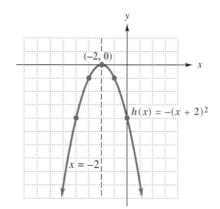

FIGURE 3.3

As shown in Example 2, the graph of $g(x) = x^2 + 2$ was shifted upward 2 units in comparison with the graph of $f(x) = x^2$. Similarly, in Example 3, the graph of $h(x) = -(x + 2)^2$ is shifted 2 units to the left when compared to

3 Graph the following.

(a) $f(x) = (x + 4)^2 - 3$

(b) $f(x) = -2(x - 3)^2 + 1$

Answers:

(a)

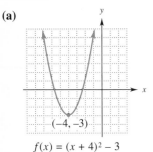

$f(x) = (x + 4)^2 - 3$

(b)

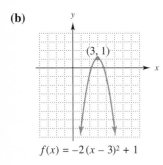

$f(x) = -2(x - 3)^2 + 1$

$f(x) = x^2$, and the parabola opens downward. Both upward (or downward) and side-to-side shifts can occur in the same parabola.

The facts about parabolas illustrated in the examples are summarized here without proof.

If f is a quadratic function defined by $y = a(x - h)^2 + k$, then the graph of the function f is a parabola having its vertex at (h, k) and axis of symmetry $x = h$.

If $a > 0$, the parabola opens upward; if $a < 0$, it opens downward.

If $0 < |a| < 1$, the parabola is "broader" than $y = x^2$, while if $|a| > 1$, the parabola is "narrower" than $y = x^2$.

▶ **EXAMPLE 4**   Graph $g(x) = 2(x - 3)^2 - 5$.

This parabola has vertex $(3, -5)$ and opens upward. The axis is the line $x = 3$. The coefficient 2 causes the values of y to increase more rapidly than in the parabola $f(x) = x^2$, so that the graph in this example is "narrower" than the graph of $f(x) = x^2$. Figure 3.4 shows a table of ordered pairs satisfying $g(x) = 2(x - 3)^2 - 5$, as well as the graph of the function. Every parabola is symmetric about its axis, so for each ordered pair on one side of the axis, symmetry can be used to find corresponding ordered pairs on the other side. Thus, the ordered pair $(1, 3)$ suggests the ordered pair $(5, 3)$ on the other side of the axis of symmetry. Similarly, $(2, -3)$ and $(4, -3)$ are symmetric with respect to the axis.   ◀   **3**

x	y
1	3
2	-3
3	-5
4	-3
5	3

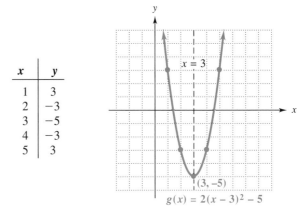

$g(x) = 2(x - 3)^2 - 5$

FIGURE 3.4

Since the vertex of a parabola is the highest or lowest point on the parabola, it is the most important point for graphing or for applications of quadratic functions. The vertex and axis of a parabola can be found quickly if the equation of the parabola is in the form

4 Complete the square for each of the following. Then graph the parabola.

(a) $f(x) = x^2 - 6x + 11$

(b) $f(x) = x^2 + 8x + 18$

Answers:

(a) $f(x) = (x - 3)^2 + 2$

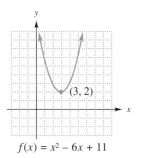

$f(x) = x^2 - 6x + 11$

(b) $f(x) = (x + 4)^2 + 2$

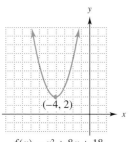

$f(x) = x^2 + 8x + 18$

$$f(x) = a(x - h)^2 + k$$

for real numbers $a \neq 0$, h and k. An equation not given in this form can be converted to it by a process called *completing the square,* first discussed in Section 1.10.

▶ **EXAMPLE 5** Graph $f(x) = x^2 - 2x + 3$ by completing the square.

Rewrite the equation in the form $f(x) = a(x - h)^2 + k$ by first writing $f(x) = x^2 - 2x + 3$ as

$$f(x) = (x^2 - 2x \quad) + 3.$$

Now complete the square for the expression in parentheses. Take half the coefficient of x, namely $(\frac{1}{2})(-2) = -1$, and square the result: $(-1)^2 = 1$. In order to complete the square we must add 1 inside the parentheses, but in order not to change the rule of the function we must also subtract 1:

$$f(x) = (x^2 - 2x + 1 - 1) + 3.$$

By the associative property,

$$f(x) = (x^2 - 2x + 1) + (-1 + 3).$$

Factor $x^2 - 2x + 1$ as $(x - 1)^2$, to get

$$f(x) = (x - 1)^2 + 2.$$

By this result, the graph of the given function is a parabola with the vertex at $(1, 2)$. Since $a = 1$ is positive, the parabola opens upward. Find a few additional points by choosing, say, $x = 2$ and $x = 3$ to get the ordered pairs $(2, 3)$ and $(3, 6)$. By symmetry, the ordered pairs $(0, 3)$ and $(-1, 6)$ are also on the graph. Use these points to get the graph shown in Figure 3.5. ◀ **4**

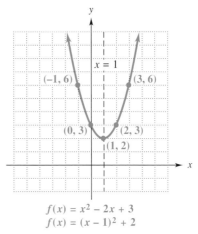

$f(x) = x^2 - 2x + 3$
$f(x) = (x - 1)^2 + 2$

FIGURE 3.5

5 Complete the square and graph the following.

(a) $f(x) = 3x^2 - 12x + 14$

(b) $f(x) - x^2 + 6x - 12$

Answers:

(a) $f(x) = 3(x - 2)^2 + 2$

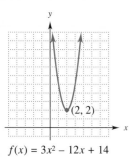

$f(x) = 3x^2 - 12x + 14$

(b) $f(x) = -(x - 3)^2 - 3$

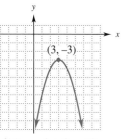

$(3, -3)$

$f(x) = -x^2 + 6x - 12$

▶**EXAMPLE 6** Graph $f(x) = -2x^2 + 12x - 19$.

Get $f(x)$ in the form $f(x) = a(x - h)^2 + k$ by first factoring -2 from $-2x^2 + 12x$.

$$f(x) = -2x^2 + 12x - 19 = -2(x^2 - 6x) - 19$$

Take half of -6: $(1/2)(-6) = -3$. Square this result: $(-3)^2 = 9$. Add and subtract 9 inside the parentheses.

$$y = -2(x^2 - 6x + 9 - 9) - 19$$
$$y = -2(x^2 - 6x + 9) + (-2)(-9) - 19 \qquad \textbf{Distributive property}$$

Simplify and factor to get

$$y = -2(x - 3)^2 - 1.$$

Since $a = -2$ is negative, the parabola opens downward. The vertex is at $(3, -1)$, and the parabola is narrower than $y = x^2$. Use these results and plot additional ordered pairs as needed to get the graph of Figure 3.6 ◀ **5**

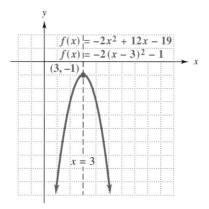

$f(x) = -2x^2 + 12x - 19$
$f(x) = -2(x - 3)^2 - 1$
$(3, -1)$
$x = 3$

FIGURE 3.6

6 Given $f(x) = x^2 - 7x + 12$, find the following.

(a) The intercepts

(b) The vertex and axis

Answers:

(a) $x = 3$, $x = 4$, $y = 12$

(b) $(7/2, -1/4)$; $x = 7/2$

FOR GRAPHERS

When you have graphed a quadratic function, the trace feature on a graphing calculator can be used to find the *approximate* coordinates of the vertex. Because the trace displays only the coordinates of points plotted by the calculator, it may not display the actual vertex, but only a point very close to it. To find the vertex exactly, use the algebraic techniques of Examples 5 and 6. To guard against mistakes, use the grapher to confirm your results.

The technique of completing the square can be used to rewrite the general equation $f(x) = ax^2 + bx + c$ in the form $f(x) = a(x - h)^2 + k$, as is shown in Exercise 36. When this is done we obtain a formula for the coordinates of the vertex.

The graph of $f(x) = ax^2 + bx + c$ is a parabola with vertex (h, k), where

$$h = \frac{-b}{2a} \quad \text{and} \quad k = f(h).$$

▶**EXAMPLE 7**   Graph $f(x) = x^2 - x - 6$.
    Since $a = 1$ and $b = -1$, the x-value of the vertex is

$$\frac{-b}{2a} = \frac{-(-1)}{2 \cdot 1} = \frac{1}{2}.$$

The y-value of the vertex is

$$f\left(\frac{1}{2}\right) = \left(\frac{1}{2}\right)^2 - \frac{1}{2} - 6 = -\frac{25}{4}.$$

The vertex is $(1/2, -25/4)$ and the axis of the parabola is $x = 1/2$. A convenient way to sketch the graph is to plot the vertex and the x- and y-intercepts. The intercepts are found by setting each variable equal to 0.

Set $f(x) = y = 0$;	Set $x = 0$:
$0 = x^2 - x - 6$	$f(x) = y = 0^2 - 0 - 6 = -6$
$0 = (x + 2)(x - 3)$	The y-intercept is -6.
$x + 2 = 0 \quad$ or $\quad x - 3 = 0$	
$x = -2 \qquad\qquad x = 3$	

The x-intercepts are -2 and 3.

Now plot the vertex and the three intercepts. Then use symmetry to complete the graph, as shown in Figure 3.7 on the next page.   ◀ 6

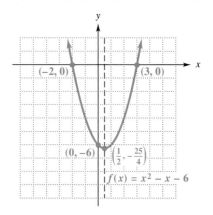

FIGURE 3.7

3.1 EXERCISES

Without graphing, determine the vertex of each of the following parabolas and state whether it opens upward or downward. (See Examples 4–7.)

1. $f(x) = 3(x - 5)^2 + 2$

2. $g(x) = -6(x - 2)^2 - 5$

3. $f(x) = -(x - 1)^2 + 2$

4. $g(x) = x^2 + 1$

5. $f(x) = x^2 + 12x + 1$

6. $g(x) = x^2 - 10x + 3$

7. $f(x) = 2x^2 + 4x + 1$

8. $g(x) = -3x^2 + 6x + 5$

Without graphing, determine the x- and y-intercepts of each of the following parabolas. (See Example 7.)

9. $f(x) = 3(x - 2)^2 - 3$

10. $f(x) = x^2 - 4x - 1$

11. $g(x) = 2x^2 + 8x + 6$

12. $g(x) = x^2 - 10x + 20$

Graph each of the following parabolas. Find the vertex and axis of symmetry of each. (See Examples 1–7.)

13. $f(x) = -x^2$

14. $f(x) = x^2 - 3$

15. $f(x) = (x + 2)^2$

16. $f(x) = -(x + 5)^2$

17. $f(x) = (x - 1)^2 - 3$

18. $f(x) = (x - 2)^2 + 1$

19. $f(x) = -(x + 4)^2 + 2$

20. $f(x) = -(x + 1)^2 - 3$

21. $f(x) = x^2 - 4x + 6$

22. $f(x) = x^2 + 6x + 3$

23. $f(x) = 2x^2 - 4x + 5$

24. $f(x) = -3x^2 + 24x - 46$

25. $f(x) = -x^2 + 6x - 6$

26. $f(x) = -x^2 + 2x + 5$

For each of the following functions, find the values of x for which f(x) is a minimum and find the minimum value of f(x).

27. $f(x) = x^2 - 2x - 15$

28. $f(x) = x^2 - 8x + 16$

In Exercises 29–34, graph the functions in parts (a)–(d) on the same set of axes; then answer part (e).

29. (a) $k(x) = x^2$ **(b)** $f(x) = 2x^2$
   (c) $g(x) = 3x^2$ **(d)** $h(x) = 3.5x^2$
   (e) Explain how the coefficient a in the function given by $f(x) = ax^2$ affects the shape of the graph, in comparison with the graph of $k(x) = x^2$, when $a > 1$.

30. (a) $k(x) = x^2$ **(b)** $f(x) = .8x^2$
   (c) $g(x) = .5x^2$ **(d)** $h(x) = .3x^2$
   (e) Explain how the coefficient a in the function given by $f(x) = ax^2$ affects the shape of the graph, in comparison with the graph of $k(x) = x^2$, when $0 < a < 1$.

31. (a) $k(x) = -x^2$ **(b)** $f(x) = -2x^2$
   (c) $g(x) = -3x^2$ **(d)** $h(x) = -3.5x^2$
   (e) Compare these graphs to the ones in Exercise 29 and explain how changing the sign of the coefficient a in the function given by $f(x) = ax^2$ effects the graph.

32. (a) $k(x) = x^2$ **(b)** $f(x) = x^2 + 2$
   (c) $g(x) = x^2 + 3$ **(d)** $h(x) = x^2 + 5$
   (e) Explain how the graph of $f(x) = x^2 + c$ (where c is a positive constant) can be obtained from the graph of $k(x) = x^2$.

33. (a) $k(x) = x^2$ **(b)** $f(x) = x^2 - 1$
   (c) $g(x) = x^2 - 2$ **(d)** $h(x) = x^2 - 4$
   (e) Explain how the graph of $f(x) = x^2 - c$ (where c is a positive constant) can be obtained from the graph of $k(x) = x^2$.

34. (a) $k(x) = x^2$ **(b)** $f(x) = (x + 2)^2$
   (c) $g(x) = (x - 2)^2$ **(d)** $h(x) = (x - 4)^2$
   (e) Explain how the graph of $f(x) = (x + c)^2$ and $f(x) = (x - c)^2$ (where c is a positive constant) can be obtained from the graph of $k(x) = x^2$.

35. Find the rule of a quadratic function whose graph is a parabola with vertex $(0, 0)$ that includes the point $(2, 12)$.

36. Verify that the right side of the equation
$$ax^2 + bx + c = a\left(x - \left(\frac{-b}{2a}\right)\right)^2 + \left(c - \frac{b^2}{4a}\right) \text{ equals}$$
the left side. Since the right side is in the form $a(x - h)^2 + k$, we conclude that the vertex of the parabola $f(x) = ax^2 + bx + c$ has x-coordinate $h = -b/2a$.

3.2 APPLICATIONS OF QUADRATIC FUNCTIONS

The fact that the vertex of a parabola $y = ax^2 + bx + c$ is the highest or lowest point on the graph can be used in applications to find a maximum or a minimum value. When $a > 0$, the graph opens upward, so the function has a minimum. When $a < 0$, the graph opens downward, producing a maximum.

▶**EXAMPLE 1**   Ms. Klein-Herring owns and operates Aunt Emma's Blueberry Pies. She has hired a consultant to analyze her business operations. The consultant tells her that her profits, $P(x)$, from the sale of x units of pies, are given by

$$P(x) = 120x - x^2.$$

How many units of pies should be sold in order to maximize profit? What is the maximum profit?

1 When a company sells x units of a product, its profit is $P(x) = -2x^2 + 40x + 280$. Find

(a) the number of units that should be sold so that maximum profit is received;

(b) the maximum profit.

Answers:

(a) 10 units

(b) $480

The profit function can be rewritten as $P(x) = -x^2 + 120x + 0$. By completing the square, we see that the graph of P is a downward opening parabola with vertex (60, 3600). For each point on the graph,

the x-coordinate is the number of units of pies;

the y-coordinate is the profit on that number of units.

Only the portion of the graph of Quadrant I (where both coordinates are positive) is relevant here because she cannot sell a negative number of pies and she is not interested in a negative profit. *Maximum* profit occurs at the point with the largest y-coordinate, namely, the vertex, as shown in Figure 3.8. The maximum profit of $3600 is obtained when 60 units of pies are sold. ◄ **1**

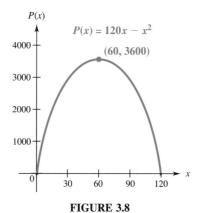

FIGURE 3.8

Supply and demand curves were introduced in Section 2.1. Here is a quadratic example.

► **EXAMPLE 2**   Suppose that the price and demand for an item are related by

$$p = 150 - 6q^2, \quad \textbf{Demand function}$$

where p is the price and q is the number of items demanded (in hundreds). The price and supply are related by

$$p = 10q^2 + 2q, \quad \textbf{Supply function}$$

where q is the number of items supplied (in hundreds). Find the equilibrium demand (and supply) and the equilibrium price.

The graphs of both of these equations are parabolas; these parabolas can be graphed as described in Section 3.1, giving the results shown in Figure 3.9. Only those portions of the graphs that lie in the first quadrant are included, because neither supply, demand, nor price can be negative.

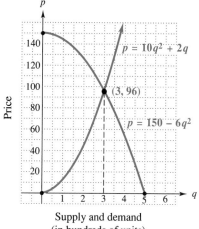

Supply and demand
(in hundreds of units)

FIGURE 3.9

The equilibrium demand and supply, by definition, occur at the same price p. Setting $150 - 6q^2$, where q is the quantity demanded, equal to $10q^2 + 2q$, where q is the quantity supplied, gives an equation that leads to the equilibrium supply and demand.

$$150 - 6q^2 = 10q^2 + 2q$$

Write this quadratic equation in standard form as follows.

$$0 = 16q^2 + 2q - 150 \qquad \text{Add } -150 \text{ and } 6q^2 \text{ to both sides}$$

$$0 = 8q^2 + q - 75 \qquad \text{Multiply both sides by } \frac{1}{2}$$

This equation can be solved by the quadratic formula given in Section 1.10. Here $a = 8$, $b = 1$, and $c = -75$.

$$q = \frac{-1 \pm \sqrt{1 - 4(8)(-75)}}{2(8)}$$

$$= \frac{-1 \pm \sqrt{1 + 2400}}{16} \qquad -4(8)(-75) = 2400$$

$$= \frac{-1 \pm 49}{16} \qquad \sqrt{1 + 2400} = \sqrt{2401} = 49$$

$$q = \frac{-1 + 49}{16} = \frac{48}{16} = 3 \quad \text{or} \quad q = \frac{-1 - 49}{16} = -\frac{50}{16} = -\frac{25}{8}$$

It is not possible to make $-25/8$ units, so discard that answer, and use only $q = 3$. Because q represents quantity in hundreds, equilibrium demand (and

2 The price and demand for an item are related by $p = 32 - x^2$, while price and supply are related by $p = x^2$. Find

(a) the equilibrium supply;

(b) the equilibrium price.

Answers:

(a) 4

(b) 16

supply) is $3 \cdot 100$, or 300 units. Find the equilibrium price by substituting 3 for q in either the supply or the demand function. Using the supply function gives

$$p = 10q^2 + 2q$$
$$p = 10 \cdot 3^2 + 2 \cdot 3 \qquad \text{Let } q = 3$$
$$= 10 \cdot 9 + 6$$
$$p = 96. \quad \blacktriangleleft \quad \boxed{2}$$

▶ **EXAMPLE 3** The rental manager of a small apartment complex with 16 units has found from experience that each $40 increase in the monthly rent results in an empty apartment. All 16 apartments will be rented at a monthly rent of $500. How many $40 increases will produce maximum monthly income for the complex?

Let x represent the number of $40 increases. Then the number of apartments rented will be $16 - x$. Also, the monthly rent per apartment will be $500 + 40x$. (There are x $40 increases for a total increase of $40x$.) The monthly income, $I(x)$, is given by the number of apartments rented times the rent per apartment, so

$$I(x) = (16 - x)(500 + 40x)$$
$$= 8000 + 640x - 500x - 40x^2$$
$$= 8000 + 140x - 40x^2.$$

The value of x that will produce maximum income occurs at the vertex, so use the methods of Section 3.1 to see that the vertex is $(1.75, 8122.50)$. Since x represents the number of $40 increases and each $40 increase causes 1 empty apartment, x must be a whole number, either 1 or 2. Try $x = 1$ and $x = 2$ in the quadratic equation to see which one gives the best result.

$$\text{If } x = 1, I(x) = -40(1)^2 + 140(1) + 8000 = 8100.$$
$$\text{If } x = 2, I(x) = -40(2)^2 + 140(2) + 8000 = 8120.$$

These results show that the best answer is to charge rents of $500 + 2(40) = 580$, or $580, and leave 2 apartments vacant. As a check, $14(580) = 8120$, as found above. ◀

3.2 EXERCISES

Work the following problems. (See Example 1.)

1. Management Shannise Cole makes and sells candy. She has found that the cost per box for making x boxes of candy is given by

$$C(x) = x^2 - 10x + 32.$$

(a) How much does it cost per box to make 2 boxes per day? 4 boxes? 10 boxes?

(b) Graph the cost function $C(x)$ and mark the points corresponding to 2, 4, and 10 boxes.

(c) What point on the graph corresponds to the number of boxes that will make the cost per box as small as possible?

(d) How many boxes should she make in order to keep the cost per box at a minimum? What is the minimum cost per box?

2. Management   George Duda sells bottled water. He has found that the average amount of time he spends with each customer is related to his weekly sales volume by the function

$$f(x) = x(60 - x),$$

where x is the number of minutes per customer and $f(x)$ is the number of cases sold per week.

(a) How many cases does he sell if he spends 10 minutes with each customer? 20 minutes? 45 minutes?

(b) Choose an appropriate scale for the axes and sketch the graph of $f(x)$. Mark the points on the graph corresponding to 10, 20, and 45 minutes.

(c) Explain what the vertex of the graph represents.

(d) How long should George spend with each customer in order to sell as many cases per week as possible? In this case, how many cases will he sell?

3. Management   French fries produce a tremendous profit (150% to 300%) for many fast food restaurants. Management, therefore, desires to maximize the number of bags sold. Suppose that a mathematical model connecting p, the profit per day from french fries (in tens of dollars), and x, the price per bag (in dimes), is $p = -2x^2 + 24x + 8$.

(a) Find the price per bag that leads to maximum profit.

(b) What is the maximum profit?

4. Natural Science   A researcher in physiology has decided that a good mathematical model for the number of impulses fired after a nerve has been stimulated is given by $y = -x^2 + 20x - 60$, where y is the number of responses per millisecond and x is the number of milliseconds since the nerve was stimulated.

(a) When will the maximum firing rate be reached?

(b) What is the maximum firing rate?

5. Physical Science   If an object is thrown upward with an initial velocity of 32 feet per second, then its height, in feet, above the ground after t seconds is given by

$$h = 32t - 16t^2.$$

Find the maximum height attained by the object. Find the number of seconds it takes for the object to hit the ground.

6. Natural Science   The number of mosquitoes in millions in a certain area, $M(x)$, depends on the June rainfall in inches, x, approximately as follows.

$$M(x) = 10x - x^2$$

Find the rainfall that will produce the maximum number of mosquitoes. What is the maximum number of mosquitoes?

7. Management   The manager of a bicycle shop has found that, at a price (in dollars) of $p(x) = 150 - \dfrac{x}{4}$ per bicycle, x bicycles will be sold.

(a) Find an expression for the total revenue from the sale of x bicycles.
(Hint: revenue = demand $\times$ price.)

(b) Find the number of bicycle sales that leads to maximum revenue.

(c) Find the maximum revenue.

8. Management   If the price (in dollars) of a videotape is $p(x) = 40 - \dfrac{x}{10}$, then x tapes will be sold.

(a) Find an expression for the total revenue from the sale of x tapes.

(b) Find the number of tapes that will produce maximum revenue.

(c) Find the maximum revenue.

9. Management   The demand for a certain type of cosmetic is given by

$$p = 500 - x,$$

where p is the price when x units are demanded.

(a) Find the revenue, $R(x)$, that would be obtained at a demand of x.

(b) Graph the revenue function, $R(x)$.

(c) From the graph of the revenue function, estimate the price that will produce the maximum revenue.

(d) What is the maximum revenue?

10. A field bounded on one side by a river is to be fenced on three sides to form a rectangular enclosure. There are 320 ft of fencing available. What should the dimensions be to have an enclosure with the maximum possible area?

11. A rectangular garden bounded on one side by a river is to be fenced on the other three sides. Fencing material for the side parallel to the river costs $30 per foot and material for the other two sides costs $10 per foot. What are the dimensions of the garden of largest possible area, if $1200 is to be spent for fencing material?

12. A trough is to be made by bending a long, flat piece of metal 10 inches wide into a rectangular shape. What depth should the trough be to have maximum possible cross-sectional area?

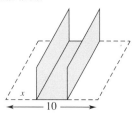

Work each of the following problems. (See Example 2.)

13. Management Suppose the price p of widgets is related to the quantity q that is demanded by

$$p = 640 - 5q^2,$$

where q is measured in hundreds of widgets. Find the price when the number of widgets demanded is
(a) 0;   **(b)** 5;   **(c)** 10.

Suppose the supply function for widgets is given by $p = 5q^2$, where q is the number of widgets (in hundreds) that are supplied at price p.

(d) Graph the demand function $p = 640 - 5q^2$ and the supply function $p = 5q^2$ on the same axes.
(e) Find the equilibrium supply.
(f) Find the equilibrium price.

Management *Suppose the supply and demand for a certain textbook are given by:*

$$supply: p = \frac{1}{5}q^2; \quad demand: p = -\frac{1}{5}q^2 + 40,$$

where p is price and q is quantity.

14. How many books are demanded at a price of
(a) 10?   **(b)** 20?   **(c)** 30?   **(d)** 40?

How many books are supplied at a price of
(e) 5?   **(f)** 10?   **(g)** 20?   **(h)** 30?
(i) Graph the supply and demand functions on the same axes.

15. Find the equilibrium demand and the equilibrium price in Exercise 14.

Work each problem. (See Example 3.)

16. Management A charter flight charges a fare of $200 per person, plus $4 per person for each unsold seat on the plane. If the plane holds 100 passengers and if x represents the number of unsold seats, find the following.

(a) An expression for the total revenue received for the flight. (Hint: multiply the number of people flying, $100 - x$, by the price per ticket.)
(b) The graph for the expression of part (a).
(c) The number of unsold seats that will produce the maximum revenue.
(d) The maximum revenue.

17. Management The revenue of a charter bus company depends on the number of unsold seats. If 100 seats are sold, the price is $50. Each unsold seat increases the cost per passenger by $1. Let x represent the number of unsold seats.

(a) Write an expression for the number of seats that are sold.
(b) Write an expression for the price per seat.
(c) Write an expression for the revenue.
(d) Find the number of unsold seats that will produce the maximum revenue.
(e) Find the maximum revenue.

18. Management Farmer Linton wants to find the best time to take her hogs to market. The current price is 88 cents per pound and her hogs weigh an average of 90 pounds. The hogs gain 5 pounds per week and the market price for hogs is falling each week by 2 cents per pound. How many weeks should Ms. Linton wait before taking her hogs to market in order to receive as much money as possible? At this time, how much money (per hog) will she get?

19. Natural Science Between the months of June and October, the percent of maximum possible chlorophyll production in a leaf is approximated by $C(x)$, where

$$C(x) = 10x + 50.$$

Here x is time in days, with $x = 0$ representing June 1, $x = 1$ representing July 1, and so on. Between November and January, $C(x)$ is approximated by

$$C(x) = -20(x - 5)^2 + 100,$$

where x is as above. Find the percent of maximum possible chlorophyll production on the first day of each of the following months.
(a) June   **(b)** July   **(c)** September
(d) October   **(e)** November   **(f)** January

20. Natural Science Use your results in Exercise 19 to sketch a graph of $y = C(x)$ from June 1 through January 1. On what day is chlorophyll production a maximum?

21. A culvert is shaped like a parabola, 18 centimeters across the top and 12 centimeters deep. How wide is the culvert 8 centimeters from the top?

22. Management Dan Lange owns a factory that manufactures souvenir key chains. His weekly profit (in hundreds of dollars) is given by $P(x) = -2x^2 + 60x - 120$, where x is the number of cases of key chains sold.

(a) What is the largest number of cases he can sell and still make a profit?

(b) Explain how is it possible for him to lose money if he sells more cases than your answer in part (a).

(c) How many cases should he make and sell in order to maximize his profits?

◁▷**Management** *Recall that profit equals revenue minus cost. In Exercises 23 and 24 find the following.*

(**a**) *The break-even point (to the nearest tenth)*

(**b**) *The x-value that makes revenue a maximum*

(**c**) *The x-value that makes profit a maximum*

(**d**) *The maximum profit*

(**e**) *For what x-values will a loss occur?*

(**f**) *For what x-values will a profit occur?*

23. $R(x) = 60x - 2x^2$; $C(x) = 20x + 80$

24. $R(x) = 75x - 3x^2$; $C(x) = 21x + 65$

---

1 Graph $f(x) = x^3 - 5$.

Answer:

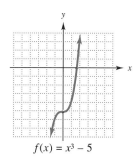

$f(x) = x^3 - 5$

3.3 POLYNOMIAL FUNCTIONS

A **polynomial function of degree n*** is a function whose rule is given by a polynomial of degree n. For example, linear functions, such as

$$f(x) = 3x - 2 \quad \text{and} \quad g(x) = 7x + 5,$$

are polynomial functions of degree 1 and quadratic functions, such as

$$f(x) = 3x^2 + 4x - 6 \quad \text{and} \quad g(x) = -2x^2 + 9$$

are polynomial functions of degree 2. Similarly, the function given by

$$f(x) = 2x^4 + 5x^3 - 6x^2 + x - 3$$

is a polynomial function of degree 4.

Perhaps the simplest polynomial functions are those defined by an expression of the form $y = x^n$.

▶**EXAMPLE 1** Graph $f(x) = x^3$.

First, find several ordered pairs belonging to the graph. Be sure to choose some negative x-values, $x = 0$, and some positive x-values to get representative ordered pairs. Find as many ordered pairs as you need in order to see the shape of the graph. Then plot the ordered pairs and draw a smooth curve through them, getting the graph in Figure 3.10 on the next page. ◀ **1**

** The degree of a polynomial was defined on page 36.

2 Graph the following.

(a) $f(x) = x^5 - 2$

(b) $f(x) = -x^4 + 3$

Answers:

(a)

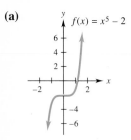

(b)

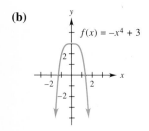

x	y
2	8
1	1
0	0
−1	−1
−2	−8

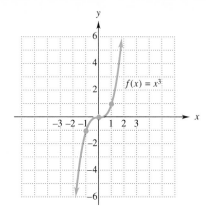

FIGURE 3.10

▶ **EXAMPLE 2**   Graph $f(x) = (3/2)x^4$.

The following table gives some typical ordered pairs.

x	−2	−1	0	1	2
y	24	3/2	0	3/2	24

The graph is shown in Figure 3.11.   ◀ **2**

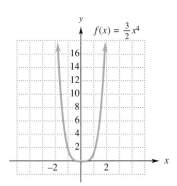

FIGURE 3.11

The domain of every polynomial function is the set of all real numbers. The range of a polynomial function of odd degree (1, 3, 5, 7, and so on) is also the set of all real numbers. Graphs typical of polynomial functions of odd degree are shown in Figure 3.12.

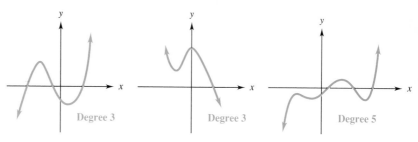

FIGURE 3.12

A polynomial function of even degree has a range that takes the form $y \leq k$ or $y \geq k$, for some real number k. Figure 3.13 shows two graphs of typical polynomial functions of even degree.

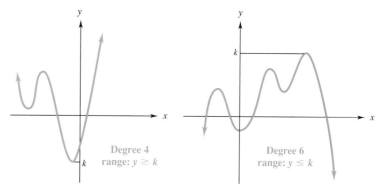

FIGURE 3.13

Since the domain of every polynomial function is the set of all real numbers, every polynomial graph extends forever to the left and the right. As the preceding figures illustrate, the graph of a typical polynomial function moves sharply away from the x-axis at the far left and right and may have several "peaks" and "valleys." For instance, the right-hand graph in Figure 3.13 has 3 peaks and 2 valleys and moves steadily downward at the far left and right. Calculus can be used to establish the following useful facts.

The total number of peaks and valleys on the graph of a polynomial function of degree n is at most $n - 1$.

The graph of every polynomial function is a smooth, unbroken curve that moves sharply away from the x-axis at the far left and the far right (that is, when $|x|$ is very large).

Identify the *x*-intercepts of each graph.

(a)

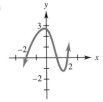

(b)

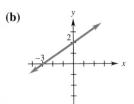

Answers:

(a) −2, 1, 2

(b) −3

When graphing polynomial functions "by hand," it is difficult to locate the peaks and valleys exactly without the use of calculus and many points may have to be plotted to obtain an accurate graph. When the polynomial can be factored, however, the general shape of its graph can often be determined quite easily by determining the *x*-intercepts and using the facts in the box above. Recall that the *x*-intercepts are found by letting $f(x) = 0$ and solving for *x*. **3**

▶ **EXAMPLE 3** Graph $f(x) = (2x + 3)(x - 1)(x + 2)$.

Multiplying out the expression on the right would show that f is a third degree polynomial, called a *cubic* polynomial. Sketch the graph of f by first finding any *x*-intercepts; do this by setting $f(x) = 0$.

$$f(x) = 0$$
$$(2x + 3)(x - 1)(x + 2) = 0$$

Solve this equation by placing each of the three factors equal to 0.

$$2x + 3 = 0 \quad \text{or} \quad x - 1 = 0 \quad \text{or} \quad x + 2 = 0$$
$$x = -\frac{3}{2} \qquad\qquad x = 1 \qquad\qquad x = -2$$

The three numbers, $-3/2$, 1, and -2, divide the *x*-axis into four regions:

$$x < -2, \quad -2 < x < -\frac{3}{2}, \quad -\frac{3}{2} < x < 1, \quad \text{and} \quad 1 < x.$$

These regions are shown in Figure 3.14.

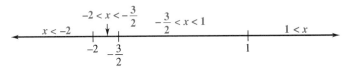

FIGURE 3.14

The sign of $f(x)$ can change from positive to negative or negative to positive only at an *x*-intercept, since $f(x) = 0$ at that point. Thus, in each of the regions in Figure 3.14 the values of $f(x)$ are either always positive or always negative. We find the sign of $f(x)$ in each region by selecting an *x*-value in each region and substituting to determine if the values of the function are positive or negative in that region. A typical selection of test numbers and the results of the tests are shown below.

Region	Test Number	Value of $f(x)$	Sign of $f(x)$	Graph
$x < -2$	−3	−12	negative	below *x*-axis
$-2 < x < -3/2$	− 7/4	11/32	positive	above *x*-axis
$-3/2 < x < 1$	0	−6	negative	below *x*-axis
$1 < x$	2	28	positive	above *x*-axis

4 Graph
$f(x) = .5(x - 1)(x + 2)(x - 3)$.

Answer:

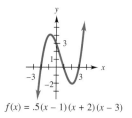

$f(x) = .5(x - 1)(x + 2)(x - 3)$

Since the graph touches the x-axis at the intercepts $x = -2$ and $x = -3/2$ and is above the x-axis between these intercepts, there must be at least one peak there. Similarly, there must be at least one valley between $x = -3/2$ and $x = 1$ because the graph is below the x-axis there. However, a polynomial function of degree 3 can have a total of at most $3 - 1 = 2$ peaks and valleys (as noted in the box before the example). So there must be exactly one peak and exactly one valley on this graph. Using these facts and plotting the x-intercepts and some additional points, leads to the graph in Figure 3.15. ◀ **4**

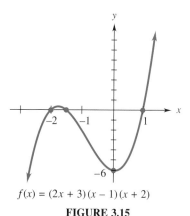

$f(x) = (2x + 3)(x - 1)(x + 2)$

FIGURE 3.15

▶**EXAMPLE 4**   Sketch the graph of $f(x) = 3x^4 + x^3 - 2x^2$.
The polynomial can be factored as follows:

$$3x^4 + x^3 - 2x^2 = x^2(3x^2 + x - 2) = x^2(3x - 2)(x + 1).$$

By letting $f(x) = 0$ we find that the x-intercepts are at $x = 0$, $x = 2/3$, and $x = -1$. These values divide the x-axis into four intervals, as indicated in the chart below. By choosing a test number in each interval, we obtain this analysis.

Region	Test Number	Value of $f(x)$	Sign of $f(x)$	Graph
$x < -1$	-2	32	positive	above x-axis
$-1 < x < 0$	$-\dfrac{1}{2}$	$-\dfrac{7}{16}$	negative	below x-axis
$0 < x < \dfrac{2}{3}$	$\dfrac{1}{2}$	$-\dfrac{3}{16}$	negative	below x-axis
$\dfrac{2}{3} < x$	1	2	positive	above x-axis

5 Graph
$$f(x) = x^3 - 4x$$

Answer:

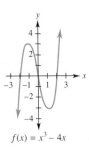

$$f(x) = x^3 - 4x$$

By plotting the x-intercepts and the points determined by the test numbers and using the fact that there can be a total of at most 3 peaks and valleys (why?), we obtain the graph in Figure 3.16.   ◀   **5**

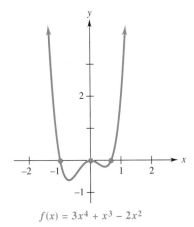

$$f(x) = 3x^4 + x^3 - 2x^2$$

FIGURE 3.16

3.3 EXERCISES

Graph each of the following polynomial functions. (See Examples 1–2.)

1. $f(x) = x^4$

2. $f(x) = x^6$

3. $f(x) = x^5$

4. $f(x) = x^7$

5. Based on your graphs in Exercises 1 and 2, what would you conjecture that the graph of $f(x) = x^n$ looks like when n is an even integer?

6. Based on your graphs in Exercises 3 and 4, what would you conjecture that the graph of $f(x) = x^n$ looks like when n is an odd integer?

In Exercises 7–10, state whether the graph could possibly be the graph of (a) some polynomial function; (b) a polynomial function of degree 3; (c) a polynomial function of degree 4. (See the box preceding Example 3.)

7.

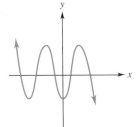

8.

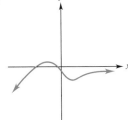

9.

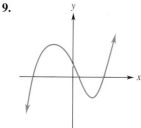

10.

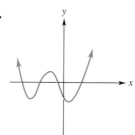

Graph each of the following polynomial functions. (See Examples 3–4.)

11. $f(x) = 2x(x - 3)(x + 2)$

12. $f(x) = -x(2x + 1)(x - 3)$

13. $f(x) = (x + 2)(x - 3)(x + 4)$

14. $f(x) = (x - 3)(x - 1)(x + 1)$

15. $f(x) = x^2(x - 2)(x + 3)$

16. $f(x) = x^2(x + 1)(x - 1)$

17. $f(x) = x^3 + x^2 - 6x$

18. $f(x) = x^3 - 2x^2 - 8x$

19. $f(x) = x^3 + 3x^2 - 4x$

20. $f(x) = x^4 - 5x^2$

21. $f(x) = x^4 - 2x^2$

22. $f(x) = x^4 - 7x^2 + 12$

In Exercises 23–26, use a calculator to evaluate the functions. Graph each function either by using a grapher or by plotting points.

23. Natural Science   The polynomial function defined by

$$A(x) = -.015x^3 + 1.058x$$

gives the approximate alcohol concentration (in tenths of a percent) in an average person's bloodstream x hours after drinking about eight ounces of 100 proof whiskey. The function is approximately valid for $0 \leq x < 8$. Find the following values.

(a) $A(1)$ (b) $A(2)$ (c) $A(4)$

(d) $A(6)$ (e) $A(8)$ (f) Graph $A(x)$.

(g) Using the graph you drew for part (f), estimate the time of maximum alcohol concentration.

(h) In one state, a person is legally drunk if the blood alcohol concentration exceeds .15%. Use the graph of part (f) to estimate the period in which this average person is legally drunk.

24. Natural Science   A technique for measuring cardiac output depends on the concentration of a dye after a known amount is injected into a vein near the heart. In a normal heart, the concentration of the dye at time x (in seconds) is given by the function defined by

$$g(x) = -.006x^4 + .140x^3 - .053x^2 + 1.79x.$$

(a) Find the following: $g(0)$; $g(1)$; $g(2)$; $g(3)$.

(b) Graph $g(x)$ for $x \geq 0$. (Hint: the only intercept is $(0, 0)$.)

25. Natural Science   The pressure of the oil in a reservoir tends to drop with time. By taking sample pressure readings for a particular oil reservoir, petroleum engineers have found that the change in pressure is given by

$$P(t) = t^3 - 18t^2 + 81t,$$

where t is time in years from the date of the first reading.

(a) Find the following: $P(0)$; $P(3)$; $P(7)$; $P(10)$.

(b) Graph $P(t)$.

(c) For what time period is the change in pressure (drop) increasing? decreasing?

26. Natural Science   During the early part of the 20th century, the deer population of the Kaibab Plateau in Arizona experienced a rapid increase because hunters had reduced the number of natural predators. The increase in population depleted the food resources and eventually caused the population to decline. For the period from 1905 to 1930, the deer population was approximated by

$$D(x) = -.125x^5 + 3.125x^4 + 4000,$$

where x is time in years from 1905.

(a) Find the following: $D(0)$; $D(5)$; $D(10)$; $D(15)$; $D(20)$; $D(25)$.

(b) Graph $D(x)$.

(c) From the graph, over what period of time (from 1905 to 1930) was the population increasing? relatively stable? decreasing?

For each of the following polynomial functions, find a viewing window that shows a complete graph (that is, a graph that includes all the peaks and valleys and indicates how the curve moves away from the x-axis at the far left and right).

27. $f(x) = x^3 + 8x^2 + 5x - 14$

28. $g(x) = -x^4 - 3x^3 + 24x^2 + 80x + 15$

29. $f(x) = x^4 + 10x^3 + 21x^2 - 40x - 80$

30. $g(x) = 84x^5 - 4x^3 + 500$

The graph of each of the following functions has exactly one peak and one valley. Graph the function and use the trace feature to find the approximate location of each peak and valley.

31. $f(x) = x^3 - x^2 - 8x + 4$

32. $g(x) = x^3 + 3x^2 - 4x - 5$

3.4 RATIONAL FUNCTIONS

A *rational function* is defined by an expression that is a quotient of two polynomials.

A function f is a **rational function** if

$$f(x) = \frac{P(x)}{Q(x)}, \qquad Q(x) \neq 0,$$

where $P(x)$ and $Q(x)$ are polynomials.

Because any values of x that make $Q(x) = 0$ are excluded from the domain of $f(x)$, a rational function usually has a graph with one or more breaks in it.

▶ **EXAMPLE 1**   Graph the rational function defined by $y = \dfrac{2}{1 + x}$.

This function is undefined for $x = -1$, since -1 leads to a 0 denominator. For this reason, the graph of this function will not intersect the vertical line $x = -1$. Since x can take on any value except -1, the values of x can approach -1 as closely as desired from either side of -1, as shown in the following table of values.

1 Graph the following.

(a) $f(x) = \dfrac{3}{5-x}$

(b) $f(x) = \dfrac{-4}{x+4}$

Answers:

(a)

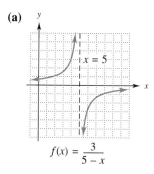

$$f(x) = \frac{3}{5-x}$$

(b)

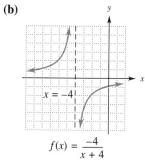

$$f(x) = \frac{-4}{x+4}$$

x approaches −1
↓

x	−1.5	−1.2	−1.1	−1.01	−.99	−.9	−.8	−.5
$1+x$	−.5	−.2	−.1	−.01	.01	.1	.2	.5
$\dfrac{2}{1+x}$	−4	−10	−20	−200	200	20	10	4

↑
$|y|$ **gets larger and larger**

The table above suggests that as x gets closer and closer to −1 from either side, the denominator $1+x$ gets closer and closer to 0, and $|2/(1+x)|$ gets larger and larger. The part of the graph in Figure 3.17 near $x = -1$ shows this behavior. The vertical line $x = -1$ that is approached by the curve is called a *vertical asymptote.*

As $|x|$ gets larger and larger so does the denominator $1+x$. Hence $y = 2/(1+x)$ gets closer and closer to 0, as shown in the table below.

x	−101	−11	−2	0	9	99
$1+x$	−100	−10	−1	1	10	100
$\dfrac{2}{1+x}$	−.02	−.2	−2	2	.2	.02

The horizontal line $y = 0$ is called a *horizontal asymptote* for this graph. Using the asymptotes and plotting the intercept and several points (shown with the figure) gives the graph of Figure 3.17. ◀ **1**

x	y
3	1/2
2	2/3
1	1
0	2
−1/2	4
−3/2	−4
−2	−2
−3	−1
−4	−2/3

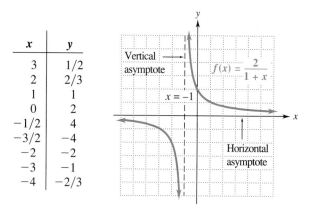

FIGURE 3.17

Example 1 suggests the following conclusion.

If a number k makes the denominator 0 but the numerator nonzero in the expression defining a rational function, then the line $x = k$ is a **vertical asymptote** for the graph of the function.

Also, wherever the values of y approach but do not equal some number k as $|x|$ gets larger and larger, the line $y = k$ is a **horizontal asymptote** for the graph.

▶**EXAMPLE 2**   Graph $f(x) = \dfrac{3x + 2}{2x + 4}$.

Find the vertical asymptote by setting the denominator equal to 0, then solving for x.

$$2x + 4 = 0$$
$$x = -2$$

Since the numerator is nonzero when $x = -2$, the line $x = -2$ is a vertical asymptote. Next, find the intercepts.

Let $x = 0$.	Let $y = 0$.
$y = \dfrac{3 \cdot 0 + 2}{2 \cdot 0 + 4} = \dfrac{1}{2}$	$0 = \dfrac{3x + 2}{2x + 4}$
	$3x + 2 = 0$
	$x = -\dfrac{2}{3}$

The y-intercept is $1/2$ and the x-intercept is $-2/3$. Now find additional ordered pairs on either side of the vertical asymptote, as shown in the table of values in Figure 3.18.

In order to see what the graph looks like when $|x|$ is very large, we rewrite the rule of the function. When $x \neq 0$, dividing both numerator and denominator by x does not change the value of the function:

$$f(x) = \frac{3x + 2}{2x + 4} = \frac{\dfrac{3x + 2}{x}}{\dfrac{2x + 4}{x}} = \frac{\dfrac{3x}{x} + \dfrac{2}{x}}{\dfrac{2x}{x} + \dfrac{4}{x}} = \frac{3 + \dfrac{2}{x}}{2 + \dfrac{4}{x}}.$$

Now when $|x|$ is very large, the fractions $2/x$ and $4/x$ are very close to 0 (for instance, when $x = 200$, $4/x = 4/200 = .02$). Therefore the numerator of $f(x)$ is very close to $3 + 0 = 3$ and the denominator is very close to $2 + 0 = 2$. Hence $f(x)$ is very close to $3/2$ when $|x|$ is large, so the line $y = 3/2$ is the horizontal asymptote of the graph, as shown in Figure 3.18. ◀

2 Graph the following.

(a) $f(x) = \dfrac{3x + 5}{x - 3}$

(b) $f(x) = \dfrac{2 - x}{x + 3}$

Answers:

(a)

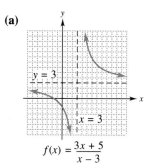

$f(x) = \dfrac{3x + 5}{x - 3}$

(b)

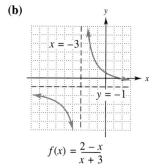

$f(x) = \dfrac{2 - x}{x + 3}$

x	y
5	1.2
3	1.1
2	1
0	1/2
−1	−1/2
−2/3	0
−5/2	11/2
−4	5/2
−6	2
−8	1.8

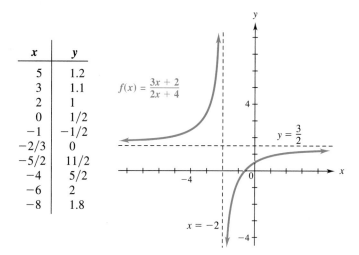

FIGURE 3.18

The argument used to find the horizontal asymptote of the function in Example 2 (which was determined by the coefficients of x in the numerator and denominator) works in the general case and leads to this conclusion.

The horizontal asymptote of $y = \dfrac{ax + b}{cx + d}$ (with $c \neq 0$) is the line $y = \dfrac{a}{c}$.

This fact makes it easy to graph this type of rational function.   **2**

▶ **EXAMPLE 3**   In many situations involving environmental pollution, much of the pollutant can be removed from the air or water at a fairly reasonable cost, but the last, small part of the pollutant can be very expensive to remove.

Cost as a function of the percentage of pollutant removed from the environment can be calculated for various percentages of removal, with a curve fitted through the resulting data points. This curve then leads to a function that approximates the situation. Rational functions often are a good choice for these **cost-benefit functions.**

For example, suppose a cost-benefit function is given by

$$f(x) = \frac{18x}{106 - x},$$

where $f(x)$ or y is the cost (in thousands of dollars) of removing x percent of a certain pollutant. The domain of x is the set of all numbers from 0 to 100, inclusive; any amount of pollutant from 0% to 100% can be removed. To remove 100% of the pollutant here would cost

$$y = \frac{18(100)}{106 - 100} = 300,$$

3 Using the function of Example 3, find the cost to remove the following percents of pollutants.

(a) 70%

(b) 85%

(c) 98%

Answers:

(a) $35,000

(b) About $73,000

(c) About $221,000

4 Rework Example 4 with the product-exchange function

$$y = \frac{60000 - 10x}{60 + x}$$

to find the maximum amount of each wine that can be produced.

Answer:
6000 tons of red, 1000 tons of white

or $300,000. Check that 95% of the pollutant can be removed for $155,000, 90% for $101,000, and 80% for $55,000. Using these points, as well as others that could be obtained from the function above, leads to the graph shown in Figure 3.19. ◀ **3**

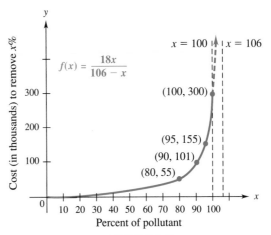

FIGURE 3.19

In management, **product-exchange functions** give the relationship between quantities of two items that can be produced by the same machine or factory. For example, an oil refinery can produce gasoline, heating oil, or a combination of the two; a winery can produce red wine, white wine, or a combination of the two. The next example discusses a product-exchange function.

▶ **EXAMPLE 4** The product-exchange function for the Golden Grape Winery for red wine x and white wine y, in tons, is

$$y = \frac{100{,}000 - 50x}{1000 + x}.$$

Graph the function in Quadrant I and find the maximum quantity of each kind of wine that can be produced.

We first find the intercepts. For $x = 0$, the y-intercept is 100. Let $y = 0$ to get $100{,}000 - 50x = 0$, from which the x-intercept is 2000. Since we are interested only in the Quadrant I portion of the graph, we can find a few more points in Quadrant I and complete the graph as shown in Figure 3.20.

The maximum value of y occurs when $x = 0$, so the maximum amount of white wine that can be produced is 100 tons, given by the y-intercept. The x-intercept gives the maximum amount of red wine that can be produced, 2000 tons. ◀ **4**

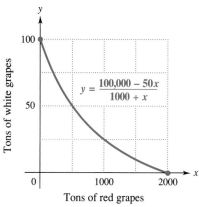

$$y = \frac{100{,}000 - 50x}{1000 + x}$$

Tons of white grapes

Tons of red grapes

FIGURE 3.20

3.4 EXERCISES

Graph each of the following. Give the equations of the vertical and horizonal asymptotes (see Examples 1–2.)

1. $f(x) = \dfrac{1}{x + 5}$

2. $g(x) = \dfrac{-7}{x - 6}$

3. $f(x) = \dfrac{-3}{2x + 5}$

4. $h(x) = \dfrac{-4}{2 - x}$

5. $f(x) = \dfrac{3x}{x - 1}$

6. $g(x) = \dfrac{x - 2}{x}$

7. $f(x) = \dfrac{x + 1}{x - 4}$

8. $f(x) = \dfrac{x - 3}{x + 5}$

9. $f(x) = \dfrac{2 - x}{x - 3}$

10. $g(x) = \dfrac{3x - 2}{x + 3}$

11. $f(x) = \dfrac{2x - 1}{4x + 2}$

12. $f(x) = \dfrac{3x - 6}{6x - 1}$

13. $f(x) = \dfrac{1 - 2x}{5x + 20}$

14. $f(x) = \dfrac{6 - 3x}{4x + 12}$

15. $f(x) = \dfrac{-x - 4}{3x + 6}$

16. $f(x) = \dfrac{-x + 8}{2x + 5}$

Find the equations of the vertical asymptotes of each of the following rational functions.

17. $f(x) = \dfrac{x - 3}{x^2 + x - 2}$

18. $g(x) = \dfrac{x + 2}{x^2 - 1}$

19. $g(x) = \dfrac{x^2 + 2x}{x^2 - 4x - 5}$

20. $f(x) = \dfrac{x^2 - 2x - 4}{x^3 - 2x^2 + x}$

---

21. Natural Science Suppose a cost-benefit model (see Example 3) is given by

$$f(x) = \frac{4.3x}{100 - x},$$

where $f(x)$ is the cost in thousands of dollars of removing x percent of a given pollutant. Find the cost of removing each of the following percents of pollutants.

(a) 50%   **(b)** 70%   **(c)** 80%   **(d)** 90%

(e) 95%   **(f)** 98%   **(g)** 99%

(h) Is it possible, according to this model to remove *all* the pollutant?

(i) Graph the function.

22. Natural Science Suppose a cost-benefit model is given by

$$f(x) = \frac{4.5x}{101 - x},$$

where $f(x)$ is the cost in thousands of dollars of removing x percent of a certain pollutant. Find the cost of removing the following percents of pollutants.

(a) 0%   **(b)** 50%   **(c)** 80%   **(d)** 90%
(e) 95%   **(f)** 99%   **(g)** 100%
(h) Graph the function.

23. Management SuperStar Cablevision Company recently began service to the city of Megapolis. Based on past experience, it estimates that the number $N(x)$ of subscribers (in thousands) at the end of x months is

$$N(x) = \frac{250x}{x + 6}.$$

Find the number of subscribers at the end of

(a) 6 months   **(b)** 18 months   **(c)** two years.
(d) Graph $N(x)$.
(e) What part of the graph is relevant to this situation?
(f) What is the horizontal asymptote of the graph? What does this suggest that the maximum possible number of subscribers will be?

24. Natural Science To calculate the dosage of the drug Naldecon for a child, pharmacists may use the formula

$$d(x) = \frac{70x}{.5x + 9} \quad (x > 0)$$

where x is the child's age in years and $d(x)$ is measured in milligrams.

(a) Graph $d(x)$.
(b) Assuming that the dosage of Naldecon should be 70 milligrams for all adults, is this formula valid for adults? (At age 18 a person is considered an adult.)
(c) What portion of this graph is relevant to a pharmacist?

25. Social Science The average waiting time in a line (or queue) before getting served is given by

$$W = \frac{S(S - A)}{A},$$

where A is the average rate that people arrive at the line and S is the average service time. At a certain fast food restaurant, the average service time is 3 minutes. Find W for each of the following average arrival times.

(a) 1 minute
(b) 2 minutes
(c) 2.5 minutes
(d) What is the vertical asymptote?
(e) Graph the equation on the interval $(0, 3]$.
(f) What happens to W when $A > 3$? What does this mean?

26. Management Antique car fans often enter their cars in a *concours d'elegance* in which a maximum of 100 points can be awarded to a particular car. Points are awarded for the general attractiveness of the car. Based on a recent article in *Business Week,* we constructed the following mathematical model for the cost, in thousands of dollars, of restoring a car so that it will win x points.

$$C(x) = \frac{10x}{49(101 - x)}$$

Find the cost of restoring a car so that it will win

(a) 99 points;   **(b)** 100 points.

27. Management At Ewing's Clothing Store, daily sales of shirts (in dollars) after x days of newspapers ads are given by

$$S = \frac{600x + 3800}{x + 1}.$$

(a) Graph the first quadrant portion of the sales function.
(b) Assuming that the ads keep running, will sales tend to level off? What is the minimum amount of sales to be expected? What feature of the graph gives this information?
(c) If newspaper ads cost $1000 per day, at what point should they be discontinued? Why?

28. Physical Science In electronics, the circuit gain is given by

$$G(R) = \frac{R}{r + R},$$

where R is the resistance of a temperature sensor in the circuit and r is a constant. Let $r = 1000$ ohms.

(a) Find any vertical asymptotes of the graph of the function.
(b) Find any horizontal asymptotes of the graph of the function.
(c) Graph $G(R)$.

Management   *Sketch the Quadrant I portion of the graph of each of the functions defined as follows, and then estimate the maximum quantities of each product that can be produced. (See Example 4.)*

29. The product-exchange function for gasoline, x, and heating oil, y, in hundreds of gallons per day, is

$$y = \frac{125,000 - 25x}{125 + 2x}.$$

30. A drug factory found the product-exchange function for a red tranquilizer, x, and a blue tranquilizer, y, is

$$y = \frac{900,000,000 - 30,000x}{x + 90,000}.$$

31. Physical Science   The failure of several O-rings in field joints was the cause of the fatal crash of the Challenger space shuttle in 1986. NASA data from 24 successful launches prior to Challenger suggested that O-ring failure was related to launch temperature by a function similar to

$$N(t) = \frac{600 - 7t}{4t - 100} \quad (50 \le t \le 85)$$

where t is the temperature (in °F) at launch and N is the approximate number of O-rings that fail. Assume that this function accurately models the number of O-ring failures that would occur at lower launch temperatures (an assumption NASA did not make).

(a) Does $N(t)$ have a vertical asymptote? At what value of t does it occur?

(b) Without graphing, what would you conjecture that the graph would look like just to the right of the vertical asymptote? What does this suggest about the number of O-ring failures that might be expected near that temperature? (The temperature at the Challenger launching was 31°.)

(c) Confirm your conjecture by graphing $N(t)$ between the vertical asymptote and $t = 85$.

32. (a) Graph $f(x) = \dfrac{x^3 + 3x^2 + x + 1}{x^2 + 2x - 1}.$

(b) Does the graph appear to have a horizontal asymptote? Does the graph appear to have some nonhorizontal straight line as an asymptote?

(c) Graph $f(x)$ and the line $y = x + 1$ on the same screen. Does this line appear to be an asymptote of the graph of $f(x)$?

33. (a) Find the vertical asymptotes of

$$f(x) = \frac{2x^2}{x^2 + x - 2}.$$

(b) Find a viewing window which clearly shows the behavior of the graph of f near the vertical asymptotes.

KEY TERMS AND SYMBOLS

	3.1 quadratic function	**3.4**	rational function
	parabola		vertical asymptote
	vertex		horizontal asymptote
	3.3 polynomial function		

KEY CONCEPTS

The **quadratic function** defined by $y = a(x - h)^2 + k$ has a graph that is a **parabola** with vertex (h, k) and axis of symmetry $x = h$. The parabola opens upward if $a > 0$, downward if $a < 0$. If the equation is in the form $f(x) = ax^2 + bx + c$, the vertex is
$$\left(-\frac{b}{2a}, f\left(-\frac{b}{2a}\right)\right).$$

A **polynomial function** of degree higher than two can be roughly sketched by finding the x-intercepts, dividing the number line into regions determined by the x-intercepts, then finding the sign of the function ($+$ or $-$) in each region.

If a number k makes the denominator of a **rational function** 0, but the numerator nonzero, then the line $x = k$ is a **vertical asymptote** for the graph.

Whenever the values of y approach, but never equal, some number k as $|x|$ gets larger and larger, the line $y = k$ is a **horizontal asymptote** for the graph.

CHAPTER 3 REVIEW EXERCISES

Without graphing, determine whether each of the following parabolas opens upward or downward and find its vertex.

1. $f(x) = 3(x - 2)^2 + 6$ **2.** $f(x) = 2(x + 3)^2 - 5$

3. $g(x) = -4(x + 1)^2 + 8$ **4.** $g(x) = -5(x - 4)^2 - 6$

Graph each of the following and label its vertex.

5. $f(x) = x^2 - 4$ **6.** $f(x) = 6 - x^2$ **7.** $f(x) = x^2 + 2x - 3$ **8.** $f(x) = -x^2 + 6x - 3$

9. $f(x) = -x^2 - 4x + 1$ **10.** $f(x) = 4x^2 - 8x + 3$ **11.** $f(x) = 2x^2 + 4x - 3$ **12.** $f(x) = -3x^2 - 12x - 8$

Determine whether each of the following functions has a minimum or a maximum value and find that value.

13. $f(x) = x^2 + 6x - 2$ **14.** $f(x) = x^2 + 4x + 5$

15. $g(x) = -4x^2 + 8x + 3$ **16.** $g(x) = -3x^2 - 6x + 3$

Solve each of the following.

17. Management The commodity market is very unstable; money can be made or lost quickly when investing in soybeans, wheat, pork bellies, and the like. Suppose that an investor kept track of her total profit, P, at time t, measured in months, after she began investing and found that $P = 4t^2 - 29t + 30$. Find the time intervals where she has been ahead. (Hint: $t > 0$ in this case.)

18. Physical Science The height h (in feet) of a rocket at time t seconds is given by $h = -16t^2 + 800t$.
(a) How long does it take the rocket to reach 3200 feet?
(b) What is the maximum height of the rocket?

19. Management The Hopkins Company's profit in thousands of dollars is given by $P = -4x^2 + 88x - 259$, where x is the number of hundreds of cases

produced. How many cases should be produced if the company is to make a profit?

20. Management The manager of a large apartment complex has found that the profit is given by $P = -x^2 + 250x - 15{,}000$, where x is the number of units rented. Decide for what values of x the complex produces a profit.

21. Find the dimensions of the rectangular region of maximum area that can be enclosed with 200 meters of fencing.

22. Find the dimensions of the rectangular region of maximum area that can be enclosed with 200 meters of fencing if no fencing is needed along one side of the region.

Graph each of the following polynomial functions.

23. $f(x) = x^3 + 1$

24. $f(x) = x^4 - 2$

25. $g(x) = x^3 - x$

26. $g(x) = x^4 - x^2$

27. $f(x) = x(x - 2)(x + 3)$

28. $f(x) = (x - 1)(x + 2)(x - 3)$

29. $f(x) = x(2x - 1)(x + 2)$

30. $f(x) = 3x(3x + 2)(x - 1)$

31. $f(x) = 2x^3 - 3x^2 - 2x$

32. $f(x) = x^3 - 3x^2 - 4x$

33. $f(x) = x^4 - 5x^2 - 6$

34. $f(x) = x^4 - 7x^2 - 8$

Graph each of the following rational functions.

35. $f(x) = \dfrac{1}{x - 3}$

36. $f(x) = \dfrac{-2}{x + 4}$

37. $f(x) = \dfrac{-3}{2x - 4}$

38. $f(x) = \dfrac{5}{3x + 7}$

39. $f(x) = \dfrac{5x + 5}{3x - 5}$

40. $f(x) = \dfrac{3 - x}{2x - 9}$

41. Management The average cost per unit of producing x cartons of cocoa is given by

$$C(x) = \frac{650}{2x + 40}.$$

Find the average cost per carton to make the following number of cartons.
(a) 10 cartons   (b) 50 cartons   (c) 70 cartons
(d) 100 cartons   (e) Graph $C(x)$

42. Management The cost and revenue functions (in dol-

lars) for a frozen yogurt shop are given by

$$C(x) = \frac{400x + 400}{x + 4} \quad \text{and} \quad R(x) = 100x,$$

where x is measured in hundreds of units.
(a) Graph $C(x)$ and $R(x)$ on the same axes.
(b) What is the break-even point for this shop?
(c) If the profit function is given by $P(x)$, does $P(1)$ represent a profit or a loss?
(d) Does $P(4)$ represent a profit or a loss?

43. Management The supply and demand functions for the yogurt shop in Exercise 42 are as follows:

$$\text{supply: } p = \frac{q^2}{4} + 25; \quad \text{demand: } p = \frac{500}{q},$$

where p is the price in dollars for q units of yogurt.
(a) Graph both functions on the same axes, and from the graph, estimate the equilibrium point.
(b) Give the q-intervals where supply exceeds demand.
(c) Give the q-intervals where demand exceeds supply.

44. Management A cost-benefit curve for pollution control is given by

$$y = \frac{9.2x}{106 - x},$$

where y is the cost in thousands of dollars of removing x percent of a specific industrial pollutant. Find y for each of the following values of x.
(a) $x = 50$ **(b)** $x = 98$
(c) What percent of the pollutant can be removed for $22,000?

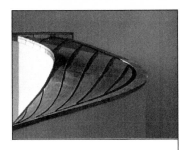

CHAPTER 4

Exponential and Logarithmic Functions

Exponential and logarithmic functions play a key role in management, economics, the social and physical sciences, and engineering. They are used to study the growth of money and organizations; learning curves; the growth of human, animal, and bacterial populations; the spread of disease; and radioactive decay. In order to work with these functions effectively, you should have a calculator equipped with keys labeled $\boxed{x^y}$ (or $\boxed{y^x}$ or $\boxed{\wedge}$), $\boxed{\log}$, $\boxed{e^x}$, and $\boxed{\ln}$ (or $\boxed{\ln x}$).

4.1 EXPONENTIAL FUNCTIONS

In polynomial functions, the variable is raised to various constant exponents; for instance, $f(x) = x^2 + 3x - 5$. In exponential functions, such as

$$f(x) = 10^x \quad \text{or} \quad g(x) = 2^{-x^2} \quad \text{or} \quad h(x) = 3^{.6x},$$

the variable is in the exponent and the base is a positive constant. We begin with the simplest type of exponential function.

If a is a positive constant other than 1, then the function defined by

$$f(x) = a^x$$

is called an **exponential function with base a.**

The exponential function with base 1 is a constant function $f(x) = 1^x = 1$. Exponential functions with negative bases are not of interest because when a is negative, a^x may not be defined for some values of x; for instance, $(-4)^{1/2} = \sqrt{-4}$ is not a real number.

1 Graph $f(x) = \left(\dfrac{1}{3}\right)^x$.

Answer:

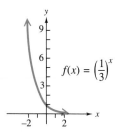

$f(x) = \left(\dfrac{1}{3}\right)^x$

▶ **EXAMPLE 1** Graph the following exponential functions.

(a) $f(x) = 2^x$

Begin by making a table of values of x and y, as shown in Figure 4.1(a). Then plot these points and draw a smooth curve through them, getting the graph shown in Figure 4.1(a). The graph approaches the negative x-axis but will never touch it, because *every* power of 2 is positive. This makes the x-axis a horizontal asymptote. The graph suggests that as x gets larger and larger, values of 2^x also get larger and larger. However, there is no vertical asymptote.

(b) $g(x) = 2^{-x}$

By the properties of exponents,

$$2^{-x} = \frac{1}{2^x} = \left(\frac{1}{2}\right)^x.$$

Construct a table of values and draw a smooth curve through the resulting points. (See Figure 4.1(b).) The graph approaches the positive x-axis and rises sharply as x takes negative values. Thus, the graph of $g(x) = (1/2)^x$ is the mirror image of the graph of $f(x) = 2^x$, with the y-axis being the mirror. ◀ **1**

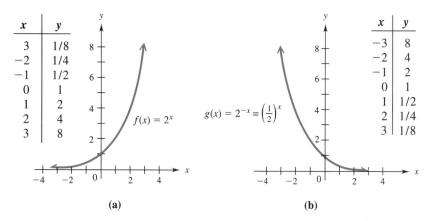

x	y
3	1/8
-2	1/4
-1	1/2
0	1
1	2
2	4
3	8

$f(x) = 2^x$

(a)

x	y
-3	8
-2	4
-1	2
0	1
1	1/2
2	1/4
3	1/8

$g(x) = 2^{-x} = \left(\dfrac{1}{2}\right)^x$

(b)

FIGURE 4.1

The graphs in Figure 4.1 illustrate a basic fact about exponential functions. The domain of $f(x) = a^x$ is the set of all real numbers and the range is the set of all positive real numbers.

The graph of $f(x) = 2^x$ in Figure 4.1(a) illustrates **exponential growth,** which is far more explosive than polynomial growth. For example, if the graph in Figure 4.1(a) were extended on the same scale to include the point $(50, 2^{50})$, the graph would be almost 3 billion *miles* high! Similarly, the graph of $g(x) = (1/2)^x$ illustrates **exponential decay.**

2 Graph $f(x) = 2^{x+1}$.

Answer:

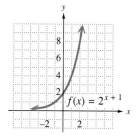

When $a > 1$, the graph of the exponential function $g(x) = a^x$ has the same basic shape as the graph of $f(x) = 2^x$ in Figure 4.1(a): it has the negative x-axis as a horizontal asymptote towards the left, crosses the y-axis at 1, and rises steeply to the right. The larger the base a is, the more steeply the graph rises, as illustrated in Figure 4.2(a).

When $0 < a < 1$, the graph of $h(x) = a^x$ has the same basic shape as the graph of $g(x) = (1/2)^x$ in Figure 4.1(b). Graphs of some typical exponential functions with $0 < a < 1$ are shown in Figure 4.2(b).

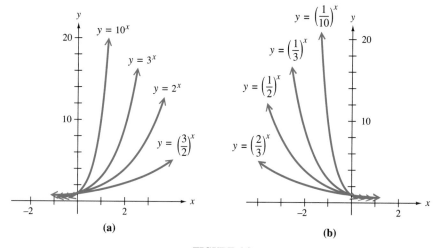

FIGURE 4.2

The graphs of exponential functions such as $f(x) = 3^{1-x}$ or $g(x) = 2^{.6x}$ or $h(x) = 3 \cdot 10^{2x+1}$ have the same general shape as the exponential graphs above. The only differences are that the graph may rise or fall at a different rate or the entire graph may be shifted vertically or horizontally.

▶ **EXAMPLE 2** Graph each of the following.

(a) $f(x) = 3^{1-x}$.

Choose values of x that make the exponent positive, zero, and negative. Plot the corresponding points and connect them with a smooth curve as in Figure 4.3(a) on the next page. Note that this graph crosses the y-axis at 3.

(b) $g(x) = 2^{.6x}$.

Use a calculator and follow the same procedure to obtain the graph of g in Figure 4.3(b). Because of the .6 in the exponent, the graph rises at a slower rate than the graph of $y = 2^x$. ◀ **2**

3 Graph $f(x) = \left(\frac{1}{2}\right)^{-x^2}$.

Answer:

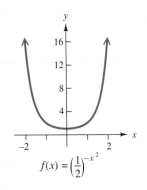

$f(x) = \left(\frac{1}{2}\right)^{-x^2}$

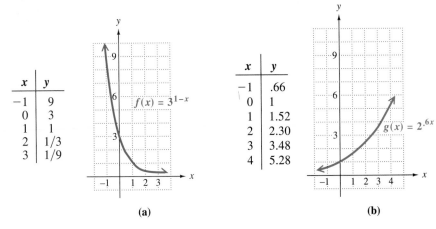

x	y
−1	9
0	3
1	1
2	1/3
3	1/9

$f(x) = 3^{1-x}$

(a)

x	y
−1	.66
0	1
1	1.52
2	2.30
3	3.48
4	5.28

$g(x) = 2^{.6x}$

(b)

FIGURE 4.3

When the exponent involves a nonlinear expression in x, an exponential graph may have a much different shape than the preceding graphs.

▶ **EXAMPLE 3**   Graph $f(x) = 2^{-x^2}$.

Using a calculator to find several ordered pairs and then plotting them gives the graph in Figure 4.4. Graphs such as this are important in probability, where the normal curve has an equation similar to the one in this example.   ◀ **3**

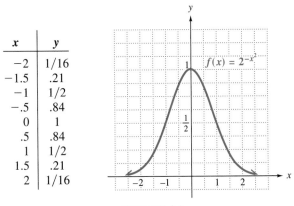

x	y
−2	1/16
−1.5	.21
−1	1/2
−.5	.84
0	1
.5	.84
1	1/2
1.5	.21
2	1/16

$f(x) = 2^{-x^2}$

FIGURE 4.4

The graphs of typical exponential functions $y = a^x$ in Figure 4.2 suggest that a given value of x leads to exactly one value of a^x and each value of a^x corresponds to exactly one value of x. Because of this, an equation with a variable in the exponent, called an **exponential equation,** can often be solved using the following property.

4 Solve each equation.

(a) $6^x = 6^4$

(b) $3^{2x} = 3^9$

(c) $5^{-4x} = 5^3$

Answers:

(a) 4

(b) 9/2

(c) −3/4

5 Solve each equation.

(a) $8^{2x} = 4$

(b) $5^{3x} = 25^4$

(c) $36^{-2x} = 6$

Answers:

(a) 1/3

(b) 8/3

(c) −1/4

6 The number of organisms present at time t is given by

$$f(t) = 75(2)^{.5t}.$$

Find the number of organisms present at

(a) $t = 0$;

(b) $t = 2$;

(c) $t = 4$.

Answers:

(a) 75

(b) 150

(c) 300

If $a > 0$, $a \neq 1$, and $a^x = a^y$, then $x = y$.

Caution   Both bases must be the same. The value $a = 1$ is excluded because $1^2 = 1^3$, for example, even though $2 \neq 3$.

As an example, solve $2^{3x} = 2^7$ using this property, as follows:

$$2^{3x} = 2^7$$
$$3x = 7$$
$$x = \frac{7}{3}. \quad \boxed{4}$$

▶**EXAMPLE 4**   Solve $9^x = 27$.

First rewrite both sides of the equation so the bases are the same. Since $9 = 3^2$ and $27 = 3^3$,

$$(3^2)^x = 3^3$$
$$3^{2x} = 3^3$$
$$2x = 3$$
$$x = \frac{3}{2}. \quad \blacktriangleleft \quad \boxed{5}$$

Exponential functions have many practical applications. For example, in situations which involve growth or decay of a population, the size of the population at a given time t often is determined by an exponential function of t.

▶**EXAMPLE 5**   The oxygen consumption of yearling salmon (in appropriate units) increases exponentially with speed of swimming according to

$$f(x) = 100(3)^{.6x},$$

where x is the speed in feet per second. Find each of the following.

(a) $f(0)$

Substitute 0 for x.

$$f(0) = 100(3)^{.6(0)} = 100(3)^0 = 100 \cdot 1 = 100$$

When the fish are still (their speed is 0) their oxygen consumption is 100 units.

(b) $f(5)$

Replace x with 5.

$$f(5) = 100(3)^{.6(5)} = 100(3)^3 = 2700$$

A speed of 5 feet per second increases the oxygen consumption to 2700 units. $\blacktriangleleft$ $\boxed{6}$

4.1 EXERCISES

Use a calculator whenever necessary to do these exercises.

Classify each of the following functions as linear, quadratic, or exponential.

1. $f(x) = 2x^2 + x - 3$

2. $g(x) = 5^{x-3}$

3. $h(x) = 3 \cdot 6^{2x^2+4}$

4. $f(x) = 3x + 1$

Without graphing, describe the shape of the graph of each of the following functions. (See Examples 1–2.)

5. $f(x) = .8^x$

6. $g(x) = 6^{-x}$

7. $f(x) = 5^{.4x}$

8. $g(x) = -(2^x)$

Graph each of the following functions. (See Examples 1–2.)

9. $f(x) = 3^x$

10. $g(x) = 3^{.5x}$

11. $f(x) = 2^{x/2}$

12. $g(x) = 4^{-x}$

13. $f(x) = (1/5)^x$

14. $g(x) = 2^{3x}$

---

15. Graph each of the following functions on the same axes.
  (a) $f(x) = 2^x$ **(b)** $g(x) = 2^{x+3}$
  (c) $h(x) = 2^{x-4}$
  (d) If c is a positive constant, explain how the graphs of $y = 2^{x+c}$ and $y = 2^{x-c}$ are related to the graph of $f(x) = 2^x$.

16. Graph each of the following functions on the same axes.
  (a) $f(x) = 3^x$ **(b)** $g(x) = 3^x + 2$
  (c) $h(x) = 3^x - 4$
  (d) If c is a positive constant, explain how the graphs of $y = 3^x + c$ and $y = 3^x - c$ are related to the graph of $f(x) = 3^x$.

Graph each of the following functions. (See Example 3.)

17. $f(x) = 2^{-x^2+2}$

18. $g(x) = 2^{x^2-2}$

19. $f(x) = x \cdot 2^x$

20. $f(x) = x^2 \cdot 2^x$

Solve each of the following equations. (See Example 4.)

21. $5^x = 25$

22. $3^x = \dfrac{1}{9}$

23. $4^x = 64$

24. $a^x = a^2 \ (a > 0)$

25. $16^x = 64$

26. $\left(\dfrac{3}{4}\right)^x = \dfrac{16}{9}$

27. $5^{-2x} = \dfrac{1}{25}$

28. $3^{x-1} = 9$

29. $16^{-x+1} = 8$

30. $25^{-2x} = 3125$

31. $81^{-2x} = 3^{x-1}$

32. $7^{-x} = 49^{x+3}$

33. $2^{|x|} = 16$

34. $5^{-|x|} = \dfrac{1}{25}$

35. $2^{x^2-4x} = \dfrac{1}{16}$

36. $5^{x^2+x} = 1$

37. $8^{x^2} = 2^{5x}$

38. $9^x = 3^{x^2}$

39. $3^{x^2-4x} = 243$

40. $5^{x^2} = 25 \cdot 5^{-x}$

41. Use a calculator and the graph of $f(x) = 2^x$ to explain why the solution of $2^x = 12$ must be a number between 3.5 and 3.6.

42. Explain why the equation $4^{x^2+1} = 2$ has no solutions.

Work the following exercises.

43. Management If $1 is deposited into an account paying 6% per year compounded annually, then after t years the account will contain

$$y = (1 + .06)^t = (1.06)^t$$

dollars.

(a) Use a calculator to complete the following table.

t	0	1	2	3	4	5	6	7	8	9	10
y	1					1.34					1.79

(b) Graph $y = (1.06)^t$.

44. Management If money loses value at the rate of 3% per year, the value of $1 in t years is given by

$$y = (1 - .03)^t = (.97)^t.$$

(a) Use a calculator to complete the following table.

t	0	1	2	3	4	5	6	7	8	9	10
y	1					.86					.74

(b) Graph $y = (.97)^t$.

45. Management If money loses value, then it takes more dollars to buy the same item. Use the results of Exercise 44(a) to answer the following questions.

(a) Suppose a house costs $65,000 today. Estimate the cost of a similar house in 10 years. (Hint: solve the equation $(.74t = \$65,000.)$

(b) Estimate the cost of a $30 textbook in 8 years.

46. Social Science Under certain conditions, the number of individuals of a species that is newly introduced into an area can double every year. That is, if t represents the number of years since the species was introduced into the area and y represents the number of individuals, then

$$y = 6 \cdot 2^t$$

if 6 animals were introduced into the area originally.

(a) Complete the following table.

t	0	1	2	3	4	5	6	7	8	9	10
y	6					192					6144

(b) Graph $y = 6 \cdot 2^t$.

47. Social Science If the world population continues to grow at the present rate, the population in billions in year t will be given by the function $P(t) = 4.2(2^{.0285(t-1980)})$.

(a) There were fewer than a billion people on earth when Thomas Jefferson died in 1826. What was the world population in 1980?

If this model is accurate, what will the population of the world be in

(b) 2000   **(c)** 2010   **(d)** 2040

48. Social Science By using the function in Exercise 47 and experimenting with a calculator, estimate how many years it will take for the world population in 2000 to triple. Is it likely that you will be alive then?

49. Management An eccentric billionaire offers you a job for the month of September. You may choose either to be paid $300,000 per day or to be paid 2 cents on the first day, 4 cents on the second day, 8 cents on the third day, and so on, with your pay doubling on each successive day.

(a) Write the rule of the function that gives your salary (in cents) on day t if you choose the second option.

(b) Which pay option will give you the larger total income for the month? (*Hint:* under the second option, how much will you be paid on September 30?)

50. Natural Science The amount of plutonium remaining from one kilogram after x years is given by the function $W(x) = 2^{-x/24360}$. How much will be left after

(a) 1000 years?   **(b)** 10,000 years?

(c) 15,000 years?

(d) Estimate how long it will take for the one kilogram to decay to half its original weight. This may help to explain why nuclear waste disposal is a serious problem.

Management   *The scrap value of a machine is the value of the machine at the end of its useful life. By one method of calculating scrap value, where it is assumed a constant percentage of value is lost annually, the scrap value S is given by*

$$S = C(1 - r)^n,$$

where C is the original cost, n is the useful life of the machine in years, and r is the constant annual percentage of value lost. Find the scrap value for each of the following machines.

51. Orginal cost, $54,000; life, 8 years; annual rate of value loss, 12%

52. Original cost, $178,000; life, 11 years; annual rate of value loss, 14%

53. **Management**   Midwest Creations finds that its total sales $T(x)$, in thousands, from the distribution of x catalogs, where x is measured in thousands, is approximated by

$$T(x) = \frac{2500}{1 + 24 \cdot 2^{-x/4}}.$$

Find the total sales resulting from the distribution of
(a) 0 catalogs;   **(b)** 5000 catalogs;
(c) 24,000 catalogs;   **(d)** 49,000 catalogs.

54. **Natural Science**   *Escherichia coli* is a strain of bacteria that occurs naturally in many different situations. Under certain conditions, the number of these bacteria present in a colony is given by

$$E(t) = E_0 \cdot 2^{t/30},$$

where $E(t)$ is the number of bacteria present t minutes after the beginning of an experiment, and E_0 is the number present when $t = 0$. Let $E_0 = 1,000,000$ and use a calculator to find the number of bacteria at the following times.
(a) $t = 5$ **(b)** $t = 10$ **(c)** $t = 15$
(d) $t = 20$ **(e)** $t = 60$ **(f)** $t = 120$

55. **Social Science**   A person learning certain skills involving repetition tends to learn quickly at first. Then learning tapers off and approaches some upper limit. Suppose the number of characters per minute a typesetter can produce is given by

$$p(t) = 250 - 120(2.8)^{-.5t},$$

where t is the number of months the typesetter has been in training. Find each of the following.

(a) $p(2)$ **(b)** $p(4)$ **(c)** $p(10)$ **(d)** Graph $p(t)$.

Natural Science   *Generally speaking, the larger an organism, the greater its complexity (as measured by counting the number of different types of cells present). The figure shows an estimate of the maximum number of cells (or the largest volume) in various organisms plotted against the number of types of cells found in those organisms.** *

56. Use the graph to estimate the maximum number of cells and the corresponding volume for each of the following organisms.
(a) whale   **(b)** sponge   **(c)** green alga

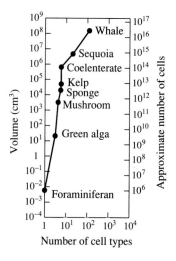

57. From the graph, estimate the number of cell types in each of the following organisms.
(a) mushroom   **(b)** kelp   **(c)** sequoia

58. Use the graph of $f(x) = 2^x$ (not a calculator) to explain why $2^x + 2^{-x}$ is approximately equal to 2^x when x is very large.

---

*From *On Size and Life* by Thomas A. McMahon and John Tyler Bonner. Copyright © 1983 by Thomas. A McMahon and John Tyler Bonner. Reprinted by permission of W. H. Freeman and Company.

59. Natural Science   The population of fish in a certain lake at time t months is given by the function

$$p(t) = \frac{20,000}{1 + 24(2^{-.36t})}.$$

(a) Graph the population function from $t = 0$ to $t = 48$ (a four-year period).

(b) What was the population at the beginning?

(c) Use the graph to estimate the one-year period in which the population grew most rapidly.

(d) When do you think the population will reach 25,000? What factors in nature might explain your answer?

60. Natural Science   The number of fruit flies present after t days of a laboratory experiment is given by $p(t) = 100 \cdot 3^{t/10}$. How many flies will be present after

(a) 15 days?   **(b)** 25 days?

(c) When will the fly population reach 3500? (*Hint:* Where does the graph of p meet the horizontal line $y = 3500$? What is the relevance of this point?)

---

1 Use a calculator to find each of the following.

(a) $e^{.06}$

(b) $e^{-.06}$

(c) $e^{2.30}$

(d) $e^{-2.30}$

Answers:

(a) 1.06184

(b) .94176

(c) 9.97418

(d) .10026

4.2 APPLICATIONS OF EXPONENTIAL FUNCTIONS

A certain irrational number, denoted e, arises naturally in a variety of mathematical situations (much the same way that the number π appears when you consider the area of a circle).* To nine decimal places,

$$e = 2.718281828.$$

Perhaps the single most useful exponential function is the function defined by $f(x) = e^x$.

A calculator should be used to evaluate powers of e. To find $e^{.14}$ on a scientific calculator, enter .14 and then press the $\boxed{e^x}$ key (or press $\boxed{\text{Inv}}$ $\boxed{\ln}$, depending on the calculator). On a graphing calculator, press the e^x key first, then enter .14 and press the $\boxed{\text{ENTER}}$ key (labeled $\boxed{\text{EXE}}$ on some calculators). In either case the display will show (to five places) 1.15027.† Most calculators allow you to use negative exponents with the e^x key, but if yours doesn't, remember that for any positive k, $e^{-k} = 1/e^k$. **1**

Note   To display the decimal expansion of e on your calculator screen, calculate e^1.

In Figure 4.5 on the next page, the functions defined by

$$g(x) = 2^x, \qquad f(x) = e^x, \quad \text{and} \quad h(x) = 3^x$$

are graphed for comparison.

---

* See pages 230–231 for an example.

† If these directions don't work with your calculator, consult the instruction booklet.

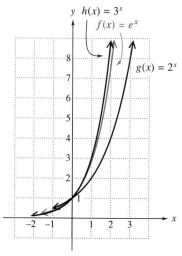

FIGURE 4.5

In many situations in biology, economics, and the social sciences, a quantity changes at a rate proportional to the quantity present. In such cases, the amount present at time t is a function of t called the *exponential growth function*.

Exponential Growth Function

Under normal conditions, growth is described by the function

$$f(t) = y_0 e^{kt},$$

where $f(t)$ is the amount or quantity present at time t, y_0 is the amount present at time $t = 0$, and k is a constant that depends on the situation.

It is understood in this context that "growth" can involve either growing larger or growing smaller. Here is an example of growing larger.

▶ **EXAMPLE 1** The approximate population of Midville is projected to be

$$P(t) = 10,000e^{.04t},$$

where the time t is measured in years. This is an example of an exponential growth function with $k = .04$ and $y_0 = 10,000$. You can see that the initial population at time $t = 0$ is, in fact, 10,000 by evaluating $P(0)$:

$$P(0) = 10,000e^{(.04)0} = 10,000e^0 = 10,000 \cdot 1 = 10,000.$$

The population in year $t = 5$ will be

$$P(5) = 10,000e^{(.04 \cdot 5)} = 10,000e^{.2}.$$

2 In Example 1, find the population of the city after

(a) 1 year;

(b) 3 years;

(c) 7 years;

(d) 10 years.

Answers:

(a) About 10,400

(b) About 11,300

(c) About 13,200

(d) About 14,900

3 Suppose the number of bacteria in a culture at time t is

$$y = 500e^{.4t},$$

where t is measured in hours.

(a) How many bacteria are present at $t = 0$?

(b) How many bacteria are present at $t = 10$?

Answers:

(a) 500

(b) About 27,300

Using a calculator to compute this last product, we find that

$$P(5) = 10{,}000e^{.2} \approx 12{,}214.$$

Thus the population of the city in 5 years will be about 12,200. ◀ **2**

When the constant k in the exponential growth function is negative, the quantity will *decrease* over time, that is, grow smaller.

▶ **EXAMPLE 2**   Suppose the amount, y, of a certain radioactive substance present at time t is given by

$$y = 1000e^{-.1t},$$

where t is measured in days and y is measured in grams.

(a) How much of the substance will be present at time $t = 0$?

This is the exponential growth function with $k = -.1$ and $y_0 = 1000$. So there are 1000 grams present at time $t = 0$. This fact can also be determined by evaluating the function when $t = 0$.

(b) How much will be present at time $t = 5$?

$$y = 1000e^{(-.1)5} = 1000e^{-.5} \approx 606.53.$$

Thus after 5 days, the amount has decreased to 606.53 grams. ◀ **3**

The exponential growth function deals with quantities that eventually either grow very large or decrease practically to 0. Other exponential functions are needed to deal with different patterns of growth.

▶ **EXAMPLE 3**   Sales of a new product often grow rapidly at first and then begin to level off with time. For example, suppose the sales, $S(x)$, in some appropriate unit, of a new model calculator are approximated by

$$S(x) = 1000 - 800e^{-x},$$

where x represents the number of years the calculator has been on the market. Calculate S_0, $S(1)$, $S(2)$, and $S(4)$. Graph $S(x)$.

Find S_0 by letting $x = 0$.

$$S_0 = S(0) = 1000 - 800 \cdot 1 = 200 \qquad e^{-x} = e^{-0} = 1$$

Using one of these calculators,

$$S(1) = 1000 - 800e^{-1} \approx 1000 - 294.3 = 705.7,$$

which rounds to 706.

In the same way, verify that $S(2) \approx 892$ and $S(4) \approx 985$. Plotting several such points leads to the graph shown in Figure 4.6. Notice that as x increases, e^{-x} decreases and approaches 0. Thus, $S(x) = 1000 - 800e^{-x}$ approaches $S(x) = 1000 - 0 = 1000$. This means that the graph approaches the horizon-

4 Suppose that sales of a product are given by $S(x) = 1200 - 800e^{-.5x}$. Find

(a) $S(0)$;

(b) $S(2)$;

(c) $S(4)$;

(d) $S(10)$.

(e) Find an equation of the horizontal asymptote for the graph.

(f) Graph $y = S(x)$.

Answers:

(a) 400

(b) 906

(c) 1092

(d) 1195

(e) $y = 1200$

(f)

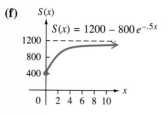

tal asymptote $y = 1000$. As the graph suggests, sales will tend to level off with time and gradually approach a level of 1000 units. ◀ **4**

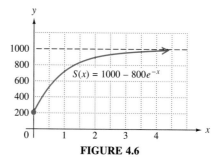

FIGURE 4.6

▶**EXAMPLE 4** Assembly line operations tend to have a high turnover of employees, forcing the companies involved to spend much time and effort in training new workers. It has been found that a worker new to the operation of a certain task on the assembly line will produce items according to

$$P(x) = 25 - 25e^{-.3x}$$

where $P(x)$ is the number of items produced by the worker on day x. How many items will be produced by a new worker on day 8?

Evaluate $P(8)$.

$$P(8) = 25 - 25e^{-.3(8)} \approx 25 - 2.3 = 22.7.$$

On the eighth day, the worker can be expected to produce about 23 items. Plotting several such points leads to the graph of $P(x)$ shown in Figure 4.7. ◀ **5**

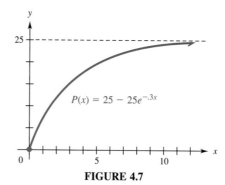

FIGURE 4.7

5 Suppose the value of the assets of a certain company at time t are given by

$$f(t) = 100{,}000 - 75{,}000e^{-.2t},$$

where t is measured in years. Find $f(t)$ for the following values of t.

(a) $t = 0$

(b) $t = 5$

(c) $t = 10$

(d) $t = 25$

(e) Find the horizontal asymptote.

(f) Graph $f(t)$.

Answers:

(a) 25,000

(b) 72,410

(c) 89,850

(d) 99,500

(e) $y = 100{,}000$

(f)

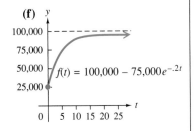

6 In Example 5, find

(a) $N(10)$

(b) $N(12)$

(c) $N(14)$

Answers:

(a) 5.00

(b) 1.68

(c) .39

Graphs such as the one in Figure 4.7 are called **learning curves.** According to such a graph, a new worker tends to learn quickly at first; then learning tapers off and approaches some upper limit. This is characteristic of the learning of certain types of skills involving the repetitive performance of the same task. Learning curves are examples of *limited growth functions.*

▶ **EXAMPLE 5**   Psychologists have measured people's ability to remember facts that they have memorized. In one such experiment, it was found that the number of facts, $N(t)$, remembered after t days was given by

$$N(t) = 10\left(\frac{1 + e^{-8}}{1 + e^{.8t-8}}\right).$$

The graph of this function is called a **forgetting curve.** Graph $y = N(t)$.
Plot some points corresponding to various values of t. If $t = 0$, then

$$N(0) = 10\left(\frac{1 + e^{-8}}{1 + e^{-8}}\right) = 10.$$

So 10 facts were known at the beginning. After 7 days the number remembered was

$$N(7) = 10\left(\frac{1 + e^{-8}}{1 + e^{(.8)(.7)-8}}\right) = 10\left(\frac{1 + e^{-8}}{1 + e^{-2.4}}\right) = 9.17.$$

Plotting several such points (see the problem at the side) leads to Figure 4.8.   ◀   **6**

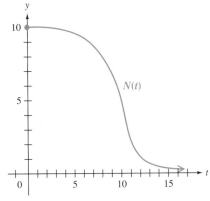

FIGURE 4.8

4.2 EXERCISES

1. Natural Science The number of bacteria present at time t hours is given by

$$f(t) = 5000e^{.4t}.$$

(a) How many bacteria are present at the beginning of the experiment? How many were present after

(b) 3 hours?   (c) 12 hours?   (d) one day?

2. Social Science The population of Metropolis is given by

$$P(t) = 1,500,000e^{.02t},$$

where t is measured in years. If year $t = 0$ corresponds to 1995, find the population in

(a) 1995   (b) 2000   (c) 2010   (d) 2030

3. Natural Science A sealed box contains radium. The number of grams present at time t is given by

$$Q(t) = 100e^{-.00043t},$$

where t is measured in years. Find the amount of radium in the box at the following times.

(a) $t = 0$ (b) $t = 800$ (c) $t = 1600$
(d) $t = 5000$

4. Natural Science The amount of a chemical in grams that will dissolve in a solution is given by

$$C(t) = 10e^{.02t},$$

where t is the temperature of the solution. Find each of the following.

(a) $C(0°)$ (b) $C(10°)$ (c) $C(30°)$
(d) $C(100°)$

5. Natural Science A population of lice, $L(t)$, is given by

$$L(t) = 100e^{.1t},$$

where t is measured in months. Find each of the following values.

(a) $L(0)$ (b) $L(1)$ (c) $L(6)$ (d) $L(12)$

6. Natural Science When a bactericide is introduced into a certain culture, the number of bacteria present, $D(t)$, is given by

$$D(t) = 50,000e^{-.01t}$$

where t is time measured in hours. Find the number of bacteria present at each of the following times.

(a) $t = 0$ (b) $t = 5$ (c) $t = 20$
(d) $t = 50$

7. Management In Chapter 5, it is shown that P dollars compounded continuously (every instant) at an annual rate of interest i would amount to

$$A = Pe^{ni}$$

at the end of n years. How much would \$20,000 amount to at 8% compounded continuously for the following number of years?

(a) 1 year   (b) 5 years   (c) 10 years
(d) By experimenting with your calculator, estimate how long it will take for the initial \$20,000 to triple.

8. Natural Science The number (in hundreds) of fish in a small commercial pond is given by

$$F(t) = 27 - 15e^{-.8t},$$

where t is in years. Find each of the following.

(a) $F(0)$ (b) $F(1)$ (c) $F(5)$ (d) $F(10)$

9. Natural Science The beaver population in a certain lake in year t is approximately

$$p(t) = \frac{2000}{1 + e^{-.5544t}}.$$

(a) What is the population now $(t = 0)$?

(b) What will the population be in five years?

10. Social Science The number of words per minute that an average typist can type is given by

$$W(t) = 60 - 30e^{-.5t},$$

where t is time in months after the beginning of a typing class. Find each of the following.

(a) $W(0)$ (b) $W(1)$ (c) $W(4)$ (d) $W(6)$

11. Management The estimated number of units that will be sold by the Goldstein Widget Works t months from now is given by

$$N(t) = 100,000e^{-.09t}.$$

(a) What are current sales $(t = 0)$?

(b) What will sales be in 2 months? In 6 months?

(c) Will sales ever regain their present level? (What does the graph of $N(t)$ look like?)

12. Management Sales of a new model can opener are approximated by

$$S(x) = 5000 - 4000e^{-x},$$

where x represents the number of years that the can opener has been on the market, and $S(x)$ represents sales in thousands. Find each of the following.

(a) $S(0)$ (b) $S(1)$ (c) $S(2)$
(d) $S(5)$ (e) $S(10)$
(f) Find the horizontal asymptote for the graph.
(g) Graph $y = S(x)$.

Management *The national debt has grown from 65 million dollars at the beginning of the Civil War to more than 4 trillion dollars currently. The graph of the debt, in which the horizontal axis represents time and the vertical axis dollars, has the approximate shape of an exponential growth function.* Over the past thirty years, the amount of debt (in billions of dollars) has been roughly approximated by the function*

$$D(t) = e^{.1293(t+36)} + 150,$$

where t is measured in years with t = 0 representing 1964.

13. Use this function to estimate the debt in each of the following years. Compare your results with the graph.

(a) 1972 (b) 1982 (c) 1992

14. (a) The Congressional Budget Office has projected that the national debt will be less than 6 trillion dollars at the turn of the century. How does this figure compare with the one predicted by the function D? What might explain the difference?

(b) Experiment with your calculator and various exponential functions to see if you can find a model for the national debt for the decade 1992–2002. Your model should indicate a debt of approximately 4077 billion in 1992 and approximately 6000 billion in 2002.

15. Social Science A class in elementary Tibetan is given a final exam worth 100 points. Then as part of an experiment they are tested weekly thereafter to see how much they remember. The average score on the test taken after t weeks is given by

$$T(x) = 77\left(\frac{1 + e^{-1}}{1 + e^{.08x-1}}\right).$$

(a) What was the average score on the original exam?
(b) What was the average score on the exam after 3 weeks? after 8 weeks?
(c) Do the students remember much Tibetan after a year?

* Graph from "Debt Dwarfs Deficit" by Sam Hodges as appeared in The Mobile Register. *Reprinted by permission of Newhouse News Service.*

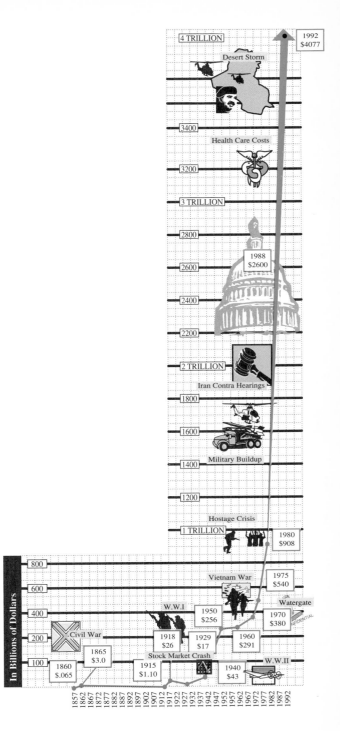

16. **Social Science**　A sociologist has shown that the fraction $y(t)$ of people in a group who have heard a rumor after t days is approximated by

$$y(t) = \frac{y_0 e^{kt}}{1 - y_0(1 - e^{kt})},$$

where y_0 is the fraction of people who have heard the rumor at time $t = 0$, and k is a constant. A graph of $y(t)$ for a particular value of k is shown in the figure.
(a) If $k = .1$ and $y_0 = .05$, find $y(10)$.
(b) If $k = .2$ and $y_0 = .10$, find $y(5)$.

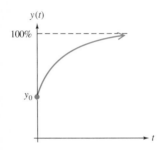

17. **Social Science**　Data from the National Highway Traffic Safety Administration for the period 1982–1992* indicates that the approximate percentage of people who use seat belts when driving is given by

$$f(t) = \frac{880}{11 + 69e^{-.3(t - 1982)}}.$$

What percentage used seat belts in
(a) 1982?　(b) 1989?　(c) 1992
Assuming this function is accurate after 1992, what percentage of people will use seat belts in
(d) 1997　(e) 2000　(f) 2005
(g) If this function remains accurate in the future, will there ever be a time when 95% of people use seat belts?

Natural Science　*Newton's Law of Cooling says that the rate at which a body cools is proportional to the difference in temperature between the body and an environ-*

ment into which it is introduced. Using calculus, the temperature $F(t)$ of the body at time t after being introduced into an environment having constant temperature T_0 is

$$F(t) = T_0 + Ce^{-kt},$$

where C and k are constants. Use this result in Exercises 18–20.

18. A piece of metal is heated to 300°C and then placed in a cooling liquid at 50°C. After 4 minutes the metal has cooled to 175°C. Find its temperature after 12 minutes.

19. Boiling water, at 100°C, is placed in a freezer at 0°C. The temperature of the water is 50°C after 24 minutes. Find the temperature of the water after 96 minutes.

20. Paisley refuses to drink coffee cooler than 95°F. She makes coffee with a temperature of 170°F in a room with a temperature of 70°F. The coffee cools to 120°F in 10 minutes. What is the longest time she can let the coffee sit before she drinks the coffee?

Natural Science　*The pressure of the atomosphere, $p(h)$, in pounds per square inch, is given by*

$$p(h) = p_0 e^{-kh},$$

where h is the height above sea level and p_0 and k are constants. The pressure at sea level is 15 pounds per square inch and the pressure is 9 pounds per square inch at a height of 12,000 feet.

21. Find the pressure at an altitude of 6000 feet.

22. What would be the pressure encountered by a spaceship at an altitude of 150,000 feet?

 23. (a) Graph the population function for Metropolis in Exercise 2. Use a viewing window that shows the entire population graph for the years 1995–2045.
(b) In what year will the population reach 2,000,000?

24. **Social Science**　The probability P percent of having an automobile accident is related to the alcohol level of the driver's blood t by the function $P(t) = e^{21.459t}$.
(a) Graph $P(t)$ in a viewing window with $0 \le t \le .2$ and $0 \le P(t) \le 100$.
(b) At what blood alcohol level is the probability of an accident at least 50 percent? What is the legal blood alcohol level in your state?

 25. **Natural Science**　When will the beaver population in Exercise 9 reach 1500?

---

*Published in *USA Today* on December 30, 1992.

1 Find each common logarithm:

(a) log 100

(b) log 1000

(c) log .1

Answers:

(a) 2

(b) 3

(c) −1

2 Find each common logarithm:

(a) log 27

(b) log 1089

(c) log .00426

Answers:

(a) 1.4314

(b) 3.0370

(c) −2.3706

4.3 LOGARITHMIC FUNCTIONS

Until the development of computers and calculators, logarithms were the only effective tool for large-scale numerical computations. They are no longer needed for this, but logarithmic functions still play a crucial role in many applications.

Logarithms are simply a *new language for old ideas*—essentially a special case of exponents.

Definition of Common (Base 10) Logarithms

$$y = \log x \quad \text{means} \quad 10^y = x.$$

"Log x", which is read "the logarithm of x", is the answer to the question

To what exponent must 10 be raised to produce x?

▶**EXAMPLE 1**   To find log 10,000 ask yourself, "To what exponent must 10 be raised to produce 10,000?" Since $10^4 = 10,000$, we see that log 10,000 = 4. Similarly,

$$\log 1 = 0 \quad \text{because} \quad 10^0 = 1;$$

$$\log .01 = -2 \quad \text{because} \quad 10^{-2} = \frac{1}{10^2} = \frac{1}{100} = .01;$$

$$\log \sqrt{10} = 1/2 \quad \text{because} \quad 10^{1/2} = \sqrt{10}. \quad ◀ \quad \boxed{1}$$

▶**EXAMPLE 2**   Log(−25) is the exponent to which 10 must be raised to produce −25. But every power of 10 is positive! So there is no exponent that will produce −25. *Logarithms of negative numbers and 0 are not defined.*   ◀

▶**EXAMPLE 3**

(a) We know that log 359 must be a number between 2 and 3 because $10^2 = 100$ and $10^3 = 1000$. Entering 359 in a scientific calculator and pressing the $\boxed{\log}$ key, we find that log 359 (to four decimal places) is 2.5551.* You can verify this by computing $10^{2.5551}$; the answer (rounded off) is 359.

(b) When 10 is raised to a negative exponent, the result is a number less than 1. Consequently, the logarithms of numbers between 0 and 1 are negative. For instance, a calculator shows that log .026 = −1.5850.   ◀ $\boxed{2}$

---

*On a graphing calculator, press the $\boxed{\log}$ key, then enter 359 and press the $\boxed{\text{ENTER}}$ (or $\boxed{\text{EXE}}$) key.

3 Find the following.

(a) ln 6.1

(b) ln 20

(c) ln .8

(d) ln .1

Answers:

(a) 1.8083

(b) 2.9957

(c) −.2231

(d) −2.3026

Common logarithms were once essential for large-scale computations but are no longer needed for such purposes because computers and calculators are faster and more accurate. Although common logarithms still have some uses (one of which is discussed in Section 4.4), the most widely used logarithms today are defined in terms of the number e (whose decimal expansion begins 2.71828· · ·) rather than 10. They have a special name and notation.

Definition of Natural (Base e) Logarithms

$$y = \ln x \quad \text{means} \quad e^y = x.$$

Thus the number **ln x** (which is sometimes read "el-en x") is the exponent to which e must be raised to produce the number x. For instance, ln 1 = 0 because $e^0 = 1$. Although logarithms to base e may not seem as "natural" as common logarithms, there are several reasons for using them, some of which are discussed in Section 4.4.

▶ EXAMPLE 4

(a) To find ln 85 on a scientific calculator, enter 85, press the ⌐ln x⌐ key, and read the result: 4.4427.* Thus 4.4427 is the exponent to which e must be raised to produce 85. You can verify this by computing $e^{4.4427}$; the answer (rounded off) is 85.

(b) A calculator shows that ln 38 = 3.6376, which means that $e^{3.6376} \approx 38$.

◀ **3**

▶ EXAMPLE 5   You don't need a calculator to find ln e^8. Just ask yourself, "To what exponent must e be raised to produce e^8?" The answer, obviously, is 8. So ln $e^8 = 8$. ◀

Example 5 is an illustration of the following fact.

$$\ln e^k = k \text{ for every real number } k.$$

The procedure used to define common and natural logarithms can be carried out with any positive number $a \neq 1$ as the base (in place of 10 or e).

---

*On a graphing calculator, press the ⌐ln⌐ key, then enter 85, and press the ⌐ENTER⌐ (or ⌐EXE⌐) key.

4 Write the logarithmic form of

(a) $5^3 = 125$;

(b) $3^{-4} = 1/81$;

(c) $8^{2/3} = 4$.

Answers:

(a) $\log_5 125 = 3$

(b) $\log_3 (1/81) = -4$

(c) $\log_8 4 = 2/3$

5 Write the exponential form of

(a) $\log_{16} 4 = 1/2$;

(b) $\log_3 (1/9) = -2$;

(c) $\log_{16} 8 = 3/4$.

Answers:

(a) $16^{1/2} = 4$

(b) $3^{-2} = 1/9$

(c) $16^{3/4} = 8$

Definition of Logarithms to the Base *a*

$$y = \log_a x \quad \text{means} \quad a^y = x.$$

Read $y = \log_a x$ as "y is the logarithm of x to the base a." For example, the exponential statement $2^4 = 16$ can be translated into the equivalent logarithmic statement $4 = \log_2 16$. Thus, **$\log_a x$** is an *exponent;* it is the answer to the question.

To what power must *a* be raised to produce *x*?

This key definition should be memorized. It is important to remember the location of the base and exponent in each part of the definition.

$$\text{logarithmic form:} \quad y = \log_{\underset{\uparrow}{a}} \overset{\text{exponent}}{x}$$
$$\text{base}$$

$$\text{exponential form:} \quad \underset{\underset{\text{base}}{\uparrow}}{a}^{\overset{\text{exponent}}{y}} = x$$

Common and natural logarithms are the special cases when $a = 10$ and when $a = e$ respectively. Both $\log u$ and $\log_{10} u$ mean the same thing. Similarly, $\ln u$ and $\log_e u$ mean the same thing.

▶ **EXAMPLE 6** This example shows several statements written in both exponential and logarithmic forms.

Exponential Form	Logarithmic Form
(a) $3^2 = 9$	$\log_3 9 = 2$
(b) $(1/5)^{-2} = 25$	$\log_{1/5} 25 = -2$
(c) $10^5 = 100{,}000$	$\log_{10} 100{,}000 = 5$
(d) $4^{-3} = 1/64$	$\log_4 (1/64) = -3$
(e) $2^{-4} = 1/16$	$\log_2 (1/16) = -4$
(f) $e^0 = 1$	$\log_e 1 = 0$

◀ **4** **5**

The usefulness of logarithmic functions depends in large part on the following *properties of logarithms.*

6 Simplify, using the properties of logarithms.

(a) $\log_a 5x + \log_a 3x^4$

(b) $\log_a 3p - \log_a 5q$

(c) $4 \log_a k - 3 \log_a m$

Answers:

(a) $\log_a 15x^5$

(b) $\log_a (3p/5q)$

(c) $\log_a (k^4/m^3)$

Properties of Logarithms

Let x and y be any positive real numbers and r be any real number. Let a be a positive real number, $a \neq 1$. Then

(a) $\log_a xy = \log_a x + \log_a y$; **(b)** $\log_a \dfrac{x}{y} = \log_a x - \log_a y$;

(c) $\log_a x^r = r \log_a x$; **(d)** $\log_a a = 1$;

(e) $\log_a 1 = 0$; **(f)** $\log_a a^y = y$;

(g) $a^{\log_a x} = x$.

Note Because these properties are so useful, they should be memorized.

To prove property (a), let

$$m = \log_a x \quad \text{and} \quad n = \log_a y.$$

Then, by the definition of logarithm,

$$a^m = x \quad \text{and} \quad a^n = y.$$

Multiply to get

$$a^m \cdot a^n = x \cdot y,$$

or, by a property of exponents,

$$a^{m+n} = xy.$$

Use the definition of logarithm to rewrite this last statement as

$$\log_a xy = m + n.$$

Replace m with $\log_a x$ and n with $\log_a y$ to get

$$\log_a xy = \log_a x + \log_a y.$$

Properties (b) and (c) can be proven in a similar way. Since $a^1 = a$ and $a^0 = 1$, properties (d) and (e) come from the definition of logarithm.

▶ **EXAMPLE 7** If all the following variable expressions represent positive numbers, then for $a > 0$, $a \neq 1$,

(a) $\log_a x + \log_a (x - 1) = \log_a x(x - 1)$;

(b) $\log_a \dfrac{x^2 + 4}{x + 6} = \log_a (x^2 + 4) - \log_a (x + 6)$;

(c) $\log_a 9x^5 = \log_a 9 + \log_a x^5 = \log_a 9 + 5 \log_a x$. ◀ **6**

Caution There is no logarithm property that allows you to simplify the logarithm of a sum, such as $\log_a (x^2 + 4)$. In particular, $\log_a (x^2 + 4)$ is *not* equal to $\log_a x^2 + \log_a 4$. Property (a) of logarithms in the box above shows that $\log_a x^2 + \log_a 4 = \log_a 4x^2$.

7 Use the properties of logarithms to rewrite and evaluate each of the following, given $\log_3 7 \approx 1.77$ and $\log_3 5 \approx 1.46$.

(a) $\log_3 35$

(b) $\log_3 7/5$

(c) $\log_3 25$

(d) $\log_3 3$

(e) $\log_3 1$

Answers:

(a) 3.23

(b) .31

(c) 2.92

(d) 1

(e) 0

▶**EXAMPLE 8**   Using the properties of logarithms, if $\log_6 7 \approx 1.09$ and $\log_6 5 \approx .9$,

(a) $\log_6 35 = \log_6 (7 \cdot 5) = \log_6 7 + \log_6 5 \approx 1.09 + .9 = 1.99;$
(b) $\log_6 5/7 = \log_6 5 - \log_6 7 \approx -.19;$
(c) $\log_6 5^3 = 3 \log_6 5 \approx 3(.9) = 2.7;$
(d) $\log_6 6 = 1;$
(e) $\log_6 1 = 0.$ ◀ **7**

In Example 8 several logarithms to base 6 were given. However, they could have been found by using a calculator and the following formula, whose proof is outlined in Exercise 83.

Change of Base Theorem

For any positive numbers b and x (with $b \neq 1$),

$$\log_b x = \frac{\ln x}{\ln b}.$$

▶**EXAMPLE 9**   To find $\log_7 3$, use the theorem with $b = 7$ and $x = 3$:

$$\log_7 3 = \frac{\ln 3}{\ln 7} \approx \frac{1.0986}{1.9459}, \approx .5646.$$

You can check this on your calculator by verifying that $7^{.5646} \approx 3.$ ◀

Equations involving logarithms are often solved by using the fact that a logarithmic equation can be rewritten (with the definition of logarithm) as an exponential equation. In other cases, the properties of logarithm may be useful in simplifying an equation involving logarithms.

▶**EXAMPLE 10**   Solve each of the following equations.
(a) $\log_x 8/27 = 3$

First, use the definition of logarithm and write the expression in exponential form.

$$x^3 = \frac{8}{27} \qquad \text{Definition of logarithm}$$

$$x^3 = \left(\frac{2}{3}\right)^3 \qquad \text{Write } \tfrac{27}{3} \text{ as a cube}$$

$$x = \frac{2}{3} \qquad \text{Set bases equal}$$

The solution is 2/3.

8 Solve each equation.

(a) $\log_x 6 = 1$

(b) $\log_5 25 = m$

(c) $\log_{27} x = 2/3$

Answers:

(a) 6

(b) 2

(c) 9

(b) $\log_4 x = 5/2$

In exponential form, the given statement becomes

$$4^{5/2} = x \qquad \text{Definition of logarithm}$$
$$(4^{1/2})^5 = x \qquad \text{Definition of rational exponent}$$
$$2^5 = x$$
$$32 = x.$$

The solution is 32. ◀ **8**

In the next example, properties of logarithms are needed to solve equations.

▶**EXAMPLE 11**   Solve each equation.

(a) $\log_2 x - \log_2 (x - 1) = 1$

By a property of logarithms, the left-hand side can be simplified as

$$\log_2 x - \log_2 (x - 1) = \log_2 \frac{x}{x - 1}.$$

The original equation then becomes

$$\log_2 \frac{x}{x - 1} = 1.$$

Use the definition of logarithm to write this last result in exponential form.

$$\frac{x}{x - 1} = 2^1 = 2$$

Solve this equation.

$$\frac{x}{x - 1} \cdot (x - 1) = 2(x - 1)$$
$$x = 2(x - 1)$$
$$x = 2x - 2$$
$$-x = -2$$
$$x = 2$$

The domain of logarithmic functions includes only positive real numbers, so it is *necessary* to check this proposed solution in the original equation.

$$\log_2 x - \log_2(x - 1) \stackrel{?}{=} 1$$
$$\log_2 2 - \log_2(2 - 1) \stackrel{?}{=} 1 \qquad \text{Let } x = 2$$
$$1 - 0 = 1 \qquad \text{Definition of logarithm}$$

The solution, 2, checks.

9 Solve each equation.

(a) $\log_5 x + 2 \log_5 x = 3$

(b) $\log_6 (a + 2)$

$-\log_6 \dfrac{a - 7}{5} = 1$

Answers:

(a) 5

(b) 52

(b) $\log x - \log 8 = 1$

Use a property of logarithms to combine the terms on the left side.

$$\log x - \log 8 = \log \frac{x}{8}$$

The equation becomes

$$\log \frac{x}{8} = 1.$$

Recall $\log x$, means $\log_{10} x$, so

$$\log \frac{x}{8} = \log_{10} \frac{x}{8} = 1$$

$$\frac{x}{8} = 10^1 = 10 \qquad \text{Definition of logarithm}$$

$$x = 80.$$

Since $x = 80$ is in the domain of $\log x$, the solution is 80. ◄ **9**

For a given *positive* value of x, the definition of logarithm leads to exactly one value of y, so that $y = \log_a x$ defines a logarithmic function of base a. (The base a must be positive, with $a \neq 1$.)

If $a > 0$ and $a \neq 1$, the **logarithmic function** with base a is defined as

$$f(x) = \log_a x.$$

The most important logarithmic function is the natural logarithmic function.

▶ **EXAMPLE 12** Graph $f(x) = \ln x$ and $g(x) = e^x$ on the same axes.

For each function, use a calculator to compute some ordered pairs. Then plot the corresponding points and connect them with a curve to obtain the graphs in Figure 4.9.

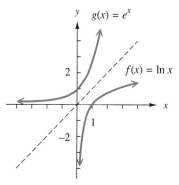

FIGURE 4.9

10 Graph $f(x) = \log x$ and $g(x) = 10^x$ on the same axes.

Answer:

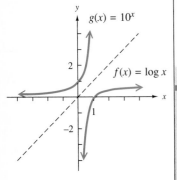

In Figure 4.9, the dashed straight line is the graph of $y = x$. Observe that the graph of $f(x) = \ln x$ is the mirror image of the graph of $g(x) = e^x$, with the line $y = x$ being the mirror. ◀ **10**

When the base $a > 1$, the graph of $f(x) = \log_a x$ has the same basic shape as the graph of the natural logarithmic function in Figure 4.9, as summarized below.

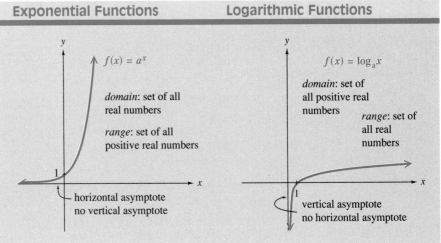

Exponential Functions	Logarithmic Functions

$f(x) = a^x$

domain: set of all real numbers

range: set of all positive real numbers

horizontal asymptote
no vertical asymptote

$f(x) = \log_a x$

domain: set of all positive real numbers

range: set of all real numbers

vertical asymptote
no horizontal asymptote

Finally, the graph of $f(x) = \log_a x$ is the mirror image of the graph of $g(x) = a^x$, with the line $y = x$ being the mirror. Functions whose graphs are related in this way are said to be **inverses** of each other. A more complete discussion of inverse functions is given in most standard college algebra books.

4.3 EXERCISES

Translate each of these logarithmic statements into an equivalent exponential statement.
(See Examples 1, 5, 6.)

1. $\log 100,000 = 5$

2. $\log .001 = -3$

3. $\ln 3 = 1.0986$

4. $\ln 10 = 2.3026$

5. $\log_3 81 = 4$

6. $\log_2(1/4) = -2$

Translate each of these exponential statements into an equivalent logarithmic statement. (See Examples 5–6.)

7. $10^{-3} = .001$

8. $10^{1.8751} = 75$

9. $e^{3.2189} = 25$

10. $e^{5.6168} = 275$

11. $3^{-2} = 1/9$

12. $16^{1/2} = 4$

Without using a calculator, evaluate each of the following. (See Examples 1, 5, 6.)

13. $\log 1000$

14. $\log 100$

15. $\log .01$

16. $\log .0001$

17. $\log_5 25$

18. $\log_9 81$

19. $\log_4 64$

20. $\log_6 216$

21. $\log_2 \dfrac{1}{4}$

22. $\log_3 \dfrac{1}{27}$

23. $\ln \sqrt{e}$

24. $\ln(1/e)$

25. $\ln e^{3.78}$

26. $\log 10^{56.9}$

Express the following as the logarithm of a single number or expression. Assume all variables represent positive numbers. (See Example 7.)

27. $\log 15 - \log 3$

28. $\log 4 + \log 8 - \log 2$

29. $3 \ln 2 + 2 \ln 3$

30. $2 \ln 5 - \frac{1}{2} \ln 25$

31. $3 \log x - 2 \log y$

32. $2 \log u + 3 \log w - 6 \log v$

33. $\ln(3x + 2) + \ln(x + 4)$

34. $2 \ln(x + 1) - \ln(x + 2)$

35. $3 \log x - 2 \log(x + 1)^2 + \frac{1}{2} \log(x + 2)$

36. $2 \log y + 3 \log z - \log(y + z)$

Assume that a is a positive number such that $\log_a 2 = .13$, $\log_a 3 = .20$, and $\log_a 5 = .30$. Use these facts and the properties of logarithms to find each of the following. (See Example 8.)

37. $\log_a 10$

38. $\log_a 15$

39. $\log_a 4$

40. $\log_a 27$

41. $\log_a(5/3)$

42. $\log_a 48$

Express each of the following in terms of u and v, where $u = \ln x$ and $v = \ln y$. For example, $\ln x^3 = 3(\ln x) = 3u$.

43. $\ln(x^2 y^5)$

44. $\ln(\sqrt{x} \cdot y^2)$

45. $\ln(x^3/y^2)$

46. $\ln(\sqrt{x}/y)$

Find each of the following. (See Example 9.)

47. $\log_6 543$

48. $\log_{20} 97$

49. $\log_{35} 6874$

50. $\log_5 50 - \log_{50} 5$

Find numerical values for b and c for which the given statement is false.

51. $\log(b + c) = \log b + \log c$

52. $\dfrac{\ln b}{\ln c} = \ln\left(\dfrac{b}{c}\right)$

Suppose $\log_b 2 = a$ and $\log_b 3 = c$. Use the properties of logarithms to find the following logarithms.

53. $\log_b 54$

54. $\log_b 144$

55. $\log_b(72b)$

56. $\log_b(4b^2)$

Solve each of the following equations. (See Examples 10 and 11.)

57. $\log_x 25 = -2$

58. $\log_x \dfrac{1}{16} = -2$

59. $\log_9 27 = m$

60. $\log_8 4 = z$

61. $\log_y 8 = \dfrac{3}{4}$

62. $\log_r 7 = \dfrac{1}{2}$

63. $\log_3(5x + 1) = 2$

64. $\log_5(9x - 4) = 1$

65. $\log x - \log(x + 3) = -1$

66. $\log m - \log(m - 4) = -2$

67. $\log_3(y + 2) = \log_3(y - 7) + \log_3 4$

68. $\log_8(z - 6) = 2 - \log_8(z + 15)$

69. $\ln(x + 9) - \ln x = 1$

70. $\ln(2x + 1) - 1 = \ln(x - 2)$

71. $\log x + \log(x - 3) = 1$

72. $\log(x - 1) + \log(x + 2) = 1$

Graph each of the following. (See Example 12.)

73. $y = \ln(x + 2)$

74. $y = \ln x + 2$

75. $y = \log(x - 3)$

76. $y = \log x - 3$

Work the following exercises.

77. Natural Science   Two people with the flu visited the campus of Big State U. The number of days T that it took for the flu virus to infect n people is given by

$$T = -1.43 \ln\left(\frac{10,000 - n}{4998n}\right).$$

How many days will it take for the virus to infect
(a) 500 people?   **(b)** 5000 people?

78. Management   The doubling function

$$D(r) = \frac{\ln 2}{\ln(1 + r)}$$

gives the number of years required to double your money when it is invested at interest rate r (expressed as a decimal), compounded annually. How long does it take to double your money at each of the following rates?
(a) 4%   **(b)** 8%   **(c)** 18%   **(d)** 36%
(e) Round each of your answers in (a)-(d) to the nearest year and compare them with these numbers: 72/4, 72/8, 72/18, 72/36. Use this evidence to state a "rule of thumb" for determining approximate doubling time without using the function D. This rule, which has long been used by bankers, is called the **rule of 72.**

79. Management   Suppose the sales of a certain product are approximated by

$$S(t) = 125 + 83 \log(5t + 1),$$

where $S(t)$ is sales in thousands of dollars t years after the product was introduced on the market. Find
(a) $S(0)$;   **(b)** $S(2)$;   **(c)** $S(4)$;
(d) $S(31)$.   **(e)** Graph $y = S(t)$.

80. Natural Science   The population of an animal species that is introduced into a certain area may grow rapidly at first but then grow more slowly as time goes on. A logarithmic function can provide an excellent description of such growth. Suppose that the population of foxes, $F(t)$, in an area t months after the foxes were introduced there is

$$F(t) = 500 \log(2t + 3).$$

Find the population of foxes at the following times.
(a) When they are first released into the area (that is, when $t = 0$)
(b) After 3 months   **(c)** After 15 months
(d) Graph $y = F(t)$.

Management   *In many applications, data is graphed on a logarithmic scale, where differences between successive measurements are not always the same. For example, the graph from the July 30, 1979, issue of* Fortune *magazine (page 58),* shows the price/performance ratio for various models of IBM computers. Notice on the vertical scale that the distance from 100 to 500 is the same as the distance from 1000 to 5000. This is characteristic of a graph drawn with logarithmic scales.*

Price vs. performance

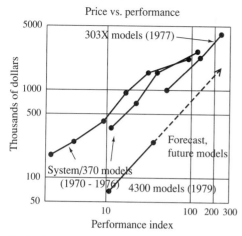

82. To locate a performance index of 50 on the horizontal scale, first find that log 50 ≈ 1.7, so that 50 would be located about .7, or 70% of the way from 10 to 100. Locate 50 on the horizontal axis, and then estimate the price for a model 4300 computer with a performance index of 50.

83. Prove the Change of Base Theorem. (*Hint:* By property (g) of logarithms $b^{\log_b x} = x$. Take the natural logarithm of each side, apply property (c), and solve the resulting equation for $\log_b x$.)

84. (a) Graph the "flu function" in Exercise 77 in a viewing window with $0 \le n \le 11,000$.
  (b) How many people are infected on the 16th day?
  (c) Explain why this model is not realistic when large numbers of people are involved.

81. Estimate the price for a 4300 model computer producing a performance index of 10. Do the same for a System/370 (use the upper graph).

4.4 APPLICATIONS OF LOGARITHMIC FUNCTIONS

This section first shows an additional method of solving exponential and logarithmic equations and then shows several applications using this method. Some of these applications depend on the following result.

Let x and y be positive numbers. Let a be a positive number, $a \ne 1$.

$$\text{If } x = y, \quad \text{then} \quad \log_a x = \log_a y.$$
$$\text{If } \log_a x = \log_a y, \quad \text{then} \quad x = y.$$

For convenience, base e is used in most of these applications.

▶ **EXAMPLE 1** Solve $3^x = 5$.

Since 3 and 5 cannot easily be written with the same base, the methods of Section 4.1 cannot be used to solve this equation. Instead, use the result given above and take natural logarithms of both sides.

$$3^x = 5$$
$$\ln 3^x = \ln 5$$
$$x \ln 3 = \ln 5 \qquad \text{Property (c) of logarithms}$$
$$x = \frac{\ln 5}{\ln 3} \approx 1.465.$$

1 Solve each equation. Round to the nearest thousandth.

(a) $2^x = 7$

(b) $5^m = 50$

(c) $3^y = 17$

Answers:

(a) 2.807

(b) 2.431

(c) 2.579

2 Solve each equation. Round to the nearest thousandth.

(a) $6^m = 3^{2m-1}$

(b) $5^{6a-3} = 2^{4a+1}$

Answers:

(a) 2.710

(b) .802

A calculator with a $\boxed{y^x}$ (or $\boxed{\wedge}$) key can be used to check this answer. Evaluate $3^{1.465}$; the result should be approximately 5. This step verifies that, to the nearest thousandth, the solution of the given equation is 1.465. ◀ **1**

Caution   There are no logarithm properties that enable you to simplify $\dfrac{\ln 5}{\ln 3}$; this number is *not* equal to $\ln\left(\dfrac{5}{3}\right)$ or $\ln 5 - \ln 3$.

▶**EXAMPLE 2**   Solve $3^{2x-1} = 4^{x+2}$.

Taking natural logarithms on both sides gives

$$\ln 3^{2x-1} = \ln 4^{x+2}.$$

Now use property (c) of logarithms and the fact that $\ln 3$ and $\ln 4$ are constants to rewrite the equation:

$$(2x - 1)(\ln 3) = (x + 2)(\ln 4)$$

$2x(\ln 3) - 1(\ln 3) = x(\ln 4) + 2(\ln 4)$ **Distributive property**

$2x(\ln 3) - x(\ln 4) = 2(\ln 4) + 1(\ln 3)$ **Collect terms with x on one side**

Factor out x on the left side to get

$$[2(\ln 3) - \ln 4]x = 2(\ln 4) + \ln 3.$$

Divide both sides by the coefficient of x:

$$x = \frac{2(\ln 4) + \ln 3}{2(\ln 3) - \ln 4}.$$

Using a calculator to evaluate this last expression, we find that

$$x = \frac{2 \ln 4 + \ln 3}{2 \ln 3 - \ln 4} \approx 4.774. \quad ◀ \;\mathbf{2}$$

Recall that $\ln e = 1$ (because 1 is the exponent to which e must be raised to produce e). This fact simplifies the solution of equations involving powers of e.

▶**EXAMPLE 3**   Solve $3e^{x^2} = 600$.

First divide each side by 3 to get

$$e^{x^2} = 200.$$

Now take natural logarithms on both sides; then use properties of logarithms.

3 Solve each equation. Round to the nearest thousandth.

(a) $e^{.1x} = 11$

(b) $e^{3+x} = .893$

(c) $e^{2x^2-3} = 9$

Answers:

(a) 23.979

(b) -3.113

(c) ± 1.612

$$e^{x^2} = 200$$
$$\ln e^{x^2} = \ln 200$$
$$x^2 \ln e = \ln 200$$
$$x^2 = \ln 200 \qquad \ln e = 1$$
$$x = \pm\sqrt{\ln 200}$$
$$x \approx \pm 2.302.$$

The solutions are ± 2.302, rounding to the nearest thousandth. (The symbol $\pm$ is used as a shortcut for writing the two solutions, 2.302 and -2.302.) ◀ **3**

The fact given at the beginning of this section, with the properties of logarithms from Section 4.3, is useful in solving logarithmic equations, as shown in the next examples.

4 Solve each equation.

(a) $\log_2(p + 9) - \log_2 p$
$= \log_2 (p + 1)$

(b) $\log_3(m + 1)$
$-\log_3 (m - 1) = \log_3 m$

Answers:

(a) 3

(b) $1 + \sqrt{2} \approx 2.414$

▶ **EXAMPLE 4** Solve $\log(x + 4) - \log(x + 2) = \log x$.

Using property (b) of logarithms, rewrite the equation as

$$\log \frac{x + 4}{x + 2} = \log x.$$

Then

$$\frac{x + 4}{x + 2} = x$$
$$x + 4 = x(x + 2)$$
$$x + 4 = x^2 + 2x$$
$$x^2 + x - 4 = 0.$$

By the quadratic formula,

$$x = \frac{-1 \pm \sqrt{1 + 16}}{2},$$

so that

$$x = \frac{-1 + \sqrt{17}}{2} \quad \text{or} \quad x = \frac{-1 - \sqrt{17}}{2}.$$

Log x cannot be evaluated for $x = (-1 - \sqrt{17})/2$, because this number is negative and not in the domain of log x. By substitution, verify that $(-1 + \sqrt{17})/2$ is a solution. ◀ **4**

5 The amount of a substance present at time t (in hours) is given by

$$A(t) = 530e^{-.2t},$$

and $A(t)$ is measured in grams. How much of the substance will remain after 5 hours? That is, find $A(5)$.

Answer:
About 195 grams

6 Find the half-life of the substance in Problem 6 above.

Answer:
About 3.5 hours

▶**EXAMPLE 5** Suppose that $A(t)$, the amount of a certain radioactive substance present at time t, is given by

$$A(t) = 1000e^{-.1t},$$

where t is measured in days and $A(t)$ in grams. At time $t = 0$,

$$A(0) = 1000e^{-.1(0)} = 1000$$

grams of the substance is present. Also,

$$A(5) = 1000e^{-.1(5)} = 1000e^{-.5} \approx 606.53,$$

so that about 607 grams are still present after 5 days. Now let us find the half-life of the substance. (The **half-life** of a radioactive substance is the time it takes for exactly half the sample to decay.)

We find the half-life by finding a value of t such that $A(t) = (1/2)(1000) = 500$ grams. That is, we find the half-life by solving the equation

$$500 = 1000e^{-.1t}.$$

First, divide both sides by 1000, obtaining

$$\frac{1}{2} = e^{-.1t}.$$

Now take natural logarithms of both sides. This gives

$$\ln \frac{1}{2} = \ln e^{-.1t}.$$

Using property (c) of logarithms.

$$\ln \frac{1}{2} = (-.1t)(\ln e),$$

and because $\ln e = 1$,

$$\ln \frac{1}{2} = -.1t,$$

or

$$t = \frac{\ln \frac{1}{2}}{-.1} = \frac{\ln .5}{-.1} \approx 6.9.$$

It will take about 6.9 days for half the sample to decay. ◀ **5** **6**

7 What is the age of a specimen in which $y = (1/3)y_0$?

Answer:
About 8880 years

▶ **EXAMPLE 6**   Carbon 14, also known as radiocarbon, is a radioactive form of carbon that is found in all living plants and animals. After a plant or animal dies, the radiocarbon disintegrates with a half-life of approximately 5600 years. Scientists can determine the age of the remains by comparing the amount of radiocarbon with the amounts present in living plants and animals. This technique is called *carbon dating*. The amount of radiocarbon present after t years is given by

$$y = y_0 e^{-(\ln 2)(1/5600)t},$$

where y_0 is the amount present in living plants and animals.

A round table hanging in Winchester Castle (England) was alleged to belong to King Arthur, who lived in the 5th century. A recent chemical analysis showed that the table had 91% of the amount of radiocarbon present in living wood. How old is the table?

The amount of radiocarbon present in the round table after y years is $.91y_0$. Therefore, in the equation

$$y = y_0 e^{-(\ln 2)(1/5600)t}$$

replace y with $.91y_0$ and solve for t.

$$.91y_0 = y_0 e^{-(\ln 2)(1/5600)t}$$
$$.91 = e^{-(\ln 2)(1/5600)t} \qquad \text{Divide both sides by } y_0$$
$$\ln .91 = \ln e^{-(\ln 2)(1/5600)t} \qquad \text{Take logarithms on both sides}$$
$$\ln .91 = -(\ln 2)(1/5600)t \qquad \text{Property (c) of logarithms and } \ln e = 1$$
$$t = \frac{(5600) \ln .91}{-\ln 2} \approx 760$$

The table is about 760 years old and therefore could not have belonged to King Arthur. ◀ **7**

Our last example uses common logarithms (base 10).

▶ **EXAMPLE 7**   The magnitude $R(i)$ of an earthquake, measured on the Richter scale, is given by

$$R(i) = \log\left(\frac{i}{i_0}\right),$$

where i is the amplitude of the ground motion of the earthquake and i_0 is the amplitude of the ground motion of the so-called *zero earthquake* (the smallest detectable earthquake, against which others are measured). The 1989 San Francisco earthquake measured 7.1 on the Richter scale.

(a) How did the ground motion of this earthquake compare with that of the zero earthquake?

8 Find the Richter scale magnitude of an earthquake whose ground motion is 100 times greater than the ground motion of the 1989 San Francisco earthquake discussed in Example 7.

Answer:
9.1

In this case $R(i) = 7.1$, that is, $\log(i/i_0) = 7.1$. Thus 7.1 is the exponent to which 10 must be raised to produce i/i_0. In other words,

$$10^{7.1} = \frac{i}{i_0}, \quad \text{or equivalently,} \quad i = 10^{7.1}i_0.$$

So this earthquake had $10^{7.1}$ (approximately 12.6 million) times more ground motion than the zero earthquake.

(b) What is the Richter scale magnitude of an earthquake with 10 times as much ground motion as the 1989 San Francisco earthquake?

Using the result from (a), the ground motion of such a quake would be

$$i = 10(10^{7.1}i_0) = 10^1 \cdot 10^{7.1}i_0 = 10^{8.1}i_0$$

so that its Richter scale magnitude would be

$$R(i) = \log\left(\frac{i}{i_0}\right) = \log\left(\frac{10^{8.1}i_0}{i_0}\right) = \log 10^{8.1} = 8.1.$$

Therefore, a ten-fold increase in ground motion increases the Richter scale magnitude by just 1. ◀ **8**

The most important applications of exponential and logarithmic functions for the fields of management and economics are considered in Chapter 5 (Mathematics of Finance).

4.4 EXERCISES

Solve each of the following exponential equations. Round to the nearest thousandth. (See Examples 1–3.)

1. $3^x = 5$

2. $5^x = 4$

3. $2^x = 3^{x-1}$

4. $4^{x+2} = 2^{x-1}$

5. $3^{1-2x} = 5^{x+5}$

6. $4^{3x-1} = 3^{x-2}$

7. $2^{1-3x} = 3^{x+1}$

8. $3^{z+3} = 2^z$

9. $e^{2x} = 5$

10. $e^{-3x} = 2$

11. $2e^{5a+2} = 8$

12. $10e^{3z-7} = 5$

13. $2^{x^2-1} = 12$

14. $3^{2-x^2} = 4$

15. $2(e^x + 1) = 10$

16. $5(e^{2x} - 2) = 15$

Solve each of the following equations for c.

17. $10^{4c-3} = d$

18. $3 \cdot 10^{2c+1} = 4d$

19. $e^{2c-1} = b$

20. $3e^{5c-7} = b$

Solve each of the following logarithmic equations. (See Example 4.)

21. $\ln(3x - 1) - \ln(2 + x) = \ln 2$

22. $\ln(8k - 7) - \ln(3 + 4k) = \ln(9/11)$

23. $\ln(5 + 4y) - \ln(3 + y) = \ln 3$

24. $\ln m + \ln(2m + 5) = \ln 7$

25. $\ln x + 1 = \ln(x - 4)$

26. $\ln(4x - 2) = \ln 4 - \ln(x - 2)$

27. $2 \ln(x - 3) = \ln(x + 5) + \ln 4$

28. $\ln(k + 5) + \ln(k + 2) = \ln 14k$

29. $\log_5(r + 2) + \log_5(r - 2) = 1$

30. $\log_4(z + 3) + \log_4(z - 3) = 1$

31. $\log_3(a - 3) = 1 + \log_3(a + 1)$

32. $\log w + \log(3w - 13) = 1$

33. $\log_2 \sqrt{2y^2 - 1} = 1/2$

34. $\log_2(\log_2 x) = 1$

35. $\log z = \sqrt{\log z}$

36. $\log x^2 = (\log x)^2$

Solve each of the following equations for c.

37. $\log (3 + b) = \log (4c - 1)$

38. $\ln (b + 7) = \ln (6c + 8)$

39. $2 - b = \log (6c + 5)$

40. $8b + 6 = \ln (2c) + \ln c$

41. Explain why the equation $3^x = -4$ has no solutions.

42. Explain why the equation $\log (-x) = -4$ does have a solution and find that solution.

Work the following exercises. (See Examples 5 and 6.)

43. Natural Science The amount of cobalt-60 (in grams) in a storage facility at time t is given by

$$C(t) = 25e^{-.14t},$$

where time is measured in years.
(a) How much cobalt-60 was present initially?
(b) What is the half-life of cobalt 60?

44. Social Science Over the past half-century, the population of a midwestern industrial city in year t has been given by

$$P(t) = 1,100,000e^{-.021t},$$

where $t = 0$ corresponds to 1950. What was the population in
(a) 1950?   **(b)** 1980?   **(c)** 1995?
(d) In what year was the population half of what it was in 1950?
(e) If this trend continues, when will the population be 250,000?

45. Natural Science An American Indian mummy was found recently. It had 73.6% of the amount of radiocarbon present in living beings. Approximately how long ago did this person die?

46. Natural Science How old is a piece of ivory that has lost 36% of its radiocarbon?

Work the following exercises. (See Example 7.)

47. Natural Science Find the Richter scale magnitude of earthquakes whose ground motion is
(a) $1000i_0$ **(b)** $100,000i_0$ **(c)** $10,000,000i_0$
(d) Fill the blank in this statement: increasing the ground motion by a factor of 10^k increases the Richter magnitude by ___ units.

48. Natural Science The great San Francisco earthquake of 1906 measured 8.3 on the Richter scale. How much greater was the ground motion in 1906 than in the 1989 earthquake, which measured 7.1 on the Richter scale?

49. Natural Science The loudness of sound is measured in units called decibels. The decibel rating of a sound is given by

$$D(i) = 10 \cdot \log\left(\frac{i}{i_0}\right),$$

where i is the intensity of the sound and i_0 is the minimum intensity detectable by the human ear (the so-called *threshold sound*). Find the decibel rating of each of the following sounds whose intensities are given. Round answers to the nearest whole number.
(a) Whisper, $115i_0$
(b) Busy street, $9,500,000i_0$
(c) Rock music, $895,000,000,000i_0$
(d) Jetliner at takeoff, $109,000,000,000,000i_0$

50. (a) How much more intense is a sound that measures 100 decibels than the threshold sound?
(b) How much more intense is a sound that measures 50 decibels than the threshold sound?
(c) How much more intense is a sound measuring 100 decibels than one measuring 50 decibels?

Natural Science *To find the maximum permitted levels of certain pollutants in fresh water, the EPA has established the functions defined in Exercises 51–52, where M(h) is the maximum permitted level of pollutant for a water hardness of h milligrams per liter. Find M(h) in each case. (These results give the maximum permitted average concentration in micrograms per liter for a 24-hour period.)*

51. Copper: $M(h) = e^r$, where $r = .65 \ln h - 1.94$ and $h = 9.7$.

52. Lead: $M(h) = e^r$, where $r = 1.51 \ln h - 3.37$ and $h = 8.4$.

Work the following exercises.

53. Social Science The number of years, $N(r)$, since two independently evolving languages split off from a common ancestral language is approximated by

$$N(r) = -5000 \ln r,$$

where r is the proportion of the words from the ancestral language that is common to both languages now. Find each of the following.
(a) $N(.9)$ **(b)** $N(.5)$ **(c)** $N(.3)$
(d) How many years have elapsed since the split if 70% of the words of the ancestral language are common to both languages today?
(e) If two languages split off from a common ancestral language about 1000 years ago, find r.

54. Natural Science A large cloud of radioactive debris from a nuclear explosion has floated over the Pacific Northwest, contaminating much of the hay supply. Consequently, farmers in the area are concerned that the cows who eat this hay will give contaminated milk. (The tolerance level for radioactive iodine in milk is 0.) The percent of the initial amount of radioactive iodine still present in the hay after t days is approximated by $P(t)$, which is given by the function

$$P(t) = 100e^{-.1t}.$$

(a) Find the percent remaining after 4 days.
(b) Find the percent remaining after 10 days.
(c) Some scientists feel that the hay is safe after the percent of radioactive iodine has declined to 10% of the original amount. Solve the equation $10 = 100e^{-.1t}$ to find the number of days before the hay may be used.
(d) Other scientists believe that the hay is not safe until the level of radioactive iodine has declined to only 1% of the original level. Find the number of days that this would take.

55. Natural Science The number of fruit flies present in an experiment at time t days is given by

$$N(t) = 100e^{.11t}.$$

Find the number of fruit flies present on day
(a) $t = 0$ **(b)** $t = 15$ **(c)** $t = 25$
(d) When will there be 2,000 fruit flies?

Natural Science *The following graphs are plotted on a logarithmic scale, where differences between successive measurements are not always the same. Data that do not plot in a linear pattern on the usual Cartesian axes often form a linear pattern when plotted on a logarithmic scale. Notice that on the vertical scale, the distance from 1 to 2 is not the same as the distance from 2 to 3, and so on. This is characteristic of a graph drawn with logarithmic scales.*

56. The graph below gives the rate of oxygen consumption for resting guinea pigs of various sizes. This rate is proportional to body mass raised to the power .67.* Estimate the oxygen consumption for a guinea pig with body mass of .3 kilograms. Do the same for one with body mass of .7.

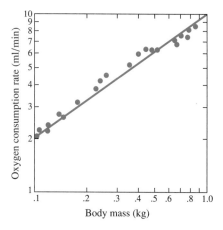

57. The graph on the next page gives the world weight-lifting records as log W_T, plotted against the logarithm of body weight. Here W_T is the total weight lifted in three lifts: the press, the snatch, and the clean-and-jerk.

---

*Figures for Exercises 56 and 57 are from *On Size and Life* by Thomas A. McMahon and John Tyler Bonner. Copyright ©1983 by Thomas A. McMahon and John Tylor Bonner. Reprinted by permission of W. H. Freeman and Company.

The numbers beside each point indicate the body weight class, in pounds.

(a) Find the record for a weight of 150 pounds and for a weight of 165 pounds. (Use base 10 logarithms.) 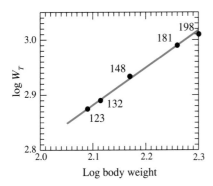 **58.**

(b) Find the body weight that corresponds to a record of 750 pounds.

58. Natural Science In the central Sierra Nevada Mountains of California, the percent of moisture that falls as snow rather than rain is approximated reasonably well by

$$p = 8.63 \ln h - 680,$$

where p is the percent of moisture as snow at an altitude of h feet (with $3000 \le h < 8500$).

(a) Graph p.

(b) At what altitude is 50 percent of the moisture snow?

CHAPTER 4   SUMMARY

KEY TERMS AND SYMBOLS

4.1 exponential function
exponential growth and decay
exponential equation

4.2 the number $e \approx 2.71828 \ldots$
exponential growth function
learning curve
forgetting curve

4.3 $\log x$: common logarithm (base 10 logarithm) of x

$\ln x$: natural logarithm (base e logarithm) of x

$\log_a x$: base a logarithm of x

properties of logarithms

logarithmic functions and their graphs

inverses

4.4 half-life

Richter scale

KEY CONCEPTS

An important application of exponents is the **exponential growth function,** defined as $f(t) = y_0 e^{kt}$, where y_0 is the amount of a quantity present at time $t = 0$, $e \approx 2.71828$, and k is a constant.

The **logarithm** of x to the base a is defined as follows. For $a > 0$ and $a \neq 1$, $y = \log_a x$ means $a^y = x$. Thus, $\log_a x$ is an *exponent,* the power to which a must be raised to produce x.

Properties of Logarithms

Let x, y, and a be positive real numbers, $a \neq 1$, and let r be any real number.

$$\log_a xy = \log_a x + \log_a y \qquad \log_a \frac{x}{y} = \log_a x - \log_a y$$

$$\log_a x^r = r \log_a x \qquad \log_a 1 = 0$$

$$\log_a a = 1 \qquad a^{\log_a x} = x$$

$$\log_a a^y = y$$

Solving Exponential and Logarithmic Equations

Let $a > 0$, $a \neq 1$.
If $a^x = a^y$, then $x = y$ and if $x = y$, then $a^x = a^y$.
If $x = y$, then $\log_a x = \log_a y$, $x > 0$, $y > 0$.
If $\log_a x = \log_a y$, then $x = y$, $x > 0$, $y > 0$.

CHAPTER 4   REVIEW EXERCISES

Solve each of the following equations.

1. $2^{3x} = \dfrac{1}{8}$

2. $\left(\dfrac{9}{16}\right)^x = \dfrac{3}{4}$

3. $9^{2y-1} = 27^y$

4. $\dfrac{1}{2} = \left(\dfrac{b}{4}\right)^{1/4}$

Graph each of the following.

5. $f(x) = 4^x$

6. $g(x) = 4^{-x}$

7. $f(x) = \ln x + 5$

8. $g(x) = \log x - 3$

---

9. Management   A company finds that its new workers produce

$$P(x) = 100 - 100e^{-.8x}$$

items per day, after x days on the job. Find each of the following.
(a) $P(0)$ **(b)** $P(1)$ **(c)** $P(5)$
(d) How many items per day would you expect an experienced worker to produce?

10. Natural Science   The amount of a certain radioactive material, in grams, present after t days, is given by

$$A(t) = 800e^{-.04t}.$$

Find $A(t)$ if
(a) $t = 0$; **(b)** $t = 5$.

Translate each of the following exponential statements into an equivalent logarithmic one.

11. $10^{1.6721} = 47$

12. $5^4 = 625$

13. $e^{3.6636} = 39$

14. $5^{1/2} = \sqrt{5}$

Translate each of the following logarithmic statements into an equivalent exponential one.

15. $\log 1000 = 3$

16. $\log 16.6 = 1.2201$

17. $\ln 95.4 = 4.5581$

18. $\log_2 64 = 6$

Evaluate each expression without using a calculator.

19. $\ln e^3$

20. $\log \sqrt{10}$

21. $10^{\log 7.4}$

22. $\ln e^{4k}$

23. $\log_8 16$

24. $\log_{25} 5$

Use a calculator to find the following to four decimal places.

25. $\log_8 88$

26. $\log_4 100 + \log_{16} 100$

Write each expression as a single logarithm. Assume all variables represent positive quantities.

27. $\log 4k + \log 5k^3$

28. $4 \log x - 2 \log x^3$

29. $2 \log b - 3 \log c$

30. $4 \ln x - 2(\ln x^3 + 4 \ln x)$

Solve each equation. Round to the nearest thousandth.

31. $8^p = 19$

32. $3^z = 11$

33. $5 \cdot 2^{-m} = 35$

34. $2 \cdot 15^{-k} = 18$

35. $e^{-5-2x} = 5$

36. $e^{3x-1} = 12$

37. $10^{2r-3} = 17$

38. $8^{9y-4} = 15$

39. $6^{2-m} = 2^{3m+1}$

40. $5^{3r-1} = 6^{2r+5}$

41. $(1 + .003)^k = 1.089$

42. $(1 + .094)^z = 2.387$

43. $4 \cdot 3^{x^2} = 15$

44. $3 \cdot 5^{p^2} = 28$

45. $\log(m + 2) = 1$

46. $\log x^2 = 2$

47. $\log_2(3k - 2) = 4$

48. $\log_5\left(\dfrac{5z}{z - 2}\right) = 2$

49. $\log x + \log(x + 3) = 1$

50. $\log_2 r + \log_2(r - 2) = 3$

51. $\log(p - 1) = 1 + \log p$

52. $\ln(m + 3) - \ln m = \ln 2$

53. $2 \ln(y + 1) = \ln(y^2 - 1) + \ln 5$

54. $\log_3 k + \log_3(k + 2) = \log_3 8$

55. $3 + 2 \log_4(2x - 1) = 1$

56. $\log_2(1 - 3x) - 4 = 2$

---

57. Social Science   The height, in meters, of the members of a certain tribe is approximated by

$$h = .5 + \log t,$$

where t is the tribe member's age in years, $1 \leq t \leq 20$. Find the height of the tribe member of age
(a) 2 years;   **(b)** 5 years;
(c) 10 years;   **(d)** 20 years.

58. Social Science   The turnover of legislators is a problem of interest to political scientists. One model of legislative turnover in the U.S. House of Representatives is given by

$$M = 434e^{-.08t},$$

where M is the number of continuously serving members at time t.* This model is based on the 1965 mem-

bership of the House. Thus, 1965 corresponds to $t = 0$, 1966 to $t = 1$, and so on. Find the number of continuously serving members in each of the following years.
(a) 1969   **(b)** 1973   **(c)** 1979

59. Natural Science   The amount of polonium (in grams) present after t days is given by

$$A(t) = 10e^{-.00495t}.$$

(a) How much polonium was present initially?
(b) What is the half-life of polonium?
(c) How long will it take for the polonium to decay to 3 grams?

60. Natural Science   One earthquake measures 4.6 on the Richter scale. A second earthquake has ground motion 1000 times greater than the first. What does the second one measure on the Richter scale?

---

*Excerpt from "Exponential Models of Legislative Turnover" by Thomas W. Casstevens. Reprinted by permission of COMAP, Arlington, MA.

The monkeyface prickleback (*Cebidichthys violaceus*), known to anglers as the monkeyface "eel," is found in rocky intertidal and subtidal habitats ranging from San Quintin Bay, Baja California, to Brookings, Oregon. Pricklebacks are prime targets of the few sports anglers who "poke pole" in the rocky intertidal zone at low tide. Little is known about the life history of this species. The results of a study of the length, weight, and age of this species is discussed in this case.

Data on standard length (*SL*) and total length (*TL*) were collected. Early in the study only *TL* was measured, so a conversion to *SL* was necessary. The equation relating the two lengths, calculated from 177 observations for which both lengths had been measured, is

$$SL = TL(.931) + 1.416.$$

Ages (determined by standard aging techniques) were used to estimate parameters of the von Bertalanffy growth model

$$L_t = L_x(1 - e^{-kt}) \qquad \textbf{(1)}$$

where L_t = length at age t,
$\quad L_x$ = asymptotic age of the species,
$\quad k$ = growth completion rate, and
$\quad t_0$ = theoretical age at zero length.

The constants a and b in the model

$$W = aL^b \qquad \textbf{(2)}$$

where W = weight in g,
$\quad L$ = standard length in cm,

were determined using 139 fish ranging from 27 cm and 145 g to 60 cm and 195 g.

Growth curves giving length as a function of age are shown in Figure 1. For the data marked opercle, the lengths were computed from the ages using equation (1).

* Characteristics of the Monkeyface Prickleback, by William H. Marshall and Tina Wyllie Echeverria as appeared in *California Fish & Game,* Vol. 78, Spring 1992, Number 2.

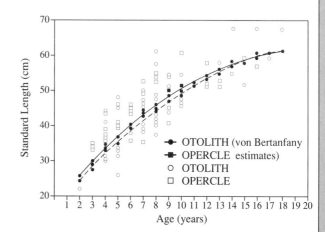

FIGURE 1

Estimated length from equation (1) at a given age was larger for males than females after age eight. See the table. Weight/length relationships found with Equation (2) are shown in Figure 2, along with data from other studies.

Structure /Sex	Age (yr)	Length (cm)	L_x	k	t_0	n
Otolith						
Est.	2–18	23–67	72	.10	−1.89	91
S.D.			8	.03	1.08	
Opercle						
Est.	2–18	23–67	71	.10	−2.63	91
S.D.			8	.04	1.31	
Opercle-Females						
Est.	0–18	15–62	62	.14	−1.95	115
S.D.			2	.02	.28	
Opercle-Males						
Est.	0–18	13–67	70	.12	−1.91	74
S.D.			5	.02	.29	

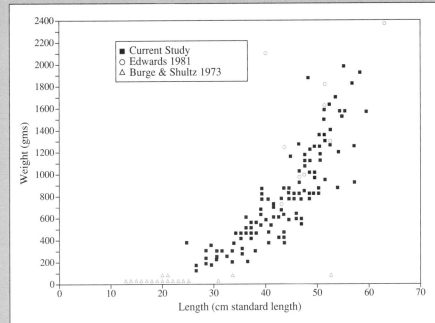

FIGURE 2

EXERCISES

1. Use Equation (1) to estimate the lengths at ages 4, 11, and 17. Let $L_x = 71.5$ and $k = .1$. Compare your answers with the results in Figure 1. What do you find?

2. Use Equation (2) with $a = .01289$ and $b = 2.9$ to estimate the weights for lengths of 25 cm, 40 cm, and 60 cm. Compare with the results in Figure 2. Are your answers reasonable compared to the curve?

After the liberation of Belgium at the end of World War II, officials began a search for Nazi collaborators. One person arrested as a collaborator was a minor painter, H. A. Van Meegeren; he was charged with selling a valuable painting by the Dutch artist Vermeer (1632–75) to the Nazi Hermann Goering. He defended himself from the very serious charge of collaboration by claiming that the painting was a fake—he had forged it himself.

He also claimed that the beautiful and famous painting "Disciples at Emmaus," as well as several other supposed Vermeers, was his own work. To prove this, he did another "Vermeer" in his prison cell. An international panel of experts was assembled, which pronounced as forgeries all the "Vermeers" in question.

Many people would not accept the verdict of this panel for the painting "Disciples at Emmaus"; it was felt to be too beautiful to be the work of a minor talent such as Van Meegeren. In fact, the painting was declared genuine by a noted art scholar and sold for $170,000. The question of the authenticity of this painting continued to trouble art historians, who began to insist on conclusive proof one way or the other. This proof was given by a group of scientists at Carnegie-Mellon University, using the idea of radioactive decay.

The dating of objects is based on radioactivity; the atoms of certain radioactive elements are unstable, and within a given time period a fixed fraction of such atoms will spontaneously disintegrate, forming atoms of a new element. If t_0 represents some initial time, N_0 represents the number of atoms present at time t_0, and N represents the number present at some later time t, then it can be shown (using physics and calculus) that

$$t - t_0 = \frac{1}{\lambda} \cdot \ln \frac{N_0}{N},$$

where λ is a "decay constant" that depends on the radioactive substance under consideration.

If t_0 is the time that the substance was formed or made, then $t - t_0$ is the age of the item. Thus, the age of an item is given by

$$\frac{1}{\lambda} \cdot \ln \frac{N_0}{N}.$$

The decay constant λ can be readily found, as can N, the number of atoms present now. The problem is N_0—we can't find a value for this variable. However, it is possible to get reasonable ranges for the values of N_0. This is done by studying the white lead in the painting. This pigment has been used by artists for over 2000 years. It contains a small amount of the radioactive substance lead-210 and an even smaller amount of radium-226.

Radium-226 disintegrates through a series of intermediate steps to produce lead-210. The lead-210, in turn, decays to form polonium-210. This last process, lead-210 to polonium-210, has a half-life of 22 years. That is, in 22 years half the initial quantity of lead-210 will decay to polonium-210.

When lead ore is processed to form white lead, most of the radium is removed with other waste products. Thus, most of the supply of lead-210 is cut off, with the remainder beginning to decay very rapidly. This process continues until the lead-210 in the white lead is once more in equilibrium with the small amount of radium then present. Let $y(t)$ be the amount of lead-210 per gram of white lead present at time t since manufacture of the pigment. Let r represent the number of disintegrations of radium-226 per minute per gram of white lead. (Actually, r is a function of time, but the half life of radium-226 is so long in comparison to the time interval in question that we assume it to be a constant.) If λ is the decay constant for lead-210, then it can be shown that

$$y(t) = \frac{r}{\lambda}\left[1 - e^{-\lambda(t - t_0)}\right] + y_0 e^{-\lambda(t - t_0)}. \tag{1}$$

*From "The Van Meegeren Art Forgeries" by Martin Braun from *Applied Mathematical Sciences,* Vol. 15. Copyright ©1975. Published by Springer-Verlag New York, Inc. Reprinted by permission.

All variables in this result can be evaluated except y_0, the amount of lead-210 present originally. To get around this problem, we use the fact that the original amount of lead-210 was in radioactive equilibrium with the larger amount of radium-226 in the ore from which the metal was extracted. We therefore take samples of different ores and compute the rate of disintegration of radium-226. The results are as shown in the table.

Location of Ore	Disintegrations of Radium-226 per Minute per gram of White Lead
Oklahoma	4.5
S.E. Missouri	.7
Idaho	.18
Idaho	2.2
Washington	140.0
British Columbia	.4

The numbers in the table vary from .18 to 140—quite a range. Since the number of disintegrations is proportional to the amount of lead-210 present originally, we must conclude that y_0 also varies over a tremendous range. Thus, equation (1) cannot be used to obtain even a crude estimate of the age of a painting. However, we want to distinguish only between a modern forgery and a genuine painting that would be 300 years old.

To do this, we observe that if the painting is very old compared to the 22 year half-life of lead-210, then the amount of radioactivity from the lead-210 will almost equal the amount of radioactivity from the radium-226. On the other hand, if the painting is modern, then the amount of radioactivity from the lead-210 will be much greater than the amount from the radium 226.

We want to know if the painting is modern or 300 years old. To find out, let $t - t_0 = 300$ in equation (1), getting

$$\lambda y_0 = \lambda \cdot y(t) \cdot e^{300\lambda} - r(e^{300\lambda} - 1) \tag{2}$$

after some rearrangement of terms. If the painting is modern, then λy_0 should be a very large number; λy_0 represents the number of disintegrations of the lead-210 per minute per gram of white lead at the time of manufacture. By studying samples of white lead, we can conclude that λy_0 should never be anywhere near as large as 30,000. Thus, we use equation (2) to calculate λy_0; if our result is greater than 30,000 we conclude that the painting is a modern forgery. The details of this calculation are left for the exercises.

EXERCISES

1. To calculate λ, use the formula

$$\lambda = \frac{\ln 2}{\text{half-life}}.$$

Find λ for lead-210, whose half-life is 22 years.

2. For the painting "Disciples at Emmaus," the current rate of disintegration of the lead-210 was measured and found to be $\lambda \cdot y(t) = 8.5$. Also, r was found to be .8. Use this information and equation (2) to calculate λy_0. Based on your results, what do you conclude about the age of the painting?

The table below lists several other possible forgeries. Decide which of them must be modern forgeries.

Title	$\lambda \cdot y(t)$	r
3. "Washing of Feet"	12.6	.26
4. "Lace Maker"	1.5	.4
5. "Laughing Girl"	5.2	.6
6. "Woman Reading Music"	10.3	.3

CHAPTER 5

Mathematics of Finance

TECHNOLOGY RESOURCES
*Explorations in Finite
Mathematics,* Schneider

*Topics in Finite Mathematics:
An Introduction to the
Electronic Spreadsheet,*
Spero

Whether you are in a position to invest money or to borrow money, it is important for both business managers and consumers to understand *interest*. The formulas for interest are developed in this chapter. A calculator with a y^x key will be useful throughout the chapter.

5.1 SIMPLE INTEREST AND DISCOUNT

Interest is the fee paid for the use of someone else's money. For example, you might pay interest to a bank for money you borrow or the bank might pay you interest on the money in your savings account. The amount of money that is borrowed or deposited is called the **principal.** The fee paid as interest depends on the interest **rate** and the length of **time** for which you have the use of the money. Unless stated otherwise, the time, t, is measured in years and the interest rate, r, is in percent per year, expressed as a decimal; for instance, $8\% = .08$ or $9.5\% = .095$.

There are two common ways of computing interest, the first of which we study in this section. **Simple interest** is interest paid only on the amount deposited and not on past interest.

> The **simple interest,** I, for t years on an amount of P dollars at a rate of interest r per year for t years is
>
> $$I = Prt.$$

d

1 Find the simple interest for the following.

(a) $1000 at 8% for 2 years

(b) $5500 at 10.5% for $1\frac{1}{2}$ years

Answers:

(a) $160

(b) $866.25

2 Find the maturity value of each loan.

(a) $10,000 at 10% for 6 months

(b) $8970 at 11% for 9 months

(c) $95,106 at 9.8% for 76 days

Answers:

(a) $10,500

(b) $9710.03

(c) $97,073.64

▶ **EXAMPLE 1** To buy furniture for a new apartment, Fontaine Evaldo borrowed $5000 at 11% simple interest for 11 months. How much interest will she pay?

From the formula, $I = Prt$, with $P = 5000$, $r = .11$, and $t = 11/12$ (in years). The total interest she will pay is

$$I = 5000(.11)(11/12) = 504.17,$$

or $504.17. ◀ **1**

A deposit of P dollars today at a rate of interest r for t years produces interest of $I = Prt$. The interest, added to the original principal P, gives

$$P + Prt = P(1 + rt).$$

This amount is called the **future value** of P dollars at an interest rate r for time t in years. When loans are involved, the future value is often called the **maturity value** of the loan. This idea is summarized as follows.

The **future value** or **maturity value,** A, of P dollars for t years all at a rate of interest r per year is

$$A = P(1 + rt).$$

▶ **EXAMPLE 2** Find the maturity value for each of the following loans at simple interest.

(a) A loan of $2500 to be repaid in 8 months with interest of 12.1%

The loan is for 8 months, or $8/12 = 2/3$ of a year. The maturity value is

$$A = P(1 + rt)$$

$$A = 2500\left[1 + .121\left(\frac{2}{3}\right)\right]$$

$$\approx 2500[1 + .08067] \approx 2701.67,$$

or $2701.67. (Interest is rounded to the nearest cent, as is customary in financial problems). Because the maturity value is the sum of principal and interest, the interest paid on this loan is

$$\$2701.67 - \$2500 = \$201.67.$$

(b) A loan of $11,280 for 85 days at 11% interest

It is common to assume 360 days in a year when working with simple interest. We shall usually make such an assumption in this book. The maturity value in this example is

$$A = 11,280\left[1 + .11\left(\frac{85}{360}\right)\right] \approx 11,280[1.0259722] = 11,572.97,$$

or $11,572.97. ◀ **2**

3 Find the present value of the following future amounts. Assume 12% interest.

(a) $7500 in 1 year

(b) $89,000 in 5 months

(c) $164,200 in 125 days

Answers:

(a) $6696.43

(b) $84,761.90

(c) $157,632.00

Part (b) of Example 2 assumed 360 days in a year. Interest found using 360 days is called **ordinary interest,** while interest found using 365 days is **exact interest.**

PRESENT VALUE   A sum of money that can be deposited today to yield some larger amount in the future is called the **present value** of that future amount. Present value refers to the principal to be invested or loaned, so we use the same variable P as we did for principal. In interest problems, P always represents the amount at the beginning of the time period, and A always represents the amount at the end of the time period. To find a formula for P, we begin with the future value formula

$$A = P(1 + rt).$$

Dividing each side by $1 + rt$ gives the following formula for present value.

$$P = \frac{A}{1 + rt}$$

The **present value** P of a future amount of A dollars at a simple interest rate r for t years is

$$P = \frac{A}{1 + rt}.$$

▶**EXAMPLE 3**   Find the present value of $32,000 in 4 months at 9% interest.

$$P = \frac{32,000}{1 + (.09)\left(\frac{4}{12}\right)} = \frac{32,000}{1.03} = 31,067.96$$

A deposit of $31,067.96 today, at 9% interest, would produce $32,000 in 4 months. These two sums, $31,067.96 today and $32,000.00 in 4 months, are equivalent (at 9%) because the first amount becomes the second amount in 4 months.   ◀ **3**

▶**EXAMPLE 4**   Because of a court settlement, Charlie Dawkins owes $5000 to Arnold Parker. The money must be paid in 10 months, with no interest. Suppose Dawkins wishes to pay the money today. What amount should Parker be willing to accept? Assume an interest rate of 5%.

4 Curt Reynolds is owed $19,500 by Cyndi Keen. The money will be paid in 11 months, with no interest. If the current interest rate is 10%, how much should Reynolds be willing to accept today in settlement of the debt?

Answer:

$17,862.60

The amount that Parker should be willing to accept is given by the present value:

$$P = \frac{5000}{1 + (.05)\left(\frac{10}{12}\right)} = 4800.00$$

Parker should be willing to accept $4800.00 today in settlement of the obligation. ◀ **4**

▶**EXAMPLE 5** Suppose you borrow $40,000 today and are required to pay $41,400 in 4 months to pay off the loan and interest. What is the simple interest rate?

We can use the future value formula, with $P = 40,000$, $A = 41,400$, and $t = 4/12 = 1/3$, and solve for r.

$$A = P(1 + rt).$$

$$41,400 = 40,000\left(1 + r \cdot \frac{1}{3}\right)$$

$$41,400 = 40,000 + \frac{40,000r}{3}$$

$$1400 = \frac{40,000r}{3}$$

$$40,000r = 3 \cdot 1400 = 4200$$

$$r = \frac{4200}{40,000} = .105$$

Therefore, the interest rate is 10.5%. ◀

SIMPLE DISCOUNT NOTES   The loans discussed up to this point are called **simple interest notes,** where interest on the face value of the loan is added to the loan and paid at maturity. Another common type of note, called a **simple discount note,** has the interest deducted in advance from the amount of a loan before giving the *balance* to the borrower. The *full* value of the note must be paid at maturity. The money that is deducted is called the **bank discount** or just the **discount,** and the money actually received by the borrower is called the **proceeds.**

For example, consider a loan of $3000 at 6% interest for 9 months. We can compare the two types of loan arrangements as follows.

	Simple Interest Note	Bank Discount Note
Interest on the note	3000(.06)(9/12) = $135	3000(.06)(9/12) = $135
Borrower receives	$3000	$2865
Borrower pays back	$3135	$3000

5 Kelly Bell signs an agreement at her bank to pay the bank $25,000 in 5 months. The bank charges a 13% discount rate. Find the amount of the discount and the amount Bell actually receives.

Answer:

$1354.17; $23,645.83

6 Refer to Problem 5 at the side above, and find the actual rate of interest paid by Bell.

Answer:

$13.7% (to the nearest tenth)

▶ **EXAMPLE 6**   Theresa DePalo needs a loan from her bank and agrees to pay $8500 to her banker in 9 months. The banker subtracts a discount of 12% and gives the balance to DePalo. Find the amount of the discount and the proceeds.

As shown above, the discount is found in the same way that simple interest is found, except that it is based on the amount to be repaid.

$$\text{Discount} = 8500(.12)\left(\frac{9}{12}\right) = 765.00$$

The proceeds are found by subtracting the discount from the original amount.

$$\text{Proceeds} = \$8500 - \$765.00 = \$7735.00 \quad ◀ \quad \boxed{5}$$

In Example 6, the borrower was charged a discount of 12%. However, 12% is *not* the interest rate paid, since 12% applies to the $8500, while the borrower actually received only $7735. In the next example, we find the rate of interest actually paid by the borrower.

▶ **EXAMPLE 7**   Find the actual rate of interest paid by DePalo in Example 6.
Use the formula for simple interest, $I = Prt$, with r the unknown. Since the borrower received only $7735 and must repay $8500, $I = 8500 - 7735 = 765$. Here, $P = 7735$ and $t = 9/12 = .75$. Substitute these values into $I = Prt$.

$$I = Prt$$
$$765 = 7735(r)(.75)$$
$$\frac{765}{7735(.75)} = r$$
$$.132 \approx r$$

The actual interest rate paid by the borrower is about 13.2%.   ◀ $\boxed{6}$

Let D represent the amount of discount on a loan. Then $D = Art$, where A is the maturity value of the loan (the amount borrowed plus interest), and r is the stated rate of interest. The amount actually received, the proceeds, can be written as $P = A - D$, or $P = A - Art = A(1 - rt)$.
The formulas for discount are summarized below.

Discount

If D is the discount on a loan having a maturity value A at a rate of interest r for t years, and if P represents the proceeds, then

$$P = A - D \quad \text{or} \quad P = A(1 - rt).$$

7 A firm accepts a $21,000 note due in 7 months with interest of 10.5%. Suppose the firm discounts the note at a bank 75 days before it is due. Find the amount the firm would receive if the bank charges a 12.4% discount rate. (Use 360 days in a year.)

Answer:

$21,710.52

▶ **EXAMPLE 8** John Young owes $4250 to Meg Holden. The loan is payable in 1 year at 10% interest. Holden needs cash to buy a new car, so 3 months before the loan is payable she goes to the bank to have the loan discounted. The bank charges an 11% discount fee. Find the amount of cash she will receive from the bank.

First find the maturity value of the loan, the amount (with interest) Young must pay Holden. By the formula for maturity value,

$$A = P(1 + rt)$$
$$= 4250[1 + (.10)(1)]$$
$$= 4250(1.10) = 4675$$

or $4675.

The bank applies its discount rate to this total:

$$\text{Amount of discount} = 4675(.11)(3/12) = 128.56.$$

(Remember that the loan was discounted 3 months before it was due.) Holden actually receives

$$\$4675 - \$128.56 = \$4546.44$$

in cash from the bank. Three months later, the bank will get $4675.00 from Young. ◀ **7**

5.1 EXERCISES

1. What factors determine the amount of interest earned on a fixed principal?

Find the simple interest in Exercises 2–5. (See Example 1.)

2. $25,000 at 7% for 9 mo

3. $3850 at 9% for 8 mo

4. $1974 at 6.3% for 7 mo

5. $3724 at 8.4% for 11 mo

Find the simple interest. Assume a 360-day year and a 30-day month. (See Example 2(b).)

6. $5147.18 at 10.1% for 58 days

7. $2930.42 at 11.9% for 123 days

8. $7980 at 10%; loan made on May 7 and due September 19

9. $5408 at 12%; loan made on August 16 and due December 30

Find the simple interest. Assume 365 days in a year, and use the exact number of days in a month. (Assume 28 days in February.)

10. $7800 at 11%; made on July 7 and due October 25

11. $11,000 at 10%; made on February 19 and due May 31

12. $2579 at 9.6%; made on October 4 and due March 15

13. $37,098 at 11.2%; made on September 12 and due July 30

14. In your own words, describe the *maturity value* of a loan.

15. What is meant by the *present value* of money?

Find the present value of each of the future amounts in Exercises 16–19. Assume 360 days in a year. (See Example 3.)

16. $15,000 for 8 mo; money earns 6%

17. $48,000 for 9 mo; money earns 5%

18. $15,402 for 125 days; money earns 6.3%

19. $29,764 for 310 days; money earns 7.2%

Find the proceeds for the amounts in Exercises 20–23. Assume 360 days in a year. (See Example 6.)

20. $7150; discount rate 12%; length of loan 11 mo

21. $9450; discount rate 10%; length of loan 7 mo

22. $35,800; discount rate 9.1%; length of loan 183 days

23. $50,900; discount rate 8.2%; length of loan 238 days

24. Why is the discount rate charged on a simple discount note different from the actual interest rate paid on the proceeds?

Find the interest rate to the nearest tenth on the proceeds for the following simple discount notes. (See Example 7.)

25. $6200; discount rate 10%; length of loan 8 mo

26. $5000; discount rate 8.1%; length of loan 6 mo

27. $58,000; discount rate 10.8%; length of loan 9 mo

28. $43,000; discount rate 9%; length of loan 4 mo

Management   *Work the following applied problems.*

29. Michelle Beese borrowed $25,900 from her father to start a flower shop. She repaid him after 11 mo, with interest of 8.4%. Find the total amount she repaid.

30. An accountant for a corporation forgot to pay the firm's income tax of $725,896.15 on time. The government charged a penalty of 12.7% interest for the 34 days the money was late. Find the total amount (tax and penalty) that was paid. (Use a 365-day year.)

31. A $100,000 certificate of deposit held for 60 days is worth $101,133.33. To the nearest tenth of a percent, what interest rate was earned?

32. Tuition of $1769 will be due when the spring term begins in 4 mo. What amount should a student deposit today, at 6.25%, to have enough to pay the tuition?

33. A firm of accountants has ordered 7 new IBM computers at a cost of $5104 each. The machines will not be delivered for 7 mo. What amount could the firm deposit in an account paying 6.42% to have enough to pay for the machines?

34. Sun Kang needs $5196 to pay for remodeling work on his house. He plans to repay the loan in 10 mo. His bank loans money at a discount rate of 13%. Find the amount of his loan.

35. Joan McKee decides to go back to college. To get to school, she buys a small car for $6100. She decides to borrow the money from a bank that charges an 11.8% discount rate. If she will repay the loan in 7 months, find the amount of the loan.

36. John Matthews signs a $4200 note at the bank. The bank charges a 12.2% discount rate. Find the net proceeds if the note is for 10 mo. Find the actual interest rate (to the nearest hundredth) charged by the bank.

37. A stock that sold for $22 at the beginning of the year was selling for $24 at the end of the year. If the stock paid a dividend of $.50 per share, what is the simple interest rate on an investment in this stock? (*Hint:* Consider the interest to be the increase in value plus the dividend.)

38. A bond with a face value of $10,000 in 10 yr can be purchased now for $5988.02. What is the simple interest rate?

39. A building contractor gives a $13,500 note to a plumber. The note is due in 9 mo, with interest of 9%. Three months after the note is signed, the plumber discounts it at the bank. The bank charges a 10.1% discount rate. How much will the plumber receive? Will it be enough to pay a bill for $13,582?

40. Maria Lopez owes $7000 to the Eastside Music Shop. She has agreed to pay the amount in 7 mo at an interest rate of 10%. Two months before the loan is due, the store needs $7350 to pay a wholesaler's bill. The bank will discount the note at a rate of 10.5%. How much will the store receive? Is it enough to pay the bill?

41. Fay, Inc., received a $30,000, six-month, 12% interest-bearing note from a customer.* The note was discounted the same day by Carr National Bank at 15%. The amount of cash received by Fay from the bank was
   a. $30,000
   b. $29,550
   c. $29,415
   d. $27,750

---

* Uniform CPA Examination, May, 1989, American Institute of Certified Public Accountants.

1 Use the formula

$$A = P(1 + rt)$$

to find the amount in the account after 2 years at 5% simple interest.

Answer:

$1100

5.2 COMPOUND INTEREST

Simple interest is normally used for loans or investments of a year or less. For longer periods **compound interest** is used. With **compound interest,** interest is charged (or paid) on interest as well as on principal. For example, if $1000 is deposited at 5% compound interest, then the interest for the first year is $1000(.05) = $50, just as with simple interest, so that the account balance is $1050 at the end of the year. During the second year, interest is paid on the entire $1050 (not just on the original $1000 as with simple interest), so the amount in the account at the end of the second year is $1050 + $1050(.05) = $1102.50. This is more than simple interest would produce. **1**

To find a formula for compound interest, suppose that P dollars are deposited at interest rate r per year. The amount A on deposit after 1 year is found by the simple interest formula, with $t = 1$.

$$A = P[1 + r(1)] = P(1 + r)$$

If the deposit earns compound interest, the interest for the second year is paid on the total amount on deposit at the end of the first year, $P(1 + r)$. Using the formula $A = P(1 + rt)$ again, with $P = P(1 + r)$ and $t = 1$ gives the total amount on deposit at the end of the second year.

$$A = [P(1 + r)](1 + r \cdot 1) = P(1 + r)^2$$

In the same way, the total amount on deposit at the end of the third year is

$$P(1 + r)^3.$$

Continuing in this way, the total amount on deposit after t years is

$$A = P(1 + r)^t,$$

called the **compound amount.**

Note   Compare this formula for compound interest with the formula for simple interest from the previous section.

$$\text{Compound interest } A = P(1 + r)^t$$

$$\text{Simple interest } A = P(1 + rt)$$

The important distinction between the two formulas is that in the compound interest formula, the number of years t is an *exponent,* so that money grows much more rapidly when interest is compounded.

Interest can be compounded more than once a year. Common **compounding periods** include *semiannually* (two periods per year), *quarterly* (four periods per year), *monthly* (twelve periods) and *daily* (usually 365 periods per year.) To find the *interest rate per period, i,* we *divide* the annual interest rate r by the

2 Suppose $17,000 is deposited at 4% compounded semiannually for 11 years.

(a) Find the compound amount.

(b) Find the amount of interest earned.

Answers:

(a) $26,281.65

(b) $9,281.65

3 Find the compound amount.

(a) $10,000 at 8% compounded quarterly for 7 years

(b) $36,000 at 6% compounded monthly for 2 years

Answers:

(a) $17,410.24

(b) $40,577.75

number of compounding periods per year. The total number of compounding periods, n, is found by *multiplying* the number of years t by the number of compounding periods per year. Then the general formula can be derived in much the same way as the formula given above.

If P dollars are deposited for n compounding periods at a rate of interest i per period, the **compound amount** A is

$$A = P(1 + i)^n.$$

▶**EXAMPLE 1** Suppose $1000 is deposited for 6 years in an account paying 8% per year compounded semiannually.
(a) Find the compound amount.
  In the formula above $P = 1000$, $i = 8\%/2 = .04$, and $n = 6 \cdot 2 = 12$. The compound amount is

$$A = P(1 + i)^n = 1000(1.04)^{12}.$$

The quantity $(1.04)^{12}$ can be found with a calculator, as shown in Section 4.1 : $(1.04)^{12} \approx 1.601032$ and so

$$A = \$1000(1.601032) = \$1601.03.$$

(b) Find the actual amount of interest earned.
  Subtract the initial deposit from the compound amount.

$$\text{Amount of interest} = \$1601.03 - \$1000 = \$601.03 \quad ◀ \quad \boxed{2}$$

▶**EXAMPLE 2** Find the amount of interest earned by a deposit of $1000 for 6 years at 6% compounded quarterly.
  Interest compounded quarterly is compounded four times a year. In 6 years there are $4 \cdot 6 = 24$ quarters, or 24 periods. Thus $n = 24$. Interest of 6% per year is $6\%/4$, or 1.5%, per quarter, so $i = .015$. The compound amount is

$$1000(1 + .015)^{24} = 1000(1.015)^{24}.$$

Using a calculator,

$$A = 1000(1.015)^{24} = 1429.50.$$

The compound amount is $1429.50 and the interest earned is $1429.50 − $1000 = $429.50. ◀ **3**

  The more often interest is compounded within a given time period, the more interest will be earned. Surprisingly, however, there is a limit on the amount of interest, no matter how often it is compounded. To see this, suppose that $1 is invested at 100% interest per year, compounded n times per year. Then the interest rate (in decimal form) is 1.00 and the interest rate per period is $1/n$. According to the formula (with $P = 1$), the compound amount at the end of 1 year will be $A = \left(1 + \dfrac{1}{n}\right)^n$. Using a computer with double precision accuracy gives the following results for various values of n.

Interest Is Compounded	n	$\left(1 + \dfrac{1}{n}\right)^n$
Annually	1	$\left(1 + \dfrac{1}{1}\right)^1 = 2$
Semiannually	2	$\left(1 + \dfrac{1}{2}\right)^2 = 2.25$
Quarterly	4	$\left(1 + \dfrac{1}{4}\right)^4 \approx 2.4414$
Monthly	12	$\left(1 + \dfrac{1}{12}\right)^{12} \approx 2.6130$
Daily	365	$\left(1 + \dfrac{1}{365}\right)^{365} \approx 2.71457$
Hourly	8760	$\left(1 + \dfrac{1}{8760}\right)^{8760} \approx 2.718127$
Every minute	525,600	$\left(1 + \dfrac{1}{525,600}\right)^{525,600} \approx 2.7182792$
Every second	31,536,000	$\left(1 + \dfrac{1}{31,536,000}\right)^{31,536,000} \approx 2.7182818$

Because interest is rounded to the nearest penny, the compound amount never exceeds $2.72, no matter how big n is.

The table above suggests that as n takes larger and larger values, then the corresponding values of $\left(1 + \dfrac{1}{n}\right)^n$ get closer and closer to a specific real number, whose decimal expansion begins $2.71828 \cdots$. This is indeed the case, as is shown in calculus, and the number $2.71828 \cdots$ is denoted e.*

The preceding example is typical of what happens when interest is compounded n times per year, with larger and larger values of n. It can be shown that no matter what interest rate or principal is used, there is always an upper limit on the compound amount, which is called the compound amount from **continuous compounding.** Not surprisingly, it involves the number e.

---

*Applications of the exponential function $f(x) = e^x$ are discussed in Section 4.2.

4 Find the compound amount, assuming continuous compounding.

(a) $12,000 at 10% for 5 years

(b) $22,867 at 7.2% for 9 years

Answers:

(a) $19,784.66

(b) $43,732.36

5 Find the effective rate corresponding to a nominal rate of

(a) 12% compounded monthly;

(b) 8% compounded quarterly.

Answers:

(a) 12.68%

(b) 8.24%

Continuous Compounding

The compound amount A for a deposit of P dollars at interest rate r per year compounded continuously for t years is given by

$$A = Pe^{rt}.$$

Many calculators have an e^x key for computing powers of e. See Section 4.2 for more details on using a calculator to evaluate e^x.

▶ **EXAMPLE 3** Suppose $5000 is invested at an annual rate of 4% compounded continuously for 5 years. Find the compound amount.

In the formula for continuous compounding, let $P = 5000$, $r = .04$ and $t = 5$. Then a calculator with an e^x key shows that

$$A = 5000e^{(.04)5} = 5000e^{.2} = \$6107.01$$

You can readily verify that daily compounding would have produced a compound amount about 6¢ less. ◀ **4**

EFFECTIVE RATE   If $1 is deposited at 4% compounded quarterly, a calculator can be used to find that at the end of one year, the compound amount is $1.0406, an increase of 4.06% over the original $1. The actual increase of 4.06% in the money is somewhat higher than the stated increase of 4%. To differentiate between these two numbers, 4% is called the **nominal** or **stated rate** of interest, while 4.06% is called the **effective rate.*** To avoid confusion between stated rates and effective rates, we shall continue to use r for the stated rate and we will use r_e for the effective rate.

▶ **EXAMPLE 4**   Find the effective rate corresponding to a nominal rate of 6% compounded semiannually.

A calculator shows that $100 at 6% compounded semiannually will grow to

$$A = 100\left(1 + \frac{.06}{2}\right)^2 = 100(1.03)^2 = \$106.09.$$

Thus, the actual amount of compound interest is $106.09 − $100 = $6.09. Now if you earn $6.09 interest on $100 in 1 year with annual compounding, your rate is $6.09/100 = .0609 = 6.09\%$. Thus, the effective rate is $r_e = 6.09\%$. ◀ **5**

In the preceding example we found the effective rate by dividing compound interest for 1 year by the original principal. The same thing can be done with any principal P and rate r compounded m times per year.

_____

* When applied to consumer finance, the effective rate is called the annual percentage rate, or APR.

6 Find the effective rate corresponding to a nominal rate of

(a) 15% compounded monthly;

(b) 10% compounded quarterly.

Answers:

(a) 16.08%

(b) 10.38%

$$\text{Effective rate} = \frac{\text{compound interest}}{\text{principal}}$$

$$r_e = \frac{\text{compound amount} - \text{principal}}{\text{principal}}$$

$$= \frac{P\left(1 + \frac{r}{m}\right)^m - P}{P} = \frac{P\left[\left(1 + \frac{r}{m}\right)^m - 1\right]}{P}$$

$$r_e = \left(1 + \frac{r}{m}\right)^m - 1.$$

The **effective rate** corresponding to a stated rate of interest r per year compounded m times per year is

$$r_e = \left(1 + \frac{r}{m}\right)^m - 1.$$

▶**EXAMPLE 5**   Find the effective rate of a loan with interest of 9% compounded monthly.

Use the formula in the box, with $r = .09$ and $m = 12$. The effective rate is

$$r_e = \left(1 + \frac{.09}{12}\right)^{12} - 1 = (1.0075)^{12} - 1 = .0938,$$

or 9.38%.   ◀ **6**

▶**EXAMPLE 6**   Bank A is now lending money at 13.2% interest compounded annually. The rate at Bank B is 12.6% compounded monthly and the rate at Bank C is 12.7% compounded quarterly. If you need to borrow money, at which bank will you pay the least interest?

Compare the effective rates.

$$\text{Bank A: } \left(1 + \frac{.132}{1}\right)^1 - 1 = .132 = 13.2\%$$

$$\text{Bank B: } \left(1 + \frac{.126}{12}\right)^{12} - 1 \approx .13354 = 13.354\%$$

$$\text{Bank C: } \left(1 + \frac{.127}{4}\right)^4 - 1 \approx .13318 = 13.318\%$$

The lowest effective interest rate is at Bank A, which has the highest nominal rate.   ◀

7 Find P in Example 7 if the interest rate is

(a) 6%;

(b) 10%.

Answers:

(a) $4483.55

(b) $3725.53

PRESENT VALUE WITH COMPOUND INTEREST The formula for compound interest, $A = P(1 + i)^n$, has four variables, A, P, i, and n. Given the values of any three of these variables, the value of the fourth can be found. In particular, if A (the future amount), i, and n are known, then P can be found. Here P is the amount that should be deposited today to produce A dollars in n periods.

▶ **EXAMPLE 7** Joan Nakamura must pay a lump sum of $6000 in 5 years. What amount deposited today at 4% compounded annually will grow to $6000 in 5 years?

Here $A = 6000$, $i = .04$, $n = 5$, and P is unknown. Substituting these values into the formula for the compound amount gives

$$6000 = P(1.04)^5$$

and

$$P = \frac{6000}{(1.04)^5} \approx \$4931.56$$

If Nakamura deposits $4931.56 for 5 years in an account paying 4% interest compounded annually, she will have $6000 when she needs it. ◀ **7**

As the last example shows, $6000 in 5 years is the same as $4931.56 today (if money can be deposited at 4% compounded annually). Recall from the first section that an amount that can be deposited today to yield a given sum in the future is called the *present value* of this future sum.

Generalizing from Example 7, we get this result.

The **present value** of A dollars compounded at an interest rate i per period for n periods is

$$P = \frac{A}{(1 + i)^n} \quad \text{or} \quad P = A(1 + i)^{-n}.$$

Compare this with the present value of an amount at simple interest r for t years, given in the previous section:

$$P = \frac{A}{1 + rt}.$$

▶ **EXAMPLE 8** Find the present value of $16,000 in 9 years if money can be deposited at 6% compounded semiannually.

In 9 years there are $2 \cdot 9 = 18$ semiannual periods. A rate of 6% per year is 3% in each semiannual period. Apply the formula with $A = 16,000$, $i = .03$, and $n = 18$.

8 Find the present value in Example 8 if money is deposited at 10% compounded semiannually.

Answer:

$6648.33

9 Using a calculator, estimate the number of years it will take for $500 to increase to $750 in an account paying 6% interest compounded semiannually.

Answer:

About 7 years

$$P = \frac{A}{(1 + i)^n} = \frac{16{,}000}{(1 + .03)^{18}} \approx \frac{16{,}000}{1.702433} \approx 9398.31$$

A deposit of $9398.31 today, at 6% compounded semiannually, will produce a total of $16,000 in 9 years. ◄ **8**

The formula for compound amount can also be solved for n.

▶**EXAMPLE 9**   Suppose the general level of inflation in the economy averages 8% per year. Find the number of years it would take for the overall level of prices to double.

To find the number of years it will take for $1 worth of goods or services to cost $2, find n in the equation

$$2 = 1(1 + .08)^n,$$

where $A = 2$, $P = 1$, and $i = .08$. This equation simplifies to

$$2 = (1.08)^n.$$

By trying various values of n, we find that $n = 9$ is approximately correct, because $1.08^9 = 1.99900 \approx 2$. The exact value of n can be found quickly by using logarithms as in Chapter 4. ◄ **9**

When interest is compounded continuously, the present value can be found by solving the continuous compounding formula $A = Pe^{rt}$ for P.

▶**EXAMPLE 10**   How much must be deposited today in an account paying 7.5% interest compounded continuously in order to have $10,000 in 4 years?

Here the future value is $A = 10{,}000$, the interest rate is $r = .075$, and the number of years is $t = 4$. The present value P is found as follows.

$$A = Pe^{rt}$$
$$10{,}000 = Pe^{(.075)4}$$
$$10{,}000 = Pe^{.3}$$
$$P = \frac{10{,}000}{e^{.3}} \approx \$7408.18 \quad ◄$$

At this point, it seems helpful to summarize the notation and the most important formulas for simple and compound interest. We use the following variables.

P = principal or present value
A = future or maturity value
r = annual (stated or nominal) interest rate
t = number of years

m = number of compounding periods per year
i = interest rate per period ($i = r/m$)
r_e = effective rate
n = total number of compounding periods ($n = tm$)

Simple Interest

$$A = P(1 + rt)$$

$$P = \frac{A}{1 + rt}$$

Compound Interest

$$A = P(1 + i)^n$$

$$P = \frac{A}{(1 + i)^n} = A(1 + i)^{-n}$$

$$r_e = \left(1 + \frac{r}{m}\right)^m - 1$$

5.2 EXERCISES

1. Explain the difference between simple interest and compound interest.

2. As the number of compounding periods per year increases, what happens to the interest for the year?

Find the compound amount for each of the following deposits. (See Examples 1–2.)

3. $1000 at 6% compounded annually for 8 yr

4. $1000 at 7% compounded annually for 10 yr

5. $470 at 10% compounded semiannually for 12 yr

6. $15,000 at 6% compounded semiannually for 11 yr

7. $6500 at 12% compounded quarterly for 6 yr

8. $9100 at 8% compounded quarterly for 4 yr

Find the amount of interest earned by each of the following deposits. (See Examples 1–2.)

9. $26,000 at 7% compounded annually for 5 years

10. $32,000 at 5% compounded annually for 10 years

11. $8000 at 4% compounded semiannually for 6.4 years

12. $2500 at 4.5% compounded semiannually for 8 years

13. $5124.98 at 6.3% compounded quarterly for 5.2 years

14. $27,630.35 at 7.1% compounded quarterly for 3.7 years

Find the compound amount if $25,000 is invested at 6% compounded continuously for the following number of years. (See Example 3.)

15. 1

16. 5

17. 10

18. 15

19. How do the nominal or stated interest rate and the effective interest rate differ?

20. If interest is compounded more than once per year, which rate is higher, the stated rate or the effective rate?

Find the effective rate corresponding to the following nominal rates. (See Examples 4–6.)

21. 4% compounded semiannually

22. 8% compounded quarterly

23. 8% compounded semiannually

24. 10% compounded semiannually

25. 12% compounded semiannually

26. 12% compounded quarterly

Find the present value of the following amounts. (See Examples 7 and 8.)

27. $12,000 at 5% compounded annually for 6 years

28. $8500 at 6% compounded annually for 9 years

29. $4253.91 at 6.8% compounded semiannually for 4 years

30. $27,692.53 at 4.6% compounded semiannually for 5 years

31. $17,230 at 4% compounded quarterly for 10 years

32. $5240 at 8% compounded quarterly for 8 years

33. If money can be invested at 8% compounded quarterly, which is larger: $1000 now or $1210 in 5 years?

34. If money can be invested at 6% compounded annually, which is larger: $10,000 now or $15,000 in 10 years?

Find the present value of $17,200 for the following number of years, if money can be deposited at 11.4% compounded continuously.

35. 2   **36.** 4   **37.** 7   **38.** 10

Under certain conditions, Swiss banks pay negative interest—they charge you. (You didn't think all that secrecy was free?) Suppose a bank "pays" −2.4% interest compounded annually. Use a calculator and find the compound amount for a deposit of $150,000 after the following.

39. 2 years   **40.** 4 years   **41.** 8 years   **42.** 12 years

Management *Work the following applied problems.*

43. When Lindsay Branson was born, her grandfather made an initial deposit of $3000 in an account for her college education. Assuming an interest rate of 6% compounded quarterly, how much will the account be worth in 18 yr?

44. Frank Capek has $10,000 in an Individual Retirement Account (IRA). Because of new tax laws, he decides to make no further deposits. The account earns 6% interest compounded semiannually. Find the amount on deposit in 15 yr.

45. Suppose the average price of a house nationally is $78,000. Housing prices are increasing at a rate of 3% per year. Assuming prices continue to increase at the same rate, what will be the average price of a house in 12 yr?

46. Natural gas rates are rising by 5% per year. If the average monthly bill is now $52, what will the average monthly bill be in 10 yr?

47. A small business borrows $50,000 for expansion at 12% compounded monthly. The loan is due in 4 yr. How much interest will the business pay?

48. A developer needs $80,000 to buy land. He is able to borrow the money at 10% per year compounded quarterly. How much will the interest amount to if he pays off the loan in 5 yr?

49. A company has agreed to pay $2.9 million in 5 yr to settle a lawsuit. How much must they invest now in an account paying 8% compounded monthly to have that amount when it is due?

50. Kent Merrill wants to have $20,000 available in 5 yr for a down payment on a house. He has inherited $15,000. How much of the inheritance should he invest now to accumulate the $20,000, if he can get an interest rate of 8% compounded quarterly?

51. Two partners agree to invest equal amounts in their business. One will contribute $10,000 immediately. The other plans to contribute an equivalent amount in 3 yr, when she expects to acquire a large sum of money. How much should she contribute at that time to match her partner's investment now, assuming an interest rate of 6% compounded semiannually?

52. As the prize in a contest, you are offered $1000 now or $1210 in 5 yr. If money can be invested at 6% compounded annually, which is larger?

53. The consumption of electricity has increased historically at 6% per year. If it continues to increase at this rate indefinitely, find the number of years before the electric utilities will need to double their generating capacity.

54. Suppose a conservation campaign coupled with higher rates causes the demand for electricity to increase at only 2% per year, as it has recently. Find the number of years before the utilities will need to double generating capacity.

Management *Use the approach in Example 9 to find the time it would take for the general level of prices in the economy to double at each of the following average annual inflation rates.*

55. 4%   **56.** 5%

57. On January 1, 1980, Jack deposited $1000 into Bank X to earn interest at the rate of j per annum compounded semiannually. On January 1, 1985, he transferred his account to Bank Y to earn interest at the rate of k per annum compounded quarterly. On January 1, 1988, the balance at Bank Y was $1990.76. If Jack could have earned interest at the rate of k per annum compounded quarterly from January 1, 1980, through January 1, 1988, his balance would have been $2203.76. Which of the following represents the ratio k/j?*

(a) 1.25   (b) 1.30   (c) 1.35   (d) 1.40
(e) 1.45

---

*Problem from "Course 140 Examination, Mathematics of Compound Interest" of the Education and Examination Committee of The Society of Actuaries. Reprinted by permission of The Society of Actuaries.

58. On January 1, 1987, Tone Company exchanged equipment for a $200,000 noninterest bearing note due on January 1, 1990. The prevailing rate of interest for a note of this type at January 1, 1987 was 10%. The present value of $1 at 10% for three periods is 0.75. What amount of interest revenue should be included in Tone's 1988 income statement?**

(a) $7,500   (b) $15,000   (c) $16,500
(d) $20,000

** Uniform CPA Examination, May, 1989, American Institute of Certified Public Accountants.

5.3 ANNUITIES

So far in this chapter, only lump sum deposits and payments have been discussed. But many financial situations involve a sequence of equal payments at regular intervals, such as weekly deposits in a savings account or monthly payments on a mortgage or car loan. In order to develop formulas to deal with periodic payments like these, we must first discuss sequences.

GEOMETRIC SEQUENCES   If a and r are fixed nonzero numbers, then the infinite list of numbers, $a, ar, ar^2, ar^3, ar^4, \ldots$ is called a **geometric sequence.** For instance, if $a = 5$ and $r = 2$, we have the sequence

$$5, 5 \cdot 2, 5 \cdot 2^2, 5 \cdot 2^3, 5 \cdot 2^4, \ldots,$$

or

$$5, 10, 20, 40, 80, \ldots.$$

In the sequence $a, ar, ar^2, ar^3, ar^4, \ldots$, the number a is the first term of the sequence, ar the second term, ar^2 the third term, ar^3 the fourth term, and so on. Thus, for any $n \geq 1$,

$$ar^{n-1} \text{ is the } n\text{th term of the sequence.}$$

Each term in the sequence is r times the preceding term. The number r is called the **common ratio** of the sequence.

1 Write the first four terms of the geometric sequence with $a = 5$ and $r = -2$. Then find the seventh term.

Answer:
$5, -10, 20, -40; 320$

▶**EXAMPLE 1** Find the seventh term of the geometric sequence 6, 24, 96, ... Here $a = 6$ and $r = 24/6 = 4$. Using ar^{n-1}, with $n = 7$, the **seventh** term is

$$ar^6 = 6(4)^6 = 6(4096) = 24,576. \quad ◀$$

▶**EXAMPLE 2** Write the first five terms of the geometric sequence with $a = 100$ and $r = .1$.

The first five terms are

$$100, 100(.1), 100(.1)^2, 100(.1)^3, 100(.1)^4,$$

or

$$100, 10, 1, .1, .01. \quad ◀ \quad \boxed{1}$$

Next we find the sum S_n of the first n terms of a geometric sequence. That is, we find S_n, where

$$S_n = a + ar + ar^2 + ar^3 + \cdots + ar^{n-1}. \tag{1}$$

If $r = 1$, this is easy because

$$S_n = \underbrace{a + a + a + \cdots + a}_{n \text{ terms}} = na.$$

If $r \neq 1$, multiply both sides of equation (1) by r, obtaining

$$rS_n = ar + ar^2 + ar^3 + ar^4 + \cdots + ar^n. \tag{2}$$

Now subtract corresponding sides of equation (1) from equation (2):

$$
\begin{aligned}
rS_n &= \qquad\quad ar + ar^2 + ar^3 + \cdots + ar^{n-1} + ar^n \\
-S_n &= -(a + ar + ar^2 + ar^3 + \cdots + ar^{n-1}) \\
\hline
rS_n - S_n &= ar^n - a \\
S_n(r - 1) &= a(r^n - 1) \\
S_n &= \frac{a(r^n - 1)}{r - 1}.
\end{aligned}
$$

Hence, we have this useful formula.

If a geometric sequence has first term a and common ratio r, then the **sum of the first n terms** is given by

$$S_n = \frac{a(r^n - 1)}{r - 1}, \quad r \neq 1.$$

2 (a) Find S_4 and S_7 for the geometric sequence 5, 15, 45, 135, ...

(b) Find S_5 for the geometric sequence having $a = -3$ and $r = -5$.

Answers:

(a) $S_4 = 200$, $S_7 = 5465$

(b) $S_5 = -1563$

▶**EXAMPLE 3** Find the sum of the first six terms of the geometric sequence 3, 12, 48, ...

Here $a = 3$ and $r = 4$. Find S_6 by the result above.

$$S_6 = \frac{3(4^6 - 1)}{4 - 1} \qquad \text{Let } n = 6, a = 3, r = 4$$

$$= \frac{3(4096 - 1)}{3}$$

$$= 4095 \quad ◀ \quad \boxed{2}$$

ORDINARY ANNUITIES A sequence of equal payments made at equal periods of time is called an **annuity.** If the payments are made at the end of the time period, and if the frequency of payments is the same as the frequency of compounding, the annuity is called an **ordinary annuity.** The time between payments is the **payment period,** and the time from the beginning of the first payment period to the end of the last period is called the **term of the annuity.** The **future value of the annuity,** the final sum on deposit, is defined as the sum of the compound amounts of all the payments, compounded to the end of the term.

For example, suppose $1500 is deposited at the end of the year for the next 6 years in an account paying 8% per year compounded annually. Figure 5.1 shows this annuity schematically.

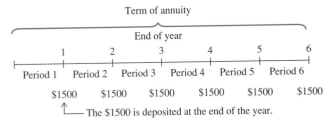

FIGURE 5.1

To find the future value of this annuity, look separately at each of the $1500 payments. The first of these payments will produce a compound amount of

$$1500(1 + .08)^5 = 1500(1.08)^5.$$

Use 5 as the exponent instead of 6 since the money is deposited at the *end* of the first year and earns interest for only 5 years. The second payment of $1500 will produce a compound amount of $1500(1.08)^4$. As shown in Figure 5.2, the future value of the annuity is

$$1500(1.08)^5 + 1500(1.08)^4 + 1500(1.08)^3 + 1500(1.08)^2$$
$$+ 1500(1.08)^1 + 1500.$$

3 Complete these steps for an annuity of $2000 at the end of the each year for 3 years. Assume interest of 8% compounded annually.

(a) The first deposit of $2000 produces a total of _____ .

(b) The second deposit becomes _____ .

(c) No interest is earned on the third deposit, so the total in the account is _____ .

Answers:

(a) $2332.80

(b) $2160.00

(c) $6492.80

(The last payment earns no interest at all.)

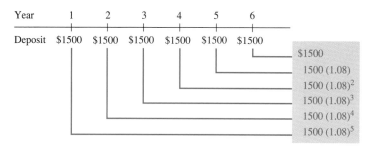

FIGURE 5.2

Reading this in reverse order, we see that it is just the sum of the first six terms of a geometric sequence with $a = 1500$, $r = 1.08$, and $n = 6$. Therefore, the sum is

$$\frac{a(r^n - 1)}{r - 1} = \frac{1500[(1.08)^6 - 1]}{1.08 - 1} = \$11,003.89. \quad \blacksquare \; 3$$

To generalize this result, suppose that payments of R dollars each are deposited into an account at the *end of each period* for *n periods,* at a rate of interest *i per period.* The first payment of R dollars will produce a compound amount of $R(1 + i)^{n-1}$ dollars, the second payment will produce $R(1 + i)^{n-2}$ dollars, and so on; the final payment earns no interest and contributes just R dollars to the total. If S represents the future value of the annuity, then (as shown in Figure 5.3),

$$S = R(1 + i)^{n-1} + R(1 + i)^{n-2} + R(1 + i)^{n-3} + \cdots + R(1 + i) + R$$

or, written in reverse order,

$$S = R + R(1 + i)^1 + R(1 + i)^2 + \cdots + R(1 + i)^{n-1}.$$

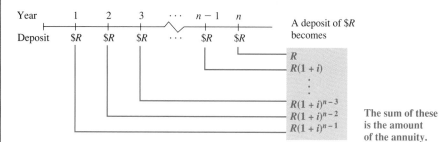

FIGURE 5.3

4 Johnson Building Materials deposits $2500 at the end of each year into an account paying 8% per year compounded annually. Find the total amount on deposit after

(a) 6 years;

(b) 10 years.

Answers:

(a) $18,339.82

(b) $36,216.41

5 Helen's Dry Goods deposits $5800 at the end of each quarter for 4 years. Find the final amount on deposit if the deposits earn

(a) 8% compounded quarterly;

(b) 6% compounded quarterly.

Answers:

(a) $108,107.85

(b) $104,007.75

This result is the sum of the first n terms of the geometric sequence having first term R and common ratio $1 + i$. Using the formula for the sum of the first n terms of a geometric sequence,

$$S = \frac{R[(1 + i)^n - 1]}{(1 + i) - 1} = \frac{R[(1 + i)^n - 1]}{i} = R\left[\frac{(1 + i)^n - 1}{i}\right].$$

The quantity in brackets is commonly written $s_{\overline{n}|i}$ (read "s-angle-n at i"), so that

$$S = R \cdot s_{\overline{n}|i}.$$

We can summarize this work as follows.

The **future value S of an ordinary annuity** of n payments of R dollars each at the end of consecutive interest periods with interest compounded at a rate of interest i per period is

$$S = R\left[\frac{(1 + i)^n - 1}{i}\right] \quad \text{or} \quad S = R \cdot s_{\overline{n}|i}.$$

▶ **EXAMPLE 4** Glenn Russell is an athlete who feels that his playing career will last 7 years. To prepare for his future, he deposits $22,000 at the end of each year for 7 years in an account paying 6% compounded annually. How much will he have on deposit after 7 years?

His payments form an ordinary annuity with $R = 22,000$, $n = 7$, and $i = .06$. The future value of this annuity (by the formula above) is

$$S = 22,000\left[\frac{(1.06)^7 - 1}{.06}\right]. = \$184,664.43 \quad ◀ \quad \boxed{4}$$

▶ **EXAMPLE 5** Suppose $1000 is deposited at the end of each 6-month period for 5 years in an account paying 10% compounded semiannually. Find the future value of the annuity.

Interest of $10\%/2 = 5\%$ is earned semiannually. In 5 years there are $5 \cdot 2 = 10$ semiannual periods. The future value is

$$S = 1000\left(\frac{(1.05)^{10} - 1}{.05}\right) \approx 1000(12.57789) = \$12,577.89 \quad ◀ \quad \boxed{5}$$

ANNUITIES DUE The formula developed above is for *ordinary annuities*—those with payments at the *end* of each time period. These results can be modified slightly to apply to **annuities due**—annuities where payments are made at the *beginning* of each time period.

To find the future value of an annuity due, treat each payment as if it were made at the end of the *preceding* period. that is, find $S_{\overline{n}|i}$ for *one additional period;* to compensate for this, subtract the amount of one payment. Thus, the **future value of an annuity due** of n payments of R dollars each at the beginning

6 Ms. Black deposits $800 at the beginning of each 6-month period for 5 years. Find the final amount if the account pays

(a) 6% compounded semiannually;

(b) 8% compounded semiannually.

Answers:

(a) $9446.24

(b) $9989.08

of consecutive interest periods with interest compounded at the rate of i per period is

$$S = R\left[\frac{(1 + i)^{n+1} - 1}{i}\right] - R \quad \text{or} \quad S = R \cdot s_{\overline{n+1}|i} - R.$$

▶ **EXAMPLE 6**   Find the future value of an annuity due if payments of $500 are made at the beginning of each quarter for 7 years in an account paying 12% compounded quarterly.

In 7 years there are 28 quarterly periods. Add one period to get $n = 29$, and use the formula with $i = 12\%/4 = 3\%$.

$$S = 500\left[\frac{(1.03)^{29} - 1}{.03}\right] - 500$$

$$S \approx 500(45.21885) - 500 = \$22,109.43$$

The account will contain a total of $22,109.43 after 7 years. ◀ **6**

SINKING FUNDS   A fund set up to receive periodic payments is called a **sinking fund.** The periodic payments, together with the interest earned by the payments, are designed to produce a given sum at some time in the future. For example, a sinking fund might be set up to receive money that will be needed to pay off the principal on a loan at some future time. If the payments are all the same amount and are made at the end of a regular time period, they form an ordinary annuity.

▶ **EXAMPLE 7**   The Chinns are close to retirement. They agree to sell an antique urn to the local museum for $17,000. Their tax adviser suggests that they defer receipt of this money until they retire, 5 years in the future. (At that time, they might well be in a lower tax bracket.) Find the amount of each payment the museum must make into a sinking fund so that it will have the necessary $17,000 in 5 years. Assume that the museum can earn 6% compounded annually on its money. Also, assume that the payments are made annually.

These payments are the periodic payments into an ordinary annuity. The annuity will amount to $17,000 in 5 years at 6% compounded annually. Using the formula and a calculator,

$$17,000 = R \cdot s_{\overline{5}|.06} = R\left[\frac{(1.06)^5 - 1}{.06}\right]$$

$$17,000 \approx R(5.637093)$$

$$R \approx \frac{17,000}{5.637093} = 3015.74$$

or $3015.74. If the museum management deposits $3,015.74 at the end of each year for 5 years in an account paying 6% compounded annually, it will have the

7 Francisco Arce needs $8000 in 6 years so that he can go on an archeological dig. He wants to deposit equal payments at the end of each quarter so that he will have enough to go on the dig. Find the amount of each payment if the bank pays

(a) 12% compounded quarterly;

(b) 8% compounded quarterly.

Answers:

(a) $232.38

(b) $262.97

total amount needed. This result is shown in the following sinking fund table. In these tables the last payment may differ slightly from the others because of rounding errors.

Payment Number	Amount of Deposit	Interest Earned	Total in Account
1	$3015.74	$0	$3015.74
2	3015.74	180.94	6212.42
3	3015.74	372.75	9600.91
4	3015.74	576.05	13,192.70
5	3015.74	791.56	17,000.00

To construct the table, notice that the first payment does not earn interest until the second payment is made. Line 2 of the table shows the second payment, 6% interest of $180.94 on the first payment, and the sum of these amounts added to the total in line 1. Line 3 shows the third payment, 6% interest of $372.75 on the total from line 2, and the new total found by adding these amounts to the total in line 2. This procedure is continued to complete the table. ◄ **7**

5.3 EXERCISES

Find the fourth term of each of the following geometric sequences. (See Example 1.)

1. $a = 5, r = 3$

2. $a = 20, r = 2$

3. $a = 48, r = .5$

4. $a = 80, r = .1$

5. $a = 2000, r = 1.05$

6. $a = 10,000, r = 1.01$

Find the sum of the first four terms for each of the following geometric sequences. (See Example 3.)

7. $a = 1, r = 2$

8. $a = 3, r = 3$

9. $a = 5, r = .2$

10. $a = 6, r = .5$

11. $a = 128, r = 1.1$

12. $a = 100, r = 1.05$

Find each of the following values.

13. $s_{\overline{12}|.05}$

14. $s_{\overline{20}|.06}$

15. $s_{\overline{16}|.04}$

16. $s_{\overline{40}|.02}$

17. $s_{\overline{20}|.01}$

18. $s_{\overline{18}|.015}$

19. Explain the difference between an ordinary annuity and an annuity due.

Find the future value of the following ordinary annuities. Payments are made and interest is compounded as given. (See Examples 4 and 5.)

20. $R = \$1500$, 4% interest compounded annually for 8 years

21. $R = \$680$, 5% interest compounded annually for 6 years

22. $R = \$12,000$, 6.2% interest compounded annually for 10 years

23. $R = \$20,000$, 5.5% interest compounded annually for 12 years

24. $R = \$865$, 6% interest compounded semiannually for 8 years

25. $R = \$7300$, 9% interest compounded semiannually for 6 years

26. $R = \$1200$, 8% interest compounded quarterly for 10 years

27. $R = \$20,000$, 6% interest compounded quarterly for 12 years

Find the future value of each annuity due. (See Example 6.)

28. Payments of $500 for 10 years at 5% compounded annually

29. Payments of $1050 for 6 years at 3.5% compounded annually

30. Payments of $16,000 for 8 years at 4.7% compounded annually

31. Payments of $25,000 for 12 years at 6% compounded annually

32. Payments of $1000 for 9 years at 8% compounded semiannually

33. Payments of $750 for 15 years at 6% compounded semiannually

34. Payments of $100 for 7 years at 12% compounded quarterly

35. Payments of $1500 for 11 years at 12% compounded quarterly

Find the periodic payment that will amount to the given sums under the given conditions, if payments are made at the end of each period. (See Example 7.)

36. $S = \$14,500$, interest is 5% compounded semiannually for 8 years

37. $S = \$43,000$, interest is 6% compounded semiannually for 5 years

38. $S = \$62,000$, interest is 8% compounded quarterly for 6 years

39. $S = \$12,800$, interest is 6% compounded monthly for 4 years

40. What is meant by a sinking fund? Give an example of a sinking fund.

Find the amount of each payment to be made into a sinking fund to accumulate the following amounts. Payments are made at the end of each period. (See Example 7.)

41. $11,000, money earns 6% compounded semiannaully, for 6 years

42. $75,000, money earns 6% compounded semiannually, for $4\frac{1}{2}$ years

43. $50,000, money earns 10% compounded quarterly, for $2\frac{1}{2}$ years

44. $25,000, money earns 12% compounded quarterly, for $3\frac{1}{2}$ years

45. $6000, money earns 8% compounded monthly, for 3 years

46. $9000, money earns 12% compounded monthly, for $2\frac{1}{2}$ years

Management *Work the following applied problems.*

47. Charlie Dawkins is saving for a computer. At the end of each month he puts $60 in a savings account that pays 4% interest compounded monthly. How much is in the account after 3 years?

48. Pat Quinlan deposits $6000 at the end of each year for 4 years in an account paying 5% interest compounded annually.
(a) Find the final amount she will have on deposit.
(b) Pat's brother-in-law works in a bank that pays 4.8% compounded annually. If she deposits her money in this bank instead of the one above, how much will she have in her account?
(c) How much would Quinlan lose over 5 years by using her brother-in-law's bank?

49. A father opened a savings acount for his daughter on the day she was born, depositing $1000. Each year on her birthday he deposits another $1000, making the last deposit on her twenty-first birthday. If the account pays 9.5% interest compounded annually, how much is in the account at the end of the day on the daughter's twenty-first birthday?

50. A 45-year-old man puts $1000 in a retirement account at the end of each quarter until he reaches the age of 60 and makes no further deposits. If the account pays 11% interest compounded quarterly, how much will be in the account when the man retires at age 65?

51. At the end of each quarter a 50-year-old woman puts $1200 in a retirement account that pays 7% interest compounded quarterly. When she reaches age 60, she withdraws the entire amount and places it in a mutual fund that pays 9% interest compounded monthly. From then on she deposits $300 in the mutual fund at the end of each month. How much is in the account when she reaches age 65?

52. Jasspreet Kaur deposits $2435 at the beginning of each semiannual period for 8 years in an account paying 6% compounded semiannually. She then leaves that money alone, with no further deposits, for an additional 5 years. Find the final amount on deposit after the entire 13-year period.

53. Chuck Hickman deposits $10,000 at the beginning of each year for 12 years in an account paying 5% compounded annually. He then puts the total amount on deposit in another account paying 6% compounded semiannually for another 9 years. Find the final amount on deposit after the entire 21-year period.

54. David Horwitz needs $10,000 in 8 years.
   (a) What amount should he deposit at the end of each quarter at 8% compounded quarterly so that he will have his $10,000?
   (b) Find Horwitz's quarterly deposit if the money is deposited at 6% compounded quarterly.

55. Harv's Meats knows that it must buy a new deboner machine in 4 years. The machine costs $12,000. In order to accumulate enough money to pay for the machine, Harv decides to deposit a sum of money at the end of each 6 months in an account paying 6% compounded semiannually. How much should each payment be?

56. Karin Sandberg wants to buy an $18,000 car in 6 years. How much money must she deposit at the end of each quarter in an account paying 12% compounded quarterly so that she will have enough to pay for her car?

57. In a recent state lottery, the jackpot was $27 million. An Australian investment firm tried to buy all possible combinations of numbers, which would have cost $7 million. (In fact, the firm ran out of time and was unable to buy all combinations.) Suppose the investment firm had accomplished its goal and held the only winning ticket in the lottery. Assume that the firm would receive the jackpot in payments of $1.35 million paid at the end of each year for 20 years.
   (a) Suppose each jackpot payment received by the firm is invested at 8% interest compounded annually. How many years would it take until the value of this investment would be greater than the amount the firm would have if it had simply invested its original $7 million at the same rate (instead of buying lottery tickets)? (*Hint:* Experiment with different values of n, the number of years.)
   (b) How many years would it take in part (a) at an interest rate of 12%?

58. Diane Gray sells some land in Nevada. She will be paid a lump sum of $60,000 in 7 yr. Until then, the buyer pays 8% simple interest quarterly.
   (a) Find the amount of each quarterly interest payment.
   (b) The buyer sets up a sinking fund so that enough money will be present to pay off the $60,000. The buyer wants to make semiannual payments into the sinking fund; the account pays 6% compounded semiannually. Find the amount of each payment into the fund.
   (c) Prepare a table showing the amount in the sinking fund after each deposit.

59. Joe Seniw bought a rare stamp for his collection. He agreed to pay a lump sum of $4000 after 5 yr. Until then, he pays 6% simple interest semiannually.
   (a) Find the amount of each semiannual interest payment.
   (b) Seniw sets up a sinking fund so that enough money will be present to pay off the $4000. He wants to make annual payments into the fund. The account pays 8% compounded annually. Find the amount of each payment.
   (c) Prepare a table showing the amount in the sinking fund after each deposit.

5.4 PRESENT VALUE OF AN ANNUITY; AMORTIZATION

Suppose that at the end of each year, for the next 10 years, $500 is deposited in a savings account paying 7% interest compounded annually. This is an example of an ordinary annuity. The **present value** of this annuity is the amount that would have to be deposited in one lump sum today (at the same compound interest rate) in order to produce exactly the same balance at the end of 10 years. We can find a formula for the present value of an annuity as follows.

Suppose deposits of R dollars are made at the end of each period for n periods at interest rate i per period. Then the amount in the account after n periods is the future value of this annuity:

$$S = R \cdot s_{\overline{n}|i} = R\left[\frac{(1 + i)^n - 1}{i}\right].$$

On the other hand, if P dollars are deposited today at the same compound interest rate i, then at the end of n periods, the amount in the account is $P(1 + i)^n$. This amount must be the same as the amount S in the formula above, that is,

$$P(1 + i)^n = R\left[\frac{(1 + i)^n - 1}{i}\right].$$

To solve this equation for P, multiply both sides by $(1 + i)^{-n}$.

$$P = R(1 + i)^{-n}\left[\frac{(1 + i)^n - 1}{i}\right]$$

Use the distributive property; also recall that $(1 + i)^{-n}(1 + i)^n = 1$.

$$P = R\left[\frac{(1 + i)^{-n}(1 + i)^n - (1 + i)^{-n}}{i}\right]$$

$$P = R\left[\frac{1 - (1 + i)^{-n}}{i}\right]$$

The amount P is called the **present value of the annuity.** The quantity in brackets is abbreviated as $a_{\overline{n}|i}$, (read "a-angle-n at i"), so

$$a_{\overline{n}|i} = \frac{1 - (1 + i)^{-n}}{i}.$$

Compare this quantity with $s_{\overline{n}|i}$ in the previous section.
Here is a summary of what we have done.

Present Value of an Annuity

The present value P of an annuity of n payments of R dollars each at end of consecutive interest periods with interest compounded at a rate of interest i per period is

$$P = R\left[\frac{1 - (1 + i)^{-n}}{i}\right] \quad \text{or} \quad P = R \cdot a_{\overline{n}|i}.$$

Caution   Don't confuse the formula for the present value of an annuity with the one for the future value of an annuity. Notice the difference: the numerator of the fraction in the present value formula is $1 - (1 + i)^{-n}$, but in the future value formula, it is $(1 + i)^n - 1$.

1 What lump sum deposited today would be equivalent to equal payments of

(a) $650 at the end of each year for 9 years at 4% compounded annually?

(b) $1000 at the end of each quarter for 4 years at 4% compounded quarterly?

Answers:

(a) $4832.97

(b) $14,717.87

2 Kelly Erin buys a small business for $174,000. She agrees to pay off the cost in payments at the end of each semiannual period for 7 years, with interest of 10% compounded semiannually on the unpaid balance. Find the amount of each payment.

Answer:

$17,578.17

▶ **EXAMPLE 1** Edison Diest and Maria Gonzalez are both graduates of the Forestvire Institute of Technology. They both agree to contribute to the endowment fund of FIT. Diest says that he will give $500 at the end of each year for 9 years. Gonzalez prefers to give a lump sum today. What lump sum can she give that will equal the present value of Diest's annual gifts, if the endowment fund earns 7.5% compounded annually?

Here $R = 500$, $n = 9$, and $i = .075$ and we have

$$P = R \cdot a_{\overline{9}|.075} = 500\left[\frac{1 - (1.075)^{-9}}{.075}\right]$$

$$\approx 500(6.37889) = 3189.44.$$

Therefore, Gonzalez must donate a lump sum of $3189.44 today. ◀ **1**

▶ **EXAMPLE 2** A car costs $12,000. After a downpayment of $2000, the balance will be paid off in 36 monthly payments with interest of 9% per year on the unpaid balance. Find the amount of each payment.

A single lump sum payment of $10,000 today would pay off the loan. So $10,000 is the present value of an annuity of 36 monthly payments with interest of 9%/12 = .75% per month. Thus $P = 10,000$, $n = 36$, $i = .0075$, and we must find the monthly payment R in the formula

$$P = R\left[\frac{1 - (1 + i)^{-n}}{i}\right]$$

$$10,000 = R\left[\frac{1 - (1.0075)^{-36}}{.0075}\right] \approx R(31.446805)$$

$$R = \frac{10,000}{31.446805} = 318.00.$$

A monthly payment of $318.00 will be needed. ◀ **2**

AMORTIZATION A loan is **amortized** if both the principal and interest are paid by a sequence of equal periodic payments. In Example 2 above, a loan of $10,000 at 9% interest compounded monthly could be amortized by paying $318.00 per month for 36 months.

The periodic payment needed to amortize a loan may be found, as in Example 2, by solving the present value equation for R.

A loan of P dollars at interest rate i per period may be amortized in n equal periodic payments of R dollars made at the end of each period, where

$$R = \frac{P}{a_{\overline{n}|i}} = \frac{P}{\left[\dfrac{1 - (1 + i)^{-n}}{i}\right]} = \frac{Pi}{1 - (1 + i)^{-n}}.$$

3 If the mortgage in Example 3 runs for 15 years, find

(a) the monthly payment;

(b) the total amount of interest paid.

Answers:

(a) $819.21

(b) $69,457.80

▶ **EXAMPLE 3** The Beckenstein family buys a house for $94,000 with a down payment of $16,000. They take out a 30-year mortgage for $78,000 at an annual interest rate of 9.6%.

(a) Find the amount of the monthly payment needed to amortize this loan.

Here $P = 78,000$ and the monthly interest rate is $9.6\%/12 = .096/12 = .008.$* The number of monthly payments is $12 \cdot 30 = 360$. Therefore,

$$R = \frac{78,000}{a_{\overline{360}|.008}} = \frac{78,000}{\left[\dfrac{1 - (1.008)^{-360}}{.008}\right]} \approx \frac{78,000}{117.90229} = 661.56.$$

Monthly payments of $661.56 are required to amortize the loan.

(b) Find the total amount of interest paid when the loan is amortized over 30 years.

The Beckenstein family makes 360 payments of $661.56 each, for a total of $238,161.60. Since the amount of the loan was $78,000, the total interest paid is

$$\$238,161.60 - \$78,000 = \$160,161.60.$$

This large amount of interest is typical of what happens with a long mortgage. A 15-year mortgage would have higher payments but would involve significantly less interest. **3**

(c) Find the part of the first payment that is interest and the part that is applied to reducing the debt.

As we saw in part (a), the monthly interest rate is .008. During the first month, the entire $78,000 is owed. Interest on this amount for 1 month is found by the formula for simple interest.

$$I = Prt = 78,000(.008)(1) = 624$$

At the end of the month, a payment of $661.56 is made; since $624 of this is interest, a total of

$$\$661.56 - \$624 = \$37.56$$

is applied to the reduction of the original debt. ◀

AMORTIZATION SCHEDULES In the preceding example, 360 payments are made to amortize a $78,000 loan. The loan balance after the first payment is reduced by only $37.56, which is much less than $(1/360)(78,000) \approx \$216.67$. Therefore, even though equal *payments* are made to amortize a loan, the loan *balance* does not decrease in equal steps. This fact is very important if a loan is paid off early.

---

*Mortgage rates are quoted in terms of annual interest, but it is always understood that the monthly rate is 1/12 of the annual rate and that interest is compounded monthly.

4 Find the following for a car loan of $10,000 for 48 months at 9% interest compounded monthly.

(a) the monthly payment

(b) How much is still due after 12 payments have been made?

Answers:

(a) $248.85

(b) $7825.54 or $7825.56

▶ **EXAMPLE 4** Jill Stuart borrows $1000 for 1 year at 12% annual interest compounded monthly.

(a) What is her monthly loan payment?

The monthly interest rate is $12\%/12 = 1\% = .01$, so that her payment is

$$R = \frac{1000}{a_{\overline{12}|.01}} = \frac{1000}{\left[\dfrac{1 - (1.01)^{-12}}{.01}\right]} \approx \frac{1000}{11.25508} = \$88.85.$$

(b) After making three payments, she decides to pay off the remaining balance all at once. How much must she pay?

Since nine payments remain to be paid, they can be thought of as an annuity consisting of nine payments of $88.25 at 1% interest per period. The present value of this annuity is

$$88.85\left[\frac{1 - (1.01)^{-9}}{.01}\right] = 761.09.$$

So Jill's remaining balance, computed by this method, is $761.09.

An alternative method of figuring the balance is to consider the payments already made as an annuity of three payments. At the beginning, the present value of this annuity was

$$88.85\left[\frac{1 - (1.01)^{-3}}{.01}\right] = 261.31.$$

So she still owes the difference $1000 - \$261.31 = \$738.69$. Furthermore, she owes the interest on this amount for 3 months, for a total of

$$(738.69)(1.01)^3 = \$761.07.$$

This balance due differs from the one obtained by the first method by 2¢ because the monthly payment and the other calculations were rounded to the nearest penny. ◀ **4**

Although most people wouldn't quibble about a 2¢ difference in the balance due in Example 4, the difference in other cases (larger amounts or longer terms) might be more than that. A bank or business must keep its books accurately to the nearest penny, so it must determine the balance due in cases like this unambiguously and exactly. This is done by means of an **amortization schedule,** which lists how much of each payment is interest, how much goes to reduce the balance, and how much is owed after *each* payment.

▶ **EXAMPLE 5** Determine the exact amount Jill Stuart in Example 4 owes after three monthly payments.

An amortization table for the loan is shown below. It is obtained as follows. The annual interest rate is 12% compounded monthly, so the interest rate per month is $12\%/12 = 1\% = .01$. When the first payment is made, 1 month's interest, namely .01(1000) = $10, is owed. Subtracting this from the $88.85 payment leaves $78.85 to be applied to repayment. Hence, the principal at the end of the first payment period is 1000 − 78.85 = $921.15, as shown in the "payment 1" line of the chart.

When payment 2 is made, 1 month's interest on $921.15 is owed, namely .01(921.15) = $9.21. Subtracting this from the $88.85 payment leaves $79.64 to reduce the principal. Hence, the principal at the end of payment 2 is 921.15 − 79.64 = $841.51. The interest portion of payment 3 is based on this amount and the remaining lines of the table are found in a similar fashion.

Payment Number	Amount of Payment	Interest for Period	Portion to Principal	Principal at End of Period
0	—	—	—	$1000.00
1	$88.85	$10.00	$78.85	921.15
2	88.85	9.21	79.64	841.51
3	88.85	8.42	80.43	761.08
4	88.85	7.61	81.24	679.84
5	88.85	6.80	82.05	597.79
6	88.85	5.98	82.87	514.92
7	88.85	5.15	83.70	431.22
8	88.85	4.31	84.54	346.68
9	88.85	3.47	85.38	261.30
10	88.85	2.61	86.24	175.06
11	88.85	1.75	87.10	87.96
12	88.84	.88	87.96	0

The schedule shows that after three payments, she still owes $761.08, an amount that differs slightly from that obtained by either method in Example 4. ◀

The amortization schedule in Example 5 is typical. In particular, note that all payments are the same except the last one. It is often necessary to adjust the amount of the final payment to account for rounding off earlier and to insure that the final balance is exactly 0.

An amortization schedule also shows how the periodic payments are applied to interest and principal. The amount going to interest decreases with each payment, while the amount going to reduce the principal owed increases with each payment.

5.4 EXERCISES

1. Which of the following is represented by $a_{\overline{n}|i}$?

(a) $\dfrac{(1 + i)^{-n} - 1}{i}$ (b) $\dfrac{(1 + i)^{n} - 1}{i}$ (c) $\dfrac{1 - (1 + i)^{-n}}{i}$ (d) $\dfrac{1 - (1 + i)^{n}}{i}$

2. Which of the choices in Exercise 1 represents $s_{\overline{n}|i}$?

Find each of the following.

3. $a_{\overline{15}|.06}$

4. $a_{\overline{10}|.03}$

5. $a_{\overline{18}|.04}$

6. $a_{\overline{30}|.01}$

7. $a_{\overline{16}|.01}$

8. $a_{\overline{32}|.02}$

9. Explain the difference between the present value of an annuity and the future value of an annuity. For a given annuity, which is larger? Why?

Find the present value of each ordinary annuity (See Example 1.)

10. Payments of $5200 are made annually for 6 years at 5% compounded annually.

11. Payments of $1250 are made annually for 8 years at 4% compounded annually.

12. Payments of $675 are made semiannually for 10 years at 6% compounded semiannually.

13. Payments of $750 are made semiannually for 8 years at 6% compounded semiannually.

14. Payments of $16,908 are made quarterly for 3 years at 4.4% compounded quarterly.

15. Payments of $12,125 are made quarterly for 5 years at 5.6% compounded quarterly.

Find the lump sum deposited today that will yield the same total amount as payments of $10,000 at the end of each year for 15 yr, at each of the given interest rates. (See Example 1.)

16. 4% compounded annually

17. 5% compounded annually

18. 6% compounded annually

19. What does it mean to amortize a loan?

Find the payment necessary to amortize each of the following loans. (See Example 2.)

20. $800, 10 annual payments at 5%

21. $12,000, 8 quarterly payments at 6%

22. $25,000, 8 quarterly payments at 6%

23. $35,000, 12 quarterly payments at 4%

24. $5000, 36 monthly payments at 12%

25. $472, 48 monthly payments at 12%

Find the monthly house payment necessary to amortize the following loans. Interest is calculated on the unpaid balance. (See Examples 3 and 4.)

26. $49,560 at 10.75% for 25 years

27. $70,892 at 11.11% for 30 years

28. $53,762 at 12.45% for 30 years

29. $96,511 at 10.57% for 25 years

Use the amortization table in Example 5 to answer the questions in Exercises 30–33.

30. How much of the fifth payment is interest?

31. How much of the tenth payment is used to reduce the debt?

32. How much interest is paid in the first 5 months of the loan?

33. How much interest is paid in the last 5 months of the loan?

34. What sum deposited today at 5% compounded annually for 8 yr will provide the same amount as $1000 deposited at the end of each year for 8 yr at 6% compounded annually?

35. What lump sum deposited today at 8% compounded quarterly for 10 yr will yield the same final amount as deposits of $4000 at the end of each six-month period for 10 yr at 6% compounded semiannually?

Management *Work the following applied problems.*

36. Stereo Shack sells a stereo system for $600 down and monthly payments of $30 for the next 3 yr. If the interest rate is 1.25% per month on the unpaid balance, find
(a) the cost of the stereo system;
(b) the total amount of interest paid.

37. Hong Le buys a car costing $6000. He agrees to make payments at the end of each monthly period for 4 years. He pays 12% interest, compounded monthly.
(a) What is the amount of each payment?
(b) Find the total amount of interest Le will pay.

38. A speculator agrees to pay $15,000 for a parcel of land; this amount, with interest, will be paid over 4 yr, with semiannual payments, at an interest rate of 10% compounded semiannually. Find the amount of each payment.

39. In the Million-Dollar Lottery, a winner is paid a million dollars at the rate of $50,000 per year for 20 yr. Assume that these payments form an ordinary annuity, and that the lottery managers can invest money at 6% compounded annually. Find the lump sum that the management must put away to pay off the million-dollar winner.

40. The Adams family bought a house for $81,000. They paid $20,000 down and took out a 30-year mortgage for the balance at 11%. Use one of the methods in Example 4 to estimate the mortgage balance after 100 payments.

41. After making 180 payments on their mortgage, the Adams family of Exercise 40 sells their house for $136,000. They must pay closing costs of $3700 plus 2.5% of the sale price of the house. Approximately how much money will they receive at the close of the sale after their current mortgage is paid off and closing costs are deducted?

Management   *In Exercises 42–45, prepare an amortization schedule showing the first 4 payments for each loan.*

42. An insurance firm pays $4000 for a new printer for its computer. It amortizes the loan for the printer in 4 annual payments at 8% compounded annually.

43. Large semitrailer trucks cost $72,000 each. Ace Trucking buys such a truck and agrees to pay for it by a loan that will be amortized with 9 semiannual payments at 6% compounded semiannually.

44. One retailer charges $1048 for a computer monitor. A firm of tax accountants buys 8 of these monitors. They make a down payment of $1200 and agree to amortize the balance with monthly payments at 12% compounded monthly for 4 yr.

45. When Teresa Flores opened her law office, she bought $14,000 worth of law books and $7200 worth of office furniture. She paid $1200 down and agreed to amortize the balance with semiannual payments for 5 yr, at 7% compounded semiannually.

46. Don Cole buys a house for $285,000. He pays $60,000 down and takes out a mortgage at 9.5% on the balance. Find his monthly payment and the total amount of interest he will pay if the length of the mortgage is
(**a**) 15 yr;   (**b**) 20 yr;   (**c**) 25 yr.

47. Mary Yaroslavsky has received $25,000 in a divorce settlement. She deposits the money at 6% compounded annually. She wants to make annual withdrawals from the account so that the money (principal and interest) is gone in exactly 8 yr.
(**a**) Find the amount of each withdrawal.
(**b**) Find the amount of each withdrawal if the money must last 12 yr.

48. The trustees of a college have accepted a gift of $150,000. The donor has directed the trustees to deposit the money in an account paying 6% per year, compounded semiannually. The trustees may make equal withdrawals at the end of each six-month period; the money must last 5 yr.
(**a**) Find the amount of each withdrawal.
(**b**) Find the amount of each withdrawal if the money must last 6 yr.

Management   *Use a computer or graphing calculator to prepare an amortization schedule for each of the following loans.*

49. A loan of $37,497.50 with interest at 8.5% compounded annually, to be paid with equal annual payments over 10 years

50. A loan of $4835.80 at 9.25% interest compounded semiannually, to be repaid in 5 years in equal semiannual payments

5.5 APPLYING FINANCIAL FORMULAS

We have presented a lot of new formulas in this chapter. By answering the following questions, you can decide which formula to use for a particular problem.

1.   Is simple or compound interest involved?
     Simple interest is normally used for investments or loans of a year or less; compound interest is normally used in all other cases.

2. If simple interest is being used, what is being sought: interest amount, future value, present value, or discount?
3. If compound interest is being used, does it involve a lump sum (single payment) or an annuity (sequence of payments)?
   (a) For a lump sum,
      (i) Is ordinary compound interest or continuous interest involved?
      (ii) What is being sought: present value, future value, number of periods at interest, or effective rate?
   (b) For an annuity,
      (i) Is it an ordinary annuity (payment at the end of each period) or an annuity due (payment at the beginning of each period)?
      (ii) What is being sought: present value, future value, or payment amount?

Once you have answered these questions, choose the appropriate formula from the Chapter Summary, as in the following examples.

▶ **EXAMPLE 1**   Karen La Bonte must pay $27,000 in settlement of an obligation in 3 years. What amount can she deposit today, at 4% compounded monthly, to have enough?

In this problem, the future value of $27,000 is known, and the present value must be found. Because the time is more than one year, we should use the formula for the present value of compound interest.   ◀

▶ **EXAMPLE 2**   A bond sells for $800 and pays interest of 7.5%. How much interest is earned in 6 months?

The time period is for less than a year, so use the simple interest formula. The cost of the bond is the principal P, so $P = 800$, $i = .075$, and $t = 6/12 = 1/2$ year.   ◀

▶ **EXAMPLE 3**   A new car is is priced at $11,000. The buyer must pay $3000 down and and pay the balance in 48 equal monthly payments at 12% compounded monthly. Find the payment.

The payments form an ordinary annuity with a present value of $11,000 - 3000 = 8000$. Use the formula for the present value of an annuity with $P = 8000$, $i = .01$, and $n = 48$ to find R, the amount of the payment.   ◀

▶ **EXAMPLE 4**   Hassi is paid on the first of each month and $80 is automatically deducted from his pay and deposited in a savings account. If the account pays 4.5% interest compounded monthly, how much will be in the account after 3 years and 9 months?

The time period is for more than a year, so we use compound interest. There will be $3(12) + 9 = 45$ monthly deposits of $80, so this is an annuity due (the payments are at the first of each month). We want the future value of the annuity. Use the formula for an annuity due with $R = 80$, $i = .045$, and $n = 45$.   ◀

After you have analyzed the situation and chosen the correct formula, as in the preceding examples, work the problem. As a final step, consider whether the answer you get makes sense. For instance, present value should always be less than future value. Similarly, the future value of an annuity (which includes interest) should be larger than the sum of the payments.

5.5 EXERCISES

For the exercises in this section, round money amounts to the nearest cent, time to the nearest day, and rates to the nearest tenth of a percent.

Find the present value of $82,000 for the following number of years, if the money can be deposited at 12% compounded quarterly.

1. 5 years   **2.** 7 years

Find the amount of the payment necessary to amortize the following amounts.

3. $4250, 13 quarterly payments at 4%

4. $58,000, 23 quarterly payments at 6%

Find the compound amount for each of the following.

5. $4792.35 at 4% compounded semiannually for $5\frac{1}{2}$ years

6. $2500 at 6% compounded quarterly for $3\frac{3}{4}$ years

Find the simple interest for each of the following. Assume 360 days in a year.

7. $42,500 at 5.75% for 10 months

8. $32,662 at 6.882% for 225 days

Find the amount of interest earned by each of the following deposits.

9. $22,500 at 6% compounded quarterly for $5\frac{1}{4}$ years

10. $53,142 at 8% compounded monthly for 32 months

Find the compound amount and the amount of interest earned by a deposit of $32,750 at 5% compounded continuously for the following number of years.

11. $7\frac{1}{2}$ years

12. 9.2 years

Find the present value of the following future amounts. Assume 360 days in a year and use simple interest.

13. $17,320 for 9 months, money earns 3.5%

14. $122,300 for 138 days, money earns 4.75%

Find the proceeds for the following. Assume 360 days in a year and use simple interest.

15. $23,561 for 112 days, discount rate 4.33%

16. $267,100 for 271 days, discount rate 5.72%

Find the future value of each of the following annuities.

17. $2500 is deposited at the end of each semiannual period for $5\frac{1}{2}$ years, money earns 7% compounded semiannually

18. $800 is deposited at the end of each month for $1\frac{1}{4}$ years, money earns 8% compounded monthly

19. $250 is deposited at the end of each quarter for $7\frac{1}{4}$ years, money earns 4% compounded quarterly

20. $100 is deposited at the beginning of each year for 5 years, money earns 8% compounded yearly

Find the amount of each payment to be made to a sinking fund to accumulate the indicated amounts.

21. $10,000, money earns 5% compounded annually, 7 annual payments

22. $42,000, money earns 4% compounded quarterly, 13 quarterly payments

23. $100,000, money earns 6% compounded semiannually, 9 semiannual payments

24. $53,000, money earns 6% compounded monthly, 35 monthly payments

Find the present value of each ordinary annuity.

25. Payments of $1200 are made annually for 7 years at 5% compounded annually

26. Payments of $500 are made semiannually at 4% compounded semiannually for $3\frac{1}{2}$ years

27. Payments of $1500 are made quarterly for $5\frac{1}{4}$ years at 6% compounded quarterly

28. Payments of $905.43 are made monthly for $2\frac{11}{12}$ years at 8% compounded monthly

 Use a computer or graphing calculator to prepare an amortization schedule for each of the following loans. Interest is calculated on the unpaid balance.

29. $8500 loan repaid in semiannual payments for $3\frac{1}{2}$ years at 8%

30. $40,000 loan repaid in quarterly payments for $1\frac{3}{4}$ years at 9%

Management *Solve the following applied problems.*

31. A 10-month loan of $42,000 at simple interest produces $1785 interest. Find the interest rate.

32. Willa Burke deposits $803.47 at the end of each quarter for $3\frac{3}{4}$ years in an account paying 4% compounded quarterly. Find the final amount in the account and the amount of interest earned.

33. Makarim Wibison owes $7850 to a relative. He has agreed to pay the money in 5 months at simple interest of 7%. One month before the loan is due, the relative discounts the loan at the bank. The bank charges a 9.2% discount rate. How much money does the relative receive?

34. A small resort must add a swimming pool to compete with a new resort built nearby. The pool will cost $28,000. The resort borrows the money and agrees to repay it with equal payments at the end of each quarter for $6\frac{1}{2}$ years at an interest rate of 6% compounded quarterly. Find the amount of each payment.

35. According to the terms of a divorce settlement, one spouse must pay the other a lump sum of $2800 in 17 months. What lump sum can be invested today, at 6% compounded monthly, so that enough will be available for the payment?

36. An accountant loans $28,000 at simple interest to her business. The loan is at 9% and earns $3255 interest. Find the time of the loan in months.

37. A firm of attorneys deposits $5000 of profit sharing money at the end of each semiannual period for $7\frac{1}{2}$ years. Find the final amount in the account if the deposit earns 5% compounded semiannually. Find the amount of interest earned.

38. Find the principal that must be invested at 5.25% simple interest to earn $937.50 interest in 8 months.

39. In 3 years Ms. Thompson must pay a pledge of $7500 to her college's building fund. What lump sum can she deposit today, at 4% compounded semiannually, so that she will have enough to pay the pledge?

40. The owner of Eastside Hallmark borrows $48,000 to expand the business. The money will be repaid in equal payments at the end of each year for 7 years. Interest is 8%. Find the amount of each payment.

41. To buy a new computer, Mark Nguyen borrows $3250 from a friend at 9% interest compounded annually for 4 years. Find the compound amount he must pay back at the end of the 4 years.

42. A small business invests some spare cash for 3 months at 5.2% simple interest and earns $244 in interest. Find the amount of the investment.

43. When the Lee family bought their home, they borrowed $115,700 at 10.5% compounded monthly for 25 years. If they make all 300 payments, repaying the loan on schedule, how much interest will they pay? Assume the last payment is the same as the previous ones.

44. Suppose $84,720 is deposited for 7 months and earns $2,372.16 in interest. Find the rate of interest.

CHAPTER 5 SUMMARY

KEY TERMS AND SYMBOLS

5.1
interest
principal
rate
time
simple interest
future value (maturity value)
ordinary interest
exact interest
present value
discount (bank discount)
proceeds

5.2 compound interest
compound amount
compounding period
continuous compounding
nominal rate (stated rate)
effective rate

5.3 geometric sequence
common ratio
annuity
ordinary annuity
payment period
term of an annuity
future value of an annuity
annuity due
sinking fund

5.4 present value of an annuity
amortize a loan
amortization schedule

KEY CONCEPTS

Simple Interest

The **simple interest** I on an amount of P dollars for t years at interest rate r per year is $I = Prt$.

The **future value** A of P dollars at simple interest rate r for t years is $A = P(1 + rt)$.

The **present value** P of a future amount of A dollars at simple interest rate r for t years is

$$P = \frac{A}{1 + rt}.$$

If D is the **discount** on a loan having maturity value A at simple interest rate r for t years, then $D = Art$. If D is the discount and P the **proceeds** of a loan having maturity value A at simple interest rate r for t years, then $P = A - D$ or $P = A(1 - rt)$.

Compound Interest

If P dollars is deposited for n time periods at compound interest rate i per period, the **compound amount (future value)** A is $A = P(1 + i)^n$.

The **present value** P of A dollars at compound interest rate i per period for n periods is

$$P = \frac{A}{(1 + i)^n} = A(1 + i)^{-n}.$$

The **effective rate** corresponding to a stated interest rate r per year, compounded m times per year, is

$$r_e = \left[1 + \frac{r}{m}\right]^m - 1.$$

Continuous Compound Interest

If P dollars is deposited for t years at interest rate r per year, compounded continuously, the **compound amount (future value)** A is $A = Pe^{rt}$.

The **present value** P of A dollars at interest rate r per year compounded continuously for t years is

$$P = \frac{A}{e^{rt}}.$$

Annuties

The **future value S of an ordinary annuity** of n payments of R dollars each at the end of consecutive interest periods with interest compounded at rate i per period is

$$S = R\left[\frac{(1 + i)^n - 1}{i}\right] \quad \text{or} \quad S = R \cdot s_{\overline{n}|i}.$$

The **present value P of an ordinary annuity** of n payments of R dollars each at the end of consecutive interest periods with interest compounded at rate i per period is

$$P = R\left[\frac{1 - (1 + i)^{-n}}{i}\right] \quad \text{or} \quad P = R \cdot a_{\overline{n}|i}.$$

To find the payment, solve the formula for R.

The **future value S of an annuity due** of n payments of R dollars each at the beginning of consecutive interest periods with interest compounded at rate i per period is

$$S = R\left[\frac{(1 + i)^{n+1} - 1}{i}\right] - R \quad \text{or} \quad S = R \cdot s_{\overline{n+1}|i} - R.$$

CHAPTER 5   REVIEW EXERCISES

Find the simple interest for the following loans.

1. $4902 at 9.5% for 11 months

2. $42,368 at 15.22% for 5 months

3. $3478 at 7.4% for 88 days (assume a 360-day year)

4. $2390 at 18.7% from May 3 to July 28 (assume 365 days in a year)

5. What is meant by the present value of an amount A?

Find the present value of the following future amounts. Assume 360 days in a year; use simple interest.

6. $459.57 in 7 months, money earns 8.5%

7. $80,612 in 128 days, money earns 6.77%

8. Explain what happens when a borrower is charged a discount. What are the proceeds?

Find the proceeds in Exercises 9 and 10. Assume 360 days in a year.

9. $802.34; discount rate 8.6%; length of loan 11 mo

10. $12,000; discount rate 7.09%; length of loan 145 days

11. For a given amount of money at a given interest rate for a given time period greater than 1 year, does simple interest or compound interest produce more interest? Explain.

Find the compound amount and the amount of interest earned in each of the following.

12. $2800 at 6% compounded annually for 10 yr

13. $57,809.34 at 4% compounded quarterly for 5 yr

14. $12,903.45 at 6.37% compounded quarterly for 29 quarters

15. $4677.23 at 4.57% compounded monthly for 32 mo

Find the present value of the following amounts.

16. $42,000 in 7 yr, 12% compounded monthly

17. $17,650 in 4 yr, 8% compounded quarterly

18. $1347.89 in 3.5 yr, 6.77% compounded semiannually

19. $2388.90 in 44 mo, 5.93% compounded monthly

20. Write the first five terms of the geometric sequence with $a = 2$ and $r = 3$.

21. Write the first four terms of the geometric sequence with $a = 4$ and $r = 1/2$.

22. Find the sixth term of the geometric sequence with $a = -3$ and $r = 2$.

23. Find the fifth term of the geometric sequence with $a = -2$ and $r = -2$.

24. Find the sum of the first four terms of the geometric sequence with $a = -3$ and $r = 3$.

25. Find the sum of the first five terms of the geometirc sequence with $a = 8000$ and $r = -1/2$.

26. Find $s_{\overline{30}|.01}$.

27. What is meant by the future value of an annuity?

Find the future value of each annuity.

28. $1288 deposited at the end of each year for 14 yr; money earns 8% compound annually

29. $4000 deposited at the end of each quarter for 7 yr; money earns 6% compounded quarterly

30. $233 deposited at the end of each month for 4 yr; money earns 12% compounded monthly

31. $672 deposited at the beginning of each quarter for 7 yr; money earns 8% compounded quarterly

32. $11,900 deposited at the beginning of each month for 13 mo; money earns 12% compounded monthly

33. What is the purpose of a sinking fund?

Find the amount of each payment that must be made into a sinking fund to accumulate the following amounts.

34. $6500; money earns 5% compounded annually; 6 annual payments

35. $57,000; money earns 6% compounded semiannually for $8\frac{1}{2}$ yr

36. $233,188; money earns 5.7% compounded quarterly for $7\frac{3}{4}$ yr

37. $56,788 money earns 6.12% compounded monthly for $4\frac{1}{2}$ yr

Find the present value of each ordinary annuity.

38. Payments of $850 annually for 4 yr at 5% compounded annually

39. Payments of $1500 quarterly for 7 yr at 8% compounded quarterly

40. Payments of $4210 semiannually for 8 yr at 8.6% compounded semiannually

41. Payments of $877.34 monthly for 17 mo at 6.4% compounded monthly

42. Give two examples of the types of loans that are commonly amortized.

Find the amount of the payment necessary to amortize each of the following loans.

43. $32,000 at 9.4% compounded quarterly, 10 quarterly payments

44. $5607 at 7.6% compounded monthly, 32 monthly payments

Find the monthly house payments for the following mortgages.

45. $56,890 at 10.74% for 25 yr

46. $77,110 at 8.45% for 30 yr

A portion of an amortization table is given below for a $127,000 loan at 8.5% interest compounded monthly for 25 yr.

Payment Number	Amount of Payment	Interest for Period	Portion to Principal	Principal at End of Period
0	—	—	—	$127,000.00
1	$1022.64	$899.58	$123.06	126,876.94
2	1022.64	898.71	123.93	126,753.02
3	1022.64	897.83	124.80	126,628.21
4	1022.64	896.95	125.69	126,502.53
5	1022.64	896.06	126.58	126,375.95
6	1022.64	895.16	127.48	126,248.47

Use the table to answer the following questions.

47. How much of the fifth payment is interest?

48. How much of the sixth payment is used to reduce the debt?

49. How much interest is paid in the first 3 months of the loan?

50. How much has the debt been reduced at the end of the first 6 months?

Management   *Work the following applied problems.*

51. Larry DiCenso needs $9812 to buy new equipment for his business. The bank charges a discount rate of 12%. Find the amount of DiCenso's loan if he borrows the money for 7 mo.

52. A florist borrows $1400 at simple interest to pay taxes. The loan is at 11.5% and costs $120.75 in interest. Find the time of the loan in months.

53. A developer deposits $84,720 for 7 mo and earns $4055.46 in interest. Find the interest rate.

54. Tom Wilson owes $5800 to his mother. He has agreed to pay back the money in 10 mo, at an interest rate of 10%. Three months before the loan is due, Tom's mother discounts the loan at the bank to get cash for $6000 worth of new furniture. The bank charges a 13.45% discount rate. How much money does Tom's mother receive? Is it enough to buy the furniture?

55. In 3 yr Ms. Flores must pay a pledge of $7500 to her favorite charity. What lump sum can she deposit today, at 10% compounded semiannually, so that she will have enough to pay the pledge?

56. Each year a firm must set aside enough funds to provide employee retirement benefits of $52,000 in 20 years. If the firm can invest, money at 7.5% compounded monthly, what amount must be invested at the end of each month for this purpose?

57. Chalon Bridges deposits semiannual payments of $3200, received in payment of a debt, in an ordinary annuity at 6.8% compounded semiannually. Find the final amount in the account and the interest earned at the end of 3.5 years.

58. To finance the $15,000 cost of their kitchen remodeling, the Chews will make equal payments at the end of each month for 36 months. They will pay interest at the rate of 7.2% compounded monthly. Find the amount of each payment.

59. To expand her business, the owner of a small restaurant borrows $40,000. She will repay the money in equal payments at the end of each semiannual period for 8 years at 9% interest compounded semiannually. What payments must she make?

60. The Taggart family bought a house for $91,000. They paid $20,000 down and took out a 30-yr mortgage for the balance at 9%.
  (a) Find their monthly payment.
  (b) How much of the first payment is interest?

After 180 payments, the family sells their house for $136,000. They must pay closing costs of $3700 plus 2.5% of the sale price.
  (c) Estimate the current mortgage balance at the time of the sale using one of the methods from Example 4 in Section 5.4.
  (d) Find the total closing costs.
  (e) Find the amount of money they receive from the sale after paying off the mortgage.

61. The proceeds of a $10,000 death benefit are left on deposit with an insurance company for 7 yr at an annual effective interest rate of 5%.* The balance at the end of 7 yr is paid to the beneficiary in 120 equal monthly payments of X, with the first payment made immediately. During the payout period, interest is credited at an annual effective interest rate of 3%. Calculate X.
  (a) 117 **(b)** 118 **(c)** 129 **(d)** 135 **(e)** 158

————————

*Problem from "Course 140 Examination, Mathematics of Compound Interest" of the Education and Examination Committee of The Society of Actuaries. Reprinted by permission of The Society of Actuaries.

A *time line* is often helpful for evaluating complex investments. For example, suppose you buy a $1000 CD at time t_0. After one year $2500 is added to the CD at t_1. By time t_2, after another year, your money has grown to $3851 with interest. What rate of interest, called *yield to maturity* (*YTM*), did your money earn? A time line for this situation is shown in Figure 1.

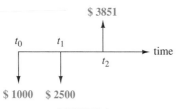

FIGURE 1

Assuming interest is compounded annually at a rate i, and using the compound interest formula gives the following description of the YTM.

$$1000(1 + i)^2 + 2500(1 + i) = 3851$$

To determine the yield to maturity, we must solve this equation for i. Since the quantity $1 + i$ is repeated, let $x = 1 + i$ and first solve the second-degree (quadratic) polynomial equation for x.

$$1000x^2 + 2500x - 3851 = 0$$

We can use the quadratic formula with $a = 1000$, $b = 2500$, and $c = -3851$.

$$x = \frac{-2500 \pm \sqrt{2500^2 - 4(1000)(-3851)}}{2(1000)}$$

We get $x = 1.0767$ and $x = -3.5767$. Since $x = 1 + i$, the two values for i are $.0767 = 7.67\%$ and $-4.5767 = -457.67\%$. We reject the negative value because the final accumulation is greater than the sum of the deposits. In some applications, however, negative rates may be meaningful. By

---

*From *Time, Money, and Polynomials*, COMAP

checking in the first equation, we see that the yield to maturity for the CD is 7.67%.

Now let us consider a more complex but realistic problem. Suppose Curt Reynolds has contributed for four years to a retirement fund. He contributed $6000 at the beginning of the first year. At the beginning of the next three years, he contributed $5840, $4000, and $5200, respectively. At the end of the fourth year, he had $29,912.38 in his fund. The interest rate earned by the fund varied between 21% and -3%, so Reynolds would like to know the $YTM = i$ for his hard-earned retirement dollars. From a time line (see Figure 2), we set up the following equation in $1 + i$ for Reynolds' savings program.

$$6000(1 + i)^4 + 5840(1 + i)^3 + 4000(1 + i)^2$$
$$+ 5200(1 + i) = 29,912.38$$

Let $x = 1 + i$. We need to solve the fourth-degree polynomial equation

$$f(x) = 6000x^4 + 5840x^3 + 4000x^2 + 5200x$$
$$- 29,912.38 = 0.$$

There is no simple way to solve a fourth-degree polynomial equation, so we will use a guess and check method.

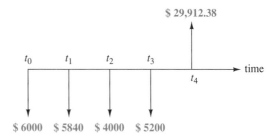

FIGURE 2

We expect that $0 < i < 1$, so that $1 < x < 2$. Let us calculate $f(1)$ and $f(2)$. If there is a change of sign, we will know that there is a solution to $f(x) = 0$ between 1 and 2. We find that

$$f(1) = -8872.38 \text{ and } f(2) = 139,207.62.$$

There is a change in sign as expected. Now we find $f(1.1)$, $f(1.2), f(1.3)$, and so on, and look for another change in sign. (A computer or calculator program would help here to try values of x between 1 and 2.) Right away we find

$$f(1.1) = -2794.74 \quad \text{and} \quad f(1.2) = 4620.74.$$

This process can be repeated for values of x between 1.1 and 1.2, to get $f(1.11) = -2116.59$, $f(1.12) = -1424.88$, $f(1.13) = -719.42$, and $f(1.14) = 0$. We were lucky; the solution for $f(x) = 0$ is exactly 1.14, so $i = YTM = .14 = 14\%$.

EXERCISES

1. Lucinda Turley received $50 on her 16th birthday, and $70 on her 17th birthday, both of which she immediately invested in the bank with interest compounded annually. On her 18th birthday, she had $127.40 in her account. Draw a time line, set up a polynomial equation, and calculate the *YTM*.

2. At the beginning of the year, Jay Beckenstein invested $10,000 at 5% for the first year. At the beginning of the second year, he added $12,000 to the account. The total account earned 4.5% for the second year.
   (a) Draw a time line for this investment.
   (b) How much was in the fund at the end of the second year?
   (c) Set up and solve a polynomial equation and determine the *YTM*. What do you notice about the *YTM*?

3. On January 2 each year for three years, Earl Karn deposited bonuses of $1025, $2200, and $1850, respectively, in an account. He received no bonus the following

year, so he made no deposit. At the end of the fourth year, there was $5864.17 in the account.
   (a) Draw a time line for these investments.
   (b) Write a polynomial equation in x ($x = 1 + i$) and use the guess and check method to find the *YTM* for these investments.

4. Pat Kelley invested yearly in a fund for his children's college education. At the beginning of the first year, he invested $1000; at the beginning of the second year, $2000; at the third through the sixth, $2500 each year; and at the beginning of the seventh, he invested $5000. At the beginning of the eighth year, there was $21,259 in the fund.
   (a) Draw a time line for this investment program.
   (b) Write a seventh-degree polynomial equation in $1 + i$ that gives the *YTM* for this investment program.
    (c) Use a grapher to show that the *YTM* is less than 5.07% and greater than 5.05%.
   (d) Use a grapher to calculate the solution for $1 + i$ and find the *YTM*.

5. People often lose money on investments. Jim Carlson invested $50 at the beginning of each of two years in a mutual fund, and at the end of two years his investment was worth $90.
   (a) Draw a time line and set up a polynomial equation in $1 + i$. Solve for i.
   (c) Examine each negative solution (rate of return on the investment) to see if it has a reasonable interpretation in the context of the problem. To do this, use the compound interest formula on each value of i to trace each $50 payment to maturity.

CHAPTER 6

Systems of Linear Equations and Matrices

Many applications of mathematics require finding the solution of a *system* of first-degree equations. This chapter presents methods for solving such systems, including matrix methods. Matrix algebra and other applications of matrices are also discussed.

6.1 SYSTEMS OF LINEAR EQUATIONS

This section deals with **linear** (or **first-degree) equations** such as

$$2x + 3y = 14 \quad \text{linear equation in two variables,}$$
$$4x - 2y + 5z = 8 \quad \text{linear equation in three variables,}$$

and so on. A **solution** of such an equation is an ordered set of numbers that when substituted for the variables in the order they appear produces a true statement. For instance, $(1, 4)$ is a solution of the equation $2x + 3y = 14$ because substituting $x = 1$ and $y = 4$ produces the true statement $2(1) + 3(4) = 14$. Similarly, $(0, -4, 0)$ is a solution of $4x - 2y + 5z = 8$ because $4(0) - 2(-4) + 5(0) = 8$.

Many applications involve **systems of linear equations,** such as these two:

Two equations in two variables	Three equations in four variables
$5x - 3y = 7$	$2x + y + z = 3$
$2x + 4y = 8$	$x + y + z + w = 5$
	$-4x + z + w = 0$

1 Which of $(-8, 3, 0)$, $(8, -2, -2)$, $(-6, 5, -2)$ are solutions of the system

$$x + 2y + 3z = -2$$
$$2x + 6y + z = 2$$
$$3x + 3y + 10z = -2 ?$$

Answer:
Only $(8, -2, -2)$.

A **solution of a system** is a solution that satisfies *all* the equations in the system. For instance, in the right-hand system of equations above, $(1, 0, 1, 3)$ is a solution of all three equations (check it) and hence is a solution of the system. On the other hand, $(1, 1, 0, 3)$ is a solution of the first two equations but not the third. Hence $(1, 1, 0, 3)$ is not a solution of the system. **1**

SYSTEMS OF TWO EQUATIONS IN TWO VARIABLES The graph of a linear equation in two variables is a straight line (see Section 2.1). The coordinates of each point on the graph represent a solution of the equation. Thus, the solution of a system of two such equations is represented by the point or points where the two lines intersect. There are exactly three geometric possibilities for two lines: they intersect at a single point, or they coincide, or they are distinct and parallel. As illustrated in Figure 6.1, each of these geometric possibilities corresponds to an algebraic outcome for which special terminology is used.

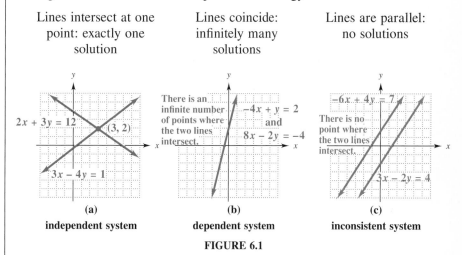

FIGURE 6.1

In theory, every system of two equations in two variables can be solved by graphing the lines and finding their intersection points (if any). In practice, however, algebraic techniques are usually needed to determine the solutions precisely. Such systems can be solved by the **elimination method**, as illustrated in the following examples.

▶**EXAMPLE 1** Solve the system

$$3x - 4y = 1 \tag{1}$$
$$2x + 3y = 12. \tag{2}$$

To eliminate one variable by addition of the two equations, the coefficients of either x or y in the two equations must be additive inverses. For example, let

2 Solve the system of equations

$$3x + 2y = -1$$
$$5x - 3y = 11.$$

Draw the graph of each equation on the same axes.

Answer:
$(1, -2)$

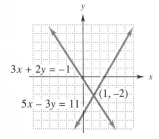

3 Solve the following system.

$$3x - 4y = 13$$
$$12x - 16y = 52$$

Answer:
All ordered pairs that satisfy the equation $3x - 4y = 13$ (or $12x - 16y = 52$).

us choose to eliminate x. We multiply both sides of equation (1) by 2 and both sides of equation (2) by -3 to get

$$6x - 8y = 2 \tag{3}$$
$$-6x - 9y = -36. \tag{4}$$

Adding these equations gives a new equation with just one variable, y. This equation can be solved for y.

$$\begin{array}{r} 6x - 8y = 2 \\ -6x - 9y = -36 \\ \hline -17y = -34 \\ y = 2 \end{array}$$ Variable x is eliminated.

To find the corresponding value of x, substitute 2 for y in either of equations (1) or (2). We choose equation (1).

$$3x - 4(2) = 1$$
$$3x - 8 = 1$$
$$x = 3$$

Therefore, the solution of the system is $(3, 2)$. The graphs of both equations of the system are shown in Figure 6.1(a). They intersect at the point $(3, 2)$, the solution of the equation. ◀ **2**

▶**EXAMPLE 2** Solve the system

$$-4x + y = 2$$
$$8x - 2y = -4.$$

Eliminate x by multiplying both sides of the first equation by 2 and adding the results to the second equation.

$$\begin{array}{r} -8x + 2y = 4 \\ 8x - 2y = -4 \\ \hline 0 = 0 \end{array}$$ Both variables eliminated.

Although both variables have been eliminated, the resulting statement "$0 = 0$" is true, which is the algebraic indication that the two equations have the same graph, as shown in Figure 6.1(b). Therefore the system is dependent and has an infinite number of solutions: every ordered pair that is a solution of the equation $-4x + y = 2$ is a solution of the system. ◀ **3**

▶**EXAMPLE 3** Solve the system

$$3x - 2y = 4$$
$$-6x + 4y = 7.$$

4 Solve the system
$$x - y = 4$$
$$2x - 2y = 3.$$

Draw the graph of each equation on the same axes.

Answer:
No solution.

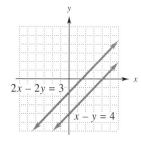

5 Verify that $x = 2$, $y = 1$ is the solution of the system
$$x - 3y = -1$$
$$3x + 2y = 8.$$

(a) Replace the second equation by the sum of itself and -3 times the first equation.

(b) What is the solution of the system in part (a)?

Answers:

(a) The system becomes
$$x - 3y = -1$$
$$11y = 11 \cdot$$

(b) $x = 2$, $y = 1$

The graphs of these equations are parallel lines (each has slope $3/2$), as shown in Figure 6.1(c). Therefore the system has no solutions. However, you do not need the graphs to discover this fact. If you try to solve the system algebraically by multiplying both sides of the first equation by 2 and adding the results to the second equation, you obtain

$$
\begin{aligned}
6x - 4y &= 8 \\
-6x + 4y &= 7 \\
\hline
0 &= 15.
\end{aligned}
$$

The false statement "$0 = 15$" is the algebraic signal that the system is inconsistent and has no solutions. ◀ **4**

LARGER SYSTEMS OF LINEAR EQUATIONS   Two systems of equations are said to be **equivalent** if they have the same solutions. The basic procedure for solving a large system of equations is to use properties of algebra to transform the system into a simpler, equivalent system and then solve this simpler system.

Performing any one of the following **elementary operations** on a system of linear equations produces an equivalent system.

1. Interchange any two equations.
2. Multiply both sides of an equation by a nonzero constant.
3. Replace an equation by the sum of itself and a constant multiple of another equation in the system.

Performing either of the first two elementary operations produces an equivalent system because rearranging the order of the equations or multiplying both sides of an equation by a constant does not affect the solutions of the individual equations, and hence does not affect the solutions of the system. No formal proof will be given here that performing the third elementary operation produces an equivalent system, but Side Problem 5 illustrates this fact. **5**

Example 4 shows how to use elementary operations on a system to eliminate certain variables and produce an equivalent system that is easily solved. The italicized statements provide an outline of the procedure.

▶**EXAMPLE 4**   Use the elimination method to solve the system

$$
\begin{aligned}
2x + y - z &= 2 \\
x + 3y + 2z &= 1 \\
x + y + z &= 2.
\end{aligned}
$$

First, *use elementary operations to produce an equivalent system in which 1 is the coefficient of x in the first equation.* One way to do this is to interchange the first two equations (another would be to multiply both sides of the first equation by 1/2):

$$x + 3y + 2z = 1 \qquad \text{Interchange } R_1, R_2$$
$$2x + y - z = 2$$
$$x + y + z = 2$$

Here and below we use R_1 to denote the first equation in a system, R_2 the second equation, and so on.

Next, *use elementary operations to produce an equivalent system in which the x term has been eliminated from the second and third equations.* To eliminate the x term from the second equation above, replace the second equation by the sum of itself and -2 times the first equation:

$$
\begin{array}{rl}
-2R_1 & -2x - 6y - 4z = -2 \\
R_2 & \underline{2x + y - z = 2} \\
-2R_1 + R_2 & -5y - 5z = 0
\end{array}
$$

$$x + 3y + 2z = 1$$
$$-5y - 5z = 0 \qquad -2R_1 + R_2 \longleftarrow$$
$$x + y + z = 2.$$

To eliminate the x term from the third equation of this last system, replace the third equation by the sum of itself and -1 times the first equation:

$$
\begin{array}{rl}
-1R_1 & -x - 3y - 2z = -1 \\
R_3 & \underline{x + y + z = 2} \\
-1R_1 + R_3 & -2y - z = 1
\end{array}
$$

$$x + 3y + 2z = 1$$
$$-5y - 5z = 0$$
$$-2y - z = 1 \qquad -1R_1 + R_3 \longleftarrow$$

Now that x has been eliminated from all but the first equation, we ignore the first equation and work on the remaining ones. *Use elementary operations to produce an equivalent system in which 1 is the coefficient of y in the second equation.* This can be done by multiplying the second equation in the system above by $-1/5$:

$$x + 3y + 2z = 1$$
$$y + z = 0 \qquad -\frac{1}{5}R_2$$
$$-2y - z = 1$$

6 Use the elimination method to solve each system.

(a) $2x + y = -1$

$\quad x + 3y = 2$

(b) $2x - y + 3z = 2$

$\quad x + 2y - z = 6$

$\quad -x - y + z = -5$

Answers:

(a) $(-1, 1)$

(b) $(3, 1, -1)$

Then *use elementary operations to obtain an equivalent system in which y has been eliminated from the third equation:* replace the third equation by the sum of itself and 2 times the second equation:

$$x + 3y + 2z = 1$$
$$y + z = 0$$
$$z = 1. \quad 2R_2 + R_3$$

The solution of the third equation is obvious: $z = 1$. Now work backwards in the system. Substitute 1 for z in the second equation and solve for y, obtaining $y = -1$. Finally, substitute 1 for z and -1 for y in the first equation and solve for x, obtaining $x = 2$. This process is known as **back substitution.** When it is finished we have the solution of the original system, namely, $(2, -1, 1)$. ◄

The procedure used in Example 4 to eliminate variables and produce a system in which back substitution works can be carried out with any system, as summarized below. In this summary, the first variable that appears in an equation with nonzero coefficient is called the **leading variable** of that equation and its nonzero coefficient is called the **leading coefficient.**

The Elimination Method for Solving Large Systems of Linear Equations

Use elementary operations to transform the given system into an equivalent one in which the leading coefficient in each equation is 1 and the leading variable in each equation does not appear in any later equations. This can be done systematically as follows:

1. Make the leading coefficient of the first equation, 1.
2. Eliminate the leading variable of the first equation from each later equation by replacing the later equation by the sum of itself and a suitable multiple of the first equation.
3. Repeat Steps 1 and 2 for the second equation: make its leading coefficient 1 and eliminate its leading variable from each later equation by replacing the later equation by the sum of itself and a suitable multiple of the second equation.
4. Repeat Steps 1 and 2 for the third equation, fourth equation, and so on, until it is not possible to go further.

Then solve the resulting system by back substitution.

At various stages in the elimination process, you may have a choice of elementary operations that can be used. As long as the final result is a system in which back substitution can be used, the choice does not matter. To avoid unnecessary errors, choose elementary operations that minimize the amount of computation and, as far as possible, avoid complicated fractions. **6**

DEPENDENT AND INCONSISTENT SYSTEMS The possible number of solutions of a system with more than two variables or equations is the same as for smaller systems. Such a system has exactly one solution (**independent system**); or infinitely many solutions (**dependent system**); or no solutions (**inconsistent system**). The elimination method will always produce the unique solution of an independent system. It also provides a useful way of describing the infinitely many solutions of a dependent system, as we now see.

▶**EXAMPLE 5**   Solve the system

$$2x - 3y + 4z = 6$$
$$x - 2y + z = 9.$$

Use the steps of the elimination method as far as possible. Exchange the two equations so that the coefficient of x is 1 in the first equation of the system. This gives

$$x - 2y + z = 9 \quad \text{\textbf{Interchange } } R_1 \text{ \textbf{and} } R_2 \qquad \textbf{(1)}$$
$$2x - 3y + 4z = 6. \qquad\qquad\qquad\qquad\qquad\qquad \textbf{(2)}$$

To eliminate the x term from equation (2), multiply both sides of equation (1) by -2 and add it to equation (2). The new system is:

$$x - 2y + z = 9 \qquad\qquad\qquad\qquad \textbf{(1)}$$
$$y + 2z = -12 \quad -2R_1 + R_2 \qquad \textbf{(3)}$$

Since there are only two equations, it is not possible to continue the elimination process. We have a system in which one variable (namely, z) is not the leading variable of an equation. This indicates a dependent system. Its solutions can be found as follows. Solve equation (3) for y:

$$y = -2z - 12$$

Now substitute the result for y in equation (1), and solve for x.

$$x - 2y + z = 9$$
$$x - 2(-2z - 12) + z = 9$$
$$x + 4z + 24 + z = 9$$
$$x + 5z = -15$$
$$x = -5z - 15$$

Each choice of a value for z leads to values for x and y. For example,

if $z = 1$,
then $x = -5(1) - 15 = -20$ and $y = -2(1) - 12 = -14$;
if $z = -6$,
then $x = 15$ and $y = 0$;
if $z = 0$,
then $x = -15$ and $y = -12$.

7 Use the following values of z to find additional solutions for the system of Example 5.

(a) $z = 7$

(b) $z = -14$

(c) $z = 5$

Answers:

(a) $(-50, -26, 7)$

(b) $(55, 16, -14)$

(c) $(-40, -22, 5)$

There are infinitely many solutions for the original system, since z can take on infinitely many values. The solutions are all triples

$$(-5z - 15, -2z - 12, z),$$

where z is any real number. ◀ **7**

Since both x and y in Example 5 were expressed in terms of z, the variable z is called a **parameter.** If we solved the system in a different way, x or y could be the parameter. The system in Example 5 had one more variable than equations. If there are two more variables than equations, there usually will be two parameters, and so on.

Whenever there are more variables than equations, as in Example 5, then the system cannot have a unique solution. It must be either dependent (infinitely many solutions) or inconsistent (no solutions).

When a system is inconsistent, the elimination process will indicate this fact, as in the next example.

▶**EXAMPLE 6**   Solve the system

$$
\begin{aligned}
4x + 12y + 8z &= -4 \\
2x + 8y + 5z &= 0 \\
3x + 9y + 6z &= 2 \\
3x + 2y - z &= 6
\end{aligned}
$$

Multiply both sides of the first equation by $1/4$ to make the coefficient of x equal to 1:

$$
\begin{aligned}
x + 3y + 2z &= -1 &\quad& \tfrac{1}{4}R_1 \\
2x + 8y + 5z &= 0 \\
3x + 9y + 6z &= 2 \\
3x + 2y - z &= 6
\end{aligned}
$$

Now eliminate the x term from the last three equations by adding suitable multiples of the first equation to each of them, as indicated here:

$$
\begin{aligned}
x + 3y + 2z &= -1 \\
2y + z &= 2 &\quad& -2R_1 + R_2 \\
3x + 9y + 6z &= 2 \\
3x + 2y - z &= 6
\end{aligned}
$$

$$
\begin{aligned}
x + 3y + 2z &= -1 \\
2y + z &= 2 \\
0x + 0y + 0z &= 5 &\quad& -3R_1 + R_3 \\
3x + 2y - z &= 6
\end{aligned}
$$

Stop! The third equation has no solutions because its left side is always 0 and the right side is 5, resulting in the false statement "0 = 5". Therefore the entire system cannot have any solutions and is inconsistent. ◀

Caution It is possible for the elimination method to result in an equation that is 0 on *both* sides, such as $0x + 0y + 0z = 0$. Unlike the situation in Example 6, such an equation has infinitely many solutions, so a system that contains it may also have solutions.

As a general rule, there is no way to determine in advance whether a system is independent, dependent, or inconsistent. So you should carry out the elimination process as far as necessary. If you obtain an equation of the form $0 = c$ for some nonzero c (as in Example 6), then the system is inconsistent and has no solution. Otherwise, the system will either be independent with a unique solution (Example 4) or dependent with infinitely many solutions (Example 5).

APPLICATIONS The mathematical techniques in this text will be useful to you only if you are able to apply them to practical problems. To do this, always begin by reading the problem carefully. Next, identify what must be found. Let each unknown quantity be represented by a variable. (It is a good idea to *write down* exactly what each variable represents.) Now reread the problem, looking for all necessary data. Write that down, too. Finally, look for one or more sentences that lead to equations or inequalities.

▶**EXAMPLE 7** An animal feed is to be made from corn, soybeans, and cottonseed. Determine how many units of each ingredient are needed to make a feed that supplies 1800 units of fiber, 2800 units of fat, and 2200 units of protein, given that one unit of each ingredient provides the numbers of units shown in the table below. The table states, for example, that a unit of corn provide 10 units of fiber, 30 units of fat, and 20 units of protein.

	Corn	Soybeans	Cottonseed
Units of Fiber	10	20	30
Units of Fat	30	20	40
Units of Protein	20	40	25

Let x represent the required number of units of corn, y the number of units of soybeans, and z the number of units of cottonseed. Since the total amount of fiber is to be 1800,

$$10x + 20y + 30z = 1800.$$

The feed must supply 2800 units of fat, so

$$30x + 20y + 40z = 2800.$$

8 List a sequence of elementary operations that will transform system **(1)** in Example 7 into system **(2)**.

Answer:
Many sequences are possible, including this one:

replace R_1 by $\frac{1}{10}R_1$;

replace R_2 by $\frac{1}{10}R_2$;

replace R_3 by $\frac{1}{5}R_3$;

replace R_2 by $-3R_1 + R_2$;
replace R_3 by $-4R_1 + R_3$;

replace R_2 by $-\frac{1}{4}R_2$;

replace R_3 by $-\frac{1}{7}R_3$.

Finally, since 2200 units of protein are required,

$$20x + 40y + 25z = 2200.$$

Thus we must solve this system of equations:

$$\begin{aligned}10x + 20y + 30z &= 1800 \\ 30x + 20y + 40z &= 2800 \\ 20x + 40y + 25z &= 2200\end{aligned} \qquad (1)$$

The elimination method leads to the following equivalent system, as shown in problem 8 at the side. **8**

$$\begin{aligned}x + 2y + 3z &= 180 \\ y + \frac{5}{4}z &= 65 \\ z &= 40\end{aligned} \qquad (2)$$

Back substitution now shows that $z = 40$,

$$y = 65 - \frac{5}{4}(40) = 15 \quad \text{and} \quad x = 180 - 2(15) - 3(40) = 30.$$

Thus, the feed should contain 30 units of corn, 15 units of soybeans, and 40 units of cottonseed. ◀

▶**EXAMPLE 8** Kelly Karpet Kleaners sells rug cleaning machines. The EZ model weighs 10 pounds and comes in a 10 cubic ft box. The compact model weighs 20 pounds and comes in an 8 cubic ft box. The commercial model weights 60 pounds and comes in a 28 cubic ft box. Each of their delivery vans has 248 cubic feet of space and can hold a maximum of 440 pounds. In order for a van to be fully loaded, how many of each model should it carry?

Let x be the number of EZ, y the number of compact, and z the number of commercial models carried by a van. Then we can summarize the information in this chart:

Model	Number	Weight	Volume
EZ	x	10	10
Compact	y	20	8
Commercial	z	60	28
Total for van load		$10x + 20y + 60z$	$10x + 8y + 28z$

9 List a sequence of elementary operations that will transform system **(1)** in Example 8 into system **(2)**.

Answer:
Many sequences are possible, including this one:

replace R_1 by $\frac{1}{10}R_1$;

replace R_2 by $\frac{1}{2}R_2$;

replace R_2 by $-5R_1 + R_2$;

replace R_2 by $-\frac{1}{6}R_2$.

Since a fully loaded van can carry 440 pounds and 248 cubic feet, we must solve this system of equations:

$$10x + 20y + 60z = 440 \quad \text{weight equation}$$
$$10x + 8y + 28z = 248 \quad \text{volume equation} \tag{1}$$

As shown in problem 9 at the side, the elimination method produces the following equivalent system. **9**

$$x + 2y + 6z = 44$$
$$y + \frac{8}{3}z = 16 \tag{2}$$

Solving this dependent system by back substitution, we have

$$y = 16 - \frac{8}{3}z$$

$$x = 44 - 2y - 6z = 44 - 2\left(16 - \frac{8}{3}z\right) - 6z = 12 - \frac{2}{3}z.$$

so that all solutions of the system are given by $\left(12 - \frac{2}{3}z, 16 - \frac{8}{3}z, z\right)$. The only solutions that apply in this situation, however, are those given by $z = 0, 3$ or 6 because all other values of z lead to fractions or negative numbers (you can't deliver a negative number of boxes, or part of a box). Hence, there are three ways to have a fully loaded van:

Solution	*Van Load*
(12, 16, 0)	12 EZ, 16 compact, 0 commercial
(10, 8, 3)	10 EZ, 8 compact, 3 commercial
(8, 0, 6)	8 EZ, 0 compact, 6 commercial ◀

6.1 EXERCISES

Determine whether the given ordered set of numbers is a solution of the system of equations.

1. $(-1, 3)$

$$2x + y = 1$$
$$-3x + 2y = 9$$

2. $(2, 1.5, -.5)$

$$3x + 4y - 2z = -.5$$
$$.5x + 8z = -3$$
$$x - 3y + 5z = -5$$

Solve each of the following systems of two equations in two variables. (See Examples 1–3.)

3. $x - 2y = 5$
$2x + y = 3$

4. $3x - y = 1$
$-x + 2y = 4$

5. $2x - 2y = 12$
$-2x + 3y = 10$

6. $3x + 2y = -4$
$4x - 2y = -10$

7. $x + 3y = -1$
$\quad 2x - y = 5$

8. $4x - 3y = -1$
$\quad x + 2y = 19$

9. $2x + 3y = 15$
$\quad 8x + 12y = 40$

10. $2x + 5y = 8$
$\quad 6x + 15y = 18$

11. $2x - 8y = 2$
$\quad 3x - 12y = 3$

12. $3x - 2y = 4$
$\quad 6x - 4y = 8$

13. $3x + 2y = 5$
$\quad 6x + 4y = 8$

14. $\quad 9x - 5y = 1$
$\quad -18x + 10y = 1$

In Exercises 15–18, multiply both sides of each equation by a common denominator to eliminate the fractions. Then solve the system.

15. $\dfrac{x}{2} + \dfrac{y}{3} = 8$

$\dfrac{2x}{3} + \dfrac{3y}{2} = 17$

16. $\dfrac{x}{5} + 3y = 31$

$2x - \dfrac{y}{5} = 8$

17. $\dfrac{x}{2} + y = \dfrac{3}{2}$

$\dfrac{x}{3} + y = \dfrac{1}{3}$

18. $x + \dfrac{y}{3} = -6$

$\dfrac{x}{5} + \dfrac{y}{4} = -\dfrac{7}{4}$

In Exercises 19–24, carry out the elimination method as far as necessary to determine whether the system is independent, dependent, or inconsistent. (See Examples 4–6.)

19. $x + 2y \quad = 0$
$\quad y - z = 2$
$\quad x + y + z = -2$

20. $x + 2y + z = 0$
$\quad y + 2z = 0$
$\quad x + y - z = 0$

21. $x + 2y + 4z = 6$
$\quad y + z = 1$
$\quad x + 3y + 5z = 10$

22. $\quad x + y + 2z + 3w = 1$
$\quad 2x + y + 3z + 4w = 1$
$\quad 3x + y + 4z + 5w = 2$

23. $\quad a - 3b - 2c = -3$
$\quad 3a + 2b - c = 12$
$\quad -a - b + 4c = 3$

24. $2x + 2y + 2z = 6$
$\quad 3x - 3y - 4z = -1$
$\quad x + y + 3z = 11$

Use the elimination method to solve each of the following systems. If the system is dependent, express the solutions in terms of the parameter z or w (whichever is the last variable). (See Examples 4–6.)

25. $x + y + z = 2$
$\quad 2x + y - z = 5$
$\quad x - y + z = -2$

26. $2x + y + z = 9$
$\quad -x - y + z = 1$
$\quad 3x - y + z = 9$

27. $x + 3y + 4z = 14$
$\quad 2x - 3y + 2z = 10$
$\quad 3x - y + z = 9$

28. $4x - y + 3z = -2$
$\quad 3x + 5y - z = 15$
$\quad -2x + y + 4z = 14$

29. $x + 2y + 3z = 8$
$\quad 3x - y + 2z = 5$
$\quad -2x - 4y - 6z = 5$

30. $3x - 2y - 8z = 1$
$\quad 9x - 6y - 24z = -2$
$\quad x - y + z = 1$

31. $2x - 4y + z = -4$
$\quad x + 2y - z = 0$
$\quad -x + y + z = 6$

32. $4x - 3y + z = 9$
$\quad 3x + 2y - 2z = 4$
$\quad x - y + 3z = 5$

33. $5x + 3y + 4z = 19$
$\quad 3x - y + z = -4$

34. $3x + y - z = 0$
$\quad 2x - y + 3z = -7$

35. $x + 2y + 3z = 11$
$\quad 2x - y + z = 2$

36. $-x + y - z = -7$
$\quad 2x + 3y + z = 7$

37. $11x + 10y + 9z = 5$
$\quad x + 2y + 3z = 1$
$\quad 3x + 2y + z = 1$

38. $-x + 2y - 3z + 4w = 8$
$\quad 2x - 4y + z + 2w = -3$
$\quad 5x - 4y + z + 2w = -3$

39. $x + y = 3$
$\quad 5x - y = 3$
$\quad 9x - 4y = 1$

40. $x + y + z + w = 10$
$\quad x + y + 2z = 11$
$\quad x - 3y + w = -14$
$\quad y + 3z - w = 7$

41. Find constants a, b, c such that the points $(2, 3), (-1, 0)$, and $(-2, 2)$ lie on the graph of the equation $y = ax^2 + bx + c$. (*Hint:* Since $(2, 3)$ is on the graph, we must have $3 = a(2^2) + b(2) + c$, that is, $4a + 2b + c = 3$. Similarly, the other two points lead to two more equations. Solve the resulting system for a, b, c.)

42. (a) Find the equation of the straight line through $(1, 2)$ and $(3, 4)$.
   (b) Find the equation of the line through $(-1, 1)$ with slope 3.
   (c) Find a point that lies on both the lines in (a) and (b).

43. Graph the equations in the following system. Then explain why the graphs show that the system is inconsistent.

$$2x + 3y = 8$$
$$x - y = 4$$
$$5x + y = 7$$

44. Explain why a system with more variables than equations cannot have a unique solution (that is, be an independent system). (*Hint:* When you apply the elimination method to such a system, what must necessarily happen?)

Work the following problems by writing and solving a system of two equations in two variables. (See Examples 7–8.)

45. Management Shirley Cicero has $16,000 invested in Boeing and GE stock. The Boeing stock currently sells for $30 a share and the GE stock for $70 a share. If GE stock triples in value and Boeing stock goes up 50%, her stock will be worth $34,500. How many shares of each stock does she own?

46. Management The price at which grapefruit is sold affects the consumer demand for grapefruit according to the equation

$$2p = -.2q + 5,$$

where p is the price per pound (in dollars) at which consumers will demand q thousand pounds of grapefruit. The amount q of grapefruit that producers will supply at price p is governed by the equation $5p = .3q + 5.3$. Find the equilibrium quantity and the equilibrium price. (That is, find the quantity and price at which supply equals demand, or the values of p and q that satisfy both equations.)

47. Management If 20 pounds of rice and 10 pounds of potatoes cost $16.20 and 30 pounds of rice and 12 pounds of potatoes cost $23.04, how much will 10 pounds of rice and 50 pounds of potatoes cost?

48. Management An apparel shop sells skirts for $45 and blouses for $35. Its entire stock is worth $51,750. But sales are slow and only half the skirts and two thirds of the blouses are sold, for a total of $30,600. How many skirts and blouses are left in the store?

49. Management A theater charges $4 for main floor seats and $2.50 for balcony seats. If all seats are sold, the ticket income is $2100. At one show, 25% of the main floor seats and 40% of the balcony seats were sold and ticket income was $600. How many seats are on the main floor and how many are in the balcony?

50. Management A company produces two models of bicycles, model 201 and model 301. Model 201 requires 2 hours of assembly time and model 301 requires 3 hours of assembly time. The parts for model 201 cost $25 per bike and the parts for model 301 cost $30 per bike. If the company has a total of 34 hours of assembly time and $365 available per day for these two models, how many of each can be made in a day?

Work the following problems by writing and solving a suitable system of equations. (See Examples 7–8.)

51. Management Juanita invests $10,000, received from her grandmother, in three ways. With one part, she buys mutual funds which offer a return of 8% per year. The second part, which amounts to twice the first, is used to buy government bonds at 9% per year. She puts the rest in the bank at 5% annual interest. The first year her investments bring a return of $830. How much did she invest in each way?

52. Management To get the necessary funds for a planned expansion, a small company took out three loans totaling $25,000. The company was able to borrow some of the money at 16%. They borrowed $2,000 more than one half the amount of the 16% loan at 20%, and the rest at 18%. The total annual interest was $4440. How much did they borrow at each rate?

53. Natural Science Three brands of fertilizer are available that provide nitrogen, phosphoric acid, and soluble potash to the soil. One bag of each brand provides the following units of each nutrient.

Brand	Nutrient		
	Nitrogen	Phosphoric Acid	Potash
A	1	3	2
B	2	1	0
C	3	2	1

For ideal growth, the soil in a certain country needs 18 units of nitrogen, 23 units of phosphoric acid, and 13 units of potash per acre. How many bags of each brand of fertilizer should be used per acre for ideal growth?

54. Management   A furniture manufacturer has 1950 machine hours available each week in the cutting department, 1490 hours in the assembly department, and 2160 in the finishing department. Manufacturing a chair requires .2 hours of cutting, .3 hours of assembly, and .1 hours of finishing. A cabinet requires .5 hours of cutting, .4 hours of assembly, and .6 hours of finishing. A buffet requires .3 hours of cutting, .1 hours of assembly, and .4 hours of finishing. How many chairs, cabinets, and buffets should be produced in order to use all the available production capacity?

55. Management   A company produces three color television sets: models X, Y, and Z. Each model X set requires 2 hours of electronics work, 2 hours of assembly time, and 1 hour of finishing time. Each model Y requires 1, 3, and 1 hours of electronics, assembly, and finishing time, respectively. Each model Z requires 3, 2, and 2 hours of the same work, respectively. There are 100 hours available for electronics, 100 hours available for assembly, and 65 hours available for finishing per week. How many of each model should be produced each week if all available time must be used?

56. Management   Turley Tailor Inc. makes long-sleeve, short-sleeve, and sleeveless blouses. A sleeveless blouse requires .5 hours of cutting and .6 hours of sewing. A short-sleeve blouse requires 1 hour of cutting and .9 hours of sewing. A long-sleeve blouse requires 1.5 hours of cutting and 1.2 hours of sewing. There are 380 hours of labor available in the cutting department each day and 330 hours in the sewing department. If the plant is to run at full capacity, how many of each type of blouse should be made each day?

57. Management   Felsted Furniture makes dining room furniture. A buffet requires 30 hours for construction and 10 hours for finishing. A chair requires 10 hours for construction and 10 hours for finishing. A table requires 10 hours for construction and 30 hours for finishing. The construction department has 350 hours of labor and the finishing department 150 hours of labor available each week. How many pieces of each type of furniture should be produced each week if the factory is to run at full capacity?

58. Natural Science   **(a)** A hospital dietician is planning a special diet for a certain patient. The total amount per meal of food groups A, B, and C must equal 400 grams. The diet should include one-third as much of group A as of group B, and the sum of the amounts of group A and group C should equal twice the amount of group B. How many grams of each food group should be included?

(b) Suppose we drop the requirement that the diet include one-third as much of group A as of group B. Describe the set of all possible solutions.

(c) Suppose that, in addition to the conditions given in part (a), foods A and B cost 2 cents per gram and food C costs 3 cents per gram, and that a meal must cost $8. Is a solution possible?

59. Management   At a pottery factory, fuel consumption for heating the kilns varies with the size of the order being fired. In the past, the company recorded these figures.

x = Number of Platters	y = Fuel Cost per Platter
6	$2.80
8	2.48
10	2.24

(a) Find an equation of the form $y = ax^2 + bx + c$ whose graph contains the three points corresponding to the data in the table. (See Exercise 41.)

(b) How many platters should be fired at one time in order to minimize the fuel cost per platter? What is the minimum fuel cost per platter?

60. Management   The business analyst for Melcher Manufacturing wants to find an equation that can be used to project sales of a relatively new product. For the years 1992, 1993, and 1994, sales were $15,000, $32,000, and $123,000, respectively.

(a) Graph the sales for the years 1992, 1993, and 1994, letting the year 1992 equal 0 on the x-axis. Let the values on the vertical axis be in thousands. (For example, the point (1993, 32,000) will be graphed as (1, 32).)

(b) Find the equation of the straight line $ax + by = c$ through the points for 1992 and 1994.

(c) Find the equation of the parabola $y = ax^2 + bx + c$ through the three given points.

(d) Find the projected sales for 1997 first by using the equation from part (b) and then by using the equation from part (c). If you were to estimate sales of the product in 1997, which result would you choose? Why?

1 (a) Write the augmented matrix of this system:

$$4x - 2y + 3z = 4$$
$$3x + 5y + z = -7$$
$$5x - y + 4z = 6$$

(b) Write the system of equations associated with this augmented matrix:

$$\begin{bmatrix} 2 & -2 & | & -2 \\ 1 & 1 & | & 4 \\ 3 & 5 & | & 8 \end{bmatrix}$$

Answers:

(a) $$\begin{bmatrix} 4 & -2 & 3 & | & 4 \\ 3 & 5 & 1 & | & -7 \\ 5 & -1 & 4 & | & 6 \end{bmatrix}$$

(b)
$$2x - 2y = -2$$
$$x + y = 4$$
$$3x + 5y = 8$$

6.2 THE GAUSS-JORDAN METHOD

Since the variables in a system of linear equations remain unchanged during the solution process, we really need to keep track of just the coefficients and the constants. For instance, consider the system in Example 4 of the previous section:

$$2x + y - z = 2$$
$$x + 3y + 2z = 1$$
$$x + y + z = 2.$$

This system can be written in an abbreviated form as

$$\begin{bmatrix} 2 & 1 & -1 & 2 \\ 1 & 3 & 2 & 1 \\ 1 & 1 & 1 & 2 \end{bmatrix}.$$

Such a rectangular array of numbers, consisting of horizontal **rows** and vertical **columns,** is called a **matrix** (plural **matrices**). Each number in the array is an **element** or **entry.** To separate the constants in the last column of the matrix from the coefficients of the variables, we use a vertical line, producing the following **augmented matrix.**

$$\begin{bmatrix} 2 & 1 & -1 & | & 2 \\ 1 & 3 & 2 & | & 1 \\ 1 & 1 & 1 & | & 2 \end{bmatrix}$$ **1**

The rows of the augmented matrix can be transformed in the same way as the equations of the system, since the matrix is just a shortened form of the system. The following **row operations** on the augmented matrix correspond to the elementary operations used on systems of equations in the previous section.

Performing any one of the following **row operations** on the augmented matrix of a system of linear equations produces the augmented matrix of an equivalent system.

1. Interchange any two rows.
2. Multiply each element of a row by a nonzero constant.
3. Replace a row by the sum of itself and a constant multiple of another row of the matrix.

2 Perform the following row operations on the matrix

$$\begin{bmatrix} -1 & 5 \\ 3 & -2 \end{bmatrix}.$$

(a) Interchange R_1 and R_2.

(b) $2R_1$

(c) Replace R_2 by $-3R_1 + R_2$.

(d) Replace R_1 by $2R_2 + R_1$.

Answers:

(a) $\begin{bmatrix} 3 & -2 \\ -1 & 5 \end{bmatrix}$

(b) $\begin{bmatrix} -2 & 10 \\ 3 & -2 \end{bmatrix}$

(c) $\begin{bmatrix} -1 & 5 \\ 6 & -17 \end{bmatrix}$

(d) $\begin{bmatrix} 5 & 1 \\ 3 & -2 \end{bmatrix}$

Row operations on a matrix are denoted by the same notation used earlier for elementary operations on a system of equations. For example, $2R_3 + R_1$ indicates the sum of 2 times row 3 and row 1. **2**

THE GAUSS-JORDAN METHOD   The **Gauss-Jordan method** for solving a system of equations is an extension of the elimination method. It uses row operations on the augmented matrix, as illustrated in the next example.

▶**EXAMPLE 1**   Solve the system

$$\begin{aligned} x - 2y &= 6 - 4z \\ x + 13z &= 6 - y \\ -2x + 6y - z &= -10 \end{aligned}$$

First, put the system in proper form, with the constants on the right side of the equal sign and the terms with variables on the left side of the equal sign in the same order in each equation. Then write the augmented matrix of the system:

System of Equations	*Augmented Matrix*

$$\begin{aligned} x - 2y + 4z &= 6 \\ x + y + 13z &= 6 \\ -2x + 6y - z &= -10 \end{aligned} \qquad \begin{bmatrix} 1 & -2 & 4 & 6 \\ 1 & 1 & 13 & 6 \\ -2 & 6 & -1 & -10 \end{bmatrix}$$

The first part of the Gauss-Jordan method is the same as the elimination method, except that row operations are used on the augmented matrix instead of elementary operations on the corresponding system of equations, as shown in this side-by-side comparison.

Elimination Method

Replace the second equation by the sum of itself and -1 times the first equation.

$$\begin{aligned} x - 2y + 4z &= 6 \\ 3y + 9z &= 0 \\ -2x + 6y - z &= -10 \end{aligned}$$

Replace the third equation by the sum of itself and 2 times the first equation.

$$\begin{aligned} x - 2y + 4z &= 6 \\ 3y + 9z &= 0 \\ 2y + 7z &= 2 \end{aligned}$$

Gauss-Jordan Method

Replace the second row by the sum of itself and -1 times the first row.

$$0 \leftarrow -1R_1 + R_2 \rightarrow \begin{bmatrix} 1 & -2 & 4 & 6 \\ 0 & 3 & 9 & 0 \\ -2 & 6 & -1 & -10 \end{bmatrix}$$

Replace the third row by the sum of itself and 2 times the first row.

$$\leftarrow 2R_1 + R_3 \rightarrow \begin{bmatrix} 1 & -2 & 4 & 6 \\ 0 & 3 & 9 & 0 \\ 0 & 2 & 7 & 2 \end{bmatrix}$$

Multiply both sides of the second equation by 1/3.

Multiply each element of row 2 by 1/3.

$$
\begin{aligned}
x - 2y + 4z &= 6 \\
y + 3z &= 0 \\
2y + 7z &= 2
\end{aligned}
\qquad \xleftarrow{\;\frac{1}{3}R_2\;}
\qquad
\begin{bmatrix}
1 & -2 & 4 & \bigm| & 6 \\
0 & 1 & 3 & \bigm| & 0 \\
0 & 2 & 7 & \bigm| & 2
\end{bmatrix}
$$

Replace the third equation by the sum of itself and -2 times the second equation.

Replace the third row by the sum of itself and -2 times the second row.

$$
\begin{aligned}
x - 2y + 4z &= 6 \\
y + 3z &= 0 \\
z &= 2
\end{aligned}
\qquad \xleftarrow{\;-2R_2 + R_3\;}
\qquad
\begin{bmatrix}
1 & -2 & 4 & \bigm| & 6 \\
0 & 1 & 3 & \bigm| & 0 \\
0 & 0 & 1 & \bigm| & 2
\end{bmatrix}
$$

At this point, back substitution could be used to solve the system. In the Gauss-Jordan method, however, additional elimination of variables replaces back substitution, as follows.

Replace the second equation by the sum of itself and -3 times the third equation.

Replace the second row by the sum of itself and -3 times the third row.

$$
\begin{aligned}
x - 2y + 4z &= 6 \\
y &= -6 \\
z &= 2
\end{aligned}
\qquad \xleftarrow{\;-3R_3 + R_2\;}
\qquad
\begin{bmatrix}
1 & -2 & 4 & \bigm| & 6 \\
0 & 1 & 0 & \bigm| & -6 \\
0 & 0 & 1 & \bigm| & 2
\end{bmatrix}
$$

Replace the first equation by the sum of itself and -4 times the third equation.

Replace the first row by the sum of itself and -4 times the third row.

$$
\begin{aligned}
x - 2y &= -2 \\
y &= -6 \\
z &= 2
\end{aligned}
\qquad \xleftarrow{\;-4R_3 + R_1\;}
\qquad
\begin{bmatrix}
1 & -2 & 0 & \bigm| & -2 \\
0 & 1 & 0 & \bigm| & -6 \\
0 & 0 & 1 & \bigm| & 2
\end{bmatrix}
$$

Replace the first equation by the sum of itself and 2 times the second equation.

Replace the first row by the sum of itself and 2 times the second row.

$$
\begin{aligned}
x &= -14 \\
y &= -6 \\
z &= 2
\end{aligned}
\qquad \xleftarrow{\;2R_2 + R_1\;}
\qquad
\begin{bmatrix}
1 & 0 & 0 & \bigm| & -14 \\
0 & 1 & 0 & \bigm| & -6 \\
0 & 0 & 1 & \bigm| & 2
\end{bmatrix}
$$

The solution of the system is now obvious; it is $(-14, -6, 2)$. Note that this solution is the last column of the final augmented matrix. ◀

3 Use the Gauss-Jordan method to solve the system

$$x + 2y = 11$$
$$-4x + y = -8,$$

as follows. Give the shorthand notation and the new matrix in (b)–(d).

(a) Set up the augmented matrix.

(b) Get 0 in row two, column one.

(c) Get 1 in row two, column two.

(d) Finally, get 0 in row one, column two.

(e) The solution for the system is _____ .

Answers:

(a) $\begin{bmatrix} 1 & 2 & | & 11 \\ -4 & 1 & | & -8 \end{bmatrix}$

(b) $4R_1 + R_2; \begin{bmatrix} 1 & 2 & | & 11 \\ 0 & 9 & | & 36 \end{bmatrix}$

(c) $\frac{1}{9}R_2; \begin{bmatrix} 1 & 2 & | & 11 \\ 0 & 1 & | & 4 \end{bmatrix}$

(d) $-2R_2 + R_1; \begin{bmatrix} 1 & 0 & | & 3 \\ 0 & 1 & | & 4 \end{bmatrix}$

(e) $(3, 4)$

In the Gauss-Jordan method row operations can be performed in any order, provided they eventually lead to the augmented matrix of a system in which the leading variable of each equation does not appear in any other equation of the system. When there is a unique solution, as in Example 1, this final system will be of the form $x =$ constant, $y =$ constant, $z =$ constant, and so on. **3**

It is best to transform the matrix systematically. Either follow the procedure in Example 1 (which first puts the system in a form where back substitution could be used and then eliminates additional variable terms) or work column by column from left to right, as in the next example.

▶**EXAMPLE 2** Use the Gauss-Jordan method to solve the system

$$\begin{aligned} x \quad\quad + 5z &= -6 + y \\ 3x + 3y \quad\quad &= 10 + z \\ x + 3y + 2z &= 5. \end{aligned}$$

The system must first be rewritten in proper form as follows.

$$\begin{aligned} x - y + 5z &= -6 \\ 3x + 3y - z &= 10 \\ x + 3y + 2z &= 5 \end{aligned}$$

Begin the solution by writing the augmented matrix of the linear system.

$$\begin{bmatrix} 1 & -1 & 5 & | & -6 \\ 3 & 3 & -1 & | & 10 \\ 1 & 3 & 2 & | & 5 \end{bmatrix}$$

The first element in column one is already 1. Get 0 for the second element in column one by multiplying each element in the first row by -3 and adding the results to the corresponding elements in row two.

$$\begin{bmatrix} 1 & -1 & 5 & | & -6 \\ 0 & 6 & -16 & | & 28 \\ 1 & 3 & 2 & | & 5 \end{bmatrix} \quad -3R_1 + R_2$$

Now, change the first element in row three to 0 by multiplying each element of the first row by -1 and adding the results to the corresponding elements of the third row.

$$\begin{bmatrix} 1 & -1 & 5 & | & -6 \\ 0 & 6 & -16 & | & 28 \\ 0 & 4 & -3 & | & 11 \end{bmatrix} \quad -1R_1 + R_3$$

4 Continue the solution of the system in Example 2 as follows. Give the shorthand notation and the matrix for each step.

(a) Get 1 in row two, column two.

(b) Get 0 in row one, column two.

(c) Now get 0 in row three, column two.

Answers:

(a) $\frac{1}{6}R_2$;

$$\begin{bmatrix} 1 & -1 & 5 & | & -6 \\ 0 & 1 & -\frac{8}{3} & | & \frac{14}{3} \\ 0 & 4 & -3 & | & 11 \end{bmatrix}$$

(b) $1R_2 + R_1$;

$$\begin{bmatrix} 1 & 0 & \frac{7}{3} & | & -\frac{4}{3} \\ 0 & 1 & -\frac{8}{3} & | & \frac{14}{3} \\ 0 & 4 & -3 & | & 11 \end{bmatrix}$$

(c) $-4R_2 + R_3$;

$$\begin{bmatrix} 1 & 0 & \frac{7}{3} & | & -\frac{4}{3} \\ 0 & 1 & -\frac{8}{3} & | & \frac{14}{3} \\ 0 & 0 & \frac{23}{3} & | & -\frac{23}{3} \end{bmatrix}$$

(Solution continued in the text.)

5 Use the Gauss-Jordan method to solve

$$\begin{aligned} x + y - z &= 6 \\ 2x - y + z &= 3 \\ -x + y + z &= -4. \end{aligned}$$

Answer:
$(3, 1, -2)$

This transforms the first column. Transform the second column in a similar manner. **4**

Complete the solution by transforming the third column.

$$\begin{bmatrix} 1 & 0 & \frac{7}{3} & | & -\frac{4}{3} \\ 0 & 1 & -\frac{8}{3} & | & \frac{14}{3} \\ 0 & 0 & 1 & | & -1 \end{bmatrix} \quad \frac{3}{23}R_3$$

$$\begin{bmatrix} 1 & 0 & 0 & | & 1 \\ 0 & 1 & -\frac{8}{3} & | & \frac{14}{3} \\ 0 & 0 & 1 & | & -1 \end{bmatrix} \quad -\frac{7}{3}R_3 + R_1$$

$$\begin{bmatrix} 1 & 0 & 0 & | & 1 \\ 0 & 1 & 0 & | & 2 \\ 0 & 0 & 1 & | & -1 \end{bmatrix} \quad \frac{8}{3}R_3 + R_2$$

The linear system associated with this last augmented matrix is

$$\begin{aligned} x \quad &= 1 \\ y \quad &= 2 \\ z &= -1, \end{aligned}$$

and the solution is $(1, 2, -1)$. ◀

Note Notice that the first two row operations are used to get the ones and the third row operation is used to get the zeros. **5**

▶ **EXAMPLE 3** Use the Gauss-Jordan method to solve the system

$$\begin{aligned} 2x + 4y &= 4 \\ 3x + 6y &= 8 \\ 2x + y &= 7. \end{aligned}$$

Write the augmented matrix and perform row operations to obtain a first column whose entries (from top to bottom) are 1, 0, 0:

$$\begin{bmatrix} 2 & 4 & | & 4 \\ 3 & 6 & | & 8 \\ 2 & 1 & | & 7 \end{bmatrix}$$

$$\begin{bmatrix} 1 & 2 & | & 2 \\ 3 & 6 & | & 8 \\ 2 & 1 & | & 7 \end{bmatrix} \quad \frac{1}{2}R_1$$

6 Solve each system.

(a)
$$x - y = 4$$
$$-2x + 2y = 1$$

(b) $3x - 4y = 0$
$$2x + y = 0$$

Answers:

(a) No solution

(b) $(0, 0)$

7 Use the Gauss-Jordan method to solve the system

$$x + 3y = 4$$
$$4x + 8y = 4$$
$$6x + 12y = 6$$

Answer:
$(-5, 3)$

$$\begin{bmatrix} 1 & 2 & | & 2 \\ 0 & 0 & | & 2 \\ 2 & 1 & | & 7 \end{bmatrix} \quad -3R_1 + R_2$$

Stop! The second row of this augmented matrix denotes the equation $0x + 0y = 2$. Since the left side of this equation is always 0 and the right side is 2, it has no solutions. Hence, the original system has no solutions. ◀ **6**

Whenever the Gauss-Jordan method produces a row whose elements are all 0 except the last one, such as $[0 \ 0 \ | \ 2]$ in Example 3, the system is inconsistent and has no solutions. On the other hand, if a row with *every* element 0 is produced, the system may have solutions. In that case, continue carrying out the Gauss-Jordan method. **7**

▶ **EXAMPLE 4**   Use the Gauss-Jordan method to solve the system

$$x + 2y - z = 0$$
$$3x - y + z = 6$$

Start with the augmented matrix and use row operations to obtain a first column whose entries (from top to bottom) are 1, 0:

$$\begin{bmatrix} 1 & 2 & -1 & | & 0 \\ 3 & -1 & 1 & | & 6 \end{bmatrix}$$

$$\begin{bmatrix} 1 & 2 & -1 & | & 0 \\ 0 & -7 & 4 & | & 6 \end{bmatrix} \quad -3R_1 + R_2$$

Now use row operations to obtain a second column whose entries (from top to bottom) are 0, 1:

$$\begin{bmatrix} 1 & 2 & -1 & | & 0 \\ 0 & 1 & -\frac{4}{7} & | & -\frac{6}{7} \end{bmatrix} \quad -\frac{1}{7}R_2$$

$$\begin{bmatrix} 1 & 0 & \frac{1}{7} & | & \frac{12}{7} \\ 0 & 1 & -\frac{4}{7} & | & -\frac{6}{7} \end{bmatrix} \quad -2R_2 + R_1$$

This last matrix is the augmented matrix of the system

$$x + \frac{1}{7}z = \frac{12}{7}$$

$$y - \frac{4}{7}z = -\frac{6}{7}.$$

8 Use the Gauss-Jordan method to solve the following.

(a) $3x + 9y = -6$
     $-x - 3y = 2$

(b) $2x + 9y = 12$
    $4x + 18y = 5$

Answers:

(a) y arbitrary,
$x = -3y - 2$
or $(-3y - 2, y)$

(b) No solution

Solving the first equation for x and the second for y gives the solution

$$z \text{ arbitrary}$$

$$y = -\frac{6}{7} + \frac{4}{7}z$$

$$x = \frac{12}{7} - \frac{1}{7}z,$$

or $(12/7 - z/7, -6/7 + 4z/7, z)$. ◀ **8**

The techniques used in Examples 1–4 can be summarized as follows.

The Gauss-Jordan Method for Solving a System of Linear Equations

1. Arrange the equations with the variable terms in the same order on the left of the equal sign and the constants on the right.
2. Write the augmented matrix of the system.
3. Use row operations to transform the augmented matrix into this form:
   (a) The rows consisting entirely of zeros are grouped together at the bottom of the matrix.
   (b) In each row that does not consist entirely of zeros, the leftmost nonzero element is 1 (called a *leading* 1).
   (c) Each column that contains a leading 1 has zeros in all other entries.
   (d) The leading 1 in any row is to the left of any leading 1's in the rows below it.
4. Stop the process in step 3 if you obtain a row whose elements are all zero except the last one. In that case, the system is inconsistent and has no solutions. Otherwise, finish Step 3 and read the solutions of the system from the final matrix.

When doing Step 3, try to choose row operations so that as few fractions as possible are carried through the computation. This makes calculation easier when working by hand and avoids introducing round-off errors when using a calculator or computer.

▶ **EXAMPLE 5** The U-Drive Rent-a-Truck Company plans to spend 3 million dollars on 200 new trucks. Each van will cost $10,000, each small truck, $15,000, and each large truck, $25,000. Past experience shows that they need twice as many vans as small trucks. How many of each kind of vehicle can they buy?

Let x be the number of vans, y the number of small trucks, and z the number of large trucks. Then $x + y + z = 200$. The cost of x vans at $10,000 each is $10,000x$, the cost of y small trucks is $15,000y$, and the cost of z large trucks is $25,000z$, so that $10,000x + 15,000y + 25,000z = 3,000,000$. Dividing this

9 In Example 5, suppose the U-Drive Company can spend only 2 million dollars on 150 new trucks, and that they need three times as many vans as small trucks. Write a system of equations to express these conditions.

Answer:

$$x + y + z = 150$$
$$2x + 3y + 5z = 400$$
$$x - 3y = 0$$

equation on each side by 5000 makes it $2x + 3y + 5z = 600$. Finally, the number of vans is twice the number of small trucks: $x = 2y$, or equivalently, $x - 2y = 0$.

To solve the system

$$x + y + z = 200$$
$$2x + 3y + 5z = 600$$
$$x - 2y = 0,$$

we form the augmented matrix and use the indicated row operations.

$$\begin{bmatrix} 1 & 1 & 1 & | & 200 \\ 2 & 3 & 5 & | & 600 \\ 1 & -2 & 0 & | & 0 \end{bmatrix}$$

$$\begin{bmatrix} 1 & 1 & 1 & | & 200 \\ 0 & 1 & 3 & | & 200 \\ 0 & -3 & -1 & | & -200 \end{bmatrix} \quad \begin{matrix} -2R_1 + R_2 \\ -R_1 + R_3 \end{matrix}$$

$$\begin{bmatrix} 1 & 0 & -2 & | & 0 \\ 0 & 1 & 3 & | & 200 \\ 0 & 0 & 8 & | & 400 \end{bmatrix} \quad \begin{matrix} -R_2 + R_1 \\ \\ 3R_2 + R_3 \end{matrix}$$

$$\begin{bmatrix} 1 & 0 & -2 & | & 0 \\ 0 & 1 & 3 & | & 200 \\ 0 & 0 & 1 & | & 50 \end{bmatrix} \quad \frac{1}{8}R_3$$

$$\begin{bmatrix} 1 & 0 & 0 & | & 100 \\ 0 & 1 & 0 & | & 50 \\ 0 & 0 & 1 & | & 50 \end{bmatrix} \quad \begin{matrix} 2R_3 + R_1 \\ -3R_3 + R_2 \end{matrix}$$

The final matrix corresponds to the system

$$x = 100$$
$$y = 50$$
$$z = 50.$$

Therefore, U-Drive should buy 100 vans, 50 small trucks, and 50 large trucks. ◀ **9**

The elimination method and the Gauss-Jordan method can be used with systems of equations of any size and are very suitable for use with computers. One or the other is usually the best way to solve such systems. However, there are other solution techniques for certain systems, one of which is considered in Section 6.6.

> ### FOR GRAPHERS
> Many graphing calculators allow you to enter matrices and perform row operations on them. The Gauss-Jordan method can be used with such calculators to solve systems of equations. Check your instruction manual. Calculators that handle matrices, but do not perform row operations, can be used to carry out the alternative solution technique presented in Section 6.6.

6.2 EXERCISES

Write the augmented matrix of each of the following systems. Do not solve the systems.

1. $2x + y + z = 3$
$3x - 4y + 2z = -5$
$x + y + z = 2$

2. $3x + 4y - 2z - 3w = 0$
$x - 3y + 7z + 4w = 9$
$2x + 5z - 6w = 0$

Write the system of equations associated with the following augmented matrices. Do not solve the systems.

3. $\begin{bmatrix} 2 & 3 & 8 & | & 20 \\ 1 & 4 & 6 & | & 12 \\ 0 & 3 & 5 & | & 10 \end{bmatrix}$

4. $\begin{bmatrix} 3 & 2 & 6 & | & 18 \\ 2 & -2 & 5 & | & 7 \\ 1 & 0 & 5 & | & 20 \end{bmatrix}$

Use the indicated row operation to transform each matrix.

5. Interchange R_2 and R_3.

$\begin{bmatrix} 1 & 2 & 3 & | & -1 \\ 6 & 5 & 4 & | & 6 \\ 2 & 0 & 7 & | & -4 \end{bmatrix}$

6. Replace R_3 by $-3R_1 + R_3$.

$\begin{bmatrix} 1 & 5 & 2 & 0 & | & -1 \\ 8 & 5 & 4 & 6 & | & 6 \\ 3 & 0 & 7 & 1 & | & -4 \end{bmatrix}$

7. Replace R_2 by $2R_1 + R_2$.

$\begin{bmatrix} -4 & -3 & 1 & -1 & | & 2 \\ 8 & 2 & 5 & 0 & | & 6 \\ 0 & -2 & 9 & 4 & | & 5 \end{bmatrix}$

8. Replace R_3 by $\dfrac{1}{4} R_3$.

$\begin{bmatrix} 2 & 5 & 1 & | & -1 \\ -4 & 0 & 4 & | & 6 \\ 6 & 0 & 8 & | & -4 \end{bmatrix}$

Use the Gauss-Jordan method to solve each of the following systems of equations. (See Examples 1–4.)

9. $x + 2y = 5$
$2x + y = -2$

10. $3x - 2y = 4$
$3x + y = -2$

11. $x + 3y - 6z = 7$
$2x - y + 2z = 0$
$x + y + 2z = -1$

12. $x = 1 - y$
$2x = z$
$2z = -2 - y$

13. $3x + 5y - z = 0$
$4x - y + 2z = 1$
$-6x - 10y + 2z = 0$

14. $x + y = -1$
$y + z = 4$
$x + z = 1$

15. $x + y - z = 6$
$2x - y + z = -9$
$x - 2y + 3z = 1$

16. $y = x - 1$
$y = 6 + z$
$z = -1 - x$

17. $x - 2y + z = 5$
$2x + y - z = 2$
$-2x + 4y - 2z = 2$

18. $2x + 3y + z = 9$
$4x + y - 3z = -7$
$6x + 2y - 4z = -8$

19. $-8x - 9y = 11$
$24x + 34y = 2$
$16x + 11y = -57$

20. $2x + y = 7$
$x - y = 3$
$x + 3y = 4$

21. $\begin{aligned} x + y - z &= -20 \\ 2x - y + z &= 11 \end{aligned}$

22. $\begin{aligned} 4x + 3y + z &= 1 \\ -2x - y + 2z &= 0 \end{aligned}$

23. $\begin{aligned} 2x + y + 3z - 2w &= -6 \\ 4x + 3y + z - w &= -2 \\ x + y + z + w &= -5 \\ -2x - 2y + 2z + 2w &= -10 \end{aligned}$

24. $\begin{aligned} x + y + z + w &= -1 \\ -x + 4y + z - w &= 0 \\ x - 2y + z - 2w &= 11 \\ -x - 2y + z + 2w &= -3 \end{aligned}$

25. $\begin{aligned} x + 2y - z &= 3 \\ 3x + y + w &= 4 \\ 2x - y + z + w &= 2 \end{aligned}$

26. $\begin{aligned} x - 2y - z - 3w &= -3 \\ -x + y + z &= 2 \\ 4y + 3z - 6w &= -2 \end{aligned}$

27. Management McFrugal Snack Shops plan to hire two public relations firms to survey 500 customers by phone, 750 by mail, and 250 by in-person interviews. The Garcia firm has personnel to do 10 phone surveys, 30 mail surveys, and 5 interviews per hour. The Wong firm can handle 20 phone surveys, 10 mail surveys, and 10 interviews per hour. For how many hours should each firm be hired to produce the exact number of surveys needed?

28. Management A knitting shop ordered yarn from three suppliers, I, II, and III. One month the shop ordered a total of 100 units of yarn from these suppliers. The delivery costs were $80, $50, and $65 per unit for the orders from suppliers I, II, and III, respectively, with total delivery costs of $5990. The shop ordered the same amount from suppliers I and III. How many units were ordered from each supplier? Use the Gauss-Jordan method to find the solution.

29. Management An electronics company produces three models of stereo speakers, models A, B, and C, and can deliver them by truck, van, or station wagon. A truck holds 2 boxes of model A, 1 of model B, and 3 of model C. A van holds 1 box of model A, 3 boxes of model B, and 2 boxes of model C. A station wagon holds 1 box of model A, 3 boxes of model B, and 1 box of model C. If 15 boxes of model A, 20 boxes of model B, and 22 boxes of model C are to be delivered, how many vehicles of each type should be used so that all operate at full capacity?

30. Management Pretzels cost $3 per pound, dried fruit $4 per pound, and nuts $8 per pound. How many pounds of each should be used to produce 140 pounds of trail mix costing $6 per pound in which there are twice as many pretzels (by weight) as dried fruit?

31. Management A manufacturer purchases a part for use at both of its two plants—one in Canoga Park, Califor-

nia, the other in Wooster, Ohio. The part is available in limited quantities from two suppliers. Each supplier has 75 units available. The Canoga Park plant needs 40 units and the Wooster plant requires 75 units. The first supplier charges $70 per unit delivered to Canoga Park and $90 per unit delivered to Wooster. Corresponding costs from the second supplier are $80 and $120. The manufacturer wants to order a total of 75 units from the first, less expensive, supplier, with the remaining 40 units to come from the second supplier. If the company spends $10,750 to purchase the required number of units for the two plants, find the number of units that should be purchased from each supplier for each plant as follows.

(a) Assign variables to the four unknowns.

(b) Write a system of five equations with the four variables. (Not all equations will involve all four variables.)

(c) Use the Gauss-Jordan method to solve the system of equations.

32. Management An auto manufacturer sends cars from two plants, I and II, to dealerships A and B located in a midwestern city. Plant I has a total of 28 cars to send, and plant II has 8. Dealer A needs 20 cars, and dealer B needs 16. Transportation costs based on the distance of each dealership from each plant are $220 from I to A, $300 from I to B, $400 from II to A, and $180 from II to B. The manufacturer wants to limit transportation costs to $10,640. How many cars should be sent from each plant to each of the two dealerships? Use the Gauss-Jordan method to find the solution.

33. Management The electronics company in Exercise 29 no longer makes model C. Each kind of delivery vehicle can now carry one more box of model B than previously and the same number of boxes of model A. If 16 boxes of model A and 22 boxes of model B are to be delivered, how many vehicles of each type should be used so that all operate at full capacity?

34. Natural Science An animal breeder can buy four types of tiger food. Each case of Brand A contains 25 units of fiber, 30 units of protein, and 30 units of fat. Each case of Brand B contains 50 units of fiber, 30 units of protein, and 20 units of fat. Each case of Brand C contains 75 units of fiber, 30 units of protein, and 20 units of fat. Each case of Brand D contains 100 units of fiber, 60 units of protein, and 30 units of fat. How many cases of each brand should the breeder mix together to obtain a food that provides 1200 units of fiber, 600 units of protein, and 400 units of fat?

Set up a system of equations and use a graphing calculator with matrix (and row operation) capabilities or appropriate computer software to solve the system in Exercises 35–36.

35. Natural Science Three species of bacteria are fed three foods, I, II, and III. A bacterium of the first species consumes 1.3 units each of foods I and II and 2.3 units of food III each day. A bacterium of the second species consumes 1.1 units of food I, 2.4 units of food II, and 3.7 units of food III each day. A bacterium of the third species consumes 8.1 units of I, 2.9 units of II, and 5.1 units of III each day. If 16,000 units of I, 28,000 units of II, and 44,000 units of III are supplied each day, how many of each species can be maintained in this environment?

36. Management A company produces three combinations of mixed vegetables which sell in 1 kilogram packages. Italian style combines .3 kilograms of zucchini, .3 of broccoli, and .4 of carrots. French style combines .6 kilograms of broccoli and .4 of carrots. Oriental style combines .2 kilograms of zucchini, .5 of broccoli, and .3 of carrots. The company has a stock of 16,200 kilograms of zucchini, 41,400 kilograms of broccoli, and 29,400 kilograms of carrots. How many packages of each style should they prepare to use up their supplies?

6.3 BASIC MATRIX OPERATIONS

Until now we have used matrices only as a convenient shorthand for dealing with systems of equations. However, matrices are also important in the fields of management, natural science, engineering, and social science because they provide a convenient way to organize data, as Example 1 demonstrates.

▶ **EXAMPLE 1** The EZ Life Company manufactures sofas and armchairs in three models, A, B, and C. The company has regional warehouses in New York, Chicago, and San Francisco. In its August shipment, the company sends 10 model A sofas, 12 model B sofas, 5 model C sofas, 15 model A chairs, 20 model B chairs, and 8 model C chairs to each warehouse.

This data might be organized by first listing it as follows.

Sofas	10 model A	12 model B	5 model C
Chairs	15 model A	20 model B	8 model C

Alternatively, we might tabulate the data in a chart.

		Model		
		A	B	C
Furniture	Sofa	10	12	5
	Chair	15	20	8

1 Rewrite this information in a matrix with three rows and two columns.

Answer:

$$\begin{bmatrix} 10 & 15 \\ 12 & 20 \\ 5 & 8 \end{bmatrix}$$

2 Give the size of each of the following matrices.

(a) $\begin{bmatrix} 2 & 1 & -5 & 6 \\ 3 & 0 & 7 & -4 \end{bmatrix}$

(b) $\begin{bmatrix} 1 & 2 & 3 \\ 4 & 5 & 6 \\ 9 & 8 & 7 \end{bmatrix}$

Answers:

(a) 2×4

(b) 3×3

3 Use the numbers 2, 5, −8, 4 to write

(a) a row matrix;

(b) a column matrix;

(c) a square matrix.

Answers:

(a) $\begin{bmatrix} 2 & 5 & -8 & 4 \end{bmatrix}$

(b) $\begin{bmatrix} 2 \\ 5 \\ -8 \\ 4 \end{bmatrix}$

(c) $\begin{bmatrix} 2 & 5 \\ -8 & 4 \end{bmatrix}$ or $\begin{bmatrix} 2 & -8 \\ 5 & 4 \end{bmatrix}$

(Other answers are possible.)

With the understanding that the numbers in each row refer to the furniture type (sofa, chair) and the numbers in each column refer to the model (A, B, C), the same information can be given by a matrix, as follows.

$$M = \begin{bmatrix} 10 & 12 & 5 \\ 15 & 20 & 8 \end{bmatrix} \quad \blacktriangleleft \quad \boxed{1}$$

Matrices often are named with capital letters, as in Example 1. A matrix is classified by its size, that is, by the number of horizontal rows and vertical columns that it contains. For example, matrix M above has two rows and three columns. This matrix is called a 2×3 (read "2 by 3") matrix. By definition, a matrix with m rows and n columns is size $m \times n$. The number of rows is always given first.

▶ **EXAMPLE 2**

(a) The matrix $\begin{bmatrix} 6 & 5 \\ 3 & 4 \\ 5 & -1 \end{bmatrix}$ is a 3×2 matrix.

(b) $\begin{bmatrix} 5 & 8 & 9 \\ 0 & 5 & -3 \\ -4 & 0 & 5 \end{bmatrix}$ is a 3×3 matrix.

(c) $\begin{bmatrix} 1 & 6 & 5 & -2 & 5 \end{bmatrix}$ is a 1×5 matrix.

(d) $\begin{bmatrix} 3 \\ -5 \\ 0 \\ 2 \end{bmatrix}$ is a 4×1 matrix. $\quad \blacktriangleleft \quad \boxed{2}$

A matrix with only one row, as in Example 2(c), is called a **row matrix** or **row vector.** A matrix with only one column, as in Example 2(d), is called a **column matrix** or **column vector.** A matrix with the same number of rows as columns is called a **square matrix.** The matrix in Example 2(b) above is a square matrix, as are

$$A = \begin{bmatrix} -5 & 6 \\ 8 & 3 \end{bmatrix} \quad \text{and} \quad B = \begin{bmatrix} 0 & 0 & 0 & 0 \\ -2 & 4 & 1 & 3 \\ 0 & 0 & 0 & 0 \\ -5 & -4 & 1 & 8 \end{bmatrix}. \quad \boxed{3}$$

When a matrix is denoted by a single letter, such as the matrix A above, then the element in row i and column j is denoted a_{ij}. For example, $a_{21} = 8$ (the element in row 2, column 1). Similarly, in matrix B above, $b_{42} = -4$ (the element in row 4, column 2).

Equality of matrices is defined as follows.

4 Give the values of the variables that make each of the following statements true.

(a) $\begin{bmatrix} x & 2 \\ 5 & y \end{bmatrix} = \begin{bmatrix} 6 & p \\ q & -1 \end{bmatrix}$

(b)
$[1 \quad 2 \quad x] = \begin{bmatrix} y \\ 2 \\ 8 \end{bmatrix}$

Answers:

(a) $x = 6$, $y = -1$, $p = 2$, $q = 5$

(b) Can never be true

Two matrices are **equal** if they are the same size and if corresponding elements are equal.

Using this definition, the matrices

$$\begin{bmatrix} 2 & 1 \\ 3 & -5 \end{bmatrix} \text{ and } \begin{bmatrix} 1 & 2 \\ -5 & 3 \end{bmatrix}$$

are not equal (even though they contain the same elements and are the same size) because corresponding elements differ.

▶**EXAMPLE 3** **(a)** From the definition of matrix equality given above, the only way that the statement

$$\begin{bmatrix} 2 & 1 \\ p & q \end{bmatrix} = \begin{bmatrix} x & y \\ -1 & 0 \end{bmatrix}$$

can be true is if $2 = x$, $1 = y$, $p = -1$, and $q = 0$.
(b) The statement

$$\begin{bmatrix} x \\ y \end{bmatrix} = \begin{bmatrix} 1 \\ 4 \\ 0 \end{bmatrix}$$

can never be true, because the two matrices are different sizes. (One is 2×1 and the other is 3×1.) ◀ **4**

ADDITION The matrix given in Example 1,

$$M = \begin{bmatrix} 10 & 12 & 5 \\ 15 & 20 & 8 \end{bmatrix},$$

shows the August shipment from the EZ Life plant to each of its warehouses. If matrix *N* below gives the September shipment to the New York warehouse, what is the total shipment for each item of furniture to the New York warehouse for these two months?

$$N = \begin{bmatrix} 45 & 35 & 20 \\ 65 & 40 & 35 \end{bmatrix}$$

If 10 model A sofas were shipped in August and 45 in September, then altogether 55 model A sofas were shipped in the 2 months. Adding the other corresponding entries gives a new matrix, *Q*, that represents the total shipment for the 2 months.

5 Find each sum when possible.

(a) $\begin{bmatrix} 2 & 5 & 7 \\ 3 & -1 & 4 \end{bmatrix}$

$\quad + \begin{bmatrix} -1 & 2 & 0 \\ 10 & -4 & 5 \end{bmatrix}$

(b) $\begin{bmatrix} 1 \\ 2 \\ 3 \end{bmatrix} + \begin{bmatrix} 2 & -1 \\ 4 & 5 \\ 6 & 0 \end{bmatrix}$

(c) $[5 \quad 4 \quad -1] + [-5 \quad 2 \quad 3]$

Answers:

(a) $\begin{bmatrix} 1 & 7 & 7 \\ 13 & -5 & 9 \end{bmatrix}$

(b) Not possible

(c) $[0 \quad 6 \quad 2]$

6 From the result of Example 5, find the total number of the following shipped to the three warehouses.

(a) Model A chairs

(b) Model B sofas

(c) Model C chairs

Answers:

(a) 139

(b) 92

(c) 100

$$Q = \begin{bmatrix} 55 & 47 & 25 \\ 80 & 60 & 43 \end{bmatrix}$$

It is convenient to refer to Q as the "sum" of M and N.

The way these two matrices were added illustrates the following definition of addition of matrices.

The **sum** of two $m \times n$ matrices X and Y is the $m \times n$ matrix $X + Y$ in which each element is the sum of the corresponding elements of X and Y.

It is important to remember that only matrices that are the same size can be added.

▶**EXAMPLE 4** Find each sum if possible.

(a) $\begin{bmatrix} 5 & -6 \\ 8 & 9 \end{bmatrix} + \begin{bmatrix} -4 & 6 \\ 8 & -3 \end{bmatrix} = \begin{bmatrix} 5 + (-4) & -6 + 6 \\ 8 + 8 & 9 + (-3) \end{bmatrix} = \begin{bmatrix} 1 & 0 \\ 16 & 6 \end{bmatrix}$

(b) The matrices

$$A = \begin{bmatrix} 5 & 8 \\ 6 & 2 \end{bmatrix} \quad \text{and} \quad B = \begin{bmatrix} 3 & 9 & 1 \\ 4 & 2 & 5 \end{bmatrix}$$

are different sizes, so it is not possible to find the sum $A + B$. ◀ **5**

▶**EXAMPLE 5** The September shipments from the EZ Life Company to the New York, San Francisco, and Chicago warehouses are given in matrices N, S, and C below.

$$N = \begin{bmatrix} 45 & 35 & 20 \\ 65 & 40 & 35 \end{bmatrix}, \quad S = \begin{bmatrix} 30 & 32 & 28 \\ 43 & 47 & 30 \end{bmatrix}, \quad C = \begin{bmatrix} 22 & 25 & 38 \\ 31 & 34 & 35 \end{bmatrix}$$

What was the total amount shipped to the three warehouses in September?

The total of the September shipments is represented by the sum of the three matrices N, S, and C.

$$N + S + C = \begin{bmatrix} 45 & 35 & 20 \\ 65 & 40 & 35 \end{bmatrix} + \begin{bmatrix} 30 & 32 & 28 \\ 43 & 47 & 30 \end{bmatrix} + \begin{bmatrix} 22 & 25 & 38 \\ 31 & 34 & 35 \end{bmatrix}$$

$$= \begin{bmatrix} 97 & 92 & 86 \\ 139 & 121 & 100 \end{bmatrix}$$

For example, from this sum the total number of model C sofas shipped to the three warehouses in September was 86. ◀ **6**

As mentioned in Section 1.1, the additive inverse of the real number a is $-a$; a similar definition is given for the additive inverse of a matrix.

The **additive inverse** (or *negative*) of a matrix X is the matrix $-X$ in which each element is the additive inverse of the corresponding element of X.

If

$$A = \begin{bmatrix} 1 & 2 & 3 \\ 0 & -1 & 5 \end{bmatrix} \quad \text{and} \quad B = \begin{bmatrix} -2 & 3 & 0 \\ 1 & -7 & 2 \end{bmatrix},$$

then by the definition of the additive inverse of a matrix,

$$-A = \begin{bmatrix} -1 & -2 & -3 \\ 0 & 1 & -5 \end{bmatrix} \quad \text{and} \quad -B = \begin{bmatrix} 2 & -3 & 0 \\ -1 & 7 & -2 \end{bmatrix}.$$

By the definition of matrix addition, for each matrix X, the sum $X + (-X)$ is a **zero matrix,** O, whose elements are all zeros. There is an $m \times n$ zero matrix for each pair of values of m and n. Zero matrices have the following *identity property:*

If O is the $m \times n$ zero matrix, and A is any $m \times n$ matrix, then

$$A + O = O + A = A.$$

Compare this with the identity property for real numbers: for any real number a, $a + 0 = 0 + a = a$.

SUBTRACTION The **subtraction** of matrices can be defined in a manner comparable to subtraction for real numbers.

For two $m \times n$ matrices X and Y, the **difference** of X and Y is the $m \times n$ matrix $X - Y$ in which each element is the difference of the corresponding elements of X and Y, or, equivalently,

$$X - Y = X + (-Y).$$

According to this definition, matrix subtraction can be performed by subtracting corresponding elements. For example, using A, B, and $-B$ as defined above,

7 Find each of the following differences when possible.

(a) $\begin{bmatrix} 2 & 5 \\ -1 & 0 \end{bmatrix} - \begin{bmatrix} 6 & 4 \\ 3 & -2 \end{bmatrix}$

(b) $\begin{bmatrix} 1 & 5 & 6 \\ 2 & 4 & 8 \end{bmatrix} - \begin{bmatrix} 2 & 1 \\ 10 & 3 \end{bmatrix}$

(c) $[5 \quad -4 \quad 1] - [6 \quad 0 \quad -3]$

Answers:

(a) $\begin{bmatrix} -4 & 1 \\ -4 & 2 \end{bmatrix}$

(b) Not possible

(c) $[-1 \quad -4 \quad 4]$

$$A - B = \begin{bmatrix} 1 & 2 & 3 \\ 0 & -1 & 5 \end{bmatrix} - \begin{bmatrix} -2 & 3 & 0 \\ 1 & -7 & 2 \end{bmatrix}$$

$$= \begin{bmatrix} 1 - (-2) & 2 - 3 & 3 - 0 \\ 0 - 1 & -1 - (-7) & 5 - 2 \end{bmatrix}$$

$$= \begin{bmatrix} 3 & -1 & 3 \\ -1 & 6 & 3 \end{bmatrix}.$$

▶ **EXAMPLE 6**

(a) $[8 \quad 6 \quad -4] - [3 \quad 5 \quad -8] = [5 \quad 1 \quad 4]$

(b) The matrices

$$\begin{bmatrix} -2 & 5 \\ 0 & 1 \end{bmatrix} \quad \text{and} \quad \begin{bmatrix} 3 \\ 5 \end{bmatrix}$$

are different sizes and cannot be subtracted. ◀ **7**

 FOR GRAPHERS

A graphing calculator with matrix capabilities can be used to add or subtract matrices, and to find the additive inverse of a matrix. Check your instruction manual.

▶ **EXAMPLE 7** During September the Chicago warehouse of the EZ Life Company shipped out the following numbers of each model.

$$K = \begin{bmatrix} 5 & 10 & 8 \\ 11 & 14 & 15 \end{bmatrix}$$

What was the Chicago warehouse inventory on October 1, taking into account only the number of items received and sent out during the month?

The number of each kind of item received during September is given by matrix *C* from Example 5; the number of each model sent out during September is given by matrix *K* above. The October 1 inventory will be represented by the matrix *C* − *K* as shown below.

$$\begin{bmatrix} 22 & 25 & 38 \\ 31 & 34 & 35 \end{bmatrix} - \begin{bmatrix} 5 & 10 & 8 \\ 11 & 14 & 15 \end{bmatrix} = \begin{bmatrix} 17 & 15 & 30 \\ 20 & 20 & 20 \end{bmatrix}$$ ◀

▶ **EXAMPLE 8** A drug company is testing 200 patients to see if Painoff (a new headache medicine) is effective. Half the patients receive Painoff and half receive a placebo. The data on the first 50 patients is summarized in this matrix:

Pain Relief Obtained

Yes   No

Patient took Painoff $\begin{bmatrix} 22 & 3 \\ 8 & 17 \end{bmatrix}$
Patient took placebo

For example, row 2 shows that of the people who took the placebo, 8 got relief, but 17 did not. The test was repeated on three more groups of 50 patients each, with the results summarized by these matrices.

$$\begin{bmatrix} 21 & 4 \\ 6 & 19 \end{bmatrix} \begin{bmatrix} 19 & 6 \\ 10 & 15 \end{bmatrix} \begin{bmatrix} 23 & 2 \\ 3 & 22 \end{bmatrix}$$

The total results of the test can be obtained by adding these four matrices.

$$\begin{bmatrix} 22 & 3 \\ 8 & 17 \end{bmatrix} + \begin{bmatrix} 21 & 4 \\ 6 & 19 \end{bmatrix} + \begin{bmatrix} 19 & 6 \\ 10 & 15 \end{bmatrix} + \begin{bmatrix} 23 & 2 \\ 3 & 22 \end{bmatrix} = \begin{bmatrix} 85 & 15 \\ 27 & 73 \end{bmatrix}$$

Because 85 of 100 patients got relief with Painoff and only 27 of 100 with the placebo, it appears that Painoff is effective. ◄

6.3 EXERCISES

Find the size of each of the following. Identify any square, column, or row matrices. (See Example 2.) Give the additive inverse of each matrix.

1. $\begin{bmatrix} 7 & -8 & 4 \\ 0 & 13 & 9 \end{bmatrix}$

2. $\begin{bmatrix} -7 & 23 \\ 5 & -6 \end{bmatrix}$

3. $\begin{bmatrix} -3 & 0 & 11 \\ 1 & \frac{1}{4} & -7 \\ 5 & -3 & 9 \end{bmatrix}$

4. $\begin{bmatrix} 6 & -4 & \frac{2}{3} & 12 & 2 \end{bmatrix}$

5. $\begin{bmatrix} 7 \\ 11 \end{bmatrix}$

6. $\begin{bmatrix} -5 \end{bmatrix}$

7. If A is a 5×3 matrix and $A + B = A$, what do you know about B?

8. If C is a 3×3 matrix and D is a 3×4 matrix, then $C + D$ is _____ .

Mark each of the following statements as true or false. If false, tell why.

9. $\begin{bmatrix} 1 & 3 \\ 5 & 7 \end{bmatrix} = \begin{bmatrix} 1 & 5 \\ 3 & 7 \end{bmatrix}$

10. $\begin{bmatrix} 1 \\ 2 \\ 3 \end{bmatrix} = \begin{bmatrix} 1 & 2 & 3 \end{bmatrix}$

11. $\begin{bmatrix} x \\ y \end{bmatrix} = \begin{bmatrix} 3 \\ 5 \end{bmatrix}$ if $x = 3$ and $y = 5$.

12. $\begin{bmatrix} 3 & 5 & 2 & 8 \\ 1 & -1 & 4 & 0 \end{bmatrix}$ is a 4×2 matrix.

Find the values of the variables in each of the following. (See Example 3.)

13. $\begin{bmatrix} x-4 & y+1 \\ 5 & -2 \end{bmatrix} = \begin{bmatrix} 5 & -3 \\ 5 & z \end{bmatrix}$

14. $\begin{bmatrix} 2 & 3 \\ 0 & a \end{bmatrix} = \begin{bmatrix} r+2 & s-10 \\ 0 & 12 \end{bmatrix}$

15. $\begin{bmatrix} -7+z & 4r & 8s \\ 6p & 2 & 5 \end{bmatrix} + \begin{bmatrix} -9 & 8r & 3 \\ 2 & 5 & 4 \end{bmatrix} = \begin{bmatrix} 2 & 36 & 27 \\ 20 & 7 & 12a \end{bmatrix}$

16. $\begin{bmatrix} -a+2 & 3z+1 & 5m \\ 4k & 0 & 3 \end{bmatrix} + \begin{bmatrix} 3a & 2z & 5m \\ 2k & 5 & 6 \end{bmatrix} = \begin{bmatrix} 10 & -14 & 80 \\ 10 & 5 & 9 \end{bmatrix}$

Perform the indicated operations where possible. (See Examples 4 and 6.)

17. $\begin{bmatrix} 1 & 2 & 5 & -1 \\ 3 & 0 & 2 & -4 \end{bmatrix} + \begin{bmatrix} 8 & 10 & -5 & 3 \\ -2 & -1 & 0 & 0 \end{bmatrix}$

18. $\begin{bmatrix} 1 & 5 \\ 2 & -3 \\ 3 & 7 \end{bmatrix} + \begin{bmatrix} 2 & 3 \\ 8 & 5 \\ -1 & 9 \end{bmatrix}$

19. $\begin{bmatrix} 1 & 5 & 7 \\ 2 & 2 & 3 \end{bmatrix} + \begin{bmatrix} 4 & 8 & -7 \\ 1 & -1 & 5 \end{bmatrix}$

20. $\begin{bmatrix} 2 & 4 \\ -8 & 1 \end{bmatrix} + \begin{bmatrix} 9 & -3 \\ 8 & 5 \end{bmatrix}$

21. $\begin{bmatrix} 4 & -2 & 5 \\ 3 & 7 & 0 \end{bmatrix} - \begin{bmatrix} 1 & 5 & -2 \\ -1 & 3 & 8 \end{bmatrix}$

22. $\begin{bmatrix} 9 & 1 \\ 0 & -3 \\ 4 & 10 \end{bmatrix} - \begin{bmatrix} 1 & 9 & -4 \\ -1 & 1 & 0 \end{bmatrix}$

Using matrices

$$O = \begin{bmatrix} 0 & 0 \\ 0 & 0 \end{bmatrix}, P = \begin{bmatrix} m & n \\ p & q \end{bmatrix}, T = \begin{bmatrix} r & s \\ t & u \end{bmatrix}, \text{ and}$$

$$X = \begin{bmatrix} x & y \\ z & w \end{bmatrix},$$

verify that the statements in Exercises 23–28 are true.

23. $X + T$ is a 2×2 matrix. (See Example 2.)

24. $X + T = T + X$ (Commutative property of addition of matrices)

25. $X + (T + P) = (X + T) + P$ (Associative property of addition of matrices)

26. $X + (-X) = O$ (Inverse property of addition of matrices)

27. $P + O = P$ (Identity property of addition of matrices)

28. Which of the above properties are valid for matrices that are not square?

29. **Management**   An investment group planning a shopping center decided to include a market, a barber shop, a variety store, a drug store, and a bakery. They estimated the initial cost and the guaranteed rent (both in dollars per square foot) for each type of store, respectively, as follows. Initial cost: 18, 10, 8, 10, and 10; guaranteed rent: 2.7, 1.5, 1.0, 2.0, and 1.7. Write this information first as a 5×2 matrix and then as a 2×5 matrix. (See Example 1.)

30. **Natural Science**   A dietician prepares a diet specifying the amounts a patient should eat of four basic food groups: group I, meats; group II, fruits and vegetables; group III, breads and starches; group IV, milk products. Amounts are given in "exchanges" which represent 1 ounce (meat), 1/2 cup (fruits and vegetables), 1 slice (bread), 8 ounces (milk), or other suitable measurements.

(a) The number of "exchanges" for breakfast for each of the four food groups, respectively, are 2, 1, 2, and 1; for lunch, 3, 2, 2, and 1; and for dinner, 4, 3, 2, and 1. Write a 3×4 matrix using this information.

(b) The amounts of fat, carbohydrates, and protein in each food group respectively are as follows.

Fat: 5, 0, 0, 10

Carbohydrates: 0, 10, 15, 12

Protein: 7, 1, 2, 8

Use this information to write a 4×3 matrix.

(c) There are 8 calories per unit of fat, 4 calories per unit of carbohydrates, and 5 calories per unit of protein; summarize this data in a 3×1 matrix.

31. Natural Science At the beginning of a laboratory experiment, five baby rats measured 5.6, 6.4, 6.9, 7.6, and 6.1 centimeters in length, and weighed 144, 138, 149, 152, and 146 grams, respectively.

(a) Write a 2 × 5 matrix using this information.

(b) At the end of 2 weeks, their lengths were 10.2, 11.4, 11.4, 12.7, and 10.8 centimeters, and they weighed 196, 196, 225, 250, and 230 grams. Write a 2 × 5 matrix with this information.

(c) Use matrix subtraction and the matrices found in (a) and (b) to write a matrix that gives the amount of change in length and weight for each rat. (See Examples 5, 7, and 8.)

(d) The following week the rats gained as shown in the matrix below.

$$\begin{array}{c} \text{Length} \\ \text{Weight} \end{array} \begin{bmatrix} 1.8 & 1.5 & 2.3 & 1.8 & 2.0 \\ 25 & 22 & 29 & 33 & 20 \end{bmatrix}$$

What were their lengths and weights at the end of this week?

32. Management There are three convenience stores in Gambier. This week, Store I sold 88 loaves of bread, 48 quarts of milk, 16 jars of peanut butter, and 112 pounds of cold cuts. Store II sold 105 loaves of bread, 72 quarts of milk, 21 jars of peanut butter, and 147 pounds of cold cuts. Store III sold 60 loaves of bread, 40 quarts of milk, no peanut butter, and 50 pounds of cold cuts.

(a) Use a 3 × 4 matrix to express the sales information for the three stores.

(b) During the following week, sales on these products at Store I increased by 25%; sales at Store II increased by 1/3; and sales at Store III increased by 10%. Write the sales matrix for that week.

(c) Write a matrix that represents total sales over the two-week period.

33. Management A toy company has plants in Boston, Chicago, and Seattle that manufacture toy rockets and robots. The matrix below gives the production costs (in dollars) for each item at the Boston plant:

$$\begin{array}{c} \\ \text{Material} \\ \text{Labor} \end{array} \begin{array}{cc} \text{Rockets} & \text{Robots} \\ \begin{bmatrix} 4.27 & 6.94 \\ 3.45 & 3.65 \end{bmatrix} \end{array}$$

(a) In Chicago, a rocket costs $4.05 for materials and $3.27 for labor; a robot costs $7.01 for material and $3.51 for labor. In Seattle, material costs are $4.40 for rockets and $6.90 for robots; labor costs are $3.54 for rockets and $3.76 for robots. Write the production cost matrices for Chicago and Seattle.

(b) Assume each plant makes the same number of each item. Write a matrix that expresses the average production costs for all three plants.

(c) Suppose labor costs increase by .11 per item in Chicago and material costs there increase by .37 for a rocket and .42 for a robot. What is the new production cost matrix for Chicago?

(d) After the Chicago cost increases, the Boston plant is closed and production divided evenly among the other two plants. What is the matrix that now expresses the average production costs for the entire country?

34. Social Sciences The following table gives the educational attainment of the U.S. population 25 years and older.*

	Male		Female	
	Four Years of High School or More	*Four Years of College or More*	*Four Years of High School or More*	*Four Years of College or More*
1940	22.7%	5.5%	26.3%	3.8%
1950	32.6	7.3	36.0	5.2
1959	42.2	10.3	45.2	6.0
1970	55.0	14.1	55.4	8.2
1980	69.1	20.8	68.1	13.5
1987	76.0	23.6	75.3	16.5
1991	78.5	24.3	78.3	18.8

(a) Write a matrix for the educational attainment of males.

(b) Write a matrix for the educational attainment of females.

(c) Use the matrices from (a) and (b) to write a matrix showing how much more (or less) education males have attained than females.

* "Educational Attainment by Percentage of Population 25+Years, 1940–91" from "The Universal Almanac, 1993," John W. Wright, General Editor, Kansas City, New York: Andrews and McMeel.

6.4 MULTIPLICATION OF MATRICES

Suppose one of the EZ Life Company warehouses receives the following order, written in matrix form, where the entries have the same meaning as in the previous section.

$$\begin{bmatrix} 5 & 4 & 1 \\ 3 & 2 & 3 \end{bmatrix}$$

Later, the store that sent the order asks the company to send five more of the same order. The five new orders can be written as one matrix by multiplying each element in the matrix by 5, giving the product

$$5\begin{bmatrix} 5 & 4 & 1 \\ 3 & 2 & 3 \end{bmatrix} = \begin{bmatrix} 25 & 20 & 5 \\ 15 & 10 & 15 \end{bmatrix}.$$

In work with matrices, a real number, like the 5 in the product above, is called a **scalar.**

The **product** of a scalar k and a matrix X is the matrix kX, each of whose elements is k times the corresponding element of X.

For example,

$$(-3)\begin{bmatrix} 2 & -5 \\ 1 & 7 \end{bmatrix} = \begin{bmatrix} -6 & 15 \\ -3 & -21 \end{bmatrix}.$$

Next we shall define the product of two matrices. To understand the reasoning behind the definition of matrix multiplication, look again at the EZ Life Company. Suppose sofas and chairs of the same model are often sold as sets with matrix W showing the number of each model set in each warehouse.

$$
\begin{array}{c}
\\
\text{New York} \\
\text{Chicago} \\
\text{San Francisco}
\end{array}
\begin{array}{ccc}
\text{A} & \text{B} & \text{C} \\
\end{array}
\begin{bmatrix}
10 & 7 & 3 \\
5 & 9 & 6 \\
4 & 8 & 2
\end{bmatrix} = W
$$

If the selling price of a model A set is \$800, of a model B set \$1000, and of a model C set \$1200, find the total value of the sets in the New York warehouse as follows.

1 In this example of the EZ Life Company, find the total value of the New York sets if model A sofas sell for $1200, model B for $1600, and model C for $1300.

Answer:
$27,100

Type	Number of Sets		Price of Set		Total
A	10	×	$ 800	=	$ 8000
B	7	×	1000	=	7000
C	3	×	1200	=	3600
				Total for New York	$18,600

The total value of the three kinds of sets in New York is $18,600. **1**
The work done in the table above is summarized as follows:

$$10(\$800) + 7(\$1000) + 3(\$1200) = \$18,600.$$

In the same way, the Chicago sets have a total value of

$$5(\$800) + 9(\$1000) + 6(\$1200) = \$20,200,$$

and in San Francisco, the total value of the sets is

$$4(\$800) + 8(\$1000) + 2(\$1200) = \$13,600.$$

The selling prices can be written as a column matrix, *P*, and the total value in each location as a column matrix, *V*.

$$\begin{bmatrix} 800 \\ 1000 \\ 1200 \end{bmatrix} = P \quad \text{and} \quad \begin{bmatrix} 18,600 \\ 20,200 \\ 13,600 \end{bmatrix} = V$$

Consider how the first row of the matrix *W* and the single column *P* lead to the first entry of *V*.

Similarly, adding the products of corresponding entries in the second row of *W* and the column *P* produces the second entry in *V*. The third entry in *V* is obtained in the same way by using the third row of *W* and column *P*. This suggests that it is reasonable to *define* the product *WP* to be *V*.

$$WP = \begin{bmatrix} 10 & 7 & 3 \\ 5 & 9 & 6 \\ 4 & 8 & 2 \end{bmatrix} \begin{bmatrix} 800 \\ 1000 \\ 1200 \end{bmatrix} = \begin{bmatrix} 18,600 \\ 20,200 \\ 13,600 \end{bmatrix} = V$$

2 Matrix A is 4 × 6 and matrix B is 2 × 4.

(a) Can AB be found? If so, give its size.

(b) Can BA be found? If so, give its size.

Answers:

(a) No

(b) Yes; 2 × 6

Note the sizes of the matrices here: the product of a 3 × 3 matrix and a 3 × 1 matrix is a 3 × 1 matrix.

This example provides a model for the definition of matrix multiplication. We first define the **product of a row and a column** (with the same number of entries in each) to be the *number* obtained by multiplying the corresponding entries (first by first, second by second, and so on) and adding the results. For instance,

$$[3 \quad -2 \quad 1] \cdot \begin{bmatrix} 4 \\ 5 \\ 0 \end{bmatrix} = 3 \cdot 4 + (-2) \cdot 5 + 1 \cdot 0 = 12 - 10 + 0 = 2.$$

The product of a row and a column is sometimes called the **dot product** of the row vector and column vector. Now **matrix multiplication** is defined as follows.

> Let A be an $m \times n$ matrix and let B be an $n \times k$ matrix. The **product matrix AB** is the $m \times k$ matrix whose entry in the i-th row and j-th column is
>
> the product of the i-th row of A and the j-th column of B.

Caution Be careful when multiplying matrices. Remember that the number of *columns* of A must equal the number of *rows* of B in order to get the product matrix AB. The final product will have as many rows as A and as many columns as B.

▶ **EXAMPLE 1** Suppose matrix A is 2 × 2 and matrix B is 2 × 4. Can the product AB be calculated? What is the size of the product?

The following diagram helps decide the answers to these questions.

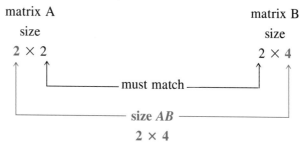

The product of A and B can be calculated because A has two columns and B has two rows. The product will be a 2 × 4 matrix. ◀ **2**

3 Find the product AB given

$$A = \begin{bmatrix} 2 & 4 \\ 5 & 6 \end{bmatrix} \text{ and }$$

$$B = \begin{bmatrix} -3 \\ 4 \end{bmatrix}.$$

Answer:

$$AB = \begin{bmatrix} 10 \\ 9 \end{bmatrix}$$

▶ **EXAMPLE 2** Find the product AB given

$$A = \begin{bmatrix} 2 & 3 & -1 \\ 4 & 2 & 2 \end{bmatrix} \text{ and } B = \begin{bmatrix} 1 \\ 8 \\ 6 \end{bmatrix}.$$

Since matrix A is 2×3 and matrix B is 3×1, matrix AB can be found and will be a 2×1 matrix.

Step 1 Multiply the elements of the first row of A and the corresponding elements of the column of B.

$$\begin{bmatrix} \mathbf{2} & \mathbf{3} & \mathbf{-1} \\ 4 & 2 & 2 \end{bmatrix} \begin{bmatrix} \mathbf{1} \\ \mathbf{8} \\ \mathbf{6} \end{bmatrix} \qquad \mathbf{2 \cdot 1} + \mathbf{3 \cdot 8} + \mathbf{(-1) \cdot 6} = 20$$

Therefore, 20 is the first row entry of the product matrix AB.

Step 2 Multiply the elements of the second row of A and the corresponding elements of B.

$$\begin{bmatrix} 2 & 3 & -1 \\ \mathbf{4} & \mathbf{2} & \mathbf{2} \end{bmatrix} \begin{bmatrix} \mathbf{1} \\ \mathbf{8} \\ \mathbf{6} \end{bmatrix} \qquad \mathbf{4 \cdot 1} + \mathbf{2 \cdot 8} + \mathbf{2 \cdot 6} = 32$$

Step 3 Write the product using the two entries we found above.

$$AB = \begin{bmatrix} 2 & 3 & -1 \\ 4 & 2 & 2 \end{bmatrix} \begin{bmatrix} 1 \\ 8 \\ 6 \end{bmatrix} = \begin{bmatrix} 20 \\ 32 \end{bmatrix} \quad ◀ \; \mathbf{3}$$

▶ **EXAMPLE 3** Find the product CD given

$$C = \begin{bmatrix} -3 & 4 & 2 \\ 5 & 0 & 4 \end{bmatrix} \text{ and } D = \begin{bmatrix} -6 & 4 \\ 2 & 3 \\ 3 & -2 \end{bmatrix}.$$

Here matrix C is 2×3 and matrix D is 3×2, so matrix CD can be found and will be 2×2.

Step 1

$$\begin{bmatrix} \mathbf{-3} & \mathbf{4} & \mathbf{2} \\ 5 & 0 & 4 \end{bmatrix} \begin{bmatrix} \mathbf{-6} & 4 \\ \mathbf{2} & 3 \\ \mathbf{3} & -2 \end{bmatrix} \qquad \mathbf{(-3) \cdot (-6)} + \mathbf{4 \cdot 2} + \mathbf{2 \cdot 3} = 32$$

4 Find the product CD given

$$C = \begin{bmatrix} 1 & 3 & 5 \\ 2 & -4 & -1 \end{bmatrix}$$

and

$$D = \begin{bmatrix} 2 & -1 \\ 4 & 3 \\ 1 & -2 \end{bmatrix}.$$

Answer:

$$CD = \begin{bmatrix} 19 & -2 \\ -13 & -12 \end{bmatrix}$$

5 Give the size of each of the following products that can be found.

(a) $\begin{bmatrix} 2 & 4 \\ 6 & 8 \end{bmatrix}\begin{bmatrix} 1 & 2 & 3 \\ 0 & -1 & 2 \end{bmatrix}$

(b) $\begin{bmatrix} 1 & 2 \\ 5 & 10 \\ 12 & 7 \end{bmatrix}\begin{bmatrix} 2 & 4 \\ 3 & 6 \\ 9 & 1 \end{bmatrix}$

(c) $\begin{bmatrix} 5 \\ 2 \\ 4 \end{bmatrix}[1 \quad 0 \quad 6]$

Answers:

(a) 2×3

(b) Not possible

(c) 3×3

Step 2

$$\begin{bmatrix} \mathbf{-3} & \mathbf{4} & \mathbf{2} \\ 5 & 0 & 4 \end{bmatrix}\begin{bmatrix} -6 & \mathbf{4} \\ 2 & \mathbf{3} \\ 3 & \mathbf{-2} \end{bmatrix} \quad \mathbf{(-3)} \cdot \mathbf{4} \cdot + \mathbf{4} \cdot \mathbf{3} + \mathbf{2} \cdot \mathbf{(-2)} = -4$$

Step 3

$$\begin{bmatrix} -3 & 4 & 2 \\ \mathbf{5} & \mathbf{0} & \mathbf{4} \end{bmatrix}\begin{bmatrix} \mathbf{-6} & 4 \\ \mathbf{2} & 3 \\ \mathbf{3} & -2 \end{bmatrix} \quad \mathbf{5} \cdot \mathbf{(-6)} + \mathbf{0} \cdot \mathbf{2} + \mathbf{4} \cdot \mathbf{3} = -18$$

Step 4

$$\begin{bmatrix} -3 & 4 & 2 \\ \mathbf{5} & \mathbf{0} & \mathbf{4} \end{bmatrix}\begin{bmatrix} -6 & \mathbf{4} \\ 2 & \mathbf{3} \\ 3 & \mathbf{-2} \end{bmatrix} \quad \mathbf{5} \cdot \mathbf{4} + \mathbf{0} \cdot \mathbf{3} + \mathbf{4} \cdot \mathbf{(-2)} = 12$$

Step 5 The product is

$$CD = \begin{bmatrix} -3 & 4 & 2 \\ 5 & 0 & 4 \end{bmatrix}\begin{bmatrix} -6 & 4 \\ 2 & 3 \\ 3 & -2 \end{bmatrix} = \begin{bmatrix} 32 & -4 \\ -18 & 12 \end{bmatrix}. \quad \blacktriangleleft \quad \boxed{4}$$

► EXAMPLE 4 Find BA given

$$A = \begin{bmatrix} 1 & -3 \\ 7 & 2 \end{bmatrix} \quad \text{and} \quad B = \begin{bmatrix} 1 & 0 & -1 \\ 3 & 1 & 4 \end{bmatrix}.$$

Since B is a 2×3 matrix and A is a 2×2 matrix, the product BA cannot be found. ◄

Note In Example 4, although BA cannot be found, matrix AB can. Since A is 2×2 and B is 2×3, AB will be a 2×3 matrix. **5**

FOR GRAPHERS

A graphing calculator with matrix capabilities can be used to multiply matrices or to find the product of a scalar and a matrix. Check your instruction manual.

Matrix multiplication has some similarities with the multiplication of numbers.

For any matrices A, B, C, such that all the indicated sums and products exist, matrix multiplication is associative and distributive.

$$A(BC) = (AB)C \quad A(B + C) = AB + AC \quad (B + C)A = BA + CA$$

However, there are important differences between matrix multiplication and multiplication of numbers. (See Exercises 29–32 at the end of this section.) In particular, matrix multiplication is *not* commutative.

If A and B are matrices such that the products AB and BA exist,

$$AB \text{ may not equal } BA.$$

▶**EXAMPLE 5** A contractor builds three kinds of houses, models A, B, and C, with a choice of two styles, Spanish or contemporary. Matrix P shows the number of each kind of house planned for a new 100-home subdivision.

$$
\begin{array}{c}
 \\
\text{Model A} \\
\text{Model B} \\
\text{Model C}
\end{array}
\begin{array}{c}
\text{Spanish} \quad \text{Contemporary} \\
\left[
\begin{array}{cc}
0 & 30 \\
10 & 20 \\
20 & 20
\end{array}
\right] = P
\end{array}
$$

The amounts for each of the exterior materials used depend primarily on the style of the house. These amounts are shown in matrix Q. (Concrete is in cubic yards, lumber in units of 1000 board feet, brick in 1000s, and shingles in units of 100 square feet.)

$$
\begin{array}{c}
\\
\text{Spanish} \\
\text{Contemporary}
\end{array}
\begin{array}{c}
\text{Concrete} \quad \text{Lumber} \quad \text{Brick} \quad \text{Shingles} \\
\left[
\begin{array}{cccc}
10 & 2 & 0 & 2 \\
50 & 1 & 20 & 2
\end{array}
\right] = Q
\end{array}
$$

Matrix R gives the cost for the each kind of material

$$
\begin{array}{c}
\\
\text{Concrete} \\
\text{Lumber} \\
\text{Brick} \\
\text{Shingles}
\end{array}
\begin{array}{c}
\text{Cost per Unit} \\
\left[
\begin{array}{c}
20 \\
180 \\
60 \\
25
\end{array}
\right] = R
\end{array}
$$

(a) What is the total cost for each model house?

First find PQ. The product PQ shows the amount of each material needed for each model house.

6 (Reference on next page.)
Let matrix A be

$$
\begin{array}{c}
\\
\\
Brand
\end{array}
\begin{array}{c}
Vitamin \\
\begin{array}{ccc} C & E & K \end{array} \\
\begin{array}{c} X \\ Y \end{array}
\begin{bmatrix} 2 & 7 & 5 \\ 4 & 6 & 9 \end{bmatrix}
\end{array}
$$

and matrix B be

$$
\begin{array}{c}
\\
Vitamin
\end{array}
\begin{array}{c}
Cost \\
\begin{array}{cc} X & Y \end{array} \\
\begin{array}{c} C \\ E \\ K \end{array}
\begin{bmatrix} 12 & 14 \\ 18 & 15 \\ 9 & 10 \end{bmatrix}.
\end{array}
$$

(a) What quantities do matrices A and B represent?

(b) What quantities does the product AB represent?

(c) What quantities does the product BA represent?

Answers:

(a) A = brand/vitamin,
B = vitamin/cost

(b) AB = brand/cost

(c) Not meaningful, although the product BA can be found

$$
PQ = \begin{bmatrix} 0 & 30 \\ 10 & 20 \\ 20 & 20 \end{bmatrix}
\begin{bmatrix} 10 & 2 & 0 & 2 \\ 50 & 1 & 20 & 2 \end{bmatrix}
$$

$$
= \begin{array}{c}
\begin{array}{cccc} Concrete & Lumber & Brick & Shingles \end{array} \\
\begin{bmatrix} 1500 & 30 & 600 & 60 \\ 1100 & 40 & 400 & 60 \\ 1200 & 60 & 400 & 80 \end{bmatrix}
\begin{array}{c} Model\ A \\ Model\ B \\ Model\ C \end{array}
\end{array}
$$

Now multiply PQ and R, the cost matrix, to get the total cost for each model house.

$$
\begin{bmatrix} 1500 & 30 & 600 & 60 \\ 1100 & 40 & 400 & 60 \\ 1200 & 60 & 400 & 80 \end{bmatrix}
\begin{bmatrix} 20 \\ 180 \\ 60 \\ 25 \end{bmatrix}
= \begin{array}{c}
Cost \\
\begin{bmatrix} 72,900 \\ 54,700 \\ 60,800 \end{bmatrix}
\begin{array}{c} Model\ A \\ Model\ B \\ Model\ C \end{array}
\end{array}
$$

(b) How much of each of the four kinds of material must be ordered?

The totals of the columns of matrix PQ will give a matrix whose elements represent the total amounts of each material needed for the subdivision. Call this matrix T, and write it as a row matrix.

$$
T = \begin{bmatrix} 3800 & 130 & 1400 & 200 \end{bmatrix}
$$

(c) What is the total cost for material?

Find the total cost of all the materials by taking the product of matrix T, the matrix showing the total amounts of each material, and matrix R, the cost matrix. (To multiply these and get a 1×1 matrix, representing total cost, we must multiply a 1×4 matrix by a 4×1 matrix. This is why T was written as a row matrix in (b) above.)

$$
TR = \begin{bmatrix} 3800 & 130 & 1400 & 200 \end{bmatrix}
\begin{bmatrix} 20 \\ 180 \\ 60 \\ 25 \end{bmatrix}
= \begin{bmatrix} 188,400 \end{bmatrix}
$$

(d) Suppose the contractor builds the same number of homes in five subdivisions. Calculate the total amount of each material for each model for all five subdivisions.

Multiply PQ by the scalar 5, as follows.

$$
5 \begin{bmatrix} 1500 & 30 & 600 & 60 \\ 1100 & 40 & 400 & 60 \\ 1200 & 60 & 400 & 80 \end{bmatrix}
= \begin{bmatrix} 7500 & 150 & 3000 & 300 \\ 5500 & 200 & 2000 & 300 \\ 6000 & 300 & 2000 & 400 \end{bmatrix} \blacktriangleleft
$$

We can introduce a notation to help keep track of the quantities a matrix represents. For example, we can say that matrix P, from Example 5, represents models/styles, matrix Q represents styles/materials, and matrix R represents materials/cost. In each case, the meaning of the rows is written first and the columns second. When we found the product PQ in Example 5, the rows of the matrix represented models and the columns represented materials. Therefore, we can say the matrix product PQ represents models/materials. The common quantity, styles, in both P and Q was eliminated in the product PQ. Do you see that the product $(PQ)R$ represents models/cost?

In practical problems this notation helps decide in what order to multiply two matrices so that the results are meaningful. In Example 5(c) we could have found either product RT or product TR. However, since T represents subdivisions/materials and R represents materials/cost, the product TR gives subdivisions/cost. **6** (See page 303.)

6.4 EXERCISES

Find the sizes of the product AB and the product BA, whenever these products exist. Do not compute any products. (See Examples 1 and 4.)

1.
$$A = \begin{bmatrix} 3 & 6 & 7 \\ 8 & 0 & 1 \end{bmatrix}, B = \begin{bmatrix} 2 & 5 & 9 & 1 \\ 7 & 0 & 0 & 6 \\ -1 & 3 & 8 & 7 \end{bmatrix}$$

2.
$$A = \begin{bmatrix} -1 & -2 & -5 \\ 9 & 2 & -1 \\ 10 & 34 & 5 \end{bmatrix}, B = \begin{bmatrix} 17 & -9 \\ -6 & 12 \\ 3 & 5 \end{bmatrix}$$

3.
$$A = \begin{bmatrix} 1 & 0 \\ 1 & 1 \\ 0 & 1 \end{bmatrix}, B = \begin{bmatrix} 5 & 6 & 11 \\ 7 & 8 & 15 \end{bmatrix}$$

4.
$$A = \begin{bmatrix} 1 & -5 & 7 \\ 2 & 4 & 8 \\ 1 & -1 & 2 \end{bmatrix}, B = \begin{bmatrix} -2 & 4 & 9 \\ 13 & -2 & 1 \\ 5 & 25 & 0 \end{bmatrix}$$

5.
$$A = \begin{bmatrix} -4 & 15 \\ 3 & -7 \\ 2 & 10 \end{bmatrix}, B = \begin{bmatrix} 1 & 2 \\ 3 & 4 \end{bmatrix}$$

6.
$$A = \begin{bmatrix} 10 & 12 \\ -6 & 0 \\ 1 & 23 \\ -4 & 3 \end{bmatrix}, B = \begin{bmatrix} 1 & 2 & 3 \\ 3 & 2 & 1 \end{bmatrix}$$

Let $A = \begin{bmatrix} -2 & 4 \\ 0 & 3 \end{bmatrix}$ and $B = \begin{bmatrix} -6 & 2 \\ 4 & 0 \end{bmatrix}$. Find each of the following.

7. $2A$ **8.** $-3B$ **9.** $-4B$ **10.** $5A$ **11.** $-4A + 5B$ **12.** $3A - 10B$

Find each of the following matrix products where possible. (See Examples 1–4.)

13. $\begin{bmatrix} 1 & 2 \\ 3 & 4 \end{bmatrix} \begin{bmatrix} -1 \\ 7 \end{bmatrix}$

14. $\begin{bmatrix} -1 & 5 \\ 7 & 0 \end{bmatrix} \begin{bmatrix} 6 \\ 2 \end{bmatrix}$

15. $\begin{bmatrix} 2 & 2 & -1 \\ 3 & 0 & 1 \end{bmatrix} \begin{bmatrix} 0 & 2 \\ -1 & 4 \\ 0 & 2 \end{bmatrix}$

16. $\begin{bmatrix} -9 & 2 & 1 \\ 3 & 0 & 0 \end{bmatrix} \begin{bmatrix} 2 \\ -1 \\ 4 \end{bmatrix}$

17. $\begin{bmatrix} -4 & 1 \\ 2 & -3 \end{bmatrix} \begin{bmatrix} 1 & 0 \\ 0 & 1 \end{bmatrix}$

18. $\begin{bmatrix} 1 & 0 \\ 0 & 1 \end{bmatrix} \begin{bmatrix} 3 & -2 \\ 1 & -5 \end{bmatrix}$

19. $\begin{bmatrix} 1 & 0 & 0 \\ 0 & 1 & 0 \\ 0 & 0 & 1 \end{bmatrix} \begin{bmatrix} 3 & -5 & 7 \\ -2 & 1 & 6 \\ 0 & -3 & 4 \end{bmatrix}$

20. $\begin{bmatrix} -8 & 9 \\ 3 & -4 \\ -1 & 6 \end{bmatrix} \begin{bmatrix} 1 & 0 & 0 \\ 0 & 1 & 0 \end{bmatrix}$

21. $\begin{bmatrix} 1 & 2 & 3 \\ 4 & 5 & 6 \\ 7 & 8 & 9 \end{bmatrix} \begin{bmatrix} -1 & 5 \\ 7 & 0 \\ 1 & 2 \end{bmatrix}$

22. $\begin{bmatrix} -2 & 0 & 3 \\ 5 & -3 & -1 \end{bmatrix} \begin{bmatrix} 2 & 0 & -1 & 3 \\ 0 & 1 & 0 & -1 \\ 4 & 2 & 5 & -4 \end{bmatrix}$

Let $A = \begin{bmatrix} 1 & -2 \\ 3 & 4 \end{bmatrix}$, $B = \begin{bmatrix} -1 & 0 \\ 1 & 2 \end{bmatrix}$, $C = \begin{bmatrix} 1 & 0 \\ 0 & 1 \end{bmatrix}$, $D = \begin{bmatrix} 1 & -2 & 3 \\ 0 & 4 & 1 \end{bmatrix}$, and $E = \begin{bmatrix} 1 & 2 \\ -2 & 1 \\ 0 & 5 \end{bmatrix}$.

Find each of the following.

23. $-3B + 4C$

24. $A - 2B$

25. $AD + BD$

26. $EA + EB$

Let $A = \begin{bmatrix} 1 & -2 \\ 4 & 3 \end{bmatrix}$ and $B = \begin{bmatrix} 2 & -1 \\ 0 & 5 \end{bmatrix}$. *Find a matrix X satisfying the given equation.*

27. $2X = 2A + 3B$

28. $3X = A - 3B$

In Exercises 29–31, use the matrices

$$A = \begin{bmatrix} -3 & -9 \\ 2 & 6 \end{bmatrix} \quad \text{and} \quad B = \begin{bmatrix} 4 & 6 \\ 2 & 3 \end{bmatrix}.$$

29. Show that $AB \neq BA$. Hence, matrix multiplication is not commutative.

30. Show that $(A + B)^2 \neq A^2 + 2AB + B^2$.

31. Show that $(A + B)(A - B) \neq A^2 - B^2$.

32. Show that $D^2 = D$, where

$$D = \begin{bmatrix} 1 & 0 & 0 \\ \frac{1}{2} & 0 & \frac{1}{2} \\ 0 & 0 & 1 \end{bmatrix}.$$

Given matrices

$$P = \begin{bmatrix} m & n \\ p & q \end{bmatrix}, \quad X = \begin{bmatrix} x & y \\ z & w \end{bmatrix}, \quad T = \begin{bmatrix} r & s \\ t & u \end{bmatrix},$$

verify that the statements in Exercises 33–36 are true.

33. $(PX)T = P(XT)$ (Associative property)

34. $P(X + T) = PX + PT$ (Distributive property)

35. $k(X + T) = kX + kT$ for any real number k

36. $(k + h)P = kP + hP$ for any real numbers k and h

37. **Management** Burger Barn's three locations sell hamburgers, fries, and soft drinks. Barn I sells 900 burgers, 600 orders of fries, and 750 soft drinks each day. Barn II sells 1500 burgers a day and Barn III sells 1150. Soft drink sales number 900 a day at Barn II and 825 a day at Barn III. Barn II sells 950 and Barn III sells 800 orders of fries per day.

(a) Write a 3×3 matrix S that displays daily sales figures for all locations.

(b) Burgers cost $1.50 each, fries $.90 an order and soft drinks $.60 each. Write a 1×3 matrix P that displays the prices.

(c) What matrix product displays the daily revenue at each of the three locations?

(d) What is the total daily revenue from all locations?

38. **Management** The four departments of Stagg Enterprises need to order the following amounts of the same products.

	Paper	Tape	Printer Ribbon	Memo Pads	Pens
Department 1	10	4	3	5	6
Department 2	7	2	2	3	8
Department 3	4	5	1	0	10
Department 4	0	3	4	5	5

The unit price (in dollars) of each product is given below for two suppliers.

	Supplier A	Supplier B
Paper	2	3
Tape	1	1
Printer Ribbon	4	3
Memo Pads	3	3
Pens	1	2

(a) Use matrix multiplication to get a matrix showing the comparative costs for each department for the products from the two suppliers.

(b) Find the total cost to buy products from each supplier. From which supplier should the company make the purchase?

39. **Management** The Perulli Candy Company makes three types of chocolate candy: Cheery Cherry, Mucho Mocha, and Almond Delight. The company produces its products in San Diego, Mexico City, and Managua using two main ingredients: chocolate and sugar.

(a) Each kilogram of Cheery Cherry requires .5 kg of sugar and .2 kg of chocolate; each kilogram of Mucho Mocha requires .4 kg of sugar and .3 kg of chocolate; and each kilogram of Almond Delight requires .3 kg of sugar and .3 kg of chocolate. Put this information into a 2 × 3 matrix, labeling the rows and columns.

(b) The cost of 1 kg of sugar is $3 in San Diego, $2 in Mexico City, and $1 in Managua. The cost of 1 kg of chocolate is $3 in San Diego, $3 in Mexico City, and $4 in Managua. Put this information into a matrix in such a a way that when you multiply it with your matrix from part (a), you get a matrix representing the ingredient cost of producing each type of candy in each city.

(c) Multiply the matrices in parts (a) and (b), labeling the product matrix.

(d) From part (c) what is the combined sugar-and-chocolate cost to produce 1 kg of Mucho Mocha in Managua?

(e) Perulli Candy needs to quickly produce a special shipment of 100 kg of Cheery Cherry, 200 kg of Mucho Mocha, and 500 kg of Almond Delight, and it decides to select one factory to fill the entire order. Use matrix multiplication to determine in which city the total sugar-and-chocolate cost to produce the order is the smallest.

40. **Social Sciences** The average birth and death rates per million for several regions and the world population (in millions) by region are given below.*

	Births	Deaths
Asia	.027	.009
Latin America	.030	.007
North America	.015	.009
Europe	.013	.011
Soviet Union	.019	.011

	Asia	*Latin America*	*North America*	*Europe*	*USSR*
1960	1596	218	199	425	214
1970	1996	286	226	460	243
1980	2440	365	252	484	266
1990	2906	455	277	499	291

(a) Write the information in each table as a matrix.

(b) Use the matrices from part (a) to find the total number (in millions) of births and deaths in each year.

(c) Using the results of part (b), compare the number of births in 1960 and in 1990. Also compare the birth rates from part (a). Which gives better information?

(d) Using the results of part (b), compare the number of deaths in 1980 and in 1990. Discuss how this comparison differs from comparing death rates from part (a).

41. In Exercise 30, Section 6.3, label the matrices found in parts (a), (b), and (c) respectively *X*, *Y*, and *Z*.

(a) Find the product matrix *XY*. What do the entries of this matrix represent?

(b) Find the product matrix *YZ*. What do the entries represent?

42. Explain why the system of equations

$$x - 3y = 4$$
$$2x + y = 1$$

is equivalent to the matrix equation $AX = B$, where

$A = \begin{bmatrix} 1 & -3 \\ 2 & 1 \end{bmatrix}$, $X = \begin{bmatrix} x \\ y \end{bmatrix}$, and $B = \begin{bmatrix} 4 \\ 1 \end{bmatrix}$. (*Hint*: what is the product AX?)

---

*"Vital Events and Rates by Region and Development Category, 1987" and "World Population by Region and Development Category, 1950–2025" from U.S. Bureau of the Census, *World Population Profile: 1987*.

Solve the matrix equation AX = B. (See Exercise 42.)

43.
$$A = \begin{bmatrix} 1 & 2 \\ 5 & -4 \end{bmatrix}, X = \begin{bmatrix} x \\ y \end{bmatrix}, B = \begin{bmatrix} 3 \\ -6 \end{bmatrix}.$$

44.
$$A = \begin{bmatrix} 1 & 2 & 3 \\ 2 & 6 & 1 \\ 3 & 3 & 10 \end{bmatrix}, X = \begin{bmatrix} x \\ y \\ z \end{bmatrix}, B = \begin{bmatrix} -2 \\ 2 \\ -2 \end{bmatrix}.$$

 Use a graphing calculator with matrix capabilities or appropriate computer software to find the matrix products in Exercises 45–50.

$$A = \begin{bmatrix} 2 & 3 & -1 & 5 & 10 \\ 2 & 8 & 7 & 4 & 3 \\ -1 & -4 & -12 & 6 & 8 \\ 2 & 5 & 7 & 1 & 4 \end{bmatrix} \quad B = \begin{bmatrix} 9 & 3 & 7 & -6 \\ -1 & 0 & 4 & 2 \\ -10 & -7 & 6 & 9 \\ 8 & 4 & 2 & -1 \\ 2 & -5 & 3 & 7 \end{bmatrix}$$

$$C = \begin{bmatrix} -6 & 8 & 2 & 4 & -3 \\ 1 & 9 & 7 & -12 & 5 \\ 15 & 2 & -8 & 10 & 11 \\ 4 & 7 & 9 & 6 & -2 \\ 1 & 3 & 8 & 23 & 4 \end{bmatrix} \quad D = \begin{bmatrix} 5 & -3 & 7 & 9 & 2 \\ 6 & 8 & -5 & 2 & 1 \\ 3 & 7 & -4 & 2 & 11 \\ 5 & -3 & 9 & 4 & -1 \\ 0 & 3 & 2 & 5 & 1 \end{bmatrix}$$

45. *CD*

48. *DB*

46. *AC*

49. Is *AC = CA*?

47. *CA*

50. Is *CD = DC*?

---

1 Let $A = \begin{bmatrix} 3 & -2 \\ 4 & -1 \end{bmatrix}$

and $I = \begin{bmatrix} 1 & 0 \\ 0 & 1 \end{bmatrix}$.

Find *IA* and *AI*.

Answer:

$$IA = \begin{bmatrix} 3 & -2 \\ 4 & -1 \end{bmatrix} = A$$

and

$$AI = \begin{bmatrix} 3 & -2 \\ 4 & -1 \end{bmatrix} = A$$

6.5 MATRIX INVERSES

In Section 6.3, a zero matrix was defined with properties similar to those of the real number zero, the identity for addition. Recall from Section 1.1 that the real number 1 is the identity element for multiplication of real numbers: for any real number a, $a \cdot 1 = 1 \cdot a = a$. In this section, an **identity matrix I** is defined that has properties similar to those of the number 1. This identity matrix is then used to find the multiplicative inverse of any square matrix that has an inverse.

If I is to be the identity matrix, the products AI and IA must both equal A. The 2 × 2 identity matrix that satisfies these conditions is

$$I = \begin{bmatrix} 1 & 0 \\ 0 & 1 \end{bmatrix}. \quad \mathbf{1}$$

To check that I is really the 2 × 2 identity matrix, let

$$A = \begin{bmatrix} a & b \\ c & d \end{bmatrix}.$$

Then AI and IA should both equal A.

$$AI = \begin{bmatrix} a & b \\ c & d \end{bmatrix} \begin{bmatrix} 1 & 0 \\ 0 & 1 \end{bmatrix} = \begin{bmatrix} a(1) + b(0) & a(0) + b(1) \\ c(1) + d(0) & c(0) + d(1) \end{bmatrix} = \begin{bmatrix} a & b \\ c & d \end{bmatrix} = A$$

$$IA = \begin{bmatrix} 1 & 0 \\ 0 & 1 \end{bmatrix} \begin{bmatrix} a & b \\ c & d \end{bmatrix} = \begin{bmatrix} 1(a) + 0(c) & 1(b) + 0(d) \\ 0(a) + 1(c) & 0(b) + 1(d) \end{bmatrix} = \begin{bmatrix} a & b \\ c & d \end{bmatrix} = A$$

This verifies that I has been defined correctly. (It can also be shown that I is the only 2×2 identity matrix.)

The identity matrices for 3×3 matrices and 4×4 matrices, respectively, are

$$I = \begin{bmatrix} 1 & 0 & 0 \\ 0 & 1 & 0 \\ 0 & 0 & 1 \end{bmatrix} \quad \text{and} \quad I = \begin{bmatrix} 1 & 0 & 0 & 0 \\ 0 & 1 & 0 & 0 \\ 0 & 0 & 1 & 0 \\ 0 & 0 & 0 & 1 \end{bmatrix}.$$

By generalizing, an identity matrix can be found for any n by n matrix: this identity matrix will have 1's on the main diagonal from upper left to lower right, with all other entries equal to 0.

Recall that for every nonzero real number a, the equation $ax = 1$ has a solution, namely, $x = 1/a$. Similarly, for a square matrix A it is natural to consider the matrix equation $AX = I$. This equation does not always have a solution, but when it does, we use special terminology. If there is a matrix A^{-1} satisfying

$$AA^{-1} = I$$

(that is A^{-1} is a solution of $AX = I$), then A^{-1} is the called the **inverse matrix** of A. In this case it can be proved that $A^{-1}A = I$ and that A^{-1} is unique (that is, a square matrix has no more than one inverse). When a matrix has an inverse, it can be found by using the row operations of Section 6.2, as we shall see below.

Caution　Only square matrices have inverses, but not every square matrix has one. Note that the symbol A^{-1} (read A-inverse) does *not* mean $1/A$ or I/A; the symbol A^{-1} is just the notation for the inverse of matrix A. There is no such thing as matrix division.

▶ **EXAMPLE 1**　Given matrices A and B below, decide if they are inverses.

$$A = \begin{bmatrix} 2 & 3 \\ 1 & 8 \end{bmatrix} \quad B = \begin{bmatrix} -1 & 3 \\ 1 & -2 \end{bmatrix}$$

The matrices are inverses if AB and BA both equal I.

$$AB = \begin{bmatrix} 2 & 3 \\ 1 & 8 \end{bmatrix} \begin{bmatrix} -1 & 3 \\ 1 & -2 \end{bmatrix} = \begin{bmatrix} 1 & 0 \\ 7 & -13 \end{bmatrix} \neq I$$

2 Given

$$A = \begin{bmatrix} 1 & 2 \\ 4 & 6 \end{bmatrix}$$

and

$$B = \begin{bmatrix} -3 & 1 \\ 2 & -\frac{1}{2} \end{bmatrix},$$

decide if they are inverses.

Answer:
Yes because $AB = BA = I$.

Since $AB \neq I$, the two matrices are not inverses of each other. ◀ **2**

As an example of finding the multiplicative inverse of a matrix, let us look for the inverse of

$$A = \begin{bmatrix} 2 & 4 \\ 1 & -1 \end{bmatrix}.$$

Let the unknown inverse matrix be

$$A^{-1} = \begin{bmatrix} x & y \\ z & w \end{bmatrix}.$$

By the definition of matrix inverse, $AA^{-1} = I$, or

$$AA^{-1} = \begin{bmatrix} 2 & 4 \\ 1 & -1 \end{bmatrix} \begin{bmatrix} x & y \\ z & w \end{bmatrix} = \begin{bmatrix} 1 & 0 \\ 0 & 1 \end{bmatrix}.$$

Use matrix multiplication to get

$$\begin{bmatrix} 2x + 4z & 2y + 4w \\ x - z & y - w \end{bmatrix} = \begin{bmatrix} 1 & 0 \\ 0 & 1 \end{bmatrix}.$$

Setting corresponding elements equal gives the system of equations

$$2x + 4z = 1 \tag{1}$$
$$2y + 4w = 0 \tag{2}$$
$$x - z = 0 \tag{3}$$
$$y - w = 1. \tag{4}$$

Since equations (1) and (3) involve only x and z, while equations (2) and (4) involve only y and w, these four equations lead to two systems of equations,

$$\begin{array}{cc} 2x + 4z = 1 & 2y + 4w = 0 \\ \quad\quad\text{and} & \\ x - z = 0 & y - w = 1. \end{array}$$

Writing the two systems as augmented matrices gives

$$\begin{bmatrix} 2 & 4 & | & 1 \\ 1 & -1 & | & 0 \end{bmatrix} \quad \text{and} \quad \begin{bmatrix} 2 & 4 & | & 0 \\ 1 & -1 & | & 1 \end{bmatrix}.$$

Each of these systems can be solved by the Gauss-Jordan method. However, since the elements to the left of the vertical bar are identical, the two systems can be combined into one matrix

$$\begin{bmatrix} 2 & 4 & | & 1 & 0 \\ 1 & -1 & | & 0 & 1 \end{bmatrix} \tag{5}$$

3 (a) Find A^{-1} if

$$A = \begin{bmatrix} 2 & 2 \\ 4 & 1 \end{bmatrix}.$$

(b) Check your answer by finding AA^{-1}.

Answers:

(a) $\begin{bmatrix} -\frac{1}{6} & \frac{1}{3} \\ \frac{2}{3} & -\frac{1}{3} \end{bmatrix}$

(b) $\begin{bmatrix} 1 & 0 \\ 0 & 1 \end{bmatrix}$

and solved simultaneously as follows. Interchange the two rows to get a 1 in the upper left corner.

$$\begin{bmatrix} 1 & -1 & \bigm| & 0 & 1 \\ 2 & 4 & \bigm| & 1 & 0 \end{bmatrix} \qquad \text{Interchange } R_1, R_2$$

Multiply row one by -2 and add the results to row two to get

$$\begin{bmatrix} 1 & -1 & \bigm| & 0 & 1 \\ 0 & 6 & \bigm| & 1 & -2 \end{bmatrix}. \qquad -2R_1 + R_2$$

Now, to get a 1 in the second row, second column position, multiply row two by $1/6$.

$$\begin{bmatrix} 1 & -1 & \bigm| & 0 & 1 \\ 0 & 1 & \bigm| & \frac{1}{6} & -\frac{1}{3} \end{bmatrix} \qquad \frac{1}{6}R_2$$

Finally, add row two to row one to get a 0 in the second column above the 1.

$$\begin{bmatrix} 1 & 0 & \bigm| & \frac{1}{6} & \frac{2}{3} \\ 0 & 1 & \bigm| & \frac{1}{6} & -\frac{1}{3} \end{bmatrix} \qquad R_2 + R_1 \tag{6}$$

The left half of the augmented matrix (6) is the identity matrix, so the Gauss-Jordan process is finished and the solutions can be read from the right half of the augmented matrix. The numbers in the first column to the right of the vertical bar give the values of x and z. The second column to the right of the bar gives the values of y and w. That is,

$$\begin{bmatrix} 1 & 0 & \bigm| & x & y \\ 0 & 1 & \bigm| & z & w \end{bmatrix} = \begin{bmatrix} 1 & 0 & \bigm| & \frac{1}{6} & \frac{2}{3} \\ 0 & 1 & \bigm| & \frac{1}{6} & -\frac{1}{3} \end{bmatrix}$$

so that

$$A^{-1} = \begin{bmatrix} x & y \\ z & w \end{bmatrix} = \begin{bmatrix} \frac{1}{6} & \frac{2}{3} \\ \frac{1}{6} & -\frac{1}{3} \end{bmatrix}.$$

Thus the original augmented matrix (5) has A as its left half and the identity matrix as its right half, while the final augmented matrix (6), at the end of the Gauss-Jordan process, has the identity matrix as its left half and the inverse matrix A^{-1} as its right half.

Check by multiplying A and A^{-1}. The result should be I.

$$AA^{-1} = \begin{bmatrix} 2 & 4 \\ 1 & -1 \end{bmatrix} \begin{bmatrix} \frac{1}{6} & \frac{2}{3} \\ \frac{1}{6} & -\frac{1}{3} \end{bmatrix} = \begin{bmatrix} \frac{1}{3} + \frac{2}{3} & \frac{4}{3} - \frac{4}{3} \\ \frac{1}{6} - \frac{1}{6} & \frac{2}{3} + \frac{1}{3} \end{bmatrix} = \begin{bmatrix} 1 & 0 \\ 0 & 1 \end{bmatrix} = I. \quad \textbf{3}$$

This procedure for finding the inverse of a matrix can be generalized as follows.

4 (a) Complete this step.

(b) Write this row transformation as _____.

Answers:

(a)

$$\begin{bmatrix} 1 & 0 & 1 & | & 1 & 0 & 0 \\ 0 & -2 & -3 & | & -2 & 1 & 0 \\ 0 & 0 & -3 & | & -3 & 0 & 1 \end{bmatrix}$$

(b) $-3R_1 + R_3$

5 (a) Complete this step.

(b) Write this row transformation as _____.

Answers:

(a)

$$\begin{bmatrix} 1 & 0 & 0 & | & 0 & 0 & \frac{1}{3} \\ 0 & 1 & \frac{3}{2} & | & 1 & -\frac{1}{2} & 0 \\ 0 & 0 & 1 & | & 1 & 0 & -\frac{1}{3} \end{bmatrix}$$

(b) $-1R_3 + R_1$

To obtain an **inverse matrix** A^{-1} for any $n \times n$ matrix A for which A^{-1} exists, follow these steps.

1. Form the augmented matrix $[A \mid I]$ where I is the $n \times n$ identity matrix.
2. Perform row operations on $[A \mid I]$ to get a matrix of the form $[I \mid B]$.
3. Matrix B is A^{-1}.

▶**EXAMPLE 2**   Find A^{-1} if $A = \begin{bmatrix} 1 & 0 & 1 \\ 2 & -2 & -1 \\ 3 & 0 & 0 \end{bmatrix}$.

First write the augmented matrix $[A \mid I]$.

$$[A \mid I] = \begin{bmatrix} 1 & 0 & 1 & | & 1 & 0 & 0 \\ 2 & -2 & -1 & | & 0 & 1 & 0 \\ 3 & 0 & 0 & | & 0 & 0 & 1 \end{bmatrix}$$

The augmented matrix already has 1 in the upper left-hand corner as needed, so begin by selecting the row operation which will result in a 0 for the first element in row two. Multiply row one by -2 and add the result to row two. This gives

$$\begin{bmatrix} 1 & 0 & 1 & | & 1 & 0 & 0 \\ 0 & -2 & -3 & | & -2 & 1 & 0 \\ 3 & 0 & 0 & | & 0 & 0 & 1 \end{bmatrix}. \qquad -2R_1 + R_2$$

Get 0 for the first element in row three by multiplying row one by -3 and adding to row three. **4**

Get 1 for the second element in row two by muliplying row two of the matrix found in Problem 4 at the side by $-1/2$, obtaining the new matrix

$$\begin{bmatrix} 1 & 0 & 1 & | & 1 & 0 & 0 \\ 0 & 1 & \frac{3}{2} & | & 1 & -\frac{1}{2} & 0 \\ 0 & 0 & -3 & | & -3 & 0 & 1 \end{bmatrix}. \qquad -\frac{1}{2}R_2$$

Get 1 for the third element in row three by multiplying row three by $-1/3$, with the result

$$\begin{bmatrix} 1 & 0 & 1 & | & 1 & 0 & 0 \\ 0 & 1 & \frac{3}{2} & | & 1 & -\frac{1}{2} & 0 \\ 0 & 0 & 1 & | & 1 & 0 & -\frac{1}{3} \end{bmatrix}. \qquad -\frac{1}{3}R_3$$

Now get 0 for the third element in row one by multiplying row three by -1 and adding to row one. **5**

6

(a) Find A^{-1} if
$$A = \begin{bmatrix} 2 & 8 \\ -1 & -5 \end{bmatrix}.$$

(b) Check this answer by finding AA^{-1}.

Answers:

(a) $\begin{bmatrix} \frac{5}{2} & 4 \\ -\frac{1}{2} & -1 \end{bmatrix}$

(b) $\begin{bmatrix} 1 & 0 \\ 0 & 1 \end{bmatrix}.$

---

7 Find the inverse, if it exists, of each matrix below.

(a) $\begin{bmatrix} 8 & 4 \\ 6 & 3 \end{bmatrix}$ **(b)** $\begin{bmatrix} 0 & 1 \\ 1 & 0 \end{bmatrix}$

(c) $\begin{bmatrix} 6 & -1 & 2 \\ 4 & 1 & 3 \\ -3 & \frac{1}{2} & -1 \end{bmatrix}$

Answers:

(a) and **(c)** have no inverse;

(b) is its own inverse.

Get 0 for the third element in row two by multiplying row three of the matrix found in Problem 5 at the side by $-3/2$ and adding to row two.

$$\begin{bmatrix} 1 & 0 & 0 & | & 0 & 0 & \frac{1}{3} \\ 0 & 1 & 0 & | & -\frac{1}{2} & -\frac{1}{2} & \frac{1}{2} \\ 0 & 0 & 1 & | & 1 & 0 & -\frac{1}{3} \end{bmatrix} \quad -\frac{3}{2}R_3 + R_2$$

This last transformation gives the desired inverse:

$$A^{-1} = \begin{bmatrix} 0 & 0 & \frac{1}{3} \\ -\frac{1}{2} & -\frac{1}{2} & \frac{1}{2} \\ 1 & 0 & -\frac{1}{3} \end{bmatrix}.$$

Confirm this by computing the product AA^{-1}. It should equal I. ◄ **6**

▶ **EXAMPLE 3** Find A^{-1} if $A = \begin{bmatrix} 2 & -4 \\ 1 & -2 \end{bmatrix}$.

Using row operations to transform the first column of the augmented matrix

$$\begin{bmatrix} 2 & -4 & | & 1 & 0 \\ 1 & -2 & | & 0 & 1 \end{bmatrix}$$

results in the following matrices.

$$\begin{bmatrix} 1 & -2 & | & \frac{1}{2} & 0 \\ 1 & -2 & | & 0 & 1 \end{bmatrix} \quad \frac{1}{2}R_1$$

$$\begin{bmatrix} 1 & -2 & | & \frac{1}{2} & 0 \\ 0 & 0 & | & -\frac{1}{2} & 1 \end{bmatrix} \quad -1R_1 + R_2$$

At this point, the matrix should be changed so that the second element of row two will be 1. Since that element is now 0, there is no way to complete the desired transformation. What is wrong? Remember, near the beginning of this section we mentioned that some matrices do not have inverses. Matrix A is an example of a matrix that has no inverse. In this case, there is no matrix A^{-1} such that $AA^{-1} = I$. ◄ **7**

FOR GRAPHERS

Only one or two keystrokes are needed to find the inverse of a matrix with a graphing calculator that has matrix capabilities. Check your instruction manual. If you attempt to find the inverse of a matrix A that does not have one, you should get an error message. However, because of round-off error, the calculator may sometimes display a matrix that it says is A^{-1}. As an accuracy check you should multiply A by A^{-1} to see if the product is the identity matrix. If it is not, then A does not have an inverse.

6.5 EXERCISES

Determine whether the given matrices are inverses of each other by computing their product. (See Example 1.)

1. $\begin{bmatrix} 5 & 2 \\ 3 & -1 \end{bmatrix}$ and $\begin{bmatrix} -1 & 2 \\ 3 & -4 \end{bmatrix}$

2. $\begin{bmatrix} 0 & 1 \\ 1 & 0 \end{bmatrix}$ and $\begin{bmatrix} 3 & 5 \\ 7 & 9 \end{bmatrix}$

3. $\begin{bmatrix} 3 & -1 \\ -4 & 2 \end{bmatrix}$ and $\begin{bmatrix} 1 & \frac{1}{2} \\ 2 & \frac{3}{2} \end{bmatrix}$

4. $\begin{bmatrix} 1 & 1 \\ .1 & .2 \end{bmatrix}$ and $\begin{bmatrix} 2 & -10 \\ -1 & 10 \end{bmatrix}$

5. $\begin{bmatrix} 1 & 1 & 1 \\ 2 & 3 & 0 \\ 1 & 2 & 1 \end{bmatrix}$ and $\begin{bmatrix} 1.5 & .5 & -1.5 \\ -1 & 0 & 1 \\ .5 & -2 & 2 \end{bmatrix}$

6. $\begin{bmatrix} 2 & 5 & 4 \\ 1 & 4 & 3 \\ 1 & 3 & 2 \end{bmatrix}$ and $\begin{bmatrix} 1 & 2 & 1 \\ -5 & 8 & 2 \\ 7 & -11 & -3 \end{bmatrix}$

7. Does a square matrix with a row of all zeros have an inverse? Explain why.

8. If A is a matrix that has an inverse A^{-1}, then what does $(A^{-1})^{-1}$ equal? (*Hint:* Experiment with a few matrices to see what you get.)

Find the inverse, if it exists, for each of the following matrices. (See Examples 2 and 3.)

9. $\begin{bmatrix} 2 & 3 \\ 1 & 2 \end{bmatrix}$

10. $\begin{bmatrix} -1 & 2 \\ 1 & -1 \end{bmatrix}$

11. $\begin{bmatrix} 2 & 4 \\ 3 & 6 \end{bmatrix}$

12. $\begin{bmatrix} -3 & -5 \\ 6 & 10 \end{bmatrix}$

13. $\begin{bmatrix} 2 & 6 \\ 1 & 4 \end{bmatrix}$

14. $\begin{bmatrix} 1 & 2 \\ 3 & 4 \end{bmatrix}$

15. $\begin{bmatrix} 1 & -1 & 1 \\ 0 & 2 & -1 \\ 2 & 3 & 0 \end{bmatrix}$

16. $\begin{bmatrix} 1 & 2 & 3 \\ 1 & 1 & 2 \\ 0 & 1 & 2 \end{bmatrix}$

17. $\begin{bmatrix} 1 & 4 & 3 \\ 1 & -3 & -2 \\ 2 & 5 & 4 \end{bmatrix}$

18. $\begin{bmatrix} 1 & 2 & 0 \\ 3 & -1 & 2 \\ -2 & 3 & -2 \end{bmatrix}$

19. $\begin{bmatrix} 1 & -3 & 4 \\ 2 & -5 & 7 \\ 0 & -1 & 1 \end{bmatrix}$

20. $\begin{bmatrix} 5 & 0 & 2 \\ 2 & 2 & 1 \\ -3 & 1 & -1 \end{bmatrix}$

21. $\begin{bmatrix} 2 & 4 & 6 \\ -1 & -4 & -3 \\ 0 & 1 & -1 \end{bmatrix}$

22. $\begin{bmatrix} 2 & 2 & -4 \\ 2 & 6 & 0 \\ -3 & -3 & 5 \end{bmatrix}$

23. $\begin{bmatrix} 1 & -2 & 3 & 0 \\ 0 & 1 & -1 & 1 \\ -2 & 2 & -2 & 4 \\ 0 & 2 & -3 & 1 \end{bmatrix}$

24. $\begin{bmatrix} 1 & 1 & 0 & 2 \\ 2 & -1 & 1 & -1 \\ 3 & 3 & 2 & -2 \\ 1 & 2 & 1 & 0 \end{bmatrix}$

In Exercises 25–28, let $A = \begin{bmatrix} 2 & 1 \\ 3 & 2 \end{bmatrix}$ and $B = \begin{bmatrix} 5 & -1 \\ 9 & -2 \end{bmatrix}$.

25. Find the product AB and its inverse matrix $(AB)^{-1}$.

26. (a) Find A^{-1} and B^{-1} and the products $A^{-1}B^{-1}$ and $B^{-1}A^{-1}$.
(b) Is it true that $(AB)^{-1} = A^{-1}B^{-1}$?
(c) Is it true that $(AB)^{-1} = B^{-1}A^{-1}$?

27. Does the matrix $A + B$ have an inverse? If so, what is it?

28. Does the matrix $A - B$ have an inverse? If so, what is it?

29. Let $A = \begin{bmatrix} a & 0 \\ 0 & d \end{bmatrix}$, where a and d are nonzero constants and find A^{-1}.

30. Based on Exercise 29, what matrix do you think is the inverse of $A = \begin{bmatrix} a & 0 & 0 & 0 \\ 0 & b & 0 & 0 \\ 0 & 0 & c & 0 \\ 0 & 0 & 0 & d \end{bmatrix}$ (where a, b, c, d are nonzero?) Confirm your conjecture by calculating AA^{-1}.

In Exercises 31–32, let $A = \begin{bmatrix} a & b \\ c & d \end{bmatrix}$, where a, b, c, d are constants. Assume $ad - bc \neq 0$.

31. Find A^{-1} and show that $AA^{-1} = I$.

32. Show that $A^{-1}A = I$.

33. Show that the matrix equation $X^2 = \begin{bmatrix} 1 & 0 \\ 0 & 1 \end{bmatrix}$ has at least four different solutions. $\left(\textit{Hint: consider } \begin{bmatrix} -1 & 0 \\ 0 & 1 \end{bmatrix}.\right)$

34. Find two solutions for the equation in Exercise 33, in addition to the four you found there.

Use a graphing calculator with matrix capabilities or suitable computer software to find the inverses in Exercises 35–40.

$$C = \begin{bmatrix} -6 & 8 & 2 & 4 & -3 \\ 1 & 9 & 7 & -12 & 5 \\ 15 & 2 & -8 & 10 & 11 \\ 4 & 7 & 9 & 6 & -2 \\ 1 & 3 & 8 & 23 & 4 \end{bmatrix}$$

$$D = \begin{bmatrix} 5 & -3 & 7 & 9 & 2 \\ 6 & 8 & -5 & 2 & 1 \\ 3 & 7 & -4 & 2 & 11 \\ 5 & -3 & 9 & 4 & -1 \\ 0 & 3 & 2 & 5 & 1 \end{bmatrix}$$

35. C^{-1}

36. $(CD)^{-1}$

37. D^{-1}

38. Is $C^{-1}D^{-1} = (CD)^{-1}$?

39. Find $C^{-1}C$. This product *should* equal the identity matrix I, but doesn't *quite* equal I, because of round-off error.

40. Find DD^{-1}. Does this product exactly equal I?

6.6 APPLICATIONS OF MATRICES

This section gives a variety of applications of matrices.

SOLVING SYSTEMS WITH MATRICES   Consider this system of linear equations.

$$2x - 3y = 4$$
$$x + 5y = 2$$

Let

$$A = \begin{bmatrix} 2 & -3 \\ 1 & 5 \end{bmatrix}, \quad X = \begin{bmatrix} x \\ y \end{bmatrix}, \quad B = \begin{bmatrix} 4 \\ 2 \end{bmatrix}.$$

Since

$$AX = \begin{bmatrix} 2 & -3 \\ 1 & 5 \end{bmatrix}\begin{bmatrix} x \\ y \end{bmatrix} = \begin{bmatrix} 2x - 3y \\ x + 5y \end{bmatrix} \quad \text{and} \quad B = \begin{bmatrix} 4 \\ 2 \end{bmatrix},$$

the original system is equivalent to the single matrix equation $AX = B$. Similarly, any system of linear equations can be written as a matrix equation $AX = B$. The matrix A is called the **coefficient matrix.**

A matrix equation $AX = B$ can be solved if A^{-1} exists. Assuming A^{-1} exists and using the facts that $A^{-1}A = I$ and $IX = X$ along with the associative property of multiplication of matrices gives

$$AX = B$$
$$A^{-1}(AX) = A^{-1}B \qquad \text{Multiply both sides by } A^{-1}$$
$$(A^{-1}A)X = A^{-1}B \qquad \text{Associative property}$$
$$IX = A^{-1}B \qquad \text{Inverse property}$$
$$X = A^{-1}B. \qquad \text{Identity property}$$

When multiplying by matrices on both sides of a matrix equation, be careful to multiply in the same order on both sides of the equation, since multiplication of matrices is not commutative (unlike multiplication of real numbers). This discussion is summarized below.

A system of equations $AX = B$, where A is the matrix of coefficients, X is the matrix of variables, and B is the matrix of constants, is solved by first finding A^{-1}. Then, if A^{-1} exists, $X = A^{-1}B$.

This method is most practical in cases where A^{-1} is known or easily found.

▶**EXAMPLE 1** Use the inverse of the coefficient matrix to solve the following systems.

(a)
$$-x - 2y + 2z = 9$$
$$2x + y - z = -3$$
$$3x - 2y + z = -6$$

(b)
$$-x - 2y + 2z = 3$$
$$2x + y - z = 3$$
$$3x - 2y + z = 7$$

(c)
$$-x - 2y + 2z = 12$$
$$2x + y - z = 0$$
$$3x - 2y + z = 18$$

Notice that the three systems all have the same matrix of coefficients and the same matrix of variables.

$$A = \begin{bmatrix} -1 & -2 & 2 \\ 2 & 1 & -1 \\ 3 & -2 & 1 \end{bmatrix} \quad \text{and} \quad X = \begin{bmatrix} x \\ y \\ z \end{bmatrix}$$

Find A^{-1} first, then use it to solve all three systems.

To find A^{-1}, we start with matrix

1 (a) Write the matrix of coefficients, the matrix of variables, and the matrix of constants for the system

$$2x + 6y = -14$$
$$-x - 2y = 3.$$

(b) Solve the system in part (a) by using the inverse of the coefficient matrix.

Answers:

(a) $A = \begin{bmatrix} 2 & 6 \\ -1 & -2 \end{bmatrix}$,

$X = \begin{bmatrix} x \\ y \end{bmatrix}$,

$B = \begin{bmatrix} -14 \\ 3 \end{bmatrix}$,

(b) $(5, -4)$

$$[A \mid I] = \begin{bmatrix} -1 & -2 & 2 & \bigm| & 1 & 0 & 0 \\ 2 & 1 & -1 & \bigm| & 0 & 1 & 0 \\ 3 & -2 & 1 & \bigm| & 0 & 0 & 1 \end{bmatrix}$$

and use row operations to get $[I \mid A^{-1}]$, from which

$$A^{-1} = \begin{bmatrix} \frac{1}{3} & \frac{2}{3} & 0 \\ \frac{5}{3} & \frac{7}{3} & -1 \\ \frac{7}{3} & \frac{8}{3} & -1 \end{bmatrix}.$$

Now we can solve each of the three systems by using $X = A^{-1}B$.

(a) Here the matrix of coefficients is

$$B = \begin{bmatrix} 9 \\ -3 \\ -6 \end{bmatrix}.$$

Since $X = A^{-1}B$,

$$X = \begin{bmatrix} \frac{1}{3} & \frac{2}{3} & 0 \\ \frac{5}{3} & \frac{7}{3} & -1 \\ \frac{7}{3} & \frac{8}{3} & -1 \end{bmatrix} \begin{bmatrix} 9 \\ -3 \\ -6 \end{bmatrix} = \begin{bmatrix} 1 \\ 14 \\ 19 \end{bmatrix}.$$

From this result, $x = 1$, $y = 14$, and $z = 19$, and the solution is $(1, 14, 19)$.

(b) The martix of coefficients is $B = \begin{bmatrix} 3 \\ 3 \\ 7 \end{bmatrix}$.

$$X = A^{-1}B = \begin{bmatrix} \frac{1}{3} & \frac{2}{3} & 0 \\ \frac{5}{3} & \frac{7}{3} & -1 \\ \frac{7}{3} & \frac{8}{3} & -1 \end{bmatrix} \begin{bmatrix} 3 \\ 3 \\ 7 \end{bmatrix} = \begin{bmatrix} 3 \\ 5 \\ 8 \end{bmatrix}.$$

The solution is $(3, 5, 8)$.

(c) This time, $B = \begin{bmatrix} 12 \\ 0 \\ 18 \end{bmatrix}$.

$$X = A^{-1}B = \begin{bmatrix} \frac{1}{3} & \frac{2}{3} & 0 \\ \frac{5}{3} & \frac{7}{3} & -1 \\ \frac{7}{3} & \frac{8}{3} & -1 \end{bmatrix} \begin{bmatrix} 12 \\ 0 \\ 18 \end{bmatrix} = \begin{bmatrix} 4 \\ 2 \\ 10 \end{bmatrix}.$$

This solution is $(4, 2, 10)$. ◀ **1**

2 Write a 2 × 2 technological matrix in which 1 unit of electricity requires 1/2 unit of water and 1/3 unit of electricity, while 1 unit of water requires no water but 1/4 unit of electricity.

Answer:

$$\begin{array}{cc} & \begin{array}{cc} \text{Elec.} & \text{Water} \end{array} \\ \begin{array}{c} \text{Elec.} \\ \text{Water} \end{array} & \begin{bmatrix} \frac{1}{3} & \frac{1}{4} \\ \frac{1}{2} & 0 \end{bmatrix} \end{array}$$

FOR GRAPHERS
A graphing calculator with matrix capabilities is ideal for implementing the matrix inverse method of solving systems of linear equations.

INPUT-OUTPUT ANALYSIS   An interesting application of matrix theory to economics was developed by Nobel Prize winner Wassily Leontief. His application of matrices to the interdependencies in an economy is called **input-output** analysis. In practice, input-output analysis is very complicated with many variables. We shall discuss only simple examples with a few variables.

Input-output models are concerned with the production and flow of goods (and perhaps services). In an economy with n basic commodities (or sectors), the production of each commodity uses some (perhaps all) of the commodities in the economy as inputs. The amounts of each commodity used in the production of 1 unit of each commodity can be written as an $n \times n$ matrix A, called the **technological** or **input-output matrix** of the economy.

▶**EXAMPLE 2**   Suppose a simplified economy involves just three commodity categories: agriculture, manufacturing, and transportation, all in appropriate units. Production of 1 unit of agriculture requires 1/2 unit of manufacturing and 1/4 unit of transportation. Production of 1 unit of manufacturing requires 1/4 unit of agriculture and 1/4 unit of transportation; while production of 1 unit of transportation requires 1/3 unit of agriculture and 1/4 unit of manufacturing. Write the input-output matrix of this economy.

The matrix is shown below.

$$\begin{array}{c} & & \textit{Output} \\ & & \begin{array}{ccc} \text{Agricul-} & \text{Manufac-} & \text{Trans-} \\ \text{ture} & \text{turing} & \text{portation} \end{array} \\ \textit{Input} \begin{array}{c} \textbf{Agriculture} \\ \textbf{Manufacturing} \\ \textbf{Transportation} \end{array} & \begin{bmatrix} 0 & \frac{1}{4} & \frac{1}{3} \\ \frac{1}{2} & 0 & \frac{1}{4} \\ \frac{1}{4} & \frac{1}{4} & 0 \end{bmatrix} = A \end{array}$$

The first column of the input-output matrix represents the amount of each of the three commodities consumed in the production of 1 unit of agriculture. The second column gives the corresponding amounts required to produce 1 unit of manufacturing, and the last column gives the amounts needed to produce 1 unit of transportation. (Although it is unrealistic, perhaps, that production of 1 unit of a commodity requires none of that commodity, the simpler matrix involved is useful for our purposes.) ◀ **2**

Another matrix used with the input-output matrix is a matrix giving the amount of each commodity produced, called the **production matrix,** or the **vector of gross output.** In an economy producing n commodities, the production matrix can be represented by a column matrix x with entries x_1, x_2, x_3, . . . ,x_n.

3 (a) Write a 2 × 1 matrix X to represent gross production of 9000 units of electricity and 12,000 units of water.

(b) Find AX using A from the last problem.

(c) Find D using $D = X - AX$.

Answers:

(a) $\begin{bmatrix} 9000 \\ 12,000 \end{bmatrix}$

(b) $\begin{bmatrix} 6000 \\ 4500 \end{bmatrix}$

(c) $\begin{bmatrix} 3000 \\ 7500 \end{bmatrix}$

▶ **EXAMPLE 3** In Example 2, suppose the production matrix is

$$X = \begin{bmatrix} 60 \\ 52 \\ 48 \end{bmatrix}.$$

Then 60 units of agriculture, 52 units of manufacturing, and 48 units of transportation are produced. As 1/4 unit of agriculture is used for each unit of manufacturing produced, $1/4 \times 52 = 13$ units of agriculture must be used up in the "production" of manufacturing. Similarly, $1/3 \times 48 = 16$ units of agriculture will be used up in the "production" of transportation. Thus $13 + 16 = 29$ units of agriculture are used for production in the economy. Look again at the matrices A and X. Since X gives the number of units of each commodity produced and A gives the amount (in units) of each commodity used to produce 1 unit of the various commodities, the matrix product AX gives the amount of each commodity used up in production.

$$AX = \begin{bmatrix} 0 & \frac{1}{4} & \frac{1}{3} \\ \frac{1}{2} & 0 & \frac{1}{4} \\ \frac{1}{4} & \frac{1}{4} & 0 \end{bmatrix} \begin{bmatrix} 60 \\ 52 \\ 48 \end{bmatrix} = \begin{bmatrix} 29 \\ 42 \\ 28 \end{bmatrix}$$

This product shows that 29 units of agriculture, 42 units of manufacturing, and 28 units of transportation are used to produce 60 units of agriculture, 52 units of manufacturing, and 48 units of transportation. ◀

We have seen that the matrix product AX represents the amount of each commodity used in the production process. The remainder (if any) must be enough to satisfy the demand for the various commodities from outside the production system. In an n-commodity economy, this demand can be represented by a **demand matrix** D with entries $d_1, d_2, \ldots, d_n$. The difference between the production matrix, X, and the amount, AX, used in the production process must equal the demand, D, or

$$D = X - AX.$$

In Example 3,

$$D = \begin{bmatrix} 60 \\ 52 \\ 48 \end{bmatrix} - \begin{bmatrix} 29 \\ 42 \\ 28 \end{bmatrix} = \begin{bmatrix} 31 \\ 10 \\ 20 \end{bmatrix}.$$

This result shows that production of 60 units of agriculture, 52 units of manufacturing, and 48 units of transportation would satisfy a demand of 31, 10, and 20 units of each, respectively. ◀ **3**

5 Write the message *"when"* using 2 × 1 matrices.

Answer:

$$\begin{bmatrix} 23 \\ 8 \end{bmatrix}, \begin{bmatrix} 5 \\ 14 \end{bmatrix}$$

6 Use the matrix given below to find the 2 × 1 matrices to be transmitted for the message you encoded in Problem 5 at the side.

$$\begin{bmatrix} 2 & 1 \\ 5 & 0 \end{bmatrix}$$

Answer:

$$\begin{bmatrix} 54 \\ 115 \end{bmatrix}, \begin{bmatrix} 24 \\ 25 \end{bmatrix}$$

$$\begin{bmatrix} 13 \\ 1 \\ 20 \end{bmatrix}.$$

The coded message then consists of the 3 × 1 column matrices:

$$\begin{bmatrix} 13 \\ 1 \\ 20 \end{bmatrix}, \begin{bmatrix} 8 \\ 5 \\ 13 \end{bmatrix}, \begin{bmatrix} 1 \\ 20 \\ 9 \end{bmatrix}, \begin{bmatrix} 3 \\ 19 \\ 27 \end{bmatrix}, \begin{bmatrix} 9 \\ 19 \\ 27 \end{bmatrix}, \begin{bmatrix} 6 \\ 15 \\ 18 \end{bmatrix}, \begin{bmatrix} 27 \\ 20 \\ 8 \end{bmatrix}, \begin{bmatrix} 5 \\ 27 \\ 2 \end{bmatrix}, \begin{bmatrix} 9 \\ 18 \\ 4 \end{bmatrix}, \begin{bmatrix} 19 \\ 27 \\ 27 \end{bmatrix}. \quad \textbf{5}$$

We can further complicate the code by choosing a matrix that has an inverse (in this case a 3 × 3 matrix, call it *M*) and finding the products of this matrix and each of the above column matrices. The size of each group, the assignment of numbers to letters, and the choice of matrix *M* must all be predetermined.

Suppose we choose

$$M = \begin{bmatrix} 1 & 3 & 3 \\ 1 & 4 & 3 \\ 1 & 3 & 4 \end{bmatrix}.$$

If we find the products of *M* and the column matrices above, we have a new set of column matrices,

$$\begin{bmatrix} 1 & 3 & 3 \\ 1 & 4 & 3 \\ 1 & 3 & 4 \end{bmatrix} \begin{bmatrix} 13 \\ 1 \\ 20 \end{bmatrix} = \begin{bmatrix} 76 \\ 77 \\ 96 \end{bmatrix}, \text{ and so on.}$$

The entries of these matrices can then be transmitted to an agent as the message 76, 77, 96, and so on. **6**

When the agent receives the message, it is divided into groups of numbers with each group formed into a column matrix. After multiplying each column matrix by the matrix M^{-1}, the message can be read.

Although this type of code is relatively simple, it is actually difficult to break. Many complications are possible. For example, a long message might be placed in groups of 20, thus requiring a 20 × 20 matrix for coding and decoding. Finding the inverse of such a matrix would require an impractical amount of time if calculated by hand. For this reason some of the largest computers are used by government agencies involved in coding.

ROUTING The diagram in Figure 6.2 shows the roads connecting four cities. Another way of representing this information is shown in matrix *A*, where the entries represent the number of roads connecting two cities without passing through another city.* For example, from the diagram we see that there are two roads connecting city 1 to city 4 without passing through either city 2 or 3. This information is entered in row one, column four and again in row four, column one of matrix A,

$$A = \begin{bmatrix} 0 & 1 & 2 & 2 \\ 1 & 0 & 1 & 0 \\ 2 & 1 & 0 & 1 \\ 2 & 0 & 1 & 0 \end{bmatrix}$$

Note that there are zero roads connecting each city to itself. Also, there is one road connecting cities 3 and 2.

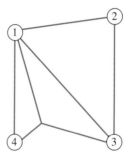

FIGURE 6.2

How many ways are there to go from city 1 to city 2, for example, by going through exactly one other city? Because we must go through one other city, we must go through either city 3 or city 4. On the diagram in Figure 6.2, we see that we can go from city 1 to city 2 through city 3 in two ways. We can go from city 1 to city 3 in two ways and then from city 3 to city 2 in one way, so there are $2 \cdot 1 = 2$ ways to get from city 1 to city 2 through city 3. It is not possible to go from city 1 to city 2 through city 4, because there is no direct route between cities 4 and 2.

The matrix A^2 gives the number of ways to travel between any two cities by passing through exactly one other city. Multiply matrix A by itself, to get A^2. Let the first row, second column entry of A^2 be b_{12}. (We use a_{ij} to denote the entry in the i-th row and j-th column of matrix A.) The entry b_{12} is found as follows.

$$b_{12} = a_{11}a_{12} + a_{12}a_{22} + a_{13}a_{32} + a_{14}a_{42}$$
$$= 0 \cdot 1 + 1 \cdot 0 + 2 \cdot 1 + 2 \cdot 0$$
$$= 2$$

*From *Matrices with Applications,* section 3.2, example 5, by Hugh G. Campbell. Copyright © 1968, pp. 50–51. Adapted by permission of Prentice-Hall, Englewood Cliffs, New Jersey.

The first product $0 \cdot 1$ in the calculations above represents the number of ways to go from city 1 to city 1, (0), and then from city 1 to city 2, (1). The 0 result indicates that such a trip does not involve a third city. The only nonzero product $(2 \cdot 1)$ represents the two routes from city 1 to city 3 and the one route from city 3 to city 2 which result in the $2 \cdot 1$ or 2 routes from city 1 to city 2 by going through city 3.

Similarly, A^3 gives the number of ways to travel between any two cities by passing through exactly two cities. Also, $A + A^2$ represents the total number of ways to travel between two cities with at most one intermediate city.

The diagram can be given many other interpretations. For example, the lines could represent lines of mutual influence between people or nations, or they could represent communication lines such as telephone lines.

6.6 EXERCISES

Solve the matrix equation $AX = B$ for X. (See Example 1.)

1. $A = \begin{bmatrix} 1 & -1 \\ 5 & -6 \end{bmatrix}$, $B = \begin{bmatrix} 2 \\ 4 \end{bmatrix}$

2. $A = \begin{bmatrix} 3 & -2 \\ -1 & 1 \end{bmatrix}$, $B = \begin{bmatrix} -3 \\ 5 \end{bmatrix}$

3. $A = \begin{bmatrix} 3 & 1 \\ 4 & 2 \end{bmatrix}$, $B = \begin{bmatrix} 3 & 4 \\ 5 & 6 \end{bmatrix}$

4. $A = \begin{bmatrix} 7 & -3 \\ -2 & 1 \end{bmatrix}$, $B = \begin{bmatrix} 0 & 8 \\ 4 & 1 \end{bmatrix}$

5. $A = \begin{bmatrix} 1 & -2 & -3 \\ -1 & 4 & 6 \\ 1 & -1 & -2 \end{bmatrix}$, $B = \begin{bmatrix} 2 \\ 7 \\ 4 \end{bmatrix}$

6. $A = \begin{bmatrix} 3 & -1 & 0 \\ 0 & 1 & 2 \\ 6 & 0 & 5 \end{bmatrix}$, $B = \begin{bmatrix} -6 \\ 12 \\ 15 \end{bmatrix}$

Use matrix algebra to solve the following matrix equations for X. Then use the given matrices to find X and check your work.

7. $N = X - MX$, $N = \begin{bmatrix} 8 \\ -12 \end{bmatrix}$, $M = \begin{bmatrix} 0 & 1 \\ -2 & 1 \end{bmatrix}$

8. $A = BX + X$, $A = \begin{bmatrix} 4 & 6 \\ -2 & 2 \end{bmatrix}$, $B = \begin{bmatrix} -2 & -2 \\ 3 & 3 \end{bmatrix}$

Use the inverse of the coefficient matrix to solve each system of equations. The inverses for Exercises 11–16 were found in Exercises 17 and 20–24 of Section 6.5. (See Example 1.)

9.
$$\begin{aligned} x + 2y + 3z &= 5 \\ 2x + 3y + 2z &= 2 \\ -x - 2y - 4z &= -1 \end{aligned}$$

10.
$$\begin{aligned} x + y - 3z &= 4 \\ 2x + 4y - 4z &= 8 \\ -x + y + 4z &= -3 \end{aligned}$$

11.
$$\begin{aligned} x + 4y + 3z &= -12 \\ x - 3y - 2z &= 0 \\ 2x + 5y + 4z &= 7 \end{aligned}$$

12.
$$\begin{aligned} 5x \quad\ + 2z &= 3 \\ 2x + 2y + z &= 4 \\ -3x + y - z &= 5 \end{aligned}$$

13.
$$\begin{aligned} 2x + 4y + 6z &= 4 \\ -x - 4y - 3z &= 8 \\ y - z &= -4 \end{aligned}$$

14.
$$\begin{aligned} 2x + 2y - 4z &= 12 \\ 2x + 6y &= 16 \\ -3x - 3y + 5z &= -20 \end{aligned}$$

15.
$$\begin{aligned} x - 2y + 3z \quad &= 4 \\ y - z + w &= -8 \\ -2x + 2y - 2z + 4w &= 12 \\ 2y - 3z + w &= -4 \end{aligned}$$

16.
$$\begin{aligned} x + y \quad\ + 2w &= 3 \\ 2x - y + z - w &= 3 \\ 3x + 3y + 2z - 2w &= 5 \\ x + 2y + z \quad &= 3 \end{aligned}$$

Find the production matrix given the following input-output and demand matrices. (See Examples 4 and 5.)

17. $A = \begin{bmatrix} \frac{1}{2} & \frac{2}{5} \\ \frac{1}{4} & \frac{1}{5} \end{bmatrix}$, $D = \begin{bmatrix} 2 \\ 4 \end{bmatrix}$

18. $A = \begin{bmatrix} \frac{1}{5} & \frac{1}{25} \\ \frac{3}{5} & \frac{1}{20} \end{bmatrix}$, $D = \begin{bmatrix} 3 \\ 10 \end{bmatrix}$

19. $A = \begin{bmatrix} .1 & .03 \\ .07 & .6 \end{bmatrix}$, $D = \begin{bmatrix} 5 \\ 10 \end{bmatrix}$

20. $A = \begin{bmatrix} .01 & .03 \\ .05 & .05 \end{bmatrix}$, $D = \begin{bmatrix} 100 \\ 200 \end{bmatrix}$

21. $A = \begin{bmatrix} .4 & 0 & .3 \\ 0 & .8 & .1 \\ 0 & .2 & .4 \end{bmatrix}$, $D = \begin{bmatrix} 1 \\ 3 \\ 2 \end{bmatrix}$

22. $A = \begin{bmatrix} .1 & .5 & 0 \\ 0 & .3 & .4 \\ .1 & .2 & .1 \end{bmatrix}$, $B = \begin{bmatrix} 10 \\ 4 \\ 2 \end{bmatrix}$

Solve the following problems by using the inverse of the coefficient matrix to solve a system of equations.

23. Management The Badgett Bakery sells three types of cake, each requiring the amounts of the basic ingredients shown in the following matrix.

		Type of Cake	
		I	II	III
	Flour (in cups)	2	4	2
Ingredient	Sugar (in cups)	2	1	2
	Eggs	2	1	3

To fill its daily order for these three kinds of cake, the bakery uses 72 cups of flour, 48 cups of sugar, and 60 eggs.
(a) Write a 3×1 matrix for the amounts used daily.
(b) Let the number of daily orders for cakes be a 3×1 matrix X with entries x_1, x_2, and x_3. Write a matrix equation that can be solved for X, using the given matrix and the matrix from part (a).
(c) Solve the equation from part (b) to find the number of daily orders for each type of cake.

24. Management An electronics company produces transistors, resistors, and computer chips. Each transistor requires 3 units of copper, 1 unit of zinc, and 2 units of glass. Each resistor requires 3, 2, and 1 units of the three materials, and each computer chip requires 2, 1, and 2 units of these materials, respectively. How many of each product can be made with the following amounts of materials?
(a) 810 units of copper, 410 units of zine, and 490 units of glass
(b) 765 units of copper, 385 units of zinc, and 470 units of glass
(c) 1010 units of copper, 500 units of zinc, and 610 units of glass

25. Management An investment firm recommends that a client invest in AAA, A, and B rated bonds. The average yield on AAA bonds is 6%, on A bonds 7%; and on B bonds 10%. The client wants to invest twice as much in AAA bonds as in B bonds. How much should be invested in each type of bond under the following conditions?
(a) The total investment is $25,000, and the investor wants an annual return of $1810 on the three investments.
(b) The values in part (a) are changed to $30,000 and $2150, respectively.
(c) The values in part (a) are changed to $40,000 and $2900, respectively.

Exercises 26 and 27 refer to Example 5.

26. Management If the demand is changed to 690 metric tons of wheat and 920 metric tons of oil, how many units of each commodity should be produced?

27. Management Change the technological matrix so that production of 1 metric ton of wheat requires 1/5 metric ton of oil (and no wheat), and the production of 1 metric ton of oil requires 1/3 metric ton of wheat (and no oil). To satisfy the same demand matrix, how many units of each commodity should be produced?

28. Management A simplified economy has only two industries, the electric company and the gas company. Each dollar's worth of the electric company's output requires .40 of its own output and .50 of the gas company's output. Each dollar's worth of the gas company's output requires .25 of its own output and .60 of the electric company's output. What should the production of electricity and gas be (in dollars) if there is a $12 million demand for gas and a $15 million demand for electricity?

29. Management A two-segment economy consists of manufacturing and agriculture. To produce one unit of manufacturing output requires .40 unit of its own output and .20 unit of agricultural output. To produce one unit of agricultural output requires .30 unit of its own output and .40 unit of manufacturing output. If there is a demand of 240 units of manufacturing and 90 units of agriculture, what should be the output of each segment?

30. Management A primitive economy depends on two basic goods, yams and pork. Production of 1 bushel of yams requires 1/4 bushel of yams and 1/2 of a pig. To produce 1 pig requires 1/6 bushel of yams. Find the amount of each commodity that should be produced to get
(a) 1 bushel of yams and 1 pig;
(b) 100 bushels of yams and 70 pigs.

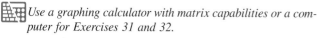 *Use a graphing calculator with matrix capabilities or a computer for Exercises 31 and 32.*

31. Management A simplified economy is based on agriculture, manufacturing, and transportation. Each unit of agricultural output requires .4 unit of its own output, .3 unit of manufacturing, and .2 unit of transportation output. One unit of manufacturing output requires .4 unit of its own output, .2 unit of agricultural, and .3 unit of transportation output. One unit of transportation output requires .4 unit of its own output, .1 unit of agricultural, and .2 unit of manufacturing output. There is demand for 35 units of agricultural, 90 units of manufacturing, and 20 units of transportation output. How many units should each segment of the economy produce?

32. Management How many units of each segment should the economy in Exercise 31 produce, if the demand is for 55 units of agricultural, 20 units of manufacturing, and 10 units of transportation output?

33. Use the method discussed in the text to encode the message

Anne is home.

Break the message into groups of two letters and use the matrix

$$M = \begin{bmatrix} 1 & 3 \\ 2 & 7 \end{bmatrix}.$$

34. Use the matrix of Exercise 33 to encode the message

Head for the hills!

35. Decode the following message, which was encoded by using the matrix M of Exercise 33.

$$\begin{bmatrix} 90 \\ 207 \end{bmatrix}, \begin{bmatrix} 39 \\ 87 \end{bmatrix}, \begin{bmatrix} 26 \\ 57 \end{bmatrix}, \begin{bmatrix} 66 \\ 145 \end{bmatrix}, \begin{bmatrix} 61 \\ 142 \end{bmatrix}, \begin{bmatrix} 89 \\ 205 \end{bmatrix}.$$

36. Use matrix A in the discussion on routing in the text to find A^2. Then answer the following questions. How many ways are there to travel from
(a) City 1 to city 3 by passing through exactly one city?
(b) City 2 to city 4 by passing through exactly one city?
(c) City 1 to city 3 by passing through at most one city?
(d) City 2 to city 4 by passing through at most one city?

37. Find A^3. (See Exercise 36.) Then answer the following questions.
(a) How many ways are there to travel between cities 1 and 4 by passing through exactly two cities?
(b) How many ways are there to travel between cities 1 and 4 by passing through at most two cities?

38. Management A small telephone system connects three cities. There are four lines between cities 3 and 2, three lines connecting city 3 with city 1, and two lines between cities 1 and 2.
(a) Write a matrix B to represent this information.
(b) Find B^2.
(c) How many lines which connect cities 1 and 2 go through exactly one other city (city 3)?
(d) How many lines which connect cities 1 and 2 go through at most one other city?

39. Management The figure shows four southern cities served by Supersouth Airlines.

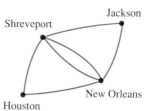

(a) Write a matrix to represent the number of nonstop routes between cities.
(b) Find the number of one-stop flights between Houston and Jackson.
(c) Find the number of flights between Houston and Shreveport which require at most one stop.
(d) Find the number of one-stop flights between New Orleans and Houston.

40. Natural Science The figure shows a food web. The arrows indicate the food sources of each population. For example, cats feed on rats and on mice.

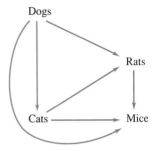

Dogs

Rats

Cats — Mice

(a) Write a matrix C in which each row and corresponding column represents a population in the food chain. Enter a 1 when the population in a given row feeds on the population in the given column.

(b) Calculate and interpret C^2.

CHAPTER 6 SUMMARY

KEY TERMS AND SYMBOLS

6.1 linear equations
systems of linear equations
solution of a system
independent system
dependent system
inconsistent system
equivalent systems
elementary operations
elimination method
parameter

6.2 matrix (matrices)
row
column
element
entry
augmented matrix
row operations
Gauss-Jordan method

6.3 row matrix (row vector)
column matrix (column vector)

square matrix
equal matrices
sum of matrices
additive inverse of a matrix
zero matrix
subtraction (difference) of matrices

6.4 scalar
product of a scalar and a matrix
product of matrices

6.5 identity matrix I
inverse matrix

6.6 coefficient matrix
input-output model
technological matrix (input-output matrix)
production matrix (vector of gross output)
demand matrix
code theory
routing theory

KEY CONCEPTS

Solving Systems of Equations

The following **elementary operations** are used to transform a system of equations into an equivalent system with a simpler solution.

1. Interchange any two equations.
2. Multiply both sides of an equation by a nonzero constant.
3. Replace an equation by the sum of itself and a constant multiple of another equation in the system.

The **elimination method** is a systematic way of using elementary operations to transform a system into an equivalent system that can be solved by **back substitution.** See Section 6.1 for details.

The following **row operations** on a matrix correspond to elementary operations on a system of equations.

1. Interchange any two rows.
2. Multiply each element of a row by a nonzero constant.
3. Replace a row by the sum of itself and a constant multiple of another row in the matrix.

The **Gauss-Jordan method** is an extension of the elimination method for solving a system of linear equations. It uses row operations on the augmented matrix of the system. See Section 6.2 for details.

Operations on Matrices

The **sum** of two $m \times n$ matrices X and Y is the $m \times n$ matrix $X + Y$ in which each element is the sum of the corresponding elements of X and Y. The **difference** of two $m \times n$ matrices X and Y is the $m \times n$ matrix $X - Y$ in which each element is the difference of the corresponding elements of X and Y.

The **product** of a scalar k and a matrix X is the matrix kX, with each element k times the corresponding element of X.

The **product matrix** AB of an $m \times n$ matrix A and an $n \times k$ matrix B is the $m \times k$ matrix whose entry in the i-th row and j-th column is the product of the i-th row of A and the j-th column of B.

The **inverse matrix** A^{-1} for any $n \times n$ matrix A for which A^{-1} exists is found as follows. Form the augmented matrix $[A \,|\, I]$; perform row operations on $[A \,|\, I]$ to get the matrix $[I \,|\, A^{-1}]$.

Use the elimination method to solve each of the following systems. Identity any dependent or inconsistent systems.

1. $-5x - 3y = 4$
$2x + y = -3$

2. $3x - y = 6$
$2x + 3y = 7$

3. $3x - 5y = 10$
$4x - 3y = 6$

4. $\dfrac{1}{4}x - \dfrac{1}{3}y = -\dfrac{1}{4}$
$\dfrac{1}{10}x + \dfrac{2}{5}y = \dfrac{2}{5}$

5. $x - 2y = 1$
$4x + 4y = 2$
$10x + 8y = 4$

6. $x + y - 4z = 0$
$2x + y - 3z = 2$

7. $3x + y - z = 13$
$x \qquad + 2z = 9$
$-3x - y + 2z = 9$

8. $4x - y - 2z = 4$
$x - y - \dfrac{1}{2}z = 1$
$2x - y - z = 8$

---

9. Management   An office supply manufacturer makes two kinds of paper clips, standard and extra large. To make 1000 standard paper clips requires 1/4 hr on a cutting machine and 1/2 hr on a machine that shapes the clips. One thousand extra large paper clips require 1/3 hr on each machine. The manager of paper clip production has 4 hr per day available on the cutting machine and 6 hr per day on the shaping machine. How many of each kind of clip can he make?

10. Management   Gretchen Schmidt plans to buy shares of two stocks. One costs $32 per share and pays dividends of $1.20 per share. The other costs $23 per share and pays dividends of $1.40 per share. She has $10,100 to spend and wants to earn dividends of $540. How many shares of each stock should she buy?

11. Management   Joyce Pluth has money in two investment funds. Last year the first fund paid a dividend of 8% and the second a dividend of 2% and Joyce received a total of $780. This year the first fund paid a 10% dividend and the second only 1% and Joyce received $810. How much does she have invested in each fund?

12. You are given $144 in one, five, and ten dollar bills. There are 35 bills. There are two more ten dollar bills

than five dollar bills. How many bills of each type are there?

13. Social Science   A social service agency provides counseling, meals, and shelter to clients referred by sources I, II, and III. Clients from source I require an average of $100 for food, $250 for shelter, and no counseling. Source II clients require an average of $100 for counseling, $200 for food, and nothing for shelter. Source III clients require an average of $100 for counseling, $150 for food, and $200 for shelter. The agency has funding of $25,000 for counseling, $50,000 for food, and $32,500 for shelter. How many clients from each source can be served?

14. Management   The Waputi Indians make woven blankets, rugs, and skirts. Each blanket requires 24 hr for spinning the yarn, 4 hr for dying the yarn, and 15 hr for weaving. Rugs require 30, 5, and 18 hr and skirts 12, 3, and 9 hr, respectively. If there are 306, 59, and 201 hr available for spinning, dying, and weaving, respectively, how many of each item can be made? (*Hint:* Simplify the equations you write, if possible, before solving the system.)

Use the Gauss-Jordan method to solve the following systems.

15. $\begin{aligned} x - z &= -3 \\ y + z &= 6 \\ 2x - 3z &= -9 \end{aligned}$

16. $\begin{aligned} 2x - y + 4z &= -1 \\ -3x + 5y - z &= 5 \\ 2x + 3y + 2z &= 3 \end{aligned}$

17. $\begin{aligned} 5x - 8y + z &= 1 \\ 3x - 2y + 4z &= 3 \\ 10x - 16y + 2z &= 3 \end{aligned}$

18. $\begin{aligned} x - 2y + 3z &= 4 \\ 2x + y - 4z &= 3 \\ -3x + 4y - z &= -2 \end{aligned}$

19. $\begin{aligned} 3x + 2y - 6z &= 9 \\ x + y + 2z &= 4 \\ 2x + 2y + 5z &= 0 \end{aligned}$

20. Management   Each week at a furniture factory, there are 2000 work hours available in the construction department, 1400 work hours in the painting department, and 1300 work hours in the packing department. Producing a chair requires 2 hours of construction, 1 hour of painting, and 2 hours for packing. Producing a table requires 4 hours of construction, 3 hours of painting, and 3 hours for packing. Producing a chest requires 8 hours of construction, 6 hours of painting, and 4 hours for packing. If all available time is used in every department, how many of each item are produced each week?

For each of the following, find the sizes of the matrices, find the values of any variables, and identify any square, row, or column matrices.

21. $\begin{bmatrix} 2 & 3 \\ 5 & q \end{bmatrix} = \begin{bmatrix} a & b \\ c & 9 \end{bmatrix}$

22. $\begin{bmatrix} 2 & x \\ y & 6 \\ 5 & z \end{bmatrix} = \begin{bmatrix} a & -1 \\ 4 & 6 \\ p & 7 \end{bmatrix}$

23. $[m \quad 4 \quad z \quad -1] = [12 \quad k \quad -8 \quad r]$

24. $\begin{bmatrix} a + 5 & 3b & 6 \\ 4c & 2 + d & -3 \\ -1 & 4p & q - 1 \end{bmatrix} = \begin{bmatrix} -7 & b + 2 & 2k - 3 \\ 3 & 2d - 1 & 4l \\ m & 12 & 8 \end{bmatrix}$

25. $\begin{bmatrix} 6 & m \\ k & 5 - x \end{bmatrix} + \begin{bmatrix} r & m + 3 \\ k + 1 & 2x \end{bmatrix} = \begin{bmatrix} 8 & 10 \\ 12 & -2 \end{bmatrix}$

26. $\begin{bmatrix} 3p & 1 & y \\ q & 5 & z \end{bmatrix} - \begin{bmatrix} 3 + p & r & 2y + 1 \\ q + 2 & m & 10 \end{bmatrix} = \begin{bmatrix} 4p & 2r & 5y \\ 3q & 2m & -z \end{bmatrix}$

27. Natural Science   The activities of a grazing animal can be classified roughly into three categories: grazing, moving, and resting. Suppose horses spend 8 hours grazing, 8 moving, and 8 resting; cattle spend 10 grazing, 5 moving, and 9 resting; sheep spend 7 grazing, 10 moving, and 7 resting; and goats spend 8 grazing, 9 moving, and 7 resting. Write this information as a 4 × 3 matrix.

28. Management   The New York Stock Exchange reports in the daily newspapers give the dividend, price-to-earnings ratio, sales (in hundreds of shares), last price, and change in price for each company. Write the following stock reports as a 4 × 5 matrix. American Telephone & Telegraph: 5, 7, 2532, $52\frac{3}{8}$, $-\frac{1}{4}$. General Electric: 3, 9, 1464, 56, $+\frac{1}{8}$. Gulf Oil: 2.50, 5, 4974, 41, $-1\frac{1}{2}$. Sears: 1.36, 10, 1754, 18, $+\frac{1}{2}$.

Given the matrices

$$A = \begin{bmatrix} 4 & 10 \\ -2 & -3 \\ 6 & 9 \end{bmatrix}, \qquad B = \begin{bmatrix} 2 & 3 & -2 \\ 2 & 4 & 0 \\ 0 & 1 & 2 \end{bmatrix}, \qquad C = \begin{bmatrix} 5 & 0 \\ -1 & 3 \\ 4 & 7 \end{bmatrix}, \qquad D = \begin{bmatrix} 6 \\ 1 \\ 0 \end{bmatrix}, \qquad E = \begin{bmatrix} 1 & 3 & -4 \end{bmatrix},$$

$$F = \begin{bmatrix} -1 & 4 \\ 3 & 7 \end{bmatrix}, \qquad G = \begin{bmatrix} 2 & 5 \\ 1 & 6 \end{bmatrix},$$

find each of the following (if possible).

29. $-B$

30. $-D$

31. $3A - 2C$

32. $F + 3G$

33. $2B - 5C$

34. $G - 2F$

---

35. Management Refer to Exercise 28. Write a 4 × 2 matrix using the sales and price changes for the four companies. The next day's sales and price changes for the same four companies were 2310, 1258, 5061, 1812 and $-1/4$, $-1/4$, $+1/2$, $+1/2$, respectively. Write a 4 × 2 matrix using these new sales and price change figures. Use matrix addition to find the total sales and price changes for the two days.

36. Management An oil refinery in Tulsa sent 110,000 gallons of oil to a Chicago distributor, 73,000 to a Dallas distributor, and 95,000 to an Atlanta distributor. Another refinery in New Orleans sent the following amounts to the same three distributors: 85,000, 108,000, 69,000. The next month the two refineries sent the same distributors new shipments of oil as follows: from Tulsa, 58,000 to Chicago, 33,000 to Dallas, and 80,000 to Atlanta; from New Orleans, 40,000, 52,000, and 30,000, respectively.

(a) Write the monthly shipments from the two distributors to the three refineries as 3 × 2 matrices.

(b) Use matrix addition to find the total amounts sent to the refineries from each distributor.

Use the matrices given above Exercise 29 to find each of the following (if possible).

37. AG

38. EB

39. GF

40. CA

41. AGF

42. B^2D

---

43. Management An office supply manufacturer makes two kinds of paper clips, standard and extra large. To make a unit of standard paper clips requires 1/4 hour on a cutting machine and 1/2 hour on a machine that shapes the clips. A unit of extra large paper clips requires 1/3 hour on each machine.

(a) Write this information as a 2 × 2 matrix (size/machine).

(b) If 48 units of standard and 66 units of extra large clips are to be produced, use matrix multiplication to find out how many hours each machine will operate. (Hint: write the units as a 1 × 2 matrix.)

44. Management   Theresa DePalo buys shares of three stocks. Their cost per share and dividend earnings per share are $32, $23, $54, and $1.20, $1.49, and $2.10, respectively. She buys 50 shares of the first stock, 20 shares of the second, and 15 shares of the third.

(a) Write the cost per share and earnings per share of the stocks as a 3 × 2 matrix.

(b) Write the number of shares of each stock as a 1 × 3 matrix.

(c) Use matrix multiplication to find the total cost and total dividend earnings of these stocks.

45. If $A = \begin{bmatrix} 3 & 0 \\ 2 & 1 \end{bmatrix}$, find a matrix B such that both AB and BA are defined and $AB \neq BA$.

46. Is it possible to do Exercise 45 if $A = \begin{bmatrix} 4 & 0 \\ 0 & 4 \end{bmatrix}$? Explain why.

Find the inverse of each of the following matrices that has an inverse.

47. $\begin{bmatrix} -4 & 2 \\ 0 & 3 \end{bmatrix}$ **48.** $\begin{bmatrix} 2 & 1 \\ 5 & 3 \end{bmatrix}$ **49.** $\begin{bmatrix} 6 & 4 \\ 3 & 2 \end{bmatrix}$ **50.** $\begin{bmatrix} 2 & 0 \\ -1 & 5 \end{bmatrix}$

51. $\begin{bmatrix} 2 & 0 & 4 \\ 1 & -1 & 0 \\ 0 & 1 & -2 \end{bmatrix}$ **52.** $\begin{bmatrix} 2 & -1 & 0 \\ 1 & 0 & 1 \\ 1 & -2 & 0 \end{bmatrix}$ **53.** $\begin{bmatrix} 2 & 3 & 5 \\ -2 & -3 & -5 \\ 1 & 4 & 2 \end{bmatrix}$ **54.** $\begin{bmatrix} 1 & 3 & 6 \\ 4 & 0 & 9 \\ 5 & 15 & 30 \end{bmatrix}$

Refer again to the matrices given above Exercise 29 to find each of the following (if possible).

55. F^{-1} **56.** G^{-1} **57.** $(G - F)^{-1}$ **58.** $(F + G)^{-1}$ **59.** B^{-1}

60. Explain why the matrix $\begin{bmatrix} a & 0 \\ c & 0 \end{bmatrix}$, where a and c are nonzero constants, cannot possibly have an inverse.

Solve each of the following matrix equations $AX = B$ for X.

61. $A = \begin{bmatrix} 2 & 4 \\ -1 & -3 \end{bmatrix}$, $B = \begin{bmatrix} 8 \\ 3 \end{bmatrix}$

62. $A = \begin{bmatrix} 1 & 3 \\ -2 & 4 \end{bmatrix}$, $B = \begin{bmatrix} 15 \\ 10 \end{bmatrix}$

63. $A = \begin{bmatrix} 1 & 0 & 2 \\ -1 & 1 & 0 \\ 3 & 0 & 4 \end{bmatrix}$, $B = \begin{bmatrix} 8 \\ 4 \\ -6 \end{bmatrix}$

64. $A = \begin{bmatrix} 2 & 4 & 0 \\ 1 & -2 & 0 \\ 0 & 0 & 3 \end{bmatrix}$, $B = \begin{bmatrix} 72 \\ -24 \\ 48 \end{bmatrix}$

Use the method of matrix inverses to solve each of the following equations.

65. $\begin{aligned} x + y &= 4 \\ 2x + 3y &= 10 \end{aligned}$

66. $\begin{aligned} 5x - 3y &= -2 \\ 2x + 7y &= -9 \end{aligned}$

67. $\begin{aligned} 2x + y &= 5 \\ 3x - 2y &= 4 \end{aligned}$

68. $\begin{aligned} x - 2y &= 7 \\ 3x + y &= 7 \end{aligned}$

69. $\begin{aligned} x + y + z &= 1 \\ 2x - y &= -2 \\ 3y + z &= 2 \end{aligned}$

70. $\begin{aligned} x &= -3 \\ y + z &= 6 \\ 2x - 3z &= -9 \end{aligned}$

71.
$$3x - 2y + 4z = 4$$
$$4x + y - 5z = 2$$
$$-6x + 4y - 8z = -2$$

72.
$$x + 2y = -1$$
$$3y - z = -5$$
$$x + 2y - z = -3$$

Solve each of the following problems by any method.

73. Management A wine maker has two large casks of wine. One is 8% alcohol and the other is 18% alcohol. How many liters of each wine should be mixed to produce 30 liters of wine that is 12% alcohol?

74. Management A gold merchant has some 12 carat gold (12/24 pure gold), and some 22 carat gold (22/24 pure). How many grams of each could be mixed to get 25 grams of 15 carat gold?

75. Natural Science A chemist has some 40% acid solution and some 60% solution. How many liters of each should be used to get 40 liters of a 45% solution?

76. Management How many pounds of tea worth $4.60 a pound should be mixed with tea worth $6.50 a pound to get 10 pounds of a mixture worth $5.74 a pound?

77. Management A machine in a pottery factory takes 3 minutes to form a bowl and 2 minutes to form a plate. The material for a bowl costs $.25 and the material for a plate costs $.20. If the machine runs for 8 hours and exactly $44 is spent for material, how many bowls and plates can be produced?

78. A boat travels at a constant speed a distance of 57 km downstream in 3 hours, then turns around and travels 55 km upstream in 5 hours. What is the speed of the boat and of the current?

79. Management Ms. Tham invests $50,000 three ways—at 8%, $8\frac{1}{2}$%, and 11%. In total, she receives $4436.25 per year in interest. The interest from the 11% investment is $80 more than the interest on the 8% investment. Find the amount she has invested at each rate.

80. Tickets to a band concert cost $2 for children, $3 for teenagers, and $5 for adults. 570 people attended the concert and total ticket receipts were $1950. Three fourths as many teenagers as children attended. How many children, teenagers, and adults were at the concert?

Find the production matrix given the following input-output and demand matrices.

81. $A = \begin{bmatrix} .01 & .05 \\ .04 & .03 \end{bmatrix}$, $D = \begin{bmatrix} 200 \\ 300 \end{bmatrix}$

82. $A = \begin{bmatrix} .2 & .1 & .3 \\ .1 & 0 & .2 \\ 0 & 0 & .4 \end{bmatrix}$, $D = \begin{bmatrix} 500 \\ 200 \\ 100 \end{bmatrix}$

83. Given the input-output matrix $A = \begin{bmatrix} 0 & \frac{1}{4} \\ \frac{1}{2} & 0 \end{bmatrix}$ and the demand matrix $D = \begin{bmatrix} 2100 \\ 1400 \end{bmatrix}$, find each of the following.

(a) $I - A$ **(b)** $(I - A)^{-1}$
(c) the production matrix X

84. Management An economy depends on two commodities, goats and cheese. It takes 2/3 of a unit of goats to produce 1 unit of cheese and 1/2 unit of cheese to produce 1 unit of goats.

(a) Write the input-output matrix for this economy.
(b) Find the production required to satisfy a demand of 400 units of cheese and 800 units of goats.

85. Management In a simple economic model, a country has two industries: agriculture and manufacturing. To produce $1 of agricultural output requires .10 of agricultural output and .40 of manufacturing output. To produce $1 of manufacturing output requires .70 of agricultural output and .20 of manufacturing output. If agricultural demand is $60,000 and manufacturing demand is $20,000, what must each industry produce? (Round answers to the nearest whole number.)

86. Management   The matrix below represents the number of direct flights between four cities.

$$\begin{array}{c} \\ A \\ B \\ C \\ D \end{array} \begin{array}{cccc} A & B & C & D \\ \begin{bmatrix} 0 & 1 & 0 & 1 \\ 1 & 0 & 0 & 1 \\ 0 & 0 & 0 & 1 \\ 1 & 1 & 1 & 0 \end{bmatrix} \end{array}$$

(a) Find the number of one-stop flights between cities A and C.

(b) Find the total number of flights between cities B and C that are either direct or one-stop.

(c) Find the matrix that gives the number of two-stop flights between these cities.

87. (a) Use the matrix $M = \begin{bmatrix} 2 & 6 \\ 1 & 4 \end{bmatrix}$ to encode the message "leave now".

(b) What matrix should be used to decode this message?

Suppose that three people have contracted a contagious disease.* A second group of five people may have been in contact with the three infected persons. A third group of six people may have been in contact with the second group. We can form a 3×5 matrix P with rows representing the first group of three and columns representing the second group of five. We enter a one in the corresponding position if a person in the first group has contact with a person in the second group. These direct contacts are called *first-order contacts*. Similarly, we form a 5×6 matrix Q representing the first-order contacts between the second and third group. For example, suppose

$$P = \begin{bmatrix} 1 & 0 & 0 & 1 & 0 \\ 0 & 0 & 1 & 1 & 0 \\ 1 & 1 & 0 & 0 & 0 \end{bmatrix} \text{ and}$$

$$Q = \begin{bmatrix} 1 & 1 & 0 & 1 & 1 & 1 \\ 0 & 0 & 0 & 0 & 1 & 0 \\ 0 & 0 & 0 & 0 & 0 & 0 \\ 0 & 1 & 0 & 1 & 0 & 0 \\ 1 & 0 & 0 & 0 & 1 & 0 \end{bmatrix}.$$

From matrix P we see that the first person in the first group had contact with the first and fourth persons in the second group. Also, none of the first group had contact with the last person in the second group.

A *second-order contact* is an indirect contact between persons in the first and third group through some person in the second group. The product matrix PQ indicates these contacts. Verify that the second row, fourth column entry of PQ is 1. That is, there is one second-order contact between the second person in group one and the fourth person in group three. Let a_{ij} denote the element in the i-th row and j-th column of the matrix PQ. By looking at the products that form a_{24} below, we see that the common contact was with the fourth individual in group two. (The p_{ij} are entries in P, and the q_{ij} are entries in Q.)

$$a_{24} = p_{21}q_{14} + p_{22}q_{24} + p_{23}q_{34} + p_{24}q_{44} + p_{25}q_{54}$$
$$= 0 \cdot 1 + 0 \cdot 0 + 1 \cdot 0 + 1 \cdot 1 + 0 \cdot 1$$
$$= 1$$

The second person in group one and the fourth person in group three both had contact with the fourth person in group two.

This idea could be extended to third, fourth, and larger order contacts. It indicates a way to use matrices to trace the spread of a contagious disease. It could also pertain to the dispersal of ideas or anything that might pass from one individual to another.

EXERCISES

1. Find the second-order contact matrix PQ mentioned in the text.

2. How many second-order contacts were there between the second contagious person and the third person in the third group?

3. Is there anyone in the third group who has had no contacts at all with the first group?

4. The totals of the columns in PQ give the total number of second-order contacts per person, while the column totals in P and Q give the total number of first-order contacts per person. Which person(s) in the third group had the most contacts, counting first- and second-order contacts?

---

* "First and Second Order Contact to a Contagious Disease" from *Finite Mathematics with Applications to Business, Life Sciences, and Social Sciences* by Stanley Grossman.

▶ **Leontief's Model of the American Economy**

In the April 1965 issue of *Scientific American,* Wassily Leontief explained his input-output system using the 1958 American economy as an example.* He divided the economy into 81 sectors, grouped into six families of related sectors. In order to keep the discussion reasonably simple, we will treat each family of sectors as a single sector and so, in effect, work with a six-sector model. The sectors are listed in Table 1.

Table 1

Sector	Examples
Final Nonmetal (FN)	Furniture, processed food
Final Metal (FM)	Household appliances, motor vehicles
Basic Metal (BM)	Machine-shop products, mining
Basic Nonmetal (BN)	Agriculture, printing
Energy (E)	Petroleum, coal
Services (S)	Amusements, real estate

The workings of the American economy in 1958 are described in the input-output table (Table 2) based on Leontief's figures. We will demonstrate the meaning of Table 2 by considering the first left-hand column of numbers. The numbers in this column mean that 1 unit of final nonmetal production requires the consumption of .170 unit of (other) final nonmetal production, .003 unit of final metal output, .025 unit of basic metal products, and so on down the column. Since the unit of measurement that Leontief used for this table is millions of dollars, we conclude that the production of $1 million worth of final nonmetal production consumes $.170 million, or $170,000, worth of other final nonmetal products, $3000 of final metal products, $25,000 of basic metal products, and so on. Similarly, the entry in the column headed FM and opposite S of .074 means that $74,000 worth of input from the service industries goes into the production of $1 million worth of final

*Adapted from *Applied Finite Mathematics* by Robert F. Brown and Brenda W. Brown. Copyright © 1977 by Robert F. Brown and Brenda W. Brown. Reprinted by permission.

Table 2

	FN	FM	BM	BN	E	S
FN	.170	.004	0	.029	0	.008
FM	.003	.295	.018	.002	.004	.016
BM	.025	.173	.460	.007	.011	.007
BN	.348	.037	.021	.403	.011	.048
E	.007	.001	.039	.025	.358	.025
S	.120	.074	.104	.123	.173	.234

metal products, and the number .358 in the column headed E and opposite E means that $358,000 worth of energy must be consumed to produce $1 million worth of energy.

By the underlying assumption of Leontief's model, the production of n units (n = any number) of final nonmetal production consumes $.170n$ units of final nonmetal output, $.003n$ units of final metal output, $.025n$ units of basic metal production, and so on. Thus, production of $50 million worth of products from the final nonmetal section of the 1958 American economy required $(.170)(50) = 8.5$ units ($8.5 million) worth of final nonmetal input, $(.003)(50) = .15$ units of final metal input, $(.025)(50) = 1.25$ units of basic metal production, and so on.

▶ **EXAMPLE 1** According to the simplified input-output table for the 1958 American economy, how many dollars worth of final metal products, basic nonmetal products, and services are required to produce $120 million worth of basic metal products?

Each unit ($1 million worth) of basic metal products requires .018 units of final metal products because the number in the BM column of the table opposite FM is .018. Thus, $120 million, or 120 units, requires $(.108)(120) = 2.16$ units, or $2.16 million worth of final metal products. Similarly, 120 units of basic metal production uses $(.021)(120) = 2.52$ units of basic nonmetal production and $(.104)(120) = 12.48$ units of services, or $2.52 million and $12.48 million worth of basic nonmetal output and services, respectively. ◄

The Leontief model also involves a *bill of demands,* that is, a list of requirements for units of output beyond that required for its inner workings as described in the input-output table. These demands represent exports, surpluses, government and individual consumption, and the like. The bill of demands (in millions) for the simplified version of the 1958 American economy we have been using is shown below.

FN	$99,640
FM	$75,548
BM	$14,444
BN	$33,501
E	$23,527
S	$263,985

We can now use the methods developed in Section 6.6 to answer this question: how many units of output from each sector are needed in order to run the economy and fill the bill of demands? The units of output from each sector required to run the economy and fill the bill of demands is unknown, so we denote them by variables. In our example, there are six quantities which are, at the moment, unknown. The number of units of final nonmetal production required to solve the problem will be our first unknown, because this sector is represented by the first row of the input-output matrix. The unknown quantity of final nonmetal units will be represented by the symbol x_1. Following the same pattern, we represent the unknown quantities in the following manner.

x_1 = units of final nonmetal production required
x_2 = units of final metal production required
x_3 = units of basic metal production required
x_4 = units of basic nonmetal production required
x_5 = units of energy required
x_6 = units of services required

These six numbers are the quantities we are attempting to calculate, and the variables will make up the entries in our production matrix.

To find these numbers, first let A be the 6×6 matrix corresponding the input-output table.

$$A = \begin{bmatrix} .170 & .004 & 0 & .029 & 0 & .008 \\ .003 & .295 & .018 & .002 & .004 & .016 \\ .025 & .173 & .460 & .007 & .011 & .007 \\ .348 & .037 & .021 & .403 & .011 & .048 \\ .007 & .001 & .039 & .025 & .358 & .025 \\ .120 & .074 & .104 & .123 & .173 & .234 \end{bmatrix}$$

A is the input-output matrix. The bill of demands leads to a 6×1 demand matrix D, and X is the matrix of unknowns.

$$D = \begin{bmatrix} 99,640 \\ 75,548 \\ 14,444 \\ 33,501 \\ 23,527 \\ 263,985 \end{bmatrix} \quad \text{and} \quad X = \begin{bmatrix} x_1 \\ x_2 \\ x_3 \\ x_4 \\ x_5 \\ x_6 \end{bmatrix}$$

To use the formula $D = (I - A)X$, or $X = (I - A)^{-1}D$, we need to find $I - A$.

$$I - A = \begin{bmatrix} 1 & 0 & 0 & 0 & 0 & 0 \\ 0 & 1 & 0 & 0 & 0 & 0 \\ 0 & 0 & 1 & 0 & 0 & 0 \\ 0 & 0 & 0 & 1 & 0 & 0 \\ 0 & 0 & 0 & 0 & 1 & 0 \\ 0 & 0 & 0 & 0 & 0 & 1 \end{bmatrix}$$
$$- \begin{bmatrix} .170 & .004 & 0 & .029 & 0 & .008 \\ .003 & .295 & .018 & .002 & .004 & .016 \\ .025 & .173 & .460 & .007 & .011 & .007 \\ .348 & .037 & .021 & .403 & .011 & .048 \\ .007 & .001 & .039 & .025 & .358 & .025 \\ .120 & .074 & .104 & .123 & .173 & .234 \end{bmatrix}$$
$$= \begin{bmatrix} .830 & -.004 & 0 & -.029 & 0 & -.008 \\ -.003 & .705 & -.018 & -.002 & -.004 & -.016 \\ -.025 & -.173 & .540 & -.007 & -.011 & -.007 \\ -.348 & -.037 & -.021 & .597 & -.011 & -.048 \\ -.007 & -.001 & -.039 & -.025 & .642 & -.025 \\ -.120 & -.074 & -.104 & -.123 & -.173 & .766 \end{bmatrix}$$

By the method given in Section 6.5, we find the inverse (actually an approximation).

$$(I - A)^{-1} = \begin{bmatrix} 1.234 & .014 & .007 & .064 & .006 & .017 \\ .017 & 1.436 & .056 & .014 & .019 & .032 \\ .078 & .467 & 1.878 & .036 & .044 & .031 \\ .752 & .133 & .101 & 1.741 & .065 & .123 \\ .061 & .045 & .130 & .083 & .1.578 & .059 \\ .340 & .236 & .307 & .315 & .376 & 1.349 \end{bmatrix}$$

From this result,

$$x_1 = 131,033$$
$$x_2 = 120,459$$
$$x_3 = 80,861$$
$$x_4 = 178,732$$
$$x_5 = 66,929$$
$$x_6 = 431,562.$$

Therefore,

$$X = (I - A)^{-1} D =$$

$$\begin{bmatrix} 1.234 & .014 & .007 & .064 & .006 & .017 \\ .017 & 1.436 & .056 & .014 & .019 & .032 \\ .078 & .467 & 1.878 & .036 & .044 & .031 \\ .752 & .133 & .101 & 1.741 & .065 & .123 \\ .061 & .045 & .130 & .083 & 1.578 & .059 \\ .340 & .236 & .307 & .315 & .376 & 1.349 \end{bmatrix} \begin{bmatrix} 99,640 \\ 75,548 \\ 14,444 \\ 33,501 \\ 23,527 \\ 263,985 \end{bmatrix}$$

$$= \begin{bmatrix} 131,033 \\ 120,459 \\ 80,681 \\ 178,732 \\ 66,929 \\ 431,562 \end{bmatrix}.$$

In other words, it would require 131,033 units ($131,033 million worth) of final nonmetal production, 120,459 units of final metal output, 80,861 units of basic metal products, and so on to run the 1958 American economy and completely fill the stated bill of demands.

EXERCISES

Use a grapher with matrix capability to work the following problems.

1. A much simplified version of Leontief's 42-sector analysis of the 1947 American economy divides the economy into just 3 sectors: agriculture, manufacturing, and the household (i.e., the sector of the economy that produces labor). It is summarized in the following input-output table.

	Agriculture	Manufacturing	Household
Agriculture	.245	.102	.051
Manufacturing	.099	.291	.279
Household	.433	.372	.011

The bill of demands (in billions of dollars) is shown below.

Agriculture	2.88
Manufacturing	31.45
Household	30.91

(a) Write the input-output matrix A, the demand matrix D, and the matrix X.

(b) Compute $I - A$.

(c) Check that $(I - A)^{-1} = \begin{bmatrix} 1.453 & .291 & .157 \\ .532 & 1.762 & .525 \\ .836 & .790 & 1.277 \end{bmatrix}$ is an approximation to the inverse of $I - A$ by calculating $(I - A)^{-1}(I - A)$.

(d) Use the matrix of part (c) to compute X.

(e) Explain the meaning of the numbers in X in dollars.

2. An analysis of the 1958 Israeli economy* is here simplified by grouping the economy into three sectors: agriculture, manufacturing, and energy. The input-output table is given below.

*Wassily Leontief, Input-Output Economics (New York: Oxford University Press, 1966), pp. 54–57.

	Agriculture	*Manufacturing*	*Energy*
Agriculture	.293	0	0
Manufacturing	.014	.207	.017
Energy	.044	.010	.216

Exports (in thousands of Israeli pounds) were as follows.

Agriculture	138,213
Manufacturing	17,597
Energy	1,786

(a) Write the input-output matrix A and the demand (export) matrix D.

(b) Compute $I - A$.

(c) Check that $(I - A)^{-1} = \begin{bmatrix} 1.414 & 0 & 0 \\ .027 & 1.261 & .027 \\ .080 & .016 & 1.276 \end{bmatrix}$ is an approximation to the inverse of $I - A$ by calculating $(I - A)^{-1}(I - A)$.

(d) Use the matrix of part (c) to determine the number of Israeli pounds worth of agricultural products, manufactured goods, and energy required to run this model of the Israeli economy and export the stated value of products.

CHAPTER 7

Linear Programming

Many realistic problems involve inequalities. For example, a factory may have no more than 200 workers on a shift and must manufacture at least 3000 units at a cost of no more than $35 each. How many workers should it have per shift in order to produce the required units at minimal cost? *Linear programming* is a method for finding the optimal (best possible) solution for such problems, if there is one.

In this chapter we shall study two methods of solving linear programming problems: the graphical method and the simplex method. The graphical method requires a knowledge of **linear inequalities,** those involving only first-degree polynomials in x and y. So we begin with a study of such inequalities.

7.1 GRAPHING LINEAR INEQUALITIES IN TWO VARIABLES

Examples of linear inequalities in two variables include

$$x + 2y < 4, \quad 3x + 2y > 6, \quad \text{and} \quad 2x - 5y \geq 10.$$

A solution of a linear inequality is an ordered pair that satisfies the inequality. For example (4, 4) is a solution of

$$3x - 2y \leq 6.$$

(Check by substituting 4 for x and 4 for y.) A linear inequality has an infinite number of solutions, one for every choice of a value for x. The best way to show these solutions is with a graph that consists of all the points in the plane whose coordinates satisfy the inequality.

1 Graph.

(a) $2x + 5y \le 10$

(b) $x - y \ge 4$

Answers:

(a)

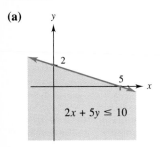

(b)

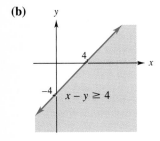

▶**EXAMPLE 1** Graph the linear inequality $3x - 2y \le 6$.

Because of the "=" portion of $\le$, the points of the line $3x - 2y = 6$ satisfy the linear inequality $3x - 2y \le 6$ and are part of its graph. As in Chapter 2, find the intercepts by first letting $x = 0$ and then letting $y = 0$; use these points to get the graph of $3x - 2y = 6$ shown in Figure 7.1.

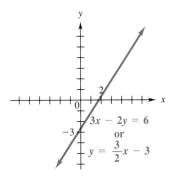

FIGURE 7.1

The points on the line satisfy "$3x - 2y$ *equals* 6." The points satisfying "$3x - 2y$ *is less than* 6" can be found by first solving $3x - 2y \le 6$ for y.

$$3x - 2y \le 6$$
$$-2y \le -3x + 6$$
$$y \ge \frac{3}{2}x - 3 \qquad \text{Multiply by } -1/2;$$
$$\text{reverse the inequality}$$

As shown in Figure 7.2, the points *above* the line $3x - 2y = 6$ satisfy

$$y > \frac{3}{2}x - 3,$$

while those below the line satisfy

$$y < \frac{3}{2}x - 3.$$

The line itself is the **boundary.** In summary, the inequality $3x - 2y \le 6$ is satisfied by all points *on or above* the line $3x - 2y = 6$. Indicate the points above the line by shading, as in Figure 7.3. The line and shaded region of Figure 7.3 make up the graph of the linear inequality $3x - 2y \le 6$. ◀ **1**

After an inequality is solved for y, the inequality symbol tells whether the points above, on, or below the boundary satisfy the inequality. An alternative method that does not require first solving the inequality for y involves the use of a test point.

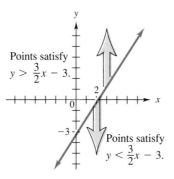

FIGURE 7.2   **FIGURE 7.3**

▶ **EXAMPLE 2**   Graph $x + 4y < 4$.

The boundary here is the line $x + 4y = 4$. Since the points on this line do not satisfy $x + 4y < 4$, the line is drawn dashed, as in Figure 7.4. We decide whether to shade the region above the line or the region below the line by choosing as a test point any point not on the boundary line. For example, we choose the point $(0, 0)$, which is not on the line $x + 4y = 4$. Substitute 0 for x and 0 for y in the given inequality.

$$x + 4y < 4$$
$$0 + 4(0) < 4 \qquad \text{Let } x = 0, y = 0$$
$$0 < 4 \qquad \text{True}$$

The result $0 < 4$ is true, so the test point $(0, 0)$ is on the side of the line where all the points satisfy $x + 4y < 4$. For this reason, shade the side containing $(0, 0)$, as in Figure 7.4.

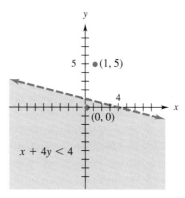

FIGURE 7.4

2 Graph.

(a) $2x + 3y > 12$

(b) $3x - 2y < 6$

Answers:

(a)

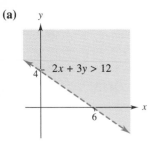

(b)

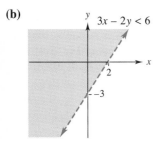

3 Graph each of the following.

(a) $x \geq 3$

(b) $y - 3 \leq 0$

Answers:

(a)

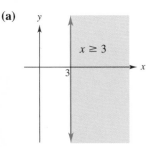

(b)

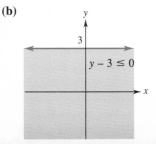

If a different test point is chosen, say $(1, 5)$, and substituted in $x + 4y < 4$, we obtain $1 + 4 \cdot 5 < 4$, which is *false*. Therefore, the side of the boundary line that does *not* contain $(1, 5)$ is the set of solutions, as shown in Figure 7.4. Thus, either test point leads to the same conclusion. ◀ **2**

Note If it is not on the boundary, $(0, 0)$ is usually the best choice for a test point. When $(0, 0)$ is on the boundary, choose a test point on one of the axes. These choices make calculations simpler, which avoids mistakes. Furthermore, it is usually easy to tell which side of the boundary such points lie on, even on a poorly drawn graph, which may not be the case for other choices of test points.

As these examples suggest, the graph of a linear inequality is a region in the plane, perhaps including the line that is the boundary of the region. Each of the shaded regions is an example of a **half-plane,** a region on one side of a line. For example, in Figure 7.5 line r divides the plane into half-planes P and Q. The points of line r belong to neither P nor Q. Line r is the boundary of each half-plane.

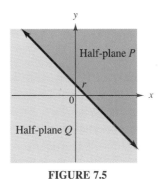

FIGURE 7.5

▶ **EXAMPLE 3** Graph each of the following.

(a) $x \leq -1$

Recall that the graph of $x = -1$ is the vertical line through $(-1, 0)$. The points where $x < -1$ lie in the half-plane to the left of the boundary line, as shown in Figure 7.6(a).

(b) $y \geq 2$

The graph of $y = 2$ is the horizontal line through $(0, 2)$. The points where $y \geq 2$ lie in the half-plane above this boundary line, as shown in Figure 7.6(b). ◀ **3**

▶ **EXAMPLE 4** Graph $x \geq 3y$.

Start by graphing the boundary, $x = 3y$. If $x = 0$, then $y = 0$, giving the point $(0, 0)$. Setting $y = 0$ would produce $(0, 0)$ again. To get a second point on the line, choose another number for x. Choosing $x = 3$ gives $y = 1$, so that $(3, 1)$ is a point on the line $x = 3y$. The points $(0, 0)$ and $(3, 1)$ lead to the line graphed in Figure 7.7. Since $(0, 0)$ is on the boundary, we choose $(6, 0)$ as a test

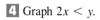 Graph $2x < y$.

Answer:

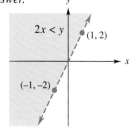

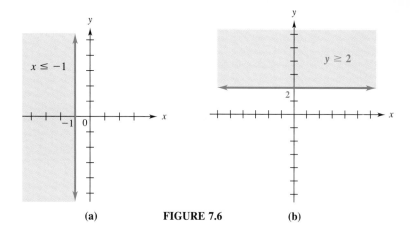

(a) **FIGURE 7.6** (b)

point to decide which half-plane to shade. (Any point that is not on the line $x = 3y$ may be used.) Replacing x with 6 and y with 0 in the original inequality gives a true statement, so the half-plane containing $(6, 0)$ is shaded, as shown in Figure 7.7. ◀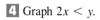

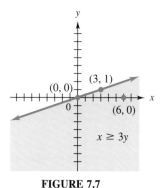

FIGURE 7.7

The steps used to graph a linear inequality are summarized below.

Graphing a Linear Inequality

1. Draw the graph of the boundary line. Make the line solid if the inequality involves $\leq$ or $\geq$; make the line dashed if the inequality involves $<$ or $>$.
2. Decide which half-plane to shade: either
   (a) solve the inequality for y, getting $y > mx + b$ or $y < mx + b$; shade the region above the line $y = mx + b$ if $y > mx + b$; shade the region below the line if $y < mx + b$; or
   (b) choose any point not on the line as a test point; shade the half-plane that includes the test point if the test point satisfies the original inequality; otherwise, shade the half-plane on the other side of the boundary line.

5 Graph the system
$x + y \leq 6$, $2x + y \geq 4$.

Answer:

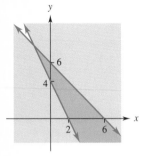

SYSTEMS OF INEQUALITIES Realistic problems often involve many inequalities. For example, a manufacturing problem might produce inequalities resulting from production requirements, as well as inequalities about cost requirements. A set of at least two inequalities is called a **system of inequalities.** The graph of a system of inequalities is made up of all those points that satisfy all the inequalities of the system at the same time. The next example shows how to graph a system of linear inequalities.

▶ **EXAMPLE 5** Graph the system

$$y < -3x + 12$$
$$x < 2y.$$

First, graph the solution of $y < -3x + 12$. The boundary, the line with equation $y = -3x + 12$, is dashed. The points below the boundary satisfy $y < -3x + 12$. Now graph the solution of $x < 2y$ on the same axes. Again, the boundary line $x = 2y$ is dashed. Use a test point to see that the region above the boundary satisfies $x < 2y$. The heavily shaded region in Figure 7.8 shows all the points that satisfy both inequalities of the system. ◀ **5**

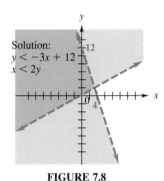

FIGURE 7.8

The heavily shaded region of Figure 7.8 is sometimes called the **region of feasible solutions,** or just the **feasible region,** since it is made up of all the points that satisfy (are feasible for) each inequality of the system.

▶ **EXAMPLE 6** Graph the feasible region for the system

$$2x - 5y \leq 10$$
$$x + 2y \leq 8$$
$$x \geq 0, \quad y \geq 0.$$

On the same axes, graph each inequality by graphing the boundary and choosing the appropriate half-plane. Graph the solid boundary line $2x - 5y = 10$ by first locating the intercepts that give the points $(5, 0)$ and $(0, -2)$. Use a

6 Graph the feasible region of the system
$x + 4y \leq 8, x - y \geq 3$
$x \geq 0, y \geq 0.$

Answer:

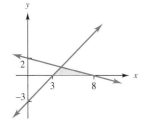

test point to see that $2x - 5y < 10$ is satisfied by the points above the boundary. In the same way, the solid boundary line $x + 2y = 8$ goes through $(8, 0)$ and $(0, 4)$. A test point will show that the graph of $x + 2y < 8$ includes all points below the boundary. The inequalities $x \geq 0$ and $y \geq 0$ restrict the feasible region to the first quadrant. Find the feasible region by locating the intersection (overlap) of *all* the half-planes. This feasible region is shaded in Figure 7.9. ◄ **6**

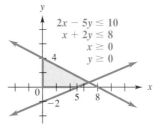

FIGURE 7.9

APPLICATIONS   As we shall see in the rest of this chapter, many realistic problems lead to systems of linear inequalities. The next example is typical of such problems.

►**EXAMPLE 7**   Midtown Manufacuturing Company makes plastic plates and cups, both of which require time on two machines. A unit of plates requires 1 hour on machine A and 2 on machine B, while a unit of cups requires 3 hours on machine A and 1 on machine B. Each machine is operated for at most 15 hours per day. Write a system of inequalities expressing these conditions and graph the feasible region.

Let x represent the number of units of plates to be made, and y represent the number of units of cups. Then make a chart that summarizes the given information.

	Number Made	Time on Machine A	Time on Machine B
Plates	x	1	2
Cups	y	3	1
Maximum time available		15	15

On machine A, x units of plates require a total of $1 \cdot x = x$ hours while y units of cups require $3 \cdot y = 3y$ hours. Since machine A is available no more than 15 hours a day,

$$x + 3y \leq 15. \quad \text{Machine A}$$

The requirement that machine B be used no more than 15 hours a day gives

$$2x + y \leq 15. \quad \text{Machine B}$$

It is not possible to produce a negative number of cups or plates, so that

$$x \geq 0 \quad \text{and} \quad y \geq 0.$$

The feasible region for this system of inequalities is shown in Figure 7.10. ◀

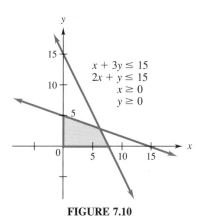

FIGURE 7.10

7.1 EXERCISES

Graph each of the following linear inequalities. (See Examples 1–4.)

1. $y < 5 - 2x$ **2.** $y > x + 3$ **3.** $3x - 2y \geq 18$ **4.** $2x + 5y \leq 10$ **5.** $2x - y \leq 4$

6. $4x - 3y \geq 24$ **7.** $y \leq -4$ **8.** $x \geq -2$ **9.** $x + 4y \leq 2$ **10.** $3x + 2y \geq 6$

11. $4x + 3y > -3$ **12.** $5x + 3y > 15$ **13.** $2x - 4y < 3$ **14.** $4x - 3y < 12$ **15.** $x \leq 5y$

16. $2x \geq y$ **17.** $-3x < y$ **18.** $-x > 6y$ **19.** $y < x$ **20.** $y > -2x$

Graph the feasible region for the following systems of inequalities. (See Examples 5 and 6.)

21. $x - y \geq 1$
$\quad x \leq 3$

22. $2x + y \leq 5$
$\quad x + 2y \leq 5$

23. $4x + y \geq 9$
$\quad 2x + 3y \leq 7$

24. $2x + y > 8$
$\quad 4x - y < 3$

25. $x + y > 5$
$\quad x - 2y < 2$

26. $3x - 4y < 6$
$\quad 2x + 5y > 15$

27. $2x - y < 1$
$\quad 3x + y < 6$

28. $x + 3y \leq 6$
$\quad 2x + 4y \geq 7$

29. $-x - y < 5$
$\quad 2x - y < 4$

30. $6x - 4y > 8$
$\quad 3x + 2y > 4$

31. $3x + y \geq 6$
$\quad x + 2y \geq 7$
$\quad x \geq 0$
$\quad y \geq 0$

32. $2x + 3y \geq 12$
$\quad x + y \geq 4$
$\quad x \geq 0$
$\quad y \geq 0$

33. $-2 < x < 3$
$\quad -1 \leq y \leq 5$
$\quad 2x + y < 6$

34. $-2 < x < 2$
$\quad y > 1$
$\quad x - y > 0$

35. $2y - x \geq -5$
$\quad y \leq 3 + x$
$\quad x \geq 0$
$\quad y \geq 0$

36. $2x + 3y \le 12$
$2x + 3y > -6$
$3x + y < 4$
$x \ge 0$
$y \ge 0$

37. $3x + 4y > 12$
$2x - 3y < 6$
$0 \le y \le 2$
$x \ge 0$

38. $0 \le x \le 9$
$x - 2y \ge 4$
$3x + 5y \le 30$
$y \ge 0$

In Exercises 39–40, find a system of inequalities whose feasible region is the interior of the given polygon.

39. Rectangle with vertices $(2, 3)$, $(2, -1)$, $(7, 3)$, $(7, -1)$.

40. Triangle with vertices $(2, 4)$, $(-4, 0)$, $(2, -1)$.

41. Management Lillie Chalmers and Brett Sullivan produce handmade shawls and afghans. They spin the yarn, dye it, and then weave it. A shawl requires 1 hour of spinning, 1 hour of dyeing, and 1 hour of weaving. An afghan needs 2 hours of spinning, 1 of dyeing, and 4 of weaving. Together, they spend at most 8 hours spinning, 6 hours dyeing, and 14 hours weaving.
(a) Complete the following chart.

	Number	Hours Spinning	Hours Dyeing	Hours Weaving
Shawls	x			
Afghans	y			
Maximum number of hours available		8	6	14

(b) Use the chart to write a system of inequalities that describe the situation.
(c) Graph the feasible region of this system of inequalities.

42. Management An electric shaver manufacturer makes two models, the regular and the flex. Because of demand, the number of regular shavers made is never more than half the number of flex shavers. The factory's production cannot exceed 1200 shavers per week.
(a) Write a system of inequalities that describe the possibilities for making x regular and y flex shavers per week.
(b) Graph the feasible region of this system of inequalities.

In each of the following, write a system of inequalities that describes all the possibilities and graph the feasible region of the system.

43. Management Southwestern Oil supplies two distributors located in the Northwest. One distributor needs at least 3000 barrels of oil, and the other needs at least 5000 barrels. Southwestern can send out at most 10,000 barrels. Let $x =$ the number of barrels of oil sent to distributor 1 and $y =$ the number sent to distributor 2.

44. Management The California Almond Growers have 2400 boxes of almonds to be shipped from their plant in Sacramento to Des Moines and San Antonio. The Des Moines market needs at least 1000 boxes, while the San Antonio market must have at least 800 boxes. Let $x =$ the number of boxes to be shipped to Des Moines and $y =$ the number of boxes to be shipped to San Antonio.

45. Management A cement manufacturer produces at least 3.2 million barrels of cement annually. He is told by the Environmental Protection Agency that his operation emits 2.5 pounds of dust for each barrel produced. The EPA has ruled that annual emissions must be reduced to 1.8 million pounds. To do this, the manufacturer plans to replace the present dust collectors with two types of electronic precipitators. One type would reduce emissions to .5 pounds per barrel and would cost 16¢ per barrel. The other would reduce the dust to .3 pounds per barrel and would cost 20¢ per barrel. The manufacturer does not want to spend more than .8 million dollars on the precipitators. He needs to know how many barrels he should produce with each type. Let $x =$ the number of barrels in millions produced with the first type and $y =$ the number of barrels in millions produced with the second type.

46. Natural Science A dietician is planning a snack package of fruit and nuts. Each ounce of fruit will supply 1 unit of protein, 2 units of carbohydrates, and 1 unit of fat. Each ounce of nuts will supply 1 unit of protein, 1 unit of carbohydrates, and 1 unit of fat. Every package must provide at least 7 units of protein, and at least 10 units of carbohydrates, and no more than 9 units of fat. Let x be the ounces of fruit and y the ounces of nuts to be used in each package.

7.2 LINEAR PROGRAMMING: THE GRAPHICAL METHOD

Many problems in business, science, and economics involve finding the optimal value of a function (for instance, the maximum value of the profit function or the minimum value of the cost function) subject to various **constraints** (such as transportation costs, environmental protection laws, availability of parts, interest rates, etc.). **Linear programming** deals with such situations in which the function to be optimized, called the **objective function,** is linear and the constraints are given by linear inequalities. Linear programming problems that involve only two variables can be solved by the graphical method, which is explained in Example 1.

▶ **EXAMPLE 1**   Find the maximum and minimum values of the objective function $z = 2x + 5y$, subject to the following constraints.

$$3x + 2y \leq 6$$
$$-2x + 4y \leq 8$$
$$x + y \geq 1$$
$$x \geq 0, \quad y \geq 0$$

First, graph the feasible region of the system of inequalities (Figure 7.11). The points in this region are the only ones that satisfy all the constraints. However, each such point may produce a different value of the objective function. For instance, the points $(.5, 1)$ and $(1, 0)$ in the feasible region lead to the values

$$z = 2(.5) + 5(1) = 6 \quad \text{and} \quad z = 2(1) + 5(0) = 2.$$

We must find the points that produce the maximum and minimum value of z.

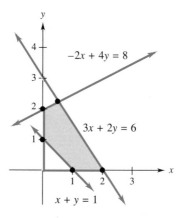

FIGURE 7.11

1 Suppose the objective function in Example 1 is changed to $z = 5x + 2y$.

(a) Sketch the graphs of the objective function when $z = 0$, $z = 5$, and $z = 10$ on the region of feasible solutions given in Figure 7.11.

(b) From the graph, decide what values of x and y will maximize the objective function.

Answers:

(a)

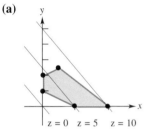

$z = 0$ $z = 5$ $z = 10$

(b) $(2, 0)$

To find the maximum value, consider various possible values for z. For instance, when $z = 0$, then the objective function is $0 = 2x + 5y$, whose graph is a straight line. Similarly, when z is 5, 10, and 15, the objective function becomes (in turn)

$$5 = 2x + 5y, \qquad 10 = 2x + 5y, \qquad 15 = 2x + 5y.$$

These four lines are graphed in Figure 7.12. (All the lines are parallel because they have the same slope.) The figure shows that z cannot take on the value 15 because the graph for $z = 15$ is entirely outside the feasible region. The maximum possible value of z will be obtained from a line parallel to the others and between the lines representing the objective function when $z = 10$ and $z = 15$. The value of z will be as large as possible and all constraints will be satisfied if this line just touches the feasible region. This occurs at point A.

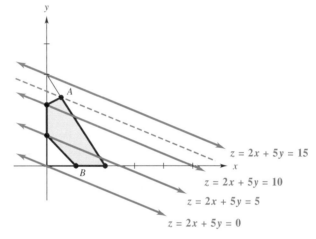

$z = 2x + 5y = 15$

$z = 2x + 5y = 10$

$z = 2x + 5y = 5$

$z = 2x + 5y = 0$

FIGURE 7.12

The coordinates of point A are found by solving the system

$$3x + 2y = 6$$
$$-2x + 4y = 8$$

to get $(1/2, 9/4)$. The value of z at this point is

$$z = 2x + 5y$$
$$= 2\left(\frac{1}{2}\right) + 5\left(\frac{9}{4}\right)$$
$$z = 12\frac{1}{4}.$$

The maximum possible value of z is $12\frac{1}{4}$. Similarly, the minimum value of z occurs at point B, which has coordinates $(1, 0)$. The minimum value of z is $2(1) + 5(0) = 2$. ◀ **1**

Points such as A and B in Example 1 are called corner points. A **corner point** is a point in the feasible region where the boundary lines of two constraints cross. The feasible region in Figure 7.11 is **bounded,** because the region is enclosed by boundary lines on all sides. Linear programming problems with bounded regions always have solutions. However, if Example 1 did not include the constraint $3x + 2y \le 6$, the feasible region would be **unbounded,** and there would be no way to *maximize* the value of the objective function.

Some general conclusions can be drawn from the method of solution used in Example 1. Figure 7.13 shows various feasible regions and the lines that result from various values of z. (Figure 7.13 shows the situation in which the lines are in order from left to right as z increases.) In part (a) of the figure, the objective function takes on its minimum value at corner point Q and its maximum value at P. The minimum is again at Q in part (b), but the maximum occurs at P_1 or P_2, or any point on the line segment connecting them. Finally, in part (c), the minimum value occurs at Q, but the objective function has no maximum value because the feasible region is unbounded.

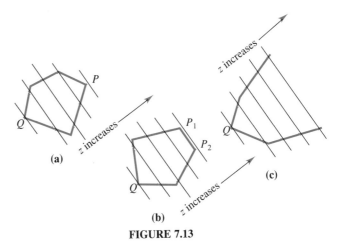

(a)   (b)   (c)

FIGURE 7.13

The preceding discussion suggests the truth of the **corner point theorem.**

CORNER POINT THEOREM

If the feasible region is bounded, then the objective function has both a maximum and a minimum value and each occurs at one or more corner points.

If the feasible region is unbounded, the objective function may not have a maximum or minimum. But if a maximum or minimum value exists, it will occur at one or more corner points.

2

(a) Identify the corner points in the graph.

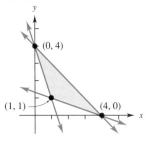

(b) Which corner point would minimize $z = 2x + 3y$?

Answers:

(a) (0, 4), (1, 1), (4, 0)

(b) (1, 1)

This theorem simplifies the job of finding an optimum value: First, graph the feasible region and find all corner points. Then test each point in the objective function. Finally, identify the corner point producing the optimum solution.

With the theorem, the problem in Example 1 could have been solved by identifying the five corner points of Figure 7.11: (0, 1), (0, 2), (1/2, 9/4), (2, 0), and (1, 0). Then, substituting each of these points into the objective function $z = 2x + 5y$ would identify the corner points that produce the maximum or the minimum value of z.

Corner Point	Value of $z = 2x + 5y$
(0, 1)	$2(0) + 5(1) = 5$
(0, 2)	$2(0) + 5(2) = 10$
$\left(\dfrac{1}{2}, \dfrac{9}{4}\right)$	$2\left(\dfrac{1}{2}\right) + 5\left(\dfrac{9}{4}\right) = 12\dfrac{1}{4}$ (maximum)
(2, 0)	$2(2) + 5(0) = 4$
(1, 0)	$2(1) + 5(0) = 2$ (minimum)

From these results, the corner point (1/2, 9/4) yields the maximum value of $12\frac{1}{4}$ and the corner point (1, 0) gives the minimum value of 2. These are the same values found earlier. **2**

A summary of the steps in solving a linear programming problem by the graphical method is given here.

Solving a Linear Programming Problem Graphically

1. Write the objective function and all necessary constraints.
2. Graph the feasible region.
3. Determine the coordinates of each of the corner points.
4. Find the value of the objective function at each corner point.
5. If the feasible region is bounded, the solution is given by the corner point producing the optimum value of the objective function.
6. If the feasible region is an unbounded region in the first quadrant and both coefficients of the objective function are positive,* then the minimum value of the objective function occurs at a corner point and there is no maximum value.

*This is the only case of an unbounded region that occurs in the applications considered here.

3 Use the region of feasible solutions in the sketch to find the following.

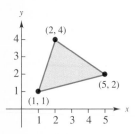

(a) The values of x and y that maximize $z = 2x - y$.

(b) The maximum value of $z = 2x - y$.

(c) The values of x and y that minimize $z = 4x + 3y$.

(d) The minimum value of $z = 4x + 3y$.

Answers:

(a) $(5, 2)$

(b) 8

(c) $(1, 1)$

(d) 7

▶ **EXAMPLE 2** Sketch the feasible region for the following set of constraints:

$$3y - 2x \geq 0$$
$$y + 8x \leq 52$$
$$y - 2x \leq 2$$
$$x \geq 3.$$

Then find the maximum and minimum values of the objective function $z = 5x + 2y$.

The graph in Figure 7.14 shows that the feasible region is bounded. The corner points were found by solving systems of two equations using the methods of Chapter 6. For example, the point $(5, 12)$ is the intersection of the lines corresponding to the equations $y + 8x = 52$ and $y - 2x = 2$, and the solution of the system

$$y + 8x = 52$$
$$y - 2x = 2$$

is the ordered pair $(5, 12)$. Also, the corner point $(6, 4)$ was found by solving the system

$$y + 8x = 52$$
$$3y - 2x = 0.$$

Use the corner points from the graph to find the maximum and minimum values of the objective function.

Corner Point	Value of $z = 5x + 2y$
$(3, 2)$	$5(3) + 2(2) = 19$ (minimum)
$(6, 4)$	$5(6) + 2(4) = 38$
$(5, 12)$	$5(5) + 2(12) = 49$ (maximum)
$(3, 8)$	$5(3) + 2(8) = 31$

The minimum value of $z = 5x + 2y$ is 19 at the corner point $(3, 2)$. The maximum value is 49 at $(5, 12)$. ◀ **3**

▶ **EXAMPLE 3** Solve the following linear programming problem.

$$\text{Minimize} \quad z = x + 2y$$
$$\text{subject to:} \quad x + y \leq 10$$
$$3x + 2y \geq 6$$
$$x \geq 0, \ y \geq 0.$$

The feasible region is shown in Figure 7.15. From the figure, the corner points are $(0, 3), (0, 10), (10, 0),$ and $(2, 0)$. These corner points give the following values of z.

4 The sketch shows a feasible region. Let $z = x + 3y$. Use the sketch to find the values of x and y that

(a) minimize z;

(b) maximize z.

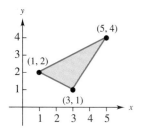

Answers:

(a) $(3, 1)$

(b) $(5, 4)$

5 The sketch below shows a region of feasible solutions. From the sketch decide what ordered pair would minimize, $z = 2x + 4y$.

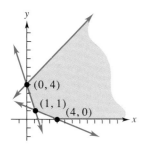

Answer:
$(1, 1)$

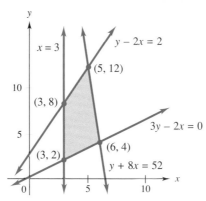

FIGURE 7.14

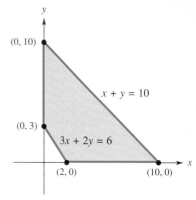

FIGURE 7.15

Corner Point	Value of $z = x + 2y$
$(0, 3)$	$0 + 2(3)\ = 6$
$(0, 10)$	$0 + 2(10) = 20$
$(10, 0)$	$10 + 2(0)\ = 10$
$(2, 0)$	$2 + 2(0)\ = 2$ **(minimum)**

The minimum value of z is 2; it occurs at $(2, 0)$. ◀ **4**

▶ **EXAMPLE 4** Solve the following linear programming problem.

$$\text{Minimize} \qquad z = 2x + 4y$$
$$\text{subject to:} \qquad x + 2y \geq 10$$
$$3x + y \geq 10$$
$$x \geq 0, \quad y \geq 0.$$

Figure 7.16 on the next page shows the feasible region with corner points $(0, 10)$, $(2, 4)$, and $(10, 0)$. Find the value of z for each point.

Corner Point	Value of $z = 2x + 4y$
$(0, 10)$	$2(0) + 4(10) = 40$
$(2, 4)$	$2(2) + 4(4)\ = 20$ **(minimum)**
$(10, 0)$	$2(10) + 4(0) = 20$ **(minimum)**

In this case, both $(2, 4)$ and $(10, 0)$, as well as all the points on the boundary line between them, give the same optimum value of z. There is an infinite number of equally "good" values of x and y which give the same minimum value of the objective function $z = 2x + 4y$. The minimum value is 20. ◀ **5**

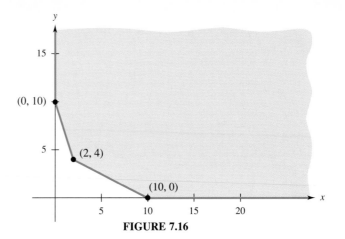

FIGURE 7.16

7.2 EXERCISES

Exercises 1–6 show regions of feasible solutions. Use these regions to find maximum and minimum values of each given objective function. (See Examples 1–2.)

1. $z = 3x + 5y$

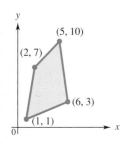

2. $z = 6x + y$

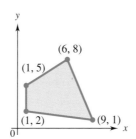

3. $z = .40x + .75y$

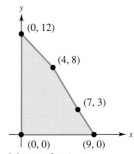

4. $z = .35x + 1.25y$

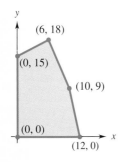

5. **(a)** $z = 4x + 2y$
(b) $z = 2x + 3y$
(c) $z = 2x + 4y$
(d) $z = x + 4y$

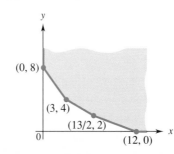

6. **(a)** $z = 4x + y$
(b) $z = 5x + 6y$
(c) $z = x + 2y$
(d) $z = x + 6y$

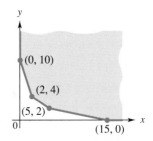

Use graphical methods to solve Exercises 7–12. (See Examples 2–4.)

7. Maximize $z = 5x + 2y$
subject to: $2x + 3y \le 6$
$4x + y \le 6$
$x \ge 0, \ y \ge 0.$

8. Minimize $z = x + 3y$
subject to: $2x + y \le 10$
$5x + 2y \ge 20$
$-x + 2y \ge 0$
$x \ge 0, \ y \ge 0.$

9. Minimize $z = 2x + y$
subject to: $3x - y \ge 12$
$x + y \le 15$
$x \ge 2, y \ge 3.$

10. Maximize $z = x + 3y$
subject to: $2x + 3y \le 100$
$5x + 4y \le 200$
$x \ge 10, \ y \ge 20.$

11. Maximize $z = 4x + 2y$
subject to: $x - y \le 10$
$5x + 3y \le 75$
$x \ge 0, y \ge 0.$

12. Maximize $z = 4x + 5y$
subject to: $10x - 5y \le 100$
$20x + 10y \ge 150$
$x \ge 0, \ y \ge 0.$

Find the minimum and maximum value of $z = 3x + 4y$ (if possible) for each of the following sets of constraints. (See Examples 2–4.)

13. $3x + 2y \ge 6$
$x + 2y \ge 4$
$x \ge 0, y \ge 0$

14. $2x + y \le 20$
$10x + y \ge 36$
$2x + 5y \ge 36$

15. $x + y \le 6$
$-x + y \le 2$
$2x - y \le 8$

16. $-x + 2y \le 6$
$3x + y \ge 3$
$x \ge 0, \ y \ge 0$

17. Find values of $x \ge 0$ and $y \ge 0$ which maximize $z = 10x + 12y$ subject to each of the following sets of constraints.
  (a) $x + y \le 20$ **(b)** $3x + y \le 15$
  $x + 3y \le 24$ $x + 2y \le 18$
  (c) $x + 2y \ge 10$
  $2x + y \ge 12$
  $x - y \le 8$

18. Find values of $x \ge 0$ and $y \ge 0$ that minimize $z = 3x + 2y$ subject to each of the following sets of constraints.
  (a) $10x + 7y \le 42$ **(b)** $6x + 5y \ge 25$
  $4x + 10y \ge 35$ $2x + 6y \ge 15$
  (c) $2x + 5y \ge 22$
  $4x + 3y \le 28$
  $2x + 2y \le 17$

19. Explain why it is impossible to maximize the function $z = 3x + 4y$ subject to the constraints:
$$x + y \ge 8, \quad 2x + y \le 10, \quad x + 2y \le 8,$$
$$x \ge 0, \quad y \ge 0.$$

20. You are given the following linear programming problem:[*]

Maximize $z = c_1 x_1 + c_2 x_2$
subject to: $2x_1 + x_2 \le 11$
$-x_1 + 2x_2 \le 2$
$x_1 \ge 0, x_2 \ge 0.$

If $c_2 > 0$, determine the range of c_1/c_2 for which $(x_1, x_2) = (4, 3)$ is an optimal solution.
  (a) $[-2, 1/2]$ **(b)** $[-1/2, 2]$
  (c) $[-11, -1]$ **(d)** $[1, 11]$ **(e)** $[-11, 11]$

*Problem from "Course 130 Examination Operations Research" of the *Education and Examination Committee of The Society of Actuaries.* Reprinted by permission of The Society of Actuaries.

7.3 APPLICATIONS OF LINEAR PROGRAMMING

In this section we show several applications of linear programming problems with two variables.

▶**EXAMPLE 1**   A 4-H Club member raises only geese and pigs. She wants to raise no more than 16 animals including no more than 10 geese. She spends \$15 to raise a goose and \$45 to raise a pig, and has \$540 available for this project. Find the maximum profit she can make if each goose produces a profit of \$6 and each pig a profit of \$20.

The total profit is determined by the number of geese and pigs. So let x be the number of geese to be produced, and let y be the number of pigs. Then summarize the information of the problem in a table.

	Number	*Cost to Raise*	*Profit Each*
Geese	x	\$ 15	\$ 6
Pigs	y	45	20
Maximum available	16	\$540	

Use this table to write the necessary constraints. Since the total number of animals cannot exceed 16, the first contraint is

$$x + y \leq 16.$$

"No more than 10 geese" leads to

$$x \leq 10.$$

The cost to raise x geese at \$15 per goose is $15x$ dollars, while the cost for y pigs at \$45 each is $45y$ dollars. Only \$540 is available, so

$$15x + 45y \leq 540.$$

Dividing both sides by 15 gives the equivalent inequality

$$x + 3y \leq 36.$$

The number of geese and pigs cannot be negative, so

$$x \geq 0, \quad y \geq 0.$$

The 4-H Club member wants to know the number of geese and the number of pigs that should be raised for maximum profit. Each goose produces a profit of \$6, and each pig, \$20. If z represents total profit, then

$$z = 6x + 20y$$

1 Find the corner points P and Q in Figure 7.17.

Answer:

$P = (6, 10)$

$Q = (10, 6)$

is the objective function, which is to be maximized.

We must solve the following linear programming problem.

Maximize $z = 6x + 20y$ **Objective function**

subject to: $x + y \leq 16$
$x \leq 10$
$x + 3y \leq 36$ **Constraints**
$x \geq 0, \quad y \geq 0.$

Using the methods of the previous section, graph the feasible region for the system of inequalities given by the constraints as in Figure 7.17.

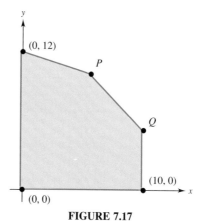

FIGURE 7.17

The corner points $(0, 12)$, $(0, 0)$, and $(10, 0)$ can be read directly from the graph. The coordinates of each of the other corner points can be found by solving a system of linear equations. **1**

Test each corner point in the objective function to find the maximum profit.

Corner Point	$z = 6x + 20y$
$(0, 12)$	$6(0) + 20(12) = 240$ (maximum)
$(6, 10)$	$6(6) + 20(10) = 236$
$(10, 6)$	$6(10) + 20(6) = 180$
$(10, 0)$	$6(10) + 20(0) = 60$
$(0, 0)$	$6(0) + 20(0) = 0$

The maximum of 240 occurs at $(0, 12)$. Thus, 12 pigs and no geese will produce a maximum profit of $240. ◀

▶**EXAMPLE 2** An office manager needs to purchase new filing cabinets. He knows that Ace cabinets cost $40 each, require 6 square feet of floor space, and hold 8 cubic feet of files. On the other hand, each Excello cabinet costs $80, requires 8 square feet of floor space, and holds 12 cubic feet. His budget permits him to spend no more than $560 on files, while the office has room for no more than 72 square feet of cabinets. The manager desires the greatest storage capacity within the limitations imposed by funds and space. How many of each type cabinet should he buy?

Let x represent the number of Ace cabinets to be bought and let y represent the number of Excello cabinets. The information given in the problem can be summarized as follows.

	Number	*Cost of Each*	*Space Required*	*Storage Capacity*
Ace	x	$ 40	6 sq ft	8 cu ft
Excello	y	$ 80	8 sq ft	12 cu ft
Maximum available		$560	72 sq ft	

Write the constraints imposed by cost and space.

$$40x + 80y \le 560 \qquad \text{Cost}$$
$$6x + 8y \le 72 \qquad \text{Floor space}$$

The number of cabinets cannot be negative, so $x \ge 0$ and $y \ge 0$. The objective function to be maximized gives the amount of storage capacity provided by some combination of Ace and Excello cabinets. From the information in the chart, the objective function is

$$\textbf{Storage space} = z = 8x + 12y.$$

In summary, the given problem has produced the following linear programming problem.

$$\text{Maximize} \qquad z = 8x + 12y$$
$$\text{subject to:} \qquad 40x + 80y \le 560$$
$$6x + 8y \le 72$$
$$x \ge 0, \quad y \ge 0.$$

A graph of the feasible region is shown in Figure 7.18. Three of the corner points can be identified from the graph as $(0, 0)$, $(0, 7)$, and $(12, 0)$. The fourth corner point, labeled Q in the figure, can be found by solving the system of equations

$$40x + 80y = 560$$
$$6x + 8y = 72.$$

2 Find the corner point labeled *Q* on the region of feasible solutions given below.

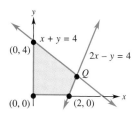

Answer: (8/3, 4/3)

3 A popular cereal combines oats and corn. At least 27 tons of the cereal is to be made. For the best flavor, the amount of corn should be at least twice the amount of oats. Corn costs $200 per ton and oats cost $300 per ton. How much of each grain should be used to minimize the cost?

(a) Make a chart to organize the information given in the problem.

(b) Write an equation for the objective function.

(c) Write four inequalities for the constraints.

Answers:

(a)

	Number of Tons	Cost/Ton
Oats	x	$300
Corn	y	200
	27	

(b) $z = 300x + 200y$

(c) $x + y \geq 27$
$\quad\quad y \geq 2x$
$\quad\quad x \geq 0$
$\quad\quad y \geq 0$

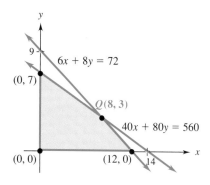

FIGURE 7.18

By solving this system, we find that Q is the point $(8, 3)$. **2**

Use the corner point theorem to find the maximum value of z.

Corner Point	Value of $z = 8x + 12y$
(0, 0)	0
(0, 7)	84
(12, 0)	96
(8, 3)	100 (maximum)

The objective function, which represents storage space, is maximized when $x = 8$ and $y = 3$. The manager should buy 8 Ace cabinets and 3 Excello cabinets. ◀ **3**

▶ **EXAMPLE 3** Certain laboratory animals must have at least 30 grams of protein and at least 20 grams of fat per feeding period. These nutrients come from food A, which costs 18¢ per unit and supplies 2 grams of protein and 4 of fat, and food B, with 6 grams of protein and 2 of fat, costing 12¢ per unit. Food B is bought under a long-term contract requiring that at least 2 units of B be used per serving. How much of each food must be bought to produce minimum cost per serving?

Let x represent the amount of food A needed, and y the amount of food B. Use the given information to produce the following table.

Food	Number of Units	Grams of Protein	Grams of Fat	Cost
A	x	2	4	18¢
B	y	6	2	12¢
Minimum required		30	20	

4 Use the information in side problem 3 above to do the following.

(a) Graph the feasible region and find the corner points.

(b) Determine the minimum value of the objective function and the point where it occurs.

(c) Is there a maximum cost? Explain your answer.

Answers:

(a)

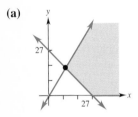

Corner points: (27, 0), (9, 18)

(b) $6300 at (9, 18)

(c) No

The linear programming problem can be stated as follows.

$$\text{Minimize} \quad z = .18x + .12y$$
$$\text{subject to:} \quad 2x + 6y \geq 30$$
$$4x + 2y \geq 20$$
$$y \geq 2$$
$$x \geq 0, y \geq 0.$$

(The constraint $y \geq 0$ is redundant because of the constraint $y \geq 2$.) A graph of the feasible region with the corner points identified is shown in Figure 7.19.

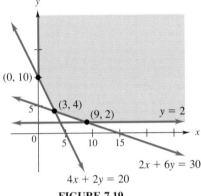

FIGURE 7.19

Find the corner points by solving the systems of equations for each pair of intersecting boundary lines. Use the corner point theorem to find the minimum value of z as shown in the chart below.

Corner Points	$z = .18x + .12y$
(0, 10)	$.18(0) + .12(10) = 1.20$
(3, 4)	$.18(3) + .12(4) = 1.02$ (minimum)
(9, 2)	$.18(9) + .12(2) = 1.86$

The minimum value of 1.02 occurs at (3, 4). Thus, 3 units of A and 4 units of B will produce a minimum cost of $1.02 per serving. ◀ **4**

The feasible region in Figure 7.19 is an unbounded feasible region—the region extends indefinitely to the upper right. With this region it would not be possible to *maximize* the objective function, because the total cost of the food could always be increased by encouraging the animals to eat more.

7.3 EXERCISES

Write the constraints in Exercises 1–5 as linear inequalities and identify all variables used. In some instances, not all of the information is needed to write the constraint. (See Examples 1–3.)

1. A canoe requires 6 hours of fabrication and a rowboat 4 hours. The fabrication department has at most 90 hours of labor available each week.

2. Doug Gilbert needs at least 2400 mg of Vitamin C per day. Each Supervite pill provides 250 mg and each Vitahealth pill provides 350 mg.

3. A candidate can afford to spend no more than $8500 on radio and TV advertising. Each radio spot costs $150 and each TV ad costs $750.

4. A hospital dietician has two meal choices, one for patients on solid food that costs $2.25, and one for patients on liquids that costs $3.75. There are a maximum of 400 patients in the hopsital.

5. Cashews costing $8 per pound are to be mixed with peanuts costing $3 per pound to obtain at least 30 pounds of mixed nuts.

6. An agricultural advisor looks at the results of Example 1 and claims that it cannot possibly be correct. He says that since the 4-H Club member could raise 16 animals, but is only raising 12, she could earn more profit by raising all 16 animals. How would you respond?

Solve the following linear programming problems. (See Examples 1–3.)

7. **Management** Audio City Corporation has warehouses in Meadville and Cambridge. It has 80 stereo systems stored in Meadville and 70 in Cambridge. Superstore orders 35 systems and ValueHouse orders 60. It costs $8 to ship a system from Meadville to Superstore and $12 to ship one to ValueHouse. It costs $10 to ship a system from Cambridge to Superstore and $13 to ship one to ValueHouse. How should the orders be filled to keep shipping costs as low as possible? What is the minimum shipping cost? (*Hint:* If x systems are shipped from Meadville to Superstore, then $35 - x$ systems are shipped from Cambridge to Superstore.)

8. **Management** A company is considering two insurance plans with the types of coverage (in thousands of dollars) and premiums per thousand dollars as shown below.

		Coverage		
		Fire/Theft	Liability	Premium
Policy	A	$10	$ 80	$50
	B	$15	$120	$40

The company wants at least $100,000 fire/theft insurance and at least $1,000,000 liability insurance from these plans. How many units should be purchased from each plan to minimize the cost of the premiums?

9. **Management** A manufacturer of refrigerators must ship at least 100 refrigerators to its two West Coast warehouses. Each warehouse holds a maximum of 100 refrigerators. Warehouse A holds 25 refrigerators already, and warehouse B has 20 on hand. It costs $12 to ship a refrigerator to warehouse A and $10 to ship one to warehouse B. How many refrigerators should be shipped to each warehouse to minimize costs? What is the minimum cost?

10. **Management** Hotnews Magazine publishes a U.S. and a Canadian edition each week. There are 30,000 subscribers in the United States and 20,000 in Canada. Other copies are sold at newsstands. Postage and shipping costs average $80 per thousand copies for the U.S. and $60 per thousand copies for Canada. Surveys show that no more than 120,000 copies of each issue can be sold (including subscriptions) and that the number of copies of the Canadian edition should not exceed twice the number of copies of the U.S. edition. The publisher can spend at most $8400 a month on postage and shipping. If the profit is $200 for each thousand copies of the U.S. edition and $150 for each thousand copies of the Canadian edition, how many copies of each version should be printed to earn as large a profit as possible? What is that profit?

11. **Management** The manufacturing process requires that oil refineries must manufacture at least 2 gallons of gasoline for every gallon of fuel oil. To meet the winter demand for fuel oil, at least 3 million gallons a day must be produced. The demand for gasoline is no more than 6.4 million gallons per day. If the refinery sells gasoline for $1.25 per gallon and fuel oil for $1 per gallon, how much of each should be produced to maximize revenue? Find the maximum revenue.

12. Management A machine shop manufactures two types of bolts. Each type of bolt requires three machines for its manufacture. The time required on each machine is shown in the table below.

		Machine Group		
		I	II	III
Bolts	Type 1	.1 min	.1 min	.1 min
	Type 2	.1 min	.4 min	.5 min

Production schedules are made up one day at a time. In a day there are 240, 720, and 160 minutes available, respectively, on these machines. Type 1 bolts sell for 10¢ and type 2 bolts for 12¢. How many of each type of bolt should be manufactured per day to maximize revenue? What is the maximum revenue?

13. Natural Science Kim Walrath has a nutritional deficiency and is told to take at least 2400 mg of iron, 2100 mg of Vitamin B-1, and 1500 mg of Vitamin B-2. One Maxivite pill contains 40 mg of iron, 10 mg of B-1, and 5 mg of B-2, and costs 6¢. One Healthovite pill provides 10 mg of iron, 15 mg of B-1, and 15 mg of B-2, and costs 8¢. What combination of Maxivite and Healthovite pills will meet the requirement at lowest cost?

14. Management A candy company has 100 kilograms of chocolate-covered nuts and 125 kilograms of chocolate-covered raisins to be sold as two different mixes. One mix will contain 1/2 nuts and 1/2 raisins and will sell for $6 per kilogram. The other mix will contain 1/3 nuts and 2/3 raisins and will sell for $4.80 per kilogram. How many kilograms of each mix should the company prepare for maximum revenue? Find the maximum revenue.

15. Management A small country can grow only two crops for export, coffee and cocoa. The country has 500,000 hectares of land available for the crops. Long-term contracts require that at least 100,000 hectares be devoted to coffee and at least 200,000 hectares to cocoa. Cocoa must be processed locally, and production bottlenecks limit cocoa to 270,000 hectares. Coffee requires two workers per hectare, with cocoa requiring five. No more than 1,750,000 people are available for working with these crops. Coffee produces a profit of $220 per hectare and cocoa a profit of $310 per hectare. How many hectares should the country devote to each crop in order to maximize profit? Find the maximum profit.

16. Management A greeting card manufacturer has 400 copies of a particular card in warehouse I and 500 copies of the same card in warehouse II. A greeting card shop in San Jose orders 350 copies of the card, and another shop in Memphis orders 300 copies. The shipping costs per copy to these shops from the two warehouses are shown in the following table.

		Destination	
		San Jose	Memphis
Warehouse	I	.25	.15
	II	.20	.30

How many copies should be shipped to each city from each warehouse to minimize shipping costs? What is the minimum cost? (Hint: use x, $350 - x$, y, and $300 - y$ as the variables.)

17. Management 60 pounds of chocolates and 100 pounds of mints are available to make up 5-pound boxes of candy. A regular box has 4 pounds of chocolates and 1 pound of mints and sells for $10. A deluxe box has 2 pounds of chocolates and 3 pounds of mints and sells for $16. How many boxes of each kind should be made to maximize revenue?

18. Management A pension fund manager decides to invest at most $40 million in U.S. Treasury Bonds paying 12% annual interest and in mutual funds paying 8% annual interest. He plans to invest at least $20 million in bonds and at least $15 million in mutual funds. How much should be invested in each to maximize annual interest? What is the maximum annual interest?

19. Natural Science A certain predator requires at least 10 units of protein and 8 units of fat per day. One prey of Species I provides 5 units of protein and 2 units of fat; one prey of Species II provides 3 units of protein and 4 units of fat. Capturing and digesting each Species II prey requires 3 units of energy, and capturing and digesting each Species I prey requires 2 units of energy. How many of each prey would meet the predator's daily food requirements with the least expenditure of energy? Are the answers reasonable? How could they be interpreted?

20. Management In a small town in South Carolina, zoning rules require that the window space (in square feet) in a house be at least one-sixth of the space used up by solid walls. The monthly cost to heat the house is 20¢ for each square foot of solid walls and 80¢ for each square foot of windows. Find the maximum total area (windows plus walls) if $160 is available to pay for heat.

21. Social Science Students at Upscale U are rquired to take at least 3 humanities and 4 science courses. The maximum allowable number of science courses is 12. Each humanities course carries 4 credits and each science course 5 credits. The total number of credits in science and humanities cannot exceed 80. Quality points for each course are assigned in the usual way: the number of credit hours times 4 for an A grade; times 3 for a B grade; times 2 for a C grade. Susan Katz expects to get B's in all her science courses. She expects to get C's in half her humanities courses, B's in one-fourth of them, and A's in the rest. Under these assumptions, how many courses of each kind should she take in order to earn the maximum possible number of quality points?

22. Social Science In Exercise 21, find Susan's grade point average (the total number of quality points divided by the total number of credit hours) at each corner point of the feasible region. Does the distribution of courses that produces the highest number of quality points also yield the highest grade point average? Is this a contradiction?

*The importance of linear programming is shown by the inclusion of linear programming problems on most qualification examinations for Certified Public Accountants. Exercises 23–25 are reprinted from one such examination.**

The Random Company manufactures two products, Zeta and Beta. Each product must pass through two processing operations. All materials are introduced at the start of Process No. 1. There are no work-in-process inventories. Random may produce either one product exclusively or various combinations of both products subject to the following constraints.

---

*Material from Uniform CPA Examinations and Unofficial Answers, copyright © 1973, 1974, 1975 by the American Institute of Certified Public Accountants, Inc., is reprinted with permission.

	Process No. I	Process No. 2	Contribution Margin Per Unit
Hours required to produce 1 unit of:			
Zeta	1 hour	1 hour	$4.00
Beta	2 hours	3 hours	5.25
Total capacity in hours per day	1000 hours	1275 hours	

A shortage of technical labor has limited Beta production to 400 units per day. There are no constraints on the production of Zeta other than the hour constraints in the above schedule. Assume that all the relationships between capacity and production are linear.

23. Given the objective to maximize total contribution margin, what is the production constraint for Process No. 1?
(a) Zeta + Beta ≤ 1000
(b) Zeta + 2 Beta ≤ 1000
(c) Zeta + Beta ≥ 1000
(d) Zeta + 2 Beta ≥ 1000

24. Given the objective to maximize total contribution margin, what is the labor constraint for production of Beta?
(a) Beta ≤ 400   **(b)** Beta ≥ 400
(c) Beta ≤ 425   **(d)** Beta ≥ 425

25. What is the objective function of the data presented?
(a) Zeta + 2 Beta = $9.25
(b) $4.00 Zeta + 3($5.25) Beta = total contribution margin
(c) $4.00 Zeta + $5.25 Beta = total contribution margin
(d) 2($4.00) Zeta + 3($5.25) Beta = total contribution margin

7.4 THE SIMPLEX METHOD: MAXIMIZATION

For linear programming problems with more than two variables or with two variables and many constraints, the graphical method is usually too inefficient, so the **simplex method** is used. The simplex method, which is introduced here, was developed for the U.S. Air Force by George B. Danzig in 1947. It was used successfully during the Berlin airlift in 1948–49 to maximize the amount of cargo delivered under very severe constraints and is widely used today in a variety of industries.

Because the simplex method is used for problems with many variables, it usually is not convenient to use letters such as x, y, z, or w as variable names. Instead, the symbols x_1 (read "x-sub-one"), x_2, x_3, and so on, are used. In the simplex method, all constraints must be expressed in the linear form

$$a_1 x_1 + a_2 x_2 + a_3 x_3 + \cdots \leq b,$$

where x_1, x_2, x_3, . . . are variables, a_1, a_2, . . . , a_n are coefficients, and b is a constant.

We first discuss the simplex method for linear programming problems in *standard maximum form*.

Standard Maximum Form

A linear programming problem is in **standard maximum form** if

1. the objective function is to be maximized;
2. all variables are nonnegative ($x_i \geq 0$, $i = 1, 2, 3, . . .$);
3. all constraints involve $\leq$;
4. the constants on the right side in the constraints are all nonnegative ($b \geq 0$).

Problems that do not meet all of these conditions are considered in Sections 7.6 and 7.7.

The "mechanics" of the simplex method are demonstrated in Examples 1–5. Although the procedures to be followed will be made clear, as will the fact that they result in an optimal solution, the reasons why these procedures are used may not be immediately apparent. Examples 6 and 7 will supply these reasons and explain the connection between the simplex method and the graphical method used in Section 7.3.

SETTING UP THE PROBLEM   The first step is to convert each constraint, a linear inequality, into a linear equation. This is done by adding a nonnegative variable, called a **slack variable,** to each constraint. For example, convert the inequality $x_1 + x_2 \leq 10$ into an equation by adding the slack variable x_3, to get

$$x_1 + x_2 + x_3 = 10, \quad \text{where } x_3 \geq 0.$$

The inequality $x_1 + x_2 \leq 10$ says that the sum $x_1 + x_2$ is less than or perhaps equal to 10. The variable x_3 "takes up any slack" and represents the amount by

1 Rewrite the following set of constraints as equations by adding nonnegative slack variables.

$$x_1 + x_2 + x_3 \leq 12$$
$$2x_1 + 4x_2 \qquad \leq 15$$
$$x_2 + 3x_3 \leq 10$$

Answer:

$$x_1 + x_2 + x_3 + x_4 = 12$$
$$2x_1 + 4x_2 \qquad + x_5 = 15$$
$$x_2 + 3x_3 + x_6 = 10$$

2 Set up the initial simplex tableau for the following linear programming problem:

Maximize $\quad z = 2x_1 + 3x_2$
subject to: $\quad x_1 + 2x_2 \leq 85$
$$2x_1 + x_2 \leq 92$$
$$x_1 + 4x_2 \leq 104$$
with $x_1 \geq 0, \quad x_2 \geq 0.$

Answer:

$$
\begin{array}{cccccc}
x_1 & x_2 & x_3 & x_4 & x_5 & z
\end{array}
$$
$$
\left[
\begin{array}{cccccc|c}
1 & 2 & 1 & 0 & 0 & 0 & 85 \\
2 & 1 & 0 & 1 & 0 & 0 & 92 \\
1 & 4 & 0 & 0 & 1 & 0 & 104 \\
\hline
-2 & -3 & 0 & 0 & 0 & 1 & 0
\end{array}
\right]
$$

which $x_1 + x_2$ fails to equal 10. For example, if $x_1 + x_2$ equals 8, then x_3 is 2. If $x_1 + x_2 = 10$, the value of x_3 is 0.

Caution A different slack variable must be used for each constraint.

▶**EXAMPLE 1** Restate the following linear programming problem by introducing slack variables.

$$\text{Maximize} \qquad z = 2x_1 + 3x_2 + x_3$$
$$\text{subject to} \qquad x_1 + x_2 + 4x_3 \leq 100$$
$$x_1 + 2x_2 + x_3 \leq 150$$
$$3x_1 + 2x_2 + x_3 \leq 320$$
$$\text{with} \quad x_1 \geq 0, x_2 \geq 0, x_3 \geq 0.$$

Rewrite the three constraints as equations by introducing nonnegative slack variables x_4, x_5, and x_6, one for each constraint. Then the problem can be restated as

$$\text{Maximize} \qquad z = 2x_1 + 3x_2 + x_3$$
$$\text{subject to} \qquad x_1 + x_2 + 4x_3 + x_4 \qquad\qquad = 100$$
$$x_1 + 2x_2 + x_3 \qquad + x_5 \qquad = 150$$
$$3x_1 + 2x_2 + x_3 \qquad\qquad + x_6 = 320$$
$$\text{with} \quad x_1 \geq 0, x_2 \geq 0, x_3 \geq 0, x_4 \geq 0, x_5 \geq 0, x_6 \geq 0. \quad ◀ \quad \boxed{1}$$

Adding slack variables to the constraints converts a linear programming problem into a system of linear equations. These equations should have all variables on the left of the equals sign and all constants on the right. All the equations of Example 1 satisfy this condition except for the objective function, $z = 2x_1 + 3x_2 + x_3$, which may be written with all variables on the left as

$$-2x_1 - 3x_2 - x_3 + z = 0.$$

Now the equations of Example 1 can be written as the following augmented matrix.

$$
\begin{array}{ccccccc}
x_1 & x_2 & x_3 & x_4 & x_5 & x_6 & z
\end{array}
$$
$$
\left[
\begin{array}{ccccccc|c}
1 & 1 & 4 & 1 & 0 & 0 & 0 & 100 \\
1 & 2 & 1 & 0 & 1 & 0 & 0 & 150 \\
3 & 2 & 1 & 0 & 0 & 1 & 0 & 320 \\
\hline
-2 & -3 & -1 & 0 & 0 & 0 & 1 & 0
\end{array}
\right]
$$
$$\underbrace{\qquad\qquad\qquad\qquad}_{\text{indicators}}$$

This matrix is the initial **simplex tableau.** Except for the last entry on the right end, the numbers in the bottom row of a simplex tableau are called **indicators.** **2**

This simplex tableau represents a system of 4 linear equations in 7 variables. Since there are more variables than equations, the system is dependent and has infinitely many solutions. Our goal is to find a solution in which all the variables are nonnegative and z is as large as possible. This will be done by using row operations to replace the given system by an equivalent one in which certain variables are eliminated from some of the equations. The process will be repeated until the optimum solution can be read from the matrix, as explained below.

SELECTING THE PIVOT Recall how row operations are used to eliminate variables in the Gauss-Jordan method. A particular nonzero entry in the matrix is chosen and changed to a 1; then all other entries in that column are changed to zeros. A similar process is used in the simplex method. The chosen entry is called the **pivot.** The procedure for selecting the appropriate pivot in the simplex method is explained in the next example. The reason why this procedure is used is discussed in Example 7.

▶**EXAMPLE 2** Determine the pivot in the simplex tableau for the problem in Example 1.

Look at the indicators (the last row of the tableau) and choose the most negative one:

$$
\begin{array}{ccccccc}
x_1 & x_2 & x_3 & x_4 & x_5 & x_6 & z \\
\end{array}
$$

$$
\left[
\begin{array}{ccccccc|c}
1 & 1 & 4 & 1 & 0 & 0 & 0 & 100 \\
1 & 2 & 1 & 0 & 1 & 0 & 0 & 150 \\
3 & 2 & 1 & 0 & 0 & 1 & 0 & 320 \\
\hline
-2 & -3 & -1 & 0 & 0 & 0 & 1 & 0 \\
\end{array}
\right]
$$

most negative indicator

The most negative indicator identifies the variable that is to be eliminated from all but one of the equations (rows), in this case x_2. The column containing the most negative indicator is called the **pivot column.** Now for each *positive* entry in the pivot column, divide the number in the far right column of the same row by the positive number in the pivot column:

$$
\begin{array}{ccccccc}
x_1 & x_2 & x_3 & x_4 & x_5 & x_6 & z \\
\end{array}
$$

$$
\left[
\begin{array}{ccccccc|c}
1 & 1 & 4 & 1 & 0 & 0 & 0 & 100 \\
1 & 2 & 1 & 0 & 1 & 0 & 0 & 150 \\
3 & 2 & 1 & 0 & 0 & 1 & 0 & 320 \\
\hline
-2 & -3 & -1 & 0 & 0 & 0 & 1 & 0 \\
\end{array}
\right]
$$

quotients
$100/1 = 100$
$150/2 = 75 \leftarrow$ smallest
$320/2 = 160$

The row with the smallest quotient (in this case, the second row) is called the **pivot row.** The entry in the pivot row and pivot column is the pivot:

3 Find the pivot for the following tableau.

$$\begin{bmatrix} x_1 & x_2 & x_3 & x_4 & x_5 & z & \\ 0 & 1 & 1 & 0 & 0 & 0 & 50 \\ -2 & 3 & 0 & 1 & 0 & 0 & 78 \\ 2 & 4 & 0 & 0 & 1 & 0 & 65 \\ \hline -5 & -3 & 0 & 0 & 0 & 1 & 0 \end{bmatrix}$$

Answer:
2 (in first column)

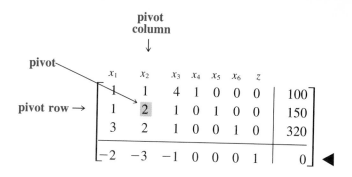

Caution   In some simplex tableaus the pivot column may contain zeros or negative entries. Only the positive entries in the pivot column should be used to form the quotients and determine the pivot row. If there are no positive entries in the pivot column (so that a pivot row cannot be chosen), then no maximum solution exists.   **3**

PIVOTING   Once the pivot has been selected, row operations are used to replace the initial simplex tableau by another simplex tableau in which the pivot column variable is eliminated from all but one of the equations. Since this new tableau is obtained by row operations, it represents an equivalent system of equations (that is, a system with the same solutions as the original system). This process, which is called **pivoting,** is explained in the next example.

▶**EXAMPLE 3**   Use the indicated pivot, 2, to perform the pivoting on the simplex tableau of Example 2:

$$\begin{bmatrix} x_1 & x_2 & x_3 & x_4 & x_5 & x_6 & z & \\ 1 & 1 & 4 & 1 & 0 & 0 & 0 & 100 \\ 1 & 2 & 1 & 0 & 1 & 0 & 0 & 150 \\ 3 & 2 & 1 & 0 & 0 & 1 & 0 & 320 \\ \hline -2 & -3 & -1 & 0 & 0 & 0 & 1 & 0 \end{bmatrix}.$$

Start by multiplying each entry of row two by $1/2$ in order to change the pivot to 1.

$$\begin{bmatrix} x_1 & x_2 & x_3 & x_4 & x_5 & x_6 & z & \\ 1 & 1 & 4 & 1 & 0 & 0 & 0 & 100 \\ \frac{1}{2} & 1 & \frac{1}{2} & 0 & \frac{1}{2} & 0 & 0 & 75 \\ 3 & 2 & 1 & 0 & 0 & 1 & 0 & 320 \\ \hline -2 & -3 & -1 & 0 & 0 & 0 & 1 & 0 \end{bmatrix} \quad \frac{1}{2}R_2$$

4 For the simplex tableau below,

(a) Find the pivot.

(b) Perform the pivoting and write the new tableau.

$$\begin{bmatrix} x_1 & x_2 & x_3 & x_4 & x_5 & z & \\ 1 & 2 & 6 & 1 & 0 & 0 & 16 \\ 1 & 3 & 0 & 0 & 1 & 0 & 25 \\ \hline -1 & -4 & -3 & 0 & 0 & 1 & 0 \end{bmatrix}$$

Answers:

(a) 2

(b)

$$\begin{bmatrix} x_1 & x_2 & x_3 & x_4 & x_5 & z & \\ \frac{1}{2} & 1 & 3 & \frac{1}{2} & 0 & 0 & 8 \\ -\frac{1}{2} & 0 & -9 & -\frac{3}{2} & 1 & 0 & 1 \\ \hline 1 & 0 & 9 & 2 & 0 & 1 & 32 \end{bmatrix}$$

Now use row operations to make the entry in row one, column two a 0.

$$\begin{bmatrix} x_1 & x_2 & x_3 & x_4 & x_5 & x_6 & z & \\ \frac{1}{2} & 0 & \frac{7}{2} & 1 & -\frac{1}{2} & 0 & 0 & 25 \\ \frac{1}{2} & 1 & \frac{1}{2} & 0 & \frac{1}{2} & 0 & 0 & 75 \\ 3 & 2 & 1 & 0 & 0 & 1 & 0 & 320 \\ \hline -2 & -3 & -1 & 0 & 0 & 0 & 1 & 0 \end{bmatrix} \quad -R_2 + R_1$$

Change the 2 in row three, column two to a 0 by a similar process.

$$\begin{bmatrix} x_1 & x_2 & x_3 & x_4 & x_5 & x_6 & z & \\ \frac{1}{2} & 0 & \frac{7}{2} & 1 & -\frac{1}{2} & 0 & 0 & 25 \\ \frac{1}{2} & 1 & \frac{1}{2} & 0 & \frac{1}{2} & 0 & 0 & 75 \\ 2 & 0 & 0 & 0 & -1 & 1 & 0 & 170 \\ \hline -2 & -3 & -1 & 0 & 0 & 0 & 1 & 0 \end{bmatrix} \quad -2R_2 + R_3$$

Finally, add 3 times row 2 to the last row in order to change the indicator -3 to 0.

$$\begin{bmatrix} x_1 & x_2 & x_3 & x_4 & x_5 & x_6 & z & \\ \frac{1}{2} & 0 & \frac{7}{2} & 1 & -\frac{1}{2} & 0 & 0 & 25 \\ \frac{1}{2} & 1 & \frac{1}{2} & 0 & \frac{1}{2} & 0 & 0 & 75 \\ 2 & 0 & 0 & 0 & -1 & 1 & 0 & 170 \\ \hline -\frac{1}{2} & 0 & \frac{1}{2} & 0 & \frac{3}{2} & 0 & 1 & 225 \end{bmatrix} \quad 3R_2 + R_4$$

The pivoting is now complete because the pivot column variable x_2 has been eliminated from all equations except the one represented by the pivot row. The initial simplex tableau has been replaced by a new simplex tableau, which represents an equivalent system of equations. ◀ **4**

Caution During pivoting, do not interchange rows of the matrix. Make the pivot entry 1 by multiplying the pivot row by an appropriate constant, as in Example 3.

When at least one of the indicators in the last row of a simplex tableau is negative (as is the case with the tableau obtained in Example 3), the simplex method requires that a new pivot be selected and the pivoting be performed again. This procedure is repeated until a simplex tableau with no negative indicators in the last row is obtained or a tableau is reached in which no pivot row can be chosen.

▶**EXAMPLE 4**   In the simplex tableau obtained in Example 3, select a new pivot and perform the pivoting.

First, locate the pivot column by finding the most negative indicator in the last row. Then locate the pivot row by computing the necessary quotients and finding the smallest one, as shown here.

$$
\begin{array}{c}
\text{pivot row} \rightarrow
\end{array}
\begin{array}{ccccccc}
x_1 & x_2 & x_3 & x_4 & x_5 & x_6 & z \\
\end{array}
$$

	x_1	x_2	x_3	x_4	x_5	x_6	z			Quotients
pivot row →	$\frac{1}{2}$	0	$\frac{7}{2}$	1	$-\frac{1}{2}$	0	0		25	$\dfrac{25}{1/2} = 50$ smallest
	$\frac{1}{2}$	1	$\frac{1}{2}$	0	$\frac{1}{2}$	0	0		75	$\dfrac{75}{1/2} = 150$
	2	0	0	0	-1	1	0		170	$170/2 = 85$
	$-\frac{1}{2}$	0	$\frac{1}{2}$	0	$\frac{3}{2}$	0	1		225	

pivot column

So the pivot is the number $1/2$ in row one, column one. Begin the pivoting by multiplying every entry in row one by 2. Then continue as indicated below, to obtain the following simplex tableau.

x_1	x_2	x_3	x_4	x_5	x_6	z			
1	0	7	2	-1	0	0		50	$2R_1$
0	1	-3	-1	1	0	0		50	$-\frac{1}{2}R_1 + R_2$
0	0	-14	-4	1	1	0		70	$-2R_1 + R_3$
0	0	4	1	1	0	1		250	$\frac{1}{2}R_1 + R_4$

Since there are no negative indicators in the last row, no further pivoting is necessary and we call this the **final simplex tableau.** ◀

READING THE SOLUTION   The next example shows how to read an optimal solution of the original linear programming problem from the final simplex tableau.

▶**EXAMPLE 5**   Solve the linear programming problem introduced in Example 1.

Look at the final simplex tableau for this problem, which was obtained in Example 4:

x_1	x_2	x_3	x_4	x_5	x_6	z		
1	0	7	2	-1	0	0		50
0	1	-3	-1	1	0	0		50
0	0	-14	-4	1	1	0		70
0	0	4	1	1	0	1		250

5 A linear programming problem with slack variables x_4 and x_5 has the final simplex tableau shown below. What is the optimal solution?

$$\begin{array}{ccccccc} x_1 & x_2 & x_3 & x_4 & x_5 & z \\ \left[\begin{array}{cccccc|c} 0 & 3 & 1 & 5 & 2 & 0 & 9 \\ 1 & -2 & 0 & 4 & 1 & 0 & 6 \\ 0 & 5 & 0 & 1 & 0 & 1 & 21 \end{array}\right] \end{array}$$

Answer:
$z = 21$ when $x_1 = 6$, $x_2 = 0$, $x_3 = 9$.

The last row of this matrix represents the equation

$$4x_3 + x_4 + x_5 + z = 250, \quad \text{or equivalently,} \quad z = 250 - 4x_3 - x_4 - x_5.$$

If x_3, x_4, and x_5 are all 0, then the value of z is 250. If any one of x_3, x_4, or x_5 is positive, then z will have a smaller value than 250 (why?). Consequently, since we want a solution for this system in which all the variables are nonnegative and z is as large as possible, we must have

$$x_3 = 0, \quad x_4 = 0, \quad x_5 = 0.$$

When these values are substituted in the first equation (represented by the first row of the final simplex tableau), the result is

$$x_1 + 7 \cdot 0 + 2 \cdot 0 - 1 \cdot 0 = 50, \quad \text{that is,} \quad x_1 = 50.$$

Similarly, substituting 0 for x_3, x_4, and x_5 in the last three equations represented by the final simplex tableau shows that

$$x_2 = 50, \quad x_6 = 70, \quad z = 250.$$

Therefore, the maximum value of $z = 2x_1 + 3x_2 + x_3$ occurs when

$$x_1 = 50, \quad x_2 = 50, \quad x_3 = 0,$$

in which case $z = 2 \cdot 50 + 3 \cdot 50 + 0 = 250$. (The values of the slack variables are irrelevant in stating the solution of the original problem.) ◀ **5**

In any simplex tableau, some columns look like columns of an identity matrix (one entry is 1, the rest are 0). The variables corresponding to these columns are called **basic variables** and the variables corresponding to the other columns **nonbasic variables.** In the tableau of Example 5, for instance, the basic variables are x_1, x_2, x_6, and z (shown in color below), and the nonbasic variables are x_3, x_4, x_5.

$$\begin{array}{ccccccc} x_1 & x_2 & x_3 & x_4 & x_5 & x_6 & z \\ \left[\begin{array}{ccccccc|c} 1 & 0 & 7 & 2 & -1 & 0 & 0 & 50 \\ 0 & 1 & -3 & -1 & 1 & 0 & 0 & 50 \\ 0 & 0 & -14 & -4 & 1 & 1 & 0 & 70 \\ \hline 0 & 0 & 4 & 1 & 1 & 0 & 1 & 250 \end{array}\right] \end{array}$$

The optimal solution in Example 5 was obtained from the final simplex tableau by setting the nonbasic variables equal to 0 and solving for the basic variables. Furthermore, the values of the basic variables are easy to read off the matrix: find the 1 in the column representing a basic variable; the last entry in that row is the value of that basic variable in the optimal solution. In particular, *the entry in the lower right hand corner of the final simplex tableau is the maximum value of z.*

6 A linear programming problem has the initial tableau given below. Use the simplex method to solve the problem.

$$
\begin{array}{ccccc|c}
x_1 & x_2 & x_3 & x_4 & z & \\
\hline
1 & 1 & 1 & 0 & 0 & 40 \\
2 & 1 & 0 & 1 & 0 & 24 \\
\hline
-300 & -200 & 0 & 0 & 1 & 0
\end{array}
$$

Answer:
$x_1 = 0$, $x_2 = 24$, $x_3 = 16$, $x_4 = 0$, $z = 4800$

Caution   If there are two identical columns in a tableau, each of which is a column in an identity matrix, only one of the variables corresponding to these columns can be a basic variable. The other is treated as a nonbasic variable. You may choose either one to be the basic variable, unless one of them is z, in which case z must be the basic variable.

The steps involved in solving a standard maximum linear programming problem by the simplex method have been illustrated in Examples 1–5 and are summarized here.

Simplex Method

1. Determine the objective function.
2. Write all necessary constraints.
3. Convert each constraint into an equation by adding slack variables.
4. Set up the initial simplex tableau.
5. Locate the most negative indicator. If there are two such indicators, choose one. This indicator determines the pivot column.
6. Use the positive entries in the pivot column to form the quotients necessary for determining the pivot. If there are no positive entries in the pivot column, no maximum solution exists. If two quotients are equally the smallest, let either determine the pivot.*
7. Multiply every entry in the pivot row by the reciprocal of the pivot to change the pivot to 1. Then use row operations to change all other entries in the pivot column to 0 by adding suitable multiples of the pivot row to the other rows.
8. If the indicators are all positive or 0, this is the final tableau. If not, go back to Step 5 above the repeat the process until a tableau with no negative indicators is obtained.**
9. Determine the basic and nonbasic variables and read the solution from the final tableau. The maximum value of the objective function is the number in the lower right-hand corner of the final tableau.

The solution found by the simplex method may not be unique, especially when choices are possible in steps 5, 6, or 9. There may be other solutions that produce the same maximum value of the objective function. (See Exercises 37 and 38.) **6**

---

*It may be that the first choice of a pivot does not produce a solution. In that case try the other choice.

**Some special circumstances are noted at the end of Section 7.7.

GEOMETRIC INTERPRETATION OF THE SIMPLEX METHOD   Although it may not be immediately apparent, the simplex method is based on the same geometrical considerations as the graphical method. This can be seen by looking at a problem that can be readily solved by both methods.

▶ **EXAMPLE 6**   In Example 2 of Section 7.3 the following problem was solved graphically (using x, y instead of x_1, x_2):

$$\begin{aligned} \text{Maximize} \quad & z = 8x_1 + 12x_2 \\ \text{subject to:} \quad & 40x_1 + 80x_2 \le 560 \\ & 6x_1 + 8x_2 \le 72 \\ & x_1 \ge 0, \quad x_2 \ge 0. \end{aligned}$$

Graphing the feasible region and evaluating z at each corner point shows that the maximum value of z occurs at $(8, 3)$.

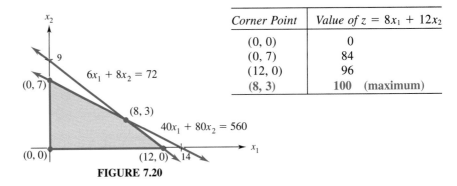

Corner Point	Value of $z = 8x_1 + 12x_2$
$(0, 0)$	0
$(0, 7)$	84
$(12, 0)$	96
$(8, 3)$	100 (maximum)

FIGURE 7.20

To solve the same problem by the simplex method, add a slack variable to each constraint.

$$\begin{aligned} 40x_1 + 80x_2 + x_3 \quad\quad\; &= 560 \\ 6x_1 + \; 8x_2 \quad\quad + x_4 &= \; 72 \end{aligned}$$

Then write the initial simplex tableau.

$$\begin{array}{ccccc} x_1 & x_2 & x_3 & x_4 & z \\ \left[\begin{array}{ccccc|c} 40 & 80 & 1 & 0 & 0 & 560 \\ 6 & 8 & 0 & 1 & 0 & 72 \\ \hline -8 & -12 & 0 & 0 & 1 & 0 \end{array}\right] \end{array}$$

In this tableau the basic variables are x_3, x_4, and z (why?). By setting the nonbasic variables (namely, x_1 and x_2) equal to 0 and solving for the basic varibles, we obtain the following solution (which will be called a **basic feasible solution**):

$$x_1 = 0, \quad x_2 = 0, \quad x_3 = 560, \quad x_4 = 72, \quad z = 0.$$

Since $x_1 = 0$ and $x_2 = 0$, this solution corresponds to the corner point at the origin in the graphical solution (Figure 7.20). The value $z = 0$ at the origin is obviously not maximal and pivoting in the simplex method is designed to improve it.

The most negative indicator in the initial tableau is -12 and the necessary quotients are

$$\frac{560}{80} = 7 \quad \text{and} \quad \frac{72}{8} = 9.$$

The smaller quotient is 7, giving 80 as the pivot. Performing the pivoting leads to this tableau:

$$
\begin{array}{ccccc}
x_1 & x_2 & x_3 & x_4 & z
\end{array}
$$
$$
\left[
\begin{array}{ccccc|c}
\frac{1}{2} & 1 & \frac{1}{80} & 0 & 0 & 7 \\
2 & 0 & -\frac{1}{10} & 1 & 0 & 16 \\
\hline
-2 & 0 & \frac{3}{20} & 0 & 1 & 84
\end{array}
\right]
\begin{array}{l}
\frac{1}{80}R_1 \\[4pt]
-8R_1 + R_2 \\[4pt]
12R_1 + R_3
\end{array}
$$

The basic variables here are x_2, x_4, and z and the basic feasible solution (found by setting the nonbasic variables equal to 0 and solving for the basic variables) is

$$x_1 = 0, \quad x_2 = 7, \quad x_3 = 0, \quad x_4 = 16, \quad z = 84,$$

which corresponds to the corner point $(0, 7)$ in Figure 7.20. Note that the new value of the pivot variable x_2 is precisely the smallest quotient, 7, that was used to select the pivot row. Although this value of z is better, further improvement is possible.

Now the most negative indicator is -2 and the quotients are

$$\frac{7}{1/2} = 14 \quad \text{and} \quad \frac{16}{2} = 8.$$

Since 8 is smaller, the pivot is the number 2 in row two, column one. Pivoting produces the final tableau.

$$
\begin{array}{ccccc}
x_1 & x_2 & x_3 & x_4 & z
\end{array}
$$
$$
\left[
\begin{array}{ccccc|c}
0 & 1 & \frac{3}{80} & -\frac{1}{4} & 0 & 3 \\
1 & 0 & -\frac{1}{20} & \frac{1}{2} & 0 & 8 \\
\hline
0 & 0 & \frac{1}{20} & 1 & 1 & 100
\end{array}
\right]
\begin{array}{l}
-\frac{1}{2}R_2 + R_1 \\[4pt]
\frac{1}{2}R_2 \\[4pt]
2R_2 + R_3
\end{array}
$$

Here the basic feasible solution is

$$x_1 = 8, \quad x_2 = 3, \quad x_3 = 0, \quad x_4 = 0, \quad z = 100,$$

which corresponds to the corner point (8, 3) in Figure 7.20. Once again, the new value of the pivot variable x_1 is the smallest quotient, 8, that was used to select the pivot. From the graphical method we know that this solution provides the maximum value of the objective function. This fact can also be seen algebraically by using an algebraic argument similar to the one used in Example 5. Thus there is no need to move to another corner point and the simplex method ends. ◀

As illustrated in Example 6, the basic feasible solution obtained from a simplex tableau corresponds to a corner point of the feasible region. Pivoting, which replaces one tableau with another, is a systematic way of moving from one corner point to another, each time improving the value of the objective function. The simplex method ends when a corner point that produces the maximum value of the objective function is reached (or when it becomes clear that the problem has no maximum solution).

When there are three or more variables in a linear programming problem, it may be difficult or impossible to draw a picture, but it can be proved that the optimal value of the objective function occurs at a basic feasible solution (corresponding to a corner point in the two variable case). The simplex method provides a means of moving from one basic feasible solution to another until one that produces the optimal value of the objective function is reached.

EXPLANATION OF PIVOTING   The rules for selecting the pivot in the simplex method can be understood by examining how the first pivot was chosen in Example 6.

▶**EXAMPLE 7**   The initial simplex tableau of Example 6 provides a basic feasible solution with $x_1 = 0$, $x_2 = 0$.

$$
\begin{array}{ccccc}
x_1 & x_2 & x_3 & x_4 & z
\end{array}
$$
$$
\left[
\begin{array}{ccccc|c}
40 & 80 & 1 & 0 & 0 & 560 \\
6 & 8 & 0 & 1 & 0 & 72 \\
\hline
-8 & -12 & 0 & 0 & 1 & 0
\end{array}
\right]
$$

This solution certainly does not give a maximum value for the objective function $z = 8x_1 + 12x_2$. Since x_2 has the largest coefficient, z will be increased most if x_2 is increased. In other words, the most negative indicator in the tableau (which corresponds to the largest coefficient in the objective function) identifies the variable that will provide the greatest change in the value of z.

To determine how much x_2 can be increased without leaving the feasible region, look at the first two equations

$$
\begin{aligned}
40x_1 + 80x_2 + x_3 \quad\quad\ &= 560 \\
6x_1 + 8x_2 \quad\quad + x_4 &= 72
\end{aligned}
$$

and solve for the basic variables x_3 and x_4:

$$x_3 = 560 - 40x_1 - 80x_2$$
$$x_4 = 72 - 6x_1 - 8x_2$$

Now x_2 is to be increased while x_1 is to keep the value 0. Hence

$$x_3 = 560 - 80x_2$$
$$x_4 = 72 - 8x_2$$

Since $x_3 \geq 0$ and $x_4 \geq 0$, we must have:

$$0 \leq 560 - 80x_2 \quad \text{and} \quad 0 \leq 72 - 8x_2$$
$$80x_2 \leq 560 \qquad\qquad\qquad 8x_2 \leq 72$$
$$x_2 \leq \frac{560}{80} = 7 \qquad\qquad x_2 \leq \frac{72}{8} = 9$$

The right sides of these last inequalities are the quotients used to select the pivot row. Since x_2 must satisfy both inequalities, x_2 can be at most 7. In other words, the smallest quotient formed from positive entries in the pivot column identifies the value of x_2 that produces the largest change in z while remaining in the feasible region. By pivoting with the pivot determined in this way, we obtain the second tableau and a basic feasible solution in which $x_2 = 7$, as was shown in Example 6. ◄

An analysis similar to that in Example 7 applies to each occurrence of pivoting in the simplex method. The idea is to improve the value of the objective function by adjusting one variable at a time. The most negative indicator identifies the variable that will account for the largest increase in z. The smallest quotient determines the largest value of that variable that will produce a feasible solution. Pivoting leads to a solution in which the selected variable has this largest value.

The simplex method is easily implemented on a computer and even on some graphing calculators. A computer is essential for its use in any situation where there are a large number of variables and constraints (and hence an enormous number of corner points to check). A computer disk that includes the simplex method is available to users of this text (see the preface).

7.4 EXERCISES

In Exercises 1–4, (a) determine the number of slack variables needed; (b) name them; (c) use the slack variables to convert each constraint into a linear equation. (See Example 1.)

1. Maximize $z = 32x_1 + 9x_2$
subject to: $4x_1 + 2x_2 \leq 20$
$5x_1 + x_2 \leq 50$
$2x_1 + 3x_2 \leq 25$
$x_1 \geq 0, \quad x_2 \geq 0.$

2. Maximize $z = 3.7x_1 + 4.3x_2$
subject to: $2.4x_1 + 1.5x_2 \leq 10$
$1.7x_1 + 1.9x_2 \leq 15$
$x_1 \geq 0, \quad x_2 \geq 0.$

3. Maximize $z = 8x_1 + 3x_2 + x_3$
   subject to:
   $$3x_1 - x_2 + 4x_3 \leq 95$$
   $$7x_1 + 6x_2 + 8x_3 \leq 118$$
   $$4x_1 + 5x_2 + 10x_3 \leq 220$$
   $$x_1 \geq 0, \quad x_2 \geq 0, \quad x_3 \geq 0.$$

4. Maximize $z = 12x_1 + 15x_2 + 10x_3$
   subject to:
   $$2x_1 + 2x_2 + x_3 \leq 8$$
   $$x_1 + 4x_2 + 3x_3 \leq 12$$
   $$x_1 \geq 0, \quad x_2 \geq 0, \quad x_3 \geq 0.$$

Introduce slack variables as necessary and then write the initial simplex tableau for each of these linear programming problems.

5. Maximize $z = 5x_1 + x_2$
   subject to:
   $$2x_1 + 3x_2 \leq 6$$
   $$4x_1 + x_2 \leq 6$$
   $$5x_1 + 2x_2 \leq 15$$
   $$x_1 \geq 0, \quad x_2 \geq 0.$$

6. Maximize $z = 5x_1 + 3x_2 + 4x_3$
   subject to:
   $$4x_1 + 3x_2 + 2x_3 \leq 60$$
   $$3x_1 + 4x_2 + x_3 \leq 24$$
   $$x_1 \geq 0, \quad x_2 \geq 0, \quad x_3 \geq 0.$$

7. Maximize $z = x_1 + 5x_2 + 10x_3$
   subject to:
   $$x_1 + 2x_2 + 3x_3 \leq 10$$
   $$2x_1 + x_2 + x_3 \leq 8$$
   $$3x_1 + 2x_3 \leq 6$$
   $$x_1 \geq 0, \quad x_2 \geq 0, \quad x_3 \geq 0.$$

8. Maximize $z = 5x_1 - x_2 + 3x_3$
   subject to:
   $$3x_1 + 2x_2 + x_3 \leq 36$$
   $$x_1 + 4x_2 + x_3 \leq 24$$
   $$x_1 - x_2 - x_3 \leq 32$$
   $$x_1 \geq 0, \quad x_2 \geq 0, \quad x_3 \geq 0.$$

Find the pivot in each of the following simplex tableaus. (See Example 2.)

9.

x_1	x_2	x_3	x_4	x_5	z	
2	2	0	3	1	0	15
3	4	1	6	0	0	20
-2	-1	0	1	0	1	10

10.

x_1	x_2	x_3	x_4	x_5	z	
0	2	1	1	3	0	5
1	-5	0	1	2	0	8
0	-2	0	-1	1	1	10

11.

x_1	x_2	x_3	x_4	x_5	x_6	z	
6	2	1	3	0	0	0	8
0	2	0	1	0	1	0	7
2	1	0	3	1	0	0	6
-3	-2	0	2	0	0	1	12

12.

x_1	x_2	x_3	x_4	x_5	x_6	z	
0	2	0	1	2	2	0	3
0	3	1	0	1	2	0	2
1	4	0	0	3	5	0	5
0	-4	0	0	4	-3	1	20

In Exercises 13–16, use the indicated entry as the pivot and perform the pivoting. (See Examples 3 and 4.)

13.

x_1	x_2	x_3	x_4	x_5	z	
1	2	4	1	0	0	56
2	**2**	1	0	1	0	40
-1	-3	-2	0	0	1	0

14.

x_1	x_2	x_3	x_4	x_5	x_6	z	
2	2	**1**	1	0	0	0	12
1	2	3	0	1	0	0	45
3	1	1	0	0	1	0	20
-2	-1	-3	0	0	0	1	0

15.

x_1	x_2	x_3	x_4	x_5	x_6	z	
1	1	1	1	0	0	0	60
3	1	**2**	0	1	0	0	100
1	2	3	0	0	1	0	200
-1	-1	-2	0	0	0	1	0

16.

x_1	x_2	x_3	x_4	x_5	x_6	z	
4	2	3	1	0	0	0	22
2	2	**5**	0	1	0	0	28
1	3	2	0	0	1	0	45
-3	-2	-4	0	0	0	1	0

*For each simplex tableau in Exercises 17–20, (**a**) list the basic and the nonbasic variables;
(**b**) find the basic feasible solution determined by setting the nonbasic variables equal to 0;
(**c**) decide whether this is a maximum solution. (See Examples 5 and 6.)*

17.

$$
\begin{array}{ccccccc|c}
x_1 & x_2 & x_3 & x_4 & x_5 & z & & \\
3 & 2 & 0 & -3 & 1 & 0 & & 29 \\
4 & 0 & 1 & -2 & 0 & 0 & & 16 \\
\hline
-5 & 0 & 0 & -1 & 0 & 1 & & 11
\end{array}
$$

18.

$$
\begin{array}{ccccccc|c}
x_1 & x_2 & x_3 & x_4 & x_5 & x_6 & z & \\
-3 & 0 & \frac{1}{2} & 1 & -2 & 0 & 0 & 22 \\
2 & 0 & -3 & 0 & 1 & 1 & 0 & 10 \\
4 & 1 & 4 & 0 & \frac{3}{4} & 0 & 0 & 17 \\
\hline
-1 & 0 & 0 & 0 & 1 & 0 & 1 & 120
\end{array}
$$

19.

$$
\begin{array}{ccccccc|c}
x_1 & x_2 & x_3 & x_4 & x_5 & x_6 & z & \\
1 & 0 & 2 & \frac{1}{2} & 0 & \frac{1}{3} & 0 & 6 \\
0 & 1 & -1 & 5 & 0 & -1 & 0 & 13 \\
0 & 0 & 1 & \frac{3}{2} & 1 & -\frac{1}{3} & 0 & 21 \\
\hline
0 & 0 & 2 & \frac{1}{2} & 0 & 3 & 1 & 18
\end{array}
$$

20.

$$
\begin{array}{cccccccc|c}
x_1 & x_2 & x_3 & x_4 & x_5 & x_6 & x_7 & z & \\
-1 & 0 & 0 & 1 & 0 & 3 & -2 & 0 & 47 \\
2 & 0 & 1 & 0 & 0 & 2 & -\frac{1}{2} & 0 & 37 \\
3 & 0 & 0 & 0 & 1 & -1 & 6 & 0 & 43 \\
\hline
4 & 1 & 0 & 0 & 0 & 6 & 0 & 1 & 86
\end{array}
$$

Use the simplex method to solve Exercises 21–36.

21. Maximize $z = x_1 + 3x_2$
subject to: $x_1 + x_2 \le 10$
$5x_1 + 2x_2 \le 20$
$x_1 + 2x_2 \le 36$
$x_1 \ge 0, \quad x_2 \ge 0.$

22. Maximize $z = 5x_1 + x_2$
subject to: $2x_1 + 3x_2 \le 8$
$4x_1 + 8x_2 \le 12$
$5x_1 + 2x_2 \le 30$
$x_1 \ge 0, \quad x_2 \ge 0.$

23. Maximize $z = 2x_1 + x_2$
subject to: $x_1 + 3x_2 \le 12$
$2x_1 + x_2 \le 10$
$x_1 + x_2 \le 4$
$x_1 \ge 0, \quad x_2 \ge 0.$

24. Maximize $z = 4x_1 + 2x_2$
subject to: $-x_1 - x_2 \le 12$
$3x_1 - x_2 \le 15$
$x_1 \ge 0, \quad x_2 \ge 0.$

25. Maximize $z = 5x_1 + 4x_2 + x_3$
subject to: $-2x_1 + x_2 + 2x_3 \le 3$
$x_1 - x_2 + x_3 \le 1$
$x_1 \ge 0, \quad x_2 \ge 0, \quad x_3 \ge 0.$

26. Maximize $z = 3x_1 + 2x_2 + x_3$
subject to: $2x_1 + 2x_2 + x_3 \le 10$
$x_1 + 2x_2 + 3x_3 \le 15$
$x_1 \ge 0, \quad x_2 \ge 0, \quad x_3 \ge 0.$

27. Maximize $z = 2x_1 + x_2 + x_3$
subject to: $x_1 - 3x_2 + x_3 \le 3$
$x_1 - 2x_2 + 2x_3 \le 12$
$x_1 \ge 0, \quad x_2 \ge 0, \quad x_3 \ge 0.$

28. Maximize $z = 4x_1 + 5x_2 + x_3$
subject to: $x_1 + 2x_2 + 4x_3 \le 10$
$2x_1 + 2x_2 + x_3 \le 10$
$x_1 \ge 0, \quad x_2 \ge 0, \quad x_3 \ge 0.$

29. Maximize $z = 2x_1 + 2x_2 - 4x_3$
subject to: $3x_1 + 3x_2 - 6x_3 \le 51$
$5x_1 + 5x_2 + 10x_3 \le 99$
$x_1 \ge 0, \quad x_2 \ge 0, \quad x_3 \ge 0.$

30. Maximize $z = 4x_1 + x_2 + 3x_3$
subject to: $x_1 + 3x_3 \le 6$
$6x_1 + 3x_2 + 12x_3 \le 40$
$x_1 \ge 0, \quad x_2 \ge 0, \quad x_3 \ge 0.$

31. Maximize $z = 300x_1 + 200x_2 + 100x_3$
subject to: $x_1 + x_2 + x_3 \le 100$
$2x_1 + 3x_2 + 4x_3 \le 320$
$2x_1 + x_2 + x_3 \le 160$
$x_1 \ge 0, \quad x_2 \ge 0, \quad x_3 \ge 0.$

32. Maximize $z = x_1 + 5x_2 - 10x_3$
subject to: $8x_1 + 4x_2 + 12x_3 \le 18$
$x_1 + 6x_2 + 2x_3 \le 45$
$5x_1 + 7x_2 + 3x_3 \le 60$
$x_1 \ge 0, \quad x_2 \ge 0, \quad x_3 \ge 0.$

33. Maximize $z = 4x_1 - 3x_2 + 2x_3$
subject to: $\quad 2x_1 - x_2 + 8x_3 \le 40$
$\quad\quad\quad 4x_1 - 5x_2 + 6x_3 \le 60$
$\quad\quad\quad 2x_1 - 2x_2 + 6x_3 \le 24$
$\quad\quad x_1 \ge 0, \quad x_2 \ge 0, \quad x_3 \ge 0.$

34. Maximize $z = 3x_1 + 2x_2 - 4x_3$
subject to: $\quad x_1 - x_2 + x_3 \le 10$
$\quad\quad\quad 2x_1 - x_2 + 2x_3 \le 30$
$\quad\quad\quad -3x_1 + x_2 + 3x_3 \le 40$
$\quad\quad x_1 \ge 0, \quad x_2 \ge 0, \quad x_3 \ge 0$

35. Maximize $z = x_1 + 2x_2 + x_3 + 5x_4$
subject to: $\quad x_1 + 2x_2 + x_3 + x_4 \le 50$
$\quad\quad\quad 3x_1 + x_2 + 2x_3 + x_4 \le 100$
$\quad\quad x_1 \ge 0, \quad x_2 \ge 0, \quad x_3 \ge 0, \quad x_4 \ge 0.$

36. Maximize $z = x_1 + x_2 + 4x_3 + 5x_4$
subject to: $\quad x_1 + 2x_2 + 3x_3 + x_4 \le 115$
$\quad\quad\quad 2x_1 + x_2 + 8x_3 + 5x_4 \le 200$
$\quad\quad\quad x_1 + x_3 \le 50$
$\quad\quad x_1 \ge 0, \quad x_2 \ge 0, \quad x_3 \ge 0, \quad x_4 \ge 0.$

---

37. The initial simplex tableau of a linear programming problem is given below.

$$\begin{array}{cccccc} x_1 & x_2 & x_3 & x_4 & x_5 & z \\ \left[\begin{array}{cccccc|c} 1 & 1 & 1 & 1 & 0 & 0 & 12 \\ 2 & 1 & 2 & 0 & 1 & 0 & 30 \\ \hline -2 & -2 & -1 & 0 & 0 & 1 & 0 \end{array}\right] \end{array}$$

(a) Use the simplex method to solve the problem, with column one as the first pivot column.

(b) Now use the simplex method to solve the problem, with column two as the first pivot column.

(c) Does this problem have a unique maximum solution? Why?

38. The final simplex tableau of a linear programming problem is given here.

$$\begin{array}{ccccc} x_1 & x_2 & x_3 & x_4 & z \\ \left[\begin{array}{ccccc|c} 1 & 1 & 2 & 0 & 0 & 24 \\ 2 & 0 & 2 & 1 & 0 & 8 \\ \hline 4 & 0 & 0 & 0 & 1 & 40 \end{array}\right] \end{array}$$

(a) What is the solution given by this tableau?

(b) Even though all the indicators are nonnegative, perform one more round of pivoting on this tableau, using column three as the pivot column and choosing the pivot row by forming quotients in the usual way.

(c) Show that there is more than one solution to the linear programming problem by comparing your answer in part (a) to the basic feasible solution given by the tableau found in part (b). Does it give the same value of z as the solution in part (a)?

7.5 MAXIMIZATION APPLICATIONS

Appplications of linear programming that use the simplex method are considered in this section. First, however, we make a slight change in notation. You have probably noticed that the column representing z in a simplex tableau never changes during pivoting. Furthermore, the value of z in the basic feasible solution associated with the tableau is the number in the lower right-hand corner. Consequently, the z column is unnecessary and will be omitted from all simplex tableaus hereafter.

▶**EXAMPLE 1** A farmer has 100 acres of available land he wishes to plant with a mixture of potatoes, corn, and cobbage. It costs him $400 to produce an acre of potatoes, $160 to produce an acre of corn, and $280 to produce an acre of cabbage. He has a maximum of $20,000 to spend. He makes a profit of $120 per acre of potatoes, $40 per acre of corn, and $60 per acre of cabbage. How many acres of each crop should he plant to maximize his profit?

Begin by summarizing the given information as follows.

Crop	Number of Acres	Cost per Acre	Profit per Acre
Potatoes	x_1	$400	$120
Corn	x_2	160	40
Cabbage	x_3	280	60
Maximum available	100	$20,000	

If the number of acres allotted to each of the three crops is represented by x_1, x_2, and x_3, respectively then the constraints can be expressed as

$$x_1 + x_2 + x_3 \leq 100 \quad \textbf{Number of acres}$$
$$400x_1 + 160x_2 + 280x_3 \leq 20{,}000 \quad \textbf{Production costs}$$

where x_1, x_2, and x_3 are all nonnegative. The first of these constraints says that $x_1 + x_2 + x_3$ is less than or perhaps equal to 100. Use x_4 as the slack variable, giving the equation

$$x_1 + x_2 + x_3 + x_4 = 100.$$

Here x_4 represents the amount of the farmer's 100 acres that will not be used. (x_4 may be 0 or any value up to 100.)

In the same way, the constraint $400x_1 + 160x_2 + 280x_3 \leq 20{,}000$ can be converted into an equation by adding a slack variable, x_5:

$$400x_1 + 160x_2 + 280x_3 + x_5 = 20{,}000.$$

The slack variable x_5 represents any unused portion of the farmer's $20,000 capital. (Again, x_5 may have any value from 0 to 20,000.)

The farmer's profit on potatoes is the product of the profit per acre ($120) and the number x_1 of acres, that is, $120x_1$. His profit on corn and cabbage is computed similarly. Hence his total profit is given by

$$z = \text{profit on potatoes} + \text{profit on corn} + \text{profit on cabbage}$$
$$z = 120x_1 + 40x_2 + 60x_3.$$

The linear programming problem can now be stated as follows:

Maximize $\quad z = 120x_1 + 40x_2 + 60x_3$

subject to:
$$x_1 + x_2 + x_3 + x_4 = 100$$
$$400x_1 + 160x_2 + 280x_3 + x_5 = 20{,}000$$

with $x_1 \geq 0, \quad x_2 \geq 0, \quad x_3 \geq 0, \quad x_4 \geq 0, \quad x_5 \geq 0.$

The initial simplex tableau (without the z column) is

$$
\begin{array}{ccccc}
x_1 & x_2 & x_3 & x_4 & x_5 \\
\end{array}
$$

$$
\left[
\begin{array}{ccccc|c}
1 & 1 & 1 & 1 & 0 & 100 \\
400 & 160 & 280 & 0 & 1 & 20{,}000 \\
\hline
-120 & -40 & -60 & 0 & 0 & 0
\end{array}
\right]
$$

The most negative indicator is -120; column one is the pivot column. The quotients needed to determine the pivot row are $100/1 = 100$ and $20{,}000/400 = 50$. So the pivot is 400 in row two, column one. Multiplying row two by $1/400$ and completing the pivoting leads to the final simplex tableau:

$$
\begin{array}{ccccc}
x_1 & x_2 & x_3 & x_4 & x_5 \\
\end{array}
$$

$$
\left[
\begin{array}{ccccc|c}
0 & .6 & .3 & 1 & -.0025 & 50 \\
1 & .4 & .7 & 0 & .0025 & 50 \\
\hline
0 & 8 & 24 & 0 & .3 & 6000
\end{array}
\right]
\quad
\begin{array}{l}
-1R_2 + R_1 \\[4pt]
\dfrac{1}{400}R_2 \\[6pt]
120R_2 + R_3
\end{array}
$$

Setting the nonbasic variables x_2, x_3, and x_5 equal to 0, solving for the basic variables x_1 and x_4, and remembering that the value of z is in the lower right hand corner leads to this maximum solution:

$$x_1 = 50, \quad x_2 = 0, \quad x_3 = 0, \quad x_4 = 50, \quad x_5 = 0, \quad z = 6000.$$

Therefore, the farmer will make a maximum profit of $6000 by planting 50 acres of potatoes, and no corn or cabbage. Thus 50 acres are left unplanted (represented by x_4, the slack variable for potatoes). It may seem strange that leaving assets unused can produce a maximum profit, but such results occur quite often. ◀

▶ **EXAMPLE 2** Set up the initial simplex tableau for the following problem.
Irene Mills, who is a candidate for the state legislature, has $96,000 to buy TV advertising time. Ads cost $400 per minute on a local cable channel, $4000 per minute on a regional independent channel, and $12,000 per minute on a national network channel. Because of existing contracts the TV stations can provide at most 30 minutes of advertising time, with a maximum of 6 minutes on the national network channel. At any given time during the evening, approximately 100,000 people watch the cable channel, 200,000 the independent channel, and 600,000 the network channel. To get maximum exposure, how much time should Irene buy from each station?
Let x_1 be the number of minutes of ads on the cable channel, x_2 the number of minutes on the independent channel, and x_3 the number of minutes on the network channel. Exposure is measured in viewer-minutes. For instance, 100,000 people watching x_1 minutes of ads on the cable channel produces

1 What is the optimal solution in Example 2?

Answer:
Buying 20 minutes of time on the cable channel, 4 minutes on the independent channel, and 6 minutes on the network channel produces the maximum of 6,400,000 viewer-minutes.

$100,000x_1$ viewer-minutes. The amount of exposure is given by the total number of viewer-minutes for all three channels, namely,

$$100,000x_1 + 200,000x_2 + 600,000x_3.$$

Since 30 minutes are available,

$$x_1 + x_2 + x_3 \le 30.$$

The fact that only 6 minutes can be used on the network channel means that

$$x_3 \le 6.$$

Expenditures are limited to $96,000, so

$$\text{Cable cost} + \text{independent cost} + \text{network cost} \le 96,000$$
$$400x_1 + 4000x_2 + 12,000x_3 \le 96,000.$$

Therefore, Irene must solve the following linear programming problem:

$$\text{Maximize } z = 100,000x_1 + 200,000x_2 + 600,000x_3$$
$$\text{Subject to:} \qquad x_1 + x_2 + x_3 \le 30$$
$$x_3 \le 6$$
$$400x_1 + 4000x_2 + 12,000x_3 \le 96,000$$
$$\text{with} \quad x_1 \ge 0, \quad x_2 \ge 0, \quad x_3 \ge 0.$$

Introducing slack variables x_4, x_5, and x_6 (one for each constraint), rewriting the constraints as equations, and expressing the objective function as

$$-100,000x_1 - 200,000x_2 - 600,000x_3 + z = 0,$$

leads to the initial simplex tableau:

$$
\begin{bmatrix}
\begin{array}{ccccccc|c}
x_1 & x_2 & x_3 & x_4 & x_5 & x_6 & \\
1 & 1 & 1 & 1 & 0 & 0 & 30 \\
0 & 0 & 1 & 0 & 1 & 0 & 6 \\
400 & 4000 & 12,000 & 0 & 0 & 1 & 96,000 \\
\hline
-100,000 & -200,000 & -600,000 & 0 & 0 & 0 & 0
\end{array}
\end{bmatrix}.
$$

A computer program can be used to find the final simplex tableau:

$$
\begin{bmatrix}
\begin{array}{cccccc|c}
x_1 & x_2 & x_3 & x_4 & x_5 & x_6 & \\
1 & 0 & 0 & \frac{10}{9} & \frac{20}{9} & \frac{-25}{90,000} & 20 \\
0 & 0 & 1 & 0 & 1 & 0 & 6 \\
0 & 1 & 0 & -\frac{1}{9} & -\frac{29}{9} & \frac{25}{90,000} & 4 \\
\hline
0 & 0 & 0 & \frac{800,000}{9} & \frac{1,600,000}{9} & \frac{250}{9} & 6,400,000
\end{array}
\end{bmatrix}
$$

◀ **1**

7.5 EXERCISES

Set up the initial simplex tableau for each of the following problems.

1. **Management** A cat breeder has the following amounts of cat food: 90 units of tuna, 80 units of liver, and 50 units of chicken. To raise a Siamese cat, the breeder must use 2 units of tuna, 1 of liver, and 1 of chicken per day, while raising a Persian cat requires 1, 2, and 1 units, respectively, per day. If a Siamese cat sells for $12 while a Persian cat sells for $10, how many of each should be raised in order to obtain maximum gross income? What is the maximum gross income?

2. **Management** Banal, Inc., produces art for motel rooms. Its painters can turn out mountain scences, seascapes, and pictures of clowns. Each painting is worked on by three different artists, T, D, and H. Artist T works only 25 hours per week, while D and H work 45 and 40 hours per week, respectively. Artist T spends 1 hour on a mountain scene, 2 hours on a seascape, and 1 hour on a clown. Corresponding times for D and H are 3, 2, and 2 hours, and 2, 1, and 4 hours respectively. Banal makes $20 on a mountain scene, $18 on a seascape, and $22 on a clown. The head painting packer can't stand clowns, so that no more than 4 clown paintings may be done in a week. Find the number of each type of painting that should be made weekly in order to maximize profit. Find the maximum possible profit.

3. **Management** A manufacturer makes two products, toy trucks and toy fire engines. Both are processed in four different departments, each of which has a limited capacity. The sheet metal department can handle at least $1\frac{1}{2}$ times as many trucks as fire engines. The truck assembly department can handle at most 6700 trucks per week, while the fire engine assembly department assembles at most 5500 fire engines weekly. The painting department, which finishes both toys, has a maximum capacity of 12,000 per week. If the profit is $8.50 for a toy truck and $12.10 for a toy fire engine, how many of each item should the company produce to maximize profit?

4. **Natural Science** A lake is stocked each spring with three species of fish, A, B, and C. The average weights of the fish are 1.62, 2.12, and 3.01 kilograms for species A, B, and C, respectively. Three foods, I, II, and III, are available in the lake. Each fish of species A requires 1.32 units of food I, 2.9 units of food II, and 1.75 units of food III on the average each day. Species B fish each require 2.1 units of food I, .95 units of food II, and .6 units of food III daily. Species C fish require .86, 1.52, and 2.01 units of I, II, and III per day, respectively. If 490 units of food I, 897 units of food II, and 653 units of food III are available daily, how should the lake be stocked to maximize the weight of the fish supported by the lake?

Use the simplex method to solve the following problems.

5. **Management** A manufacturer of bicycles builds one-, three-, and ten-speed models. The bicycles need both aluminum and steel. The company has available 91,800 units of steel and 42,000 units of aluminum. The one-, three-, and ten-speed models need, respectively, 20, 30, and 40 units of steel and 12, 21, and 16 units of aluminum. How many of each type of bicycle should be made in order to maximize profit if the company makes $8 per one-speed bike, $12 per three-speed, and $24 per ten-speed? What is the maximum possible profit?

6. **Social Science** Jayanta is working to raise money for the homeless by sending information letters and making follow-up calls to local labor organizations and church groups. She discovered that each church group requires 2 hours of letter writing and 1 hour of follow-up, while for each labor union she needs 2 hours of letter writing and 3 hours of follow-up. Jayanta can raise $100 from each church group and $200 from each union local, and she has a maximum of 16 hours of letter-writing time and a maximum of 12 hours of follow-up time available per month. Determine the most profitable mixture of groups she should contact and the most money she can raise in a month.

7. **Management** A baker has 60 units of flour, 132 units of sugar, and 102 units of raisins. A loaf of raisin bread requires 1 unit of flour, 1 unit of sugar, and 2 units of raisins, while a raisin cake needs 2, 4, and 1 units respectively. If raisin bread sells for $3 a loaf and a raisin cake for $4, how many of each should be baked so that the gross income is maximized? What is the maximum gross income?

8. **Management** Mellow Sounds Inc. produces three types of compact discs: easy listening, jazz, and rock. Each easy listening disc requires 6 hours of recording, 12 hours of mixing, and 2 hours of editing. Each jazz disc requires 8 hours of recording, 8 hours of mixing, and 4

hours of editing. Each rock disc requires 3 hours of recording, 6 hours of mixing, and 1 hour of editing. Each week 288 hours of recording studio time are available, 312 hours of mixing board time are available, and editors are available for at most 124 hours. Mellow Sounds receives $6 for each easy listening and rock disc and $8 for each jazz disc. How many of each type of disc should the company process each week to maximize their income? What is the maximum income.

9. **Management**   The Cut-Right Company sells sets of kitchen knives. The Basic Set consists of 2 utility knives and 1 chef's knife. The Regular Set consists of 2 utility knives, 1 chef's knife, and 1 slicer. The Deluxe Set consists of 3 utility knives, 1 chef's knife, and 1 slicer. Their profit is $30 on a Basic Set, $40 on a Regular Set, and $60 on a Deluxe Set. The factory has on hand 800 utility knives, 400 chef's knives, and 200 slicers. Assuming that all sets will be sold, how many of each type should be made up in order to maximize profit? What is the maximum profit?

10. **Management**   Super Souvenir Company makes paper weights, plaques, and ornaments. Each paperweight requires 8 units of plastic, 3 units of metal, and 2 units of paint. Each plaque requires 4 units of plastic, and 1 unit each of metal and paint. Each ornament requires 2 units each of plastic and metal, and 1 unit of paint. They make a profit of $3 on each paperweight and each ornament, and $4 on each plaque. If 36 units of plastic, 24 units of metal, and 30 units of paint are available today, how many of each kind of souvenir should be made in order to maximize profit?

11. **Management**   The Fancy Fashions Store has $8000 available each month for advertising. Newspaper ads cost $400 each and no more than 20 can be run per month. Radio ads cost $200 each and no more than 30 can run per month. TV ads cost $1200 each, with a maximum of 6 available each month. Approximately 2000 women will see each newspaper ad, 1200 will hear each radio commercial, and 10,000 will see each TV ad. How much of each type of advertising should be used if the store wants to maximize its ad exposure?

12. **Management**   Caroline's Quality Candy Confectionery is famous for fudge, chocolate cremes, and pralines. Its candy-making equipment is set up to make 100-pound batches at a time. Currently there is a chocolate shortage and the company can get only 120 pounds of chocolate in the next shipment. On a week's run, the confectionery's cooking and processing equipment is available for a total of 42 machine hours. During the same period the employees have a total of 56 work hours available for packaging. A batch of fudge requires 20 pounds of chocolate while a batch of cremes uses 25 pounds of chocolate. The cooking and processing take 120 minutes for fudge, 150 minutes for chocolate cremes, and 200 minutes for pralines. The packaging times measured in minutes per 1-pound box are 1, 2, 3, respectively, for fudge, cremes, and pralines. Determine how many batches of each type of candy the confectionery should make, assuming that the profit per pound box is 50¢ on fudge, 40¢ on chocolate cremes, and 45¢ on pralines. Also, find the maximum profit for the week.

Management   *The next two problems come from past CPA examinations.* Select the appropriate answer for each question.*

13. The Ball Company manufactures three types of lamps, labeled A, B, and C. Each lamp is processed in two departments, I and II. Total available man-hours per day for departments I and II are 400 and 600, respectively. No additional labor is available. Time requirements and profit per unit for each lamp type are as follows.

	A	*B*	*C*
Man-hours in I	2	3	1
Man-hours in II	4	2	3
Profit per unit	$5	$4	$3

The company has assigned you as the accounting member of its profit planning committee to determine the numbers of types of A, B, and C lamps that it should produce in order to maximize its total profit from the sale of lamps. The following questions relate to a linear programming model that your group has developed.

(a) The coefficients of the objective function would be
  (**1**) 4, 2, 3;   (**2**) 2, 3, 1;
  (**3**) 5, 4, 3;   (**4**) 400,600.

(b) The constraints in the model would be
  (**1**) 2, 3, 1;   (**2**) 5, 4, 3;
  (**3**) 4, 2, 3;   (**4**) 400,600.

---

*Material from *Uniform CPA Examination Questions and Unofficial Answers,* copyright © 1973, 1974, 1975 by the American Institute of Certified Public Accountants, Inc., is reprinted with permission.

(c) The constraint imposed by the available man-hours in department I could be expressed as

(1) $4X_1 + 2X_2 + 3X_3 \leq 400$;

(2) $4X_1 + 2X_2 + 3X_3 \geq 400$;

(3) $2X_1 + 3X_2 + 1X_3 \leq 400$;

(4) $2X_1 + 3X_2 + 1X_3 \geq 400$.

14. The Golden Hawk Manufacturing Company wants to maximize the profits on products A, B, and C. The contribution margin for each product follows.

Product	Contribution Margin
A	$2
B	5
C	4

The production requirements and departmental capacities, by departments, are as follows.

Department	Production Requirements by Product (hours)			Departmental Capacity (total hours)
	A	*B*	*C*	
Assembling	2	3	2	30,000
Painting	1	2	2	38,000
Finishing	2	3	1	28,000

(a) What is the profit-maximization formula for the Golden Hawk Company?

(1) $\$2A + \$5B + \$4C = X$ (where X = profit)

(2) $5A + 8B + 5C \leq 96,000$

(3) $\$2A + \$5B + \$4C \leq X$

(4) $\$2A + \$5B + \$4C = 96,000$

(b) What is the constraint for the Painting Department of the Golden Hawk Company?

(1) $1A + 2B + 2C \geq 38,000$

(2) $\$2A + \$5B + \$4C \geq 38,000$

(3) $1A + 2B + 2C \leq 38,000$

(4) $2A + 3B + 2C \leq 30,000$

15. Solve the problem in Exercise 1.

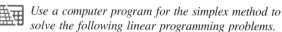 *Use a computer program for the simplex method to solve the following linear programming problems.*

16. Exercise 2. Your final answer should consist of whole numbers (Banal can't sell half a painting).

17. Exercise 3.

18. Exercise 4.

7.6 THE SIMPLEX METHOD: DUALITY AND MINIMIZATION

In this section the simplex method is extended to linear programming problems satisfying the following conditions:

1. The objective function is to be *minimized.*
2. All the coefficients of the objective function are nonnegative.
3. All constraints involve $\geq$.
4. All variables are nonnegative.

The method of solving minimization problems presented here is based on an interesting connection between maximizing and minimizing problems: any solution of a maximizing problem produces the solution of an associated minimizing problem, or vice-versa. Each of the associated problems is called the **dual** of the other. Thus, duals enable us to solve minimization problems of the type de-

1 Use the corner points in Figure 7.21(a) on the next page to find the minimum value of $w = 8y_1 + 16y_2$ and where it occurs.

Answer:
48 when $y_1 = 4$, $y_2 = 1$

2 Use Figure 7.21(b) to find the maximum value of $z = 9x_1 + 10x_2$ and where it occurs.

Answer:
48 when $x_1 = 2$, $x_2 = 3$

scribed above by the simplex method introduced in Section 7.4. (An alternative approach for solving minimization problems is given in the next section.)

When dealing with minimization problems, we use y_1, y_2, y_3, etc. as variables and denote the objective function by w. An example will explain the idea of a dual.

▶**EXAMPLE 1**   Minimize $w = 8y_1 + 16y_2$

subject to: $y_1 + 5y_2 \geq 9$

$2y_1 + 2y_2 \geq 10$

$y_1 \geq 0, \quad y_2 \geq 0.$

Without considering slack variables just yet, write the augmented matrix of the system of inequalities, and include the coefficients of the objective function (not their negatives) as the last row in the matrix.

$$\begin{bmatrix} 1 & 5 & | & 9 \\ 2 & 2 & | & 10 \\ \hline 8 & 16 & | & 0 \end{bmatrix}$$

Look now at the following new matrix, obtained from the one above by interchanging rows and columns.

$$\begin{bmatrix} 1 & 2 & | & 8 \\ 5 & 2 & | & 16 \\ \hline 9 & 10 & | & 0 \end{bmatrix}$$

The *rows* of the first matrix (for the minimizing problem) are the *columns* of the second matrix.

The entries in this second matrix could be used to write the following maximizing problem in standard form (again ignoring the fact that the numbers in the last row are not negative).

Maximize $z = 9x_1 + 10x_2$

subject to: $x_1 + 2x_2 \leq 8$

$5x_1 + 2x_2 \leq 16$

$x_1 \geq 0, \quad x_2 \geq 0.$

Figure 7.21(a) shows the region of feasible solutions for the minimization problem given above, while Figure 7.21(b) shows the region of feasible solutions for the maximization problem produced by exchanging rows and columns.   ◀ **1** **2**

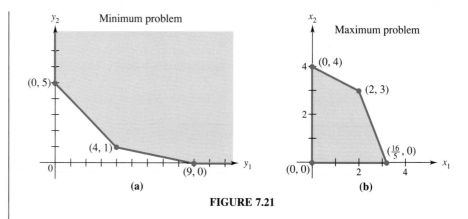

FIGURE 7.21

The two feasible regions in Figure 7.21 are different and the corner points are different, but the values of the objective functions found in Problems 1 and 2 at the side are equal—both are 48. An even closer connection between the two problems is shown by using the simplex method to solve the maximization problem given above.

Maximization problem

$$
\begin{array}{cccc}
x_1 & x_2 & x_3 & x_4 \\
\end{array}
$$

$$
\left[
\begin{array}{cccc|c}
1 & 2 & 1 & 0 & 8 \\
5 & 2 & 0 & 1 & 16 \\
\hline
-9 & -10 & 0 & 0 & 0 \\
\end{array}
\right]
$$

$$
\begin{array}{cccc}
x_1 & x_2 & x_3 & x_4 \\
\end{array}
$$

$$
\left[
\begin{array}{cccc|c}
\frac{1}{2} & 1 & \frac{1}{2} & 0 & 4 \\
4 & 0 & -1 & 1 & 8 \\
\hline
-4 & 0 & 5 & 0 & 40 \\
\end{array}
\right]
\begin{array}{l}
\frac{1}{2}R_1 \\
-2R_1 + R_2 \\
10R_1 + R_3 \\
\end{array}
$$

$$
\begin{array}{cccc}
x_1 & x_2 & x_3 & x_4 \\
\end{array}
$$

$$
\left[
\begin{array}{cccc|c}
0 & 1 & \frac{5}{8} & -\frac{1}{8} & 3 \\
1 & 0 & -\frac{1}{4} & \frac{1}{4} & 2 \\
\hline
0 & 0 & 4 & 1 & 48 \\
\end{array}
\right]
\begin{array}{l}
-\frac{1}{2}R_2 + R_1 \\
\frac{1}{4}R_2 \\
4R_2 + R_3 \\
\end{array}
$$

The maximum is 48 when $x_1 = 2$, $x_2 = 3$.

3 Give the transpose of each matrix.

(a) $\begin{bmatrix} 2 & 4 \\ 6 & 3 \\ 1 & 5 \end{bmatrix}$

(b) $\begin{bmatrix} 4 & 7 & 10 \\ 3 & 2 & 6 \\ 5 & 8 & 12 \end{bmatrix}$

Answers:

(a) $\begin{bmatrix} 2 & 6 & 1 \\ 4 & 3 & 5 \end{bmatrix}$

(b) $\begin{bmatrix} 4 & 3 & 5 \\ 7 & 2 & 8 \\ 10 & 6 & 12 \end{bmatrix}$

Notice that the solution to the *minimization problem* (namely, $y_1 = 4$, $y_2 = 1$) is found in the bottom row and slack variable columns of the final simplex tableau for the maximizaton problem. This result suggests that a minimalization problem can be solved by forming the dual maximization problem, solving it by the simplex method, and then reading the solution for the minimalization problem from the bottom row of the final simples tableau.

Before using this method to solve a minimization problem, let us find the duals of some typical linear programming problems. The process of exchanging the rows and columns of a matrix, which is used to find the dual, is called **transposing** the matrix, and each of the two matrices is the **transpose** of the other.

▶ **EXAMPLE 2** Find the transpose of each matrix.

(a) $A = \begin{bmatrix} 2 & -1 & 5 \\ 6 & 8 & 0 \\ -3 & 7 & -1 \end{bmatrix}$

Write the rows of matrix A as the columns of the transpose.

$$\text{Transpose of A} = \begin{bmatrix} 2 & 6 & -3 \\ -1 & 8 & 7 \\ 5 & 0 & -1 \end{bmatrix}$$

(b) The transpose of $\begin{bmatrix} 1 & 2 & 4 & 0 \\ 2 & 1 & 7 & 6 \end{bmatrix}$ is $\begin{bmatrix} 1 & 2 \\ 2 & 1 \\ 4 & 7 \\ 0 & 6 \end{bmatrix}$. ◀ **3**

▶ **EXAMPLE 3** Write the dual of the following minimization linear programming problems.

(a) Minimize $w = 10y_1 + 8y_2$
  subject to: $\quad y_1 + 2y_2 \geq 2$
  $\quad\quad\quad\quad\quad y_1 + y_2 \geq 5$
  $\quad\quad y_1 \geq 0, \quad y_2 \geq 0.$

Begin by writing the augmented matrix for the given problem.

$$\begin{bmatrix} 1 & 2 & | & 2 \\ 1 & 1 & | & 5 \\ \hline 10 & 8 & | & 0 \end{bmatrix}$$

Form the transpose of this matrix to get

$$\begin{bmatrix} 1 & 1 & | & 10 \\ 2 & 1 & | & 8 \\ \hline 2 & 5 & | & 0 \end{bmatrix}$$

4 Write the dual of the following linear programming problem. Minimize $w = 2y_1 + 5y_2 + 6y_3$ subject to:

$$2y_1 + 3y_2 + y_3 \geq 15$$
$$y_1 + y_2 + 2y_3 \geq 12$$
$$5y_1 + 3y_2 \qquad \geq 10$$
$$y_1 \geq 0, \quad y_2 \geq 0, \quad y_3 \geq 0.$$

Answer:
Maximize $z = 15x_1 + 12x_2 + 10x_3$ subject to:

$$2x_1 + x_2 + 5x_3 \leq 2$$
$$3x_1 + x_2 + 3x_3 \leq 5$$
$$x_1 + 2x_2 \qquad \leq 6$$
$$x_1 \geq 0, \quad x_2 \geq 0, \quad x_3 \geq 0.$$

The dual problem is stated from this second matrix as follows (using x instead of y.)

$$\text{Maximize} \quad z = 2x_1 + 5x_2$$
$$\text{subject to:} \qquad x_1 + x_2 \leq 10$$
$$2x_1 + x_2 \leq 8$$
$$x_1 \geq 0, \quad x_2 \geq 0.$$

(b) Minimize $w = 7y_1 + 5y_2 + 8y_3$
subject to:
$$3y_1 + 2y_2 + y_3 \geq 10$$
$$y_1 + y_2 + y_3 \geq 8$$
$$4y_1 + 5y_2 \qquad \geq 25$$
$$y_1 \geq 0, \quad y_2 \geq 0, \quad y_3 \geq 0.$$

The dual problem is stated as follows.

$$\text{Maximize} \quad z = 10x_1 + 8x_2 + 25x_3$$
$$\text{subject to:} \quad 3x_1 + x_2 + 4x_3 \leq 7$$
$$2x_1 + x_2 + 5x_3 \leq 5$$
$$x_1 + x_2 \qquad \leq 8$$
$$x_1 \geq 0, \quad x_2 \geq 0, \quad x_3 \geq 0. \quad \blacktriangleleft \quad \boxed{4}$$

In Example 3, all the constraints of the minimization problems were $\geq$ inequalities, while all those in the dual maximization problems were $\leq$ inequalities. This is generally the case; inequalities are reversed when the dual problem is stated.

The following table shows the close connection between a problem and its dual.

Given Problem	*Dual Problem*
m variables	n variables
n constraints	m constraints
Coefficients from objective function	Constants
Constants	Coefficients from objective function

The next theorem, whose proof requires advanced methods, guarantees that a minimization problem can be solved by forming a dual maximization problem.

Theorem of Duality

The objective function w of a minimizing linear programming problem takes on a minimum value if and only if the objective function z of the corresponding dual maximizing problem takes on a maximum value. The maximum value of z equals the minimum value of w.

This method is illustrated in the following example.

▶ **EXAMPLE 4**   Minimize $\quad w = 3y_1 + 2y_2$

$$\text{subject to:} \quad \begin{aligned} y_1 + 3y_2 &\geq 6 \\ 2y_1 + y_2 &\geq 3 \\ y_1 \geq 0, \quad y_2 &\geq 0. \end{aligned}$$

Use the given information to write the matrix.

$$\begin{bmatrix} 1 & 3 & | & 6 \\ 2 & 1 & | & 3 \\ \hline 3 & 2 & | & 0 \end{bmatrix}$$

Transpose to get the following matrix for the dual problem.

$$\begin{bmatrix} 1 & 2 & | & 3 \\ 3 & 1 & | & 2 \\ \hline 6 & 3 & | & 0 \end{bmatrix}$$

Write the dual problem from this matrix, as follows.

$$\text{Maximize} \quad z = 6x_1 + 3x_2$$

$$\text{subject to:} \quad \begin{aligned} x_1 + 2x_2 &\leq 3 \\ 3x_1 + x_2 &\leq 2 \\ x_1 \geq 0, \quad x_2 &\geq 0. \end{aligned}$$

Solve this standard maximization problem using the simplex method. Start by introducing slack variables to give the system

$$\begin{aligned} x_1 + 2x_2 + x_3 \qquad\qquad &= 3 \\ 3x_1 + x_2 \qquad + x_4 \quad &= 2 \\ -6x_1 - 3x_2 - 0x_3 - 0x_4 + z &= 0 \end{aligned}$$

with $x_1 \geq 0, \quad x_2 \geq 0, \quad x_3 \geq 3, \quad x_4 \geq 0.$

5 Minimize $w = 10y_1 + 8y_2$ subject to:

$$y_1 + 2y_2 \geq 2$$
$$y_1 + y_2 \geq 5$$
$$y_1 \geq 0, \quad y_2 \geq 0.$$

Answer:
$y_1 = 0$, $y_2 = 5$, for a minimum of 40

The first tableau for this system is given below, with the pivot as indicated.

$$\begin{array}{cccc} x_1 & x_2 & x_3 & x_4 \end{array} \qquad \text{Quotients}$$

$$\left[\begin{array}{cccc|c} 1 & 2 & 1 & 0 & 3 \\ 3 & 1 & 0 & 1 & 2 \\ \hline -6 & -3 & 0 & 0 & 0 \end{array}\right] \begin{array}{l} 3/1 = 3 \\ 2/3 \end{array}$$

The simplex method gives the following final tableau.

$$\begin{array}{cccc} x_1 & x_2 & x_3 & x_4 \end{array}$$

$$\left[\begin{array}{cccc|c} 0 & 1 & \frac{3}{5} & -\frac{1}{5} & \frac{7}{5} \\ 1 & 0 & -\frac{1}{5} & \frac{2}{5} & \frac{1}{5} \\ \hline 0 & 0 & \frac{3}{5} & \frac{9}{5} & \frac{27}{5} \end{array}\right]$$

The last row of this final tableau shows that the solution of the given minimization problem is as follows:

The minimum value of $w = 3y_1 + 2y_2$, subject to the given constraints, is $27/5$ and occurs when $y_1 = 3/5$ and $y_2 = 9/5$.

The minimum value of w, $27/5$, is the same as the maximum value of z. ◀ **5**

A minimizing problem that meets the conditions listed at the beginning of the section can be solved by the method of duals, as illustrated in Examples 1 and 4 and summarized here.

Solving Minimum Problems with Duals

1. Find the dual standard maximum problem.*
2. Solve the maximum problem using the simplex method.
3. The minimum value of the objective function w is the maximum value of the objective function z.
4. The optimum solution is given by the entries in the bottom row of the columns corresponding to the slack variables.

FURTHER USES OF THE DUAL The dual is useful not only in solving minimization problems, but also in seeing how small changes in one variable will affect the value of the objective function. For example, suppose an animal breeder needs at least 6 units per day of nutrient A and at least 3 units of nutrient B and that the breeder can choose between two different feeds, feed 1 and feed

---

*The coefficients of the objective function in the minimization problem are the constants on the right side of the constraints in the dual maximization problem. So when all these coefficients are nonnegative (condition 2), the dual problem is in standard maximum form.

2. Find the minimum cost for the breeder if each bag of feed 1 costs $3 and provides 1 unit of nutrient A and 2 units of B, while each bag of feed 2 costs $2 and provides 3 units of nutrient A and 1 of B.

If y_1 represents the number of bags of feed 1 and y_2 represents the number of bags of feed 2, the given information leads to

$$\text{Minimize} \quad w = 3y_1 + 2y_2$$
$$\text{subject to:} \quad y_1 + 3y_2 \geq 6$$
$$2y_1 + y_2 \geq 3$$
$$y_1 \geq 0, \quad y_2 \geq 0.$$

This minimization linear programming problem is the one we solved in Example 4 of this section. In that example, we formed the dual and reached the following final tableau.

$$
\begin{array}{cccc}
x_1 & x_2 & x_3 & x_4 \\
\end{array}
$$
$$
\left[
\begin{array}{cccc|c}
0 & 1 & \frac{3}{5} & -\frac{1}{5} & \frac{7}{5} \\
1 & 0 & -\frac{1}{5} & \frac{2}{5} & \frac{1}{5} \\
\hline
0 & 0 & \frac{3}{5} & \frac{9}{5} & \frac{27}{5} \\
\end{array}
\right]
$$

This final tableau shows that the breeder will obtain minimum feed costs by using 3/5 bag of feed 1 and 9/5 bag of feed 2 per day, for a daily cost of 27/5 = 5.40 dollars.

Now look at the data from the feed problem shown in the table below.

	Units of Nutrient (per Bag)		Cost per Bag
	A	B	
Feed 1	1	2	$3
Feed 2	3	1	$2
Minimum nutrient needed	6	3	

If x_1 and x_2 are the cost per unit of nutrients A and B, the constraints of the dual problem can be stated as follows.

$$\text{Cost of feed 1:} \quad x_1 + 2x_2 \leq 3$$
$$\text{Cost of feed 2:} \quad 3x_1 + x_2 \leq 2$$

The solution of the dual problem, which maximizes nutrients, can be read from the final tableau:

$$x_1 = \frac{1}{5} = .20 \quad \text{and} \quad x_2 = \frac{7}{5} = 1.40,$$

6 The final tableau of the dual of the problem about filing cabinets in Example 2, Section 7.3 and Example 6, Section 7.4 is given below.

$$\begin{array}{cccc} x_1 & x_2 & x_3 & x_4 \\ \begin{bmatrix} 0 & 1 & \frac{1}{2} & -\frac{1}{4} & 1 \\ 1 & 0 & -\frac{1}{20} & -\frac{1}{80} & \frac{1}{20} \\ 0 & 0 & 8 & 3 & 100 \end{bmatrix} \end{array}$$

(a) What are the imputed amounts of storage for each unit of cost and floor space?

(b) What are the shadow values of the cost and the floor space?

Answers:

(a) Cost: 28 sq ft
floor space: 72 sq ft

(b) $\frac{1}{20}$, 1

which means that a unit of nutrient A costs 1/5 of a dollar = \$.20, while a unit of nutrient B costs 7/5 dollars = \$1.40. The minimum daily cost, \$5.40, is found by the following procedure.

$$(\$.20 \text{ per unit of A}) \times (6 \text{ units of A}) = \$1.20$$
$$\underline{+ (\$1.40 \text{ per unit of B}) \times (3 \text{ units of B}) = \$4.20}$$
$$\text{Minimum daily cost} = \$5.40$$

The numbers .20 and 1.40 are called the **shadow costs** of the nutrients. These two numbers from the dual, \$.20 and \$1.40, also allow the breeder to estimate feed costs for "small" changes in nutrient requirements. For example, an increase of 1 unit in the requirement for each nutrient would produce a total cost of

\$5.40	6 units of A, 3 of B
.20	1 extra unit of A
1.40	1 extra unit of B
\$7.00	Total cost per day **6**

7.6 EXERCISES

Find the transpose of each matrix. (See Example 2.)

1. $\begin{bmatrix} 3 & -4 & 5 \\ 1 & 10 & 7 \\ 0 & 3 & 6 \end{bmatrix}$

2. $\begin{bmatrix} 3 & -5 & 9 & 4 \\ 1 & 6 & -7 & 0 \\ 4 & 18 & 11 & 9 \end{bmatrix}$

3. $\begin{bmatrix} 3 & 0 & 14 & -5 & 3 \\ 4 & 17 & 8 & -6 & 1 \end{bmatrix}$

4. $\begin{bmatrix} 15 & -6 & -2 \\ 13 & -1 & 11 \\ 10 & 12 & -3 \\ 24 & 1 & 0 \end{bmatrix}$

State the dual problem for each of the following, but do not solve it. (See Example 3.)

5. Minimize $w = 3y_1 + 5y_2$
subject to: $3y_1 + y_2 \geq 4$
$-y_1 + 2y_2 \geq 6$
$y_1 \geq 0, \quad y_2 \geq 0.$

6. Minimize $w = 4y_1 + 7y_2$
subject to: $y_1 + y_2 \geq 17$
$3y_1 + 6y_2 \geq 21$
$2y_1 + 4y_2 \geq 19$
$y_1 \geq 0, \quad y_2 \geq 0.$

7. Minimize $w = 2y_1 + 8y_2$
subject to: $y_1 + 7y_2 \geq 18$
$4y_1 + y_2 \geq 15$
$5y_1 + 3y_2 \geq 20$
$y_1 \geq 0, \quad y_2 \geq 0.$

8. Minimize $w = 5y_1 + y_2 + 3y_3$
subject to: $7y_1 + 6y_2 + 8y_3 \geq 18$
$4y_1 + 5y_2 + 10y_3 \geq 20$
$y_1 \geq 0, \quad y_2 \geq 0, \quad y_3 \geq 0.$

9. Minimize $w = y_1 + 2y_2 + 6y_3$
subject to: $3y_1 + 4y_2 + 6y_3 \geq -8$
$y_1 + 5y_2 + 2y_3 \geq 12$
$y_1 \geq 0, \quad y_2 \geq 0, \quad y_3 \geq 0.$

10. Minimize $w = 4y_1 + 3y_2 + y_3$
subject to: $y_1 + 2y_2 + 3y_3 \geq 115$
$2y_1 + y_2 + 8y_3 \geq 200$
$y_1 \qquad - \quad y_3 \geq 50$

11. Minimize $w = 8y_1 + 9y_2 + 3y_3$
subject to: $y_1 + y_2 + y_3 \geq 5$
$y_1 + y_2 \qquad \geq 4$
$2y_1 + y_2 + 3y_3 \geq 15$
$y_1 \geq 0, \quad y_2 \geq 0, \quad y_3 \geq 0.$

12. Minimize $w = y_1 + 2y_2 + y_3 + 5y_4$
subject to: $y_1 + y_2 + y_3 + y_4 \geq 50$
$3y_1 + y_2 + 2y_3 + y_4 \geq 100$
$y_1 \geq 0, \quad y_2 \geq 0, \quad y_3 \geq 0, y_4 \geq 0.$

Use duality to solve the following problems. (See Example 4.)

13. Minimize $w = 2y_1 + y_2 + 3y_3$
subject to: $y_1 + y_2 + y_3 \geq 100$
$2y_1 + y_2 \qquad \geq 50$
$y_1 \geq 0, \quad y_2 \geq 0, \quad y_3 \geq 0.$

14. Minimize $w = 2y_1 + 4y_2$
subject to: $4y_1 + 2y_2 \geq 10$
$4y_1 + y_2 \geq 8$
$2y_1 + y_2 \geq 12.$
$y_1 \geq 0, \quad y_2 \geq 0$

15. Minimize $w = 3y_1 + y_2 + 4y_3$
subject to: $2y_1 + y_2 + y_3 \geq 6$
$y_1 + 2y_2 + y_3 \geq 8$
$2y_1 + y_2 + 2y_3 \geq 12$
$y_1 \geq 0, \quad y_2 \geq 0, \quad y_3 \geq 0.$

16. Minimize $w = y_1 + y_2 + 3y_3$
subject to: $2y_1 + 6y_2 + y_3 \geq 8$
$y_1 + 2y_2 + 4y_3 \geq 12$
$y_1 \geq 0, \quad y_2 \geq 0, \quad y_3 \geq 0.$

17. Minimize $w = 6y_1 + 4y_2 + 2y_3$
subject to: $2y_1 + 2y_2 + y_3 \geq 2$
$y_1 + 3y_2 + 2y_3 \geq 3$
$y_1 + y_2 + 2y_3 \geq 4$
$y_1 \geq 0, \quad y_2 \geq 0, \quad y_3 \geq 0.$

18. Minimize $w = 12y_1 + 10y_2 + 7y_3$
subject to: $2y_1 + y_2 + y_3 \geq 7$
$y_1 + 2y_2 + y_3 \geq 4$
$y_1 \geq 0, \quad y_2 \geq 0, \quad y_3 \geq 0.$

19. Minimize $w = 20y_1 + 12y_2 + 40y_3$
subject to: $y_1 + y_2 + 5y_3 \geq 20$
$2y_1 + y_2 + y_3 \geq 30$
$y_1 \geq 0, \quad y_2 \geq 0, \quad y_3 \geq 0.$

20. Minimize $w = 4y_1 + 5y_2$
subject to: $10y_1 + 5y_2 \geq 100$
$20y_1 + 10y_2 \geq 150$
$y_1 \geq 0, \quad y_2 \geq 0.$

21. Minimize $w = 4y_1 + 2y_2 + y_3$
subject to: $y_1 + y_2 + y_3 \geq 4$
$3y_1 + y_2 + 3y_3 \geq 6$
$y_1 + y_2 + 3y_3 \geq 5$
$y_1 \geq 0, \quad y_2 \geq 0, \quad y_3 \geq 0.$

22. Minimize $w = 3y_1 + 2y_2$
subject to: $2y_1 + 3y_2 \geq 60$
$y_1 + 4y_2 \geq 40$
$y_1 \geq 0, \quad y_2 \geq 0.$

23. Glenn Russell, who is dieting, requires two food supplements, I and II. He can get these supplements from two different products, A and B, as shown in the following table.

	Supplement (grams per serving)	
	I	**II**
Product A	4	2
Product B	2	5

Glenn's physician has recommended that he include at least 20 grams of supplement I and 18 grams of supplement II in his diet. If product A costs 24¢ per serving and product B costs 40¢ per serving, how can he satisfy these requirements most economically?

24. Management   An animal food must provide at least 54 units of vitamins and 60 calories per serving. One gram of soybean meal provides 2.5 units of vitamins and 5 calories. One gram of meat byproducts provides

4.5 units of vitamins and 3 calories. One gram of grain provides 5 units of vitamins and 10 calories. If a gram of soybean meal costs 8¢, a gram of meat byproducts 9¢, and a gram of grain 10¢, what mixture of these three ingredients will provide the required vitamins and calories at minimum cost?

25. **Management** A furniture company makes tables and chairs. Union contracts require that the total number of tables and chairs produced must be at least 60 per week. Sales experience has shown that at least 1 table must be made for every 3 chairs that are made. If it costs $152 to make a table and $40 to make a chair, how many of each should be produced each week to minimize the cost. What is the minimum cost?

26. **Management** Brand X canners produce canned corn, beans, and carrots. Labor contracts require them to produce at least 1000 cases per month. Based on past sales, they should produce at least twice as many cases of corn as of beans. At least 340 cases of carrots must be produced to fulfill commitments to a major distributor. It costs $10 to produce a case of beans, $15 to produce a case of corn, and $25 to produce a case of carrots. How many cases of each vegetable should be produced to minimize costs?

27. Refer to the end of this Section, to the text on minimizing the daily cost of feeds.
    (a) Find a combination of feeds that will cost $7.00 and give 7 units of A and 4 units of B.
    (b) Use the dual variables to predict the daily cost of feed if the requirements change to 5 units of A and 4 units of B. Find a combination of feeds to meet these requirements at the predicted price.

28. **Management** A small toy manufacturing firm has 200 squares of felt, 600 ounces of stuffing, and 90 feet of trim available to make two types of toys, a small bear and a monkey. The bear requires 1 square of felt and 4 ounces of stuffing. The monkey requires 2 squares of felt, 3 ounces of stuffing, and 1 foot of trim. The firm makes $1 profit on each bear and $1.50 profit on each monkey. The linear program to maximize profit is

$$\text{Maximize} \quad x_1 + 1.5x_2 = z$$
$$\text{subject to:} \quad x_1 + 2x_2 \leq 200$$
$$4x_1 + 3x_2 \leq 600$$
$$x_2 \leq 90$$
$$x_1 \geq 0, \quad x_2 \geq 0.$$

The final simplex tableau is

$$\begin{bmatrix} 0 & 1 & .8 & -.2 & 0 & | & 40 \\ 1 & 0 & -.6 & .4 & 0 & | & 120 \\ 0 & 0 & -.8 & .2 & 1 & | & 50 \\ \hline 0 & 0 & .6 & .1 & 0 & | & 180 \end{bmatrix}$$

(a) What is the corresponding dual problem?
(b) What is the optimal solution to the dual problem?
(c) Use the shadow values to estimate the profit the firm will make if their supply of felt increases to 210 squares.
(d) How much profit will the firm make if their supply of stuffing is cut to 590 ounces and their supply of trim is cut to 80 feet?

29. Refer to Example 1 in Section 7.5.
    (a) Give the dual problem.
    (b) Use the shadow values to estimate the farmer's profit if land is cut to 90 acres but capital increases to $21,000.
    (c) Suppose the farmer has 110 acres but only $19,000. Find the optimum profit and the planting strategy that will produce this profit.

Use duality and a computer program for doing the simplex method to solve these problems.

30. **Management** Natural Brand plant food is made from three chemicals. In a batch of the plant food there must be at least 81 kilograms of the first chemical and the other two chemicals must be in the ratio of at most 4 to 3. If the three chemicals cost $1.09, $.87, and $.65 per kilogram, respectively, how much of each should be used to minimize the cost of producing at least 750 kilograms of the plant food?

7.7 THE SIMPLEX METHOD: NONSTANDARD PROBLEMS

So far we have solved only linear programming problems in standard maximum form or problems whose duals were in standard maximum form. In this section we present a two-stage method for solving certain nonstandard problems (those with negative coefficients, mixed $\leq$ and $\geq$ constraints, and those with only $\geq$ constraints).

The first step in applying the two-stage method to a maximization problem is to write each constraint so that the constant on the right side is nonnegative. For instance, the inequality

$$4x_1 + 5x_2 - 12x_3 \leq -30$$

can be replaced by the equivalent one obtained by multiplying both sides by -1 and reversing the direction of the inequality sign:

$$-4x_1 - 5x_2 + 12x_3 \geq 30.$$

The next step is to write each constraint as an equation. Recall that constraints involving $\leq$ are converted to equations by adding a nonnegative slack variable. Similarly, constraints involving $\geq$ are converted to equations by *subtracting* a nonnegative **surplus variable.** For example, the inequality $2x_1 - x_2 + 5x_3 \geq 12$ means that

$$2x_1 - x_2 + 5x_3 - x_4 = 12$$

for some nonnegative x_4. The surplus variable x_4 represents the amount by which $2x_1 - x_2 + 5x_3$ exceeds 12.

▶ **EXAMPLE 1**   Restate the following problem in terms of equations and write its initial simplex tableau.

$$\text{Maximize} \quad z = 4x_1 + 10x_2 + 6x_3$$
$$\text{subject to:} \quad x_1 + 4x_2 + 4x_3 \geq 8$$
$$x_1 + 3x_2 + 2x_3 \leq 6$$
$$3x_1 + 4x_2 + 8x_3 \leq 22$$
$$x_1 \geq 0, \quad x_2 \geq 0, \quad x_3 \geq 0.$$

In order to write the constraints as equations, subtract a surplus variable from each $\geq$ constraint and add a slack variable to each $\leq$ constraint. So the problem becomes

$$\text{Maximize} \quad z = 4x_1 + 10x_2 + 6x_3$$
$$\text{subject to:} \quad x_1 \ 1 \ 4x_2 \ 1 \ 4x_3 \ - \ x_4 \qquad\qquad = \ 8$$
$$x_1 \ 1 \ 3x_2 \ 1 \ 2x_3 \qquad 1 \ x_5 \qquad = \ 6$$
$$3x_1 \ 1 \ 4x_2 \ 1 \ 8x_3 \qquad\qquad 1 \ x_6 = \ 22$$

1 **(a)** Restate this problem in terms of equations:

$$\text{Maximize } z = 3x_1 - 2x_2$$

$$\text{subject to: } 2x_1 + 3x_2 \leq 8$$
$$6x_1 - 2x_2 \geq 3$$
$$x_1 + 4x_2 \geq 1$$
$$x_1 \geq 0, \quad x_2 \geq 0.$$

(b) Write the initial simplex tableau.

Answers:

(a) Maximize $z = 3x_1 - 2x_2$
subject to:
$$2x_1 + 3x_2 + x_3 \qquad = 8$$
$$6x_1 - 2x_2 \qquad - x_4 \qquad = 3$$
$$x_1 + 4x_2 \qquad\qquad - x_5 = 1$$
$x_1 \geq 0, x_2 \geq 0, x_3 \geq 0, x_4 \geq 0,$
$x_5 \geq 0.$

(b)
$$\begin{bmatrix} 2 & 3 & 1 & 0 & 0 & 8 \\ 6 & -2 & 0 & -1 & 0 & 3 \\ 1 & 4 & 0 & 0 & -1 & 1 \\ \hline -3 & 2 & 0 & 0 & 0 & 0 \end{bmatrix}$$

2 State the basic solution given by each tableau. Is it feasible?

(a)
$$\begin{bmatrix} 3 & -5 & 1 & 0 & 0 & 12 \\ 4 & 7 & 0 & 1 & 0 & 6 \\ 1 & 3 & 0 & 0 & -1 & 5 \\ \hline -7 & 4 & 0 & 0 & 0 & 0 \end{bmatrix}$$

(b)
$$\begin{bmatrix} 9 & 8 & -1 & 1 & 0 & 12 \\ -5 & 3 & 0 & 0 & 1 & 7 \\ \hline 4 & 2 & 3 & 0 & 0 & 0 \end{bmatrix}$$

Answers:

(a) $x_1 = 0, x_2 = 0, x_3 = 12,$
$x_4 = 6, x_5 = -5$; no.

(b) $x_1 = 0, x_2 = 0, x_3 = 0,$
$x_4 = 12, x_5 = 7$; yes.

Write the objective function as $z - 4x_1 - 10x_2 - 6x_3 = 0$ and use the coefficients of the four equations to write the initial simplex tableau (omitting the z column):

$$\begin{array}{cccccc} x_1 & x_2 & x_3 & x_4 & x_5 & x_6 \end{array}$$
$$\begin{bmatrix} 1 & 4 & 4 & -1 & 0 & 0 & 8 \\ 1 & 3 & 2 & 0 & 1 & 0 & 6 \\ 3 & 4 & 8 & 0 & 0 & 1 & 22 \\ \hline -4 & -10 & -6 & 0 & 0 & 0 & 0 \end{bmatrix}. \quad \blacktriangleleft \; \boxed{1}$$

The tableau in Example 1 resembles those that have appeared previously, and similar terminology is used. The variables whose columns have one entry ± 1 and the rest 0 will be called **basic variables;** the other variables are nonbasic. A solution obtained by setting the nonbasic variables equal to 0 and solving for the basic variables (by looking at the constants in the right-hand column) will be called a **basic solution.** A basic solution that is feasible is called a **basic feasible solution.** In the tableau of Example 1, for instance, the basic variables are x_4, x_5, and x_6, and the basic solution is:

$$x_1 = 0, \quad x_2 = 0, \quad x_3 = 0, \quad x_4 = -8, \quad x_5 = 6, \quad x_6 = 22.$$

However, because one variable is negative, this solution is not feasible. **2**

Stage I of the two-stage method for nonstandard maximization problems consists of finding a basic *feasible* solution that can be used as the starting point for the simplex method. (This stage is unnecessary in a standard maximization problem because the solution given by the initial tableau is always feasible.) There are many systematic ways of finding a feasible solution, all of which depend on the fact that row operations (such as pivoting) produce a tableau that represents a system with the same solutions as the original one. One such technique is explained in the next example. Since the immediate goal is to find a feasible solution, not necessarily an optimal one, the procedures for choosing pivots differ from those in the ordinary simplex method.

▶**EXAMPLE 2** Find a basic feasible solution for the problem in Example 1, whose initial tableau is

$$\begin{array}{cccccc} x_1 & x_2 & x_3 & x_4 & x_5 & x_6 \end{array}$$
$$\begin{bmatrix} 1 & 4 & 4 & -1 & 0 & 0 & 8 \\ 1 & 3 & 2 & 0 & 1 & 0 & 6 \\ 3 & 4 & 8 & 0 & 0 & 1 & 22 \\ \hline -4 & -10 & -6 & 0 & 0 & 0 & 0 \end{bmatrix}.$$

In the basic solution given by this tableau, x_4 has a negative value. The only nonzero entry in its column is the -1 in row one. Choose any *positive* entry in row one except the entry on the far right. The column that the chosen entry is in will be the pivot column. We choose the first positive entry in row one, the 1 in column one. The pivot row is determined in the usual way by considering quotients (constant at the right end of the row divided by the positive entry in the pivot column) in each row except the objective row:

$$8/1 = 8, \quad 6/1 = 6, \quad 22/3 = 7\frac{1}{3}.$$

The smallest quotient is 6, so the pivot is the 1 in row two, column one. Pivoting in the usual way leads to this tableau

$$
\begin{array}{c}
\begin{array}{cccccc}
x_1 & x_2 & x_3 & x_4 & x_5 & x_6
\end{array} \\
\left[
\begin{array}{cccccc|c}
0 & 1 & 2 & -1 & -1 & 0 & 2 \\
1 & 3 & 2 & 0 & 1 & 0 & 6 \\
0 & -5 & 2 & 0 & -3 & 1 & 4 \\
\hline
0 & 2 & 2 & 0 & 4 & 0 & 24
\end{array}
\right]
\begin{array}{l}
-R_2 + R_1 \\
\\
-3R_2 + R_3 \\
4R_2 + R_4
\end{array}
\end{array}
$$

and the basic solution

$$x_1 = 6, \quad x_2 = 0, \quad x_3 = 0, \quad x_4 = -2, \quad x_5 = 0, \quad x_6 = 4.$$

Since the basic variable x_4 is negative, this solution is not feasible. So we repeat the pivoting process described above. The column of x_4 has a -1 in row one, so we choose a positive entry in that row, namely, the 1 in row one, column two. This choice makes column two the pivot column. The pivot row is determined by the quotients $2/1 = 2$ and $6/3 = 2$ (negative entries in the pivot column and the entry in the objective row are not used). Since there is a tie, we can choose either row one or row two. We choose row one and use the 1 in row one, column 2 as the pivot. Pivoting produces this tableau

$$
\begin{array}{c}
\begin{array}{cccccc}
x_1 & x_2 & x_3 & x_4 & x_5 & x_6
\end{array} \\
\left[
\begin{array}{cccccc|c}
0 & 1 & 2 & -1 & -1 & 0 & 2 \\
1 & 0 & -4 & 3 & 4 & 0 & 0 \\
0 & 0 & 12 & -5 & -8 & 1 & 14 \\
\hline
0 & 0 & -2 & 2 & 6 & 0 & 20
\end{array}
\right]
\begin{array}{l}
\\
-3R_1 + R_2 \\
5R_1 + R_3 \\
-2R_1 + R_4
\end{array}
\end{array}
$$

and the basic *feasible* solution

$$x_1 = 0, \quad x_2 = 2, \quad x_3 = 0, \quad x_4 = 0, \quad x_5 = 0, \quad x_6 = 14. \quad \blacktriangleleft$$

Once a basic feasible solution has been found, Stage I is ended. The procedures used in Stage I are summarized below.*

---

*Except in rare cases that do not occur in this book, this method eventually produces a basic feasible solution or shows that one does not exist. The *two-phase method* using artificial variables, which is discussed in more advanced texts, works in all cases and often is more efficient.

3 The first tableau of a maximization problem is given below. Use column one as the pivot column for carrying out Stage I and state the basic feasible solution that results.

$$\begin{bmatrix} 1 & 3 & 1 & 0 & | & 70 \\ 2 & 4 & 0 & -1 & | & 50 \\ \hline -8 & -10 & 0 & 0 & | & 0 \end{bmatrix}$$

Answer:

$$\begin{bmatrix} 0 & 1 & 1 & \frac{1}{2} & | & 45 \\ 1 & 2 & 0 & -\frac{1}{2} & | & 25 \\ \hline 0 & 6 & 0 & -4 & | & 200 \end{bmatrix}$$

$x_1 = 25$, $x_2 = 0$, $x_3 = 45$, $x_4 = 0$.

Finding a Basic Feasible Solution

1. If any basic variable has a negative value, locate the -1 in that variable's column and note the row it is in.
2. In the row determined in Step 1, choose a positive entry (other than the one at the far right) and note the column it is in. This is the pivot column.
3. Use the positive entries in the pivot column (except in the objective row), to form quotients and select the pivot.
4. Pivot as usual, which results in the pivot column's having one entry 1 and the rest 0.
5. Repeat Steps 1–4 until every basic variable is nonnegative, so that the basic solution given by the tableau is feasible. If it ever becomes impossible to continue, then the problem has no feasible solution.

One way to make the required choices systematically is to choose the first possibility in each case (going from the top for rows or from the left for columns). However, any choice meeting the required conditions may be used. For maximum efficiency, it is usually best to choose the pivot column in Step 2 so that the pivot is in the same row chosen in Step 1, if this is possible. **3**

In Stage II, the simplex method is applied as usual to the tableau that produced the basic feasible solution in Stage I. Just as in Section 7.4, each round of pivoting replaces the basic feasible solution of one tableau with the basic feasible solution of a new tableau in such a way that the value of the objective function is increased, until an optimal value is obtained (or it becomes clear that no optimal solution exists).

▶ **EXAMPLE 3** Solve the linear programming problem in Example 1.

A basic feasible solution for this problem was found in Example 2 by using the tableau shown below. However, this solution is not maximal because there is a negative indicator in the objective row. So we use the simplex method: the most negative indicator determines the pivot column and the usual quotients determine that the number 2 in row one, column three is the pivot.

	x_1	x_2	x_3	x_4	x_5	x_6		Quotients
	0	1	**2**	-1	-1	0	2	2/2 ← smallest
	1	0	-4	3	4	0	0	
	0	0	12	-5	-8	1	14	14/12
	0	0	-2	2	6	0	20	

most negative indicator

4 Complete Stage II and find an optimal solution for side problem 3 above. What is the optimal value of the objective function z?

Answer:
The optimal value $z = 560$ occurs when $x_1 = 120$, $x_2 = 0$, $x_3 = 0$, $x_4 = 90$.

Pivoting leads to the final tableau.

$$
\begin{array}{c}
\begin{array}{cccccc}
x_1 & x_2 & x_3 & x_4 & x_5 & x_6
\end{array} \\
\left[
\begin{array}{cccccc|c}
0 & \frac{1}{2} & 1 & -\frac{1}{2} & -\frac{1}{2} & 0 & 1 \\
1 & 0 & -4 & 3 & 4 & 0 & 0 \\
0 & 0 & 12 & -5 & -8 & 1 & 14 \\
\hline
0 & 0 & -2 & 2 & 6 & 0 & 20
\end{array}
\right]
\end{array}
\qquad \frac{1}{2}R_1
$$

$$
\begin{array}{c}
\begin{array}{cccccc}
x_1 & x_2 & x_3 & x_4 & x_5 & x_6
\end{array} \\
\left[
\begin{array}{cccccc|c}
0 & \frac{1}{2} & 1 & -\frac{1}{2} & -\frac{1}{2} & 0 & 1 \\
1 & 2 & 0 & 1 & 2 & 0 & 4 \\
0 & -6 & 0 & 1 & -2 & 1 & 2 \\
\hline
0 & 1 & 0 & 1 & 5 & 0 & 22
\end{array}
\right]
\end{array}
\qquad
\begin{array}{l}
\\
4R_1 + R_2 \\
-12R_1 + R_3 \\
2R_1 + R_4
\end{array}
$$

Therefore, the maximum value of z occurs when $x_1 = 4$, $x_2 = 0$, and $x_3 = 1$, in which case $z = 22$. ◀ **4**

The two-stage method for maximization problems illustrated in Examples 1–3 also provides a means of solving minimization problems. To see why, consider this simple fact: when a number t gets smaller, then $-t$ gets larger, and vice-versa. For instance, if t goes from 6 to 1 to 0 to -8, then $-t$ goes from -6 to -1 to 0 to 8. Thus, if w is the objective function of a linear programming problem, the feasible solution that produces the minimum value of w also produces the maximum value of $-w$, and vice-versa. Therefore, to solve a minimization problem with objective function w, we need only solve the maximization problem with the same constraints and objective function $z = -w$.

▶ **EXAMPLE 4**   Minimize $w = 2y_1 + y_2 - y_3$

subject to: $-y_1 - y_2 + y_3 \le -4$
$$y_1 + 3y_2 + 3y_3 \ge 6$$
$$y_1 \ge 0, \quad y_2 \ge 0, \quad y_3 \ge 0.$$

Make the constant in the first constraint positive by multiplying both sides by -1. Then solve this maximization problem:

Maximize $z = -w = -2y_1 - y_2 + y_3$
subject to: $y_1 + y_2 - y_3 \ge 4$
$$y_1 + 3y_2 + 3y_3 \ge 6$$
$$y_1 \ge 0, \quad y_2 \ge 0, \quad y_3 \ge 0.$$

Convert the constraints to equations by subtracting surplus variables, and set up the first tableau.

$$\begin{array}{ccccc} y_1 & y_2 & y_3 & y_4 & y_5 \\ \left[\begin{array}{ccccc|c} 1 & 1 & -1 & -1 & 0 & 4 \\ 1 & 3 & 3 & 0 & -1 & 6 \\ \hline 2 & 1 & -1 & 0 & 0 & 0 \end{array}\right] \end{array}$$

The basic solution given by this tableau, $y_1 = 0$, $y_2 = 0$, $y_3 = 0$, $y_4 = -4$, $y_5 = -6$ is not feasible, so the procedures of Stage I must be used to find a basic feasible solution. In the column of the negative basic variable y_4, there is a -1 in row one; we choose the first positive entry in that row, so that column one will be the pivot column. The quotients $4/1 = 4$ and $6/1 = 6$ show that the pivot is the 1 in row one, column one. Pivoting produces this tableau:

$$\begin{array}{ccccc} y_1 & y_2 & y_3 & y_4 & y_5 \\ \left[\begin{array}{ccccc|c} 1 & 1 & -1 & -1 & 0 & 4 \\ 0 & 2 & 4 & 1 & -1 & 2 \\ \hline 0 & -1 & 1 & 2 & 0 & -8 \end{array}\right] \end{array} \quad \begin{array}{l} \\ -R_1 + R_2 \\ -2R_1 + R_3 \end{array}$$

The basic solution $y_1 = 4$, $y_2 = 0$, $y_3 = 0$, $y_4 = 0$, $y_5 = -2$ is not feasible because y_5 is negative, so we repeat the process. We choose the first positive entry in row two (the row containing the -1 in the y_5 column), which is in column two, so that column two is the pivot column. The relevant quotients are $4/1 = 4$ and $2/2 = 1$, so the pivot is the 2 in row two, column two. Pivoting produces a new tableau.

$$\begin{array}{ccccc} y_1 & y_2 & y_3 & y_4 & y_5 \\ \left[\begin{array}{ccccc|c} 1 & 1 & -1 & -1 & 0 & 4 \\ 0 & 1 & 2 & \frac{1}{2} & -\frac{1}{2} & 1 \\ \hline 0 & -1 & 1 & 2 & 0 & -8 \end{array}\right] \end{array} \quad \begin{array}{l} \\ \frac{1}{2}R_2 \\ \\ \end{array}$$

$$\begin{array}{ccccc} y_1 & y_2 & y_3 & y_4 & y_5 \\ \left[\begin{array}{ccccc|c} 1 & 0 & -3 & -\frac{3}{2} & \frac{1}{2} & 3 \\ 0 & 1 & 2 & \frac{1}{2} & -\frac{1}{2} & 1 \\ \hline 0 & 0 & 3 & \frac{5}{2} & -\frac{1}{2} & -7 \end{array}\right] \end{array} \quad \begin{array}{l} -R_2 + R_1 \\ \\ R_2 + R_3 \end{array}$$

The basic solution $y_1 = 3$, $y_2 = 1$, $y_3 = 0$, $y_4 = 0$, $y_5 = 0$, is feasible, so Stage I is complete. However, this solution is not optimal because the objective row contains the negative indicator $-1/2$ in column five. According to the simplex method, column five is the next pivot column. The only positive ratio $3/\frac{1}{2} = 6$ is in row one, so the pivot is $1/2$ in row one, column five. Pivoting produces the final tableau.

5 Minimize $w = 2y_1 + 3y_2$ subject to:

$$y_1 + y_2 \geq 10$$
$$2y_1 + y_2 \geq 16$$
$$y_1 \geq 0, \quad y_2 \geq 0.$$

Answer:
$y_1 = 10, y_2 = 0; w = 20$

$$
\begin{array}{ccccc}
y_1 & y_2 & y_3 & y_4 & y_5 \\
\end{array}
$$

$$
\left[
\begin{array}{ccccc|c}
2 & 0 & -6 & -3 & 1 & 6 \\
0 & 1 & 2 & \frac{1}{2} & -\frac{1}{2} & 1 \\
\hline
0 & 0 & 3 & \frac{5}{2} & -\frac{1}{2} & -7 \\
\end{array}
\right]
\begin{array}{l}
2R_1 \\
\\
\\
\end{array}
$$

$$
\begin{array}{ccccc}
y_1 & y_2 & y_3 & y_4 & y_5 \\
\end{array}
$$

$$
\left[
\begin{array}{ccccc|c}
2 & 0 & -6 & -3 & 1 & 6 \\
1 & 1 & -1 & -1 & 0 & 4 \\
\hline
1 & 0 & 0 & 1 & 0 & -4 \\
\end{array}
\right]
\begin{array}{l}
\\
\frac{1}{2}R_1 + R_2 \\
\\
\frac{1}{2}R_1 + R_3 \\
\end{array}
$$

Since there are no negative indicators, the solution given by this tableau ($y_1 = 0$, $y_2 = 4$, $y_3 = 0$, $y_4 = 0$, $y_5 = 6$) is optimal. The maximum value of $z = -w$ is -4. Therefore, the minimum value of the original objective function w is $-(-4) = 4$, which occurs when $y_1 = 0$, $y_2 = 4$, $y_3 = 0$. ◄ **5**

Here is a summary of the two-stage method that was illustrated in Examples 1–4.

Solving Nonstandard Problems

1. If necessary, write each constraint with a positive constant and convert the problem to a maximum problem by letting $z = -w$.
2. Add slack variables and subtract surplus variables as needed to convert the constraints into equations.
3. Write the initial simplex tableau.
4. Find a basic feasible solution for the problem, if one exists (Stage I).
5. When a basic feasible solution is found, use the simplex method to solve the problem (Stage II).

Note It may happen that the tableau that gives the basic feasible solution in Stage I has no negative indicators in its last row. In this case, the solution found is already optimal and Stage II is not necessary.

► **EXAMPLE 5** A college textbook publisher has received orders from two colleges, C_1 and C_2. C_1 needs at least 500 books, and C_2 needs at least 1000. The publisher can supply the books from either of two warehouses. Warehouse W_1

has 900 books available and warehouse W_2 has 700. The costs to ship a book from each warehouse to each college are given below.

		To	
		C_1	C_2
From	W_1	$1.20	1.80
	W_2	$2.10	1.50

How many books should be sent from each warehouse to each college to minimize the shipping costs?

To begin, let

$$y_1 = \text{the number of books shipped from } W_1 \text{ to } C_1;$$
$$y_2 = \text{the number of books shipped from } W_2 \text{ to } C_1;$$
$$y_3 = \text{the number of books shipped from } W_1 \text{ to } C_2;$$
$$y_4 = \text{the number of books shipped from } W_2 \text{ to } C_2.$$

C_1 needs at least 500 books, so

$$y_1 + y_2 \geq 500.$$

Similarly,

$$y_3 + y_4 \geq 1000.$$

Since W_1 has 900 books available and W_2 has 700 available,

$$y_1 + y_3 \leq 900 \quad \text{and} \quad y_2 + y_4 \leq 700.$$

The company wants to minimize shipping costs, so the objective function is

$$w = 1.20y_1 + 2.10y_2 + 1.80y_3 + 1.50y_4.$$

Now write the problem as a system of linear equations, adding slack or surplus variables as needed, and let $z = -w$.

$$
\begin{aligned}
y_1 + y_2 \qquad\qquad\qquad - y_5 \qquad\qquad\qquad &= 500 \\
y_3 + y_4 \qquad - y_6 \qquad\qquad &= 1000 \\
y_1 \qquad + y_3 \qquad\qquad + y_7 \qquad &= 900 \\
y_2 \qquad + y_4 \qquad\qquad + y_8 &= 700 \\
1.20y_1 + 2.10y_2 + 1.80y_3 + 1.50y_4 \qquad\qquad + z &= 0
\end{aligned}
$$

Set up the first simplex tableau.

$$
\begin{array}{cccccccc}
y_1 & y_2 & y_3 & y_4 & y_5 & y_6 & y_7 & y_8 \\
\end{array}
$$

$$
\left[
\begin{array}{cccccccc|c}
1 & 1 & 0 & 0 & -1 & 0 & 0 & 0 & 500 \\
0 & 0 & 1 & 1 & 0 & -1 & 0 & 0 & 1000 \\
1 & 0 & 1 & 0 & 0 & 0 & 1 & 0 & 900 \\
0 & 1 & 0 & 1 & 0 & 0 & 0 & 1 & 700 \\
\hline
1.20 & 2.10 & 1.80 & 1.50 & 0 & 0 & 0 & 0 & 0
\end{array}
\right]
$$

The indicated solution is

$$y_5 = -500, \quad y_6 = -1000, \quad y_7 = 900, \quad y_8 = 700,$$

which is not feasible since y_5 and y_6 are negative.

There is a -1 in row one of the y_5 column. We choose the first positive entry in row one, which makes column one the pivot column. The quotients $500/1 = 500$ and $900/1 = 900$ show that the pivot is the 1 in row one, column one. Pivoting produces this tableau.

$$
\begin{array}{cccccccc}
y_1 & y_2 & y_3 & y_4 & y_5 & y_6 & y_7 & y_8 \\
\end{array}
$$

$$
\left[
\begin{array}{cccccccc|c}
1 & 1 & 0 & 0 & -1 & 0 & 0 & 0 & 500 \\
0 & 0 & 1 & 1 & 0 & -1 & 0 & 0 & 1000 \\
0 & -1 & 1 & 0 & 1 & 0 & 1 & 0 & 400 \\
0 & 1 & 0 & 1 & 0 & 0 & 0 & 1 & 700 \\
\hline
0 & .9 & 1.8 & 1.5 & 1.2 & 0 & 0 & 0 & -600
\end{array}
\right]
$$

The basic variable y_6 is negative and there is a -1 in row two of its column. We choose the first positive entry in that row, which makes column three the pivot column. The quotients $1000/1 = 1000$ and $400/1 = 400$ show that the pivot is the 1 in row three, column three. Pivoting leads to the next tableau.

$$
\begin{array}{cccccccc}
y_1 & y_2 & y_3 & y_4 & y_5 & y_6 & y_7 & y_8 \\
\end{array}
$$

$$
\left[
\begin{array}{cccccccc|c}
1 & 1 & 0 & 0 & -1 & 0 & 0 & 0 & 500 \\
0 & 1 & 0 & 1 & -1 & -1 & -1 & 0 & 600 \\
0 & -1 & 1 & 0 & 1 & 0 & 1 & 0 & 400 \\
0 & 1 & 0 & 1 & 0 & 0 & 0 & 1 & 700 \\
\hline
0 & 2.7 & 0 & 1.5 & -.6 & 0 & -1.8 & 0 & -1320
\end{array}
\right]
$$

The basic variable y_6 is still negative, so we must choose a positive entry in row two (the row containing the -1 in the y_6 column). If we choose the first positive entry, as we usually have done, then the quotients for determining the pivot will be $500/1$ and $700/1$ so that the pivot will be in row one. However, it is usually more efficient to have the pivot in the same row as the -1 of the basic variable (in this case row two). So we choose the 1 in row two, column four. Then the quotients for determining the pivot are $600/1$ and $700/1$ and the pivot is the 1 in row two, column four. Pivoting produces this tableau

$$
\begin{array}{cccccccc}
y_1 & y_2 & y_3 & y_4 & y_5 & y_6 & y_7 & y_8 \\
\end{array}
$$

$$
\left[
\begin{array}{cccccccc|c}
1 & 1 & 0 & 0 & -1 & 0 & 0 & 0 & 500 \\
0 & 1 & 0 & 1 & -1 & -1 & -1 & 0 & 600 \\
0 & -1 & 1 & 0 & 1 & 0 & 1 & 0 & 400 \\
0 & 0 & 0 & 0 & 1 & 1 & 1 & 1 & 100 \\
\hline
0 & 1.2 & 0 & 0 & .9 & 1.5 & -.3 & 0 & -2220
\end{array}
\right]
$$

and the basic feasible solution $y_1 = 500$, $y_2 = 0$, $y_3 = 400$, $y_4 = 600$, $y_5 = 0$, $y_6 = 0$, $y_7 = 0$, $y_8 = 100$. Hence Stage I is ended. This solution is not optimal because there is a negative indicator in the objective row, so we proceed to Stage II and the usual simplex method. Column seven has the only negative indicator, $-.3$, and row four the smallest quotient, $100/1 = 100$. Pivoting on the 1 in row four, column seven produces the final simplex tableau.

$$
\begin{array}{cccccccc}
y_1 & y_2 & y_3 & y_4 & y_5 & y_6 & y_7 & y_8 \\
\end{array}
$$

$$
\left[
\begin{array}{cccccccc|c}
1 & 1 & 0 & 0 & -1 & 0 & 0 & 0 & 500 \\
0 & 1 & 0 & 1 & 0 & 0 & 0 & 1 & 700 \\
0 & -1 & 1 & 0 & 0 & -1 & 0 & -1 & 300 \\
0 & 0 & 0 & 0 & 1 & 1 & 1 & 1 & 100 \\
\hline
0 & 1.2 & 0 & 0 & 1.2 & 1.8 & 0 & .3 & -2190
\end{array}
\right]
$$

Since there are no negative indicators, the solution given by this tableau ($y_1 = 500$, $y_3 = 300$, $y_4 = 700$) is optimal. The publisher should ship 500 books from W_1 to C_1, 300 books from W_1 to C_2, and 700 books from W_2 to C_2 for a minimum shipping cost of $2190 (remember that the optimal value for the original minimization problem is the negative of the optimal value for the associated maximization problem). ◀

Although they will not occur in this book, various complications can arise in using the simplex method. Some of the possible difficulties, which are treated in more advanced texts, include the following:

1. Some of the constraints may be *equations* instead of inequalities. In this case, *artificial variables* must be used.
2. Occasionally, a transformation will cycle—that is, produce a "new" solution which was an earlier solution in the process. These situations are known as *degeneracies* and special methods are available for handling them.
3. It may not be possible to convert a nonfeasible basic solution to a feasible basic solution. In that case, no solution can satisfy all the constraints. Graphically, this means there is no region of feasible solutions.

Two linear programming models in actual use, one on merit pay, the other on making ice cream, are presented at the end of this chapter. These models illustrate the usefulness of linear programming. In most real applications, the number of variables is so large that these problems could not be solved without the use of a method, like the simplex method, that can be adapted to a computer.

7.7 EXERCISES

In Exercises 1–4, (a) restate the problem in terms of equations by introducing slack and surplus variables; (b) write the initial simplex tableau. (See Example 1.)

1. Maximize $z = 5x_1 + 2x_2 - x_3$
   subject to: $2x_1 + 3x_2 + 5x_3 \geq 8$
   $4x_1 - x_2 + 3x_3 \leq 7$
   $x_1 \geq 0, \quad x_2 \geq 0, \quad x_3 \geq 0.$

2. Maximize $z = x_1 + 4x_2 + 6x_3$
   subject to: $5x_1 + 8x_2 - 5x_3 \leq 10$
   $6x_1 + 2x_2 + 3x_3 \geq 7$
   $x_1 \geq 0, \quad x_2 \geq 0, \quad x_3 \geq 0.$

3. Maximize $z = 2x_1 - 3x_2 + 4x_3$
   subject to: $x_1 + x_2 + x_3 \leq 100$
   $x_1 + x_2 + x_3 \geq 75$
   $x_1 + x_2 \geq 27$
   $x_1 \geq 0, \quad x_2 \geq 0, \quad x_3 \geq 0.$

4. Maximize $z = -x_1 + 5x_2 + x_3$
   subject to: $2x_1 + x_3 \leq 40$
   $x_1 + x_2 \geq 18$
   $x_1 + x_3 \geq 20$
   $x_1 \geq 0, \quad x_2 \geq 0, \quad x_3 \geq 0.$

Convert Exercises 5–8 into maximization problems with positive constants on the right side of each constraint and write the initial simplex tableau. (See Example 4.)

5. Minimize $w = 2y_1 + 5y_2 - 3y_3$
   subject to: $y_1 + 2y_2 + 3y_3 \geq 115$
   $2y_1 + y_2 + y_3 \leq 200$
   $y_1 + y_3 \geq 50$
   $y_1 \geq 0, \quad y_2 \geq 0, \quad y_3 \geq 0.$

6. Minimize $w = 7y_1 + 6y_2 + y_3$
   subject to: $y_1 + y_2 + y_3 \geq 5$
   $-y_1 + y_2 \leq -4$
   $2y_1 + y_2 + 3y_3 \geq 15$
   $y_1 \geq 0, \quad y_2 \geq 0, \quad y_3 \geq 0.$

7. Minimize $w = y_1 - 4y_2 + 2y_3$
   subject to: $-7y_1 + 6y_2 - 8y_3 \leq -18$
   $4y_1 + 5y_2 + 10y_3 \geq 20$
   $y_1 \geq 0, \quad y_2 \geq 0, \quad y_3 \geq 0.$

8. Minimize $w = y_1 + 2y_2 + y_3 + 5y_4$
   subject to: $-y_1 + y_2 + y_3 + y_4 \leq -50$
   $3y_1 + y_2 + 2y_3 + y_4 \geq 100$
   $y_1 \geq 0, \quad y_2 \geq 0, \quad y_3 \geq 0, \quad y_4 \geq 0.$

Use the two-stage method to solve Exercises 9–18. (See Examples 1–4.)

9. Maximize $z = 12x_1 + 10x_2$
   subject to: $x_1 + 2x_2 \geq 24$
   $x_1 + x_2 \leq 40$
   $x_1 \geq 0, \quad x_2 \geq 0.$

10. Find $x_1 \geq 0, x_2 \geq 0,$ and $x_3 \geq 0$ such that
   $x_1 + x_2 + x_3 \leq 150$
   $x_1 + x_2 + x_3 \geq 100$
   and $z = 2x_1 + 5x_2 + 3x_3$ is maximized.

11. Find $x_1 \geq 0, x_2 \geq 0,$ and $x_3 \geq 0$ such that
   $x_1 + x_2 + 2x_3 \leq 38$
   $2x_1 + x_2 + x_3 \geq 24$
   and $z = 3x_1 + 2x_2 + 2x_3$ is maximized.

12. Maximize $z = 6x_1 + 8x_2$
   subject to: $3x_1 + 12x_2 \geq 48$
   $2x_1 + 4x_2 \leq 60$
   $x_1 \geq 0, \quad x_2 \geq 0.$

13. Find $x_1 \geq 0$ and $x_2 \geq 0$ such that
   $x_1 + 2x_2 \leq 18$
   $x_1 + 3x_2 \geq 12$
   $2x_1 + 2x_2 \leq 30$
   and $z = 5x_1 + 10x_2$ is maximized.

14. Find $x_1 \geq 0$ and $x_2 \geq 0$ such that
   $x_1 + x_2 \leq 100$
   $x_1 + x_2 \geq 50$
   $2x_1 + x_2 \leq 110$
   and $z = 2x_1 + 3x_2$ is maximized.

15. Find $y_1 \geq 0$, $y_2 \geq 0$ such that
$$10y_1 + 5y_2 \geq 100$$
$$20y_1 + 10y_2 \geq 160$$
and $w = 4y_1 + 5y_2$ is minimized.

16. Minimize $\quad w = 3y_1 + 2y_2$
subject to: $\quad 2y_1 + 3y_2 \geq 60$
$$y_1 + 4y_2 \geq 40$$
$$y_1 \geq 0, \quad y_2 \geq 0.$$

17. Minimize $\quad w = 3y_1 + 4y_2$
subject to: $\quad y_1 + 2y_2 \geq 10$
$$y_1 + y_2 \geq 8$$
$$2y_1 + y_2 \leq 22$$
$$y_1 \geq 0, \quad y_2 \geq 0.$$

18. Minimize $\quad w = 4y_1 + 2y_2$
subject to: $\quad y_1 + y_2 \geq 20$
$$y_1 + 2y_2 \geq 25$$
$$-5y_1 + y_2 \leq 4$$
$$y_1 \geq 0, \quad y_2 \geq 0.$$

In Exercises 19–22, set up the initial simplex tableau, but do not solve the problem.

19. **Management** The manufacturer of a popular personal computer has orders from two dealers. Dealer D_1 wants at least 32 computers, and dealer D_2 wants at least 20 computers. The manufacturer can fill the orders from either of two warehouses, W_1 or W_2. W_1 has 25 of the computers on hand, and W_2 has 30. The costs (in dollars) to ship one computer to each dealer from each warehouse are given below.

		To	
		D_1	D_2
From	W_1	14	22
	W_2	12	10

How should the orders be filled to minimize shipping costs?

20. **Management** Natural Brand plant food is made from three chemicals. In a batch of the plant food there must be at least 81 kilograms of the first chemical and the other two chemicals must be in the ratio of at most 4 to 3. If the three chemicals cost $1.09, $.87, and $.65 per kilogram, respectively, how much of each should be used to minimize the cost of producing at least 750 kilograms of the plant food?

21. **Management** A company is developing a new additive for gasoline. The additive is a mixture of three liquid ingredients, I, II, and III. For proper performance, the total amount of additive must be at least 10 ounces per gallon of gasoline. However, for safety reasons, the amount of additive should not exceed 15 ounces per gallon of gasoline. At least 1/4 ounce of ingredient I must be used for every ounce of ingredient II and at least 1 ounce of ingredient III must be used for every ounce of ingredient I. If the cost of I, II, and III is $.30, $.09, and

$.27 per ounce, respectively, find the mixture of the three ingredients that produces the minimum cost of the additive. How much of the additive should be used per gallon of gasoline?

22. **Management** A popular soft drink called Sugarlo, which is advertised as having a sugar content of no more than 10%, is blended from five ingredients, each of which has some sugar content. Water may also be added to dilute the mixture. The sugar content of the ingredients and their costs per gallon are given below.

	Ingredient					
	1	2	3	4	5	Water
Sugar content(%)	.28	.19	.43	.57	.22	0
Cost ($/gal.)	.48	.32	.53	.28	.43	.04

At least .01 of the content of Sugarlo must come from ingredients 3 or 4, .01 must come from ingredients 2 or 5, and .01 from ingredients 1 or 4. How much of each ingredient should be used in preparing at least 15,000 gallons of Sugarlo to minimize the cost?

Use the two-stage method to solve Exercises 23–30.

23. **Management** Southwestern Oil supplies two distributors in the Northwest from two outlets. Distributor D_1 needs at least 3000 barrels of oil, and distributor D_2 needs at least 5000 barrels. The two outlets can each furnish 5000 barrels of oil. The costs per barrel to send the oil are given below.

		To	
		D_1	D_2
From	S_1	$30	$20
	S_2	$25	$22

How should the oil be supplied to minimize shipping costs?

24. Mark, who is ill, takes vitamin pills. Each day he must have at least 16 units of vitamin A, 5 units of vitamin B_1, and 20 units of vitamin C. He can choose between pill #1 which costs 10¢ and contains 8 units of A, 1 of B_1, and 2 of C, and pill #2 which costs 20¢ and contains 2 units of A, 1 of B_1, and 7 of C. How many of each pill should he buy in order to minimize his cost?

25. Management A bank has set aside a maximum of $25 million for commercial and home loans. The bank's policy is to loan at least four times as much for home loans as for commerical loans. Because of prior commitments, at least $10 million will be used for these two types of loans. The bank earns 12% on home loans and 10% on commercial loans. What amount of money should be loaned out for each type of loan to maximize the interest income?

26. Sam, who is dieting, requires two food supplements, I and II. He can get these supplements from two different products, A and B, as shown in the following table.

		Supplement (Grams per Serving)	
		I	II
Product	A	3	2
	B	2	4

Sam's physician has recommended that he include at least 15 grams of supplement I but no more than 12 grams of II in his daily diet. If product A costs 25¢ per serving and product B costs 40¢ per serving, how can he satisfy his requirements most economically?

27. Management Brand X Canners produce canned whole tomatoes and tomato sauce. This season, they have available 3,000,000 kilograms of tomatoes for these two products. To meet the demands of regular customers, they must produce at least 80,000 kilograms of sauce and 800,000 kilograms of whole tomatoes. The cost per kilogram is $4 to produce canned whole tomatoes and $3.25 to produce tomato sauce. How many kilograms of tomatoes should they use for each product to minimize cost?

28. Managment A brewery produces regular beer and a lower-carbohydrate "light" beer. Steady customers of the brewery buy 12 units of regular beer and 10 units of light beer. While setting up the brewery to produce the beers, the management decides to produce extra beer, beyond that needed to satisfy the steady customers. The cost per unit of regular beer is $36,000 and the cost per unit of light beer is $48,000. The number of units of light beer should not exceed twice the number of units of regular beer. At least 20 additional units of beer can be sold. How much of each type beer should be made so as to minimize total production costs?

29. The chemistry department at a local college decides to stock at least 800 small test tubes and 500 large test tubes. It wants to buy at least 1500 test tubes to take advantage of a special price. Since the small tubes are broken twice as often as the larger, the department will order at least twice as many small tubes as large. If the small test tubes cost 15¢ each and the large ones, made of a cheaper glass, cost 12¢ each, how many of each size should they order to minimize cost?

30. Management Topgrade Turf lawn seed mixture contains three types of seeds: bluegrass, rye, and bermuda. The costs per pound of the three types of seed are 20¢, 15¢ and 5¢. In each mixture there must be at least 20% bluegrass seed and the amount of bermuda must be no more than the amount of rye. To fill current orders, the company must make at least 5000 pounds of the mixture. How much of each kind of seed should be used minimize cost?

KEY TERMS AND SYMBOLS

	linear inequality	indicator
7.1	boundary	pivot and pivoting
	half-plane	basic variables
	system of inequalities	nonbasic variables
	region of feasible solutions	basic feasible solution
	(feasible region)	**7.6** dual
7.2	linear programming	transpose of a matrix
	objective function	theorem of duality
	constraints	shadow costs
	corner point	**7.7** surplus variable
	bounded feasible region	basic variables
	unbounded feasible region	basic solution
	corner point theorem	basic feasible solution
7.4	standard maximum form	Stage I
	slack variable	Stage II
	simplex tableau	minimization problems

KEY CONCEPTS

Graphing a Linear Inequality

Graph the boundary line as a solid line if the inequality includes "or equal", a broken line otherwise. Shade the half-plane that includes a test point that makes the inequality true. The graph of a system of inequalities, called the **region of feasible solutions,** includes all points that satisfy all the inequalities of the system at the same time.

Solving Linear Programming Problems

Graphically: Determine the objective function and all necessary constraints. Graph the region of feasible solutions. The maximum or minimum value will occur at one or more of the corner points of this region.

Simplex Method: Determine the objective function and all necessary constraints. Convert each constraint into an equation by adding slack variables. Set up the initial simplex tableau. Locate the most negative indicator. Form the quotients to determine the pivot. Use row operations to change the pivot to 1 and all other numbers in that column to 0. If the indicators are all positive or 0, this is the final tableau. If not, choose a new pivot and repeat the process until no indicators are negative. Read the solution from the final tableau. The optimum value of the objective function is the number in the lower right corner of the final tableau. For problems with **mixed constraints,** add surplus variables as well as slack variables. In stage I, use row operations to transform the matrix until the solution is feasible. In stage II, use the simplex method as described above. For **minimum** problems, let the objective function be w and set $-w = z$. Then proceed as with mixed constraints.

Solving Minimum Problems With Duals

Find the dual maximum problem. Solve the dual using the simplex method. The minimum value of the objective function w is the maximum value of the dual objective function z. The optimal solution is found in the entries in the bottom row of the columns corresponding to the slack variables.

CHAPTER 7 REVIEW EXERCISES

Graph each of the following linear inequalities.

1. $y \le 3x + 2$

2. $2x - y \ge 6$

3. $4x + 3y \ge 12$

4. $y \le x$

5. $y \le 4$

6. $4x - 2y \ge 10$

Graph the solution of each of the following systems of inequalities.

7. $x + y \le 6$
$2x - y \ge 3$

8. $4x + y \ge 8$
$2x - 3y \le 6$

9. $-4 \le x \le 2$
$-1 \le y \le 3$
$x + y \le 4$

10. $2 \le x \le 5$
$1 \le y \le 7$
$x - y \le 3$

11. $x + 3y \ge 6$
$4x - 3y \le 12$
$x \ge 0$
$y \ge 0$

12. $x + 2y \le 4$
$2x - 3y \le 6$
$x \ge 0$
$y \ge 0$

Set up a system of inequalities for each of the following problems; then graph the region of feasible solutions.

13. A bakery makes both cakes and cookies. Each batch of cakes requires 2 hours in the oven and 3 hours in the decorating room. Each batch of cookies needs $1\frac{1}{2}$ hours in the oven and 2/3 of an hour in the decorating room. The oven is available no more than 15 hours a day, while the decorating room can be used no more than 13 hours a day.

14. A company makes two kinds of pizza, basic and plain. Basic contains cheese and beef, while plain contains onions and beef. The company sells at least 3 units a day of basic, and at least 2 units of plain. The beef costs $5 per unit for basic, and $4 per unit for plain. They can spend no more than $50 per day on beef. Dough for basic is $2 per unit, while dough for plain is $1 per unit. The company can spend no more than $16 per day on dough.

Use the given regions to find the maximum and minimum values of the objective function $z = 2x + 4y$.

15.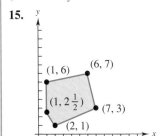

16.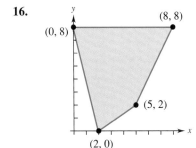

Use the graphical method to solve Exercises 17–22.

17. Maximize $z = 3x + 2y$
  subject to: $2x + 7y \leq 14$
          $2x + 3y \leq 10$
          $x \geq 0, \quad y \geq 0.$

18. Find $x \geq 0$ and $y \geq 0$ such that
          $8x + 9y \geq 72$
          $6x + 8y \geq 72$
  and $w = 4x + 12y$ is minimized.

19. Find $x \geq 0$ and $y \geq 0$ such that
          $x + \quad y \leq 50$
          $2x + \quad y \geq 20$
          $x + 2y \geq 30$
  and $w = 8x + 3y$ is minimized.

20. Maximize $z = 2x - 5y$
  subject to: $3x + 2y \leq 12$
          $5x + \quad y \geq 5$
          $x \geq 0, \quad y \geq 0.$

21. How many batches of cakes and cookies should the bakery of Exercise 13 make in order to maximize profits if cookies produce a profit of $20 per batch and cakes produce a profit of $30 per batch?

22. How many units of each kind of pizza should the company of Exercise 14 make in order to maximize profits if basic sells for $20 per unit and plain for $15 per unit?

For Exercises 23–26, (a) select appropriate variables, (b) write the objective function, (c) write the constraints as inequalities.

23. Roberta Hernandez sells three items, A, B, and C, in her gift shop. Each unit of A costs her $2 to buy, $1 to sell, and $2 to deliver. For each unit of B, the costs are $3, $2, and $2, respectively, and for each unit of C the costs are $6, $2, and $4, respectively. The profit on A is $4, on B it is $3, and on C, $3. How many of each should she get to maximize her profit if she can spend $1200 to buy, $800 on selling costs, and $500 on delivery costs?

24. An investor is considering three types of investment: a high-risk venture into oil leases with a potential return of 15%, a medium-risk investment in bonds with a 9% return, and a relatively safe stock investment with a 5% return. He has $50,000 to invest. Because of the risk, he will limit his investment in oil leases and bonds to 30% and his investment in oil leases and stock to 50%. How much should he invest in each to maximize his return, assuming investment returns are as expected?

25. The Aged Wood Winery makes two white wines, Fruity and Crystal, from two kinds of grapes and sugar. The wines require the following amounts of each ingredient per gallon and produce a profit per gallon as shown below.

	Grape A (bushels)	Grape B (bushels)	Sugar (pounds)	Profit (dollars)
Fruity	3	2	2	12
Crystal	1	2	1	15

The winery has available 110 bushels of grape A, 120 bushels of grape B, and 90 pounds of sugar. How much of each wine should be made to maximize profit?

26. A company makes three sizes of plastic bags: 5 gallon, 10 gallon and 20 gallon. The production time in hours for cutting, sealing, and packaging a unit of each size is shown below.

Size	Cutting	Sealing	Packaging
5 gallon	1	1	2
10 gallon	1.1	1.2	3
20 gallon	1.5	1.3	4

There are at most 8 hours available each day for each of the three operations. If the profit on a unit of 5-gallon bags is $1, 10-gallon bags is $.90, and 20-gallon bags is $.95, how many of each size should be made per day to maximize the profit?

For each of the following problems, (**a**) add slack variables and (**b**) set up the initial simplex tableau.

27. Maximize $z = 2x_1 + 7x_2$
subject to:
$$3x_1 + 5x_2 \le 47$$
$$x_1 + x_2 \le 25$$
$$5x_1 + 2x_2 \le 35$$
$$2x_1 + x_2 \le 30$$
$$x_1 \ge 0, \quad x_2 \ge 0.$$

28. Maximize $z = 15x_1 + 10x_2$
subject to:
$$2x_1 + 5x_2 \le 50$$
$$x_1 + 3x_2 \le 25$$
$$4x_1 + x_2 \le 18$$
$$x_1 + x_2 \le 12$$
$$x_1 \ge 0, \quad x_2 \ge 0.$$

29. Maximize $z = 4x_1 + 6x_2 + 3x_3$
subject to:
$$x_1 + x_2 + x_3 \le 100$$
$$2x_1 + 3x_2 \le 500$$
$$x_1 + 2x_3 \le 350$$
$$x_1 \ge 0, \quad x_2 \ge 0, \quad x_3 \ge 0.$$

30. Maximize $z = x_1 + 4x_2 + 2x_3$
subject to:
$$x_1 + x_2 + x_3 \le 90$$
$$2x_1 + 5x_2 + x_3 \le 120$$
$$x_1 + 3x_2 \le 80$$
$$x_1 \ge 0, \quad x_2 \ge 0, \quad x_3 \ge 0.$$

For each of the following, use the simplex method to solve the maximizing linear programming problems with initial tableaus as given.

31.

x_1	x_2	x_3	x_4	x_5	
1	2	3	1	0	28
2	4	8	0	1	32
-5	-2	-3	0	0	0

32.

x_1	x_2	x_3	x_4	
2	1	1	0	10
9	3	0	1	15
-2	-3	0	0	0

33.

x_1	x_2	x_3	x_4	x_5	x_6	
1	2	2	1	0	0	50
4	24	0	0	1	0	20
1	0	2	0	0	1	15
-5	-3	-2	0	0	0	0

34.

x_1	x_2	x_3	x_4	x_5	
1	-2	1	0	0	38
1	-1	0	1	0	12
2	1	0	0	1	30
-1	-2	0	0	0	0

Convert the following problems into maximization problems without using duals.

35. Minimize $w = 18y_1 + 10y_2$
subject to:
$$y_1 + y_2 \ge 17$$
$$5y_1 + 8y_2 \ge 42$$
$$y_1 \ge 0, \quad y_2 \ge 0.$$

36. Minimize $w = 12y_1 + 20y_2 - 8y_3$
subject to:
$$y_1 + y_2 + 2y_3 \ge 48$$
$$y_1 + y_2 \ge 12$$
$$y_3 \ge 10$$
$$3y_1 + y_3 \ge 30$$
$$y_1 \ge 0, \quad y_2 \ge 0, \quad y_3 \ge 0.$$

37. Minimize $w = 6y_1 - 3y_2 + 4y_3$
subject to:
$$2y_1 + y_2 + y_3 \ge 112$$
$$y_1 + y_2 + y_3 \ge 80$$
$$y_1 + y_2 \ge 45$$
$$y_1 \ge 0, \quad y_2 \ge 0, \quad y_3 \ge 0.$$

Use the simplex method to solve the following mixed constraint problems.

38. Maximize $z = 2x_1 + 4x_2$
subject to:
$$3x_1 + 2x_2 \le 12$$
$$5x_1 + x_2 \ge 5$$
$$x_1 \ge 0, \quad x_2 \ge 0.$$

39. Minimize $w = 4y_1 - 8y_2$
subject to:
$$y_1 + y_2 \le 50$$
$$2y_1 - 4y_2 \ge 20$$
$$y_1 - y_2 \le 22$$
$$y_1 \ge 0, \quad y_2 \ge 0.$$

The following tableaus are the final tableaus of minimizing problems solved by letting $w = -z$. Give the solution and the minimum value of the objective function for each problem.

40.
$$\begin{bmatrix} 0 & 1 & 0 & 2 & 5 & 0 & | & 17 \\ 0 & 0 & 1 & 3 & 1 & 1 & | & 25 \\ 1 & 0 & 0 & 4 & 2 & \frac{1}{2} & | & 8 \\ \hline 0 & 0 & 0 & 2 & 5 & 0 & | & -427 \end{bmatrix}$$

41.
$$\begin{bmatrix} 0 & 0 & 2 & 1 & 0 & 6 & 6 & | & 92 \\ 1 & 0 & 3 & 0 & 0 & 0 & 2 & | & 47 \\ 0 & 1 & 0 & 0 & 0 & 1 & 0 & | & 68 \\ 0 & 0 & 4 & 0 & 1 & 0 & 3 & | & 35 \\ \hline 0 & 0 & 5 & 0 & 0 & 2 & 9 & | & -1957 \end{bmatrix}$$

The following tableaus are the final tableaus of minimizing problems solved by the method of duals. State the solution and the minimum value of the objective function for each problem.

42.
$$\begin{bmatrix} 1 & 0 & 0 & 3 & 1 & 2 & | & 12 \\ 0 & 0 & 1 & 4 & 5 & 3 & | & 5 \\ 0 & 1 & 0 & -2 & 7 & -6 & | & 8 \\ \hline 0 & 0 & 0 & 5 & 7 & 3 & | & 172 \end{bmatrix}$$

43.
$$\begin{bmatrix} 0 & 0 & 1 & 6 & 3 & 1 & | & 2 \\ 1 & 0 & 0 & 4 & -2 & 2 & | & 8 \\ 0 & 1 & 0 & 10 & 7 & 0 & | & 12 \\ \hline 0 & 0 & 0 & 9 & 5 & 8 & | & 62 \end{bmatrix}$$

44.
$$\begin{bmatrix} 1 & 0 & 7 & -1 & | & 100 \\ 0 & 1 & 1 & 3 & | & 27 \\ \hline 0 & 0 & 7 & 2 & | & 640 \end{bmatrix}$$

45. Solve Exercise 23.

46. Solve Exercise 24.   **47.** Solve Exercise 25.   **48.** Solve Exercise 26.

Solve the following minimization problems.

49. A contractor builds boathouses in two basic models, the atlantic and pacific. Each atlantic model requires 1000 feet of framing lumber, 3000 cubic feet of concrete, and $2000 for advertising. Each pacific model requires 2000 feet of framing lumber, 3000 cubic feet of concrete, and $3000 for advertising. Contracts call for using at least 8000 feet of framing lumber, 18,000 cubic feet of concrete, and $15,000 worth of advertising. If the total spent on each atlantic model is $3000 and the total spent on each pacific model is $4000, how many of each model should be built to minimize costs?

50. A steel company produces two types of alloys. A run of type I requires 3000 pounds of molybdenum and 2000 tons of iron ore pellets as well as $2000 in advertising. A run of type II requires 3000 pounds of molybdenum and 1000 tons of iron ore pellets as well as $3000 in advertising. Total costs are $15,000 on a run of type I and $6000 on a run of type II. Because of various contracts, the company must use at least 18,000 pounds of molybdenum and 7000 tons of iron ore pellets and spend at least $14,000 on advertising. How much of each type should be produced to minimize costs?

Individuals doing the same job within the management of a company often receive different salaries. These salaries may differ because of the length of service of an employee, the productivity of an individual worker, and so on. However, for each job there is usually an established minimum and maximum salary.

Many companies make annual reviews of the salary of each of their management employees. At these reviews, an employee may receive a general cost of living increase, an increase based on merit, both, or neither.

In this case, we look at a mathematical model for distributing merit increases in an optimum way. An individual who is due for salary review may be described as shown in the figure below. Here i represents the number of the employee whose salary is being reviewed.

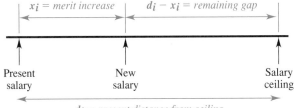

Here the salary ceiling is the maximum salary for the job classification, x_i is the merit increase to be awarded to the individual ($x_i \geq 0$), d_i is the present distance of the current salary from the salary ceiling, and the difference, $d_i - x_i$, is the remaining gap.

We let w_i be a measure of the relative worth of the individual to the company. This is the most difficult variable of the model to actually calculate. One way to evaluate w_i is to give a rating

sheet to a number of co-workers and supervisors of employee i. An average rating can be obtained and then divided by the highest rating received by any employee with that same job. This number then gives the worth of employee i relative to all other employees with that job.

The best way to allocate money available for merit pay increases is to minimize

$$\sum_{i=1}^{n} w_i(d_i - x_i).$$

This sum is found by multiplying the relative worth of employee i and the salary gap of employee i after any merit increase. Here n represents the total number of employees who have this job. One constraint here is that the total of all merit increases cannot exceed P, the total amount available for merit increases. That is,

$$\sum_{i=1}^{n} x_i \leq P.$$

Also, the increases for an employee must not put that employee over the maximum salary. That is, for employee i,

$$x_i \leq d_i.$$

We can simplify the objective function, using rules from algebra.

$$\sum_{i=1}^{n} w_i(d_i - x_i) = \sum_{i=1}^{n} (w_i d_i - w_i x_i)$$

$$= \sum_{i=1}^{n} w_i d_i - \sum_{i=1}^{n} w_i x_i$$

---

*Based on part of a paper by Jack Northam, Head, Mathematical Services Department, The Upjohn Company, Kalamazoo, Michigan.

For a given individual, w_i and d_i are constant. Therefore, the sum of the $w_i d_i$ is some constant, say Z, and

$$\sum_{i=i}^{n} w_i(d_i - x_i) = Z - \sum_{i=1}^{n} w_i x_i.$$

We want to minimize the sum on the left; we can do so by *maximizing* the sum on the right. (Why?) Thus, the original model simplifies to maximizing

$$\sum_{i=1}^{n} w_i x_i,$$

subject to

$$\sum_{i=1}^{n} x_i \le P \quad \text{and} \quad x_i \le d_i$$

for each i.

EXERCISES

Here are current salary information and job evaluation averages for six employees who have the same job. The salary ceiling is $1700 per month.

Employee Number	Evaluation Average	Current Salary
1	570	$1600
2	500	1550
3	450	1500
4	600	1610
5	520	1530
6	565	1420

1. Find w_i for each employee by dividing that employee's evaluation average by the highest evaluation average.

2. Use the simplex method and a computer to find the merit increase for each employee. Assume that P is 400.

3. What are some of the limitations and advantages of solving this problem by a strictly mathematical approach?

The first step in the commercial manufacture of ice cream is to blend several ingredients (such as dairy products, eggs, and sugar) to obtain a mix that meets the necessary minimum quality restrictions regarding butterfat content, serum solids, and so on.

Usually, many different combinations of ingredients may be blended to obtain a mix of the necessary quality. Within this range of possible substitutes, the firm desires the combination that produces minimum total cost. This problem is quite suitable for solution by linear programming methods.

A "mid-quality" line of ice cream requires the following minimum percentages of constituents by weight.

Fat	16.00%
Serum solids	8.00
Sugar solids	16.00
Egg solids	.35
Stabilizer	.25
Emulsifier	.15
Total	40.75%

The balance of the mix is water. A batch of mix is made by blending a number of ingredients, each of which contains one or more of the necessary constituents, and nothing else. The chart on the next page shows the possible ingredients and their costs.

In setting up the mathematical model, use $c_1, c_2, \ldots, c_{14}$ as the cost per unit of ingredients, and $x_1, x_2, \ldots, x_{14}$ for the quantities of ingredients. The table shows the composition of each ingredient. For example, 1 pound of ingredient 1 (the 40% cream) contains .400 pounds of fat and .054 pounds of serum solids, with the balance being water.

* From *Linear Programming in Industry: Theory and Application, An Introduction* by Sven Danø, Fourth revised and enlarged edition. Copyright © 1974 by Springer-Verlag/Wein. Reprinted by permission of Springer-Verlag New York, Inc.

The ice cream mix is made up in batches of 100 pounds at a time. For a batch to contain 16% fat, at least $100(.16) = 16$ pounds of fat is necessary. Fat is contained in ingredients 1, 2, 3, 4, 5, 6, 10, and 11. The requirement of at least 16 pounds of fat produces the constraint

$$.400x_1 + .230x_2 + .805x_3 + .800x_4 + .998x_5 + .040x_6 + .500x_{10} + .625x_{11} \geq 16.$$

Similar constraints can be obtained for the other constituents. Because of the minimum requirements, the total of the ingredients will be at least 40.75 pounds, or

$$x_1 + x_2 + \cdots + x_{13} \geq 40.75.$$

The balance of the 100 pounds is water, making $x_{14} \leq 59.25$.

The table shows that ingredients 12 and 13 must be used—there is no alternate way of getting these consituents into the final mix. For a 100-pound batch of ice cream, $x_{12} = .25$ pound and $x_{13} = .15$ pound. Removing x_{12} and x_{13} as variables permits a substantial simplification of the model; the problem is now reduced to the following system of four constraints:

$$.400x_1 + .230x_2 + .805x_3 + .800x_4 + .998x_5 + .040x_6 + .500x_{10} + .625x_{11} \geq 16$$

$$.054x_1 + .069x_2 + .025x_4 + .078x_6 + .280x_7 + .970x_8 \geq 8$$

$$.707x_9 + .100x_{10} \geq 16$$

$$.350x_{10} + .315x_{11} \geq .35.$$

The objective function, which is to be minimized, is

$$c = .298x_1 + .174x_2 + \cdots + 1.090x_{11}.$$

As usual, $x_1 \geq 0$, $x_2 \geq 0$, . . . , $x_{14} \geq 0$.

Solving this linear programming problem with the simplex method gives

$$x_2 = 67.394, \quad x_8 = 3.459, \quad x_9 = 22.483, \quad x_{10} = 1.000.$$

The total cost of these ingredients is $14.094. To this, we must add the cost of ingredients 12 and 13, producing a total cost of

$$14.094 + (.600)(.25) + (.420)(.15) = 14.307,$$

or $14.307.

		Ingredients								
		1	2	3	4	5	6	7	8	9
		40% Cream	23% Cream	Butter	Plastic Cream	Butter Oil	4% Milk	Skim Condensed Milk	Skim Milk Powder	Liquid Sugar
		Cost ($/lb)								
		.298	.174	.580	.576	.718	.045	.052	.165	.061
Constituents	1. Fat	.400	.230	.805	.800	.998	.040			
	2. Serum solids	.054	.069		.025		.078	.280	.970	
	3. Sugar solids									.707
	4. Egg solids									
	5. Stabilizer									
	6. Emulsifier									

		Ingredients					
		10	11	12	13	14	
		Sugared Egg Yolk	Powdered Egg Yolk	Stabilizer	Emulsifier	Water	
		Cost ($/lb)					
		.425	1.090	.600	.420	0	Requirements
Constituents	1. Fat	.500	.625				16.00
	2. Serum solids						8.00
	3. Sugar solids	.100					16.00
	4. Egg solids	.350	.315				.35
	5. Stabilizer			1			.25
	6. Emulsifier				1		.15

EXERCISES

1. Find the mix of ingredients needed for a 100-pound batch of the following grades of ice cream.
   (a) "Generic," sold in a white box, with at least 14% fat and at least 6% serum solids (all other ingredients the same)
   (b) "Premium," sold with a funny foreign name, with at least 17% fat and at least 16.5% sugar solids (all other ingredients the same)

2. Solve the problem given in this case by using a computer.

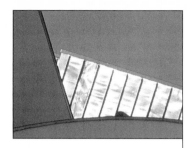

CHAPTER 8

Sets and Probability

Federal officials cannot predict exactly how traffic deaths are affected by the trends toward fewer drunken drivers and increased use of seat belts. Economists cannot tell exactly how stricter federal regulations on bank loans affect the U.S. economy. The number of traffic deaths and the growth of the economy are subject to many factors that cannot be predicted precisely.

Probability theory enables us to deal with uncertainty. The basic concepts of probability are discussed in this chapter and applications of probability are discussed in the next chapter. Sets and set operations are the basic tools for the study of probability and so we begin with them.

8.1 SETS

Think of a **set** as a collection of objects. For example, a set of coins might contain one of each type of coin now put out by the United States government. Another set might include all the students in a class. The set containing the numbers 3, 4, and 5 is written

$$\{3, 4, 5\},$$

where **set braces**, { }, are used to enclose the numbers belonging to the set. The numbers, 3, 4, and 5 are called the **elements** or **members** of this set. To show

1 Write *true* or *false*.

(a) $9 \in \{8, 4, -3, -9, 6\}$

(b) $4 \notin \{3, 9, 7\}$

(c) If $M = \{0, 1, 2, 3, 4\}$, then $0 \in M$.

Answers:

(a) False

(b) True

(c) True

that 4 is an element of the set $\{3, 4, 5\}$, we use the symbol $\in$ and write

$$4 \in \{3, 4, 5\},$$

read "4 is an element of the set containing 3, 4, and 5."
Also, $5 \in \{3, 4, 5\}$. Place a slash through the symbol $\in$ to show that 8 is *not* an element of this set.

$$8 \notin \{3, 4, 5\}$$

This is read "8 is not an element of the set $\{3, 4, 5\}$."
Sets are often named with capital letters, so that if

$$B = \{5, 6, 7\},$$

then, for example, $6 \in B$ and $10 \notin B$. **1**

Sometimes a set has no elements. Some examples are the set of woman presidents of the United States, the set of counting numbers less than 1, and the set of men more than 10 feet tall. A set with no elements is called the **empty set.** The symbol $\emptyset$ is used to represent the empty set.

Caution Be careful to distinguish between the symbols 0, $\emptyset$ and $\{0\}$. The symbol 0 represents a *number;* $\emptyset$ represents a *set* with no elements; and $\{0\}$ represents a *set* with one element, the number 0. Do not confuse the empty set symbol $\emptyset$ with the zero $\emptyset$ on a computer screen or printout.

Two sets are **equal** if they contain exactly the same elements. The sets $\{5, 6, 7\}$, $\{7, 6, 5\}$, and $\{6, 5, 7\}$ all contain exactly the same elements and are equal. In symbols,

$$\{5, 6, 7\} = \{7, 6, 5\} = \{6, 5, 7\}.$$

This means that the ordering of the elements in a set is unimportant. Sets that do not contain exactly the same elements are *not equal.* For example, the sets $\{5, 6, 7\}$ and $\{5, 6, 7, 8\}$ do not contain exactly the same elements and are not equal. We show this by writing

$$\{5, 6, 7\} \neq \{5, 6, 7, 8\}.$$

Sometimes we describe a set by a common property of its elements rather than by a list of elements. This common property can be expressed with **set-builder notation;** for example,

$$\{x \mid x \text{ has property } P\}$$

(read "the set of all elements x such that x has property P") represents the set of all elements x having some property P.

2 List the elements in the following sets.

(a) $\{x \mid x$ is a counting number more than 5 and less than 8$\}$

(b) $\{x \mid x$ is an integer, $-3 < x \le 1\}$

Answers:

(a) $\{6, 7\}$

(b) $\{-2, -1, 0, 1\}$

3 Write *true* or *false*.

(a) $\{3, 4, 5\} \subseteq \{2, 3, 4, 6\}$

(b) $\{x \mid x$ is an automobile$\}$ $\subseteq \{x \mid x$ is a motor vehicle$\}$

(c) $\{3, 6, 9, 10\}$ $\subseteq \{3, 9, 11, 13\}$

Answers:

(a) False

(b) True

(c) False

▶**EXAMPLE 1**   List the elements belonging to each of the following sets.
(a) $\{x \mid x$ is a natural number less than 5$\}$
   The natural numbers less than 5 make up the set $\{1, 2, 3, 4\}$.
(b) $\{x \mid x$ is a state that touches Florida$\} = \{$Alabama, Georgia$\}$ ◀ **2**

The **universal set** in a particular discussion is a set that contains all of the objects being discussed. In grade-school arithmetic, for example, the set of whole numbers might be the universal set, whereas in a college calculus class the universal set might be the set of all real numbers. When it is necessary to consider the universal set being used, it will be clearly specified or easily understood from the context of the problem.

Sometimes every element of one set also belongs to another set. For example, if

$$A = \{3, 4, 5, 6\}$$

and

$$B = \{2, 3, 4, 5, 6, 7, 8\},$$

then every element of A is also an element of B. This is an example of the following definition.

A set A is a **subset** of a set B (written $A \subseteq B$) provided that every element of A is also an element of B.

▶**EXAMPLE 2**   Decide whether or not $M \subseteq N$.
(a) M is the set of all daughters in a family and N is the set of all children in the family.
   Each daughter in a family is a child in the family, so $M \subseteq N$.
(b) M is the set of all fourth-grade students in a school at the end of the school year, and N is the set of all nine-year-old students in the school at the end of the school year.
   By the end of the school year, some fourth-grade students are ten years old, so there are elements in M that are not in N. Thus, M is not a subset of N, written $M \nsubseteq N$.   ◀ **3**

Every set A is a subset of itself because the statement "every element of A is also an element of A" is always true. It is also true that the empty set is a subset of every set.*

———————————
   * This fact is not intuitively obvious to most people. If you wish, you can think of it as a convention that we agree to adopt in order to simplify the statements of several results later.

4 List all subsets of
{w, x, y, z}.

Answer:
∅, {w}, {x}, {y}, {z}, {w, x},
{w, y}, {w, z}, {x, y}, {x, z},
{y, z}, {w, x, y}, {w, x, z},
{w, y, z}, {x, y, z}, {w, x, y, z}

For any set A,

$$\emptyset \subseteq A \quad \text{and} \quad A \subseteq A.$$

▶**EXAMPLE 3** List all possible subsets for each of the following sets.

(a) {7, 8}

There are 4 subsets of {7, 8}:

$$\emptyset, \quad \{7\}, \quad \{8\}, \quad \{7, 8\}.$$

(b) {a, b, c}

There are 8 subsets of {a, b, c}:

$$\emptyset, \quad \{a\}, \quad \{b\}, \quad \{c\}, \quad \{a, b\}, \quad \{a, c\}, \quad \{b, c\}, \quad \{a, b, c\}. \quad ◀$$

In Example 3, the subsets of {7, 8} and the subsets of {a, b, c} were found by trial and error. An alternative method uses a **tree diagram,** a systematic way of listing all the subsets of a given set. Figures 8.1(a) and (b) show tree diagrams for finding the subsets of {7, 8} and {a, b, c}. **4**

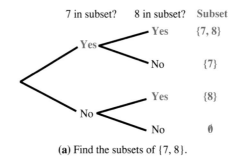

(a) Find the subsets of {7, 8}.

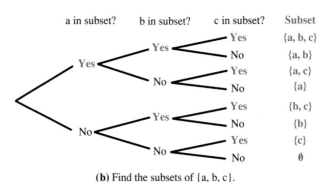

(b) Find the subsets of {a, b, c}.

FIGURE 8.1

5 Find the number of subsets for each of the following sets.

(a) $\{x \mid x$ is a season of the year$\}$

(b) $\{-6, -5, -4, -3, -2, -1, 0\}$

(c) $\{6\}$

Answers:

(a) 16   **(b)** 128   **(c)** 2

6 Refer to sets A, B, C, and U in the diagram.

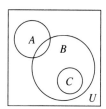

(a) Is $A \subseteq B$?

(b) Is $C \subseteq B$?

(c) Is $C \subseteq U$?

(d) Is $\emptyset \subseteq A$?

Answers:

(a) No

(b) Yes

(c) Yes

(d) Yes

By using the fact that there are two possibilities for each element (either it is in the subset or it is not) we have found that a set with 2 elements has 4 ($=2^2$) subsets and a set with 3 elements has 8 ($=2^3$) subsets. Similar arguments work for any finite set and lead to this conclusion.

A set of n distinct elements has 2^n subsets.

▶**EXAMPLE 4**   Find the number of subsets for each of the following sets.
(a) $\{3, 4, 5, 6, 7\}$
   Since this set has 5 elements, it has 2^5 or 32 subsets.
(b) $\{x \mid x$ is a day of the week$\}$
   This set has 7 elements and therefore has $2^7 = 128$ subsets.
(c) $\emptyset$
   Since the empty set has 0 elements, it has $2^0 = 1$ subset, $\emptyset$ itself.   ◀

Figure 8.2 shows a set A, which is a subset of a set B, because A is entirely in B. (The areas of the regions are not meant to be proportional to the size of the corresponding sets.) The rectangle represents the universal set, U. Such diagrams, called **Venn diagrams,** are used to illustrate relationships among sets.   **6**

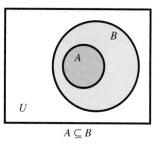

$A \subseteq B$

FIGURE 8.2

We can combine sets to get other sets, just as operations on numbers, such as addition, subtraction, and multiplication produce other numbers. For example, given a set A and a universal set U, the set of all elements of U which do *not* belong to A is called the **complement** of set A. For example, if set A is the set of all the female students in your class and U is the set of all students in the class, then the complement of A would be the set of all male students in the class. The

7 Let $U = \{a, b, c, d, e, f, g\}$, with $K = \{c, d, f, g\}$ and $R = \{a, c, d, e, g\}$. Find

(a) K';

(b) R'.

Answers:

(a) $\{a, b, e\}$

(b) $\{b, f\}$

complement of set A is written A' (read "A-prime"). The Venn diagram of Figure 8.3 shows a set B. Its complement, B', is shown in color.

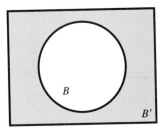

FIGURE 8.3

▶ **EXAMPLE 5**   Let $U = \{1, 2, 3, 4, 5, 6, 7\}$, $A = \{1, 3, 5, 7\}$, and $B = \{3, 4, 6\}$. Find the following sets.

(a) A'

Set A' contains the elements of U that are not in A.

$$A' = \{2, 4, 6\}$$

(b) $B' = \{1, 2, 5, 7\}$

(c) $\emptyset' = U$ and $U' = \emptyset$ ◀ **7**

Given two sets A and B, the set of all elements belonging to *both* set A and set B is called the **intersection** of the two sets, written $A \cap B$. For example, the elements that belong to both $A = \{1, 2, 4, 5, 7\}$ and $B = \{2, 4, 5, 7, 9, 11\}$ are 2, 4, 5, and 7, so that

$$A \cap B = \{1, 2, 4, 5, 7\} \cap \{2, 4, 5, 7, 9, 11\} = \{2, 4, 5, 7\}.$$

The Venn diagram of Figure 8.4 shows two sets A and B with their intersection, $A \cap B$, shown in color.

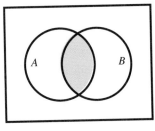

$A \cap B$

FIGURE 8.4

8 Find the following.

(a) $\{1, 2, 3, 4\} \cap \{3, 5, 7, 9\}$

(b) Suppose set K is the set of all blue-eyed blondes in a class and J is the set of all blue-eyed brunettes in the class. If the class has only blondes or brunettes, describe $K \cap J$ in words.

(c) Let $P = \{x \mid x$ is a brown-eyed redhead$\}$. Describe $K \cap P$.

Answers:

(a) $\{3\}$

(b) All members of the class with blue eyes

(c) $\emptyset$

▶**EXAMPLE 6**

(a) $\{9, 15, 25, 36\} \cap \{15, 20, 25, 30, 35\} = \{15, 25\}$

The elements 15 and 25 are the only ones belonging to both sets.

(b) $\{x \mid x$ is a teen-ager$\} \cap \{x \mid x$ is a senior citizen$\}$ is an empty set. ◀

Two sets that have no elements in common are called **disjoint sets.** For example, there are no elements common to both $\{50, 51, 54\}$ and $\{52, 53, 55, 56\}$, so that these two sets are disjoint, and

$$\{50, 51, 54\} \cap \{52, 53, 55, 56\} = \emptyset.$$

The result of this example can be generalized:

For any sets A and B,

if A and B are disjoint sets then $A \cap B = \emptyset$.

Figure 8.5 is a Venn diagram of disjoint sets.

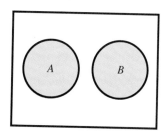

A and *B* are disjoint sets.

FIGURE 8.5

The set of all elements belonging to set A, to set B, or to both sets is called the **union** of the two sets, written $A \cup B$. For example,

$$\{1, 3, 5\} \cup \{3, 5, 7, 9\} = \{1, 3, 5, 7, 9\}.$$

The Venn diagram of Figure 8.6 shows two sets A and B, with their union $A \cup B$ shown in color.

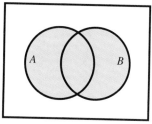

$A \cup B$

FIGURE 8.6

9 Find the following.

(a) {a, b, c} ∪ {a, c, e}

(b) Describe $K \cup J$ in words for the sets given in Problem 8(b).

Answers:

(a) {a, b, c, e}

(b) All members of the class with blue eyes.

▶ **EXAMPLE 7**

(a) Find the union of {1, 2, 5, 9, 14} and {1, 3, 4, 8}.

Begin by listing the elements of the first set, {1, 2, 5, 9, 14}. Then include any elements from the second set *that are not already listed.* Doing this gives

$$\{1, 2, 5, 9, 14\} \cup \{1, 3, 4, 8\} = \{1, 2, 3, 4, 5, 8, 9, 14\}.$$

(b) {terriers, spaniels, chows, dalmatians} ∪ {spaniels, collies, bulldogs} = {terriers, spaniels, chows, dalmatians, collies, bulldogs} ◀ **9**

Finding the complement of a set, the intersection of two sets, or the union of two sets are examples of *set operations.*

The set operations are summarized below.

Operations on Sets

Let A and B be any sets with U the universal set. Then

the **complement** of A, written A', is

$$A' = \{x \,|\, x \notin A \text{ and } x \in U\};$$

the **intersection** of A and B is

$$A \cap B = \{x \,|\, x \in A \text{ and } x \in B\};$$

the **union** of A and B is

$$A \cup B = \{x \,|\, x \in A \text{ or } x \in B\}.$$

Caution As shown in the definitions given above, an element is in the *intersection* of sets A and B if it is in *both A and B.* On the other hand, an element is in the *union* of sets A and B if it is in *A or B or both.*

10

If $U =$
$$\{-3, -2, -1, 0, 1, 2, 3\},$$
$A = \{-1, -2, -3\}$, and
$B = \{-2, 0, 2\}$, find

(a) $A \cup B$; **(b)** B';

(c) $A \cap B$.

Answers:

(a) $\{-3, -2, -1, 0, 2\}$

(b) $\{-3, -1, 1, 3\}$ **(c)** $\{-2\}$

▶**EXAMPLE 8** The table below gives recent dividends, price to earnings ratios (the quotient of the price per share and the annual earnings per share), and price changes at the end of a day for six companies, as listed on the New York Stock Exchange.

Stock	*Dividend*	*Price to Earnings Ratio*	*Price Change*
ATT	1.32	18	−1 3/8
Gen Mill	1.68	21	−3/4
Hershey	1.08	18	+1/8
IBM	2.16	none	−3/4
Mobil	3.20	21	−1/8
PepsiCo	.52	25	0

Let set A include all stocks with a dividend greater than 2: B all stocks with a price to earnings ratio of at least 20, and C all stocks with a positive price change. Find the following.

(a) A'

Set A' contains all the listed stocks outside set A, those with a dividend less than or equal to 2, so

$$A' = \{\text{ATT, Gen Mill, Hershey, PepsiCo}\}.$$

(b) $A \cap B$

The intersection of A and B will contain those stocks that offer a dividend greater than 2 *and* have a price to earnings ratio of at least 20.

$$A \cap B = \{\text{Mobil}\}$$

(c) $A \cup C$

The union of A and C contains all stocks with a dividend greater than 2 or a positive price change (or both).

$$A \cup C = \{\text{Hershey, IBM, Mobil}\} \quad ◀ \quad \boxed{10}$$

8.1 EXERCISES

Write true or false for each statement.

1. $3 \in \{2, 5, 7, 9, 10\}$

2. $6 \in \{-2, 6, 9, 5\}$

3. $9 \notin \{2, 1, 5, 8\}$

4. $3 \notin \{7, 6, 5, 4\}$

5. $\{2, 5, 8, 9\} = \{2, 5, 9, 8\}$

6. $\{3, 7, 12, 14\} = \{3, 7, 12, 14, 0\}$

7. {all whole numbers greater than 7 and less than 10} = {8, 9}

8. {all counting numbers not greater than 3} = {0, 1, 2}

9. $\{x \mid x$ is an odd integer, $6 \le x \le 18\} =$ {7, 9, 11, 15, 17}

10. $\{x \mid x$ is a vowel$\} = \{$a, e, i, o, u$\}$

11. The elements of a set may be sets themselves, as in $\{1, \{1, 3\}, \{2\}, 4\}$. Explain why the set $\{\emptyset\}$ is not the same set as $\{0\}$.

12. What is set-builder notation? Give an example.

Let $A = \{-3, 0, 3\}$, $B = \{-2, -1, 0, 1, 2\}$, $C = \{-3, -1\}$, $D = \{0\}$, $E = \{-2\}$, and $U = \{-3, -2, -1, 0, 1, 2, 3\}$. Insert $\subseteq$ or $\not\subseteq$ to make the following statements true.

13. $A \_\_ U$

14. $E \_\_ A$

15. $A \_\_ E$

16. $B \_\_ C$

17. $\emptyset \_\_ A$

18. $\{0, 2\} \_\_ D$

19. $D \_\_ B$

20. $A \_\_ C$

Find the number of subsets for each set.

21. $\{A, B, C\}$

22. {red, yellow, blue, black, white}

23. $\{x \mid x$ is an integer between 0 and 7$\}$

24. $\{x \mid x$ is a whole number less than 4$\}$

25. Describe the intersection and union of sets. How do they differ?

Insert $\cap$ or $\cup$ to make each statement true. (See Examples 6 and 7.)

26. $\{5, 7, 9, 19\} \_\_ \{7, 9, 11, 15\} = \{7, 9\}$

27. $\{8, 11, 15\} \_\_ \{8, 11, 19, 20\} = \{8, 11\}$

28. $\{2, 1, 7\} \_\_ \{1, 5, 9\} = \{1\}$

29. $\{6, 12, 14, 16\} \_\_ \{6, 14, 19\} = \{6, 14\}$

30. $\{3, 5, 9, 10\} \_\_ \emptyset = \emptyset$

31. $\{3, 5, 9, 10\} \_\_ \emptyset = \{3, 5, 9, 10\}$

32. $\{1, 2, 4\} \_\_ \{1, 2, 4\} = \{1, 2, 4\}$

33. $\{1, 2, 4\} \_\_ \{1, 2\} = \{1, 2, 4\}$

34. Is it possible for two nonempty sets to have the same intersection and union? If so, give an example.

Let $U = \{2, 3, 4, 5, 7, 9\}$; $X = \{2, 3, 4, 5\}$; $Y = \{3, 5, 7, 9\}$; and $Z = \{2, 4, 5, 7, 9\}$. List the members of each of the following sets, using set braces. (See Example 5.)

35. $X \cap Y$

36. $X \cup Y$

37. X'

38. Y'

39. $X' \cap Y'$

40. $X' \cap Z$

41. $X \cup (Y \cap Z)$

42. $Y \cap (X \cup Z)$

Let $U = \{$all students in this school$\}$; $M = \{$all students taking this course$\}$; $N = \{$all students taking accounting$\}$; $P = \{$all students taking zoology$\}$.

Describe each of the following sets in words.

43. M'

44. $M \cup N$

45. $N \cap P$

46. $N' \cap P'$

47. Refer to the sets listed in the directions for Exercises 13–20. Which pairs of sets are disjoint?

48. Refer to the sets listed in the directions for Exercises 35–42. Which pairs are disjoint?

Refer to Example 8 in the text. Describe each of the sets in Exercises 49–52 in words; then list the elements of each set.

49. B'

50. $A \cup B$

51. $(A \cap B)'$

52. $(A \cup C)'$

Management   *A department store classifies credit applicants by sex, marital status, and employment status. Let the universal set be the set of all applicants, M be the set of male applicants, S be the set of all single applicants, and E be the set of employed applicants. Describe the following sets in words.*

53. $M \cap E$

54. $M' \cap S$

55. $M' \cup S'$

Social Science   *The top five pay-cable services in 1988 are listed below.* Use this information for Exercises 56–61.*

Network	Subscribers (in thousands)	Content
Home Box Office (HBO)	16,500	Movies, variety, sports, documentaries, etc.
Showtime	6,100	Movies, variety, comedy specials
Cinemax	5,100	Movies, comedy, music specials
The Disney Channel	4,000	Original and classic movies, cartoons, etc.
The Movie Channel	2,500	Movies, film festivals, etc.

List the elements of the following sets.

56. F, the set of networks with more than 5000 subscribers

57. G, the set of networks that show comedy (not cartoons)

58. H, the set of networks that show movies

59. $F \cup G$

60. $H \cap G$

61. G'

---

* "Top Five Pay-Cable Services, 1988" from *The Universal Almanac, 1990*, edited by John W. Wright (Kansas City and New York: Andrews and McMeel, 1990).

Natural Science   *The table below shows some symptoms of an overactive thyroid and an underactive thyroid.*

Underactive Thyroid	Overactive Thyroid
Sleepiness, *s*	Insomnia, *i*
Dry hands, *d*	Moist hands, *m*
Intolerance of cold, *c*	Intolerance of heat, *h*
Goiter, *g*	Goiter, *g*

Let U be the smallest possible set that includes all the symptoms listed, N be the set of symptoms for an underactive thyroid, and O be the set of symptoms for an overactive thyroid. Find each of the following sets.

62. O' **63.** N' **64.** $N \cap O$ **65.** $N \cup O$ **66.** $N \cap O'$

8.2 APPLICATIONS OF VENN DIAGRAMS

Venn diagrams were used in the last section to illustrate set union and intersection. The rectangular region in a Venn diagram represents the universal set, U. Including only a single set, A, inside the universal set, as in Figure 8.7, divides U into two nonoverlapping regions. Region 1 represents those elements outside set A, while region 2 represents those elements belonging to set A. (The numbering of these regions is arbitrary.)

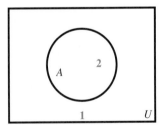

One set leads to 2 regions
(numbering is arbitrary).

FIGURE 8.7

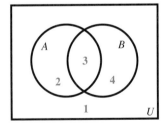

Two sets lead to 4 regions
(numbering is arbitrary).

FIGURE 8.8

The Venn diagram of Figure 8.8 shows two sets inside U. These two sets divide the universal set into four nonoverlapping regions. As labeled in Figure 8.8, region 1 includes those elements outside both set A and set B. Region 2 includes those elements belonging to A and not to B. Region 3 includes those elements belonging to both A and B. Which elements belong to region 4? (Again, the numbering is arbitrary.)

1 Draw Venn diagrams for the following.

(a) $A \cup B'$

(b) $A' \cap B'$

Answers:

(a)

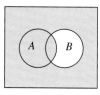

$A \cup B'$

(b)

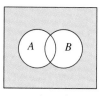

$A' \cap B'$

▶**EXAMPLE 1**   Draw Venn diagrams similar to Figure 8.8 and shade the regions representing the following sets.

(a) $A' \cap B$

Set A' contains all the elements outside set A. As labeled in Figure 8.8, A' is represented by regions 1 and 4. Set B is represented by the elements in regions 3 and 4. The intersection of sets A' and B, the set $A' \cap B$, is given by the region common to regions 1 and 4 and regions 3 and 4. The result, region 4, is shaded in Figure 8.9.

(b) $A' \cup B'$

Again, set A' is represented by regions 1 and 4, and set B' by regions 1 and 2. To find $A' \cup B'$, identify the region that represents the set of all elements in A', B', or both. The result, which is shaded in Figure 8.10, includes regions 1, 2, and 4.   ◀ **1**

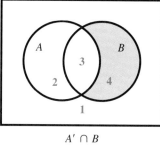

$A' \cap B$

FIGURE 8.9

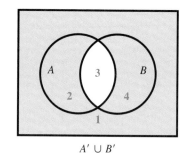

$A' \cup B'$

FIGURE 8.10

Venn diagrams also can be drawn with three sets inside U. These three sets divide the universal set into eight nonoverlapping regions which can be numbered (arbitrarily) as in Figure 8.11.

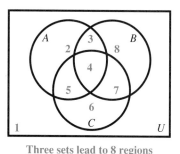

Three sets lead to 8 regions

FIGURE 8.11

$A' \cup (B \cap C')$

FIGURE 8.12

▶**EXAMPLE 2**   Shade $A' \cup (B \cap C')$ on a Venn diagram.

First find $B \cap C'$. See Figure 8.12. Set B is represented by regions 3, 4, 7, and 8, and C' by regions 1, 2, 3, and 8. The overlap of these regions, regions 3

2 Draw a Venn diagram for the following.

(a) $B' \cap A$

(b) $(A \cup B)' \cap C$

Answers:

(a)

$B' \cap A$

(b)

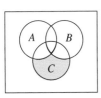

$(A \cup B)' \cap C$

3 **(a)** Place numbers in the regions on a Venn diagram if the data on the 100 households showed

29 home computers

63 VCRs

20 with both.

(b) How many have a VCR but not a computer?

Answers:

(a)

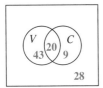

(b) 43

and 8, represents the set $B \cap C'$. Set A' is represented by regions 1, 6, 7, and 8. The union of regions 3 and 8 and regions 1, 6, 7, and 8 includes regions 1, 3, 6, 7, and 8, which are shaded in Figure 8.12. ◄ **2**

Venn diagrams can be used to solve problems that result from surveying groups of people. As an example, suppose a researcher collecting data on 100 households finds that

21 have a home computer;

56 have a videocassette recorder (VCR); and

12 have both.

The researcher wants to answer the following questions.

(a) How many do not have a VCR?

(b) How many have neither a computer nor a VCR?

(c) How many have a computer but not a VCR?

A Venn diagram like the one in Figure 8.13 will help sort out the information. In Figure 8.13(a), we put the number 12 in the region common to both a VCR and a computer, because 12 households have both. Of the 21 with a home computer, $21 - 12 = 9$ have no VCR, so in Figure 8.13(b) we put 9 in the region for a computer but no VCR. Similarly, $56 - 12 = 44$ households have a VCR but not a computer, so we put 44 in that region. Finally, the diagram shows that $100 - 44 - 12 - 9 = 35$ households have neither a VCR nor a computer. Now we can answer the questions:

(a) $35 + 9 = 44$ do not have a VCR;

(b) 35 have neither,

(c) 9 have a computer but not a VCR. ◄ **3**

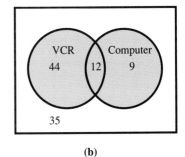

FIGURE 8.13

▶ **EXAMPLE 3** A group of 60 freshman business students at a large university was surveyed, with the following results.

19 of the students read *Business Week;*

18 read *The Wall Street Journal;*

50 read *Fortune;*

13 read *Business Week* and *The Journal;*
11 read *The Journal* and *Fortune;*
13 read *Business Week* and *Fortune;*
 9 read all three.

Use these data to answer the following questions.

(a) How many students read none of the publications?
(b) How many read only *Fortune?*
(c) How many read *Business Week* and *The Journal,* but not *Fortune?*

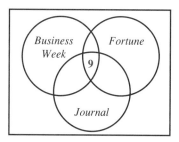

FIGURE 8.14(a)

Once again, use a Venn diagram to represent the data. Since 9 students read all three publications, begin by placing 9 in the area in Figure 8.14(a) that belongs to all three regions.

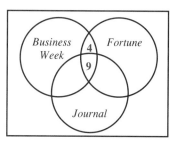

FIGURE 8.14(b)

Of the 13 students who read *Business Week* and *Fortune,* 9 also read *The Journal.* Therefore only $13 - 9 = 4$ students read just *Business Week* and *Fortune.* So place a 4 in the region common only to *Business Week* and *Fortune* readers, as in Figure 8.14(b).

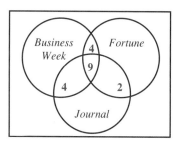

FIGURE 8.14(c)

In the same way, place a 4 in the region of Figure 8.14(c) common only to *Business Week* and *The Journal,* and 2 in the region common only to *Fortune* and *The Journal.*

4 In the example about the three publications, how many students read exactly

(a) 1;

(b) 2 of the publications?

Answers:

(a) 40

(b) 10

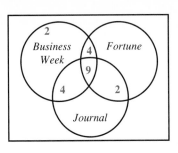

FIGURE 8.14(d)

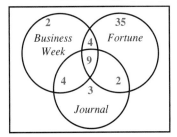

FIGURE 8.14(e)

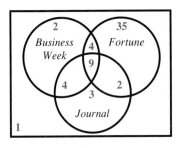

FIGURE 8.14(f)

The data show that 19 students read *Business Week*. However, $4 + 9 + 4 = 17$ readers have already been placed in the *Business Week* region. The balance of this region in Figure 8.14(d) will contain only $19 - 17 = 2$ students. These 2 students read *Business Week* only—not *Fortune* and not *The Journal*.

In the same way, 3 students read only *The Journal* and 35 read only *Fortune,* as shown in Figure 8.14(e).

A total of $2 + 4 + 3 + 4 + 9 + 2 + 35 = 59$ students are placed in the various regions of Figure 8.13(f). Since 60 students were surveyed, $60 - 59 = 1$ student reads none of the three publications and 1 is placed outside the other regions in Figure 8.14(f).

Figure 8.14(f) can now be used to answer the questions asked above.

(a) Only 1 student reads none of the publications.
(b) There are 35 students who read only *Fortune*.
(c) The overlap of the regions representing *Business Week* and *The Journal* shows that 4 students read *Business Week* and *The Journal* but not *Fortune*. ◀ **4**

▶**EXAMPLE 4**   Jeff Friedman is a section chief for an electric utility company. The employees in his section cut down tall trees, climb poles, and splice wire. Friedman reported the following information to the management of the utility.

Of the 100 employees in my section,
45 can cut tall trees;

5 In Example 4, suppose 46 employees can cut tall trees. Then how many

(a) can only cut tall trees?

(b) can cut trees or climb poles?

(c) can cut trees or climb poles or splice wire?

Answers:

(a) 4

(b) 68

(c) 91

50 can climb poles;
57 can splice wire;
28 can cut trees and climb poles;
20 can climb poles and splice wire;
25 can cut trees and splice wire;
11 can do all three;
  9 can't do any of the three (management trainees).

The data supplied by Friedman lead to the numbers shown in Figure 8.15. Add the numbers from all the regions to get the total number of Friedman's employees.

$$9 + 3 + 14 + 23 + 11 + 9 + 17 + 13 = 99$$

Friedman claimed to have 100 employees, but his data indicate only 99. The management decided that Friedman didn't qualify as a section chief, and reassigned him as a nightshift meter reader in Guam. (Moral: he should have taken this course.) ◀ **5**

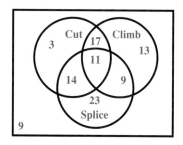

FIGURE 8.15

Caution   Note that in all the examples above, we started in the innermost region with the intersection of the three categories. This is usually the best way to begin solving problems of this type.

We use the symbol $n(A)$ to denote the *number* of elements in A. For instance, if $A = \{w, x, y, z\}$, then $n(A) = 4$. The following useful fact is proved below.

Union Rule For Counting

$$n(A \cup B) = n(A) + n(B) - n(A \cap B)$$

For example, if $A = \{r, s, t, u, v\}$ and $B = \{r, t, w\}$, then $A \cap B = \{r, t\}$, so that $n(A) = 5$, $n(B) = 3$, and $n(A \cap B) = 2$. By the formula in the box, $n(A \cup B) = 5 + 3 - 2 = 6$, which is certainly true since $A \cup B = \{r, s, t, u, v, w\}$.

6 If $n(A) = 10$, $n(B) = 7$, and $n(A \cap B) = 3$, find $n(A \cup B)$.

Answer:
14

Here is a proof of the statement in the box: let x be the number of elements in A that are not in B, y the number of elements in $A \cap B$, and z the number of elements in B that are not in A, as indicated in Figure 8.16. That diagram shows that $n(A \cup B) = x + y + z$. It also shows that $n(A) = x + y$ and $n(B) = y + z$, so that

$$n(A) + n(B) - n(A \cap B) = (x + y) + (z + y) - y$$
$$= x + y + z.$$
$$= n(A \cup B).$$

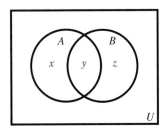

FIGURE 8.16

▶ **EXAMPLE 5** A group of 10 students meet to plan a school function. All are majors in accounting or economics or both. Five of the students are economics majors and 7 are majors in accounting. How many major in both subjects?

Use the union rule, with $n(A) = 5$, $n(B) = 7$, and $n(A \cup B) = 10$. We must find $n(A \cap B)$.

$$n(A \cup B) = n(A) + n(B) - n(A \cap B)$$
$$10 = 5 + 7 - n(A \cap B),$$

so

$$n(A \cap B) = 5 + 7 - 10 = 2. \quad ◀ \quad \boxed{6}$$

8.2 EXERCISES

Sketch a Venn diagram like the one to the right and use shading to show each of the following sets. (See Example 1.)

1. $B \cap A'$

2. $A \cup B'$

3. $A' \cup B$

4. $A' \cap B'$

5. $B' \cup (A' \cap B')$

6. $(A \cap B) \cup B'$

7. U'

8. $\emptyset'$

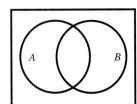

Sketch a Venn diagram like the one shown, and use shading to show each of the following sets. (See Example 2.)

9. $(A \cap B) \cap C$

10. $(A \cap C') \cup B$

11. $A \cap (B \cup C')$

12. $A' \cap (B \cap C)$

13. $(A' \cap B') \cap C$

14. $(A \cap B') \cup C$

15. $(A \cap B') \cap C$

16. $A' \cap (B' \cup C)$

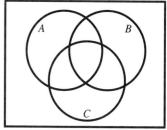

Use Venn diagrams to answer the following questions. (See Examples 3 and 4.)

17. Social Science A survey of people attending a Lunar New Year celebration in Chinatown yielded the following results:

> 120 were women;
> 150 spoke Cantonese;
> 170 lit firecrackers;
> 108 of the men spoke Cantonese;
> 100 of the men did not light firecrackers;
> 18 of the non-Cantonese-speaking women lit firecrackers;
> 78 non-Cantonese-speaking men did not light firecrackers;
> 30 of the women who spoke Cantonese lit firecrackers.

(a) How many attended?

(b) How many of those who attended did not speak Cantonese?

(c) How many women did not light firecrackers?

(d) How many of those who lit firecrackers were Cantonese-speaking men?

18. Management Jeff Friedman, of Example 4 in the text, was again reassigned, this time to the home economics department of the electric utility. He interviewed 140 people in a suburban shopping center to find out some of their cooking habits. He obtained the following results. Should he be reassigned yet one more time?

> 58 use microwave ovens
> 63 use electric ranges;
> 58 use gas ranges;
> 19 use microwave ovens and electric ranges;
> 17 use microwave ovens and gas ranges;
> 4 use both gas and electric ranges;
> 1 uses all three;
> 2 cook only with solar energy.

19. Social Science A recent nationwide survey revealed the following data concerning unemployment in the United States. Out of the total population of people who are working or seeking work,

> 9% are unemployed;
> 11% are black;
> 30% are youths;
> 1% are unemployed black youths;
> 5% are unemployed youths;
> 7% are employed black adults;
> 23% are nonblack employed youths.

What percent are
(a) not black?
(b) employed adults?
(c) unemployed adults?
(d) unemployed black adults?
(e) unemployed nonblack youths?

20. Country-western songs emphasize three basic themes: love, prison, and trucks. A survey of the local country-western radio station produced the following data.

> 12 songs were about a truck driver who was in love while in prison;
> 13 about a prisoner in love;
> 28 about a person in love;
> 18 about a truck driver in love;
> 3 about a truck driver in prison who was not in love;
> 2 about a prisoner who was not in love and did not drive a truck;
> 8 about a person out of jail who was not in love, and did not drive a truck;
> 16 about truck drivers who were not in prison.

(a) How many songs were surveyed?

Find the number of songs about

(b) truck drivers;

(c) prisoners;

(d) truck drivers in prison;

(e) people not in prison;

(f) people not in love.

21. Natural Science   After a genetics experiment, the number of pea plants having certain characteristics was tallied, with the results as follows.

> 22 were tall;
> 25 had green peas;
> 39 had smooth peas;
> 9 were tall and had green peas;
> 17 were tall and had smooth peas;
> 20 had green peas and smooth peas;
> 6 had all three characteristics;
> 4 had none of the characteristics.

(a) Find the total number of plants counted.

(b) How many plants were tall and had peas that were neither smooth nor green?

(c) How many plants were not tall but had peas that were smooth and green?

22. Natural Science   Human blood can contain either no antigens, the A antigen, the B antigen, or both the A and B antigens. A third antigen, called the Rh antigen, is important in human reproduction, and again may or may not be present in an individual. Blood is called type A-positive if the individual has the A and Rh, but not the B antigen. A person having only the A and B antigens is said to have type AB-negative blood. A person having only the Rh antigen has type O-positive blood. Other blood types are defined in a similar manner. Identify the blood type of the individuals in regions (a)–(g) of the Venn diagram.

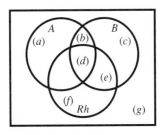

23. Natural Science   (Use the diagram from Exercise 22.) In a certain hospital, the following data was recorded.

> 25 patients had the A antigen;
> 17 had the A and B antigens;
> 27 had the B antigen;
> 22 had the B and Rh antigens;
> 30 had the Rh antigen;
> 12 had none of the antigens;
> 16 had the A and Rh antigens;
> 15 had all three antigens.

How many patients

(a) were represented?

(b) had exactly one antigen?

(c) had exactly two antigens?

(d) had O-positive blood?

(e) had AB-positive blood?

(f) had B-negative blood?

(g) had O-negative blood?

(h) had A-positive blood?

24. At a pow-wow in Arizona, Native Americans from all over the Southwest came to participate in the ceremonies. A coordinator of the pow-wow took a survey and found that

> 15 families brought food, costumes, and crafts;
> 25 families brought food and crafts;
> 42 families brought food;
> 20 families brought costumes and food;
> 6 families brought costumes and crafts, but not food;
> 4 families brought crafts, but neither food nor costumes;
> 10 families brought none of the three items;
> 18 families brought costumes but not crafts.

(a) How many families were surveyed?

(b) How many families brought costumes?

(c) How many families brought crafts, but not costumes?

(d) How many families did not bring crafts?

(e) How many families brought food or costumes?

25. Management   The following table shows the number of people in a certain small town in Georgia who fit in the given categories.

Age	Drink Coke (V)	Drink Pepsi (B)	Drink 7-up (G)	Totals
20–25 (Y)	40	15	15	70
26–35 (M)	30	30	20	80
Over 35 (O)	10	50	10	70
Totals	80	95	45	220

Using the letters given in the table, find the number of people in each of the following sets.
(a) $Y \cap V$ (b) $M \cap B$
(c) $M \cup (B \cap Y)$ (d) $Y' \cap (B \cup G)$
(e) $O' \cup G$ (f) $M' \cap (V' \cap G')$

27. Restate the union rule in words.

26. Management   The following table shows the results of a survey in a medium-sized town in Tennessee. The survey asked questions about the investment habits of local citizens.

Age	Stocks (S)	Bonds (B)	Savings Accounts (A)	Totals
18–29 (Y)	6	2	15	23
30–49 (M)	14	5	14	33
50 or over (O)	32	20	12	64
Totals	52	27	41	120

Using the letters given in the table, find the number of people in each of the following sets.
(a) $Y \cap B$ (b) $M \cup A$
(c) $Y \cap (S \cup B)$ (d) $O' \cup (S \cup A)$
(e) $(M' \cup O') \cap B$

Use Venn diagrams to answer the following questions. (See Example 5.)

28. If $n(A) = 5$, $n(B) = 8$, and $n(A \cap B) = 4$, what is $n(A \cup B)$?

29. If $n(A) = 12$, $n(B) = 27$, and $n(A \cup B) = 30$, what is $n(A \cap B)$?

30. Suppose $n(B) = 7$, $n(A \cap B) = 3$, and $n(A \cup B) = 20$. What is $n(A)$?

31. Suppose $n(A \cap B) = 5$, $n(A \cup B) = 35$, and $n(A) = 13$. What is $n(B)$?

Draw a Venn diagram and use the given information to fill in the number of elements for each region.

32. $n(U) = 38$, $n(A) = 16$, $n(A \cap B) = 12$, $n(B') = 20$

33. $n(A) = 26$, $n(B) = 10$, $n(A \cup B) = 30$, $n(A') = 17$

34. $n(A \cup B) = 17$, $n(A \cap B) = 3$, $n(A) = 8$, $n(A' \cup B') = 21$

35. $n(A') = 28$, $n(B) = 25$, $n(A' \cup B') = 45$, $n(A \cap B) = 12$

36. $n(A) = 28$, $n(B) = 34$, $n(C) = 25$, $n(A \cap B) = 14$, $n(B \cap C) = 15$, $n(A \cap C) = 11$, $n(A \cap B \cap C) = 9$, $n(U) = 59$

37. $n(A) = 54$, $n(A \cap B) = 22$, $n(A \cup B) = 85$, $n(A \cap B \cap C) = 4$, $n(A \cap C) = 15$, $n(B \cap C) = 16$, $n(C) = 44$, $n(B') = 63$

38. $n(A \cap B) = 6$, $n(A \cap B \cap C) = 4$, $n(A \cap C) = 7$, $n(B \cap C) = 4$, $n(A \cap C') = 11$, $n(B \cap C') = 8$, $n(C) = 15$, $n(A' \cap B' \cap C') = 5$

39. $n(A) = 13$, $n(A \cap B \cap C) = 4$, $n(A \cap C) = 6$, $n(A \cap B') = 6$, $n(B \cap C) = 6$, $n(B \cap C') = 11$, $n(B \cup C) = 22$, $n(A' \cap B' \cap C') = 5$

In Exercises 40–43, prove that the statements are true by drawing Venn diagrams and shading the regions representing the sets on each side of the equals signs.*

40. $(A \cup B)' = A' \cap B'$

41. $(A \cap B)' = A' \cup B'$

42. $A \cap (B \cup C) = (A \cap B) \cup (A \cap C)$

43. $A \cup (B \cap C) = (A \cup B) \cap (A \cup C)$

*The statements in Exercises 40 and 41 are known as De Morgan's laws. They are named for the English mathematician Augustus De Morgan (1806–71).

8.3 PROBABILITY

If you go to a supermarket and buy five pounds of peaches at 54¢ per pound, you can easily find the *exact* price of your purchase: $2.70. On the other hand, the produce manager of the market is faced with the problem of ordering peaches. The manager may have a good estimate of the number of pounds of peaches that will be sold during the day, but there is no way to know *exactly*. The number of pounds that customers will purchase during a day cannot be predicted exactly.

Many problems that come up in applications of mathematics involve phenomena for which exact prediction is impossible. The best that we can do is to determine the *probability* of the possible outcomes. In this section and the next we introduce some of the terminology and basic premises of probability theory.

In probability, an **experiment** is an activity or occurrence with an observable result. Each repetition of an experiment is called a **trial.** The possible results of each trial are called **outcomes.** An example of a probability experiment is the tossing of a coin. Each trial of the experiment (each toss) has two possible outcomes: heads, *h*, and tails, *t*. If the outcomes *h* and *t* are equally likely to occur, then the coin is not "loaded" to favor one side over the other. Such a coin is called **fair.** For a coin that is not loaded, this "equally likely" assumption is made for each trial.

Because *two* equally likely outcomes are possible, *h* and *t*, theoretically a coin tossed many, many times should come up heads approximately 1/2 of the time. Also, the more times the coin is tossed, the closer the occurrence of heads should be to 1/2. Therefore, it is reasonable to define the probability of a head coming up to be 1/2. Similarly, we define the probability of tails coming up to be 1/2.

For now we shall concentrate on experiments, like coin tossing, in which each outcome is equally likely and use the probabilities suggested by this fact.

If there are *n* equally likely outcomes of an experiment, then the probability of any one outcome is 1/*n*.

▶**EXAMPLE 1**　Suppose a spinner like the one shown in Figure 8.17 is spun.

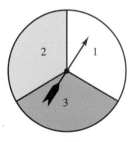

FIGURE 8.17

1 Find the probability that the spinner of Figure 8.17

(a) points to 3;

(b) does not point to 1.

Answers:

(a) 1/3

(b) 2/3

2 A jar holds a red marble, a blue marble, and a green marble. If one marble is drawn at random, what is the probability it is

(a) green?

(b) red?

Answers:

(a) 1/3

(b) 1/3

(a) Find the probability that the spinner will point to 1.

It is natural to assume that if this spinner were spun many, many times, it would point to 1 about 1/3 of the time, so that the required probability is 1/3.

(b) Find the probability that the spinner will point to 2.

Since 2 is one of three possible outcomes, the probability is 1/3. ◄ **1**

An ordinary die is a cube whose six faces show the following numbers of dots: 1, 2, 3, 4, 5, and 6. If the die is not "loaded" to favor certain faces over others (a fair die), then any one of the six faces is equally likely to come up when the die is rolled.

▶**EXAMPLE 2** **(a)** If a single fair die is rolled, find the probability of rolling the number 4.

One out of six faces is a 4, so the required probability is 1/6.

(b) Using the same die, find the probability of rolling a 6.

One out of the six faces shows a 6, so the probability is 1/6. ◄ **2**

SAMPLE SPACES   The universal set, the set of all possible outcomes for an experiment, is called the **sample space** for the experiment. A sample space for the experiment of tossing a coin has two outcomes: heads, h, and tails, t. If S represents this sample space, then

$$S = \{h, t\}.$$

▶**EXAMPLE 3**   Give the sample space for each experiment.

(a) A spinner like the one in Figure 8.17 is spun.

The three outcomes 1, 2, or 3 are equally likely, so the sample space is

$$\{1, 2, 3\}.$$

(b) For the purposes of a public opinion poll, people are classified as young, middle-aged, or older, and as male or female.

A sample space for this poll could be written as a set of ordered pairs of these types.

{(young, male), (young, female), (middle-aged, male),

(middle-aged, female), (older, male), (older, female)}

(c) An experiment consists of studying the number of boys and girls in families with exactly 3 children. Let b represent *boy* and g represent *girl*.

A 3-child family can have 3 boys, written *bbb,* 3 girls, *ggg,* or various combinations, such as *bgg.* A sample space with four outcomes (not eqaully likely) is

$$S_1 = \{3 \text{ boys, 2 boys and 1 girl, 1 boy and 2 girls, 3 girls}\}.$$

Another sample space that considers the ordering of the births of the 3 children with, for example, *bgg* different from *gbg* or *ggb,* is

$$S_2 = \{bbb, bbg, bgb, gbb, bgg, gbg, ggb, ggg\}.$$

3 **(a)** Write an equally likely sample space for the experiment of rolling a single fair die.

(b) Write an equally likely sample space for the experiment of tossing 2 fair coins.

(c) Write a sample space for the experiment of tossing 2 fair coins, if we are interested only in the number of heads. Is this sample space equally likely?

Answers:

(a) {1, 2, 3, 4, 5, 6}

(b) {hh, ht, th, tt}

(c) {0, 1, 2}; no

4 Suppose a die is tossed. Write the following events.

(a) The number showing is less than 3.

(b) The number showing is 5.

(c) The number showing is 8.

Answers:

(a) {1, 2}

(b) {5}

(c) $\emptyset$

The second sample space, S_2, has equally likely outcomes; S_1 does not, since there is more than one way to get a family with 2 boys and 1 girl or a family of 2 girls and 1 boy, but only one way to get 3 boys or 3 girls. ◄

Caution An experiment may have more than one sample space, as shown in Example 3(c). The most useful sample spaces have equally likely outcomes, but it is not always possible to choose such a sample space. **3**

EVENTS An **event** is a subset of outcomes from a sample space. If the sample space for tossing a coin is $S = \{h, t\}$, then one event is the subset $\{h\}$, which represents the outcome "heads." For the sample space of rolling a single fair die, $S = \{1, 2, 3, 4, 5, 6\}$, some possible events are listed below.

The die shows an even number: $E_1 = \{2, 4, 6\}$.
The die shows a 1: $E_2 = \{1\}$.
The die shows a number less than 5: $E_3 = \{1, 2, 3, 4\}$.
The die shows a multiple of 3: $E_4 = \{3, 6\}$.

►**EXAMPLE 4** For the sample space S_2 in Example 3(c), write the following events.
(a) Event H: the family has exactly 2 girls.
Families can have exactly 2 girls with either bgg, gbg, or ggb, so that event H is

$$H = \{bgg, gbg, ggb\}.$$

(b) Event K: the 3 children are the same sex.
Two outcomes satisfy this condition, all boys or all girls.

$$K = \{bbb, ggg\}.$$

(c) Event J: the family has 3 girls.
Only ggg satisfies this condition, so

$$J = \{ggg\}. \quad ◄ \quad \boxed{4}$$

In Example 4(c), event J had only one possible outcome, *ggg*. Such an event, with only one possible outcome, is a **simple event.** If an event E equals the sample space S, then E is a **certain event.** If event $E = \emptyset$, then E is an **impossible event.**

►**EXAMPLE 5** Suppose a die is rolled. As shown in Problem 3(a), the sample space is {1, 2, 3, 4, 5, 6}.
(a) The event "the die shows a 4," {4}, has only one possible outcome. It is a simple event.

5 Which of the events listed in Problem 4 at the side is

(a) simple?

(b) certain?

(c) impossible?

Answers:

(a) Part (b)

(b) None

(c) Part (c)

6 A fair die is rolled. Find the probability of rolling

(a) an odd number;

(b) 2, 4, 5, or 6;

(c) a number greater than 5;

(d) the number 7.

Answers:

(a) 1/2

(b) 2/3

(c) 1/6

(d) 0

(b) The event "the number showing is less than 10" equals the sample space $S = \{1, 2, 3, 4, 5, 6\}$. This event is a certain event; if a die is rolled the number showing (either 1, 2, 3, 4, 5, or 6), must be less than 10.

(c) The event "the die shows a 7" is the empty set, $\emptyset$; this is an impossible event. ◀ **5**

PROBABLITY   For sample spaces with *equally likely* outcomes, the *probability of an event* is defined as follows.

Basic Probability Principle

Suppose event E is a subset of a sample space S. Then the **probability that event E occurs,** written $P(E)$, is

$$P(E) = \frac{n(E)}{n(S)}.$$

This definition means that the probability of an event is a number that indicates the relative likelihood of the event.

▶ **EXAMPLE 6**   Suppose a single die is rolled. Use the sample space $S = \{1, 2, 3, 4, 5, 6\}$ and give the probability of each of the following events.

(a) E: the die shows an even number.

Here, $E = \{2, 4, 6\}$, a set with three elements. Because S contains six elements,

$$P(E) = \frac{3}{6} = \frac{1}{2}.$$

(b) F: the die shows a number greater than 4.

Since F contains two elements, 5 and 6,

$$P(F) = \frac{2}{6} = \frac{1}{3}.$$

(c) G: the die shows a number less than 10.

Event G is a certain event, with

$$G = \{1, 2, 3, 4, 5, 6\}, \quad \text{so} \quad P(G) = \frac{6}{6} = 1.$$

(d) H: the die shows an 8.

This event is impossible, so

$$P(H) = \frac{0}{6} = 0. \quad ◀ \quad \boxed{6}$$

7 A single playing card is drawn at random from an ordinary 52-card deck. Find the probability of drawing

(a) a queen;

(b) a diamond;

(c) a red card.

Answers:

(a) 1/13

(b) 1/4

(c) 1/2

▶**EXAMPLE 7** If a single card is drawn at random from an ordinary 52-card bridge deck (shown in Figure 8.18), find the probability of each of the following events.

(a) Drawing an ace

There are 4 aces in the deck. The event "drawing an ace" is

{heart ace, diamond ace, club ace, spade ace}.

Therefore,

$$P(\text{ace}) = \frac{4}{52} = \frac{1}{13}.$$

(b) Drawing a face card

Since there are 12 face cards,

$$P(\text{face card}) = \frac{12}{52} = \frac{3}{13}.$$

(c) Drawing a spade

The deck contains 13 spades, so

$$P(\text{spade}) = \frac{13}{52} = \frac{1}{4}.$$

(d) Drawing a spade or a heart

Besides the 13 spades, the deck contains 13 hearts, so

$$P(\text{spade or heart}) = \frac{26}{52} = \frac{1}{2}. \quad ◀ \quad 7$$

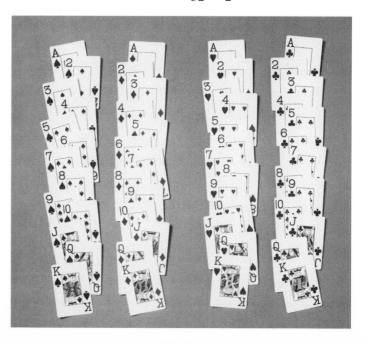

FIGURE 8.18

The probability of each event in the preceding examples was a number between 0 and 1. The same thing is true in the general case. If an event E consists of m of the possible n outcomes, then $0 \le m \le n$ and hence, $0 \le m/n \le 1$. Since $P(E) = m/n$, $P(E)$ is between 0 and 1.

For any event E, $\quad 0 \le P(E) \le 1$.

For any sample space S, $P(S) = 1$ and $P(\emptyset) = 0$.

8.3 EXERCISES

1. What is meant by a "fair" coin or die?

2. What is meant by the sample space for an experiment?

Write sample spaces for the experiments in Exercises 3–8. (See Example 3.)

3. A day in June is chosen for a picnic.

4. At a sleep center, a person is asked the number of hours (to the nearest hour) he slept yesterday.

5. The manager of a small company must decide whether to buy office space or lease it.

6. A student must decide which of three courses that interest her to take. (Assume she can take more than one.)

7. A CPA chooses a month as the end of the fiscal year for a new corporation.

8. A coin is tossed and a die is rolled.

9. Define an event.

10. What is a simple event?

For the experiments in Exercises 11–16, write out an equally likely sample space, and then write the indicated events in set notation. (See Examples 3–5.)

11. A marble is drawn at random from a bowl containing 3 yellow, 4 white, and 8 blue marbles.
    (a) A yellow marble is drawn.
    (b) A blue marble is drawn.
    (c) A white marble is drawn.
    (d) A black marble is drawn.

12. Slips of paper marked with the numbers 1, 2, 3, 4, and 5 are placed in a box. After being mixed, two slips are drawn.
    (a) Both slips are marked with even numbers.

    (b) One slip is marked with an odd number and the other is marked with an even number.
    (c) Both slips are marked with the same number.

13. An unprepared student takes a three-question true/false quiz in which he guesses the answers to all three questions.
    (a) The student gets three answers wrong.
    (b) The student gets exactly two answers correct.
    (c) The student gets only the first answer corrcet.

14. A die is tossed twice, with the tosses recorded as ordered pairs.
    (a) The first die shows a 3.
    (b) The sum of the numbers showing is 8.
    (c) The sum of the numbers showing is 13.

15. One urn contains four balls, labeled 1, 2, 3, and 4. A second urn contains five balls, labeled 1, 2, 3, 4, and 5. An experiment consists of taking one ball from the first urn, and then taking a ball from the second urn.
    (a) The number on the first ball is even.
    (b) The number on the second ball is even.
    (c) The sum of the numbers on the two balls is 5.
    (d) The sum of the numbers on the two balls is 1.

16. From 5 employees, Strutz, Martin, Hampton, Williams, and Ewing, 2 are selected to attend a conference.
    (a) Hampton is selected.
    (b) Strutz and Martin are not both selected.
    (c) Both Williams and Ewing are selected.

A single fair die is rolled. Find the probabilities of the following events. (See Example 6.)

17. Getting a 5

18. Getting a number less than 4

19. Getting a number greater than 4

20. Getting a 2 or a 5

A card is drawn from a well-shuffled deck of 52 cards. Find the probability of drawing each of the following. (See Example 7.)

21. A 9

22. A black card

23. A black 9

24. A heart

25. The 9 of hearts

26. A 2 or a queen

27. A black 7 or a red 8

28. A red face card

A jar contains 5 red, 4 black, 7 purple, and 9 green marbles. If a marble is drawn at random, what is the probability that the marble is

29. red?

30. black?

31. green?

32. purple?

33. not black?

34. not purple?

35. red or black?

36. black or purple?

37. **Management** According to an article in *Business Week*, in 1989 funding for university research in the United States totaled $15 billion. Support came from various sources, as shown in the table.*

1989 Funding for University Research

Source	Amount (in billions of dollars)
Federal government	9.0
State and local government	1.2
Institutional	2.7
Industry	1.0
Other	1.1

Find the probability that funds for a particular project came from each of the following sources.

(a) Federal government

(b) Industry

(c) The institution

38. **Social Science** The population of the United States by race in 1990, and the projected population by race for the year 2000 are given below.*

U.S. Population by Race (in thousands)

Race	1990	2000
White	191,594	197,634
Black	30,915	35,440
Hispanic	21,854	30,295
Other	7,930	11,110

Find the probability of a randomly selected person being each of the following.

(a) Hispanic in 1990

(b) White in 2000

(c) Black in 2000

---

*"Projections of the Population, by Race and Hispanic Origin, 1990–2000" from U.S. Bureau of the Census, *Projections of the Population of the U.S. by Age, Sex, and Race, 1983–2080.*

---

*"University Research: The Squeeze Is On," reprinted from May 20, 1991 issue of *Business Week* by special permission. Copyright © 1991 by McGraw-Hill, Inc.

1 Give the following events for the experiment of Example 1 if $E = \{1, 3\}$ and $F = \{2, 3, 4, 5\}$.

(a) $E \cap F$

(b) $E \cup F$

(c) E'

Answers:

(a) $\{3\}$

(b) $\{1, 2, 3, 4, 5\}$

(c) $\{2, 4, 5, 6\}$

8.4 BASIC CONCEPTS OF PROBABILITY

We discuss the probability of more complex events in this section. Since events are sets, we can use set operations to find unions, intersections, and complements of events.

▶**EXAMPLE 1**   Suppose a die is tossed. Let E be the event "the die shows a number greater than 3," and let F be the event "the die shows an even number." Then

$$E = \{4, 5, 6\} \quad \text{and} \quad F = \{2, 4, 6\}.$$

Find each of the following.

(a) The die shows an even number greater than 3.

We are looking for an even number *and* a number that is greater than 3. The event $E \cap F$ includes the outcomes common to *both* E and F. Here,

$$E \cap F = \{4, 6\}.$$

(b) The die shows an even number or a number greater than 3. We want numbers that are even, greater than 3, *or* both even and greater than 3, which is the event $E \cup F$.

$$E \cup F = \{2, 4, 5, 6\}$$

(c) The die shows a number less than or equal to 3.

This set, the complement of E, includes numbers less than or equal to 3.

$$E' = \{1, 2, 3\}$$

The Venn diagrams of Figure 8.19 show the events $E \cap F$, $E \cup F$, and E'. ◀ **1**

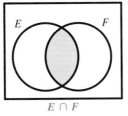

$E \cap F$

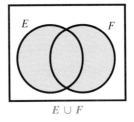

$E \cup F$

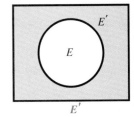

E'

FIGURE 8.19

Two events that cannot both occur at the same time, such as getting both a head and a tail on the same toss of a coin, are called **mutually exclusive events.**

Events A and B are mutually exclusive events if $A \cap B = \emptyset$.

2 In Example 2, let $F = \{2, 4, 6\}$ and $K = \{1, 3, 5\}$. Are the following events mutually exclusive?

(a) F and K

(b) F and G

Answers:

(a) Yes

(b) No

For any event E, E and E' are mutually exclusive. See Figure 8.19. By definition, mutually exclusive events are disjoint sets.

▶ **EXAMPLE 2**   Let $S = \{1, 2, 3, 4, 5, 6\}$, the sample space for tossing a die. Let $E = \{4, 5, 6\}$, and let $G = \{1, 2\}$. Then E and G are mutually exclusive events because they have no outcomes in common; $E \cap G = \emptyset$. See Figure 8.20.

◀ **2**

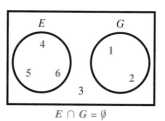

$E \cap G = \emptyset$

FIGURE 8.20

A summary of the set operations for events is given below.

Let E and F be events for a sample space S.

$E \cap F$ occurs when both E and F occur.
$E \cup F$ occurs when either E or F or both occur.
E' occurs when E does not.
E and F are mutually exclusive if $E \cap F = \emptyset$.

To determine the probability of the union of the two events E and F in a sample space S, we use the union rule for counting from Section 8.2:

$$n(E \cup F) = n(E) + n(F) - n(E \cap F).$$

Dividing both sides by $n(S)$ shows that

$$\frac{n(E \cup F)}{n(S)} = \frac{n(E)}{n(S)} + \frac{n(F)}{n(S)} - \frac{n(E \cap F)}{n(S)}$$

$$P(E \cup F) = P(E) + P(F) - P(E \cap F).$$

This discussion is summarized below.

Union Rule For Probability

For any events E and F from a sample space S,

$$\boldsymbol{P(E \cup F) = P(E) + P(F) - P(E \cap F).}$$

3 A single card is drawn from an ordinary deck. Find the probability that it is black or a 9.

Answer:
7/13

▶ **EXAMPLE 3** If a single card is drawn from an ordinary deck of cards, find the probability that it will be red or a face card.

Let R represent the event "red card" and F the event "face card." There are 26 red cards in the deck, so $P(R) = 26/52$. There are 12 face cards in the deck, so $P(F) = 12/52$. Since there are 6 red face cards in the deck, $P(R \cap F) = 6/52$. By the union rule, the probability that the card will be red or a face card is

$$P(R \cup F) = P(R) + P(F) - P(R \cap F)$$

$$= \frac{26}{52} + \frac{12}{52} - \frac{6}{52} = \frac{32}{52} = \frac{8}{13}. \quad ◀ \ 3$$

Caution Recall from Section 8.1, the word "or" always indicates a *union*.

▶ **EXAMPLE 4** Suppose two fair dice are rolled. Find each of the following probabilities.

(a) The first die shows a 2 or the sum of the results is 6 or 7.

The sample space for the throw of two dice is shown in Figure 8.21, where 1-1 represents the event "the first die shows a 1 and the second die shows a 1", 1-2 represents "the first die shows a 1 and the second die shows a 2," and so on. Let A represent the event "the first die shows a 2" and B represent the event "the sum of the results is 6 or 7." These events are indicated in Figure 8.21. From the diagram, event A has 6 elements, B has 11 elements, and the sample space has 36 elements. Thus,

$$P(A) = \frac{6}{36}, \quad P(B) = \frac{11}{36}, \quad \text{and} \quad P(A \cap B) = \frac{2}{36}.$$

By the union rule,

$$P(A \cup B) = P(A) + P(B) - P(A \cap B),$$

$$P(A \cup B) = \frac{6}{36} + \frac{11}{36} - \frac{2}{36} = \frac{15}{36} = \frac{5}{12}.$$

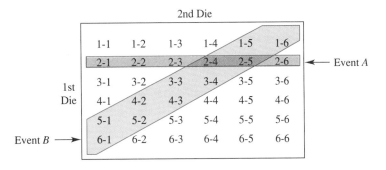

FIGURE 8.21

4 In the experiment of Example 4, find the following probabilities.

(a) The sum is 5 or the second die shows a 3.

(b) Both dice show the same number or the sum is at least 11.

(c) Explain why we did not use the sample space {2, 3, 4, 5, 6, 7, 8, 9, 10, 11, 12} in Example 4.

Answers:

(a) 1/4

(b) 2/9

(c) The events are not equally likely.

5 In Example 5, find the probability of no more than 2 girls.

Answer:
7/8

(b) The sum is 11 or the second die is 5.
P(sum is 11) = 2/36, P(second die is 5) = 6/36, and P(sum is 11 and second die is 5) = 1/36, so

$$P(\text{sum is 11 or second die is 5}) = \frac{2}{36} + \frac{6}{36} - \frac{1}{36} = \frac{7}{36}. \blacktriangleleft \boxed{4}$$

When events E and F are mutually exclusive, then $E \cap F = \emptyset$ by definition; hence, $P(E \cap F) = 0$. The converse of this fact can be used to determine that two events are mutually exclusive.

For any events E and F, if $P(E \cap F) = 0$, then E and F are mutually exclusive.

Applying the union rule to mutually exclusive events yields a special case.

For mutually exclusive events E and F,
$$P(E \cup F) = P(E) + P(F).$$

▶**EXAMPLE 5**   Assume that the probability of a couple having a boy is the same as the probability of their having a girl. If the couple has 3 children, find the probability that at least 2 of them are girls.

The event of having at least 2 girls is the union of the mutually exclusive events E = "the family has exactly 2 girls" and F = "the family has exactly 3 girls." Using the equally likely sample space

{**ggg, ggb, gbg, bgg**, gbb, bgb, bbg, bbb},

we see that $P(2 \text{ girls}) = 3/8$ and $P(3 \text{ girls}) = 1/8$. Therefore,

$$P(\text{at least 2 girls}) = P(2 \text{ girls}) + P(3 \text{ girls})$$
$$= \frac{3}{8} + \frac{1}{8} = \frac{1}{2}. \blacktriangleleft \boxed{5}$$

By definition of E', for any event E from a sample space S,
$$E \cup E' = S \quad \text{and} \quad E \cap E' = \emptyset.$$
Because $E \cap E' = \emptyset$, events E and E' are mutually exclusive, so that
$$P(E \cup E') = P(E) + P(E').$$

6 **(a)** Let $P(K) = 2/3$. Find $P(K')$.

(b) If $P(X') = 3/4$, find $P(X)$.

Answers:

(a) $1/3$

(b) $1/4$

7 In Example 7, find the probability that the sum of the numbers rolled is at least 5.

Answer:
$5/6$

However, $E \cup E' = S$, the sample space, and $P(S) = 1$. Thus
$$P(E \cup E') = P(E) + P(E') = 1.$$
Rearranging these terms gives the following useful rule.

Complement Rule

For any event E,
$$P(E') = 1 - P(E) \quad \text{and} \quad P(E) = 1 - P(E').$$

▶ **EXAMPLE 6** If a fair die is rolled, what is the probability that any number but 5 will come up?

If E is the event that 5 comes up, then E' is the event that any number but 5 comes up. $P(E) = 1/6$, so we have $P(E') = 1 - 1/6 = 5/6$. ◀ **6**

▶ **EXAMPLE 7** If two fair dice are rolled, find the probability that the sum of the numbers showing is greater than 3.

To calculate this probability directly, we must find the probabilities that the sum is 4, 5, 6, 7, 8, 9, 10, 11, or 12 and then add them. It is much simpler to first find the probability of the complement, the event that the sum is less than or equal to 3.

$$P(\text{sum} \le 3) = P(\text{sum is } 2) + P(\text{sum is } 3)$$
$$= \frac{1}{36} + \frac{2}{36} = \frac{3}{36} = \frac{1}{12}$$

Now use the fact that $P(E) = 1 - P(E')$ to get

$$P(\text{sum} > 3) = 1 - P(\text{sum} \le 3) = 1 - \frac{1}{12} = \frac{11}{12}. \quad ◀ \; \boxed{7}$$

Venn diagrams are useful in finding probabilities, as shown in the next example.

▶ **EXAMPLE 8** Susan is a college student who receives heavy sweaters from her aunt at the first sign of cold weather. Suppose the probability that a sweater is the wrong size is .47, the probability that it is a loud color is .59, and the probability that it is both the wrong size and a loud color is .31.

(a) Find the probability that the sweater is the correct size and not a loud color.

Let W be the event "wrong size" and L be the event "loud color." Draw a Venn diagram and determine the probabilities of the mutually exclusive events

8 Find the probability that the sweater in Example 8 is

(a) not a loud color;

(b) the correct size *and* a loud color.

Answers:

(a) .41

(b) .28

I–IV in Figure 8.22 as follows. We are given $P(\text{II}) = P(W \cap L) = .31$. Because regions I and II are mutually exclusive, $P(W) = P(\text{I}) + P(\text{II})$, so that $P(\text{I}) = P(W) - P(\text{II}) = .47 - .31 = .16$. Similarly, $P(\text{III}) = P(L) - P(\text{II})$ $=.59 - .31 = .28$. Finally, since IV is the complement of $W \cup L$, we have

$$P(\text{IV}) = 1 - P(W \cup L) = 1 - (.16 + .31 + .28) = .25.$$

The event "correct size *and* not a loud color" is region IV, the area outside of both W and L. Therefore, its probability is .25.

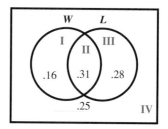

FIGURE 8.22

(b) Find the probability that the sweater is the correct size *or* not a loud color.
The event "correct size *or* not a loud color" is the region outside of region II (in II every sweater is both the wrong size and a loud color), that is, the complement of II. Hence, its probability is $1 - .31 = .69$. ◀ **8**

In many realistic problems, it is not possible to establish exact probabilities for events. Useful approximations, however, often can be found by using past experience as a guide to the future, as in the next example.

▶ **EXAMPLE 9**   The manager of a store has decided to make a study of the amounts of money spent by people coming into the store. To begin, he chooses a day that seems fairly typical and gathers the following data. (Purchases have been rounded to the nearest dollar, with sales tax ignored.)

Amount Spent	Number of Customers
$0	158
$1–10	94
$11–20	203
$21–50	126
$51–100	47
$101–200	38
$201 and over	53

First, the manager might add the numbers of customers to find that 719 people came into the store that day. Of these 719 people, $126/719 \approx .175$

9 A traffic engineer gathered the following data on the number of accidents at four busy intersections.

Intersection	Number of Accidents
1	25
2	32
3	28
4	35

If an accident occurs at one of these four intersections, what is the probability that it occurs at intersections 1 or 3?

Answer:
.44

made a purchase of at least $21 but no more than $50. Also, $53/719 \approx .074$ of the customers spent $201 or more. For this day, the probability that a customer entering the store will spend from $21 to $50 is .175, and the probability that the customer will spend $201 or more is .074. Probabilities for the other purchase amounts can be assigned in the same way, giving the results below.

Amount Spent	Probability
$0	.220
$1–10	.131
$11–20	.282
$21–50	.175
$51–100	.065
$101–200	.053
$201 and over	.074
	Total 1.000

The categories in the table are mutually exclusive simple events. Thus, by the union rule,

$$P(\text{spending} < \$21) = .220 + .131 + .282 = .633.$$

From the table, .282 of the customers spend from $11 to $20, inclusive. Since this price range attracts more than a quarter of the customers, perhaps the store's advertising should emphasize items in, or near, this price range.

The manager should use this table of probabilities to help in predicting the results on other days only if the manager is reasonably sure that the day when the measurements were made is fairly typical of the other days the store is open. For example, on the last few days before Christmas the probabilities might be quite different. ◀ **9**

In this section and the previous one, we have used more than one way to assign probabilities of events. There are three types of probability: *theoretical* (or *classical*); *empirical;* and *subjective.* We use theoretical probability when we assign probabilities for experiments with a coin or with dice, as in Examples 1–7. Probabilities determined by the data from experiments, using the frequency definition, $P(E) = n(E)/n(S)$, are empirical probabilities. See Examples 9 and 10. Subjective probabilities are assigned when no experimental results or long-run frequencies are available. Since these probabilities are based on individual experience, different people may assign different probabilities to the same event. Example 8 gives an example of subjective probability. In the next chapter the use of subjective probability is discussed in more detail. No matter how probabilities are assigned, they must satisfy the following properties.

10 Suppose $P(E) = 9/10$. Find the odds

(a) in favor of E;

(b) against E.

Answers:

(a) 9 to 1

(b) 1 to 9

Properties of Probability

Let S be a sample space consisting of n distinct outcomes $s_1, s_2, \ldots, s_n$. An acceptable probability assignment consists of assigning to each outcome s_i a number p_i (the probability of s_i) according to these rules.

1. The probability of each outcome is a number between 0 and 1.

$$0 \leq p_1 \leq 1, \quad 0 \leq p_2 \leq 1, \ldots, \quad 0 \leq p_n \leq 1$$

2. The sum of the probabilities of all possible outcomes is 1.

$$p_1 + p_2 + p_3 + \cdots + p_n = 1.$$

ODDS Sometimes probability statements are given in terms of **odds,** a comparison of $P(E)$ with $P(E')$. For example, suppose $P(E) = \frac{4}{5}$. Then $P(E') = 1 - \frac{4}{5} = \frac{1}{5}$. These probabilities predict that E will occur 4 out of 5 times and E' will occur 1 out of 5 times. Then we say the **odds in favor** of E are 4 to 1 or 4:1. This ratio may be found from the fraction $P(E)/P(E') = \frac{4}{5}/\frac{1}{5} = 4/1$.

Odds

The **odds in favor** of an event E are defined as the ratio of $P(E)$ to $P(E')$, or

$$\frac{P(E)}{P(E')}, \quad P(E') \neq 0.$$

▶ **EXAMPLE 10** Suppose the weather forecaster says that the probability of rain tomorrow is $1/3$. Find the odds in favor of rain tomorrow.

Let E be the event "rain tomorrow." Then E' is the event "no rain tomorrow." Since $P(E) = 1/3$, $P(E') = 2/3$. By the definition of odds, the odds in favor of rain are

$$\frac{1/3}{2/3} = \frac{1}{2}, \quad \text{written 1 to 2} \quad \text{or} \quad 1:2.$$

On the other hand, the odds that it will *not* rain, or the odds *against* rain, are

$$\frac{2/3}{1/3} = \frac{2}{1}, \quad \text{written 2 to 1.} \quad ◀ \quad \boxed{10}$$

If the odds in favor of an event are, say, 3 to 5, then the probability of the event is 3/8, while the probability of the complement of the event is 5/8. (Odds of 3 to 5 indicate 3 outcomes in favor of the event out of a total of 8 outcomes.) This example suggests the following generalization.

11 If the odds in favor of event E are 1 to 5, find

(a) $P(E)$;

(b) $P(E')$.

Answers:

(a) 1/6

(b) 5/6

If the odds favoring event E are m to n, then

$$P(E) = \frac{m}{m + n} \quad \text{and} \quad P(E') = \frac{n}{m + n}.$$

▶ **EXAMPLE 11**   The odds that a particular bid will be the low bid are 4 to 5.

(a) Find the probability that the bid will be the low bid.

Odds of 4 to 5 show 4 favorable chances out of $4 + 5 = 9$ chances altogether.

$$P(\text{bid will be low bid}) = \frac{4}{4 + 5} = \frac{4}{9}$$

(b) Find the odds against that bid being the low bid.

There is a 5/9 chance that the bid will not be the low bid, so the odds against a low bid are

$$\frac{P(\text{bid will not be low})}{P(\text{bid will be low})} = \frac{5/9}{4/9} = \frac{5}{4},$$

or $5 : 4$ ◀ **11**

8.4 EXERCISES

Decide whether the events in Exercises 1–6 are mutually exclusive. (See Example 2.)

1. Being 15 and being a teenager

2. Wearing jogging shoes and wearing sandals

3. Being male and being a dancer

4. Owning a bicycle and owning a car

5. Being a U.S. Senator and being a U.S. Congressman concurrently

6. Being female and owning a truck

7. If the probability of an event is .857, what is the probability that the event will not occur?

8. Given $P(A) = .5$, $P(B) = .35$, $P(A \cup B) = .85$. Can this be correct? Explain under what conditions it is correct and under what conditions it is not correct.

Find the probabilities in Exercises 9–16. (See Examples 3–7.)

9. If a marble is drawn from a bag containing 2 yellow, 5 red, and 3 blue marbles, what are the probabilities of the following?

(a) The marble is red.

(b) The marble is either yellow or blue.

(c) The marble is yellow or red.

(d) The marble is green.

10. The law firm of Alam, Bartolini, Chinn, Dickinson, and Ellsberg has two senior partners, Alam and Bartolini. Two of the attorneys are to be selected to attend a conference. Assuming that all are equally likely to be selected, find the following probabilities.

(a) Chinn is selected.

(b) Ellsberg is not selected.

(c) Alam and Dickinson are selected.

(d) At least one senior partner is selected.

11. Ms. Bezzone invites 10 relatives to a party: her mother, two uncles, three brothers, and four cousins. If the chances of any one guest arriving first are equally likely, find the following probabilities.

(a) The first guest is an uncle or a cousin.

(b) The first guest is a brother or a cousin.

(c) The first guest is an uncle or her mother.

12. A card is drawn from a well-shuffled deck of 52 cards. Find the probability that the card is the following.
    (a) a queen
    (b) red
    (c) a black 3
    (d) a club or red

13. In Exercise 12, find the probability of the following.
    (a) a face card (K, Q, J of any suit)
    (b) red or a 3
    (c) less than a four (consider aces as 1's)

14. Two dice are rolled. Find the probability of the following.
    (a) The sum of the points is at least 10.
    (b) The sum of the points is either 7 or at least 10.
    (c) The sum of the points is 3 or the dice both show the same number.

15. **Management** The management of a firm wants to survey its workers, who are classified as follows for the purpose of an interview: 30% have worked for the company more than 5 years; 28% are female; 65% contribute to a voluntary retirement plan; 1/2 of the female workers contribute to the retirement plan. Find the following probabilities.
    (a) A male worker is selected.
    (b) A worker with less than 5 years in the company is selected.
    (c) A worker who contributes to the retirement plan or a female worker is selected.

16. The numbers 1, 2, 3, 4, and 5 are written on five slips of paper, and two slips are drawn at random without replacement. Find the probability of each of the following.
    (a) Both numbers are even.
    (b) One of the numbers is even or greater than 3.
    (c) The sum of the two numbers is 5 or the second number is 2.

Use Venn diagrams to work Exercises 17–22. (See Example 8.)

17. **Social Science** In a refugee camp in southern Mexico, it was found that 90% of the refugees came to escape political oppression, 80% came to escape abject poverty, and 70% came to escape both. What is the probability that a refugee in the camp was not poor nor seeking political asylum?

18. **Social Science** A study on body types gave the following results: 45% were short, 25% were short and overweight, and 24% were tall and not overweight. Find the probability that a person is
    (a) overweight;
    (b) short, but not overweight;
    (c) tall and overweight.

19. **Social Science** A teacher found that 85% of the students in her math class had passed a course in algebra, 60% had passed a course in geometry, and 55% had passed both courses. Find the probability that a student selected randomly from the math class has passed at least one of the two courses.

20. **Social Science** The following data were gathered for 130 adult U.S. workers: 55 were women, 3 women earned more than $40,000, 62 men earned less than $40,000. Find the probability that an individual is
    (a) a woman earning less than $40,000;
    (b) a man earning more than $40,000;
    (c) a man or is earning more than $40,000;
    (d) a woman or is earning less than $40,000.

21. **Management** Suppose than 8% of a certain batch of calculators have a defective case, and that 11% have defective batteries. Also, 3% have both a defective case and defective batteries. A calculator is selected from the batch at random. Find the probability that the calculator has a good case and good batteries.

22. **Social Science** Fifty students in a Texas school were interviewed with the following results: 45 spoke Spanish, 10 spoke Vietnamese, and 8 spoke both languages. Find the probability that a randomly selected student from this group
    (a) speaks both languages;
    (b) speaks neither language;
    (c) speaks only one of the two languages.

Work Exercises 23–29 on odds. (See Examples 10–11)

23. A single die is rolled. Find the odds in favor of getting the following results.
    (a) 3
    (b) 5 or 6
    (c) a number greater than 3
    (d) a number less than 2

24. A marble is drawn from a box containing 3 yellow, 4 white, and 8 blue marbles. Find the odds in favor of drawing the following.
(a) A yellow marble
(b) A blue marble
(c) A white marble

25. The probability that a company will make a profit this year is .74. Find the odds against the company making a profit.

26. If the odds that a given candidate will win an election are 3 to 2, what is the probability that the candidate will lose?

27. Social Science   A nationwide survey showed that the odds that an individual uses a seat belt when in a car's front seat are 17:8.* What is the probability that such an individual does not use a seat belt?

28. Social Science   The survey mentioned in Exercise 27 also showed that the odds that a driver does not drink or does not drive after drinking are 4:1.* What is the probability that a driver was drinking?

29. On page 134 of Roger Staubach's autobiography, *First Down, Lifetime to Go*, Staubach makes the following statement regarding his experience in Vietnam: "Odds against a direct hit are very low but when your life is in danger, you don't worry too much about the odds." Is this wording consistent with our definition of odds for and against? How could it have been said so as to be technically correct?

An experiment is conducted for which the sample space is $S = \{s_1, s_2, s_3, s_4, s_5\}$. Which of the probability assignments in Exercises 30–35 is possible for this experiment? If an assignment is not possible, tell why.

30.

Outcomes	s_1	s_2	s_3	s_4	s_5
Probabilities	.02	.27	.35	.22	.14

31.

Outcomes	s_1	s_2	s_3	s_4	s_5
Probabilities	.50	.30	.10	.08	.02

32.

Outcomes	s_1	s_2	s_3	s_4	s_5
Probabilities	1/8	1/6	1/5	1/3	1/2

33.

Outcomes	s_1	s_2	s_3	s_4	s_5
Probabilities	1/10	1/8	1/5	1/5	1/4

34.

Outcomes	s_1	s_2	s_3	s_4	s_5
Probabilities	.23	.17	.32	.38	−.10

35.

Outcomes	s_1	s_2	s_3	s_4	s_5
Probabilities	.3	.4	−.4	.4	.3

Work the following problems. (See Example 9.)

36. Social Science   A consumer survey randomly selects 2,000 people and asks them about their income and their TV-watching habits. The results are shown in this table.

Annual Income	Hours of TV Watched per Week				
	0–8	9–15	16–22	23–30	More than 30
Less than $12,000	11	15	8	14	120
$12,000–$24,999	10	19	28	96	232
$25,000–$39,999	18	32	88	327	189
$40,000–$59,999	31	85	160	165	100
$60,000 or more	73	60	52	55	12

As a reward for participation, each person questioned is given a small prize. Then the names of all participants are placed in a box and one name is randomly drawn to receive the grand prize of a free vacation trip. What is the probability that the winner of the grand prize

(a) watches TV at least 16 hours per week?
(b) has an income of at least $25,000?
(c) has an income of at least $40,000 and watches TV at least 16 hours per week?
(d) has an income of $0–$24,999 and watches TV more than 30 hours per week?

---

*Exercises 27 and 28 based on an article in the *Sacramento Bee*, June 4, 1992.

37. Social Science The table shows the probability of a person accumulating credit card charges over a 12-month period.

Charges	Probability
Under $100	.31
$100–$499	.18
$500–$999	.18
$1000–$1999	.13
$2000–$2999	.08
$3000–$4999	.05
$5000–$9999	.06
$10000 or more	.01

Find the probability that a person's total charges during the period are
(a) $500 or more;
(b) less than $1000;
(c) $500 to $2999;
(d) $3000 or more.

38. Social Science The table below gives the probability that a person has life insurance in the indicated range.

Amount of Insurance	Probability
None	.17
Less than $10,000	.20
$10,000–$24,999	.17
$25,000–$49,999	.14
$50,000–$99,999	.15
$100,000–$199,999	.12
$200,000 or more	.05

Find the probability that an individual has the following amounts of life insurance.
(a) Less than $10,000
(b) $10,000 to $99,999
(c) $50,000 or more
(d) Less than $50,000 or $100,000 or more

39. Natural Science The results of a study relating blood cholesterol level to coronary disease are given below.

Cholesterol Level	Probability of Coronary Disease
Under 200	.10
200–219	.15
220–239	.20
240–259	.26
Over 259	.29

Find the probability of coronary disease if the cholesterol level is
(a) less than 240;
(b) 220 or more;
(c) from 200 to 239;
(d) from 220 to 259.

Natural Science *Color blindness is an inherited characteristic which is sex-linked, so that it is more common in males than in females. If M represents male and C represents red-green color blindness, using the relative frequencies of the incidence of males and red-green color blindness as probabilities, P(C) = .049, P(M $\cap$ C) = .042, P(M $\cup$ C) = .534. Find the following. (Hint: use a Venn diagram with two circles labeled M and C.)*

40. $P(C')$

41. $P(M)$

42. $P(M')$

43. $P(M' \cap C')$

44. $P(C \cap M')$

45. $P(C \cup M')$

Natural Science *Gregor Mendel, an Austrian monk, was the first to use probability in the study of genetics. In an effort to understand the mechanisms of character transmittal from one generation to the next in plants, he counted the number of occurrences of various characteristics. Mendel found that the flower color in certain pea plants obeyed this scheme:*

Pure red crossed with pure white produces red.

The red offspring received from its parents genes for both red (R) and white (W), but in this case red is dominant *and white* recessive, *so the offspring exhibits the color red. However, the offspring still carries both genes, and when two such offspring are crossed, several things can happen in the third generation, as shown in the table below, which is called a* Punnet *square.*

		2nd Parent	
		R	W
1st parent	R	RR	RW
	W	WR	WW

This table shows the possible combinations of genes. Use the fact that red is dominant over white to find

46. P(red);

47. P(white).

Natural Science *Mendel found no dominance in snapdragons, with one red gene and one white gene producing pink-flowering offspring. These second-generation pinks, however, still carry one red and one white gene, and when they are crossed, the next generation still yields the Punnet square above. Find*

48. P(red);

49. P(pink);

50. P(white.)

(Mendel verified these probability ratios experimentally with large numbers of observations, and did the same for many character units other than flower color. The importance of his work, published in 1866, was not recognized until 1900.)

Natural Science *In most animals and plants, it is very unusual for the number of main parts of the organism (arms, legs, toes, flower petals, etc.) to vary from generation to generation. Some species, however, have meristic variability, in which the number of certain body parts varies from generation from generation. One researcher* studied the front feet of certain guinea pigs and produced the probabilities shown below.*

$$P(\text{only four toes, all perfect}) = .77$$
$$P(\text{one imperfect toe and four good ones}) = .13$$
$$P(\text{exactly five good toes}) = .10$$

Find the probability of having the following.

51. No more than four good toes

52. Five toes, whether perfect or not

---

*From "An Analysis of Variability in Guinea Pigs" by J. R. Wright, from *Genetics*, Vol. 19, 1934, pp. 506–36. Reprinted by permission.

*One way to solve a probability problem is to repeat the experiment (or a simulation of the experiment) many times, keeping track of the results. Then the probability can be approximated using the basic definition of the probability of an event E: P(E) = m/n, where m favorable outcomes occur in n trials of an experiment. This is called the **Monte Carlo method** of approximating probabilities.*

Use a calculator with a random number generator or appropriate computer software and the Monte Carlo method to simulate the experiments in Exercises 53–58.

Approximate the probabilities in Exercises 53 and 54 if five coins are tossed. Then calculate the theoretical probabilities using the methods of the text and compare the results.

53. P(4 heads)

54. P(2 heads, 1 tail, 2 heads) (in the order given)

Approximate the following probabilities if 4 cards are drawn from a 52-card deck.

55. P(any 2 cards and then 2 kings)

56. P(exactly 2 kings)

Approximate the probabilities in Exercises 57 and 58

57. A jeweler received 8 indentical watches each in a box marked with the series number of the watch. An assistant, who does not know that the boxes are marked, is told to polish the watches and then put them back in the boxes. She puts them in the boxes at random. What is the probability that she gets at least one watch in the right box?

58. A check room attendant has 10 hats but has lost the numbers identifying them. If he gives them back randomly, what is the probability that at least 2 of the hats are given back correctly?

1 Use the data in the table to find

(a) $P(B)$,

(b) $P(A')$,

(c) $P(B')$.

Answers:

(a) .45

(b) .4

(c) .55

8.5 CONDITIONAL PROBABILITY; INDEPENDENT EVENTS

The training manager for a large stockbrokerage firm has noticed that some of the firm's brokers use the firm's research advice, while other brokers tend to go with their own beliefs of which stocks will go up. To see whether the research department performs better than the beliefs of the brokers, the manager conducted a survey of 100 brokers, with results as shown in the following table.

	Picked a Stock that Went Up	*Didn't Pick a Stock that Went Up*	*Totals*
Used Research	30	15	45
Didn't Use Research	30	25	55
Totals	60	40	100

Letting A be the event "picked a stock that went up" and letting B be the event "used research," $P(A)$, $P(A')$, $P(B)$, and $P(B')$ can be found. For example, the chart shows that a total of 60 brokers picked stocks that went up, so $P(A) = 60/100 = .6$. **1**

Suppose we want to find the probability that a broker using research will pick a stock that goes up. From the table above, of the 45 brokers who use research, there are 30 who picked stocks that went up, so

$$P(\text{broker who uses research picks stocks that go up}) = \frac{30}{45} \approx .667$$

This is a different number than the probability that a broker picks a stock that goes up, .6, because *we have additional information* (the broker uses research) *that has reduced the sample space*. In other words, we found the probability that a broker picks a stock that goes up, A, given the additional information that the broker uses research, B. This is called the *conditional probability* of event A, given that event B has occurred, written $P(A \mid B)$. ($P(A \mid B)$ may also be read as "the probability of A given B.")

In the example above,

$$P(A \mid B) = \frac{30}{45}.$$

If we divide the numerator and denominator by 100, this can be written as

$$P(A \mid B) = \frac{\dfrac{30}{100}}{\dfrac{45}{100}} = \frac{P(A \cap B)}{P(B)},$$

where $P(A \cap B)$ represents, as usual, the probability that both A and B will occur.

To generalize this result, assume that E and F are two events for a particular experiment. Assume that the sample space S for this experiment has n possible equally likely outcomes. Suppose event F has m elements, and $E \cap F$ has k elements $(k \leq m)$. Using the fundamental principle of probability,

$$P(F) = \frac{m}{n} \quad \text{and} \quad P(E \cap F) = \frac{k}{n}.$$

We now want $P(E \mid F)$, the probability that E occurs given that F has occurred. Since we assume F has occurred, reduce the sample space to F: look only at the m elements inside F. (See Figure 8.23). Of these m elements, there are k elements where E also occurs, because $E \cap F$ has k elements. This makes

$$P(E \mid F) = \frac{k}{m}.$$

Divide numerator and denominator by n to get

$$P(E \mid F) = \frac{k/n}{m/n} = \frac{P(E \cap F)}{P(F)}.$$

This last result gives the definition of conditional probability.

The **conditional probability** of an event E given event F, written $P(E \mid F)$, is

$$P(E \mid F) = \frac{P(E \cap F)}{P(F)}, \quad P(F) \neq 0.$$

This definition tells us that, for equally likely outcomes, conditional probability is found by *reducing the sample space to event F*, and then finding the number of outcomes in F that are also in event E. Thus,

$$P(E \mid F) = \frac{n(E \cap F)}{n(F)}.$$

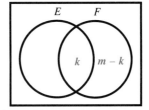

Event F has a total of m elements
FIGURE 8.23

2 The table shows the results of a survey of a buffalo herd.

	Males	Females	Totals
Adults	500	1300	1800
Calves	520	500	1020
Totals	1020	1800	2820

Let M represent "male" and A represent "adult." Find each of the following.

(a) $P(M \mid A)$

(b) $P(M' \mid A)$

(c) $P(A \mid M')$

(d) $P(A' \mid M)$

(e) State the probability in part (d) in words.

Answers:

(a) 5/18

(b) 13/18

(c) 13/18

(d) 26/51

(e) The probability that a buffalo is a calf given that it is a male.

▶ **EXAMPLE 1**   Use the chart in the stockbroker's problem at the beginning of this section to find the following probabilities, where A is the event "picked a stock that went up" and B is the event "used research."

(a) $P(B \mid A)$

This represents the probability that the broker used research, given that the broker picked a stock that went up. Reduce the sample space to A. Then find $n(A \cap B)$ and $n(A)$.

$$P(B \mid A) = \frac{P(B \cap A)}{P(A)} = \frac{n(A \cap B)}{n(A)} = \frac{30}{60} = \frac{1}{2}$$

If a broker picked a stock that went up, then the probability is 1/2 that the broker used research.

(b) $P(A' \mid B)$

In words, this is the probability that a broker picks a stock that does not go up, even though he used research.

$$P(A' \mid B) = \frac{n(A' \cap B)}{n(B)} = \frac{15}{45} = \frac{1}{3}$$

(c) $P(B' \mid A')$

Here, we want the probability that a broker who picked a stock that did not go up did not use research.

$$P(B' \mid A') = \frac{n(B' \cap A')}{n(A')} = \frac{25}{40} = \frac{5}{8} \qquad \blacktriangleleft \quad \boxed{2}$$

Venn diagrams can be used to illustrate problems in conditional probability. A Venn diagram for Example 1, in which the probabilities are used to indicate the number in the set defined by each region, is shown in Figure 8.24. In the diagram, $P(B \mid A)$ is found by *reducing the sample space to just set A*. Then $P(B \mid A)$ is the ratio of the number in that part of set B which is also in A to the number in set A, or $.3/.6 = .5$.

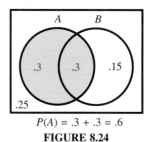

$P(A) = .3 + .3 = .6$

FIGURE 8.24

3 Find $P(F \mid E)$ if $P(E) = .3$, $P(F) = .4$, and $P(E \cup F) = .6$.

Answer:
1/3

▶**EXAMPLE 2** Given $P(E) = .4$, $P(F) = .5$, and $P(E \cup F) = .7$, find $P(E \mid F)$.

Find $P(E \cap F)$ first. Then use a Venn diagram to find $P(E \mid F)$. By the union rule,

$$P(E \cup F) = P(E) + P(F) - P(E \cap F)$$
$$.7 = .4 + .5 - P(E \cap F)$$
$$P(E \cup F) = .2.$$

Now use the probabilities to indicate the number in each region of the Venn diagram in Figure 8.25. $P(E \mid F)$ is the ratio of the probability of that part of E which is in F to the probability of F or

$$P(E \mid F) = \frac{P(E \cap F)}{P(F)} = \frac{.2}{.5} = \frac{2}{5} = .4. \quad ◀ \quad \boxed{3}$$

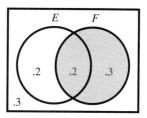

FIGURE 8.25

▶**EXAMPLE 3** Two fair coins were tossed, and it is known that at least one was a head. Find the probability that both were heads.

The sample space has four equally likely outcomes, $S = \{hh, ht, th, tt\}$. Define two events:

$$E_1 = \text{at least 1 head} = \{hh, ht, th\}$$

and

$$E_2 = 2 \text{ heads} = \{hh\}.$$

Because there are four equally likely outcomes, $P(E_1) = 3/4$. Also, $P(E_1 \cap E_2) = 1/4$. We want the probability that both were heads, given that at least one was a head: that is, we want to find $P(E_2 \mid E_1)$. Because of the condition that at least one coin was a head, the reduced sample space is

$$\{hh, ht, th\}.$$

Since only one outcome in this reduced sample space is 2 heads,

$$P(E_2 \mid E_1) = \frac{1}{3}.$$

4 In Example 3, find the probability that exactly one coin showed a head, given that at least one was a head.

Answer:
2/3

5 In a litter of puppies, 3 were female and 4 were male. Half the males were black. Find the probability that a puppy chosen at random from the litter would be a black male.

Answer:
2/7

Alternatively, use the definition given above.

$$P(E_2 \mid E_1) = \frac{P(E_2 \cap E_1)}{P(E_1)} = \frac{1/4}{3/4} = \frac{1}{3} \quad \blacktriangleleft \quad \boxed{4}$$

PRODUCT RULE   If $P(E) \neq 0$ and $P(F) \neq 0$, then the definition of conditional probability shows that

$$P(E \mid F) = \frac{P(E \cap F)}{P(F)} \quad \text{and} \quad P(F \mid E) = \frac{P(F \cap E)}{P(E)}.$$

Using the fact that $P(E \cap F) = P(F \cap E)$, and solving each of these equations for $P(E \cap F)$, we obtain the following rule.

Product Rule of Probability

If E and F are events, then $P(E \cap F)$ may be found by either of these formulas.

$$P(E \cap F) = P(F) \cdot P(E \mid F) \quad \text{or} \quad P(E \cap F) = P(E) \cdot P(F \mid E).$$

The **product rule** gives a method for finding the probability that events E and F both occur, as illustrated by the next few examples.

▶ **EXAMPLE 4**   In a class with 2/5 women and 3/5 men, 25% of the women are business majors. Find the probability that a student chosen at random from the class is a female business major.

Let B and W represent the events "business major" and "women," respectively. We want to find $P(B \cap W)$. By the product rule,

$$P(B \cap W) = P(W) \cdot P(B \mid W).$$

From the given information, $P(W) = 2/5 = .4$ and the probability that a woman is a business major is $P(B \mid W) = .25$. Then

$$P(B \cap W) = .4(.25) = .10. \quad \blacktriangleleft \quad \boxed{5}$$

In Section 8.1 we used a tree diagram to find the number of subsets of a given set. By including the probabilities for each branch of a tree diagram, we convert it to a **probability tree.** The next examples show how conditional probability is used with probability trees.

▶ **EXAMPLE 5**   A company needs to hire a new director of advertising. It has decided to try to hire either person A or person B, who are assistant advertising directors for its major competitor. To decide between A and B, the company does research on the campaigns managed by A or B (none are managed by both), and finds that A is in charge of twice as many advertising campaigns as B. Also,

A's campaigns have satisfactory results three out of four times, while B's campaigns have satisfactory results only two out of five times. Suppose one of the competitor's advertising campaigns (managed by A or B) is selected randomly.

We can represent this situation schematically as follows. Let A denote the event "Person A does the job" and B the event "Person B does the job." Let S be the event "satisfactory results" and U the event "unsatisfactory results." Then the given information can be summarized in the probability tree in Figure 8.26. Since A does twice as many jobs as B, $P(A) = 2/3$ and $P(B) = 1/3$, as noted on the first stage branches of the tree. When A does a job, the probability of satisfactory results is $3/4$ and of unsatisfactory results $1/4$, as noted on the second-stage branches. Similarly, the probabilities when B does the job are noted on the remaining second-stage branches. The composite branches labeled 1–4 represent the four mutually exclusive possibilities for the running and outcome of the campaign.

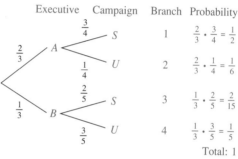

FIGURE 8.26

(a) Find the probability that A is in charge of a campaign that produces satisfactory results.

We are asked to find $P(A \cap S)$. We know that when A does the job, the probability of success is $3/4$, that is, $P(S \mid A) = 3/4$. Hence, by the product rule,

$$P(A \cap S) = P(A) \cdot P(S \mid A) = \frac{2}{3} \cdot \frac{3}{4} = \frac{1}{2}.$$

The event $A \cap S$ is represented by branch 1 of the tree, and as we have just seen, its probability is the product of the probabilities of the pieces that make up the branch.

(b) Find the probability that B runs a campaign that produces satisfactory results.

We must find $P(B \cap S)$. The event is represented by branch 3 of the tree and, as before, its probability is the product of the probabilities of the pieces of that branch:

$$P(B \cap S) = P(B) \cdot P(S \mid B) = \frac{1}{3} \cdot \frac{2}{5} = \frac{2}{15}.$$

6 Find each of the following probabilities for Example 5.

(a) $P(U \mid A)$

(b) $P(U \mid B)$

Answers:

(a) 1/4

(b) 3/5

7 Find the probability of drawing a green marble and then a white marble.

Answer:
1/5

(c) What is the probability that the campaign is satisfactory?

The event S is the union of the mutually exclusive events $A \cap S$ and $B \cap S$, which are represented by branches 1 and 3 of the tree diagram. By the union rule,

$$P(S) = P(A \cap S) + P(B \cap S) = \frac{1}{2} + \frac{2}{15} = \frac{19}{30}.$$

Thus, the probability of an event that appears on several branches is the sum of the probabilities of each of these branches.

(d) What is the probability that the campaign is unsatisfactory?

$P(U)$ can be read from branches 2 and 4 of the tree.

$$P(U) = \frac{1}{6} + \frac{1}{5} = \frac{11}{30}$$

Alternatively, because U is the complement of S,

$$P(U) = 1 - P(S) = 1 - \frac{19}{30} = \frac{11}{30}.$$

(e) Find the probability that either A runs the campaign or the results are satisfactory (or possibly both).

Event A combines branches 1 and 2, while event S combines branches 1 and 3, so use branches 1, 2, and 3.

$$P(A \cup S) = \frac{1}{2} + \frac{1}{6} + \frac{2}{15} = \frac{4}{5} \quad \blacktriangleleft \quad \boxed{6}$$

▶ **EXAMPLE 6**   From a box containing 1 red, 3 white, and 2 green marbles, two marbles are drawn one at a time without replacing the first before the second is drawn. Find the probability that one white and one green marble are drawn.

A probability tree showing the various possible outcomes is given in Figure 8.27. In this diagram, W represents the event "drawing a white marble" and G represents "drawing a green marble." On the first draw, $P(W$ on the 1st $) = 3/6 = 1/2$ because three of the six marbles in the box are white. On the second draw, $P(G$ on the 2nd $\mid W$ on the 1st$) = 2/5$. One white marble has been removed, leaving 5, of which 2 are green.

We want to find the probability of drawing exactly one white marble and exactly one green marble. Two events satisfy this condition: drawing a white marble first and then a green one (branch 2 of the tree), or drawing a green marble first and then a white one (branch 4). For branch 2,

$$P(W \text{ on 1st}) \cdot P(G \text{ on 2nd} \mid W \text{ on 1st}) = \frac{1}{2} \cdot \frac{2}{5} = \frac{1}{5}. \quad \boxed{7}$$

8 In Example 6, find the probability of drawing 1 white and 1 red marble.

Answer:
1/5

For branch 4, where the green marble is drawn first,

$$P(G \text{ first}) \cdot P(W \text{ second} \mid G \text{ first}) = \frac{1}{3} \cdot \frac{3}{5} = \frac{1}{5}.$$

Since these two events are mutually exclusive, the final probability is the sum of the two probabilities.

$$P(\text{one } W, \text{ one } G) = P(W \text{ on 1st}) \cdot P(G \text{ on 2nd} \mid W \text{ on 1st})$$

$$+ P(G \text{ on 1st}) \cdot P(W \text{ on 2nd} \mid G \text{ on 1st}) = \frac{2}{5} \quad \blacktriangleleft \quad \boxed{8}$$

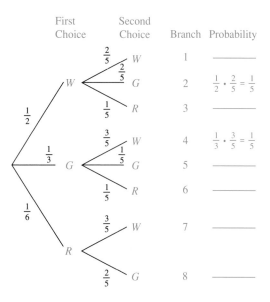

FIGURE 8.27

The product rule is often used with *stochastic processes,* where the outcome of an experiment depends on the outcomes of previous experiments. For example, the outcome of a draw of a card from a deck depends on any cards previously drawn. (Stochastic processes are studied further in Section 9.4.)

▶**EXAMPLE 7** Two cards are drawn without replacement from an ordinary deck (52 cards). Find the probability that the first card is a heart and the second card is red.

9 Find the probability of drawing a heart on the first draw and a black card on the second, if two cards are drawn without replacement.

Answer:
13/102 or .1275

10 Use the tree in Example 8 to find the probability that exactly one of the cards is red.

Answer:
13/34 ≈ .382

Start with the probability tree of Figure 8.28. (You may wish to refer to the deck of cards shown in Figure 8.18.) On the first draw, since there are 13 hearts in the 52 cards, the probability of drawing a heart first is $13/52 = 1/4$. On the second draw, since a (red) heart has been drawn already, there are 25 red cards in the remaining 51 cards. Thus the probability of drawing a red card on the second draw, given that the first is a heart, is $25/51$. By the product rule of probability

$$P(\text{heart on 1st and red on 2nd})$$
$$= P(\text{heart on 1st}) \cdot P(\text{red on 2nd} \mid \text{heart on 1st})$$
$$= \frac{1}{4} \cdot \frac{25}{51} = \frac{25}{204} = .1225. \quad \blacktriangleleft \quad \boxed{9}$$

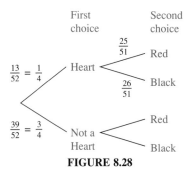

FIGURE 8.28

▶**EXAMPLE 8** Three cards are drawn, without replacement, from an ordinary deck. Find the probability that exactly 2 of the cards are red.

Here we need a probability tree with three stages, as shown in Figure 8.29. The three branches indicated with arrows produce exactly 2 red cards from the draws. Multiply the probabilities along each of these branches and then add.

$$P(\text{exactly 2 red cards}) = \frac{26}{52} \cdot \frac{25}{51} \cdot \frac{26}{50} + \frac{26}{52} \cdot \frac{26}{51} \cdot \frac{25}{50} + \frac{26}{52} \cdot \frac{26}{51} \cdot \frac{25}{50}$$
$$= \frac{50,700}{132,600} = \frac{13}{34} = .382 \quad \blacktriangleleft \quad \boxed{10}$$

INDEPENDENT EVENTS Suppose a fair coin is tossed and shows heads. The probability of heads on the next toss is still $1/2$; the fact that heads was the result on a given toss has no effect on the outcome of the next toss. Coin tosses are **independent events,** since the outcome of one toss does not help decide the

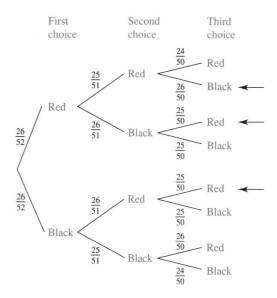

First choice
Second choice
Third choice

FIGURE 8.29

outcome of the next toss. Rolls of a fair die are independent events; the fact that a 2 came up on one roll does not increase our knowledge of the outcome of the next roll. On the other hand, the events "today is cloudy" and "today is rainy" are **dependent events;** if it is cloudy, then there is an increased chance of rain. Similarly, in the example at the beginning of this section, the events A (broker picked a stock that went up) and B (broker used research) are dependent events, because information about the use of research affects the probability of picking a stock that goes up. That is, $P(A \mid B)$ is different from $P(A)$.

If events E and F are independent, then the knowledge that E has occurred gives no (probability) information about the occurrence or nonoccurrence of event F. That is, $P(F)$ is exactly the same as $P(F \mid E)$, or

$$P(F \mid E) = P(F).$$

In fact, this is the formal definition of independent events.

E and F are **independent events** if
$$P(F \mid E) = P(F) \quad \text{or} \quad P(E \mid F) = P(E).$$

11 Find the probability of getting 4 successive heads on 4 tosses of a fair coin.

Answer:
1/16

When E and F are independent events, then $P(F \mid E) = P(F)$ and the product rule becomes

$$P(E \cap F) = P(E) \cdot P(F \mid E) = P(E) \cdot P(F).$$

Conversely, if this equation holds, then it follows that $P(F) = P(F \mid E)$. Consequently, we have this useful fact

Product Rule for Independent Events

E and F are independent events if and only if

$$P(E \cap F) = P(E) \cdot P(F).$$

▶ **EXAMPLE 9**   A calculator requires a key-stroke assembly and a logic circuit. Assume that 99% of the key-stroke assemblies are satisfactory and 97% of the logic circuits are satisfactory. Find the probability that a finished calculator will be satisfactory.

If the failure of a key-stroke assembly and the failure of a logic circuit are independent events, then

P(satisfactory calculator)

$= P$(satisfactory key-stroke assembly) $\cdot P$(satisfactory logic circuit)

$= (.99)(.97) \approx .96.$

The probability of a defective calculator is $1 - .96 = .04.$ ◀   **11**

Caution   It is common for students to confuse the ideas of *mutually exclusive* events and *independent* events. Events E and F are mutually exclusive if $E \cap F = \emptyset$. For example, if a family has exactly one child, the only possible outcomes are $B = \{\text{boy}\}$ and $G = \{\text{girl}\}$. These two events are mutually exclusive. However, the events are *not* independent, since $P(G \mid B) = 0$ (if a family with only one child has a boy, the probability it has a girl is then 0). Since $P(G \mid B) \neq P(G)$, the events are not independent. Of all the families with exactly *two* children, the events $G_1 = \{\text{first child is a girl}\}$ and $G_2 = \{\text{second child is a girl}\}$ are independent, because $P(G_2 \mid G_1)$ equals $P(G_2)$. However, G_1 and G_2 are not mutually exclusive, since $G_1 \cap G_2 = \{\text{both children are girls}\} \neq \emptyset$.

To show that two events E and F are independent, we can show that $P(F \mid E) = P(F)$ or that $P(E \mid F) = P(E)$ or that $P(E \cap F) = P(E) \cdot P(F)$. Another way is to observe that knowledge of one outcome does not influence the probability of the other outcome, as we did for coin tosses.

12 In the U.S. population, the probability of being Hispanic is .11, the probability of living in California is .12, and the probability of being a Hispanic living in California is .04. Are the events being Hispanic and living in California independent?

Answer:
No

▶**EXAMPLE 10**   On a typical January day in Manhattan the probability of snow is .10, the probability of a traffic jam is .80, and the probability of snow or a traffic jam (or both) is .82. Are the event "it snows" and the event "a traffic jam occurs" independent?

Let S represent the event "it snows" and T represent the event "a traffic jam occurs." We must determine whether

$$P(T \mid S) = P(T) \quad \text{or} \quad P(S \mid T) = P(S).$$

We know $P(S) = .10$, $P(T) = .8$, and $P(S \cup T) = .82$. We can use the union rule (or a Venn diagram) to find $P(S \cap T) = .08$, $P(T \mid S) = .8$, and $P(S \mid T) = .1$. Since

$$P(T \mid S) = P(T) = .8 \quad \text{and} \quad P(S \mid T) = P(S) = .1,$$

the events "it snows" and "a traffic jam occurs" are independent. ◀ **12**

Although we showed $P(T \mid S) = P(T)$ and $P(S \mid T) = P(S)$ in Example 10, only one of these results is needed to establish independence.

8.5 EXERCISES

If a single fair die is rolled, find the probability of rolling the following.

1. 3, given that the number rolled was odd

2. 5, given that the number rolled was even

3. An odd number, given that the number rolled was 3

If two fair dice are rolled (recall the 36-outcome sample space), find the probability of rolling the following.

4. A sum of 8, given the sum was greater than 7

5. A sum of 6, given the roll was a "double" (two identical numbers)

6. A double, given that the sum was 9

If two cards are drawn without replacement from an ordinary deck (see Example 7), find the following probabilities.

7. The second is a heart, given that the first is a heart.

8. The second is black, given that the first is a spade.

9. The second is a face card, given that the first is a jack.

10. The second is an ace, given that the first is not an ace.

11. A jack and a 10 are drawn.

12. An ace and a 4 are drawn.

13. Two black cards are drawn.

14. Two hearts are drawn.

15. In your own words, explain how to find the conditional probability $P(E \mid F)$.

16. Your friend asks you to explain how the product rule for independent events differs from the product rule for dependent events. How would you respond?

17. Another friend asks you to explain how to tell whether two events are dependent or independent. How would you reply? (Use your own words.)

18. A student reasons that the probability in Example 3 of both coins being heads is just the probability that the other coin is a head, that is, 1/2. Explain why this reasoning is wrong.

Use a tree diagram or Venn diagram in Exercises 19–30. (See Examples 2 and 5–8.)

Social Science   *Marriages between cousines are very common in some countries, where about 50% of marriages are consanguineous (between first cousins or people even more closely related). A recent study in Pakistan has shown that 16% of children from unrelated marriages died by age 10, while 21% of children from consanguineous marriages died by age 10. Find the following probabilities.*

19. a child survives

20. a child from a consanguineous marriage survives

Social Science   *A recent survey showed that 46% of all U.S. public school students used a computer at school, and 24% of those students also used a computer at home. Only 11% of the remaining students used a computer at home. Find the following probabilities.*

21. A student uses a computer at home.

22. A student uses a computer at school and at home.

Social Science   *A survey has shown that 52% of the women in a certain community work outside the home. Of these women, 64% are married, while 86% of the women who do not work outside the home are married. Find the probability that a woman in that community is*

23. married;

24. a single woman working outside the home.

Two-thirds of the population is on a diet at least occasionally. Of this group, 4/5 drink diet soft drinks, while 1/2 of the rest of the population drinks diet soft drinks. Find the probability that a person

25. drinks diet soft drinks;

26. diets but does not drink diet soft drinks.

Social Science   *Two of the most popular syndicated TV shows are "Wheel of Fortune," seen by 16.5% of all households, and "The Oprah Winfrey Show," viewed by 14% of households. If 85% of "Wheel of Fortune's" viewers are women, and 83% of Oprah's viewers are women, find the following probabilities.* *

27. A man watches "The Oprah Winfrey Show."

28. A man watches either show.

Management   *A shop that produces custom kitchen cabinets has two employees, Sitlington and Capek. 95% of Capek's work is satisfactory and 10% of Sitlington's work is unsatisfactory. 60% of the shop's cabinets are made by Capek (the rest by Sitlington). Find the following probabilities.*

29. An unsatisfactory cabinet was made by Capek.

30. A finished cabinet is unsatisfactory.

Management   *The table below shows employment figures for managerial/professional occupations in 1991 for U.S. civilians with four or more years of college.* **

	White	Black	
Women	6813	617	7430
Men	9453	435	9888
	16266	1052	17318

Letting A represent white, B represent black, C represent women, and D represent men, express each of the following probabilities in words and find its value. (See Example 1.)

31. $P(A \mid D)$ **32.** $P(C \mid A)$ **33.** $P(B \mid C)$ **34.** $P(D \cap A)$

Natural Science   *The following table shows frequencies for red-green color blindness, where M represents male and C represents color-blind.*

	M	M'	Totals
C	.042	.007	.049
C'	.485	.466	.951
Totals	.527	.473	1.000

Use this table to find the following probabilities.

35. $P(M)$ **36.** $P(C)$

37. $P(M \cap C)$ **38.** $P(M \cup C)$

39. $P(M \mid C)$ **40.** $P(M' \mid C)$

41. Are the events C and M described above dependent? (Recall that two events E and F are dependent if $P(E \mid F) \neq P(E)$.)

---

*Source: Nielsen Media Research, February, 1991.
**U.S. Bureau of the Census, Statistical Abstract of the United States: 1992 (112th edition). Washington, DC, 1992

42. Natural Science A scientist wishes to determine if there is any dependence between color blindness (C) and deafness (D). Given the probabilities listed in the table below, what should his findings be? (See Exercises 35–41.)

	D	D'	Totals
C	.0004	.0796	.0800
C'	.0046	.9154	.9200
Totals	.0050	.9950	1.0000

Social Science *The Motor Vehicle Department has found that the probability of a person passing the test for a driver's license on the first try is .75. The probability that an individual who fails on the first test will pass on the second try is .80, and the probability that an individual who fails the first and second tests will pass the third time is .70. Find the probability that an individual*

43. fails both the first and second tests;

44. will fail three times in a row;

45. will require at least two tries to pass the test.

Natural Science *Four different medications, C, D, E, and F, may be used to control high blood pressure. A physician usually prescribes C first because it is least likely to cause side effects. If blood pressure remains high, the patient is switched to D. If this fails to work, the patient is switched to E, and if necessary to F. The probability that C will work is .7. If C fails, the probability that D will work is .8. If D fails, the probability that E will work is .62, If E fails, the probability that F will work is .45 Find the probability that*

46. a patient's blood pressure will not be reduced by any of the medications.

47. a patient will have to take at least two medications and will have his or her blood pressure reduced.

48. If medications C and D fail, what is the probability that a patient's blood pressure will be reduced by medication E or F?

Management *The number of vehicles (in thousands) on the road from the major worldwide producers in selected years is shown in the following table.*

	U.S.	Europe	Japan	Canada
1975	4495	6737	3471	712
1980	6008	11,584	8282	1031
1985	9322	12,767	9817	1546
1990	9291	17,683	12,812	1801

Find the following probabilities for a vehicle selected at random.

49. It was made in the U.S.

50. It was made in 1990.

51. It was made in Japan in 1985.

52. It was made in 1980, given that it was made in Europe.

The probability that the first record by a singing group will be a hit is .32. If their first record is a hit, so are all their subsequent records. If their first record is not a hit, the probability of their second record and all subsequent ones being hits is .16. If the first two records are not hits, the probability that the third is a hit is .08. The probability of a hit continues to decrease by half with each successive nonhit record. Find the probability that

53. the group will have at least one hit in their first four records.

54. the group will have exactly one hit in their first three records.

55. the group will have a hit in their first six records if the first three are not hits.

Work the following problems on independent events. (See Examples 9 and 10.)

56. Management Corporations such as banks, where a computer is essential to day-to-day operations, often have a second, backup computer in case of failure by the main computer. Suppose that there is a .003 chance that the main computer will fail in a given time period and a .005 chance that the backup computer will fail while the main computer is being repaired. Assume these failures represent independent events, and find the fraction of the time that the corporation can assume it will have computer service. How realistic is our assumption of independence?

57. Management   According to a booklet put out by East-west Airlines, 98% of all scheduled Eastwest flights actually take place. (The other flights are canceled due to weather, equipment problems, and so on.) Assume that the event that a given flight takes place is independent of the event that another flight takes place.

(a) Elisabeta Guervara plans to visit her company's branch offices; her journey requires 3 separate flights on Eastwest Airlines. What is the probability that all of these flights will take place?

(b) Based on the reasons we gave for a flight to be canceled, how realistic is the assumption of independence that we made?

58. In one area, 4% of the population drive luxury cars. However, 17% of the CPAs drive luxury cars. Are the events "person drives a luxury car" and "person is a CPA" independent?

59. The probability that a key component of a space rocket will fail is .03.

(a) How many such components must be used as backups to ensure that the probability of at least one of the components' working is .999999?

(b) Is it reasonable to assume independence here?

60. Natural Science   A medical experiment showed that the probability that a new medicine is effective is .75, the probability that a patient will have a certain side effect is .4, and the probability that both events occur is .3. Decide whether these events are dependent or independent.

61. Social Science   A teacher has found that the probability that a student studies for a test is .6, the probability that a student gets a good grade on a test is .7, and the probability that both occur is .52. Are these events independent?

8.6 BAYES' FORMULA

Suppose the probability that a person gets lung cancer, given that the person smokes a pack or more of cigarettes daily, is known. For a research project, it might be necessary to know the probability that a person smokes a pack or more of cigarettes daily, given that the person has lung cancer. More generally, if $P(E \mid F)$ is known for two events E and F, can $P(F \mid E)$ be found? The answer is yes, we can find $P(F \mid E)$ using the formula to be developed in this section. To develop this formula, we can use a probability tree to find $P(F \mid E)$. Since $P(E \mid F)$ is known, the first outcome is either F or F'. Then for each of these outcomes, either E or E' occurs, as shown in Figure 8.30.

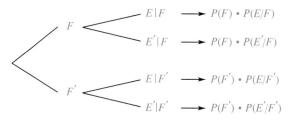

FIGURE 8.30

1 Use the special case of
Bayes' formula to find $P(F \mid E)$
if $P(F) = .2$, $P(E \mid F) = .1$, and
$P(E \mid F') = .3$. (Hint: $P(F') = 1 - P(F)$.)

Answer:
$1/13 \approx .077$

2 In Example 1, find
$P(F' \mid E)$.

Answer:
$6/7 \approx .857$

The four cases have the probabilities shown on the right. Notice that $P(E)$ is the sum of the first and third cases. By the definition of conditional probability and the product rule

$$P(F \mid E) = \frac{P(F \cap E)}{P(E)} = \frac{P(F) \cdot P(E \mid F)}{P(F) \cdot P(E \mid F) + P(F') \cdot P(E \mid F')}.$$

We have proved a special case of Bayes' theorem, which is generalized later in this section.

Bayes' Formula (Special Case)

$$P(F \mid E) = \frac{P(F) \cdot P(E \mid F)}{P(F) \cdot P(E \mid F) + P(F') \cdot P(E \mid F')}. \quad \boxed{1}$$

▶ **EXAMPLE 1**   For a fixed length of time, the probability of worker error on a certain production line is .1, the probability that an accident will occur when there is a worker error is .3, and the probability that an accident will occur when there is no worker error is .2. Find the probability of a worker error if there is an accident.

Let E represent the event of an accident, and let F represent the event of worker error. From the information above,

$$P(F) = .1, \quad P(E \mid F) = .3, \quad \text{and} \quad P(E \mid F') = .2.$$

These probabilities are shown on the probability tree in Figure 8.31.

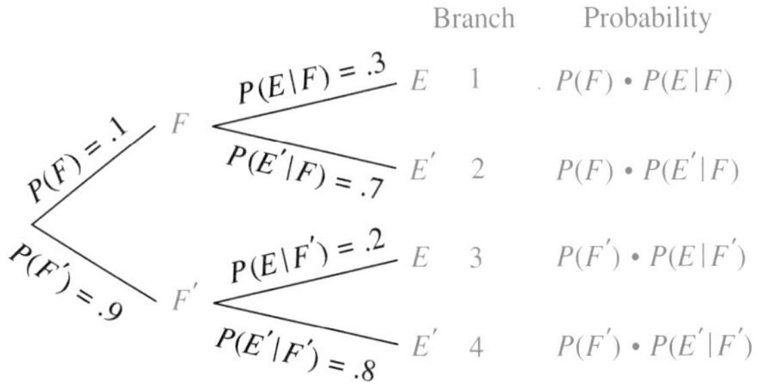

FIGURE 8.31

Find $P(F \mid E)$ using the tree or Bayes' theorem.

$$P(F \mid E) = \frac{P(F) \cdot P(E \mid F)}{P(F) \cdot P(E \mid F) + P(F') \cdot P(E \mid F')}$$

$$= \frac{(.1)(.3)}{(.1)(.3) + (.9)(.2)} \approx .143 \quad ◀ \boxed{2}$$

Bayes' formula can be generalized to more than two possibilities with the probability tree of Figure 8.32. This diagram shows the paths that can produce an event E. We assume that events $F_1, F_2, \ldots, F_n$ are pairwise mutually exclusive events (that is, events which, taken two at a time, are disjoint), whose union is the sample space, and E is an event that has occurred. See Figure 8.33.

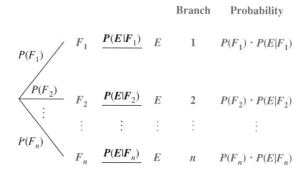

FIGURE 8.32

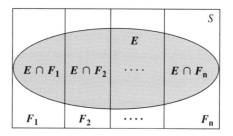

FIGURE 8.33

The probability $P(F_i \mid E)$, where $1 \leq i \leq n$, can be found by dividing the probability for the branch containing $P(E \mid F_i)$ by the sum of the probabilities of all the branches producing event E.

Bayes' Formula

$$P(F_i \mid E) = \frac{P(F_i) \cdot P(E \mid F_i)}{P(F_1) \cdot P(E \mid F_1) + \cdots + P(F_n) \cdot P(E \mid F_n)}.$$

This result is known as **Bayes' formula,** after the Reverend Thomas Bayes (1702–61), whose paper on probability was published about two hundred years ago.

The statement of Bayes' formula can be daunting. Actually, it is easier to remember the formula by thinking of the probability tree that produced it. Go through the following steps.

Using Bayes' Formula

Step 1 Start a probability tree with branches representing events F_1, $F_2, \ldots, F_n$. Label each branch with its corresponding probability.

Step 2 From the end of each of these branches, draw a branch for event E. Label this branch with the probability of getting to it, or $P(E \mid F_i)$.

Step 3 There are now n different paths that result in event E. Next to each path, put its probability—the product of the probabilities that the first branch occurs, $P(F_i)$, and that the second branch occurs, $P(E \mid F_i)$: that is, $P(F_i) \cdot P(E \mid F_i)$.

Step 4 $P(F_i \mid E)$ is found by dividing the probability of the branch for F_i by the sum of the probabilities of all the branches producing event E.

▶**EXAMPLE 2**   Based on past experience, a company knows that an experienced machine operator (one or more years of experience) will produce a defective item 1% of the time. Operators with some experience (up to one year) have a 2.5% defect rate, while new operators have a 6% defect rate. At any one time, the company has 60% experienced employees, 30% with some experience, and 10% new employees. Find the probability that a particular defective item was produced by a new operator.

Let E represent the event "item is defective," with F_1 representing "item was made by an experienced operator," F_2 "item was made by an operator with some experience," and F_3 "item was made by a new operator." Then

$$P(F_1) = .60 \qquad P(E \mid F_1) = .01$$
$$P(F_2) = .30 \qquad P(E \mid F_2) = .025$$
$$P(F_3) = .10 \qquad P(E \mid F_3) = .06.$$

We need to find $P(F_3 \mid E)$, the probability that an item was produced by a new operator, given that it is defective. First, draw a probability tree using the given information, as in Figure 8.34 on the next page. The steps leading to event E are shown.

3 In Example 2, find

(a) $P(F_1 \mid E)$;

(b) $P(F_2 \mid E)$.

Answers:

(a) $4/13 \approx .3075$

(b) $5/13 \approx .385$

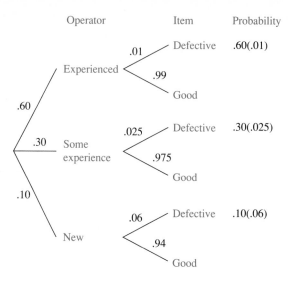

FIGURE 8.34

Find $P(F_3 \mid E)$ using the bottom branch of the tree in Figure 8.34: divide the probability for this branch by the sum of the probabilities of all the branches leading to E.

$$P(F_3 \mid E) = \frac{.10(.06)}{.60(.01) + .30(.025) + .10(.06)} = \frac{.006}{.0195} \approx .3075 \quad \blacktriangleleft \; \boxed{3}$$

After working Problem 3 at the side, check that $P(F_1 \mid E) + P(F_2 \mid E) + P(F_3 \mid E) = 1$. (That is, the defective item was made by *someone*.)

▶ **EXAMPLE 3** A manufacturer buys items from six different suppliers. The fraction of the total number of items obtained from each supplier, along with the probability that an item purchased from that supplier is defective, is shown in the following table.

Supplier	*Fraction of Total Supplied*	*Probability of Defective*
1	.05	.04
2	.12	.02
3	.16	.07
4	.23	.01
5	.35	.03
6	.09	.05

Find the probability that a defective item came from supplier 5.

Let F_1 be the event that an item came from supplier 1, with F_2, F_3, F_4, F_5, and F_6 defined in a similar manner. Let E be the event that an item is defective.

4 In Example 3, find the probability that the defective item came from

(a) supplier 3;

(b) supplier 6.

Answers:

(a) .340

(b) .137

We want to find $P(F_5 \mid E)$. Use the probabilities in the table above to prepare a probability tree. By Bayes' formula,

$$P(F_5 \mid E) =$$

$$\frac{(.35)(.03)}{(.05)(.04) + (.12)(.02) + (.16)(.07) + (.23)(.01) + (.35)(.03) + (.09)(.05)}$$

$$= \frac{.0105}{.0329} \approx .319.$$

There is about a 32% chance that a defective item came from supplier 5. ◀ **4**

8.6 EXERCISES

For two events, M and N, $P(M) = .4$, $P(N \mid M) = .3$, and $P(N \mid M') = .4$. Find each of the following. (See Example 1.)

1. $P(M \mid N)$

2. $P(M' \mid N)$

For mutually exclusive events R_1, R_2, R_3, $P(R_1) = .05$, $P(R_2) = .6$, and $P(R_3) = .35$, Also, $P(Q \mid R_1) = .40$, $P(Q \mid R_2) = .30$, and $P(Q \mid R_3) = .60$. Find each of the following. (See Example 2.)

3. $P(R_1 \mid Q)$

4. $P(R_2 \mid Q)$

5. $P(R_3 \mid Q)$

6. $P(R_1' \mid Q)$

Suppose three jars have the following contents: 2 black balls and 1 white ball in the first; 1 black ball and 2 white balls in the second; 1 black ball and 1 white ball in the third. If the probability of selecting one of the three jars is 1/2, 1/3, and 1/6, respectively, find the probability that if a white ball is drawn, it came from the

7. second jar;

8. third jar.

Social Science *In 1990, 1% of the United States population was Native American. The probability that a Native American lives in the West or South is .76. The probability is .56 that a person who is not Native American lives in the West or South. Find the following probabilities.*

9. a person living in the West or South is Native American

10. a person who does not live in the West or South is not Native American

Social Science *A federal study showed that in 1990, 49% of all those involved in a fatal car crash wore seat belts. Of those in a fatal crash who wore seat belts, 44% were injured and 27% were killed. For those not wearing seat belts, the comparable figures were 41% and 50%, respectively.* *

11. Find the probability that a randomly selected person who was killed in a car crash was wearing a seat belt.

12. Find the probability that a randomly selected person who was unharmed in a fatal crash was not wearing a seat belt.

* National Highway Traffic Safety Administration, Office of Driver and Pedestrian Research: "Occupant Protection Trends in 19 Cities" (November 1989) and "Use of Automatic Safety Belt Systems in 19 Cities" (February 1991).

Management *The probability that a customer of a local department store will be a "slow pay" is .02. The probability that a "slow pay" will make a large down payment when buying a referigerator is .14. The probability that a person who is not a "slow pay" will make a large down payment when buying a refrigerator is .50. Suppose a customer makes a large down payment on a refrigerator. Find the probability that the customer is*

13. a "slow pay";

14. not a "slow pay."

Management *Companies A , B, and C produce 15%, 40%, and 45%, respectively, of the major appliances sold in a certain area. In that area, 1% of the Company A appliances, $1\frac{1}{2}$% of the Company B appliances, and 2% of the Company C appliances need service within the first year. Suppose an appliance that needs service within the first year is chosen at random; find the probability that it was manufactured by Company*

15. A;

16. B.

Management *On a given weekend in the fall, a tire company can buy television advertising time for a college football game, a baseball game, or a professional football game. If the company sponsors the college game, there is a 70% chance of a high rating, a 50% chance if they sponsor a baseball game, and a 60% chance if they sponsor a professional football game. The probability of the company sponsoring these various games is .5, .2, and .3, respectively. Suppose the company does get a high rating; find the probability that it sponsored*

17. a college game;

18. a professional football game.

Management *According to a business publication, there is a 50% chance of a booming economy next summer, a 20% chance of a mediocre economy, and a 30% chance of a recession. The probabilities that a particular investment strategy will produce a high profit under each of these possibilities are .1, .6, and .3, respectively. Suppose it turns out that the strategy does produce huge profits; find the probability that the economy was*

19. booming;

20. in recession.

21. Management A manufacturing firm finds that 70% of its new hires turn out to be good workers and 30% poor workers. All current workers are given a reasoning test. Of the good workers, 80% pass it; 40% of the poor workers pass it. Assume that these figures will hold true in the future. If the company makes the test part of its hiring procedure and only hires people who meet the previous requirements and pass the test, what percent of the new hires will turn out to be good workers?

22. Management A bank finds that the relationship between mortgage defaults and the size of the down payment is given by this table.

Down Payment (%)	*5%*	*10%*	*20%*	*25%*
Number of mortgages with this down payment	1260	700	560	280
Probability of default	.05	.03	.02	.01

If a default occurs, what is the probability that it is on a mortgage with a 5% down payment? (See Examples 2 and 3.)

23. Management The following information pertains to three shipping terminals operated by Krag Corp.:*

Terminal	*Percentage of Cargo Handled*	*Percentage of Error*
Land	50	2
Air	40	4
Sea	10	14

Krag's internal auditor randomly selects one set of shipping documents, ascertaining that the set selected contains an error. Which of the following gives the probability that the error occurred in the Land Terminal?
(a) .02   **(b)** .10   **(c)** .25   **(d)** .50

*Uniform CPA Examination, November 1989

Natural Science   *In a test for toxemia, a disease that affects pregnant women, the woman lies on her left side and then rolls over on her back. The test is considered positive if there is a 20 mm rise in her blood pressure within one minute. The results have produced the following probabilities, where T represents having toxemia at some time during the pregnancy, and N represents a negative test.*

$$P(T' \mid N) = .90 \quad \text{and} \quad P(T \mid N') = .75$$

Assume that $P(N') = .11$, and find each of the following.

24. $P(N \mid T)$

25. $P(N' \mid T)$

26. Natural Science   The probability that a person with certain symptoms has hepatitis is .8. The blood test used to confirm this diagnosis gives positive results for 90% of those who have the disease and 5% of those without the disease. What is the probability that an individual with the symptoms who reacts positively to the test has hepatitis?

27. Natural Science   Suppose the probability that an individual has AIDS is .01, the probability of a person testing positive if he or she has AIDS is .95, and the probability of a person testing positive if he or she does not have AIDS is .05.

(a) Find the probability that a person who tests positive has AIDS.

(b) It has been argued that everyone should be tested for AIDS. Based on the results of part (a), how useful would the results of such testing be?

Social Science   *The following table gives the proportions of adult men and women in the U.S. population, and the proportions of adult men and women who have never married, in a recent year.[*]*

Age	Men Proportion of Men	Proportion Never Married
18–24	.151	.875
25–29	.126	.433
30–34	.126	.250
35–39	.110	.140
40 or over	.487	.054

Age	Women Proportion of Women	Proportion Never Married
18–24	.142	.752
25–29	.117	.295
30–34	.116	.161
35–39	.103	.090
40 or over	.522	.033

28. Find the probability that a randomly selected man who has never married is between 30 and 34 years old (inclusive).

29. Find the probability that a randomly selected woman who has been married is between 18 and 24 (inclusive).

30. Find the probability that a randomly selected woman who has never been married is between 35 and 39 (inclusive).

[*] From the *Sacramento Bee,* September 10, 1987.

KEY TERMS AND SYMBOLS

{ }	set braces
$\in$	is an element of
$\notin$	is not an element of
$\subseteq$	is a subset of
$\not\subseteq$	is not a subset of
A'	complement of set A
$\cap$	set intersection
$\cup$	set union

8.1 set
element (member) of a set
empty set
set-builder notation
universal set
equal sets
subset
tree diagram
Venn diagram
complement
intersection
disjoint sets
union

8.2 Union Rule for Counting

8.3 $P(E)$ probability of event E
experiment
trial

outcome
fair coin
sample space
event
simple event
certain event
impossible event
basic probability principle
probability of an event

8.4 mutually exclusive events
Union Rule for Probability
Complement Rule
odds
empirical probability
probability distribution

8.5 $P(F \mid E)$ probability of F, given that E has occurred
conditional probability
Product Rule of Probability
probability tree
independent events
dependent events
Product Rule for Independent Events

8.6 Bayes' formula

KEY CONCEPTS

Sets

Set A is a **subset** of set B if every element of A is also an element of B.

A set of n elements has 2^n subsets.

Let A and B be any sets with universal set U.

The **complement** of A is $A' = \{x \mid x \notin A \text{ and } x \in U\}$.

The **intersection** of A and B is $A \cap B = \{x \mid x \in A \text{ and } x \in B\}$.

The **union** of A and B is $A \cup B = \{x \mid x \in A \text{ or } x \in B \text{ or both }\}$.
$n(A \cup B) = n(A) + n(B) - n(A \cap B)$

Probability

If $n(S) = n$ and $n(E) = m$, where S is the sample space, then $P(E) = \dfrac{m}{n}$.

The probability of any outcome is a number between 0 and 1, inclusive. The sum of the probabilities of all possible distinct outcomes in a sample space is 1.

Let E and F be events from a sample space S.

$P(E') = 1 - P(E)$ and $P(E) = 1 - P(E')$.

$P(E \cup F) = P(E) + P(F) - P(E \cap F)$.

$P(E \mid F) = \dfrac{P(E \cap F)}{P(F)}, \quad P(F) \neq 0$.

$P(E \cap F) = P(F) \cdot P(E \mid F) \quad \text{or} \quad P(E \cap F) = P(E) \cdot P(F \mid E)$.

Odds: The odds in favor of event E are given by the ratio of $P(E)$ to $P(E')$.

Events E and F are **mutually exclusive** if $E \cap F = \emptyset$. In that case,
$P(E \cup F) = P(E) + P(F)$.

Events E and F are **independent events** if $P(F \mid E) = P(F)$ or $P(E \mid F) = P(E)$. In that case, $P(E \cap F) = P(E) \cdot P(F)$.

Bayes' Formula: $P(F_i \mid E) = \dfrac{P(F_i) \cdot P(E \mid F_i)}{P(F_1) \cdot P(E \mid F_1) + \ldots + P(F_n) \cdot P(E \mid F_n)}$.

CHAPTER 8 REVIEW EXERCISES

Write true or false for each of the following.

1. $9 \in \{8, 4, -3, -9, 6\}$

2. $4 \in \{3, 9, 7\}$

3. $2 \notin \{0, 1, 2, 3, 4\}$

4. $0 \notin \{0, 1, 2, 3, 4\}$

5. $\{3, 4, 5\} \subseteq \{2, 3, 4, 5, 6\}$

6. $\{1, 2, 5, 8\} \subseteq \{1, 2, 5, 10, 11\}$

7. $\emptyset \subseteq \{1\}$

8. $0 \subseteq \emptyset$

List the elements in the following sets.

9. $\{x \mid x$ is a national holiday$\}$

10. $\{x \mid x$ is an integer, $-3 \le x < 1\}$

11. $\{$all counting numbers less than 5$\}$

12. $\{x \mid x$ is a leap year between 1989 and 2001$\}$

Let $U = \{$Vitamins A, B_1, B_2, B_3, B_6, B_{12}, C, D, $E\}$, $M = \{A$, C, D, $E\}$, *and* $N = \{A$, B_1, B_2, C, $E\}$. *Find the following.*

13. M' **14.** N' **15.** $M \cap N$ **16.** $M \cup N$ **17.** $M \cup N'$ **18.** $M' \cap N$

Let $U = \{$all students in a class$\}$, $A = \{$all male students$\}$, $B = \{$all A students$)$, $C = \{$all students with red hair$\}$, *and* $D = \{$all students younger than 21$\}$. *Describe the following sets in words.*

19. $A \cap C$ **20.** $B \cap D$ **21.** $A \cup D$ **22.** $A' \cap D$ **23.** $B' \cap C'$

Draw Venn diagrams for Exercises 24–27.

24. $B \cup A'$ **25.** $A' \cap B$ **26.** $A' \cap (B' \cap C)$ **27.** $(A \cup B)' \cap C$

Social Science *A survey of a group of military personnel revealed the following information.*

20 officers
27 minorities
19 women
5 women officers
8 minority women
10 minority officers
3 women minority officers
6 Caucasian male enlisted personnel

28. How many were interviewed?

29. How many were enlisted minority women?

30. How many were male minority officers?

Write sample spaces for the following.

31. A die is rolled and the number of points showing is noted.

32. A card is drawn from a deck containing only 4 aces.

33. A color is selected from the set $\{$red, blue, green$\}$, and then a number is chosen from the set $\{$10, 20, 30$\}$.

A jar contains 5 discs labeled 2, 4, 6, 8, 10, and another jar contains 2 blue and 3 yellow balls. One disc is drawn and then a ball is drawn. Give the following.

34. The sample space

35. Event F, the ball is blue.

36. Event E, the disc shows a number greater than 5.

37. Are the outcomes in this sample space equally likely?

Management *A company sells typewriters and copiers. Let E be the event "a customer buys a typewriter," and let F be the event "a customer buys a copier." In Exercises 38–39, write each of the following using $\cap$, $\cup$, or ', as necessary.*

38. A customer buys neither.

39. A customer buys at least one.

40. A student gives the answer to a probability problem as 6/5. Explain why this answer must be incorrect.

41. Describe what is meant by disjoint sets and give an example.

42. Describe what is meant by mutually exclusive events and give an example.

43. How are disjoint sets and mutually exclusive events related?

A single card is drawn from an ordinary deck. Find the probability of drawing each of the following.

44. a black king

45. a face card

46. a red card or a face card

47. a black card, given it is a 2

48. a jack, given it is a face card

49. a face card, given it is a jack

Find the odds in favor of drawing the following.

50. a spade

51. a red queen

52. a black face card or a 7

Management *A sample shipment of five electric motors is chosen at random. The probability of exactly 0, 1, 2, 3, 4, or 5 motors being defective is given in the following table.*

Number defective	0	1	2	3	4	5
Probability	.31	.25	.18	.12	.08	.06

Find the probability that

53. no more than 3 are defective.

54. at least 3 are defective.

Natural Science *The square shows the four possible (equally likely) combinations when both parents are carriers of the sickle cell anemia trait. Each carrier parent has normal cells (N) and trait cells (T).*

		2nd Parent	
		N_2	T_2
1st parent	N_1		$N_1 T_2$
	T_1		

55. Complete the table.

56. If the disease occurs only when two trait cells combine, find the probability that a child born to these parents will have sickle cell anemia.

57. The child will carry the trait but not have the disease if a normal cell combines with a trait cell. Find this probability.

58. Find the probability that the child is neither a carrier nor has the disease.

Find the probabilities for the following sums when two fair dice are rolled.

59. 8

60. At least 10

61. No more than 5

62. Odd and greater than 8

63. 12, given the sum is greater than 10

64. 7, given that at least one die shows a 4

Suppose $P(E) = .51$, $P(F) = .37$, and $P(E \cap F) = .22$. Find each of the following probabilities.

65. $P(E \cup F)$

66. $P(E \cap F')$

67. $P(E' \cup F)$

68. $P(E' \cap F')$

69. For events E and F, $P(E) = .2$, $P(F \mid E) = .3$, and $P(F \mid E') = .2$. Find each of the following.
 (a) $P(E \mid F)$ **(b** $P(E \mid F')$

70. Define independent events and give an example.

71. Are independent events always mutually exclusive? Are they ever mutually exclusive? Give examples.

Management *Of the appliance repair shops listed in the phone book, 80% are competent and 20% are not. A competent shop can repair an appliance correctly 95% of the time; an incompetent shop can repair an appliance correctly 60% of the time. Suppose an appliance was repaired correctly. Find the probability that it was repaired by*

72. a competent shop;

73. an incompetent shop.

Suppose an appliance was repaired incorrectly. Find the probability that it was repaired by

74. a competent shop;

75. an incompetent shop.

76. Four red and one orange slips of paper are placed in a box. Two red and three orange slips are placed in a second box. A box is chosen at random, and a slip of paper is selected from it. The probability of choosing the first box is 3/8. If the selected slip of paper is orange, what is the probability that it came from the first box?

77. Find the probability that the slip of paper in Exercise 76 came from the second box, given that it is red.

78. Management A manufacturer buys items from four different suppliers. The fraction of the total number of items that is obtained from each supplier, along with the probability that an item purchased from that supplier is defective, is shown in the following table.

Supplier	Fraction of Total Supplied	Probability of Defective
1	.17	.04
2	.39	.02
3	.35	.07
4	.09	.03

(a) Find the probability that a defective item came from supplier 4.

(b) Find the probability that a defective item came from supplier 2.

79. Management The table below shows the results of a survey of buyers of a certain model of car.

Car Type	Satisfied	Not Satisfied	Totals
New	300	100	
Used	450		600
Totals		250	

(a) Complete the table:

(b) How many buyers were surveyed?

(c) How many bought a new car and were satisfied?

(d) How many were not satisfied?

(e) How many bought used cars?

(f) How many of those who were not satisfied had bought a used car?

(g) Rewrite the event stated in part (f) using the expression "given that."

(h) Find the probability of the outcome in parts (f) and (g).

(i) Find the probability that a used-car buyer is not satisfied.

(j) You should have different answers in parts (h) and (i). Explain why.

When a patient is examined, information, typically incomplete, is obtained about his or her state of health. Probability theory provides a mathematical model appropriate for this situation, as well as a procedure for quantitatively interpreting such partial information to arrive at a reasonable diagnosis.*

To do this, list the states of health that can be distinguished in such a way that the patient can be in one and only one state at the time of the examination. Each state of health H is associated with a number $P(H)$ between 0 and 1 such that the sum of all these numbers is 1. This number $P(H)$ represents the probability, before examination, that a patient is in the state of health H, and $P(H)$ may be chosen subjectively from medical experience, using any information available prior to the examination. The probability may be most conveniently established from clinical records; that is, a mean probability is established for patients in general, although the number would vary from patient to patient. Of course, the more information that is brought to bear in establishing $P(H)$, the better the diagnosis.

For example, limiting the discussion to the condition of a patient's heart, suppose there are exactly 3 states of health, with probabilities as follows.

	State of Health H	P(H)
H_1	patient has a normal heart	.8
H_2	patient has minor heart irregularities	.15
H_3	patient has a severe heart condition	.05

Having selected $P(H)$, the information of the examination is processed. First, the results of the examination must be classified. The examination itself consists of observing the state of a number of characteristics of the patient. Let us assume that the examination for a heart condition consists of a stethoscope examination and a cardiogram. The outcome of such an examination, C, might be one of the following:

C_1 —stethoscope shows normal heart and cardiogram shows normal heart;

C_2 —stethoscope shows normal heart and cardiogram shows minor irregularities;

and so on.

*From "Probabilistic Medical Diagnosis," Roger Wright, *Some Mathematical Models in Biology,* Robert M. Thrall, ed., (The University of Michigan, 1967), by permission of Robert M. Thrall.

It remains to assess for each state of health H the conditional probability $P(C \mid H)$ of each examination outcome C using only the knowledge that a patient is in a given state of health. (This may be based on the medical knowledge and clinical experience of the doctor.) The conditional probabilities $P(C \mid H)$ will not vary from patient to patient, so that they may be built into a diagnostic system, although they should be reviewed periodically.

Suppose the result of the examination is C_1. Let us assume the following probabilities.

$$P(C_1 \mid H_1) = .9$$
$$P(C_1 \mid H_2) = .4$$
$$P(C_1 \mid H_3) = .1$$

Now, for a given patient, the appropriate probability associated with each state of health H, after examination, is $P(H \mid C)$ where C is the outcome of the examination. This can be calculated by using Bayes' theorem. For example, to find $P(H_1 \mid C_1)$—that is, the probability that the patient has a normal heart given that the examination showed a normal stethoscope examination and a normal cardiogram—we use Bayes' theorem as follows.

$$P(H_1 \mid C_1)$$
$$= \frac{P(C_1 \mid H_1)P(H_1)}{P(C_1 H_1)P(H_1) + P(C_1 \mid H_2)P(H_2) + P(C_1 \mid H_3)P(H_3)}$$
$$= \frac{(.9)(.8)}{(.9)(.8) + (.4)(.15) + (.1)(.05)} \approx .92$$

Hence, the probability is about .92 that the patient has a normal heart on the basis of the examination results. This means that in 8 out of 100 patients, some abnormality will be present and not be detected by the stethoscope or the cardiogram.

EXERCISES

1. Find $P(H_2 \mid C_1)$.

2. Assuming the following probabilities, find $P(H_1 \mid C_2)$:
$$P(C_2 \mid H_1) = .2, \quad P(C_2 \mid H_2) = .8, \quad P(C_2 \mid H_3) = .3.$$

3. Assuming the probabilities of Exercise 2, find $P(H_3 \mid C_2)$.

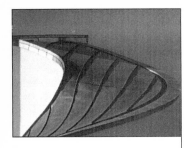

CHAPTER 9

Further Topics in Probability

A recent survey by *Money* magazine found that supermarket scanners are overcharging customers at 30% of stores. If you shop at 3 supermarkets that use scanners, what is the probability that you will be overcharged? Techniques for finding this probability and solving other special kinds of probability problems are introduced in this chapter.

9.1 PERMUTATIONS AND COMBINATIONS

Up to this point we have simply listed the outcomes in a sample space *S* and an event *E* in order to find *P(E)*. However, when *S* has many outcomes, listing them becomes very tedious. In this section we discuss counting methods that do not require listing.

Let us begin with a simple example. If there are 3 roads from town A to town B and 2 roads from town B to town C, in how many ways can someone travel from A to C by way of B? For each of the 3 roads from A there are 2 different routes leading from B to C, making $3 \cdot 2 = 6$ different trips, as shown in Figure 9.1.

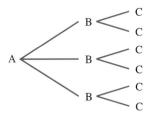

FIGURE 9.1

This example illustrates a general principle of counting called the *multiplication principle*.

1 **(a)** In how many ways can 6 business tycoons line up their golf carts at the country club?

(b) How many ways can 4 pupils be seated in a row with 4 seats?

Answers:

(a) $6 \cdot 5 \cdot 4 \cdot 3 \cdot 2 \cdot 1 = 720$

(b) $4 \cdot 3 \cdot 2 \cdot 1 = 24$

Multiplication Principle

Suppose n choices must be made, with

$$m_1 \text{ ways to make choice 1,}$$

and for each of these,

$$m_2 \text{ ways to make choice 2,}$$

and so on, with

$$m_n \text{ ways to make choice } n.$$

Then there are

$$m_1 \cdot m_2 \cdots m_n$$

different ways to make the entire sequence of choices.

▶ **EXAMPLE 1**   A combination lock can be set to open to any 3-letter sequence. How many such sequences are possible?

Since there are 26 letters in the alphabet, there are 26 choices for each of the 3 letters, and, by the multiplication principle, $26 \cdot 26 \cdot 26 = 17,576$ different sequences. ◀

▶ **EXAMPLE 2**   Morse code uses a sequence of dots and dashes to represent letters and words. How many sequences are possible with at most 3 symbols?

"At most 3" means "1 or 2 or 3." Each symbol may be either a dot or a dash. Thus, the following number of sequences are possible in each case.

Number of Symbols	Number of Sequences
1	2
2	$2 \cdot 2 = 4$
3	$2 \cdot 2 \cdot 2 = 8$

Altogether, $2 + 4 + 8 = 14$ different sequences of at most 3 symbols are possible. Because there are 26 letters in the alphabet, some letters must be represented by 4 symbols in Morse code. ◀

▶ **EXAMPLE 3**   A teacher has 5 different books to be arranged side by side. How many different arrangements are possible?

Five choices will be made, 1 for each space that will hold a book. Any of the 5 possible books could be chosen for the first space. There are 4 possible choices for the second space, since 1 book has already been placed in the first space, 3 possible choices for the third space, and so on. By the multiplication principle, the number of different possible arrangements (sequence of choices) is $5 \cdot 4 \cdot 3 \cdot 2 \cdot 1 = 120$. ◀ **1**

2 Evaluate:

(a) 4!

(b) 6!

(c) 1!

(d) 6!/4!

Answers:

(a) 24

(b) 720

(c) 1

(d) 30

3 In how many ways can 3 of 7 items be arranged?

Answer:
$7 \cdot 6 \cdot 5 = 210$

The use of the multiplication principle often leads to products such as $5 \cdot 4 \cdot 3 \cdot 2 \cdot 1$, the product of all the natural numbers from 5 down to 1. If n is a natural number, the symbol $n!$ (read *"n factorial"*) denotes the product of all the natural numbers from n down to 1. If $n = 1$, this formula is understood to give $1! = 1$.

n-factorial

For any natural number n,
$$n! = n(n - 1)(n - 2) \cdots (3)(2)(1).$$
Also, 0! is defined to be the number 1.

With this symbol, the product $5 \cdot 4 \cdot 3 \cdot 2 \cdot 1$ can be written as 5!. Also, $3! = 3 \cdot 2 \cdot 1 = 6$. The definition of $n!$ could be used to show that $n! = n \cdot (n - 1)!$ for all natural numbers $n \geq 2$. It is helpful if this result also holds for $n = 1$. This can only happen if 0! equals 1, as defined above. **2**

Some calculators have an $\boxed{n!}$ key. A calculator with a 10-digit display and scientific notation capability will usually give the exact value of $n!$ for $n \leq 13$ and approximate values of $n!$ for $14 \leq n \leq 69$.

▶**EXAMPLE 4** Suppose the teacher in Example 3 wishes to place only 3 of the 5 books on his desk. How many arrangements of 3 books are possible?

The teacher again has 5 ways to fill the first space, 4 ways to fill the second space, and 3 ways to fill the third. Because he wants to use only 3 books, there are only 3 spaces to be filled giving $5 \cdot 4 \cdot 3 = 60$ arrangements. ◀ **3**

PERMUTATIONS The answer 60 in Example 4 is called the number of *permutations* of 5 things taken 3 at a time. A **permutation** of r elements (where $r \geq 1$) from a set of n elements is any arrangement, *without repetition*, of the r elements. The number of permutations of n things taken r at a time (with $r \leq n$) is written $_nP_r$.* Based on the work in Example 4,

$$_5P_3 = 5 \cdot 4 \cdot 3 = 60.$$

Factorial notation can be used to express this product as follows.

$$5 \cdot 4 \cdot 3 = 5 \cdot 4 \cdot 3 \cdot \frac{2 \cdot 1}{2 \cdot 1} = \frac{5 \cdot 4 \cdot 3 \cdot 2 \cdot 1}{2 \cdot 1} = \frac{5!}{2!} = \frac{5!}{(5 - 3)!}$$

---

*Another notation that is sometimes used is $P(n, r)$.

4 Find the number of permutations of

(a) 5 things taken 2 at a time;

(b) 9 things taken 3 at a time.

Find each of the following.

(c) $_3P_1$

(d) $_7P_3$

(e) $_{12}P_2$

Answers:

(a) 20

(b) 504

(c) 3

(d) 210

(e) 132

5 Find the number of permutations of the letters C, O, D, and E

(a) using all the letters;

(b) using 2 of the 4 letters.

(c) using 3 of the 4 letters. (Can you find this without calculating?)

Answers:

(a) 24

(b) 12

(c) 24 (yes)

This example illustrates the general rule of permutations, which is stated below.

Permutations

If $_nP_r$ (where $r \le n$) is the number of permutations of n elements taken r at a time, then

$$_nP_r = \frac{n!}{(n - r)!}.$$

Many calculators have a key (often labeled $_nP_r$) for finding permutations.

To find $_nP_r$, we can use either the rule above or direct application of the multiplication principle, as the following example shows.

▶ **EXAMPLE 5**   Find the number of permutations of 8 elements taken 3 at a time. Since there are 3 choices to be made, the multiplication principle gives $_8P_3 = 8 \cdot 7 \cdot 6 = 336$. Alternatively, by the formula for $_nP_r$,

$$_8P_3 = \frac{8!}{(8 - 3)!} = \frac{8!}{5!} = \frac{8 \cdot 7 \cdot 6 \cdot 5 \cdot 4 \cdot 3 \cdot 2 \cdot 1}{5 \cdot 4 \cdot 3 \cdot 2 \cdot 1} = 8 \cdot 7 \cdot 6 = 336. \ ◀ \ \boxed{4}$$

▶ **EXAMPLE 6**   Find each of the following.
(a) The number of permutations of the letters A, B, and C
By the formula for $_nP_r$ with both n and r equal to 3,

$$_3P_3 = \frac{3!}{(3 - 3)!} = \frac{3!}{0!} = \frac{3!}{1} = 3! = 3 \cdot 2 \cdot 1 = 6.$$

The 6 permutations (or arrangements) are

$$\text{ABC, ACB, BAC, BCA, CAB, CBA.}$$

(b) The number of permutations possible using just 2 of the letters A, B, and C. Find $_3P_2$:

$$_3P_2 = \frac{3!}{(3 - 2)!} = \frac{3!}{1!} = 3! = 6.$$

This result is exactly the same answer as in part (a). This is because, in the case of $_3P_3$, after the first 2 choices are made, the third is already determined, as shown in the table below.

First two letters	AB	AC	BA	BC	CA	CB
Third letter	C	B	C	A	B	A

◀ **5**

6 A collection of 3 paintings by one artist and 2 by another is to be displayed. In how many ways can the paintings be shown

(a) in a row?

(b) if the works of the artists are to be alternated?

Answers:

(a) 120

(b) 12

▶ **EXAMPLE 7** A televised talk show will include 4 women and 3 men as panelists.

(a) In how many ways can the panelists be seated in a row of 7 chairs?

Find $_7P_7$, the total number of ways to seat 7 panelists in 7 chairs.

$$_7P_7 = \frac{7!}{(7-7)!} = \frac{7!}{0!} = \frac{7!}{1} = 7 \cdot 6 \cdot 5 \cdot 4 \cdot 3 \cdot 2 \cdot 1 = 5040$$

There are 5040 ways to seat the 7 panelists.

(b) In how many ways can the panelists be seated if the men and women are to be alternated?

In order to alternate men and women, a woman must be seated in the first chair (since there are 4 women and only 3 men), any of the men next, and so on. Thus, there are 4 ways to fill the first seat, 3 ways to fill the second seat, 3 ways to fill the third seat (with any of the 3 remaining women), and so on. Use the multiplication principle. There are

$$4 \cdot 3 \cdot 3 \cdot 2 \cdot 2 \cdot 1 \cdot 1 = 144$$

ways to seat the panelists. ◀ **6**

COMBINATIONS In Example 4, we found that there are 60 ways that a teacher can arrange 3 of 5 different books on a desk. That is, there are 60 permutations of 5 things taken 3 at a time. Suppose now that the teacher does not wish to arrange the books on his desk, but rather wishes to choose, at random, any 3 of the 5 books to give to a book sale to raise money for his school. In how many ways can he do this?

At first glance, we might say 60 again, but this is incorrect. The number 60 counts all possible *arrangements* of 3 books chosen from 5. However, the following arrangements would all lead to the same set of 3 books being given to the book sale.

mystery-biography-textbook   biography-textbook-mystery
mystery-textbook-biography   textbook-biography-mystery
biography-mystery-textbook   textbook-mystery-biography

The list shows 6 different *arrangements* of 3 books, but only one subset of 3 books selected from the 5 books for the book sale. A subset of items selected *without regard to order* is called a **combination.** The number of combinations of 5 things taken 3 at a time is written $\binom{5}{3}$ (read "5 over 3"). Since they are subsets, combinations are *not ordered.*

To evaluate $\binom{5}{3}$, start with the $5 \cdot 4 \cdot 3$ *permutations* of 5 things taken 3 at a time. Combinations are unordered, therefore, find the number of combinations by dividing the number of permutations by the number of ways each group of 3 can be ordered—that is, by 3!.

7 Evaluate $\frac{_nP_r}{r!}$ for the following values.

(a) $n = 6, r = 2$

(b) $n = 8, r = 4$

(c) $n = 7, r = 0$

Answers:

(a) 15

(b) 70

(c) 1

$$\binom{5}{3} = \frac{5 \cdot 4 \cdot 3}{3!} = \frac{5 \cdot 4 \cdot 3}{3 \cdot 2 \cdot 1} = 10$$

There are 10 ways that the teacher can choose 3 books at random for the book sale.

Generalizing this discussion gives the formula for the number of combinations of n elements taken r at a time, written $\binom{n}{r}$.*

$$\binom{n}{r} = \frac{_nP_r}{r!}$$

$$= \frac{n!}{(n-r)!} \cdot \frac{1}{r!} \qquad \text{Definition of } _nP_r$$

$$= \frac{n!}{(n-r)!r!}$$

This last form is the most useful for setting up the calculation. **7**

Combinations

If $\binom{n}{r}$ denotes the number of combinations of n elements taken r at a time, where $r \leq n$, then

$$\binom{n}{r} = \frac{n!}{(n-r)! \, r!}.$$

Many calculators have a key (often labeled $_nC_r$) for finding combinations.

Replacing r by $n - r$ in the combinations formula gives

$$\binom{n}{n-r} = \frac{n!}{(n-[n-r])! \, (n-r)!}$$

$$= \frac{n!}{r! \, (n-r)!} = \frac{n!}{(n-r)! \, r!}.$$

Therefore,

$$\binom{n}{r} = \binom{n}{n-r}.$$

For example, by this result,

$$\binom{5}{3} = \binom{5}{2} \quad \text{and} \quad \binom{10}{4} = \binom{10}{6}.$$

---

* Other comon notations for $\binom{n}{r}$ are $_nC_r$ and $C(n, r)$.

8 Use $\dfrac{n!}{(n-r)!\,r!}$ to evaluate $\dbinom{n}{r}$.

(a) $\dbinom{6}{2}$

(b) $\dbinom{8}{4}$

(c) $\dbinom{7}{0}$

Compare your answers with the answers for Problem 7.

Answers:

(a) 15

(b) 70

(c) 1

▶ **EXAMPLE 8** How many committees of 3 people can be formed from a group of 8 people?

A committee is an unordered group, so find $\dbinom{8}{3}$. By the formula for combinations,

$$\binom{8}{3} = \frac{8!}{5!\,3!} = \frac{8 \cdot 7 \cdot 6 \cdot 5 \cdot 4 \cdot 3 \cdot 2 \cdot 1}{5 \cdot 4 \cdot 3 \cdot 2 \cdot 1 \cdot 3 \cdot 2 \cdot 1} = \frac{8 \cdot 7 \cdot 6}{3 \cdot 2 \cdot 1} = 56. \quad ◀ \; \boxed{8}$$

▶ **EXAMPLE 9** Three secretaries are to be selected from a group of 30 to work on a special project.

(a) In how many different ways can the secretaries be selected?

Here we wish to know the number of 3-element combinations that can be formed from a set of 30 elements. (We want combinations and not permutations, since order within the group of 3 does not matter.)

$$\binom{30}{3} = \frac{30!}{27!\,3!} = \frac{30(29)(28)(27) \cdots (3)(2)(1)}{27(26)(25) \cdots (2)(1)(3)(2)(1)}$$

$$= \frac{30(29)(28)}{3(2)(1)} = 4060$$

There are 4060 ways to select the project group.

(b) In how many ways can the group of 3 be selected if a certain secretary must work on the project?

Since 1 secretary has already been selected for the project, the problem is reduced to selecting 2 more from the remaining 29 secretaries.

$$\binom{29}{2} = \frac{29!}{27!\,2!} = \frac{29 \cdot 28 \cdot 27!}{27! \cdot 2 \cdot 1} = \frac{29 \cdot 28}{2 \cdot 1}$$

$$= 29 \cdot 14 = 406$$

In this case, the project group can be selected in 406 ways.

(c) In how many ways can a nonempty group of at most 3 secretaries be selected from these 30 secretaries.

The group is to be nonempty; therefore, "at most 3" means "1 or 2 or 3." Find the number of ways for each case.

Case	Number of Ways		
1	$\dbinom{30}{1} = \dfrac{30!}{29!\,1!}$	$= \dfrac{30 \cdot 29!}{29!\,(1)!}$	$= 30$
2	$\dbinom{30}{2} = \dfrac{30!}{28!\,2!}$	$= \dfrac{30 \cdot 29 \cdot 28!}{28! \cdot 2 \cdot 1}$	$= 435$
3	$\dbinom{30}{3} = \dfrac{30!}{27!\,3}$	$= \dfrac{30 \cdot 29 \cdot 28 \cdot 27!}{27! \cdot 3 \cdot 2 \cdot 1}$	$= 4060$

9 Five orchids from a collection of 20 are to be selected for a flower show.

(a) In how many ways can this be done?

(b) In how many different ways can the group of 5 be selected if 2 particular orchids must be included?

(c) In how many ways can at least 1 and at most 5 orchids be selected? (Hint: Use a calculator or refer to Table 1 in Appendix B.)

Answers:

(a) $\binom{20}{5} = 15,504$

(b) $\binom{18}{3} = 816$

(c) 21,699

10 Solve the problems in Example 10.

Answers:

(a) 5040

(b) 455

(c) 28

(d) 360

The total number of ways to select at most 3 secretaries will be the sum

$$30 + 435 + 4060 = 4525. \quad \blacktriangleleft \quad \boxed{9}$$

The formulas for permutations and combinations given in this section will be very useful in solving probability problems in later sections. Any difficulty in using these formulas usually comes from being unable to differentiate between them. Both permutations and combinations give the number of ways to choose r objects from a set of n objects. The differences between permutations and combinations are outlined below.

Permutations	**Combinations**
Different orderings or arrangements of the r objects are different permutations.	Each choice or subset of r objects gives 1 combination. Order within the r objects does not matter.
$$_nP_r = \frac{n!}{(n-r)!}$$	$$\binom{n}{r} = \frac{n!}{(n-r)!\,r!}$$
Clue words: Arrangement, Schedule, Order	Clue words: Group, Committee, Sample

In the next examples, concentrate on recognizing which of the formulas should be applied.

▶**EXAMPLE 10** For each of the following problems, tell whether permutations or combinations should be used to solve the problem.

(a) How many 4-digit code numbers are possible if no digits are repeated?

Since changing the order of the 4 digits results in a different code, we use permutations.

(b) A sample of 3 light bulbs is randomly selected from a batch of 15 items. How many different samples are possible?

The order in which the 3 light bulbs items are selected is not important. The sample is unchanged if the items are rearranged, so combinations should be used.

(c) In a basketball tournament with 8 teams, how many games must be played so that each team plays every other team exactly once?

Selection of 2 teams for a game is an *unordered* subset of 2 from the set of 8 teams. Use combinations again.

(d) In how many ways can 4 patients be assigned to 6 hospital rooms so that each patient has a private room?

The room assignments are an *ordered* selection of 4 rooms from the 6 rooms. Exchanging the rooms of any 2 patients within a selction of 4 rooms gives a different assignment, so permutations should be used. ◀ **10**

11 A salesman has the names of 6 prospects.

(a) In how many ways can he arrange his schedule if he calls on all 6?

(b) In how many ways can he do it if he decides to call on only 4 of the 6?

Answers:

(a) 720

(b) 360

12 In how many ways can 4 aces and any other card be dealt?

Answer:
48

▶ **EXAMPLE 11**   A manager must select 4 employees for promotion: 12 employees are eligible.

(a) In how many ways can the 4 be chosen?

Because there is no reason to differentiate among the 4 who are selected, we use combinations.

$$\binom{12}{4} = \frac{12!}{4! \; 8!} = 495$$

(b) In how many ways can 4 employees be chosen (from 12) to be placed in 4 different jobs?

In this case, once a group of 4 is selected, they can be assigned in many different ways (or arrangements) to the 4 jobs. Therefore, this problem requires permutations.

$$_{12}P_4 = \frac{12!}{8!} = 11,880 \quad ◀ \; \boxed{11}$$

The following problems involve a standard deck of 52 playing cards, shown in Figure 8.18 on page 442.

▶ **EXAMPLE 12**   In how many ways can a full house of aces and eights (3 aces and 2 eights) be dealt in 5-card poker?

The arrangement of the 3 aces or the 2 eights does not matter, so use combinations and the multiplication principle. There are $\binom{4}{3}$ ways to get 3 aces from the 4 aces in the deck, and $\binom{4}{2}$ ways to get 2 eights. By the multiplication principle, the number of ways to get 3 aces *and* 2 eights is

$$\binom{4}{3} \cdot \binom{4}{2} = 4 \cdot 6 = 24. \quad ◀ \; \boxed{12}$$

▶ **EXAMPLE 13**   Five cards are dealt from a standard 52-card deck.

(a) How many such hands have all face cards?

The face cards are the king, queen, and jack of each suit. There are 4 suits, so there are 12 face cards. The arrangement of the 5 cards is not important, so use combinations to get

$$\binom{12}{5} = \frac{12!}{5! \; 7!} = 792.$$

(b) How many 5-card hands have all cards of the same suit?

13 In how many ways can 5 red cards be dealt?

Answer:
65,780

The arrangement of the 5 cards is not important, so use combinations. The total number of ways that 5 cards of a particular suit of 13 cards can be dealt is $\binom{13}{5}$.

Since there are 4 different suits, by the multiplication principle, there are

$$4 \cdot \binom{13}{5} = 4 \cdot 1287 = 5148$$

ways to deal 5 cards of the same suit. ◀ **13**

As Examples 12 and 13 show, often both combinations and the multiplication principle must be used in the same problem.

▶**EXAMPLE 14**   To illustrate the differences between permutations and combinations in another way, suppose 2 cans of soup are to be selected from 4 cans on a shelf: noodle (N), bean (B), mushroom (M), and tomato (T). As shown in Figure 9.2 (a), there are 12 ways to select 2 cans from the 4 cans if the order matters (if noodle first and bean second is considered different from bean, then noodle, for example). On the other hand, if order is unimportant, then there are 6 ways to choose 2 cans of soup from the 4, as illustrated in Figure 9.2(b). ◀

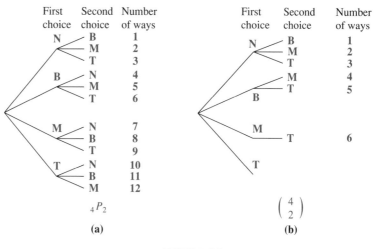

FIGURE 9.2

Caution   It should be stressed that not all counting problems lend themselves to either permutations or combinations. Whenever a tree diagram or the multiplication principle can be used directly, as in the example at the beginning of this section, then use it.

9.1 EXERCISES

Evaluate the following factorials, permutations, and combinations.

1. $_4P_2$ **2.** $3!$ **3.** $\binom{8}{3}$ **4.** $7!$ **5.** $_8P_1$ **6.** $\binom{8}{1}$

7. $4!$ **8.** $_4P_4$ **9.** $\binom{12}{5}$ **10.** $\binom{10}{8}$ **11.** $_{13}P_2$ **12.** $_{12}P_3$

Use a calculator to find values for Exercises 13–20.

13. $_{25}P_5$ **14.** $_{38}P_4$ **15.** $_{14}P_5$ **16.** $_{17}P_8$

17. $\binom{21}{10}$ **18.** $\binom{34}{25}$ **19.** $\binom{25}{16}$ **20.** $\binom{30}{15}$

Use the multiplication principle to solve the following problems. (See Examples 1–4.)

21. A social security number has 9 digits. How many social security numbers are there? The United States population in 1989 was about 248 million. Is it possible for every U.S. resident to have a unique social security number? (Assume no restrictions.)

22. The United States Postal Service currently uses 5-digit zip codes in most areas. How many zip codes are possible if there are no restrictions on the digits used? How many would be possible if the first number could not be 0?

23. The Postal Service is encouraging the use of 9-digit zip codes in some areas, adding 4 digits after the usual 5-digit code. How many such zip codes are possible with no restrictions?

24. How many different types of homes are available if a builder offers a choice of 5 basic plans, 3 roof styles, and 2 exterior finishes?

25. An auto manufacturer produces 7 models, each available in 6 different colors, with 4 different upholstery fabrics, and 5 interior colors. How many varieties of the auto are available?

26. How many different 4-letter radio station call letters can be made
   (a) if the first letter must be K or W and no letter may be repeated?

(b) if repeats are allowed (but the first letter is K or W)?

(c) How many of the 4-letter call letters (starting with K or W) with no repeats end in R?

27. For many years, the state of California used 3 letters followed by 3 digits on its automobile license plates.
   (a) How many different license plates are possible with this arrangement?
   (b) When the state ran out of new numbers, the order was reversed to 3 digits followed by 3 letters. How many new license plate numbers were then possible?
   (c) Several years ago, the numbers described in (b) were also used up. The state then issued plates with 1 letter followed by 3 digits and then 3 letters. How many new license plate numbers will this provide?

The United States is rapidly running out of telephone numbers. In large cities, telephone companies introduced new area codes as numbers were used up; as a result, only three unallocated area codes are left: 210, 810 and 910.

28. **(a)** At present, all area codes have a 0 or a 1 as the middle digit and the first digit cannot be 0 or 1. How many area codes are there with this arrangement? How many telephone numbers does the current 7-digit sequence permit per area code? (The 3-digit sequence that follows the area code cannot start with 0 or 1. Assume there are no other restrictions.)

   (b) The actual number of area codes available is 152. Explain the discrepancy between this number and your answer to part (a).

29. How many area codes would be possible if all restrictions on the second digit were removed?

30. A problem with the plan in Exercise 29 is that the second digit in the area code now tells the phone company equipment that a long-distance call is being made. To avoid changing all equipment, an alternative plan proposes a 4-digit area code and restricting the first and second digits as before. How many area codes would this plan provide?

31. Still another alternative solution is to increase the local dialing sequence to 8 digits instead of 7. How many additional numbers would this plan create? (Assume the same restrictions.)

32. Define permutation in your own words.

Use permutations to solve each of the following problems. (See Examples 5–7.)

33. A baseball team has 20 players. How many 9-player batting orders are possible?

34. In a game of musical chairs, 12 children will sit in 11 chairs arranged in a row (one will be left out). In how many ways can the 11 children find seats?

35. From a carton of 12 cans of a soft drink, 2 are to be selected for testing. In how many ways can this be done?

36. In an election with 3 candidates for one office and 5 candidates for another office, how many different ballots may be printed?

37. From a pool of 7 secretaries, 3 are selected to be assigned to 3 managers. In how many ways can they be selected?

38. A chapter of union Local 715 has 35 members. In how many different ways can the chapter select a president, a vice-president, a treasurer, and a secretary?

39. The television schedule for a certain evening shows 8 choices from 8 to 9 P.M., 5 choices from 9 to 10 P.M., and 6 choices from 10 to 11 P.M.. In how many different ways could a person schedule that evening of television viewing from 8 to 11 P.M.? (Assume each program that is selected is watched for an entire hour.)

40. In a club with 15 members, how many ways can a slate of 3 officers consisting of president, vice-president, and secretary/treasurer be chosen?

Use combinations to solve each of the following problems. (See Examples 8–9 and 12–13.)

41. Management Five items are to be randomly selected from the first 50 items on an assembly line to determine the defect rate. How many different samples of 5 items can be chosen?

42. Social Science A group of 3 students is to be selected from a group of 12 students to take part in a class in cell biology.
  (a) In how many ways can this be done?
  (b) In how many ways can the group which will *not* take part be chosen?

43. Natural Science From a group of 16 smokers and 20 nonsmokers, a researcher wants to randomly select 8 smokers and 8 nonsmokers for a study. In how many ways can the study group be selected?

44. Five cards are drawn from an ordinary deck. In how many ways is it possible to draw
  (a) all queens;
  (b) all face cards (face cards are the Jack, Queen, and King);
  (c) no face card;
  (d) exactly 2 face cards;
  (e) 1 heart, 2 diamonds, and 2 clubs.

Exercises 45–64 are mixed problems that may require permutations, combinations, or the multiplication principle. (See Examples 10, 11, and 14.)

45. Use a tree diagram to find the number of ways 2 letters can be chosen from the set {L, M, N} if order is important and
  (a) if repetition is allowed;
  (b) if no repeats are allowed.
  (c) Find the number of combinations of 3 elements taken 2 at a time. Does this answer differ from (a) or (b)?

46. Repeat Exercise 45 using the set {L, M, N, P}.

47. Explain the difference between a permutation and a combination.

48. Padlocks with digit dials are often referred to as "combination locks." According to the mathematical definition of combination, is this an accurate description? Explain.

49. Social Science A legislative committee consists of 5 Democrats and 4 Republicans. A delegation of 3 is to be selected to visit a small Pacific Island republic.
(a) How many different delegations are possible?
(b) How many delegations would have all Democrats?
(c) How many delegations would have 2 Democrats and 1 Republican?
(d) How many delegations would include at least 1 Republican?

50. Natural Science In an experiment on plant hardiness, a researcher gathers 6 wheat plants, 3 barley plants, and 2 rye plants. She wishes to select 4 plants at random.
(a) In how many ways can this be done?
(b) In how many ways can this be done if 2 wheat plants must be included?

51. Baskin-Robbins advertises that it has 31 flavors of ice cream.
(a) How many different double-scoop cones can be made?
(b) How many different triple-scoop cones can be made?

52. A concert to raise money for an economics prize is to consist of 5 works: 2 overtures, 2 sonatas, and a piano concerto.
(a) In how many ways can the program be arranged?
(b) In how many ways can the program be arranged if an overture must come first?

53. A state lottery game requires that you pick 6 different numbers from 1 to 99. If you pick all 6 winning numbers, you win $1 million.
(a) How many ways are there to choose 6 numbers if order is not important?
(b) How many ways are there to choose 6 numbers if order matters?

54. From 10 names on a ballot, 4 will be elected to a political party committee. In how many ways can the committee of 4 be formed if each person will have a different responsibility?

55. In Exercise 53, if you pick 5 of the 6 numbers correctly, you win $250,000. In how many ways can you pick exactly 5 of the 6 winning numbers without regard to order?

56. The coach of the Morton Valley Softball Team has 6 good hitters and 8 poor hitters. He chooses 3 hitters at random.
(a) In how many ways can he choose 2 good hitters and 1 poor hitter?
(b) In how many ways can he choose all good hitters?
(c) In how many ways can he choose at least 2 good hitters?

57. In how many ways can 5 out of 9 plants be arranged in a row on a window sill?

58. A bag contains 5 black, 1 red, and 3 yellow jelly beans; you take 3 at random. How many samples are possible in which the jelly beans are
(a) all black;
(b) all red;
(c) all yellow;
(d) 2 black, 1 red;
(e) 2 black, 1 yellow;
(f) 2 yellow, 1 black;
(g) 2 red, 1 yellow.

59. In Example 6, there are six 3-letter permutations of the letters A, B, and C. How many 3-letter subsets (unordered groups of letters) are there?

60. In Example 6, how many unordered 2-letter subsets of the letters A, B, and C are there?

61. Natural Science Eleven drugs have been found to be effective in the treatment of a disease. It is believed that the sequence in which the drugs are administered is important in the effectiveness of the treatment. In how many orders can 5 of the 11 drugs be administered?

62. Natural Science A biologist is attempting to classify 52,000 species of insects by assigning 3 initials to each species. Is it possible to classify all the species in this way? If not, how many initials should be used?

63. One play in a state lottery consists of choosing 6 numbers from 1 to 44. If your 6 numbers are drawn (in any order), you win the jackpot.
(a) How many possible ways are there to draw the 6 numbers?
(b) If you get 2 plays for a dollar, how much would it cost to guarantee that 1 of your choices would be drawn?
(c) Assuming that you work alone and can fill out a betting ticket (for 2 plays) every second and the lotto drawing will take place 3 days from now, can you place enough bets to guarantee that 1 of your choices will be drawn?

64. Powerball is a lottery game played in 15 states across the United States. For $1 a ticket, a player selects five numbers from 1 to 45 and one Powerball number from 1 to 45. A match of all six numbers wins the jackpot. How many different selections are possible?

*If the n objects in a permutations problem are not all distinguishable, that is, there are n_1 of type 1, n_2 of type 2, and so on for r different types, then the number of **distinguishable permutations** is*

$$\frac{n!}{n_1! \, n_2! \ldots n_r!}.$$

Example *In how many ways can you arrange the letters in the word Mississippi?*

This word contains 1 m, 4 i's, 4 s's, and 2 p's. To use the formula, let $n = 11$, $n_1 = 1$, $n_2 = 4$, $n_3 = 4$, $n_4 = 2$ to get

$$\frac{11!}{1! \, 4! \, 4! \, 2!} = 34{,}650$$

arrangements. The letters in a word with 11 different letters can be arranged in $11! = 39{,}916{,}800$ ways.

65. Find the number of distinguishable permutations of the letters in each of the following words.
  (a) initial   **(b)** little   **(c)** decreed

66. A printer has 5 A's, 4 B's, 2 C's, and 2 D's. How many different "words" are possible which use all these letters? (A "word" does not have to have any meaning here.)

67. Mike has 4 blue, 3 green, and 2 red books to arrange on a shelf.
  (a) In how many ways can this be done if they can be arranged in any order?
  (b) In how many ways if books of the same color are identical and must be grouped together?
  (c) In how many distinguishable ways if books of the same color are identical but need not be grouped together?

68. A child has a set of different shaped plastic objects. There are 3 pyramids, 4 cubes, and 7 spheres.
  (a) In how many ways can she arrange them in a row if they are all different colors?
  (b) In how many ways if the same shapes must be grouped?
  (c) In how many distinguishable ways can they be arranged in a row if objects of the same shape are also the same color, but need not be grouped?

9.2 APPLICATIONS OF COUNTING

Many of the probability problems involving *dependent* events that were solved with probability trees in Chapter 8 can also be solved by using combinations. Combinations are especially helpful when the numbers involved would require a tree with a large number of branches.

The use of combinations to solve probability problems depends on the basic probability principle introduced in Section 8.3 and repeated here.

If event E is a subset of sample space S, then the probability that event E occurs, written $P(E)$, is

$$P(E) = \frac{n(E)}{n(S)}.$$

It is also helpful to keep in mind that in probability statements

"and" corresponds to multiplication,
"or" corresponds to addition.

To compare the method of using combinations with the method of probability trees used in Section 8.5, the first example repeats Example 6 from that section.

▶ **EXAMPLE 1**   From a box containing 3 white, 2 green, and 1 red marble, 2 marbles are drawn one at a time without replacement. Find the probability that 1 white and 1 green marble are drawn.

In Example 6 of Section 8.5, it was necessary to consider the order in which the marbles were drawn. With combinations, it is not necessary. Simply count the number of ways in which 1 white and 1 green marble can be drawn from the given selection. The white marble can be drawn from the 3 white marbles in $\binom{3}{1}$ ways, and the green marble can be drawn from the 2 green marbles in $\binom{2}{1}$ ways. By the multiplication principle, both results can occur in

$$\binom{3}{1} \cdot \binom{2}{1} \text{ ways,}$$

giving the numerator of the probability fraction, $P(E) = m/n$. For the denominator, 2 marbles are to be drawn from a total of 6 marbles. This can occur in $\binom{6}{2}$ ways. The required probability is

$$P(1 \text{ white and } 1 \text{ green}) = \frac{\binom{3}{1}\binom{2}{1}}{\binom{6}{2}} = \frac{\dfrac{3!}{2!\,1!} \cdot \dfrac{2!}{1!\,1!}}{\dfrac{6!}{4!\,2!}} = \frac{6}{15} = \frac{2}{5}.$$

This agrees with the answer found earlier.   ◀

▶ **EXAMPLE 2**   From a group of 22 nurses, 4 are to be selected to present a list of grievances to management.

(a) In how many ways can this be done?

Four nurses from a group of 22 can be selected in $\binom{22}{4}$ ways. (Use combinations, since the group of 4 is an unordered set.)

$$\binom{22}{4} = \frac{22!}{4!\,18!} = \frac{22(21)(20)(19)}{4(3)(2)(1)} = 7315.$$

There are 7315 ways to choose 4 people from 22.

(b) One of the nurses is Michael Branson. Find the probability that Branson will be among the 4 selected.

The probability that Branson will be selected is given by m/n, where m is the number of ways the chosen group includes him, and n is the total number of ways

1 A jar contains 1 white and 4 red jelly beans.

(a) What is the probability that out of 2 jelly beans selected at random from the jar, 1 will be white?

(b) What is the probability of choosing 3 red jelly beans?

Answers:

(a) 2/5

(b) 2/5

the group of 4 can be chosen. If Branson must be one of the 4 selected, the problem reduces to finding the number of ways that the 3 additional nurses can be chosen. The 3 are chosen from 21 nurses; this can be done in

$$\binom{21}{3} = \frac{21!}{3!\,18!} = 1330$$

ways, so $m = 1330$. Since n is the number of ways 4 nurses can be selected from 22,

$$n = \binom{22}{4} = 7315.$$

The probability that Branson will be one of the 4 chosen is

$$P(\text{Branson is chosen}) = \frac{1330}{7315} \approx .182.$$

(b) Find the probability that Branson will not be selected.

The probability that he will not be chosen is $1 - .182 = .818$. ◀ **1**

▶ **EXAMPLE 3** When shipping diesel engines abroad, it is common to pack 12 engines in 1 container which is then loaded on a railcar and sent to a port. Suppose that a company has received complaints from its customers that many of the engines arrive in nonworking condition. To help solve this problem, the company decides to make a spot check of containers after loading—the company will test 3 engines from a container at random; if any of the 3 is nonworking, the container will not be shipped until each engine in it is checked. Suppose a given container has 2 nonworking engines. Find the probability that the container will not be shipped.

The container will not be shipped if the sample of 3 engines contains 1 or 2 defective engines. Thus, letting $P(1 \text{ defective})$ represent the probability of exactly 1 defective engine in the sample,

$$P(\text{not shipping}) = P(1 \text{ defective}) + P(2 \text{ defectives}).$$

There are $\binom{12}{3}$ ways to choose the 3 engines for testing:

$$\binom{12}{3} = \frac{12!}{3!\,9!} = \frac{12(11)(10)}{3(2)(1)} = 220.$$

There are $\binom{2}{1}$ ways of choosing 1 defective engine from the 2 in the container, and for each of these ways, there are $\binom{10}{2}$ ways of choosing 2 good engines from

2 Calculate $\binom{2}{1}\binom{10}{2}$.

Answer:
90

3 Calculate $\binom{2}{2}\binom{10}{1}$.

Answer:
10

4 In Example 3, if a sample of 2 engines is tested, what is the probability that the container will not be shipped?

Answer:
.318

among the 10 in the container. By the multiplication principle, there are

$$\binom{2}{1}\binom{10}{2}$$

ways of choosing a sample of 3 engines containing 1 defective. **2**
   Using the result from Problem 2 at the side,

$$P(1 \text{ defective}) = \frac{90}{220}.$$

There are $\binom{2}{2}$ ways of choosing 2 defective engines from the 2 defective engines in the container, and $\binom{10}{1}$ ways of choosing 1 good engine from among the 10 good engines, giving

$$\binom{2}{2}\binom{10}{1}$$

ways of choosing a sample of 3 engines containing 2 defectives. **3**
   Then, using the result from Problem 3 at the side,

$$P(2 \text{ defectives}) = \frac{10}{220}$$

and

$$P(\text{not shipping}) = P(1 \text{ defective}) + P(2 \text{ defectives})$$
$$= \frac{90}{220} + \frac{10}{220} = \frac{100}{220} = \frac{5}{11} \approx .455$$

The probability is $1 - .455 = .545$ that the container *will* be shipped, even though it has 2 defective engines. The management must decide if this probability is acceptable; if not, it may be necessary to test more than 3 engines from a container. ◄

   Instead of finding the sum $P(1 \text{ defective}) + P(2 \text{ defectives})$, the result in Example 3 could be found by calculating $1 - P(\text{no defectives})$.

$$P(\text{not shipping}) = 1 - P(\text{no defectives in sample})$$

$$= 1 - \frac{\binom{2}{0}\binom{10}{3}}{\binom{12}{3}} = 1 - \frac{1(120)}{220}$$

$$= 1 - \frac{120}{220} = \frac{100}{220} \approx .455 \quad \boxed{4}$$

5 Find the probability of being dealt a poker hand (5 cards) with 4 kings.

Answer:
.00001847

▶ **EXAMPLE 4**   In a common form of the card game *poker,* a hand of 5 cards is dealt to each player from a deck of 52 cards. There are a total of

$$\binom{52}{5} = \frac{52!}{5!47!} = 2,598,960$$

such hands possible. Find each of the following probabilities.

(a) A hand containing only hearts, called a *heart flush*
There are 13 hearts in a deck; there are

$$\binom{13}{5} = \frac{13!}{5!8!} = \frac{13(12)(11)(10)(9)}{5(4)(3)(2)(1)} = 1287$$

different hands containing only hearts. The probability of a heart flush is

$$P(\text{heart flush}) = \frac{1287}{2,598,960} = \frac{33}{66,640} \approx .000495.$$

(b) A flush of any suit (5 cards, all from 1 suit)
There are 4 suits to a deck, so

$$P(\text{flush}) = 4 \cdot P(\text{heart flush}) = 4 \cdot \frac{33}{66,640} \approx .00198.$$

(c) A full house of aces and eights (3 aces and 2 eights)
There are $\binom{4}{3}$ ways to choose 3 aces from among the 4 in the deck, and $\binom{4}{2}$ ways to choose 2 eights.

$$P(3 \text{ aces, } 2 \text{ eights}) = \frac{\binom{4}{3} \cdot \binom{4}{2}}{2,598,960} = \frac{1}{108,290} = \approx .00000923$$

(d) Any full house (3 cards of one value, 2 of another)
There are 13 values in a deck, so there are 13 choices for the first value mentioned, leaving 12 choices for the second value (order *is* important here, since a full house of aces and eights, for example, is not the same as a full house of eights and aces). Because, from part (c), the probability of any particular full house is 1/108,290,

$$P(\text{full house}) = 13 \cdot 12 \cdot \left(\frac{1}{108,290}\right) = \frac{156}{108,290} \approx .00144. \quad ◀ \quad \boxed{5}$$

6 An office manager must select 4 employees, A, B, C, and D, to work on a project. Each employee will be responsible for a specific task and 1 will be the coordinator.

(a) If the manager assigns the tasks randomly, what is the probability that B is the coordinator?

(b) What is the probability of one specific assignment of tasks?

Answers:

(a) 1/4

(b) 1/24

7 Evaluate

$$1 - \frac{{}_{365}P_n}{(365)^n}$$

for

(a) $n = 3$;

(b) $n = 6$.

Answers:

(a) .008

(b) .040

▶**EXAMPLE 5**   A music teacher has 3 violin pupils, Fred, Carl, and Helen. For a recital, the teacher selects a first violinist and a second violinist. The third pupil will play with the others, but not solo. If the teacher selects randomly, what is the probability that Helen is first violinist, Carl is second violinist, and Fred does not solo?

Use *permutations* to find the number of arrangements in the sample space.

$${}_3P_3 = 3! = 6$$

The 6 arrangements are equally likely, since the teacher will select randomly. Thus, the required probability is 1/6.   ◀   **6**

▶**EXAMPLE 6**   Suppose a group of 5 people is in a room. Find the probability that at least 2 of the people have the same birthday.

"Same birthday" refers to the month and the day, not necessarily the same year. Also, ignore leap years, and assume that each day in the year is equally likely as a birthday. First find the probability that *no 2 people* among 5 people have the same birthday. There are 365 different birthdays possible for the first of the 5 people, 364 for the second (so that the people have different birthdays), 363 for the third, and so on. The number of ways the 5 people can have different birthdays is thus the number of permutations of 365 things (days) taken 5 at a time or

$${}_{365}P_5 = 365 \cdot 364 \cdot 363 \cdot 362 \cdot 361.$$

The number of ways that the 5 people can have the same or different birthdays is

$$365 \cdot 365 \cdot 365 \cdot 365 \cdot 365 = (365)^5.$$

Finally, the *probability* that none of the 5 people have the same birthday is

$$\frac{{}_{365}P_5}{(365)^5} = \frac{365 \cdot 364 \cdot 363 \cdot 362 \cdot 361}{365 \cdot 365 \cdot 365 \cdot 365 \cdot 365} \approx .973.$$

The probability that at least 2 of the 5 people *do* have the same birthday is $1 - .973 = .027.$ ◀

Example 6 can be extended for more than 5 people. In general, the probability that no 2 people among n people have the same birthday is

$$\frac{{}_{365}P_n}{(365)^n}.$$

The probability that at least 2 of the n people *do* have the same birthday is

$$1 - \frac{{}_{365}P_n}{(365)^n}.$$ **7**

8 Set up (do not calculate) the probability that at least 2 of the 7 astronauts in the Mercury project have the same birthday.

Answer:
$1 - {_{365}P_7}/365^7$

The following table shows this probability for various values of n.

Number of People, n	Probability that 2 Have the Same Birthday
5	.027
10	.117
15	.253
20	.411
22	.476
23	.507
25	.569
30	.706
35	.814
40	.891
50	.970
366	1

The probability that 2 people among 23 have the same birthday is .507, a little more than half. Many people are surprised at this result—somehow it seems that a larger number of people should be required. **8**

9.2 EXERCISES

Management *Refer to Example 3. The management feels that the probability of .545 that a container will be shipped even though it contains 2 defectives is too high. They decide to increase the sample size chosen. Find the probability that a container will be shipped even though it contains 2 defectives if the sample size is increased to*

1. 4. **2.** 5.

Management *A shipment of 9 computers contains 2 with defects. Find the probability that a sample of the following size, drawn from the 9, will not contain a defective. (See Example 3.)*

3. 1 **4.** 2 **5.** 3 **6.** 4

A grab-bag contains 10 $1 prizes, 6 $5 prizes, and 2 $20 prizes. Three prizes are randomly drawn. Find the following probabilities. (See Examples 1 and 2.)

7. all $1 prizes

8. all $5 prizes

9. two $20 prizes

10. one prize of each kind

11. at least one $20 prize

12. no $1 prize

Two cards are drawn at random from an ordinary deck of 52 cards. (See Example 4.)

13. How many 2-card hands are possible?

Find the probability that the two-card hand in Exercise 13 contains the following.

14. 2 queens

15. no aces

16. 2 face cards

17. different suits

18. at least 1 black card

19. no more than 1 heart

20. Discuss the relative merits of using probability trees versus combinations to solve probability problems. When would each approach be most appropriate?

21. Several examples in this section used the rule $P(E') = 1 - P(E)$. Explain the advantage (especially in Example 6) of using this rule.

A bridge hand consists of 13 cards from a deck of 52. Set up the probabilities that a bridge hand includes each of the following.

22. 6 face cards

23. 2 aces and 3 kings

24. 7 cards of one suit and 6 of another

25. In Exercise 53 in the previous section, we found the number of ways to pick 6 different numbers from 1 to 99 in a state lottery.
(a) Assuming order is unimportant, what is the probability of picking all 6 numbers correctly to win the big prize?
(b) What is the probability, if order matters?

26. In Exercise 25, what is the probability of picking exactly 5 of the 6 numbers correctly?

27. In December 1993, Percy Ray Pridgen, a District of Columbia resident, and Charles Gill of Richmond, Virginia shared a $90 million Powerball jackpot. The winning numbers were 1, 3, 13, 15, and 29. The Powerball number was 12. The answer to Exercise 64 in Section 9.1 gives the number of ways to win a Powerball jackpot as about 55 million. What is the probability of two individuals independently selecting the winning numbers?

28. Management A cellular phone manufacturer randomly selects 5 of every 50 phones from the assembly line and tests them. If at least 4 of the 5 pass inspection, the batch of 50 is considered acceptable. Find the probability that a batch is considered acceptable if it contains the following.
(a) 2 defective phones
(b) no defective phones   (c) 3 defective phones

29. Social Science Of the 16 members of President Clinton's cabinet, 4 are women. Suppose the president randomly selected 4 advisors from the cabinet for a meeting. Find the probability that the group of 4 would be composed as follows.
(a) 2 women and 2 men
(b) all men
(c) at least 1 woman

30. A car dealer has 8 red, 11 gray, and 9 blue cars in stock. Ten cars are randomly chosen to be displayed in front of the dealership. Find the probability that
(a) 4 are red and the others are blue.
(b) 3 are red, 3 are blue, and 4 are gray.
(c) exactly 5 are gray and none are blue.
(d) all 10 are gray.

31. Social Science The teacher of a multicultural first-grade class on the first day of school found that in her class of 25 students, 10 spoke only English, 6 spoke only Spanish, 4 spoke only Russian, 3 spoke only Vietnamese, and 2 spoke only Hmong. The children were assigned randomly to groups of 5. Find the probability that a group included the following.
(a) 2 English-speaking and 3 Russian-speaking children
(b) all English-speaking children
(c) no English-speaking children
(d) at least two childen who spoke Vietnamese or Hmong.

For Exercises 32–34 refer to Example 6 in this section.

32. Set up the probability that at least 2 of the 42 men who have served as president of the United States have had the same birthday.*

33. Set up the probability that at least 2 of the 100 U.S. Senators have the same birthday.

34. What is the probability that 2 of the 435 members of the House of Representatives have the same birthday?

35. An elevator has 4 passengers and stops at 7 floors. It is equally likely that a person will get off at any one of the 7 floors. Find the probability that no 2 passengers leave at the same floor.

---

*In fact, James Polk and Warren Harding were both born on November 2.

36. Show that the probability that in a group of n people *exactly one pair* have the same birthday is

$$\binom{n}{2} \cdot \frac{_{365}P_{n-1}}{(365)^n}.$$

37. On National Public Radio, the "Weekend Edition" program on Sunday, September 7, 1991, posed the following probability problem: Given a certain number of balls, of which some are blue, pick 5 at random. The probability that all 5 are blue is $1/2$. Determine the original number of balls and decide how many were blue.

38. The Big Eight Conference during the 1988 college football season ended the season in a "perfect progression", as shown in the table below.*

Won	Lost	Team
7	0	Nebraska (NU)
6	1	Oklahoma (OU)
5	2	Oklahoma State (OSU)
4	3	Colorado (CU)
3	4	Iowa State (ISU)
2	5	Missouri (MU)
1	6	Kansas (KU)
0	7	Kansas State (KSU)

*From Richard Madsen. "On the Probability of a Perfect Progression," *The American Statistician,* August 1991, vol. 45, no. 3, p. 214.

Someone wondered what the probability of such an outcome might be.

(a) Assuming no ties and that each team had an equally likely probability of winning each game, find the probability of the perfect progression shown above.

(b) Find a general expression for the probability of a perfect progression in an n-team league with the same assumptions.

39. Use a computer or a graphing calculator and the Monte Carlo method with $n = 50$ to estimate the probabilities of the following hands at poker. Assume aces are either high or low. Since each hand has 5 cards, you will need $50 \cdot 5 = 250$ random numbers to "look at" 50 hands. Compare these experimental results with the theoretical results.

(a) A pair of aces

(b) Any two cards of the same value

(c) Three of a kind

40. Use a computer or a graphing calculator and the Monte Carlo method with $n = 20$ to estimate the probabilities of the following 13-card bridge hands. Since each hand has 13 cards, you will need $20 \cdot 13 = 260$ random numbers to "look at" 20 hands.

(a) No aces

(b) 2 kings and 2 aces

(c) No cards of any one suit—that is, only 3 suits represented

9.3 BERNOULLI TRIALS

Many probability problems are concerned with experiments in which an event is repeated many times. Some examples include finding the probability of getting 7 heads in 8 tosses of a coin, of hitting a target 6 times out of 6, and of finding 1 defective item in a sample of 15 items. Probability problems of this kind are called **Bernoulli trials** problems, or **Bernoulli processes,** named after the Swiss mathematician Jakob Bernoulli (1654–1705). In each case, some outcome is designated a success, and any other outcome is considered a failure. Thus, if the probability of a success in a single trial is p, the probability of failure will be $1 - p$. A Bernoulli trials problem, or **binomial experiment** must satisfy the following conditions.

Bernoulli Trials

1. The same experiment is repeated several times.
2. There are only two possible outcomes, success or failure.
3. The repeated trials are independent so that the probability of each outcome remains the same for each trial.

Consider a typical problem of this type: find the probability of getting 5 ones on 5 rolls of a die. Here, the experiment, rolling a die, is repeated 5 times. If getting a one is designated as a success, then getting any other outcome is a failure. The 5 trials are independent; the probability of success (getting a one) is $1/6$ on each trial, while the probability of a failure (any other result) is $5/6$. Thus, the required probability is

$$P(5 \text{ ones on 5 rolls}) = P(1) \cdot P(1) \cdot P(1) \cdot P(1) \cdot P(1) = \left(\frac{1}{6}\right)^5$$

$$\approx .00013.$$

Now suppose the problem is changed to that of finding the probability of getting a one exactly 4 times in 5 rolls of the die. Again, a success on any roll of the die is defined as getting a one, and the trials are independent, with the same probability of success each time. The desired outcome for this experiment can occur in more than one way, as shown below, where s represents a success (getting a one), and f represents a failure (getting any other result), and each row represents a different outcome.

$$
\begin{array}{ccccc}
s & s & s & s & f \\
s & s & s & f & s \\
s & s & f & s & s \\
s & f & s & s & s \\
f & s & s & s & s \\
\end{array}
$$

Using combinations, the total number of ways in which 4 successes (and 1 failure) can occur is $\dbinom{5}{4} = 5$.

1 Find the probability of rolling a one

(a) exactly twice in 5 rolls of a die;

(b) exactly once in 5 rolls of a die.

(c) Do the probabilities found in the text, plus the two probabilities found here cover all possible outcomes for this experiment? If not, what other outcomes are possible?

Answers:

(a) .161

(b) .402

(c) No; 0 ones

The probability of each of these 5 outcomes is

$$\left(\frac{1}{6}\right)^4\left(\frac{5}{6}\right).$$

Since the 5 outcomes represent mutually exclusive alternative events, add the 5 probabilities, or multiply by $\binom{5}{4} = 5$.

$$P(4 \text{ ones in 5 rolls}) = \binom{5}{4}\left(\frac{1}{6}\right)^4\left(\frac{5}{6}\right)^{5-4} = 5\left(\frac{1}{6}\right)^4\left(\frac{5}{6}\right)^1 = \frac{5^2}{6^5} \approx .0032$$

The probability of rolling a one exactly 3 times in 5 rolls of a die can be computed in the same way. The probability of any one way of achieving 3 successes and 2 failures will be

$$\left(\frac{1}{6}\right)^3\left(\frac{5}{6}\right)^2.$$

Again the desired outcome can occur in more than one way. The number of ways in which 3 successes (and 2 failures) can occur, using combinations, is $\binom{5}{3} = 10$.

$$P(3 \text{ ones in 5 rolls}) = \binom{5}{3}\left(\frac{1}{6}\right)^3\left(\frac{5}{6}\right)^{5-3} = 10\left(\frac{1}{6}\right)^3\left(\frac{5}{6}\right)^2 = \frac{250}{6^5} \approx .032. \quad \blacksquare \;\; \mathbf{1}$$

A similar argument works in the general case.

Binomial Probability

If p is the probability of success in a single trial of a binomial experiment, the probability of x successes and $n - x$ failures in n independent repeated trials of the experiment is

$$\binom{n}{x}p^x(1 - p)^{n-x}.$$

This formula plays an important role in the study of *binomial distributions*, which are discussed in Section 10.4.

2 Eighty percent of all students at a certain school ski. If a sample of 5 students at this school is selected, and if their responses are independent, find the probability that exactly

(a) 1 of the 5 students skis;

(b) 4 of the 5 students ski.

Answers:

(a) $\binom{5}{1}(.8)^1(.2)^4 = .0064$

(b) $\binom{5}{4}(.8)^4(.2)^1 = .4096$

3 Find the probability of getting 2 fours in 8 tosses of a die.

Answer:

$\binom{8}{2}\left(\frac{1}{6}\right)^2\left(\frac{5}{6}\right)^6 \approx .2605$

▶ **EXAMPLE 1**   The advertising agency that handles the Diet Supercola account believes that 40% of all consumers prefer this product over its competitors. Suppose a random sample of 6 people is chosen. Assume that all responses are independent of each other. Find the probability of the following.

(a) Exactly 4 of the 6 people prefer Diet Supercola.

   We can think of the 6 responses as 6 independent trials. A success occurs if a person prefers Diet Supercola. Then this is a binomial experiment with $p = P(\text{success}) = P(\text{prefer Diet Supercola}) = .4$. The sample is made up of 6 people, so $n = 6$. To find the probability that exactly 4 people prefer this drink, we let $x = 4$ and use the result in the box.

$$P(\text{exactly } 4) = \binom{6}{4}(.4)^4(1 - .4)^{6-4}$$
$$= 15(.4)^4(.6)^2$$
$$= 15(.0256)(.36) \qquad \text{Use a calculator}$$
$$= .13824$$

(b) None of the 6 people prefers Diet Supercola.

   Let $x = 0$.

$$P(\text{exactly } 0) = \binom{6}{0}(.4)^0(1 - .4)^6$$
$$= 1(1)(.6)^6$$
$$\approx .0467 \quad ◀ \; \boxed{2}$$

▶ **EXAMPLE 2**   Find the probability of getting exactly 7 heads in 8 tosses of a fair coin.

   The probability of success, getting a head in a single toss, is 1/2, so the probability of failure, getting a tail, is 1/2.

$$P(7 \text{ heads in } 8 \text{ tosses}) = \binom{8}{7}\left(\frac{1}{2}\right)^7\left(\frac{1}{2}\right)^1 = 8\left(\frac{1}{2}\right)^8 = .03125 \quad ◀ \; \boxed{3}$$

▶ **EXAMPLE 3**   Assuming that selection of items for a sample can be treated as independent trials, and that the probability that any 1 item is defective is .01, find the following.

(a) The probability of 1 defective item in a random sample of 15 items from a production line

   Here, a "success" is a defective item. Selecting each item for the sample is assumed to be an independent trial, so the binomial probability formula applies.

4 Five percent of the clay pots fired in a certain way are defective. Find the probability of getting exactly 2 defective pots in a sample of 12.

Answer:

$\binom{12}{2}(.05)^2(.95)^{10} \approx .0988$

5 In Example 4, find the probability that

(a) you are overcharged in more than 1 store.

(b) you are overcharged in less than 1 store.

Answers:

(a) .216

(b) .343

The probability of success (a defective item) is .01, while the probability of failure (an acceptable item) is .99. This makes

$$P(1 \text{ defective in } 15 \text{ items}) = \binom{15}{1}(.01)^1(.99)^{14}$$
$$= 15(.01)(.99)^{14}$$
$$\approx .130.$$

(b) The probability of at most 1 defective item in a random sample of 15 items from a production line

"At most 1" means 0 defective items or 1 defective item. Since 0 defective items is equivalent to 15 acceptable items,

$$P(0 \text{ defective}) = (.99)^{15} \approx .860.$$

Use the union rule, noting that 0 defective and 1 defective are mutually exclusive events, to get

$$P(\text{at most } 1 \text{ defective}) = P(0 \text{ defective}) + P(1 \text{ defective})$$
$$\approx .860 + .130$$
$$= .990. \quad \blacktriangleleft \quad \boxed{4}$$

In the next example we return to the problem posed at the beginning of this chapter.

▶**EXAMPLE 4**   A recent survey by *Money* magazine found that supermarket scanners are overcharging customers at 30% of stores.* If you shop at 3 supermarkets that use scanners what is the probability that you will be overcharged in at least one store?

We can treat this as a binomial experiment, letting $n = 3$ and $p = .3$. At least 1 of 3 means 1 or 2 or 3. It will be simpler here to find the probability of being overcharged in none of the 3 stores, that is, $P(0 \text{ overcharges})$, and then find $1 - P(0 \text{ overcharges})$.

$$P(0 \text{ overcharges}) = \binom{3}{0}(.3)^0(.7)^3$$
$$= 1(1)(.343)$$
$$= .343$$
$$P(\text{at least } 1) = 1 - P(0 \text{ overcharges})$$
$$= 1 - .343$$
$$= .657 \quad \blacktriangleleft \quad \boxed{5}$$

---

* "Don't Get Cheated by Supermarket Scanners," pp. 132–138, by Vanessa O'Connell reprinted from the April 1993 issue of *Money* by special permission (reprinted in the *Chicago Tribune* as "Beware, When Scanners Err, You Pay Price"). Copyright © 1993, Time Inc.

9.3 EXERCISES

Social Science *In 1990 12.7% of the U.S. population were 65 or older, an increase from 4.1% in 1900. Find the probabilities that the following number of persons selected at random from 20 U.S. residents in 1990 were 65 or older. (See Examples 1–3.)*

1. exactly 5

2. exactly 1

3. none

4. all

5. at least 1

6. at most 3

A die is rolled 12 times. Find the probability of rolling

7. exactly 12 ones;

8. exactly 6 ones;

9. exactly 1 one;

10. exactly 2 ones;

11. no more than 3 ones;

12. no more than 1 one.

A coin is tossed 5 times. Find the probability of getting

13. all heads;

14. exactly 3 heads;

15. no more than 3 heads;

16. at least 3 heads.

17. How do you identify a probability problem that involves a binomial experiment?

18. Why do combinations occur in the binomial probability formula?

Management *The survey discussed in Example 4 also found the following:*

> *Customers overpay for 1 out of every 10 items, on average. Scanner errors account for more than half of supermarket profits. In 7% of the stores customers were undercharged.*

Suppose a customer purchases 15 items. Find the following probabilities.

19. A customer overpays on 3 items.

20. A customer is not overcharged for any item.

21. A customer underpays on 3 items.

22. A customer underpays on no items.

23. A customer underpays on at least one item.

24. A customer overpays on no more than one item.

Management *The insurance industry has found that the probability is .9 that a life insurance applicant will qualify at the regular rates. Find the probabilities that of the next 10 applicants for life insurance, the following numbers will qualify at the regular rates.*

25. Exactly 10

26. Exactly 9

27. At least 9

28. Less than 8

Natural Science *The probability that a birth will result in twins is .012. Assuming independence (perhaps not a valid assumption), what are the probabilities that out of 100 births in a hospital, there will be the following numbers of sets of twins?*

29. Exactly 2 sets of twins

30. At most 2 sets of twins

Natural Science *Six mice from the same litter, all suffering from a vitamin A deficiency, are fed a certain dose of carrots. If the probability of recovery under such treatment is .70, find the probabilities of the following results.*

31. None of the mice recover.

32. Exactly 3 of the 6 mice recover.

33. All of the mice recover.

34. No more than 3 mice recover.

35. **Natural Science** In an experiment on the effects of a radiation dose on cells, a beam of radioactive particles is aimed at a group of 10 cells. Find the probability that 8 of the cells will be hit by the beam if the probability that any single cell will be hit is .6. (Assume independence.)

36. **Natural Science** The probability of a mutation of a given gene under a dose of 1 roentgen of radiation is approximately 2.5×10^{-7}. What is the probability that in 10,000 genes, at least 1 mutation occurs?

37. Natural Science   A new drug being tested causes a serious side effect in 5 out of 100 patients. What is the probability that no side effects occur in a sample of 10 patients taking the drug?

38. Natural Science   Children in a certain school are monitored for a year and it is found that 30% of those who do not brush their teeth regularly don't have a cavity during the year. A child who brushes regularly has a 70% chance of having no cavities during the year.
(a) If 10 children brush regularly, what is the probability that at least 9 will have no cavities during the year?
(b) 150 children who brush regularly are divided into 15 groups of 10 each. What is the probability that in at least one group of 10, at least 9 will have no cavities during the year?

39. Social Science   A study recently released by the College Board and the Western Interstate Commission for Higher Education predicted that by 1995, one-third of U.S. public elementary and secondary school students would belong to ethnic minorities. Assuming independence, find the probability that of 5 students selected at random in 1995 from schools throughout the nation, no more than 3 belonged to ethnic minorities.

Social Science   *Educators have always been aware of the problem of guessing the answers to multiple choice questions on tests. Suppose an unprepared student guesses the answers to each of 10 questions on a multiple choice test with 5 choices for each item. To get some idea of the likelihood of guessing, find the following probabilities.*

40. 5 questions are answered correctly.

41. No questions are answered correctly.

42. At least 3 questions are answered correctly.

43. At most 2 questions are answered correctly.

44. Social Science   According to the state of California, 33% of all state community college students belong to ethnic minorities. Find the probabilities of the following results in a random sample of 10 California community college students.
(a) Exactly 2 belong to an ethnic minority.
(b) Three or fewer belong to an ethnic minority.
(c) Exactly 5 do not belong to an ethnic minority.
(d) Six or more do not belong to an ethnic minority.

 Use a calculator or computer to calculate each of the following probabilities.

45. Natural Science   A flu vaccine has a probability of 80% of preventing a person who is inoculated from getting flu. A county health office inoculates 134 people. What is the probability that
(a) exactly 10 of them get the flu?
(b) no more than 10 get the flu?
(c) none of them get the flu?

46. Natural Science   The probability that a male will be color-blind is .042. What is the probability that in a group of 53 men
(a) exactly 5 are color-blind?
(b) no more than 5 are color-blind?
(c) at least 1 is color-blind?

47. Management   The probability that a certain machine turns out a defective item is .05. What is the probability that in a run of 75 items
(a) exactly 5 defectives are produced?
(b) no defectives are produced?
(c) at least 1 defective is produced?

48. Management   A company is taking a survey to find out if people like their product. Their last survey indicated that 70% of the population like their product. Based on that, of a sample of 58 people, what is the probability that
(a) all 58 like the product?
(b) from 28 to 30 (inclusive) like the product?

9.4 MARKOV CHAINS

In Section 8.5, we touched on **stochastic processes,** mathematical models in which the outcome of an experiment depends on the outcomes of previous experiments. In this section we study a special kind of stochastic process called a **Markov chain,** where the outcome of an experiment depends only on the outcome of the previous experiment. In other words, the next **state** of the system depends only on the present state, not on preceding states. Such experiments are common enough in applications to make their study worthwhile. Markov chains are named after the Russian mathematician A. A. Markov, 1856–1922, who started the theory of stochastic processes. To see how Markov chains work, we look at an example.

▶ **EXAMPLE 1**   A small town has only two dry cleaners, Johnson and North-Clean. Johnson's manager hopes to increase the firm's market share by an extensive advertising campaign. After the campaign, a market research firm finds that there is a probability of .8 that a Johnson customer will bring his next batch of dirty items to Johnson, and a .35 chance that a NorthClean customer will switch to Johnson for his next batch. Assume that the probability that a customer comes to a given cleaners depends only on where the last load of clothes was taken. If there is an .8 chance that a Johnson customer will return to Johnson, then there must be a $1 - .8 = .2$ chance that the customer will switch to NorthClean. In the same way, there is a $1 - .35 = .65$ chance that a NorthClean customer will return to NorthClean. If an individual bringing a load to Johnson is said to be in state 1 and an individual bringing a load to NorthClean is said to be in state 2, then these probabilities of change from one cleaner to the other are as shown in the following table.

		Second Load	
	State	1	2
First	1	.8	.2
Load	2	.35	.65

◀

The information from the table can be written in other forms. Figure 9.3 is a **transition diagram** that shows the two states and the probabilities of going from one state to another.

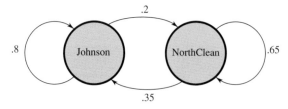

FIGURE 9.3

In a **transition matrix,** the states are indicated at the side and top, as follows.

1 Given the transition matrix

$$
\begin{array}{c}
\quad\quad \textit{State} \\
\quad\quad 1 \quad\ 2 \\
\textit{State}\ \begin{array}{c} 1 \\ 2 \end{array} \begin{bmatrix} .3 & .7 \\ .1 & .9 \end{bmatrix}
\end{array}
$$

(a) what is the probability of changing from state 1 to state 2?

(b) what does the number .1 represent?

(c) Draw a transition diaagram for this information.

Answers:

(a) .7

(b) The probability of changing from state 2 to state 1

(c)

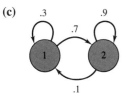

		Second Load	
		Johnson	NorthClean
First Load	Johnson	.8	.2
	NorthClean	.35	.65

1

A **transition matrix** has the following features.

1. It is square, since all possible states must be used both as rows and as columns.
2. All entries are between 0 and 1, inclusive, because all entries represent probabilities.
3. The sum of the entries in any row must be 1, as the numbers in the row give the probability of changing from the state at the left to one of the states indicated across the top.

▶**EXAMPLE 2** Suppose that when the new promotional campaign began, Johnson had 40% of the market and NorthClean had 60%. Use the probability tree in Figure 9.4 to find how these proportions would change after another week of advertising.

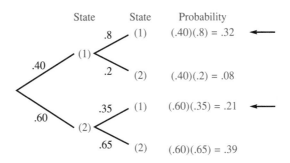

State 1: taking cleaning to Johnson
State 2: taking cleaning to North Clean

FIGURE 9.4

Add the numbers indicated with arrows to find the proportion of people taking their cleaning to Johnson after 1 week.

$$.32 + .21 = .53$$

Similarly, the proportion taking their cleaning to NorthClean is

$$.08 + .39 = .47.$$

The initial distribution of 40% and 60% becomes, after 1 week, 53% and 47%. ◀

2 Find the product

$$[.53 \quad .47]\begin{bmatrix} .8 & .2 \\ .35 & .65 \end{bmatrix}.$$

Answer:
[.59 .41] (rounded)

3 Find each cleaner's market share after 3 weeks.

Answer:
[.62 .38]

These distributions can be written as the *probability vectors*

$$[.40 \quad .60] \quad \text{and} \quad [.53 \quad .47].$$

A **probability vector** is a matrix of only one row, having nonnegative entries with the sum of the entries equal to 1.

The results from the probability tree above are exactly the same as the result of multiplying the initial probability vector by the transition matrix. (Multiplication of matrices was discussed in Section 6.4.)

$$[.4 \quad .6]\begin{bmatrix} .8 & .2 \\ .35 & .65 \end{bmatrix} = [.53 \quad .47]$$

Find the market share after 2 weeks by multiplying the vector [.53 .47] and the transition matrix. **2**

The product from Problem 2 at the side (with the numbers rounded to the nearest hundredth) shows that after 2 weeks of advertising, Johnson's share of the market has increased to 59%. To get the share after 3 weeks, multiply the probability vector [.59 .41] and the transition matrix. Continuing this process gives each cleaner's share of the market after any number of weeks. **3**

The following table gives the market share for each cleaner at the end of week n for several values of n.

Week	Johnson	NorthClean
Start	.4	.6
1	.53	.47
2	.59	.41
3	.62	.38
4	.63	.37
5	.63	.37
12	.64	.36

The results seem to approach the probability vector [.64 .36]. This vector is called the **equilibrium vector** or **fixed vector** for the given transition matrix. The equilibrium vector gives a long-range prediction—the shares of the market will stabilize (under the same conditions) at 64% for Johnson and 36% for NorthClean.

Starting with some other initial probability vector and going through the steps above would give the same equilibrium vector. In fact, the long-range trend is the same no matter what the initial vector is. The long-range trend depends only on the transition matrix, not on the initial distribution.

One of the many applications of Markov chains is in finding these long-range predictions. It is not possible to make long-range predictions with all transition matrices, but there is a large set of transition matrices for which long-range predictions *are* possible. Such predictions are always possible with *regular transition matrices*. A transition matrix is **regular** if some power of the matrix contains all positive entries. A Markov chain is a **regular Markov chain** if its transition matrix is regular. A regular Markov chain always has an equilibrium vector.

▶ **EXAMPLE 3**   Decide if the following transition matrices are regular.

(a)
$$A = \begin{bmatrix} .75 & .25 & 0 \\ 0 & .5 & .5 \\ .6 & .4 & 0 \end{bmatrix}$$

Square the matrix A by multiplying it by itself.

$$A^2 = \begin{bmatrix} .5625 & .3125 & .125 \\ .3 & .45 & .25 \\ .45 & .35 & .2 \end{bmatrix}$$

All entries in A^2 are positive, so that matrix A is regular.

(b)
$$B = \begin{bmatrix} .5 & 0 & .5 \\ 0 & 1 & 0 \\ 0 & 0 & 1 \end{bmatrix}$$

Find various powers of B.

$$B^2 = \begin{bmatrix} .25 & 0 & .75 \\ 0 & 1 & 0 \\ 0 & 0 & 1 \end{bmatrix} \quad B^3 = \begin{bmatrix} .125 & 0 & .875 \\ 0 & 1 & 0 \\ 0 & 0 & 1 \end{bmatrix} \quad B^4 = \begin{bmatrix} .0625 & 0 & .9375 \\ 0 & 1 & 0 \\ 0 & 0 & 1 \end{bmatrix}$$

Notice that all of the powers of B shown here have zeros in the same locations. Thus, further powers of B will still give the same zero entries, so that no power of matrix B contains all positive entries. For this reason, B is not regular.   ◀ **4**

The equilibrium vector can be found without doing all the work shown above. Let $V = \begin{bmatrix} v_1 & v_2 \end{bmatrix}$ represent the desired vector. Let P represent the transition matrix. By definition, V is the fixed probability vector if $VP = V$. In our example,

$$\begin{bmatrix} v_1 & v_2 \end{bmatrix} \begin{bmatrix} .8 & .2 \\ .35 & .65 \end{bmatrix} = \begin{bmatrix} v_1 & v_2 \end{bmatrix}.$$

5 Solve the equation for v_2. Round to the nearest thousandth.

Answer:
.364

Multiply on the left to get

$$[.8v_1 + .35v_2 \quad .2v_1 + .65v_2] = [v_1 \quad v_2].$$

Set corresponding entries from the two matrices equal to get

$$.8v_1 + .35v_2 = v_1 \quad .2v_1 + .65v_2 = v_2.$$

Simplify each of these equations.

$$-.2v_1 + .35v_2 = 0 \quad .2v_1 - .35v_2 = 0$$

These last two equations are really the same. (The equations in the system obtained from $VP = V$ are always dependent.) To find the values of v_1 and v_2, recall that $V = [v_1 \quad v_2]$ is a probability vector, so that

$$v_1 + v_2 = 1.$$

Find v_1 and v_2 by solving the system

$$-.2v_1 + .35v_2 = 0$$
$$v_1 + \quad v_2 = 1.$$

From the second equation, $v_1 = 1 - v_2$. Substitute for v_1 in the first equation:

$$-.2(1 - v_2) + .35v_2 = 0. \quad \blacksquare 5$$

Since $v_2 = .364$ (from Problem 5 at the side) and $v_1 = 1 - v_2$, then $v_1 = 1 - .364 = .636$ and the equilibrium vector is $[.636 \quad .364]$.

▶ **EXAMPLE 4** Find the equilibrium vector for the transition matrix

$$P = \begin{bmatrix} \frac{1}{3} & \frac{2}{3} \\ \frac{3}{4} & \frac{1}{4} \end{bmatrix}.$$

Look for a vector $[v_1 \quad v_2]$ such that

$$[v_1 \quad v_2] \begin{bmatrix} \frac{1}{3} & \frac{2}{3} \\ \frac{3}{4} & \frac{1}{4} \end{bmatrix} = [v_1 \quad v_2].$$

Multiply on the left, obtaining the two equations

$$\frac{1}{3}v_1 + \frac{3}{4}v_2 = v_1 \quad \frac{2}{3}v_1 + \frac{1}{4}v_2 = v_2.$$

Simplify each of these equations.

$$-\frac{2}{3}v_1 + \frac{3}{4}v_2 = 0 \quad \frac{2}{3}v_1 - \frac{3}{4}v_2 = 0$$

6 Find the equilibrium vector for the transition matrix

$$p = \begin{bmatrix} .3 & .7 \\ .5 & .5 \end{bmatrix}.$$

Answer:
[5/12   7/12]

We know that $v_1 + v_2 = 1$, so $v_1 = 1 - v_2$. Substitute this into either of the two given equations. Using the first equation gives

$$-\frac{2}{3}(1 - v_2) + \frac{3}{4}v_2 = 0$$

$$-\frac{2}{3} + \frac{2}{3}v_2 + \frac{3}{4}v_2 = 0$$

$$-\frac{2}{3} + \frac{17}{12}v_2 = 0$$

$$\frac{17}{12}v_2 = \frac{2}{3}$$

$$v_2 = \frac{8}{17},$$

and $v_1 = 1 - 8/17 = 9/17$. The equilibrium vector is $[9/17 \quad 8/17]$. ◀ **6**

▶ **EXAMPLE 5**   The probability that a complex assembly line works correctly depends on whether or not the line worked correctly the last time it was used. The various probabilities are as given in the following transition matrix.

	Works Properly Now	Does Not
Worked Properly Before	.9	.1
Did Not	.7	.3

Find the long-range probability that the assembly line will work properly.
Begin by finding the equilibrium vector $[v_1 \quad v_2]$, where

$$\begin{bmatrix} v_1 & v_2 \end{bmatrix}\begin{bmatrix} .9 & .1 \\ .7 & .3 \end{bmatrix} = \begin{bmatrix} v_1 & v_2 \end{bmatrix}.$$

Multiplying on the left and setting corresponding entries equal gives the equations

$$.9v_1 + .7v_2 = v_1 \quad \text{and} \quad .1v_1 + .3v_2 = v_2$$

or

$$-.1v_1 + .7v_2 = 0 \quad \text{and} \quad .1v_1 - .7v_2 = 0.$$

Substitute $v_1 = 1 - v_2$ in the first of these equations to get

$$-.1(1 - v_2) + .7v_2 = 0$$

$$-.1 + .1v_2 + .7v_2 = 0$$

$$-.1 + .8v_2 = 0$$

$$.8v_2 = .1$$

$$v_2 = \frac{1}{8},$$

7 In Example 5, suppose the company modifies the line so that the transition matrix becomes

$$\begin{bmatrix} .95 & .05 \\ .8 & .2 \end{bmatrix}.$$

Find the long-range probability that the assembly line will work properly.

Answer:
16/17

and $v_1 = 1 - 1/8 = 7/8$. The equilibrium vector is $[7/8 \quad 1/8]$. In the long run, the company can expect the assembly line to run properly 7/8 of the time. ◄ **7**

The examples suggest the following generalization. If a regular Markov chain has transition matrix P and initial probability vector v, the products

$$vP$$
$$vPP = vP^2$$
$$vP^2P = vP^3$$
$$\vdots$$
$$vP^n$$

lead to the equilibrium vector V. The results of this section can be summarized as follows.

Suppose a regular Markov chain has a transition matrix P.

1. For any initial probability vector v, the products vP^n approach a unique vector V as n gets larger and larger. Vector V is called the equilibrium or fixed vector.
2. Vector V has the property that $VP = V$.
3. Vector V is found by solving a system of equations obtained from the matrix equation $VP = V$ and from the fact that the sum of the entries of V is 1.

9.4 EXERCISES

Decide which of the following could be a probability vector.

1. $\begin{bmatrix} \dfrac{1}{2} & \dfrac{2}{3} \end{bmatrix}$

2. $\begin{bmatrix} \dfrac{3}{4} & \dfrac{1}{2} \end{bmatrix}$

3. $[0 \quad 1]$

4. $[.4 \quad .2 \quad 0]$

5. $[.3 \quad -.1 \quad .6]$

6. $\begin{bmatrix} \dfrac{1}{4} & \dfrac{1}{8} & \dfrac{5}{8} \end{bmatrix}$

Decide which of the following could be a transition matrix. Sketch a transition diagram for any transition matrices.

7. $\begin{bmatrix} .7 & .1 \\ .5 & .5 \end{bmatrix}$

8. $\begin{bmatrix} \dfrac{1}{4} & \dfrac{3}{4} \\ 0 & 1 \end{bmatrix}$

9. $\begin{bmatrix} \dfrac{2}{3} & \dfrac{1}{3} \\ \dfrac{1}{5} & \dfrac{4}{5} \end{bmatrix}$

10. $\begin{bmatrix} 0 & 1 & 0 \\ \dfrac{1}{3} & \dfrac{1}{3} & \dfrac{1}{3} \\ 1 & 0 & 0 \end{bmatrix}$

11. $\begin{bmatrix} \dfrac{1}{2} & \dfrac{1}{4} & 1 \\ \dfrac{2}{3} & 0 & \dfrac{1}{3} \\ \dfrac{1}{3} & 1 & 0 \end{bmatrix}$

12. $\begin{bmatrix} .2 & .3 & .5 \\ 0 & 0 & 1 \\ .1 & .9 & 0 \end{bmatrix}$

In Exercises 13-15, write any transition diagrams as transition matrices.

13.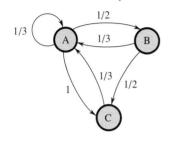

14.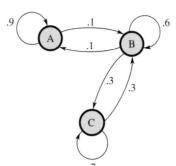

15.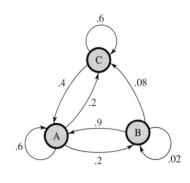

Decide whether the following transition matrices are regular. (See Example 3.)

16. $\begin{bmatrix} 1 & 0 \\ .6 & .4 \end{bmatrix}$

17. $\begin{bmatrix} .55 & .45 \\ 0 & 1 \end{bmatrix}$

18. $\begin{bmatrix} .2 & .8 \\ .9 & .1 \end{bmatrix}$

19. $\begin{bmatrix} .3 & .5 & .2 \\ 1 & 0 & 0 \\ .5 & .1 & .4 \end{bmatrix}$

20. $\begin{bmatrix} 0 & 1 & 0 \\ .4 & .2 & .4 \\ 1 & 0 & 0 \end{bmatrix}$

21. $\begin{bmatrix} .22 & .78 & 0 \\ 0 & .43 & .57 \\ 0 & 0 & 1 \end{bmatrix}$

Find the equilibrium vector for each of the following transition matrices. (See Examples 4 and 5.)

22. $\begin{bmatrix} .3 & .7 \\ .4 & .6 \end{bmatrix}$

23. $\begin{bmatrix} .8 & .2 \\ .1 & .9 \end{bmatrix}$

24. $\begin{bmatrix} \frac{1}{4} & \frac{3}{4} \\ \frac{1}{2} & \frac{1}{2} \end{bmatrix}$

25. $\begin{bmatrix} \frac{2}{3} & \frac{1}{3} \\ \frac{1}{8} & \frac{7}{8} \end{bmatrix}$

26. $\begin{bmatrix} .25 & .35 & .4 \\ .1 & .3 & .6 \\ .55 & .4 & .05 \end{bmatrix}$

27. $\begin{bmatrix} .16 & .28 & .56 \\ .43 & .12 & .45 \\ .86 & .05 & .09 \end{bmatrix}$

28. $\begin{bmatrix} .1 & .1 & .8 \\ .4 & .4 & .2 \\ .1 & .2 & .7 \end{bmatrix}$

29. $\begin{bmatrix} .5 & .2 & .3 \\ .1 & .4 & .5 \\ .2 & .2 & .6 \end{bmatrix}$

30. Social Science The probability that a homeowner will become a renter in five years is .03. The probability that a renter will become a homeowner in five years is .1. In 1989 56% of Americans were homeowners and the rest were renters.* Assume that these figures continue to apply. Set up a transition matrix with this information and find the long-range probability that an American will be a homeowner.

31. One of Markov's own applications was a 1913 study of how often a vowel is followed by another vowel or a consonant by another consonant in Russian text. A similar study of a passage of English text revealed the following transition matrix.

	Vowel	Consonant
Vowel	.12	.88
Consonant	.54	.46

Find the percent of letters in English text that are expected to be vowels.

32. Management The probability that a complex assembly line works correctly depends on whether the line worked correctly the last time it was used. There is a .95 chance that the line will work correctly if it worked correctly the time before, and a .7 chance that it will work correctly if it did *not* work correctly the time before. Set up a transition matrix with this information and find the long-run probability that the line will work correctly. (See Example 5.)

*Some of these data are from *The Universal Almanac,1993,* John W. Wright, General Editor, Kansas City, New York: Andrews and McMeel p. 294.

33. Management Suppose improvements are made in the assembly line of Exercise 32, so that the transition matrix becomes

	Works	Doesn't Work
Works	.95	.05
Doesn't Work	.85	.15

Find the new long-run probability that the line will work properly.

34. Natural Science In Exercises 46–47 of Section 8.4 we discussed the effect on flower color of cross-pollinating pea plants. As shown there, since the gene for red is dominant and the gene for white is recessive, 75% of these pea plants have red flowers and 25% have white flowers, because plants with 1 red and 1 white gene appear red. If a red-flowerd plant is crossed with a red flowered plant known to have 1 red and 1 white gene, then 75% of the offspring will be red and 25% will be white. Crossing a red-flowered plant that has 1 red and 1 white gene with a white-flowered plant produces 50% red-flowered offspring and 50% white-flowered offspring.

(a) Write a transition matrix using this information.
(b) Write a probability vector for the initial distribution of colors.
(c) Find the distribution of colors after 4 generations.
(d) Find the long-range distribution of colors.

35. Natural Science Snapdragons with 1 red gene and 1 white gene produce pink-flowered offspring. If a red snapdragon is crossed with a pink snapdragon, the probabilities that the offspring will be red, pink, or white are $1/2$, $1/2$, and 0, respectively. If 2 pink snapdragons are crossed, the probabilites of red, pink, or white offspring are $1/4$, $1/2$, and $1/4$, respectively. For a cross between a white and a pink snapdragon, the corresponding probabilities are $0, 1/2$, and $1/2$. Set up a transition matrix and find the long-range prediction for the fraction of red, pink, and white snapdragons.

36. Many phenomena can be viewed as examples of a random walk. Consider the following simple example. A security guard can stand in front of any one of three doors 20 ft apart in front of a building, and every minute he decides whether to move to another door chosen at random. If he is at the middle door, he is equally likely to stay where he is, move to the door to the left, or move to the door to the right. If he is at the door on either end, he is equally likely to stay where he is or move to the middle door.

(a) Verify that the transition matrix is given by

$$
\begin{array}{cc}
 & \begin{array}{ccc} 1 & 2 & 3 \end{array} \\
\begin{array}{c} 1 \\ 2 \\ 3 \end{array} &
\left[\begin{array}{ccc}
\frac{1}{2} & \frac{1}{2} & 0 \\
\frac{1}{3} & \frac{1}{3} & \frac{1}{3} \\
0 & \frac{1}{2} & \frac{1}{2}
\end{array} \right]
\end{array}
$$

(b) Find the long-range trend for the fraction of time the guard spends in front of each door.

37. The figures in Exercise 30 ignore the homeless. Suppose the proportions in the population are 56% homeowners (O), 43.5% renters (R), and .5% homeless (H), with the following transition matrix.*

	O	R	H
O	.97	.03	0
R	.1	.899	.001
H	0	.4	.6

Find the long-range probabilities for the three categories.

38. Management An insurance company classifies its drivers into three groups: G_0 (no accidents), G_1 (one accident), and G_2 (more than one accident). The probability that a driver in G_0 will stay in G_0 after 1 year is .85, that he will become a G_1 is .10, and that he will become a G_2 is .05. A driver in G_1 cannot move to G_0 (this insurance company has a long memory). There is an .80 probability that a G_1 driver will stay in G_1 and a .20 probability that he will become a G_2. A driver in G_2 must stay in G_2.

(a) Write a transition matrix using this information. Suppose that the company accepts 50,000 new policyholders, all of whom are in group G_0. Find the number in each group
(b) after 1 year;   (c) after 2 years;
(d) after 3 years;   (e) after 4 years.
(f) Find the equilibrium vector here. Interpret your result.

39. Management The difficulty with the mathematical model of Exercise 38 is that no "grace period" is provided; there should be a certain probability of moving from G_1 or G_2 back to G_0 (say, after 4 years with no accidents). A new system with this feature might produce the following transition matrix.

$$
\left[\begin{array}{ccc}
.85 & .10 & .05 \\
.15 & .75 & .10 \\
.10 & .30 & .60
\end{array} \right]
$$

*Some of these data are from *The Universal Almanac*, 1993, John W. Wright, General Editor, Kansas City, New York: Andrews and McMeel, p. 294.

Suppose that when this new policy is adopted, the company has 50,000 policyholders in group G_0. Find the number of these in each group

(a) after 1 year;   **(b)** after 2 years;   **(c)** after 3 years.
(d) Find the equilibrium vector here. Interpret your result.

40. Management   Research done by the Gulf Oil Corporation* produced the following transition matrix for the probability that a person with one form of home heating would switch to another.

$$\begin{array}{c} \\ \textit{Now Has} \end{array} \begin{array}{c} \\ \text{Oil} \\ \text{Gas} \\ \text{Electric} \end{array} \begin{array}{c} \textit{Will Switch to} \\ \begin{array}{ccc} \text{Oil} & \text{Gas} & \text{Electric} \end{array} \\ \begin{bmatrix} .825 & .175 & 0 \\ .060 & .919 & .021 \\ .049 & 0 & .951 \end{bmatrix} \end{array}$$

The current share of the market held by these three types of heat is given by $[.26 \quad .60 \quad .14]$. Find the share of the market held by each type of heat after

(a) 1 year;   **(b)** 2 years;   **(c)** 3 years.
(d) What is the long-range prediction?

41. The results of cricket matches between England and Australia have been found to be modeled by a Markov chain.** The probability that England wins, loses, or draws is based on the result of the previous game, with the following transition matrix:

$$\begin{array}{c} \\ \text{Wins} \\ \text{Loses} \\ \text{Draws} \end{array} \begin{array}{c} \begin{array}{ccc} \text{Wins} & \text{Loses} & \text{Draws} \end{array} \\ \begin{bmatrix} .443 & .364 & .193 \\ .277 & .436 & .287 \\ .266 & .304 & .430 \end{bmatrix} \end{array}$$

(a) Compute the transition matrix for the game after the next one, based on the result of the last game.
(b) Use your answer from part (a) to find the probability that, if England won the last game, England will win the game after the next one.

(c) Use your answer from part (a) to find the probability that, if Australia won the last game, England will win the game after the next one.

42. Social Science   At one liberal arts college, students are classified as humanities majors, science majors, or undecided. There is a 20% chance that a humanities major will change to a science major from one year to the next, and a 45% chance that a humanities major will change to undecided. A science major will change to humanities with probability .15, and to undecided with probability .35. An undecided will switch to humanities or science with probabilities of .5 and .3, respectively. Find the long-range prediction for the fraction of students in each of these three majors.

43. Management   In the queuing chain, we assume that people are queuing up to be served by, say, a bank teller. For simplicity, let us assume that once two people are in line, no one else can enter the line. Let us further assume that one person is served every minute, as long as someone is in line. Assume further that in any minute, there is a probability of $\frac{1}{2}$ that no one enters the line, a probability of $\frac{1}{3}$ that exactly one person enters the line, and a probability of $\frac{1}{6}$ that exactly two people enter the line, assuming there is room. If there is not enough room for two people then the probability that one person enters the line is $\frac{1}{2}$. Let the state be given by the number of people in line.

(a) Verify that the transition matrix is

$$\begin{array}{c} \\ 0 \\ 1 \\ 2 \end{array} \begin{array}{c} \begin{array}{ccc} 0 & 1 & 2 \end{array} \\ \begin{bmatrix} \frac{1}{2} & \frac{1}{3} & \frac{1}{6} \\ \frac{1}{2} & \frac{1}{3} & \frac{1}{6} \\ 0 & \frac{1}{2} & \frac{1}{2} \end{bmatrix} \end{array}.$$

(b) Find the transition matrix for a two-minute period.
(c) Use your result from part (b) to find the probability that a queue with no one in line has two people in line two minutes later.

Exercises 44–49 are computer or calculator problems. For each of the following transition matrices, find the first five powers of the matrix. Then find the probability that state 2 changes to state 4 after 5 repetitions of the experiment

44. $\begin{bmatrix} .1 & .2 & .2 & .3 & .2 \\ .2 & .1 & .1 & .2 & .4 \\ .2 & .1 & .4 & .2 & .1 \\ .3 & .1 & .1 & .2 & .3 \\ .1 & .3 & .1 & .1 & .4 \end{bmatrix}$

45. $\begin{bmatrix} .3 & .2 & .3 & .1 & .1 \\ .4 & .2 & .1 & .2 & .1 \\ .1 & .3 & .2 & .2 & .2 \\ .2 & .1 & .3 & .2 & .2 \\ .1 & .1 & .4 & .2 & .2 \end{bmatrix}$

* From "Forecasting Market Shares of Alternative Home-Heating Units by Markov Process Using Transition Probabilities Estimates from Aggregate Time Series Data" by Ali Ezzati, from *Management Science*, Vol. 21, No. 4, December 1974. Copyright © 1974 The Institute of Management Sciences. Reprinted by permission.

** From Derek Colwell, Brian Jones, and Jack Gillett, "A Markov Chain in Cricket," *The Mathematical Gazette*, June 1991.

46. Management A company with a new training program classified each employee in one of four states: s_1, never in the program; s_2, currently in the program; s_3, discharged; s_4, completed the program. The transition matrix for this company is given below

$$\begin{array}{c} \\ s_1 \\ s_2 \\ s_3 \\ s_4 \end{array} \begin{array}{c} \begin{array}{cccc} s_1 & s_2 & s_3 & s_4 \end{array} \\ \left[\begin{array}{cccc} .4 & .2 & .05 & .35 \\ 0 & .45 & .05 & .5 \\ 0 & 0 & 1.0 & 0 \\ 0 & 0 & 0 & 1.0 \end{array} \right] \end{array}$$

(a) What percent of employees who had never been in the program (state s_1) completed the program (state s_4) after the program had been offered five times?

(b) If the initial percent of employees in each state was [.5 .5 0 0], find the corresponding percents after the program had been offered four times.

Management *Find the equilibrium vector for the following transition matrices by taking powers of the matrix.*

47. Exercise 44

48. Exercise 45

49. Management Find the long-range prediction for the percent of employees in each state for the company training program from Exercise 46.

9.5 PROBABILITY DISTRIBUTIONS; EXPECTED VALUE

In this section we shall see that the *expected value* of a probability distribution is a kind of average. Probability distributions were introduced briefly in Chapter 8. A probability distribution depends on the idea of a *random variable*, so we begin with that.

Suppose a bank is interested in improving its services to the public. The manager decides to begin by finding the amount of time tellers spend on each transaction, rounded to the nearest minute. To each transaction, then, will be assigned one of the numbers 0, 1, 2, 3, 4, ... That is, if t represents an outcome of the experiment of timing a transaction, then t may take on any of the values from the list 0, 1, 2, 3, 4, ... The value that t takes on for a particular transaction is random, so t is called a random variable.

A **random variable** is a function that assigns a real number to each outcome of an experiment.

PROBABILITY DISTRIBUTIONS Suppose that the bank manager in our example records the times for 75 different transactions, with results as shown in Table 1. As the table shows, the shortest transaction time was 1 minute, with 3 transactions of 1-minute duration. The longest time was 10 minutes. Only 1 transaction took that long.

In Table 1, the first column gives the 10 values assumed by the random variable t, and the second column shows the number of occurences corresponding to each of these values, the *frequency* of that value. Table 1 is an example of a **frequency distribution,** a table listing the frequencies for each value a random variable may assume.

Table 1

Time	Frequency
1	3
2	5
3	9
4	12
5	15
6	11
7	10
8	6
9	3
10	1
	Total: 75

Now suppose that several weeks after starting new procedures to speed up transactions, the manager takes another survey. This time she observes 57 transactions, and she records their times as shown in the frequency distribution in Table 2.

Table 2

Time	Frequency
1	4
2	5
3	8
4	10
5	12
6	17
7	0
8	1
9	0
10	0
	Total: 57

Do the results in Table 2 indicate an improvement? It is hard to compare the two tables, since one is based on 75 transactions and the other on 57. To make them comparable, we can add a column to each table that gives the *relative*

1 Complete the probability distribution below. (*f* represents frequency and *p* represents probability.)

x	f	p
1	3	
2	7	
3	9	
4	3	
5	2	

Answer:

x	f	p
1	3	.13
2	7	.29
3	9	.38
4	3	.13
5	2	.08
	Total: 24	

frequency of each transaction time. These results are shown in Tables 3 and 4. Where necessary, decimals are rounded to the nearest hundredth. To find a **relative frequency,** divide each frequency by the total of the frequencies. Here the individual frequencies are divided by 75 or 57. Many of these frequencies have been rounded.

Table 3

Time	Frequency	Relative Frequency
1	3	$3/75 = .04$
2	5	$5/75 \approx .07$
3	9	$9/75 = .12$
4	12	$12/75 = .16$
5	15	$15/75 = .20$
6	11	$11/75 \approx .15$
7	10	$10/75 \approx .13$
8	6	$6/75 = .08$
9	3	$3/75 = .04$
10	1	$1/75 \approx .01$

Table 4

Time	Frequency	Relative Frequency
1	4	$4/57 \approx .07$
2	5	$5/57 \approx .09$
3	8	$8/57 \approx .14$
4	10	$10/57 \approx .18$
5	12	$12/57 \approx .21$
6	17	$17/57 \approx .30$
7	0	$0/57 = 0$
8	1	$1/57 \approx .02$
9	0	$0/57 = 0$
10	0	$0/57 = 0$

The results shown in the two tables are easier to compare using the relative frequencies. We will return to this problem later in this section.

The relative frequencies in Tables 3 and 4 can be considered as probabilities. Such a table that lists all the possible values of a random variable, together with the corresponding probabilities, is called a **probability distribution.** The sum of the probabilities in a probability distribution must always be 1. (The sum in an actual distribution may vary slightly from 1 because of rounding.) **1**

▶ **EXAMPLE 1** Many plants have seed pods with variable numbers of seeds. One variety of green beans has no more than 6 seeds per pod. Suppose that examination of 30 such bean pods gives the results shown in Table 5. Here the random variable *x* tells the number of seeds per pod. Give a probability distribution for these results.

The probabilities are found by computing the relative frequencies. A total of 30 bean pods were examined, so each frequency should be divided by 30 to

Table 5

x	Frequency
0	3
1	4
2	6
3	8
4	5
5	3
6	1
	Total: 30

Table 6

x	Frequency	Probability
0	3	$3/30 = .10$
1	4	$4/30 \approx .13$
2	6	$6/30 = .20$
3	8	$8/30 \approx .27$
4	5	$5/30 \approx .17$
5	3	$3/30 = .10$
6	1	$1/30 \approx .03$
	Total: 30	

2 In Example 1, what is

(a) $P(x = 0)$?

(b) $P(x = 1)$?

Answers:

(a) .10

(b) .13

get the probability distribution shown as Table 6. Some of the results have been rounded to the nearest hundredth.

As shown in Table 6, the probability that the random variable x takes on the value 2 is 6/30, or .20. This is often written as

$$P(x = 2) = .20.$$

Also, $P(x = 5) = .10$, and $P(x = 6) \approx .03$. ◀ **2**

Instead of writing the probability distribution in Example 1 as a table, we could write the same information as a set of ordered pairs:

$$\{(0, .10), (1, .13), (2, .20), (3, .27), (4, .17), (5, .10), (6, .03)\}$$

There is just one probability for each value of the random variable. Thus, a probability distribution defines a function, called a **probability distribution function** or simply a **probability function.** We shall use the terms "probability distribution" and "probability function" interchangeably.

The information in a probability distribution is often displayed graphically as a special kind of bar graph called a **histogram.** The bars of a histogram all have the same width, usually 1. The heights of the bars are determined by the frequencies. A histogram for the data in Table 3 is given in Figure 9.5. A histogram shows important characteristics of a distribution that may not be readily apparent in tabular form, such as the relative sizes of the probabilities and any symmetry in the distribution.

The area of the bar above $t = 1$ in Figure 9.5 is the product of 1 and .04, or .04 × 1 = .04. Since each bar has a width of 1, its area is equal to the probability that corresponds to that value of t. The probability that a particular value will occur is thus given by the area of the appropriate bar of the graph. For example, the probability that a transaction will take less than 4 minutes is the sum of the areas for $t = 1$, $t = 2$, and $t = 3$. This area, shown in color in Figure 9.6, corresponds to 23% of the total area, since

$$P(t < 4) = P(t = 1) + P(t = 2) + P(t = 3)$$
$$= .04 + .07 + .12 = .23.$$

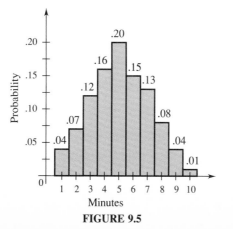

FIGURE 9.5

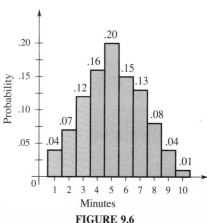

FIGURE 9.6

3 **(a)** Prepare a histogram for the probability distribution in Problem 1 at the side.

(b) Find $P(x \le 2)$.

Answers:

(a)

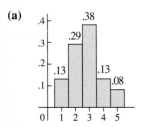

(b) .42

4 **(a)** Give the probability distribution for the number of heads showing when 3 coins are tossed.

(b) Find the probability that no more than 1 coin shows heads.

Answers:

(a)

x	$P(x)$
0	1/8
1	3/8
2	3/8
3	1/8

(b) 1/2

▶**EXAMPLE 2** Construct a histogram for the probability distribution in Example 1. Find the area that gives the probability that the number of seeds will be more than 4.

A histogram for this distribution is shown in Figure 9.7. The portion of the histogram in color represents

$$P(x > 4) = P(x = 5) + P(x = 6)$$
$$= .10 + .03 = .13,$$

or 13% of the total area. ◀

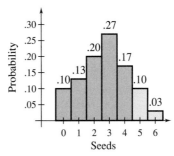

FIGURE 9.7

▶**EXAMPLE 3**

(a) Give the probability distribution for the number of heads showing when 2 coins are tossed.

Let x represent the outcome "number of heads," Then x can take on the values 0, 1, or 2. Now find the probability of each outcome. The results are shown in Table 7.

Table 7

x	$P(x)$
0	1/4
1	1/2
2	1/4

(b) Draw a histogram for the distribution in Table 7. Find the probability that at least 1 coin comes up heads.

The histogram is shown in Figure 9.8. The portion in color represents

$$P(x \ge 1) = P(x = 1) + P(x = 2)$$
$$= \frac{3}{4}. \quad ◀ \boxed{4}$$

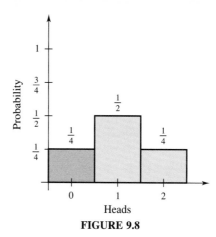

FIGURE 9.8

In working with experimental data, it is often useful to have a typical or "average" number that represents the entire set of data. For example, we compare our heights and weights to those of the typical or "average" person on weight charts. Students are familiar with the "class average" and their own "average" in a given course.

In a recent year, a citizen of the United States could expect to complete about 13 years of school, to be a member of a household earning $35,400 per year, and to live in a household of 2.6 people. What do we mean here by "expect"? Many people have completed less than 13 years of school; many others have completed more. Many households have less income than $35,400 per year; many others have more. The idea of a household of 2.6 people is a little hard to swallow. The numbers all refer to *averages*. When the term "expect" is used in this way, it refers to *mathematical expectation,* which is a kind of average.

EXPECTED VALUE What about an average value for a random variable? As an example, let us find the average number of offspring for a certain species of pheasant, given the probability distribution in Table 8.

Table 8

Number of Offspring	Frequency	Probability
0	8	.08
1	14	.14
2	29	.29
3	32	.32
4	17	.17
	Total: 100	

5 Using Table 4, find the expected transaction time for the second survey.

Answer:
4.4

We might be tempted to find the typical number of offspring by averaging the numbers 0, 1, 2, 3, and 4, which represent the numbers of offspring possible. This won't work, however, since the various numbers of offspring do not occur with equal probability: for example, a brood with 3 offspring is much more common than one with 0 or 1 offspring. The differing probabilities of occurrence can be taken into account with a **weighted average,** found by multiplying each of the possible numbers of offspring by its corresponding probability, as follows.

$$\text{Typical number of offspring} = 0(.08) + 1(.14) + 2(.29) + 3(.32) + 4(.17)$$
$$= 0 + .14 + .58 + .96 + .68$$
$$= \mathbf{2.36}$$

Based on the data given above, a typical brood of pheasants includes 2.36 offspring.

It is certainly not possible for a pair of pheasants to produce 2.36 offspring. If the numbers of offspring produced by many different pairs of pheasants are found, however, the average of these numbers will be about 2.36.

We can use the results of this example to define the mean, or *expected value,* of a probability distribution as follows. (We will have more to say about the mean in Chapter 10.)

Expected Value

Suppose the random variable x can take on the n values $x_1, x_2, x_3, \ldots, x_n$. Also, suppose the probabilities that these values occur are respectively p_1, $p_2, p_3, \ldots, p_n$ Then the **expected value** of the random variable is

$$E(x) = x_1p_1 + x_2p_2 + x_3p_3 + \cdots + x_np_n.$$

As in the example above, the expected value of a random variable may be a number that can never occur in any one trial of the experiment.

Physically, the expected value of a probability distribution represents a balance point. Figure 9.9 shows a histogram for the distribution of the pheasant offspring. If the histogram is thought of as a series of weights with magnitudes represented by the heights of the bars, then the system would balance if supported at the point corresponding to the expected value.

▶**EXAMPLE 4** Find the expected transaction time for each of the survey results from the bank example at the beginning of this section. What can we conclude from the results?

Refer to Tables 3 and 4. From Table 3, using the relative frequencies as probabilities, the expected transaction time is

$$1(.04) + 2(.07) + 3(.12) + 4(.16) + 5(.20) + 6(.15) + 7(.13)$$
$$+ 8(.08) + 9(.04) + 10(.01) = 5.09. \quad \mathbf{5}$$

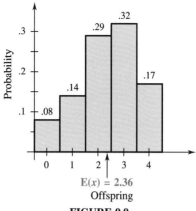

FIGURE 9.9

From Problem 5 at the side, the expected transaction time for the second survey is 4.4. The expected values give us a useful way to compare the results of the two surveys. The second survey had a smaller expected transaction time, so the new procedures may have sped up transactions. ◀

▶**EXAMPLE 5**   Suppose a local church decides to raise money by raffling a microwave oven worth $400. A total of 2000 tickets are sold at $1 each. Find the expected value of winning for a person who buys 1 ticket in the raffle.

Here the random variable represents the possible amounts of net winnings, where net winnings = amount won − cost of ticket. The net winnings of the person winning the oven are $400 (amount won) − $1 (cost of ticket) = $399. The net winnings for each losing ticket are $0 − $1 = −$1.

The probability of winning is 1 in 2000, or 1/2000, while the probability of losing is 1999/2000. See Table 9.

Table 9

Outcome (net winnings)	Probability
$399	$\dfrac{1}{2000}$
−$1	$\dfrac{1999}{2000}$

The expected winnings for a person buying 1 ticket are

$$399\left(\frac{1}{2000}\right) + (-1)\left(\frac{1999}{2000}\right) = \frac{399}{2000} - \frac{1999}{2000}$$

$$= -\frac{1600}{2000}$$

$$= -.80.$$

6 Suppose you buy 1 of 1000 tickets at 10¢ each in a lottery where the first prize is $50. What are your expected net winnings? What does this answer mean?

Answer:
−5¢. On the average you lose 5¢ per ticket purchased.

7 Find Mary's expected winnings. If Mary tosses and Donna calls, is it still a fair game?

Answer:
0; yes

On the average, a person buying 1 ticket in the raffle will lose $.80, or 80¢.

It is not possible to lose 80¢ in this raffle—either you lose $1, or you win a $400 prize. If you bought tickets in many such raffles over a long period of time, however, you would lose 80¢ per ticket, on the average. ◀ **6**

▶ **EXAMPLE 6** Each day Donna and Mary toss a coin to see who buys the coffee (60¢ a cup). One tosses and the other calls the outcome. If the person who calls the outcome is correct, the other buys the coffee; otherwise the caller pays. Find Donna's expected winnings.

Assume that an honest coin is used, that Mary tosses the coin, and that Donna calls the outcome. The possible results and corresponding probabilities are shown below.

	Possible Results			
Result of loss	Heads	Heads	Tails	Tails
Call	Heads	Tails	Heads	Tails
Caller Wins?	Yes	No	No	Yes
Probability	1/4	1/4	1/4	1/4

Donna wins a 60¢ cup of coffee whenver the results and calls match, and she loses a 60¢ cup when there is no match. Her expected winnings are

$$(.60)\left(\frac{1}{4}\right) + (-.60)\left(\frac{1}{4}\right) + (-.60)\left(\frac{1}{4}\right) + (.60)\left(\frac{1}{4}\right) = 0.$$

On the average, over the long run. Donna neither wins nor loses. ◀ **7**

A game with an expected value of 0 (such as the one in Example 6) is called a **fair game.** Casinos do not offer fair games. If they did, they would win (on the average) $0, and have a hard time paying the help! Casino games have expected winnings for the house that vary from 1.5¢ per dollar to 60¢ per dollar. Exercises 37–42 at the end of the section ask you to find the expected winnings for certain games of chance.

The idea of expected value can be very useful in decision making, as shown by the next example.

▶ **EXAMPLE 7** At age 50, Earl Karn receives a letter from the Mutual of Mauritania Insurance Company. According to the letter, he must tell the company immediately which of the following two options he will choose: take $20,000 at age 60 (if he is alive, $0 otherwise) or $30,000 at age 70 (again, if he is alive, $0 otherwise). Based *only* on the idea of expected value, which should he choose?*

---

*Other considerations might affect the decision, such as the rate at which Karn might invest the $20,000 at age 60 for 10 years.

8 A person can take one of two jobs. With job A, there is a 50% chance of making $60,000 per year after 5 years and a 50% chance of making $30,000. With job B, there is a 30% chance of making $90,000 per year after 5 years and a 70% chance of making $20,000. Based strictly on expected value, which job should be taken?

Answer:
Job A has an expected salary of $45,000; job B, $41,000. Take job A.

Life insurance companies have constructed elaborate tables showing the probability of a person living a given number of years in the future. From a recent such table, the probability of living from age 50 to age 60 is .88, while the probability of living from age 50 to 70 is .64. The expected value of the two options are given below.

First Option: $(20,000)(.88) + (0)(.12) = 17,600$
Second Option: $(30,000)(.64) + (0)(.36) = 19,200$

Based strictly on expected values, Karn should choose the second option ◄ **8**

9.5 EXERCISES

*In Exercises 1–6 (**a**) give the probability distribution, and (**b**) sketch its histogram. (See Examples 1–3.)*

1. The number of accidents at a certain intersection in a metropolitan area were counted each day for 7 days with the following results.

Number of Accidents	Frequency
0	2
1	3
2	1
3	0
4	1

2. A teacher polled a finite mathematics class each semester for 10 semesters to see how many students were business majors. The results are given in the table.

Number of Business Majors	Frequency
13	1
14	0
15	3
16	3
17	2
18	1

3. The winning team's scores in the seven no-hit games in 1991 are given in the table.

Winning Scores	Frequency
1	1
2	3
3	1
4	0
5	0
6	0
7	2

4. At a training program for police officers, each member of a class of 25 took 6 shots at a target. The numbers of bullseyes shot are shown in the table.

Number of Bullseyes	Frequency
0	0
1	1
2	0
3	4
4	10
5	8
6	2
	Total: 25

5. The telephone company kept track of the calls for the correct time during a 24-hour period for 2 weeks. The results are shown in the table.

Number of Calls	Frequency
28	1
29	1
30	2
31	3
32	2
33	2
34	2
35	1
	Total: 14

6. Five people are given a pain reliever. After a certain time period, the number who experience relief is noted. The experiment is repeated 20 times with the results shown in the table.

Number Who Experience Relief	Frequency
0	3
1	5
2	6
3	3
4	2
5	1
	Total: 20

For each of the experiments below, let x represent a random variable, and use your knowledge of probability to prepare a probability distribution.

7. The numbers 1 to 5 are written on cards, placed in a basket, mixed up, and then 2 cards are drawn. The number of even numbers are counted.

8. A bowl contains 5 yellow, 8 orange, and 3 white jelly beans. Two jelly beans are drawn at random and the number of white jelly beans is counted.

9. Two dice are rolled, and the total number of points is recorded.

10. Three cards are drawn from a deck. The number of aces is counted.

11. Two balls are drawn from a bag in which there are 4 white balls and 2 black balls. The number of black balls is counted.

12. Five cards are drawn from a deck. The number of black threes is counted.

Draw a histogram for each of the following, and shade the region that gives the indicated probability. (See Example 2.)

13. Exercise 9; $P(x \geq 11)$

14. Exercise 10; P (at least one ace)

15. Exercise 11; P(at least one black ball)

16. Exercise 12; $P(x = 1 \text{ or } x = 2)$

17. Exercise 7; $P(1 \leq x \leq 2)$

18. Exercise 12; $P(0 \leq x \leq 1)$

Find the expected value for each random variable. (See Example 4.)

19.

x	5	6	7	8
$P(x)$	.3	.1	.4	.2

20.

y	10	11	12	13	14
$P(y)$	.1	.05	.15	.4	.3

21.

z	22	23	24	25	26
$P(z)$	.15	.23	.34	.20	.08

22.

x	0	1	2	3	4
$P(x)$	.31	.35	.20	.11	.03

23. Refer to Example 6. Suppose one day Mary brings a two-headed coin and uses it to toss for the coffee. Since Mary tosses, Donna calls.
 (a) Is this still a fair game?
 (b) What is Donna's expected gain if she calls heads?
 (c) What is it if she calls tails?

Find the expected values for the random variables x having the probability functions graphed below.

24.

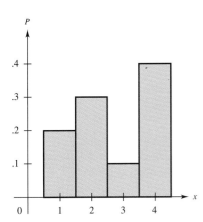

25.

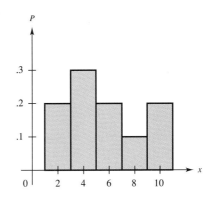

26.

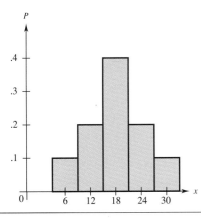

27.

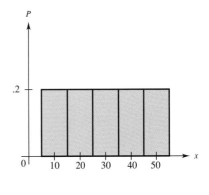

28. A raffle offers a first prize of $100, and two second prizes of $40 each. One ticket costs $1, and 500 tickets are sold. Find the expected winnings for a person who buys 1 ticket. (See Example 5.)

29. A raffle offers a first prize of $1000, two second prizes of $300 each, and twenty prizes of $10 each. If 10,000 tickets are sold at 50¢ each, find the expected winnings for a person buying 1 ticket.

Many of the following exercises use the idea of combinations discussed in Section 9.1.

30. If 3 marbles are drawn from a bag containing 3 yellow and 4 white marbles, what is the expected number of yellow marbles in the sample?

31. From a group of 2 women and 5 men, a delegation of 2 is selected. Find the expected number of women in the delegation.

32. In a club with 20 senior and 10 junior members, what is the expected number of junior members on a 3-member committee?

33. If 2 cards are drawn at one time from a deck of 52 cards, what is the expected number of diamonds?

34. Suppose someone offers to pay you $5 if you draw 2 diamonds in the game of Exercise 33. He says that you should pay 50¢ for the chance to play. Is this a fair game?

The following chart shows the number of times in the last 52 weeks that each possible number was chosen in any of the four digit positions in the Pick-4 game of the Illinois lottery.

No.	1st	2d	3d	4th	Any
0	38	29	32	34	133
1	32	35	39	40	146
2	35	37	39	39	150
3	46	43	35	42	166
4	33	50	38	40	161
5	31	35	28	37	131
6	46	35	36	32	149
7	41	37	42	33	153
8	26	28	43	35	132
9	35	34	31	31	131

Find each of the following expected values.

35. the expected number in first position

36. the expected number in any position

Find the expected winnings for each for the following games of chance.

37. A state lottery requires you to choose 1 heart, 1 club, 1 diamond, and 1 spade in that order from the 13 cards in each suit. If all four choices are correct, you win $5000. It costs $1 to play.

38. If exactly 3 of the 4 selections in Exercise 37 are correct, the player wins $200. (It still costs $1 to play.)

39. In one form of roulette, you bet $1 on "even". If 1 of the 18 even numbers comes up, you get your dollar back, plus another one. If 1 of the 20 noneven (18 odd, 0, and 00) numbers comes up, you lose.

40. In another form of roulette, there are only 19 noneven numbers (no 00).

41. Numbers is an illegal game in which you bet $1 on any 3-digit number from 000 to 999. If your number comes up, you get $500.

42. In Keno, the house has a pot containing 80 balls, each marked with a different number from 1 to 80. You buy a ticket for $1 and mark 1 of the 80 numbers on it. The house then selects 20 numbers at random. If your number is among the 20, you get $3.20 (for a net winning of $2.20).

43. Your friend missed class the day probability distributions were discussed. How would you explain to him what a probability distribution is?

44. How is the relative frequency of a random variable found?

45. The expected value of a random variable is
  (a) the most frequent value of the random variable;
  (b) the average of all possible values of the random variable;
  (c) the sum of the products formed by multiplying each value of the random variable by its probability;
  (d) the product of the sum of all values of the random variable and the sum of the probabilities.

46. **Management** An insurance company has written 100 policies of $10,000, 500 of $5000, and 1000 polices of $1000 on people of age 20. If experience shows that the probability of dying during the 20th year of life is .001, how much can the company expect to pay out during the year the policies were written?

47. Jack must choose at age 40 to inherit $25,000 at age 50 (if he is still alive) or $30,000 at age 55 (if he is still alive). If the probabilities for a person of age 40 to live to be 50 and 55 are .90 and .85, respectively, which choice gives him the larger expected inheritance? (See Example 7.)

48. **Management** A magazine distributor offers a first prize of $100,000, two second prizes of $40,000 each, and two third prizes of $10,000 each. A total of 2,000,000 entries are received in the contest. Find the expected winnings if you submit 1 entry to the contest. If it would cost you 25¢ in time, paper, and stamps to enter, would it be worth it?

49. Management Levi Strauss and Company* use expected value to help its salespeople rate their accounts. For each account, a salesperson estimates potential additional volume and the probability of getting it. The product of these gives the expected value of the potential, which is added to the existing volume. The totals are then classified as A, B, or C as follows: below $40,000, class C; between $40,000 and $55,000, class B; above $55,000, class A. Complete the chart below for one of its salespeople.

Account Number	Existing Volume	Potential Additional Volume	Probability of Additional Volume	Expected Value of Potential	Existing Volume + Expected Volume of Potential	Class
1	$15,000	$10,000	.25	$2,500	$17,500	C
2	40,000	0	—	—	40,000	C
3	20,000	10,000	.20			
4	50,000	10,000	.10			
5	5,000	50,000	.50			
6	0	100,000	.60			
7	30,000	20,000	.80			

50. A recent McDonald's contest offered cash prizes and probabilities of winning on one visit, as shown in the following table. Suppose you spend $1 to buy a bus pass that lets you go to 25 different McDonald's and pick up entry forms. Find the expected value of winning a prize.

Prize	Probability
$100,000	$\dfrac{1}{176,402,500}$
25,000	$\dfrac{1}{39,200,556}$
5000	$\dfrac{1}{17,640,250}$
1000	$\dfrac{1}{1,568,022}$
100	$\dfrac{1}{282,244}$
5	$\dfrac{1}{7056}$
1	$\dfrac{1}{588}$

---

* This example was supplied by James McDonald, Levi Strauss and Company, San Francisco.

51. Natural Science One of the few methods that can be used in an attempt to cut the severity of a hurricane is to *seed* the storm. In this process, silver iodide crystals are dropped into the storm. Unfortunately, silver iodide crystals sometimes cause the storm to *increase* its speed. Wind speeds may also increase or decrease even with no seeding. The probabilities and amounts of property damage shown in the probability tree are from an article by R. A. Howard, J. E. Matheson, and D. W. North, "The Decision to Seed Hurricanes."*

(a) Find the expected amount of damage under each option, "seed" and "do not seed."

(b) To minimize total expected damage, what option should be chosen?

---

*Figure, "The Decision to Seed Hurricanes," by R. A. Howard from *Science,* Vol. 176, pp. 1191–1202, Copyright 1972 by the American Association for the Advancement of Science and the author.

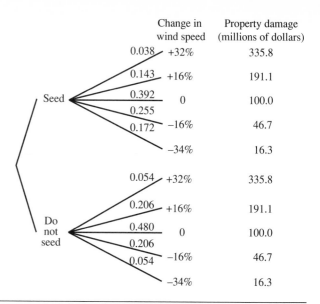

	Change in wind speed	Property damage (millions of dollars)
	+32%	335.8
	+16%	191.1
Seed	0	100.0
	−16%	46.7
	−34%	16.3
	+32%	335.8
	+16%	191.1
Do not seed	0	100.0
	−16%	46.7
	−34%	16.3

9.6 DECISION MAKING

John F. Kennedy once remarked he had assumed that as president it would be difficult to choose between distinct, opposite alternatives when a decision needed to be made. Actually, however, he found that such decisions were easy to make; the hard decisions came when he was faced with choices that were not as clear-cut. Most decisions fall in this last category—decisions which must be made under conditions of uncertainty. In the previous section we saw how to use expected values to help make a decision. These ideas are extended in this section, where we consider decision making in the face of uncertainty. Let us begin with an example.

Freezing temperatures are endangering the orange crop in central California. A farmer can protect his crop by burning smudge pots—the heat from the pots keeps the oranges from freezing. However, burning the pots is expensive; the cost is $4000. The farmer knows that if he burns smudge pots he will be able to sell his crop for a net profit (after smudge pot costs are deducted) of $10,000, provided that the freeze does develop and wipes out other orange crops in California. If he does nothing he will either lose $2000 already invested in the

1 Explain how each of the following payoffs in the matrix were obtained.

(a) −$2000

(b) $10,000

Answers:

(a) If it freezes and smudge pots are not used, the farmer's profit is −$2000 for labor costs.

(b) If it freezes and smudge pots are used, the farmer makes a profit of $10,000.

2 What should the farmer do if the probability of a freeze is .6? What is his expected profit?

Answer:
Use smudge pots; $8240

crop if it does freeze, or make a profit of $9600 if it does not freeze. (If it does not freeze, there will be a large supply of oranges, and thus his profit will be lower than if there was a small supply.)

What should the farmer do? He should begin by carefully defining the problem. First, he must decide on the **states of nature,** the possible alternatives over which he has no control. Here there are two: freezing temperatures, or no freezing temperatures. Next, the farmer should list the things he can control— his actions or **strategies.** He has two possible strategies: to use smudge pots or not. The consequences of each action under each state of nature, called **payoffs,** are summarized in a **payoff matrix,** as shown below. The payoffs in this case are the profits for each possible combination of events.

$$
\begin{array}{c}
 & & \text{\textit{States of Nature}} \\
 & & \text{Freeze} \qquad \text{No Freeze} \\
\textit{Strategies of Farmer} & \begin{array}{l}\text{Use Smudge Pots} \\ \text{Do not Use Pots}\end{array} & \begin{bmatrix} \$10,000 & \$5600 \\ -2000 & \$9600 \end{bmatrix}
\end{array}
$$

To get the $5600 entry in the payoff matrix, use the profit if there is no freeze, $9600, and subtract the $4000 cost of using the pots.   **1**

Once the farmer makes the payoff matrix, what then? The farmer might be an optimist (some might call him a gambler); in this case he might assume that the best will happen and go for the biggest number of the matrix ($10,000). For this profit, he must adopt the strategy "use smudge pots."

On the other hand, if the farmer is a pessimist, he would want to minimize the worst thing that could happen. If he uses smudge pots, the worst thing that could happen to him would be profit of $5600, which will result if there is no freeze. If he does not use smudge pots, he might face a loss of $2000. To minimize the worst, he once again should adopt the strategy "use smudge pots."

Suppose the farmer decides that he is neither an optimist nor a pessimist, but would like further information before choosing a strategy. For example, he might call the weather forecaster and ask for the probability of a freeze. Suppose the forecaster says that this probability is only .1. What should the farmer do? He should recall the discussion of expected value from the previous section and work out the expected profit for each of his two possible strategies. If the probability of a freeze is .1, then the probability that there is no freeze is .9. This information leads to the following expected values.

If smudge pots are used: $\quad 10,000(.1) + 5600(.0) = 6040$

If no smudge pots are used: $\quad -2000(.1) + 9600(.9) = 8440$

Here the maximum expected profit, $8440, is obtained if smudge pots are not used.   **2**

As the example shows, the farmer's beliefs about the probabilities of a freeze affect his choice of strategies.

3 Suppose the manufacturer reads another article which gives the following predictions: 35% chance of a weak economy, 25% chance of an in-between economy, and a 40% chance of a strong economy. What is the best strategy now? What is the expected profit?

Answer:
Modern; $83,250

▶**EXAMPLE 1**   A small manufacturer of Christmas cards must decide in February about the type of cards she should emphasize in her fall line. She has three possible strategies: emphasize modern cards, emphasize old-fashioned cards, or a mixture of the two. Her success is dependent on the state of the economy in December. If the economy is strong, she will do well with her modern cards, while in a weak economy people long for the old days and buy old-fashioned cards. In an in-between economy, her mixture of lines would do the best. She first prepares a payoff matrix for all three possibilities. The numbers in the matrix represent her profits in thousands of dollars.

	States of Nature		
	Weak Economy	In-Between	Strong Economy
Modern	40	85	120
Old-fashioned	106	46	83
Mixture	72	90	68

Strategies

(a) If the manufacturer is an optimist, she should aim for the biggest number on the matrix, 120 (representing $120,000 in profit). Her strategy in this case would be to produce modern cards.

(b) A pessimistic manufacturer wants to find the best of the worst of all bad things that can happen. If she produces modern cards, the worst that can happen is a profit of $40,000. For old-fashioned cards, the worst is a profit of $46,000, while the worst that can happen from a mixture is a profit of $68,000. Her strategy here is to use a mixture.

(c) Suppose the manufacturer reads in a business magazine that leading experts believe there is a 50% chance of a weak economy at Christmas, a 20% chance of an in-between economy, and a 30% chance of a strong economy. The manufacturer can now find her expected profit for each possible strategy.

Modern: $40(.50) + 85(.20) + 120(.30) = 73$

Old fashioned: $106(.50) + 46(.20) + 83(.30) = 87.1$

Mixture: $72(.50) + 90(.20) + 68(.30) = 74.4$

Here the best strategy is old-fashioned cards; the expected profit is 87.1, or $87,100. ◀ **3**

9.6 EXERCISES

1. **Management**   A developer has $100,000 to invest in land. He has a choice of two parcels (at the same price), one on the highway and one on the coast. With both parcels, his ultimate profit depends on whether he faces light opposition from environmental groups or heavy opposition. He estimates that the payoff matrix is as follows (the numbers represent his profit).

	Opposition	
	Light	Heavy
Highway	$70,000	$30,000
Coast	$150,000	−$40,000

What should the developer do if he is
(a) an optimist?   **(b)** a pessimist?

(c) Suppose the probability of heavy opposition is .8. What is his best strategy? What is the expected profit?

(d) What is the best strategy if the probability of heavy opposition is only .4?

2. Hillsdale College has sold out all tickets for a jazz concert to be held in the stadium. If it rains, the show will have to be moved to the gym, which has a much smaller seating capacity. The dean must decide in advance whether to set up the seats and the stage in the gym or in the stadium, or both, just in case. The payoff matrix below shows the net profit in each case.

$$
\begin{array}{c}
 & & \text{States of Nature} \\
 & & \text{Rain} \quad\quad \text{No Rain} \\
 & \text{Set up in Stadium} & \begin{bmatrix} -\$1550 & \$1500 \\ \$1000 & \$1000 \\ \$750 & \$1400 \end{bmatrix} \\
\text{Strategies} & \text{Set up in Gym} & \\
 & \text{Set up Both} &
\end{array}
$$

What strategy should the dean choose if she is
(a) an optimist?   (b) a pessimist?
(c) If the weather forecaster predicts rain with a probability of .6, what strategy should she choose to maximize expected profit? What is the maximum expected profit?

3. **Management**   An analyst must decide what fraction of the items produced by a certain machine are defective. He has already decided that there are three possibilites for the fraction of defective items: .01, .10, and .20. He may recommend two courses of action: repair the machine or make no repairs. The payoff matrix below represents the *costs* to the company in each case, in hundreds of dollars.

$$
\begin{array}{c}
 & & \text{States of Nature} \\
 & & .01 \quad .10 \quad .20 \\
\text{Strategies} & \text{Repair} & \begin{bmatrix} 130 & 130 & 130 \\ 25 & 200 & 500 \end{bmatrix} \\
 & \text{No Repair} &
\end{array}
$$

What strategy should the analyst recommend if he is
(a) an optimist?   (b) a pessimist?
(c) Suppose the analyst is able to estimate probabilities for the three states of nature as follows.

Fraction of Defectives	Probability
.01	.70
.10	.20
.20	.10

Which strategy should he recommend? Find the expected cost to the company if this strategy is chosen.

4. **Management**   The research department of the Allied Manufacturing Company has developed a new process which it believes will result in an improved product. Management must decide whether or not to go ahead and market the new product. The new product may be better than the old or it may not be better. If the new product is better and the company decides to market it, sales should increase by $50,000. If it is not better and they replace the old product with the new product on the market, they will lose $25,000 to competitors. If they decide not to market the new product they will lose $40,000 if it is better, and research costs of $10,000 if it is not.

(a) Prepare a payoff matrix.
(b) If management believes the probability that the new product is better to be .4, find the expected profits under each strategy and determine the best action.

5. **Management**   A businessman is planning to ship a used machine to his plant in Nigeria. He would like to use it there for the next 4 years. He must decide whether or not to overhaul the machine before sending it. The cost of overhaul is $2600. If the machine fails when in operation in Nigeria, it will cost him $6000 in lost production and repairs. He estimates the probability that it will fail at .3 if he does not overhaul it, and .1 if he does overhaul it. Neglect the possibility that the machine might fail more than once in the 4 years.

(a) Prepare a payoff matrix.
(b) What should the businessman do to minimize his expected costs?

6. **Management**   A contractor prepares to bid on a job. If all goes well, his bid should be $30,000, which will cover his costs plus his usual profit margin of $4500. However, if a threatened labor strike actually occurs, his bid should be $40,000 to give him the same profit. If there is a strike and he bids $30,000, he will lose $5500. If his bid is too high, he may lose they job entirely, while if it is too low, he may lose money.

(a) Prepare a payoff matrix.
(b) If the contractor believes that the probability of a strike is .6, how much should he bid?

7. **Natural Science** A community is considering an anti-smoking campaign.* The city council will choose one of three possible strategies: a campaign for everyone over age 10 in the community, a campaign for youths only, or no campaign at all. The two states of nature are a true cause-effect relationship between smoking and cancer and no cause-effect relationship. The costs to the community (including loss of life and productivity) in each case are as shown below.

	States of Nature	
Strategies	Cause-Effect Relationship	No Cause-Effect Relationship
Campaign for All	$100,000	$800,000
Campaign for Youth	$2,820,000	$20,000
No Campaign	$3,100,100	$0

What action should the city council choose if it is
(a) optimistic?   (b) pessimistic?
(c) If the Director of Public Health estimates that the probability of a true cause-effect relationship is .8, which strategy should the city council choose?

8. An investor has $20,000 to invest in stocks. She has two possible strategies: buy conservative blue-chip stocks or buy highly speculative stocks. There are two states of nature: the market goes up or the market goes down. The following payoff matrix shows the net amounts she will have under the various circumstances.

	Market Up	Market Down
Buy Blue-Chip	$25,000	$18,000
Buy Speculative	$30,000	$11,000

What should the investor do if she is
(a) an optimist?   (b) a pessimist?
(c) Suppose there is a .7 probability of the market going up. What is the best strategy? What is the expected profit?
(d) What is the best strategy if the probability of a market rise is .2?

---

*This problem is based on an article by B. G. Greenberg in the September 1969 issue of the *Journal of the American Statistical Association.*

Sometimes the numbers (or payoffs) in a payoff matrix do not represent money (profits or costs, for example), but utility. A **utility** is a number that measures the satisfaction (or lack of it) that results from a certain action. The numbers must be assigned by each individual, depending on how he or she feels about a situation. For example, one person might assign a utility of +20 for a week's vacation in San Francisco, with −6 being assigned if the vacation were moved to Sacramento. Work the following problems in the same way as those above.

9. **Social Science** A politician must plan her reelection strategy. She can emphasize jobs or she can emphasize the environment. The voters can be concerned about jobs or about the environment. A payoff matrix showing the utility of each possible outcome is shown below.

		Voters	
		Jobs	Environment
Candidate	Jobs	+25	−10
	Environment	−15	+30

The political analysts feel that there is a .35 chance that the voters will emphasize jobs. What strategy should the candidate adopt? What is its expected utility?

10. In an accounting class, the instructor permits the students to bring a calculator or a reference book (but not both) to an examination. The examination itself can emphasize either problems or definitions. In trying to decide which aid to take to an examination, a student first decides on the utilities shown in the following payoff matrix.

		Exam Emphasizes	
		Numbers	Definitions
Student Chooses	Calculator	+50	0
	Book	+10	+40

(a) What strategy should the student choose if the probability that the examination will emphasize numbers is .6? What is the expected utility in this case?
(b) Suppose the probability that the examination emphasizes numbers is .4. What strategy should be chosen by the student?

KEY TERMS AND SYMBOLS

9.1 $n!$ n factorial
multipication principle
permutations
combinations

9.3 Bernoulli trials

9.4 stochastic processes
Markov chain
state
transition diagram
transition matrix
probability vector
equilibrium vector (fixed vector)
regular transition matrix
regular Markov chain

9.5 random variable
frequency distribution
relative frequency
probability distribution
probability function (probability distribution function)
histogram
weighted average
expected value
fair game

9.6 states of nature
strategies
payoffs
payoff matrix

KEY CONCEPTS

Multiplication Principle: If there are m_1 ways to make a first choice, m_2 ways to make a second choice, and so on, where none of the choices depend on any of the others, then there are $m_1 m_2 \cdots m_n$ different ways to make the entire sequence of choices.

The number of **permutations** of n elements taken r at a time is $_nP_r = \dfrac{n!}{(n-r)!}$.

The number of **combinations** of n elements taken r at a time is

$$\binom{n}{r} = \frac{n!}{(n-r)!\, r!}.$$

Binomial experiments have the following characteristics: The same experiment is repeated several times. There are only *two* outcomes, labeled success and failure. The trials are independent so that the probability of success is the same for each trial. If the probability of success in a single trial is p, then the probability of x successes in n trials is

$$\binom{n}{x} p^x (1-p)^{n-x}.$$

Markov Chains: A **transition matrix** must be square, with all entries between 0 and 1 inclusive, and the sum of the entries in any row must be 1. A Markov chain is *regular* if some power of its transition matrix P contains all positive entries. The long-range probabilities for a regular Markov chain are given by the **equilibrium or fixed vector** V, where for any initial probability vector v, the products vP^n approach V as n gets larger and larger, and $VP = V$. To find V, solve the system of equations formed by $VP = V$ and the fact that the sum of the entries of V is 1.

Expected Value of a Probability Distribution: For a random variable x with values $x_1, x_2, \ldots, x_n$ and probabilities $p_1, p_2, \ldots, p_n$, respectively, the expected value is

$$E(x) = x_1 p_1 + x_2 p_2 + \cdots + x_n p_n.$$

Decision Making: A **payoff matrix** which includes all available strategies and states of nature is used in decision making to define the problem and the possible solutions. The expected value of each strategy can help to determine the best course of action.

CHAPTER 9 REVIEW EXERCISES

1. In how many ways can 5 shuttle vans line up at the airport?

2. How many variations in first-, second-, and third-place finishes are possible in a 100-yard dash with 7 runners?

3. In how many ways can a sample of 3 pears be taken from a basket containing a dozen pears?

4. If 2 of the pears in Exercise 3 are spoiled, in how many ways can the sample of 3 include the following?
   (a) 1 spoiled pear   (b) no spoiled pears
   (c) at most 1 spoiled pear

5. In how many ways can 3 pictures, selected from a group of 6 pictures, be arranged in a row on a wall?

6. In how many ways can the 6 pictures in Exercise 5 be arranged in a row, if a certain one must be first?

7. In how many ways can the 6 pictures in Exercise 5 be arranged if 3 are landscapes and 3 are puppies, and if
   (a) like types must be kept together?
   (b) landscapes and puppies are alternated?

8. A representative is to be selected from each of 3 departments in a large company. There are 7 people in the first department, 5 in the second department, and 8 in the third department.
   (a) How many different groups of 3 representative are possible?
   (b) How many groups are possible, if any number (at least 1) up to 3 representatives can form a group?

9. Explain under what circumstances a permutation should be used in a probability problem, and under what circumstances a combination should be used.

10. Discuss under what circumstances the binomial probability formula should be used in a probability problem.

Suppose a family plans to have 4 children, and the probability that a particular child is a boy is 1/2. Find the probability that the family will have the following.

11. exactly 2 boys

12. all girls

13. at least 2 boys

14. no more than 3 girls

Suppose 2 cards are drawn without replacement from an ordinary deck of 52 cards. Find the probabilities of the following results.

15. Both cards are black.

16. Both cards are hearts.

17. Exactly 1 is a face card.

18. At most 1 is an ace.

A collection of golf balls contains 4 yellow, 2 blue, and 6 white balls. A golfer selects 3 balls at random. Find the probability that the selection includes the following.

19. all white balls

20. all yellow balls

21. at least 1 blue ball

22. 1 ball of each color

Decide whether each of the following is a regular transition matrix.

23. $\begin{bmatrix} 0 & 1 \\ .8 & .2 \end{bmatrix}$

24. $\begin{bmatrix} -.1 & .4 \\ .3 & .7 \end{bmatrix}$

25. $\begin{bmatrix} .3 & 0 & .7 \\ .5 & .1 & .4 \\ 1 & 0 & 0 \end{bmatrix}$

26. $\begin{bmatrix} .2 & .3 & .5 \\ .4 & .4 & .2 \\ .5 & .1 & .4 \end{bmatrix}$

Management *A bottle capping machine has an error rate of .01. A random sample of 20 bottles is selected. Find the following probabilities.*

27. Exactly 4 bottles are improperly capped.

28. No more than 3 bottles are improperly capped.

29. At least 1 bottle is improperly capped.

30. Management A credit card company classifies its customers in three groups: nonusers in a given month, light users, and heavy users. The transition matrix for these states is

$$\begin{array}{c} \\ \text{Nonuser} \\ \text{Light} \\ \text{Heavy} \end{array} \begin{array}{ccc} \text{Nonuser} & \text{Light} & \text{Heavy} \\ \begin{bmatrix} .8 & .15 & .05 \\ .25 & .55 & .2 \\ .04 & .21 & .75 \end{bmatrix} \end{array}$$

Suppose the initial distribution for the three states is $[.4 \quad .4 \quad .2]$. Find the distribution after

(a) 1 month; **(b)** 2 months.

(c) What is the long-range prediction for the distribution of users?

31. Management Savmor Investments starts a heavy advertising campaign. At the start of the campaign, Savmor sells 35% of all mutual funds sold in the area, while 65% are sold by the Highrate Company. The campaign produces the following transition matrix.

$$\begin{array}{cc} & \begin{array}{cc} \textit{After Campaign} \\ \text{Savmor} \quad\quad \text{Highrate} \end{array} \\ \textit{Before Campaign} \begin{array}{c} \text{Savmor} \\ \text{Highrate} \end{array} & \begin{bmatrix} .8 & .2 \\ .4 & .6 \end{bmatrix} \end{array}$$

(a) Find the market share for each company after the campaign.

(b) Find the share of the market for each company after three such campaigns.

(c) Predict the long-range market share for Savmor.

In Exercises 32–35, (a) give a probability distribution, and (b) sketch its histogram.

32.

x	6	7	8	9	10
frequency	3	7	9	3	2

33.

x	1	2	3	4	5	6
frequency	1	0	2	5	8	4

34. A coin is tossed 3 times and the number of heads is recorded.

35. A pair of dice are rolled and the sum of the results for each roll is recorded.

36. Natural Science Patients in groups of 5 are given a new cancer treatment. The experiment is repeated 10 times with the following results.

Number with Significant Improvement	*Frequency*
0	1
1	1
2	2
3	3
4	3
5	0
	Total: $\overline{10}$

In Exercises 37 and 38, give the probability that corresponds to the shaded region of each histogram.

37.

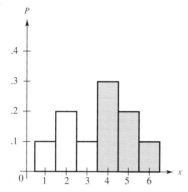

38.

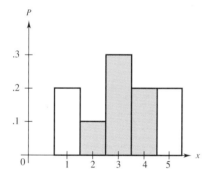

Solve the following problems.

39. Suppose someone offers to pay you $100 if you draw 3 cards from a standard deck of 52 cards and all the cards are clubs. What should you pay for the chance to win if it is a fair game?

40. You pay $6 to play in a game where you will roll a die, with payoffs as follows: $8 for a 6, $7 for a 5, and $4 for any other results. What are your expected winnings? Is the game fair?

41. A lottery has a first prize of $5000, two second prizes of $1000 each, and two $100 third prizes. A total of 10,000 tickets is sold, at $1 each. Find the expected winnings of a person buying 1 ticket.

42. Find the expected number of girls in a family of 5 children.

43. Three cards are drawn from a standard deck of 52 cards.
 (a) What is the expected number of aces?
 (b) What is the expected number of clubs?

44. Management In labor-management relations, both labor and management can adopt either a friendly or a hostile attitude. The results are shown in the following payoff matrix. The numbers give the wage gains made by an average worker.

		Management	
		Friendly	Hostile
Labor	Friendly	$600	$800
	Hostile	$400	$950

 (a) Suppose the chief negotiator for labor is an optimist. What strategy should he choose?
 (b) What strategy should he choose if he is a pessimist?
 (c) The chief negotiator for labor feels that there is a 70% chance that the company will be hostile. What strategy should he adopt? What is the expected payoff?
 (d) Just before negotiations begin, a new management is installed in the company. There is only a 40% chance that the new management will be hostile. What strategy should be adopted by labor?

45. Social Science A candidate for city council can come out in favor of a new factory, be opposed to it, or waffle on the issue. The change in votes for the candidate depends on what her opponent does, with payoffs as shown.

		Opponent		
		Favors	Waffles	Opposes
	Favors	0	−1000	−4000
Candidate	Waffles	1000	0	−500
	Opposes	5000	2000	0

 (a) What should the candidate do if she is an optimist?
 (b) What should she do if she is a pessimist?
 (c) Suppose the candidate's campaign manager feels there is a 40% chance that the opponent will favor the plant, and a 35% chance that he will waffle. What strategy should the candidate adopt? What is the expected change in the number of votes?
 (d) The opponent conducts a new poll which shows strong opposition to the new factory. This changes the probability he will favor the factory to 0 and the probability he will waffle to .7. What strategy should our candidate adopt? What is the expected change in the number of votes now?

Exercises 46 and 47 are taken from actuarial examinations given by the Society of Actuaries. *

46. Management A company is considering the introduction of a new product that is believed to have probability .5 of being successful and probability .5 of being unsuccessful. Successful products pass quality control 80% of the time. Unsuccessful products pass quality control 25% of the time. If the product is successful, the net profit to the company will be $40 million: if unsuccessful, the net loss will be $15 million. Determine the expected net profit if the product passes quality control.
(a) $23 million   (b) $24 million
(c) $25 million
(d) $26 million   (e) $27 million

47. Management A merchant buys boxes of fruit from a grower and sells them. Each box of fruit is either Good or Bad. A Good box contains 80% excellent fruit and will earn $200 profit on the retail market. A Bad box contains 30% excellent fruit and will produce a loss of $1000. The a priori probability of receiving a Good box of fruit is .9. Before the merchant decides to put the box on the market, he can sample one piece of fruit to test whether it is excellent. Based on that sample, he has the option of rejecting the box without paying for it. Determine the expected value of the right to sample. (*Hint:* If the merchant samples the fruit, what are the probabilities of accepting a Good box, accepting a Bad box, and not

accepting the box? What are these probabilities if he does not sample the fruit?)

(a) 0   (b) $16   (c) $34   (d) $72   (e) $80

 48. Management The March 1982 issue of *Mathematics Teacher* included "Overbooking Airline Flights," an article by Joe Dan Austin. In this article, Austin developed a model for the expected income for an airline flight. With appropriate assumptions, the probability that exactly x of n people with reservations show up at the airport to buy a ticket is given by the binomial probability formula. Assume the following: 6 reservations have been accepted for 3 seats, $p = .6$ is the probability that a person with a reservation will show up, a ticket costs $100, and the airline must pay $100 to anyone with a reservation who does not get a ticket. Complete the following table.

Number Who Show Up (x)	0	1	2	3	4	5	6
Airline's Income							
$P(x)$							

(a) Use the table to find $E(I)$, the expected airline income from the 3 seats.
(b) Find $E(I)$ for $n = 3$, $n = 4$, and $n = 5$. Compare these answers with $E(I)$ for $n = 6$. For these values of n, how many reservations should the airline book for the 3 seats in order to maximize the expected revenue?

---

*Problem from "Course 130 Examination, Operations Research," of the *Education and Examination Committee of the Society of Actuaries*. Reprinted by permission of the Society of Actuaries.

K. H. Lu of the University of Oregon Dental School has applied Markov chains to analyze the progress of teeth from a healthy state to having cavities to being restored through a filling or becoming so decayed as to require removal.*

Dental records of 184 grade and high school students were observed at six-month intervals (a total of 1080 records.) The state of the upper second premolar of each student was recorded as a five-digit number. To interpret this number, consider the tooth being roughly shaped like a box, with one side fastened to the gum. There are five sides of the tooth where cavities can occur, and each digit of the five-digit number refers to one of these sides. The first digit referred to the occlusal (or chewing) surface, the second to the mesial surface (the surface closest to the front of the mouth), the third to the

distal surface (the surface farthest from the front of the mouth), the fourth to the lingual surface (next to the tongue), and the fifth to the buccal surface (next to the cheek). A 0 was used to signify a healthy surface, a 1 for a decayed surface, and a 2 for a restored surface, in which the decayed area has been filled. Thus, (00000) signifies a healthy tooth and (20100) a tooth with a filling on the occlusal surface and a cavity on the distal surface, with the other three surfaces healthy.

The total number of possible states in this Markov chain is $3^5 = 243$, but only 18 of these states occured among the students. In the initial analysis, all 9 states with a 2 in one of the digits were grouped together under the heading "restoration."

Lu found the following transition matrix.

	00000	00100	01000	01100	10000	10100	11000	11100	Restoration	11111
00000	.84853	.07003	.03420	.01140	.01791	.00326	0	.00326	.01141	0
00100	0	.74725	0	.18681	0	.02198	0	0	.04396	0
01000	0	0	.70312	.26563	0	0	.03125	0	0	0
01100	0	0	0	.90510	0	0	0	.03650	.05840	0
10000	0	0	0	0	.71430	.17857	0	.03571	.03571	.03571
10100	0	0	0	0	0	.75000	0	.25000	0	0
11000	0	0	0	0	0	0	.50000	.50000	0	0
11100	0	0	0	0	0	0	0	.90000	.05000	.05000
Restoration	0	0	0	0	0	0	0	0	1	0
11111	0	0	0	0	0	0	0	0	0	1

EXERCISE

1. Interpret the states 00100, 10100, and 11111.

---

*From K. H. Lu, "A Markov Chain Analysis of Caries Process with Consideration for the Effect of Restoration," *Archives of Oral Biology* 13, 1968, pp. 1119–32.

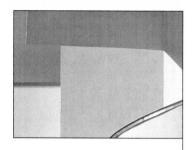

CHAPTER 10

Introduction to Statistics

TECHNOLOGY RESOURCES

Explorations in Finite Mathematics, Schneider

StatExplorer

The Electronic Spreadsheet, Spero

Statistics is a branch of mathematics that deals with the collection and summarization of data. Methods of statistical analysis make it possible to draw conclusions about a population based on data from a sample of the population. Statistical models have become increasingly useful in manufacturing, government, agriculture, medicine, and the social sciences, and in all types of research. An Indianapolis race-car team is using statistics to improve performance by gathering data on each run around the track. They sample data 300 times a second, and use computers to process the data. In this chapter we give a brief introduction to some of the key topics from statistical theory.

In the previous chapter, we saw that a frequency distribution can be transformed into a probability distribution by using the relative frequency of each value of the random variable as its probability. Sometimes it is convenient to work directly with a frequency distribution.

10.1 FREQUENCY DISTRIBUTIONS; MEASURES OF CENTRAL TENDENCY

Often, a researcher wishes to learn something about a characteristic of a population, but because the population is very large or mobile, it is not possible to examine all of its elements. Instead, a limited sample drawn from the population is studied to determine the characteristics of the population. For example, a book by Frances Cerra Whittelsey. *Why Women Pay More,* published by Ralph Nader's Center for Responsive Law, documents how women get bad deals. In the studies cited in this book, the population is U.S. women. The studies involved data collected from a sample of U.S. women.

For these inferences to be correct, the sample chosen must be a **random sample.** Random samples are representative of the population because they are chosen so that every element of the population is equally likely to be selected. A hand dealt from a well-shuffled deck of cards is a random sample.

1 An accounting firm selected 24 complex tax returns prepared by a certain tax preparer. The number of errors per return were as follows.

```
8  12   0   6  10   8   0  14
8  12  14  16   4  14   7  11
9  12   7  15  11  21  22  19
```

Prepare a grouped frequency distribution for this data. Use intervals 0–4, 5–9, and so on.

Answer:

Interval	Frequency
0–4	3
5–9	7
10–14	9
15–19	3
20–24	2
	Total: 24

After a sample has been chosen and all data of interest are collected, the data must be organized so that conclusions may be more easily drawn. One method of organization is to group the data into intervals; equal intervals are usually chosen.

▶**EXAMPLE 1** A survey asked 30 business executives how many college units in management each had. The results are shown below. Group the data into intervals and find the frequency of each interval.

```
 3  25  22  16   0   9  14   8  34  21
15  12   9   3   8  15  20  12  28  19
17  16  23  19  12  14  29  13  24  18
```

The highest number in the list is 34 and the lowest is 0; one convenient way to group the data is in intervals of size 5, starting with 0–4 and ending with 30–34. This gives an interval for each number in the list and results in seven equal intervals of a convenient size. Too many intervals of smaller size would not simplify the data enough, while too few intervals of larger size would conceal information that the data might provide. A rule of thumb is to use from six to fifteen intervals.

First tally the number of college units falling into each interval. Then total the tallies in each interval, as in the table below. This table is an example of a **grouped frequency distribution.**

College Units	Tally	Frequency
0–4	III	3
5–9	IIII	4
10–14	JHI I	6
15–19	JHI III	8
20–24	JHI	5
25–29	III	3
30–34	I	1
	Total:	30

◀ **1**

The frequency distribution in Example 1 shows information about the data that might not have been noticed before. For example, the interval with the largest number of units is 15–19, and 19 executives (more than half) had between 9 and 25 units. Also, the frequency in each interval increases rather evenly (up to 8) and then decreases at about the same pace. However, some information has been lost; for example, we no longer know how many executives had 12 units.

The information in a grouped frequency distribution can be displayed in a histogram similar to the histograms for probability distributions in the previous chapter. The intervals determine the widths of the bars; if equal intervals are

2 In Problem 1 at the side, the following grouped frequency distribution was found.

Interval	Frequency
0–4	3
5–9	7
10–14	9
15–19	3
20–24	2

Make a histogram and a frequency polygon for this distribution.

Answer:

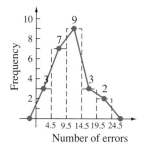

used, all the bars have the same width. The heights of the bars are determined by the frequencies.

A **frequency polygon** is another form of graph that illustrates a grouped frequency distribution. The polygon is formed by joining consecutive midpoints of the tops of the histogram bars with straight line segments. The midpoints of the first and last bars are joined to endpoints on the horizontal axis where the next midpoint would appear.

▶**EXAMPLE 2**   A grouped frequency distribution of college units was found in Example 1. Draw a histogram and a frequency polygon for this distribution.

First, draw a histogram, shown in black in Figure 10.1. To get a frequency polygon, connect consecutive midpoints of the tops of the bars. The frequency polygon is shown in color. ◀ **2**

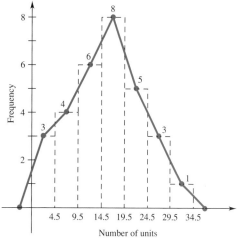

FIGURE 10.1

FOR GRAPHERS

Many graphing calculators can display the histogram and frequency polygon for a frequency distribution. Methods vary with the calculator, so consult your manual for specific instructions.

MEAN   The average value of a probability distribution is the expected value of the distribution. Three measures of central tendency, or "averages," are used with frequency distributions: the mean, the median, and the mode. The most important of these is the mean, which is similar to the expected value of a probability distribution. The **mean** (the arithmetic average) of a set of numbers is the sum of the numbers, divided by the total number of numbers. We write the sum of n numbers $x_1, x_2, x_3, \ldots, x_n$ in a compact way using **summation notation,** also called **sigma notation.** With the Greek letter Σ (sigma), the sum

$$x_1 + x_2 + x_3 + \cdots + x_n$$

is written

$$x_1 + x_2 + x_3 + \cdots + x_n = \sum_{i=1}^{n} x_i.$$

In statistics, $\sum_{i=1}^{n} x_i$ is often abbreviated as just $\Sigma\, x$. The symbol $\overline{x}$ (read x-bar) is used to represent the mean of a sample.

Mean

The mean of the n numbers $x_1, x_2, x_3, \ldots, x_n$ is

$$\overline{x} = \frac{\Sigma\, x}{n}.$$

▶**EXAMPLE 3**   The number of bankruptcy petitions (in thousands) filed in the United States in the fiscal years 1985–1990 are given in the table below.* Find the mean number of bankruptcy petitions filed annually during this period.

Year	Petitions Field
1985	365
1986	478
1987	561
1988	594
1989	643
1990	725

Let $x_1 = 365, x_2 = 478$, and so on. Here, $n = 6$, since there are 6 numbers.

$$\overline{x} = \frac{365 + 478 + 561 + 594 + 643 + 725}{6}$$

$$\overline{x} = \frac{3366}{6} = 561$$

---

* From *The World Almanac and Book of Facts, 1992.*

3 Find the mean of the following list of sales at a boutique.

$25.12	$42.58
$76.19	$32
$81.11	$26.41
$19.76	$59.32
$71.18	$21.03

Answer:
$45.47

4 Find $\overline{x}$ for the following frequency distribution.

Value	Frequency
7	2
9	3
11	6
13	4
15	1
17	4

Answer:
$\overline{x} = 12.1$

The mean number of bankruptcy petitions filed during the given years is 561,000. ◀ **3**

The mean of data that have been arranged in a frequency distribution is found in a similar way. For example, suppose the following data are collected.

Value	Frequency
84	2
87	4
88	7
93	4
99	3
Total:	20

The value 84 appears twice, 87 four times, and so on. To find the mean, first add 84 two times, 87 four times, and so on; or get the same result faster by multiplying 84 by 2, 87 by 4, and so on, and then by adding the results. Dividing the sum by 20, the total of the frequencies, gives the mean.

$$\overline{x} = \frac{(84 \cdot 2) + (87 \cdot 4) + (88 \cdot 7) + (93 \cdot 4) + (99 \cdot 3)}{20}$$

$$= \frac{168 + 348 + 616 + 372 + 297}{20} = \frac{1801}{20}$$

$$\overline{x} = 90.05$$

▶ **EXAMPLE 4** Find the mean for the data shown in the following frequency distribution.

Value	Frequency	Value × Frequency
30	6	30 · 6 = 180
32	9	32 · 9 = 288
33	7	33 · 7 = 231
37	12	37 · 12 = 444
42	6	42 · 6 = 252
	Total: 40	Total: 1395

A new column, "Value × Frequency," has been added to the frequency distribution. Adding the products from this column gives a total of 1395. The total from the frequency column is 40. The mean is

$$\overline{x} = \frac{1395}{40} = 34.875. \quad ◀ \quad \boxed{4}$$

5 Find the mean of the following grouped frequency distribution.

Interval	Frequency
0–4	6
5–9	4
10–14	7
15–19	3

Answer:
8.75

Note The mean in Example 4 is found in the same way as was the expected value of a probability distribution in the previous chapter. In fact, the words *mean* and *expected value* are often used interchangeably.

The mean of grouped data is found in a similar way. For grouped data, intervals are used rather than single values. To calculate the mean, it is assumed that all these values are located at the midpoint of the interval. The letter x is used to represent the midpoints and f represents the frequencies, as shown in the next example.

▶**EXAMPLE 5** Find the mean for the following grouped frequency distribution.

Interval	Midpoint, x	Frequency, f	Product, xf
40–49	44.5	2	89
50–59	54.5	4	218
60–69	64.5	7	451.5
70–79	74.5	9	670.5
80–89	84.5	5	422.5
90–99	94.5	3	283.5
		Total: 30	Total: 2135

A column for the midpoint of each intervals has been added. The numbers in this column are found by adding the endpoints of each interval and dividing by 2. For the interval 40–49, the midpoint is $(40 + 49)/2 = 44.5$. The numbers in the product column on the right are found by multiplying frequencies and corresponding midpoints. Finally, we divide the total of the product column by the total of the frequency column to get

$$\overline{x} = \frac{2135}{30} = 71.2 \text{ (to the nearest tenth).} \quad ◀$$

The formula for the **mean of a grouped frequency distribution** is given below.

Mean of a Grouped Distribution

The mean of a distribution where x represents the midpoints, f the frequencies, and $n = \Sigma(f)$, is

$$\overline{x} = \frac{\Sigma(xf)}{n}.$$

Caution Note that in the formula above, n is the sum of the frequencies in the entire data set, not the number of intervals. **5**

Many calculators have built in programs that calculate the mean and other statistical measures. In most cases, you input the data and the frequencies and the calculator gives you $\bar{x}$, Σx, Σx^2, n, and measures of variation, discussed in the next section.

MEDIAN   Asked by a reporter to give the average height of the players on his team the Little League coach lined up his 15 players by increasing height. He picked out the player in the middle and pronounced this player to be of average height. This kind of average, called the **median,** is defined as the middle entry in a set of data arranged in either increasing or decreasing order. If there is an even number of entries, the median is defined to be the mean of the two center entries. The following table shows how to find the median for two sets of data: {8, 7, 4, 3, 1} and {2, 3, 4, 7, 9, 12}.

Odd Number of Entries	*Even Number of Entries*
8	2
7	3
Median = 4	4
3	7 $\Big\}$ Median $= \dfrac{4 + 7}{2} = 5.5$
1	9
	12

Note   As shown in the table above, when there are an even number of entries, the median is not always equal to one of the data entries.

The procedure for finding the median of a grouped frequency distribution is more complicated. We omit it here because it is more common to find the mean when working with grouped frequency distributions.

▶ **EXAMPLE 6**   Find the median for the following lists of numbers.
(a) 11,   12,   17,   20,   23,   28,   29
The median is the middle number, in this case 20. (Note that the numbers are aleady arranged in numerical order.) In this list, three numbers are smaller than 20 and three are larger.
(b) 15,   13,   7,   11,   19,   30,   39,   5,   10
First arrange the numbers in numerical order, from smallest to largest.

$$5, \quad 7, \quad 10, \quad 11, \quad 13, \quad 15, \quad 19, \quad 30, \quad 39$$

The middle number can now be determined: the median is 13.
(c) 47,   59,   32,   81,   74,   153
Write the numbers in numerical order.

$$32, \quad 47, \quad 59, \quad 74, \quad 81, \quad 153$$

6 Find the median for each of the following lists of numbers.

(a) 12, 15, 17, 19, 35, 42, 58

(b) 28, 68, 7, 15, 47, 59, 13, 74, 32, 25

Answers:

(a) 19

(b) 30

7 Find the mode for each of the following lists of numbers.

(a) 29, 35, 29, 18, 29, 56, 48

(b) 13, 17, 19, 20, 20, 13, 25, 27, 13, 20

(c) 512, 546, 318, 729, 854, 253

Answers:

(a) 29

(b) 13 and 20

(c) No mode

There are six numbers here; the median is the mean of the two middle numbers, or

$$\text{Median} = \frac{59 + 74}{2} = \frac{133}{2} = 66\frac{1}{2}. \quad \blacktriangleleft$$

Caution Remember, the data must be arranged in numerical order before locating the median. **6**

In some situations, the median gives a truer representative or typical element of the data than the mean. For example, suppose in an office there are 10 salespersons, 4 secretaries, the sales manager, and Ms. Daly, who owns the business. Their annual salaries are as follows: secretaries, $15,000 each; salespersons, $25,000 each; manager, $35,000; and owner, $200,000. The mean salary is

$$\bar{x} = \frac{(15,000)4 + (25,000)10 + 35,000 + 200,000}{16} = \$34,062.50.$$

However, since 14 people earn less than $34,062.50 and only 2 earn more, this does not seem very representative. The median salary is found by ranking the salaries by size: $15,000, $15,000, $15,000, $15,000, $25,000, $25,000, . . . , $200,000. There are 16 salaries (an even number) in the list, so the mean of the 8th and 9th entries will give the value of the median. The 8th and 9th entries are both $25,000, so the median is $25,000. In this example, the median is more representative of the distribution than the mean.

MODE Sue's scores on ten class quizzes include one 7, two 8's, six 9's and one 10. She claims that her average grade on quizzes is 9, because most of her scores are 9's. This kind of "average," found by selecting the most frequent entry, is called the **mode.**

▶ **EXAMPLE 7** Find the mode for each list of numbers.

(a) 57, 38, **55**, **55**, 80, 87, 98

The number 55 occurs more often than any other, so it is the mode. It is not necessary to place the numbers in numerical order when looking for the mode.

(b) 182, **185**, 183, **185**, **187**, **187**, 189

Both 185 and 187 occur twice. This list has *two* modes,

(c) 10,708, 11,519, 10,972, 17,546, 13,905, 12,182

No number occurs more than once. This list has no mode. ◀ **7**

The mode has the advantages of being easily found and not being influenced by data that are very large or very small compared to the rest of the data. It is often used in samples where the data to be "averaged" are not numerical. The major disadvantage to the mode is that there may be more than one, in case of ties, or there may be no mode at all when all entries occur with the same frequency.

The mean is the most commonly used measure of central tendency. Its advantages are that it is easy to compute, it takes all the data into consideration, and it is reliable—that is, repeated samples are likely to give very similar means. A disadvantage of the mean is that it is influenced by extreme values, as illustrated in the salary example above.

The median can be easy to compute and is influenced very little by extremes. Like the mode, the median can be found in situations where the data are not numerical. A disadvantage of the median is the need to rank the data in order; this can be tedious when the number of items is large.

10.1 EXERCISES

For Exercises 1–4, (a) *group the data as indicated;* (b) *prepare a frequency distribution with a column for intervals and frequencies;* (c) *construct a histogram;* (d) *construct a frequency polygon;*

1. Use six intervals, starting with 0–24.

74	133	4	127	20	30
103	27	139	118	138	121
149	132	64	141	130	76
42	50	95	56	65	104
4	140	12	88	119	64

2. Use seven intervals, starting with 30–39.

79	71	78	87	69	50	63	51	60	46
65	65	56	88	94	56	74	63	87	62
84	76	82	67	59	66	57	81	93	93
54	88	55	69	78	63	63	48	89	81
98	42	91	66	60	70	64	70	61	75
82	65	68	39	77	81	67	62	73	49
51	76	94	54	83	71	94	45	73	95
72	66	71	77	48	51	54	57	69	87

3. Use 70–74 as the first interval.

79	84	88	96	102	104	110	108	106	106
104	99	97	92	94	90	82	74	72	83
84	92	100	99	101	107	111	102	97	94	92

4. Use 140–149 as the first interval.

174	190	172	182	179	186	171	152	174	185
180	170	160	173	163	177	165	157	149	167
169	182	178	158	182	169	181	173	183	176
170	162	159	147	150	192	179	165	148	188

5. How does a frequency polygon differ from a histogram?

6. Discuss the advantages and disadvantages of the mean as a measure of central tendency.

Find the mean for each list of numbers. Round to the nearest tenth.

7. 8, 10, 16, 21, 25

8. 44, 41, 25, 36, 67, 51

9. 21,900, 22,850, 24,930, 29,710, 28,340, 40,000

10. 38,500, 39,720, 42,183, 21,982, 43,250

11. 9.4, 11.3, 10.5, 7.4, 9.1, 8.4, 9.7, 5.2, 1.1, 4.7

12. 30.1, 42.8, 91.6, 51.2, 88.3, 21.9, 43.7, 51.2

Find the mean for each of the following. Round to the nearest tenth.

13.

Value	Frequency
3	4
5	2
9	1
12	3

14.

Value	Frequency
9	3
12	5
15	1
18	1

15.

Value	Frequency
12	4
13	2
15	5
19	3
22	1
23	5

16.

Value	Frequency
25	1
26	2
29	5
30	4
32	3
33	5

Find the median for each of the following lists of numbers.

17. 12, 18, 32, 51, 58, 92, 106

18. 596, 604, 612, 683, 719

19. 100, 114, 125, 135, 150, 172

20. 1072, 1068, 1093, 1042, 1056, 1005, 1009

21. 28.4, 9.1, 3.4, 27.6, 59.8, 32.1, 47.6, 29.8

22. .6, .4, .9, 1.2, .3, 4.1, 2.2, .4, .7, .1

Find the mode or modes for each of the following lists of numbers.

23. 4, 9, 8, 6, 9, 2, 1, 3

24. 21, 32, 46, 32, 49, 32, 49

25. 74, 68, 68, 68, 75, 75, 74, 74, 70

26. 158, 162, 165, 162, 165, 157, 163

27. 6.8, 6.3, 6.3, 6.9, 6.7, 6.4, 6.1, 6.0

28. 12.75, 18.32, 19.41, 12.75, 18.30, 19.45, 18.33

29. When is the median the most appropriate measure of central tendency?

30. Under what circumstances would the mode be an appropriate measure of central tendency?

For grouped data, the modal class *is the interval containing the most data values. Give the mean and modal class for each of the following collections of grouped data.*

31. The distribution in Exercise 3.

32. The distribution in Exercise 4.

For each set of ungrouped data, (a) Find the mean, median, and mode. (b) Discuss which of the three measures best represents the data and why.

33. The weight gains of 10 experimental rats fed on a special diet were −1, 0, −3, 7, 1, 1, 5, 4, 1, 4.

34. A sample of 7 measurements of the thickness of a copper wire were .010, .010, .009, .008, .007, .009, .008.

35. The times in minutes that 12 patients spent in a doctor's office were 20, 15, 18, 22, 10, 12, 16, 17, 19, 21, 23, 13.

36. The scores on a 10-point botany quiz were 6, 6, 8, 10, 9, 7, 6, 5, 6, 8, 3.

37. **Management** A firm took a random sample of the number of days absent in a year for 40 employees, with results as shown below.

Days Absent	Frequency
0–2	23
3–5	11
6–8	5
9–11	0
12–14	1

Sketch a histogram and a frequency polygon for the data.

38. Natural Science The size of the home ranges (in square kilometers) of several pandas were surveyed over a year's time, with the following results.

Home Range	Frequency
.1–.5	11
.6–1.0	12
1.1–1.5	7
1.6–2.0	6
2.1–2.5	2
2.6–3.0	1
3.1–3.5	1

Sketch a histogram and frequency polygon for the data.

39. Social Science The histogram below shows the percent of the U.S. population in each age group in 1980.* What percent of the population was in each of the following age groups?
(a) 10–19 **(b)** 60–69
(c) What age group had the largest percent of the population?

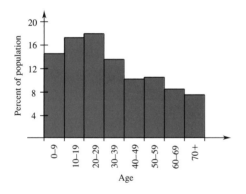

40. Social Science The histogram in the next column shows the estimated percent of the U.S. population in each age group in the year 2000.* What percent of the population is estimated to be in each of the following age groups then?
(a) 20–29 **(b)** 70+
(c) What age group will have the largest percent of the population?

(d) Compare the histogram in Exercise 39 with the histogram below. What seems to be true of the U.S. population?

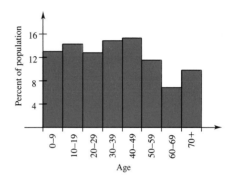

Management *U.S. wheat prices and production figures for a recent decade are given below.*

Year	Price ($ per bushel)	Production (millions of bushels)
1982	2.70	2200
1983	2.30	2000
1984	2.95	1750
1985	3.80	2200
1986	3.90	2400
1987	3.60	2800
1988	3.55	2800
1989	3.50	2450
1990	3.35	2600
1991	3.20	2750

Find the mean for each of the following.

41. Price per bushel of wheat

42. Wheat production

---

*Data from Census Bureau statistics.

43. Social Science The table shows the median age at first marriage in the U.S. over the last 50 years.

Year	Male	Female
1990	26.1	23.9
1980	24.7	22.0
1970	23.2	20.8
1960	22.8	20.3
1950	22.8	20.3
1940	24.3	21.5

(a) Find the mean of the median ages for males.
(b) Find the mean of the median ages for females.

44. Social Science The number of nations participating in the winter Olympic games, from 1968 to 1992, is given below.

Year	Nations Participating
1968	37
1972	37
1976	37
1980	37
1984	49
1988	57
1992	65

Find the following measures for the data.
(a) Mean   **(b)** Median   **(c)** Mode
(d) Which of these measures best represents the data? Explain your reasoning.

10.2 MEASURES OF VARIATION

The mean gives a measure of central tendency of a list of numbers, but tells nothing about the *spread* of the numbers in the list. For example, look at the following three samples.

I	3	5	6	3	3
II	4	4	4	4	4
III	10	1	0	0	9

Each of these three samples has a mean of 4, and yet they are quite different; the amount of dispersion or variation within the samples is different. Therefore, in addition to a measure of central tendency, another kind of measure is needed that describes how much the numbers vary.

The largest number in sample I is 6, while the smallest is 3, a difference of 3. In sample II this difference is 0; in sample III it is 10. The difference between the largest and smallest number in a sample is called the **range,** one example of a measure of variation. The range of sample I is 3, of sample II is 0, and of sample III is 10. The range has the advantage of being very easy to compute and gives a rough estimate of the variation among the data in the sample. However, it depends only on the two extremes and tells nothing about how the other data are distributed between the extremes.

▶**EXAMPLE 1**   Find the range for each list of numbers.
(a) 12,   27,   6,   19,   38,   9,   42,   15

1 Find the range for the numbers 159, 283, 490, 390, 375, 297.

Answer:
331

2 Find the deviations from the mean for each set of numbers.

(a) 19, 25, 36, 41, 52, 61

(b) 6, 9, 5, 11, 3, 2

Answers:

(a) Mean is 39; deviations are −20, −14, −3, 2, 13, 22

(b) Mean is 6; deviations from the mean are 0, 3, −1, 5, −3, −4

The highest number here is 42; the lowest is 6. The range is the difference of these numbers, or

$$42 - 6 = 36.$$

(b) 74, 112, 59, 88, 200, 73, 92, 175

$$\text{Range} = 200 - 59 = 141 \quad \blacktriangleleft \quad \boxed{1}$$

The most useful measure of variation is the *standard deviation*. Before defining it, however, we must find the **deviations from the mean,** the differences found by subtracting the mean from each number in a distribution.

▶ **EXAMPLE 2** Find the deviations from the mean for the numbers

$$32, \quad 41, \quad 47, \quad 53, \quad 57.$$

Adding these numbers and dividing by 5 gives a mean of 46. To find the deviations from the mean, subtract 46 from each number in the list. For example, the first deviation from the mean is $32 - 46 = -14$; the last is $57 - 46 = 11$.

Number	Deviation from Mean
32	−14
41	−5
47	1
53	7
57	11

To check your work, find the sum of these deviations. It should always equal 0. (The answer is always 0 because the positive and negative numbers cancel each other.) ◀ **2**

To find a measure of variation, we might be tempted to use the mean of the deviations. However, as mentioned above, this number is always 0, no matter how widely the data are dispersed. One way to solve this problem is to use absolute value and find the mean of the absolute values of the deviations from the mean.

$$\frac{|-14| + |-5| + |1| + |7| + |11|}{5} = \frac{38}{5} = 7.6$$

Absolute value is awkward to work with algebraically, however.

Another way to get a list of positive numbers is to square each deviation as shown in the table below and then find the mean.

Number	Deviation from Mean	Square of Deviation
32	−14	196
41	−5	25
47	1	1
53	7	49
57	11	121

To find the mean of the squared deviations of a sample, statisticians prefer to divide by $n - 1$, rather than n, to create what is known as an *unbiased estimator,* a concept that is beyond the scope of this text. (Note that, if n is large, it does not affect the result much.) For the distribution above, this gives

$$\frac{196 + 25 + 1 + 49 + 121}{5 - 1} = \frac{392}{4} = 98.$$

This number 98 is called the **variance** of the distribution. Because it is found by averaging a list of squares, the variance of a sample is represented by s^2.

For a sample of n numbers $x_1, x_2, x_3, \ldots, x_n$, with mean $\bar{x}$, the variance is

$$s^2 = \frac{\Sigma(x - \bar{x})^2}{n - 1}.$$

The following shortcut formula for the variance can be derived algebraically from the formula given above.

Variance

The variance of a sample of n numbers, $x_1, x_2, x_3, \ldots, x_n$, with mean $\bar{x}$, is

$$s^2 = \frac{\Sigma x^2 - n\bar{x}^2}{n - 1}.$$

To find the variance, we square the deviations from the mean, so the variance is in squared units. To return to the same units as the data, we use the *square root* of the variance, called the **standard deviation,** denoted s.

Standard Deviation

The standard deviation of the n numbers, $x_1, x_2, x_3, \ldots, x_n$, with mean $\overline{x}$, is

$$s = \sqrt{\frac{\Sigma(x - \overline{x})}{n - 1}} = \sqrt{\frac{\Sigma x^2 - n\overline{x}^2}{n - 1}}.$$

As its name indicates, the standard deviation is the most commonly used measure of variation. The standard deviation is a measure of the variation from the mean. The size of the standard deviation indicates how spread out the data are from the mean.

▶ **EXAMPLE 3** Find the standard deviation of the numbers

$$7, \quad 9, \quad 18, \quad 22, \quad 27, \quad 29, \quad 32, \quad 40.$$

The mean of the numbers is

$$\frac{7 + 9 + 18 + 22 + 27 + 29 + 32 + 40}{8} = 23.$$

Arrange the work in columns, as shown in the table.

Number	Square of the Number
7	49
9	81
18	324
22	484
27	729
29	841
32	1024
40	1600
	Total: 5132

The total of the second column gives $\Sigma x^2 = 5132$. Now find the quantity under the radical. The variance is

$$s^2 = \frac{\Sigma x^2 - n\overline{x}^2}{n - 1}$$

$$= \frac{5132 - 8(23)^2}{8 - 1}$$

$$= 128.6 \quad \text{(rounded)},$$

3 Find the standard deviation of each set of numbers. The deviations from the mean were found in Problem 2 at the side.

(a) 19, 25, 36, 41, 52, 61

(b) 6, 9, 5, 11, 3, 2

Answers:

(a) 15.9

(b) 3.5

and the standard deviation is

$$s \approx \sqrt{128.6} \approx 11.3 \quad \blacktriangleleft$$

Caution Be careful to divide by $n - 1$, not n, when calculating the standard deviation of a sample. **3**

One way to interpret the standard deviation uses the fact that, for many populations, most of the data are within three standard deviations of the mean. (See Section 10.3.) This implies that, in Example 3, most of the population from which this sample is taken is between

$$\bar{x} - 3s = 23 - 3(11.3) = -10.9$$

and

$$\bar{x} + 3s = 23 + 3(11.3) = 56.9.$$

This has important implications for quality control. If the sample in Example 3 represents measurements of a product that the manufacturer wants to be between 5 and 45, the standard deviation is too large, even though all the numbers are within these bounds.

For data in a grouped frequency distribution, a slightly different formula for the standard deviation is used.

Standard Deviation for a Grouped Distribution

The standard deviation for a distribution with mean $\bar{x}$, where x is an interval midpoint with frequency f, and $n = \Sigma f$, is

$$s = \sqrt{\frac{\Sigma f x^2 - n\bar{x}^2}{n - 1}}.$$

The formula indicates that the product fx^2 is to be found for each interval. Then these products are summed, n times the square of the mean is subtracted, and the difference is divided by one less than the total frequency; that is, by $n - 1$. The square root of this result is s, the standard deviation. The standard deviation found by this formula may (probably will) differ somewhat from the standard deviation found from the original data.

Caution In calculating the standard deviation for either a grouped or ungrouped distribution, using a rounded value for the mean or variance may produce an inaccurate value.

▶ **EXAMPLE 4** Find s for the grouped data of Example 5, Section 10.1.
Begin by including columns for x^2 and fx^2 in the table.

4 Find the standard deviation for the grouped data that follows. (Hint: $\bar{x} = 28.5$)

Value	Frequency
20–24	3
25–29	2
30–34	4
35–39	1

Answer:
5.3

Interval	f	x	x^2	fx^2
40–49	2	44.5	1980.25	3,960.50
50–59	4	54.5	2970.25	11,881.00
60–69	7	64.5	4160.25	29,121.75
70–79	9	74.5	5550.25	49,952.25
80–89	5	84.5	7140.25	35,701.25
90–99	3	94.5	8930.25	26,790.75
Total: 30				Total: 157,407.50

Recall from Section 10.1 that $\bar{x} = 71.2$. Use the formula above with $n = 30$ to find s.

$$s = \sqrt{\frac{\Sigma fx^2 - n\bar{x}^2}{n - 1}}$$

$$= \sqrt{\frac{157,407.50 - 30(71.2)^2}{30 - 1}}$$

$$\approx 13.5 \quad \blacktriangleleft \quad \boxed{4}$$

Note A calculator is almost a necessity for finding a standard deviation. A good procedure to follow is to first calculate $\bar{x}$. Then for each x, square that number, then multiply the result by the appropriate frequency. If your calculator has a key that accumulates a sum, use it to accumulate the total in the last column of the table.*

10.2 EXERCISES

A calculator will be helpful with many of the exercises in this set.

1. How are the variance and the standard deviation related?

2. Why can't we use the sum of the deviations from the mean as a measure of dispersion of a distribution?

Find the range and standard deviation for each of the following sets of numbers. (See Examples 1 and 3.)

3. 6, 8, 9, 10, 12

4. 12, 15, 19, 23, 26

5. 7, 6, 12, 14, 18, 15

6. 4, 3, 8, 9, 7, 10, 1

7. 42, 38, 29, 74, 82, 71, 35

8. 122, 132, 141, 158, 162, 169, 180

9. 241, 248, 251, 257, 252, 287

10. 51, 58, 62, 64, 67, 71, 74, 78, 82, 93

*Some calculators are equipped with statistical keys that compute the variance and standard deviation of a list of numbers directly, without your having to do any arithmetic or taking square roots. Some of these calculators use $n - 1$ and others use n for these computations. Others may give you a choice. Check the instruction book before using a statistical calculator for the exercises.

Find the standard deviation for the following grouped data. (See Example 4.)

11. (From Exercise 1, Section 10.1)

College Units	Frequency
0–24	4
25–49	3
50–74	6
75–99	3
100–124	5
125–149	9

12. (From Exercise 2, Section 10.1)

Scores	Frequency
30–39	1
40–49	6
50–59	13
60–69	22
70–79	17
80–89	13
90–99	8

13. The data of Exercise 3, Section 10.1

14. The data of Exercise 4, Section 10.1

An application of standard deviation is given by **Chebyshev's theorem.** *(PL. Chebyshev was a Russian mathematician who lived from 1821 to 1894). This theorem applies to any distribution of numbers. It states:*

For any distribution of numbers, at least $1 - 1/k^2$ of the numbers lie within k standard deviations of the mean.

Example *For any distribution, at least*

$$1 - \frac{1}{3^2} = 1 - \frac{1}{9} = \frac{8}{9}$$

of the numbers lie within 3 standard deviations of the mean.
Find the fraction of all the numbers of a data set lying within the following numbers of standard deviations from the mean.

15. 2

16. 4

17. 5

In a certain distribution of numbers, the mean is 50 with a standard deviation of 6. Use Chebyshev's theorem to tell what percent of the numbers are

18. between 32 and 68;

19. between 26 and 74;

20. less than 38 or more than 62;

21. less than 32 or more than 68;

22. less than 26 or more than 74.

---

23. Management The Britelite Company conducted tests on the life of its light bulbs and those of a competitor (Brand X) with the following results for samples of 10 bulbs of each brand.

	Hours of Use (in 100s)									
Britelite	20	22	22	25	26	27	27	28	30	35
Brand X	15	18	19	23	25	25	28	30	34	38

Compute the mean and standard deviation for each sample. Compare the means and standard deviations of the two brands and then answer the questions below.

(a) Which bulbs have a more uniform life in hours?

(b) Which bulbs have the highest average life in hours?

24. Management The weekly wages of the six employees of Harold's Hardware Store are $300, $320, $380, $420, $500, and $2000.

(a) Find the mean and standard deviation of this distribution.

(b) How many of the employees earn within one standard deviation of the mean? How many earn within two standard deviations of the mean?

25. Social Science   The number of unemployed workers in the United States in recent years (in millions) is given below.*

Year	Number Unemployed
1984	8.54
1985	8.31
1986	8.24
1987	7.43
1988	6.70
1989	6.53
1990	6.87

(a) Find the mean number unemployed (in millions) in this period. Which year has unemployment closest to the mean?

(b) Find the standard deviation for the data.

(c) In how many of these years is unemployment within 1 standard deviation of the mean?

26. Social Science   In an article comparing national mathematics examinations in the U.S. and some European countries, a researcher found the results shown in the histogram for the number of minutes allowed for each open-ended question.**

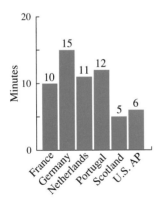

*Excerpts from "Employment and Unemployment in the U.S." from *The World Almanac and Book of Facts 1992.* Copyright © 1991 by Pharos Books. Reprinted by permission of World Almanac, an imprint of Pharos Books, a Scripps Howard Company.

** Information from article comparing national mathematics examinations in the U.S. and some European countries as appeared in FOCUS, June 1993. Reprinted by permission of the Mathematical Association of America.

(a) Find the mean and standard deviation. (*Hint:* Do not use the rounded value of the mean to find the standard deviation.)

(b) How many standard deviations from the mean is the largest number of minutes?

(c) How many standard deviations from the mean is the U.S. AP examination?

27. Social Science   For all questions, the researcher in Exercise 26 found the following results for Scotland and the United States, the countries with the least rigorous examinations.

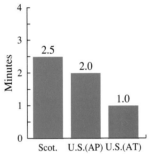

(a) Find the mean and standard deviation.

(b) How many standard deviations from the mean is the AT (achievement test)?

(c) How many standard deviations from the mean is the Scottish test?

28. Social Science   Recently Germany has endured rioting and disruption because of the increasing numbers of immigrants. The nine top countries of origin of German immigrants are shown in the histogram.

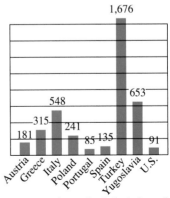

(a) Find the mean and standard deviation of this data.

(b) How many standard deviations from the mean is the largest number of immigrants? the smallest? Which country of origin is closest to the mean?

29. Social Science The numbers of immigrants to the United States (in thousands) from selected parts of the world in 1990 are shown in the table.*

Region	Immigrants
Europe	124.0
Asia	321.9
Canada	24.6
Mexico	680.2
Other America	346.2

(a) Find the mean number of immigrants from these regions. Which region produces the number of immigrants closest to the mean?

(b) Find the standard deviation. Are any of the data more than 2 standard deviations from the mean? If so, which ones? Are any of the data more than 1 standard deviation from the mean? If so, which ones?

30. Management The Quaker Oats Company conducted a survey to determine if a proposed premium, to be included in their cereal, was appealing enough to generate new sales.** Four cities were used as test markets, where the cereal was distributed with the premium, and four cities as control markets, where the cereal was distributed without the premium. The eight cities were chosen on the basis of their similarity in terms of population, per capita income, and total cereal purchase volume. The results were as follows.

		Percent Change in Average Market Shares per Month
Test Cities	1	+18
	2	+15
	3	+7
	4	+10
Control Cities	1	+1
	2	−8
	3	−5
	4	0

(a) Find the mean of the change in market share for the four test cities.

(b) Find the mean of the change in market share for the four control cities.

(c) Find the standard deviation of the change in market share for the test cities.

(d) Find the standard deviation of the change in market share for the control cities.

(e) Find the difference between the mean of part (a) and the mean of part (b). This represents the estimate of the percent change in sales due to the premium.

(f) The two standard deviations from part (c) and part (d) were used to calculate an "error" of ± 7.95 for the estimate in part (e). With this amount of error what is the smallest and largest estimate of the increase in sales? (Hint: use the answer to part (e).)

On the basis of the results of Exercise 30, the company decided to mass produce the premium and distribute it nationally.

▷ **31. Management** The following table gives 10 samples, of three measurements, made during a production run.

				Sample Number					
1	2	3	4	5	6	7	8	9	10
2	3	−2	−3	−1	3	0	−1	2	0
−2	−1	0	1	2	2	1	2	3	0
1	4	1	2	4	2	2	3	2	2

(a) Find the mean $\bar{x}$ for each sample of three measurements.

(b) Find the standard deviation s for each sample of three measurements.

(c) Find the mean $\bar{X}$ of the sample means.

(d) Find the mean $\bar{s}$ of the sample standard deviations.

(e) The upper and lower control limits of the sample means here are $\bar{X} \pm 1.954\bar{s}$. Find these limits. If any of the measurements are outside these limits, the process is out of control. Decide if this production process is out of control.

32. Discuss what the standard deviation tells us about a distribution.

---

*Table reprinted with permission from *The World Almanac and Book of Facts 1992*. Copyright © 1991. All rights reserved. The World Almanac is an imprint of Funk and Wagnalls Corporation.

**This example was supplied by Jeffery S. Berman, Senior Analyst, Marketing Information, Quaker Oats Company.

Use a computer or a calculator to solve the problems in Exercises 33–35.

33. Natural Science Twenty-five laboratory rats, used in an experiment to test the food value of a new product, made the following weight gains in grams.

5.25	5.03	4.90	4.97	5.03
5.12	5.08	5.15	5.20	4.95
4.90	5.00	5.13	5.18	5.18
5.22	5.04	5.09	5.10	5.11
5.23	5.22	5.19	4.99	4.93

Find the mean gain and the standard deviation of the gains.

34. Management An assembly-line machine turns out washers with the following thicknesses (in millimeters).

1.20	1.01	1.25	2.20	2.58	2.19	1.29	1.15
2.05	1.46	1.90	2.03	2.13	1.86	1.65	2.27
1.64	2.19	2.25	2.08	1.96	1.83	1.17	2.24

Find the mean and standard deviation of these thicknesses.

35. Natural Science A medical laboratory tested 21 samples of human blood for acidity on the pH scale, with the following results.

7.1	7.5	7.3	7.4	7.6	7.2	7.3
7.4	7.5	7.3	7.2	7.4	7.3	7.5
7.5	7.4	7.4	7.1	7.3	7.4	7.4

Find the mean and standard deviation.

Social Science *The reading scores of a second-grade class given individualized instruction are shown below. The table also shows the reading scores of a second-grade class given traditional instruction in the same school.*

Scores	Individualized Instruction	Traditional Instruction
50–59	2	5
60–69	4	8
70–79	7	8
80–89	9	7
90–99	8	6

36. Find the mean and standard deviation for the individualized instruction scores.

37. Find the mean and standard deviation for the traditional instruction scores.

38. Discuss a possible interpretation of the differences in the means and the standard deviation in Exercises 36 and 37.

10.3 NORMAL DISTRIBUTIONS

The general idea of a probability distribution was first discussed in Section 9.5. In this section, we consider an important specific type of probability distribution.

Figure 10.2(a) on the next page shows a histogram and frequency polygon for the bank transaction example in Section 9.5. The heights of the bars are the probabilities. The transaction times in the example were given to the nearest minute. Theoretically, at least, they could have been timed to the nearest tenth of a minute, or hundredth of a minute, or even more precisely. In each case, a histogram and frequency polygon could be drawn. If the times are measured with smaller and smaller units, there are more bars in the histogram, and the frequency polygon begins to look more and more like the curve in Figure 10.2(b) instead of a polygon. Actually it is possible for the transaction times to take on any real number value greater than 0. A distribution in which the outcomes can take any real number value within some interval is a **continuous distribution.** The graph of a continuous distribution is a curve.

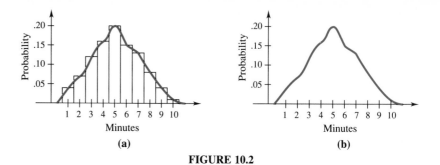

FIGURE 10.2

The distribution of heights (in inches) of college women is another example of a continuous distribution, since these heights include infinitely many possible measurements, such as 53, 58.5, 66.3, 72.666, . . . , and so on. Figure 10.3 shows the continuous distribution of heights of college women. Here the most frequent heights occur near the center of the interval shown.

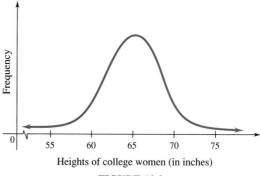

Heights of college women (in inches)

FIGURE 10.3

Another continuous curve, which approximates the distribution of yearly incomes in the United States, is given in Figure 10.4. The graph shows that the most frequent incomes are grouped near the low end of the interval. This kind of distribution, where the peak is not at the center, is called **skewed.**

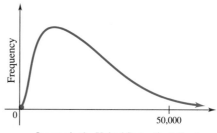

Income in the United States (in dollars)

FIGURE 10.4

Many natural and social phenomena produce continuous probability distributions whose graphs are bell-shaped curves, such as those shown in Figure 10.5. Such distributions are called **normal distributions** and their graphs are called **normal curves.** Examples of normal distributions are the heights of college women and the errors made in filling 1-pound cereal boxes. We use the Greek letters μ (mu) to denote the mean and σ (sigma) to denote the standard deviation of a normal distribution.

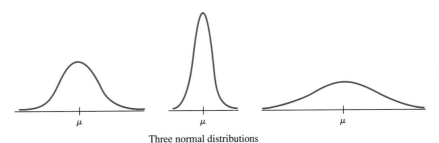

Three normal distributions

FIGURE 10.5

There are many normal distributions. Some of the corresponding normal curves are tall and thin and others short and wide, as shown in Figure 10.5. But every normal curve has the following properties.

1. Its peak occurs directly above the mean μ.
2. The curve is symmetric about the vertical line through the mean (that is, if you fold the page along this line, the left half of the graph will fit exactly on the right half).
3. The curve never touches the x-axis—it extends indefinitely in both directions.
4. The area under the curve (and above the horizontal axis) is 1. (As is shown in calculus, this is a consequence of the fact that the sum of the probabilities in any distribution is 1.)

It can be shown that a normal distribution is completely determined by its mean μ and standard deviation σ.* A small standard deviation leads to a tall, narrow curve like the one in the center of Figure 10.5, because most of the data are close to the mean. A large standard deviation means the data are very spread out, producing a flat wide curve like the one on the right in Figure 10.5.

---

*As is shown in more advanced courses, its graph is the graph of the function

$$f(x) = \frac{1}{\sigma\sqrt{2\pi}}e^{-(x-\mu)^2/(2\sigma^2)},$$

where $e \approx 2.71828$ is the real number discussed in Section 4.2.

Since the area under a normal curve is 1, parts of this area can be used to determine certain probabilities. For instance, Figure 10.6(a) is the probability distribution of the annual rainfall in a certain region. The probability that the annual rainfall will be between 25 and 35 inches is the area under the curve from 25 to 35. The general case, shown in Figure 10.6(b), can be stated as follows.

The area of the shaded region under the normal curve from a to b is the probability that an observed data value will be between a and b.

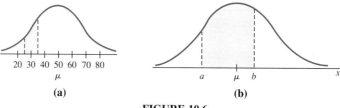

(a)   (b)

FIGURE 10.6

To use normal curves effectively we must be able to calculate areas under portions of these curves. These calculations have already been done for the normal curve with mean $\mu = 0$ and standard deviation $\sigma = 1$ (which is called the **standard normal curve**) and are available in Table 2 at the back of the book. The following examples demonstrate how to use Table 2 to find such areas. Later we shall see how the standard normal curve may be used to find areas under any normal curve.

The horizontal axis of the standard normal curve is usually labeled z. Since the standard deviation of the standard normal curve is 1, the numbers along the horizontal axis (the z-values) measure the number of standard deviations above or below the mean $z = 0$.

▶ **EXAMPLE 1**   Find the following areas under the standard normal curve.
(a) The area between $z = 0$ and $z = 1$, the shaded region in Figure 10.7 is 34.13% of the total area under the normal curve.

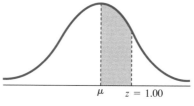

FIGURE 10.7

Find the entry 1 in the z-column of Table 2. The entry next to it in the A-column is .3413, which means that the area between $z = 0$ and $z = 1$ is .3413. Since the total area under the curve is 1, the shaded area in Figure 10.7 is 34.13% of the total area under the normal curve.

1 Find the percent of area between the mean and

(a) $z = 1.51$;

(b) $z = -2.04$.

(c) Find the percent of area in the shaded region.

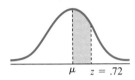

Answers:

(a) 43.45%

(b) 47.93%

(c) 26.42%

(b) The area between $z = -2.43$ and $z = 0$.

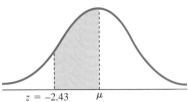

FIGURE 10.8

Table 2 lists only positive values of z. But the normal curve is symmetric around the mean $z = 0$, so the area between $z = 0$ and $z = -2.43$ is the same as the area between $z = 0$ and $z = 2.43$. Find 2.43 in the z-column of Table 2. The entry next to it in the A-column shows that the area is .4925. Hence, the shaded area in Figure 10.8 is 49.25% of the total area under the curve. ◀ **1**

▶ **EXAMPLE 2** Find the percent of the total area for the following areas under the standard normal curve.

(a) The area between 1.41 standard deviations *below* the mean and 2.25 standard deviations *above* the mean (that is, between $z = -1.41$ and $z = 2.25$)

First, draw a sketch showing the desired area, as in Figure 10.9. From Table 2, the area between the mean and 1.41 standard deviations below the mean is .4207. Also, the area from the mean to 2.25 standard deviations above the mean is .4878. As the figure shows, the total desired area can be found by *adding* these numbers.

$$\begin{array}{r} .4207 \\ +\ .4878 \\ \hline .9085 \end{array}$$

The shaded area in Figure 10.9 represents 90.85% of the total area under the normal curve.

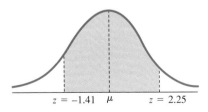

FIGURE 10.9

2 Find the following standard normal curve areas as percents of the total area.

(a) Between .31 standard deviations below the mean and 1.01 standard deviations above the mean

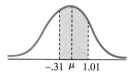

−.31 μ 1.01

(b) Between .38 and 1.98 standard deviations below the mean

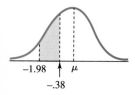

−1.98 μ
−.38

(c) To the right of 1.49 standard deviations above the mean

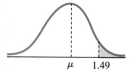

μ 1.49

(d) What percent of the area is within 1 standard deviation of the mean? within 2 standard deviations of the mean? within 3 standard deviations of the mean? What can you conclude from the last answer?

Answers:

(a) 46.55%

(b) 32.82%

(c) 6.81%

(d) 68.3%, 95.47%, 99.7%. Almost all the data lies within 3 standard deviations of the mean.

(b) The area between .58 standard deviations above the mean and 1.94 standard deviations above the mean

Figure 10.10 shows the desired area. The area between the mean and .58 standard deviations above the mean is .2190. The area between the mean and 1.94 standard deviations above the mean is .4738. As the figure shows, the desired area is found by *subtracting* the two areas.

$$\begin{array}{r} .4738 \\ - \ .2190 \\ \hline .2548 \end{array}$$

The shaded area of Figure 10.10 represents 25.48% of the total area under the normal curve.

(c) The area to the right of 2.09 standard deviations above the mean

The total area under a normal curve is 1. Thus, the total area to the right of the mean is 1/2, or .5000. From Table 2, the area from the mean to 2.09 standard deviations above the mean is .4817. The area to the right of 2.09 standard deviations is found by substracting .4817 from .5000.

$$\begin{array}{r} .5000 \\ - \ .4817 \\ \hline .0183 \end{array}$$

A total of 1.83% of the total area is to the right of 2.09 standard deviations above the mean. Figure 10.11 shows the desired area. ◀ **2**

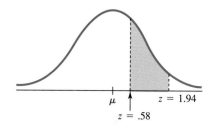

μ $z = 1.94$
$z = .58$

FIGURE 10.10

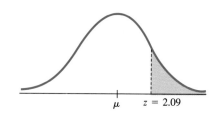

μ $z = 2.09$

FIGURE 10.11

The key to finding areas under *any* normal curve is to express each number x on the horizontal axis in terms of standard deviations above or below the mean. The **z-score** for x is the number of standard deviations that x lies from the mean (positive if x is above the mean, negative if x is below the mean).

▶**EXAMPLE 3**   If a normal distribution has mean 50 and standard deviation 4, find the following z-scores.

(a) The z-score for $x = 46$

Since 46 is 4 units below 50 and the standard deviation is 4, 46 is 1 standard deviation below the mean. So, its z-score is −1.

3 Find each z-score using the information in Example 3.

(a) $x = 36$

(b) $x = 55$

Answers:

(a) -3.5

(b) 1.25

(b) The z-score for $x = 60$

The z-score is 2.5 because 60 is 10 units above the mean (since $60 - 50 = 10$) and 10 units is 2.5 standard deviations (since $10/4 = 2.5$). ◀ **3**

In Example 3(b) we found the z-score by taking the difference between 60 and the mean and dividing this difference by the standard deviation. The same procedure works in the general case.

> If a normal distribution has mean μ and standard deviation σ, then the z-score for the number x is
>
> $$z = \frac{x - \mu}{\sigma}.$$

The importance of z-scores is the following fact, whose proof is omitted.

Area Under a Normal Curve

> The area under a normal curve between $x = a$ and $x = b$ is the same as the area under the standard normal curve between the z-score for a and the z-score for b.

Therefore, by converting to z-scores and using Table 2 for the standard normal curve, we can find areas under any normal curve. Since these areas are probabilities (as explained earlier), we can now handle a variety of applications.

FOR GRAPHERS

Graphers, computer programs, and CAS programs (such as DERIVE) can be used to find areas under the normal curve and hence, probabilities. The equation of the standard normal curve, with $\mu = 0$ and $\sigma = 1$, is $f(x) = (1/\sqrt{2\pi})e^{-x^2/2}$. A good approximation of the area under this curve (and above $y = 0$) can be found by using the x-interval $[-4, 4]$. See Chapter 13 for more information on finding such areas.

▶ **EXAMPLE 4** Dixie Office Supplies finds that its sales force drives an average of 1200 miles per month per person, with a standard deviation of 150 miles. Assume that the number of miles driven by a salesperson is closely approximated by a normal distribution.

(a) Find the probability that a salesperson drives between 1200 and 1600 miles per month.

4 The heights of female sophomore college students at one school have $\mu = 172$ centimeters, with $\sigma = 10$ centimeters. Find the probability that the height of such a student is

(a) between 172 cm and 185 cm;

(b) between 160 cm and 180 cm;

(c) less than 165 cm.

Answers:

(a) .4032

(b) .6730

(c) .2420

Here $\mu = 1200$ and $\sigma = 150$, and we must find the area under the normal distribution curve between $x = 1200$ and $x = 1600$. We begin by finding the z-score for $x = 1200$.*

$$z = \frac{x - \mu}{\sigma} = \frac{1200 - 1200}{150} = \frac{0}{150} = 0.$$

The z-score for $x = 1600$ is

$$z = \frac{x - \mu}{\sigma} = \frac{1600 - 1200}{150} = \frac{400}{150} = 2.67.$$

So the area under the curve from $x = 1200$ to $x = 1600$ is the same as the area under the standard normal curve from $z = 0$ to $z = 2.67$, as indicated in Figure 10.12. Table 2 shows that this area is .4962. Therefore, the probability that a salesperson drives between 1200 and 1600 miles per month is .4962.

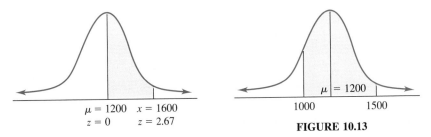

$\mu = 1200$ $x = 1600$
$z = 0$ $z = 2.67$

FIGURE 10.12

$\mu = 1200$
1000 1500

FIGURE 10.13

(b) Find the probability that a salesperson drives between 1000 and 1500 miles per month.

As shown in Figure 10.13, z-scores for both $x = 1000$ and $x = 1500$ are needed.

For $x = 1000$,

$$z = \frac{1000 - 1200}{150}$$

$$= \frac{-200}{150}$$

$$= -1.33$$

For $x = 1500$,

$$z = \frac{1500 - 1200}{150}$$

$$= \frac{300}{150}$$

$$= 2.00$$

From the table $z = 1.33$ leads to an area of .4082, while $z = 2.00$ corresponds to .4773. A total of $.4082 + .4773 = .8855$, or 88.55%, of all drivers travel between 1000 and 1500 miles per month. From this, the probability that a driver travels between 1000 and 1500 miles per month is .8855. ◀ **4**

---

*All z-scores here are rounded to two decimal places.

▶**EXAMPLE 5**   A tire store finds that the tread life of its tires is normally distributed, with a mean of 26,640 miles and a standard deviation of 4000 miles. The store sold 9000 tires this month. How many of them can be expected to last more than 35,000 miles?

Here $\mu = 26,640$ and $\sigma = 4000$. The probability that a tire will last more than 35,000 miles is the area under the normal curve to the right of $x = 35,000$. The $z =$ score for $x = 35,000$ is

$$z = \frac{x - \mu}{\sigma} = \frac{35,000 - 26,640}{4000} = \frac{8360}{4000} = 2.09.$$

Example 2(c) and Figure 10.11 show that the area to the right of $z = 2.09$ is .0183, which is 1.83% of the total area under the curve. Therefore, 1.83% of the tires can be expected to last more than 35,000 miles. Thus,

$$1.83\% \text{ of } 9000 = .0183 \cdot 9000 = 164.7,$$

or approximately 165 tires can be expected to last more than 35,000 miles. ◀

10.3 EXERCISES

1. In your own words, discuss the characteristics of a normal curve, mentioning the area under the curve, the location of the peak, and the end-behavior.

2. Explain how z-scores are found for normal distributions where
$$\mu \neq 0 \quad \text{or} \quad \sigma \neq 1.$$

3. Explain how the standard normal curve is used to find probabilities for other normal distributions.

Find the percent of the area under a normal curve between the mean and the following number of standard deviations from the mean. (See Example 3.)

4. 1.75   **5.** .23   **6.** −.43   **7.** −2.1   **8.** 3.1

Find the percent of the total area under the standard normal curve between the following z-scores. (See Examples 1 and 2.)

9. $z = 1.41$ and $z = 2.83$
10. $z = .64$ and $z = 2.11$
11. $z = -2.48$ and $z = -.05$
12. $z = -1.74$ and $z = -1.02$
13. $z = -3.11$ and $z = 1.44$
14. $z = -2.94$ and $z = -.43$

Find a z-score satisfying each of the following conditions. (Hint: use Table 2 backwards.)

15. 5% of the total area is to the right of z.

16. 1% of the total area is to the left of z.

17. 15% of the total area is to the left of z.

18. 25% of the total area is to the right of z.

Assume the following distributions are all normal, and use the areas under the normal curve given in Table 2 to answer the questions. (See Example 4.)

19. Management   According to the label, a regular can of Campbell's soup holds an average of 305 grams with a standard deviation of 4.2 grams. What is the probability that a can will be sold that holds more than 306 grams.?

20. **Management**   A jar of Adams Old Fashioned Peanut Butter contains 453 grams with a standard deviation of 10.1 grams. Find the probability that one of these jars contains less than 450 grams.

21. **Management**   A General Electric soft white 3-way bulb has an average life of 1200 hours with a standard deviation of 50 hours. Find the probability that the life of one of these bulbs will be between 1150 and 1300 hours.

22. **Management**   A 100-watt light bulb has an average brightness of 1640 lumens with a standard deviation of 62 lumens. What is the probability that a 100-watt bulb will have a brightness between 1600 and 1700 lumens?

23. **Social Science**   The scores on a standardized test in a suburban high school have a mean of 80 with a standard deviation of 12. What is the probability that a student will have a score less than 60?

24. **Social Science**   The average time to complete the test in Exercise 23 is 40 minutes with a standard deviation of 5.2 minutes. Find the probability that a student will require more than 50 minutes to finish the test.

25. A 120-minute video tape has a standard deviation of .7 minutes. Find the probability that a 118-minute movie will not be entirely copied.

26. **Natural Science**   The distribution of low temperatures for a city has an average of 44° with a standard deviation of 6.7°. What is the probability that the low temperature will be above 55°?

Social Science   *New studies by Federal Highway Administration traffic engineers suggest that speed limits on many thoroughfares are set arbitrarily and often are artificially low. According to traffic engineers, the ideal limit should be the "85th percentile speed." This means the speed at or below which 85 percent of the traffic moves. Assuming speeds are normally distributed, find the 85th percentile speed for roads with the following conditions.*

27. The mean speed is 50 mph with a standard deviation of 10 mph.

28. The mean speed is 30 mph with a standard deviation of 5 mph.

Social Science   *One professor uses the following system for assigning letter grades in a course.*

Grade	Total Points
A	Greater than $\mu + \frac{3}{2}\sigma$
B	$\mu + \frac{1}{2}\sigma$ to $\mu + \frac{3}{2}\sigma$
C	$\mu - \frac{1}{2}\sigma$ to $\mu + \frac{1}{2}\sigma$
D	$\mu - \frac{3}{2}\sigma$ to $\mu - \frac{1}{2}\sigma$
F	Below $\mu - \frac{3}{2}\sigma$

What percent of the students receive the following grades?

29. A   **30.** B   **31.** C

32. Do you think the system in Exercise 31 would be more likely to be fair in a large freshmen class in psychology or in a graduate seminar of five students? Why?

Natural Science   *In nutrition, the recommended daily allowance of vitamins is a number set by the government as a guide to an individual's daily vitamin intake. Actually, vitamin needs vary drastically from person to person, but the needs are very closely approximated by a normal curve. To calculate the recommended daily allowance, the government first finds the average need for vitamins among people in the population and then the standard deviation. The recommended daily allowance is defined as the mean plus 2.5 times the standard deviation.*

33. What percentage of the population will receive adequate amounts of vitamins under this plan?

Find the recommended daily allowance for the following vitamins.

34. Mean = 1800 units, standard deviation = 140 units

35. Mean = 159 units, standard deviation = 12 units

36. Mean = 1200 units, standard deviation = 92 units

Social Science   *The mean performance score of a large group of fifth-grade students on a math achievement test is 88. The scores are known to be normally distributed. What percent of the students had scores as follows?*

37. More than 1 standard deviation above the mean

38. More than 2 standard deviations above the mean

Social Science *A teacher gives a test to a large group of students. The results are closely approximated by a normal curve. The mean is 74, with a standard deviation of 6. The teacher wishes to give A's to the top 8% of the students and F's to the bottom 8%. A grade of B is given to the next 15%, with D's given similarly. All other students get C's. Find the bottom cutoff (rounded to the nearest whole number) for the following grades. (Hint: use Table 2 backwards.)*

39. A   **40.** B

41. C   **42.** D

Social Science *Studies have shown that women are charged an average of $500 more than men for cars.* Assume a normal distribution of overcharges with a mean of $500 and a standard deviation of $60. Find the probability of a woman's paying the following additional amounts for a car.*

43. Less than $500

44. At least $600

45. Between $400 and $600

Social Science *Women earn an average of $.74 for every $1 earned by a man, a difference of $.26. Assume differences in pay are normally distributed with a mean of $.26 and a standard deviation of $.08. Find the probability that a woman earns the following amounts less than a man in the same job.*

46. Between $.20 and $.30

47. More than $.19

48. At most $.26

Management *At a ranch in the Sacramento valley, the tomato crop yields an average of 25 tons per acre with a standard deviation of 2.2 tons per acre. Find the probability of each of the following yields in tons per acre.*

49. At least 24

50. At most 25.7

51. Between 24.5 and 25.5

Management *According to the manufacturer, a certain automobile averages 28.2 miles per gallon of gasoline with a standard deviation of 2.1 miles per gallon. Find the probability of each of the following miles per gallon.*

52. Between 27 and 29

53. At least 28

54. At most 29

 Use a graphing calculator or a computer to find the following probabilities by finding the comparable area under a standard normal curve.

55. $P(1.372 \le z \le 2.548)$

56. $P(-2.751 \le z \le 1.693)$

57. $P(z > -2.476)$

58. $P(z < 1.692)$

Use a graphing calculator or a computer to find the following probabilities for a distribution with a mean of 35.693 and a standard deviation of 7.104.

59. $P(12.275 < x < 28.432)$

60. $P(x > 38.913)$

61. $P(x < 17.462)$

62. $P(17.462 \le x \le 53.106)$

---

* "From repair shops to cleaners, women pay more," by Bob Dart as appeared in *The Chicago Tribune*, May 27, 1993. Reprinted by permission of the author.

10.4 BINOMIAL DISTRIBUTIONS

Many practical experiments have only two possible outcomes, *success* or *failure.* Such experiments are called *binomial experiments* or *Bernoulli trials* and were first studied in Section 9.3. Examples of binomial experiments include flipping a coin (with heads being a success, for instance, and tails a failure) or testing TVs coming off the assembly line to see whether or not they are defective.

A **binomial distribution** is a probability distribution that satisfies the following conditions.

1. The experiment is a series of repeated independent trials.
2. There are only two possible outcomes for each trial, success or failure.
3. The probability of success is constant from trial to trial.

For example, suppose a fair die is tossed 5 times, with 1 or 2 considered a success and 3, 4, 5, or 6 a failure. Each trial is independent of the others and the probability of success in each trial is $p = \frac{2}{6} = \frac{1}{3}$. In 5 tosses there can be any number of successes from 0 through 5. But these 6 possible results are not equally likely, as shown in the chart at the side, which was obtained by using the following formula from Section 9.3.

$$P(x) = \binom{n}{x} p^x (1 - p)^{n-x},$$

x	$P(x)$
0	$\binom{5}{0}\left(\frac{1}{3}\right)^0\left(\frac{2}{3}\right)^5 = \dfrac{32}{243}$
1	$\binom{5}{1}\left(\frac{1}{3}\right)^1\left(\frac{2}{3}\right)^4 = \dfrac{80}{243}$
2	$\binom{5}{2}\left(\frac{1}{3}\right)^2\left(\frac{2}{3}\right)^3 = \dfrac{80}{243}$
3	$\binom{5}{3}\left(\frac{1}{3}\right)^3\left(\frac{2}{3}\right)^2 = \dfrac{40}{243}$
4	$\binom{5}{4}\left(\frac{1}{3}\right)^4\left(\frac{2}{3}\right)^1 = \dfrac{10}{243}$
5	$\binom{5}{5}\left(\frac{1}{3}\right)^5\left(\frac{2}{3}\right)^0 = \dfrac{1}{243}$

where n is the number of trials (here, $n = 5$), x is the number of successes, p is the probability of success in a single trial (here, $p = \frac{1}{3}$), and $P(x)$ is the probability that exactly x of the n trials result in success.

The expected value of this experiment (that is, the expected number of successes in 5 trials) is the sum

$$0 \cdot P(0) + 1 \cdot P(1) + 2 \cdot P(2) + 3 \cdot P(3) + 4 \cdot P(4) + 5 \cdot P(5)$$

$$= 0\left(\frac{32}{243}\right) + 1\left(\frac{80}{243}\right) + 2\left(\frac{80}{243}\right) + 3\left(\frac{40}{243}\right) + 4\left(\frac{10}{243}\right) + 5\left(\frac{1}{243}\right)$$

$$= \frac{405}{243} = 1\frac{2}{3} \quad \text{or} \quad \frac{5}{3}.$$

This result agrees with our intuition because there is, on average, 1 success in 3 trials; so we would expect 2 successes in 6 trials, and a bit less than 2 in 5 trials. From another point of view, since the probability of success is 1/3 in each trial and 5 trials were performed, the expected number of successes should be $5 \cdot (1/3) = 5/3$. More generally, we can show the following.

The expected number of successes in n binomial trials is np, where p is the probability of success in a single trial.

1 Find μ and σ for a binomial distribution having $n = 120$ and $p = 1/6$.

Answer:
$\mu = 20$; $\sigma = 4.08$

The expected value of any probability distribution is actually its mean. To show this, suppose an experiment has possible numerical outcomes x_1, x_2, ..., x_k, which occur with frequencies $f_1, f_2, \ldots, f_k$, respectively. If n is the total number of items, then (as shown in Section 10.1) the mean of this grouped distribution is

$$\mu = \frac{x_1 f_1 + x_2 f_2 + \cdots + x_k f_k}{n} = \frac{x_1 f_1}{n} + \frac{x_2 f_2}{n} + \cdots + \frac{x_k f_k}{n}$$

$$= x_1\left(\frac{f_1}{n}\right) + x_2\left(\frac{f_2}{n}\right) + \cdots + x_k\left(\frac{f_k}{n}\right).$$

But in each case, the fraction f_i/n (frequency over total number) is the probability p_i of outcome x_i. Hence,

$$\mu = x_1 p_1 + x_2 p_2 + \cdots + x_k p_k,$$

which is the expected value.

We saw above that the expected value (mean) of a binomial distribution is np. It can be shown that the variance and standard deviation of a binomial distribution are also given by convenient formulas.

$$\sigma^2 = np(1 - p) \quad \text{and} \quad \sigma = \sqrt{np(1 - p)}$$

Substituting the appropriate values for n and p from the example into this new formula gives

$$\sigma^2 = 5\left(\frac{1}{3}\right)\left(\frac{2}{3}\right) = \frac{10}{9},$$

and $\sigma = \sqrt{10/9} \approx 1.05$. **1**

A summary of these results follows.

Binomial Distribution

Suppose an experiment is a series of n independent repeated trials, where the probability of a success in a single trial is always p. Let x be the number of successes in the n trials. Then the probability that exactly x successes will occur in n trials is given by

$$\binom{n}{x} p^x (1 - p)^{n-x}.$$

The mean μ and variance σ^2 of a binomial distribution are, respectively,

$$\mu = np \quad \text{and} \quad \sigma^2 = np(1 - p).$$

The standard deviation is

$$\sigma = \sqrt{np(1 - p)}.$$

2 The probability that a can of soda from a certain plant is defective is .005. A sample of 4 cans is selected at random. Write a distribution for the number of defective cans in the sample, and give its mean and standard deviation.

Answer:

x	$P(x)$
0	.98
1	.0197
2	.00015
3	.0000005
4	.00000000

$\mu = .02$, $\sigma = .14$

▶**EXAMPLE 1**   The probability that a plate picked at random from the assembly line in a china factory will be defective is .01. A sample of 3 is to be selected. Write the distribution for the number of defective plates in the sample, and give its mean and standard deviation.

Since 3 plates will be selected, the possible number of defective plates ranges from 0 to 3. Here, n (the number of trials) is 3; p (the probability of selecting a defective on a single trial) is .01. The distribution and the probability of each outcome are shown below.

x	$P(x)$
0	$\binom{3}{0}(.01)^0(.99)^3 = .970$
1	$\binom{3}{1}(.01)(.99)^2 = .029$
2	$\binom{3}{2}(.01)^2(.99) = .0003$
3	$\binom{3}{3}(.01)^3(.99)^0 = .000001$

The mean of the distribution is

$$\mu = np = 3(.01) = .03.$$

The standard deviation is

$$\sigma = \sqrt{np(1 - p)} = \sqrt{3(.01)(.99)} = \sqrt{.0297} \approx .17. \quad ◀ \ \boxed{2}$$

Binomial distributions are very useful, but the probability calculations become difficult if n is large. However, the standard normal distribution of the previous section (which shall be referred to as the normal distribution from now on) can be used to get a good approximation of binomial distributions.

As an example, to show how the normal distribution is used, consider the distribution for the expected number of heads if 1 coin is tossed 15 times. This is a binomial distribution with $p = 1/2$ and $n = 15$. The distribution is shown with Figure 10.14. The mean of the distribution is

$$\mu = np = 15\left(\frac{1}{2}\right) = 7.5.$$

The standard deviation is

$$\sigma = \sqrt{np(1 - p)} = \sqrt{15\left(\frac{1}{2}\right)\left(1 - \frac{1}{2}\right)}$$

$$= \sqrt{15\left(\frac{1}{2}\right)\left(\frac{1}{2}\right)} = \sqrt{3.75} \approx 1.94.$$

In Figure 10.14 the normal curve with $\mu = 7.5$ and $\sigma = 1.94$ has been super-imposed over the histogram of the distribution.

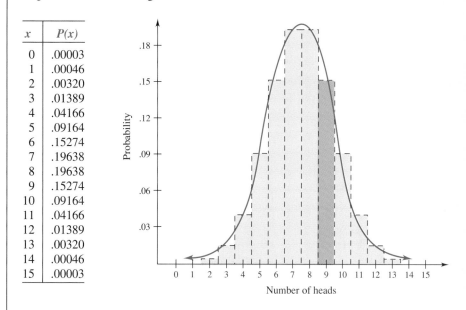

x	$P(x)$
0	.00003
1	.00046
2	.00320
3	.01389
4	.04166
5	.09164
6	.15274
7	.19638
8	.19638
9	.15274
10	.09164
11	.04166
12	.01389
13	.00320
14	.00046
15	.00003

FIGURE 10.14

As shown in the distribution, the probability of getting exactly 9 heads in the 15 tosses is approximately .153. This answer is about the same fraction that would be found by dividing the area of the bar in color in Figure 10.14 by the total area of all 16 bars in the graph. (Some of the bars at the extreme left and right ends of the graph are too short to show up.)

As the graph suggests, the area in color is approximately equal to the area under the normal curve from $x = 8.5$ to $x = 9.5$. The normal curve is higher than the top of the bar in the left half but lower in the right half.

To find the area under the normal curve from $x = 8.5$ to $x = 9.5$, first find z-scores, as in the previous section. Use the mean and the standard deviation for the distribution, which we have already calculated, to get z-scores for $x = 8.5$ and $x = 9.5$.

$$\text{For } x = 8.5, \qquad\qquad \text{For } x = 9.5,$$

$$z = \frac{8.5 - 7.5}{1.94} \qquad\qquad z = \frac{9.5 - 7.5}{1.94}$$

$$= \frac{1.00}{1.94} \qquad\qquad\quad = \frac{2.00}{1.94}$$

$$z \approx .52 \qquad\qquad\qquad z \approx 1.03$$

From Table 2, $z = .52$ gives an area of .1985, while $z = 1.03$ gives .3485.

3 Use the normal distribution to find the probability of getting exactly the following number of heads in 15 tosses of a coin.

(a) 7

(b) 10

Answers:

(a) .1985

(b) .0909

4 About 9% of the transistors produced by a certain factory are defective. Find the probability that in a sample of 200, the following numbers of transistors will be defective. (Hint: $\sigma = 4.05$)

(a) Exactly 11

(b) 16 or fewer

(c) More than 14

Answers:

(a) .0226

(b) .3557

(c) .8051

Find the required result by substracting these two numbers.

$$.3485 - .1985 = .1500$$

This answer, .1500, is close to the answer, .153, found above. **3**

▶**EXAMPLE 2** About 6% of the bolts of cloth produced by a certain machine have defects.

(a) Find the probability that in a sample of 100 bolts, 3 or fewer have defects.

This problem satisfies the conditions of the definition of a binomial distribution, so the normal curve approximation can be used. First find the mean and the standard deviation using $n = 100$ and $p = 6\% = .06$.

$$\mu = 100(.06) \qquad \sigma = \sqrt{100(.06)(1 - .06)}$$
$$\mu = 6 \qquad \qquad = \sqrt{100(.06)(.94)}$$
$$= \sqrt{5.64}$$
$$\sigma \approx 2.37$$

As the graph of Figure 10.15 shows, the area to the left of $x = 3.5$ (since we want 3 or fewer bolts with defects) must be found. For $x = 3.5$,

$$z = \frac{3.5 - 6}{2.37} = \frac{-2.5}{2.37} \approx -1.05.$$

From Table 8, $z = -1.05$ corresponds to an area of .3531. Find the area to the left of z by subtracting .3531 from .5000 to get .1469. The probability of 3 or fewer bolts with defects in a sample of 100 bolts is .1469, or 14.69%. **4**

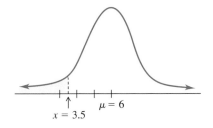

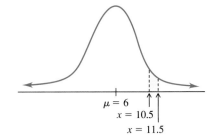

FIGURE 10.15   **FIGURE 10.16**

(b) Find the probability of exactly 11 bolts with defects in a sample of 100 bolts.

As Figure 10.16 shows, the area between $x = 10.5$ and $x = 11.5$ must be found.

$$\text{If } x = 10.5, \text{ then } z = \frac{10.5 - 6}{2.37} = 1.90.$$

$$\text{If } x = 11.5, \text{ then } z = \frac{11.5 - 6}{2.37} = 2.32.$$

Look in Table 2; $z = 1.90$ gives an area of .4713, while $z = 2.32$ yields, .4898.

The final answer is the difference of these numbers, or

$$.4898 - .4713 = .0185.$$

The probability of having exactly 11 bolts with defects is .0185. ◄

Caution   The normal curve approximation to a binomial distribution is quite accurate *provided that n is large and p is not close to 0 or 1.* As a rule of thumb, the normal curve approximation can be used as long as both np and $n(1 - p)$ are at least 5.

10.4 EXERCISES

1. What three conditions must be satisfied for a distribution to be binomial?

2. What must be known to find the mean and standard deviation of a binomial distribution?

In Exercises 3–6, binomial experiments are described. For each one (a) write the distribution; (b) find the mean; (c) find the standard deviation. (See Example 1.)

3. Management   According to a recent survey, fifty percent of cereals are bought with a coupon. Six cereal purchases are selected randomly and checked for use of a coupon. Write the distribution for the number purchased with a coupon.*

4. Management   Private labels (not nationally advertised) account for 7% of cold cereal sales. The label on each of six randomly selected purchases is noted. Write the distribution of the number with a private label.*

5. Social Science   In the United States 14% of people aged five and older speak a language other than English at home. Five people are interviewed. Write the distribution of the number speaking a language other than English at home.

6. Management   To maintain quality control on the production line, the Bright Lite Company randomly selects three light bulbs each day for testing. Experience has shown a defective rate of .02. Write the distribution for the number of defectives in the daily samples.

Natural Science   *Work the following exercises involving binomial experiments.*

7. The probability that an infant will die in the first year of life is about .025. In a group of 500 babies, what are the mean and standard deviation of the number of babies who can be expected to die in their first year of life?

8. The probability that a particular kind of mouse will have a brown coat is $1/4$. In a litter of 8, assuming independence, how many could be expected to have a brown coat? With what standard deviation?

9. A certain drug is effective 80% of the time. Give the mean and standard deviation of the number of patients using the drug who recover, out of a group of 64 patients.

10. The probability that a newborn infant will be a girl is .49. If 50 infants are born on Susan B. Anthony's birthday, how many can be expected to be girls? With what standard deviation?

11. What is the rule of thumb for using the normal distribution to approximate a binomial distribution?

For the remaining exercises, use the normal curve approximation to a binomial distribution. (See Example 2.)

Suppose 16 coins are tossed. Find the probability of getting exactly

12. 8 heads;   13. 7 heads;   14. 10 tails.

Suppose 1000 coins are tossed. Find the probability of getting each of the following.

15. Exactly 500 heads

16. Exactly 510 heads

17. 480 heads or more

18. Fewer than 470 tails

19. Fewer than 518 heads

20. More than 550 tails

---

*"Promotion costs soften cereal profit," by George Lazarus from *The Chicago Tribune,* © Copyright May 27, 1993, Chicago Tribune Company. All rights reserved. Used with permission.

A die is tossed 120 times. Find the probability of getting each of the following. (Hint: $\sigma = 4.08$)

21. Exactly 20 fives

22. Exactly 24 sixes

23. Exactly 17 threes

24. Exactly 22 twos

25. More than 18 threes

26. Fewer than 22 sixes

Management *Two percent of the quartz heaters produced in a certain plant are defective. Suppose the plant produced 10,000 such heaters last month. Find the probability that among these heaters, the following numbers were defective.*

27. Fewer than 170

28. More than 222

29. **Natural Science** For certain bird species, with appropriate assumptions, the number of nests escaping predation has a binomial distribution.* Suppose the probability of success (that is, a nest escaping predation) is .3. Find the probability that at least half of 26 nests escape predation.

30. **Natural Science** Under certain appropriate assumptions, the probability of a competing young animal eating x units of food is binomially distributed, with n equal to the maximum number of food units the animal can acquire, and p equal to the probability per time unit that an animal eats a unit of food.** Suppose $n = 120$ and $p = .6$.
(a) Find the probability that an animal consumes 80 units of food.
(b) Suppose the animal must consume at least 70 units of food to survive. What is the probability that this happens?

31 **Natural Science** An experimental drug causes a rash in 15% of all people taking it. If the drug is given to 12,000 people, find the probability that more than 1700 people will get the rash.

32. **Social Science** In one state, 55% of the voters expect to vote for Edison Diest. Suppose 1400 people are asked the name of the person for whom they expect to vote. Find the probability that at least 750 people will say that they expect to vote for Diest.

Use a computer and the normal curve to approximate the following binomial probabilities. Compare these answers with those obtained by other methods in Exercises 45–48 of Section 9.3.

33. **Natural Science** A flu vaccine has a probability of 80% of preventing a person who is inoculated from getting the flu. A county health office inoculates 134 people. Find the probabilities of the following.
(a) Exactly 10 of the people inoculated get the flu.
(b) No more than 10 of the people inoculated get the flu.
(c) None of the people inoculated get the flu.

34. **Natural Science** The probability that a male will be color-blind is .042. Find the probabilities that in a group of 53 men, the following will be true.
(a) Exactly 5 are color-blind.
(b) No more than 5 are color-blind.
(c) At least 1 is color-blind.

35. **Management** The probability that a certain machine turns out a defective item is .05. Find the probabilities that in a run of 75 items, the following results are obtained.
(a) Exactly 5 defectives   (b) No defectives
(c) At least 1 defective

36. **Management** A company is taking a survey to find out whether people like its product. Their last survey indicated that 70% of the population like the product. Based on that, of a sample of 58 people, find the probabilities of the following.
(a) All 58 like the product.
(b) From 28 to 30 (inclusive) like the product.

---

* From H. M. Wilbur, *American Naturalist,* vol. 111.

** From G. deJong, *American Naturalist,* vol. 110.

CHAPTER 10 SUMMARY

KEY TERMS AND SYMBOLS

10.1 Σ summation (sigma) notation

$\bar{x}$ sample mean

random sample

grouped frequency distribution

frequency polygon

mean (arithmetic average)

median

mode

10.2 s^2 sample variance

s sample standard deviation

range

deviations from the mean

10.3 μ population mean

σ population standard deviation

Chebyshev's Theorem

continuous distribution

skewed distribution

normal distributions

normal curves

standard normal curve

z-score

10.4 binomial distribution

KEY CONCEPTS

To organize the data from a sample, we use a **grouped frequency distribution,** a set of intervals with their corresponding frequencies. The same information can be displayed with a **histogram,** a bar graph with a bar for each interval. Each bar has width 1 and height equal to the probability of the corresponding interval. Another way to display this information is with a **frequency polygon,** which is formed by connecting the midpoints of consecutive bars of the histogram.

The **mean** $\bar{x}$ of a frequency distribution is the expected value.

For n numbers $x_1, x_2, \ldots, x_n$

$$\bar{x} = \frac{\Sigma(x)}{n}.$$

For a grouped distribution

$$\bar{x} = \frac{\Sigma(xf)}{n}.$$

The **median** is the middle entry in a set of data arranged in either increasing or decreasing order.

The **mode** is the most frequent entry in a set of numbers.

The **range** of a distribution is the difference between the largest and smallest numbers in the distribution.

The **standard deviation** s is the square root of the **variance.**

For n numbers

$$s = \sqrt{\frac{\Sigma x^2 - n\bar{x}^2}{n - 1}}.$$

For a grouped distribution

$$s = \sqrt{\frac{\Sigma fx^2 - n\bar{x}^2}{n - 1}}.$$

A **normal distribution** is a continuous distribution with the following properties: The highest frequency is at the mean; the graph is symmetric about a vertical line through the mean; the total area under the curve, above the x-axis, is 1. If a normal distribution has mean μ and standard deviation σ, then the z-score for the number x is $z = \dfrac{x - \mu}{\sigma}$.

Area Under a Normal Curve   The area under a normal curve between $x = a$ and $x = b$ gives the probability that an observed data value will be between a and b.

The **binomial distribution** is a distribution with the following properties: For n independent repeated trials, where the probability of success in a single trial is p, the probability of x successes is $\dbinom{n}{x}p^x(1 - p)^{n-x}$. The mean is $\mu = np$ and the standard deviation is

$$\sigma = \sqrt{np(1 - p)}.$$

CHAPTER 10   REVIEW EXERCISES

1. Discuss some reasons for organizing data into a grouped frequency distribution.

2. What is the rule of thumb for an appropriate interval in a grouped frequency distribution?

In Exercises 3 and 4 (a) write a frequency distribution; (b) draw a histogram; (c) draw a frequency polygon.

3. The following numbers give the sales in dollars for the lunch hour at a local hamburger store for the last twenty Fridays. (Use intervals 450–474, 475–499, and so on.)

480 451 501 478 512 473 509 515 458 566
516 535 492 558 488 547 461 475 492 471

4. The number of units carried in one semester by the students in a business mathematics class was as follows. (Use intervals of 9–10, 11–12, 13–14, 15–16.)

10 9 16 12 13 15 13 16 15 11 13
12 12 15 12 14 10 12 14 15 15 13

Find the mean for each of the following.

5. 41, 60, 67, 68, 72, 74, 78, 83, 90, 97

6. 105, 108, 110, 115, 106, 110, 104, 113, 117

7.

Interval	Frequency
10–19	6
20–29	12
30–39	14
40–49	10
50–59	8

8.

Interval	Frequency
40–44	2
45–49	5
50–54	7
55–59	10
60–64	4
65–69	1

9. What do the mean, median, and mode of a distribution have in common? How do they differ? Describe each in a sentence or two.

Find the median and the mode (or modes) for each of the following.

10. 32, 35, 36, 44, 46, 46, 59

11. 38, 36, 42, 44, 38, 36, 48, 35

For grouped data the modal class is the interval containing the most data values. Find the modal class for the distributions of

12. Exercise 7 above;

13. Exercise 8 above.

14. What is meant by the range of a distribution?

15. How are the variance and the standard deviation of a distribution related? What is measured by the standard deviation?

Find the range and standard deviation for each of the following distributions.

16. 14, 17, 18, 19, 32

17. 26, 43, 51, 29, 37, 56, 29, 82, 74, 93

Find the standard deviation for the following.

18. Exercise 7 above

19. Exercise 8 above

20. Describe the characteristics of a normal distribution.

21. What is meant by a skewed distribution?

22. A probability distribution has mean of 28 and a standard deviation of 4. Use Chebyshev's theorem to decide what percent of the distribution is as follows.

(a) Between 20 and 36

(b) Less than 23.2 or greater than 32.8

 23. (a) Find the percent of the area under a normal curve within 2.5 standard deviations of the mean.

(b) Compare your answer for part (a) with the result found by using Chebyshev's theorem.

Find the following areas under the standard normal curve.

24. Between $z = 0$ and $z = 1.27$

25. To the left of $z = .41$

26. Between $z = -1.88$ and $z = 2.10$

27. Between $z = 1.53$ and $z = 2.82$

28. Find a z-score such that 8% of the area under the curve is to the right of z.

29. Why is the normal distribution not a good approximation of a binomial distribution that has a value of p close to 0 or 1?

30. Management The annual returns of two stocks for three years are given below.

	1988	1989	1990
Stock I	11%	−1%	14%
Stock II	9%	5%	10%

(a) Find the mean and standard deviation for each stock over the three-year period.

(b) If you are looking for security with an 8% return, which of these two stocks would you choose?

31. Natural Science   The weight gains of two groups of 10 rats fed on two different experimental diets were as follows.

	Weight Gains									
Diet A	1	0	3	7	1	1	5	4	1	4
Diet B	2	1	1	2	3	2	1	0	1	0

Compute the mean and standard deviation for each group and compare them to answer the questions below.
(a) Which diet produced the greatest mean gain?
(b) Which diet produced the most consistent gain?

32. Social Science   Between 1980 and 1990 HMO enrollment in the United States increased as shown below.

Location	Gains
South	7%
Midwest	10%
Northeast	13%
West	12%

(a) Compute the mean and standard deviation of the gains.
(b) How many standard deviations from the mean is the greatest gain? the smallest gain?

Social Science   *On standard IQ tests, the mean is 100, with a standard deviation of 15. The results are very close to fitting a normal curve. Suppose an IQ test is given to a very large group of people. Find the percent of people whose IQ score is*

33. more than 130;

34. less than 85;

35. between 85 and 115.

36. Management   A machine that fills quart orange juice cartons is set to fill them with 32.1 oz. If the actual contents of the cartons vary normally, with a standard deviation of .1 oz, what percent of the cartons contain less than a quart (32 oz)?

Management   *About 60% of the blended coffee produced by a certain company contains cheap coffee. Assume a binomial distribution and use the normal distribution to find the probability that, in a sample of 500 bags of coffee, the following is true.*

37. 310 or fewer contain cheap coffee.

38. Exactly 300 contain cheap coffee.

An area infested with fruit flies is to be sprayed with a chemical which is known to be 98% effective for each application. A sample of 100 flies is checked. Assume a binomial distribution and use the normal distribution to approximate the following probabilities.

39. 95% of the flies are killed in one application.

40. At least 95% of the flies are killed in one application.

41. At least 90% of the flies are killed in one application.

42. All the flies are killed in one application. Compare your answer with the probability found using the binomial distribution.

In Exercises 43–47, assume a binomial distribution and use the normal distribution approximation.

Natural Science   *Twenty percent of third-world women would like to avoid pregnancy but are not using birth control. For 50 such women, use the normal curve approximation to find the probability of each of the following.*

43. Exactly 10 are not using birth control.

44. At least 20 are using birth control.

45. Social Science   About 19% of the U.S. population older than 100 are African-Americans. Find the probability that in a group of 10 Americans older than 100, 4 are African-American.

46. Social Science   In Japan about 58% of married women work full- or part-time outside the home. Find the probability that at least 15 of a group of 20 married Japanese women work outside the home.

47. Social Science   Only 51.4% of Americans drank coffee in 1991 (compared with 74.7% in 1962). Find the probability that at most 10 of a group of 16 people at a dinner party drink coffee.

48. Much of our work in Chapters 9 and 10 is interrelated. Note the similarities in the following parallel treatments of a frequency distribution and a probability distribution.

Frequency Distribution

Complete the table below for the following data. (Recall that x is the midpoint of the interval.)

14, 7, 1, 11, 2, 3, 11, 6, 10, 13, 11, 11, 16, 12, 9, 11, 9, 10, 7, 12, 9, 6, 4, 5, 9, 16, 12, 12, 11, 10, 14, 9, 13, 10, 15, 11, 11, 1, 12, 12, 6, 7, 8, 2, 9, 12, 10, 15, 9, 3

Interval	Tally	x	f	$x \cdot f$
1–3				
4–6				
7–9				
10–12				
13–15				
16–18				

Probability Distribution

A binomial distribution has $n = 10$ and $p = .5$. Complete the table below.

x	$P(x)$	$x \cdot P(x)$
0	.001	
1	.010	
2	.044	
3	.117	
4		
5		
6		
7		
8		
9		
10		

(a) Find the mean (or expected value) for each distribution.

(b) Find the standard deviation for each distribution.

(c) For both distributions, use Chebyshev's theorem to find the interval that contains 75% of the distribution.

(d) Use the normal approximation of the binomial probability distribution to find the interval that contains 95.44% of that distribution.

(e) Why can't we use the normal distribution to answer probability questions about the frequency distribution?

Although farming has been an industry that has relied on the wisdom in almanacs and the whims of nature to bear fruit, it is becoming more of a science that constantly incorporates new technology. A Global Positioning System (GPS), which is a satellite-based compass, and a computerized soil analysis system that depends on sampling work together to yield a high-tech harvest.

The GPS uses 24 satellites orbiting the earth to give instant latitude and longitude coordinates by radio waves to a receiver, which a farmer may install in any farm machinery. Three GPS satellites are used to pin-point a farmer's location. The time it takes for a GPS receiver's signal to reach the satellites is used along with the position of the satellites to determine the location of the signal. An antenna on top of the tractor receives new coordinates from the satellites every three seconds, and the information is recorded on an onboard computer.

Farmers spend billions of dollars a year on fertilizers to prepare their fields for planting. By using satellites and a soil analysis map indicating nutrient levels, farmers can apply fertilizer only where it is needed. Money is saved and fewer chemicals are released into the environment. Variable-rate technology, or VRT, is a soil-analysis and mapping system that relies on GPS's exact positioning to apply site-specific amounts of fertilizer.

A field is measured and divided into 3.3-acre squares. Five soil samples are taken from the center of each square. The five samples are combined and analyzed for content. A color-coded map is generated from the soil analysis showing nutrient levels and how much fertilizer should be applied for each color. A farmer matches the tractor's GPS coordinates with the VRT map to control the flow of fertilizer as it is released over each color-coded area.

---

* "Harvesting from the heavens," as appeared in *The Sacramento Bee,* November 26, 1993. Reprinted by permission of Tribune Media Services.

Errors in survey data can be caused by respondents themselves (through intentional or accidental misrepresentation) or by other sources (mistakes in recording, coding, and so on). Evidence, including the data analyzed here, indicates that the frequency of such errors can be substantial. Errors can have a significant impact on inferences about consumer behavior. Also, the effective size of a sample can be reduced greatly with an error-free sample.

Let p represent the proportion of consumers who purchase a given product or brand. If purchases or lack of same are reported and recorded accurately in a survey, then $p = r/n$, where r is the number who purchased the product and n is the total number in the sample. However, errors in the data modify this formula. Let e_1 denote the probability that a purchaser is erroneously recorded as a nonpurchaser and e_2 denote the probability that a nonpurchaser is recorded as a purchaser. Then the probability that an individual is recorded as a purchaser is

$$q = p(1 - e_1) + (1 - p)e_2, \quad \text{and}$$
$$1 - q = pe_1 + (1 - p)(1 - e_2)$$

is the probability that an individual is recorded as a nonpurchaser. The *recording* (as opposed to the actual status) of purchasers and nonpurchasers follows a binomial process, so the probability that there are r recorded purchasers and $n - r$ nonpurchasers is

$$q^r(1 - q)^{n-r} \quad \text{or}$$
$$[p(1 - e_1) + (1 - p)e_2]^r [pe_1 + (1 - p)(1 - e_2)]^{n-r}. \quad (1)$$

If the error rates e_1 and e_2 are known, then the most likely estimate of p is

$$\bar{p} = \frac{(r/n) - e_2}{1 - e_1 - e_2} \quad (2)$$

Note how this differs from r/n, the most likely estimate, if we ignore the possibility of errors in the data.

One area in which some past data are available is purchase recall. To illustrate the model, we examine a study that involved both diary panel records and purchase recall data for one sample, and in addition, purchase recall data alone for a second (matched) sample. The product we focus on from this study is instant coffee. The proportions of interview claims (purchased/no) and panel records (buyers/nonbuyers) are shown below.

		Panel Records		
		Buyers	Nonbuyers	
Interview	Purchased	.55	.21	.76
Claims	No	.02	.22	.24
		.57	.43	1.00

The estimates of error rates can be calculated from the chart.

$$e_1 \approx .02/.57 = .04 \quad \text{and} \quad e_2 \approx .21/.43 = .49.$$

Note that nonpurchasers are reported as purchasers much more often than vice versa.

Next we consider the data from the matched sample of nonpanel housewives, in which $r = 724$ of $n = 940$ claimed they purchased instant coffee. From Equation (1) the likelihood is

$$[p(1 - e_1) + (1 - p)e_2]^{724}[pe_1 + (1 - p)(1 - e_2)]^{216}.$$

Using the estimates of e_1 and e_2 from above, the most likely estimate of p for the nonpanel housewives, from (2), would be

$$\bar{p} = \frac{(724/940) - .49}{1 - .04 - .49} = .60.$$

On the other hand, if we do not consider e_1 and e_2, $\bar{p} = 724/940 = .77$.

In conclusion, since errors lead to considerable loss of information, they should be taken into account in the design of surveys to gather data. Errors cause drastic reductions in the value of information without corresponding reductions in its

*Reprinted by permission of Robert L. Winkler and Anil Gaba, "Implications of Errors in Survey Data: a Bayesian Model," *Management Science,* Volume 38, Number 7, July 1992. Copyright © 1992, The Institute of Management Sciences, 290 Westminster Street, Providence, RI 02903.

cost. In other words, much larger samples are needed to achieve a given level of precision than in the case of an error-free process.

EXERCISES

For another product, suppose the proportions of interview claims and panel records are as follows.

Panel Records

		Buyers	**Nonbuyers**	
Interview	**Purchased**	.60	.21	.81
Claims	**No**	.05	.14	.19
		.65	.35	1.00

A matched sample of nonpanel shoppers showed that 622 of 850 claimed they purchased the product.

1. Calculate $\bar{p}$ without using the error estimates.
2. Calculate e_1 and e_2.
3. Calculate $\bar{p}$ using the error estimates.
4. Compare your answers for Exercises 1 and 3. Are they what would be expected?

CHAPTER 11

Differential Calculus

TECHNOLOGY RESOURCES
GraphExplorer

Visual Calculus, Schneider

Electronic Spreadsheet, Spero

The algebraic problems considered in earlier chapters dealt with *static* situations.

What is the revenue when *x* items are sold?
How much interest is earned in 2 years?
What is the equilibrium price?

Calculus, on the other hand, deals with *dynamic* situations.

At what rate is the economy growing?
How fast is a rocket going at any instant after lift-off?
How quickly can production be increased without adversely affecting profits?

The techniques of calculus will allow us to answer many questions like these that deal with rates of change.

The key idea underlying the development of calculus is the concept of limit. So we begin by studying limits.

11.1 LIMITS

It is easy to see that the value of the function

$$f(x) = \frac{x^2 - 4}{x - 2}$$

when $x = 1$ is the number $f(1) = -3/-1 = 3$. It is also true that when x is a number *very close* to 1 (on either side of 1), then $f(x)$ is a number *very close* to 3 as shown in the following table.

x	.99	.9999	1	1.0001	1.01
$f(x)$	2.99	2.9999	3	3.0001	3.01

1 Use a calculator to find $\lim\limits_{x \to 4} \dfrac{x^2 - 4}{x - 2}$ by completing the following table.

x	$f(x)$
3.8	
3.9	
3.99	
3.999	
4.001	
4.1	
4.2	

Answer:
5.8; 5.9; 5.99; 5.999; 6.001; 6.1; 6.2; limit is 6

At $x = 2$ the situation is different: $f(2)$ is *not defined* because substituting 2 in the rule of the function makes the denominator 0. But we can still ask, what happens to $f(x)$ when x is a number *very close* to (but not equal to) 2? The following table provides an answer.

x	1.9	1.99	1.9999	2	2.00001	2.001	2.05	2.1
$f(x)$	3.9	3.99	3.9999		4.00001	4.001	4.05	4.1

The table suggests that

as x gets closer and closer to 2 from either direction, the corresponding value of $f(x)$ gets closer and closer to 4.

In fact, by experimenting with a calculator you can convince yourself that the values of $f(x)$ can be made *as close as you want* to 4 by taking values of x close enough to 2. This situation is usually described by saying that "the *limit* of $f(x)$ as x approaches 2 is the number 4," which is written symbolically as

$$\lim_{x \to 2} f(x) = 4.$$

Similarly, we write

$$\lim_{x \to 1} f(x) = 3,$$

which is read "the limit of $f(x)$ as x approaches 1 is 3." This means that

as x takes values closer and closer to 1 from either direction, the corresponding value of $f(x)$ gets closer and closer to 3

and that

the value of $f(x)$ can be made *arbitrarily close* (as close as you want) to 3 by taking values of x close enough to 1. **1**

In the general case, we have this informal definition.

Limit of a Function

Let f be a function and let a and L be real numbers. Suppose that

as x takes values closer and closer (but not equal) to a (on both sides of a), the corresponding values of $f(x)$ get closer and closer (and possibly are equal) to L;

and that

the values of $f(x)$ can be made arbitrarily close to L by taking values of x close enough to a.

Then L is the **limit** of the function $f(x)$ as x approaches a, written:

$$\lim_{x \to a} f(x) = L.$$

This definition is *informal* because the expressions "closer and closer" and "arbitrarily close" have not been precisely defined. In particular, the charts used above provide strong intuitive confirmation, but not a rigorous proof, that the limits are what was claimed.

Two observations about limits are in order. First, the concept of limit deals with the behavior of a function *f near* a number *a*, without regard to whether or not $f(a)$ is defined. Second, the definition of limit implies that *if* a limit exists, it is necessarily unique (because the values of $f(x)$ cannot be arbitrarily close to two different numbers).

▶**EXAMPLE 1**   If $f(x) = x^2 + x + 1$, then $\lim\limits_{x \to 3} f(x)$ can be written $\lim\limits_{x \to 3} (x^2 + x + 1)$. Find this limit.
To find this limit, we make a chart showing the values of the function at numbers very close to 3.

	x approaches 3 from the left → 3 ← *x* approaches 3 from the right						
x	2.9	2.99	2.9999	3	3.0001	3.01	3.1
$f(x)$	12.31	12.9301	12.9993 . . .		13.0007 . . .	13.0701	13.71

The chart strongly suggests that as *x* approaches 3 from either direction, $f(x)$ gets closer and closer to 13 and, hence, that $\lim\limits_{x \to 3} (x^2 + x + 1) = 13$. ◀

For Graphers

The computation time needed in Example 1 can be avoided on a graphing calculator. Graph the function in a very small viewing window (for instance, $2.9 \le x \le 3.1$ and $12.9 \le y \le 13.1$) and use the trace feature to move along the graph. The calculator will display the coordinates of points on the graph and you will be able to see that when the *x*-coordinate is very close to 3, the corresponding *y*-coordinate (the value of $f(x)$) is very close to 13.

The function in Example 1 is defined at 3 and $f(3) = 3^2 + 3 + 1 = 13$. So the limit of $f(x)$ as *x* approaches 3 is $f(3)$, the value of the function at 3; that is

$$\lim_{x \to 3} f(x) = f(3).$$

It can be shown that what happend in Example 1 always occurs for *polynomial* functions.

If $f(x)$ is a polynomial function and *a* is a real number, then

$$\lim_{x \to a} f(x) = f(a).$$

2 If $f(x) = 2x^4 - 4x^3 + 3x$, find

(a) $\lim_{x \to 2} f(x)$;

(b) $\lim_{x \to -1} f(x)$.

Answers:

(a) 6

(b) 3

▶**EXAMPLE 2**   According to the fact in the box,

$$\lim_{x \to -1} (x^3 - 2x^2 - 8x - 3) = (-1)^3 - 2(-1)^2 - 8(-1) - 3 = 2. \quad ◀ \quad \boxed{2}$$

Similarly, the limit of any constant polynomial as x approaches any constant a is the value of that constant.

$$\lim_{x \to a} 5 = 5 \quad \text{and} \quad \lim_{x \to a} (-7) = -7$$

Limits of a function can often be found graphically, provided that the graph of the function is known.

▶**EXAMPLE 3**   The graph of a function f is known for all values of x except $x = 2$ (Figure 11.1). Find $\lim_{x \to 2} f(x)$.

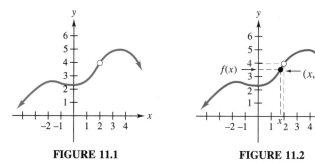

FIGURE 11.1   **FIGURE 11.2**

A typical point on the graph is of the form $(x, f(x))$, where the first coordinate is a number on the horizontal axis and the second coordinate is a number on the vertical axis, as indicated by the dashed lines in Figure 11.2 above. As x takes values closer and closer to 2 (along the horizontal axis), the corresponding values of $f(x)$ (along the vertical axis) get closer and closer to 4. Therefore, $\lim_{x \to 2} f(x) = 4$.   ◀

When $f(x)$ is not a polynomial function and $f(a)$ is defined, it is possible that the limit of $f(x)$ as x approaches a may not be $f(a)$.

▶**EXAMPLE 4**   Let f be the function whose rule is

$$f(x) = \begin{cases} 0 \text{ if } x \text{ is an integer} \\ 1 \text{ if } x \text{ is not an integer} \end{cases}$$

and whose graph is shown in Figure 11.3.

3 Let $f(x) = \dfrac{x^2 + 9}{x - 3}$. Find the following.

(a) $\lim\limits_{x \to 3} f(x)$

(b) $\lim\limits_{x \to 0} f(x)$

Answers:

(a) Does not exist

(b) -3

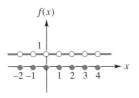

FIGURE 11.3

The graph shows that whenever x is a number very close to (but not equal to) 3, then $f(x) = 1$. Therefore, $\lim\limits_{x \to 3} f(x) = 1$. But since 3 is an integer, $f(3) = 0$. Therefore, $\lim\limits_{x \to 3} f(x) \neq f(3)$. ◀

It is possible that $\lim\limits_{x \to a} f(x)$ may not exist, that is, there may be no number L satisfying the definition. This can happen in several ways.

▶ **EXAMPLE 5**   Let $g(x) = \dfrac{x^2 + 4}{x - 2}$ and find $\lim\limits_{x \to 2} g(x)$.

Make a table of values.

	x approaches 2 from the left $\to$ 2 $\leftarrow$ x approaches 2 from the right							
x	1.8	1.9	1.99	1.999	2	2.001	2.01	2.05
$g(x)$	-36.2	-76.1	-796	-7996		8004	804	164
	$g(x)$ gets smaller and smaller					$g(x)$ gets larger and larger		

The table above and the graph of $g(x)$ in Figure 11.4 show that as x approaches 2 from the left, $g(x)$ gets smaller and smaller, but as x approaches 2 from the right, $g(x)$ gets larger and larger. Since $g(x)$ does not get closer and closer to a single real number as x approaches 2 from either side,

$$\lim\limits_{x \to 2} \frac{x^2 + 4}{x - 2} \text{ does not exist.} \quad ◀ \quad \boxed{3}$$

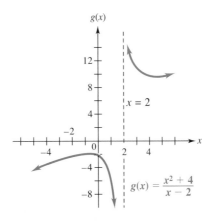

FIGURE 11.4

▶**EXAMPLE 6** What is $\lim\limits_{x \to 0} \dfrac{|x|}{x}$?

The function $f(x) = |x|/x$ is not defined when $x = 0$. When $x > 0$, then the definition of absolute value shows that $f(x) = |x|/x = x/x = 1$. When $x < 0$, then $|x| = -x$ and $f(x) = -x/x = -1$. The graph of f is shown in Figure 11.5. As x approaches 0 from the right, x is always positive and the corresponding value of $f(x)$ is 1. But as x approaches 0 from the left, x is always negative and the corresponding value of $f(x)$ is -1. Thus, as x approaches 0 from *both* sides, the corresponding values of $f(x)$ do not get closer and closer to a *single* real number. Therefore, the limit does not exist.* ◀

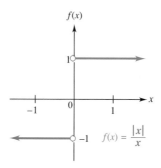

FIGURE 11.5

The preceding examples illustrate the following facts.

Existence of Limits

1. The limit of a function f as x approaches a may fail to exist.
   (a) If $f(x)$ becomes infinitely large or infinitely small as x approaches a from either side, then the limit does not exist (Example 5).
   (b) If $f(x)$ gets closer and closer to L as x approaches a from the left and $f(x)$ gets closer and closer to M as x approaches a from the right and $L \ne M$, then the limit does not exist (Example 6).
2. If the limit does exist, it may not be equal to $f(a)$ (Example 4). In fact, $f(a)$ may not even be defined (Example 3).
3. But for a polynomial function $p(x)$, the limit as x approaches a is always defined and is equal to $p(a)$ (Examples 1 and 2).

The function f whose graph is shown in Figure 11.6 illustrates the facts listed in the box above.

---

* In a situation like this, one sometimes says that -1 is the *limit of $f(x)$ from the left* and that 1 is the *limit of $f(x)$ from the right*. The concept of limit as we have defined it is sometimes called a *two-sided limit*.

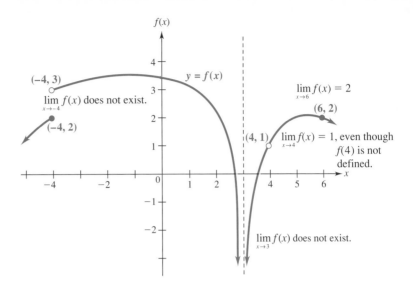

FIGURE 11.6

RULES FOR LIMITS As shown by the examples above, tables and graphs can be used to find limits. However, it is usually more efficient to find limits algebraically by using the following rules for limits. The proofs of these rules are omitted since they require a rigorous definition of limit, which we have not given.

Rules for Limits

Let a, k, n, A, and B be real numbers, and let f and g be functions such that

$$\lim_{x \to a} f(x) = A \quad \text{and} \quad \lim_{x \to a} g(x) = B.$$

1. If k is a constant, then (a) $\lim_{x \to a} k = k$ and (b) $\lim_{x \to a} k \cdot f(x) = k \cdot \lim_{x \to a} f(x)$.

2. $\lim_{x \to a} [f(x) \pm g(x)] = \lim_{x \to a} f(x) \pm \lim_{x \to a} g(x) = A \pm B$

   (The limit of a sum or difference is the sum or difference of the limits.)

3. If $p(x)$ is a polynomial, then $\lim_{x \to a} p(x) = p(a)$.

4. $\lim_{x \to a} [f(x) \cdot g(x)] = \left[\lim_{x \to a} f(x)\right] \cdot \left[\lim_{x \to a} g(x)\right] = A \cdot B$

   (The limit of a product is the product of the limits.)

5. $\lim_{x \to a} \dfrac{f(x)}{g(x)} = \dfrac{\lim_{x \to a} f(x)}{\lim_{x \to a} g(x)} = \dfrac{A}{B}$ if $B \neq 0$.

   (The limit of a quotient is the quotient of the limits, provided the limit of the denominator is not zero.)

6. For any real number n for which A^n exists,
   $\lim_{x \to a} [f(x)]^n = \left[\lim_{x \to a} f(x)\right]^n = A^n$.

7. If $f(x) = g(x)$ for all $x \neq a$, then $\lim_{x \to a} f(x) = \lim_{x \to a} g(x)$.

4 Use the limit rules to find the following.

(a) $\lim_{x \to 4} (3x - 9)$

(b) $\lim_{x \to -1} (2x^2 - 4x + 1)$

(c) $\lim_{x \to 2} \dfrac{x - 1}{3x + 2}$

(d) $\lim_{x \to 2} \sqrt{3x + 3}$

Answers:

(a) 3

(b) 7

(c) 1/8

(d) 3

This list may seem imposing, but most limit problems have solutions that agree with your common sense. Algebraic techniques and tables of values can also help resolve questions about limits, as some of the following examples will show.

▶ **EXAMPLE 7** Find each limit.

(a) $\lim_{x \to 2} [(x^2 + 1) + (x^3 - x + 3)]$

$\lim_{x \to 2} [(x^2 + 1) + (x^3 - x + 3)]$

$= \lim_{x \to 2} (x^2 + 1) + \lim_{x \to 2} (x^3 - x + 3)$ Rule 2

$= (2^2 + 1) + (2^3 - 2 + 3) = 5 + 9 = 14$ Rule 3

(b) $\lim_{x \to -1} (x^3 + 4x)(2x^2 - 3x)$

$\lim_{x \to -1} (x^3 + 4x)(2x^2 - 3x)$

$= \lim_{x \to -1} (x^3 + 4x) \cdot \lim_{x \to -1} (2x^2 - 3x)$ Rule 4

$= [(-1)^3 + 4(-1)] \cdot [2(-1)^2 - 3(-1)]$ Rule 3

$= (-1 - 4)(2 + 3) = -25$

(c) $\lim_{x \to -1} 5(3x^2 + 2)$

$\lim_{x \to -1} 5(3x^2 + 2) = 5 \cdot \lim_{x \to -1} (3x^2 + 2)$ Rule 1

$= 5[3(-1)^2 + 2]$ Rule 3

$= 25$

(d) $\lim_{x \to 4} \dfrac{x}{x + 2}$

$\lim_{x \to 4} \dfrac{x}{x + 2} = \dfrac{\lim_{x \to 4} x}{\lim_{x \to 4} (x + 2)}$ Rule 5

$= \dfrac{4}{4 + 2} = \dfrac{2}{3}$ Rule 3

(e) $\lim_{x \to 9} \sqrt{4x - 11}$

As $x \to 9$, the expression $4x - 11$ approaches $4 \cdot 9 - 11 = 25$. Using Rule 6, with $n = 1/2$, gives

$\lim_{x \to 9} (4x - 11)^{1/2} = [\lim_{x \to 9} (4x - 11)]^{1/2} = \sqrt{25} = 5.$ ◀ **4**

Rule 7 for limits reflects the fact that the limit as x approaches a involves only the values of the function *near a*, but not *at a*. So two functions that agree for all values of x except $x = a$ will necessarily have the same limit at a. This fact and some algebraic manipulation can be used to evaluate many limits involving quotients.

5 Use the limit rules to find
$$\lim_{x \to 1} \frac{2x^2 + x - 3}{x - 1}.$$

Answer:
5

6 Find the following.

(a) $\lim\limits_{x \to 1} \dfrac{\sqrt{x} - 1}{x - 1}$.

(b) $\lim\limits_{x \to 9} \dfrac{\sqrt{x} - 3}{x - 9}$

Answers:

(a) 1/2

(b) 1/6

▶ **EXAMPLE 8** Find $\lim\limits_{x \to 2} \dfrac{x^2 + x - 6}{x - 2}$.

Rule 5 cannot be used here, because
$$\lim_{x \to 2} (x - 2) = 0.$$
We can, however, simplify the function by rewriting the fraction as
$$\frac{x^2 + x - 6}{x - 2} = \frac{(x + 3)(x - 2)}{x - 2}.$$
When $x \neq 2$, the quantity $x - 2$ is nonzero and may be cancelled, so that
$$\frac{x^2 + x - 6}{x - 2} = x + 3 \quad \text{for all} \quad x \neq 2.$$
Now Rule 7 can be used.
$$\lim_{x \to 2} \frac{x^2 + x - 6}{x - 2} = \lim_{x \to 2} (x + 3) = 2 + 3 = 5 \quad ◀ \boxed{5}$$

▶ **EXAMPLE 9** Find $\lim\limits_{x \to 4} \dfrac{\sqrt{x} - 2}{x - 4}$.

As $x \to 4$, the numerator approaches 0 and the denominator also approaches 0, giving the meaningless expression $0/0$. To change the form of the expression, algebra can be used to rationalize the numerator by multiplying both the numerator and the denominator by $\sqrt{x} + 2$. This gives
$$\frac{\sqrt{x} - 2}{x - 4} = \frac{\sqrt{x} - 2}{x - 4} \cdot \frac{\sqrt{x} + 2}{\sqrt{x} + 2} = \frac{\sqrt{x} \cdot \sqrt{x} - 2\sqrt{x} + 2\sqrt{x} - 4}{(x - 4)(\sqrt{x} + 2)}$$
$$= \frac{x - 4}{(x - 4)(\sqrt{x} + 2)} = \frac{1}{\sqrt{x} + 2}$$
for all $x \neq 4$. Now use rules for limits.
$$\lim_{x \to 4} \frac{\sqrt{x} - 2}{x - 4} = \lim_{x \to 4} \frac{1}{\sqrt{x} + 2} = \frac{1}{\sqrt{4} + 2} = \frac{1}{2 + 2} = \frac{1}{4} \quad ◀ \boxed{6}$$

▶ **EXAMPLE 10** Find $\lim\limits_{x \to 1} \dfrac{x + 1}{x^2 - 1}$.

Again, Rule 5 cannot be used since $\lim\limits_{x \to 1} x^2 - 1 = 0$. However,
$$\frac{x + 1}{x^2 - 1} = \frac{x + 1}{(x + 1)(x - 1)} = \frac{1}{x - 1} \quad \text{for all} \quad x \neq 1.$$
Hence,
$$\lim_{x \to 1} \frac{x + 1}{x^2 - 1} = \lim_{x \to 1} \frac{1}{x - 1}$$

7 Find $\lim\limits_{x \to 2} \dfrac{2}{3x - 6}$.

Answer:
Does not exist

by Rule 7. None of the rules can be used to find

$$\lim_{x \to 1} \frac{1}{x - 1}.$$

The table below shows that the values of $1/(x - 1)$ are increasing positive numbers as x approaches 1 from the right and decreasing negative numbers as x approaches 1 from the left.

x	.9	.99	.999	1	1.001	1.01	1.1
$\dfrac{1}{x - 1}$	-10	-100	-1000		1000	100	10

Therefore $\lim\limits_{x \to 1} \dfrac{1}{x - 1}$ does not exist. ◀ **7**

11.1 EXERCISES

In each of the following, use the graph to determine the value of the indicated limits. (See Examples 3–5 and Figure 11.6.)

1. (a) $\lim\limits_{x \to 3} f(x);$ **(b)** $\lim\limits_{x \to -1.5} f(x)$

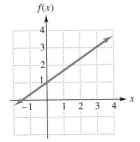

2. (a) $\lim\limits_{x \to 2} F(x)$ **(b)** $\lim\limits_{x \to -1} F(x)$

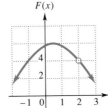

3. (a) $\lim\limits_{x \to -2} f(x)$ **(b)** $\lim\limits_{x \to 1} f(x)$

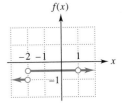

4. (a) $\lim\limits_{x \to -1} g(x)$ **(b)** $\lim\limits_{x \to 3} g(x)$

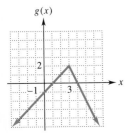

5. (a) $\lim\limits_{x \to 0} f(x)$ **(b)** $\lim\limits_{x \to -1} f(x)$

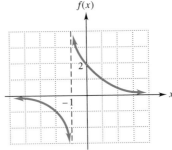

6. (a) $\lim\limits_{x \to 1} h(x)$ **(b)** $\lim\limits_{x \to 2} h(x)$

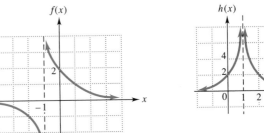

7. (a) $\lim\limits_{x \to 1} g(x)$ **(b)** $\lim\limits_{x \to -1} g(x)$

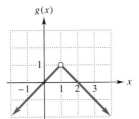

8. (a) $\lim\limits_{x \to 3} f(x)$ **(b)** $\lim\limits_{x \to 0} f(x)$

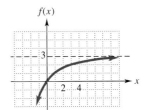

▷**9.** Explain why $\lim\limits_{x \to 2} F(x)$ in Exercise 2(a) exists, but $\lim\limits_{x \to -2} f(x)$ in Exercise 3(a) does not.

▷**10.** In Exercise 7(a), why does $\lim\limits_{x \to 1} g(x)$ exist even though $g(1)$ is not defined?

In each of the following, use a calculator to construct a table and use it to estimate the limit. (See Examples 1 and 5.)

11. $\lim\limits_{x \to 1} \dfrac{\ln x}{x - 1}$

12. $\lim\limits_{x \to 3} \dfrac{\ln x - \ln 3}{x - 3}$

13. $\lim\limits_{x \to 0} \dfrac{e^{2x} - 1}{x}$

14. $\lim\limits_{x \to 0}(x \cdot \ln |x|)$

Suppose $\lim\limits_{x \to 4} f(x) = 16$ and $\lim\limits_{x \to 4} g(x) = 8$. Use the limit rules to find the following limits.

15. $\lim\limits_{x \to 4}[f(x) - g(x)]$

16. $\lim\limits_{x \to 4}[g(x) \cdot f(x)]$

17. $\lim\limits_{x \to 4} \dfrac{f(x)}{g(x)}$

18. $\lim\limits_{x \to 4}[3 \cdot f(x)]$

19. $\lim\limits_{x \to 4} \sqrt{f(x)}$

20. $\lim\limits_{x \to 4}[g(x)]^3$

21. $\lim\limits_{x \to 4} \dfrac{f(x) + g(x)}{2g(x)}$

22. $\lim\limits_{x \to 4} \dfrac{5g(x) + 2}{1 - f(x)}$

▷**23. (a)** Graph the function f whose rule is

$$f(x) = \begin{cases} 3 - x & \text{if } x < -2 \\ x + 2 & \text{if } -2 \le x < 2. \\ 1 & \text{if } x \ge 2 \end{cases}$$

Use the graph in part (a) to find these limits.
(b) $\lim\limits_{x \to -2} f(x)$ **(c)** $\lim\limits_{x \to 1} f(x)$ **(d)** $\lim\limits_{x \to 2} f(x)$

▷**24. (a)** Graph the function g whose rule is

$$g(x) = \begin{cases} x^2 & \text{if } x < -1 \\ x + 2 & \text{if } -1 \le x < 1. \\ 3 - x & \text{if } x \ge 1 \end{cases}$$

Use the graph in part (a) to find these limits.
(b) $\lim\limits_{x \to -1} g(x)$ **(c)** $\lim\limits_{x \to 0} g(x)$ **(d)** $\lim\limits_{x \to 1} g(x)$

Use algebra and the rules for limits as needed to find the following limits. If the limit does not exist, say so. (See Examples 7–10.)

25. $\lim\limits_{x \to 2}(2x^3 + 5x^2 + 2x + 1)$

26. $\lim\limits_{x \to -1}(4x^3 - x^2 + 3x - 1)$

27. $\lim\limits_{x \to 3} \dfrac{5x - 6}{2x + 1}$

28. $\lim\limits_{x \to -2} \dfrac{2x + 1}{3x - 4}$

29. $\lim\limits_{x \to 3} \dfrac{x^2 - 9}{x - 3}$

30. $\lim\limits_{x \to -2} \dfrac{x^2 - 4}{x + 2}$

31. $\lim\limits_{x \to -2} \dfrac{x^2 - x - 6}{x + 2}$

32. $\lim\limits_{x \to 5} \dfrac{x^2 - 3x - 10}{x - 5}$

33. $\lim\limits_{x \to 2} \dfrac{x^2 - 5x + 6}{x^2 - 6x + 8}$

34. $\lim\limits_{x \to -2} \dfrac{x^2 + 3x + 2}{x^2 - x - 6}$

35. $\lim\limits_{x \to 4} \dfrac{(x + 4)^2(x - 5)}{(x - 4)(x + 4)^2}$

36. $\lim\limits_{x \to -3} \dfrac{(x + 3)(x - 3)(x + 4)}{(x + 8)(x + 3)(x - 4)}$

37. $\lim\limits_{x \to 3} \sqrt{x^2 - 4}$

38. $\lim\limits_{x \to 3} \sqrt{x^2 - 5}$

39. $\lim\limits_{x \to 4} \dfrac{-6}{(x - 4)^2}$

40. $\lim\limits_{x \to -2} \dfrac{3x}{(x + 2)^3}$

41. $\lim\limits_{x \to 0} \dfrac{[1/(x + 3)] - 1/3}{x}$

42. $\lim\limits_{x \to 0} \dfrac{[-1/(x + 2)] + 1/2}{x}$

43. $\lim\limits_{x \to 25} \dfrac{\sqrt{x} - 5}{x - 25}$

44. $\lim\limits_{x \to 36} \dfrac{\sqrt{x} - 6}{x - 36}$

45. $\lim\limits_{x \to 5} \dfrac{\sqrt{x} - \sqrt{5}}{x - 5}$

46. $\lim\limits_{x \to 8} \dfrac{\sqrt{x} - \sqrt{8}}{x - 8}$

---

47. Management The cost for manufacturing a particular videotape is

$$c(x) = 20{,}000 + 5x,$$

where x is the number of tapes produced. The average cost per tape, denoted by $\overline{c}(x)$, is found by dividing $c(x)$ by x. Find the following.
(a) $\overline{c}(1000)$ **(b)** $\overline{c}(100{,}000)$
(c) $\lim\limits_{x \to 10{,}000} \overline{c}(x)$

49. Management When the price of an essential commodity (such as gasoline) rises rapidly, consumption drops slowly at first. If the price continues to rise, however, a "tripping" point may be reached, at which consumption takes a sudden, substantial drop. Suppose the accompanying graph shows the consumption of gasoline, $G(t)$, in millions of gallons, in a certain area. We assume that the price is rising rapidly. Here t is time in months after the price began rising. Use the graph to find the following.
(a) $\lim\limits_{t \to 12} G(t)$ **(b)** $\lim\limits_{t \to 16} G(t)$ **(c)** $G(16)$
(d) The tipping point (in months)

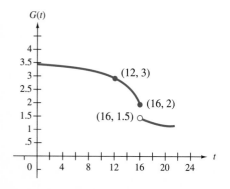

48. Management A company training program has determined that a new employee can do an average of $P(s)$ pieces of work per day after s days of on-the job training, where

$$P(s) = \frac{90s}{s + 6}.$$

Find the following.
(a) $P(1)$ **(b)** $P(11)$ **(c)** $\lim\limits_{s \to 11} P(s)$

50. Management The graph below shows the profit from the daily production of x thousand kilograms of an industrial chemical. Use the graph to find the following limits.
(a) $\lim\limits_{x \to 6} P(x)$ **(b)** $\lim\limits_{x \to 10} P(x)$ **(c)** $\lim\limits_{x \to 15} P(x)$
(d) Use the graph to estimate the number of units of the chemical that must be produced before the second shift is beneficial.

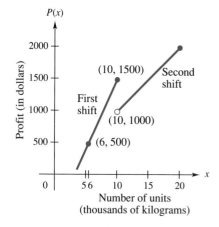

51. Natural Science   The concentration of a drug in a patient's bloodstream h hours after it was injected is given by

$$A(h) = \frac{.2h}{h^2 + 2}.$$

Find the following.
(a) $A(.5)$ **(b)** $A(1)$ **(c)** $\lim\limits_{h \to 1} A(h)$

52. Management   Weekly sales (in dollars) at Sam's Shoppe x weeks after the end of an advertising campaign are given by

$$S(x) = 5000 + \frac{3600}{x + 2}.$$

Find the following.
(a) $S(2)$ **(b)** $\lim\limits_{x \to 5} S(x)$ **(c)** $\lim\limits_{x \to 16} S(x)$

Use a very small viewing window and the trace feature on a graphing calculator to estimate the following limits.

53. $\lim\limits_{x \to 0} (x \ln |x|)$

54. $\lim\limits_{x \to 0} \dfrac{x}{e^x - 1}$

55. $\lim\limits_{x \to 3} \dfrac{x^3 - 3x^2 - x + 3}{x - 3}$

56. $\lim\limits_{x \to 4} \dfrac{.1x^4 - .8x^3 + 1.6x^2 + 2x - 8}{x - 4}$

57. $\lim\limits_{x \to -2} \dfrac{x^4 + 2x^3 - x^2 + 3x + 1}{x + 2}$

58. $\lim\limits_{x \to 0} \dfrac{e^{2x} + e^x - 2}{e^x - 1}$

11.2 RATES OF CHANGE

One of the main applications of calculus is determining how one variable changes in relation to another. A person in business wants to know how profit changes with respect to advertising, while a person in medicine wants to know how a patient's reaction to a drug changes with respect to the dose.

We begin the discussion with a familiar situation. A driver makes the 168-mile trip from Cleveland to Columbus, Ohio, in 3 hours. The following table shows how far the driver has traveled from Cleveland at various times.

Time (in hours)	0	.5	1	1.5	2	2.5	3
Distance (in miles)	0	22	52	86	118	148	168

If f is the function whose rule is

$$f(x) = \text{distance from Cleveland at time } x,$$

then the table shows, for example, that $f(2) = 118$ and $f(3) = 168$. So the distance traveled from time $x = 2$ to $x = 3$ is $168 - 118$, that is, $f(3) - f(2)$. In a similar fashion, we obtain the other entries in the following chart.

1 Find the average speed

(a) from $t = 1.5$ to $t = 2$;

(b) from $t = s$ to $t = r$.

Answers:

(a) 64 mph

(b) $\dfrac{f(r) - f(s)}{r - s}$

Time Interval	Distance Traveled
$x = 2$ to $x = 3$	$f(3) - f(2) \ = 168 - 118 = 50$
$x = 1$ to $x = 3$	$f(3) - f(1) \ = 168 - 52 = 116$
$x = 0$ to $x = 2.5$	$f(2.5) - f(0) \ = 148 - 0 = 148$
$x = .5$ to $x = 2$	$f(2) - f(.5) \ = 118 - 22 = 96$
$x = a$ to $x = b$	$f(b) - f(a)$

The last line of the chart shows how to find the distance traveled in any time interval ($0 \le a < b \le 3$).

Since distance = average speed × time,

$$\text{Average speed} = \frac{\text{distance traveled}}{\text{time interval}}.$$

In the chart above, you can compute the length of each time interval by taking the difference between the two times. Thus, for example, from $x = 1$ to $x = 3$ is a time interval of length $3 - 1 = 2$ hours and, hence, the average speed over this interval is $116/2 = 58$ mph. Similarly, we have the following information.

Time Interval	Average speed $= \dfrac{\text{Distance Traveled}}{\text{Time Interval}}$
$x = 2$ to $x = 3$	$\dfrac{f(3) - f(2)}{3 - 2} = \dfrac{168 - 118}{3 - 2} = \dfrac{50}{1} = 50$ mph
$x = 1$ to $x = 3$	$\dfrac{f(3) - f(1)}{3 - 1} = \dfrac{168 - 52}{3 - 1} = \dfrac{116}{2} = 58$ mph
$x = 0$ to $x = 2.5$	$\dfrac{f(2.5) - f(0)}{2.5 - 0} = \dfrac{148 - 0}{2.5 - 0} = \dfrac{148}{2.5} = 59.2$ mph
$x = .5$ to $x = 2$	$\dfrac{f(2) - f(.5)}{2 - .5} = \dfrac{118 - 22}{2 - .5} = \dfrac{96}{1.5} = 64$ mph
$x = a$ to $x = b$	$\dfrac{f(b) - f(a)}{b - a}$ mph

The last line of the chart shows how to compute the average speed over any time interval ($0 \le a < b \le 3$). **1**

Now speed (miles per hour) is simply the *rate of change* of distance with respect to time and what was done for the distance function f in the preceding discussion can be done with any function.

2 Find the average rate of change of $f(x)$ in Example 1 when x changes from

(a) 0 to 4;

(b) 2 to 7.

Answers:

(a) 8

(b) 13

Quantity	*Meaning for the Distance Function*	*Meaning for an Arbitrary Function f*
$b - a$	time interval = change in time from $x = a$ to $x = b$	change in x from $x = a$ to $x = b$
$f(b) - f(a)$	distance traveled = corresponding change in distance as time changes from a to b	corresponding change in $f(x)$ as x changes from a to b
$\dfrac{f(b) - f(a)}{b - a}$	average speed = average rate of change of distance with respect to time as time changes from a to b	**average rate of change** of $f(x)$ with respect to x as x changes from a to b (where $a < b$)

▶**EXAMPLE 1**   If $f(x) = x^2 + 4x + 5$, find the average rate of change of $f(x)$ with respect to x as x changes from -2 to 3.

This is the situation described in the last line of the chart above, with $a = -2$ and $b = 3$. The average rate of change is

$$\frac{f(3) - f(-2)}{3 - (-2)} = \frac{26 - 1}{5} = \frac{25}{5} = 5. \quad ◀ \;\; \boxed{2}$$

▶**EXAMPLE 2**   The graph in Figure 11.7 shows the profit $P(x)$, in hundreds of thousands of dollars, from a popular new computer program x months after its introduction on the market.

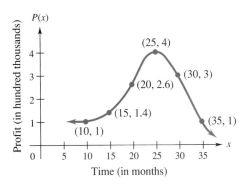

FIGURE 11.7

3 Use Figure 11.7 and find the average rate of change of profit with respect to time if time changes from

(a) $x = 15$ to $x = 20$;

(b) $x = 20$ to $x = 25$;

(c) $x = 25$ to $x = 35$;

(d) $x = 30$ to $x = 35$.

Answers:

(a) .24 hundred thousand, or $24,000

(b) .28 hundred thousand, or $28,000

(c) $-.3$ hundred thousand, or $-$30,000

(d) $-.4$ hundred thousand, or $-$40,000

(a) The graph shows that $P(25) = 4$ and $P(15) = 1.4$. So the average rate of change of profit from the 15th to the 25th month is

$$\frac{P(25) - P(15)}{25 - 15} = \frac{4 - 1.4}{10} = \frac{2.6}{10} = .26$$

Thus, the average rate of change is .26 hundred thousand dollars, that is $26,000 per month.

(b) From 25 months to 30 months, the average rate of change is

$$\frac{P(30) - P(25)}{30 - 25} = \frac{3 - 4}{5} = \frac{-1}{5} = -.20,$$

with each month in this interval showing an average *decline* in profits of .20 hundred thousand dollars, or $20,000.

(c) From 10 to 35 months, the average rate of change of profit is

$$\frac{P(35) - P(10)}{35 - 10} = \frac{1 - 1}{25} = \frac{0}{25} = 0. \quad \blacktriangleleft \quad \boxed{3}$$

INSTANTANEOUS RATE OF CHANGE   Suppose a car is stopped at a traffic light. When the light turns green, the car begins to move along a straight road. Assume that the distance traveled by the car is given by the function

$$s(t) = 2t^2 \quad (0 \le t \le 30),$$

where time t is measured in seconds and the distance $s(t)$ at time t is measured in feet. We know how to find the *average* speed of the car over any time interval, so we now turn to a different problem: determining the *exact* speed of the car at a particular instant, say $t = 10$.*

The intuitive idea is that the exact speed at $t = 10$ is very close to the average speed over a very short time interval near $t = 10$. If we take shorter and shorter time intervals near $t = 10$, the average speeds over these intervals should get closer and closer to the exact speed at $t = 10$. In other words,

the exact speed at $t = 10$ is the limit of the average speeds over shorter and shorter time intervals near $t = 10$.

---

*As distance is measured in feet and time in seconds here, speed is measured in feet per second. It may help to know that 15 mph is equivalent to 22 ft/sec and 60 mph to 88 ft/sec.

The following chart illustrates this idea.

Interval	Average speed
$t = 10$ to $t = 10.1$	$\dfrac{s(10.1) - s(10)}{10.1 - 10} = \dfrac{204.02 - 200}{.1} = 40.2$
$t = 10$ to $t = 10.01$	$\dfrac{s(10.01) - s(10)}{10.01 - 10} = \dfrac{200.4002 - 200}{.01} = 40.02$
$t = 10$ to $t = 10.001$	$\dfrac{s(10.001) - s(10)}{10.001 - 10} = \dfrac{200.040002 - 200}{.001} = 40.002$

The chart suggests that the exact speed at $t = 10$ is 40 ft/sec. We can confirm this intutition by computing the average speed from $t = 10$ to $t = 10 + h$, where h is any very small nonzero number. (The chart does this for $h = .1$, $h = .01$, and $h = .001$). The average speed from $t = 10$ to $t = 10 + h$ is

$$\frac{s(10 + h) - s(10)}{(10 + h) - 10} = \frac{s(10 + h) - s(10)}{h}$$

$$= \frac{2(10 + h)^2 - 2 \cdot 10^2}{h}$$

$$= \frac{2(100 + 20h + h^2) - 200}{h}$$

$$= \frac{200 + 40h + 2h^2 - 200}{h}$$

$$= \frac{40h + 2h^2}{h} = \frac{h(40 + 2h)}{h}$$

$$= 40 + 2h.$$

Saying that the time interval gets shorter and shorter is equivalent to saying h gets closer and closer to 0. Hence, the exact speed at $t = 10$ is the limit as h approaches 0 of the average speeds over the intervals from $t = 10$ to $t = 10 + h$; that is,

$$\lim_{h \to 0} \frac{s(10 + h) - s(10)}{h} = \lim_{h \to 0}(40 + 2h)$$

$$= 40 \text{ ft/sec.}$$

The preceding example can easily be generalized. Suppose an object is moving in a straight line with its position (distance from some fixed point) at time t given by the function $s(t)$. The speed of the object is called its **velocity** and its exact speed at time t is called the **instantaneous velocity at time t** (or just velocity at time t).

Let t_0 be a constant. By replacing 10 by t_0 in the discussion above, we see that the average velocity of the object from time $t = t_0$ to time $t = t_0 + h$ is the

4 In Example 3, if $s(t) = s^2 + 3$, find

(a) the average velocity from 1 second to 5 seconds;

(b) the instantaneous velocity at 5 seconds.

Answers:

(a) 6 ft per second

(b) 10 ft per second

quotient

$$\frac{s(t_0 + h) - s(t_0)}{(t_0 + h) - t_0} = \frac{s(t_0 + h) - s(t_0)}{h}.$$

The instantaneous velocity at time t is the limit of this quotient as h approaches 0.

Velocity

If an object moves along a straight line, with position $s(t)$ at time t, then the velocity of the object at $t = t_0$ is

$$\lim_{h \to 0} \frac{s(t_0 + h) - s(t_0)}{h},$$

provided this limit exists.

▶**EXAMPLE 3**   The distance in feet of an object from a starting point is given by $s(t) = 2t^2 - 5t + 40$, where t is time in seconds.

(a) Find the average velocity of the object from 2 seconds to 4 seconds.
     The average velocity is

$$\frac{s(4) - s(2)}{4 - 2} = \frac{52 - 38}{2} = \frac{14}{2} = 7$$

feet per second.

(b) Find the instantaneous velocity at 4 seconds.
     For $t = 4$, the instantaneous velocity is

$$\lim_{h \to 0} \frac{s(4 + h) - s(4)}{h}$$

feet per second. We have

$$
\begin{aligned}
s(4 + h) &= 2(4 + h)^2 - 5(4 + h) + 40 \\
&= 2(16 + 8h + h^2) - 20 - 5h + 40 \\
&= 32 + 16h + 2h^2 - 20 - 5h + 40 \\
&= 2h^2 + 11h + 52
\end{aligned}
$$

and

$$s(4) = 2(4)^2 - 5(4) + 40 = 52.$$

Thus,

$$s(4 + h) - s(4) = (2h^2 + 11h + 52) - 52 = 2h^2 + 11h$$

and the instantaneous velocity at $t = 4$ is

$$\lim_{h \to 0} \frac{2h^2 + 11h}{h} = \lim_{h \to 0} \frac{h(2h + 11)}{h}$$

$$= \lim_{h \to 0} (2h + 11) = 11 \text{ ft/sec.} \quad ◀ \quad \boxed{4}$$

5 Repeat Example 4 with $s(t) = .3t - 2$.

Answer:
The velocity is .3 millimeters per second.

▶ **EXAMPLE 4** The velocity of blood cells is of interest to physicians; a slower velocity than normal might indicate a constriction, for example. Suppose the position of a red blood cell in a capillary is given by

$$s(t) = 1.2t + 5,$$

where $s(t)$ gives the position of a cell in millimeters from some reference point and t is time in seconds. Find the velocity of this cell at time t.

Evaluate the limit given above. To find $s(t + h)$, substitute $t + h$ for the variable t in $s(t) = 1.2t + 5$.

$$s(t + h) = 1.2(t + h) + 5$$

Now use the definition of velocity.

$$v(t) = \lim_{h \to 0} \frac{s(t + h) - s(t)}{h}$$

$$= \lim_{h \to 0} \frac{1.2(t + h) + 5 - (1.2t + 5)}{h}$$

$$= \lim_{h \to 0} \frac{1.2t + 1.2h + 5 - 1.2t - 5}{h} = \lim_{h \to 0} \frac{1.2h}{h} = 1.2$$

The velocity of the blood cell is a constant 1.2 millimeters per second. ◀ **5**

The ideas underlying the concept of the velocity of a moving object can be extended to any function $f(x)$. In place of average velocity at time t, we have the average rate of change of $f(x)$ with respect to x as x changes from one value to another.

The **instantaneous rate of change** for a function f when $x = x_0$ is

$$\lim_{h \to 0} \frac{f(x_0 + h) - f(x_0)}{h},$$

provided this limit exists.

▶ **EXAMPLE 5** A company determines that the cost in hundreds of dollars of manufacturing x units of a certain item is

$$C(x) = -.2x^2 + 8x + 40 \quad (0 \le x \le 20).$$

(a) Find the average rate of change of cost per item for manufacturing between 5 and 10 items.

Use the formula for average rate of change. The cost to manufacture 5 items is

$$C(5) = -.2(5^2) + 8(5) + 40 = 75$$

or $7500. The cost to manufacture 10 items is

$$C(10) = -.2(10^2) + 8(10) + 40 = 100,$$

or $10,000. The average rate of change of cost is

$$\frac{C(10) - C(5)}{10 - 5} = \frac{100 - 75}{5} = 5.$$

Thus, on the average, cost is increasing at the rate of $500 per item when production is increased from 5 to 10 units.

(b) Find the instantaneous rate of change with respect to the number of items produced when 5 items are produced.

The instantaneous rate of change when $x = 5$ is given by

$$\lim_{h \to 0} \frac{C(5 + h) - C(5)}{h}$$

$$= \lim_{h \to 0} \frac{[-.2(5 + h)^2 + 8(5 + h) + 40] - [-.2(5^2) + 8(5) + 40]}{h}$$

$$= \lim_{h \to 0} \frac{[-5 - 2h - .2h^2 + 40 + 8h + 40] - [75]}{h}$$

$$= \lim_{h \to 0} \frac{6h - .2h^2}{h} \quad \text{Combine terms}$$

$$= \lim_{h \to 0} (6 - .2h) \quad \text{Divide by } h$$

$$= 6. \quad \text{Calculate the limit}$$

When 5 items are manufactured, the cost is increasing at the rate of $600 per item.

(c) Find the instantaneous rate of change of cost when 10 items are made.

The instantaneous rate of change when $x = 10$ is given by

$$\lim_{h \to 0} \frac{C(10 + h) - C(10)}{h}$$

$$= \lim_{h \to 0} \frac{[-.2(10 + h)^2 + 8(10 + h) + 40] - [-.2(10^2) + 8(10) + 40]}{h}$$

$$= \lim_{h \to 0} \frac{[-20 - 4h - .2h^2 + 80 + 8h + 40] - [100]}{h}$$

$$= \lim_{h \to 0} \frac{4h - .2h^2}{h} = \lim_{h \to 0} (4 - .2h) = 4.$$

When 10 items are manufactured, the cost is increasing at the rate of only $400 per item. This reflects the fact that cost per item is usually smaller when more items are manufactured. ◀

The rate of change of the cost function is called the **marginal cost.*** Similarly, **marginal revenue** and **marginal profit** are the rates of change of the revenue and profit functions, respectively.

Part (b) of Example 5 shows that the marginal cost when 5 items are manufactured is $600 and part (c) shows that the marginal cost when 10 items are manufactured is $400. Notice that as the number of items produced goes up, the marginal cost goes down.

_____

*Marginal cost for linear cost functions was discussed in Section 2.4.

11.2 EXERCISES

Find the average rate of change for the following functions over the given closed intervals. (See Example 1.)

1. $f(x) = x^2 + 2x$ between $x = 0$ and $x = 5$

2. $f(x) = -4x^2 - 6$ between $x = 2$ and $x = 5$

3. $f(x) = 2x^3 - 4x^2 + 6x$ between $x = -1$ and $x = 2$

4. $f(x) = -3x^3 + 2x^2 - 4x + 1$ between $x = 0$ and $x = 1$

5. $f(x) = \sqrt{x}$ between $x = 1$ and $x = 4$

6. $f(x) = \sqrt{3x - 2}$ between $x = 1$ and $x = 3$

7. $f(x) = \dfrac{1}{x - 1}$ between $x = -2$ and $x = 0$

8. $f(x) = \dfrac{-5}{2x - 3}$ between $x = 2$ and $x = 4$

_____

Exercises 9–11 deal with a car moving along a straight road, as discussed on pages 610–611. At time t seconds the distance of the car (in feet) from the starting point is $s(t) = 2t^2$. Find the instantaneous velocity (speed) of the car at

9. $t = 5$;

10. $t = 20$;

11. What was the average speed of the car during the first 30 seconds?

_____

An object moves along a straight line; its distance (in feet) from a fixed point at time t seconds is $s(t) = t^2 + 5t + 2$. Find the instantaneous velocity of the object at the following times. (See Example 3.)

12. $t = 6$

13. $t = 1$

14. $t = 10$

_____

15. Use the graph of the function f to find each of the following. (See Example 2.)

(a) $f(1)$ **(b)** $f(3)$ **(c)** $f(5) - f(1)$

(d) The average rate of change of $f(x)$ as x changes from 1 to 5.

(e) The average rate of change of $f(x)$ as x changes from 3 to 5.

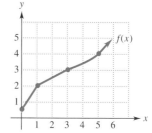

16. Use the graph of the function g to answer these questions.
   (a) Between which pair of consecutive points (P, Q, R, S, T) is the average rate of change of $g(x)$ the smallest?
   (b) Between which pair of consecutive points is the average rate of change of $g(x)$ the largest?
   (c) Between which pair of consecutive points is the average rate of change of $g(x)$ closest to 0?

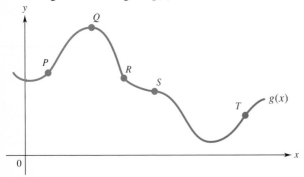

17. Explain the difference between the average rate of change of y as x changes from a to b, and the instantaneous rate of change of y at $x = a$.

18. If the instantaneous rate of change of $f(x)$ with respect to x is positive when $x = 1$, is f increasing or decreasing there?

19. **Management** The graph shows the total sales in thousands of dollars from the distribution of x thousand catalogs. Find and interpret the average rate of change of sales with respect to the number of catalogs distributed for the following changes in x.
   (a) 10 to 20   **(b)** 20 to 30   **(c)** 30 to 40
   (d) What is happening to the average rate of change of sales as the number of catalogs distributed increases?
   (e) Explain why (d) might happen.

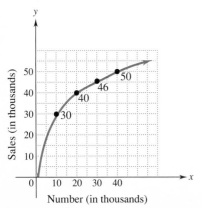

20. **Natural Science** The graph shows the temperature T in degrees Celsius as a function of the altitude h in feet when an inversion layer is over Southern California. (An inversion layer is formed when air at a higher altitude, say 3000 ft, is warmer than air at sea level, even though air normally is cooler with increasing altitude.) Estimate and interpret the average rate of change in temperature for the following changes in altitude.
   (a) 1000 to 3000 ft   **(b)** 1000 to 5000 ft
   (c) 3000 to 9000 ft   **(d)** 5000 to 9000 ft
   (e) At what altitude at or below 7000 ft is the temperature highest? lowest? How would your answer change if 7000 ft is changed to 10,000 ft?
   (f) At what altitude is the temperature the same as it is at 1000 ft?

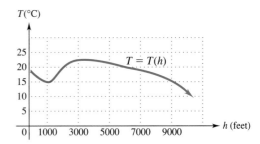

21. **Management** The graph shows annual sales (in appropriate units) of a computer game. Find the average annual rate of change in sales for the following changes in years.
   (a) 1 to 4   **(b)** 4 to 7   **(c)** 7 to 12
   (d) What do your answers for (a) to (c) tell you about the sales of this product?
   (e) Give an example of another product that might have such a sales curve.

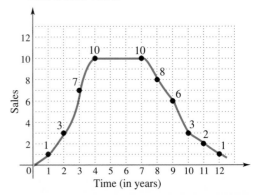

22. Management   IBM Eruope steadily lost its share of the market from 1985 to 1990, as shown in the figure.* Estimate the average drop in market share (in percent) over the following intervals.

(a) 1985 to 1986   **(b)** 1986 to 1988
(c) 1988 to 1989   **(d)** 1989 to 1990
(e) From your answers to parts (a)–(d), did the decline in market share seem to be increasing or tapering off?
(f) In which one-year period was the decline the greatest?

23. Management   The figure* shows the profit margin (in percent) for IBM Europe from 1985 to 1990. Estimate the average rate of change over the following intervals.

(a) 1985 to 1987   **(b)** 1987 to 1988
(c) 1988 to 1989   **(d)** 1989 to 1990

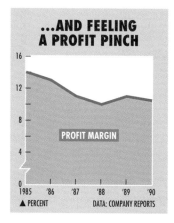

---

* "IBM Europe Is Losing Market Share . . . and Feeling a Profit Pinch" reprinted from May 6, 1991 issue of *Business Week* by special permission, copyright © 1991 by McGraw-Hill, Inc.

(e) In what time period(s) did the profit margin increase?
(f) Do the results in parts (a)–(d) indicate an improving, stable, or declining profit margin?

24. Management   Profit figures for a small firm over a 7-year period are shown in the following table.

Year	1988	1989	1990	1991	1992	1993	1994
Profit (in dollars)	5000	6000	6500	6800	7200	7500	8000

(a) Find the average rate of change of profit from 1988 to 1994.
(b) Find the average rate of change of profit from 1989 to 1993.

25. Physical Science   A car is moving along a straight test track. The position in feet of the car, $s(t)$, at various times t is measured, with the following results.

t (seconds)	0	2	6	10	12	18	20
$s(t)$ (feet)	0	10	14	18	30	36	40

Find the average velocity (velocity equals distance divided by time) as the time in seconds changes from
(a) 0 to 6;   **(b)** 2 to 10;   **(c)** 2 to 12;
(d) 6 to 10;   **(e)** 6 to 18;   **(f)** 12 to 20.

26. Management   The revenue (in thousands of dollars) from producing x units of an item is

$$R = 10x - .002x^2.$$

(a) Find the average rate of change of revenue when production is increased from 1000 to 1001 units.
(b) Find the marginal revenue when 1000 units are produced.
(c) Find the additional revenue if production is increased from 1000 to 1001 units.
(d) Compare your answers for parts (a) and (c). What do you find?

27. Management   Suppose customers in a hardware store are willing to buy $N(p)$ boxes of nails at p dollars per box, as given by

$$N(p) = 80 - 5p^2, \quad 1 \le p \le 4.$$

(a) Find the average rate of change of demand for a change in price from $2 to $3.
(b) Find the instantaneous rate of change of demand when the price is $2.
(c) Find the instantaneous rate of change of demand when the price is $3.
(d) As the price is increased from $2 to $3, how is demand changing? Is the change to be expected?

28. Natural Science The graph shows the population in millions of bacteria *t* minutes after a bactericide is introduced into a culture. Find and interpret the average rate of change of population with respect to time for the following time intervals.

(a) 1 to 2 **(b)** 2 to 3 **(c)** 3 to 4 **(d)** 4 to 5

(e) How long after the bactericide was introduced did the population begin to decrease?

(f) At what time did the rate of decrease of the population begin to slow down?

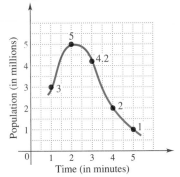

29. Management The graph* shows the amount spent by the Federal government and state Medicaid on prescription drugs during an eleven-year period. Spending went from 1.6 billion dollars in 1981 to 4.4 billion in 1990 to 6.7 billion in 1992. What was the average rate of change in spending (in billions of dollars per year) from

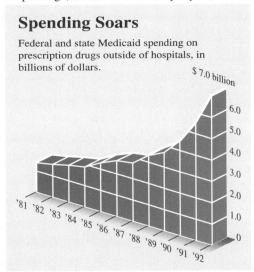

Spending Soars

Federal and state Medicaid spending on prescription drugs outside of hospitals, in billions of dollars.

(a) 1981 to 1984? **(b)** 1984 to 1987?

(c) 1987 to 1990? **(d)** 1990 to 1992?

(e) How does the rate of change in the first three years of the period shown on the graph compare with the rate during the last two years?

30. Social Science The graph** shows the amounts paid to the U.S. Treasury in fines by polluters during the 80's.

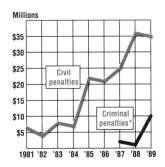

*Accurate data not available before 1987.

Source: EPA; art by Mark Holmes, NGS staff

(a) Find the average rate of change for civil penalties from 1987 to 1988, and from 1988 to 1989.

(b) Find the average rate of change of criminal penalties from 1987 to 1988, and from 1988 to 1989.

(c) Compare your results for parts (a) and (b). What do they tell you? What might account for the differences?

(d) Find the average rate of change for civil penalties from 1981 to 1989. What was the general trend over this eight-year period? What might explain this trend?

31. Social Science In an experiment, teenagers memorize a list of 15 words and then listen to rock music for certain periods of time. After *t* minutes of listening to rock music, the typical subject can remember approximately

$$R(t) = -.03t^2 + 15$$

of the words, where $0 \leq t \leq 20$. Find the instantaneous

---

*"Spending Soars" from *The New York Times,* July 7, 1993. Copyright © 1993 by the New York Times Company. Reprinted by permission.

**Graph entitled "Higher EPA fines Make Pollution Costly" by Mark Holmes from *National Geographic,* February 1991. Copyright © 1991 by National Geographic Society. Reprinted by permission.

rate at which the typical student remembers the words at the following times.

(a) $t = 5$ **(b)** $t = 15$

32. Physical Science The distance (in ft) of a particle from some fixed point is given by

$$s(t) = t^2 + 5t + 2,$$

where t is time measured in seconds. Find the average velocity of the particle over each of the following intervals.

(a) 4 to 6 sec   **(b)** 4 to 5 sec

(c) Complete the following table.

h	1	.1	.01	.001	.0001
$4 + h$	5	4.1	4.01	4.001	4.0001
$s(4 + h)$	52	39.31	38.1301	38.013001	38.00130001
$s(4)$	38	38	38	38	38
$s(4 + h) - s(4)$	14	1.31	.1301	.013001	.00130001
$\dfrac{s(4 + h) - s(4)}{h}$	14	13.1			

(d) Find $\lim\limits_{h \to 0} \dfrac{s(4 + h) - s(4)}{h}$, and give the instantaneous velocity of the particle when $x = 4$.

11.3 DEFINITION OF THE DERIVATIVE

In the previous section, the formula

$$\lim_{h \to 0} \frac{f(x_0 + h) - f(x_0)}{h}$$

was used to calculate the instantaneous rate of change of a function f at the point where $x = x_0$. Now we will give a geometric interpretation of this limit in terms of tangent lines.

In geometry, a tangent line to a circle is defined as a line that touches the circle at only one point, as at the point P in Figure 11.8 (which shows only the

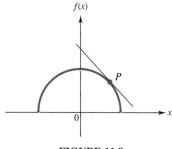

FIGURE 11.8

top half of the circle). If you think of this half-circle as a road on which you are driving at night, then the tangent line indicates the direction of the light beam from your headlights as you pass through the point P. Intuitively, the tangent line to an arbitrary curve at a point P on the curve ought to have these same properties if possible.

1. It should touch the curve at *P*, but not at any nearby points;
2. It should indicate the "direction" of the curve as you move left to right through the point *P*.

For example, the lines through P_1 and P_3 in Figure 11.9 meet these intuitive criteria for tangent lines, but the lines through P_2 and P_4 do not.

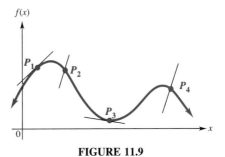

FIGURE 11.9

We can use these ideas to develop a precise definition of the tangent line to the graph of a function *f* at the point *R*. Let *R* have coordinates $(x_0, f(x_0))$ as in Figure 11.10. Choose a different point *S* on the graph and draw the line through *R* and *S*: this line is called a **secant line.** If *S* has coordinates $(x_0 + h, f(x_0 + h))$, then by the definition of slope, the slope of the secant line *RS* is

$$\text{Slope of secant} = \frac{f(x_0 + h) - f(x_0)}{x_0 + h - x_0} = \frac{f(x_0 + h) - f(x_0)}{h}.$$

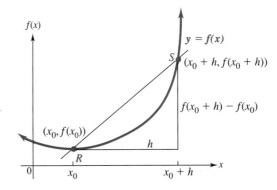

FIGURE 11.10

Calculus at Work

This section shows you that calculus is not just a body of theorems, concepts, and exercises without any real human dimension; there are practicing mathematicians, scientists, and engineers using this knowledge. The following essays were written by professionals who work with mathematics daily to create products and services we commonly use.

W. Weston Meyer
Mathematics Department, Research Laboratories
General Motors Corporation
"Perfecting the Skin of an Automobile"

Richard E. Klabunde, Ph.D.
Senior Cardiovascular Group Leader, Department of Pharmacology
Abbott Laboratories
Roberto Mercado, Ph.D.
Manager, Nonclinical Statistics, Pharmaceutical Products Division
Abbott Laboratories
"Drugs, Derivatives, and Heart Disease"

Jane Gillum (Summer Intern), Alan Opsahl, Allen Blaurock
Technology Center
Kraft General Foods
"Crystals, Integration, and . . . Margarine!"

Henry Robinson
Meteorologist, National Weather Service
National Oceanic and Atmospheric Administration
"Calculus in a Cornfield"

When you read these testaments, you may appreciate more the phenomenal versatility of calculus–how it is useful, in fact indispensable, in virtually countless aspects of our lives: from the margarine you spread on your breakfast toast, to the car you drive, to predicting the weather you experience.

The fact that calculus works as well as it does in as many ways as it does leads us to the following conclusion: learning about calculus is learning about an amazing pattern, one repeated in the most disparate and surprising places in our world.

Perfecting the Skin of an Automobile

W. Weston Meyer

Mathematics Department, Research Laboratories / General Motors Corporation, Warren, Michigan

As far as we know, it was Jacob Bernoulli (1654 – 1705) who first published a formula to measure curvature. Three centuries later, the same formula is put to uses Bernoulli could never have imagined. Designing the "skin" of an automobile — its outer panel surface — is one such use.

Bernoulli's formula, given below, provides an expression for the radius R of the circle whose curvature is identical to the curvature of the curve $y = f(x)$ at a fixed point (see Figure 1).

$$R(x) = \frac{\left[1 + f'(x)^2\right]^{3/2}}{\left|f''(x)\right|}$$

Since curvature of a circle is equal to the inverse of its radius, the curvature of f at a fixed point is the inverse of the expression for $R(x)$.

Traditionally, the outer panels of an automobile materialized from scores of longitudinal and transverse section lines painstakingly inscribed on parchment by drafting artisans, then transferred to a clay mass by means of wooden templates. The clay was then smoothed between section lines by hand to give an iconic model.

Today, the design process is largely computer-aided, the artisan sitting before a graphic display console and choosing control parameters to generate section lines, plus the surface that interpolates them, as mathematical entities. Feedback information is immediate and visual, but a computer image does not always convey enough information for the designer to make an aesthetic judgment. A line or surface that seems impeccable on the computer screen may have very

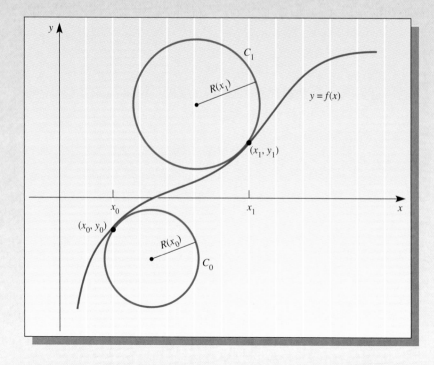

Figure 1 The curvature of circle C_0 is identical to the curvature of $y = f(x)$ at x_0; the curvature of circle C_1 is identical to the curvature of $y = f(x)$ at x_1. In each case the radius R is a function of x:

$$R(x) = \frac{\left[1 + f'(x)^2\right]^{3/2}}{\left|f''(x)\right|}$$

minuscule undulations, flat spots, or depressions that become apparent only after full-scale manufacturing.

Curvature measurements can be especially helpful in detecting and eliminating such flaws at the design level. Figures 2(a) and 2(b) show a line "before and after" adjustment, with emanating quills to represent, by their lengths, the local curvatures according to Bernoulli. Before adjustment, the line has invisible irregularities, as evidenced by occasional spikes among the quills. After adjustment, the quill pattern is quite smooth.

Calculus at a somewhat higher level provides techniques to measure curvature of surfaces. Figure 3(a) shows a typical segment of an automobile outer panel. Superimposed on the surface in Figure 3(b) are curvature measurements, communicated by means of color coding. The presence of blue, in contrast to yellow, indicates a local flattening that needs to be eliminated. Figure 3(c) shows the result of a minor mathematical adjustment: the flattening is gone.

Figure 3 On the surface of an automobile's outer panel (a), curvature measurements coded by color indicate flattenings that need to be eliminated (b). In (c), the surface appears smooth after minor mathematical adjustments are made.

General Motors Research

(a)

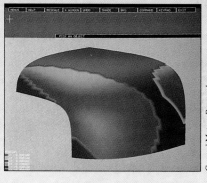

General Motors Research

(b)

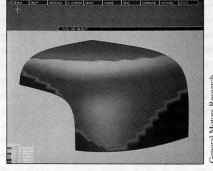

General Motors Research

(c)

Figure 2 The length of each quill emanating from the curve in (a) represents the local curvature according to Bernoulli's formula. In (b), the quills are quite regular, after adjustments have fixed the unseen undulations in the curve. (Figure from "The Porcupine Technique: Principle, Application, and Algorithm" by Klaus-Peter Beier, October 1987. Reprinted by permission of Klaus-Peter Beier.)

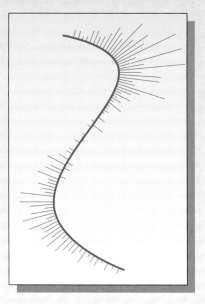

(a)

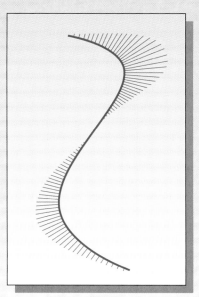

(b)

Drugs, Derivatives, and Heart Disease

Richard E. Klabunde, Ph.D.

Senior Cardiovascular Group Leader, Department of Pharmacology
Abbott Laboratories, Abbott Park, Illinois

Roberto Mercado, Ph.D.

Manager, Nonclinical Statistics, Pharmaceutical Products Division
Abbott Laboratories, Abbott Park, Illinois

(Dr. Alan Sewell served as a consultant for this article.)

A healthy heart is a flexible muscle, one that contracts vigorously to force blood out its left ventricle, then relaxes easily to repeat the process. The pressure exerted by blood leaving the heart is measured as a function of time (see Figure 1). The *rate* of change of pressure with respect to time measures the *contractility* of the heart — that is, how quickly the heart contracts and relaxes. A heart with poor contractility will demonstrate a consistently slow rate of change of pressure. Drugs are used to improve this rate.

Mathematically, we can describe the pressure of blood leaving the left ventricle as $p(t)$, where p is the pressure at a given time t. The rate of change of pressure with respect to time, then, can be described as the derivative of $p(t)$, or dp/dt. Pharmacologists are interested in both the positive rate of change, measuring the contraction of

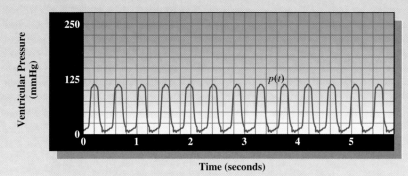

Figure 1 Left ventricular pressure is measured in mmHg as a function of time.

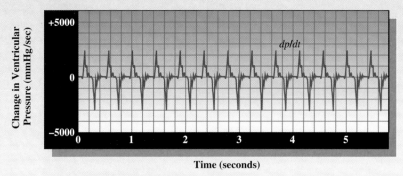

Figure 2 Change in left ventricular pressure is measured in mmHg / sec as a function of time. *dp/dt* is computed using an analog differentiator.

the heart, and the negative rate of change, measuring the relaxation of the heart. The positive rate of change is denoted by $+dp/dt$, while the negative rate of change is denoted by $-dp/dt$. The critical measurements are the maximum positive rate of change and the maximum negative rate of change: the heart's "best efforts" in contracting and relaxing.

As is often the case when we use mathematics to model the real world, an explicit equation describing $p(t)$ does not exist. Pressure is clearly a function of time, and can be measured accurately. But we cannot express the relationship except in the form of data and graphs.

Since no equation explicitly describes $p(t)$, no equation explicitly describes dp/dt. In practice, dp/dt is determined by inserting a fluid-filled plastic tube into the ventricular cavity. The tube is hooked up to a gauge and an amplifier, and pressure is measured directly. A special type of computer, called an *analog differentiator,* receives the output signal of the amplifier and numerically approximates values for $+dp/dt$ and $-dp/dt$, particularly the maximum values (see Figure 2). With these results, a physician can judge the contractility of the heart and, when a drug is being used, the effectiveness of the drug in improving contractility.

Abbott Laboratories

Pharmacologist making amplifier adjustments to the analog differentiator.

Crystals, Integration, and . . . Margarine!

Jane Gillum (Summer Intern), Alan Opsahl, Allen Blaurock

Technology Center / Kraft General Foods, Glenview, Illinois

Crystals of gems are valuable commodities, but in the manufacturing of certain foods, it is crystals of fat that are valuable. Makers of peanut butter, for example, add a small amount of a crystallizable fat — "partially hydrogenated vegetable oil" — so that a separate layer of oil doesn't form on top.

In making margarine, the rate fat crystals form is very important, because if the crystals form too rapidly, the margarine solidifies and plugs the pipes on its way to the machine where the margarine is shaped into sticks. Of course, sticks won't form at all if fat crystals form too slowly.

Crystallization is an orderly process involving two key events: (1) the creation of a crystal "seed" and (2) the growth of the crystal as oil molecules add on to the seed. The number n of crystal seeds depends on temperature, time, and the volume of oil. The size g, in meters, of the growing crystal seed depends on tem-

Figure 1 *Beta* crystals formed by very slow cooling of trilaurin, an edible oil. The feathery growth pattern started from a *single* crystal seed. Polarized-light micrograph (x 62.5).

perature and time. Given a constant temperature and a fixed volume of oil, however, both n and g are simple, constant functions of time: $n(t) = Nt$ and $g(t) = Gt$ (in meters). The rate of seed creation can thus be given as $n'(t) = N$, a constant, and the rate of size growth as $g'(t) = G$, also a constant.

The volume of the crystals is a function of the rate of seed creation, the rate of size growth, and time. It is, in fact, an integral function. Fixing the amount of oil and the temperature, the volume V, in cubic meters, of crystals can be expressed as

$$V(t) = \left(8NG^3\right)\int_0^t x^3 \, dx$$

Finding the antiderivative and evaluating,

$$V(t) = 8NG^3\left[x^4/4\right]_0^t.$$
$$= 2NG^3t^4.$$

Keen-eyed readers will recognize that this expression for V cannot be valid for arbitrarily large values of t, for then the volume of the crystals would grow arbitrarily large, too. This growth does not in fact occur, however, since the amount of oil, after all, is fixed.

Thus, the expression for V does not hold true for large values of t. It is valid only for the early period of crystallization, that is, when t is small.

Since the rate N at which crystal seeds appear varies with temperature, margarine makers alter temperature to create varying textures. Lower temperatures mean more crystal seeds and a finer texture; higher temperatures mean fewer crystal seeds and a coarser texture. Spreadability and "mouth-feel" are controlled, then, by adjusting temperature and, hence, texture.

Figure 2 *Beta prime* (dim circular regions) and *alpha* crystals (brighter regions) formed by more rapid cooling. Each region ("Maltese crosses") started with a single crystal seed. Polarized-light micrograph (x 125).

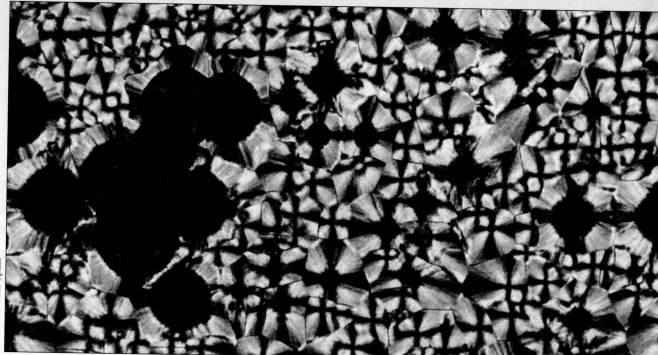

Calculus in a Cornfield

Henry Robinson

Meteorologist, National Weather Service

National Oceanic and Atmospheric Administration, Silver Springs, Maryland

Grant Heilman / Grant Heilman Photography

Predicting the weather is important for almost everybody, but for farmers it often means the difference between success and failure.

For example, agriculture specialists have found that the amount of time a crop needs to reach peak maturity is closely related to the temperature during its growing time. Predicting the daily temperature after planting can help a farmer know the optimal harvest day and schedule a manageable workload.

The relationship between temperature and growing time is expressed in terms of *growing degree days*, determined as follows: when the average temperature on a day is one degree above a certain threshold (specific to each crop), the day amounts to one growing degree day; when the average temperature on a day is two degrees above threshold, the day amounts to two growing degree days; and so on. For peas, the threshold temperature has

been found to be 39° F. Thus, if a day's average temperature is 72° F, the day amounts to 33 growing degree days for peas.

The total number of growing degree days D between the time t_p of planting and the time t_h of optimal harvest is thus a sum, expressible as an integral:

$$D = \int_{t_p}^{t_h} [T(t) - r]\, dt,$$

where $T(t)$ is the temperature at any time t and r is the threshold temperature.

Predicting t_h, the time of optimal harvest, for a crop involves knowing the threshold temperature and total number of growing degree days for that crop, and finding an explicit expression for $T(t)$, the temperature at any time t.

For virtually all cash crops, experience and experimentation has allowed agricultural specialists to determine the

threshold temperature and the total number of growing degree days. This information has not gone unnoticed by investors who deal in crop futures on the various worldwide exchanges. The missing piece of information is an explicit expression for $T(t)$.

One formulation is the Cosine Day of the Year Model for Temperature. Patterns of temperature change for individual locales tend to resemble a cosine curve, as the red line in the figure shows. Modifying the cosine function by appropriate factors results in an explicit expression of temperature as a function of time (the blue line). In general the model used is

$$T(t) = a \cdot \cos(t - b) + c,$$

where T is temperature, t is the day of the year (1 to 365), and a, b, c are constants.

Using this model allows the integral expression for the total number of growing degree days D to be evaluated, and hence the time of optimal harvest t_h to be approximated.

On the national and international scene, this information helps to predict potential shortages or oversupplies of crops. Farmers can use a simpler technique to determine the optimal harvest time, since they can reasonably consider the change in time dt to be a single day. By taking each day's temperature and subtracting the threshold, the farmer keeps a running total of growing degree days. When the total gets close to D, the farmer predicts the growing degree days on the basis of the weather forecast, thereby identifying the day for harvesting.

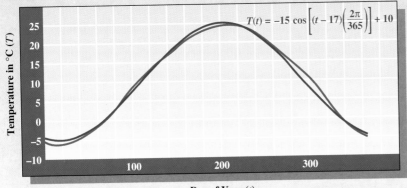

$$T(t) = -15 \cos\left[(t - 17)\left(\frac{2\pi}{365}\right)\right] + 10$$

Day of Year (t)

The Cosine Day of the Year Model for Temperature approximates the climatology of temperature for Des Moines, Iowa, with little error.

The slope corresponds to the average rate of change of y with respect to x over the interval from x_0 to $x_0 + h$. As h approaches 0, point S will slide along the curve, getting closer and closer to the fixed point R. See Figure 11.11, which shows successive positions S_1, S_2, S_3, S_4 of the point S. If the slopes of the corresponding secant lines approach a limit as h approaches 0, then this limit is defined to be the slope of the **tangent line** at point R.

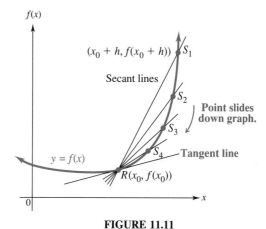

FIGURE 11.11

Tangent Line

The tangent line to the graph of $y = f(x)$ at the point $(x_0, f(x_0))$ is the line through this point having slope

$$\lim_{h \to 0} \frac{f(x_0 + h) - f(x_0)}{h},$$

provided this limit exists. If this limit does not exist, then there is no tangent at the point.

The slope of this line at a point is also called the **slope of the curve** at the point and corresponds to the instantaneous rate of change of y with respect to x at the point.

In certain applications of mathematics it is necessary to determine the equation of a line tangent to the graph of a function at a given point, as in the next example.

▶**EXAMPLE 1**   Find the slope of the tangent line to the graph of $y = x^2 + 2$ when $x = -1$. Find the equation of the tangent line.

1 Rework Example 1 for $x = 2$.

Answer:
Slope is 4; equation is $y = 4x - 2$.

Use the definition above, with $f(x) = x^2 + 2$ and $x_0 = -1$. The slope of the tangent line is

$$\text{Slope of tangent} = \lim_{h \to 0} \frac{f(x_0 + h) - f(x_0)}{h}$$

$$= \lim_{h \to 0} \frac{[(-1 + h)^2 + 2] - [(-1)^2 + 2]}{h}$$

$$= \lim_{h \to 0} \frac{[1 - 2h + h^2 + 2] - [1 + 2]}{h}$$

$$= \lim_{h \to 0} \frac{-2h + h^2}{h} = \lim_{h \to 0} (-2 + h) = -2.$$

The slope of the tangent line at $(-1, f(-1)) = (-1, 3)$ is -2. The equation of the tangent line can be found with the point-slope form of the equation of a line from Chapter 2.

$$y - y_1 = m(x - x_1)$$
$$y - 3 = -2[x - (-1)]$$
$$y - 3 = -2(x + 1)$$
$$y - 3 = -2x - 2$$
$$y = -2x + 1$$

Figure 11.12 shows a graph of $f(x) = x^2 + 2$, along with a graph of the tangent line at $x = -1$. ◀ **1**

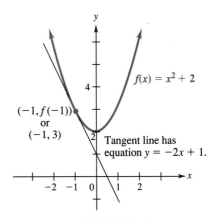

FIGURE 11.12

THE DERIVATIVE   If $y = f(x)$ is a function and x_0 is a number in its domain, then we shall use the symbol $f'(x_0)$ to denote the special limit

$$\lim_{h \to 0} \frac{f(x_0 + h) - f(x_0)}{h},$$

provided that it exists. In other words, to each number x_0, we can assign the number $f'(x_0)$ obtained by calculating this limit. This process defines an important new function.

The **derivative** of the function f is the function denoted f' whose value at the number x is defined to be the number

$$f'(x) = \lim_{h \to 0} \frac{f(x + h) - f(x)}{h},$$

provided this limit exists.

The derivative function f' has as its domain all the points at which the specified limit exists, and the value of the derivative function at the number x is the number $f'(x)$. Using x instead of x_0 here is similar to the way that $g(x) = 2x$ denotes the function that assigns to each number x_0 the number $2x_0$.

If $y = f(x)$ is a function, then its derivative is denoted either by f' or by y'. If x is a number in the domain of $y = f(x)$ such that $y' = f'(x)$ is defined, then the function f is said to be **differentiable** at x. The process that produces the function f' from the function f is called **differentiation.**

The derivative function may be interpreted in many ways, two of which were discussed above.

1. The derivative function f' gives the *slope* of the graph of f at any point. If the derivative is evaluated at $x = x_0$, then $f'(x_0)$ is the slope of the tangent line to the curve at the point $(x_0, f(x_0))$.
2. The derivative function f' gives the *instantaneous rate of change* of $y = f(x)$ with respect to x. This instantaneous rate of change can be interpreted as marginal cost, revenue, or profit (if the original function represents cost, revenue, or profit) or as velocity (if the original function represents displacement along a line). From now on we will use "rate of change" to mean "instantaneous rate of change."

The next few examples show how to use the definition to find the derivative of a function by means of a four-step procedure.

▶**EXAMPLE 2**   Let $f(x) = x^2$.
(a) Find the derivative.
By definition,

$$f'(x) = \lim_{h \to 0} \frac{f(x + h) - f(x)}{h}.$$

In attempting to evaluate this limit when it exists, x is fixed and h is the variable. Use the following sequence of steps to evaluate the limit.

2 Let $f(x) = -2x^2 + 7$. Find the following.

(a) $f(x + h)$

(b) $f(x + h) - f(x)$

(c) $\dfrac{f(x + h) - f(x)}{h}$

(d) $f'(x)$

(e) $f'(4)$

(f) $f'(0)$

Answers:

(a) $-2x^2 - 4xh - 2h^2 + 7$

(b) $-4xh - 2h^2$

(c) $-4x - 2h$

(d) $-4x$

(e) -16

(f) 0

Step 1 Find $f(x + h)$.
Replace x with $x + h$ in the equation for $f(x)$. Simplify the result.

$$f(x) = x^2$$
$$f(x + h) = (x + h)^2$$
$$= x^2 + 2xh + h^2$$

Step 2 Find $f(x + h) - f(x)$.
Since $f(x) = x^2$,

$$f(x + h) - f(x) = x^2 + 2xh + h^2 - x^2$$
$$= 2xh + h^2.$$

Step 3 Form and simplify the quotient $\dfrac{f(x + h) - f(x)}{h}$.

$$\frac{f(x + h) - f(x)}{h} = \frac{2xh + h^2}{h} = \frac{h(2x + h)}{h} = 2x + h$$

Step 4 Find the limit of the result in step 3 as h approaches 0.

$$f'(x) = \lim_{h \to 0} \frac{f(x + h) - f(x)}{h} = \lim_{h \to 0} (2x + h) = 2x + 0 = 2x$$

From the last step, the derivative of $f(x) = x^2$ is $f'(x) = 2x$.

(b) Calculate and interpret $f'(3)$.
The procedure in part (a) works for *every* x, Hence, when $x = 3$,

$$f'(3) = 2(3) = 6.$$

The number 6 is the slope of the tangent line to the graph of $f(x) = x^2$ at the point where $x = 3$, that is, at $(3, f(3)) = (3, 9)$. See Figure 11.13. ◀ **2**

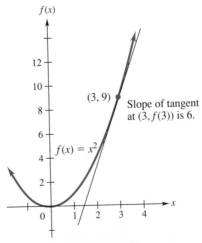

FIGURE 11.13

Caution

1. In Example 2(a), note that $f(x + h)$ is *not* equal to $f(x) + h$ because
    $$f(x + h) = (x + h)^2 + x^2 + 2xh + h^2,$$
    but
    $$f(x) + h = x^2 + h.$$

2. In Example 2(b), do not confuse $f(3)$ and $f'(3)$. The value $f(3)$ is found by substituting 3 for x in the rule of the original function $f(x) = x^2$.
    $$f(3) = 3^2 = 9$$
    On the other hand, $f'(3)$ is the slope of the tangent line to the curve $y = f(x)$ at $x = 3$. It is found by substituting 3 for x in the rule of the derivative function $f'(x) = 2x$.
    $$f'(3) = 2 \cdot 3 = 6$$

The four steps used when finding the derivative of $f(x)$ are summarized below.

Finding $f'(x)$ from the Definition of the Derivative

Step 1   Find $f(x + h)$.

Step 2   Find $f(x + h) - f(x)$.

Step 3   Divide by h to get $\dfrac{f(x + h) - f(x)}{h}$.

Step 4   Let $h \to 0$; $f'(x) = \lim\limits_{h \to 0} \dfrac{f(x + h) - f(x)}{h}$ if this limit exists.

▶ **EXAMPLE 3**   Let $f(x) = 2x^3 + 4x$. Find $f'(x)$. Then find $f'(2)$ and $f'(-3)$. Go through the four steps used above.

Step 1   Find $f(x + h)$ by replacing x with $x + h$.
$$\begin{aligned} f(x + h) &= 2(x + h)^3 + 4(x + h) \\ &= 2(x^3 + 3x^2h + 3xh^2 + h^3) + 4(x + h) \\ &= 2x^3 + 6x^2h + 6xh^2 + 2h^3 + 4x + 4h \end{aligned}$$

Step 2 $f(x + h) - f(x) = 2x^3 + 6x^2h + 6xh^2 + 2h^3 + 4x + 4h$
$$-2x^3 - 4x$$
$$= 6x^2h + 6xh^2 + 2h^3 + 4h$$

3 Let $f(x) = -3x^3 + 2x$. Find the following.

(a) $f(x + h)$

(b) $f(x + h) - f(x)$

(c) $\dfrac{f(x + h) - f(x)}{h}$

(d) $f'(x)$

(e) $f'(-1)$

Answers:

(a) $-3x^3 - 9x^2h - 9xh^2 - 3h^3 + 2x + 2h$

(b) $-9x^2h - 9xh^2 - 3h^3 + 2h$

(c) $-9x^2 - 9xh - 3h^2 + 2$

(d) $-9x^2 + 2$

(e) -7

4 Let $f(x) = -5/x$. Find the following.

(a) $f(x + h)$

(b) $f(x + h) - f(x)$

(c) $\dfrac{f(x + h) - f(x)}{h}$

(d) $f'(x)$

(e) $f'(-1)$

Answers:

(a) $\dfrac{-5}{x + h}$

(b) $\dfrac{5h}{x(x + h)}$

(c) $\dfrac{5}{x(x + h)}$

(d) $\dfrac{5}{x^2}$

(e) 5

Step 3 $\dfrac{f(x + h) - f(x)}{h} = \dfrac{6x^2h + 6xh^2 + 2h^3 + 4h}{h}$

$$= \dfrac{h(6x^2 + 6xh + 2h^2 + 4)}{h}$$

$$= 6x^2 + 6xh + 2h^2 + 4$$

Step 4 Now use the rules for limits to get

$$f'(x) = \lim_{h \to 0} \dfrac{f(x + h) - f(x)}{h} = \lim_{h \to 0} (6x^2 + 6xh + 2h^2 + 4)$$

$$= 6x^2 + 6x(0) + 2(0)^2 + 4$$

$$= 6x^2 + 4.$$

Use this result to find $f'(2)$ and $f'(-3)$.

$$f'(2) = 6 \cdot 2^2 + 4 = 28$$

$$f'(-3) = 6 \cdot (-3)^2 + 4 = 58 \quad \blacktriangleleft \quad \boxed{3}$$

▶ **EXAMPLE 4** Let $f(x) = 4/x$. Find $f'(x)$.

Step 1 $f(x + h) = \dfrac{4}{x + h}$

Step 2 $f(x + h) - f(x) = \dfrac{4}{x + h} - \dfrac{4}{x}$

$$= \dfrac{4x - 4(x + h)}{x(x + h)} \qquad \text{Find a common denominator}$$

$$= \dfrac{4x - 4x - 4h}{x(x + h)} \qquad \text{Simplify the numerator}$$

$$= \dfrac{-4h}{x(x + h)}$$

Step 3 $\dfrac{f(x + h) - f(x)}{h} = \dfrac{\dfrac{-4h}{x(x + h)}}{h}$

$$= \dfrac{-4h}{x(x + h)} \cdot \dfrac{1}{h} \qquad \text{Invert and multiply}$$

$$= \dfrac{-4}{x(x + h)}$$

Step 4 $f'(x) = \lim_{h \to 0} \dfrac{f(x + h) - f(x)}{h} = \lim_{h \to 0} \dfrac{-4}{x(x + h)}$

$$= \dfrac{-4}{x(x + 0)} = \dfrac{-4}{x(x)} = \dfrac{-4}{x^2} \quad \blacktriangleleft \quad \boxed{4}$$

▶**EXAMPLE 5** Let $f(x) = \sqrt{x}$. Find $f'(x)$.

Step 1 $f(x + h) = \sqrt{x + h}$

Step 2 $f(x + h) - f(x) = \sqrt{x + h} - \sqrt{x}$

Step 3 $\dfrac{f(x + h) - f(x)}{h} = \dfrac{\sqrt{x + h} - \sqrt{x}}{h}$

At this point, in order to be able to divide by h, multiply both numerator and denominator by $\sqrt{x + h} + \sqrt{x}$; that is, rationalize the *numerator*.

$$\frac{f(x + h) - f(x)}{h} = \frac{\sqrt{x + h} - \sqrt{x}}{h} \cdot \frac{\sqrt{x + h} + \sqrt{x}}{\sqrt{x + h} + \sqrt{x}}$$

$$= \frac{(\sqrt{x + h})^2 - (\sqrt{x})^2}{h(\sqrt{x + h} + \sqrt{x})}$$

$$= \frac{x + h - x}{h(\sqrt{x + h} + \sqrt{x})} = \frac{1}{\sqrt{x + h} + \sqrt{x}}$$

Step 4 $f'(x) = \lim\limits_{h \to 0} \dfrac{1}{\sqrt{x + h} + \sqrt{x}} = \dfrac{1}{\sqrt{x} + \sqrt{x}} = \dfrac{1}{2\sqrt{x}}$ ◀

FOR GRAPHERS

The approximate value of the derivative of a function at a particular number can be found on many graphing calculators, even if the formula for the derivative is not known, by using the "NDer" key (labeled "NDeriv" or "d/dx" on some calculators), which is usually located on the "math" or the "calc" menu. This capability is particularly useful in applications such as the following examples. Since procedures vary depending on the calculator, check your instruction manual under "derivative" or "numerical derivative."

▶**EXAMPLE 6** The cost in dollars to manufacture x graphic calculators is given by $C(x) = -.005x^2 + 20x + 150$ ($0 \le x \le 2000$). Find the rate of change of cost with respect to the number manufactured when 100 calculators are made and when 1000 calculators are made.

The rate of change of cost is given by the derivative of the cost function,

$$C'(x) = \lim_{h \to 0} \frac{C(x + h) - C(x)}{h}.$$

Going through the steps for finding $C'(x)$ gives

$$C'(x) = -.01x + 20.$$

When $x = 100$, $\qquad C'(100) = -.01(100) + 20 = 19.$

When 1000 calculators are made, the marginal cost is

$$C'(1000) = -.01(1000) + 20 = 10. \quad ◀$$

5 Go through the four steps to find $s'(t)$, the velocity of the car at any time t.

Answer:
$s'(t) = -15t^2 + 60t$

▶**EXAMPLE 7**   A sales representative for a textbook publishing company frequently makes a 4-hour drive from her home in a large city to a university in another city. If $s(t)$ represents her distance (in miles) from home t hours into the trip, then $s(t)$ is given by

$$s(t) = -5t^3 + 30t^2.$$

(a) How far from home will she be after 1 hour? After $1\frac{1}{2}$ hours?

Her distance from home after 1 hour is

$$s(1) = -5(1)^3 + 30(1)^2 = 25,$$

or 25 miles. After $1\frac{1}{2}$ (or $3/2$) hours, it is

$$s\left(\frac{3}{2}\right) = -5\left(\frac{3}{2}\right)^3 + 30\left(\frac{3}{2}\right)^2 = \frac{405}{8} = 50.625,$$

or 50.625 miles.

(b) How far apart are the two cities?

Since the trip takes 4 hours and the distance is given by $s(t)$, the university city is $s(4) = 160$ miles from her home.

(c) How fast is she driving 1 hour into the trip? $1\frac{1}{2}$ hours into the trip?

Velocity (or speed) is the instantaneous rate of change in position with respect to time. We need to find the value of the derivative $s'(t)$ at $t = 1$ and $t = 1\frac{1}{2}$. **5**

From Problem 5 at the side, $s'(t) = -15t^2 + 60t$. At $t = 1$, the velocity is

$$s'(1) = -15(1)^2 + 60(1) = 45,$$

or 45 miles per hour. At $t = 1\frac{1}{2}$, the velocity is

$$s'\left(\frac{3}{2}\right) = -15\left(\frac{3}{2}\right)^2 + 60\left(\frac{3}{2}\right) = 56.25,$$

about 56 miles per hour.

(d) Does she ever exceed the speed limit of 65 miles per hour on the trip?

To find the maximum velocity, notice that the graph of the velocity function $s'(t) = -15t^2 + 60t$ is a parabola opening downward. The maximum velocity will occur at the vertex. Verify that the vertex of the parabola is (2, 60). Thus, her maximum velocity during the trip is 60 miles per hour, so she never exceeds the speed limit. ◀

EXISTENCE OF THE DERIVATIVE   The definition of the derivative included the phrase "provided this limit exists." If the limit used to define $f'(x)$ does not exist, then of course the derivative does not exist at that x. For example, a derivative cannot exist at a point where the function itself is not defined. If there is no function value for a particular value of x, there can be no tangent line for that value. This was the case in Example 4—there was no tangent line (and no derivative) when $x = 0$.

Derivatives also do not exist at "corners" or "sharp points" on a graph. For example, the function graphed in Figure 11.14 is the *absolute value function,* defined by

$$f(x) = \begin{cases} x & \text{if} \quad x \geq 0 \\ -x & \text{if} \quad x < 0 \end{cases}$$

and written $f(x) = |x|$.

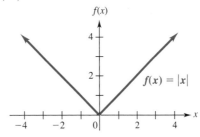

FIGURE 11.14

By the definition of derivative, the derivative at any value of x is given by

$$\lim_{h \to 0} \frac{f(x + h) - f(x)}{h},$$

provided this limit exists. To find the derivative at 0 for $f(x) = |x|$, replace x with 0 and $f(x)$ with $|0|$ to get

$$\lim_{h \to 0} \frac{|0 + h| - |0|}{h} = \lim_{h \to 0} \frac{|h|}{h}.$$

In Example 6 of Section 11.1 (with x in place of h) we showed that

$$\lim_{h \to 0} \frac{|h|}{h} \text{ does not exist.}$$

Therefore, there is no derivative at 0. However, the derivative of $f(x) = |x|$ *does* exist for all values of x other than 0.

Since a vertical line has an undefined slope, the derivative cannot exist at any point where the tangent line is vertical, as at x_5 in Figure 11.15 on the next page. Figure 11.15 summarizes the various ways that a derivative can fail to exist.

FOR GRAPHERS

Be careful when using the "NDer" key on a grapher to evaluate derivatives. Because of the approximation methods used, this key may produce a value at a point where the derivative is not defined. For instance, most calculators do this when asked for the derivative of $f(x) = |x|$ at $x = 0$, even though the derivative is not defined there, as shown above.

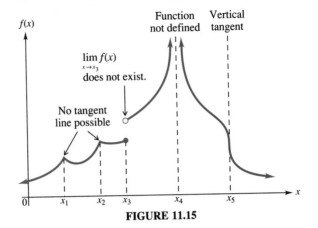

FIGURE 11.15

11.3 EXERCISES

Find the slope of the tangent line to each of the following curves when x has the given value.
(Hint for Exercise 5: in step 3, multiply numerator and denominator by $\sqrt{16+h} + \sqrt{16}$.)
(See Example 1.)

1. $f(x) = -4x^2 + 11x;\ x = -3$

2. $f(x) = 6x^2 - 4x;\ x = -2$

3. $f(x) = -\dfrac{2}{x};\ x = 4$

4. $f(x) = \dfrac{6}{x};\ x = -1$

5. $f(x) = \sqrt{x};\ x = 16$

6. $f(x) = -3\sqrt{x};\ x = 1$

Find the equation of the tangent line to each of the following curves when x has the given value. (See Example 1.)

7. $f(x) = x^2 + 2x;\ x = 2$

8. $f(x) = 6 - x^2;\ x = -2$

9. $f(x) = \dfrac{5}{x};\ x = 2$

10. $f(x) = -\dfrac{3}{x+1};\ x = 1$

11. $f(x) = 4\sqrt{x};\ x = 9$

12. $f(x) = \sqrt{x};\ x = 25$

Estimate the slope of the tangent line at the given point (x, y) in each of the following graphs.

13.

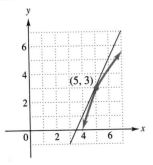

14.

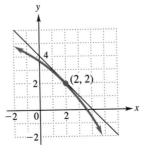

15.

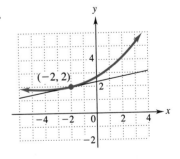

16. (a) Sketch the graph of $g(x) = \sqrt[3]{x}$ for $-1 \le x \le 1$.
   (b) Explain why the derivative of $g(x)$ is not defined at $x = 0$.
   (*Hint:* What is the slope of the tanget line at $x = 0$?)

Find $f'(x)$ for each function defined as follows. (Many of these derivatives were found in Exercises 1–6.) Then find $f'(2)$, $f'(0)$, and $f'(-3)$. (See Examples 2–5).

17. $f(x) = -4x^2 + 11x$ **18.** $f(x) = 6x^2 - 4x$ **19.** $f(x) = 8x + 6$ **20.** $f(x) = x^3 + 3x$

21. $f(x) = -\dfrac{2}{x}$ **22.** $f(x) = \dfrac{6}{x}$ **23.** $f(x) = \dfrac{4}{x - 1}$ **24.** $f(x) = \sqrt{x}$

Find all points where the functions whose graphs are shown do not have derivatives.

25.

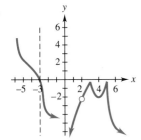

26.

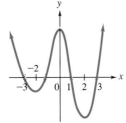

27.

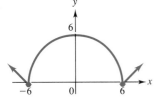

28.

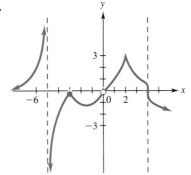

---

29. In the graph of the function f, at which of the labeled x-values is
   (a) $f(x)$ the largest?
   (b) $f(x)$ the smallest?
   (c) $f'(x)$ the smallest?
   (d) $f'(x)$ the closest to 0?

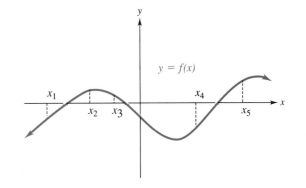

30. Draw the graph of a function f which has the following properties. (Many correct answers are possible.)
(a) $f'(x) > 0$ for $x < -2$
(b) $f'(x) = 0$ at $x = -2$
(c) $f'(x) < 0$ for $-2 < x < 3$
(d) $f'(x) = 0$ at $x = -3$
(e) $f'(x) > 0$ for $x > 3$

32. Sketch the graph of a function g with the property that $g'(x) > 0$ for $x < 0$ and $g'(x) < 0$ for $x > 0$. Many correct answers are possible.

31. Sketch the graph of the *derivative* of the function g whose graph is shown. (*Hint:* consider the slope of the tangent line at each point along the graph of g. Are there any points where there is no tangent line?)

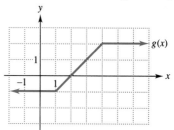

In Exercises 33–34, tell which graph, (a) or (b), represents velocity and which represents distance from a starting point. (Hint: Consider where the derivative is zero, positive, or negative.)

33. (a)

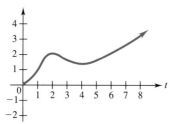

(b)

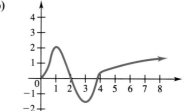

34. (a)

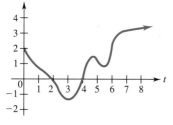

(b)

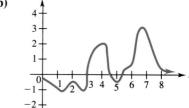

35. Management Suppose the demand for a certain item is given by $D(p) = -2p^2 + 4p + 6$, where p represents the price of the item in dollars.
(a) Given $D'(p) = -4p + 4$, find the rate of change of demand with respect to price.
(b) Find and interpret the rate of change of demand when the price is $10.

36. Management The revenue generated from the sale of x picnic tables is given by

$$R(x) = 20x - \frac{x^2}{500}.$$

(a) Find the marginal revenue when $x = 1000$ units.
(b) Determine the actual revenue from the sale of the 1001st item.
(c) Compare the answers to parts (a) and (b). How are they related?

37. Management The cost of producing x tacos is $C(x) = 1000 + .24x^2$, $0 \le x \le 30,000$.
(a) Find the marginal cost, $C'(x)$.
(b) Find and interpret $C'(100)$.
(c) Find the exact cost to produce the 101st taco.
(d) Compare the answers to parts (b) and (c). How are they related?

38. Management The profit (in dollars) from the expenditure of x thousand dollars on advertising is given by $P(x) = 1000 + 32x - 2x^2$, with $P'(x) = 32 - 4x$. Find the marginal profit at the following expenditures. In each case, decide whether the firm should increase the expenditure.
(a) $8000   (b) $6000   (c) $12,000
(d) $20,000

39. Natural Science A biologist has estimated that if a bactericide is introduced into a culture of bacteria, the number of bacteria, $B(t)$, present at time t (in hours) is given by $B(t) = 1000 + 50t - 5t^2$ million. If $B'(t) = 50 - 10t$, find the rate of change of the number of bacteria with respect to time after each of the following numbers of hours:
(a) 3   (b) 5   (c) 6.
(d) When does the population of bacteria start to decline?

40. Natural Science In one research study, the population of a certain shellfish in an area at time t was closely approximated by the graph below. Estimate and interpret the derivative at each of the marked points.

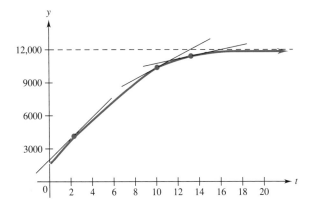

41. The figure gives the percent of outstanding credit card loans that were at least 30 days past due from June 1989 to March 1991.[*] Assume the function changes smoothly.

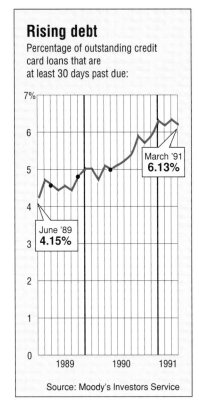

Rising debt
Percentage of outstanding credit card loans that are at least 30 days past due:

March '91
6.13%

June '89
4.15%

Source: Moody's Investors Service

(a) Estimate and interpret the rate of change in the debt level at the indicated points on the curve.
(b) During which months in 1990 did the rate of change indicate a decreasing debt level?
(c) The overall direction of the curve is upward. What does this say about the average rate of change of outstanding loans over the time period shown in the figure?

---

[*] "Credit Card Defaults Pose Worry for Banks" as appeared in *The Sacramento Bee*, June 1991. Reprinted by permission of The Associated Press.

42. Natural Science The graph shows the relationship between the speed of the arctic tern in flight and the required power expended by its flight muscles.* Several significant flight speeds are indicated on the curve.

(a) The speed V_{mp} minimizes energy costs per unit of time. What is the slope of the line tangent to the curve at the point corresponding to V_{mp}? What is the physical significance of the slope at that point?

(b) The speed V_{mr} minimizes the energy costs per unit of distance covered. Estimate the slope of the curve at the point corresponding to V_{mr}. Give the significance of the slope at that point.

(c) The speed V_{opt} minimizes the total duration of the migratory journey. Estimate the slope of the curve at the point corresponding to V_{opt}. Relate the significance of this slope to the slopes found in (a) and (b).

(d) By looking at the shape of the curve, describe how the power level decreases and increases for various speeds.

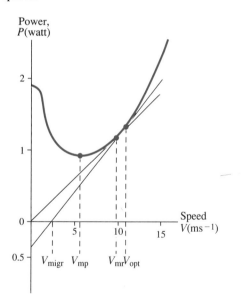

Power, P(watt)

Speed V(ms^{-1})

V_{migr} V_{mp} $V_{mr}V_{opt}$

---

*"Bird Flight and Optimal Migration" by Thomas Alerstam from *Trends in Ecology and Evolution,* July 1991, Volume 6, number 7. Reprinted by permission of Elsevier Trends Journals and Thomas Alerstam.

43. Natural Science The graph shows the temperature in degrees Celsius as a function of the altitude h in feet when an inversion layer is over Southern California. (See Exercise 20 in the previous section.) Estimate and interpret the derivatives of $T(h)$ at the marked points.

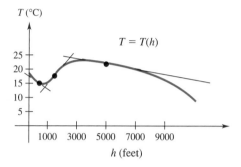

T (°C)

$T = T(h)$

h (feet)

44. Natural Science Penicillin is often injected into the muscle so that it will be absorbed slowly. At the beginning, there is no penicillin in the bloodstream, then the amount slowly increases to a maximum and slowly decays to zero again.

(a) Let $f(t)$ be the amount of penicillin in the bloodstream at time t. Sketch a possible graph of f based on the information given above.

(b) Using your graph from part (a), explain in words how the *rate* at which penicillin is entering or leaving the blood stream changes over time.

In Exercises, 45–48, use the fact that directing a grapher to graph "NDer($f(x),x$)" will produce the (approximate) graph of the derivative function of $f(x)$. *

45. (a) Graph the derivative of $f(x) = .5x^5 - 2x^3 + x^2 - 3x + 2$ for $-3 \le x \le 3$.

(b) Graph $g(x) = 2.5x^4 - 6x^2 + 2x - 3$ on the same screen.

(c) How do the graphs of $f'(x)$ and $g(x)$ compare? What does this suggest that the derivative of $f(x)$ is?

46. Repeat Exercise 45 for $f(x) = (x^2 + x + 1)^{1/3}$ (with $-6 \le x \le 6$) and $g(x) = \dfrac{2x + 1}{3(x^2 + x + 1)^{2/3}}$.

---

*On some graphing calculators, you must use either NDer $(f(x), x, x)$ or $d/dx(f(x))$; check your instruction manual.

47. By comparing graphs, as in Exercises 45 and 46, decide which of the following functions could *possibly* be the derivative of $y = \dfrac{4x^2 + x}{x^2 + 1}$?

(a) $f(x) = \dfrac{2x + 1}{2x}$ **(b)** $g(x) = \dfrac{x^2 + x}{2x}$

(c) $h(x) = \dfrac{2x + 1}{x^2 + 1}$ **(d)** $k(x) = \dfrac{-x^2 + 8x + 1}{(x^2 + 1)^2}$

48. (a) Use NDer to find the slope of the tangent line to the graph of $f(x) = .1x^3 - .2x^2 - 2x + 2$ at the point where $x = -3$.

(b) Graph $f(x)$ and the tangent line where $x = -3$ on the same screen.

11.4 TECHNIQUES FOR FINDING DERIVATIVES

In the previous section, the derivative of a function was defined as a special limit. The mathematical process of finding this limit, called *differentiation,* resulted in a new function that was interpreted in several different ways. Using the definition to calculate the derivative of a function is a very involved process even for simple functions. In this section we develop rules that make the calculation of derivatives much easier. Keep in mind that even though the process of finding a derivative will be greatly simplified with these rules, *the interpretation of the derivative will not change.*

In addition to y' and $f'(x)$, there are several other commonly used notations for the derivative.

Notations for the Derivative

The derivative of the function $y = f(x)$ may be denoted in any of the following ways.

$$f'(x), \quad y', \quad \frac{dy}{dx}, \quad \frac{d}{dx}[f(x)], \quad D_x y, \quad D_x[f(x)]$$

The dy/dx notation for the derivative is sometimes referred to as *Leibniz notation,* named after one of the co-inventors of calculus, Gottfried Wilhelm Leibniz (1646–1716). (The other was Sir Isaac Newton (1642-1727).)

1 Use the results of some of Exercises 20–24 in the previous section to find each of the following.

(a) $\dfrac{d}{dx}(x^3 + 3x)$

(b) $\dfrac{d}{dx}\left(-\dfrac{2}{x}\right)$

(c) $D_x\left(\dfrac{4}{x - 1}\right)$

(d) $D_x(\sqrt{x})$

Answers:

(a) $3x^2 + 3$

(b) $\dfrac{2}{x^2}$

(c) $\dfrac{-4}{(x - 1)^2}$

(d) $\dfrac{1}{\sqrt{x}}$

For example, the derivative of $y = 2x^3 + 4x$, which we found in Example 3 of the last section to be $y' = 6x^2 + 4$, can also be written

$$\frac{dy}{dx} = 6x^2 + 4$$

$$\frac{d}{dx}(2x^3 + 4x) = 6x^2 + 4$$

$$D_x(2x^3 + 4x) = 6x^2 + 4. \quad \boxed{1}$$

A variable other than x may be used as the independent variable. For example, if $y = f(t)$ gives population growth as a function of time, then the derivative of y with respect to t could be written

$$f'(t), \quad \frac{dy}{dt}, \quad \frac{d}{dt}[f(t)], \quad \text{or} \quad D_t[f(t)].$$

In this section the definition of the derivative,

$$f'(x) = \lim_{h \to 0} \frac{f(x + h) - f(x)}{h},$$

is used to develop some rules for finding derivatives more easily than by the four-step process of the previous section.

The first rule tells how to find the derivative of a constant function $f(x) = k$, where k is a constant real number. Since $f(x + h)$ is also k, by definition $f'(x)$ is

$$f'(x) = \lim_{h \to 0} \frac{f(x + h) - f(x)}{h}$$

$$= \lim_{h \to 0} \frac{k - k}{h} = \lim_{h \to 0} \frac{0}{h} = \lim_{h \to 0} 0 = 0,$$

establishing the following rule.

Constant Rule

If $f(x) = k$ where k is any real number, then

$$f'(x) = 0.$$

(The derivative of a constant function is 0.)

Figure 11.16 illustrates the constant rule; it shows a graph of the horizontal line $y = k$. At any point P on this line, the tangent line at P is the line itself. Since a horizontal line has a slope of 0, the slope of the tangent line is 0. This agrees with the result above: the derivative of a constant is 0.

2 Find the derivatives of the following.

(a) $y = -4$

(b) $f(x) = \pi^3$

(c) $y = 0$

Answers:

(a) 0

(b) 0

(c) 0

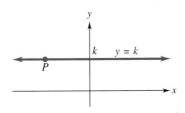

FIGURE 11.16

▶ **EXAMPLE 1**

(a) If $f(x) = 9$, then $f'(x) = 0$.

(b) If $y = \pi$, then $y' = 0$.   **(c)** If $y = 2^3$, then $dy/dx = 0$.   ◀ **2**

Functions of the form $y = x^n$, where n is a real number, are very common in applications. To get a rule for finding the derivative of such a function, we can use the definition to work out the derivatives for various special values of n. This was done in the previous section in Example 2 showing that for $f(x) = x^2$, $f'(x) = 2x$.

For $f(x) = x^3$, the derivative is found as follows.

$$f'(x) = \lim_{h \to 0} \frac{f(x + h) - f(x)}{h}$$

$$= \lim_{h \to 0} \frac{(x + h)^3 - x^3}{h}$$

$$= \lim_{h \to 0} \frac{(x^3 + 3x^2h + 3xh^2 + h^3) - x^3}{h}$$

$$= \lim_{h \to 0} \frac{3x^2h + 3xh^2 + h^3}{h}$$

$$= \lim_{h \to 0} (3x^2 + 3xh + h^2) = 3x^2$$

The results in the table below were found in a similar way, using the definition of the derivative. (These results are modifications of some of the examples and exercises from the previous section.)

Function	*n*	*Derivative*
$y = x^2$	2	$y' = 2x = 2x^1$
$y = x^3$	3	$y' = 3x^2$
$y = x^4$	4	$y' = 4x^3$
$y = x^{-1}$	-1	$y' = -1 \cdot x^{-2} = \dfrac{-1}{x^2}$
$y = x^{1/2}$	$\dfrac{1}{2}$	$y' = \dfrac{1}{2}x^{-1/2} = \dfrac{1}{2x^{1/2}}$

These results suggest the following rule.

3 Find the following.

(a) $y = x^4$; find y'

(b) $y = x^{17}$; find y'

(c) $y = x^{-2}$; find dy/dx

(d) $y = t^{-5}$; find dy/dt

(e) $D_t(t^{3/2})$

Answers:

(a) $y' = 4x^3$

(b) $y' = 17x^{16}$

(c) $dy/dx = -2/x^3$

(d) $dy/dt = -5/t^6$

(e) $D_t(t^{3/2}) = (3/2)t^{1/2}$

Power Rule

If $f(x) = x^n$ for any nonzero real number n, then
$$f'(x) = nx^{n-1}.$$

(The derivative of $f(x) = x^n$ is found by multiplying by the exponent n and decreasing the exponent on x by 1.)

For a proof of the power rule, see Exercise 67 at the end of this section.

▶ **EXAMPLE 2**

(a) If $y = x^6$, find y'.
Multiply x^{6-1} by 6.
$$y' = 6 \cdot x^{6-1} = 6x^5$$

(b) If $y = x = x^1$, find y'.
$$y' = 1 \cdot x^{1-1} = 1 \cdot x^0 = 1 \cdot 1 = 1.$$
(Recall that $x^0 = 1$ if $x \neq 0$.)

(c) If $y = t^{-3}$, find dy/dt.
$$\frac{dy}{dt} = -3 \cdot t^{-3-1} = -3 \cdot t^{-4} = -\frac{3}{t^4}.$$

(d) Find $D_t(t^{4/3})$.
$$D_t t^{(4/3)} = \frac{4}{3} t^{(4/3)-1} = \frac{4}{3} t^{1/3}.$$

(e) If $y = \sqrt{x}$, find dy/dx.
Replace $\sqrt{x}$ by the equivalent expression $x^{1/2}$. Then
$$\frac{dy}{dx} = \frac{1}{2} \cdot x^{(1/2)-1} = \frac{1}{2} x^{-1/2} = \frac{1}{2x^{1/2}} = \frac{1}{2\sqrt{x}}. \quad ◀ \; \boxed{3}$$

The next rule shows how to find the derivative of the product of a constant and a function.

Constant Times a Function

Let k be a real number. If $g'(x)$ exists, then the derivative of $f(x) = k \cdot g(x)$ is
$$f'(x) = k \cdot g'(x).$$

(The derivative of a constant times a function is the constant times the derivative of the function.)

4 Find the derivatives of the following.

(a) $y = 12x^3$

(b) $f(t) = 30t^7$

(c) $y = -35t$

(d) $y = 5\sqrt{x}$

(e) $y = -10/t$

Answers:

(a) $36x^2$

(b) $210t^6$

(c) -35

(d) $(5/2)x^{-1/2}$ or $5/(2\sqrt{x})$

(e) $10t^{-2}$ or $10/t^2$

▶ **EXAMPLE 3**

(a) If $y = 8x^4$, find y'.

Since the derivative of $g(x) = x^4$ is $g'(x) = 4x^3$ and $y = 8x^4 = 8g(x)$,

$$y' = 8g'(x) = 8(4x^3) = 32x^3.$$

(b) If $y = -\dfrac{3}{4}t^{12}$, find dy/dt.

$$\frac{dy}{dt} = -\frac{3}{4}\left[\frac{dy}{dt}(t^{12})\right] = -\frac{3}{4}(12t^{11}) = -9t^{11}$$

(c) Find $D_x(15x)$.

$$D_x(15x) = 15 \cdot D_x(x) = 15(1) = 15$$

(d) Find $D_x(10x^{3/2})$.

$$D_x(10x^{3/2}) = 10\left(\frac{3}{2}x^{1/2}\right) = 15x^{1/2}$$

(e) If $y = \dfrac{6}{x}$, find y'.

Replace $\dfrac{6}{x}$ with $6 \cdot \dfrac{1}{x}$, or $6x^{-1}$. Then

$$y' = 6(-1x^{-2}) = -6x^{-2} = -\frac{6}{x^2}. \quad ◀ \boxed{4}$$

The final rule in this section is for the derivative of a function that is a sum or difference of functions.

Sum or Difference Rule

If $f(x) = u(x) + v(x)$, and if $u'(x)$ and $v'(x)$ exist, then

$$f'(x) = u'(x) + v'(x).$$

If $f(x) = u(x) - v(x)$, then

$$f'(x) = u'(x) - v'(x).$$

(The derivative of a sum or difference of two functions is the sum or difference of the derivatives of the functions.)

For a proof of this rule, see Exercise 68 at the end of this section. This rule can be generalized to sums and differences with more than two terms.

5 Find the derivatives of the following.

(a) $y = -4x^5 - 8x + 6$

(b) $y = 32x^5 - 100x^2 + 12x$

(c) $y = 8t^{3/2} + 2t^{1/2}$

(d) $f(t) = -\sqrt{t} + 6/t$

Answers:

(a) $y' = -20x^4 - 8$

(b) $y' = 160x^4 - 200x + 12$

(c) $y' = 12t^{1/2} + t^{-1/2}$ or $12t^{1/2} + 1/t^{1/2}$

(d) $f'(t) = -1/(2\sqrt{t}) - 6/t^2$

▶ **EXAMPLE 4** Find the derivatives of the following functions.

(a) $y = 6x^3 + 15x^2$

Let $u(x) = 6x^3$ and $v(x) = 15x^2$; then $y = u(x) + v(x)$. Since $u'(x) = 18x^2$ and $v'(x) = 30x$,

$$\frac{dy}{dx} = 18x^2 + 30x.$$

(b) $p(t) = 8t^4 - 6\sqrt{t} + \dfrac{5}{t}$

Rewrite $p(t)$ as $p(t) = 8t^4 - 6t^{1/2} + 5t^{-1}$; then $p'(t) = 32t^3 - 3t^{-1/2} - 5t^{-2}$, which also may be written as $p'(t) = 32t^3 - \dfrac{3}{\sqrt{t}} - \dfrac{5}{t^2}$.

(c) $f(x) = 5\sqrt[3]{x^2} + 4x^{-2} + 7$

Rewrite $f(x)$ as $f(x) = 5x^{2/3} + 4x^{-2} + 7$. Then

$$D_x[f(x)] = \frac{10}{3}(x^{-1/3}) - 8x^{-3},$$

or

$$D_x[f(x)] = \frac{10}{3\sqrt[3]{x}} - \frac{8}{x^3}. \quad \blacktriangleleft \quad \boxed{5}$$

The rules developed in this section make it possible to find the derivative of a function more directly, so that applications of the derivative can be dealt with more effectively. The following examples illustrate some business applications.

MARGINAL ANALYSIS In business and economics the rates of change of such variables as cost, revenue, and profit are important considerations. Economists use the word marginal to refer to rates of change: for example, *marginal cost* refers to the rate of change of cost. Since the derivative of a function gives the rate of change of the function, a marginal cost (or revenue, or profit) function is found by taking the derivative of the cost (or revenue, or profit) function. Roughly speaking, the marginal cost at some level of production x is the cost to produce the $(x + 1)$st item, as we now show. (Similar statements could be made for revenue or profit.)

Look at Figure 11.17, where $C(x)$ represents the cost of producing x units of some item. Then the cost of producing $x + 1$ units is $C(x + 1)$. The cost of the $(x + 1)$st unit is, therefore, $C(x + 1) - C(x)$. This quantity is shown on the graph in Figure 11.17.

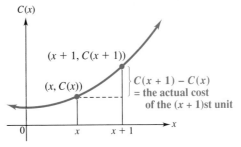

FIGURE 11.17

Now if $C(x)$ is the cost function, then the marginal cost $C'(x)$ represents the slope of the tangent line at any point $(x, C(x))$. The graph in Figure 11.18 shows the cost function $C(x)$ and the tangent line at point $P = (x, C(x))$. We know that the slope of the tangent line is $C'(x)$ and that the slope can be computed using the triangle PQR in Figure 11.18.

$$C'(x) = \text{slope} = \frac{QR}{PR} = \frac{QR}{1} = QR$$

So the length of the line segment QR is the number $C'(x)$.

Superimposing the graphs from Figures 11.17 and 11.18 as in Figure 11.19 shows that $C'(x)$ is indeed very close to $C(x + 1) - C(x)$. The two values are closest when $C'(x)$ is very large, so that 1 unit is relatively small.

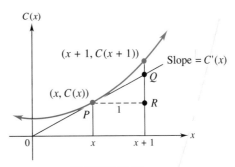

FIGURE 11.18

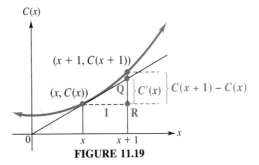

FIGURE 11.19

6 The cost in dollars to produce x units of wheat is given by
$C(x) = 5000 + 20x + 10\sqrt{x}$.
Find the marginal cost when

(a) $x = 9$;

(b) $x = 16$;

(c) $x = 25$.

(d) As more wheat is produced, what happens to the marginal cost?

Answers:

(a) $65/3 \approx \$21.67$

(b) $\$85/4 \approx \$21.25$

(c) $\$21$

(d) It decreases and approaches $20.

▶**EXAMPLE 5** Suppose that the total cost in hundreds of dollars to produce x thousand barrels of a beverage is given by

$$C(x) = 4x^2 + 100x + 500 \; (0 \le x \le 50).$$

Find the marginal cost for the following values of x.

(a) $x = 5$

To find the marginal cost, first find $C'(x)$, the derivative of the total cost function.

$$C'(x) = 8x + 100$$

When $x = 5$,

$$C'(5) = 8(5) + 100 = 140.$$

After 5 thousand barrels of the beverage have been produced, the cost to produce 1 thousand more barrels will be *approximately* 140 hundred dollars, or \$14,000.

The *actual* cost to produce 1 thousand more barrels is $C(6) - C(5)$.

$$C(6) - C(5) = (4 \cdot 6^2 + 100 \cdot 6 + 500) - (4 \cdot 5^2 + 100 \cdot 5 + 500)$$
$$= 1244 - 1100$$
$$= 144$$

The actual cost is 144 hundred dollars, or \$14,400.

(b) $x = 30$

After 30 thousand barrels have been produced, the cost to produce 1 thousand more barrels will be approximately

$$C'(30) = 8(30) + 100 = 340,$$

or \$34,000. Notice that the cost to produce an additional thousand barrels of beverage has increased by approximately \$20,000 at a production level of 30 thousand barrels compared to a production level of 5 thousand barrels. Management must be careful to keep track of marginal costs. If the marginal cost of producing an extra unit exceeds the revenue received from selling it, then the company will lose money on that unit. ◀ **6**

DEMAND FUNCTIONS The **demand function,** defined by $p = f(x)$, relates the number of units x of an item that consumers are willing to purchase at the price p. (Demand functions were also discussed in Chapter 2.) The total revenue $R(x)$ is related to price per unit and the amount demanded (or sold) by the equation

$$R(x) = xp = x \cdot f(x).$$

▶**EXAMPLE 6** The demand function for a certain product is given by

$$p = \frac{50,000 - x}{25,000}.$$

Find the marginal revenue when $x = 10,000$ units and p is in dollars.

7 Suppose the demand function for x units of an item is

$$p = 5 - \frac{x}{1000},$$

Where x is the price in dollars. Find

(a) the marginal revenue;

(b) marginal revenue at $x = 500$;

(c) marginal revenue at $x = 1000$.

Answers:

(a) $R'(x) = 5 - \dfrac{x}{500}$

(b) $4

(c) $3

From the given function for p, the revenue function is given by

$$R(x) = xp$$
$$= x\left(\frac{50,000 - x}{25,000}\right)$$
$$= \frac{50,000x - x^2}{25,000} = 2x - \frac{1}{25,000}x^2.$$

The marginal revenue is

$$R'(x) = 2 - \frac{2}{25,000}x.$$

When $x = 10,000$, the marginal revenue is

$$R'(10,000) = 2 - \frac{2}{25,000}(10,000) = 1.2,$$

or $1.20 per unit. Thus, the next unit sold (at sales of 10,000) will produce additional revenue of about $1.20 per unit. ◀ **7**

In economics, the demand function is written in the form $p = f(x)$, as shown above. From the perspective of a consumer, it is probably more reasonable to think of the quantity demanded as a function of price. Mathematically, these two viewpoints are equivalent. In Example 6, the demand function could have been written from the consumer's viewpoint as

$$x = 50,000 - 25,000p.$$

▶ **EXAMPLE 7**   Suppose that the cost function for the product in Example 6 is given by

$$C(x) = 2100 + .25x \quad \text{where } 0 \le x \le 30,000.$$

Find the marginal profit from the production of the following numbers of units.
(a) 15,000
     From Example 6, the revenue from the sale of x units is

$$R(x) = 2x - \frac{1}{25,000}x^2.$$

Since profit, P, is given by $P = R - C$,

$$P(x) = R(x) - C(x)$$
$$= \left(2x - \frac{1}{25,000}x^2\right) - (2100 + .25x)$$
$$= 2x - \frac{1}{25,000}x^2 - 2100 - .25x$$
$$= 1.75x - \frac{1}{25,000}x^2 - 2100.$$

8 For a certain product, the cost is $C(x) = 1250 + .75x$, and the revenue is

$$R(x) = 5x - \frac{x^2}{10,000} \text{ for } x \text{ units.}$$

(a) Find the profit $P(x)$.

(b) Find $P'(20,000)$.

(c) Find $P'(30,000)$.

(d) Interpret the results of (b) and (c).

Answers:

(a) $P(x) = 4.25x - x^2/10,000$
$\qquad\qquad\qquad - 1250$

(b) .25

(c) -1.75

(d) Profit is increasing by $.25 per unit at 20,000 units in (b) and decreasing by $1.75 per unit at 30,000 units in (c).

The marginal profit from the sale of x units is

$$P'(x) = 1.75 - \frac{2}{25,000}x = 1.75 - \frac{1}{12,500}x.$$

At $x = 15,000$, the marginal profit is

$$P'(15,000) = 1.75 - \frac{1}{12,500}(15,000) = .55,$$

or $.55 per unit.

(b) 21,875

When $x = 21,875$, the marginal profit is

$$P'(21,875) = 1.75 - \frac{1}{12,500}(21,875) = 0.$$

(c) 25,000

When $x = 25,000$, the marginal profit is

$$P'(25,000) = 1.75 - \frac{1}{12,500}(25,000) = -.25,$$

or $-$.25 per unit.

As shown by parts (b) and (c), if more than 21,875 units are sold, the marginal profit is negative. This indicates that increasing production beyond that level will *reduce* profit. ◀ **8**

The final example shows a medical application of the derivative as the rate of change of a function.

▶**EXAMPLE 8** A tumor has the approximate shape of a cone. See Figure 11.20. The radius of the tumor is fixed by the bone structure at 2 centimeters, but the tumor is growing along the height of the cone. The formula for the volume of a cone is $V = \frac{1}{3}\pi r^2 h$, where r is the radius of the base and h is the height of the cone. Find the rate of change in the volume of the tumor with respect to the height.

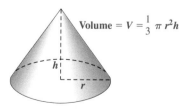

Volume $= V = \frac{1}{3}\pi r^2 h$

FIGURE 11.20

9 A balloon is spherical. The formula for the volume of a sphere is $V = (4/3)\pi r^3$, where r is the radius of the sphere. Find the following.

(a) dV/dr

(b) The rate of change of the volume when $r = 3$ inches

Answers:

(a) $4\pi r^2$

(b) 36π cubic inches per inch

To emphasize that the rate of change of the volume is found with respect to the height, use the symbol dV/dh for the derivative. For this tumor, r is fixed at 2 cm. By substituting 2 for r,

$$V = \frac{1}{3}\pi r^2 h \quad \text{becomes} \quad V = \frac{1}{3}\pi \cdot 2^2 \cdot h \quad \text{or} \quad V = \frac{4}{3}\pi h.$$

Since $4\pi/3$ is constant,

$$\frac{dV}{dh} = \frac{4\pi}{3} \approx 4.2 \text{ cu cm per cm.}$$

For each additional centimeter that the tumor grows in height, its volume will increase approximately 4.2 cubic centimeters. ◀ **9**

11.4 EXERCISES

Find the derivatives of the following functions. (See Examples 1–4.)

1. $f(x) = 4x^2 - 6x + 5$

2. $g(x) = 8x^2 + x - 12$

3. $y = 2x^3 + 3x^2 - 5x + 2$

4. $y = 4x^3 + 4x + 4$

5. $g(x) = x^4 + 3x^3 - 6x - 7$

6. $f(x) = 5x^6 - 3x^4 + x^3 - 3x + 9$

7. $f(x) = 6x^{1.5} - 4x^{.5}$

8. $f(x) = -2x^{2.5} + 8x^{.5}$

9. $y = -15x^{3.2} + 2x^{1.9}$

10. $y = 18x^{1.6} - 4x^{3.1}$

11. $y = 24t^{3/2} + 4t^{1/2}$

12. $y = -24t^{5/2} - 6t^{1/2}$

13. $y = 8\sqrt{x} + 6x^{3/4}$

14. $y = -100\sqrt{x} - 11x^{2/3}$

15. $g(x) = 6x^{-5} - x^{-1}$

16. $y = 4x^{-3} + x^{-1} + 5$

17. $y = 10x^{-2} + 3x^{-4} - 6x$

18. $y = x^{-5} - x^{-2} + 5x^{-1}$

19. $f(t) = \dfrac{6}{t} - \dfrac{8}{t^2}$

20. $f(t) = \dfrac{4}{t} + \dfrac{2}{t^3}$

21. $y = \dfrac{9 - 8x + 2x^3}{x^4}$

22. $y = \dfrac{3 + x - 7x^4}{x^6}$

23. $g(x) = 8x^{-1/2} - 5x^{1/2} + x$

24. $f(x) = -12x^{-1/2} + 12x^{1/2} - 12x$

25. $y = 4x^{-3/2} + 8x^{-1/2} + x^2$

26. $y = 2x^{1/2} + 5 + 2x^{-1/2} + x^{-3/2}$

27. $y = \dfrac{6}{\sqrt[4]{x}}$

28. $y = \dfrac{-2}{\sqrt[3]{x}}$

29. $y = \dfrac{-5t}{\sqrt[3]{t^2}}$

30. $g(t) = \dfrac{9t}{\sqrt{t^3}}$

Find each of the following.

31. $\dfrac{dy}{dx}$ if $y = 8x^{-5} - 9x^{-4}$

32. $\dfrac{dy}{dx}$ if $y = -3x^{-2} - 4x^{-5}$

33. $D_x\left(9x^{-1/2} + \dfrac{2}{x^{3/2}}\right)$

34. $D_x\left(\dfrac{8}{\sqrt[4]{x}} - \dfrac{3}{\sqrt{x^3}}\right)$

35. $f'(-2)$ if $f(x) = 6x^2 - 4x$

36. $f'(3)$ if $f(x) = 9x^3 - 8x^2$

37. $f'(4)$ if $f(t) = 2\sqrt{t} - \dfrac{3}{\sqrt{t}}$

38. $f'(8)$ if $f(t) = -5\sqrt[3]{t} + \dfrac{6}{\sqrt[3]{t}}$

39. If $f(x) = \dfrac{(3x^2 + x)^2}{7}$, which of the following is *closest* to $f'(1)$?

(a) -12 (b) -9 (c) -6

(d) -3 (e) 0 (f) 3

40. If $g(x) = -3x^{3/2} + 4x^2 - 9x$, which of the following is closest to $g'(4)$?

(a) 3 (b) 6 (c) 9

(d) 12 (e) 15 (f) 18

Find the slope and the equation of the tangent line to the graph of each function at the given value of x.

41. $f(x) = x^4 - 2x^2 + 1; \; x = 1$

42. $g(x) = -x^5 + 4x^2 - 2x + 2; \; x = 2$

43. $y = 4x^{1/2} + 2x^{3/2} + 1; \; x = 4$

44. $y = -x^{-3} + 5x^{-1} + x; \; x = 2$

---

45. What is the y-intercept of the tangent line to the graph of $y = x^3 - 2x^2 - 3$ at the point where $x = 1$?

(a) -12 (b) -9 (c) -6

(d) -3 (e) 0 (f) 3

46. If the cost of mining x tons of coal is $f(x)$ dollars, what does it mean to say that $f'(3000) = 75$?

Work the following exercises. (See Examples 5–8.)

47. Management The profit in dollars from the sale of x expensive cassette recorders is

$$P(x) = x^3 - 5x^2 + 7x + 10.$$

Find the marginal profit for the following values of x.

(a) $x = 4$ (b) $x = 8$ (c) $x = 10$

(d) $x = 12$

48. Management The total cost to produce x handcrafted weathervanes is

$$C(x) = 100 + 8x - x^2 + 4x^3.$$

Find the marginal cost for the following values of x.

(a) $x = 0$ (b) $x = 4$ (c) $x = 6$

(d) $x = 8$

49. Management The cost (in thousands of dollars) of manufacturing x sailboats is given by

$$C(x) = 600 + x + 42x^{2/3} \; (0 \le x \le 100).$$

(a) Find the marginal cost function.

(b) What is the marginal cost at $x = 40$?

(c) What is the actual cost of manufacturing the 41st sailboat?

(d) Is the marginal cost at $x = 40$ a reasonable approximation of the actual cost of making the 41st sailboat?

50. Management Often sales of a new product grow rapidly at first and then level off with time. This is the case with the sales represented by the function

$$S(t) = 100 - 100t^{-1},$$

where t represents time in years. Find the rate of change of sales for the following values of t.

(a) $t = 1$ (b) $t = 10$

51. The revenue from selling x wallets is given by

$$R(x) = 201\sqrt[3]{x} + 2x \quad (4 \le x \le 80).$$

The cost of manufacturing x wallets is given by

$$C(x) = .1x^2 + 5x + 40.$$

(a) Find the profit function.

(b) What is the profit on selling 10 wallets? 20 wallets? 30 wallets? 50 wallets?

(c) Find the marginal profit function.

(d) What is the marginal profit at $x = 10$? at $x = 20$? at $x = 30$? at $x = 50$?

(e) What is the relationship between your answers in parts (b) and (d)?

52. Management An analyst has found that a company's costs and revenues for one product are given by

$$C(x) = 2x \quad \text{and} \quad R(x) = 6x - \frac{x^2}{1000},$$

respectively, where x is the number of items produced.

(a) Find the marginal cost function.

(b) Find the marginal revenue function.

(c) Using the fact that profit is the difference between revenue and costs, find the marginal profit function,

(d) What value of x makes marginal profit equal 0?

(e) Find the profit when the marginal profit is 0.

(As we shall see in the next chapter, this process is used to find *maximum* profit.)

53. Natural Science Suppose $P(t) = 100/t$ represents the percent of acid in a chemical solution after t days of

exposure to an ultraviolet light source. Find the percent of acid in the solution after the following number of days.
(a) 1 day (b) 100 days
(c) Find and interpret $P'(100)$.

54. **Management** If the price of a product is given by

$$P(x) = \frac{1000}{x} + 1000,$$

where x represents the demand for the product, find the rate of change of price when the demand is 10.

55. **Social Science** Based on population trends from 1990 to 1995, it is predicted that the population of a certain city in year t will be given by

$$P(t) = 4t^2 + 2000\sqrt{t} + 50{,}000 \quad (0 \le t \le 30),$$

where $t = 0$ corresponds to 1990.
(a) At what rate (in people per year) is the population changing in year t?
(b) Use part (a) to explain why the population is growing during this 30-year period.
(c) What will the population growth rate be in the years 1996, 2000, and 2010?
(d) Does the population of the city in 1990 affect the growth rate in subsequent years? Why?

56. **Natural Science** A short length of blood vessel has a cylindrical shape. The volume of a cylinder is given by $V = \pi r^2 h$. Suppose an experimental device is set up to measure the volume of blood in a blood vessel of fixed length 80 mm as the radius changes.
(a) Find dV/dr.
Suppose a drug is administered which causes the blood vessel to expand. Evaluate dV/dr for the following values of r and interpret your answers.
(b) 4 mm (c) 6 mm (d) 8 mm

57. **Management** Assume that a demand equation is given by $x = 5000 - 100p$. Find the marginal revenue for the following production levels (values of x). (Hint: solve the demand equation for p and use $R(x) = xp$.)
(a) 1000 units (b) 2500 units (c) 3000 units

58. **Management** Suppose that for the situation in Exercise 57, the cost of producing x units is given by $C(x) = 3000 - 20x + .03x^2$. Find the marginal profit for each of the following production levels.
(a) 500 units (b) 815 units (c) 1000 units

59. **Natural Science** In an experiment testing methods of sexually attracting male insects to sterile females, equal numbers of males and females of a certain species are permitted to intermingle. Assume that

$$M(t) = 4t^{3/2} + 2t^{1/2}$$

approximates the number of matings observed among the insects in an hour, where t is the temperature in degrees Celsius. (This formula is only valid for certain temperature ranges.) Find each of the following.
(a) $M(16)$ (b) $M(25)$
(c) The rate of change of M when $t = 16$
(d) Interpret your answer for part (c).

60. **Social Science** Living standards are defined by the total output of goods and services divided by the total population. In the United States during the 1980s, living standards were closely approximated by

$$f(x) = -.023x^3 + .3x^2 - .4x + 11.6,$$

where $x = 0$ corresponds to 1981. Find the derivative of f. Use the derivative to find the rate of change in living standards in the following years.
(a) 1981 (b) 1983 (c) 1988
(d) 1989 (e) 1990
(f) What do your answers to parts (a)–(e) tell you about living standards in those years?

Physical Science *We saw earlier that the velocity of a particle moving in a straight line is given by*

$$\lim_{h \to 0} \frac{s(t + h) - s(t)}{h},$$

where s(t) gives the position of the particle at time t. This limit is the derivative of s(t), so the velocity of a particle is given by s'(t). If v(t) represents velocity at time t, then v(t) = s'(t). For each of the following position functions, find (a) v(t); (b) the velocity when t = 0, t = 5, and t = 10.

61. $s(t) = 8t^2 + 3t + 1$

62. $s(t) = 10t^2 - 5t + 6$

63. $s(t) = 2t^3 + 6t^2$

64. $s(t) = -t^3 + 3t^2 + t - 1$

65. Physical Science If a rock is dropped from a 144-ft building, its position (in feet above the ground) is given by $s(t) = -16t^2 + 144$, where t is the time in seconds since it was dropped.
(a) What is its velocity 1 second after being dropped? 2 seconds after being dropped?
(b) When will it hit the ground?
(c) What is its velocity upon impact?

66. Physical Science A ball is thrown vertically upward from the ground at a velocity of 64 feet per second. Its distance from the ground at t seconds is given by $s(t) = -16t^2 + 64t$.
(a) How fast is the ball moving 2 seconds after being thrown? 3 seconds after being thrown?
(b) How long after the ball is thrown does it reach its maximum height?
(c) How high will it go?

67. Perform each step and give reasons for your results in the following proof that the derivative of $y = x^n$ is $y' = n \cdot x^{n-1}$. (We prove this result only for positive integer values of n, but it is valid for all values of n.)

(a) Recall the binomial theorem from algebra:
$$(p + q)^n = p^n + n \cdot p^{n-1}q$$
$$+ \frac{n(n-1)}{2}p^{n-2}q^2 + \cdots + q^n.$$
Let $y = (x + h)^n$, and evaluate $(x + h)^n$.
(b) Find the quotient $\dfrac{(x + h)^n - x^n}{h}$.
(c) Use the definition of derivative to find y'.

68. Perform each step and give reasons for your result in the proof that the derivative of $y = f(x) + g(x)$ is $y' = f'(x) + g'(x)$.
(a) Let $s(x) = f(x) + g(x)$. Show that
$$s'(x) =$$
$$\lim_{h \to 0} \frac{[f(x + h) + g(x + h)] - [f(x) + g(x)]}{h}.$$
(b) Show that $s'(x)$ equals
$$\lim_{h \to 0} \left[\frac{f(x + h) - f(x)}{h} + \frac{g(x + h) - g(x)}{h} \right].$$
(c) Finally, show that $s'(x) = f'(x) + g'(x)$.

 For each of the following, use a computer or a graphing calculator to graph each function and its derivative on the same screen. Determine the values of x where the derivative is **(a)** *positive,* **(b)** *zero, and* **(c)** *negative.* **(d)** *What is true of the graph of the function in each case?*

69. $g(x) = 6 - 4x + 3x^2 - x^3$

70. $k(x) = 2x^4 - 3x^3 + x$

11.5 DERIVATIVES OF PRODUCTS AND QUOTIENTS

In the previous section we developed several rules for finding derivatives. We develop two additional rules in this section, again using the definition of the derivative.

The derivative of a sum of two functions is found from the sum of the derivatives. What about products? Is the derivative of a product equal to the product of the derivatives? For example, if

$$u(x) = 2x + 3 \quad \text{and} \quad v(x) = 3x^2;$$

then

$$u'(x) = 2 \quad \text{and} \quad v'(x) = 6x.$$

Let $f(x)$ be the product of u and v; that is, $f(x) = (2x + 3)(3x^2) = 6x^3 + 9x^2$. By the rules of the preceding section, $f'(x) = 18x^2 + 18x$. On the other hand, $u'(x) = 2$ and $v'(x) = 6x$, with the product $u'(x) \cdot v'(x) = 2(6x) = 12x \neq f'(x)$. In this example, the derivative of a product is *not* equal to the product of the derivatives, nor is this usually the case.

The rule for finding derivatives of products is given below.

Product Rule

If $f(x) = u(x) \cdot v(x)$, and if both $u'(x)$, and $v'(x)$ exist, then

$$f'(x) = u(x) \cdot v'(x) + v(x) \cdot u'(x).$$

(The derivative of a product of two functions is the first function times the derivative of the second, plus the second function times the derivative of the first.)

To sketch the method used to prove the product rule, let

$$f(x) = u(x) \cdot v(x).$$

Then $f(x + h) = u(x + h) \cdot v(x + h)$, and, by definition, $f'(x)$ is given by

$$f'(x) = \lim_{h \to 0} \frac{f(x + h) - f(x)}{h}$$

$$= \lim_{h \to 0} \frac{(x + h) \cdot v(x + h) - u(x) \cdot v(x)}{h}.$$

Now subtract and add $u(x + h) \cdot v(x)$ in the numerator, giving

$$f'(x) = \lim_{h \to 0} \frac{u(x + h) \cdot v(x + h) - u(x + h) \cdot v(x) + u(x + h) \cdot v(x) - u(x) \cdot v(x)}{h}$$

$$= \lim_{h \to 0} \frac{u(x + h)[v(x + h) - v(x)] + v(x)[u(x + h) - u(x)]}{h}$$

$$= \lim_{h \to 0} u(x + h)\left[\frac{v(x + h) - v(x)}{h}\right] + \lim_{h \to 0} v(x)\left[\frac{u(x + h) - u(x)}{h}\right]$$

$$= \lim_{h \to 0} u\,(x + h) \cdot \lim_{h \to 0} \frac{v(x + h) - v(x)}{h} + \lim_{h \to 0} v(x) \cdot \lim_{h \to 0} \frac{u(x + h) - u(x)}{h} \qquad (*)$$

If u' and v' both exist, then

$$\lim_{h \to 0} \frac{u(x + h) - u(x)}{h} = u'(x) \quad \text{and} \quad \lim_{h \to 0} \frac{v(x + h) - v(x)}{h} = v'(x).$$

The fact that u' exists can be used to prove

$$\lim_{h \to 0} u(x + h) = u(x),$$

and since no h is involved in $v(x)$,

$$\lim_{h \to 0} v(x) = v(x).$$

1 Use the product rule to find the derivatives of the following.

(a) $f(x) = (5x^2 + 6)(3x)$

(b) $g(x) = (8x)(4x^2 + 5x)$

Answers:

(a) $45x^2 + 18$

(b) $96x^2 + 80x$

2 Find the derivatives of the following.

(a) $f(x) = (x^2 - 3)(\sqrt{x} + 5)$

(b) $g(x) = (\sqrt{x} + 4)(5x^2 + x)$

Answers:

(a) $\dfrac{5}{2}x^{3/2} + 10x - \dfrac{3}{2}x^{-1/2}$

(b) $\dfrac{25}{2}x^{3/2} + 40x$

$+ \dfrac{3}{2}x^{1/2} + 4$

Substituting these results into equation (*) gives

$$f'(x) = u(x) \cdot v'(x) + v(x) \cdot u'(x),$$

the desired result.

▶ **EXAMPLE 1** Let $f(x) = (2x + 3)(3x^2)$. Use the product rule to find $f'(x)$.

Here f is given as the product of $u(x) = 2x + 3$ and $v(x) = 3x^2$. By the product rule and the fact that $u'(x) = 2$ and $v'(x) = 6x$,

$$f'(x) = u(x) \cdot v'(x) + v(x) \cdot u'(x)$$
$$= (2x + 3)(6x) + (3x^2)(2)$$
$$= 12x^2 + 18x + 6x^2 = 18x^2 + 18x.$$

This result is the same as that found at the beginning of the section. ◀ **1**

▶ **EXAMPLE 2** Find the derivative of $y = (\sqrt{x} + 3)(x^2 - 5x)$.

Let $u(x) = \sqrt{x} + 3 = x^{1/2} + 3$, and $v(x) = x^2 - 5x$. Then

$$y' = u(x) \cdot v'(x) + v(x) \cdot u'(x)$$

$$= (x^{1/2} + 3)(2x - 5) + (x^2 - 5x)\left(\frac{1}{2}x^{-1/2}\right).$$

$$= 2x^{3/2} + 6x - 5x^{1/2} - 15 + \frac{1}{2}x^{3/2} - \frac{5}{2}x^{1/2}$$

$$= \frac{5}{2}x^{3/2} + 6x - \frac{15}{2}x^{1/2} - 15. ◀ **2**$$

We could have found the derivatives above by multiplying out the original functions. The product rule then would not have been needed. In the next section, however, we shall see products of functions where the product rule is essential.

What about *quotients* of functions? To find the derivative of the quotient of two functions, use the next rule.

Quotient Rule

If $f(x) = \dfrac{u(x)}{v(x)}$, if all indicated derivatives exist, and if $v(x) \neq 0$, then

$$f'(x) = \frac{v(x) \cdot u'(x) - u(x) \cdot v'(x)}{[v(x)]^2}.$$

(The derivative of a quotient is the denominator times the derivative of the numerator, minus the numerator times the derivative of the denominator, all divided by the square of the denominator.)

3 Find the derivatives of the following.

(a) $f(x) = \dfrac{3x + 7}{5x + 8}$

(b) $g(x) = \dfrac{2x + 11}{5x - 1}$

Answers:

(a) $\dfrac{-11}{(5x + 8)^2}$

(b) $\dfrac{-57}{(5x - 1)^2}$

4 Find each derivative. Write answers with positive exponents.

(a) $D_x\left(\dfrac{x^{-2} - 1}{x^{-1} + 2}\right)$

(b) $D_x\left(\dfrac{2 + x^{-1}}{x^{-3} + 1}\right)$

Answers:

(a) $\dfrac{-1 - 4x - x^2}{x^2 + 4x^3 + 4x^4}$

(b) $\dfrac{2x + 6x^2 - x^4}{1 + 2x^3 + x^6}$

The proof of the quotient rule is similar to that of the product rule and is omitted here.

▶ **EXAMPLE 3**　Find $f'(x)$ if $f(x) = \dfrac{2x - 1}{4x + 3}$.

Let $u(x) = 2x - 1$, with $u'(x) = 2$. Also, let $v(x) = 4x + 3$, with $v'(x) = 4$. Then, by the quotient rule,

$$f'(x) = \frac{v(x) \cdot u'(x) - u(x) \cdot v'(x)}{[v(x)]^2}$$

$$= \frac{(4x + 3)(2) - (2x - 1)(4)}{(4x + 3)^2}$$

$$= \frac{8x + 6 - 8x + 4}{(4x + 3)^2}$$

$$f'(x) = \frac{10}{(4x + 3)^2}. \quad ◀ \;\; \boxed{3}$$

▶ **EXAMPLE 4**　Find $D_x\left(\dfrac{x^{-1} - 2}{x^{-2} + 4}\right)$.

Use the quotient rule.

$$D_x\left(\frac{x^{-1} - 2}{x^{-2} + 4}\right) = \frac{(x^{-2} + 4) \cdot D_x(x^{-1} - 2) - (x^{-1} - 2) \cdot D_x(x^{-2} + 4)}{(x^{-2} + 4)^2}$$

$$= \frac{(x^{-2} + 4)(-x^{-2}) - (x^{-1} - 2)(-2x^{-3})}{(x^{-2} + 4)^2}$$

$$= \frac{-x^{-4} - 4x^{-2} + 2x^{-4} - 4x^{-3}}{(x^{-2} + 4)^2} = \frac{x^{-4} - 4x^{-3} - 4x^{-2}}{(x^{-2} + 4)^2}$$

When derivatives are used to solve practical problems, the work is easier if the derivatives are first simplified. This derivative, for example, can be simplified if the negative exponents are removed by multiplying numerator and denominator by x^4, and using $(x^{-2} + 4)^2 = x^{-4} + 8x^{-2} + 16$.

$$\frac{x^4(x^{-4} - 4x^{-3} - 4x^{-2})}{x^4(x^{-2} + 4)^2} = \frac{1 - 4x - 4x^2}{x^4(x^{-4} + 8x^{-2} + 16)} = \frac{1 - 4x - 4x^2}{1 + 8x^2 + 16x^4} \quad ◀ \;\; \boxed{4}$$

▶ **EXAMPLE 5**　Find $D_x\left(\dfrac{(3 - 4x)(5x + 1)}{7x - 9}\right)$.

This function has a product within a quotient. Instead of multiplying the factors in the numerator first (which is an option), we can use the quotient rule together with the product rule, as follows. Use the quotient rule first to get

$$D_x\left(\frac{(3 - 4x)(5x + 1)}{7x - 9}\right)$$

$$= \frac{(7x - 9)[D_x(3 - 4x)(5x + 1)] - [(3 - 4x)(5x + 1)D_x(7x - 9)]}{(7x - 9)^2}$$

5 Find each derivative.

(a) $D_x\left(\dfrac{(3x-1)(4x+2)}{2x}\right)$

(b) $D_x\left(\dfrac{5x^2}{(2x+1)(x-1)}\right)$

Answers:

(a) $\dfrac{6x^2+1}{x^2}$

(b) $\dfrac{-5x^2-10x}{(2x+1)^2(x-1)^2}$

Now use the product rule to find $D_x(3-4x)(5x+1)$ in the numerator.

$$= \frac{(7x-9)[(3-4x)5 + (5x+1)(-4)] - (3+11x-20x^2)(7)}{(7x-9)^2}$$

$$= \frac{(7x-9)(15-20x-20x-4) - (21+77x-140x^2)}{(7x-9)^2}$$

$$= \frac{(7x-9)(11-40x) - 21 - 77x + 140x^2}{(7x-9)^2}$$

$$= \frac{-280x^2 + 437x - 99 - 21 - 77x + 140x^2}{(7x-9)^2}$$

$$= \frac{-140x^2 + 360x - 120}{(7x-9)^2} \quad \blacktriangleleft \boxed{5}$$

AVERAGE COST Suppose $y = C(x)$ gives the total cost to manufacture x items. As mentioned earlier, the average cost per item is found by dividing the total cost by the number of items. The rate of change of average cost, called the *marginal average cost*, is the derivative of the average cost.

Average Cost

If the total cost to manufacture x items is given by $C(x)$, then the **average cost per item** is

$$\overline{C}(x) = \frac{C(x)}{x}.$$

The **marginal average cost** is the derivative of the average cost function, $\overline{C}'(x)$.

A company naturally would be interested in making the average cost as small as possible. We will see in the next chapter that this can be done by using the derivative of $C(x)/x$. The derivative often can be found with the quotient rule, as in the next example.

▶**EXAMPLE 6**   The total cost in thousands of dollars to manufacture x electrical generators is given by $C(x)$, where

$$C(x) = -x^3 + 15x^2 + 1000.$$

(a) Find the average cost per generator.

The average cost is given by the total cost divided by the number of items, or

$$\frac{C(x)}{x} = \frac{-x^3 + 15x^2 + 1000}{x}.$$

6 The total revenue in thousands of dollars from the sale of x dozen CB radios is given by

$$R(x) = 32x^2 + 7x + 80.$$

(a) Find the average revenue.

(b) Find the marginal average revenue.

Answers:

(a) $\dfrac{32x^2 + 7x + 80}{x}$

(b) $\dfrac{32x^2 - 80}{x^2}$

7 If the cost in Example 7 is given by

$$C(x) = x^2 + 10x + 16,$$

find the production level at which the marginal average cost is zero.

Answer:
400 items

(b) Find the marginal average cost.

The marginal average cost is the derivative of the average cost function. Using the quotient rule.

$$\frac{d}{dx}\left(\frac{C(x)}{x}\right) = \frac{x(-3x^2 + 30x) - (-x^3 + 15x^2 + 1000)(1)}{x^2}$$

$$= \frac{-3x^3 + 30x^2 + x^3 - 15x^2 - 1000}{x^2}$$

$$= \frac{-2x^3 + 15x^2 - 1000}{x^2}. \quad \blacktriangleleft \boxed{6}$$

▶ **EXAMPLE 7** Suppose the cost in dollars of manufacturing x hundred items is given by

$$C(x) = 3x^2 + 7x + 12.$$

(a) Find the average cost.

The average cost is

$$\overline{C}(x) = \frac{C(x)}{x} = \frac{3x^2 + 7x + 12}{x} = 3x + 7 + \frac{12}{x}.$$

(b) Find the marginal average cost.

The marginal average cost is

$$\frac{d}{dx}(\overline{C}(x)) = \frac{d}{dx}\left(3x + 7 + \frac{12}{x}\right) = 3 - \frac{12}{x^2}.$$

(c) Find the marginal cost.

The marginal cost is

$$\frac{d}{dx}(C(x)) = \frac{d}{dx}(3x^2 + 7x + 12) = 6x + 7.$$

(d) Find the level of production at which the marginal average cost is zero.

Set the derivative $\overline{C}'(x) = 0$ and solve for x.

$$3 - \frac{12}{x^2} = 0$$

$$\frac{3x^2 - 12}{x^2} = 0$$

$$3x^2 - 12 = 0$$

$$x^2 = 4$$

$$x = 2$$

Since x is in hundreds, production of 200 items will produce a marginal average cost of zero dollars. $\blacktriangleleft \boxed{7}$

11.5 EXERCISES

Use the product rule to find the derivatives of the following functions. (See Examples 1 and 2.) Hint for Exercises 9–12: Write the quantity as a product.

1. $y = (x^2 - 2)(3x + 1)$

2. $y = (2x^2 + 3)(4x + 5)$

3. $y = (6x^3 + 2)(5x - 3)$

4. $y = (x^2 + 3x)(x^2 + 2)$

5. $y = (7x^4 + 2x)(x^2 - 4)$

6. $y = (x^3 + 4x^2 + 2x)(x^2 - 1)$

7. $y = (2x^2 + 4x - 3)(5x^3 + x + 2)$

8. $y = (x^4 - 2x^3 + 2x)(4x^2 + x - 3)$

9. $y = (3x - 2)^2$

10. $y = (6x^2 + 4x)^2$

11. $y = (x^2 - 1)^2$

12. $y = (3x^3 + x^2)^2$

13. $y = (2x - 3)(\sqrt{x} - 1)$

14. $y = (5\sqrt{x} - 1)(2\sqrt{x} + 1)$

15. $y = (-3\sqrt{x} + 6)(4\sqrt{x} - 2)$

Use the quotient rule to find the derivatives of the following functions. (See Examples 3 and 4.)

16. $y = \dfrac{x + 1}{2x - 1}$

17. $y = \dfrac{3x - 5}{x - 4}$

18. $f(x) = \dfrac{7x + 1}{3x + 8}$

19. $y = \dfrac{2}{3x - 5}$

20. $y = \dfrac{9 - 7x}{1 - x}$

21. $f(t) = \dfrac{t^2 + t}{t - 1}$

22. $f(t) = \dfrac{t^2 - 4t}{t + 3}$

23. $y = \dfrac{4x + 11}{x^2 - 3}$

24. $g(x) = \dfrac{3x^2 + x}{2x^3 - 1}$

25. $k(x) = \dfrac{-x^2 + 6x}{4x^3 + 1}$

26. $y = \dfrac{x^2 - 4x + 2}{x + 3}$

27. $y = \dfrac{x^2 + 7x - 2}{x - 2}$

28. $r(t) = \dfrac{\sqrt{t}}{2t + 3}$

29. $y = \dfrac{5x + 6}{\sqrt{x}}$

30. $y = \dfrac{9x - 8}{\sqrt{x}}$

Find the derivative of each of the following. (See Example 5.)

31. $f(p) = \dfrac{(2p + 3)(4p - 1)}{3p + 2}$

32. $g(t) = \dfrac{(5t - 2)(2t + 3)}{t - 4}$

33. $g(x) = \dfrac{x^3 + 1}{(2x + 1)(5x + 2)}$

34. $f(x) = \dfrac{x^3 - 4}{(2x + 1)(3x - 2)}$

---

35. Find the error in the following work.

$$D_x\left(\frac{2x + 5}{x^2 - 1}\right) = \frac{(2x + 5)(2x) - (x^2 - 1)2}{(x^2 - 1)^2}$$

$$= \frac{4x^2 + 10x - 2x^2 + 2}{(x^2 - 1)^2}$$

$$= \frac{2x^2 + 10x + 2}{(x^2 - 1)^2}$$

36. Find the error in the following work.

$$D_x\left(\frac{x^2 - 4}{x^3}\right) = x^3(2x) - (x^2 - 4)(3x^2)$$

$$= 2x^4 - 3x^4 + 12x^2 = -x^4 + 12x^2$$

37. Find an equation of the line tangent to the graph of $f(x) = \dfrac{x}{x - 2}$ at $(3, 3)$.

38. If $f(x) = 6 - 7x$ and $g(x) = 4x^3 - 9$, then which of the following is closest to $(fg)'(1)$?

(a) 0 (b) 6 (c) 12 (d) 18
(e) 24 (f) 30

Work the following exercises. (See Examples 6 and 7.)

39. Management The total cost (in hunderds of dollars) to produce x units of perfume is

$$C(x) = \frac{3x + 2}{x + 4}.$$

Find the average cost for each of the following production levels.

(a) 10 units   **(b)** 20 units   **(c)** x units
(d) Find the marginal average cost function.

40. Management   The total profit (in tens of dollars) from selling x self-help books is

$$P(x) = \frac{5x - 6}{2x + 3}.$$

Find the average profit from each of the following sales levels.

(a) 8 books   **(b)** 15 books   **(c)** x books
(d) Find the marginal average profit function.
(e) Is this a reasonable function for profit? Why?

41. Social Science   After t hours of instruction, a typical typing student can type

$$N(t) = \frac{70t^2}{30 + t^2}$$

words per minute.

(a) Find $N'(t)$, the rate at which the student is improving after t hours.
(b) At what rate is the student improving after 3 hours? 5 hours? 7 hours? 10 hours? 15 hours?
(c) Describe the student's progress during the first 15 hours of instruction.

42. Management   A company that manufactures bicycles has determined that a new employee can assemble $M(d)$ bicycles per day after d days of on-the-job training, where

$$M(d) = \frac{200d}{3d + 10}.$$

(a) Find $M'(d)$.
(b) Find and interpret $M'(2)$ and $M'(5)$.

43. Management   Suppose you are the manager of a trucking firm, and one of your drivers reports that, according to her calculations, her truck burns fuel at the rate of

$$G(x) = \frac{1}{200}\left(\frac{800}{x} + x\right)$$

gallons per mile when traveling at x miles per hour on a smooth, dry road.

(a) If the driver tells you that she wants to travel 20 miles per hour, what should you tell her? (Hint: take the derivative of G and evaluate if for $x = 20$. Then interpret your results.)
(b) If the driver wants to go 40 miles per hour, what should you say? (Hint: find $G'(40)$).

44. Natural Science   When a certain drug is introduced into a muscle, the muscle responds by contracting. The amount of contraction, s, in millimeters, is related to the concentration of the drug, x, in milliliters, by

$$s(x) = \frac{x}{m + nx},$$

where m and n are constants.

(a) Find $s'(x)$.
(b) Evaluate $s'(x)$ when $x = 50$, $m = 10$, and $n = 3$.
(c) Intrerpret your results in part (b).

45. Natural Science   Assume that the total number (in millions) of bacteria present in a culture at a certain time t (in hours) is given by

$$N(t) = (t - 10)^2(2t) + 50.$$

(a) Find $N'(t)$.

Find the rate at which the population of bacteria is changing at each of the following times.

(b) 8 hr   **(c)** 11 hr
(d) The answer in part (b) is negative, and the answer in part (c) is positive. What does this mean in terms of the population of bacteria?

46. Social Science   Some psychologists contend that the number of facts of a certain type that are remembered after t hr is given by

$$f(t) = \frac{90t}{99t - 90}.$$

Find the rate at which the number of facts remembered is changing after the following numbers of hours.

(a) 1   **(b)** 10

1 Let $f(x) = 3x - 2$ and $g(x) = (x - 1)^5$. Find the following.

(a) $g[f(2)]$

(b) $f[g(2)]$

Answers:

(a) 243

(b) 1

11.6 THE CHAIN RULE

Many of the most useful functions for applications are created by combining simpler functions. Viewing complex functions as combinations of simpler functions often makes them easier to understand and use.

COMPOSITION OF FUNCTIONS Consider the function h whose rule is $h(x) = \sqrt{x^3}$. To compute $h(4)$, for example, you first find $4^3 = 64$ and then take the square root $\sqrt{64} = 8$. So the rule of h may be rephrased as:

First apply the function $f(x) = x^3$,

Then apply the function $g(x) = \sqrt{x}$ to the result.

The same idea can be expressed in functional notation like this:

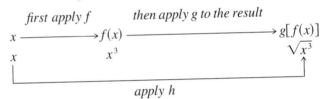

So the rule of h may be written as $h(x) = g[f(x)]$, where $f(x) = x^3$ and $g(x) = \sqrt{x}$. We can think of the functions g and f as being "composed" to create the function h. Here is a formal definition of this idea.

Composite Function

Let f and g be functions. A **composite function**, or **composition**, of g and f is the function whose values are given by $g[f(x)]$ for all x in the domain of f such that $f(x)$ is in the domain of g.

▶**EXAMPLE 1** Let $f(x) = 2x - 1$ and $g(x) = \sqrt{3x + 5}$. Find each of the following.

(a) $g[f(4)]$

Find $f(4)$ first.

$$f(4) = 2 \cdot 4 - 1 = 8 - 1 = 7$$

Then

$$g[f(4)] = g[7] = \sqrt{3 \cdot 7 + 5} = \sqrt{26}.$$

(b) $f[g(4)]$

Since $g(4) = \sqrt{3 \cdot 4 + 5} = \sqrt{17}$.

$$f[g(4)] = 2 \cdot \sqrt{17} - 1 = 2\sqrt{17} - 1.$$

(c) $f[g(-2)]$ does not exist since -2 is not in the domain of g. ◀ **1**

2 Let $f(x) = \sqrt{x + 4}$ and $g(x) = x^2 + 5x + 1$. Find $f[g(x)]$ and $g[f(x)]$.

Answers:

$f[g(x)] = \sqrt{x^2 + 5x + 5}$,
$g[f(x)] = (\sqrt{x + 4})^2$
$\qquad + 5\sqrt{x + 4} + 1$
$\qquad = x + 5 + 5\sqrt{x + 4}$

▶ **EXAMPLE 2** Let $f(x) = 4x + 1$ and $g(x) = 2x^2 + 5x$. Find each of the following.

(a) $g[f(x)]$

Using the given functions,

$$\begin{aligned}
g[f(x)] &= g[4x + 1] \\
&= 2(4x + 1)^2 + 5(4x + 1) \\
&= 2(16x^2 + 8x + 1) + 20x + 5 \\
&= 32x^2 + 16x + 2 + 20x + 5 \\
&= 32x^2 + 36x + 7.
\end{aligned}$$

(b) $f[g(x)]$

By the definition above, with f and g interchanged,

$$\begin{aligned}
f[g(x)] &= f[2x^2 + 5x] \\
&= 4(2x^2 + 5x) + 1 \\
&= 8x^2 + 20x + 1. \quad \blacktriangleleft \quad \boxed{2}
\end{aligned}$$

As Example 2 shows, $f[g(x)]$ usually is *not* equal to $g[f(x)]$. In fact, it is rare to find two functions f and g for which $f[g(x)] = g[f(x)]$.

THE CHAIN RULE   The product and quotient rules tell how to find the derivative of fg and f/g from the derivatives of f and g. Similarly, the **chain rule,** which is given below, tells how to find the derivative of the composite function $f[g(x)]$ from the derivatives of f and g. The proof of the chain rule is beyond the scope of this book, but we can illustrate the idea with some examples.

Metal expands or contracts as the temperature changes and the temperature changes over a period of time. Suppose the length of a metal bar is increasing at the rate of 2 mm for every degree increase in temperature and that temperature is increasing at the rate of 4° per hour. Then during the course of an hour, the length of the bar will increase by $2 \cdot 4 = 8$ mm. In other words, the rate of change of length with respect to time is the following product.

$$\begin{pmatrix} \text{rate of change} \\ \text{of length with} \\ \text{respect to time} \end{pmatrix} = \begin{pmatrix} \text{rate of change} \\ \text{of length with} \\ \text{respect to temperature} \end{pmatrix} \cdot \begin{pmatrix} \text{rate of change} \\ \text{of temperature with} \\ \text{respect to time} \end{pmatrix} (*)$$

If we think of length as a function of temperature, say $L = f(d)$, and temperature as a function of time, say $d = g(t)$, then the composite function $L = f(d) = f[g(t)]$ gives length as a function of time. Since the derivative of a function is its rate of change, statement (*) says

$$L'(t) = f'(d) \cdot g'(t) = f'[g(t)] \cdot g'(t).$$

We shall see that a similar result holds for any composite function. For instance, if $f(x) = x^2$ and $g(x) = x^3 + x$, then the composite function is

$$f[g(x)] = f[x^3 + x] = [x^3 + x]^2 = (x^3)^2 + 2(x^3)x + x^2 = x^6 + 2x^4 + x^2.$$

3 Use the chain rule to find $D_x\sqrt{14x + 1}$.

Answer:

$$\frac{7}{\sqrt{14x + 1}}$$

The derivative of the composite function $y = x^6 + 2x^4 + x^2$ is

$$y' = 6x^5 + 8x^3 + 2x$$

and the derivatives of $f(x) = x^2$ and $g(x) = x^3 + x$ are

$$f'(x) = 2x \quad \text{and} \quad g'(x) = 3x^2 + 1.$$

The relationship of these three derivatives can be seen when we factor.

$$\begin{aligned} y' &= 6x^5 + 8x^3 + 2x \\ &= 2(3x^5 + 4x^3 + x) \\ &= 2(x^3 + x)(3x^2 + 1) \\ &= [2 \cdot g(x)] \cdot g'(x) \end{aligned}$$

Since $f'(x) = 2x$, then $f'[g(x)] = 2g(x)$, so that the last line can be written as

$$y' = [2 \cdot g(x)] \cdot g'(x) = f'[g(x)] \cdot g'(x).$$

Although there are different letters for the variables, this is the same result as in the case of the expanding metal bar: the derivative of the composite function $y = f[g(x)]$ is the product of the derivative of f, evaluated at $g(x)$, and the derivative of g. This same relationship holds in all cases.

Chain Rule

If f and g are functions and $y = f[g(x)]$, then

$$y' = f'[g(x)] \cdot g'(x),$$

provided $f'[g(x)]$ and $g'(x)$ exist. (To find the derivative of $f[g(x)]$, find the derivative of $f(x)$, replace each x with $g(x)$, and multiply the result by the derivative of $g(x)$.)

▶ **EXAMPLE 3** Use the chain rule to find $D_x\sqrt{15x^2 + 1}$.

Write $\sqrt{15x^2 + 1}$ as $(15x^2 + 1)^{1/2}$. Let $f(x) = x^{1/2}$ and $g(x) = 15x^2 + 1$. Then $\sqrt{15x^2 - 1} = f[g(x)]$ and

$$D_x(15x^2 + 1)^{1/2} = f'[g(x)] \cdot g'(x).$$

Here $f'(x) = \frac{1}{2}x^{-1/2}$, with $f'[g(x)] = \frac{1}{2}[g(x)]^{-1/2} = \frac{1}{2}(15x^2 + 1)^{-1/2}$, and

$$\begin{aligned} D_x\sqrt{15x^2 + 1} &= \frac{1}{2}[g(x)]^{-1/2} \cdot g'(x) \\ &= \frac{1}{2}(15x^2 + 1)^{-1/2} \cdot (30x) \\ &= \frac{15x}{(15x^2 + 1)^{1/2}}. \end{aligned}$$ ◀ **3**

4 Use the chain rule to find dy/dx if $y = 10(2x^2 + 1)^4$.

Answer:
$160x(2x^2 + 1)^3$

The chain rule can also be stated using the Leibnitz notation for derivatives. If y is a function of u, say $y = f(u)$, and u is a function of x, say $u = g(x)$, then

$$f'(u) = \frac{dy}{du} \quad \text{and} \quad g'(x) = \frac{du}{dx}.$$

Now y can be considered as a function of x, namely, $y = f(u) = f(g(x))$. According to the chain rule, the derivative of y is

$$\frac{dy}{dx} = f'(g(x)) \cdot g'(x) = f'(u) \cdot g'(x) = \frac{dy}{du} \cdot \frac{du}{dx}.$$

Thus we have this alternative version of the chain rule.

The Chain Rule (Alternative Form)

If y is a function of u, say $y = f(u)$, and if u is a function of x, say $u = g(x)$, then $y = f(u) = f[g(x)]$, and

$$\frac{dy}{dx} = \frac{dy}{du} \cdot \frac{du}{dx}$$

provided dy/du and du/dx exist.

One way to remember the chain rule is to *pretend* that dy/du and du/dx are fractions, with du "canceling out."

▶**EXAMPLE 4**   Find dy/dx if $y = (3x^2 - 5x)^{1/2}$.
Let $y = u^{1/2}$, and $u = 3x^2 - 5x$. Then

$$\frac{dy}{dx} = \frac{dy}{du} \cdot \frac{du}{dx}$$

$$= \frac{1}{2} u^{-1/2} \cdot (6x - 5).$$

Replacing u with $3x^2 - 5x$ gives

$$\frac{dy}{dx} = \frac{1}{2}(3x^2 - 5x)^{-1/2}(6x - 5) = \frac{6x - 5}{2(3x^2 - 5x)^{1/2}} \quad ◀ \boxed{4}$$

While the chain rule is essential for finding the derivatives of some of the functions discussed later, the derivatives of the algebraic functions discussed so far can be found by the following *generalized power rule,* a special case of the chain rule.

5 Find dy/dx for the following.

(a) $y = (2x + 5)^6$

(b) $y = (4x^2 - 7)^3$

(c) $f(x) = \sqrt{3x^2 - x}$

(d) $g(x) = (2 - x^4)^{-3}$

Answers:

(a) $12(2x + 5)^5$

(b) $24x(4x^2 - 7)^2$

(c) $\dfrac{6x - 1}{2\sqrt{3x^2 - x}}$

(d) $\dfrac{12x^3}{(2 - x^4)^4}$

Generalized Power Rule

Let u be a function of x, and let $y = u^n$, for any real number n. Then

$$y' = n \cdot u^{n-1} \cdot u',$$

Provided that u' exists. (The derivative of $y = u^n$ is found by decreasing the exponent on u by 1 and multiplying the result by the exponent n and by the derivative of u with respect to x.)

▶ **EXAMPLE 5**

(a) Use the generalized power rule to find the derivative of $y = (3 + 5x)^2$.

Let $u = 3 + 5x$, and $n = 2$. Then $u' = 5$. By the generalized power rule,

$$y' = \frac{dy}{dx} = n \cdot u^{n-1} \cdot u'$$

$$= 2 \cdot (3 + 5x)^{2-1} \cdot \frac{d}{dx}(3 + 5x)$$

$$= 2(3 + 5x)^{2-1} \cdot 5 = 10(3 + 5x)$$

$$= 30 + 50x.$$

(b) Find y' if $y = (3 + 5x)^{-3/4}$.

Use the generalized power rule with $n = -\dfrac{3}{4}$, $u = 3 + 5x$, and $u' = 5$.

$$y' = -\frac{3}{4}(3 + 5x)^{(-3/4)-1}(5)$$

$$= -\frac{15}{4}(3 + 5x)^{-7/4}$$

This result would not have been found by any of the rules given earlier. ◀ **5**

▶ **EXAMPLE 6** Find the derivative of the following.

(a) $y = 2(7x^2 + 5)^4$

Let $u = 7x^2 + 5$. Then $u' = 14x$, and

$$y' = 2 \cdot 4(7x^2 + 5)^{4-1} \cdot \frac{d}{dx}(7x^2 + 5)$$

$$= 2 \cdot 4(7x^2 + 5)^3(14x)$$

$$= 112x(7x^2 + 5)^3.$$

6 Find dy/dx for the following.

(a) $y = 12(x^2 + 6)^5$

(b) $y = 8(4x^2 + 2)^{3/2}$

Answers:

(a) $120x(x^2 + 6)^4$

(b) $96x(4x^2 + 2)^{1/2}$

7 Find the derivatives of the following.

(a) $y = 6x(x + 2)^2$

(b) $y = -9x(2x^2 + 1)^3$

Answers:

(a) $6(x + 2)(3x + 2)$

(b) $-9(2x^2 + 1)^2(14x^2 + 1)$

8 Find the derivatives of the following.

(a) $y = \dfrac{(2x + 1)^3}{3x}$

(b) $y = \dfrac{(x - 6)^5}{3x - 5}$

Answers:

(a) $\dfrac{(2x + 1)^2(4x - 1)}{3x^2}$

(b) $\dfrac{(x - 6)^4(12x - 7)}{(3x - 5)^2}$

(b) $y = \sqrt{9x + 2}$

Write $y = \sqrt{9x + 2}$ as $y = (9x + 2)^{1/2}$. Then

$$y' = \frac{1}{2}(9x + 2)^{-1/2}(9) = \frac{9}{2}(9x + 2)^{-1/2}.$$

The derivative also can be written as

$$y' = \frac{9}{2(9x + 2)^{1/2}} \quad \text{or} \quad y' = \frac{9}{2\sqrt{9x + 2}}. \quad \blacktriangleleft \;\; \boxed{6}$$

Sometimes both the generalized power rule and either the product or quotient rule are needed to find a derivative, as the next examples show.

▶ **EXAMPLE 7** Find the derivative of $y = 4x(3x + 5)^5$.

Write $4x(3x + 5)^5$ as the product

$$4x \cdot (3x + 5)^5.$$

To find the derivative of $(3x + 5)^5$, let $u = 3x + 5$ with $u' = 3$. Now use the product rule and the generalized power rule.

derivative of $(3x + 5)^5$ · · · derivative of $4x$

$$y' = 4x\overbrace{[5(3x + 5)^4 \cdot 3]} + (3x + 5)^5(4)$$

$$= 60x(3x + 5)^4 + 4(3x + 5)^5$$

$$= 4(3x + 5)^4[15x + (3x + 5)^1] \qquad \text{Factor out the greatest common factor, } 4(3x + 5)^4$$

$$= 4(3x + 5)^4(18x + 5). \quad \blacktriangleleft \;\; \boxed{7}$$

▶ **EXAMPLE 8** Find the derivative of $y = \dfrac{(3x + 2)^7}{x - 1}$.

Use the quotient rule and the generalized power rule.

$$\frac{dy}{dx} = \frac{(x - 1)[7(3x + 2)^6 \cdot 3] - (3x + 2)^7(1)}{(x - 1)^2}$$

$$= \frac{21(x - 1)(3x + 2)^6 - (3x + 2)^7}{(x - 1)^2}$$

$$= \frac{(3x + 2)^6[21(x - 1) - (3x + 2)]}{(x - 1)^2} \qquad \text{Factor out the greatest common factor, } (3x + 2)^6$$

$$= \frac{(3x + 2)^6[21x - 21 - 3x - 2]}{(x - 1)^2} \qquad \text{Simplify inside brackets.}$$

$$\frac{dy}{dx} = \frac{(3x + 2)^6(18x - 23)}{(x - 1)^2} \quad \blacktriangleleft \;\; \boxed{8}$$

Some applications requiring the use of the chain rule or the generalized power rule are illustrated in the next examples.

9 Suppose the revenue in Example 9 is given by

$$R(n) = \frac{4500n}{n + 5}$$

and the work hours are decreasing at the rate of 4 per day. How fast is the revenue decreasing?

Answer:
About $73.47 per day

▶**EXAMPLE 9**   The revenue realized by a small city from the collection of fines from parking tickets is given by

$$R(n) = \frac{8000n}{n + 2},$$

where n is the number of work hours each day that can be devoted to parking patrol. At the outbreak of a flu epidemic, 30 work hours are used daily in parking patrol, but during the epidemic that number is decreasing at the rate of 6 work hours per day. How fast is revenue from parking fines decreasing during the epidemic?

We want to find dR/dt, the change in revenue with respect to time. By the chain rule,

$$\frac{dR}{dt} = \frac{dR}{dn} \cdot \frac{dn}{dt}.$$

First find dR/dn, as follows.

$$\frac{dR}{dn} = \frac{(n + 2)(8000) - 8000n(1)}{(n + 2)^2} = \frac{16000}{(n + 2)^2}$$

Since $n = 30$, $dR/dn = 15.625$. Also, $dn/dt = -6$. Thus,

$$\frac{dR}{dt} = (15.625)(-6) = -93.75.$$

Revenue is being lost at the rate of approximately $94 per day.   ◀ **9**

▶**EXAMPLE 10**   Suppose a sum of $500 is deposited in an account with an interest rate of r percent per year compounded monthly. At the end of 10 years, the balance in the account is given by

$$A = 500\left(1 + \frac{r}{1200}\right)^{120}.$$

Find the rate of change of A with respect to r if $r = 5, 4.2$, or 3.
First find dA/dr using the generalized power rule.

$$\frac{dA}{dr} = (120)(500)\left(1 + \frac{r}{1200}\right)^{119}\left(\frac{1}{1200}\right) = 50\left(1 + \frac{r}{1200}\right)^{119}$$

If $r = 5$,

$$\frac{dA}{dr} = 50\left(1 + \frac{5}{1200}\right)^{119} \approx 82.01,$$

or $82.01 per percentage point. If $r = 4.2$

$$\frac{dA}{dr} = 50\left(1 + \frac{4.2}{1200}\right)^{119} \approx 75.78,$$

or $75.78 per percentage point. If $r = 3$,

$$\frac{dA}{dr} = 50\left(1 + \frac{3}{1200}\right)^{119} \approx 67.30,$$

or $67.30 per percentage point.   ◀

The chain rule can be used to develop the formula for **marginal revenue product,** an economic concept that approximates the change in revenue when a manufacturer hires an additional employee. Start with $R = px$, where R is total revenue from the daily production of x units and p is the price per unit. The demand function is $p = f(x)$, as before. Also, x can be considered a function of the number of employees, n. Since $R = px$, and x and therefore p depend on n, R can also be considered a function on n. To find an expression for dR/dn, use the product rule for derivatives on the function $R = px$ to get

$$\frac{dR}{dn} = p \cdot \frac{dx}{dn} + x \cdot \frac{dp}{dn}. \tag{1}$$

By the chain rule,

$$\frac{dp}{dn} = \frac{dp}{dx} \cdot \frac{dx}{dn}.$$

Substituting for dp/dn in equation (1) gives

$$\frac{dR}{dn} = p \cdot \frac{dx}{dn} + x\left(\frac{dp}{dx} \cdot \frac{dx}{dn}\right)$$

$$= \left(p + x \cdot \frac{dp}{dx}\right)\frac{dx}{dn}. \qquad \text{Factor out } \frac{dx}{dn}$$

The expression for dR/dn gives the marginal revenue product.

▶ **EXAMPLE 11**   Find the marginal revenue product dR/dn (in dollars) when $n = 20$ if the demand function is $p = 600/\sqrt{x}$ and $x = 5n$.

As shown above,

$$\frac{dR}{dn} = \left(p + x \cdot \frac{dp}{dx}\right)\frac{dx}{dn}.$$

Find dp/dx and dx/dn. From

$$p = \frac{600}{\sqrt{x}} = 600x^{-1/2},$$

we have the derivative

$$\frac{dp}{dx} = -300x^{-3/2}.$$

Also, from $x = 5n$,

$$\frac{dx}{dn} = 5.$$

10 Find marginal revenue product at $n = 10$ if the demand function is $p = 1000/x^2$ and $x = 8n$. Interpret your answer.

Answer:
−$1.25; hiring an additional employee will produce a decrease in revenue of $1.25.

Then, by substitution,

$$\frac{dR}{dn} = \left[\frac{600}{\sqrt{x}} + x(-300x^{-3/2})\right]5 = \frac{1500}{\sqrt{x}}.$$

If $n = 20$, then $x = 100$ and

$$\frac{dR}{dn} = \frac{1500}{\sqrt{100}}$$

$$= 150.$$

This means that hiring an additional employee when production is at a level of 20 items will produce an increase in revenue of $150. ◄ **10**

11.6 EXERCISES

Let $f(x) = 2x^2 + 3x$ and $g(x) = 4x - 1$. Find each of the following. (See Example 1.)

1. $f[g(3)]$ **2.** $f[g(-4)]$ **3.** $g[f(3)]$ **4.** $g[f(-4)]$

Find $f[g(x)]$ and $g[f(x)]$ in each of the following. (See Example 2.)

5. $f(x) = 8x + 12$; $g(x) = 3x - 1$ **6.** $f(x) = -6x + 9$; $g(x) = 5x + 7$ **7.** $f(x) = -x^3 + 2$; $g(x) = 4x$

8. $f(x) = 2x$; $g(x) = 6x^2 - x^3$ **9.** $f(x) = \dfrac{1}{x}$; $g(x) = x^2$ **10.** $f(x) = \dfrac{2}{x^4}$; $g(x) = 2 - x$

11. $f(x) = \sqrt{x + 2}$; $g(x) = 8x^2 - 6$ **12.** $f(x) = 9x^2 - 11x$; $g(x) = 2\sqrt{x + 2}$

Write each function as a composition of two functions. (There may be more than one way to do this.)

13. $y = (4x + 3)^5$ **14.** $y = (x^2 + 2)^{1/3}$

15. $y = \sqrt{6 + 3x}$ **16.** $y = \sqrt{x + 3} - \sqrt[3]{x + 3}$

17. $y = \dfrac{\sqrt{x + 3}}{\sqrt{x - 3}}$ **18.** $y = \dfrac{2}{\sqrt{x + 5}}$

19. $y = (x^{1/2} - 3)^2 + (x^{1/2} - 3) + 5$ **20.** $y = (x^2 + 5x)^{1/3} - 2(x^2 + 5x)^{2/3} + 7$

Find the derivative of each of the following functions. (See Examples 3–6.)

21. $y = (3x + 4)^3$ **22.** $y = (6x - 1)^3$ **23.** $y = 6(3x + 2)^4$ **24.** $y = -5(2x - 1)^4$

25. $y = -2(8x^2 + 6)^4$ **26.** $y = -4(x^3 + 5x^2)^4$ **27.** $y = 12(2x + 5)^{3/2}$ **28.** $y = 45(3x - 8)^{3/2}$

29. $y = -7(4x^2 + 9x)^{3/2}$ **30.** $y = 11(5x^2 + 6x)^{3/2}$ **31.** $y = 8\sqrt{4x + 7}$ **32.** $y = -3\sqrt{7x - 1}$

33. $y = -2\sqrt{x^2 + 4x}$ **34.** $y = 4\sqrt{2x^2 + 3}$

Use the product or quotient rule to find the derivative of each of the following functions. (See Examples 7–8.)

35. $y = (x + 1)(x - 3)^2$ **36.** $y = (2x + 1)^2(x - 5)$ **37.** $y = 5(x + 3)^2(2x - 1)^5$ **38.** $y = -9(x + 4)^2(2x - 3)^2$

39. $y = (3x + 1)^3\sqrt{x}$ **40.** $y = (3x + 5)^2\sqrt{x}$ **41.** $y = \dfrac{1}{(x - 4)^2}$ **42.** $y = \dfrac{-5}{(2x + 1)^2}$ **43.** $y = \dfrac{(4x + 3)^2}{2x - 1}$

44. $y = \dfrac{(x-6)^2}{3x+4}$ **45.** $y = \dfrac{x^2 + 4x}{(5x+2)^3}$ **46.** $y = \dfrac{3x^2 - x}{(x-1)^2}$ **47.** $y = (x^{1/2} + 1)(x^{1/2} - 1)^{1/2}$

48. $y = (3 - x^{2/3})(x^{2/3} + 2)^{1/2}$

Consider the following table of values of the functions f and g and their derivatives at various points:

x	1	2	3	4
$f(x)$	2	4	1	3
$f'(x)$	-6	-7	-8	-9
$g(x)$	2	3	4	1
$g'(x)$	$2/7$	$3/7$	$4/7$	$5/7$

Find each of the following.

49. (a) $D_x(f[g(x)])$ at $x = 1$ **(b)** $D_x(f[g(x)])$ at $x = 2$

50. (a) $D_x(g[f(x)])$ at $x = 1$ **(b)** $D_x(g[f(x)])$ at $x = 2$

51. If $f(x) = (2x^2 + 3x + 1)^{50}$, then which of the following is closest to $f'(0)$?
(a) 1   **(b)** 50   **(c)** 100   **(d)** 150
(e) 200   **(f)** 250

52. The graphs of $f(x) = 3x + 5$ and $g(x) = 4x - 1$ are straight lines.
(a) Show that the graph of $f[g(x)]$ is also a straight line.
(b) How are the slopes of the graphs of $f(x)$ and $g(x)$ related to the slope of the graph of $f[g(x)]$?

Work the following exercises. (See Examples 9–11.)

53. Management   Suppose the demand for a certain brand of vacuum cleaner is given by

$$D(p) = \frac{-p^2}{100} + 500,$$

where p is the price in dollars. If the price, in terms of the cost c, is expressed as

$$p(c) = 2c - 10,$$

find the demand in terms of the cost.

54. Natural Science   Suppose the population P of a certain species of fish depends on the number x (in hundreds) of a smaller fish that serves as its food supply, so that

$$P(x) = 2x^2 + 1.$$

Suppose, also, that the number x (in hundreds) of the smaller species of fish depends upon the amount a (in appropriate units) of its food supply, a kind of plankton. Suppose

$$x = f(a) = 3a + 2.$$

Find $P[f(a)]$, the relationship between the population P of the large fish and the amount a of plankton available.

55. Natural Science   An oil well off the Gulf Coast is leaking, with the leak spreading oil over the surface as a circle. At any time t, in minutes, after the beginning of the leak, the radius of the circular oil slick on the surface is $r(t) = t^2$ feet. Let $A(r) = \pi r^2$ represent the area of a circle of radius r. Find and interpret $A[r(t)]$.

56. Natural Science   When there is a thermal inversion layer over a city (as happens often in Los Angeles), pollutants cannot rise vertically but are trapped below the layer and must disperse horizontally. Assume that a factory smokestack begins emitting a pollutant at 8 A.M. Assume that the pollutant disperses horizontally, forming a circle. If t represents the time, in hours, since the factory began emitting pollutants ($t = 0$ represents 8 A.M.), assume that the radius of the circle of pollution is $r(t) = 2t$ miles. Let $A(r) = \pi r^2$ represent the area of a circle of radius r. Find and interpret $A[r(t)]$.

57. Management   The Acme Company's total cost for producing x widgets is given by

$$C(x) = 600 + \sqrt{50 + 15x^2}\ \ (0 \le x \le 200).$$

Find the marginal cost function.

58. Management   Find the marginal revenue product for a manufacturer with 8 workers if the demand function is $p = 300/x^{1/3}$ and if $x = 8n$.

59. Management   Suppose the demand function for a product is $p = 200/x^{1/2}$. Find the marginal revenue product if there are 25 employees and if $x = 15n$.

60. Management A manufacturer's weekly profit from the sales of x souvenir cups is given by

$$P(x) = (x^3 + 12x + 120)^{1/3} - 200,$$

where $(0 \le x \le 2000)$.

(a) Use a calculator to find $P(50)$, $P(100)$, $P(200)$, and $P(1000)$.

(b) Explain why it is reasonable that some of the numbers found in part (a) are negative.

(c) Find the marginal profit function.

61. Management The revenue from the sale of x items is given by $R(x) = 10\sqrt{300x - 2x^2}$ $(0 \le x \le 150)$.

(a) Find the marginal revenue function.

(b) Evaluate the marginal revenue function at $x = 30$, 60, 90, and 120.

(c) Explain the significance of the answers found in part (b).

62. Management Suppose a demand function is given by

$$x = 30\left(5 - \frac{p}{\sqrt{p^2 + 1}}\right),$$

where x is the demand for a product and p is the price per item in dollars. Find the rate of change in the demand for the product (i.e., find dx/dp).

63. Social Science Studies show that after t hours on the job, the number of items a supermarket cashier can ring up per minute is given by

$$F(t) = 60 - \frac{150}{\sqrt{8 + t^2}}.$$

(a) Find $F'(t)$, the rate at which the cahsier's speed is increasing.

(b) At what rate is the cashier's speed increasing after 5 hours? 10 hours? 20 hours? 40 hours?

(c) Are your answers in part (b) increasing or decreasing with time? Is this reasonable? Explain.

64. Natural Science The total number of bacteria (in millions) present in a culture is given by

$$N(t) = 2t(5t + 9)^{1/2} + 12,$$

where t represents time in hours after the beginning of an experiment. Find the rate of change of the population of bacteria with respect to time for each of the following.

(a) $t = 0$ **(b)** $t = 7/5$ **(c)** $t = 8$ **(d)** $t = 11$.

65. Natural Science To test an individual's use of calcium, a researcher injects a small amount of radioactive calcium into the person's bloodstream. The calcium remaining in the bloodstream is measured each day for several days. Suppose the amount of the calcium remaining in the bloodstream in milligrams per cubic centimeter t days after the initial injection is approximated by

$$C(t) = \frac{1}{2}(2t + 1)^{-1/2}.$$

Find the rate of change of C with respect to time for each of the following.

(a) $t = 0$ **(b)** $t = 4$

(c) $t = 6$ (Use a calculator.)   **(d)** $t = 7.5$

66. Natural Science The strength of a person's reaction to a certain drug is given by

$$R(Q) = Q\left(C - \frac{Q}{3}\right)^{1/2},$$

where Q represents the quantity of the drug given to the patient and C is a constant.

(a) The derivative $R'(Q)$ is called the *sensitivity* to the drug. Find $R'(Q)$.

(b) Find $R'(Q)$ if $Q = 87$ and $C = 59$.

 For each of the following, use a computer or a graphing calculator to graph each function and its derivative on the same axes. Determine the values of x where the derivative is
(a) *positive,* **(b)** *zero, and* **(c)** *negative.* **(d)** *What is true of the graph of the function in each case?*

67. $G(x) = \dfrac{2x}{(x - 1)^2}$

68. $K(x) = \sqrt[3]{(2x - 1)^2}$

11.7 DERIVATIVES OF EXPONENTIAL AND LOGARITHMIC FUNCTIONS

Exponential and logarithmic functions to the base e were examined in detail in Chapter 4. (Recall that $e \approx 2.7182818$.) In this section formulas will be developed for the derivatives of $f(x) = e^x$ and $g(x) = \ln x$.

To find the derivative of $f(x) = e^x$, we use the definition of the derivative function:

$$f'(x) = \lim_{h \to 0} \frac{f(x + h) - f(x)}{h},$$

provided this limit exists. (Remember that h is the variable here and x is treated as a constant.)

For $f(x) = e^x$, we see that

$$f'(x) = \lim_{h \to 0} \frac{e^{x+h} - e^x}{h}$$

$$= \lim_{h \to 0} \frac{e^x e^h - e^x}{h} \qquad \textbf{Product property of exponents}$$

$$= \lim_{h \to 0} \frac{e^x(e^h - 1)}{h}$$

$$= \lim_{h \to 0} e^x \cdot \lim_{h \to 0} \frac{e^h - 1}{h}, \qquad \textbf{Product property of limits}$$

provided that both of these last limits exist. But h is the variable here and x is constant and therefore,

$$\lim_{h \to 0} e^x = e^x.$$

We claim that

$$\lim_{h \to 0} \frac{e^h - 1}{h} = 1.$$

Although a rigorous proof of this fact is beyond the scope of this book, the following chart (constructed with a calculator) makes it highly plausible.

	h approaches 0 from the left $\to$ 0 $\leftarrow$ h approaches 0 from the right						
h	$-.001$	$-.0001$	$-.00001$	0	$.00001$	$.0001$	$.001$
$\dfrac{e^h - 1}{h}$	.999500	.999950	.999995		1.000005	1.000050	1.000500

Therefore,

$$f'(x) = \lim_{h \to 0} e^x \cdot \lim_{h \to 0} \frac{e^h - 1}{h} = e^x \cdot 1 = e^x.$$

In other words, *the exponential function $f(x) = e^x$ is its own derivative.*

1 Differentiate the following.

(a) $(2x^2 - 1)e^x$

(b) $(1 - e^x)^{1/2}$

Answers:

(a) $(2x^2 - 1)e^x + 4xe^x$

(b) $\dfrac{-e^x}{2(1 - e^x)^{1/2}}$

2 Find each derivative.

(a) $y = 3e^{12x}$

(b) $y = -6e^{(-10x+1)}$

(c) $y = e^{-x^2}$

Answers:

(a) $y' = 36e^{12x}$

(b) $y' = 60e^{(-10x+1)}$

(c) $y' = -2xe^{-x^2}$

▶**EXAMPLE 1** Find each derivative.

(a) $y = x^3 e^x$

The product rule and the fact that $f(x) = e^x$ is its own derivative show that

$$y' = x^3 \cdot D_x(e^x) + D_x(x^3) \cdot e^x$$
$$= x^3 \cdot e^x + 3x^2 \cdot e^x = e^x(x^3 + 3x^2).$$

(b) $y = (2e^x + x)^5$

By the generalized power, sum, and constant rules,

$$y' = 5(2e^x + x)^4 \cdot D_x(2e^x + x)$$
$$= 5(2e^x + x)^4 \cdot [D_x(2e^x) + D_x(x)]$$
$$= 5(2e^x + x)^4 \cdot [2D_x(e^x) + 1]$$
$$= 5(2e^x + x)^4(2e^x + 1). \quad ◀ \quad \boxed{1}$$

▶**EXAMPLE 2** Find the derivative of $y = e^{x^2-3x}$.

Let $f(x) = e^x$ and $g(x) = x^2 - 3x$. Then

$$y = e^{x^2-3x} = e^{g(x)} = f[g(x)]$$

and $f'(x) = e^x$ and $g'(x) = 2x - 3$. By the chain rule,

$$y' = f'[g(x)] \cdot g'(x)$$
$$= e^{g(x)} \cdot (2x - 3)$$
$$= e^{x^2-3x} \cdot (2x - 3) = (2x - 3)e^{x^2-3x}. \quad ◀$$

The argument used in Example 2 can be used to find the derivative of $y = e^{g(x)}$ for any differentiable function g. By the chain rule,

$$y' = f'[g(x)] \cdot g'(x) = e^{g(x)} \cdot g'(x) = g'(x)e^{g(x)}.$$

We can summarize these results as follows.

Derivative of e^x and $e^{g(x)}$

If $y = e^x$, then $y' = e^x$.
If $y = e^{g(x)}$, then $y' = g'(x) \cdot e^{g(x)}$.

▶**EXAMPLE 3** Find derivatives of the following functions.

(a) $y = 4e^{5x}$

Let $g(x) = 5x$, with $g'(x) = 5$. Then

$$y' = 4 \cdot 5e^{5x} = 20e^{5x}.$$

(b) $y = 3e^{-4x}$

$$y' = 3(-4e^{-4x}) = -12e^{-4x}$$

(c) $y = 10e^{3x^2}$

$$y' = 6x(10e^{3x^2}) = 60xe^{3x^2} \quad ◀ \quad \boxed{2}$$

3 Find each derivative.

(a) $y = \dfrac{e^x}{1 + x}$

(b) $y = \dfrac{10,000}{1 + 2e^x}$

Answers:

(a) $y' = \dfrac{xe^x}{(1 + x)^2}$

(b) $y' = \dfrac{-20,000e^x}{(1 + 2e^x)^2}$

▶ **EXAMPLE 4**   Let $y = \dfrac{100,000}{1 + 100e^{-.3x}}$. Find y'.

Use the quotient rule.

$$y' = \frac{(1 + 100e^{-.3x})(0) - 100,000(-30e^{-.3x})}{(1 + 100e^{-.3x})^2}$$

$$= \frac{3,000,000e^{-.3x}}{(1 + 100e^{-.3x})^2} \quad ◀ \;\boxed{3}$$

To find the derivative of $g(x) = \ln x$, we use the definition and properties of natural logarithms that were developed in Section 4.3:

$$g(x) = \ln x \text{ means } e^{g(x)} = x$$

and for all $x > 0$, $y > 0$ and every real number r

$$\ln xy = \ln x + \ln y, \qquad \ln \frac{x}{y} = \ln x - \ln y, \qquad \ln x^r = r \ln x.$$

Note that $x > 0$, $y > 0$ because logarithms of negative numbers are not defined. Differentiating with respect to x on each side of $e^{g(x)} = x$ shows that

$$D_x(e^{g(x)}) = D_x(x)$$
$$e^{g(x)} \cdot g'(x) = 1.$$

Because $e^{g(x)} = x$, this last equation becomes

$$x \cdot g'(x) = 1$$
$$g'(x) = \frac{1}{x}$$
$$\frac{d}{dx}(\ln x) = \frac{1}{x} \quad \text{for all } x > 0.$$

▶ **EXAMPLE 5**   Find the derivative of $y = \ln 6x$. Assume $x > 0$.
Use the properties of logarithms and the rules for derivatives.

$$y' = \frac{d}{dx}(\ln 6x)$$

$$= \frac{d}{dx}(\ln 6 + \ln x) \qquad\qquad \text{Product rule for logarithms}$$

$$= \frac{d}{dx}(\ln 6) + \frac{d}{dx}(\ln x) = 0 + \frac{1}{x} = \frac{1}{x} \qquad \text{ln 6 is a constant.} \quad ◀$$

Caution   To find the derivative of $y = \ln 6$, remember that $\ln 6$ is a *constant* ($\ln 6 \approx 1.79$), so its derivative is 0 (*not* $1/6$).

The derivative of $y = \ln g(x)$, where $g(x) > 0$, can be found by letting $f(x) = \ln x$, so that $y = f[g(x)]$, and applying the chain rule:

4 Find y' for the following.

(a) $y = \ln(7 + x)$

(b) $y = \ln(4x^2)$

(c) $y = \ln(8x^3 - 3x)$

(d) $y = x^2 \ln x$

Answers:

(a) $y' = \dfrac{1}{7 + x}$

(b) $y' = \dfrac{2}{x}$

(c) $y' = \dfrac{24x^2 - 3}{8x^3 - 3x}$

(d) $y' = x(1 + 2 \ln x)$

$$y' = f'[g(x)] \cdot g'(x) = \frac{1}{g(x)} \cdot g'(x) = \frac{g'(x)}{g(x)}.$$

We can summarize these results as follows.

Derivative of ln x and ln $g(x)$

If $y = \ln x$, then $y' = \dfrac{1}{x}$ $(x > 0)$.

If $y = \ln g(x)$, then $y' = \dfrac{g'(x)}{g(x)}$ $(g(x) > 0)$.

▶ **EXAMPLE 6** Find the derivatives of the following functions.

(a) $y = \ln 5x$

Let $g(x) = 5x$, so that $g'(x) = 5$. From the formula above,

$$y' = \frac{g'(x)}{g(x)} = \frac{5}{5x} = \frac{1}{x}.$$

(b) $y = \ln(3x^2 - 4x)$

$$y' = \frac{6x - 4}{3x^2 - 4x}$$

(c) $y = 3x \ln x^2$

Since $3x \ln x^2$ is the product of $3x$ and $\ln x^2$, use the product rule.

$$y' = (3x)\left(\frac{d}{dx} \ln x^2\right) + (\ln x^2)\left(\frac{d}{dx} 3x\right)$$

$$= 3x\left(\frac{2x}{x^2}\right) + (\ln x^2)(3) \qquad \text{Take derivatives}$$

$$= 6 + 3 \ln x^2$$

$$= 6 + \ln (x^2)^3 \qquad\qquad \text{Property of logarithms}$$

$$y' = 6 + \ln x^6 \qquad\qquad\quad \text{Property of exponents} \quad ◀ \; \boxed{4}$$

The function $y = \ln(-x)$ is defined for all $x < 0$ (since $-x > 0$ when $x < 0$). Its derivative can be found by applying the derivative rule for $\ln g(x)$ with $g(x) = -x$.

$$y' = \frac{g'(x)}{g(x)} = \frac{-1}{-x} = \frac{1}{x}$$

This is the same as the derivative of $y = \ln x$, with $x > 0$. Since

$$|x| = \begin{cases} x & \text{if } x > 0 \\ -x & \text{if } x < 0, \end{cases}$$

we can combine two results into one, as follows.

5 Find each derivative.

(a) $y = e^{x^2} \ln |x|$

(b) $y = x^2 / \ln |x|$

Answers:

(a) $e^{x^2} \left(\dfrac{1}{x} + 2x \ln |x| \right)$

(b) $\dfrac{2x \ln |x| - x}{(\ln |x|)^2}$

6 Suppose a deer population is given by

$$f(x) = \frac{10,000}{1 + 2e^x},$$

where x is time in years. (See Problem 3(b) at the side.) Find the rate of change of the population when

(a) $x = 0$;

(b) $x = 5$.

(c) Is the population increasing or decreasing?

Answers:

(a) About -2200

(b) About -33

(c) Decreasing

$$\text{If } y = \ln |x|, \text{ then } y' = \frac{1}{x}.$$

▶ **EXAMPLE 7** Let $y = e^x \cdot \ln |x|$. Find y'.

Use the product rule.

$$y' = e^x \cdot \frac{1}{x} + \ln |x| \cdot e^x = e^x \left(\frac{1}{x} + \ln |x| \right) \quad \blacktriangleleft \; \boxed{5}$$

Often a population, or the sales of a certain product, will start growing slowly, then grow more rapidly, and then gradually level off. Such growth can often be approximated by a mathematical model of the form

$$f(x) = \frac{b}{1 + ae^{kx}}$$

for appropriate constants a, b, and k.

▶ **EXAMPLE 8** Suppose that the sales of a new product can be approximated for its first few years on the market by

$$S(x) = \frac{100,000}{1 + 100e^{-.3x}},$$

where x is time in years since the introduction of the product. Find the rate of change of the sales when $x = 4$.

The derivative was given in Example 4. Using this derivative and a calculator,

$$S'(4) = \frac{3,000,000e^{-.3(4)}}{(1 + 100e^{-.3(4)})^2} = \frac{3,000,000e^{-1.2}}{(1 + 100e^{-1.2})^2} \approx 933.$$

The rate of change of sales at time $x = 4$ is an increase of about 933 units per year. $\quad \blacktriangleleft \; \boxed{6}$

Find the derivatives of the following functions. (See Examples 1–7.)

1. $y = e^{3x}$

2. $y = e^{-4x}$

3. $f(x) = 5e^{2x}$

4. $f(x) = 4e^{-3x}$

5. $g(x) = -4e^{-5x}$

6. $g(x) = 6e^{x/2}$

7. $y = e^{x^2}$

8. $y = e^{-x^2}$

9. $f(x) = e^{x^2/2}$

10. $y = 4e^{2x^2-4}$

11. $y = -3e^{3x^2+5}$

12. $y = xe^x$

13. $y = x^2 e^{-2x}$

14. $y = (x - 3)^2 e^{2x}$

15. $y = (3x^2 - 4x)e^{-3x}$

16. $y = \ln(3 - x)$

17. $y = \ln(1 + x^2)$

18. $y = \ln(2x^2 - 7x)$

19. $y = \ln(-8x^2 + 6x)$

20. $y = \ln\sqrt{x + 5}$

21. $y = \ln\sqrt{2x + 1}$

22. $y = \ln[(3x - 1)(5x + 2)]$

23. $f(x) = \ln[(2x - 3)(x^2 + 4)]$

24. $f(x) = \ln\left(\dfrac{4x + 3}{5x - 2}\right)$

25. $y = \ln\left(\dfrac{6 - x}{3x + 5}\right)$

26. $y = \ln(x^4 + 5x^2)^{3/2}$

27. $y = \ln(5x^3 - 2x)^{3/2}$

28. $y = -3x \ln(x + 2)$

29. $y = x \ln(2 - x^2)$

30. $y = \dfrac{x^2}{e^x}$

31. $y = \dfrac{e^x}{2x + 1}$

32. $y = (2x^3 - 1) \ln |x|$

33. $y = \dfrac{\ln |x|}{x^3}$

34. $y = \dfrac{3 \ln |x|}{3x + 4}$

35. $y = \dfrac{-4 \ln |x|}{5 - 2x}$

36. $y = \dfrac{3x^2}{\ln |x|}$

37. $y = \dfrac{x^3 - 1}{2 \ln |x|}$

38. $y = [\ln(x + 1)]^4$

39. $y = \sqrt{\ln(x - 3)}$

40. $y = \dfrac{e^x}{\ln |x|}$

41. $y = \dfrac{e^x - 1}{\ln |x|}$

42. $y = \dfrac{e^x + e^{-x}}{x}$

43. $y = \dfrac{e^x - e^{-x}}{x}$

44. $y = e^{x^3} \ln |x|$

45. $f(x) = e^{3x+2} \ln (4x - 5)$

46. $f(x) = \dfrac{2400}{3 + 8e^{.2x}}$

47. $y = \dfrac{500}{7 - 10e^{.4x}}$

48. $y = \dfrac{10,000}{9 + 4e^{-.2x}}$

49. $y = \dfrac{500}{12 + 5e^{-.5x}}$

50. $y = \ln(\ln |x|)$

---

51. If $f(x) = x^2 e^{-2x}$, which of the following is closest to $f'(-1)$?

(a) 0   (b) -10 (c) -20
(d) -30 (e) -40 (f) -50

52. If $g(x) = 10e^x + 3 \ln (x + 1)$, which of the following is closest to $g'(0)$?

(a) 6   (b) 9   (c) 12
(d) 15   (e) 18   (f) 21

▶ **53.** If $f(x) = e^{2x}$, find $f'[\ln (1/4)]$

▶ **54.** If $g(x) = 3e \ln [\ln x]$, find $g'(e)$.

Work the following exercises. (See Example 8.)

55. Management Suppose the demand function for x thousand of a certain item is

$$p = 100 + \frac{50}{\ln x}, \quad x > 1.$$

where p is in dollars.
(a) Find the marginal revenue.
(b) Find the revenue from the next thousand items at a demand of 8000 ($x = 8$).

56. Management Assume that the total revenue received from the sale of x items is given by

$$R(x) = 30 \ln (2x + 1),$$

while the total cost to produce x items is $C(x) = x/2$. Find the number of items that should be manufactured so that marginal profit is 0.

57. Management If the cost function in dollars for x thousand of the item in Exercise 55 is $C(x) = 100x + 100$, find the following.
(a) The marginal cost
(b) The profit function $P(x)$
(c) The profit from the next thousand items at a demand of 8000 ($x = 8$).

58. Management The demand function for x units of a product is

$$p = 100 - 10 \ln x, \quad 1 < x < 20,000,$$

where $x = 6n$ and n is the number of employees producing the product.
(a) Find the revenue function $R(x)$.
(b) Find the marginal revenue product function. (See Example 11 in Section 11.6.)

(c) Evaluate and interpret the marginal revenue product when $x = 20$.

59. Social Science Based on data from 1980 to the present, the population of a certain city in year t is expected to be

$$P(t) = 50,000(1 + .2t)e^{-.04t} \quad (0 \le t \le 40),$$

where $t = 0$ corresponds to 1980.
(a) At what rate is the population changing in year t?
(b) Is the city gaining or losing population in the years 1985, 1995, 2005, and 2015.

60. Management Suppose $P(x) = e^{-.02x}$ represents the proportion of cars manufactured by a given company that are still free of defects after x months of use. Find the proportion of cars free of defects after
(a) 1 month; (b) 10 months; (c) 100 months.
(d) Calculate and interpret $P'(100)$.

61. Management A landowner estimates that the value of her property is given by

$$V(t) = 50,000e^{.4\sqrt{t}},$$

where t is measured in years, with $t = 0$ being 1990. What is the value of the property in each of the following years?
(a) 1995 (b) 2000 (c) 2010

At what rate is the land increasing in value in each of the following years?
(d) 1995 (e) 2000 (f) 2010

62. Natural Science Suppose that the population of a certain collection of rare Brazilian ants is given by

$$P(t) = 1000e^{.2t},$$

where t represents the time in days. Find the rate of change of the population when $t = 2$; when $t = 8$. Does the ant population ever decrease?

63. Management A timber company plans to sell a certain stand of timber and invest the proceeds. The timber is increasing in value, but waiting to sell may result in lost interest. Taking interest rates and inflation into account, the company projects that its revenue (in dollars) from selling in year t will be

$$R(t) = 600,000e^{-.07t + \sqrt{t}/2}.$$

At what rate will this revenue increase or decrease in
(a) year 5 (b) year 10 (c) year 15
(d) year 20
(e) Based on your answers to (a)–(d), approximately when should the company sell?

64. Natural Science Assume that the amount of a radioactive substance present at time t is given by

$$A(t) = 500e^{-.25t}$$

grams. Find the rate of change of the quantity present when
(a) $t = 0$ (b) $t = 4$; (c) $t = 6$;
(d) $t = 10$.
(e) Does the substance decay more slowly or more quickly as time goes on?

65. Natural Science Consider an experiment in which equal numbers of male and female insects of a certain species are permitted to intermingle. Assume that

$$M(t) = (e^{.1t} + 1) \ln \sqrt{t}$$

represents the number of matings observed among the insects in an hour, where t is the temperature in degrees Celsius. (Note: The formula is an approximation at best and holds only for specific temperature intervals.) Find
(a) $M(15)$; (b) $M(25)$.
(c) Find the rate of change of $M(t)$ when $t = 15$.

66. Natural Science Suppose that the population of a certain colony of honey bees is given by

$$P(t) = (t + 100) \ln(t + 2),$$

where t represents the time in days. Find the rates of change of the population when $t = 2$, when $t = 8$, and $t = 20$. Is the population increasing or decreasing?

67. Natural Science The concentration of pollutants, in grams per liter, in the east fork of the Big Weasel River is approximated by

$$P(x) = .04e^{-4x},$$

where x is the number of miles downstream from a paper mill that the measurement is taken. Find
(a) $P(.5)$; (b) $P(1)$ (c) $P(2)$.

Find the rate of change of the concentration with respect to distance at
(d) $x = .5$; (e) $x = 1$; (f) $x = 2$.

68. Social Science According to work by the psychologist C. L. Hull, the strength of a habit is a function of the number of times the habit is repeated. If N is the number of repetitions and $H(N)$ is the strength of the habit, then

$$H(N) = 1000(1 - e^{-kN}),$$

where k is a constant. Find $H'(N)$ if $k = .1$ and
(a) $N = 10$; (b) $N = 100$; (c) $N = 1000$.
(d) Show that $H'(N)$ is always positive. What does this mean?

11.8 CONTINUITY AND DIFFERENTIABILITY

Intuitively speaking, a function is **continuous** at a point if you can draw the graph of the function near that point without lifting your pencil from the paper. Conversely, a function is **discontinuous** at a point if the pencil *must* be lifted from the paper in order to draw the graph on both sides of the point.

Looking first at graphs having points of discontinuity will clarify the idea of continuity at a point. For example, the graph of the function in Figure 11.21(a) has an open circle at $(2, 3)$, which indicates that there is a "hole" in the graph at that point. Therefore the function is discontinuous at $x = 2$ because to draw the graph from $x = 1$ to $x = 3$, you must lift the pencil for an instant as you pass through $(2, 3)$.

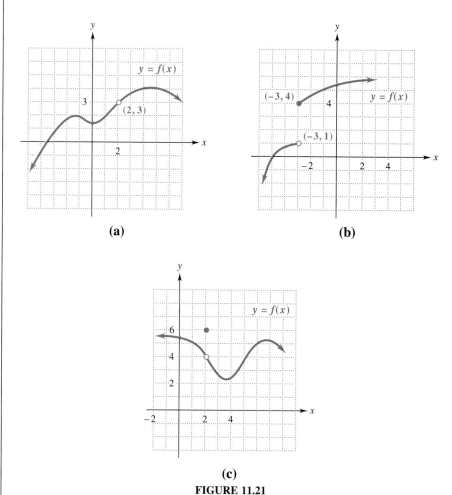

(a)

(b)

(c)

FIGURE 11.21

1 Find any points of discontinuity for the following functions.

(a)

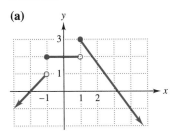

(b)

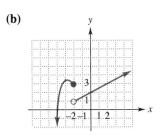

(c)

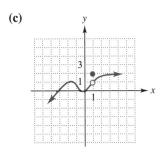

Answers:

(a) −1, 1

(b) −2

(c) 1

The function in Figure 11.21(b) is discontinuous at $x = -3$ because of the "jump" in the graph there (which necessitates lifting the pencil to draw the graph on both sides of $x = -3$). Although the function is not continuous at $x = -3$, it *is* continuous at $x = -1$ because you can draw the graph from $x = -2$ to $x = 0$, say, without lifting pencil from paper.

Finally, the function in Figure 11.21(c) is discontinuous at $x = 2$. When x is near (but not equal to) 2, all of the corresponding values of $f(x)$ are very near 4, so that on either side of $x = 2$, the graph is very near the point $(2, 4)$ and can be drawn without lifting pencil from paper. But the instant $x = 2$, you must lift the pencil to the point $(2, f(2)) = (2, 6)$. Looked at from another point of view, as x gets closer and closer to 2, $f(x)$ gets closer and closer to 4, that is, $\lim_{x \to 2} f(x) = 4$. But $f(2) = 6$ and, hence,

$$\lim_{x \to 2} f(x) \neq f(2). \quad \boxed{1}$$

Now let's consider what it means for a function f to be continuous at $x = c$. If you *can* draw the graph of f around $x = c$ without lifting pencil from paper, then at the very least, $f(c)$ must be defined (otherwise there would be a hole in the graph). But the last example shows that this is not enough to guarantee continuity: as x gets very close to c, $f(x)$ must get very close to $f(c)$ (otherwise you have to lift the pencil at $x = c$). These considerations lead to this definition.

Definition of Continuity at a Point

A function f is **continuous** at $x = c$ if

(a) $f(c)$ is defined; **(b)** $\lim_{x \to c} f(x)$ exists; **(c)** $\lim_{x \to c} f(x) = f(c)$.

If f is not continuous at $x = c$, it is **discontinuous** there.

▶ **EXAMPLE 1** Tell why the following functions are discontinuous at the indicated points.

(a) $f(x)$ in Figure 11.22 at $x = 3$

The open circle on the graph of Figure 11.22 at the point where $x = 3$ means that $f(3)$ does not exist. Because of this, part (a) of the definition fails.

(b) $h(x)$ in Figure 11.23 at $x = 0$

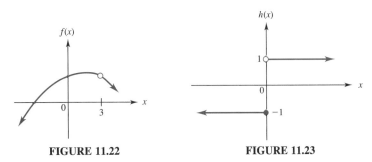

FIGURE 11.22   FIGURE 11.23

2 Tell why the following functions are discontinuous at the indicated points.

(a)

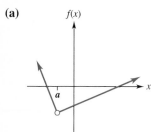

(b)

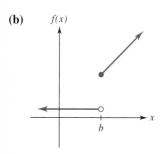

Answers:

(a) $f(a)$ does not exist

(b) $\lim_{x \to b} f(x)$ does not exist

The graph of Figure 11.23 shows that $h(0) = -1$. Also, as x approaches 0 from the left, $h(x)$ is -1. However, as x approaches 0 from the right, $h(x)$ is 1. As mentioned in Section 11.1, for a limit to exist at a particular value of x, the values of $h(x)$ must approach a single number. Since no single number is approached by the values of $h(x)$ as x approaches 0, $\lim_{x \to 0} h(x)$ does not exist, and part (b) of the definition fails.

(c) $g(x)$ at $x = 4$ in Figure 11.24

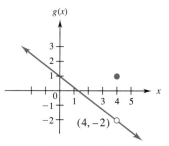

FIGURE 11.24

In Figure 11.24, the heavy dot above 4 shows that $g(4)$ is defined. In fact, $g(4) = 1$. However, the graph also shows that

$$\lim_{x \to 4} g(x) = -2,$$

so $\lim_{x \to 4} g(x) \neq g(4)$, and part (c) of the definition fails.

(d) $f(x)$ in Figure 11.25 at $x = -2$

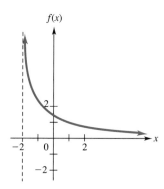

FIGURE 11.25

The function f graphed in Figure 11.25 is not defined at -2, and $\lim_{x \to -2} f(x)$ does not exist. Either of these reasons is sufficient to show that f is not continuous at -2. (Function f *is* continuous at any value of x greater than -2, however.) ◄ **2**

When discussing the continuity of a function, it is often helpful to use interval notation, which was introduced in Chapter 1. The following chart should help you recall how it is used.

 Write each of the following using interval notation.

(a)
```
-5        3
```

(b)
```
 4        7
```

(c)
```
        -1
```

Answers:

(a) $(-5, 3)$

(b) $[4, 7]$

(c) $(-\infty, -1]$

Interval	Name	Description	Interval Notation
(−2 ○—○ 3)	Open interval	$-2 < x < 3$	$(-2, 3)$
(−2 ●—● 3)	Closed interval	$-2 \le x \le 3$	$[-2, 3]$
(←— 3 ○)	Open interval	$x < 3$	$(-\infty, 3)$
(−5 ○—→)	Open interval	$x > -5$	$(-5, \infty)$

Remember, the symbol ∞ does not represent a number; ∞ is used for convenience in interval notation to indicate that the interval extends without bound in the positive direction. Also, $-\infty$ indicates no bound in the negative direction.

Continuity at a point was defined above; *continuity on an open interval* is defined as follows.

If a function is continuous at each point of an open interval, it is said to be **continuous on the open interval.**

Intuitively, the function f is continuous on the interval (a, b) if you can draw the graph between $x = a$ and $x = b$ without lifting your pencil from the paper.

▶**EXAMPLE 2** Is the function of Figure 11.26 continuous on the following x-intervals?

(a) $(-2, -1)$

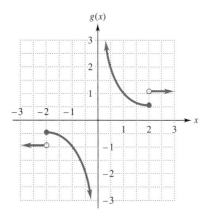

FIGURE 11.26

4 Are the functions with graphs as shown continuous on the indicated intervals?

(a) $(-4, -2)$; $(-3, 0)$

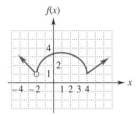

$f(x)$

(b) $(-1, 1)$; $(0, 2)$

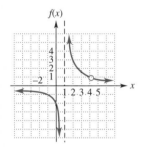

$f(x)$

Answers:

(a) Yes; no

(b) Yes; no

5 Suppose the cost is $2.25 to mail a package weighing up to 1 pound plus $.50 for each additional pound or fraction of a pound. Let $P(x)$ represent the cost of mailing a package weighing x pounds. Find any points of discontinuity for P.

Answers:

$x = 1, 2, 3, \ldots$

The function is discontinuous only at $x = -2, 0$, and 2. Thus, it is continuous at every point of the open interval $(-2, -1)$, and, therefore, is continuous on the open interval.

(b) $(1, 3)$

Since this interval includes the point of discontinuity $x = 2$, the function is not continuous on the open interval $(1, 3)$. ◄ **4**

▶ **EXAMPLE 3**   A trailer rental firm charges a flat $4 to rent a hitch. The trailer itself is rented for $11 per day or fraction of a day. Let $C(x)$ represent the cost of renting a hitch and trailer for x days.

(a) Graph C.

The charge for 1 day is $4 for the hitch and $11 for the trailer, or $15. In fact, in the interval $(0, 1]$, $C(x) = 15$. To rent the trailer for more than 1 day, but not more than 2 days, the charge is $4 + 2 \cdot 11 = 26$ dollars. For any value of x in the interval $(1, 2]$, $C(x) = 26$. Also, in $(2, 3]$ $C(x) = 37$. These results lead to the graph of Figure 11.27.

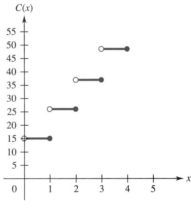

FIGURE 11.27

(b) Find any points of discontinuity for C.

As the graph suggests, C is discontinuous at 1, 2, 3, 4, and all other positive integers. ◄ **5**

CONTINUITY AND DIFFERENTIABILITY   As shown earlier in this chapter, a function fails to have a derivative at a point where the function is not defined, where the graph of the function has a "sharp point," or where the graph has a vertical tangent line. (See Figure 11.28).

The function graphed in Figure 11.28 is continuous on the interval (x_1, x_2) and has a derivative at each point on this interval. On the other hand, the function is also continuous on the interval $(0, x_2)$ but does *not* have a derivative at each point on the interval (see x_1 on the graph).

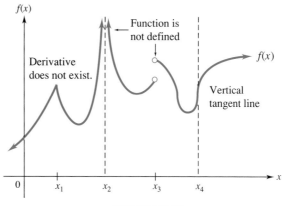

FIGURE 11.28

If the derivative of a function exists at a point, then the function is continuous at that point.

▶**EXAMPLE 4**   A nova is a star whose brightness suddenly increases and then gradually fades. The cause of the sudden increase in brightness is thought to be an explosion of some kind. The intensity of light emitted by a nova as a function of time is shown in Figure 11.29.* Notice that although the graph is a continuous curve, it is not differentiable at the point of the explosion.   ◀

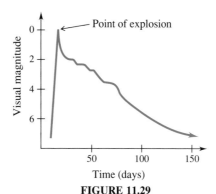

FIGURE 11.29

———————
*Reprinted with permission of Macmillan Publishing Company from *Astronomy: The Structure of the Universe* by William J. Kaufmann, III. Copyright © 1977 by William J. Kaufmann, III.

11.8 EXERCISES

Find all points of discontinuity for the functions whose graphs are shown below. (See Example 1.)

1.

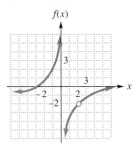

2.

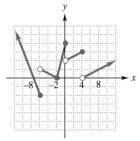

3.

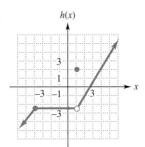

4.

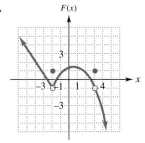

5.

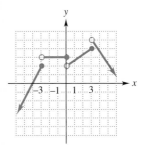

6.

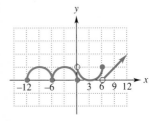

7.

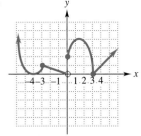

8.

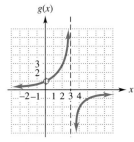

Are the following functions continuous at the given values of x?

9. $f(x) = \dfrac{4}{x - 2}$; $x = 0$, $x = 2$

10. $g(x) = \dfrac{5}{x + 5}$; $x = -5$, $x = 5$

11. $h(x) = \dfrac{1}{x(x - 3)}$; $x = 0$, $x = 3$, $x = 5$

12. $h(x) = \dfrac{-1}{(x - 2)(x + 3)}$; $x = 0$, $x = 2$, $x = 3$

13. $g(x) = \dfrac{x + 2}{x^2 - x - 2}$; $x = 1$, $x = 2$, $x = -2$

14. $h(x) = \dfrac{3x}{6x^2 + 15x + 6}$; $x = 0$, $x = -1/2$, $x = 3$

15. $g(x) = \dfrac{x^2 - 4}{x - 2}$; $x = 0$, $x = 2$, $x = -2$

16. $h(x) = \dfrac{x^2 - 25}{x + 5}$; $x = 0$, $x = 5$, $x = -5$

17. $p(x) = \dfrac{|x + 2|}{x + 2}$; $x = -2$, $x = 0$, $x = 2$

18. $r(x) = \dfrac{|5 - x|}{x - 5}$; $x = -5$, $x = 0$, $x = 5$

19. $f(x) = \begin{cases} x - 2 & \text{if } x \le 3 \\ 2 - x & \text{if } x > 3 \end{cases}$; $x = 2$, $x = 3$

20. $g(x) = \begin{cases} e^x & \text{if } x < 0 \\ x + 1 & \text{if } 0 \le x \le 3; \ x = 0, x = 3 \\ 2x - 3 & \text{if } x > 3 \end{cases}$

In Exercises 21–22, find the constant k that makes the given function continous at x = 2.

21. $f(x) = \begin{cases} x + k & \text{if } x \le 2 \\ 5 - x & \text{if } x > 2 \end{cases}$

22. $g(x) = \begin{cases} x^k & \text{if } x \le 2 \\ 2x + 4 & \text{if } x > 2 \end{cases}$

Work the following problems. (See Example 4.)

23. Social Science With certain skills (such as music) learning is rapid at first and then levels off. Sudden insights may cause learning to speed up sharply. A typical graph of such learning is shown in the figure. Where is the function discontinuous? Where is it differentiable?

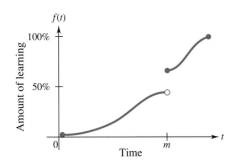

24. Natural Science Suppose a gram of ice is at a temperature of $-100°C$. The graph below shows the temperature of the ice as an increasing number of calories of heat are applied. It takes 80 calories to melt 1 gram of ice at $0°C$ into water, and 539 calories to boil 1 gram of water at $100°C$ into steam. Where is this graph discontinuous? Where is it differentiable?

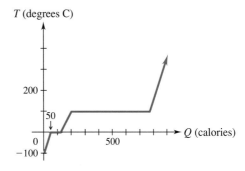

25. Like many states, California suffered a large budget deficit in 1991. As part of the solution, officials raised the sales tax by $.0125. The graph shows the California state sales tax since it was first established in 1933. Let $T(x)$ represent the sales tax in year x. Find the following.

    (a) $\lim\limits_{x \to 80} T(x)$

    (b) $\lim\limits_{x \to 73} T(x)$

    (c) List three years for which the graph indicates a discontinuity.

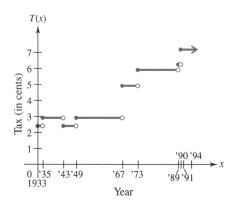

T(x)

Tax (in cents)

Year

Work the following problems. (See Example 3.)

26. **Management**   The cost to transport a mobile home depends on the distance, x, in miles that the home is moved. Let $C(x)$ represent the cost to move a mobile home x miles. One firm charges as follows.

Cost per Mile	Distance in Miles
$2	if $0 < x \le 150$
1.50	if $150 < x \le 400$
1.25	if $400 < x$

    (a) Find $C(130)$.   **(b)** Find $C(210)$.

    (c) Find $C(350)$.   **(d)** Find $C(500)$.

    (e) Graph $y = C(x)$.

    (f) For what positive values of x is C discontinuous?

27. A company charges $1.50 per pound for fertilizer on orders of less than 20 pounds. For orders of 20 pounds or more, the company charges $25 plus $1.25 for each pound over 20. On orders of more than 50 pounds, the company rebates .25 for each pound over 50. Let $F(x)$ represent the net cost (after any rebates) for buying x pounds of fertilizer. What is the net cost of buying

    (a) 10 lb   **(b)** 20 lb   **(c)** 40 lb   **(d)** 80 lb

    (e) Where is the function F discontinuous?

28. **Management**   Recently, a car rental firm charged $30 per day or portion of a day to rent a car for a period of 1 to 5 days. Days 6 and 7 were then "free," while the charge for days 8 through 12 was again $30 per day. Let $C(t)$ represent the total cost to rent the car for t days, where $0 < t \le 12$. Find the total cost of a rental for the following number of days:

    (a) 4   **(b)** 5   **(c)** 6   **(d)** 7   **(e)** 8.

    (f) Find $\lim\limits_{t \to 5} C(t)$.   **(g)** Find $\lim\limits_{t \to 6} C(t)$.

Write each of the following in interval notation.

29.
-4 6

30.
-9 15

31.
-2

32.
10

33. $\{x \mid -25 \le x \le 0\}$ **34.** $\{x \mid -5 \le x \le 14\}$ **35.** $\{x \mid \pi < x < 10\}$ **36.** $\{x \mid -7 \le x \le -3\}$

37. $\{x \mid x > -4\}$ **38.** $\{x \mid x < 3\}$ **39.** $\{x \mid x < 0\}$ **40.** $\{x \mid x > -10\}$

---

41. On which of the following intervals is the function in Exercise 5 continuous: $(-3, 0)$, $(0, 3)$, $(0, 4)$?

42. On which of the following intervals is the function in Exercise 6 continuous: $(-6, 0)$, $(0, 3)$, $(4, 8)$?

KEY TERMS AND SYMBOLS

11.1 $\lim_{x \to a} f(x)$ limit of a function as x approaches a

11.2 average rate of change

velocity

instantaneous rate of change

marginal cost, revenue, profit

11.3 y' derivative of y

$f'(x)$ derivative of $f(x)$

secant line

tangent line

slope of a curve

derivative

differentiation

11.4 $\dfrac{dy}{dx}$ derivative of $y = f(x)$

$D_x[f(x)]$ derivative of $f(x)$

$\dfrac{d}{dx}[f(x)]$ derivative of $f(x)$

11.5 $\overline{C}(x)$ average cost per item

11.6 $g[f(x)]$ composite function

marginal revenue product

11.8 continuous at a point

discontinuous

interval notation

continuous on an open interval

KEY CONCEPTS

Limit of a Function

Let f be a function and let a and L be real numbers. Suppose that as x takes values closer and closer (but not equal) to a (on both sides of a), the corresponding values of $f(x)$ get closer and closer (and possibly are equal) to L; and that the values of $f(x)$ can be made arbitrarily close to L by taking values of x close enough to a. Then L is the **limit** of f as x approaches a, written $\lim_{x \to a} f(x) = L$.

Rules for Limits

Let a, k, n, A, and B be real numbers, and let f and g be functions such that $\lim_{x \to a} f(x) = A$ and $\lim_{x \to a} g(x) = B$.

1. If k is a constant, then (a) $\lim_{x \to a} k = k$ and (b) $\lim_{x \to a} k \cdot f(x) = k \cdot \lim_{x \to a} f(x)$.

2. $\lim_{x \to a} [f(x) \pm g(x)] = \lim_{x \to a} f(x) \pm \lim_{x \to a} g(x) = A \pm B$.

3. If $p(x)$ is a polynomial, then $\lim_{x \to a} p(x) = p(a)$.

4. $\lim_{x \to a} [f(x) \cdot g(x)] = [\lim_{x \to a} f(x)] \cdot [\lim_{x \to a} g(x)] = A \cdot B$.

5. $\lim_{x \to a} \dfrac{f(x)}{g(x)} = \dfrac{\lim_{x \to a} f(x)}{\lim_{x \to a} g(x)} = \dfrac{A}{B}$ if $B \neq 0$.

6. For any real number n for which A^n exists, $\lim_{x \to a} [f(x)]^n = [\lim_{x \to a} f(x)]^n = A^n$.

7. If $f(x) = g(x)$ for all $x \neq a$, then $\lim_{x \to a} f(x) = \lim_{x \to a} g(x)$.

The **instantaneous rate of change** for a function f when $x = x_0$ is $\lim\limits_{h \to 0} \dfrac{f(x_0 + h) - f(x_0)}{h}$, provided this limit exists.

The **tangent line** to the graph of $y = f(x)$ at the point $(x_0, f(x_0))$ is the line through this point having slope $\lim\limits_{h \to 0} \dfrac{f(x_0 + h) - f(x_0)}{h}$, provided this limit exists.

The **derivative** of the function f is the function denoted f' whose value at the number x is $f'(x) = \lim\limits_{h \to 0} \dfrac{f(x + h) - f(x)}{h}$, provided this limit exists.

RULES FOR DERIVATIVES

(Assume all indicated derivatives exist.)

Constant Function

If $f(x) = k$, where k is any real number, then $f'(x) = 0$.

Power Rule

If $f(x) = x^n$, for any real number n, then $f'(x) = n \cdot x^{n-1}$.

Constant Times a Function

Let k be a real number. Then the derivative of $y = k \cdot f(x)$ is $y' = k \cdot f'(x)$.

Sum or Difference Rule

If $y = f(x) \pm g(x)$, then $y' = f'(x) \pm g'(x)$.

Product Rule

If $f(x) = g(x) \cdot k(x)$, then $f'(x) = g(x) \cdot k'(x) + k(x) \cdot g'(x)$.

Quotient Rule

If $f(x) = \dfrac{g(x)}{k(x)}$, and $k(x) \neq 0$, then $f'(x) = \dfrac{k(x) \cdot g'(x) - g(x) \cdot k'(x)}{[k(x)]^2}$.

Chain Rule

Let $y = f[g(x)]$. Then $y' = f'[g(x)] \cdot g'(x)$.

Chain Rule (alternative form)

If y is a function of u, say $y = f(u)$, and if u is a function of x, say $u = g(x)$, then $y = f(u) = f[g(x)]$, and

$$\frac{dy}{dx} = \frac{dy}{du} \cdot \frac{du}{dx}.$$

Generalized Power Rule

Let u be a function of x, and let $y = u^n$ for any real number n. Then

$$y' = n \cdot u^{n-1} \cdot u'.$$

Exponential Function

If $y = e^{g(x)}$, then $y' = g'(x) \cdot e^{g(x)}$.

Natural Logarithmic Function

If $y = \ln[g(x)]$, then $y' = \dfrac{g'(x)}{g(x)}$.

If $y = \ln|x|$, then $y' = \dfrac{1}{x}$.

A function f is **continuous** at $x = c$ if $f(c)$ is defined, $\lim\limits_{x \to c} f(x)$ exists, and $\lim\limits_{x \to c} f(x) = f(c)$.

Determine if the limits in Exercises 1–12 exist. If a limit exists, find its value.

1. $\lim\limits_{x \to -3} f(x)$

2. $\lim\limits_{x \to -1} g(x)$

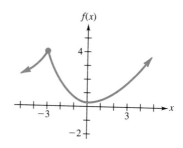

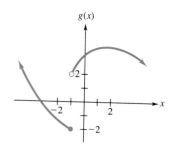

3. $\lim\limits_{x \to 2} (x^2 - 3x + 1)$

4. $\lim\limits_{x \to -1} (-2x^2 + x - 5)$

5. $\lim\limits_{x \to 4} \dfrac{3x + 1}{x - 2}$

6. $\lim\limits_{x \to 3} \dfrac{4x + 7}{x - 3}$

7. $\lim\limits_{x \to 2} \dfrac{x^2 - 4}{x - 2}$

8. $\lim\limits_{x \to -3} \dfrac{x^2 + 2x - 3}{x + 3}$

9. $\lim\limits_{x \to -4} \dfrac{2x^2 + 3x - 20}{x + 4}$

10. $\lim\limits_{x \to 3} \dfrac{3x^2 - 2x - 21}{x - 3}$

11. $\lim\limits_{x \to 9} \dfrac{\sqrt{x} - 3}{x - 9}$

12. $\lim\limits_{x \to 16} \dfrac{\sqrt{x} - 4}{x - 16}$

Use the graph to find the average rate of change of f on the following intervals.

13. $x = 0$ to $x = 4$

14. $x = 2$ to $x = 8$

15. $x = 2$ to $x = 4$

16. $x = 0$ to $x = 6$

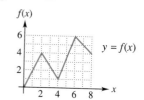

Find the average rate of change for each of the following functions.

17. $f(x) = 3x^2 - 5$, from $x = 1$ to $x = 6$

18. $g(x) = -x^3 + 2x^2 + 1$, from $x = -3$ to $x = 3$

19. $h(x) = \dfrac{6 - x}{2x + 3}$, from $x = 0$ to $x = 5$

20. $f(x) = e^{2x} + 5 \ln x$, from $x = 1$ to $x = 4$

Use the definition of the derivative to find the derivative of each of the following functions.

21. $y = 2x + 3$

22. $y = x^2 + 2x$

23. $y = 2x^2 - x + 1$

24. $y = x^3 + 5$

Find the slope of the tangent line to the given curve at the given value of x. Find the equation of each tangent line.

25. $y = x^2 - 6x$; at $x = 2$

26. $y = 8 - x^2$; at $x = 1$

27. $y = \dfrac{-2}{x + 5}$; at $x = -2$

28. $y = (3x^2 - 5x)(2x)$; at $x = -1$

29. $y = \sqrt{6x - 2}$; at $x = 3$

30. $y = -\sqrt{8x + 1}$; at $x = 3$

31. A company charges \$1.50 per pound when a certain chemical is bought in lots of 125 pounds or less, with a price per pound of \$1.35 if more than 125 pounds are purchased. Let $C(x)$ represent the cost of x pounds. Find each of the following.
  (a) $C(100)$ **(b)** $C(125)$ **(c)** $C(140)$
  (d) Graph $y = C(x)$.
  (e) Where is C discontinuous?

32. Use the information in Exercise 31 to find the average cost per pound if the following number of pounds are bought.
  (a) 100   **(b)** 125   **(c)** 140   **(d)** 200

33. Suppose the average rate of change of a function $f(x)$ from $x = 0$ to $x = 4$ is 0. Does this mean that f is constant between $x = 0$ and $x = 4$? Explain.

34. A newspaper article stated "the area continued to bleed jobs in January, but the rate has slowed from previous months . . ." What does this statement tell you about the number of jobs in the area? (Use the concepts discussed in this chapter.)

Find the derivative of each of the following.

35. $y = 5x^2 - 7x - 9$ **36.** $y = x^3 - 4x^2$

37. $y = 6x^{7/3}$

38. $y = -3x^{-2}$

39. $f(x) = x^{-3} + \sqrt{x}$ **40.** $f(x) = 6x^{-1} - 2\sqrt{x}$

41. $y = (3t^2 + 7)(t^3 - t)$

42. $y = (-5t + 4)(t^3 - 2t^2)$

43. $y = 8x^{3/4}(2x + 3)$ **44.** $y = 25x^{-3/5}(x^2 + 5)$

45. $f(x) = \dfrac{2x}{x^2 + 2}$

46. $g(x) = \dfrac{-4x^2}{3x + 4}$

47. $y = \dfrac{\sqrt{x} - 1}{x + 2}$ **48.** $y = \dfrac{\sqrt{x} + 6}{x - 3}$

49. $y = \dfrac{x^2 - x + 1}{x - 1}$

50. $y = \dfrac{2x^3 - 5x^2}{x + 2}$

51. $f(x) = (3x - 2)^4$ **52.** $k(x) = (5x - 1)^6$

53. $y = \sqrt{2t - 5}$

54. $y = -3\sqrt{8t - 1}$

55. $y = 2x(3x - 4)^3$ **56.** $y = 5x^2(2x + 3)^5$

57. $f(u) = \dfrac{3u^2 - 4u}{(2u + 3)^3}$

58. $g(t) = \dfrac{t^3 + t - 2}{(2t - 1)^5}$

59. $y = \dfrac{x^2 + 3x - 10}{x - 2}$ **60.** $y = \dfrac{x^2 - x - 6}{x - 3}$

61. $y = -6e^{2x}$

62. $y = 8e^{.5x}$

63. $y = e^{-2x^3}$ **64.** $y = -4e^{x^2}$

65. $y = 5x \cdot e^{2x}$

66. $y = -7x^2 \cdot e^{-3x}$

67. $y = \ln(x^2 + 4x + 1)$ **68.** $y = \ln(4x^3 + 2x)$

69. $y = \dfrac{\ln 4x}{x^2 - 1}$

70. $y = \dfrac{\ln(3x + 5)}{x^2 + 5x}$

Find each of the following.

71. $D_x\left(\dfrac{\sqrt{x}+1}{\sqrt{x}-1}\right)$

72. $D_x\left(\dfrac{2x+\sqrt{x}}{1-x}\right)$

73. $\dfrac{dy}{dt}$ if $y = \sqrt{t^{1/2}+t}$

74. $\dfrac{dy}{dx}$ if $y = \dfrac{\sqrt{x}-1}{x}$

75. $f'(1)$ if $f(x) = \dfrac{\sqrt{8+x}}{x+1}$

76. $f'(-2)$ if $f(t) = \dfrac{2-3t}{\sqrt{2+t}}$

---

77. Two students are working on taking the derivative of

$$f(x) = \frac{2x}{3x+4}.$$

The first one uses the quotient rule to get

$$f'(x) = \frac{(3x+4)2 - 2x(3)}{(3x+4)^2} = \frac{8}{(3x+4)^2}.$$

The second converts it into a product and uses the product rule:

$$f(x) = 2x(3x+4)^{-1}$$
$$f'(x) = 2x(-1)(3x+4)^{-2}(3) + 2(3x+4)^{-1}$$
$$= 2(3x+4)^{-1} - 6x(3x+4)^{-2}.$$

Explain the discrepancies between the two answers. Which procedure do you think is preferable?

78. Two students are working on taking the derivative of

$$f(x) = \frac{2}{(3x+1)^4}.$$

The first one uses the quotient rule as follows:

$$f'(x) = \frac{(3x+1)^4 \cdot 0 - 2 \cdot 4(3x+1)^3 \cdot 3}{(3x+1)^8}$$
$$= \frac{-24(3x+1)^3}{(3x+1)^8} = \frac{-24}{(3x+1)^5}.$$

The second rewrites the function and uses the generalized power rule as follows:

$$f(x) = 2(3x+1)^{-4}$$
$$f'(x) = (-4)2(3x+1)^{-5} \cdot 3 = \frac{-24}{(3x+1)^5}.$$

Compare the two procedures. Which procedure do you think is preferable?

Find all points of discontinuity for the following.

79.

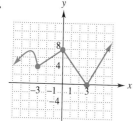

80.

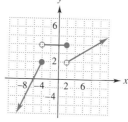

Are the following functions continuous at the given points?

81. $f(x) = \dfrac{2x-3}{2x+3}; x = -3/2, x = 0, x = 3/2$

82. $g(x) = \dfrac{2x-1}{x^3+x^2}; x = -1, x = 0, x = 1/2$

83. $h(x) = \dfrac{2-3x}{2-x-x^2}; x = -2, x = 2/3, x = 1$

84. $f(x) = \dfrac{x^2-4}{x^2-x-6}; x = 2, x = 3, x = 4$

85. $f(x) = \dfrac{x-6}{x+5}; x = 6, x = -5, x = 0$

86. $f(x) = \dfrac{x^2-9}{x+3}; x = 3, x = -3, x = 0$

87. $f(x) = x^2 + 3x - 4; x = 1, x = -4, x = 0$

88. $f(x) = 2x^2 - 5x - 3; x = -\dfrac{1}{2}, x = 3, x = 0$

Management *Work the following exercises.*

89. The profit (in dollars) from selling x specialty jackets is given by

$$P(x) = -\frac{x^2}{2} + 75x - 500 \quad (0 \leq x \leq 125).$$

Find the marginal profit when
(a) $x = 50$ **(b)** $x = 75$ **(c)** $x = 100$

90. In Exercise 89, find the average profit function and the marginal average profit function.

91. The sales of a company are related to its expenditures on research by

$$S(x) = 1000 + 50\sqrt{x} + 10x,$$

where $S(x)$ gives sales in millions when x thousand dollars is spent on research. Find and interpret dS/dx if the following amounts are spent on research.
(a) $9000 **(b)** $16,000 **(c)** $25,000
(d) As the amount spent on research increases, what happens to sales?

92. Suppose that the profit (in hundreds of dollars) from selling x units of a product is giveny by

$$P(x) = \frac{x^2}{x - 1}, \quad \text{where } x > 1.$$

Find and interpret the marginal profit when the following numbers of units are sold.
(a) 4 **(b)** 12 **(c)** 20
(d) What is happening to the marginal profit as the number sold increases?

93. A company finds that its costs are related to the amount spent on training programs by

$$T(x) = \frac{1000 + 50x}{x + 1},$$

where $T(x)$ is costs in thousands of dollars when x hundred dollars are spent on training. Find and interpret $T'(x)$ if the following amounts are spent on training.
(a) $900 **(b)** $1900
(c) Are costs per dollar spent on training always increasing or decreasing?

94. Waverly Products has found that its revenue is related to advertising expenditures by the function

$$R(x) = 5000 + 16x - 3x^2,$$

where $R(x)$ is the revenue in dollars when x hundred dollars are spent on advertising.
(a) Find the marginal revenue function.
(b) Find and interpret the marginal revenue when $1000 is spent on advertising.

95. Natural Science In a laboratory experiment the population of fruit flies on day t is given by

$$N(t) = 100e^{.11t}.$$

Find the growth rate on day
(a) 5 **(b)** 10 **(c)** 20.
(d) According to this model, does the fruit fly population increase forever? Why? Is this realistic?

▶ **Price Elasticity of Demand**

Anyone who sells a product or service is concerned with how a change in price affects demand. The sensitivity of demand to price changes varies with different items. For items such as soft drinks, pepper, and light bulbs, relatively small percentage changes in price will not change the demand for the item much. For cars, home loans, furniture, and computer equipment, however, small percentage changes in price have significant effects on demand.

One way to measure the sensitivity of demand to changes in price is by the ratio of percent change in demand to percent change in price. If q represents the quantity demanded and p the price, this ratio can be written as

$$\frac{\Delta q/q}{\Delta p/p},$$

where Δq represents the change in q and Δp represents the change in p. This ratio is always negative, because q and p are positive, while Δq and Δp have opposite signs. (An *increase* in price causes a *decrease* in demand.) If the absolute value of this ratio is large, it suggests that a relatively small increase in price causes a relatively large drop (decrease) in demand.

This ratio can be rewritten as

$$\frac{\Delta q/q}{\Delta p/p} = \frac{\Delta q}{q} \cdot \frac{p}{\Delta p} = \frac{p}{q} \cdot \frac{\Delta q}{\Delta p}.$$

Suppose $q = f(p)$. (Note that this is the inverse of the way our demand functions have been expressed so far; previously we had $p = D(q)$. Then $\Delta q = f(p + \Delta p) - f(p)$, and

$$\frac{\Delta q}{\Delta p} = \frac{f(p + \Delta p) - f(p)}{\Delta p}.$$

As $\Delta p \to 0$, this quotient becomes

$$\lim_{\Delta p \to 0} \frac{\Delta q}{\Delta p} = \lim_{\Delta p \to 0} \frac{f(p + \Delta p) - f(p)}{\Delta p} = \frac{dq}{dp},$$

and

$$\lim_{\Delta p \to 0} \frac{p}{q} \cdot \frac{\Delta q}{\Delta p} = \frac{p}{q} \cdot \frac{dq}{dp}.$$

The quantity

$$E = -\frac{p}{q} \cdot \frac{dq}{dp}$$

is positive because dq/dp is negative. E is called the **elasticity of demand** and measures the instanteous responsiveness of demand to price. For example, E may be 0.2 for medical services, but may be 1.2 for stereo equipment. The demand for essential medical services is much less responsive to price changes than is the demand for nonessential commodities, such as stereo equipment.

If $E < 1$, the relative change in demand is less than the relative change in price, and the demand is called **inelastic.** If $E > 1$, the relative change in demand is greater than the relative change in price, and the demand is called **elastic.** When $E = 1$, the percentage changes in price and demand are relatively equal and the demand is said to have **unit elasticity.**

Addiction to an illicit drug such as crack is an example of almost perfect inelastic demand. The quantity of the drug demanded by addicts does not change much no matter what the price. This fact is often used to support those who believe that increased law inforcement and longer prison terms will have little effect on the use of illicit drugs.

▶ **EXAMPLE 1** Given the demand function $q = 300 - 3p$, $0 \le p \le 100$, find the following.
(a) Calculate and interpret the elasticity of demand when $p = 25$ and when $p = 75$.

Since $q = 300 - 3p$, $dq/dp = -3$, and

$$E = -\frac{p}{q} \cdot \frac{dq}{dp}$$

$$= -\frac{p}{300 - 3p}(-3)$$

$$= \frac{3p}{300 - 3p}$$

$$= \frac{p}{100 - p}.$$

Let $p = 25$ to get

$$E = \frac{25}{100 - 25} = \frac{25}{75} = \frac{1}{3} \approx .33.$$

Since $.33 < 1$, the demand is inelastic, and a percentage change in price will result in a smaller percentage change in demand. For example, a 10% increase in price will cause a 3.3% decrease in demand.

If $p = 75$, then

$$E = \frac{75}{100 - 75} = \frac{75}{25} = 3,$$

and since $3 > 1$, demand is elastic. At this point a percentage increase in price will result in a *greater* percentage decrease in demand. Here, a 10% increase in price will cause a 30% decrease in demand.

(b) Determine the price where demand has unit elasticity. What is the significance of this price?

Demand will have unit elasticity at the price p that makes $E = 1$. Here

$$E = \frac{p}{100 - p} = 1$$

$$p = 100 - p$$

$$2p = 100$$

$$p = 50.$$

Demand has unit elasticity at a price of 50. This means that at this price the percentage changes in price and demand are about the same. ◄

The definitions from this discussion can be summarized as follows.

Elasticity of Demand

Let $q = f(p)$, where q is demand at a price p. The elasticity of demand is

$$E = -\frac{p}{q} \cdot \frac{dq}{dp}.$$

Demand is inelastic if $E < 1$.
Demand is elastic if $E > 1$.
Demand has unit elasticity if $E = 1$.

EXERCISES

1. Find the elasticity of demand (E) for the demand function $q = 400 - .2p^2$ at the indicated values of p. Is the demand elastic, inelastic, or neither in each case? Interpret your results.
   (a) $p = \$20$ **(b)** $p = \$40$

2. What must be true about the demand function if $E = 0$?

3. For some products, the demand function is actually increasing. For example, the most expensive colleges in the United States also tend to have the greatest number of applicants for each student accepted. What is true about the elasticity in this case?

CHAPTER 12

Applications of the Derivative

TECHNOLOGY RESOURCES
GraphExplorer

Visual Calculus, Schneider

The Electronic Spreadsheet, Spero

The maximum and minimum values of a function can be read from its graph. For a quadratic function (whose graph is a parabola), the maximum or minimum value can be determined without graphing by finding the vertex algebraically. For functions whose graphs are not known, other techniques are needed. In this chapter we shall see how to use derivatives to determine the maximum and minimum values of a function, as well as the intervals where the function is increasing or decreasing.

12.1 MAXIMA AND MINIMA

The graph of a typical function may have "peaks" or "valleys," as shown in Figure 12.1.

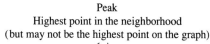

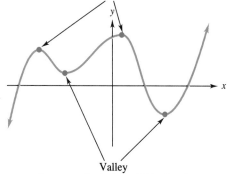

Peak
Highest point in the neighborhood
(but may not be the highest point on the graph)

Valley
Lowest point in the neighborhood
(but may not be the lowest point on the graph)

FIGURE 12.1

1 Identify the x-values of all points where these graphs have relative maxima or relative minima.

(a)

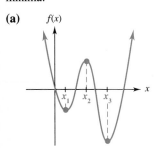

(b)

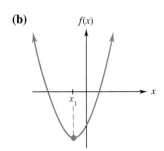

(c)

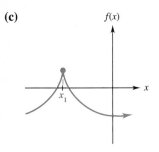

Answers:

(a) Relative maximum at x_2; relative minima at x_1 and x_3

(b) No relative maximum; relative minimum at x_1

(c) Relative maximum at x_1; no relative minimum

Suppose the graph of a function f has a peak at the point $(c, f(c))$. Then all the nearby points lie below the peak and hence have smaller y-coordinates, as shown in Figure 12.2.

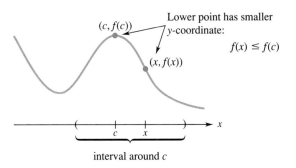

FIGURE 12.2

Analogous remarks apply to valleys and motivate the following definition in which a peak corresponds to a "relative maximum" and a valley to a "relative minimum."

Let c be a number in the domain of a function f.

1. $f(c)$ is a **relative maximum** for f if there exists an open interval (a, b) containing c such that
$$f(x) \leq f(c)$$
for all x in (a, b).
2. $f(c)$ is a **relative minimum** for f if there exists an open interval (a, b) containing c such that
$$f(x) \geq f(c)$$
for all x in (a, b).

The function f is said to have a **relative extremum** at c if it has a relative maximum or minimum there.

Note The plurals of maximum, minimum, and extremum, respectively, are maxima, minima, and extrema.

▶**EXAMPLE 1** Identify the relative extrema of the function whose graph is shown in Figure 12.3.

The function has relative maxima at x_1 and x_3 and relative minima at x_2 and x_4. ◀ **1**

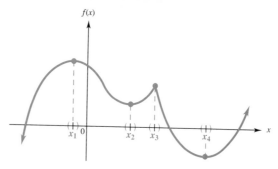

FIGURE 12.3

FOR GRAPHERS

Some graphers can approximate a relative maximum or minimum to a high degree of accuracy with a single keystroke. Check your instruction manual. On any grapher, the trace feature can be used to find the coordinates of the highest or lowest point that was plotted in a particular part of the graph. This usually gives a good approximation of the relative extremum, which can be made better by using a smaller viewing window (so that the grapher plots more points near the extremum).

The *exact* location of a relative maximum or minimum (rather than a grapher's approximation) can normally be found by using derivatives. We now develop methods for doing this.

Let f be a function and think of the graph of f as a roller coaster track, with a roller coaster car moving from left to right along the graph, as shown in Figure 12.4. As the car moves up towards a peak, its floor tilts upward. At the instant the car reaches the peak, its floor is level, but then it begins to tilt downward (perhaps very steeply) as the car rolls down toward a valley.

At any point along the graph, the floor of the car (a straight-line segment in the figure) represents the tangent line to the graph at that point. Using this

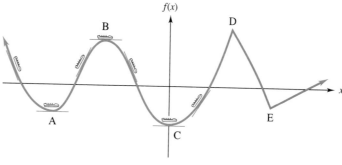

FIGURE 12.4

analogy, we see that as the car passes through the peaks and valleys at A, B, C in Figure 12.4, the tangent line is horizontal and has slope 0. At peak D and valley E, however, a real roller coaster car would have trouble. It would fly off the track at peak D and be unable to make the 90° change of direction at valley E. This corresponds to the fact that although the graph is continuous at D and E, there is no tangent line at D or E because of the sharp corners (see the discussion on the existence of the derivative in Section 11.3).

Thus, the points where a peak or a valley occurs have this property: the tangent line is horizontal and has slope 0 there *or* no tangent line is defined there. The slope of the tangent line to the graph of the function f at $(x, f(x))$ is the value of the derivative, $f'(x)$. Consequently, the following result, whose proof is omitted, should be quite plausible.

Let c be a number in the domain of a function f. If f has a relative extremum at c, then either

$$f'(c) = 0 \quad \text{or} \quad f'(c) \text{ does not exist.}$$

Because of this fact, a number c for which

$$f(c) \text{ is defined and either } f'(c) = 0 \quad \text{or} \quad f'(c) \text{ does not exist}$$

is called a **critical number** of f. The corresponding point $(c, f(c))$ on the graph of f is called a **critical point.**

Caution The result in the box above says that every relative extremum occurs at a critical number but *not* that every critical number produces a relative extremum.

▶**EXAMPLE 2** The graph of $f(x) = x^3$ is shown in Figure 12.5. The derivative, $f'(x) = 3x^2$, is 0 when $x = 0$, so that 0 is a critical number for f. But the graph shows that $f(x) = x^3$ does *not* have a relative maximum or minimum at $x = 0$ (or anywhere else, for that matter). ◀

▶**EXAMPLE 3** Find the critical numbers of the following functions.
(a) $f(x) = 2x^3 - 3x^2 - 72x + 15$

We have $f'(x) = 6x^2 - 6x - 72$, so $f'(x)$ exists for every x. Setting $f'(x) = 0$ shows that

$$6x^2 - 6x - 72 = 0$$
$$6(x^2 - x - 12) = 0$$
$$x^2 - x - 12 = 0$$
$$(x + 3)(x - 4) = 0$$
$$x + 3 = 0 \quad \text{or} \quad x - 4 = 0$$
$$x = -3 \quad \text{or} \quad x = 4.$$

2 Find the critical numbers for each of the following.

(a) $\dfrac{1}{3}x^3 - x^2 - 15x + 6$

(b) $6x^{2/3} - 4x$

Answers:

(a) $-3, 5$

(b) $0, 1$

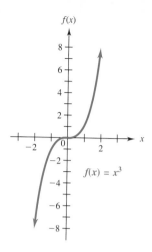

$f(x)$

$f(x) = x^3$

FIGURE 12.5

Therefore, -3 and 4 are the critical numbers of f, so the only *possibilities* for the relative maxima and minima of the function are

$$f(-3) = 2(-3)^3 - 3(-3)^2 - 72(-3) + 15 = 150 \quad \text{and}$$
$$f(4) = 2 \cdot 4^3 - 3 \cdot 4^2 - 72 \cdot 4 + 15 = -193.$$

(b) $f(x) = 3x^{4/3} - 12x^{1/3}$.

We first compute the derivative.

$$f'(x) = 3 \cdot \frac{4}{3}x^{1/3} - 12 \cdot \frac{1}{3}x^{-2/3} = 4x^{1/3} - \frac{4}{x^{2/3}}$$

$$= \frac{4x^{1/3}x^{2/3}}{x^{2/3}} - \frac{4}{x^{2/3}} = \frac{4x - 4}{x^{2/3}}$$

The derivative fails to exist when $x = 0$. Since the original function f is defined when $x = 0$, 0 is a critical number of f. If $x \neq 0$, then $f'(x)$ is 0 only when the numerator $4x - 4 = 0$; that is, when $x = 1$. So the critical numbers of f are 0 and 1; $f(0) = 0$ and $f(1) = -9$ are the *possible* relative maxima and minima. ◀ **2**

After all the critical numbers of a function have been found, as in Example 3, they must be tested to see which, if any, lead to relative extrema. One method of doing this is suggested by thinking again of the graph of f as a roller coaster track, with the cars moving from left to right, as in Figure 12.6 on the next page.

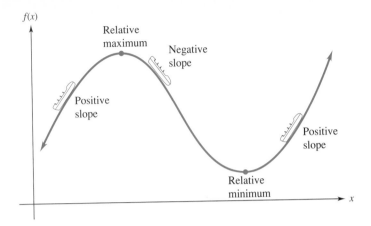

FIGURE 12.6

As the roller coaster car moves uphill from left to right toward a relative maximum, its floor (the tangent line to the graph at that point) is always tilted upward, so the slope of the tangent line is positive on the left side of a relative maximum. As the car moves downhill from the relative maximum, its floor is tilted downward, so the slope of the tangent line is negative on the right side of a relative maximum. Near a relative minimum, the opposite occurs. The tangent line has negative slope on the left side (as the car moves downhill) and positive slope on the right side (as the car moves uphill again).

A rigorous mathematical proof of this intuitive idea (which will be omitted) leads to the following test for relative extrema. Since the derivative $f'(x)$ gives the slope of f at x, our conclusions are restated in terms of $f'(x)$.

First Derivative Test

Assume that $a < c < b$ and that c is the only critical number for a function f in $[a, b]$. Assume that f is differentiable for all x in $[a, b]$ except possibly at $x = c$.

1. If $f'(a) > 0$ and $f'(b) < 0$, then $f(c)$ is a relative maximum.
2. If $f'(a) < 0$ and $f'(b) > 0$, then $f(c)$ is a relative minimum.
3. If $f'(a)$ and $f'(b)$ are both positive or both negative, then $f(c)$ is not a relative extremum.

The sketches in the following table show how the first derivative test works. Assume the same conditions on a, b, c as those stated in the box.

$f(x)$ Has:	Sign of $f'(a)$	Sign of $f'(b)$	Sketches	
Relative maximum	+	−		
Relative minimum	−	+		
No relative extrema	+	+		
No relative extrema	−	−		

▶**EXAMPLE 4**   In Example 3(a) we found that the critical points of $f(x) = 2x^3 - 3x^2 - 72x + 15$ are -3 and 4. To test $c = -3$, we can use $a = -4$ and $b = 0$ since -3 is the only critical number between -4 and 0. Many other choices of a and b are possible, but we try to select numbers that will make the computations easy. Since

$$f'(x) = 6x^2 - 6x - 72 = 6(x^2 - x - 12) = 6(x + 3)(x - 4),$$

we see that

$$f'(-4) = 6(-4 + 3)(-4 - 4) = 6(-1)(-8) > 0;$$

and

$$f'(0) = 6(0 + 3)(0 - 4) = 6(3)(-4) < 0.$$

(Note that it's not necessary to finish calculating the exact value of $f'(x)$ in order to determine its sign.) Thus, the value of the derivative is positive to the left of -3 and negative to the right of -3, as shown in Figure 12.7 on the next page. By part 1 of the first derivative test, $f(-3) = 150$ is a relative maximum.

Similarly, we can use $a = 0$ and $b = 5$ to test the critical number $c = 4$. We just saw that $f'(0) < 0$; and

$$f'(5) = 6(5 + 3)(5 - 4) = 6(8)(1) > 0.$$

3 Find all relative extrema for the following functions.

(a) $f(x) = 2x^2 - 8x + 1$

(b) $f(x) = 3 + 4x - 3x^2$

Answers:

(a) Relative minimum at $x = 2$; minimum is $f(2) = -7$

(b) Relative maximum at $x = 2/3$; maximum is $f(2/3) = 13/3$

Hence, by part 2 of the first derivative test, $f(4) = -193$ is a relative minimum. ◀ **3**

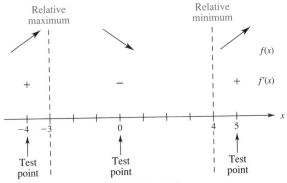

FIGURE 12.7

▶ **EXAMPLE 5** In Example 3(b) we found that 0 and 1 are the critical numbers for $f(x) = 3x^{4/3} - 12x^{1/3}$. We can use -1, $1/2$, and 2 to apply the first derivative test.

$$f'(x) = \frac{4x - 4}{x^{2/3}} = \frac{4x - 4}{\sqrt[3]{x^2}}$$

$$f'(-1) = \frac{4(-1) - 4}{\sqrt[3]{(-1)^2}} = \frac{-4 - 4}{1} < 0$$

$$f'\left(\frac{1}{2}\right) = \frac{4(1/2) - 4}{\sqrt[3]{(1/2)^2}} = \frac{-2}{\sqrt[3]{1/4}} < 0$$

$$f'(2) = \frac{4(2) - 4}{\sqrt[3]{2^2}} = \frac{4}{\sqrt[3]{4}} > 0$$

These results are shown in Figure 12.8.

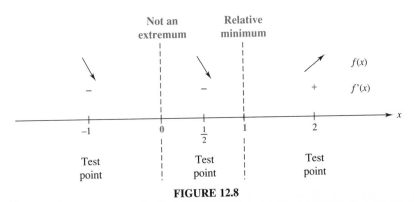

FIGURE 12.8

4 Find any relative extrema for $f(x) = x^3 + 4x^2 - 3x + 5$.

Answer:
Relative maximum at -3 of $f(-3) = 23$; relative minimum at $1/3$ of $f(1/3) = 121/27$

5 Find any relative extrema for $f(x) = x^2 e^{5x}$.

Answer:
Relative minimum of 0 at $x = 0$; relative maximum of about .02 at $x = -2/5$

6 Identify the location of any absolute extrema for these graphs.

(a)

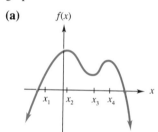

(b)

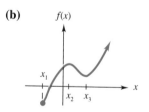

Answers:

(a) Absolute maximum at x_2; no absolute minimum

(b) Absolute minimum at x_1; no absolute maximum

Since 0 is the only critical number between -1 and $1/2$ and the derivative is negative at both -1 and $1/2$, part 3 of the first derivative test shows that there is no relative extremum at $x = 0$. The only critical number between $1/2$ and 2 is $x = 1$. Since $f'(1/2) < 0$ and $f'(2) > 0$, $f(1) = -9$ is a relative minimum. ◀ **4**

▶ **EXAMPLE 6** Find all relative maxima or relative minima for $f(x) = (2x + 1)e^{-x}$.

First take the derivative, using the product rule.

$$f'(x) = (2x + 1)(-e^{-x}) + e^{-x}(2)$$
$$= -2x \cdot e^{-x} - e^{-x} + 2e^{-x}$$
$$= -2x \cdot e^{-x} + e^{-x} = e^{-x}(-2x + 1)$$

Set the derivative equal to 0.

$$e^{-x}(-2x + 1) = 0$$

Since e^{-x} is never 0, this derivative can equal 0 only when $-2x + 1 = 0$, or when $x = 1/2$.

To decide if there is a maximum or a minimum at $x = 1/2$, use the first derivative test. Since $f'(0) > 0$ and $f'(1) < 0$, there is a relative maximum of $f(1/2) = 2/e^{1/2} \approx 1.2$ at $x = 1/2$. ◀ **5**

We can now summarize the procedure for finding relative extrema.

To find the relative extrema of a function f:

Step 1 Find the derivative $f'(x)$;

Step 2 Find every critical number c (those numbers for which $f(c)$ is defined and either $f'(c) = 0$ or $f'(c)$ does not exist);

Step 3 Use the first derivative test to identify relative maxima or minima.

ABSOLUTE MAXIMA AND MINIMA The largest possible value of a function f (if there is one) is called its **absolute maximum** and the smallest possible value (if there is one) is called its **absolute minimum.** If a function has an absolute maximum, it corresponds to the highest point on the entire graph (whereas a *relative* maximum may correspond only to a small peak). Similarly, the absolute minimum of a function, if it exists, corresponds to the lowest point on the graph. The function graphed in Figure 12.9 on the next page has an absolute maximum at x_1, a relative minimum at x_2, a relative maximum at x_3, and an absolute minimum at x_4. The function graphed in Figure 12.10 has relative extrema, but no absolute maximum and no absolute minimum. **6**

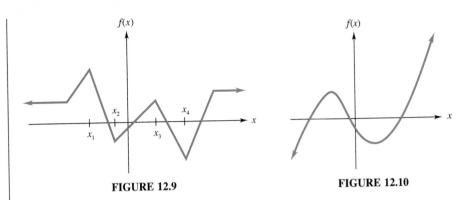

FIGURE 12.9   FIGURE 12.10

In many applications, we are interested only in the largest or smallest value of a function on a particular interval.

Let f be a function defined on some interval and let c be a number in the interval.

1. $f(c)$ is the **absolute maximum of** f **on the interval** if

$$f(x) \leq f(c)$$

for every x in the interval.
2. $f(c)$ is the **absolute minimum of** f **on the interval** if

$$f(x) \geq f(c)$$

for every x in the interval.

▶**EXAMPLE 7**   Consider the function graphed in Figure 12.11. The absolute maximum of f on the interval $[-2, 6]$ is $f(6)$, even though it is not a relative maximum nor is it the absolute maximum of the entire function f. Similarly, $f(3)$ is the absolute minimum of f on the interval $[-2, 6]$. $f(3)$ is also a relative minimum of f, but not an absolute minimum of the entire function.   ◀

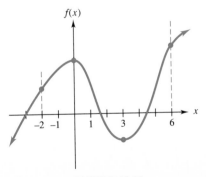

FIGURE 12.11

The absolute maximum in Example 7 occurred at $x = 6$, which is an endpoint of the interval $[-2, 6]$, and the absolute minimum occurred at $x = 3$, which is a critical number of f (why?). This is typical of the general case.

Extreme Value Theorem

If a function f is continuous on a closed interval $[a, b]$, then f has both an absolute maximum and an absolute minimum on the interval. Each of these occurs either at an endpoint of the interval or at a critical number of f.

The Extreme Value Theorem, which is proved in more advanced courses, is used as follows.

To find absolute extrema for a function f continuous on an interval $[a, b]$:

Step 1 Find f' and solve $f'(x) = 0$ to find all critical numbers c_1, $c_2, \ldots$ in (a, b). Also, identify the critical numbers where f' does not exist.

Step 2 Evaluate $f(c_1), f(c_2), \ldots$ for all critical numbers in (a, b).

Step 3 Evaluate $f(a)$ and $f(b)$ for the endpoints a and b of the interval $[a, b]$.

Step 4 The largest value found in steps 2 and 3 is the absolute maximum for f on $[a, b]$; the smallest value found is the absolute minimum.

▶ **EXAMPLE 8** Find the absolute extrema of the function given by

$$f(x) = 4x + \frac{36}{x},$$

on the interval $[1, 6]$.

First determine any critical numbers in $(1, 6)$.

$$f'(x) = 4 - \frac{36}{x^2} = 0$$

$$\frac{4x^2 - 36}{x^2} = 0$$

Since we are looking for critical numbers in $(1, 6)$, $x \neq 0$. When $f'(x) = 0$ and $x \neq 0$, then

7 Find the absolute maximum and absolute minimum values of

(a) $f(x) = -x^2 + 4x - 8$ on $[-4, 4]$;

(b) $f(x) = 2x^2 - 6x + 6$ on $[0, 5]$.

Answers:

(a) Absolute maximum of -4 at $x = 2$; absolute minimum of -40 at $x = -4$

(b) Absolute maximum of 26 at $x = 5$; absolute minimum of $3/2$ at $x = 3/2$

$$4x^2 - 36 = 0$$
$$4x^2 = 36$$
$$x^2 = 9$$
$$x = -3 \quad \text{or} \quad x = 3$$

Since -3 is not in the interval $[1, 6]$, 3 is the only critical number of interest. To find the absolute extrema, evaluate $f(x)$ at $x = 3$ and at the endpoints, where $x = 1$ and $x = 6$.

x-value	*Value of Function*	
1	40	← **Absolute maximum**
3	24	← **Absolute minimum**
6	30	

◄

Absolute extrema are particularly important in applications of mathematics. In most applications the domain is limited in some natural way that ensures the occurrence of absolute extrema.

▶ **EXAMPLE 9**   A company has found that its weekly profit from the sale of x units of an auto part is given by

$$P(x) = -.02x^3 + 600x - 20,000.$$

Production bottlenecks limit the number of units that can be made per week to no more than 120, while a long-term contract requires that at least 60 units be made each week. Find the maximum possible weekly profit that the firm can make.

Because of the restrictions, the profit function is defined only for the domain $[60, 120]$. Look first for critical numbers of the function in the open interval $(60, 120)$. Here $P'(x) = -.06x^2 + 600$. Set this derivative equal to 0 and solve for x.

$$-.06x^2 + 600 = 0$$
$$-.06x^2 = -600$$
$$x^2 = 10,000$$
$$x = 100 \quad \text{or} \quad x = -100$$

Since $x = -100$ is not in the interval $[60, 120]$, disregard it. Now evaluate the function at the remaining critical number 100 and at the endpoints of the domain, 60 and 120.

8 For a certain firm, the cost to produce x units of an item is given by

$$C(x) = x^3 - 15x^2 + 48x + 50.$$

Because of production problems, the function is defined only for the domain $[1, 6]$. Find the absolute maximum and minimum costs that the firm will face.

Answer:
Absolute maximum of 94 when $x = 2$; absolute minimum of 14 when $x = 6$

x-value	Value of Function
60	11,680
100	20,000 ← **Absolute maximum**
120	17,440

Maximum profit of $20,000 occurs when 100 units are made per week.

◀ **8**

12.1 EXERCISES

Find the location and value of all relative maxima and relative minima for the functions whose graphs are shown. (See Example 1.)

1.

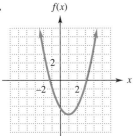

2.

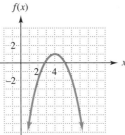

3.

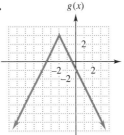

4.

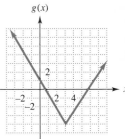

5.

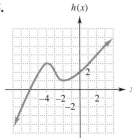

6.

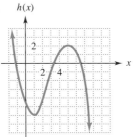

7.

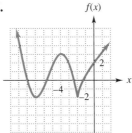

8.

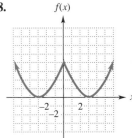

Find the x-values of all points where the given function has relative extrema. Find the value of each relative extremum. (See Examples 3–6.)

9. $f(x) = x^3 - 3x^2 + 1$

10. $f(x) = x^3 - x^2 - 5x + 1$

11. $f(x) = x^3 + 6x^2 + 9x + 2$

12. $f(x) = x^3 + 3x^2 - 24x + 2$

13. $f(x) = -\frac{4}{3}x^3 - \frac{21}{2}x^2 - 5x + 8$

14. $f(x) = -\frac{2}{3}x^3 - \frac{1}{2}x^2 + 3x - 4$

15. $f(x) = \frac{2}{3}x^3 - x^2 - 12x + 2$

16. $f(x) = \frac{4}{3}x^3 - 10x^2 + 24x - 1$

17. $f(x) = x^5 - 20x^2 + 3$

18. $f(x) = x^{11/5} - x^{6/5} + 1$

19. $f(x) = (7 - 2x)^{2/3} - 2$

20. $f(x) = -(3 - 4x)^{2/5} + 4$

21. $f(x) = x - \frac{1}{x}$

22. $f(x) = x^2 + \frac{1}{x}$

23. $f(x) = \frac{x^2}{x^2 + 1}$

24. $f(x) = \frac{x^2 - 2x + 1}{x - 3}$

25. $f(x) = -xe^x$

26. $f(x) = xe^{-x}$

27. $f(x) = e^x + e^{-x}$

28. $f(x) = -x^2 e^x$

29. $f(x) = x \cdot \ln |x|$

30. $f(x) = x - \ln |x|$

Find the values of x at which the absolute maximum and absolute minimum of the function occur on the given interval. (See Example 7.)

31. The function in Exercise 3 on $[5, 1]$

32. The function in Exercise 2 on $[2, 5]$

33. The function in Exercise 5 on $[-4, 2]$

34. The function in Exercise 6 on $[-1, 2]$

35. The function in Exercise 7 on $[-8, 0]$

36. The function in Exercise 8 on $[-4, 4]$

Find the absolute extrema of each function on the given interval. (See Example 8.)

37. $f(x) = x^4 - 32x^2 - 7; [-5, 6]$

38. $f(x) = x^4 - 18x^2 + 1; [-4, 4]$

39. $f(x) = \frac{8 + x}{8 - x}; [4, 6]$

40. $f(x) = \frac{1 - x}{3 + x}; [0, 3]$

41. $f(x) = \frac{x}{x^2 + 2}; [0, 4]$

42. $f(x) = \frac{x - 1}{x^2 + 1}; [1, 5]$

43. $f(x) = (x^2 + 18)^{2/3}; [-3, 3]$

44. $f(x) = (x^2 + 4)^{1/3}; [-2, 2]$

45. $f(x) = \frac{1}{\sqrt{x^2 + 1}}; [-1, 1]$

46. $f(x) = \frac{3}{\sqrt{x^2 + 4}}; [-2, 2]$

47. Let f be a continuous function whose graph has no horizontal segments. If c and d are the only critical numbers of f, explain why the graph of f must always rise or always fall between $x = c$ and $x = d$.

48. Let f be a continuous function which has only one critical number c. If f has a relative minimum at $x = c$, explain why this must also be the absolute minimum of f.

Work the following exercises. (See Example 9.)

49. Management The total profit $P(x)$ (in dollars) from the sale of x units of a certain prescription drug is given by

$$P(x) = -x^3 + 3x^2 + 72x.$$

(a) Find the number of units that should be sold in order to maximize the total profit.

(b) What is the maximum profit?

50. Management The demand equation for telephones at one store is

$$p = D(q) = \frac{1}{3}q^2 - \frac{25}{2}q + 100, \quad 0 \le q \le 16,$$

where p is the price (in dollars) and q is the quantity of telephones sold per week. Find the values of q and p that maximize revenue.

51. Management Suppose that the cost function for a product is given by $C(x) = .002x^3 - 9x + 4000$. Find the production level (i.e., value of x) that will produce the minimum average cost per unit $\overline{C}(x)$.

52. Management A company has found through experience that increasing its advertising also increases its sales, up to a point. The company believes that the mathematical model connecting profit in hundreds of dollars, $P(x)$, and expenditures on advertising in thousands of dollars, x, is

$$P(x) = 80 + 108x - x^3, \quad 0 \le x \le 10.$$

(a) Find the expenditure on advertising that leads to maximum profit.

(b) Find the maximum profit.

53. Management The total profit $P(x)$ (in thousands of dollars) from the sale of x hundred thousands of automobile tires is approximated by

$$P(x) = -x^3 + 9x^2 + 120x - 400, \quad 3 \le x \le 15.$$

Find the number of hundred thousands of tires that must be sold to maximize profit. Find the maximum profit.

54. Natural Science A marshy region used for agricultural drainage has become contaminated with selenium. It has been determined that flushing the area with clean water will reduce the selenium for a while, but it will then begin to build up again. A biologist has found that the percent of selenium in the soil x months after the flushing begins is given by

$$f(x) = \frac{x^2 + 36}{2x}, \quad 1 \le x \le 12.$$

When will the selenium be reduced to a minimum? What is the minimum percent?

55. Natural Science The number of salmon swimming upstream to spawn is approximated by

$$S(x) = -x^3 + 3x^2 + 360x + 5000, \quad 6 \le x \le 20,$$

where x represents the temperature of the water in degrees Celsius. Find the water temperature that produces the maximum number of salmon swimming upstream.

56. Natural Science In the summer the activity level of a certain type of lizard varies according to the time of day. A biologist has determined that the activity level is given by the function

$$a(t) = .008t^3 - .27t^2 + 2.02t + 7,$$

where t is the number of hours after 12 noon and $0 \le t \le 24$. When is the activity level highest? When is it lowest?

57. Social Science Social psychologists have found that as the discrepancy between the views of a speaker and those of an audience increases, the attitude change in the audience also increases to a point, but decreases when the discrepancy becomes too large, particularly if the communicator is viewed by the audience as having low credibility.[*] Suppose that the degree of change can be approximated by the function

$$D(x) = -x^4 + 8x^3 + 80x^2, \quad 0 \le x \le 13,$$

where x is the discrepancy between the views of the speaker and those of the audience, as measured by scores on a questionnaire. Find the amount of discrepancy the speaker should aim for to maximize the attitude change in the audience.

58. Natural Science The microbe concentration, $B(x)$, in appropriate units, of Lake Tom depends approximately on the oxygen concentration, x, again in appropriate units, according to the function

$$B(x) = x^3 - 7x^2 - 160x + 1800, \quad 0 \le x \le 20.$$

(a) Find the oxygen concentration that will lead to the minimum microbe concentration.

(b) What is the minimum concentration?

59. Social Science A group of researchers found that people prefer training films of moderate length; shorter films contain too little information, while longer films are boring. For a training film on the care of exotic birds, the researchers determined that the ratings people gave for the film could be approximated by

$$R(t) = \frac{200t}{t^2 + 100},$$

where t is the length of the film (in minutes). Find the film length that received the highest rating.

---

[*] See A. H. Eagly and K. Telaak, "Width of the Latitude of Acceptance as a Determinant of Attitude Change" in *Journal of Personality and Social Psychology*, vol. 23, 1972, pp. 388–97.

60. Suppose dots and dashes are transmitted over a telegraph line so that dots occur a fraction p of the time (where $0 \le p \le 1$) and dashes occur a fraction $1 - p$ of the time. The *information content* of the telegraph line is given by $I(p)$, where

$$I(p) = -p \cdot \ln p - (1 - p) \cdot \ln (1 - p).$$

(a) Find $I'(p)$.
(b) Let $I'(p) = 0$ and find the value of p that maximizes $I(p)$. In the long run, what fraction of the dots and dashes should be dots?

 Find the approximate location and value of all relative extrema of the following functions (accurate to two decimal places).

61. $f(x) = 3x^3 - 18.5x^2 - 4.5x - 45$

62. $g(x) = x^5 - 3x^3 + x + 1$

63. $h(x) = .1x^4 - x^3 - 12x^2 + 99x - 10$

64. $f(x) = x^5 - 12x^4 - x^3 + 232x^2 + 260x - 600$

12.2 THE SECOND DERIVATIVE TEST

A salesman says he invested $1200 in a certain mutal fund when it began two years ago and his investment is now worth $1690, a two-year increase of 41%. Over the past two years he says the price of a share of this fund (in dollars) has been given by the function

$$P(t) = 12 + t^{1/2},$$

where t is the number of months since the fund began. Since you know some calculus, he points out that the derivative of this function,

$$P'(t) = \frac{1}{2}t^{-1/2} = \frac{1}{2\sqrt{t}},$$

is always positive (because $\sqrt{t}$ is positive for $t > 0$), so the price of a share is always increasing. If you take his advice and invest, will you, too, make a huge profit?

Although the past performance of a mutual fund is not necessarily a good guide to its future price, let us assume that the salesman is right and the price function remains valid for the next two years. Now it is true that the positive derivative means the function is increasing. The catch is in *how fast* it is increasing. For example, when $t = 1$, $P'(1) = 1/2$, meaning that the price per share is increasing at a rate of 1/2 dollar (50¢) per month. At 9 months, $P'(9) = 1/6$, or about 17¢ per month. When you buy at $t = 24$ months, the price is increasing at $P'(24) \approx 10$¢ per month and the *rate* of increase appears to be steadily decreasing.

The rate of increase of the derivative function $P'(t)$ is given by *its* derivative, which is denoted $P''(t)$. Since $P'(t) = (1/2)t^{-1/2}$,

$$P''(t) = -\frac{1}{4}t^{-3/2} = \frac{-1}{4\sqrt{t^3}}.$$

1 Let $f(x) = 4x^3 - 12x^2 + x - 1$. Find

(a) $f''(0)$;

(b) $f''(4)$;

(c) $f''(-2)$.

Answers:

(a) -24

(b) 72

(c) -72

$P''(t)$ is negative whenever t is positive; therefore, the *rate* at which the price increases is always decreasing for $t > 0$. So the price of a share will continue to grow, but at a smaller and smaller rate. For instance, at $t = 24$, when you would buy, the price would be \$16.90 per share. Two years later ($t = 48$), the price would be \$18.93, a two-year increase of only 12%—hardly the "huge profit" the salesman predicts. The only investors to make huge profits on this fund would be those who got in early, when the rate of increase was much greater.

This example shows that it is important to know not only whether a function is increasing or decreasing, but also the rate at which this is occurring. As we have seen this rate is given by the derivative of the derivative of the original function. In order to deal with such situations we need some additional terminology and notation.

HIGHER DERIVATIVES   If a function f has a derivative f', then the derivative of f', if it exists, is the **second derivative** of f, written $f''(x)$. The derivative of $f''(x)$, if it exists, is called the **third derivative** of f, and so on. By continuing this process, we can find **fourth derivatives** and other higher derivatives. For example, if $f(x) = x^4 + 2x^3 + 3x^2 - 5x + 7$, then

$$f'(x) = 4x^3 + 6x^2 + 6x - 5, \quad \text{First derivative of } f$$
$$f''(x) = 12x^2 + 12x + 6, \quad \text{Second derivative of } f$$
$$f'''(x) = 24x + 12, \quad \text{Third derivative of } f$$

and $\quad f^{(4)}(x) = 24.$ Fourth derivative of f

The second derivative of $y = f(x)$ can be written with any of the following notations:

$$f''(x), \quad y'', \quad \frac{d^2y}{dx^2}, \quad \text{or} \quad D_x^2[f(x)].$$

The third derivative can be written in a similar way. For $n \geq 4$, the nth derivative is written $f^{(n)}(x)$.

▶**EXAMPLE 1**   Let $f(x) = x^3 + 6x^2 - 9x + 8$. Find the following.

(a) $f''(0)$

Here $f'(x) = 3x^2 + 12x - 9$, so that $f''(x) = 6x + 12$. Then

$$f''(0) = 6(0) + 12 = 12.$$

(b) $f''(-3) = 6(-3) + 12 = -6$ ◀ **1**

2 Find the second derivatives of the following.

(a) $y = -9x^3 + 8x^2 + 11x - 6$

(b) $y = -2x^4 + 6x^2$

(c) $y = \dfrac{x + 2}{5x - 1}$

(d) $y = e^x + \ln x$

Answers:

(a) $y'' = -54x + 16$

(b) $y'' = -24x^2 + 12$

(c) $y'' = \dfrac{110}{(5x - 1)^3}$

(d) $y'' = e^x - \dfrac{1}{x^2}$

▶**EXAMPLE 2** Find the second derivative of the following functions.

(a) $y = 8x^3 - 9x^2 + 6x + 4$

Here $y' = 24x^2 - 18x + 6$. The second derivative is the derivative of y', or

$$y'' = 48x - 18.$$

(b) $y = \dfrac{4x + 2}{3x - 1}$

Use the quotient rule to find y'.

$$y' = \frac{(3x - 1)(4) - (4x + 2)(3)}{(3x - 1)^2} = \frac{12x - 4 - 12x - 6}{(3x - 1)^2} = \frac{-10}{(3x - 1)^2}$$

Use the quotient rule again to find y''.

$$y'' = \frac{(3x - 1)^2(0) - (-10)(2)(3x - 1)(3)}{[(3x - 1)^2]^2}$$

$$= \frac{60(3x - 1)}{(3x - 1)^4} = \frac{60}{(3x - 1)^3}$$

(c) $y = xe^x$

Using the product rule gives

$$y' = x \cdot e^x + e^x \cdot 1 = xe^x + e^x.$$

Differentiate this result to get y''.

$$y'' = (xe^x + e^x) + e^x = xe^x + 2e^x = (x + 2)e^x \quad ◀ \quad \boxed{2}$$

In the previous chapter we saw that the first derivative of a function represents the rate of change of the function. The second derivative, then, represents the rate of change of the first derivative. If a function describes the position of a moving object (along a straight line) at time t, then the first derivative gives the velocity of the object. That is, if $y = s(t)$ describes the position (along a straight line) of the object at time t, than $v(t) = s'(t)$ gives the velocity at time t.

The rate of change of velocity is called **acceleration.** Since the second derivative gives the rate of change of the first derivative, the acceleration is the derivative of the velocity. Thus, if $a(t)$ represents the acceleration at time t, then

$$a(t) = \frac{d}{dt} v(t) = s''(t).$$

▶**EXAMPLE 3** Suppose that an object is moving along a straight line, with its position in feet at time t in seconds given by

$$s(t) = t^3 - 2t^2 - 7t + 9.$$

3 Rework Example 3 if
$s(t) = t^4 - t^3 + 10$.

Answers:

(a) $v(t) = 4t^3 - 3t^2$

(b) $a(t) = 12t^2 - 6t$

(c) At 0 and 1/2 second

Find the following.

(a) The velocity at any time t

The velocity is given by

$$v(t) = s'(t) = 3t^2 - 4t - 7.$$

(b) The acceleration at any time t

Acceleration is given by

$$a(t) = v'(t) = s''(t) = 6t - 4.$$

(c) The object stops when velocity is zero. For $t \geq 0$, when does that occur?
Set $v(t) = 0$.

$$3t^2 - 4t - 7 = 0$$
$$(3t - 7)(t + 1) = 0$$
$$3t - 7 = 0 \quad \text{or} \quad t + 1 = 0$$
$$t = \frac{7}{3} \qquad t = -1$$

Since we want $t \geq 0$, only $t = 7/3$ is acceptable here. The object will stop at
7/3 seconds. ◀ **3**

The first derivative of a function at any value x gives the slope of the tangent
line to the graph at that value. The second derivative (the derivative of the slope
function) gives the rate at which slope y' is changing with respect to x. When you
move from left to right along the graph near a relative minimum
(Figure 12.12(a)), the slope of the tangent line *increases* from negative to 0 to
positive. Hence the second derivative (the derivative of the slope function) is
positive. Similarly, when you move from left to right through a relative maximum (Figure 12.12(b)), the slope of the tangent line *decreases* from positive to
0 to negative, so that the second derivative is negative. These facts should make
the following result plausible.

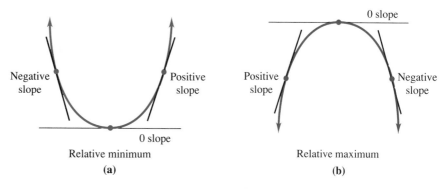

FIGURE 12.12

4 Find all relative maxima and relative minima for the following functions. Use the second derivative test.

(a) $f(x) = 6x^2 + 12x + 1$

(b) $f(x) = x^3 - 3x^2 - 9x + 8$

Answers:

(a) Relative minimum of -5 at $x = -1$

(b) Relative maximum of 13 at $x = -1$; relative minimum of -19 at $x = 3$

Second Derivative Test

Let c be a critical number of the function f such that $f'(c) = 0$ and $f''(x)$ exists for all x in some open interval containing c.

1. If $f''(c) > 0$, then $f(c)$ is a relative minimum.
2. If $f''(c) < 0$, then $f(c)$ is a relative maximum.
3. If $f''(c) = 0$, then the test gives no information about extrema.

▶**EXAMPLE 4** Find all relative extrema for

$$f(x) = 4x^3 + 7x^2 - 10x + 8.$$

First, find the critical points. Here $f'(x) = 12x^2 + 14x - 10$. Solve the equation $f'(x) = 0$.

$$12x^2 + 14x - 10 = 0$$
$$2(6x^2 + 7x - 5) = 0$$
$$2(3x + 5)(2x - 1) = 0$$
$$3x + 5 = 0 \quad \text{or} \quad 2x - 1 = 0$$
$$3x = -5 \qquad 2x = 1$$
$$x = -\frac{5}{3} \qquad x = \frac{1}{2}$$

Now use the second derivative test. The second derivative is $f''(x) = 24x + 14$. The critical points are $-5/3$ and $1/2$. Since

$$f''\left(-\frac{5}{3}\right) = 24\left(-\frac{5}{3}\right) + 14 = -40 + 14 = -26 < 0,$$

$-5/3$ leads to a relative maximum of $f(-5/3) = 691/27 \approx 25.593$. Also,

$$f''\left(\frac{1}{2}\right) = 24\left(\frac{1}{2}\right) + 14 = 12 + 14 = 26 > 0,$$

so $1/2$ gives a relative minimum of $f(1/2) = 21/4 = 5.25$. ◀ **4**

The second derivative test works only for those critical points c that make $f'(c) = 0$. This test does not work for those critical points c for which $f'(c)$ does not exist (since $f''(c)$ would not exist either). Also, the second derivative test does not work for critical points c that make $f''(c) = 0$. In both of these cases, use the first derivative test.

12.2 EXERCISES

For each of the following functions, find $f''(x)$, $f''(0)$, $f''(2)$, and $f''(-3)$. (See Examples 1 and 2.)

1. $f(x) = x^3 - 5x^2 + 1$

2. $f(x) = 2x^4 + x^3 - 3x^2 + 2$

3. $f(x) = (x + 2)^4$

4. $f(x) = \dfrac{2x + 5}{x - 3}$

5. $f(x) = \dfrac{x^2}{1 + x}$

6. $f(x) = \dfrac{-x}{1 - x^2}$

7. $f(x) = \sqrt{x + 4}$

8. $f(x) = \sqrt{2x + 9}$

9. $f(x) = 5x^{3/5}$

10. $f(x) = -2x^{2/3}$

11. $f(x) = 2e^x$

12. $f(x) = \ln(2x - 3)$

For each of the following, find $f'''(x)$, the third derivative, and $f^{(4)}(x)$, the fourth derivative.

13. $f(x) = 3x^4 + 2x^3 - x^2 + 5$

14. $f(x) = -2x^5 + 3x^3 - 4x + 7$

15. $f(x) = \dfrac{x - 1}{x}$

16. $f(x) = \dfrac{x + 2}{x - 2}$

17. $f(x) = 5e^{2x}$

18. $f(x) = 2 + e^{-x}$

19. $f(x) = \ln |x|$

20. $f(x) = \dfrac{1}{x}$

21. $f(x) = x \ln |x|$

22. $f(x) = \dfrac{\ln |x|}{x}$

Find all critical numbers of the following functions. Then use the second derivative test on each critical number to determine whether it leads to a relative maximum or minimum. (See Example 4.)

23. $f(x) = 3x^2 - 2x + 7$

24. $f(x) = 4x^3 - 3x^2 - 5$

25. $f(x) = 4x^3 - x^2 - 4x + 3$

26. $f(x) = -4x^3 - 15x^2 + 18x + 5$

27. $f(x) = -2x^3 - 3x^2 - 72x + 1$

28. $f(x) = \dfrac{2}{3}x^3 + \dfrac{1}{2}x^2 - x - \dfrac{1}{4}$

29. $f(x) = x^3 + \dfrac{3}{2}x^2 - 60x + 100$

30. $f(x) = (x - 2)^5$

31. $f(x) = x^4 - 8x^2$

32. $f(x) = x^4 - 32x^2 + 7$

33. $f(x) = x + \dfrac{3}{x}$

34. $f(x) = x - \dfrac{1}{x}$

35. $f(x) = \dfrac{x^2 + 9}{2x}$

36. $f(x) = \dfrac{x^2 + 16}{2x}$

37. $f(x) = \dfrac{2 - x}{2 + x}$

38. $f(x) = \dfrac{x + 2}{x - 1}$

Physical Science Each of the functions in Exercises 39–42 gives the distance from a starting point at time t of a particle moving along a line. Find the velocity and acceleration functions. Then find the velocity and acceleration at $t = 0$ and $t = 4$. Assume that time is measured in seconds and distance is measured in centimeters. Velocity will be in centimeters per second (cm/sec) and acceleration in centimeters per second per second (cm/sec^2). (See Example 3.)

39. $s(t) = 6t^2 + 2t$

40. $s(t) = 4t^3 - 6t^2 + 3t - 4$

41. $s(t) = 3t^3 - 4t^2 + 8t - 9$

42. $s(t) = \dfrac{-2}{3t + 4}$

In Exercises 43–44, P(t) is the price of a certain stock at time t during a particular day.

43. If the price of the stock is falling faster and faster, are $P'(t)$ and $P''(t)$ positive or negative? Explain your answer.

44. When the stock reaches its highest price of the day, are $P'(t)$ and $P''(t)$ positive or negative? Explain your answer.

In Exercises 45–46, draw the graph of a function with domain (0,10) that has the given properties. (There are many correct answers and you need not state the rule of your function.)

45. $f'(x) > 0$ and $f''(x) > 0$

46. $f'(x) > 0$ and $f''(x) < 0$

47. Management A national chain has found that advertising produces sales, but that too much advertising for a product tends to make consumers "turn off" so that sales are reduced. Based on past experience the chain expects that the number $N(x)$ of cameras sold during a week is related to the amount spent on advertising by the function

$$N(x) = -3x^3 + 135x^2 + 3600x + 12{,}000$$

where x (with $0 \leq x \leq 64$) is the amount spent on advertising in thousands of dollars. What amount of advertising results in the largest number of cameras sold?

48. Management Because of raw material shortages, it is increasingly expensive to produce fine cigars. In fact, the profit in thousands of dollars from producing x hundred thousand cigars is approximated by

$$P(x) = x^3 - 23x^2 + 120x + 60.$$

Find the levels of production ($0 \leq x \leq 16$) that will lead to maximum profit and to any minimum profit.

49. Physical Science When an object is dropped straight down, the distance in feet that it travels in t seconds is given by

$$s(t) = -16t^2.$$

Find the velocity at each of the following times.
(a) After 3 seconds
(b) after 5 seconds
(c) After 8 seconds
(d) Find the acceleration. (The answer here is a constant, the acceleration due to the influence of gravity alone.)

50. Physical Science If an object is thrown directly upward with a velocity of 256 ft/sec, its height above the ground after t seconds is given by $s(t) = 256t - 16t^2$. Find the velocity and the acceleration after t seconds. What is the maximum height the object reaches? When does it hit the ground?

51. Management A small company must hire expensive temporary help to supplement its full-time staff. It estimates that the weekly costs $C(x)$ of salaries and benefits are related to the number x of full-time employees by the function

$$C(x) = 250x + \frac{16{,}000}{x} + 1000 \quad (1 \leq x \leq 30).$$

How many full-time employees should the company have on its staff to minimize these costs?

52. Natural Science The percent of concentration of a certain drug in the bloodstream x hours after the drug is administered is given by

$$K(x) = \frac{3x}{x^2 + 4}.$$

For example, after 1 hour the concentration is given by $K(1) = 3(1)/(1^2 + 4) = (3/5)\% = .6\% = .006$.
(a) Find the time at which concentration is a maximum.
(b) Find the maximum concentration.

53. Management When a company has to pay large amounts of overtime, or build a larger factory, its profits may go down even though sales are going up. The Wizard Widget Company expects that its profit (in hundreds of dollars) during the next six months will be given by

$$P(x) = -x + 200 \sqrt{x} - 2000 \quad (0 \le x \le 35,000),$$

where x is the number of units sold. Find the number of units that produce maximum profit.

54. Natural Science   The percent of concentration of another drug in the bloodstream x hours after the drug is administered is given by

$$K(x) = \frac{4x}{3x^2 + 27}.$$

(a) Find the time at which the concentration is a maximum.

(b) Find the maximum concentration.

55. Natural Science   Assume that the number of bacteria, in millions, present in a certain culture at time t is given by

$$R(t) = t^2(t - 18) + 96t + 1000.$$

(a) At what time before 8 hours will the population be maximized?

(b) Find the maximum population.

56. Management   The revenue of a small firm, in thousands of dollars, is given by

$$R(x) = 2x^3 - 8x^2 + 4x,$$

where x is the number of units of product that the firm sells. Find the following. (See Example 3.)

(a) Velocity of revenue when $x = 8$

(b) When $x = 12$

(c) Acceleration of revenue when $x = 4$

(d) When $x = 8$

12.3 APPLICATIONS OF MAXIMA AND MINIMA

This section presents several examples showing applications of calculus to maximum and minimum problems. The following steps are used to solve these problems.

Solving Applied Problems

Step 1   Read the problem carefully. Make sure you understand what is given and what is asked for.

Step 2   If possible, sketch a diagram and label the various parts.

Step 3   Decide which variable is to be maximized or minimized. Express that variable as a function of *one* other variable. Be sure to determine the domain of this function.

Step 4   Find the critical numbers for the function in Step 3.

Step 5   If the domain is a closed interval, evaluate the function at the endpoints and at each critical number to see which yields the maximum or minimum. If the domain is an open interval, test each critical number with the first or second derivative test to see which yields a maximum or minimum.

Caution Do not skip Step 5 in the box above. If you are looking for a maximum and you find a critical number in Step 4, do not automatically assume that the maximum occurs there. It may occur at an endpoint or may not exist at all.

▶**EXAMPLE 1** An open box is to be made by cutting a square from each corner of a 12-inch by 12-inch piece of metal and then folding up the sides. The finished box must be at least 1.5 inches deep, but not deeper than 3 inches. What size square should be cut from each corner in order to produce a box of maximum volume?

Let x represent the length of a side of the square that is cut from each corner, as shown in Figure 12.13(a). The width of the box is $12 - 2x$, while the length is also $12 - 2x$. As shown in Figure 12.13(b), the depth of the box will be x inches.

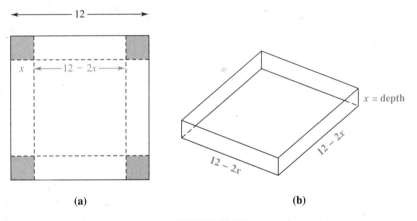

(a)   (b)

FIGURE 12.13

We must maximize the volume $V(x)$ of the box, which is given by

$$\text{Volume} = \text{length} \cdot \text{width} \cdot \text{height}$$
$$V(x) = (12 - 2x) \cdot (12 - 2x) \cdot x = 144x - 48x^2 + 4x^3$$

Since the height x must be between 1.5 and 3 inches, the domain of this volume function is the closed interval $[1.5, 3]$. First find the critical numbers by setting the derivative equal to 0:

$$V'(x) = 12x^2 - 96x + 144 = 0$$
$$12(x^2 - 8x + 12) = 0$$
$$12(x - 2)(x - 6) = 0$$
$$x - 2 = 0 \quad \text{or} \quad x - 6 = 0$$
$$x = 2 \quad \text{or} \quad x = 6$$

1 An open box is to be made by cutting squares from each corner of a 20 cm by 32 cm piece of metal and folding up the sides. Let x represent the length of the side of the square to be cut out. Find

(a) an expression for the volume of the box, $V(x)$;

(b) $V'(x)$;

(c) The value of x that leads to maximum volume. (Hint: the solutions of the equation $V'(x) = 0$ are 4 and 40/3);

(d) the maximum volume.

Answers:

(a) $V(x) = 640x - 104x^2 + 4x^3$

(b) $V'(x) = 640 - 208x + 12x^2$

(c) $x = 4$

(d) $V(4) = 1152$ cubic centimeters

Since 6 is not in the domain, the only critical number of interest here is $x = 2$. To find the depth that will maximize the volume, evaluate $V(x)$ at this critical number and at the endpoints of the domain, 1.5 and 3.

x	$V(x)$	
1.5	121.5	
2	128	← **Maximum**
3	108	

The chart indicates that the box will have maximum volume when $x = 2$ and that the maximum volume will be 128 cubic inches.   ◀ **1**

▶**EXAMPLE 2**   A truck burns fuel at the rate of

$$G(x) = \frac{1}{200}\left(\frac{800 + x^2}{x}\right) \quad (x > 0)$$

gallons per mile when traveling x miles per hour on a straight, level road. If fuel costs \$2 per gallon, find the speed that will produce the minimum total cost for a 1000-mile trip. Find the minimum total cost.

The total cost of the trip, in dollars, is the product of the number of gallons per mile, the number of miles, and the cost per gallon. If $C(x)$ represents this cost, then

$$C(x) = \left[\frac{1}{200}\left(\frac{800 + x^2}{x}\right)\right](1000)(2)$$

$$C(x) = \frac{2000}{200}\left(\frac{800 + x^2}{x}\right) = \frac{8000 + 10x^2}{x}.$$

Since x represents speed, only positive values of x make sense here. Thus, the domain of $C(x)$ is the open interval $(0, \infty)$ and there are no endpoints to check. To find the critical numbers, first find the derivative.

$$C'(x) = \frac{10x^2 - 8000}{x^2}$$

Set this derivative equal to 0 (and remember $x > 0$).

$$\frac{10x^2 - 8000}{x^2} = 0$$

$$10x^2 - 8000 = 0$$

$$10x^2 = 8000$$

$$x^2 = 800$$

Take the square root on both sides to get

$$x = \pm\sqrt{800} \approx \pm 28.3 \text{ mph.}$$

2 A diesel generator burns fuel at the rate of

$$G(x) = \frac{1}{48}\left(\frac{300}{x} + 2x\right)$$

gallons per hour when producing x thousand kilowatt hours of electricity. Suppose that fuel costs $2.25 a gallon and find the value of x that leads to minimum total cost if the generator is operated for 32 hours. Find the minimum cost.

Answer:
$x = \sqrt{150} \approx 12.2$; minimum cost is $73.50

The only critical number in the domain is $x = 28.3$. The second derivative $C''(x)$ is:

$$C''(x) = \frac{16{,}000}{x^3}$$

Since $C''(28.3) > 0$, the second derivative test shows that 28.3 leads to a minimum. (It is not necessary to calculate $C''(28.3)$; just check that it is positive.) The minimum total cost is found by the original function $C(x)$ at $x = 28.3$:

$$C(28.3) = \frac{8000 + 10(28.3)^2}{28.3} = 565.69 \text{ dollars.} \quad \blacktriangleleft \quad \boxed{2}$$

▶ **EXAMPLE 3** The U.S. Postal Service requires that boxes to be mailed have a length plus girth of no more than 108 inches, as shown in Figure 12.14. Find the dimensions of the box with largest volume that can be mailed, assuming its width and height are equal.

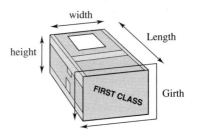

FIGURE 12.14

Let x be the width and y the length of the box. Since width and height are the same, the volume of the box is

$$V = y \cdot x \cdot x = yx^2.$$

Now express V in terms of just *one* variable. Use the facts that the girth is $x + x + x + x = 4x$ and that the length plus girth is 108, so that

$$y + 4x = 108, \quad \text{or equivalently,} \quad y = 108 - 4x.$$

Substitute for y in the expression for V to get

$$V = (108 - 4x)x^2 = 108x^2 - 4x^3.$$

Since x and y are dimensions, we must have $x > 0$ and $y > 0$. Now $y = 108 - 4x > 0$ implies that

$$4x < 108, \quad \text{or equivalently,} \quad x < 27.$$

Therefore, the domain of the volume function V (the values of x that make sense in the situation) is the open interval $(0, 27)$. Find the critical numbers for V by setting its derivative equal to 0 and solving the equation:

$$V' = 216x - 12x^2 = 0$$
$$x(216 - 12x) = 0$$
$$x = 0 \quad \text{or} \quad 12x = 216$$
$$x = 18$$

Use the second derivative test to check $x = 18$, the only critical number in the domain of V. Since $V'(x) = 216x - 12x^2$,

$$V''(x) = 216 - 24x.$$

Hence, $V''(18) = 216 - 24 \cdot 18 = -216$ and V is maximized when $x = 18$. In this case, $y = 108 - 4 \cdot 18 = 36$. Therefore, a box with dimensions 18 by 18 by 36 inches satisfies postal regulations and yields the maximum volume of $18^2 \cdot 36 = 11,664$ cubic inches. ◀

▶**EXAMPLE 4**   When Power and Money, Inc., charges $600 for a seminar on management techniques, it attracts 1000 people. For each $20 decrease in the charge, an additional 100 people will attend the seminar.
(a) Find an expression for the total revenue if there are x $20 decreases in the price.

The price charged will be

$$\text{Price per person} = 600 - 20x,$$

and the number of people in the seminar will be

$$\text{Number of people} = 1000 + 100x.$$

The total revenue, $R(x)$, is given by the product of the price and the number of people attending, or

$$R(x) = (600 - 20x)(1000 + 100x)$$
$$= 600,000 + 40,000x - 2000x^2.$$

(b) Find the value of x that maximizes revenue.
Set the derivative, $R'(x) = 40,000 - 4000x$, equal to 0.

$$40,000 - 4000x = 0$$
$$-4000x = -40,000$$
$$x = 10$$

Since $R''(x) = -4000$, $x = 10$ leads to maximum revenue.
(c) Find the maximum revenue.
Use $R(x)$ from part (a); let $x = 10$.

$$R(10) = 600,000 + 40,000(10) - 2000(10)^2 = 800,000$$

3 An investor has built a series of self-storage units near a group of apartment houses. She must now decide on the monthly rental. From past experience, she feels that 200 units will be rented for $15 per month, with 5 additional rentals for each $.25 reduction in the rental price. Let x be the number of $.25 reductions in the price and find

(a) an expression for the number of units rented;

(b) an expression for the price per unit;

(c) an expression for the total revenue;

(d) the value of x leading to maximum revenue;

(e) the maximum revenue.

Answers:

(a) $200 + 5x$

(b) $15 - .25x$

(c) $(200 + 5x)(15 - .25x) = 3000 + 25x - 1.25x^2$

(d) $x = 10$

(e) $3125

dollars. Since we want $R(x) \geq 0$, the domain is $[0, 30]$. At the endpoints, $R(0) = 600,000$ and $R(30) = 0$, so $R(10) = 800,000$ is the maximum revenue. For this level of revenue, $1000 + 100(10) = 2000$ people must attend the seminar; each person will pay $600 - 20(10) = \$400$. ◀ **3**

The preceding examples illustrate some of the factors that may affect applications in the real world. First, you must be able to find a function that models the situation. The rule of this function may be defined for values of x that do not make sense in the context of the application, so the domain must be restricted to the relevant values of x. For instance, if x represents speeds or distances, x must be positive. If x represents the number of employees on a production line, x must be restricted to the positive integers, or possibly to a few fractional values (we can conceive of a half-time employee, but probably not a 7/43-time employee).

The techniques of calculus apply to functions that are defined and continuous at every real number in some interval, so the maximum or minimum for the mathematical model (function) may not be feasible in the setting of the problem. For instance, if $C(x)$ has a minimum at $x = 80\sqrt{3}$ (≈ 138.564), where $C(x)$ is the cost of hiring x employees, then the real-life minimum occurs at either 138 or 139, whichever one leads to lower cost. Similarly, if $V(x)$ has a maximum at $x = \sqrt{5}$, where $V(x)$ is the volume of a cylindrical can of radius x, then some decimal approximation for $\sqrt{5}$ ($= 2.236067977 \cdots$) must be used. Depending on the manufacturing machinery, there may be several possibilities between 2.2 and 2.3.

FOR GRAPHERS

Approximate solutions of maximum and minimum problems can be found by graphing the function that models the situation and visually locating the highest or lowest point on the graph over a particular interval. Its coordinates can be found by using the trace feature in a very small viewing window around the point. This technique is particularly useful when the derivative is hard to compute or so complicated that the critical numbers cannot readily be found.

ECONOMIC LOT SIZE Suppose that a company manufactures a constant number of units of a product per year and that the product can be manufactured in several batches of equal size during the year. If the company were to manufacture the item only once per year, it would minimize setup costs but incur high warehouse costs. On the other hand, if it were to make many small batches, this would increase setup costs. Calculus can be used to find the number of batches per year that should be manufactured in order to minimize total cost. This number is called the **economic lot size.**

Figure 12.15 shows several of the possibilities for a product having an annual demand of 12,000 units. The top graph shows the results if only one batch of the product is made annually: in this case, an average of 6000 items will be held in a warehouse. If four batches (of 3000 each) are made at equal time intervals during a year, the average number of units in the warehouse falls to only 1500. If twelve batches are made, an average of 500 items will be in the warehouse.

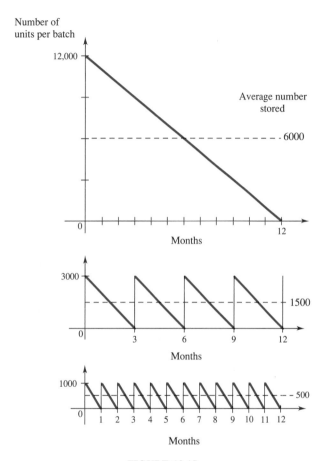

FIGURE 12.15

The following variables will be used in our discussion of economic lot size.

x = number of batches to be manufactured annually
k = cost of storing 1 unit of the product for 1 year
a = fixed setup cost to manufacture the product
b = variable cost of manufacturing a single unit of the product
M = total number of units produced annually.

The company has two types of costs associated with the production of its product: a cost associated with manufacturing the item, and a cost associated with storing the finished product.

During a year the company will produce x batches of the product, with M/x units of the product produced per batch. Each batch has a fixed cost a and a variable cost b per unit, so that the manufacturing cost per batch is

$$a + b\left(\frac{M}{x}\right).$$

There are x batches per year, so the total annual manufacturing cost is

$$\left[a + b\left(\frac{M}{x}\right)\right]x. \tag{1}$$

Each batch consists of M/x units and demand is constant; therefore, it is common to assume an average inventory of

$$\frac{1}{2}\left(\frac{M}{x}\right) = \frac{M}{2x}$$

units per year. The cost to store 1 unit of the product for a year is k, so the total storage cost is

$$k\left(\frac{M}{2x}\right) = \frac{kM}{2x}. \tag{2}$$

The total production cost is the sum of the manufacturing and storage costs, or the sum of expressions (1) and (2). If $T(x)$ is the total cost of producing x batches,

$$T(x) = \left[a + b\left(\frac{M}{x}\right)\right]x + \frac{kM}{2x} = ax + bM + \left(\frac{kM}{2}\right)x^{-1}.$$

Now find the value of x that will minimize $T(x)$. (Remember that $a, b, k,$ and M are constants.) Find $T'(x)$.

$$T'(x) = a - \frac{kM}{2}x^{-2}$$

Set this derivative equal to 0.

$$a - \frac{kM}{2}x^{-2} = 0$$

$$a = \frac{kM}{2x^2}$$

$$2ax^2 = kM$$

$$x^2 = \frac{kM}{2a}$$

$$x = \sqrt{\frac{kM}{2a}} \tag{3}$$

4 A manufacturer of business forms has an annual demand for 30,720 units of form letters to people delinquent in their payments of installment debt. It costs $5 per year to store 1 unit of the letters and $1200 to set up the machines to produce them. Find the number of batches that should be made annually to minimize total cost.

Answer:
8 batches

5 An office uses 576 cases of copy-machine paper during the year. It costs $3 per year to store 1 case. Each reorder costs $24. Find the number of orders that should be placed annually.

Answer:
6

The second derivative test can be used to show that $\sqrt{kM/(2a)}$ is the annual number of batches that gives minimum total production cost.

▶ **EXAMPLE 5**   A paint company has a steady annual demand for 24,500 cans of automobile primer. The cost accountant for the company says that it costs $2 to store 1 can of paint for 1 year and $500 to set up the plant for the production of the primer. Find the number of batches of primer that should be produced for the minimum total production cost.

Use equation (3) above.

$$x = \sqrt{\frac{kM}{2a}}$$

$$x = \sqrt{\frac{2(24,500)}{2(500)}} \qquad \text{Let } k = 2, M = 24,500, a = 500$$

$$x = \sqrt{49} = 7$$

Seven batches of primer per year will lead to minimum production costs. ◀ **4** **5**

12.3 EXERCISES

Work the following exercises. (See Examples 1–4.)

1. An open box is to be made by cutting a square from each corner of a 3 ft by 8 ft piece of cardboard and then folding up the sides. What size square should be cut from each corner in order to produce a box of maximum volume? (See Example 1.)

2. Management   A truck burns fuel at the rate of $G(x)$ gallons per mile, where

$$G(x) = \frac{1}{32}\left(\frac{64}{x} + \frac{x}{50}\right),$$

while traveling x miles per hour.
   (a) If fuel costs $1.60 per gallon, find the speed that will produce minimum total cost for a 400-mile trip. (See Example 2.)
   (b) Find the minimum total cost.

3. Management   A rock-and-roll band travels from engagement to engagement in a large bus. This bus burns fuel at the rate of $G(x)$ gallons per mile, where

$$G(x) = \frac{1}{50}\left(\frac{200}{x} + \frac{x}{15}\right),$$

while traveling x miles per hour.
   (a) If fuel costs $2 per gallon, find the speed that will produce minimum total cost for a 250-mile trip.
   (b) Find the minimum total cost.

4. Management   In Example 2 we found the speed in miles per hour that minimized cost when we considered only the cost of the fuel. Rework the problem taking into account the driver's salary of $8 per hour. (Hint: if the trip is 1000 miles at x miles per hour, the driver will be paid for $1000/x$ hours.)

5. **Management** The manager of an 80-unit apartment complex is trying to decide on the rent to charge. It is known from experience that at a rent of $200, all the units will be full. However, on the average, 1 additional unit will remain vacant for each $20 increase in rent.
   (a) Let x represent the number of $20 increases. Find an expression for the rent for each apartment. (See Example 4.)
   (b) Find an expression for the number of apartments rented.
   (c) Find an expression for the total revenue from all rented apartments.
   (d) What value of x leads to maximum revenue?
   (e) What is the maximum revenue?

6. **Management** The manager of a peach orchard is trying to decide when to have the peaches picked. If they are picked now, the average yield per tree will be 100 pounds, which can be sold for 40¢ per pound. Past experience shows that the yield per tree will increase about 5 pounds per week, while the price will decrease about 2¢ per pound per week.
   (a) Let x represent the number of weeks that the manager should wait. Find the revenue per pound.
   (b) Find the number of pounds per tree.
   (c) Find the total revenue from a tree.
   (d) When should the peaches be picked in order to produce maximum revenue?
   (e) What is the maximum revenue?

7. **Management** In planning a small restaurant, it is estimated that a profit of $5 per seat will be made if the number of seats is between 60 and 80, inclusive. On the other hand, the profit on each seat will decrease by 5¢ for each seat above 80.
   (a) Find the number of seats that will produce the maximum profit.
   (b) What is the maximum profit?

8. A local club is arranging a charter flight to Hawaii. The cost of the trip is $425 each for 75 passengers, with a refund of $5 per passenger for each passenger in excess of 75.
   (a) Find the number of passengers that will maximize the revenue received from the flight.
   (b) Find the maximum revenue.

9. A farmer has 1200 m of fencing. He want to enclose a rectangular field bordering a river, with no fencing needed along the river. (See the sketch.) Let x represent the width of the field.

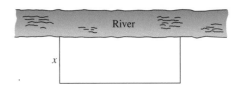

   (a) Write an expression for the length of the field.
   (b) Find the area of the field.
   (c) Find the value of x leading to the maximum area.
   (d) Find the maximum area.

10. A rectangular field is to be enclosed with a fence. One side of the field is against an existing fence, so that no fence is needed on that side. If material for the fence costs $2 per foot for the two ends and $4 per foot for the side parallel to the existing fence, find the dimensions of the field of largest area that can be enclosed for $1000.

11. A rectangular field is to be enclosed on all four sides with a fence. Fencing material costs $3 per foot for two opposite sides, and $6 per foot for the other two sides. Find the maximum area that can be enclosed for $2400.

12. A fence must be built to enclose a rectangular area of 20,000 ft^2. Fencing material costs $3 per foot for the two sides facing north and south, and $6 per foot for the other two sides. Find the cost of the least expensive fence.

13. A fence must be built in a large field to enclose a rectangular area of 15,625 m^2. One side of the area is bounded by an existing fence; no fence is needed there. Material for the fence costs $2 per meter for the two ends, and $4 per meter for the side opposite the existing fence. Find the cost of the least expensive fence.

14. **Physical Science** The altitude of a rocket (in feet) after t seconds of flight is given by
$$h(t) = -t^3 + 57t^2 + 240t + 10.$$
When does the rocket reach its maximum altitude and how high is it at that time?

15. **Management** If the price charged for a candy bar is $p(x)$ cents, then x thousand candy bars will be sold in a certain city, where
$$p(x) = 100 - \frac{x}{10}.$$

   (a) Find an expression for the total revenue from the sale of x thousand candy bars. (Hint: find the product of $p(x)$, x, and 1000.)
   (b) Find the value of x that leads to maximum revenue.
   (c) Find the maximum revenue.

16. Management The sale of cassette tapes of "lesser" performers is very sensitive to price. If a tape manufacturer charges $p(x)$ dollars per tape, where

$$p(x) = 6 - \frac{x}{8},$$

then x thousand tapes will be sold.
  (a) Find an expression for the total revenue from the sale of x thousand tapes. (Hint: find the product of $p(x)$, x, and 1000.)
  (b) Find the value of x that leads to maximum revenue.
  (c) Find the maximum revenue.

17. Management A television manufacturing firm needs to design an open-topped box with a square base. The box must hold 32 cubic inches. Find the dimensions of the box that can be built with the minimum amount of materials.

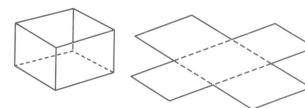

18. A closed box with a square base is to have a volume of 16,000 cubic centimeters. The material for the top and bottom of the box costs $3 per square centimeter, while the material for the sides costs $1.50 per square centimeter. Find the dimensions of the box that will lead to minimum total cost. What is the minimum total cost?

19. Management A cylindrical box will be tied up with ribbon as shown in the figure. The longest piece of ribbon available is 130 cm long, and 10 cm of that are required for the bow. Find the radius and height of the box with the largest possible volume.

20. Management A company wishes to manufacture a box with a volume of 36 cubic feet that is open on top and that is twice as long as it is wide. Find the dimensions of the box produced from the minimum amount of material.

21. Management A mathematics book is to contain 36 square inches of printed matter per page, with margins of 1 inch along the sides, and $1\frac{1}{2}$ inches along the top and bottom. Find the dimensions of the page that will lead to the minimum amount of paper being used for a page.

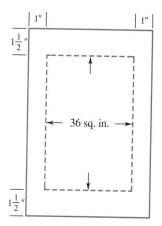

22. Management Decide what you would do if your assistant brought you the following contract to sign:

> Your firm offers to deliver 300 tables to a dealer, at $90 per table, and to reduce the price per table on the entire order by 25¢ for each additional table over 300.

Find the dollar total involved in the largest possible transaction between you and the dealer; find the smallest possible dollar amount.

23. Management A company wishes to run a utility cable from point A on the shore (see the figure on the next page) to an installation at point B on the island. The island is 6 miles from the shore (at point C) and point A is 9 miles from point C. It costs $400 per mile to run the cable on land and $500 per mile underwater. Assume that the cable starts at A and runs along the shoreline, then angles and runs underwater to the island. Find the point at which the line should begin to angle in order to yield the minimum total cost. (Hint: the length of the line underwater is $\sqrt{x^2 + 36}$.)

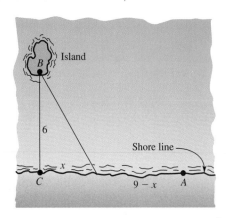

24. Natural Science Homing pigeons avoid flying over large bodies of water, preferring to fly around them instead. (One possible explanation is the fact that extra energy is required to fly over water because air pressure drops over water in the daytime.) Assume that a pigeon released from a boat 1 mi from the shore of a lake (point B in the figure) flies first to point P on the shore and then along the straight edge of the lake to reach its home at L. Assume that L is 2 mi from point A, the point on the shore closest to the boat, and that a pigeon needs 4/3 as much energy to fly over water as over land. Find the location of point P if the pigeon uses the least possible amount of energy.

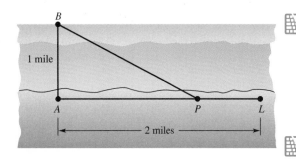

Management   *The following exercises refer to economic lot size. (See Example 5.)*

25. Find the approximate number of batches that should be produced annually if 100,000 units are to be manufactured. It costs $1 to store a unit for 1 year and it costs $500 to set up the factory to produce each batch.

26. How many units per batch will be manufactured in Exercise 25?

27. A market has a steady annual demand for 16,800 cases of sugar. It costs $3 to store 1 case for 1 year. The market pays $7 for each order that is placed. Find the number of orders for sugar that should be placed each year.

28. Find the number of cases per order in Exercise 27.

29. A bookstore has an annual demand for 100,000 copies of a best-selling book. It costs $.50 to store one copy for one year, and it costs $60 to place an order. Find the optimum number of copies per order.

30. A restaurant has an annual demand for 900 bottles of a California wine. It costs $1 to store 1 bottle for 1 year, and it costs $5 to place a reorder. Find the number of orders that should be placed annually.

31. Choose the correct answer:* The economic order quantity formula assumes that
   (a) Purchase costs per unit differ due to quantity discounts.
   (b) Costs of placing an order vary with quantity ordered.
   (c) Periodic demand for the goods is known.
   (d) Erratic usage rates are cushioned by safety stocks.

32. Management   Suppose a manufacturer's cost to produce x units is given by $c(x) = .13x^3 - 70x^2 + 10,000x$ and that no more than 300 units can be produced each week. What production level should be used in order to minimize the average cost per unit, and what is that minimum average cost? [Average cost was defined on page 652.]

33. Management   A company uses widgets throughout the year. It costs $20 every time an order for widgets is placed and $10 to store a widget until it is used. When widgets are ordered x times per year, then an average of $300/x$ widgets are in storage at any given time. How often should the company order widgets each year in order to minimize its total ordering and storage costs? [Be careful: the answer must be an integer.]

34. Management   A company makes novelty bookmarks that sell for $142 per hundred. The cost (in dollars) of making x hundred bookmarks is $x^3 - 8x^2 + 20x + 40$. Assume that the company can sell all the bookmarks it makes.

* Question from the Uniform CPA Examination of the American Institute of Certified Public Accountants, May, 1991. Reprinted by permission of the Institute of Certified Public Accountants.

(a) Because of other projects a maximum of 600 book-marks per day can be manufactured. How many should the company make per day to maximize its profits?

(b) As a result of change in other orders, as many as 1600 bookmarks can now be made each day. How many should be made in order to maximize profits?

35. *Management* A cylindrical can of volume 58 cubic inches (approximately one quart) is to be designed. For convenient handling, it must be at least one inch high and two inches in diameter. What dimensions (radius of top, height of can) will use the least amount of material?

12.4 CURVE SKETCHING

Informally, we say that a function is **increasing** on an interval if its graph is *rising* from left to right over the interval and that a function is **decreasing** on an interval if its graph is *falling* from left to right over the interval.

▶ **EXAMPLE 1** For which *x*-intervals is the function graphed in Figure 12.16 increasing? For which intervals is it decreasing?

Moving from left to right, the function is increasing up to -4, then decreasing from -4 to 0, constant (neither increasing nor decreasing) from 0 to 4, increasing from 4 to 6, and finally, decreasing from 6 on. In interval notation, the function is increasing on $(-\infty, -4)$ and $(4, 6)$, decreasing on $(-4, 0)$ and $(6, \infty)$, and constant on $(0, 4)$. ◀ **1**

1 For what values of *x* is the function graphed as follows increasing? decreasing?

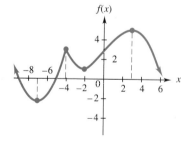

Answer:
Increasing on $(-7, -4)$ and $(-2, 3)$; decreasing on $(-\infty, -7)$, $(-4, -2)$, and $(3, \infty)$

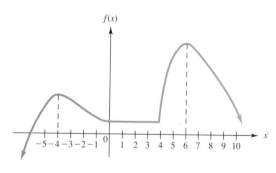

FIGURE 12.16

If *f* is a function, think of the graph of *f* as a roller coaster track, with the car moving from left to right, and the floor of the car representing the tangent line to the graph at that point as in Figure 12.17 on the next page. Using this analogy, we see that the slope of the tangent line will be *positive* when the car travels uphill (the function is *increasing*) and the slope of the tangent will be *negative* when the car travels downhill (the function is *decreasing*). Since the slope of the tangent line at the point $(x, f(x))$ is given by the derivative $f'(x)$, we have this useful test.

Suppose a function f has a derivative at each point in an open interval:

1. If $f'(x) > 0$ for each x in the interval, f is *increasing* on the interval.
2. If $f'(x) < 0$ for each x in the interval, f is *decreasing* on the interval.
3. If $f'(x) = 0$ for each x in the interval, f is *constant* on the interval.

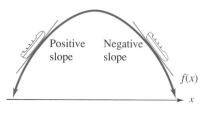

FIGURE 12.17

The derivative $f'(x)$ can change signs from positive to negative (or negative to positive) at points where $f'(x) = 0$, and also at points where $f'(x)$ does not exist. It is shown in more advanced classes that if the critical numbers of a polynomial function are used to determine open intervals on a number line, then the sign of the derivative at any point in the interval will be the same as the sign of the derivative at any other point in the interval. This suggests that the test for increasing and decreasing functions be applied as follows (assuming that no open intervals exist where the function is constant).

Step 1   Locate on a number line those values of x for which $f'(x) = 0$ or $f'(x)$ does not exist. These points determine several open intervals.

Step 2   Choose a value of x in each of the intervals determined in step 1. Use these values to decide whether $f'(x) > 0$ or $f'(x) < 0$ on that interval.

Step 3   Use the test above to decide whether f is increasing or decreasing on each interval.

▶**EXAMPLE 2**   Find the intervals in which the function defined by $f(x) = x^3 + 3x^2 - 9x + 4$ is increasing or decreasing. Locate all points where the tangent line is horizontal. Graph the function.

2 Find all intervals where the following functions are increasing or decreasing.

(a) $f(x) = 2x^2 + 5x - 4$

(b) $f(x) = 6 + 3x - x^2$

(c) $f(x) = 4x^3 + 3x^2 - 18x + 1$

Answers:

(a) Increasing on $(-5/4, \infty)$; decreasing on $(-\infty, -5/4)$

(b) Increasing on $(-\infty, 3/2)$; decreasing on $(3/2, \infty)$

(c) Increasing on $(-\infty, -3/2)$ and $(1, \infty)$; decreasing on $(-3/2, 1)$

Here $f'(x) = 3x^2 + 6x - 9$ and there are no values where $f'(x)$ fails to exist. So to find the critical numbers, we set the derivative equal to 0 and solve the equation.

$$3x^2 + 6x - 9 = 0$$
$$3(x^2 + 2x - 3) = 0$$
$$3(x + 3)(x - 1) = 0$$
$$x = -3 \quad \text{or} \quad x = 1$$

The critical numbers are $x = -3$ and $x = 1$; the tangent line is horizontal at these values of x. To determine where the function is increasing or decreasing, locate -3 and 1 on a number line as in Figure 12.18. They determine the intervals $(-\infty, -3), (-3, 1),$ and $(1, \infty)$. Choosing -4 as a test point in the first of these intervals gives

$$f'(-4) = 3(-4)^2 + 6(-4) - 9 = 15,$$

which is positive, so f is increasing on $(-\infty, -3)$. Selecting 0 from the middle interval gives $f'(0) = -9$, so f is decreasing on $(-3, 1)$. Finally, choosing 2 in the right-hand region gives $f'(2) = 15$, with f increasing on $(1, \infty)$. The intervals where f is increasing or decreasing are shown with arrows in Figure 12.18.

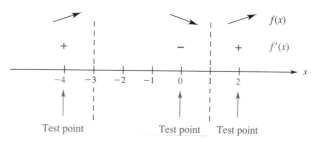

FIGURE 12.18

To graph the function, plot a point in each of the intervals determined by the critical numbers, as well as the points at the critical numbers. Use these points along with the information about where the function is increasing and decreasing to draw the graph in Figure 12.19 on the next page. ◀ **2**

It is important to note that the reverse of the test for increasing and decreasing functions is not true—it is possible for a function to be increasing on an interval even though the derivative is not positive at every point in the interval. A good example is given by $f(x) = x^3$, which is increasisng on every interval, even though $f'(x) = 0$ when $x = 0$. See Figure 12.20 on the next page.

Knowing the intervals where a function is increasing or decreasing can be important in applications, as shown by the next examples.

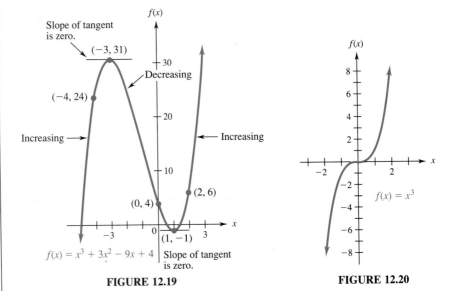

FIGURE 12.19

FIGURE 12.20

▶**EXAMPLE 3** A company sells an item with the following cost and revenue functions.

$$C(x) = 1.2x - .0006x^2, \quad 0 \le x \le 1000$$
$$R(x) = 3.6x - .003x^2, \quad 0 \le x \le 1000$$

Determine any intervals on which the profit function is increasing.

First find the profit function $P(x)$.

$$P(x) = R(x) - C(x)$$
$$= (3.6x - .003x^2) - (1.2x - .0006x^2)$$
$$= 2.4x - .0024x^2$$

To find the interval where this function is increasing, set $P'(x) = 0$.

$$P'(x) = 2.4 - .0048x = 0$$
$$x = 500$$

Use $x = 500$ to determine two intervals on a number line, as shown in Figure 12.21. Choose $x = 0$ and $x = 1000$ as test points.

$$P'(0) = 2.4 - .0048(0) = 2.4$$
$$P'(1000) = 2.4 - .0048(1000) = -2.4$$

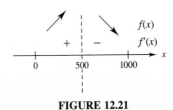

FIGURE 12.21

3 Find the intervals where profit is increasing and $P(x) > 0$ if the profit function is defined as $P(x) = 1000 + 90x - x^2$.

Answer:
(0, 45)

The test points show that the function increases on $(-\infty, 500)$ and decreases on $(500, \infty)$. Thus, the profit is increasing on the interval (0, 500) and decreasing on the interval (500, 1000), as shown in Figure 12.22.

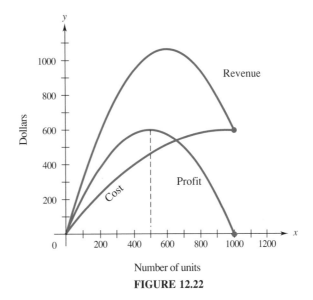

FIGURE 12.22

As the graph in Figure 12.22 shows, the profit will increase as long as the revenue function increases faster than the cost function. That is, increasing production will produce more profit as long as the marginal revenue is greater than the marginal cost. ◀ **3**

FOR GRAPHERS

The approximate intervals where a function is increasing or decreasing are easily seen on a grapher screen. A more precise approximation can be found by using the trace feature to move along the graph in a small viewing window and watching to see where the y-coordinates increase or decrease.

CONCAVITY A function is **concave downward** on an interval (a, b) if its graph lies below the tangent line at each point of the interval, as shown in Figure 12.23 on the next page. A function is **concave upward** on an interval (b, c) if its graph lies above the tangent line at each point of the interval. A point where the graph changes concavity (such as the one where $x = b$ in Figure 12.23) is called a **point of inflection.**

The left side of the graph in Figure 12.23 (up to the point of inflection) is shaped like an upside-down cup and the right side like an upright cup. This suggests a convenient informal memory hook: a graph is *concave upward* on an interval if it "holds water" there and *concave downward* if it "spills water".

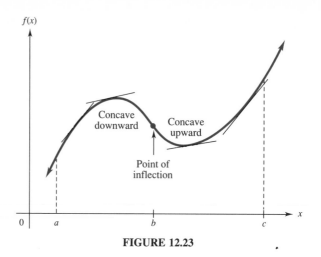

FIGURE 12.23

Just as a function can be either increasing or decreasing on an interval, it can be either concave upward or concave downward on an interval. Examples of various combinations are shown in Figure 12.24.

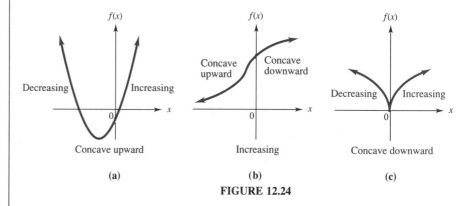

(a) **(b)** **(c)**

FIGURE 12.24

FOR GRAPHERS

The low resolution of many grapher screens (especially on calculators) makes it difficult to determine visually the intervals where a graph is concave upward or concave downward or the points of inflection.

The first derivative is used to show where a function is increasing or decreasing and where the relative extrema occur. The second derivative determines the *concavity* of the graph, as we now see. Figure 12.25 shows the graph of two functions that are concave upward on an interval (a, b). Several tangent lines are also shown. In Figure 12.25(a), the slopes of these tangent lines (moving from left to right) are first negative, then 0, and then positive. In Figure 12.25(b), the slopes are all positive but getting larger.

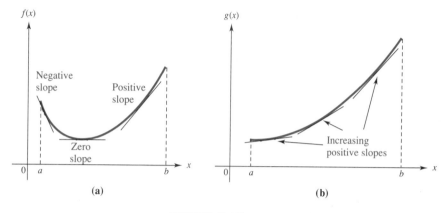

FIGURE 12.25

In both cases, the slopes are *increasing*. The slope at a point on a curve is given by the derivative. Since a function is increasing if its derivative is positive, the slope is increasing if the derivative of the function giving the slopes is positive. The derivative of a derivative is the second derivative, so a function is concave upward on an interval if its second derivative is positive at each point of the interval.

A similar result is suggested by Figure 12.26 for functions whose graphs are concave downward. In both graphs, the slopes of the tangent lines are *decreasing* as we move from left to right. Since a function is decreasing if its derivative is negative, a function is concave downward on an interval if its second derivative is negative at each point of the interval. These observations suggest the following test.

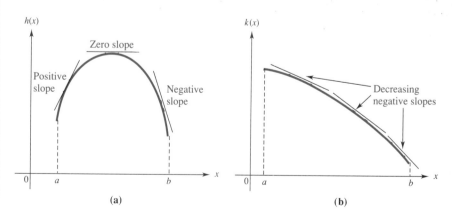

FIGURE 12.26

Let f be a function with derivatives f' and f'' existing at all ponts in an interval (a, b).

1. The function f is **concave upward** on (a, b) if $f''(x) > 0$ for all x in (a, b).
2. The function f is **concave downward** on (a, b) if $f''(x) < 0$ for all x in (a, b).

▶ **EXAMPLE 4**   Find all intervals where $f(x) = x^3 - 3x^2 + 5x - 4$ is concave upward or downward.

The first derivative is $f'(x) = 3x^2 - 6x + 5$, and the second derivative is $f''(x) = 6x - 6$. The function f is concave upward whenever $f''(x) > 0$, or

$$6x - 6 > 0$$
$$6x > 6$$
$$x > 1.$$

Also, f is concave downward if $f''(x) < 0$, or $x < 1$. In interval notation, f is concave upward on $(1, \infty)$ and concave downward on $(-\infty, 1)$, with a point of inflection at $(1, -1)$. A graph of f is shown in Figure 12.27   ◀

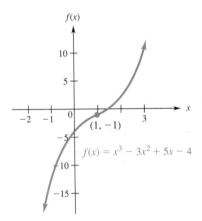

FIGURE 12.27

The graph of the function f in Figure 12.27 changes from concave downward to concave upward at $x = 1$. This means that the point $(1, f(1))$ or $(1, -1)$ in Figure 12.27 is a point of inflection. We can locate this point by finding values of x where the second derivative changes from negative to positive; that is, where the second derivative is 0. Setting the second derivative in Example 4 equal to 0 gives

$$6x - 6 = 0$$
$$x = 1.$$

4 Find the intervals where the following are concave upward. Identify any inflection points.

(a) $f(x) = 6x^3 - 24x^2 + 9x - 3$

(b) $f(x) = 2x^2 - 4x + 8$

Answers:

(a) Concave upward on $(4/3, \infty)$; point of inflection is $(4/3, -175/9)$

(b) $f''(x) = 4$, which is always positive; function is always concave upward, no inflection point.

As before, the point of inflection is $(1, f(1))$ or $(1, -1)$.

Example 4 suggests the follliowing result.

> If a function f has a point of inflection at $x = c$, then $f''(c)$ is 0 or does not exist.

Caution   The reverse of this fact is not always true. Finding a number c for which $f''(c) = 0$ does not necessarily mean that a point of inflection has been located. For example, if $f(x) = (3/2)x^4$, then $f'(x) = 6x^3$, and $f''(x) = 18x^2$, which is 0 when $x = 0$. However, the graph of f in Figure 3.11 on page 164 is always concave upward, so it has no point of inflection at $x = 0$. **4**

The **law of diminishing returns** in economics is related to the idea of concavity. The graph of the function f in Figure 12.28 shows the output y from a given input x. For instance, the input might be advertising costs and the output the corresponding revenue from sales.

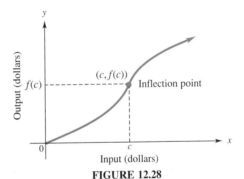

FIGURE 12.28

The graph in Figure 12.28 shows an inflection point at $(c, f(c))$. For $x < c$, the graph is concave upward, so the rate of change of the slope is increasing. This indicates that the output y is increasing at a faster rate with each additional dollar spent. When $x > c$, however, the graph is concave downward, the rate of change of the slope is decreasing, and the increase in y is smaller with each additional dollar spent. Thus, further input beyond c dollars produces diminishing returns. The point of inflection at $(c, f(c))$ is called the **point of diminishing returns.** Any investment beyond the value c is not considered a good use of capital.

▶ **EXAMPLE 5**   The revenue $R(x)$ generated from sales of a certain product is related to the amount x spent on advertising by

$$R(x) = \frac{1}{150,000}(600x^2 - x^3), \quad 0 \le x \le 600,$$

5 In Example 5,

$R(x) = \dfrac{x^3}{600} - \dfrac{x^4}{1200}$ for another

product. What is the point of diminishing returns?

Answer:

(10, .833)

where x and $R(x)$ are in thousands of dollars. Is there a point of diminishing returns for this function? If so, what is it?

Since a point of diminishing returns occurs at an inflection point, look for an x-value that makes $R''(x) = 0$. Write the function as

$$R(x) = \frac{600}{150,000}x^2 - \frac{1}{150,000}x^3 = \frac{1}{250}x^2 - \frac{1}{150,000}x^3.$$

Now find $R'(x)$ and then $R''(x)$.

$$R'(x) = \frac{2x}{250} - \frac{3x^2}{150,000} = \frac{1}{125}x - \frac{1}{50,000}x^2$$

$$R''(x) = \frac{1}{125} - \frac{1}{25,000}x$$

Set $R''(x)$ equal to 0 and solve for x.

$$\frac{1}{125} - \frac{1}{25,000}x = 0$$

$$-\frac{1}{25,000}x = -\frac{1}{125}$$

$$x = \frac{25,000}{125} = 200$$

Test a number in the interval (0, 200) to see that $R''(x)$ is positive there. Then test a number in the interval (200, 600) to find $R''(x)$ negative in that interval. Since the sign of $R''(x)$ changes from positive to negative at $x = 200$, the graph changes from concave upward to concave downward at that point, and there is a point of diminishing returns at the inflection point $(200, 106\frac{2}{3})$. Any investment in advertising beyond \$200,000 would not pay off. ◀ **5**

LIMITS AT INFINITY AND HORIZONTAL ASYMPTOTES   The function

$$f(t) = \frac{12t^2 - 15t + 12}{t^2 + 1},$$

whose graph is in Figure 12.29, gives the amount of oxygen in a small pond t weeks after some organic waste was dumped in the pond. At the beginning $(t = 0)$, there were $f(0) = 12$ units of oxygen. After two weeks, the pond contains

$$f(2) = \frac{12 \cdot 2^2 - 15 \cdot 2 + 12}{2^2 + 1} = \frac{30}{5} = 6$$

units of oxygen. The graph suggests that as time goes on the amount of oxygen slowly increases. Will it ever return to 12 units?

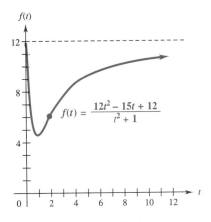

FIGURE 12.29

Consider a table of values, as t gets larger and larger.

t	10	100	1000	10,000	100,000
$f(t)$	10.5	11.85	11.985	11.9985	11.9999

The table suggests that

as t takes larger and larger values, beyond all bounds, the corresponding values of $f(t)$ get closer and closer to 12,

and that

the values of $f(t)$ can be made arbitrarily close to 12 by taking large enough values of t.

We express these facts by writing

$$\lim_{t \to \infty} f(t) = 12,$$

which is read "the limit of $f(t)$ as t approaches infinity is 12." In terms of the pond, this means that the oxygen concentration will approach 12, but never be exactly 12.

Graphically this means that as t gets larger and larger, the graph gets closer and closer to the horizontal line $y = 12$. This line is called a **horizontal asymptote** of the graph.

The preceding discussion is an example of a **limit at infinity.** The phrase "t approaches infinity" (symbolically, $t \to \infty$) is simply convenient shorthand to express the fact that t takes larger and larger values without bound. Similarly, the phrase "t approaches negative infinity" (symbolically, $t \to -\infty$) means that t takes smaller and smaller values without bound (such as $-10, -1000, -10,000$, etc.).

As indicated above, limits at infinity or negative infinity (if they exist) correspond to horizontal asymptotes of the graph of the function. For instance, the line $y = 0$ (the x-axis) is a horizontal asymptote of both the graph of $f(x) = 1/x$ (in black) and the graph of $g(x) = 1/x^2$ (in color) in Figure 12.30 and the table there suggests that

$$\lim_{x \to \infty} \frac{1}{x} = 0, \quad \lim_{x \to -\infty} \frac{1}{x} = 0, \quad \lim_{x \to \infty} \frac{1}{x^2} = 0, \quad \lim_{x \to -\infty} \frac{1}{x^2} = 0.$$

These are the cases $n = 1$ and $n = 2$ of the following rule.

Limits at Infinity

For any positive integer n,

$$\lim_{x \to \infty} \frac{1}{x^n} = 0 \quad \text{and} \quad \lim_{x \to -\infty} \frac{1}{x^n} = 0.$$

x	$\dfrac{1}{x}$	$\dfrac{1}{x^2}$
-100	$-.01$	$.0001$
-10	$-.1$	$.01$
-1	-1	1
1	1	1
10	$.1$	$.01$
100	$.01$	$.0001$

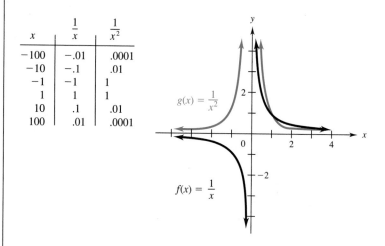

FIGURE 12.30

All the rules for limits given on page 601 in Section 11.1 (except Rule 3) remain valid when a is replaced by ∞ or $-\infty$.

▶ **EXAMPLE 6** Find each limit.

(a) $\displaystyle \lim_{x \to \infty} \frac{8x + 6}{3x - 1}$

First, we divide both numerator and denominator by x, which does not change the value of the quotient.

6 Find the following limits.

(a) $\lim\limits_{x\to\infty} \dfrac{4x^2 + 6x}{9x^2 - 3}$

(b) $\lim\limits_{x\to\infty} \dfrac{3x^2 + 4x + 5}{8x^3 + 2x + 1}$

(c) $\lim\limits_{x\to\infty} \dfrac{4x^2 + 9x + 1}{2x + 3}$

Answers:

(a) 4/9

(b) 0

(c) Does not exist

$$\lim_{x\to\infty} \frac{8x + 6}{3x - 1} = \lim_{x\to\infty} \frac{\dfrac{8x}{x} + \dfrac{6}{x}}{\dfrac{3x}{x} - \dfrac{1}{x}} = \lim_{x\to\infty} \frac{8 + 6 \cdot \dfrac{1}{x}}{3 - \dfrac{1}{x}}$$

Now apply the limit rules and the fact that $\lim\limits_{x\to\infty} 1/x^n = 0$.

$$\frac{\lim\limits_{x\to\infty}\left(8 + 6 \cdot \dfrac{1}{x}\right)}{\lim\limits_{x\to\infty}\left(3 - \dfrac{1}{x}\right)} = \frac{\lim\limits_{x\to\infty} 8 + \lim\limits_{x\to\infty} 6 \cdot \dfrac{1}{x}}{\lim\limits_{x\to\infty} 3 - \lim\limits_{x\to\infty} \dfrac{1}{x}} \qquad \text{Rules 5 and 2}$$

$$= \frac{8 + 6\left(\lim\limits_{x\to\infty} \dfrac{1}{x}\right)}{3 - \lim\limits_{x\to\infty} \dfrac{1}{x}} = \frac{8 + 6(0)}{3 - 0} = \frac{8}{3} \qquad \begin{array}{l}\text{Rule 1, limits} \\ \text{at infinity}\end{array}$$

(b) $\lim\limits_{x\to-\infty} \dfrac{4x^2 - 6x + 3}{2x^2 - x + 4}$

Divide each term of the numerator and denominator by x^2, the highest power of x, and proceed as in part (a).

$$\lim_{x\to-\infty} \frac{4x^2 - 6x + 3}{2x^2 - x + 4} = \lim_{x\to-\infty} \frac{4 - 6 \cdot \dfrac{1}{x} + 3 \cdot \dfrac{1}{x^2}}{2 - \dfrac{1}{x} + 4 \cdot \dfrac{1}{x^2}}$$

$$= \frac{4 - 0 + 0}{2 - 0 + 0} = \frac{4}{2} = 2$$

(c) $\lim\limits_{x\to\infty} \dfrac{3x + 2}{4x^3 - 1} = \lim\limits_{x\to\infty} \dfrac{3 \cdot \dfrac{1}{x^2} + 2 \cdot \dfrac{1}{x^3}}{4 - \dfrac{1}{x^3}} = \dfrac{0 + 0}{4 - 0} = \dfrac{0}{4} = 0$

Here, the highest power of x is x^3, which is used to divide each term in numerator and denominator.

(d) $\lim\limits_{x\to\infty} \dfrac{3x^2 + 2}{4x - 3} = \lim\limits_{x\to\infty} \dfrac{3 + \dfrac{2}{x^2}}{\dfrac{4}{x} - \dfrac{3}{x^2}} = \dfrac{3 + 0}{0 - 0} = \dfrac{3}{0}$

Division by 0 is undefined, so the limit does not exist. ◀ **6**

The method used in Example 6 is a useful way to rewrite expressions with fractions so that the rules for limits at infinity can be used.

Finding Limits at Infinity

If $f(x) = \dfrac{p(x)}{q(x)}$, for polynomials $p(x)$ and $q(x)$, $q(x) \neq 0$, $\lim\limits_{x \to -\infty} f(x)$ and $\lim\limits_{x \to \infty} f(x)$ can be found as follows.

1. Divide $p(x)$ and $q(x)$ by the highest power of x in either polynomial.
2. Use the rules for limits, including the rules for limits at infinity,

$$\lim_{x \to \infty} \frac{1}{x^n} = 0 \quad \text{and} \quad \lim_{x \to -\infty} \frac{1}{x^n} = 0,$$

to find the limit of the result from step 1.

 FOR GRAPHERS

Graphers generally do a good job of indicating horizontal asymptotes, particularly when a wide viewing window is used.

CURVE SKETCHING   When a grapher is not available, the tests for concavity, increasing and decreasing functions, and the concept of limits at infinity can be used to sketch the graphs of a variety of functions. This process, called **curve sketching,** uses the following steps. It may not always be feasible to carry out all the steps, but you should do as many as necessary in any convenient order to obtain a reasonable graph.

To sketch the graph of a function $y = f(x)$:

1. Find the y-intercept (if it exists) by letting $x = 0$ and computing $y = f(0)$. Find any x-intercepts by letting $y = 0$ and solving $f(x) = 0$ if this is not too difficult.
2. If f is a rational function, find any vertical asymptotes by finding the numbers for which the denominator is 0 but the numerator is not.
3. Find $f'(x)$. Locate any critical numbers by solving the equation $f'(x) = 0$ and determining where $f'(x)$ does not exist, but $f(x)$ does. Find any relative extrema and determine where f is increasing or decreasing by solving the inequalities $f'(x) > 0$ and $f'(x) < 0$.
4. Find $f''(x)$. Locate potential points of inflection by solving the equation $f''(x) = 0$ and determining where $f''(x)$ does not exist but $f(x)$ does. Determine where f is concave upward or concave downward by solving the inequalities $f''(x) > 0$ and $f''(x) < 0$.
5. Find the horizontal asymptotes (if any) by finding the limits of $f(x)$ as $x \to \infty$ and $x \to -\infty$.
6. Plot the intercepts, the critical points, the points of inflection, the asymptotes, and other points as needed.

▶**EXAMPLE 7**   Graph $f(x) = 2x^3 - 3x^2 - 12x + 1$.

Step 1   The y-intercept is $f(0) = 2 \cdot 0^3 - 3 \cdot 0^2 - 12 \cdot 0 + 1 = 1$. To find the x-intercepts, we must solve the equation $2x^3 - 3x^2 - 12x + 1 = 0$. We do not have a convenient way to do this without a grapher, so skip this part. Since $f(x)$ is a polynomial function, the graph has no vertical or horizontal asymptotes, so we can also skip Steps 2 and 5.

Step 3   The first derivative is $f'(x) = 6x^2 - 6x - 12$, which is defined for all values of x. Now $f'(x) = 0$ when

$$6(x^2 - x - 2) = 0$$
$$6(x - 2)(x + 1) = 0$$
$$x = 2 \quad \text{or} \quad x = -1.$$

These critical numbers divide the number line in Figure 12.31 into three regions. Testing a number from each region in $f'(x)$ shows that f is increasing on $(-\infty, -1)$ and $(2, \infty)$ and decreasing on $(-1, 2)$. By the first derivative test, f has a relative maximum when $x = -1$ and a relative minimum when $x = 2$. The relative maximum is $f(-1) = 8$, while the relative minimum is $f(2) = -19$.

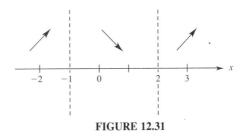

FIGURE 12.31

Step 4   Now use the second derivative to find the intervals where the function is concave upward or downward. Here

$$f''(x) = 12x - 6,$$

which is 0 when $x = 1/2$. Testing a point with x less than $1/2$, and one with x greater than $1/2$, shows that f is concave downward on $(-\infty, 1/2)$ and concave upward on $(1/2, \infty)$. The graph has an inflection point at $(1/2, f(1/2)) = (1/2, -11/2)$. This information is summarized in the chart on the next page.

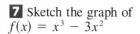

7 Sketch the graph of
$f(x) = x^3 - 3x^2$

Answer:

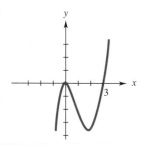

Interval	$(-\infty, -1)$	$(-1, 1/2)$	$(1/2, 2)$	$(2, \infty)$
Sign of f'	+	−	−	+
Sign of f''	−	−	+	+
f Increasing or Decreasing	Increasing	Decreasing	Decreasing	Increasing
Concavity of f	Downward	Downward	Upward	Upward
Shape of Graph	⌢	⌢	⌣	⌣

Step 6 Use the information in the chart above and plot some points to obtain the graph in Figure 12.32. ◄ **7**

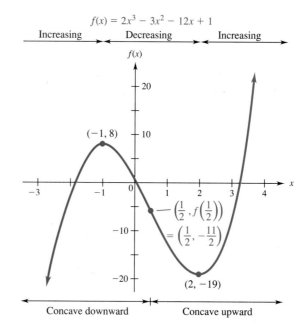

$f(x) = 2x^3 - 3x^2 - 12x + 1$

FIGURE 12.32

▶ **EXAMPLE 8** Graph $f(x) = \dfrac{3x^2}{x^2 + 5}$.

Step 1 The y-intercept is $f(0) = 0/5 = 0$. To find the x-intercepts, note that $f(x) = 0$ exactly when the numerator $3x^2 = 0$; this occurs when $x = 0$. So the point $(0, f(0)) = (0, 0)$ is both the x- and y-intercept.

Step 2   The denominator of $f(x)$ is always nonzero, so there are no vertical asymptotes.

Step 3   The first derivative is

$$f'(x) = \frac{(x^2 + 5)(6x) - (3x^2)(2x)}{(x^2 + 5)^2} = \frac{30x}{(x^2 + 5)^2}.$$

Since $x^2 + 5 \neq 0$ for all x, $f'(x)$ is always defined. $f'(x) = 0$ when its numerator $30x = 0$, which occurs when $x = 0$. The critical number 0 divides the x-axis into two regions (Figure 12.33). Testing a number in each region shows that f is decreasing on $(-\infty, 0)$ and increasing on $(0, \infty)$. By the first derivative test f has a relative minimum at $x = 0$.

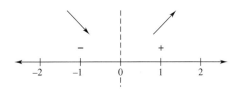

FIGURE 12.33

Step 4   The second derivative is

$$f''(x) = \frac{(x^2 + 5)^2(30) - (30x)(2)(x^2 + 5)(2x)}{(x^2 + 5)^4}$$

Factor $30(x^2 + 5)$ out of the numerator:

$$f''(x) = \frac{30(x^2 + 5)[(x^2 + 5) - (x)(2)(2x)]}{(x^2 + 5)^4}.$$

Divide a factor of $(x^2 + 5)$ out of the numerator and denominator, and simplify the numerator:

$$f''(x) = \frac{30[(x^2 + 5) - (x)(2)(2x)]}{(x^2 + 5)^3}$$

$$= \frac{30[(x^2 + 5) - (4x^2)]}{(x^2 + 5)^3}$$

$$= \frac{30(5 - 3x^2)}{(x^2 + 5)^3}.$$

The numerator of $f''(x)$ is 0 when $x = \pm \sqrt{5/3} \approx \pm 1.29$. Testing a point in each of the three intervals defined by these points shows that f is concave downward on $(-\infty, -1.29)$ and $(1.29, \infty)$, and concave upward on $(-1.29, 1.29)$. The graph has inflection points at $(\pm\sqrt{5/3}, f(\pm\sqrt{5/3})) \approx (\pm 1.29, \pm .75)$.

Interval	$(-\infty, -1.29)$	$(-1.29,0)$	$(0, 1.29)$	$(1.29, \infty)$
Sign of f'	$-$	$-$	$+$	$+$
Sign of f''	$-$	$+$	$+$	$-$
f Increasing or Decreasing	Decreasing	Decreasing	Increasing	Increasing
Concavity of f	Downward	Upward	Upward	Downward
Shape of Graph	⌒	⌣	⌣	⌒

Step 5 To find any horizontal asymptotes, divide both the numerator and denominator of $f(x)$ by x^2 and compute the limit:

$$\lim_{x \to \infty} \frac{3x^2}{x^2 + 5} = \lim_{x \to \infty} \frac{\dfrac{3x^2}{x^2}}{\dfrac{x^2}{x^2} + \dfrac{5}{x^2}} = \frac{3}{1 + 0} = 3$$

Verify that the limit of $f(x)$ as $x \to -\infty$ is also 3. Thus, the horizontal asymptote is $y = 3$.

Step 6 Plot some points (several are needed near the origin), including the intercept at the origin, and use the fact that $y = 3$ is a horizontal asymptote to obtain the graph in Figure 12.34. ◄

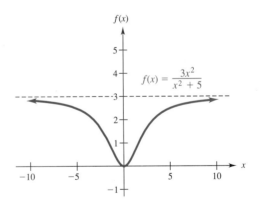

FIGURE 12.34

▶ **EXAMPLE 9** Graph $f(x) = x + 1/x$.

Step 1 Since $x = 0$ is not in the domain of the function, there is no y-intercept. To find the x-intercepts, solve $f(x) = 0$.

$$x + \frac{1}{x} = 0$$

$$x = -\frac{1}{x}$$

$$x^2 = -1$$

Since x^2 is always positive, there is also no x-intercept.

Step 2   Note that the rule of f can be written as

$$f(x) = x + \frac{1}{x} = \frac{x^2 + 1}{x}.$$

When $x = 0$, the denominator is 0, but the numerator is nonzero, so there is a vertical asymptote at $x = 0$.

Step 3   Here $f'(x) = 1 - (1/x^2)$, which is 0 when

$$\frac{1}{x^2} = 1$$

$$x^2 = 1$$

$$x = 1 \quad \text{or} \quad x = -1.$$

The derivative fails to exist at 0, where the vertical asymptote is located. Evaluating $f'(x)$ in each of the regions determined by the critical numbers and the asymptote shows that f is increasing on $(-\infty, -1)$ and $(1, \infty)$ and decreasing on $(-1, 0)$ and $(0, 1)$, as summarized in the chart on the next page. By the first derivative test, f has a relative maximum of $y = f(-1) = -2$, when $x = -1$, and a relative minimum of $y = f(1) = 2$ when $x = 1$.

Step 4   The second derivative is

$$f''(x) = \frac{2}{x^3},$$

which is never equal to 0 and does not exist when $x = 0$. (The function itself also does not exist at 0.) Because of this, there may be a change of concavity, but not an inflection point, when $x = 0$. The second derivative is negative when x is negative, making f concave downward on $(-\infty, 0)$. Also, $f''(x) > 0$ when $x > 0$, making f concave upward on $(0, \infty)$, as indicated in the chart.

Interval	*(−∞, −1)*	*(−1, 0)*	*(0, 1)*	*(1, ∞)*
sign of f'	+	−	−	+
sign of f"	−	−	+	+
f increasing or decreasing	increasing	decreasing	decreasing	increasing
concavity of f	downward	downward	upward	upward
shape of graph	⌒	⌍	⌎	⌏

Step 5 To find any horizontal asymptotes, consider the limit

$$\lim_{x \to \infty} \frac{x^2 + 1}{x} = \lim_{x \to \infty} \frac{1 + \dfrac{1}{x^2}}{1/x} = \frac{1 + 0}{0} = \frac{1}{0}.$$

Division by 0 is undefined, so the limit does not exist. Verify that $\lim_{x \to -\infty} f(x)$ also does not exist, so there are no horizontal asymptotes.

Observe that as x gets very large, the second term $(1/x)$ in $f(x)$ gets very small, so $f(x) = x + (1/x) \approx x$. The graph gets closer and closer to the straight line $y = x$ as x becomes larger and larger. This is what is known as an **oblique** or **slant asymptote.**

Step 6 Plot several points and use the information above to obtain the graph of $f(x)$ in Figure 12.35. ◀

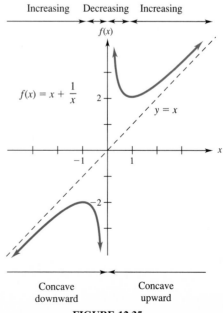

FIGURE 12.35

12.4 EXERCISES

Find the largest open intervals where the functions whose graphs are shown are increasing or decreasing. (See Example 1.)

1.

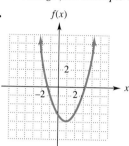

2.

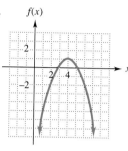

3.

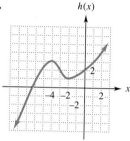

4.

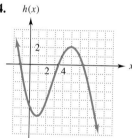

5.

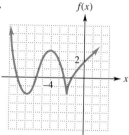

6.

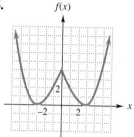

Find the largest open intervals on which each function is increasing or decreasing. (See Example 2.)

7. $f(x) = x^2 + 8x - 5$

8. $f(x) = x^2 - 5x + 7$

9. $f(x) = 2x^3 - 5x^2 - 4x + 2$

10. $f(x) = 4x^3 - 9x^2 - 12x + 7$

11. $f(x) = \dfrac{x + 1}{x + 4}$

12. $f(x) = \dfrac{x^2 + 1}{x}$

13. $f(x) = \sqrt{5 - x}$

14. $f(x) = \sqrt{x^2 + 1}$

15. $f(x) = 2x^3 - 3x^2 - 12x + 2$

16. $f(x) = 4x^3 - 15x^2 - 72x + 5$

Find the largest open intervals where the functions whose graphs are shown are concave upward or downward. Find the location of any points of inflection. (See Figure 12.23.)

17.

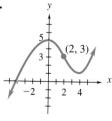

18.

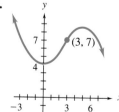

19.

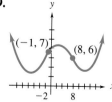

20.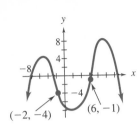

$(-2, -4)$ $(6, -1)$

21.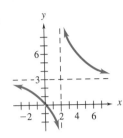

22.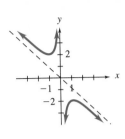

Find the largest open intervals on which each function is concave upward or concave downward and find the location of any points of inflection. (See Example 4.)

23. $f(x) = x^2 + 3x - 5$

24. $f(x) = -x^2 + 8x - 7$

25. $f(x) = x^3 + 4x^2 - 6x + 3$

26. $f(x) = 5x^3 + 12x^2 - 32x + 14$

27. $f(x) = \dfrac{2}{x - 3}$

28. $f(x) = \dfrac{-2}{x + 1}$

29. $f(x) = x^4 + 8x^3 - 30x^2 + 24x - 3$

30. $f(x) = x^4 + 8x^3 + 18x^2 + 12x - 84$

31. (a) Which of the functions in Exercises 17–22 have vertical asymptotes and where are they located?

(b) Which of the functions in Exercises 17–22 have nonvertical asymptotes and where are these asymptotes located?

Find the following limits. (See Example 6.)

32. $\lim\limits_{x \to \infty} \dfrac{3x}{5x - 1}$

33. $\lim\limits_{x \to \infty} \dfrac{5x}{3x - 1}$

34. $\lim\limits_{x \to \infty} \dfrac{2x + 3}{4x - 7}$

35. $\lim\limits_{x \to \infty} \dfrac{8x + 2}{2x - 5}$

36. $\lim\limits_{x \to \infty} \dfrac{x^2 + 2x}{2x^2 - 2x + 1}$

37. $\lim\limits_{x \to \infty} \dfrac{x^2 + 2x - 5}{3x^2 + 2}$

38. $\lim\limits_{x \to \infty} \dfrac{3x^3 + 2x - 1}{2x^4 - 3x^3 - 2}$

39. $\lim\limits_{x \to \infty} \dfrac{2x^2 + 11x - 10}{5x^3 + 3x^2 + 2x}$

40. $\lim\limits_{x \to \infty} \dfrac{2x^4 + 3x + 4}{x^2 - 4x + 1}$

In Exercises 41–44, sketch the graph of a single function that has all the properties listed. There are many correct answers and your graph need not be given by an algebraic formula.

41. (a) Continuous and differentiable for all real numbers
  (b) Increasing on $(-\infty, -3)$ and $(1, 4)$
  (c) Decreasing on $(-3, 1)$ and $(4, \infty)$
  (d) Concave downward on $(-\infty, -1)$ and $(2, \infty)$
  (e) Concave upward on $(-1, 2)$
  (f) $f'(-3) = f'(4) = 0$
  (g) Inflection points at $(-1, 3)$ and $(2, 4)$

42. (a) Continuous for all real numbers
  (b) Increasing on $(-\infty, -2)$ and $(0, 3)$
  (c) Decreasing on $(-2, 0)$ and $(3, \infty)$

  (d) Concave downward on $(-\infty, 0)$ and $(0, 5)$
  (e) Concave upward on $(5, \infty)$
  (f) $f'(-2) = f'(3) = 0$

43. (a) Continuous for all real numbers
  (b) Decreasing on $(-\infty, -6)$ and $(1, 3)$
  (c) Increasing on $(-6, 1)$ and $(3, \infty)$
  (d) Concave upward on $(-\infty, -6)$ and $(3, \infty)$
  (e) Concave downward on $(-6, 3)$
  (f) A y-intercept at $(0, 2)$

44. (a) Continuous and differentiable everywhere except at $x = 1$, where it has a vertical asymptote
  (b) Decreasing everywhere it is defined
  (c) A horizontal asymptote at $y = 2$
  (d) Concave downward on $(-\infty, 1)$ and $(2, 4)$
  (e) Concave upward on $(1, 2)$ and $(4, \infty)$

Sketch the graphs of the following functions. Identify any relative extrema and points of inflection. (See Examples 7–9.)

45. $f(x) = -x^2 - 10x - 25$

46. $f(x) = x^2 - 12x + 36$

47. $f(x) = 3x^3 - 3x^2 + 1$

48. $f(x) = 2x^3 - 4x^2 + 2$

49. $f(x) = -2x^3 - 9x^2 + 108x - 10$

50. $f(x) = -2x^3 - 9x^2 + 60x - 8$

51. $f(x) = 2x^3 + \frac{7}{2}x^2 - 5x + 3$

52. $f(x) = x^3 - \frac{15}{2}x^2 - 18x - 1$

53. $f(x) = (x + 3)^4$

54. $f(x) = x^3$

55. $f(x) = x^4 - 18x^2 + 5$

56. $f(x) = x^4 - 8x^2$

57. $f(x) = x - \frac{1}{x}$

58. $f(x) = 2x + \frac{8}{x}$

59. $f(x) = \frac{x^2 + 25}{x}$

60. $f(x) = \frac{x^2 + 4}{x}$

61. $f(x) = \frac{x - 1}{x + 1}$

62. $f(x) = \frac{x}{1 + x}$

Work the following problems. (See Examples 3 and 5.)

63. Management Suppose the total cost $C(x)$ (in dollars) to manufacture a quantity x of weed killer (in hundreds of liters) is given by

$$C(x) = x^3 - 2x^2 + 8x + 50.$$

(a) Where is $C(x)$ decreasing?

(b) Where is $C(x)$ increasing?

64. Management A county realty group estimates that the number of housing starts per year over the next 3 years will be

$$H(r) = \frac{300}{1 + .03r^2},$$

where r is the mortgage rate (in percent).

(a) Where is $H(r)$ increasing?

(b) Where is $H(r)$ decreasing?

65. Management A manufacturer sells cutlery with the following cost and revenue function, where x is the number of sets sold.

$$C(x) = 4000 + 4x, \qquad 0 \le x \le 17{,}000$$
$$R(x) = 20x - x^2/1000, \quad 0 \le x \le 17{,}000$$

Determine the intervals on which the profit function is increasing.

66. Management A manufacturer of compact disc players has determined that the profit $P(x)$ (in thousands of dollars) is related to the quantity x of players produced (in hundreds) per month by

$$P(x) = \frac{1}{3}x^3 - \frac{7}{2}x^2 + 10x - 2,$$

as long as the number of units produced is fewer than 800 per month.

(a) At what production levels is the profit increasing?

(b) At what levels is it decreasing?

67. Natural Science During a four-week long flu epidemic, the number of people $P(t)$ infected t days after the epidemic begins is approximated by

$$P(t) = t^3 - 60t^2 + 900t + 20 \ (0 \le t \le 28).$$

When will the number of people infected start to decline?

68. Natural Science The function

$$A(x) = -.15x^3 + 1.058x$$

approximates the alcohol concentration (in tenths of a percent) in an average person's bloodstream x hours after drinking 8 ounces of 100-proof whiskey. The function applies only for the interval $[0, 8]$.

(a) On what time intervals is the alcohol concentration increasing?

(b) On what intervals is it decreasing?

69. Natural Science The percent of concentration of a drug in the bloodstream x hours after the drug is administered is given by

$$K(x) = \frac{4x}{3x^2 + 27}.$$

(a) On what time intervals is the concentration of the drug increasing?

(b) On what intervals is it decreasing?

70. Natural Science Suppose a certain drug is administered to a patient, with the percent of concentration of the drug in the bloodstream t hours later given by

$$k(t) = \frac{5t}{t^2 + 1}.$$

(a) On what time intervals is the concentration of the drug increasing?

(b) On what intervals is it decreasing?

71. Natural Science The concentration of a drug in a patient's bloodstream h hours after it was injected is given by

$$A(h) = \frac{.17h}{h^2 + 2}.$$

Find and interpret $\lim\limits_{h \to \infty} A(h)$.

72. Management The cost for manufacturing a particular videotape is

$$c(x) = 15,000 + 6x,$$

where x is the number of tapes produced. The average cost per tape, denoted by $\overline{c}(x)$, is found by dividing $c(x)$ by x. Find and interpret $\lim\limits_{x \to \infty} \overline{c}(x)$.

73. Management A company training program has determined that, on the average, a new employee can do $P(s)$ pieces of work per day after s days of on-the-job training, where

$$P(s) = \frac{75s}{s + 8}.$$

Find and interpret $\lim\limits_{s \to \infty} P(s)$.

74. Social Science Members of a legislature often must vote repeatedly on the same bill. As time goes on, members may change their votes. Suppose that p_0 is the probability that an individual legislator favors an issue before the first roll call vote, and suppose that p is the probability of a change in position from one vote to the next. Then the probability that the legislator will vote "yes" on the nth roll call is given by

$$p_n = \frac{1}{2} + \left(p_0 - \frac{1}{2}\right)(1 - 2p)^n.*$$

For example, the chance of a "yes" on the third roll call vote is

$$p_3 = \frac{1}{2} + \left(p_0 - \frac{1}{2}\right)(1 - 2p)^3.$$

Suppose that there is a chance of $p_0 = .7$ that Congressman Stephens will favor the budget appropriation bill before the first roll call, but only a probability of $p = .2$ that he will change his mind on the subsequent vote. Find and interpret the following.

(a) p_2 **(b)** p_4 **(c)** p_8 **(d)** $\lim\limits_{n \to \infty} p_n$

Management *In Exercises 75 and 76, find the point of diminishing returns for the given functions, where $R(x)$ represents revenue in thousands of dollars and x represents the amount spent on advertising in thousands of dollars (See Example 5.)*

75. $R(x) = 10,000 - x^3 + 42x^2 + 800x;\ 0 \le x \le 20$

76. $R(x) = \frac{4}{27}(-x^3 + 66x^2 + 1050x - 400);$
$$0 \le x \le 25$$

 Use calculus and a grapher to find the approximate location of all relative extrema and points of inflection of these functions. Several viewing windows may be needed to see some of the graphs clearly. Be on the lookout for "hidden behavior," such as extrema that may not be obvious at first glance.

77. $f(x) = .1x^3 - .1x^2 - .005x + 1$

78. $f(x) = 2x^3 - .33x^2 - .006x + 5$

79. $f(x) = .01x^5 + x^4 - x^3 - 6x^2 + 5x + 4$

80. $f(x) = .1x^5 + 3x^4 - 4x^3 - 11x^2 + 3x + 2$

---

*Adapted excerpt from *Mathematics in the Behavioral and Social Sciences* by John W. Bishir and Donald W. Drewes, copyright © 1970 by Harcourt Brace Company. Reprinted by permission of the publisher.

KEY TERMS AND SYMBOLS

12.1   relative maximum (maxima)

relative minimum (minima)

relative extremum (extrema)

critical number

critical point

first derivative test

absolute maximum

absolute minimum

extreme value theorem

12.2 $f''(x)$ or y'' or $\dfrac{d^2y}{dx^2}$ or $D_x^2[f(x)]$

second derivative of f

$f'''(x)$ third derivative of f

$f^{(n)}(x)$ nth derivative of f

acceleration

12.3   economic lot size

12.4   increasing function

decreasing function

concave downward

concave upward

point of inflection

point of diminishing returns

limits at infinity

horizontal asymptotes

curve sketching

KEY CONCEPTS

Relative Extremum

Let c be in the interval (a, b) in the domain of function f. Then $f(c)$ is a **relative maximum** for f if $f(x) \leq f(c)$ for all x in (a, b). Also, $f(c)$ is a **relative minimum** if $f(x) \geq f(c)$ for all x in (a, b). If f has a relative extremum at c, then either $f'(c) = 0$ or $f'(c)$ does not exist.

First Derivative Test

Let f be a differentiable function for all x in $[a, b]$, except possibly at $x = c$. Assume $a < c < b$ and that c is the only critical number for f in $[a, b]$. If $f'(a) > 0$ and $f'(b) < 0$, then $f(c)$ is a relative maximum. If $f'(a) < 0$ and $f'(b) > 0$, then $f(c)$ is a relative minimum.

Absolute Extremum

Let c be in an interval where f is defined. $f(c)$ is the **absolute maximum** of f on the interval if $f(x) \leq f(c)$ for every x in the interval. $f(c)$ is the **absolute minimum** of f on the interval if $f(x) \geq f(c)$ for every x in the interval.

Second Derivative Test

Let c be a critical number of f such that $f'(c) = 0$ and $f''(x)$ exists for all x in some open interval containing c. If $f''(c) > 0$, then $f(c)$ is a relative minimum. If $f''(c) < 0$, then $f(c)$ is a relative maximum. If $f''(c) = 0$, then the test gives no information.

If $f'(x) > 0$ for each x in an interval, then f is **increasing** on the interval; if $f'(x) < 0$ for each x in the interval, then f is **decreasing** on the interval; if $f'(x) = 0$ for each x in the interval, then f is **constant** on the interval.

Let f have derivatives f' and f'' for all x in (a, b). f is **concave upward** on (a, b) if $f''(x) > 0$ for all x in (a, b). f is **concave downward** on (a, b) if $f''(x) < 0$ for all x in (a, b). f has a **point of inflection** at $x = c$ if $f''(x)$ changes sign at $x = c$.

Limits at Infinity

For any positive integer n, $\displaystyle\lim_{x \to -\infty} \frac{1}{x^n} = 0$ and $\displaystyle\lim_{x \to -\infty} \frac{1}{x^n} = 0.$

CHAPTER 12 REVIEW EXERCISES

1. When the rule of a function is given, how can you determine where it is increasing and where it is decreasing?

2. When the rule of a function is given, how can you determine where the relative extrema are located? State two ways to test whether a relative extremum is a maximum or a minimum.

3. What is the difference between a relative extremum and an absolute extremum? Can a relative extremum be an absolute extremum? Is a relative extremum necessarily an absolute extremum?

4. What information about a graph can be found from the second derivative?

Find the largest open intervals on which the following functions are increasing or decreasing.

5. $f(x) = x^2 + 7x - 9$

6. $f(x) = -3x^2 + 2x + 11$

7. $g(x) = 2x^3 - x^2 - 4x + 7$

8. $g(x) = -4x^3 + 5x^2 + 8x + 1$

9. $f(x) = \dfrac{4}{x - 3}$

10. $f(x) = \dfrac{6}{3x + 2}$

Find the locations and the values of all relative maxima and minima for each of the following functions.

11. $f(x) = 2x^3 + 3x^2 - 36x + 20$

12. $f(x) = 2x^3 + 3x^2 - 12x + 5$

13. $f(x) = x^4 + \dfrac{8}{3}x^3 - 6x^2 + 1$

14. $f(x) = x \cdot e^x$

15. $f(x) = 3x \cdot e^{-x}$

16. $f(x) = \dfrac{e^x}{x - 1}$

Find the locations and values of all absolute maxima and absolute minima for the following functions on the given intervals.

17. $f(x) = -x^2 + 5x + 1;\ [1, 4]$

18. $f(x) = 4x^2 - 8x - 3;\ [-1, 2]$

19. $f(x) = x^3 + 2x^2 - 15x + 3;\ [-4, 2]$

20. $f(x) = -2x^3 - x^2 + 4x - 1;\ [-3, 1]$

Find the second derivatives of the following functions; then find $f''(1)$ and $f''(-2)$.

21. $f(x) = 2x^5 - 4x^3 + 2x - 1$

22. $f(x) = \dfrac{3 - 2x}{x + 2}$

23. $f(x) = -5e^{4x}$

24. $f(x) = \ln|5x + 2|$

In Exercises 25–28, determine if the limit exists. If it does, find its values.

25. $\lim\limits_{x \to -\infty} g(x)$

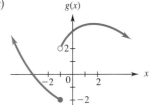

26. $\lim\limits_{x \to \infty} f(x)$

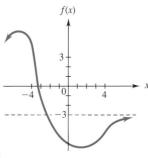

27. $\lim\limits_{x \to \infty} \dfrac{x^2 + 5}{5x^2 - 1}$

28. $\lim\limits_{x \to \infty} \dfrac{x^2 + 6x + 8}{x^3 + 2x + 1}$

Graph each of the following functions, including critical points, regions where the function is increasing or decreasing, points of inflection, regions where the function is concave up or concave down, intercepts where possible, and asymptotes where applicable.

29. $f(x) = -2x^3 - \dfrac{1}{2}x^2 + x - 3$

30. $f(x) = -\dfrac{4}{3}x^3 + x^2 + 30x - 7$

31. $f(x) = x^4 - \dfrac{4}{3}x^3 - 4x^2 + 1$

32. $f(x) = -\dfrac{2}{3}x^3 + \dfrac{9}{2}x^2 + 5x + 1$

33. $f(x) = \dfrac{x - 1}{2x + 1}$

34. $f(x) = \dfrac{2x - 5}{x + 3}$

35. $f(x) = -4x^3 - x^2 + 4x + 5$

36. $f(x) = x^3 + \dfrac{5}{2}x^2 - 2x - 3$

37. $f(x) = x^4 + 2x^2$

38. $f(x) = 6x^3 - x^4$

39. $f(x) = \dfrac{x^2 + 4}{x}$

40. $f(x) = x + \dfrac{8}{x}$

41. $f(x) = \dfrac{2x}{3 - x}$

42. $f(x) = \dfrac{-4x}{1 + 2x}$

Work the following exercises.

43. Management Suppose the profit from a product is $P(x) = 40x - x^2$, where x is the price in hundreds of dollars.
(a) At what price will the maximum profit occur?
(b) What is the maximum profit?

44. Management The total profit in hundreds of dollars from the sale of x hundred cartons of candy is given by
$$P(x) = -x^3 + 10x^2 - 12x - 4$$
(a) Find the number of cartons of candy that should be sold in order to produce maximum profit.

(b) Find the maximum profit.

45. Management The packaging department of a corporation is designing a box with a square base and top. The volume is to be 27 cubic meters. To reduce cost, the box is to have minimum surface area. What dimensions (height, length, and width) should the box have?

46. Management Another product (see Exercise 45) will be packaged in a closed cylindrical tin can with a volume of 54π cubic inches. Find the radius and height of the can if is to have minimum surface area.

47. Social Science The city park department is planning an enclosed play area in a new park. One side of the area will be against an existing building, with no fence needed there. Find the dimensions of the rectangular field of maximum area that can be enclosed with 900 meters of fence.

48. Management A company plans to package its product in a cylinder which is open at one end. The cylinder is to have a volume of 27π cubic inches. What radius should the circular bottom of the cylinder have to minimize the cost of the material? (Hint: the volume of a circular cylinder is $\pi r^2 h$ where r is the radius of the circular base and h is the height; the surface area of an open circular cylinder is $2\pi rh + \pi r^2$.)

49. Management In 1 year, a health food manufacturer produces and sells 240,000 cases of vitamins. It costs $2 to store a case for 1 year and $15 to produce each batch. Find the number of batches that should be produced annually.

50. Management A company produces 128,000 cases of soft drink annually. It costs $1 to store a case for 1 year and $10 to produce one lot. Find the number of lots that should be produced annually.

51. Social Science If the play area referred to in Exercise 47 needs fencing on all four sides, find the dimensions of the maximum rectangular area that can be made with 900 meters of fence.

▶ **The Effect of Oil Price on the Optimal Speed of Ships***

With the major increases in fuel recently, choosing an optimum speed for a large ocean-going ship has become a major concern of ship operators. Sailing slowly uses less fuel, but causes increased costs for salaries and interest on the value of the cargo.

The speed of sailing has a great effect on the amount of fuel used, the major expense in operating a ship. Past empirical studies have shown that the amount of fuel used by a ship is approximately proportional to the cube of the speed. This means that a 20% reduction in cruising speed produces about a 50% reduction in fuel consumption, since $(80\%)^3 \approx 50\%$. As an example, a medium sized ship that burns 40 tons of fuel per day at a certain speed could save 50% of this amount, or 20 tons, with a 20% reduction in speed. At a cost of $250 per ton, this 20% speed reduction would save $5000 per day in fuel costs.

To find the optimal sailing speed for one leg of a journey (a one-way trip with a given cargo between two points), let

R = income from the leg
P = profit from the leg
D_s = number of days at sea in the leg
D_p = number of days in port in the leg
L = distance of the leg in nautical miles
V = actual cruising speed
V_0 = normal cruising speed
C = daily cost of the vessel
　　　(excluding fuel for main engines)
F = actual daily fuel consumption in tons per day
F_0 = daily fuel consumption at normal cruising speed
F_c = cost of fuel in dollars per ton
V_m = minimum cruising speed.

The duration of the leg, D, is

$$D = D_s + D_p,\qquad\text{(1)}$$

*Based on "The Effect of Oil Price on the Optimal Speed of Ships" by David Ronen from *Journal of the Operations Research Society*, Vol 33, 1982. Copyright © 1982 Operational Research Society Ltd. Reprinted by permission.

where

$$D_s = \frac{L}{24\ V}.\qquad\text{(2)}$$

The profit from the leg is

$$P = R - (DC + FF_cD_s).\qquad\text{(3)}$$

As we mentioned above, the amount of fuel used is proportional to the cube of the speed, or

$$F = \left(\frac{V}{V_0}\right)^3 \cdot F_0.\qquad\text{(4)}$$

Our goal is to maximize the daily profit, Z, given by

$$Z = \frac{P}{D}.$$

Substituting equations (1), (2), and (4) into the equation for profit, (3), and dividing by D as given by equations (1) and (2) yields

$$Z = \frac{R - D_pC - \dfrac{LC}{24\ V} - \left(\dfrac{V}{V_0}\right)^3 F_0F_c\left(\dfrac{L}{24V}\right)}{D_p + \dfrac{L}{24\ V}}.\qquad\text{(5)}$$

It is not possible for a ship to sail at just any speed between 0 and its normal cruising speed, V_0. The design of the ship usually imposes some minimum cruising speed, V_m. This restriction forces V to be in the interval $[V_m, V_0]$.

To find the optimum value of V, set dZ/dV, from equation (5), equal to 0. This gives the cubic equation

$$V^3 + V^2\frac{L}{16D_p} - \frac{RV_0^3}{2F_0F_cD_p} = 0.\qquad\text{(6)}$$

Any real solutions of this equation are potential critical values. As mentioned above, these critical values must lie in the interval $[V_m, V_0]$ to be useful. To maximize Z, therefore, we must use our work with absolute extrema and check all critical values in $[V_m, V_0]$, as well as both endpoints, V_m and V_0.

EXERCISES

1. Find the optimum sailing speed for a voyage of 1536 miles, with 8 days in port, a normal cruising speed of 20 knots, a minimum speed of 10 knots, income from the leg of $86,400, fuel costs of $250 per ton, 50 tons of fuel consumed daily at normal cruising speed, and a daily cost for the vessel of $1000. (Hint: try $V = 12$.)

2. Why does the variable C not appear in equation (6) above?

▶ **A Total Cost Model for a Training Program**[*]

In this application, we set up a mathematical model for determining the total costs in setting up a training program. Then we use calculus to find the time between training programs that produces the minimum total cost. The model assumes that the demand for trainees is constant and that the fixed cost of training a batch of trainees is known. Also, it is assumed that people who are trained, but for whom no job is readily available, will be paid a fixed amount per month while waiting for a job to open up.

The model uses the following variables.

D = demand for trainees per month
N = number of trainees per batch
C_1 = fixed cost of training a batch of trainees
C_2 = variable cost of training per trainee per month
C_3 = salary paid monthly to a trainee who has not yet been given a job after training
m = time interval in months between successive batches of trainees
t = length of training program in months
$Z(m)$ = total monthly cost of program

The total cost of training a batch of trainees is given by $C_1 + NtC_2$. However, $N = mD$, so that the total cost per batch is $C_1 + mDtC_2$.

After training, personnel are given jobs at the rate of D per month. Thus, $N - D$ of the trainees will not get a job the first month, $N - 2D$ will not get a job the second month, and so on. The $N - D$ trainees who do not get a job the first month produce total costs of $(N - D)C_3$, those not getting jobs during the second month produce costs of $(N - 2D)C_3$, and so on. As $N = mD$, the costs during the first month can be written as

$$(N - D)C_3 = (mD - D)C_3 = (m - 1)DC_3,$$

while the costs during the second month are $(m - 2)DC_3$, and so on. The total cost for keeping the trainees without a job is thus $(m - 1)DC_3 + (m - 2)DC_3$
$$+ (m - 3)DC_3 + \cdots + 2DC_3 + DC_3,$$

which can be factored to give

$$DC_3[(m - 1) + (m - 2) + (m - 3) + \cdots + 2 + 1].$$

The expression in brackets is the sum of the terms of an arithmetic sequence. Using formulas for arithmetic sequences, the expression in brackets can be shown to equal $m(m - 1)/2$, so that we have

$$DC_3\left[\frac{m(m - 1)}{2}\right] \tag{1}$$

as the total cost for keeping jobless trainees.

The total cost per batch is the sum of the training cost per batch, $C_1 + mDtC_2$, and the cost of keeping trainees without a proper job, given by (1). Because we assume that a batch of trainees is trained every m months, the total cost per month, $Z(m)$, is given by

$$Z(m) = \frac{C_1 + mDtC_2}{m} + \frac{DC_3\left[\dfrac{m(m - 1)}{2}\right]}{m}$$
$$= \frac{C_1}{m} + DtC_2 + DC_3\left(\frac{m - 1}{2}\right).$$

EXERCISES

1. Find $Z'(m)$.

2. Solve the equation $Z'(m) = 0$.
   As a practical matter, it is usually required that m be a whole number. If m does not come out to be a whole number, then m^+ and m^-, the two whole numbers closest to m, must be chosen. Calculate both $Z(m^+)$ and $Z(m^-)$; the smaller of the two provides the optimum value of Z.

---

[*] Based on "A Total Cost Model for a Training Program" by P. L. Goyal and S. K. Goyal, Department of Mathematics and Computer Science, The Polytechnic of Wales, Treforest, Pontypridd. Used with permission.

3. Suppose a company finds that its demand for trainees is 3 per month, that a training program requires 12 months, that the fixed cost of training a batch of trainees is $15,000, that the variable cost per trainee per month is $100, and that trainees are paid $900 per month after training but before going to work. Use your result from Exercise 2 and find m.

4. Since m is not a whole number, find m^+ and m^-.

5. Calculate $Z(m^+)$ and $Z(m^-)$.

6. What is the optimum time interval between successive batches of trainees? How many trainees should be in a batch?

7. Write a brief essay describing other considerations, perhaps not quantifiable, that a manager might want to consider in this situation.

CHAPTER 13

Integral Calculus

TECHNOLOGY RESOURCES
Visual Calculus, Schneider
The Electronic Spreadsheet, Spero

In the previous two chapters, we studied the derivative of a function and various applications of derivatives. That material belongs to the branch of calculus called *differential calculus.* In this chapter we will study another branch of calculus, called *integral calculus.* Like the derivative of a function, the definite integral of a function is a special limit with many diverse applications. Geometrically, the derivative is related to the slope of the tangent line to a curve, while the definite integral is related to the area under a curve.

13.1 ANTIDERIVATIVES; INDEFINITE INTEGRALS

Functions used in applications in previous chapters have provided information about a *total amount* of a quantity, such as cost, revenue, profit, temperature, gallons of oil, or distance. Derivatives of these functions provided information about the rate of change of these quantities and allowed us to answer important questions about the extrema of the functions, It is not always possible to find ready-made functions that provide information about the total amount of a quantity, but it is often possible to collect enough data to come up with a function that gives the *rate* of *change* of a quantity. We know that derivatives give the rate of change when the total amount is known, is it possible, then, to reverse the process and use a known rate of change to get a function that gives the total amount of a quantity? The answer is yes; this reverse process, called *antidifferentiation,* is the topic of this section. The *antiderivative* of a function is defined as follows.

If $F'(x) = f(x)$, then $F(x)$ is an **antiderivative** of $f(x)$.

1 Find an antiderivative for each of the following.

(a) $3x^2$

(b) $5x$

(c) $8x^7$

(d) $\frac{1}{2}x^3$

Answers:
Only one possible antiderivative is given for each.

(a) x^3

(b) $\frac{5}{2}x^2$

(c) x^8

(d) $\frac{1}{8}x^4$

2 In Example 3, from the given information, could the population also be given by the function $F(x) = e^{2x} + 1000$? By $F(x) = e^{2x} + 10,000$?

Answer:
Yes: yes

▶**EXAMPLE 1**

(a) If $F(x) = 10x$, then $F'(x) = 10$, so $F(x) = 10x$ is an antiderivative of $f(x) = 10$.

(b) For $F(x) = x^2$, $F'(x) = 2x$, making $F(x) = x^2$ an antiderivative of $f(x) = 2x$. ◀

▶**EXAMPLE 2** Find an antiderivative of $f(x) = 5x^4$.

To find a function $F(x)$ whose derivative is $5x^4$, work backwards. Recall that the derivative of x^n is nx^{n-1}. If

$$nx^{n-1} \quad \text{is} \quad 5x^4,$$

then $n - 1 = 4$ and $n = 5$, so x^5 is an antiderivative of $5x^4$. ◀ **1**

▶**EXAMPLE 3** Suppose a population is growing at a rate given by $f(x) = e^{2x}$, where x is time in years from some initial date. Find a function giving the population at time x.

Let the population function be $F(x)$. Then

$$f(x) = F'(x) = e^{2x}.$$

To find the derivative of e^{2x}, we multiply e^{2x} by 2. So, to find the antiderivative of e^{2x}, we should *divide* by 2. Thus, $F(x) = e^{2x}/2$. So one population function with the given growth rate is $F(x) = e^{2x}/2$. ◀ **2**

The function from Example 1(b), defined by $F(x) = x^2$, is not the only function whose derivative is $f(x) = 2x$; for example, both $G(x) = x^2 + 2$ and $H(x) = x^2 - 4$ have $f(x) = 2x$ as a derivative. In fact, for any real number C, the function $F(x) = x^2 + C$ has $f(x) = 2x$ as its derivative. This means that there is a *family* of functions having $2x$ as an antiderivative. As the next theorem states, if two functions $F(x)$ and $G(x)$ are antiderivatives of $f(x)$, then $F(x)$ and $G(x)$ can differ only by a constant.

If $F(x)$ and $G(x)$ are both antiderivatives of $f(x)$, then there is a constant C such that

$$F(x) - G(x) = C.$$

(Two antiderivatives of a function can differ only by a constant.)

For example,

$$F(x) = x^2, \quad G(x) = x^2 + 2, \quad \text{and} \quad H(x) = x^2 - 4$$

are all antiderivatives of $f(x) = 2x$, and any two of them differ only by a constant. The derivative of a function gives the slope of the tangent line at any

x-value. The fact that these three functions have the same derivative, $f(x) = 2x$, means that their slopes at any particular value of x are the same, as shown in Figure 13.1.

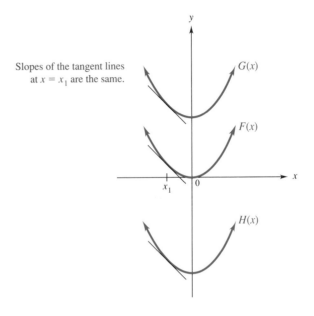

Slopes of the tangent lines at $x = x_1$ are the same.

FIGURE 13.1

The family of all antiderivatives of the function f is indicated by

$$\int f(x)\, dx.$$

The symbol $\int$ is the **integral sign,** $f(x)$ is the **integrand,** and $\int f(x)\, dx$ is called an **indefinite integral,** the most general antiderivative of f. The process of finding $\int f(x)\, dx$ is called **integration.**

Indefinite Integral

If $F'(x) = f(x)$, then

$$\int f(x)\, dx = F(x) + C,$$

for any real number C.

For example, using this notation,

$$\int 2x \, dx = x^2 + C.$$

Note The dx in the indefinite integral indicates that $\int f(x) \, dx$ is the "integral of $f(x)$ *with respct to x*" just as the symbol dy/dx denotes the "derivative of y with respect to x." For example, in the indefinite integral $\int 2ax \, dx$, dx indicates that a is to be treated as a constant and x as the variable, so that

$$\int 2ax \, dx = 2a \cdot \frac{x^2}{2} + C = ax^2 + C.$$

On the other hand,

$$\int 2ax \, da = 2x \cdot \frac{a^2}{2} + C = xa^2 + C.$$

A more complete interpretation of dx will be discussed later.

The symbol $\int f(x) \, dx$ was created by G. W. Leibniz (1646–1716) in the latter part of the seventeenth century. The $\int$ is an elongated S from *summa*, the Latin word for *sum*. The word *integral* as a term in the calculus was coined by Jakob Bernoulli (1654–1705), a Swiss mathematician who corresponded frequently with Leibniz. The relationship between sums and integrals will be clarified in Section 13.3.

Because finding an indefinite integral is the inverse of finding a derivative, each formula for derivatives leads to a rule for indefinite integrals. As mentioned above, the derivative of x^n is found by multiplying x by n and reducing the exponent on x by 1. To find an indefinite integral—that is, to undo what was done—*increase* the exponent by 1 and *divide* by the new exponent, $n + 1$.

Power Rule for Indefinite Integrals

For any real number $n \neq -1$,

$$\int x^n \, dx = \frac{1}{n + 1} x^{n+1} + C.$$

This result can be verified by differentiating the expression on the right above.

$$\frac{d}{dx}\left(\frac{1}{n + 1} x^{n+1} + C \right) = \frac{n + 1}{n + 1} x^{(n+1)-1} + 0 = x^n$$

(If $n = -1$, the expression in the denominator is 0, and the above rule cannot be used. We will see later how to find an antiderivative in this case.)

3 Find each of the following.

(a) $\displaystyle\int x^5 \, dx$

(b) $\displaystyle\int \sqrt[3]{x} \, dx$

(c) $\displaystyle\int 5 \, dx$

Answers:

(a) $\dfrac{1}{6}x^6 + C$

(b) $\dfrac{3}{4}x^{4/3} + C$

(c) $5x + C$

▶**EXAMPLE 4** Find each indefinite integral.

(a) $\displaystyle\int x^3 \, dx$

Use the power rule with $n = 3$.

$$\int x^3 \, dx = \frac{1}{3+1}x^{3+1} + C = \frac{1}{4}x^4 + C$$

(b) $\displaystyle\int \frac{1}{t^2} \, dt$

First, write $1/t^2$ as t^{-2}. Then

$$\int \frac{1}{t^2} \, dt = \int t^{-2} \, dt = \frac{1}{-2+1}t^{-2+1} = \frac{t^{-1}}{-1} + C = \frac{-1}{t} + C.$$

(c) $\displaystyle\int \sqrt{u} \, du$

Since $\sqrt{u} = u^{1/2}$,

$$\int \sqrt{u} \, du = \int u^{1/2} \, du = \frac{1}{1/2 + 1}u^{1/2+1} + C = \frac{2}{3}u^{3/2} + C.$$

To check this, differentiate $(2/3)u^{3/2} + C$; the derivative is $u^{1/2}$, the original function.

(d) $\displaystyle\int dx$

Writing dx as $1 \cdot dx$, and using the fact that $x^0 = 1$ for any nonzero number x,

$$\int dx = \int 1 \, dx = \int x^0 \, dx = \frac{1}{1}x^1 + C = x + C. \quad ◀ \text{ ▣ }$$

As shown in Chapter 11, the derivative of the product of a constant and a function is the product of the constant and the derivative of the function. A similar rule applies to indefinite integrals. Also, since derivatives of sums or differences are found term by term, indefinite integrals can also be found term by term. Proofs of these properties are given as exercises.

Properties of Indefinite Integrals: Constant Multiple Rule; Sum or Difference Rule

If all indicated integrals exist,

$$\int k \cdot f(x) \, dx = k \int f(x) \, dx, \text{ for any real number } k;$$

$$\int [f(x) \pm g(x)] \, dx = \int f(x) \, dx \pm \int g(x) \, dx.$$

4 Find each of the following.

(a) $\int (-6x^4)\, dx$

(b) $\int 9x^{2/3}\, dx$

(c) $\int \dfrac{8}{x^3}\, dx$

(d) $\int (5x^4 - 3x^2 + 6)\, dx$

(e) $\int \left(3\sqrt{x} + \dfrac{2}{x^2}\right) dx$

Answers:

(a) $-\dfrac{6}{5}x^5 + C$

(b) $\dfrac{27}{5}x^{5/3} + C$

(c) $-4x^{-2} + C$ or $-\dfrac{4}{x^2} + C$

(d) $x^5 - x^3 + 6x + C$

(e) $2x^{3/2} - \dfrac{2}{x} + C$

Caution The constant multiple rule requires that k be a *number*. The rule does not apply to a *variable*. For example,

$$\int x\sqrt{x - 1}\, dx \neq x \int \sqrt{x - 1}\, dx.$$

▶ **EXAMPLE 5** Find each of the following.

(a) $\int 2x^3\, dx$

By the constant multiple rule and the power rule,

$$\int 2x^3\, dx = 2 \int x^3\, dx = 2\left(\frac{1}{4}x^4\right) + C = \frac{1}{2}x^4 + C.$$

Since C represents any real number, it is not necessary to multiply it by 2 in next-to-last step.

(b) $\int \dfrac{12}{z^5}\, dz$

Use negative exponents.

$$\int \frac{12}{z^5}\, dz = \int 12z^{-5}\, dz$$

$$= 12 \int z^{-5}\, dz \qquad \textbf{Constant multiple rule}$$

$$= 12\left(\frac{z^{-4}}{-4}\right) + C \qquad \textbf{Power rule}$$

$$= -3z^{-4} + C$$

$$= \frac{-3}{z^4} + C.$$

(c) $\int (3z^2 - 4z + 5)\, dz$

By extending the sum or difference property given above to more than two terms,

$$\int (3z^2 - 4z + 5)\, dz = 3 \int z^2\, dz - 4 \int z\, dz + 5 \int dz$$

$$= 3\left(\frac{1}{3}z^3\right) - 4\left(\frac{1}{2}z^2\right) + 5z + C$$

$$= z^3 - 2z^2 + 5z + C.$$

Only one constant C is needed in the answer: the three constants from integrating term by term are combined. ◀ **4**

5 Find each of the following.

(a) $\displaystyle\int \frac{\sqrt{x} + 1}{x^2}\, dx$

(b) $\displaystyle\int (\sqrt{x} + 2)^2 dx$

Answers:

(a) $-\dfrac{2}{\sqrt{x}} - \dfrac{1}{x} + C$

(b) $\dfrac{x^2}{2} + \dfrac{8}{3}x^{3/2} + 4x + C$

Integration can always be checked by taking the derivative of the result. For instance, in Example 5(c) check that $z^3 - 2z^2 + 5z + C$ is the required indefinite integral by taking the derivative:

$$\frac{d}{dz}(z^3 - 2z^2 + 5z + C) = 3z^2 - 4z + 5.$$

The result is the original function to be integrated, so the work checks.

▶ **EXAMPLE 6** Find each of the following.

(a) $\displaystyle\int \frac{x^2 + 1}{\sqrt{x}}\, dx$

First rewrite the integrand as follows.

$$\int \frac{x^2 + 1}{\sqrt{x}}\, dx = \int \left(\frac{x^2}{\sqrt{x}} + \frac{1}{\sqrt{x}}\right) dx$$

$$= \int \left(\frac{x^2}{x^{1/2}} + \frac{1}{x^{1/2}}\right) dx$$

$$= \int (x^{3/2} + x^{-1/2})\, dx \qquad \textbf{Quotient rule for exponents}$$

Now find the antiderivative.

$$\int (x^{3/2} + x^{-1/2})\, dx = \frac{x^{5/2}}{5/2} + \frac{x^{1/2}}{1/2} + C$$

$$= \frac{2}{5}x^{5/2} + 2x^{1/2} + C$$

(b) $\displaystyle\int (x^2 - 1)^2\, dx$

Square the binomial first, and then find the antiderivative.

$$\int (x^2 - 1)^2\, dx = \int (x^4 - 2x^2 + 1)\, dx$$

$$= \frac{x^5}{5} - \frac{2x^3}{3} + x + C \quad ◀ \;\boxed{5}$$

As shown in Chapter 11, the derivative of $f(x) = e^x$ is $f'(x) = e^x$. Also, the derivative of $f(x) = e^{kx}$ is $f'(x) = k \cdot e^{kx}$. These results lead to the following formulas for indefinite integrals of exponential functions.

6 Find each of the following.

(a) $\int (-4e^x) \, dx$

(b) $\int e^{3x} \, dx$

(c) $\int (e^{2x} - 2e^x) \, dx$

(d) $\int (-11e^{-x}) \, dx$

Answers:

(a) $-4e^x + C$

(b) $\dfrac{1}{3} e^{3x} + C$

(c) $\dfrac{1}{2} e^{2x} - 2e^x + C$

(d) $11e^{-x} + C$

7 Find each of the following.

(a) $\int (-9/x) \, dx$

(b) $\int (8e^{4x} - 3x^{-1}) \, dx$

Answers:

(a) $-9 \cdot \ln |x| + C$

(b) $2e^{4x} - 3 \cdot \ln |x| + C$

Indefinite Integrals of Exponential Functions

If k is a real number, $k \ne 0$, then

$$\int e^x \, dx = e^x + C;$$

$$\int e^{kx} \, dx = \frac{1}{k} \cdot e^{kx} + C.$$

▶ **EXAMPLE 7**   Find each of the following.

(a) $\displaystyle\int 9e^x \, dx = 9 \int e^x \, dx = 9e^x + C$

(b) $\displaystyle\int e^{9t} \, dt = \frac{1}{9} e^{9t} + C$

(c) $\displaystyle\int 3e^{(5/4)u} \, du = 3\left(\frac{1}{5/4} e^{(5/4)u} \right) + C = 3\left(\frac{4}{5} \right) e^{(5/4)u} + C$

$$= \frac{12}{5} e^{(5/4)u} + C \quad \blacktriangleleft \boxed{6}$$

The restriction $n \ne -1$ was necessary in the formula for $\int x^n \, dx$ because $n = -1$ made the denominator of $1/(n + 1)$ equal to 0. To find $\int x^n \, dx$ when $n = -1$, that is, to find $\int x^{-1} \, dx$, recall the differentiation formula for the logarithmic function: the derivative of $f(x) = \ln |x|$, where $x \ne 0$, is $f'(x) = 1/x = x^{-1}$. This formula for the derivative of $f(x) = \ln |x|$ gives a formula for $\int x^{-1} \, dx$.

Indefinite Integral of x^{-1}

$$\int x^{-1} \, dx = \int \frac{1}{x} \, dx = \ln |x| + C, \quad \text{where } x \ne 0.$$

Caution   The domain of the logarithmic function is the set of positive real numbers. However, $y = x^{-1} = 1/x$ has as domain the set of all nonzero real numbers, so the absolute value of x *must* be used in the antiderivative.

▶ **EXAMPLE 8**   Find each of the following.

(a) $\displaystyle\int \frac{4}{x} \, dx = 4 \int \frac{1}{x} \, dx = 4 \cdot \ln |x| + C$

(b) $\displaystyle\int \left(-\frac{5}{x} + e^{-2x} \right) dx = -5 \cdot \ln |x| - \frac{1}{2} e^{-2x} + C \quad \blacktriangleleft \boxed{7}$

In all the examples above, the antiderivative family of functions was found. In many applications, however, the given information allows us to determine the value of the integration constant C. The next examples illustrates this idea.

▶**EXAMPLE 9**   According to the Cellular Telecommunications Industry Association, the rate of increase of the number of cellular phone subscribers (in millions) since service began is given by

$$S'(x) = .38x + .04,$$

where x is the number of years since 1985, when the service started. There were .25 million subscribers in 1985, year 0. Find a function that gives the number of subscribers in year x.

Since $S'(x)$ gives the rate of change in the number of subscribers,

$$S(x) = \int (.38x + .04)\, dx$$

$$= .38\frac{x^2}{2} + .04x + C$$

$$= .19x^2 + .04x + C.$$

To find the value of C, use the fact that the number of subscribers (in millions) $S(0) = .25$.

$$S(x) = .19x^2 + .04x + C$$
$$.25 = .19(0)^2 + .04(0) + C$$
$$C = .25$$

Thus, the number of subscribers (in millions) in year x is

$$S(x) = .19x^2 + .04x + .25. \quad ◀$$

▶**EXAMPLE 10**   Suppose the marginal revenue from a product is given by $50/\sqrt{x}$. Find the demand function for the product.

The marginal revenue is the derivative of the revenue function

$$\frac{dR}{dx} = \frac{50}{\sqrt{x}}.$$

$$R = \int \frac{50}{\sqrt{x}}\, dx = \int 50x^{-1/2}\, dx$$

$$= 50(2x^{1/2}) + k = 100x^{1/2} + k$$

If $x = 0$, then $R = 0$ (no items sold means no revenue), and

$$0 = 100(0)^{1/2} + k$$
$$0 = k.$$

8 The marginal cost at a level of production of x items is

$$C'(x) = 2x^3 + 6x - 5.$$

The fixed cost is $800. Find the cost function $C(x)$.

Answer:

$$C(x) = \frac{1}{2}x^4 + 3x^2 - 5x + 800$$

Thus,

$$R = 100x^{1/2}$$

gives the revenue function. Now, recall that $R = xp$, where p is the demand function.

$$100x^{1/2} = xp$$
$$100x^{-1/2} = p$$
$$\frac{100}{\sqrt{x}} = p$$

The demand function is $p = 100/\sqrt{x}$. ◀ **8**

13.1 EXERCISES

1. What must be true of $F(x)$ and $G(x)$ if both are antiderivatives of $f(x)$?

2. How is the antiderivative of a function related to the function?

3. In your own words, describe what is meant by an integrand?

4. Explain why the restriction $n \neq -1$ is necessary in the rule $\int x^n \, dx = \frac{1}{n+1}x^{n+1} + C$.

Find each of the following. (See Examples 4–8.)

5. $\int 10x \, dx$

6. $\int 25r \, dr$

7. $\int 8p^2 \, dp$

8. $\int 5t^3 \, dt$

9. $\int 100 \, dx$

10. $\int 35 \, dt$

11. $\int (5z - 1) \, dz$

12. $\int (2m + 3) \, dm$

13. $\int (z^2 - 4z + 2) \, dz$

14. $\int (2y^2 + 4y + 7) \, dy$

15. $\int (x^3 - 14x^2 + 20x + 3) \, dx$

16. $\int (x^3 + 5x^2 - 10x - 4) \, dx$

17. $\int 6\sqrt{y} \, dy$

18. $\int 8z^{1/2} \, dz$

19. $\int (6t\sqrt{t} + 3\sqrt{t}) \, dt$

20. $\int (12\sqrt{x} - x\sqrt{x}) \, dx$

21. $\int (56t^{1/2} + 18t^{7/2}) \, dt$

22. $\int (10u^{3/2} - 14u^{5/2}) \, du$

23. $\int \frac{24}{x^3} \, dx$

24. $\int \frac{-20}{x^2} \, dx$

25. $\int \left(\frac{1}{y^2} - \frac{2}{\sqrt{y}} \right) \, dy$

26. $\int \left(\frac{3}{\sqrt{u}} + \frac{2u}{\sqrt{u}} \right) \, du$

27. $\int (6x^{-3} + 4x^{-1}) \, dx$

28. $\int (3x^{-1} - 10x^{-2}) \, dx$

29. $\int 4e^{3u} \, du$

30. $\int -e^{-4x} \, dx$

31. $\int 3e^{-.2x} \, dx$

32. $\int -4e^{.2v} \, dv$

33. $\int \left(\frac{3}{x} + 4e^{-.5x} \right) \, dx$

34. $\int \left(\frac{9}{x} - 3e^{-.4x} \right) \, dx$

35. $\int \dfrac{1 + 2t^3}{t} \, dt$

36. $\int \dfrac{2y^{1/2} - 3y^2}{y} \, dy$

37. $\int \left(e^{2u} + \dfrac{u}{4} \right) du$

38. $\int \left(\dfrac{2}{v} - e^{3v} \right) dv$

39. $\int (x + 1)^2 \, dx$

40. $\int (2y - 1)^2 \, dy$

41. $\int \dfrac{\sqrt{x} + 1}{\sqrt[3]{x}} \, dx$

42. $\int \dfrac{1 - 2\sqrt[3]{z}}{\sqrt[3]{z}} \, dz$

---

43. The slope of the tangent line to a curve is given by
$$f'(x) = 6x^2 - 4x + 3.$$
If the point $(0, 1)$ is on the curve, find the equation of the curve.

44. Find the equation of the curve whose tangent line has a slope of
$$f'(x) = x^{2/3},$$
if the point $(1, 3/5)$ is on the curve.

Management   *Find the cost function for each of the following marginal cost functions. (See Example 9.)*

45. $C'(x) = .2x^2 + 5x$; fixed cost is $10

46. $C'(x) = .8x^2 - x$; fixed cost is $5

47. $C'(x) = x^{1/2}$; 16 units cost $40

48. $C'(x) = x^{2/3} + 2$; 8 units cost $58

49. $C'(x) = x^2 - 2x + 3$; 3 units cost $15

50. $C'(x) = x + \dfrac{1}{x^2}$; 2 units cost $5.50

51. $C'(x) = \dfrac{1}{x} + 2x$; 7 units cost $58.40

52. $C'(x) = 5x - \dfrac{1}{x}$; 10 units cost $94.20

53. $C'(x) = .03e^{.01x}$, no units cost $8

54. $C'(x) = 1.2e^{.02x}$, 2 units cost $9.50

Work the following problems. (See Examples 9 and 10.)

▷ **55. Management**   The marginal revenue from a product is given by
$$50 - 3x - x^2.$$
Find the demand function for the product.(*Hint:* recall $R = xp$. Also, if $x = 0$, $R = 0$.)

▷ **56. Management**   The marginal profit from the sale of x hundred items of a product is $P'(x) = 4 - 6x + 3x^2$, and the "profit" when no items are sold is $-\$40$. Find the profit function.

57. Natural Science   If the rate of excretion of a biochemical compound is given by
$$f'(t) = .01e^{-.01t},$$
the total amount excreted by time t (in minutes) is $f(t)$.

(a) Find an expression for $f(t)$.
(b) If 0 units are excreted at time $t = 0$, how many units are excreted in 10 minutes?

58. Social Science   Imports (in billions of dollars) to the United States from Canada since 1988 have changed at a rate given by $f(x) = 1.26x^2 - 5.5x + 8.33$, where x is the number of years since 1988. The United States imported $82 billion in 1988.
(a) Find a function giving the imports in year x.
(b) What was the value of imports from Canada in 1993?

59. Show that $\int k \cdot f(x) dx = k \int f(x) \, dx$, for any real number k.

60. Show that
$$\int [f(x) \pm g(x)] \, dx = \int f(x) \, dx \pm \int g(x) \, dx.$$

13.2 INTEGRATION BY SUBSTITUTION

In Section 13.1 we saw how to integrate a few simple functions. More complicated functions can sometimes be integrated by *substitution*. The technique depends on the idea of a differential. If $u = f(x)$, the **differential** of u, written

1 Find du for the following.

(a) $u = 9x$

(b) $u = 5x^3 + 2x^2$

(c) $u = e^{-2x}$

Answers:

(a) $du = 9\,dx$

(b) $du = (15x^2 + 4x)\,dx$

(c) $du = -2e^{-2x}\,dx$

du, is defined as

$$du = f'(x)\,dx.$$

For example, if $u = 6x^4$, then $du = 24x^3\,dx$. **1**

Differentials have many useful interpretations which are studied in more advanced courses. We shall only use them as a convenient notational device when finding an antiderivative such as

$$\int (3x^2 + 4)^4 6x\,dx.$$

The function $(3x^2 + 4)^4 6x$ is reminiscent of the chain rule and so we shall try to use differentials and the chain rule in *reverse* to find the antiderivative. Let $u = 3x^2 + 4$; then $du = 6x\,dx$. Now substitute u for $3x^2 + 4$ and du for $6x\,dx$ in the indefinite integral above.

$$\int (3x^2 + 4)^4 6x\,dx = \int \overbrace{(3x^2 + 4)^4}^{u}\overbrace{(6x\,dx)}^{du}$$

$$= \int u^4\,du$$

This last integral can now be found by the power rule.

$$\int u^4\,du = \frac{u^5}{5} + C$$

Finally, substitute $3x^2 + 4$ for u.

$$\int (3x^2 + 4)^4 6x\,dx = \frac{u^5}{5} + C = \frac{(3x^2 + 4)^5}{5} + C$$

We can check the accuracy of this result by using the chain rule to take the derivative.

$$\frac{d}{dx}\left[\frac{(3x^2 + 4)^5}{5} + C\right] = \frac{1}{5} \cdot 5(3x^2 + 4)^4(6x) + 0$$

$$= (3x^2 + 4)^4 6x,$$

which is the original function.

This method of integration is called **integration by substitution.** As shown above, it is simply the chain rule for derivatives in reverse. The results can always be verified by differentiation.

▶ **EXAMPLE 1** Find $\int (4x + 5)^9\,dx.$

We choose $4x + 5$ as u. Then $du = 4dx$. We are missing the constant

2 Find the following.

(a) $\displaystyle\int 8x(4x^2 - 1)^5 \, dx$

(b) $\displaystyle\int (3x - 8)^5 \, dx$

(c) $\displaystyle\int 18x^2(x^3 - 5)^{3/2} \, dx$

Answers:

(a) $\displaystyle\frac{(4x^2 - 1)^6}{6} + C$

(b) $\displaystyle\frac{(3x - 8)^6}{18} + C$

(c) $\displaystyle\frac{12(x^3 - 5)^{5/2}}{5} + C$

3 Find the following.

(a) $\displaystyle\int x(5x^2 + 6)^4 \, dx$

(b) $\displaystyle\int x\sqrt{x^2 + 16} \, dx$

Answers:

(a) $\displaystyle\frac{1}{50}(5x^2 + 6)^5 + C$

(b) $\displaystyle\frac{1}{3}(x^2 + 16)^{3/2} + C$

4. We can rewrite the integral by using the fact that $4(1/4) = 1$, as follows.

$$\int (4x + 5)^9 dx = \frac{1}{4} \cdot 4 \int (4x + 5)^9 dx$$

$$= \frac{1}{4} \int (4x + 5)^9 (4 \, dx) \qquad k \int f(x)dx = \int kf(x) \, dx$$

$$= \frac{1}{4} \int u^9 du \qquad\qquad \text{Substitute.}$$

$$= \frac{1}{4} \cdot \frac{u^{10}}{10} + C = \frac{u^{10}}{40} + C$$

$$= \frac{(4x + 5)^{10}}{40} + C \qquad\qquad \text{Substitute.} \quad \blacktriangleleft \; \boxed{2}$$

Caution When changing the x-problem to the u-problem, make sure that the change is complete; that is, that no x's are left in the u-problem.

▶**EXAMPLE 2** Find $\displaystyle\int x^2\sqrt{x^3 + 1} \, dx.$

An expression raised to a power is usually a good choice for u, so because of the square root or $1/2$ power, let $u = x^3 + 1$; then $du = 3x^2 dx$. The integrand does not contain the constant 3, which is needed for du. To take care of this, solve $du = 3x^2 \, dx$ for $x^2 \, dx$.

$$du = 3x^2 \, dx$$

$$\frac{1}{3} du = x^2 \, dx$$

Substitute $(1/3) \, du$ for $x^2 dx$.

$$\int x^2\sqrt{x^3 + 1} \, dx = \int \sqrt{x^3 + 1}(x^2 \, dx) = \int \sqrt{u} \cdot \frac{1}{3} du$$

Now use the constant multiple rule to bring the $1/3$ outside the integral sign.

$$\int x^2\sqrt{x^3 + 1} \, dx = \int \sqrt{u} \cdot \frac{1}{3} du = \frac{1}{3} \int u^{1/2} \, du$$

$$= \frac{1}{3} \cdot \frac{u^{3/2}}{3/2} + C = \frac{2}{9}u^{3/2} + C$$

Since $u = x^3 + 1,$

$$\int x^2\sqrt{x^3 + 1} \, dx = \frac{2}{9}(x^3 + 1)^{3/2} + C. \quad \blacktriangleleft \; \boxed{3}$$

4 Find the following.

(a) $\int z(z^2 + 1)^2 \, dx$

(b) $\int \dfrac{x^2 + 3}{\sqrt{x^3 + 9x}} \, dx$

Answers:

(a) $\dfrac{(z^2 + 1)^3}{6} + C$

(b) $\dfrac{2}{3} \sqrt{x^3 + 9x} + C$

Caution The substitution method given in the examples above *will not always work*. For example, we might try to find

$$\int x^3 \sqrt{x^3 + 1} \, dx$$

by substituting $u = x^3 + 1$, so that $du = 3x^2 dx$. However, there is no *constant* that can be inserted inside the integral sign to give $3x^2$. This integral, and a great many others, cannot be evaluated by substitution.

With practice, choosing u will become easy if you keep two principles in mind. First, u should equal some expression in the integral that, when replaced with u, tends to make the integral simpler. Second and most important, u must be an expression whose derivative is also present in the integral. The substitution should include as much of the integral as possible, so long as its derivative is still present. In Example 2, we could have chosen $u = 3x^2$, but $u = 3x^2 + 4$ is better, because it has the same derivative as $3x^2$ and captures more of the original integral. If we carry this reasoning further, we might try $u = (3x^2 + 4)^4$, but this is a poor choice, for $du = 4(3x^2 + 4)^3(6x) \, dx$, an expression not present in the original integral.

▶ **EXAMPLE 3** Find $\int \dfrac{x + 3}{(x^2 + 6x)^2} \, dx$.

Let $u = x^2 + 6x$, so that $du = (2x + 6) \, dx = 2(x + 3) \, dx$. The integral is missing the 2, so multiply by 2/2, putting 2 inside the integral sign and 1/2 outside.

$$\int \frac{x + 3}{(x^2 + 6x)^2} \, dx = \frac{1}{2} \int \frac{2(x + 3)}{(x^2 + 6x)^2} \, dx$$

$$= \frac{1}{2} \int \frac{du}{u^2} = \frac{1}{2} \int u^{-2} \, du$$

$$= \frac{1}{2} \cdot \frac{u^{-1}}{-1} + C = \frac{-1}{2u} + C$$

Substituting $x^2 + 6x$ for u gives

$$\int \frac{x + 3}{(x^2 + 6x)^2} \, dx = \frac{1}{2(x^2 + 6x)} + C. \quad \blacksquare \; \boxed{4}$$

Recall the formula for $\dfrac{d}{dx}(e^u)$, where $u = f(x)$.

$$\frac{d}{dx}(e^u) = e^u \frac{d}{dx}(u)$$

5 Find the following.

(a) $\int e^{5x}dx$

(b) $\int 8xe^{3x^2}\, dx$

(c) $\int 2x^3 e^{x^4-1}\, dx$

Answers:

(a) $\dfrac{1}{5}e^{5x} + C$

(b) $\dfrac{4}{3}e^{3x^2} + C$

(c) $\dfrac{1}{2}e^{x^4-1} + C$

For example, if $u = x^2$ then $\dfrac{d}{dx}(u) = \dfrac{d}{dx}(x^2) = 2x$, and

$$\frac{d}{dx}(e^{x^2}) = e^{x^2} \cdot 2x.$$

Working backwards, if $u = x^2$, then $du = 2x\, dx$, so

$$\int e^{x^2} \cdot 2x\, dx = \int e^u\, du = e^u + C$$
$$= e^{x^2} + C. \quad \blacktriangleleft$$

▶**EXAMPLE 4** Find the following.

(a) $\int e^{-11x}dx$

Choose $u = -11x$, so $du = -11dx$. Multiply the integral by $(-1/11)(-11)$, and use the rule for $\int e^u du$.

$$\int e^{-11x}x = -\frac{1}{11} \cdot -11 \int e^{-11x}dx$$
$$= -\frac{1}{11} \int e^{-11x}(-11dx)$$
$$= -\frac{1}{11} \int e^u du$$
$$= -\frac{1}{11} e^u + C$$
$$= -\frac{1}{11} e^{-11x} + C$$

(b) $\int x^2 \cdot e^{x^3}\, dx$

Let $u = x^3$, the exponent on e. Then $du = 3x^2\, dx$, and $(1/3)du = x^2\, dx$,

$$\int x^2 \cdot e^{x^3}\, dx = \int e^{x^3}(x^2\, dx)$$
$$= \int e^u\left(\frac{1}{3}\, du\right) \qquad \textbf{Substitute}$$
$$= \frac{1}{3} \int e^u\, du \qquad \textbf{Constant multiple rule}$$
$$= \frac{1}{3}e^u + C \qquad \textbf{Integrate}$$
$$= \frac{1}{3}e^{x^3} + C. \qquad \textbf{Substitute} \quad \blacktriangleleft \quad \boxed{5}$$

6 Find the following.

(a) $\displaystyle \int \frac{4\,dx}{x-3}$

(b) $\displaystyle \int \frac{(3x^2 + 8)\,dx}{x^3 + 8x + 5}$

Answers:

(a) $4 \ln|x-3| + C$

(b) $\ln|x^3 + 8x + 5| + C$

7 Find

$$\int x(x+1)^{2/3}\,dx.$$

Answer:

$$\frac{3}{8}(x+1)^{8/3} - \frac{3}{5}(x+1)^{5/3} + C$$

Recall that the antiderivative of $f(x) = 1/x$ is $\ln|x|$. The next example uses $\int x^{-1}\,dx = \ln|x| + C$, and the method of substitution.

▶ **EXAMPLE 5**   Find the following.

(a) $\displaystyle \int \frac{dx}{9x+6}$

Choose $u = 9x + 6$, so $du = 9\,dx$. Multiply by $(1/9)(9)$.

$$\int \frac{dx}{9x+6} = \frac{1}{9} \cdot 9 \int \frac{dx}{9x+6} = \frac{1}{9} \int \frac{1}{9x+6}(9\,dx)$$

$$= \frac{1}{9} \int \frac{1}{u}\,du = \frac{1}{9} \ln|u| + C = \frac{1}{9} \ln|9x+6| + C$$

(b) $\displaystyle \int \frac{(2x-3)\,dx}{x^2-3x}$.

Let $u = x^2 - 3x$, so that $du = (2x - 3)\,dx$. Then

$$\int \frac{(2x-3)\,dx}{x^2-3x} = \int \frac{du}{u} = \ln|u| + C = \ln|x^2 - 3x| + C. \quad ◀ \quad \boxed{6}$$

▶ **EXAMPLE 6**   Find $\displaystyle \int x\sqrt{1-x}\,dx$.

Let $u = 1 - x$. Then $x = 1 - u$ and $dx = -du$. Now substitute.

$$\int x\sqrt{1-x}\,dx = \int (1-u)\sqrt{u}(-du) = \int (u-1)u^{1/2}\,du$$

$$= \int (u^{3/2} - u^{1/2})\,du = \frac{2}{5}u^{5/2} - \frac{2}{3}u^{3/2} + C$$

$$= \frac{2}{5}(1-x)^{5/2} - \frac{2}{3}(1-x)^{3/2} + C. \quad ◀ \quad \boxed{7}$$

The substitution method is useful if the integral can be written in one of the following forms, where $u(x)$ is some function of x.

Substitution Method

Let $u(x)$ be some function of x.

Form of the Integral	*Form of the Antiderivative*
1. $\displaystyle \int [u(x)]^n \cdot u'(x)\,dx,\ n \neq -1$	$\displaystyle \frac{[u(x)]^{n+1}}{n+1} + C$
2. $\displaystyle \int e^{u(x)} \cdot u'(x)\,dx$	$e^{u(x)} + C$
3. $\displaystyle \int \frac{u'(x)\,dx}{u(x)}$	$\ln|u(x)| + C$

8 Sales of a new company, in thousands, are changing at a rate of

$$S'(t) = 27e^{-3t},$$

where t is time in months. Ten units were sold when $t = 0$. Find the sales function.

Answer:
$S(x) = 19 - 9e^{-3t}$

▶**EXAMPLE 7**   The research department for a hardware chain has determined that at one store the marginal price of x boxes per week of a particular type of nails is

$$p'(x) = \frac{-4000}{(2x + 15)^3}.$$

Find the demand equation if the weekly demand for this type of nails is 10 boxes when the price of a box of nails is $4.

To find the demand function $p(x)$, first integrate $p'(x)$ as follows.

$$p(x) = \int p'(x)\, dx$$

$$= \int \frac{-4000}{(2x + 15)^3}\, dx$$

Let $u = 2x + 15$. Then $du = 2\, dx$, and

$$p(x) = -2000 \int (2x + 15)^{-3}\, 2\, dx$$

$$= -2000 \int u^{-3}\, du \qquad \textbf{Substitute}$$

$$= (-2000) \frac{u^{-2}}{-2} + C \qquad \textbf{Integrate}$$

$$= \frac{1000}{u^2} + C \qquad \textbf{Simplify}$$

$$p(x) = \frac{1000}{(2x + 15)^2} + C. \qquad \textbf{Substitute} \qquad \textbf{(1)}$$

Find the value of C by using the given information that $p = 4$ when $x = 10$.

$$4 = \frac{1000}{(2 \cdot 10 + 15)^2} + C$$

$$4 = \frac{1000}{35^2} + C$$

$$4 = .82 + C$$

$$3.18 = C$$

Replacing C with 3.18 in equation (1) gives the demand function,

$$p(x) = \frac{1000}{(2x + 15)^2} + 3.18 \quad ◀ \; \boxed{8}$$

▶**EXAMPLE 8**   To determine the top 100 popular songs of each year since 1956, Jim Quirin and Barry Cohen developed a function that represents the rate of change on the charts of *Billboard* magazine required for a song to earn a "star" on the *Billboard* "Hot 100" survey.* They developed the function

$$f(x) = \frac{A}{B + x},$$

where $f(x)$ represents the rate of change in position on the charts, x is the position on the "Hot 100" survey, and A and B are appropriate constants. The function

$$F(x) = \int f(x) \, dx$$

is defined as the "Popularity Index." Find $F(x)$.

Integrating $f(x)$ gives

$$F(x) = \int f(x) \, dx$$

$$= \int \frac{A}{B + x} \, dx$$

$$= A \int \frac{1}{B + x} \, dx \qquad \text{Constant multiple rule}$$

Let $u = B + x$, so that $du = dx$. Then

$$F(x) = A \int \frac{1}{u} \, du = A \ln u + C$$

$$= A \ln(B + x) + C.$$

(Absolute value is not necessary, since $B + x$ is always positive here.) ◀

13.2 EXERCISES

1. Integration by substitution is related to what differention method? What type of integrand suggests using integration by substitution?

2. For each of the following integrals, decide what factor should be u. Then find du.

(a) $\int (3x^2 - 5)^4 \, 2x \, dx$ **(b)** $\int \sqrt{1 - x} \, dx$

(c) $\int \frac{x^2}{2x^3 + 1} \, dx$ **(d)** $\int (8x - 8)(4x^2 - 8x) \, dx$

Use substitution to find the following indefinite integrals. (See Examples 1–6.)

3. $\int 3(12x - 1)^2 \, dx$

4. $\int 5(4 - 2t)^3 \, dt$

5. $\int \frac{2}{(3t + 1)^2} \, dt$

---

*Formula for the "Popularity Index" from *Chartmasters' Rock 100,* Fourth Edition, by Jim Quirin and Barry Cohen. Copyright © 1987 by Chartmasters. Reprinted by permission.

6. $\int \dfrac{4}{\sqrt{5u-1}}\, du$

7. $\int \dfrac{x+1}{(x^2+2x-4)^{3/2}}\, dx$

8. $\int \dfrac{3x^2-2}{(2x^3-4x)^{5/2}}\, dx$

9. $\int r^2 \sqrt{r^3+3}\, dr$

10. $\int y^3 \sqrt{y^4-6}\, dy$

11. $\int (-3e^{5k})\, dk$

12. $\int (-2e^{-3z})\, dz$

13. $\int 4w^2 e^{2w^3}\, dw$

14. $\int 5ze^{-z^2}\, dz$

15. $\int (2-t)e^{4t-t^2}\, dt$

16. $\int (3-x^2)e^{9x-x^3}\, dx$

17. $\int \dfrac{e^{\sqrt{y}}}{\sqrt{y}}\, dy$

18. $\int \dfrac{e^{1/z^2}}{z^3}\, dz$

19. $\int \dfrac{-4}{2+5x}\, dx$

20. $\int \dfrac{7}{3-4x}\, dx$

21. $\int \dfrac{e^{2t}}{e^{2t}+1}\, dt$

22. $\int \dfrac{e^{w+1}}{2-e^{w+1}}\, dw$

23. $\int \dfrac{x+2}{(2x^2+8x)^3}\, dx$

24. $\int \dfrac{4y-2}{(y^2-y)^4}\, dy$

25. $\int \left(\dfrac{1}{r}+r\right)\left(1-\dfrac{1}{r^2}\right) dr$

26. $\int \left(\dfrac{2}{a}-a\right)\left(\dfrac{-2}{a^2}-1\right) da$

27. $\int \dfrac{x^2+1}{(x^3+3x)^{2/3}}\, dx$

28. $\int \dfrac{B^3-1}{(2B^4-8B)^{3/2}}\, dB$

29. $\int \dfrac{u}{\sqrt{u-1}}\, du$

30. $\int \dfrac{2x}{(x+5)^6}\, dx$

31. $\int \dfrac{x+2}{3x^2+12x+8}\, dx$

32. $\int \dfrac{x^2}{x^3+3}\, dx$

33. $\int t\sqrt{5t-1}\, dt$

34. $\int 4r\sqrt{8-r}\, dr$

35. $\int 2x(x^2+1)^3\, dx$

36. $\int y^2(y^3-4)^3\, dy$

37. $\int (\sqrt{x^2+12x})(x+6)\, dx$

38. $\int (\sqrt{x^2-6x})(x-3)\, dx$

39. $\int \dfrac{(1+\ln x)^2}{x}\, dx$

40. $\int \dfrac{1}{x(\ln x)}\, dx$

---

41. Management A company has found that the marginal cost of a new production line (in thousands) is

$$C'(x) = -100(x+10)^{-2},$$

where x is the number of years the line is in use.

(a) Find the total cost function for the production line. The fixed cost is 10 thousand dollars.

(b) The company will add the new line if the total cost is reduced to $2000 within 5 years. Should they add the new line?

42. Management The rate of growth of the profit (in millions of dollars) from a new technology is approximated by

$$P'(x) = xe^{-x^2},$$

where x represents time measured in years. The total profit in the third year that the new technology is in operation is $10,000.

(a) Find the total profit function.

(b) What happens to the total amount of profit in the long run?

43. Social Science Canadians who live near the United States border have been shopping in the United States at an increasing rate since 1992. They cross the border to shop because large savings are possible on groceries, gasoline, cigarettes, and children's clothes. The rate of change in the number of Canadian cross-border shoppers (in millions) is given by

$$S'(t) = 4.4e^{.16t},$$

where t is the number of years since 1992.

(a) Find $S(t)$ if there were 27.3 million cross-border shoppers in 1992 (year 0).

(b) In how many years will the number of Canadian cross-border shoppers double?

44. Management The rate of change of U.S. investment in Mexico (in billions of dollars) since 1987 is given by $f(x) = 1.52e^{.11x}$, where x is the number of years since 1987. In 1987, U.S. companies invested $13.8 billion in Mexico.

(a) Find a function that gives the amount invested in year x.

(b) At this rate, when would the amount invested in Mexico be double the 1987 investment?

13.3 AREA AND THE DEFINITE INTEGRAL

If the rate of change of annual maintenance charges is known, how can the total amount paid for maintenance over a 10-year period be estimated? This section introduces a method for answering such questions.

Figure 13.2 shows the region bounded by the lines $x = 0$, the x-axis, and the graph of

$$f(x) = \sqrt{4 - x^2}.$$

A very rough approximation of the area of this region can be found by using two rectangles, as in Figure 13.3. The height of the rectangle on the left is $f(0) = 2$ and the height of the rectangle on the right is $f(1) = \sqrt{3}$. The width of each rectangle is 1, making the total area of the two rectangles

$$1 \cdot f(0) + 1 \cdot f(1) = 2 + \sqrt{3} \approx 3.7321 \text{ square units.}$$

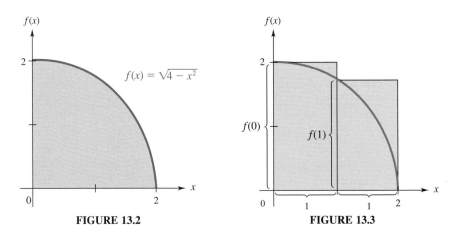

FIGURE 13.2   **FIGURE 13.3**

As Figure 13.3 suggests, this approximation is greater than the actual area. To improve the accuracy of the approximation, we could divide the interval from

1 Calculate the sum

$$\frac{1}{4} \cdot f(0) + \frac{1}{4} \cdot f\left(\frac{1}{4}\right) + \cdots +$$

$$\frac{1}{4} \cdot f\left(\frac{7}{4}\right), \text{ using this information.}$$

x	$f(x)$
0	2
1/4	1.98431
1/2	1.93649
3/4	1.85405
1	1.73205
5/4	1.56125
3/2	1.32288
7/4	.96825

Answer:
3.33982 square units

$x = 0$ to $x = 2$ into four equal parts, each of width $1/2$, as shown in Figure 13.4. As before, the height of each rectangle is given by the value of f at the left-hand side of the rectangle, and its area is the width, $1/2$, multiplied by the height. The total area of the four rectangles is

$$\frac{1}{2} \cdot f(0) + \frac{1}{2} \cdot f\left(\frac{1}{2}\right) + \frac{1}{2} \cdot f(1) + \frac{1}{2} \cdot f\left(\frac{3}{2}\right)$$

$$= \frac{1}{2}(2) + \frac{1}{2}\left(\frac{\sqrt{15}}{2}\right) + \frac{1}{2}(\sqrt{3}) + \frac{1}{2}\left(\frac{\sqrt{7}}{2}\right)$$

$$= 1 + \frac{\sqrt{15}}{4} + \frac{\sqrt{3}}{2} + \frac{\sqrt{7}}{4} \approx 3.4957 \text{ square units.}$$

This approximation looks better, but it is still greater than the actual area desired. To improve the approximation, divide the interval from $x = 0$ to $x = 2$ into eight parts with equal widths of $1/4$. (See Figure 13.5) The total area of all these rectangles is

$$\frac{1}{4} \cdot f(0) + \frac{1}{4} \cdot f\left(\frac{1}{4}\right) + \frac{1}{4} \cdot f\left(\frac{1}{2}\right) + \frac{1}{4} \cdot f\left(\frac{3}{4}\right) + \frac{1}{4} \cdot f(1) + \frac{1}{4} \cdot f\left(\frac{5}{4}\right)$$

$$+ \frac{1}{4} \cdot f\left(\frac{3}{2}\right) + \frac{1}{4} \cdot f\left(\frac{7}{4}\right). \quad \blacksquare$$

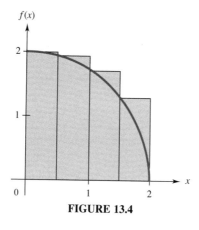

FIGURE 13.4

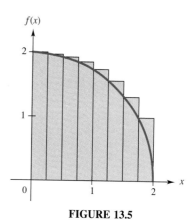

FIGURE 13.5

This process of approximating the area under a curve by using more and more rectangles to get a better and better approximation can be generalized. To do this, divide the interval from $x = 0$ to $x = 2$ into n equal parts. Each of these n intervals has width

$$\frac{2 - 0}{n} = \frac{2}{n},$$

n	Area
2	3.7321
4	3.4957
8	3.3398
10	3.3045
20	3.2285
50	3.1783
100	3.1512
500	3.1455

so each rectangle has width $2/n$ and height determined by the function-value at the left side of the rectangle. A computer was used to find approximations to the area for several values of n given in the table at the side.

The areas in the last column in the table are approximations of the area under the curve, above the x-axis, and between the lines $x = 0$ and $x = 2$. As n becomes larger and larger, the approximation is better and better, getting closer to the actual area. In this example, the exact area can be found by a formula from plane geometry. Write the given function as

$$y = \sqrt{4 - x^2},$$

then square both sides to get

$$y^2 = 4 - x^2$$
$$x^2 + y^2 = 4,$$

the equation of a circle centered at the origin with radius 2. The region in Figure 13.2 is the quarter of this circle that lies in the first quadrant. The actual area of this region is one quarter of the area of the entire circle, or

$$\frac{1}{4}\pi(2)^2 = \pi \approx 3.1416.$$

As the number of rectangles increases without bound, the sum of the areas of these rectangles gets closer and closer to the actual area of the region, π. This can be written as

$$\lim_{n\to\infty} (\text{sum of areas of } n \text{ rectangles}) = \pi.$$

(The value of π was originally approximated by a process similar to this.)*

Theoretically, this approach could be used to find the area of any region bounded by the x-axis, the lines $x = a$, and $x = b$, and the graph of a function $y = f(x)$. At this point, we need some new notation and terminology to write sums concisely. We will indicate addition (or summation) by using the Greek letter sigma, Σ, as shown in the next example.

▶**EXAMPLE 1** Find the following sums.

(a) $\sum_{i=1}^{5} i$

Replace i in turn with the integers 1 through 5 and add the resulting terms.

---

*The number π is the ratio of the circumference of a circle to its diameter. It is an example of an *irrational number* and as such it cannot be expressed as a terminating or repeating decimal. Many approximations have been used for π over the years. A passage in the Bible (I Kings 7:23) indicates a value of 3. The Egyptians used the value 3.16, and Archimedes showed that its value must be between 22/7 and 223/71. A Hindu writer, Brahmagupta, used $\sqrt{10}$ as its value in the seventh century. The search for the digits of π has continued throughout the years. Two mathematicians at Columbia University have computed the value of π to more than a billion decimal places!

2 Find each sum.

(a) $\displaystyle\sum_{i=0}^{6} i$

(b) $\displaystyle\sum_{i=2}^{7} (3i + 1)$

(c) $\displaystyle\sum_{i=1}^{4} (5x_i - 1)$, if $x_1 = 9$,
$x_2 = 8$, $x_3 = 2$, $x_4 = 7$

(d) $\displaystyle\sum_{i=3}^{6} f(x_i)\, \Delta x$, if
$f(x) = 3x^2 - 1$, $x_3 = 2$,
$x_4 = 5$, $x_5 = 1$, $x_6 = 5$, and
$\Delta x = .2$

Answers:

(a) 21

(b) 87

(c) 126

(d) 32.2

$$\sum_{i=1}^{5} i = 1 + 2 + 3 + 4 + 5 = 15$$

(b) $\displaystyle\sum_{i=1}^{4} a_i = a_1 + a_2 + a_3 + a_4$

(c) $\displaystyle\sum_{i=1}^{3} (6x_i - 2)$, if $x_1 = 2$, $x_2 = 4$, $x_3 = 6$

Letting $i = 1$, 2, and 3, respectively, gives

$$\sum_{i=1}^{3} (6x_i - 2) = (6x_1 - 2) + (6x_2 - 2) + (6x_3 - 2).$$

Now substitute the given values for x_1, x_2, and x_3.

$$\sum_{i=1}^{3} (6x_i - 2) = (6 \cdot 2 - 2) + (6 \cdot 4 - 2) + (6 \cdot 6 - 2)$$
$$= 10 + 22 + 34$$
$$= 66$$

(d) $\displaystyle\sum_{i=1}^{4} f(x_i)\, \Delta x$ if $f(x) = x^2$, $x_1 = 0$, $x_2 = 2$, $x_3 = 4$, $x_4 = 6$, and $\Delta x = 2$

$$\sum_{i=1}^{4} f(x_i)\, \Delta x = f(x_1)\, \Delta x + f(x_2)\, \Delta x + f(x_3)\, \Delta x + f(x_4)\, \Delta x$$
$$= x_1^2\, \Delta x + x_2^2\, \Delta x + x_3^2\, \Delta x + x_4^2\, \Delta x$$
$$= 0^2(2) + 2^2(2) + 4^2(2) + 6^2(2)$$
$$= 0 + 8 + 32 + 72 = 112 \quad \blacktriangleleft \quad \boxed{2}$$

Now we can generalize the method used above to find the area bounded by the curve $y = f(x)$, the x-axis, and the vertical lines $x = a$ and $x = b$, as shown in Figure 13.6. To approximate this area, we could divide the region under the curve first into ten rectangles (Figure 13.6(a)) and then into twenty rectangles

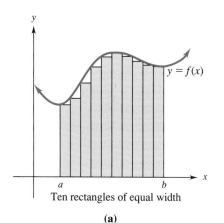

$y = f(x)$

a ∙∙∙ b

Ten rectangles of equal width

(a)

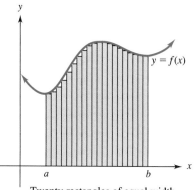

$y = f(x)$

a ∙∙∙ b

Twenty rectangles of equal width

(b)

FIGURE 13.6

(Figure 13.6 (b)). The sums of the areas of the rectangles give approximations to the area under the curve.

To get a number that can be defined as the *exact* area, begin by dividing the interval from a to b into n pieces of equal width, using each of these n pieces as the base of a rectangle. (See Figure 13.7.) The endpoints of the n intervals are labeled $x_1, x_2, x_3, \ldots, x_{n+1}$, where $a = x_1$ and $b = x_{n+1}$. In the graph of Figure 13.7, the symbol Δx is used to represent the width of each of the intervals. The darker rectangle is an arbitrary rectangle called the ith rectangle. Its area is the product of its length and width. The width of the ith rectangle is Δx and the length of the ith rectangle is given by the height $f(x_i)$, so

$$\text{Area of } i\text{th rectangle} = f(x_i) \cdot \Delta x.$$

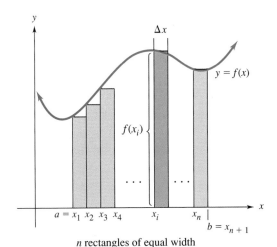

n rectangles of equal width

FIGURE 13.7

The total area under the curve is approximated by the sum of the areas of all n of the rectangles. Using the Σ symbol, the approximation to the total area becomes

$$\text{Area of all } n \text{ rectangles} = \sum_{i=1}^{n} f(x_i) \cdot \Delta x.$$

The exact area is defined to be the limit of this sum (if the limit exists) as the number of rectangles increases without bound.

$$\text{Exact area} = \lim_{n \to \infty} \sum_{i=1}^{n} f(x_i)\, \Delta x$$

This limit is called the *definite integral* of $f(x)$ from a to b and is written as follows.

The Definite Integral

If f is continuous on the interval $[a, b]$, the **definite integral** of f from a to b is given by

$$\int_a^b f(x)\,dx = \lim_{n\to\infty} \sum_{i=1}^{n} f(x_i)\,\Delta x,$$

provided the limit exists, where $\Delta x = (b - a)/n$ and x_i is *any* value of x in the ith interval.

As indicated in this definition, although the left endpoint of the ith interval has been used to find the height of the ith rectangle, any number in the ith interval can be used. (A more general definition is possible in which the rectangles do not necessarily all have the same width.) The b above the integral sign is called the **upper limit** of integration, and the a is the **lower limit** of integration. This use of the word "limit" has nothing to do with the limit of the sum; it refers to the limits, or boundaries, on x.

In the example at the beginning of this section, the area bounded by the x-axis, the curve $y = \sqrt{4 - x^2}$, the lines $x = 0$ and $x = 2$ could be written as the definite integral

$$\int_0^2 \sqrt{4 - x^2}\,dx = \pi.$$

The next section will show how antiderivatives are used in finding the definite integral and, thus, the area under a curve.

Caution   Notice that unlike the indefinite integral, which is a set of *functions*, the definite integral represents a *number*. Keep in mind that finding the definite integral of a function can be thought of as a mathematical process that gives the sum of an infinite number of individual parts (within certain limits). The definite integral represents area only if the function involved is *nonnegative* ($f(x) \ge 0$) at every x-value in the interval $[a, b]$. There are many other interpretations of the definite integral, and all of them involve this idea of approximation by appropriate sums.

▶**EXAMPLE 2**   Approximate $\int_0^4 2x\,dx$, the area of the region under the graph of $f(x) = 2x$, above the x-axis, and between $x = 0$ and $x = 4$, by using four rectangles of equal width whose heights are the values of the function at the midpoint of each rectangle.

We want to find the area of the shaded region in Figure 13.8. The heights of the four rectangles given by $f(x_i)$ for $i = 1, 2, 3,$ and 4 are as follows.

3 Divide the region of Figure 13.8 into eight rectangles of equal width whose heights are the values of the function at the midpoint of each rectangle.

(a) Complete this table.

i	x_i	$f(x)$
1	.25	
2	.75	
3	1.25	
4	1.75	
5		
6		
7		
8		

(b) Use the results from the table to approximate $\int_0^4 2x \, dx$.

Answers:

(a)

i	x_i	$f(x)$
1	.25	.50
2	.75	1.50
3	1.25	2.50
4	1.75	3.50
5	2.25	4.50
6	2.75	5.50
7	3.25	6.50
8	3.75	7.50

(b) 16

i	x_i	$f(x_i)$
1	$x_1 = \ .5$	$f(.5) = 1.0$
2	$x_2 = 1.5$	$f(1.5) = 3.0$
3	$x_3 = 2.5$	$f(2.5) = 5.0$
4	$x_4 = 3.5$	$f(3.5) = 7.0$

The width of each rectangle is $\Delta x = \dfrac{4 - 0}{4} = 1$. The sum of the areas of the four rectangles is

$$\sum_{i=1}^{4} f(x_i) \, \Delta x = f(x_1) \, \Delta x + f(x_2) \, \Delta x + f(x_3) \, \Delta x + f(x_4) \, \Delta x$$

$$= f(.5) \, \Delta x + f(1.5) \, \Delta x + f(2.5) \, \Delta x + f(3.5) \, \Delta x$$

$$= (1)(1) + (3)(1) + (5)(1) + (7)(1)$$

$$= 16. \quad \boxed{3}$$

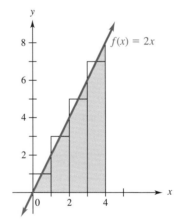

FIGURE 13.8

Using the formula for the area of a triangle, $A = (1/2)bh$, with b, the length of the base, equal to 4 and h, the height, equal to 8, gives

$$A = \frac{1}{2}bh = \frac{1}{2}(4)(8) = 16.$$

the exact value of the area. The approximation equals the exact area in this case because our use of the midpoints of each subinterval distributed the error evenly above and below the graph. ◄

FOR GRAPHERS

Most graphing calculators have keys that will produce the (approximate) value of a definite integral. See your instruction manual for details.

We now wish to connect the area under the graph of a function with rates of change. The basic idea can be seen in a simple example. Suppose a car travels along a straight road at a constant speed of 50 mph. Then the function $v(t) = 50$ is the function that gives the speed of the car at time t. The graph of this function is the straight line shown in Figure 13.9 (a).

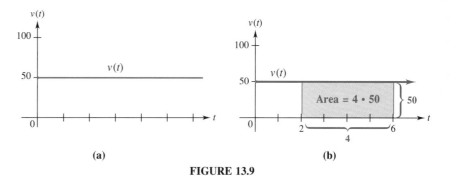

FIGURE 13.9

How far does the car travel from time $t = 2$ to $t = 6$? Since this is a 4-hour period, the answer, of course, is $4 \cdot 50 = 200$ miles. Note that this is precisely the area under the graph of the speed function $v(t)$ from $t = 2$ to $t = 6$, as shown in Figure 13.9(b) above.

As we saw in Chapter 12, the speed (or velocity) function $v(t)$ is just the rate of change of distance with respect to time, that is, the rate of change of the distance function $s(t)$ (which gives the position of the car at time t.) The distance traveled from time $t = a$ to $t = b$ is the total change in the function $s(t)$ from $t = a$ to $t = b$. In this example, therefore, the total change in $s(t)$ as t goes from $t = a$ to $t = b$ is the area under $v(t)$ from $t = a$ to $t = b$, that is, the definite integral $\int_a^b v(t)\, dt$. A more complicated argument (that is omitted here) works in the general case and leads to this useful definition.

Total Change in *F*(*x*)

Let f be a function such that f is continuous on $[a, b]$ and $f(x) \geq 0$ for all x in $[a, b]$. If $f(x)$ is the rate of change of the function $F(x)$ for x in $[a, b]$, then the **total change in *F*(*x*)** as x goes from a to b is given by

$$\int_a^b f(x)\, dx.$$

In other words, the total change in a quantity can be found from the function that gives the rate of change of the quantity, using the same methods used to approximate the area under a curve.

4 Use Figure 13.10 to estimate the maintenance charge during

(a) the first 6 years of the machine's life;

(b) the first 8 years.

Answers:

(a) $7850

(b) $14,250

▶ **EXAMPLE 3** Figure 13.10 shows the rate of change of the annual mainte-nance charges for a certain machine. Approximate the total maintenance charges over the 10-year life of the machine by using rectangles, dividing the interval from 0 to 10 into ten equal subdivisions. Each rectangle has width 1; if we use the left endpoint of each rectangle to determine the height of the rectan-gle, the approximation becomes

$$1 \cdot 0 + 1 \cdot 500 + 1 \cdot 750 + 1 \cdot 1800 + 1 \cdot 3000$$
$$+ 1 \cdot 3000 + 1 \cdot 3400 + 1 \cdot 4200 + 1 \cdot 5200 = 23,650.$$

About $23,650 will be spent on maintenance over the 10-year life of the ma-chine. ◀ **4**

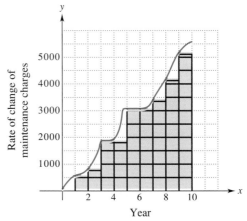

FIGURE 13.10

13.3 EXERCISES

Evaluate the following sums. (See Example 1.)

1. $\sum_{i=1}^{4} 6i$

2. $\sum_{i=1}^{5} (-4i)$

3. $\sum_{i=1}^{6} (5i - 2)$

4. $\sum_{i=1}^{8} (3i + 4)$

5. Find $\sum_{i=1}^{4} x_i$ if $x_1 = -3$, $x_2 = 1$, $x_3 = 5$, $x_4 = 10$.

6. Find $\sum_{i=1}^{5} x_i$ if $x_1 = 12$, $x_2 = 16$, $x_3 = 20$, $x_4 = 25$, $x_5 = 30$.

7. Find $\sum_{i=1}^{3} f(x_i)$ if $f(x) = 2x + 4$, $x_1 = 4$, $x_2 = 6$, $x_3 = 7$.

8. Find $\sum_{i=1}^{4} f(x_i)$ if $f(x) = .5x^2$, $x_1 = -4$, $x_2 = 0$, $x_3 = 4$, $x_4 = 8$.

9. Find $\sum_{i=1}^{4} f(x_i) \, \Delta x$ if $f(x) = \dfrac{1}{2 + x}$, $x_1 = 0$, $x_2 = 2$, $x_3 = 4$, $x_4 = 6$, $\Delta x = 2$.

10. Find $\sum_{i=1}^{4} f(x_i) \Delta x$ if $f(x) = \dfrac{1}{x}$, $x_1 = \dfrac{1}{2}$, $x_2 = 1$, $x_3 = \dfrac{3}{2}$, $x_4 = 2$, $\Delta x = \dfrac{1}{2}$.

11. The sum in Exercise 9 approximates a definite integral using rectangles. The height of each rectangle is given by the value of the function at the left endpoint. Write the definite integral that the sum approximates.

12. The sum in Exercise 10 approximates a definite integral using rectangles. The height of each rectangle is given by the value of the function at the left endpoint. Write the definite integral that the sum approximates.

Approximate the area under each given curve and above the x-axis on the given interval by using two rectangles. Let the height of the rectangle be given by the value of the function at the left side of the rectangle. Then repeat the process and approximate the area with four rectangles. (See Example 2.)

13. $f(x) = 2x + 5; [0, 4]$ **14.** $f(x) = 4 - x; [0, 4]$ **15.** $f(x) = 4 - x^2; [-2, 2]$ **16.** $f(x) = x^2 + 1; [-2, 2]$

17. $f(x) = e^x - 1; [0, 4]$ **18.** $f(x) = e^x + 1; [-2, 2]$ **19.** $f(x) = \dfrac{1}{x}; [1, 5]$ **20.** $f(x) = \dfrac{2}{x}; [1, 9]$

---

21. Explain the difference between an indefinite integral and a definite integral.

22. Complete the following statement.

$$\int_0^3 (x^2 + 2) \, dx = \lim_{n \to \infty,} \underline{\qquad}, \text{ where } \Delta x = \underline{\qquad},$$

and x_i is $\underline{\qquad}$.

Work the following exercises. (See Example 2.)

23. Consider the region below $f(x) = x/2$, above the x-axis, between $x = 0$ and $x = 4$. Let x_i be the left endpoint of the ith subinterval.
  (a) Aproximate the area of the region using four rectangles.
  (b) Approximate the area of the region using eight rectangles.
  (c) Find $\int_0^4 f(x) \, dx$ by using the formula for the area of a triangle.

24. Find $\int_0^5 (5 - x) \, dx$ by using the formula for the area of a triangle.

Estimate the area under the curve by summing the area of rectangles. Let the function value at the left side of each rectangle give the height of the rectangle. (See Example 3.)

25. Social Science   The graph shows U.S. oil production and consumption rates in millions of barrels per day. The rates for the years beyond 1991 are projected using the policy in place in 1990 (labeled "Current policy base" in the graph) and using a new policy proposed by the Bush administration in 1991 (labeled "With strategy" in the graph).* Estimate the amount of oil produced between 1990 and 2010 using the policy in place in 1990. Use rectangles with widths of 2 yr. (*Hint:* There are 365 days in a year.)

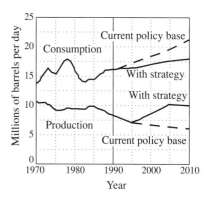

---

*Graph, "Uncle Sam's Energy Strategy" from *National Energy Strategy,* February 1991. United States Department of Energy, Washington D.C.

26. Social Science Estimate the amount of oil produced between 1990 and 2010 using the proposed policy. (See Exercise 25.) Use rectangles with widths of 2 yr.

27. Management The graph shows the average manufacturing hourly wage in the United States and Canada for 1987–1991.* All figures are in U.S. dollars. Assume that the average employee works 2000 hours per year. Estimate the total amount earned by an average U.S. worker during the four-year period from the beginning of 1987 to the beginning of 1991. Use rectangles with widths of 1 yr.

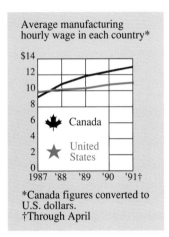

Average manufacturing hourly wage in each country*

Canada

United States

*Canada figures converted to U.S. dollars.
†Through April

28. Management Estimate the total amount earned (in U.S. dollars) by an average Canadian worker during the four-year period from the beginning of 1987 to the beginning of 1991. Use rectangles with widths of 1 yr. (See Exercise 27.)

29. Natural Science The graph below shows the rate of inhalation of oxygen by a person riding a bicycle very rapidly for 10 minutes. Estimate the total volume of oxygen inhaled in the first 20 minutes after the beginning of the ride. Use rectangles of width 1 minute.

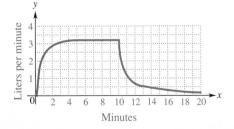

Physical Science *The next two graphs are from Road and Track magazine.** The curve shows the velocity at time t, in seconds, when the car accelerates from a dead stop. To find the total distance traveled by the car in reaching 100 miles per hour, we must estimate the definite integral*

$$\int_0^T v(t)\,dt,$$

where T represents the number of seconds it takes for the car to reach 100 mph.

Use the graphs to estimate this distance by adding the areas of rectangles with widths of 5 seconds. The last rectangle has a width of 3. To adjust your answer to miles per hour, divide by 3600 (the number of seconds in an hour). You then have the number of miles that the car traveled in reaching 100 mph. Finally, multiply by 5280 feet per mile to convert the answers to feet.

30. Estimate the distance traveled by the Porsche 928, using the graph below.

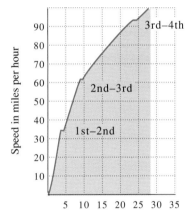

Elapsed time in seconds

* "Comparing Wages" from *The New York Times.* 1991. Copyright © 1991 by The New York Times Company. Reprinted by permission.

**From *Road & Track,* April and May, 1978. Reprinted with permission of *Road & Track.*

31. Estimate the distance traveled by the BMW 733i, using the graph below.

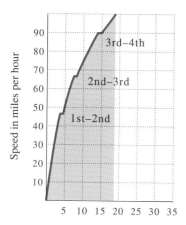

Physical Science *The graphs* in the right column show the typical heat gain, in BTUs per hour per square foot, for a window facing east and one facing south, with plain glass and with a black ShadeScreen. Estimate the total heat gain per square foot by summing the areas of the rectangles. Use rectangles with widths of 2 hr, and let the function value at the midpoint of the rectangle give the height of the rectangle.*

32. (a) Estimate the total heat gain per square foot for a plain glass window facing east.
  (b) Estimate the total heat gain per square foot for a window facing east with a ShadeScreen.

33. (a) Estimate the total heat gain per square foot for a plain glass window facing south.
  (b) Estimate the total heat gain per square foot for a window facing south with a ShadeScreen.

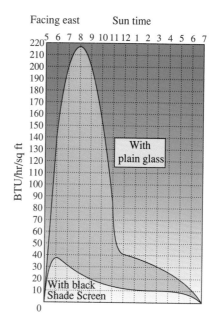

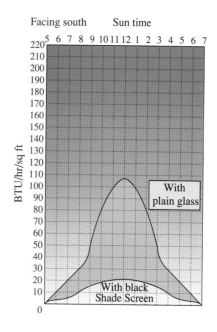

*Two graphs, "Facing east, Sun time" and "Facing south, Sun time" from Phifer Wire Products. Reprinted by permission.

For each of the following, use a grapher to approximate the definite integral that gives the area between the x-axis and the graph of the function on the indicated interval.

34. $f(x) = x \ln x$; $[1, 5]$

35. $f(x) = x^2 e^{-x}$; $[-1, 3]$

36. $f(x) = \dfrac{\ln x}{x}$; $[1, 5]$

13.4 THE FUNDAMENTAL THEOREM OF CALCULUS

In this section we develop the connection between definite integrals and antiderivatives. In the last section we saw that if $f(x) \geq 0$ for all x in $[a, b]$ and if $f(x)$ is the rate of change of the function $F(x)$, then the definite integral

$$\int_a^b f(x)\, dx \text{ is the total change of } F(x) \qquad \text{(*)}$$

as x changes from a to b.

Now to say that $f(x)$ is the rate of change of $F(x)$ means that $f(x)$ is the derivative of $F(x)$, or equivalently, $F(x)$ is an antiderivative of $f(x)$. The total change in $F(x)$ from $x = a$ to $x = b$ is the difference $F(b) - F(a)$. So we can restate (*) by saying that

$$\int_a^b f(x)\, dx = F(b) - F(a).$$

It can be proved that this relationship holds in more general situations (where $f(x)$ may not always be ≥ 0).

Fundamental Theorem of Calculus

Suppose f is continuous on the interval $[a, b]$ and F is *any* antiderivative of f. Then

$$\int_a^b f(x)\, dx = F(b) - F(a).$$

1 Let $C(x) = x^3 + 4x^2 - x + 3$. Find the following.

(a) $C(x)\Big|_1^5$

(b) $C(x)\Big|_3^4$

Answers:

(a) 216

(b) 64

Caution It is important to note that the fundamental theorem does not require $f(x) > 0$. The condition $f(x) > 0$ is necessary only when using the fundamental theorem to find area. Also, note that the fundamental theorem does not *define* the definite integral; it just provides a method for evaluating it.

When evaluating definite integrals, the symbol $F(x)\Big|_a^b$ is used to denote the number $F(b) - F(a)$. For example, if $F(x) = x^4$, then $x^4\Big|_1^2$ means $F(2) - F(1) = 2^4 - 1^4$. **1**

3. $\displaystyle\int_a^b [f(x) \pm g(x)]\, dx = \int_a^b f(x)\, dx \pm \int_a^b g(x)\, dx$

(sum or difference of functions);

4. $\displaystyle\int_a^b f(x)\, dx = \int_a^c f(x)\, dx + \int_c^b f(x)\, dx,$ for any real number c.

For $f(x) \geq 0$, because the distance from a to a is 0, the first property says that the "area" under the graph of f bounded by $x = a$ and $x = a$ is 0. Also, since $\int_a^c f(x)\, dx$ represents the darker region in Figure 13.12 and $\int_c^b f(x)\, dx$ represents the lighter region,

$$\int_a^b f(x)\, dx = \int_a^c f(x)\, dx + \int_c^b f(x)\, dx,$$

as stated in the fourth property. While the figure shows $a < c < b$, the property is true for any value of c where both $f(x)$ and $F(x)$ are defined. Of course, the properties apply even if $f(x) < 0$.

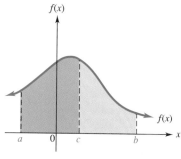

FIGURE 13.12

An algebraic proof is given here for the third property; proofs of the other properties are omitted. If $F(x)$ and $G(x)$ are antiderivatives of $f(x)$ and $g(x)$, respectively,

$$\int_a^b [f(x) + g(x)]\, dx = [F(x) + G(x)]\Big|_a^b$$

$$= [F(b) + G(b)] - [F(a) + G(a)] = [F(b) - F(a)] + [G(b) - G(a)]$$

$$= \int_a^b f(x)\, dx + \int_a^b g(x)\, dx.$$

▶**EXAMPLE 2**   Evaluate $\displaystyle\int_2^5 (6x^2 - 3x + 5)\, dx.$

Use the properties above, and the fundamental theorem, along with the power rule from Section 13.1.

3 Evaluate each definite integral.

(a) $\displaystyle\int_1^3 (x + 3x^2)\, dx$

(b) $\displaystyle\int_2^4 (6k^2 - 2k + 1)\, dk$

Answers:

(a) 30

(b) 102

4 Evaluate the following.

(a) $\displaystyle\int_0^4 e^x\, dx$

(b) $\displaystyle\int_3^5 \frac{dx}{x}$

(c) $\displaystyle\int_2^8 \frac{4}{x}\, dx$

Answers:

(a) 53.59815

(b) .51083

(c) 5.54518

5 Find $\displaystyle\int_0^2 \frac{x}{x^2 + 1}\, dx$.

Answer:

$\dfrac{1}{2} \ln 5$

$$\int_2^5 (6x^2 - 3x + 5)\, dx = 6 \int_2^5 x^2\, dx - 3 \int_2^5 x\, dx + 5 \int_2^5 dx \qquad \text{Constant multiple}$$

$$= 2x^3 \Big|_2^5 - \frac{3}{2} x^2 \Big|_2^5 + 5x \Big|_2^5 \qquad \text{Integrate}$$

$$= 2(5^3 - 2^3) - \frac{3}{2}(5^2 - 2^2) + 5(5 - 2) \qquad \begin{array}{l}\text{Evaluate}\\ \text{the limits}\end{array}$$

$$= 2(125 - 8) - \frac{3}{2}(25 - 4) + 5(3)$$

$$= 234 - \frac{63}{2} + 15 = \frac{435}{2} \quad \blacktriangleleft \; \boxed{3}$$

▶ **EXAMPLE 3** Find $\displaystyle\int_1^2 \frac{dy}{y}$.

Using a result from Section 13.1,

$$\int_1^2 \frac{dy}{y} = \ln |y| \Big|_1^2 = \ln |2| - \ln |1| \approx .6931 - 0 = .6931 \quad \blacktriangleleft \; \boxed{4}$$

▶ **EXAMPLE 4** Evaluate $\displaystyle\int_0^5 x\sqrt{25 - x^2}\, dx$.

Use substitution, Let $u = 25 - x^2$, so that $du = -2x\, dx$. With a definite integral, the limits should be changed, too. The new limits on u are found as follows.

If $x = 5$, then $u = 25 - 5^2 = 0$; if $x = 0$, then $u = 25 - 0^2 = 25$.

Then

$$\int_0^5 x\sqrt{25 - x^2}\, dx = -\frac{1}{2} \int_0^5 \sqrt{25 - x^2}\,(-2x\, dx)$$

$$= -\frac{1}{2} \int_{25}^0 \sqrt{u}\, du \qquad \text{Substitute}$$

$$= -\frac{1}{2} \int_{25}^0 u^{1/2}\, du \qquad \text{Fractional exponent}$$

$$= -\frac{1}{2} \cdot \frac{u^{3/2}}{3/2} \Big|_{25}^0 \qquad \text{Integrate}$$

$$= -\frac{1}{2} \cdot \frac{2}{3}[0^{3/2} - 25^{3/2}] \qquad \text{Evaluate limits}$$

$$= -\frac{1}{3}(-125) = \frac{125}{3}. \quad \blacktriangleleft$$

Caution Whenever you use the substitution method, be sure to replace x by its equivalent in terms of u *everywhere*. In particular, remember that the limits of integration refer to x and must be changed, too. **5**

AREA   In the previous section we saw that if $f(x) > 0$ in $[a, b]$, the definite integral $\int_a^b f(x)\, dx$ gives the area below the graph of the function $y = f(x)$, above the x-axis, and between the lines $x = a$ and $x = b$. We shall now see how to find the area between the graph of a function f and the x-axis when the graph lies below the x-axis.

Look at the graph of $f(x) = x^2 - 4$ in Figure 13.13. The area bounded by the graph of f, the x-axis, and the vertical lines $x = 0$ and $x = 2$ lies below the x-axis. Using the fundamental theorem to find this area gives

$$\int_0^2 (x^2 - 4)\, dx = \left(\frac{x^3}{3} - 4x\right)\Bigg|_0^2 = \left(\frac{8}{3} - 8\right) - (0 - 0) = \frac{-16}{3}.$$

The result is a negative number because $f(x)$ is negative for values of x in the interval $[0, 2]$. Since Δx is always positive, $f(x) < 0$ makes the product $f(x) \cdot \Delta x$ negative, so $\int_0^2 f(x)\, dx$ is negative. Area is nonnegative, so the required area is given by $\left| -16/3 \right|$ or $16/3$. Using a definite integral, the area could be written as

$$\left| \int_0^2 (x^2 - 4)\, dx \right| = \left| -\frac{16}{3} \right| = \frac{16}{3}.$$

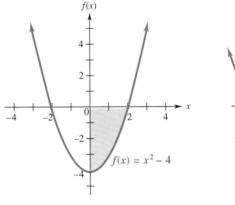

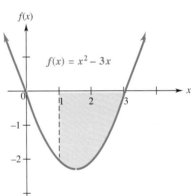

FIGURE 13.13   FIGURE 13.14

▶**EXAMPLE 5**   Find the area of the region between the x-axis and the graph of $f(x) = x^2 - 3x$ from $x = 1$ to $x = 3$.

The region is shown in Figure 13.14. The region lies below the x-axis;

6 Find each area.

(a)

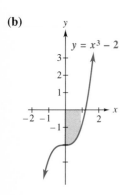

$y = x^2$

(b)

$y = x^3 - 2$

Answers:

(a) 8/3

(b) 7/4

therefore, the area is given by

$$\left| \int_1^3 (x^2 - 3x)\, dx \right|.$$

By the fundamental theorem,

$$\int_1^3 (x^2 - 3x)\, dx = \left(\frac{x^3}{3} - \frac{3x^2}{2} \right)\Big|_1^3 = \left(\frac{27}{3} - \frac{27}{2} \right) - \left(\frac{1}{3} - \frac{3}{2} \right) = -\frac{10}{3}.$$

The required area is $\left| -10/3 \right| = 10/3$ square units. ◄ **6**

▶**EXAMPLE 1** Find the area between the x-axis and the graph of $f(x) = x^2 - 4$ from $x = 0$ to $x = 4$.

Figure 13.15 shows the required region. Part of the region is below the x-axis. The definite integral over that interval will have a negative value. To find the area, integrate the negative and positive portions separately and take the absolute value of the first result before combining the two results to get the total area. Start by finding the point where the graph crosses the x-axis. This is done by solving the equation

$$x^2 - 4 = 0.$$

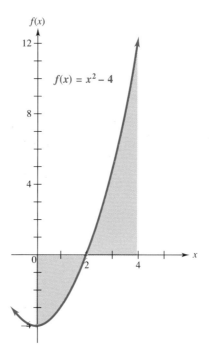

$f(x) = x^2 - 4$

FIGURE 13.15

7 Find the area between the x-axis and the graph of $f(x) = x^2 - 4$ from

(a) $x = 1$ to $x = 5$ and

(b) $x = -2$ to $x = 3$.

Answers:

(a) 86/3

(b) 13

The solutions of this equation are 2 and -2. The only solution in the interval $[0, 4]$ is 2. The total area of the region in Figure 13.15 is given by

$$\left| \int_0^2 (x^2 - 4)\, dx \right| + \int_2^4 (x^2 - 4)\, dx = \left| \left(\frac{1}{3}x^3 - 4x \right) \right|_0^2 + \left(\frac{1}{3}x^3 - 4x \right) \Big|_2^4$$

$$= \left| \frac{8}{3} - 8 \right| + \left(\frac{64}{3} - 16 \right) - \left(\frac{8}{3} - 8 \right)$$

$$= 16 \text{ square units.} \quad \blacktriangleleft$$

In Example 6, incorrectly using one integral over the entire interval to find the area would give

$$\int_0^4 (x^2 - 4)\, dx = \left(\frac{x^3}{3} - 4x \right) \Big|_0^4 = \left(\frac{64}{3} - 16 \right) - 0 = \frac{16}{3},$$

which is not the correct area. This definite integral represents no area, but is just a real number. **7**

A summary of the steps to use for finding areas follows.

Finding Area

To find the area bounded by $y = f(x)$, the vertical lines $x = a$ and $x = b$, and the x-axis, use the following steps.

Step 1   Sketch a graph.

Step 2   Find any x-intercepts in $[a, b]$. These divide the total region into subregions.

Step 3   The definite integral will be *positive* for subregions above the x-axis and *negative* for subregions below the x-axis. Use separate integrals to find the areas of the subregions.

Step 4   The total area is the sum of the areas of all of the subregions.

FOR GRAPHERS
A graphing calculator that has keys that give the (approximate) value of a definite integral can be used to find the integrals of the individual regions.

In the last section, we saw that the area under a rate of change function $f'(x)$ from $x = a$ to $x = b$ gives the total value of $f(x)$ on $[a, b]$. Now we can use the definite integral to solve these problems.

8 In Example 7, suppose a conservation campaign and higher prices cause the rate of consumption to be given by $C'(t) = \frac{1}{2}t + e^{.005t}$. Find the total amount of gas used in the nineties.

Answer:
About 35.25 trillion cubic feet.

▶ **EXAMPLE 7** The yearly rate of consumption of natural gas in trillions of cubic feet for a certain city is

$$C'(t) = t + e^{.01t},$$

where t is time in years, and $t = 0$ corresponds to 1990. At this consumption rate, what is the total amount the city will use in the 10-year period of the nineties?

To find the consumption over the 10-year period starting in 1990, use the definite integral.

$$\int_0^{10} (t + e^{.01t})\, dt = \left(\frac{t^2}{2} + \frac{e^{.01t}}{.01}\right)\Bigg|_0^{10}$$
$$= (50 + 100e^{.1}) - (0 + 100)$$
$$\approx -50 + 100(1.10517) \approx 60.5$$

Therefore, a total of about 60.5 trillion cubic feet of natural gas will be used during the nineties if the consumption rate remains the same. ◀ **8**

13.4 EXERCISES

Evaluate each of the following definite integrals. (See Examples 1–4.)

1. $\int_{-1}^{3} (6x^2 - 4x + 3)\, dx$

2. $\int_0^2 (-3x^2 + 2x + 5)\, dx$

3. $\int_0^2 3\sqrt{4u + 1}\, du$

4. $\int_3^9 \sqrt{2r - 2}\, dr$

5. $\int_0^1 2(t^{1/2} - t)\, dt$

6. $\int_0^4 -(3x^{3/2} + x^{1/2})\, dx$

7. $\int_1^4 (5y\sqrt{y} + 3\sqrt{y})\, dy$

8. $\int_4^9 (4\sqrt{r} - 3r\sqrt{r})\, dr$

9. $\int_4^6 \frac{2}{(x-3)^2}\, dx$

10. $\int_1^4 \frac{-3}{(2p+1)^2}\, dp$

11. $\int_1^5 (5n^{-1} + n^{-3})\, dn$

12. $\int_2^3 (3x^{-1} - x^{-4})\, dx$

13. $\int_2^3 \left(2e^{-.1A} + \frac{3}{A}\right)\, dA$

14. $\int_1^2 \left(\frac{-1}{B} + 3e^{.2B}\right)\, dB$

15. $\int_1^2 \left(e^{5u} - \frac{1}{u^2}\right)\, du$

16. $\int_{.5}^1 (p^3 - e^{4p})\, dp$

17. $\int_{-1}^0 y(2y^2 - 3)^5\, dy$

18. $\int_0^3 m^2(4m^3 + 2)^3\, dm$

19. $\int_1^{64} \frac{\sqrt{z} - 2}{\sqrt[3]{z}}\, dz$

20. $\int_1^8 \frac{3 - y^{1/3}}{y^{2/3}}\, dy$

21. $\int_1^2 \frac{\ln x}{x}\, dx$

22. $\int_1^3 \frac{\sqrt{\ln x}}{x}\, dx$

23. $\int_0^8 x^{1/3}\sqrt{x^{4/3} + 9}\, dx$

24. $\int_1^2 \frac{3}{x(1 + \ln x)}\, dx$

25. $\int_0^1 \frac{e^t}{(3 + e^t)^2}\, dt$

26. $\int_0^1 \frac{e^{2z}}{\sqrt{1 + e^{2z}}}\, dz$

27. $\int_1^{49} \frac{(1 + \sqrt{x})^{4/3}}{\sqrt{x}}\, dx$

28. $\int_1^8 \frac{(1 + x^{1/3})^6}{x^{2/3}}\, dx$

---

29. Suppose the function in Example 6, $f(x) = x^2 - 4$ from $x = 0$ to $x = 4$, represented the annual rate of profit of a company over a four-year period. What might the negative integral for the first two-year period indi-

cate? What integral would represent the overall profit for the four-year period?

30. In your own words describe how the fundamental theorem relates definite and indefinite integrals.

Use the definite integral to find the area between the x-axis and f(x) over the indicated inter-val. Check first to see if the graph crosses the x-asis in the given interval. (See Examples 5 and 6.)

31. $f(x) = 4 - x^2;\ [0, 3]$

32. $f(x) = x^2 - 2x - 3;\ [0, 4]$

33. $f(x) = x^3 - 1;\ [-2, 2]$

34. $f(x) = x^3 - 2x;\ [-2, 4]$

35. $f(x) = e^x - 1;\ [-1, 2]$

36. $f(x) = 1 - e^{-x};\ [-1, 2]$

37. $f(x) = \dfrac{1}{x};\ [1, e]$

38. $f(x) = \dfrac{1}{x};\ [e, e^2]$

Find the area of each shaded region.

39.

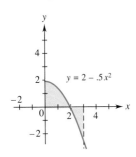

$y = 2 - .5x^2$

40.

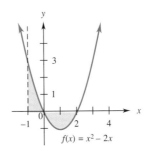

$f(x) = x^2 - 2x$

41.

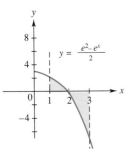

$y = \dfrac{e^2 - e^x}{2}$

42.

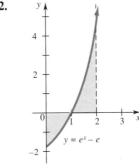

$y = e^x - e$

Use Property 4 of this section to find $\displaystyle\int_1^4 f(x)\,dx$ for the following functions.

43. $f(x) = \begin{cases} 2x + 3 & \text{if } x \le 0 \\ -.25x - 3 & \text{if } x > 0 \end{cases}$

44. $f(x) = \begin{cases} x^2 - 2 & \text{if } x \le 1 \\ -x^2 + 1 & \text{if } x > 1 \end{cases}$

45. Management   De Win Enterprises has found that its expenditure rate per day (in hundreds of dollars) on a certain type of job is given by

$$E(x) = 4x + 2,$$

where x is the number of days since the start of the job.
(a) Find the total expenditure if the job takes 10 days.
(b) How much will be spent on the job from the 10th to the 20th day?

(c) If the company wants to spend no more than $5000 on the job, in how many days must they complete it?

46. Management   De Win Enterprises (see Exercise 45) also knows that the rate of income per day (in hundreds of dollars) for the same job is

$$I(x) = 100 - x,$$

where x is the number of days since the job was started.

(a) Find the total income for the first 10 days.

(b) Find the income from the 10th to the 20th day.

(c) How many days must the job last for the total income to be at least $5000?

47. Management Karla Harby Communications, a small company of science writers, found that its rate of profits (in thousands of dollars) after t years of operation is given by

$$P'(t) = (3t + 3)(t^2 + 2t + 2)^{1/3}.$$

(a) Find the total profits in the first three years.

(b) Find the profit in the fourth year of operation.

(c) What is happening to the annual profits over the long run?

48. Management A worker new to a job will improve his efficiency with time so that it takes him fewer hours to produce an item with each day on the job, up to a certain point. Suppose the rate of change of the number of hours it takes a worker in a certain factory to produce the xth item is given by

$$H'(x) = 20 - 2x.$$

(a) What is the total number of hours required to produce the first 5 items?

(b) What is the total number of hours required to produce the first 10 items?

49. Management The rate of depreciation for a certain truck is

$$f(t) = \frac{6000(.3 + .2t)}{(1 + .3t + .1t^2)^2},$$

where t is in years and $t = 0$ is the year of purchase.

(a) Find the total depreciation at the end of 3 years.

(b) In what year will the total depreciation be at least 3000?

50. Natural Science Pollution from a factory is entering a lake. The rate of concentration of the pollutant at time t is given by

$$P'(t) = 140t^{5/2},$$

where t is the number of years since the factory started introducing pollutants into the lake. Ecologists estimate that the lake can accept a total level of pollution of 4850 units before all the fish life in the lake ends. Can the factory operate for 4 yr without killing all the fish in the lake?

51. Natural Science An oil tanker is leaking oil at a rate given in barrels per hour by

$$L'(t) = \frac{80 \ln (t + 1)}{t + 1},$$

where t is the time in hours after the tanker hits a hidden rock (when $t = 0$).

(a) Find the total number of barrels that the ship will leak on the first day.

(b) Find the total number of barrels that the ship will leak on the second day.

(c) What is happening over the long run to the amount of oil leaked per day?

52. Natural Science After long study, tree scientists conclude that a eucalyptus tree will grow at the rate of $.2 + 4t^{-4}$ feet per year, where t is time in years.

(a) Find the number of feet that the tree will grow in the second year.

(b) Find the number of feet the tree will grow in the third year.

53. Management Suppose that the rate of consumption of a natural resource is $c'(t)$, where

$$c'(t) = ke^{rt}.$$

Here t is time in years, r is a constant, and k is the consumption in the year when $t = 0$. In 1990, an oil company sold 1.2 billion barrels of oil. Assume that $r = .04$.

(a) Set up a definite integral for the amount of oil that the company will sell in the next 10 years.

(b) Evaluate the definite integral of part (a).

(c) The company has about 20 billion barrels of oil in reserve. To find the number of years that this amount will last, solve this equation for T:

$$\int_0^T 1.2e^{.04t} \, dt = 20.$$

(d) Rework part (c), assuming that $r = .02$.

54. Management In Exercise 53 the rate of consumption of oil (in billions of barrels) was given as

$$1.2e^{.04t},$$

where $t = 0$ corresponds to 1990. Find the total amount of oil used from 1990 to year T. At this rate, how much will be used in 5 years?

13.5 APPLICATIONS OF INTEGRALS

Given a function that represents the total maintenance charge for a machine from the time it is installed, the rate of maintenance at any time t is given by the derivative of the total maintenance charge. Now, the total maintenance function can be found by finding the antiderivative of the rate of maintenance function. As we have seen in this chapter, if we graph the function giving the rate of maintenance, then the total maintenance is given by the area under the curve. Example 1 shows how this works.

▶ **EXAMPLE 1**   Suppose a leasing company wants to decide on the yearly lease fee for a certain new typewriter. The company expects to lease the typewriter for 5 years, and it expects the *rate* of maintenance, $M(t)$ (in dollars), which it must supply, to be approximated by

$$M(t) = 10 + 2t + t^2,$$

where t is the number of years the typewriter has been used. How much should the company charge for maintenance on the typewriter for the 5-year period?

The definite integral can be used to find the total maintenance charge the company can expect over the life of the typewriter. Figure 13.16 shows the graph of $M(t)$. The total maintenance charge for the 5-year period will be given by the shaded area of the figure, which can be found as follows.

$$\int_0^5 (10 + 2t + t^2)\, dt = \left(10t + t^2 + \frac{t^3}{3} \right)\Big|_0^5$$

$$= 50 + 25 + \frac{125}{3} - 0$$

$$\approx 116.67$$

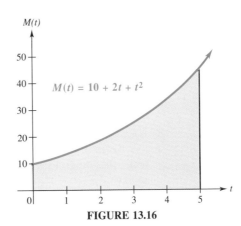

FIGURE 13.16

1 Find the total maintenance charge for a lease of

(a) 1 year;

(b) 2 years.

Answers:

(a) $11.33

(b) $26.67

2 Suppose the rate of change of cost in dollars to produce x items is given by

$$C(x) = x^2 - 8x + 15.$$

Find the total cost to produce the first 10 items.

Answer:
$83.33

3 Find the amount of savings over the first 6 years for the machine in Example 2.

Answer:
$54

The company can expect the total maintenance charge for 5 years to be about $117. Hence, the company should add about

$$\frac{\$117}{5} = \$23.40$$

to its annual lease price. ◀ **1** **2**

▶**EXAMPLE 2**   Vonalaine Crowe, who runs a factory that makes signs, has been shown a new machine that staples the signs to the handles. She estimates that the rate of savings, $S(x)$, from the machine will be approximated by

$$S(x) = 3 + 2x,$$

where x represents the number of years the stapler has been in use. If the machine costs $70, would it pay for itself in 5 years?

We need to find the area under the rate of savings curve shown in Figure 13.17 between the lines $x = 0$, $x = 5$, and the x-axis. Using a definite integral gives

$$\int_0^5 (3 + 2x)\, dx = 3x + x^2 \Big|_0^5 = 40.$$

The total savings in 5 years is $40, so the machine will not pay for itself in this time period. ◀ **3**

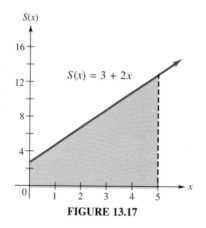

FIGURE 13.17

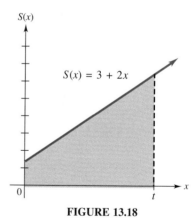

FIGURE 13.18

▶**EXAMPLE 3**   When will the machine in Example 2 pay for itself?
Because the machine in Example 2 costs a total of $70, it will pay for itself when the area under the savings curve of Figure 13.18 equals 70, or at a time t such that

$$\int_0^t (3 + 2x)\, dx = 70.$$

4 Find the number of years in which the machine in Example 2 will pay for itself if the rate of savings is $S(x) = 5 + 4x$.

Answer:
4.8 years

Evaluating the definite integral gives

$$3x + x^2 \bigg|_0^t = (3t + t^2) - (3 \cdot 0 + 0^2) = 3t + t^2.$$

Since the total savings must equal 70,

$$3t + t^2 = 70.$$

Solve this quadratic equation to verify that the solutions are 7 and -10. Since -10 cannot be used here, the machine will pay for itself in 7 years. ◀ **4**

In Section 13.3 we saw that a definite integral can be used to find the area *under* a curve. This idea can be extended to find the area *between* two curves, as shown in the next example.

▶**EXAMPLE 4**   A company is considering a new manufacturing process in one of its plants. The new process provides substantial initial savings, with the savings declining with time x according to the rate-of-savings function

$$S(x) = 100 - x^2,$$

where $S(x)$ is in thousands of dollars. At the same time, the cost of operating the new process increases with time x, according to the rate-of-cost function (in thousands of dollars)

$$C(x) = x^2 + \frac{14}{3}x.$$

(a) For how many years will the company realize savings?
   Figure 13.19 shows the graphs of the rate-of-savings and the rate-of-cost functions. The rate-of-cost (marginal cost) is increasing, while the rate-of-savings (marginal savings) is decreasing. The company should use this new process until the difference between these quantities is zero—that is, until the time at which these graphs intersect. The graphs intersect when

$$S(x) = C(x),$$

or

$$100 - x^2 = x^2 + \frac{14}{3}x.$$

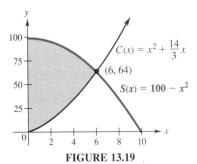

FIGURE 13.19

5 In Example 4, find the total savings if pollution control regulations permit the new process for only 4 years.

Answer:
$320,000

6 If a company's marginal savings and marginal cost functions are

$$S(x) = 150 - 2x^2$$

and

$$C(x) = x^2 + 15x,$$

where $S(x)$ and $C(x)$ give amounts in thousands of dollars, find the following.

(a) Number of years until savings equal costs

(b) Total savings

Answers:

(a) 5

(b) $437,500

Solve this equation as follows.

$$0 = 2x^2 + \frac{14}{3}x - 100$$

$$= 3x^2 + 7x - 150 \qquad \textbf{Multiply by 3/2}$$

$$= (x - 6)(3x + 25) \qquad \textbf{Factor}$$

Set each factor equal to 0 and solve both equations to get

$$x = 6 \quad \text{or} \quad x = -\frac{25}{3}.$$

Only 6 is a meaningful solution here. The company should use the new process for 6 years.

(b) What will be the net total savings during this period?

Since the total savings over the 6-year period is given by the area under the rate-of-savings curve and the total additional cost by the area under the rate-of-cost curve, the net total savings over the 6-year period is given by the difference between these areas. This difference is the shaded region between the rate-of-cost and the rate-of-savings curves and the lines $x = 0$ and $x = 6$ shown in Figure 13.19. This area can be evaluated with a definite integral as follows.

$$\text{Total savings} = \int_0^6 [S(x) - C(x)]\, dx$$

$$= \int_0^6 \left[(100 - x^2) - \left(x^2 + \frac{14}{3}x \right) \right] dx$$

$$= \int_0^6 \left(100 - \frac{14}{3}x - 2x^2 \right) dx \qquad \textbf{Combine terms}$$

$$= 100x - \frac{7}{3}x^2 - \frac{2}{3}x^3 \Big|_0^6 \qquad \textbf{Integrate}$$

$$= 100(6) - \frac{7}{3}(36) - \frac{2}{3}(216) = 372.$$

The company will save a total of $372,000 over the 6-year period. ◀ **5** **6**

▶**EXAMPLE 5** A farmer has been using a new fertilizer that gives him a better yield, but because it exhausts the soil of other nutrients he must use other fertilizers in greater and greater amounts, so that his costs increase each year. The new fertilizer produces a rate of increase in revenue (in hundreds of dollars) given by

$$R(t) = -.4t^2 + 8t + 10,$$

where t is measured in years. The rate of increase in yearly costs (also in hundreds of dollars) due to use of the fertilizer is given by

$$C(t) = 2t + 5.$$

How long can the farmer profitably use the fertilizer? What will be his net increase in revenue over this period?

The farmer should use the new fertilizer until the marginal costs equal the marginal revenue. Find this point by solving the equation $R(t) = C(t)$ as follows.

$$-.4t^2 + 8t + 10 = 2t + 5$$
$$-4t^2 + 80t + 100 = 20t + 50 \qquad \textbf{Multiply by 10}$$
$$-4t^2 + 60t + 50 = 0$$

By the quadratic formula the only positive solution is

$$t = 15.8.$$

The new fertilizer will be profitable for about 15.8 years.

To find the total amount of additional revenue over the 15.8-year period, find the area between the graphs of the rate of revenue and the rate of cost functions, as shown in Figure 13.20.

$$\text{Total savings} = \int_0^{15.8} [R(t) - C(t)] \, dt$$

$$= \int_0^{15.8} [(-.4t^2 + 8t + 10) - (2t + 5)] \, dt$$

$$= \int_0^{15.8} (-.4t^2 + 6t + 5) \, dt \qquad \textbf{Combine terms}$$

$$= \left(\frac{-.4t^3}{3} + \frac{6t^2}{2} + 5t \right) \Bigg|_0^{15.8} \qquad \textbf{Integrate}$$

$$= 302.01$$

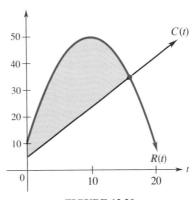

FIGURE 13.20

The total savings will amount to about $30,000 over the 15.8-year period.

It is not realistic to say that the farmer will need to use the new process for 15.8 years—he will probably have to use it for 15 years or for 16 years. In this case, when the mathematical result is not in the domain of the function, it will be necessary to find the total savings after 15 years and after 16 years and then select the best result. ◀

CONSUMERS' AND PRODUCERS' SURPLUS The· market determines the price at which a product is sold. As indicated earlier, the point of intersection of the demand curve and the supply curve for a product gives the equilibrium price. At the equilibrium price, consumers will purchase the same amount of the product that the manufacturers want to ˙sell. Some consumers, however, would be willing to spend more for an item than the equilibrium price. The total of the differences between the equilibrium price of the item and the higher prices all those individuals would be willing to pay is thought of as savings realized by those individuals and is called the **consumers' surplus.**

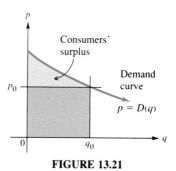

FIGURE 13.21

In Figure 13.21, the area under the demand curve is the total amount consumers are willing to spend for q_0 items. The heavily shaded area under the line $y = p_0$ shows the total amount consumers actually will spend at the equilibrium price of p_0. The lightly shaded area represents the consumers' surplus. As the figure suggests, the consumers' surplus is given by an area between the two curves $p = D(q)$ and $p = p_0$, so its value can be found with a definite integral as follows.

Consumers' Surplus

If $D(q)$ is a demand function with equilibrium price p_0 and equilibrium demand q_0, then

$$\textbf{Consumers' surplus} = \int_0^{q_0} [D(q) - p_0]\, dq.$$

Similarly, if some manufacturers would be willing to supply a product at a price *lower* than the equilibrium price p_0, the total of the differences between the equilibrium price and the lower prices at which the manufacturers would sell the product is considered added income for the manufacturers and is called the **producers' surplus.** Figure 13.22 shows the (heavily shaded) total area under the supply curve from $q = 0$ to $q = q_0$, which is the minimum total amount the manufacturers are willing to realize from the sale of q_0 items. The total area under the line $p = p_0$ is the amount actually realized. The difference between these two areas, the producers' surplus, is also given by a definite integral.

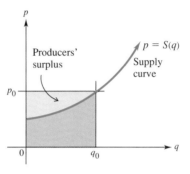

FIGURE 13.22

Producers' Surplus

If $S(q)$ is a supply function with equilibrium price p_0 and equilibrium supply q_0, then

$$\textbf{Producers' surplus} = \int_0^{q_0} [p_0 - S(q)]\, dq.$$

▶ **EXAMPLE 6**   Suppose the price (in dollars per ton) for oat bran is

$$D(q) = 900 - 20q - q^2,$$

when the demand for the product is q tons. Also, suppose the function

$$S(q) = q^2 + 10q$$

gives the price (in dollars per ton) when the supply is q tons. Find the consumers' surplus and the producers' surplus.

We begin by finding the equilibrium quantity by setting the two equations equal.

$$900 - 20q - q^2 = q^2 + 10q$$
$$0 = 2q^2 + 30q - 900$$
$$0 = q^2 + 15q - 450.$$

7 Given the demand function $D(q) = 12 - .07q$ and the supply function $S(q) = .05q$, where $D(q)$ and $S(q)$ are in dollars, find

(a) the equilibrium point;

(b) the consumers' surplus;

(c) the producers' surplus.

Answers:

(a) $x = 100$

(b) $350

(c) $250

Use the quadratic theorem or factor to see that the only positive solution of this equation is $q = 15$. At the equilibrium point where the supply and demand are both 15 tons, the price is

$$S(15) = 15^2 + 10(15) = 375,$$

or $375. Verify that the same answer is found by computing $D(15)$. The consumers' surplus, represented by the area shown in Figure 13.23, is

$$\int_0^{15} [(900 - 20q - q^2) - 375]\, dq = \int_0^{15} (525 - 20q - q^2)\, dq.$$

Evaluating this definite integral gives

$$\left(525q - 10q^2 - \frac{1}{3}q^3 \right)\Bigg|_0^{15} = \left[525(15) - 10(15)^2 - \frac{1}{3}(15)^3 \right] - 0$$

$$= 4500$$

Here, the consumers' surplus is $4500. The producers' surplus, also shown in Figure 13.23, is given by

$$\int_0^{15} [375 - (q^2 + 10q)]\, dq = \int_0^{15} (375 - q^2 - 10q)\, dq$$

$$= 325q - \frac{1}{3}q^3 - 5q^2 \Bigg|_0^{15}$$

$$= \left[375(15) - \frac{1}{3}(15)^3 - 5(15)^2 \right] - 0$$

$$= 3375.$$

The producers' surplus is $3375.   ◀ **7**

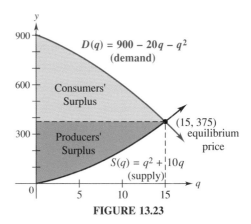

FIGURE 13.23

13.5 EXERCISES

Management   *Work the following exercises. (See Examples 1–5.)*

1. A car-leasing firm must decide how much to charge for maintenance on the cars it leases. After careful study, the firm decides that the rate of maintenance, $M(x)$, on a new car will approximate $M(x) = 60(1 + x^2)$, where x is the number of years the car has been in use. What total maintenance charge can the company expect for a 2-year lease? What amount should be added to the monthly lease payments to pay for maintenance?

2. Using the function of Exercise 1, find the maintenance charge the company can expect during the third year. Find the total charge during the first 3 years. What monthly charge should be added to cover a 3-year lease?

3. A company is considering a new manufacturing process. It knows that the rate of savings from the process, $S(t)$, will be about $S(t) = 1000(t + 2)$, where t is the number of years the process has been in use. Find the total savings during the first year. Find the total savings during the first 6 years.

4. Assume that the new process in Exercise 3 costs $16,000. About when will it pay for itself?

5. A company is introducing a new product. Production is expected to grow slowly because of difficulties in the start-up process. It is expected that the rate of production, $P(x)$, will be approximated by $P(x) = 1000e^{.2x}$, where x is the number of years since the introduction of the product. Will the company be able to supply 20,000 units during the first 4 years?

6. About when will the company of Exercise 5 be able to supply its 15,000th unit?

7. Suppose a company wants to introduce a new machine that will produce a rate of annual savings in dollars given by
$$S(x) = 150 - x^2,$$
where x is the number of years of operation of the machine, while producing a rate of annual costs in dollars of
$$C(x) = x^2 + \frac{11}{4}x.$$

   (a) For how many years will it be profitable to use this new machine?

   (b) What are the net total savings during the first year of use of the machine?

   (c) What are the net total savings over the entire period of use of the machine?

8. **Natural Science**   A new smog-control device will reduce the output of sulfur oxides from automobile exhausts. It is estimated that the rate of savings to the community from the use of this device will be approximated by
$$S(x) = -x^2 + 4x + 8,$$
where $S(x)$ is the rate of savings (in millions of dollars) after x years of use of the device. The new device cuts down on the production of sulfur oxides, but it causes an increase in the production of nitrous oxides. The rate of additional costs (in millions) to the community after x years is approximated by
$$C(x) = \frac{3}{25}x^2.$$

   (a) For how many years will it pay to use the new device?

   (b) What will be the net savings over this period of time?

9. **Management**   De Win Enterprises had an expenditure rate of $E(x) = e^{.1x}$ dollars per day and an income rate of $I(x) = 98.8 - e^{.1x}$ dollars per day on a particular job, where x was the number of days from the start of the job. The company's profit on that job will equal total income less total expenditures. Profit will be maximized if the job ends at the optimum time, which is the point where the two curves meet. Find the following.

   (a) The optimum number of days for the job to last.

   (b) The total income for the optimum number of days.

   (c) The total expenditures for the optimum number of days.

   (d) The maximum profit for the job.

10. **Management**   A factory at Harold Levinson Industries has installed a new process that will produce an increased rate of revenue (in thousands of dollars) of
$$R(t) = 104 - .4e^{t/2},$$
where t is time measured in years. The new process produces additional costs (in thousands of dollars) at the rate of
$$C(t) = .3e^{t/2}.$$

   (a) When will it no longer be profitable to use this new process?

   (b) Find the total net savings.

11. **Management**   After t years, a mine is producing at the rate of

$$P(t) = \frac{15}{t + 1}$$

tons per year. At the same time, the ore produced is consumed at a rate of $C(t) = .1t + 2$ tons per year.

(a) In how many years will the rate of consumption equal the rate of production?

(b) What is the total excess production before consumption and production are equal?

(c) Consumption equals production when $t = 0$. Why is 0 not the correct answer in part (a)?

12. **Management**   The rate of expenditure (in dollars) for maintenance of a certain machine is given by

$$M(x) = x^2 + 6x,$$

where x is time measured in years. The machine produces a rate of savings (in dollars) given by

$$S(x) = 360 - 2x^2.$$

(a) In how many years will the maintenance rate equal the savings rate?

(b) What will be the total net savings?

13. **Natural Science**   Pollution from a factory is entering a lake. The rate of concentration of the pollutant at time t (in years) is $P'(t) = 140t^4$. A cleaning substance is introduced into the lake that cleans it at a rate given by $C'(t) = 1.6t^{2.5}$. How long will it be before the total net effect is 0?

Management   *Work the following supply and demand exercises, where the price is given in dollars. (See Example 6.)*

14. Find the producers' surplus if the supply function for pork bellies is given by

$$S(q) = q^{5/2} + 2q^{3/2} + 50.$$

Assume supply and demand are in equilibrium at $q = 16$.

15. Suppose the supply function for concrete is given by

$$S(q) = 100 + 3q^{3/2} + q^{5/2},$$

and that supply and demand are in equilibrium at $q = 9$. Find the producers' surplus.

16. Find the consumers' surplus if the demand function for grass seed is given by

$$D(q) = \frac{100}{(3q + 1)^2},$$

assuming supply and demand are in equilibrium at $q = 3$.

17. Find the consumers' surplus if the demand function for extra virgin olive oil is given by

$$D(q) = \frac{16,000}{(2q + 8)^3},$$

and if supply and demand are in equilibrium at $q = 6$.

18. Suppose the supply function of a certain item is given by

$$S(q) = e^{q/2} - 1,$$

and the demand function is given by

$$D(q) = 400 - e^{q/2}.$$

(a) Graph the supply and demand curves.

(b) Find the point at which supply and demand are in equilibrium.

(c) Find the consumers' surplus.

(d) Find the producers' surplus.

19. Repeat the four steps in Exercise 18 for the supply function

$$S(q) = q^2 + \frac{11}{4}q$$

and the demand function

$$D(q) = 150 - q^2.$$

20. Find the consumers' surplus and the producers' surplus for an item having supply function

$$S(q) = \frac{q^2}{2} - 5q - 100$$

and demand function

$$D(q) = 16q - \frac{q^2}{2}.$$

21. Suppose the supply function of a certain item is given by

$$S(q) = \frac{7}{5}q$$

and the demand function is given by

$$D(q) = -\frac{3}{5}q + 10.$$

(a) Graph the supply and demand curves.
(b) Find the point at which supply and demand are in equilibrium.
(c) Find the consumers' surplus.
(d) Find the producers' surplus.

Work the following exercises.

22. Management   If a large truck has been driven x thousand miles, the rate of repair costs in dollars per mile is given by $R(x)$, where

$$R(x) = .05x^{3/2}.$$

Find the total repair costs if the truck is driven
(a) 100,000 miles;
(b) 400,000 miles.

23. Management   In a recent inflationary period, costs of a certain industrial process, in millions of dollars, were increasing according to the function

$$i(t) = .45t^{3/2},$$

where $i(t)$ is the rate of increase in costs at time t measured in years. Find the total increase in costs during the first 4 years.

24. Natural Science   From 1905 to 1920, most of the predators of the Kaibab Plateau of Arizona were killed by hunters. This allowed the deer population there to grow rapidly until they had depleted their food sources, which caused a rapid decline in population. The rate of change of this deer population during that time span is approximated by the function

$$D(t) = \frac{25}{2}t^3 - \frac{5}{8}t^4,$$

where t is the time in years $(0 \le t \le 25)$.
(a) Find the function for the deer population if there were 4000 deer in 1905 $(t = 0)$.
(b) What was the population in 1920?
(c) When was the population at a maximum?
(d) What was the maximum population?

25. Social Science   Suppose that all the people in a country are ranked according to their incomes, starting at the bottom. Let x represent the fraction of the community making the lowest income $(0 \le x \le 1)$; $x = .4$, therefore, represents the lower 40% of all income producers. Let $I(x)$ represent the proportion of the total income earned by the lowest x of all the people. Thus, $I(.4)$ represents the fraction of total income earned by the

lowest 40% of the population. Suppose

$$I(x) = .9x^2 + .1x.$$

Find and interpret the following.
(a) $I(.1)$ **(b)** $I(.5)$ **(c)** $I(.9)$

If income were distributed uniformly, we would have $I(x) = x$. The area under this line of complete equality is $1/2$. As $I(x)$ dips farther below $y = x$, there is less equality of income distribution. This inequality can be quantified by the ratio of the area between $I(x)$ and $y = x$ to $1/2$. This ratio is called the *coefficient of inequality* and equals $2 \int_0^1 (x - I(x))\, dx$.
(d) Graph $I(x) = x$ and $I(x) = .9x^2 + .1x$ for $0 \le x \le 1$ on the same axes.
(e) Find the area between the curves. What does this area represent?

26. Social Science   Repeat Exercise 25 with $I(x) = .5x^2 + .5x$.

27. Management   A worker new to a job will improve his efficiency with time so that it takes him fewer hours to produce an item with each day on the job up to a certain point. Suppose the rate of change of the number of hours it takes a worker in a certain factory to produce the xth item is given by

$$H(x) = 20 - 2x.$$

The production rate per item is a maximum when $\int_0^T H(x)\, dx$ is a maximum.
(a) How many items must be made to achieve the maximum production rate? Assume 0 items are made in 0 hours.
(b) What is the maximum production rate per item?

28. Natural Science   After long study, tree scientists conclude that a eucalyptus tree will grow at the rate of $4 + 4t^{-4}$ feet per year, where t is time in years. Find the number of feet that the tree will grow from the second to the tenth year.

29. Natural Science   The function $\int_0^T (4 + 4t^{-4})\, dt$ from Exercise 28 is not realistic. Explain why.

30. Natural Science   For a certain drug, the rate of reaction in appropriate units is given by

$$R(t) = \frac{5}{t} + \frac{2}{t^2},$$

where t is measured in hours after the drug is administered. Find the total reaction to the drug
(a) from $t = 1$ to $t = 12$;
(b) from $t = 12$ to $t = 24$.

Approximate the area between the graphs of each pair of functions on the given interval.

31. $y = \ln x$ and $y = xe^x$; $[1, 4]$

32. $y = \ln x$ and $y = 4 - x^2$; $[2, 4]$

33. $y = \sqrt{9 - x^2}$ and $y = \sqrt{x + 1}$; $[-1, 3]$

34. $y = \sqrt{4 - 4x^2}$ and $y = \sqrt{\dfrac{9 - x^2}{3}}$; $[-1, 1]$

---

1 Find the following.

(a) $\displaystyle\int \frac{4}{\sqrt{x^2 + 100}}\, dx$

(b) $\displaystyle\int \frac{-9}{\sqrt{x^2 - 4}}\, dx$

Answers:

(a) $4 \ln \left| \dfrac{x + \sqrt{x^2 + 100}}{10} \right| + C$

(b) $-9 \ln \left| \dfrac{x + \sqrt{x^2 - 4}}{2} \right| + C$

2 Find the following.

(a) $\displaystyle\int \frac{1}{x^2 - 4}\, dx$

(b) $\displaystyle\int \frac{-6}{x\sqrt{25 - x^2}}\, dx$

Answers:

(a) $\dfrac{1}{4} \ln \left| \dfrac{x - 2}{x + 2} \right| + C$

(b) $\dfrac{6}{5} \ln \left| \dfrac{5 + \sqrt{25 - x^2}}{x} \right| + C$

13.6 TABLES OF INTEGRALS

As we have noted, there are many useful functions whose antiderivatives cannot be found by the methods we have discussed. In fact, some definite integrals cannot be evaluated exactly, but can be approximated by methods similar to the method of summing rectangles discussed in Section 13.3. Many useful integrals can be found from a table of integrals such as Table 3 at the back of the book.* The next examples show how to use this table.

▶ **EXAMPLE 1** Find $\displaystyle\int \frac{1}{\sqrt{x^2 + 16}}\, dx$.

By inspecting the table, we see that if $a = 4$, this antiderivative is the same as entry 5 of the table. Entry 5 of the table is

$$\int \frac{1}{\sqrt{x^2 + a^2}}\, dx = \ln \left| \frac{x + \sqrt{x^2 + a^2}}{a} \right| + C.$$

Substituting 4 for a in this entry, we get

$$\int \frac{1}{\sqrt{x^2 + 16}}\, dx = \ln \left| \frac{x + \sqrt{x^2 + 16}}{4} \right| + C.$$

This last result could be verified by taking the derivative of the right-hand side of this last equation. ◀ **1**

▶ **EXAMPLE 2** Find $\displaystyle\int \frac{8}{16 - x^2}\, dx$.

Convert this antiderivative into the one given in entry 7 of the table by writing the 8 in front of the integral sign (permissible only with constants) and by letting $a = 4$. Doing this gives

$$8 \int \frac{1}{16 - x^2}\, dx = 8 \left[\frac{1}{2 \cdot 4} \ln \left| \frac{4 + x}{4 - x} \right| \right] + C$$

$$= \ln \left| \frac{4 + x}{4 - x} \right| + C. \quad ◀ \ \mathbf{2}$$

---

*Many computer programs are now available that can integrate functions. These will probably eventually make the use of tables obsolete.

3 Find the following.

(a) $\displaystyle\int \frac{3}{16x^2 - 1}\, dx$

(b) $\displaystyle\int \frac{-1}{100x^2 - 1}\, dx$

Answers:

(a) $\displaystyle\frac{3}{8} \ln \left| \frac{x - \dfrac{1}{4}}{x + \dfrac{1}{4}} \right| + C$

(b) $\displaystyle -\frac{1}{20} \ln \left| \frac{x - \dfrac{1}{10}}{x + \dfrac{1}{10}} \right| + C$

▶ **EXAMPLE 3** Find $\displaystyle\int \sqrt{9x^2 + 1}\, dx$.

This antiderivative seems most similar to entry 15 of the table. However, entry 15 requires that the coefficient of the x^2 term be 1. We can satisfy that requirement here by factoring out the 9.

$$\int \sqrt{9x^2 + 1}\, dx = \int \sqrt{9\left(x^2 + \frac{1}{9}\right)}\, dx$$

$$= \int 3\sqrt{x^2 + \frac{1}{9}}\, dx$$

$$= 3\int \sqrt{x^2 + \frac{1}{9}}\, dx$$

Now, use entry 15 with $a = 1/3$.

$$\int \sqrt{9x^2 + 1}\, dx = 3\left[\frac{x}{2}\sqrt{x^2 + \frac{1}{9}} + \frac{\left(\frac{1}{3}\right)^2}{2} \cdot \ln\left| x + \sqrt{x^2 + \frac{1}{9}}\right| \right] + C$$

$$= \frac{3x}{2}\sqrt{x^2 + \frac{1}{9}} + \frac{1}{6}\ln\left| x + \sqrt{x^2 + \frac{1}{9}}\right| + C \quad ◀ \boxed{3}$$

13.6 EXERCISES

Use the table of integrals to find each antiderivative. (See Examples 1–3.)

1. $\displaystyle\int \frac{-4}{\sqrt{x^2 + 36}}\, dx$

2. $\displaystyle\int \frac{9}{\sqrt{x^2 + 9}}\, dx$

3. $\displaystyle\int \frac{6}{x^2 - 9}\, dx$

4. $\displaystyle\int \frac{-12}{x^2 - 16}\, dx$

5. $\displaystyle\int \frac{-4}{x\sqrt{9 - x^2}}\, dx$

6. $\displaystyle\int \frac{3}{x\sqrt{121 - x^2}}\, dx$

7. $\displaystyle\int \frac{-2x}{3x + 1}\, dx$

8. $\displaystyle\int \frac{6x}{4x - 5}\, dx$

9. $\displaystyle\int \frac{2}{3x(3x - 5)}\, dx$

10. $\displaystyle\int \frac{-4}{3x(2x + 7)}\, dx$

11. $\displaystyle\int \frac{4}{4x^2 - 1}\, dx$

12. $\displaystyle\int \frac{-6}{9x^2 - 1}\, dx$

13. $\displaystyle\int \frac{3}{x\sqrt{1 - 9x^2}}\, dx$

14. $\displaystyle\int \frac{-2}{x\sqrt{1 - 16x^2}}\, dx$

15. $\displaystyle\int \frac{4x}{2x + 3}\, dx$

16. $\displaystyle\int \frac{4x}{6 - x}\, dx$

17. $\displaystyle\int \frac{-x}{(5x - 1)^2}\, dx$

18. $\displaystyle\int \frac{-3}{x(4x + 3)^2}\, dx$

19. $\displaystyle\int x^4 \ln |x|\, dx$

20. $\displaystyle\int 4x^2 \ln |x|\, dx$

21. $\displaystyle\int \frac{\ln |x|}{x^2}\, dx$

22. $\displaystyle\int \frac{-2 \ln |x|}{x^3}\, dx$

23. $\displaystyle\int xe^{-2x}\, dx$

24. $\displaystyle\int xe^{3x}\, dx$

Use Table 3 to solve the following problems.

25. Management The rate of change of revenue in dollars from the sale of x units of small desk calculators is

$$R'(x) = \frac{1000}{\sqrt{x^2 + 25}}.$$

Find the total revenue from the sale of the first 20 calculators.

26. Natural Science The rate of reaction to a drug is given by

$$r'(x) = 2x^2 e^{-x},$$

where x is the number of hours since the drug was administered. Find the total reaction to the drug from $x = 1$ to $x = 6$.

27. Natural Science The rate of growth of a microbe population is given by

$$m'(x) = 30xe^{2x},$$

where x is time in days. What is the total accumulated growth after 3 days?

28. Social Science The rate (in hours per item) at which a worker in a certain job produces the xth item is

$$h'(x) = \sqrt{x^2 + 16}.$$

What is the total number of hours it will take this worker to produce the first 7 items?

13.7 DIFFERENTIAL EQUATIONS

Suppose that an economist wants to develop an equation that will forecast interest rates. By studying data on previous changes in interest rates, she hopes to find a relationship between the level of interest rates and their rate of change. A function giving the rate of change of interest rates would be the derivative of the function describing the level of interest rates. A **differential equation** is an equation that involves an unknown function, $y = f(x)$, and a finite number of its derivatives. Solving the differential equation for y would give the unknown function to be used for forecasting interest rates.

Usually a solution of an equation is a *number*. A solution of a differential equation, however, is a *function*. For example, the solutions of a differential equation such as

$$\frac{dy}{dx} = 3x^2 - 2x \tag{1}$$

consist of all functions y that satisfy the equation. Since the left side of the equation is the derivative of y with respect of x, we can solve the equation for y by finding an antiderivative on each side. On the left, the antiderivative is $y + C_1$. On the right side,

$$\int (3x^2 - 2x) \, dx = x^3 - x^2 + C_2.$$

1 Find the general solution of

(a) $dy/dx = 4x$;

(b) $dy/dx = -x^3$;

(c) $dy/dx = 2x^2 - 5x$.

Answers:

(a) $y = 2x^2 + C$

(b) $y = -\dfrac{1}{4}x^4 + C$

(c) $y = \dfrac{2}{3}x^3 - \dfrac{5}{2}x^2 + C$

The solutions of equation (1) are given by

$$y + C_1 = x^3 - x^2 + C_2$$

or

$$y = x^3 - x^2 + C_2 - C_1.$$

Replacing the constant $C_2 - C_1$ with the single constant C gives

$$y = x^3 - x^2 + C. \tag{2}$$

(From now on we will add just one constant, with the understanding that it represents the difference between the two constants obtained in the two integrations.)

Each different value of C in equation (2) leads to a different solution of equation (1), showing that a differential equation can have an infinite number of solutions. Equation (2) is the **general solution** of the differential equation (1). Some of the solutions of equation (1) are graphed in Figure 13.24. **1**

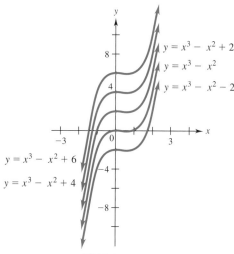

$y = x^3 - x^2 + 2$

$y = x^3 - x^2$

$y = x^3 - x^2 - 2$

$y = x^3 - x^2 + 6$

$y = x^3 - x^2 + 4$

FIGURE 13.24

The simplest kind of differential equation has the form

$$\frac{dy}{dx} = f(x).$$

Equation (1) has this form, so the solution of equation (1) suggests the following generalization.

2 Find the particular solution in Example 1 if there were 100 birds in the flock after 2 years.

Answer:
$P = 400e^{.05x} - 342$

General Solution of $dy/dx = f(x)$

The general solution of the differential equation $dy/dx = f(x)$ is

$$y = \int f(x)\, dx.$$

▶**EXAMPLE 1**   The population P of a flock of birds is growing exponentially so that

$$\frac{dP}{dx} = 20e^{.05x},$$

where x is time in years. Find P in terms of x if there were 20 birds in the flock initially.

Solve the differential equation.

$$P = \int 20e^{.05x}\, dx = \frac{20}{.05} e^{.05x} + C = 400e^{.05x} + C$$

Since P is 20 when x is 0,

$$\mathbf{20} = 400e^0 + C$$
$$-380 = C,$$

and

$$P = 400e^{.05x} - 380. \quad \blacktriangleleft \ \boxed{2}$$

In Example 1, the given information was used to produce a solution with a specific value of C. Such a solution is called a **particular solution** of the given differential equation. The given information, $P = 20$ when $t = 0$, is called an **initial condition** because $t = 0$.

Sometimes a differential equation must be rewritten in the form

$$\frac{dy}{dx} = f(x)$$

before it can be solved.

▶**EXAMPLE 2**   Find the particular solution of

$$\frac{dy}{dx} - 2x = 5,$$

given that $y = 2$ when $x = -1$.

Add $2x$ to both sides of the equation to get

$$\frac{dy}{dx} = 2x + 5.$$

3 Find the particular solution of

$$2\frac{dy}{dx} - 4 = 6x^2,$$

given $y = 4$ when $x = 1$.

Answer:
$y = x^3 + 2x + 1$

The general solution is

$$y = \frac{2x^2}{2} + 5x + C = x^2 + 5x + C.$$

Substituting 2 for y and -1 for x gives

$$2 = (-1)^2 + 5(-1) + C$$
$$C = 6.$$

The particular solution is $y = x^2 + 5x + 6$. ◄ **3**

Now let us consider solving the more general type of differential equation with the form

$$\frac{dy}{dx} = \frac{f(x)}{g(y)}.$$

Using differentials and multiplying on both sides by $g(y)\,dx$ gives

$$g(y)\,dy = f(x)\,dx.$$

In this form all terms involving y (including dy) are on one side of the equation, and all terms involving x (and dx) are on the other side. A differential equation in this form is said to be **separable,** since the variables x and y can be separated. A separable differential equation may be solved by integrating each side.

▶ **EXAMPLE 3** Find the general solution of $\dfrac{dy}{dx} = ky$, where k is a constant.

Separating variables leads to

$$\frac{1}{y}\,dy = k\,dx.$$

To solve this equation, take antiderivatives on each side.

$$\int \frac{1}{y}\,dy = \int k\,dx$$
$$\ln|y| = kx + C$$
$$|y| = e^{kx+C}. \qquad \text{Definition of logarithm}$$
$$|y| = e^{kx}e^{C}. \qquad \text{Property of exponents}$$
$$y = e^{kx}e^{C} \quad \text{or} \quad y = -e^{kx}e^{C}. \qquad \text{Definition of absolute value}$$

Since e^{C} and $-e^{C}$ are constants, replace them with the constant M, which may have any nonzero real-number value, to get the single equation

$$y = Me^{kx}.$$

This equation, $y = Me^{kx}$, defines the exponential growth or decay function that was discussed in Chapter 4. ◄

4 Find the particular solution
of $\dfrac{dy}{dx} = .05y$ if y is 2000 when
x is 0.

Answer:
$y = 2000e^{.05x}$

Recall that equations of the form $y = Me^{kx}$ arise in situations where the rate of change of a quantity is proportional to the amount present at time x; that is, where

$$\frac{dy}{dx} = ky.$$

The constant k is called the **growth rate constant,** while M represents the amount present at time $x = 0$. (A positive value of k indicates growth, while a negative value of k indicates decay.) **4**

As a model of population growth, the equation $y = Me^{kx}$ is not realistic over the long run for most populations. As shown by graphs of functions of the form $y = Me^{kx}$, with both M and k positive, growth would be unbounded. Additional factors, such as space restrictions or a limited amount of food, tend to inhibit growth of populations as time goes on. In an alternative model that assumes a maximum population of size N, the rate of growth of a population is proportional to how close the population is to that maximum. These assumptions lead to the differential equation

$$\frac{dy}{dx} = k(N - y),$$

the limited growth function mentioned in Chapter 4. Graphs of limited growth functions look like the graph in Figure 13.25, where y_0 is the initial population.

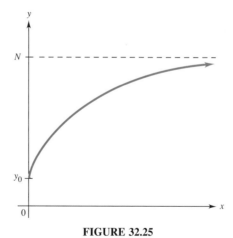

FIGURE 32.25

▶ **EXAMPLE 4**   A certain area can support no more than 4000 mountain goats. There are 1000 goats in the area at present, with a growth constant of .20.

5 An animal population is growing at a constant rate of 4%. The habitat will support no more than 10,000 animals. There are 3000 animals present now. Write an equation giving the population y in x years.

Answer:
$y = 10{,}000 - 7000e^{-.04x}$

(a) Write a differential equation for the rate of growth of this population. Let $N = 4000$ and $k = .20$. The rate of growth of the population is given by

$$\frac{dy}{dx} = .20(4000 - y).$$

To solve for y, first separate the variables.

$$\frac{dy}{4000 - y} = .2\, dx$$

$$\int \frac{dy}{4000 - y} = \int .2\, dx$$

$$-\ln(4000 - y) = .2x + C$$

$$\ln(4000 - y) = -.2x - C$$

$$4000 - y = e^{-.2x - C} = (e^{-.2x})(e^{-C})$$

The absolute value bars are not needed for $\ln(4000 - y)$ because y must be less than 4000 for this population so that $4000 - y$ is always nonnegative. Let $e^{-C} = B$. Then

$$4000 - y = Be^{-.2x}$$

$$y = 4000 - Be^{-.2x}.$$

Find B by using the fact that $y = 1000$ when $x = 0$.

$$1000 = 4000 - B$$

$$B = 3000$$

Notice that the value of B is the difference between the maximum population and the initial population. Substituting 3000 for B in the equation for y gives

$$y = 4000 - 3000e^{-.2x}.$$

(b) What will the goat population be in 5 years?
In 5 years, the population will be

$$y = 4000 - 3000e^{(-.2)(5)} = 4000 - 3000e^{-1}$$

$$= 4000 - 1103.6 = 2896.4,$$

or about 2900 goats. ◀ **5**

Marginal productivity is the rate at which production changes (increases or decreases) for a unit change in investment. Thus, marginal productivity can be expressed as the first derivative of the function which gives production in terms of investment.

6 In Example 5, if marginal productivity is changed to

$$P'(x) = 3x^2 + 2x,$$

with the same initial conditions,

(a) find an equation for production;

(b) find the increase in production if investment increases to $500,000.

Answers:

(a) $P(x) = x^3 + x^2 + 64$

(b) Production goes from 100 units to 214 units, an increase of 114 units.

▶**EXAMPLE 5** Suppose the marginal productivity of a manufacturing process is given by

$$P'(x) = 3x^2 - 10, \tag{4}$$

where x is the amount of the investment in hundreds of thousands of dollars. If the process produces 100 units per month with the present investment of $300,000 (that is, $x = 3$), by how much would production increase if the investment is increased to $500,000?

To obtain an equation for production, we can take antiderivatives on both sides of equation (4) to get

$$P(x) = x^3 - 10x + C.$$

To find C, use the given initial values: $P(x) = 100$ when $x = 3$.

$$100 = 3^3 - 10(3) + C$$
$$C = 103$$

Production is thus given by

$$P(x) = x^3 - 10x + 103,$$

and if investment is increased to $500,000, production becomes

$$P(5) = 5^3 - 10(5) + 103 = 178.$$

An increase to $500,000 in investment will increase production from 100 units to 178 units. ◀ **6**

13.7 EXERCISES

Find general solutions for the following differential equations. (See Examples 1–3.)

1. $\dfrac{dy}{dx} = -2x + 3x^2$

2. $\dfrac{dy}{dx} = 3e^{-2x}$

3. $3x^3 - 2\dfrac{dy}{dx} = 0$

4. $3x^2 - 3\dfrac{dy}{dx} = 2$

5. $y\dfrac{dy}{dx} = x$

6. $y\dfrac{dy}{dx} = x^2 - 1$

7. $\dfrac{dy}{dx} = 2xy$

8. $\dfrac{dy}{dx} = x^2y$

9. $\dfrac{dy}{dx} = 3x^2y - 2xy$

10. $(y^2 - y)\dfrac{dy}{dx} = x$

11. $\dfrac{dy}{dx} = \dfrac{y}{x}, \; x > 0$

12. $\dfrac{dy}{dx} = \dfrac{y}{x^2}$

13. $\dfrac{dy}{dx} = y - 5$

14. $\dfrac{dy}{dx} = 3 - y$

15. $\dfrac{dy}{dx} = y^2e^x$

16. $\dfrac{dy}{dx} = \dfrac{e^x}{e^y}$

Find particular solutions for the following equations. (See Examples 1–3.)

17. $\dfrac{dy}{dx} + 2x = 3x^2; \quad y = 2$ when $x = 0$

18. $\dfrac{dy}{dx} = 4x^3 - 3x^2 + x; \quad y = 0$ when $x = 1$

19. $2\dfrac{dy}{dx} = 4xe^{-x}; \quad y = 42$ when $x = 0$

20. $x\dfrac{dy}{dx} = x^2e^{3x}; \quad y = \dfrac{8}{9}$ when $x = 0$

21. $\dfrac{dy}{dx} = \dfrac{x^2}{y}; \quad y = 3$ when $x = 0$

22. $x^2\dfrac{dy}{dx} = y; \quad y = -1$ when $x = 1$

23. $(2x + 3)y = \dfrac{dy}{dx}; \quad y = 1$ when $x = 0$

24. $x\dfrac{dy}{dx} - y\sqrt{x} = 0; \quad y = 1$ when $x = 0$

25. $\dfrac{dy}{dx} = \dfrac{2x + 1}{y - 3}; \quad y = 4$ when $x = 0$

26. $\dfrac{dy}{dx} = \dfrac{x^2 + 5}{2y - 1}; \quad y = 11$ when $x = 0$

---

27. What is the difference between a general solution and a particular solution of a differential equation?

28. What is meant by a separable differential equation?

Work the following problems. (See Examples 4 and 5.)

29. Management   The marginal productivity of a process is given by

$$\frac{dy}{dx} = \frac{100}{32 - 4x},$$

where x represents the investment (in thousands of dollars). Find the productivity for each of the following expenditures if productivity is 100 units when the investement is $1000.
(a) $3000   **(b)** $5000
(c) Can expenditures ever reach $8000 according to this model? Why?

30. Natural Science   The time dating of dairy products depends on the solution of a differential equation. The rate of growth of bacteria in such products increases with time. If y is the number of bacteria (in thousands) present at a time t (in days), then the rate of growth of bacteria can be expressed as dy/dt and we have

$$\frac{dy}{dt} = kt,$$

where k is an appropriate constant. For a certain product, $k = 10$ and $y = 50$ (in thousands) when $t = 0$.
(a) Solve the differential equation for y.
(b) Suppose the maximum allowable value for y is 550 (thousand). How should the product be dated?

31. Management   Sales (in thousands) of a certain product are declining at a rate proportional to the amount of sales, with a decay constant of 25% per year.
(a) Write a differential equation to express the rate of sales decline.
(b) Find a general solution to the equation in part (a).

(c) How much time will pass before sales become 30% of their original amount?

32. Management   The rate of growth of an investment A is proportional to A, if money is compounded continuously; that is

$$\frac{dA}{dt} = kA,$$

where k is the interest rate.
(a) Solve the differential equation for A.
(b) Suppose inflation grows continuously at a rate of 4% per year. How long will it take for $1 to lose half its value?
(c) How long would it take for $1 to lose half its value at an inflation rate of 6% per year?

33. Management   Suppose that the gross national product (GNP) of a particular country increases exponentially, with a growth constant of 2% per year. Ten years ago, the GNP was 10^5 dollars. What will the GNP be in 5 years?

34. Management   In a certain area, 1500 small business firms are threatened by bankruptcy. Assume the rate of change in the number of bankruptcies is proportional to the number of small firms that are not yet bankrupt. If the growth constant is 6% and if 100 firms are bankrupt initially, how many will be bankrupt in 2 years?

35. Natural Science   The rate at which the number of bacteria in a culture is changing after the introduction of a bactericide is given by

$$\frac{dy}{dx} = 50 - y,$$

where y is the number of bacteria (in thousands) present at time x. Find the number of bacteria present at each of the following times if there were 1000 thousand bacteria present at time $x = 0$.
(a) $x = 2$ **(b)** $x = 5$ **(c)** $x = 10$

36. Natural Science The amount of a tracer dye injected into the bloodstream decreases exponentially, with a decay constant of 3% per minute. If 6 cc are present initially, how many cc are present after 10 minutes? (Here k will be negative.)

37. Physical Science The amount of a radioactive substance decreases exponentially, with a decay constant of 5% a month. There are 90 grams at the start of an experiment. Find the amount left 10 months later.

38. Social Science Suppose the rate at which a rumor spreads—that is, the number of people who have heard the rumor over a period of time—increases with the number of people who have heard it. If y is the number of people who have heard the rumor, then

$$\frac{dy}{dt} = ky,$$

where t is the time in days.
(a) If y is 1 when $t = 0$, and y is 5 when $t = 2$, find k. Using the value of k from part (a), find y for each of the following times.
(b) $t = 3$ (b) $t = 5$ (d) $t = 10$

39. Social Science A company has found that the rate at which a person new to the assembly line produces items is

$$\frac{dy}{dx} = 7.5e^{-.3y},$$

where x is the number of days the person has worked on the line. How many items can a new worker be expected to produce on the eighth day if he produces none when $x = 0$?

Management *Elasticity of demand was discussed in Exercise 47 of Section 2.5 and in Case 15. It is defined as*

$$E = -\frac{p}{q} \cdot \frac{dq}{dp},$$

for demand q and price p. Find the general demand equation $q = f(p)$ for each of the following elasticity functions. (Hint: Set each function equal to

$$-\frac{p}{q} \cdot \frac{dq}{dp}$$

then solve for q. In Exercise 41, write the constant of integration as $\ln |C|$.)

40. $E = \dfrac{4p^2}{q^2}$ **41.** $E = 2$

42. Management The sales of a new model of car are expected to follow the logistic curve

$$y = \frac{100,000}{1 + 100e^{-x}},$$

where x is the number of months since the introduction of the model. Find the following.
(a) The initial sales level
(b) The maximum sales level
(c) The month in which the maximum sales level will be achieved
(d) The rate at which sales are growing 6 months after the introduction of the model

43. Management Rapid technological advancements make many products obsolete almost overnight. The phenomenon of a technically superior new product's replacing another product can be described by the logistic equation

$$y = \frac{N}{1 + \dfrac{N - y_0}{y_0} e^{-kx}},$$

where y is the fraction of the total market share held by the new product, N is total market share, y_0 is the initial market share, and k is the growth or decay constant. Suppose $y_0 = .1$, $N = 1$, and $k = .2$.
(a) Use a grapher to graph the function for $0 \le x \le 25$, $0 \le y \le 1$.
(b) At what point does the market share of the new product appear to be growing most rapidly? (*Hint:* The most rapid growth occurs when $y = .5$.)

KEY TERMS AND SYMBOLS

13.1 $\int f(x)\,dx$ indefinite integral of f

antiderivative

integral sign

integrand

integration

power rule

constant multiple rule

sum or difference rule

integrals of e^x

integral of x^{-1}

13.2 differential

integration by substitution

13.3 $\int_a^b f(x)\,dx$ definite integral of f

Σ the sum of

limits of integration

total change in $F(x)$

13.4 $F(x)\Big|_a^b = F(b) - F(a)$

Fundamental Theorem of Calculus

13.5 consumers' surplus

producers' surplus

13.6 tables of integrals

13.7 differential equation

general solution

particular solution

initial condition

separable differential equation

KEY CONCEPTS

$F(x)$ is an antiderivative of $f(x)$ if $F'(x) = f(x)$.

Indefinite Integral

If $F'(x) = f(x)$, then $\int f(x)\,dx = F(x) + C$, for any real number C.

Properties of Integrals

$\int k \cdot f(x)\,dx = k \cdot \int f(x)\,dx$, for any real number k.

$\int [f(x) \pm g(x)]\,dx = \int f(x)\,dx \pm \int g(x)\,dx$.

Rules For Integrals

For $u = f(x)$ and $du = f'(x)\,dx$,

$$\int u^n\,du = \frac{u^{n+1}}{n+1} + C; \qquad \int e^u\,du = e^u + C; \qquad \int u^{-1}\,du = \int \frac{du}{u} = \ln|u| + C.$$

The Definite Integral

If f is continuous on $[a, b]$, the definite integral of f from a to b is

$$\int_a^b f(x)\,dx = \lim_{n \to \infty} \sum_{i=1}^{n} f(x_i)\,\Delta x,$$

provided the limit exists, where $\Delta x = \dfrac{b - a}{n}$ and x_i is any value of x in the ith interval.

Total Change in $F(x)$	Let f be continuous on $[a, b]$ and $f(x) \geq 0$ for all x in $[a, b]$. If $f(x)$ is the rate of change of $F(x)$ for x in $[a, b]$, then the **total change** in $F(x)$ as x goes from a to b is given by $$\int_a^b f(x)\, dx.$$
Fundamental Theorem of Calculus	Let f be continuous on $[a\ b]$, and let F be any antiderivative of f. Then $$\int_a^b f(x)\, dx = F(b) - F(a).$$
General Solution of $\dfrac{dy}{dx} = f(x)$	The general solution of the differential equation $dy/dx = f(x)$ is $$y = \int f(x)\, dx.$$
General Solution of $\dfrac{dy}{dx} = ky$	The general solution of the differential equation $dy/dx = ky$ is $$y = Me^{kx}.$$

CHAPTER 13   REVIEW EXERCISES

Find each indefinite integral.

1. $\displaystyle\int (x^2 - 3x + 2)\, dx$

2. $\displaystyle\int (6 - x^2)\, dx$

3. $\displaystyle\int 3\sqrt{x}\, dx$

4. $\displaystyle\int \frac{\sqrt{x}}{2}\, dx$

5. $\displaystyle\int (x^{1/2} + 3x^{-2/3})\, dx$

6. $\displaystyle\int (2x^{4/3} + x^{-1/2})\, dx$

7. $\displaystyle\int \frac{-4}{x^3}\, dx$

8. $\displaystyle\int \frac{5}{x^4}\, dx$

9. $\displaystyle\int -3e^{2x}\, dx$

10. $\displaystyle\int 5e^{-x}\, dx$

11. $\displaystyle\int \frac{2}{x-1}\, dx$

12. $\displaystyle\int \frac{-4}{x+2}\, dx$

13. $\displaystyle\int xe^{3x^2}\, dx$

14. $\displaystyle\int 2xe^{x^2}\, dx$

15. $\displaystyle\int \frac{3x}{x^2-1}\, dx$

16. $\displaystyle\int \frac{-x}{2-x^2}\, dx$

17. $\displaystyle\int \frac{x^2\, dx}{(x^3+5)^4}$

18. $\displaystyle\int (x^2 - 5x)^4(2x - 5)\, dx$

19. $\displaystyle\int \frac{4x-5}{2x^2-5x}\, dx$

20. $\displaystyle\int \frac{12(2x+9)}{x^2+9x+1}\, dx$

21. $\displaystyle\int \frac{x^3}{e^{3x^4}}\, dx$

22. $\displaystyle\int e^{3x^2+4}\, x\, dx$

23. $\displaystyle\int -2e^{-5x}\, dx$

24. $\displaystyle\int e^{-4x}\, dx$

25. Explain how rectangles are used to approximate the area under a curve.

26. Evaluate $\sum\limits_{i=1}^{4} (i^2 - i)$.

27. Let $f(x) = 3x + 1$, $x_1 = -1$, $x_2 = 0$, $x_3 = 1$, $x_4 = 2$, and $x_5 = 3$. Find $\sum\limits_{i=1}^{5} f(x_i)$.

28. Approximate the area under the graph of $f(x) = 2x + 3$ and above the x-axis from $x = 0$ to $x = 4$ using four rectangles. Let the height of each rectangle be the function value on the left side.

29. Find $\int_0^4 (2x + 3)\, dx$ by using the formula for the area of a trapezoid: $A = \dfrac{1}{2}(B + b)h$, where B and b are the lengths of the parallel sides and h is the distance between them. Compare with Exercise 28. If the answers are different, explain why.

30. Explain under what circumstances substitution is useful in integration.

31. What does the fundamental theorem of calculus state?

32. Explain why the limits of integration are changed when u is substituted for an expression in x in a definite integral.

Find each definite integral.

33. $\displaystyle\int_1^5 (3x^{-2} + x^{-3})\, dx$

34. $\displaystyle\int_2^3 (5x^{-2} + x^{-4})\, dx$

35. $\displaystyle\int_1^3 2x^{-1}\, dx$

36. $\displaystyle\int_1^6 8x^{-1}\, dx$

37. $\displaystyle\int_0^4 2e^x\, dx$

38. $\displaystyle\int_1^6 \frac{5}{2}e^{4x}\, dx$

39. $\displaystyle\int_{\sqrt{5}}^5 2x\sqrt{x^2 - 3}\, dx$

40. $\displaystyle\int_0^1 x\sqrt{5x^2 + 4}\, dx$

Find the area between the x-axis and f(x) over each of the given intervals.

41. $f(x) = e^x$; $[0, 2]$

42. $f(x) = 1 + e^{-x}$; $[0, 4]$

Management   *Find the cost function for each of the marginal cost functions in Exercises 43–46.*

43. $C'(x) = 10 - 2x$;   fixed cost is $4.

44. $C'(x) = 2x + 3x^2$;   2 units cost $12.

45. $C'(x) = 3\sqrt{2x - 1}$;   13 units cost $270.

46. $C'(x) = \dfrac{1}{x + 1}$;   fixed cost is $18.

Work the following exercises.

47. **Management**   The rate of change of sales of a new brand of tomato soup, in thousands, is given by
$$S(x) = \sqrt{x} + 2,$$
where x is the time in months that the new product has been on the market. Find the total sales after 9 months.

48. **Management**   The curve shown in the next column gives the rate that an investment accumulates income (in dollars per year). Use rectangles of width 2 units and height determined by the function value at the midpoint to find the total income accumulated over 10 yr.

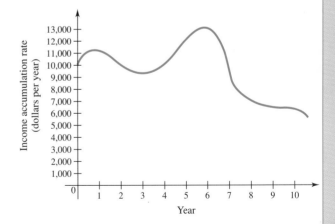

49. **Management**   A manufacturer of electronic equipment requires a certain rare metal. He has a reserve supply of 4,000,000 units that he will not be able to replace. If the rate at which the metal is used is given by
$$f(t) = 100,000e^{.03t},$$

where t is the time in years, how long will it be before he uses up the supply? (*Hint:* Find an expression for the total amount used in t years and set it equal to the known reserve supply.)

50. Management A company has installed new machinery that will produce a savings rate (in thousands of dollars) of

$$S'(x) = 225 - x^2,$$

where x is the number of years the machinery is to be used. The rate of additional costs (in thousands of dollars) to the company due to the new machinery is expected to be

$$C'(x) = x^2 + 25x + 150.$$

For how many years should the company use the new machinery? Find the net savings (in thousands of dollars) over this period.

51. Explain what consumers' surplus and producers' surplus are.

52. Management Suppose that the supply function of some commodity is

$$S(q) = q^2 + 5q + 100$$

and the demand function for the commodity is

$$D(q) = 350 - q^2.$$

(a) Find the producers' surplus.
(b) Find the consumers' surplus.

53. Natural Science The rate of infection of a disease (in people per month) is given by the function

$$I'(t) = \frac{100t}{t^2 + 1},$$

where t is the time in months since the disease broke out. Find the total number of infected people over the first four months of the disease.

Use the table of integrals to find the following.

54. $\displaystyle\int \frac{1}{\sqrt{x^2 - 64}}\, dx$

55. $\displaystyle\int \frac{5}{x\sqrt{25 + x^2}}\, dx$

56. $\displaystyle\int \frac{12}{x^2 - 9}\, dx$

57. $\displaystyle\int \frac{15x}{2x - 5}\, dx$

---

58. What is a differential equation? What is it used for?

Find general solutions for the following differential equations.

59. $\dfrac{dy}{dx} = 2x^3 + 6x$

60. $\dfrac{dy}{dx} = x^2 + 5x^4$

61. $\dfrac{dy}{dx} = \dfrac{3x + 1}{y}$

62. $\dfrac{dy}{dx} = \dfrac{e^x + x}{y - 1}$

Find particular solutions for the following differential equations.

63. $\dfrac{dy}{dx} = 5(e^{-x} - 1);\ y = 17$ when $x = 0$

64. $\dfrac{dy}{dx} = \dfrac{x}{x^2 - 3};\ y = 52$ when $x = 2$

65. $(5 - 2x)y = \dfrac{dy}{dx};\ y = 2$ when $x = 0$

66. $\sqrt{x}\dfrac{dy}{dx} = xy;\ y = 4$ when $x = 1$

---

67. Management A deposit of $10,000 is made to a savings account at 5% interest compounded continuously. Assume that continuous *withdrawals* of $1000 per year are made.
(a) Write a differential equation to describe the situation.
(b) How much will be left in the account after 1 year?

68. Management The marginal sales (in hundreds of dollars) of a computer software company are given by

$$\frac{dy}{dx} = 5e^{2x},$$

where x is the number of months the company has been in business. Assume that sales were 0 initially.
(a) Find the sales after 6 months.
(b) Find the sales after 12 months.

69. Management The rate at which a new worker in a certain factory produces items is given by

$$\frac{dy}{dx} = .2(125 - y),$$

where y is the number of items produced by the worker per day, x is the number of days worked, and the maximum production per day is 125 items. Assume the worker produced 20 items the first day on the job ($x = 0$).

(a) Find the number of items the new worker will produce in 10 days.

(b) According to the function that is the solution of the differential equation, can the worker ever produce 125 items in a day?

70. Physical Science A roast at a temperature of $40°$ is put in a $300°$ oven. After 1 hr the roast has reached a temperature of $150°$. Newton's law of cooling states that

$$\frac{dT}{dt} = k(T - T_F),$$

where T is the temperature of an object, the surrounding medium has temperature T_F at time t, and k is a constant.

(a) Use Newton's law to find the temperature of the roast after 2 hr.

(b) How long does it take for the roast to reach a temperature of $250°$?

71. Natural Science Find an equation relating x to y given the following equations, which describe the interaction of two competing species and their growth rates.

$$\frac{dx}{dt} = .2x - .5xy$$

$$\frac{dy}{dt} = -.3y + .4xy$$

Find the values of x and y for which both growth rates are 0. (Hint: solve the system of equations found by setting each growth rate equal to 0.)

72. Physical Science Acceleration is the rate of change of velocity, so $a(t) = dv/dt$. If air resistance is ignored, the acceleration of gravity is a constant g: $a(t) = g$. But air resistance cannot always be ignored, or parachutes would be of little use. In the presence of air resistance, the equation for acceleration also contains a term roughly proportional to the velocity squared. As acceleration forces a falling object downward and air resistance pushes it upward, the air resistance term is opposite in sign to the acceleration of gravity. Thus,

$$a(t) = \frac{dv}{dt} = b - kv^2,$$

where b and k are positive constants. Future calculations will be simpler if we replace b and k by the squared constants B^2 and K^2, giving

$$\frac{dv}{dt} = B^2 - K^2v^2.$$

(a) Use the table of integrals and separation of variables to solve the differential equation above. Assume $v < B/K$, which is certainly true when the object starts falling (with $v = 0$). Write your solution in the form of v as a function of t.

(b) Find $\lim\limits_{t \to \infty} v(t)$, where $v(t)$ is the solution you found in part (a). What does this tell you about a falling object in the presence of air resistance?

It is becoming more and more obvious that the earth contains only a finite quantity of minerals. The "easy and cheap" sources of minerals are being used up, forcing an ever more expensive search for new sources. For example, oil from the North Slope of Alaska would never have been used in the United States during the 1930s since there was so much Texas and California oil readily available.

Mineral usage tends to follow an exponential growth curve. Thus, if q represents the rate of consumption of a certain mineral at time t, while q_0 represents consumption when $t = 0$, then

$$q = q_0 e^{kt},$$

where k is the growth constant. For example, the world consumption of petroleum in a recent year was about 19,600 million barrels, with the value of k about 6%. Letting $t = 0$ correspond to this base year, then $q_0 = 19,600$, $k = .06$, and

$$q = 19,600 e^{.06t}$$

is the rate of consumption at time t, assuming that all present trends continue.

Based on estimates of the National Academy of Science, 2,000,000 million barrels of oil are now in provable reserves or are likely to be discovered in the future. At the present rate of consumption, how many years would be necessary to deplete these estimated reserves? Use the integral calculus of this chapter to find out.

To begin, find the total quantity of petroleum that would be used between time $t = 0$ and some future time $t = t_1$. The figure shows a typical graph of the function $q = q_0 e^{kt}$.

Following the work in Section 13.3, divide the time interval from $t = 0$ to $t = t_1$ into n subintervals. Let the ith subinterval have width Δt_i. Let the rate of consumption for the ith subinterval be approximated by q_i^*. Thus, the approximate total consumption for the subinterval is given by

$$q_i^* \cdot \Delta t_i,$$

and the total consumption over the interval from time $t = 0$ to $t = t_1$ is approximated by

$$\sum_{i=1}^{n} q_i^* \cdot \Delta t_i.$$

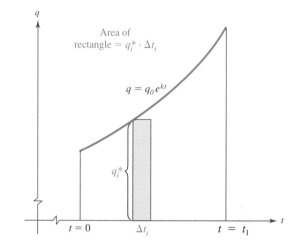

The limit of this sum as each of the Δt_i's approaches 0 gives the total consumption from time $t = 0$ to $t = t_1$. That is,

$$\text{Total consumption} = \lim_{\Delta t_i \to 0} \Sigma q_i^* \cdot \Delta t_i.$$

This limit is the definite integral of the function $q = q_0 e^{kt}$ from $t = 0$ to $t = t_1$, or

$$\text{Total consumption} = \int_0^{t_1} q_0 e^{kt} \, dt.$$

Evaluating this definite integral gives

$$\int_0^{t_1} q_0 e^{kt} \, dt = q_0 \int_0^{t_1} e^{kt} \, dt = q_0 \left(\frac{1}{k} e^{kt} \right) \Bigg|_0^{t_1}$$

$$= \frac{q_0}{k} e^{kt} \Bigg|_0^{t_1} = \frac{q_0}{k} e^{kt_1} - \frac{q_0}{k} e^0$$

$$= \frac{q_0}{k} e^{kt_1} - \frac{q_0}{k}(1)$$

$$= \frac{q^0}{k}(e^{kt_1} - 1). \tag{1}$$

Now return to the numbers given for petroleum: $q_0 = 19,600$ million barrels, where q_0 represents consumption in the base year; $k = .06$; and total petroleum reserves are estimated as 2,000,000 million barrels. Thus, using equation (1),

$$2,000,000 = \frac{19,600}{.06}(e^{.06t_1} - 1).$$

Multiply both sides of the equation by .06.

$$120,000 = 19,600(e^{.06t_1} - 1)$$

Divide both sides of the equation by 19,600.

$$6.1 = e^{.06t_1} - 1$$

Add 1 to both sides.

$$7.1 = e^{.06t_1}$$

Take natural logarithms of both sides.

$$\ln 7.1 = \ln e^{.06t_1} = .06t_1 \ln e$$
$$= .06t_1 \text{ (since } \ln e = 1)$$

Finally,

$$t_1 = \frac{\ln 7.1}{.06}.$$

A calculator gives

$$t_1 \approx 33.$$

By this result, petroleum reserves will last the world for 33 years.

The results of mathematical analyses such as this must be used with great caution. By the analysis above, the world would use all the petroleum that it wants in the thirty-second year after the base year, but there would be none at all in 34 years. This is not at all realistic. As petroleum reserves decline, the price will increase, causing demand to decline and supplies to increase.

EXERCISES

1. Find the number of years that the estimated petroleum reserves would last if used at the same rate as in the base year.

2. How long would the estimated petroleum reserves last if the growth constant was only 2% instead of 6%?

Estimate the length of time until depletion for each of the following minerals.

3. Bauxite (the ore from which aluminum is obtained), estimated reserves in base year 15,000,000 thousand tons, rate of consumption 63,000 thousand tons, growth constant 6%.

4. Bituminous coal, estimated world reserves 2,000,000 million tons, rate of consumption 2200 million tons, growth constant 4%.

This case uses some of the ideas of probability. The probability of an event is a number p, where $0 \le p \le 1$, such that p is the ratio of the number of ways that the event can happen divided by the total number of possible outcomes. For example, the probability of drawing a red card from a deck of 52 cards (of which 26 are red) is given by

$$P(\text{red card}) = \frac{26}{52} = \frac{1}{2}.$$

In the same way, the probability of drawing a black queen from a deck of 52 cards is

$$P(\text{black queen}) = \frac{2}{52} = \frac{1}{26}.$$

In this case the cost of a warranty program to a manufacturer is found. This cost depends on the quality of the products made. These variables are used.

$c =$ constant product price, per unit, including cost of warranty (We assume the price charged per unit is constant, since this price is likely to be fixed by competition.)
$m =$ expected lifetime of the product
$w =$ length of the warranty period
$N =$ size of a production lot, such as a year's production
$r =$ warranty cost per unit
$C(t) =$ pro rata customer rebate at time t
$P(t) =$ probability of product failure at any time t
$F(t) =$ number of failures occurring at time t.

Assume that the warranty is of the pro rata customer rebate type, in which the customer is paid for the proportion of the warranty left. Hence, if the product has a warranty period of w and fails at time t, then the product worked for the fraction t/w of the warranty period. Hence, the customer is reimbursed

for the unused portion of the warranty, or the fraction

$$1 - \frac{t}{w}.$$

Assuming the product cost c originally, and using $C(t)$ to represent the customer rebate at time t,

$$C(t) = c\left(1 - \frac{t}{w}\right).$$

For many different types of products, it has been shown by experience that

$$P(t) = 1 - e^{-t/m}$$

provides a good estimate of the probability of product failure at time t. The total number of failures at time t is given by the product of $P(t)$ and N, the total number of items per batch. If we use $F(t)$ to represent this total, we have

$$F(t) = N \cdot P(t) = N(1 - e^{-t/m}).$$

The total number of failures in some "tiny time interval" of width dt can be shown to be the derivative of $F(t)$,

$$F'(t) = \left(\frac{N}{m}\right)e^{-t/m},$$

while the cost for the failure in this "tiny time interval" is

$$C(t) \cdot F'(t) = c\left(1 - \frac{t}{w}\right)\left(\frac{N}{m}\right)e^{-t/m}.$$

The total cost for all failures during the warranty period is thus given by the definite integral

$$\int_0^w c\left(1 - \frac{t}{w}\right)\left(\frac{N}{m}\right)e^{-t/m}\, dt.$$

*From "Determination of Warranty Reserves," by Warren W. Menke, from *Management Science*, Vol. 15, No. 10, June 1969. Copyright © 1969 The Institute of Management Sciences. Reprinted by permission.

Using integration by parts (a method of integration not discussed), this definite integral can be shown to equal

$$Nc\left[-e^{-t/m} + \frac{t}{w} \cdot e^{-t/m} + \frac{m}{w}(e^{-t/m})\right]_0^w$$

or

$$Nc\left(1 - \frac{m}{w} + \frac{m}{w}e^{-w/m}\right).$$

This last quantity is the total warranty cost for all the units manufactured. As there are N units per batch, the warranty cost per item is

$$r = \frac{1}{N}\left[Nc\left(1 - \frac{m}{w} + \frac{m}{w}e^{-w/m}\right)\right]$$

$$= c\left(1 - \frac{m}{w} + \frac{m}{w}e^{-w/m}\right).$$

For example, suppose a product which costs $100 has an expected life of 24 months, with a 12-month warranty. Then we have $c = \$100$, $m = 24$, $w = 12$, with r, the warranty cost per unit, given by

$$r = 100(1 - 2 + 2e^{-.5})$$
$$= 100[-1 + 2(.6065)]$$
$$= 100(.2130) = 21.30.$$

EXERCISES

Find r for each of the following.

1. $c = \$50$, $m = 48$ months, $w = 24$ months

2. $c = \$1000$, $m = 60$ months, $w = 24$ months

3. $c = \$1200$, $m = 30$ months, $w = 30$ months

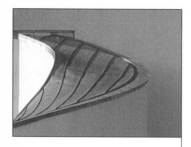

CHAPTER 14

Multivariate Calculus

Many of the ideas developed for functions of one variable also apply to functions of more than one variable. In particular, the fundamental idea of derivative generalizes in a very natural way to functions of more than one variable.

14.1 FUNCTIONS OF SEVERAL VARIABLES

If a company produces x items at a cost of \$10 per item, then the total cost $C(x)$ of producing the items is given by

$$C(x) = 10x.$$

The cost is a function of one independent variable, the number of items produced. If the company produces two products, with x of one product at a cost of \$10 each, and y of another product at a cost of \$15 each, then the total cost to the firm is a function of *two* independent variables, x and y. By generalizing $f(x)$ notation, the total cost can be written as $C(x, y)$, where

$$C(x, y) = 10x + 15y.$$

When $x = 5$ and $y = 12$, the total cost is written $C(5, 12)$, with

$$C(5, 12) = 10 \cdot 5 + 15 \cdot 12 = 230.$$

Here is a general definition.

$z = f(x, y)$ is a **function of two independent variables** if a unique value of z is obtained from each ordered pair of real numbers (x, y). The variables x and y are **independent variables;** z is the **dependent variable.** The set of all ordered pairs of real numbers (x, y) such that $f(x, y)$ is a real number is the **domain** of f; the set of all values of $f(x, y)$ is the **range.**

1

Let $f(x, y) = x^3 - 4x^2 + xy$.
Find

(a) $f(2, 4)$;

(b) $f(-2, 3)$.

Answers:

(a) 0

(b) -30

▶**EXAMPLE 1** Let $f(x, y) = 4x^2 + 2xy + 3/y$ and find each of the following.
(a) $f(-1, 3)$.
Replace x with -1 and y with 3.

$$f(-1, 3) = 4(-1)^2 + 2(-1)(3) + \frac{3}{3} = 4 - 6 + 1 = -1$$

(b) $f(2, 0)$
Because of the quotient $3/y$, it is not possible to replace y with 0, so $f(2, 0)$
is undefined. By inspection we see that the domain of the function f consists of
all ordered pairs (x, y) such that $y \neq 0$. ◀ **1**

▶**EXAMPLE 2** Let x represent the number of milliliters (ml) of carbon dioxide released by the lungs in 1 minute. Let y be the change in the carbon dioxide
content of the blood as it leaves the lungs (y is measured in ml of carbon
dioxide per 100 ml of blood). The total output of blood from the heart in one
minute (measured in ml) is given by C, where C is a function of x and y such
that

$$C = C(x, y) = \frac{100x}{y}.$$

Find $C(320, 6)$.
Replace x with 320 and y with 6 to get

$$C(320, 6) = \frac{100(320)}{6}$$

$$\approx 5333 \text{ ml of blood per minute.} ◀$$

The definition given before Example 1 was for a function of two independent
variables, but similar definitions could be given for functions of three, four, or
more independent variables. Functions of more than one independent variable
are called **multivariate functions.**

GRAPHING FUNCTIONS OF TWO INDEPENDENT VARIABLES Functions
of one independent variable are graphed by using an x-axis and a y-axis to locate
points in a plane. The plane determined by the x- and y-axes is called the
xy-plane. A third axis is needed to graph functions of two independent variables—the z-axis, which goes through the origin in the xy-plane and is perpendicular to both the x-axis and the y-axis.

Figure 14.1 on the next page shows one possible way to draw the three axes.
In Figure 14.1, the yz-plane is in the plane of the page, with the x-axis perpendicular to the plane of the page.

2 Locate $P(3, 2, 4)$ on the coordinate system below.

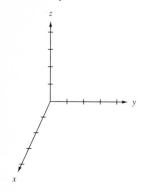

Answer:

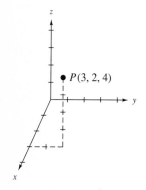

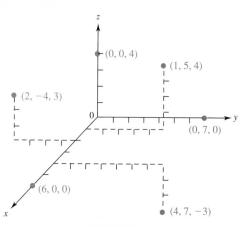

FIGURE 14.1

Just as we graphed ordered pairs earlier, we can now graph **ordered triples** of the form (x, y, z). For example, to locate the point corresponding to the ordered triple $(2, -4, 3)$, start at the origin and go 2 units along the positive x-axis. Then go 4 units in a negative direction (to the left), parallel to the y-axis. Finally, go up 3 units, parallel to the z-axis. The point representing $(2, -4, 3)$ is shown in Figure 14.1, along with several other points. The region of three-dimensional space where all coordinates are positive is called the **first octant. 2**

In Chapter 2 we saw that the graph of $ax + by = c$ (where a and b are not both 0) is a straight line. This result generalizes to three dimensions.

Plane

The graph of

$$ax + by + cz = d$$

is a plane when a, b, c, and d are real numbers, with a, b, and c not all zero.

▶**EXAMPLE 3** Graph $2x + y + z = 6$.

By the result above, the graph of this equation is a plane. Earlier, we graphed straight lines by finding x- and y-intercepts. A similar idea helps graph a plane. To find the x-intercept, the point where the graph crosses the x-axis, let $y = 0$ and $z = 0$.

$$2x + 0 + 0 = 6$$
$$x = 3$$

The point (3, 0, 0) is on the graph. Letting $x = 0$ and $z = 0$ gives the point (0, 6, 0), while $x = 0$ and $y = 0$ lead to (0, 0, 6). The plane through these three points includes the triangular surface shown in Figure 14.2. This region is the first-octant part of the plane that is the graph of $2x + y + z = 6$. ◀

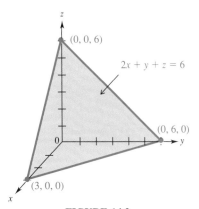

FIGURE 14.2

▶ **EXAMPLE 4**   Graph $x + z = 6$.

To find the x-intercept, let $z = 0$, giving (6, 0, 0). If $x = 0$, we get the point (0, 0, 6). Because there is no y in the equation $x + z = 6$, there can be no y-intercept. A plane that has no y-intercept is parallel to the y-axis. The first-octant portion of the graph of $x + z = 6$ is shown in Figure 14.3. ◀

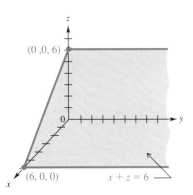

FIGURE 14.3

▶ **EXAMPLE 5**   Graph each of the following functions in two independent variables.

(a) $x = 3$.

3 Describe each graph and give any intercepts.

(a) $2x + 3y - z = 4$

(b) $x + y = 3$

(c) $z = 2$

Answers:

(a) A plane; $(2, 0, 0)$, $(0, 4/3, 0)$, $(0, 0, -4)$

(b) A plane parallel to the z-axis; $(3, 0, 0)$, $(0, 3, 0)$

(c) A plane parallel to the x-axis and the y-axis; $(0, 0, 2)$

This graph, which goes through $(3, 0, 0)$, can have no y-intercept and no z-intercept. It is, therefore, a plane parallel to the y-axis and the z-axis and, therefore, to the yz plane. The first-octant portion of the graph is shown in Figure 14.4.

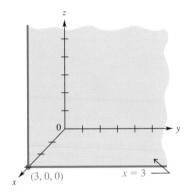

FIGURE 14.4

(b) $y = 4$.

This graph goes through $(0, 4, 0)$ and is parallel to the xz-plane. The first-octant portion of the graph is shown in Figure 14.5.

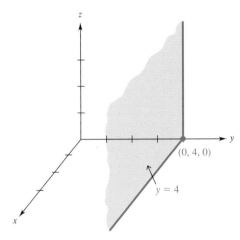

FIGURE 14.5

(c) Graph $z = 1$.

The graph is a plane parallel to the xy-plane, passing through $(0, 0, 1)$. Its first-octant portion is shown in Figure 14.6. ◀ **3**

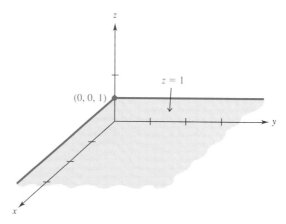

FIGURE 14.6

The graph of a function of one variable, $y = f(x)$, is a curve in the plane. If x_0 is in the domain of f, the point $(x_0, f(x_0))$ on the graph lies directly above, on, or below the number x_0 on the x-axis, as shown in Figure 14.7.

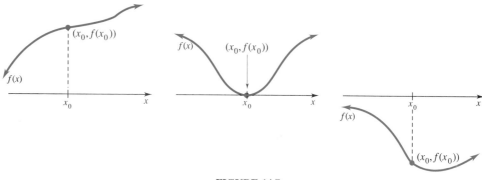

FIGURE 14.7

The graph of a function of two variables, $z = f(x, y)$ is a **surface** in three-dimensional space. If (x_0, y_0) is in the domain of f, the point $(x_0, y_0, f(x_0, y_0))$ lies directly above, on, or below the point (x_0, y_0) in the xy-plane, as shown in Figure 14.8 on the next page.

Although computer software is available for drawing the graphs of functions of two variables, you can often get a good picture of the graph without it by finding various **traces**—the curves that result when a surface is cut by a plane. The **xy-trace** is the intersection of the surface with the xy-plane. The **yz-trace** and **xz-trace** are defined similarly. You can also determine the intersection of the

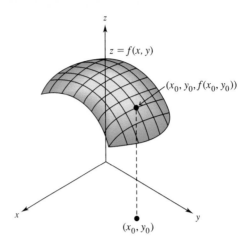

FIGURE 14.8

surface with planes parallel to the xy-plane. Such planes are of the form $z = k$, where k is a constant and the curves that result when they cut the surface are called **level curves.**

▶ **EXAMPLE 6** Graph $z = x^2 + y^2$.

The yz-plane is the plane in which every point has first coordinate 0, so its equation is $x = 0$. When $x = 0$, the equation becomes $z = y^2$, which is the equation of a parabola in the yz-plane, as shown in Figure 14.9(a). Similarly, to find the intersection of the surface with the xz-plane (whose equation is $y = 0$), let $y = 0$ in the equation. It then becomes $z = x^2$, which is the equation of a parabola in the xz-plane (shown in Figure 14.9(a)). The xy-trace (the intersection of the surface with the plane $z = 0$) is the single point $(0, 0, 0)$ because $x^2 + y^2$ is never negative, and equal to 0 only when $x = 0$ and $y = 0$.

Next, we find the level curves by intersecting the surface with the planes $z = 1, z = 2, z = 3$, etc. (all of which are parallel to the xy-plane). In each case, the result is a circle

$$x^2 + y^2 = 1, \quad x^2 + y^2 = 2, \quad x^2 + y^2 = 3,$$

and so on, as shown in Figure 14.9(b). Drawing the traces and level curves on the same set of axes suggests that the graph of $z = x^2 + y^2$ is the bowl-shaped figure, called a **paraboloid,** that is shown in Figure 14.9(c). ◀

One application of level curves in economics occurs with production functions. A **production function** $z = f(x, y)$ is a function that gives the quantity z of an item produced as a function of x and y, where x is the amount of labor and y is the amount of capital (in appropriate units) needed to produce z units.

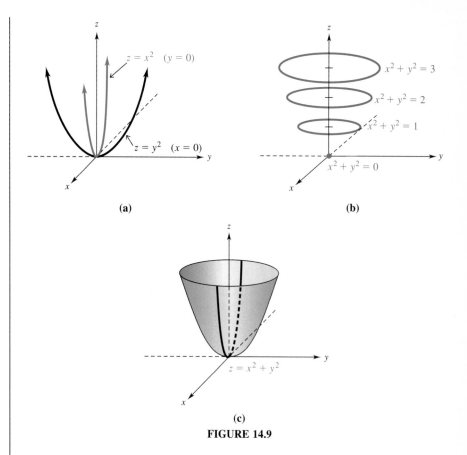

FIGURE 14.9

If the production function has the special form $z = P(x, y) = Ax^a y^{1-a}$, where A is a constant and $0 < a < 1$, the function is called a **Cobb-Douglas production function.**

▶ **EXAMPLE 7**   Find the level curve at a production of 100 items for the Cobb-Douglas production function $z = x^{2/3}y^{1/3}$.

Let $z = 100$ and solve for y to get

$$100 = x^{2/3}y^{1/3}$$

$$\frac{100}{x^{2/3}} = y^{1/3}.$$

Now cube both sides to express y as a function of x.

$$y = \frac{100^3}{x^2}$$

$$= \frac{1,000,000}{x^2}.$$

4 Find the equation of the level curve for production of 100 items if the production function is $z = 5x^{1/4}y^{3/4}$.

Answer:

$$y = \frac{20^{4/3}}{x^{1/3}}$$

The level curve of height 100 is shown graphed in three dimensions in Figure 14.10(a) and on the familiar xy-plane in Figure 14.10(b). The points of the graph correspond to those values of x and y that lead to production of 100 items. ◀ **4**

The curve in Figure 14.10 is called an *isoquant*, for *iso* (equal) and *quant* (amount). In Example 7, the "amounts" all "equal" 100.

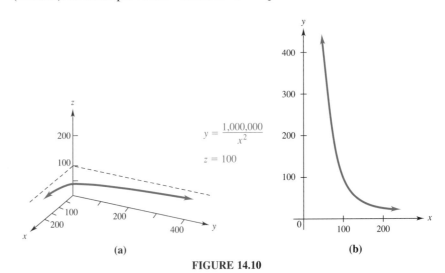

FIGURE 14.10

14.1 EXERCISES

1. What is meant by a multivariate function?

2. Describe how you would graph the function $x + y + z = 4$ and describe what the graph should look like.

For each of the following functions, find $f(2, -1), f(-4, 1), f(-2, -3),$ and $f(0, 8)$. (See Example 1.)

3. $f(x, y) = 5x + 2y - 4$ **4.** $f(x, y) = 2x^2 - xy + y^2$ **5.** $f(x, y) = \sqrt{y^2 + 2x^2}$ **6.** $f(x, y) = \dfrac{3x + 4y}{\ln |x|}$

---

7. What are xy-, xz-, and yz-traces of a graph?

8. What is a level curve?

Graph the first octant portion of each of the following planes. (See Examples 3–5.)

9. $3x + 2y + z = 12$

10. $2x + 3y + 3z = 18$

11. $x + y = 5$

12. $y + z = 3$

13. $z = 4$

14. $y = 3$

Graph the level curves in the first octant at heights of z = 0, z = 2, and z = 4 for the following equations. (See Example 6.)

15. $3x + 2y + z = 18$

16. $x + 3y + 2z = 8$

17. $y^2 - x = -z$

18. $2y - \dfrac{x^2}{3} = z$

Management *Find the level curve at a production of 500 for each of the production functions in Exercises 19–20. Graph each function on the xy-plane (See Example 7.)*

19. The production function z for the United States was once estimated as $z = x^{.7}y^{.3}$, where x stands for the amount of labor and y stands for the amount of capital.

20. If x represents the amount of labor and y the amount of capital, a production function for Canada is approximately $z = x^{.4}y^{.6}$.

The multiplier function

$$M = \frac{(1 + i)^n(1 - t) + t}{[1 + (1 - t)i]^n}$$

compares the growth of an Individual Retirement Account (IRA) with the growth of the same deposit in a regular savings account. The function M depends on the three variables n, i, and t, where n represents the number of years an amount is left at interest, i represents the interest rate in both types of accounts and t represents the income tax rate. Values of M > 1 indicate that the IRA grows faster than the savings account. Let M = f(n, i, t) and find the following.

21. Find the multiplier when funds are left for 25 years at 5% interest and the income tax rate is 28%. Which account grows faster?

22. What is the multiplier when money is invested for 25 years at 6% interest and the income tax rate is 33%? Which account grows faster?

23. Extra postage is charged for parcels sent by U.S. mail that are more than 108 inches in length and girth combined. (Girth is the distance around the parcel perpendicular to its length. See the figure.) Express the combined length and girth as a function of L, W, and H.

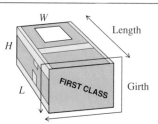

24. The holes cut in a roof for vent pipes require elliptical templates. A formula for determining the length of the major axis of the ellipse is $L = f(H, D) = \sqrt{H^2 + D^2}$, where D is the (outside) diameter of the pipe and H is the "rise" of the roof per D units of "run"; that is, the slope of the roof is H/D. (See the figure.) The width of the ellipse (minor axis) equals D. Find the length and width of the ellipse required to produce a hole for a vent pipe with a diameter of 3.75 inches in roofs with the following slopes.
(a) $3/4$ **(b)** $2/5$

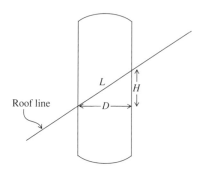

25. Let $f(x, y) = 9x^2 - 3y^2$, and find each of the following.

(a) $\dfrac{f(x + h, y) - f(x, y)}{h}$ **(b)** $\dfrac{f(x, y + h) - f(x, y)}{h}$

26. Let $f(x, y) = 7x^3 + 8y^2$, and find each of the following.

(a) $\dfrac{f(x + h, y) - f(x, y)}{h}$ **(b)** $\dfrac{f(x, y + h) - f(x, y)}{h}$

By considering traces, match each equation in Exercises 27–32 with its graph in (a)–(f).

27. $z = x^2 + y^2$

28. $z^2 - y^2 - x^2 = 1$

29. $x^2 - y^2 = z$

30. $z = y^2 - x^2$

31. $\dfrac{x^2}{16} + \dfrac{y^2}{25} + \dfrac{z^2}{4} = 1$

32. $z = 5(x^2 + y^2)^{-1/2}$

(a)

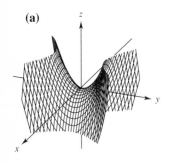

(b)

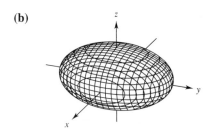

(c)

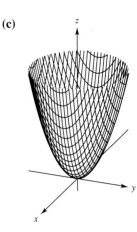

(d)

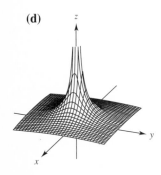

(e)

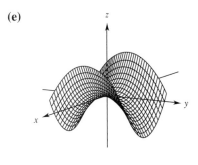

(f)

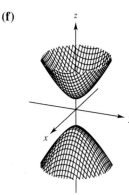

Use a graphing calculator or a computer to graph the level curves with z = 16, 36, and 64 for each of the following functions.

33. $6x + 8y + z = 64$

34. $z = 10(y - x^2)$

35. $z = 2xy$

36. $y = \sqrt{z - 4x^2}$

14.2 PARTIAL DERIVATIVES

A small firm makes only two products, radios and audiocassette recorders. The profits of the firm are given by

$$P(x, y) = 40x^2 - 10xy + 5y^2 - 80,$$

where x is the number of units of radios sold and y is the number of units of recorders sold. How will a change in x or y affect P?

Suppose that sales of radios have been steady at 10 units; only the sales of recorders vary. The management would like to find the marginal profit with respect to y, the number of recorders sold. Recall that marginal profit is given by the derivative of the profit function. Here, x is fixed at 10. Using this information, we begin by finding a new function, $f(y) = P(10, y)$. Let $x = 10$ to get

$$f(y) = P(10, y) = 40(10)^2 - 10(10)y + 5y^2 - 80$$
$$= 3920 - 100y + 5y^2.$$

The function $f(y)$ shows the profit from the sale of y recorders, assuming that x is fixed at 10 units. Find the derivative df/dy to get the marginal profit with respect to y.

$$\frac{df}{dy} = -100 + 10y$$

In this example, the derivative of the function $f(y)$ was taken with respect to y only; we assumed that x was fixed. To generalize, let $z = f(x, y)$. An intuitive definition of the *partial derivatives* of f with respect to x and y follows.

Partial Derivatives (Informal Definition)

The **partial derivative of f with respect to x** is the derivative of f obtained by treating x as a variable and y as a constant.

The **partial derivative of f with respect to y** is the derivative of f obtained by treating y as a variable and x as a constant.

The symbols $f_x(x, y)$ (no prime used), $\partial z/\partial x$, and $\partial f/\partial x$ are used to represent the partial derivative of $z = f(x, y)$ with respect to x, with similar symbols used for the partial derivative with respect to y. The symbol $f_x(x, y)$ is often abbreviated as just f_x, with $f_y(x, y)$ abbreviated f_y.

Generalizing from the definition of derivative given earlier, partial derivatives of a function $z = f(x, y)$ are formally defined as follows.

Partial Derivatives (Formal Definition)

Let $z = f(x, y)$ be a function of two variables. Then the **partial derivative of f with respect to x** is

$$f_x(x, y) = \lim_{h \to 0} \frac{f(x + h, y) - f(x, y)}{h};$$

the **partial derivative of f with respect to y** is

$$f_y(x, y) = \lim_{h \to 0} \frac{f(x, y + h) - f(x, y)}{h};$$

provided these limits exist.

1 Find f_x and f_y.

(a) $f(x, y) = -x^2y + 3xy + 2xy^2$

(b) $f(x, y) = x^3 + 2x^2y + xy$

Answers:

(a) $f_x = -2xy + 3y + 2y^2$; $f_y = -x^2 + 3x + 4xy$

(b) $f_x = 3x^2 + 4xy + y$; $f_y = 2x^2 + x$

2 Find f_x and f_y.

(a) $f(x, y) = \ln(2x + 3y)$

(b) $f(x, y) = e^{xy}$

Answers:

(a) $f_x = \dfrac{2}{2x + 3y}$; $f_y = \dfrac{3}{2x + 3y}$

(b) $f_x = ye^{xy}$; $f_y = xe^{xy}$

Similar definitions could be given for functions of more than two independent variables.

▶ **EXAMPLE 1**   Let $f(x, y) = 4x^2 - 9xy + 6y^3$. Find f_x and f_y.

To find f_x, treat y as a constant and x as a variable. The derivative of the first term, $4x^2$, is $8x$. In the second term, $-9xy$, the constant coefficient of x is $-9y$, so the derivative with x as the variable is $-9y$. The derivative of $6y^3$ is zero, since we are treating y as a constant. Thus,

$$f_x = 8x - 9y.$$

Now, to find f_y, treat y as a variable and x as a constant. Since x is a constant, the derivative of $4x^2$ is zero. In the second term, the coefficient of y is $-9x$ and the derivative of $-9xy$ is $-9x$. The derivative of the third term is $18y^2$. Thus,

$$f_y = -9x + 18y^2. \quad ◀ \;\; \boxed{1}$$

▶ **EXAMPLE 2**   Let $f(x, y) = \ln(x^2 + y)$. Find f_x and f_y.

Recall the formula for the derivative of a natural logarithm function. If $y = \ln(g(x))$, then $y' = g'(x)/g(x)$. Using this formula,

$$f_x = \frac{D_x(x^2 + y)}{x^2 + y} = \frac{2x}{x^2 + y},$$

and

$$f_y = \frac{D_y(x^2 + y)}{x^2 + y} = \frac{1}{x^2 + y} \quad ◀ \;\; \boxed{2}$$

The notation

$$f_x(a, b) \quad \text{or} \quad \frac{\partial f}{\partial x}(a, b)$$

represents the value of a partial derivative when $x = a$ and $y = b$, as shown in the next example.

▶ **EXAMPLE 3**   Let $f(x, y) = 2x^2 + 3xy^3 + 2y + 5$. Find the following.

(a) $f_x(-1, 2)$

First, find f_x by holding y constant.

$$f_x = 4x + 3y^3$$

Now let $x = -1$ and $y = 2$.

$$f_x(-1, 2) = 4(-1) + 3(2)^3 = -4 + 24 = 20$$

3 Let $f(x, y) = x^2 + xy^2 + 5y - 10$. Find the following.

(a) $f_x(2, 1)$

(b) $\dfrac{\partial f}{\partial y}(-1, 0)$

Answers:

(a) 5

(b) 5

(b) $\dfrac{\partial f}{\partial y}(-4, -3)$

Since $\partial f/\partial y = 9xy^2 + 2$,

$$\frac{\partial f}{\partial y}(-4, -3) = 9(-4)(-3)^2 + 2 = 9(-36) + 2 = -322. \quad \blacktriangleleft \quad \boxed{3}$$

The derivative of a function of one variable can be interpreted as the slope of the tangent line to the graph at that point. With some modification, the same is true of partial derivatives of functions of two variables. At a point on the graph of a function of two variables, $z = f(x, y)$, there may be many tangent lines, all of which lie in the same tangent plane, as shown in Figure 14.11.

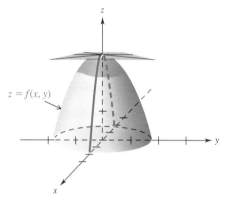

FIGURE 14.11

In any particular direction, however, there will be only one tangent line. We use partial derivatives to find the slope of the tangent lines in the x- and y-directions as follows.

Figure 14.12 shows a surface $z = f(x, y)$ and a plane that is parallel to the xz-plane. The equation of the plane is $y = a$. (This corresponds to holding y fixed).

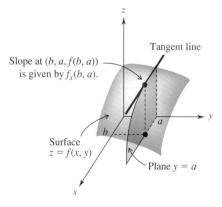

FIGURE 14.12

Because $y = a$ for points on the plane, any point on the curve that represents the intersection of the plane and the surface must have the form $(x, a, f(x, a))$. Thus, this curve can be described as $z = f(x, a)$. Since a is constant, $z = f(x, a)$ is a function of one variable. When the derivative of $z = f(x, a)$ is evaluated at $x = b$, it gives the slope of the line tangent to this curve at the point $(b, a, f(b, a))$, as shown in Figure 14.12. Thus, the partial derivative of f with respect to x, $f_x(b, a)$, gives the rate of change of the surface $z = f(x, y)$ in the x-direction at the point $(b, a, f(b, a))$. In the same way, the partial derivative with respect to y will give the slope of the line tangent to the surface in the y-direction at the point $(b, a, f(b, a))$.

RATE OF CHANGE The derivative of $y = f(x)$ gives the rate of change of y with respect to x. In the same way, if $z = f(x, y)$, then f_x gives the rate of change of z with respect to x, if y is held constant.

▶**EXAMPLE 4** Suppose that the temperature of the water at the point on a river where a nuclear power plant discharges its hot waste water is approximated by

$$T(x, y) = 2x + 5y + xy - 40.$$

Here, x represents the temperature of the river water in degrees Celsius before it reaches the power plant and y is the number of megawatts (in hundreds) of electricity being produced by the plant.
(a) Find and interpret $T_x(9, 5)$.
First, find the partial derivative T_x.

$$T_x = 2 + y$$

This partial derivative gives the rate of change of T with respect to x. Replacing x with 9 and y with 5 gives

$$T_x(9, 5) = 2 + 5 = 7.$$

Just as marginal cost is the approximate cost of one more item, this result, 7, is the approximate change in temperature of the output water if input water temperature changes by 1 degree, from $x = 9$ to $x = 9 + 1 = 10$, while y remains constant at 5 (500 megawatts of electricity produced).
(b) Find and interpret $T_y(9, 5)$.
The partial derivative T_y is

$$T_y = 5 + x.$$

This partial derivative gives the rate of change of T with respect to y, with

$$T_y(9, 5) = 5 + 9 = 14.$$

4 Use the function of Example 4 to find and interpret the following.

(a) $T_x(5, 4)$

(b) $T_y(8, 3)$

Answers:

(a) $T_x(5, 4) = 6$; the approximate increase in temperature if the input temperature increases from 5 to 6 degrees

(b) $T_y(8, 3) = 13$; the approximate increase in temperature if the production of electricity increases from 300 to 400 megawatts

This result, 14, is the approximate change in temperature resulting from a 1-unit increase in production of electricity from $y = 5$ to $y = 5 + 1 = 6$ (from 500 to 600 megawatts), while the input water temperature x remains constant at 9°C.

◀ **4**

As mentioned in the previous section, if $P(x, y)$ gives the output P produced by x units of labor and y units of capital, $P(x, y)$ is a production function. The partial derivatives of this production function have practical implications. For example, $\partial P/\partial x$ gives the marginal productivity of labor. This represents the rate at which the output is changing with respect to changes in labor for a fixed capital investment. That is, if the capital investment is held constant and labor is increased by 1 work hour, $\partial P/\partial x$ will yield the approximate change in the production level. Likewise, $\partial P/\partial y$ gives the marginal productivity of capital, which represents the rate at which the output is changing with respect to changes in capital for a fixed labor value. So if the labor force is held constant and the capital investment is increased by 1 unit, $\partial P/\partial y$ will approximate the corresponding change in the production level.

▶ **EXAMPLE 5** A company that manufactures computers has determined that its production function is given by

$$P(x, y) = 500x + 800y + 3x^2y - x^3 - \frac{y^4}{4},$$

where x is the size of the labor force (in work hours per week) and y is the amount of capital (in units of $1000) invested. Find the marginal productivity of labor and the marginal productivity of capital when $x = 50$ and $y = 20$, and interpret the results.

The marginal productivity of labor is found by taking the derivative of P with respect to x.

$$\frac{\partial P}{\partial x} = 500 + 6xy - 3x^2$$

$$\frac{\partial P}{\partial x}(50, 20) = 500 + 6(50)(20) - 3(50)^2$$

$$= -1000$$

Thus, if the capital investment is held constant at $20,000 and labor is increased from 50 to 51 work hours per week, production will decrease by about 1000 units. In the same way, the marginal productivity of capital is $\partial P/\partial y$.

$$\frac{\partial P}{\partial y} = 800 + 3x^2 - y^3$$

$$\frac{\partial P}{\partial y}(50, 20) = 800 + 3(50)^2 - (20)^3$$

$$= 300$$

5 Suppose a production function is given by $P(x, y) = 10x^2y + 100x + 400y - 5xy^2$, where x and y are defined as in Example 5. Find the marginal productivity of labor and capital when $x = 30$ and $y = 50$.

Answer:
Marginal productivity of labor is 17,600. Marginal productivity of capital is -5600.

If work hours are held constant at 50 hours per week and the capital investment is increased from \$20,000 to \$21,000, production will increase by about 300 units. ◄ **5**

SECOND-ORDER PARTIAL DERIVATIVES The second derviative of a function of one variable is very useful in determining relative maxima and minima. **Second-order partial derivatives** (partial derivatives of a partial derivative) are used in a similar way for functions of two or more variables. The situation is somewhat more complicated, however, with more independent variables. For example, $f(x, y) = 4x + x^2y + 2y$ has two first-order partial derivatives,

$$f_x = 4 + 2xy \quad \text{and} \quad f_y = x^2 + 2.$$

Because each of these has two partial derivatives, one with respect to y and one with respect to x, there are *four* second-order partial derivatives of function f. The notations for these four second-order partial derivatives are given below.

Second-Order Partial Derivatives

For a function $z = f(x, y)$, if all indicated partial derivatives exist, then

$$\frac{\partial}{\partial x}\left(\frac{\partial z}{\partial x}\right) = \frac{\partial^2 z}{\partial x^2} = f_{xx} \qquad \frac{\partial}{\partial y}\left(\frac{\partial z}{\partial y}\right) = \frac{\partial^2 z}{\partial y^2} = f_{yy}$$

$$\frac{\partial}{\partial y}\left(\frac{\partial z}{\partial x}\right) = \frac{\partial^2 z}{\partial y \,\partial x} = f_{xy} \qquad \frac{\partial}{\partial x}\left(\frac{\partial z}{\partial y}\right) = \frac{\partial^2 z}{\partial x \,\partial y} = f_{yx}.$$

As seen above, f_{xx} is used as an abbreviation for $f_{xx}(x, y)$, with f_{yy}, f_{xy}, and f_{yx} used in a similar way. The symbol f_{xx} is read "the partial derivative of f_x with respect to x," and f_{xy} is read "the partial derivative of f_x with respect to y." Also, the symbol $\partial^2 z/\partial y^2$ is read "the partial derivative of $\partial z/\partial y$ with respect to y."

Note For most functions found in applications and all the functions in this book, the second-order partial derivatives f_{xy} and f_{yx} are equal. Therefore, it is not necessary for us to be particular about the order in which these derivatives are found.

► **EXAMPLE 6** Find all second-order partial derivatives for

$$f(x, y) = -4x^3 - 3x^2y^3 + 2y^2.$$

First find f_x and f_y.

$$f_x = -12x^2 - 6xy^3 \quad \text{and} \quad f_y = -9x^2y^2 + 4y$$

To find f_{xx}, take the partial derivative of f_x with respect to x.

$$f_{xx} = -24x - 6y^3$$

6 Let $f(x, y) = 4x^2y^2 - 9xy + 8x^2 - 3y^4$. Find all second partial derivatives.

Answer:
$f_{xx} = 8y^2 + 16$
$f_{yy} = 8x^2 - 36y^2$
$f_{xy} = 16xy - 9$
$f_{yx} = 16xy - 9$

7 Let $f(x, y) = 4e^{x+y} + 2x^3y$. Find all second partial derivatives.

Answer:
$f_{xx} = 4e^{x+y} + 12xy$
$f_{yy} = 4e^{x+y}$
$f_{xy} = 4e^{x+y} + 6x^2$
$f_{yx} = 4e^{x+y} + 6x^2$

8 Let $f(x, y, z) = xyz + x^2yz + xy^2z^3$. Find f_x, f_y, f_z, and f_{xz}.

Answer:
$f_x = yz + 2xyz + y^2z^3$
$f_y = xz + x^2z + 2xyz^3$
$f_z = xy + x^2y + 3xy^2z^2$
$f_{xz} = y + 2xy + 3y^2z^2$

Take the partial derivative of f_y with respect to y; this gives f_{yy}.

$$f_{yy} = -18x^2y + 4$$

Find f_{xy} by starting with f_x, then taking the partial derivative of f_x with respect to y.

$$f_{xy} = -18xy^2$$

Finally, find f_{yx} by starting with f_y; take its partial derivative with respect to x.

$$f_{yx} = -18xy^2 \quad \blacktriangleleft \quad \boxed{6}$$

▶**EXAMPLE 7** Let $f(x, y) = 2e^x - 8x^3y^2$. Find all second-order partial derivatives.

Here $f_x = 2e^x - 24x^2y^2$ and $f_y = -16x^3y$. (Recall: if $g(x) = e^x$, then $g'(x) = e^x$.) Now find the second partial derivatives.

$$f_{xx} = 2e^x - 48xy^2 \qquad f_{xy} = -48x^2y$$
$$f_{yy} = -16x^3 \qquad\qquad f_{yx} = -48x^2y \quad \blacktriangleleft \quad \boxed{7}$$

Partial derivatives of multivariate functions with more than two independent variables are found in a way similar to that for functions with two independent variables. For example, to find f_x for $w = f(x, y, z)$ treat y and z as constants and differentiate with respect to x.

▶**EXAMPLE 8** Let $f(x, y, z) = xy^2z + 2x^2y - 4xz^2$. Find f_x, f_y, f_z, f_{xy}, and f_{yz}.

$$f_x = y^2z + 4xy - 4z^2$$
$$f_y = 2xyz + 2x^2$$
$$f_z = xy^2 - 8xz$$

To find f_{xy}, differentiate f_x with respect to y.

$$f_{xy} = 2yz + 4x$$

In the same way, differentiate f_y with respect to z to get

$$f_{yz} = 2xy. \quad \blacktriangleleft \quad \boxed{8}$$

14.2 EXERCISES

For each of the following functions, find

(a) $\dfrac{\partial z}{\partial x}$ **(b)** $\dfrac{\partial z}{\partial y}$ **(c)** $f_x(2, 3)$ **(d)** $f_y(1, -2)$

1. $z = f(x, y) = 8x^3 - 4x^2y + 9y^2$ **2.** $z = f(x, y) = -3x^2 - 2xy^2 + 5y^3$

In Exercises 3–14, find f_x and f_y. Then find $f_x(2, -1)$ and $f_y(-4, 3)$. Leave the answers in terms of e in Exercises 5–8 and 13–14. (See Examples 1–3.)

3. $f(x, y) = -x^2y + 3x^4 + 8$

4. $f(x, y) = 5y^2 - 6xy^2 + 7$

5. $f(x, y) = e^{2x+y}$

6. $f(x, y) = -4e^{x-y}$

7. $f(x, y) = \dfrac{-2}{e^{x+2y}}$

8. $f(x, y) = \dfrac{6}{e^{4x-y}}$

9. $f(x, y) = \dfrac{x + 3y^2}{x^2 + y^3}$

10. $f(x, y) = \dfrac{8x^2y}{x^3 - y}$

11. $f(x, y) = \ln|2x - x^2y|$

12. $f(x, y) = \ln|4xy^2 + 3y|$

13. $f(x, y) = x^2e^{2xy}$

14. $f(x, y) = ye^{5x+2y}$

Find all second-order partial derivatives. (See Examples 6 and 7.)

15. $f(x, y) = 10x^2y^3 - 5x^3 + 3y$

16. $g(x, y) = 8x^3y + 2x^4 + 6y^3$

17. $h(x, y) = -3y^2 - 4x^2y^2 + 7xy^2$

18. $P(x, y) = -16x^3 + 3xy^2 - 12x^4y^2$

19. $R(x, y) = \dfrac{3y}{2x + y}$

20. $C(x, y) = \dfrac{8x}{x - 4y}$

21. $z = 4xe^y$

22. $z = -3ye^x$

23. $r = \ln(x + y)$

24. $k = \ln(5x - 7y)$

25. $z = x\ln(xy)$

26. $z = (y + 1)\ln(x^3y)$

In Exercises 27 and 28 evaluate $f_{xy}(2, 1)$ and $f_{yy}(1, 2)$.

27. $f(x, y) = x\ln(xy)$

28. $f(x, y) = (y + 1)\ln(x^3y)$

For the functions defined as follows, find values of x and y such that both $f_x(x, y) = 0$ and $f_y(x, y) = 0$.

29. $f(x, y) = 6x^2 + 6y^2 + 6xy + 36x - 5$

30. $f(x, y) = 50 + 4x - 5y + x^2 + y^2 + xy$

31. $f(x, y) = 9xy - x^3 - y^3 - 6$

32. $f(x, y) = 2200 + 27x^3 + 72xy + 8y^2$

Find f_x, f_y, f_z, and f_{yz} for the functions defined as follows. In Exercises 33 and 34, also find $f_y(2, -1, 3)$ and $f_{yz}(-1, 1, 0)$. (See Example 8.)

33. $f(x, y, z) = x^2 + yz + z^4$

34. $f(x, y, z) = 3x^5 - x^2 + y^5$

35. $f(x, y, z) = \dfrac{6x - 5y}{4z + 5}$

36. $f(x, y, z) = \dfrac{2x^2 + xy}{yz - 2}$

37. $f(x, y, z) = \ln(x^2 - 5xz^2 + y^4)$

38. $f(x, y, z) = \ln(8xy + 5yz - x^3)$

39. How many partial derivatives does a function with three independent variables have? How many second-order partial derivatives? Explain why.

40. Suppose $z = f(x, y)$ describes the cost to build a certain structure, where x represents the labor costs and y represents the cost of materials. Describe what f_x and f_y represent.

41. Social Science A developmental mathematics instructor at a large university has determined that a student's probability of success in the university's pass/fail remedial algebra course is a function of s, n, and a, where s is the student's score on the departmental placement exam, n is the number of semesters of mathematics passed in high school, and a is the student's mathematics

SAT score. She estimates that p, the probability of passing the course (in percent), will be

$$p = f(s, n, a) = .003a + .1(sn)^{1/2}$$

for $200 \le a \le 800$, $0 \le s \le 10$, and $0 \le n \le 8$. Assuming that the above model has some merit, find the following.

(a) If a student scores 8 on the placement exam, has taken 6 semesters of high-school math, and has an SAT score of 450, what is the probability of passing the course?

(b) Find p for a student with 3 semesters of high school mathematics, a placement score of 3, and an SAT score of 320.

(c) Find and interpret $f_n(3, 3, 320)$ and $f_a(3, 3, 320)$.

42. A weight-loss counselor has prepared a program of diet and exercise for a client. If the client sticks to the program, the weight loss that can be expected (in pounds per week) is given by

$$\text{Weight loss} = f(n, c) = \frac{1}{8}n^2 - \frac{1}{5}c + \frac{1937}{8},$$

where c is the average daily calorie intake for the week and n is the number of 40-min aerobic workouts per week.

(a) How many pounds can the client expect to lose by eating an average of 1200 cal per day and participating in four 40-min workouts in a week?

(b) Find and interpret $\partial f / \partial n$.

(c) The client currently averages 1100 cal per day and does three 40-minute workouts each week. What would be the approximate impact on weekly weight loss of adding a fourth workout per week?

43. Management   A car dealership estimates that the total weekly sales of its most popular model is a function of the car's list price, p, and the interest rate in percent, i, offered by the manufacturer. The approximate weekly sales are given by

$$f(p, i) = 132p - 2pi - .01p^2.$$

(a) Find the weekly sales if the average list price is $9400 and the manufacturer is offering an 8% interest rate.

(b) Find and interpret f_p and f_i.

(c) What would be the effect on weekly sales if the price is $9400 and interest rates rise from 8% to 9%?

44. Management   Suppose the production function of a company is given by

$$P(x, y) = 100\sqrt{x^2 + y^2},$$

where x represents units of labor and y represents units of capital. (See Example 5.) Find the following when $x = 4$ and $y = 3$.

(a) The marginal productivity of labor

(b) The marginal productivity of capital

45. Management   A manufacturer estimates that production (in hundreds of units) is a function of the amounts x and y of labor and capital used, as follows.

$$f(x, y) = \left[\frac{1}{3}x^{-1/3} + \frac{2}{3}y^{-1/3} \right]^{-3}$$

(a) Find the number of units produced when 27 units of labor and 64 units of capital are utilized.

(b) Find and interpret $f_x(27, 64)$ and $f_y(27, 64)$.

(c) What would be the approximate effect on production of increasing labor by 1 unit?

46. Management   A manufacturer of automobile batteries estimates that his total production in thousands of units is given by

$$f(x, y) = 3x^{1/3}y^{2/3},$$

where x is the number of units of labor and y is the number of units of capital utilized.

(a) Find and interpret $f_x(64, 125)$ and $f_y(64, 125)$ if the current level of production uses 64 units of labor and 125 units of capital.

(b) What would be the approximate effect on production of increasing labor to 65 units while holding capital at the current level?

(c) Suppose that sales have been good and management wants to increase either capital or labor by 1 unit. Which option would result in a larger increase in production?

47. Management   The production function z for the United States was once estimated as

$$z = x^{.7}y^{.3},$$

where x stands for the amount of labor and y stands for the amount of capital. Find the marginal productivity of labor (find $\partial z / \partial x$) and of capital.

48. Management   A similar production function for Canada is

$$z = x^{.4}y^{.6},$$

with x, y, and z as in Exercise 47. Find the marginal productivity of labor and of capital.

49. **Natural Science** In one method of computing the quantity of blood pumped through the lungs in 1 minute, a researcher first finds each of the following (in milliliters).

b = quantity of oxygen used by body in 1 minute

a = quantity of oxygen per liter of blood that has just gone through the lungs

v = quantity of oxygen per liter of blood that is about to enter the lungs

In 1 minute,

Amount of oxygen used

     = amount of oxygen per liter

         × number of liters of blood pumped.

If C is the number of liters pumped through the blood in 1 minute, then

$$b = (a - v) \cdot C \quad \text{or} \quad C = \frac{b}{a - v}.$$

(a) Find C if $a = 160$, $b = 200$, and $v = 125$.

(b) Find C if $a = 180$, $b = 260$, and $v = 142$. Find the following partial derivatives.

(c) $\partial C/\partial b$ **(d)** $\partial C/\partial v$

50. **Natural Science** The reaction to x units of a drug t hr after it was administered is given by

$$R(x, t) = x^2(a - x)t^2 e^{-t},$$

for $0 \le x \le a$ (where a is a constant). Find the following.

(a) $\dfrac{\partial R}{\partial x}$ **(b)** $\dfrac{\partial R}{\partial t}$ **(c)** $\dfrac{\partial^2 R}{\partial x^2}$ **(d)** $\dfrac{\partial^2 R}{\partial x \, \partial t}$

(e) Interpret your answers to parts (a) and (b).

51. **Physical Science** The gravitational attraction F on a body a distance r from the center of the earth, where r is greater than the radius of the earth, is a function of its mass m and the distance r as follows:

$$F = \frac{mgR^2}{r^2},$$

where R is the radius of the earth and g is the force of gravity—about 32 feet per second per second (ft/sec^2).

(a) Find and interpret F_m and F_r.

(b) Show that $F_m > 0$ and $F_r < 0$. Why is this reasonable?

14.3 EXTREMA OF FUNCTIONS OF SEVERAL VARIABLES

One of the most important applications of calculus is in finding maxima and minima for functions. Earlier, we studied this idea extensively for functions of a single independent variable; now we shall see that extrema can be found for functions of two variables. In particular, an extension of the second derivative test can be derived and used to identify maxima or minima. We begin with the definitions of relative maxima and minima.

Relative Maxima and Minima

Let (a, b) be in the domain of a function f.

1. $f(a, b)$ is a **relative maximum** if there exists a circular region containing (a, b) such that

$$f(a, b) \ge f(x, y)$$

for all points (x, y) in the circular region.

2. $f(a, b)$ is a **relative minimum** if there exists a circular region containing (a, b) such that

$$f(a, b) \le f(x, y)$$

for all points (x, y) in the circular region.

As before, the word *extremum* is used for either a relative maximum or a relative minimum. Examples of a relative maximum and a relative minimum are given in Figures 14.13 and 14.14.

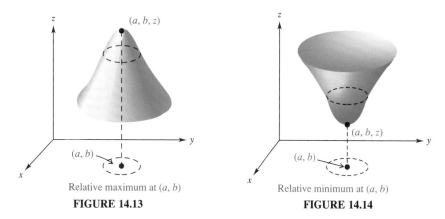

Relative maximum at (a, b)

FIGURE 14.13

Relative minimum at (a, b)

FIGURE 14.14

Note   With functions of a single variable, we made a distinction between relative extrema and absolute extrema. The methods for finding absolute extrema are quite involved for functions of two independent variables, so we will discuss only relative extrema.

As suggested by Figure 14.15 on the next page, at a relative maximum, the tangent line parallel to the x-axis has a slope of 0, as does the tangent line parallel to the y-axis. (Notice the similarity to functions of one variable.) That is, if the function $z = f(x, y)$ has a relative extremum at (a, b), then $f_x(a, b) = 0$ and $f_y(a, b) = 0$, as stated in the following theorem.

Location of Extrema

If a function $z = f(x, y)$ has a relative maximum or relative minimum at the point (a, b), and $f_x(a, b)$ and $f_y(a, b)$ both exist, then

$$f_x(a, b) = 0 \quad \text{and} \quad f_y(a, b) = 0.$$

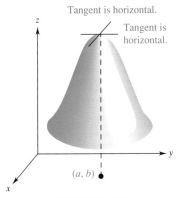

FIGURE 14.15

Just as with functions of one variable, the fact that the slopes of the tangent lines are 0 is no guarantee that a relative extremum has been located. For example, Figure 14.16 shows the graph of $z = f(x, y) = x^2 - y^2$. Both $f_x(0, 0) = 0$ and $f_y(0, 0) = 0$, and yet $(0, 0)$ leads to neither a relative maximum nor a relative minimum for the function. The point $(0, 0, 0)$ on the graph of this function is called a **saddle point;** it is a minimum when approached from one direction but a maximum when approached from another direction. A saddle point is neither a maximum nor a minimum.

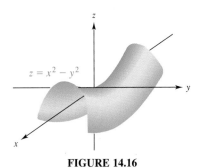

FIGURE 14.16

The theorem on location of extrema suggests a useful strategy for finding extrema. First, locate all points (a, b) where $f_x(a, b) = 0$ and $f_y(a, b) = 0$. Then test each of these points separately, using the test given after the next example. For a function $f(x, y)$, the points (a, b) such that $f_x(a, b) = 0$ and $f_y(a, b) = 0$ are called **critical points.**

▶**EXAMPLE 1** Find all critical points for
$$f(x, y) = 6x^2 + 6y^2 + 6xy + 36x - 5.$$

1 Find all critical points for the following.

(a) $f(x, y) = 4x^2 + 3xy + 2y^2 + 7x - 6y - 6$

(b) $f(x, y) = e^{-2x} + xy$

Answers:

(a) $(-2, 3)$

(b) $(0, 2)$

We must find all points (a, b) such that $f_x(a, b) = 0$ and $f_y(a, b) = 0$. Here

$$f_x = 12x + 6y + 36 \quad \text{and} \quad f_y = 12y + 6x.$$

Set each of these two partial derivatives equal to 0.

$$12x + 6y + 36 = 0 \quad \text{and} \quad 12y + 6x = 0$$

These two equations make up a system of linear equations. We can use the substitution method to solve this system. First, rewrite $12y + 6x = 0$ as follows.

$$12y + 6x = 0$$
$$x = -2y$$

Now substitute $-2y$ for x in the other equation.

$$12x + 6y + 36 = 0$$
$$12(-2y) + 6y + 36 = 0$$
$$-24y + 6y + 36 = 0$$
$$-18y + 36 = 0$$
$$y = 2$$

The equation $x = -2y$ leads to $x = -2(2) = -4$. The solution of the system of equations is $(-4, 2)$. This is the only solution of the system, so $(-4, 2)$ is the only critical point for the given function. By the theorem above, if the function has a relative extremum, it will occur at $(-4, 2)$. ◀ **1**

The results of the next theorem can be used to decide whether $(-4, 2)$ in Example 1 leads to a relative maximum, a relative minimum, or neither. The proof of this theorem is beyond the scope of this course.

Test for Relative Extrema

For a function $z = f(x, y)$, assume f_{xx}, f_{yy}, and f_{xy} all exist. Let (a, b) be a point for which

$$f_x(a, b) = 0 \quad \text{and} \quad f_y(a, b) = 0.$$

Define the number M by

$$M = f_{xx}(a, b) \cdot f_{yy}(a, b) - [f_{xy}(a, b)]^2.$$

1. If $M > 0$ and $f_{xx}(a, b) < 0$, then $f(a, b)$ is a **relative maximum.**
2. If $M > 0$ and $f_{xx}(a, b) > 0$, then $f(a, b)$ is a **relative minimum.**
3. If $M < 0$, the graph of f has a **saddle point** at (a, b). ($f(a, b)$ is neither a maximum nor a minimum).
4. If $M = 0$, the test gives **no information.**

2 Find any relative maxima or minima for the functions defined in Problem 1 at the side.

Answers:

(a) Relative minimum at $(-2, 3)$

(b) Neither a relative minimum nor a relative maximum at $(0, 2)$

The chart below summarizes the conclusions of the theorem.

	$f_{xx}(a, b) < 0$	$f_{xx}(a, b) > 0$
$M > 0$	Relative maximum	Relative Minimum
$M = 0$	No information	
$M < 0$	Saddle point	

▶**EXAMPLE 2** The previous example showed that the only critical point for the function

$$f(x, y) = 6x^2 + 6y^2 + 6xy + 36x - 5$$

is $(-4, 2)$. Does $(-4, 2)$ lead to a relative maximum, a relative minimum, or neither?

We can find out by using the test above. From Example 1,

$$f_x(-4, 2) = 0 \quad \text{and} \quad f_y(-4, 2) = 0.$$

Now find the various second partial derivatives used in finding M. From $f_x = 12x + 6y + 36$ and $f_y = 12y + 6x$,

$$f_{xx} = 12, \quad f_{yy} = 12, \quad \text{and} \quad f_{xy} = 6.$$

(If these second-order partial derivatives had not all been constants, we would have had to evaluate them at the point $(-4, 2)$.) Now

$$M = f_{xx}(-4, 2) \cdot f_{yy}(-4, 2) - [f_{xy}(-4, 2)]^2 = 12 \cdot 12 - 6^2 = 108.$$

Since $M > 0$ and $f_{xx}(-4, 2) = 12 > 0$, Part 2 of the theorem applies, showing that $f(x, y) = 6x^2 + 6y^2 + 6xy + 36x - 5$ has a relative minimum at $(-4, 2)$. This relative minimum is $f(-4, 2) = -77$. ◀ **2**

▶**EXAMPLE 3** Find all points where the function

$$f(x, y) = 9xy - x^3 - y^3 - 6$$

has any relative maxima or relative minima.

First find any critical points. Here

$$f_x = 9y - 3x^2 \quad \text{and} \quad f_y = 9x - 3y^2.$$

Set each of these partial derivatives equal to 0.

$$\begin{array}{cc}
f_x = 0 & f_y = 0 \\
9y - 3x^2 = 0 & 9x - 3y^2 = 0 \\
9y = 3x^2 & 9x = 3y^2 \\
3y = x^2 & 3x = y^2
\end{array}$$

In the first equation $(3y = x^2)$, notice that since $x^2 \geq 0$, $y \geq 0$. Also, in the second equation $(3x = y^2)$, $y^2 \geq 0$, so $x \geq 0$.

The substitution method can be used again to solve the system of equations

$$3y = x^2$$
$$3x = y^2.$$

The first equation, $3y = x^2$, can be rewritten as $y = x^2/3$. Substitute this into the second equation to get

$$3x = y^2 = \left(\frac{x^2}{3}\right)^2$$

$$3x = \frac{x^4}{9}$$

Solve this equation as follows.

$$27x = x^4$$
$$x^4 - 27x = 0$$

$$x(x^3 - 27) = 0 \qquad \text{Factor}$$

$$x = 0 \quad \text{or} \quad x^3 - 27 = 0 \qquad \text{Set each factor equal to 0}$$

$$x^3 = 27$$

$$x = 3 \qquad \text{Take the cube root on each side}$$

Use these values of x, along with the equation $3x = y^2$, to find y.

If $x = 0$	If $x = 3$
$3x = y^2$	$3x = y^2$
$3(0) = y^2$	$3(3) = y^2$
$0 = y^2$	$9 = y^2$
$0 = y$	$3 = y \quad \text{or} \quad -3 = y.$

The points $(0, 0)$, $(3, 3)$ and $(3, -3)$ appear to be critical points; however, $(3, -3)$ does not have $y \geq 0$. The only possible relative extrema for $f(x, y) = 9xy - x^3 - y^3 - 6$ occur at the critical points $(0, 0)$ or $(3, 3)$. To identify any extrema, use the test. Here

$$f_{xx} = -6x, \quad f_{yy} = -6y, \quad \text{and} \quad f_{xy} = 9.$$

Test each of the critical points.

For $(0, 0)$:

$$f_{xx}(0, 0) = -6(0) = 0 \quad f_{yy}(0, 0) = -6(0) = 0 \quad f_{xy}(0, 0) = 9,$$

so that $M = 0 \cdot 0 - 9^2 = -81$. Since $M < 0$, there is a saddle point at $(0, 0)$.

3 Find any relative extrema for
$$f(x, y) = \frac{2\sqrt{2}}{3}x^3 - xy + \frac{1}{3}y^3 - 10.$$

Answer:

Relative minimum at $\left(\frac{1}{2}, \frac{\sqrt{2}}{2}\right)$

For (3, 3):

$$f_{xx}(3, 3) = -6(3) = -18 \quad f_{yy}(3, 3) = -6(3) = -18 \quad f_{xy}(3, 3) = 9,$$

so that $M = (-18)(-18) - 9^2 = 243$. Since $M > 0$, and $f_{xx}(3, 3) = -18 < 0$, there is a relative maximum at (3, 3). ◀ **3**

▶ **EXAMPLE 4**   A company is developing a new soft drink. The cost in dollars to produce a batch of the drink is approximated by

$$C(x, y) = 2200 + 27x^3 - 72xy + 8y^2,$$

where x is the number of kilograms of sugar per batch and y is the number of grams of flavoring per batch.

(a) Find the amounts of sugar and flavoring that result in minimum cost for a batch.

Start with the following partial derivatives.

$$C_x = 81x^2 - 72y \quad \text{and} \quad C_y = -72x + 16y$$

Set each of these equal to 0 and solve for y.

$$81x^2 - 72y = 0 \qquad\qquad -72x + 16y = 0$$
$$-72y = -81x^2 \qquad\qquad 16y = 72x$$
$$y = \frac{9}{8}x^2 \qquad\qquad\qquad y = \frac{9}{2}x$$

From the equation on the left, $y \geq 0$. Since $(9/8)x^2$ and $(9/2)x$ both are equal to y, they are equal to each other. Set $(9/8)x^2$ and $(9/2)x$ equal and solve the resulting equation for x.

$$\frac{9}{8}x^2 = \frac{9}{2}x$$
$$9x^2 = 36x$$
$$9x^2 - 36x = 0$$
$$9x(x - 4) = 0$$
$$9x = 0 \quad \text{or} \quad x - 4 = 0$$

The equation $9x = 0$ leads to $x = 0$, which is not a useful answer for our problem. Substitute $x = 4$ from $x - 4 = 0$ into $y = (9/2)x$ to find y.

$$y = \frac{9}{2}x = \frac{9}{2}(4) = 18$$

Now check to see if the critical point (4, 18) leads to a relative minimum. For (4, 18),

$$C_{xx} = 162x = 162(4) = 648, \quad C_{yy} = 16, \quad \text{and} \quad C_{xy} = -72.$$

Also,

$$M = (648)(16) - (-72)^2 = 5184.$$

Since $M > 0$ and $C_{xx}(4, 18) > 0$, the cost at $(4, 18)$ is a minimum.

(b) What is the minimum cost?

To find the minimum cost, go back to the cost function and evaluate $C(4, 18)$.

$$C(x, y) = 2200 + 27x^3 - 72xy + 8y^2$$
$$C(4, 18) = 2200 + 27(4)^3 - 72(4)(18) + 8(18)^2 = 1336$$

The minimum cost for a batch is $1336.00. ◀

14.3 EXERCISES

1. Compare and contrast the way critical points are found for functions with one independent variable and functions with more than one independent variable.

2. Compare and contrast the second derivative test for $y = f(x)$ and the test for relative extrema for $z = f(x, y)$.

Find all points where the functions defined as follows have any relative extrema. Give the values of any relative extrema. Identify any saddle points. (See Examples 1–3.)

3. $f(x, y) = 2x^2 + 4xy + 6y^2 - 8x - 10$

4. $f(x, y) = x^2 + xy + y^2 - 6x - 3$

5. $f(x, y) = x^2 - xy + 2y^2 + 2x + 6y + 8$

6. $f(x, y) = 3x^2 + 6xy + y^2 - 6x - 3y$

7. $f(x, y) = 2x^2 + 4xy + y^2 - 4x + 4y$

8. $f(x, y) = 4xy - 4x^2 - 2y^2 + 4x - 8y - 7$

9. $f(x, y) = 4xy - 10x^2 - 4y^2 + 8x + 8y + 9$

10. $f(x, y) = x^2 + xy + 3x + 2y - 6$

11. $f(x, y) = x^2 + xy - 2x - 2y + 2$

12. $f(x, y) = x^2 + xy + y^2 - 3x - 5$

13. $f(x, y) = x^2 - y^2 - 2x + 4y - 7$

14. $f(x, y) = 4x + 2y - x^2 + xy - y^2 + 3$

15. $f(x, y) = 2x^3 + 2y^2 - 12xy + 15$

16. $f(x, y) = 2x^2 + 4y^3 - 24xy + 18$

17. $f(x, y) = 3x^2 + 6y^3 - 36xy + 27$

18. $f(x, y) = 2y^3 + 5x^2 + 60xy + 25$

19. $f(x, y) = e^{xy}$

20. $f(x, y) = x^2 + e^y$

21. $f(x, y) = 3xy - x^3 - y^3 + \dfrac{1}{8}$

22. $f(x, y) = \dfrac{3}{2}x - \dfrac{1}{2}x^3 - xy^2 + \dfrac{1}{16}$

23. $f(x, y) = x^4 - 2x^2 + y^2 + \dfrac{17}{16}$

24. $f(x, y) = 2y^3 - 3x^4 - 6x^2y + \dfrac{1}{16}$

25. $f(x, y) = x^4 - y^4 - 2x^2 + 2y^2 + \dfrac{1}{16}$

26. $f(x, y) = -x^4 + 4xy - 2y^2 + \dfrac{1}{16}$

Work the following exercises. (See Example 4.)

27. **Management** Suppose that the profit of a certain firm is approximated by

$$P(x, y) = 1000 + 24x - x^2 + 80y - y^2,$$

where x is the cost of a unit of labor and y is the cost of a unit of goods. Find values of x and y that maximize profit. Find the maximum profit.

28. **Management** The labor cost in dollars for manufacturing a precision camera can be approximated by

$$L(x, y) = \dfrac{3}{2}x^2 + y^2 - 2x - 2y - 2xy + 68,$$

where x is the number of hours required by a skilled craftsperson and y is the number of hours required by a semi-skilled person. Find values of x and y that minimize the labor charge. Find the minimum labor charge.

29. Management The profit (in thousands of dollars) from the sales of graphing calculators is approximated by $P(x, y) = 800 - 2x^3 + 12xy - y^2$, where x is the cost of a unit of chips and y is the cost of a unit of labor. Find the maximum profit and the cost of chips and labor that produce the maximum profit.

30. Management The total profit from one acre of a certain crop depends on the amount spent on fertilizer, x, and hybrid seed, y, according to the model

$$P(x, y) = -x^2 + 3xy + 160x - 5y^2$$
$$+ 200y + 2,600,000.$$

Find values of x and y that lead to maximum profit. Find the maximum profit.

31. Management The total cost to produce x units of electical tape and y units of packing tape is given by

$$C(x, y) = 2x^2 + 3y^2 - 2xy + 2x - 126y + 3800.$$

Find the number of units of each kind of tape that should be produced so that the total cost is a minimum. Find the minimum total cost.

32. Management The total revenue in thousands of dollars from the sale of x spas and y solar heaters is approximated by

$$R(x, y) = 12 + 74x + 85y - 3x^2 - 5y^2 - 5xy.$$

Find the number of each that should be sold to produce maximum revenue. Find the maximum revenue.

33. Management A rectangular closed box is to be built at minimum cost to hold 27 cubic meters. Since the cost will depend on the surface area, find the dimensions that will minimize the surface area of the box.

34. Management Find the dimensions that will minimize the surface area (and, hence, the cost) of a rectangular fish aquarium, open on top, with a volume of 32 cubic feet.

35. The U.S. Postal Service requires that any box sent through the mail must have a length plus girth (distance around) totaling no more than 108 inches. (See Example 3 on page 716.) Find the dimensions of the box with maximum volume that can be sent.

36. Find the dimensions of the circular tube with maximum volume that can be sent through the U.S. Mail. See Exercise 35.

37. The monthly revenue in hundreds of dollars from production of x thousand tons of grade A iron ore and y thousand tons of grade B iron ore is given by

$$R(x, y) = 2xy + 2y + 12,$$

and the corresponding cost in hundreds of dollars is given by

$$C(x, y) = 2x^2 + y^2.$$

Find the amount of each grade of ore that will produce maximum profit.

38. Suppose the revenue and cost in thousands of dollars from the manufacture of x units of one product and y units of another is

$$R(x, y) = 6xy + 3 - x^2 \quad \text{and} \quad C(x, y) = x^2 + 3y^3.$$

How many units of each product will produce maximum profit? What is the maximum profit?

CHAPTER 14 SUMMARY

KEY TERMS AND SYMBOLS

14.1 $z = f(x, y)$ function of two independent variables
      multivariate function
      ordered triple
      first octant
      xy-plane
      surface
      trace
      level curves
      paraboloid
      production function
      Cobb-Douglas production function

14.2 f_x or $\dfrac{\partial f}{\partial x}$ partial derivative of f with respect to x

$\dfrac{\partial}{\partial x}\left(\dfrac{\partial z}{\partial x}\right)$ or $\dfrac{\partial^2 z}{\partial x^2}$ or f_{xx}

Second-order partial derivative of $\partial z/\partial x$ (or f_x) with respect to x

$\dfrac{\partial}{\partial y}\left(\dfrac{\partial z}{\partial x}\right)$ or $\dfrac{\partial^2 z}{\partial y\, \partial x}$ or f_{xy}

Second-order partial derivative of $\partial z/\partial x$ (or f_x) with respect to y

14.3 saddle point

KEY CONCEPTS

The graph of $ax + by + cz = d$ is a **plane.**

The graph of $z = f(x, y)$ is a **surface** in three-dimensional space.

The graph of $z = ax^2 + by^2$ is a **paraboloid.**

The **partial derivative of f with respect to x** is the derivative of f found by treating x as a variable and y as a constant.

The **partial derivative of f with respect to y** is the derivative of f found by treating y as a variable and x as a constant.

Second-Order Partial Derivatives

For a function $z = f(x, y)$, if all indicated partial derivatives exist, then

$$\frac{\partial}{\partial x}\left(\frac{\partial z}{\partial x}\right) = \frac{\partial^2 z}{\partial x^2} = f_{xx} \qquad \frac{\partial}{\partial y}\left(\frac{\partial z}{\partial y}\right) = \frac{\partial^2 z}{\partial y^2} = f_{yy}$$

$$\frac{\partial}{\partial y}\left(\frac{\partial z}{\partial x}\right) = \frac{\partial^2 z}{\partial y\, \partial x} = f_{xy} \qquad \frac{\partial}{\partial x}\left(\frac{\partial z}{\partial y}\right) = \frac{\partial^2 z}{\partial x\, \partial y} = f_{yx}.$$

Let (a, b) be the center of a circular region in the xy plane. For a function $z = f(x, y)$, defined for every (x, y) in the region, $f(a, b)$ is a **relative maximum** if $f(a, b) \geq f(x, y)$ for all (x, y) in the circular region. $f(a, b)$ is a **relative minimum** if $f(a, b) \leq f(x, y)$ for all (x, y) in the circular region.

Location of Extrema	If $f(a, b)$ is a relative extremum, then $f_x(a, b) = 0$ and $f_y(a, b) = 0$.

Test for Relative Extrema	Let f_{xx}, f_{yy}, and f_{xy} all exist. Let (a, b) be a point for which $f_x(a, b) = 0$ and $f_y(a, b) = 0$. Let $M = f_{xx}(a, b) \cdot f_{yy}(a, b) - [f_{xy}(a, b)]^2$.

If $M > 0$ and $f_{xx}(a, b) < 0$, then $f(a, b)$ is a **relative maximum.**

If $M > 0$ and $f_{xx}(a, b) > 0$, then $f(a, b)$ is a **relative minimum.**

If $M < 0$, the graph of f has a **saddle point** at (a, b). ($f(a, b)$ is neither a maximum nor a minimum).

If $M = 0$, the test gives **no information.**

CHAPTER 14 REVIEW EXERCISES

Find $f(-1, 2)$ and $f(6, -3)$ for each of the following.

1. $f(x, y) = 6y^2 - 5xy + 2x$

2. $f(x, y) = -3x + 2x^2y^2 + 5y$

3. $f(x, y) = \dfrac{2x - 4}{x + 3y}$

4. $f(x, y) = x\sqrt{x^2 + y^2}$

5. Describe the graph of $2x + y + 4z = 12$.

6. Describe the graph of $y = 2$ on a three-dimensional grid.

Graph the first-octant portion of each plane.

7. $x + 2y + 4z = 4$

8. $3x + 2y = 6$

9. $4x + 5y = 20$

10. $x = 6$

11. Let $z = f(x, y) = -2x^2 + 5xy + y^2$. Find the following.

   (a) $\dfrac{\partial z}{\partial x}$ (b) $\dfrac{\partial z}{\partial y}(-1, 4)$ (c) $f_{xy}(2, -1)$

13. What is the difference between $\dfrac{\partial z}{\partial x}$ and $\dfrac{\partial z}{\partial y}$?

12. Let $z = f(x, y) = \dfrac{2y + x^2}{3y - x}$. Find the following.

   (a) $\dfrac{\partial z}{\partial y}$ (b) $\dfrac{\partial z}{\partial x}(0, 2)$ (c) $f_{yy}(-1, 0)$

Find f_x and f_y.

14. $f(x, y) = 3y - 7x^2y^3$

15. $f(x, y) = 4x^3y + 10xy^4$

16. $f(x, y) = \sqrt{3x^2 + 2y^2}$

17. $f(x, y) = \dfrac{3x - 2y^2}{x^2 + 4y}$

18. $f(x, y) = x^3e^{3y}$

19. $f(x, y) = (y + 1)^2e^{2x+y}$

20. $f(x, y) = \ln|x^2 - 4y^3|$

21. $f(x, y) = \ln|1 + x^3y^2|$

22. What is the difference between f_{xx} and f_{xy}?

Find f_{xx} and f_{xy}.

23. $f(x, y) = 4x^3y^2 - 8xy$

24. $f(x, y) = -6xy^4 + x^2y$

25. $f(x, y) = \dfrac{2x}{x - 2y}$

26. $f(x, y) = \dfrac{3x + y}{x - 1}$

27. $f(x, y) = x^2e^y$

28. $f(x, y) = ye^{x^2}$

29. $f(x, y) = \ln(2 - x^2y)$

30. $f(x, y) = \ln(1 + 3xy^2)$

Find all points where the functions defined below have any relative extrema. Find any saddle points.

31. $z = x^2 + 2y^2 - 4y$

32. $z = x^2 + y^2 + 9x - 8y + 1$

33. $f(x, y) = x^2 + 5xy - 10x + 3y^2 - 12y$

34. $z = x^3 - 8y^2 + 6xy + 4$

35. $z = x^3 + y^2 + 2xy - 4x - 3y - 2$

36. $f(x, y) = 7x^2 + y^2 - 3x + 6y - 5xy$

37. Write a function in terms of L, W, and H that gives the total material required to build the closed box shown in the figure.

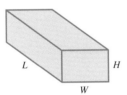

38. **Management**   The manufacturing cost in dollars for a medium-sized business computer is given by

$$c(x, y) = 2x + y^2 + 4xy + 25,$$

where x is the memory capacity of the computer in megabytes and y is the number of hours of labor required. Find each of the following.

(a) $\dfrac{\partial c}{\partial x}(64, 6)$ **(b)** $\dfrac{\partial c}{\partial y}(128, 12)$

39. **Management**   The total cost in dollars to manufacture x solar cells and y solar collectors is

$$c(x, y) = x^2 + 5y^2 + 4xy - 70x - 164y + 1800.$$

(a) Find values of x and y that produce minimum total cost.

(b) Find the minimum total cost.

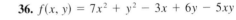**40.** **Management**   The total profit from 1 acre of a certain crop depends on the amount spent on fertilizer, x, and on hybrid seed, y, according to the model

$$P(x, y) = .01(-x^2 + 3xy + 160x - 5y^2 + 200y + 2600).$$

The budget for fertilizer and seed is limited to $280.

(a) Use the budget constraint to express one variable in terms of the other. Then substitute into the profit function to get a function with one independent variable. Use the method shown in Chapter 12 to find the amounts spent on fertilizer and seed that will maximize profit. What is the maximum profit per acre? (*Hint:* Throughout this problem you may ignore the coefficient of .01 until you need to find the maximum profit.)

(b) Find the amounts spent on fertilizer and seed that will maximize profit using the method shown in this chapter. (*Hint:* You will not need to use the budget constraint.)

(c) Discuss any relationships between these methods.

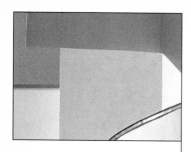

APPENDIX A

Graphing Calculators*

INTRODUCTION

Graphing calculators are a result of the amazingly rapid evolution in computer technology toward packaging more power into smaller "boxes." These machines have powerful graphing capabilities in addition to the full range of features found on programmable scientific calculators. Instead of a one-line display, graphing calculators typically can show up to eight lines of text. This makes it much easier to keep track of the steps of your work, whether you are doing routine computations, entering a long mathematical function, or writing a program. Like programmable scientific calculators, graphing calculators are capable of doing many things we have previously come to expect only from computers. Programs can be written relatively easily, and after they are stored in memory, the programs are always available. Like computers, graphing calculators can be programmed to include graphic displays as part of the program.

It takes some study to learn how to use graphing calculators, but they are much easier to master than most computers—and they can go wherever you do! New models with added features, more memory, greater ease of use, and other improvements are frequently being introduced. The most popular brand names at this time are CASIO, SHARP, Texas Instruments, and Hewlett-Packard.

GRAPHING CALCULATOR FEATURES

Every graphing calculator has keys for the usual operations of arithmetic and all commonly used functions (square root, x^2, log, ln, and so on). All of them can graph functions of the form $y = f(x)$ and have programming capabilities. Except for the cheapest ones, graphing calculators may have a variety of additional

---

*Prepared by Jim Eckerman of *American River College*.

features. Before buying one, you should consider which features you are most likely to need. The following chart may be helpful. Many of the features mentioned in the table are discussed in the next part of this appendix.

Feature	Used in
Function memory	Chapters 1–4, 11–13
Table creation capability	Chapters 1–5
Simultaneous trace feature for two or more graphs	Chapters 2–4, 11–12
Root finder (equation solver)	Chapters 1–5, 11–13
Matrix inverses	Chapter 6
Matrix row operations	Chapters 6–7
System of linear equations solver	Chapter 6
Statistical operations	Chapters 8–10
Maximum/minimum finder	Chapters 11–12
Numerical derivatives	Chapters 11–12
Numerical integration	Chapter 13

Many graphing calculators have additional features that do not play a role in this book (such as polar coordinates, parametric graphing, complex numbers, and vectors). Hewlett-Packard graphing calculators can also perform some symbolic operations (such as factoring simple polynomials and finding derivatives). However, these HP calculators are more expensive and more difficult to learn to use because of their increased complexity and the fact that they do not use standard algebraic order of operations.

ADVICE ON USING A GRAPHING CALCULATOR

1. BASICS Graphing calculators have forty-nine or more keys. Most modern desk-top computers have 101 keys on their keyboards. With fewer keys, each key must be used for more actions, so you will find special mode-changing keys such as "**2nd**," "**shift**," "**alpha**," and "**mode**." Become familiar with the capabilities of the machine, the layout of the keyboard, how to adjust the screen contrast, and so on.

2. EDITING When keying in expressions, you can pause at any time and use the **arrow keys,** located at the upper right of the keyboard, to move the cursor to any point in the text. You can then make changes by using the "**DEL**" key to delete and the "**INS**" key to insert material. ["INS" is the "second function" of "DEL."] After an expression has been entered or a calculation made, it can still be edited by using the **edit/replay** feature, available on almost every calculator model. On TI calculators (except TI-81), use "2nd, ENTER" to return to the previously entered expression. On Casio calculators, use the left or right arrow key, and on Sharp models use "2nd, up arrow."

3. SCIENTIFIC NOTATION   Learn how to enter and read data in **scientific notation** form. This form is used when the numbers become too large or too small (too many zeros between the decimal point and the first significant digit) for the machine's display.

4. FUNCTION GRAPHING

A. Setting the Range   Learn to set "**RANGE**" values to delineate a window that is appropriate for the function you are graphing before using the "**Graph**" key. This involves keying in the minimum and maximum values of x and y that will be displayed on the screen, along with the distance to be used between "tic" marks along the axes. If you do not do this, you will often find your graph screen blank! Usually one can quickly find a point on the graph by substitution of either zero or one for x. Use the "**RANGE**" command to set the x-values to the left and right of the x-coordinate of this point and set the y-values above and below the y-coordinate of this point. Then the "**zoom out**" feature can be used to see more of the graph.

B. Connected or Dot Mode for Graphs   Some functions have "trouble spots" such as at $x = 2$ in the function $f(x) = 3/(x - 2)$. A graphing calculator set in "connected" mode will sometimes draw a line from the last point plotted on the left side of this value of x to the first point plotted on the right side. This results in a misleading graph! For this example, the function is not defined at $x = 2$, and there should be no line crossing this x-value. This type of problem can be eliminated by switching from "connected" mode to "dot" mode, then redrawing the graph. Figures 1 and 2 below illustrate this situation.

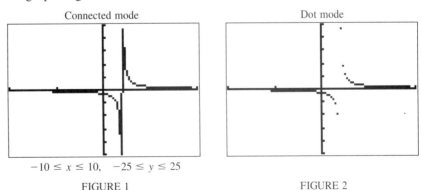

Connected mode   Dot mode

$-10 \le x \le 10, \quad -25 \le y \le 25$

FIGURE 1   FIGURE 2

C. Using the Function Memory   Learn how to redraw graphs without reentering the function. Often you will need to change the "**RANGE**" settings several times before you get the "window" that is most appropriate for your function. All graphing calculators allow you to do this without reentering the function. If you plan to graph a particular function often, then you should either store it in the **function memory** (which is labeled "$y = $" on TI and "EQTN" on Sharp models) or store it in **program memory**. Of course, you have to write a program with your function as part of the program in order to make use of the program memory. Once entered into the machine's memory, this function can be used at any time.

D. Using the Trace Feature With the "**Trace**" feature, the left/right arrows can be used to move the cursor along the last curve plotted, and the values of x and y will be displayed for each point plotted on the screen. On most calculators, if more than one graph was plotted, one can move the cursor vertically between the different graphs by using the up/down arrows.

E. Using the Zoom Feature The "**zoom**" feature allows a quick redrawing of your graph using smaller ranges of values for x and y ("**zoom in**") or larger ranges of values ("**zoom out**"). Thus one can easily examine the behavior of a function within the close vicinity of a particular point or the general behavior as seen from farther away.

F. Using the Plot Feature With "**Plot**" or "**Draw**" you can move the cursor to any point on the screen and either have the machine plot a single point or have it display the "screen coordinates" of the point. For example, on the Casio machines the command "Plot 2,3" will cause this point to be plotted on the graph screen. However, the "screen coordinates" will usually be approximations of the specified values.

SOLVING EQUATIONS GRAPHICALLY

Some mathematical procedures can be quite difficult or even impossible to do algebraically but can be done easily and to a very high degree of accuracy using a graphing calculator. Listed below are some examples.

1. SOLVING EQUATIONS OF THE FORM $f(x) = k$. The quickest way to solve this type of equation is to form the new equation $y = f(x) - k$ and then use the built-in Equation Solver or Root Finder feature. If your calculator does not have this capability or if you prefer to see the solution graphically, you can find the roots of $y = f(x) - k$ by zooming-in on the points where the graph crosses the x-axis. The "**box zoom**" is very effective for this type of situation. As an example, consider trying to find the annual interest rate, x, that is required in order to have monthly deposits of \$100 grow to \$15,000 in value in ten years if the interest is compounded monthly. The equation

$$15000 = 100\left[\frac{\left(1 + \dfrac{x}{12}\right)^{120} - 1}{\dfrac{x}{12}}\right]$$

cannot be solved algebraically. (See Section 5.3, Future Value of an Ordinary Annuity). However, if we graph the function

$$y = 100\left[\frac{\left(1 + \dfrac{x}{12}\right)^{120} - 1}{\dfrac{x}{12}}\right] - 15000$$

and then **zoom in** several times on the point of intersection with the x-axis, we can find the interest rate to any desired degree of accuracy. Using values of x from 0 to 0.1 and values of y from 0 to 20000, the picture you will see should look like Figure 3. After zooming in on the point of intersection repeatedly (with the graph of the curve exiting through the vertical sides of the "box" each time you zoom, you will see a picture similar to Figure 4.

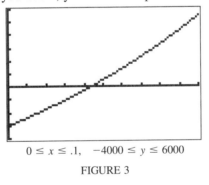

$0 \leq x \leq .1, \quad -4000 \leq y \leq 6000$

FIGURE 3

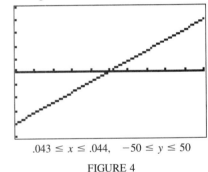

$.043 \leq x \leq .044, \quad -50 \leq y \leq 50$

FIGURE 4

Notice that no units are shown along the axes. You must look at the "**Range**" screen to determine what the tic marks represent numerically. In Figure 3 we have $0 \leq x \leq 0.1$ and $-4000 \leq y \leq 6000$. In Figure 4, $0.043 \leq x \leq 0.044$ and $-50 \leq y \leq 50$, respectively. By using the "**Trace**" feature along the curve, the point of intersection was found to be between the points

$$(.04350, -.04508) \text{ and } (.04351, .74991).$$

Thus the required annual interest rate is 4.35%, with accuracy to within one thousandth of a percent!

2. SOLVING SYSTEMS OF TWO EQUATIONS IN TWO UNKNOWNS. Some graphing calculators have built-in programs for solving systems of linear equations, but all of them can be used to solve systems of two equations (linear or not) by finding intersection points. If both equations can be solved explicitly for one variable in terms of the other, then the two equations can be put into the forms $y = f(x)$ and $y = g(x)$. Then we can graph both in the same screen and zoom in on the point or points where the two curves intersect. (See Sections 6.1 and 7.1).

3. FINDING MAXIMUM AND MINIMUM FUNCTION VALUES. The most recent graphing calculators can do this automatically. If yours does not have this feature, arrange the function in the form $y = f(x)$, graph it over its meaningful domain, and zoom in on the highest or lowest point. As an example, suppose y represents the profit in millions of dollars to be made on the production of x widgets. Suppose also that y is related to x according to the formula

$$y = 5 - 0.000000167(x - 6000)^2 + 0.000001\left(\frac{x - 6000}{100}\right)^3$$

The graph of this function for $0 \le x \le 13000$ and for $-1 \le y \le 6$ is shown in Figure 5. The rectangle is what you will see when using a "box zoom." This is the best way to find maximum and minimum points graphically. Make sure that the branches of the curve exit the "box" through either the bottom (as shown) or the top. If this is not done, the graph will become so flat after a few "zoom ins" that you will not be able to tell where the highest (or lowest) point is located. After zooming in repeatedly with the "box" placed so that the branches of the curve always exit the bottom of the box, a graph that appears very similar to the one in Figure 5 can be obtained, but with "**Range**" values close to

$$5999.8 \le x \le 6000.2 \quad \text{and} \quad 4.99999999 \le y \le 5.00000001.$$

Thus, the maximum profit (y) is 5 million dollars when 6000 widgets are produced and sold. (See Sections 3.2 and 12.3.)

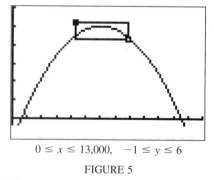

$$0 \le x \le 13,000, \quad -1 \le y \le 6$$

FIGURE 5

PROGRAMMING

Many formulas are used often enough to justify automating the process of evaluating them. The Quadratic Formula and the formulas encountered in the mathematics of finance are good examples of such formulas. (See Sections 1.10, 3.1, and all of Chapter 5.) Programs that will ask for values of the variables, evaluate the formula, and then start over again can be written very easily. Other situations where programs can be very helpful include graphing a particular function repeatedly, simulations of business situations, and processes that involve successive iterations such as the step-by-step amortization of a loan. There may be example programs included in the user's manual that will be useful to you. In general, any set of procedures done on one of these machines can be programmed into the machine as long as there is enough memory available.

SOME SUGGESTIONS FOR REDUCING FRUSTRATION

We all find ways to make even the simplest machines do the wrong things without even trying. One of the more common problems with graphing calculators is getting a blank screen when a graph was expected. This usually results from not setting the "**Range**" values appropriately before graphing the function, although it could also easily result from incorrectly entering the function.

A common problem that is particularly annoying is interpreting cryptic error messages such as "**Syn ERROR**" and "**Ma ERROR**." (Keep that manual handy!) The most common mistakes are made entering formulas and using special functions. For example, on the Casio machines "**Syn ERROR**" means that a mistake was made when entering the function or operation, such as entering "Graph $Y = \log x^y$" instead of "Graph $Y = \log x^2$".

Another common error is having more right parentheses than left parentheses. To confuse us further, these same machines think it is perfectly OK to have more left parentheses than right parentheses. For instance, the expression $5(3 - 4(2 + 7)$ has two left parentheses and one right parenthesis, but it will be evaluated as $5(3 - 4(2 + 7))$. The message "**Ma ERROR**" appears when a number is too large or when a number is not allowed. If you try to find the 1000th power of ten or divide a number by zero you will most certainly see some kind of error message. By pressing one of the "**cursor**" keys on the Casio you will see the cursor blinking at the location of the error in your expression. When the Texas Instruments models detect an error, they display a special menu that lists a code number and a name for the type of error. For certain types of errors the choice "**Go to error**" is offered. The TI-85 will display the number 9.99999999 E999 but shows "ERROR 01 OVERFLOW" for 10 E999 (which means 10 times the 999th power of ten). The latter and other more advanced machines display "(0,2)" when asked to find the square root of -4. The "(0,2)" represents the complex number $0 + 2i$.

When graphing **rational functions** you sometimes will see strange little "blips" in an otherwise smooth graph. This usually means there is a **vertical asymptote** at that location due to a division by zero, but you cannot see the actual behavior there because your "**window**" is too large. "**Zoom in**" on the graph in the region of the irregularity to obtain a better view. Also, as mentioned earlier in this article, one can switch to **dot mode** to eliminate unwanted lines in the graph. (See Section 3.4.)

SOME FINAL COMMENTS

While studying mathematics it is important to learn the mathematical concepts well enough to make intelligent decisions about when to use and when not to use "high tech" aids such as computers and graphing calculators. These machines make it easy to experiment with graphs of mathematical relations. One can learn much about the behavior of different types of functions by playing "what if" games with the formulas. However, in a timed test situation you may find yourself spending too much time working with the graphing calculator when a quick algebraic solution and a rough sketch with pencil and paper are more appropriate.

To get the most return on your investment, learn to use as many features of your machine as possible. Of course, some of the features may not be of use to you, so feel free to ignore them. A first session of two or three hours with your graphing calculator and your user's manual is essential. Be sure to keep your manual handy, referring to it when needed.

A final word of caution: These machines are fun to use, but they can be addictive. So set time limits for yourself, or you may find that your graphing calculator has been more of a detriment than a help!

APPENDIX B

Tables

Table 1 Combinations $\dbinom{n}{r} = \dfrac{n!}{(n-r)!\, r!}$

n	$\dbinom{n}{0}$	$\dbinom{n}{1}$	$\dbinom{n}{2}$	$\dbinom{n}{3}$	$\dbinom{n}{4}$	$\dbinom{n}{5}$	$\dbinom{n}{6}$	$\dbinom{n}{7}$	$\dbinom{n}{8}$	$\dbinom{n}{9}$	$\dbinom{n}{10}$
0	1										
1	1	1									
2	1	2	1								
3	1	3	3	1							
4	1	4	6	4	1						
5	1	5	10	10	5	1					
6	1	6	15	20	15	6	1				
7	1	7	21	35	35	21	7	1			
8	1	8	28	56	70	56	28	8	1		
9	1	9	36	84	126	126	84	36	9	1	
10	1	10	45	120	210	252	210	120	45	10	1
11	1	11	55	165	330	462	462	330	165	55	11
12	1	12	66	220	495	792	924	792	495	220	66
13	1	13	78	286	715	1287	1716	1716	1287	715	286
14	1	14	91	364	1001	2002	3003	3432	3003	2002	1001
15	1	15	105	455	1365	3003	5005	6435	6435	5005	3003
16	1	16	120	560	1820	4368	8008	11440	12870	11440	8008
17	1	17	136	680	2380	6188	12376	19448	24310	24310	19448
18	1	18	153	816	3060	8568	18564	31824	43758	48620	43758
19	1	19	171	969	3876	11628	27132	50388	75582	92378	92378
20	1	20	190	1140	4845	15504	38760	77520	125970	167960	184756

For $r > 10$, it may be necessary to use the identity

$$\binom{n}{r} = \binom{n}{n-r}$$

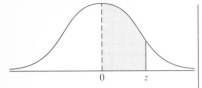

Table 2 Areas under the Normal Curve

The column under A gives the proportion of the area under the entire curve that is between $z = 0$ and a positive value of z.

z	A	z	A	z	A	z	A
.00	.0000	.49	.1879	.98	.3365	1.47	.4292
.01	.0040	.50	.1915	.99	.3389	1.48	.4306
.02	.0080	.51	.1950	1.00	.3413	1.49	.4319
.03	.0120	.52	.1985	1.01	.3438	1.50	.4332
.04	.0160	.53	.2019	1.02	.3461	1.51	.4345
.05	.0199	.54	.2054	1.03	.3485	1.52	.4357
.06	.0239	.55	.2088	1.04	.3508	1.53	.4370
.07	.0279	.56	.2123	1.05	.3531	1.54	.4382
.08	.0319	.57	.2157	1.06	.3554	1.55	.4394
.09	.0359	.58	.2190	1.07	.3577	1.56	.4406
.10	.0398	.59	.2224	1.08	.3599	1.57	.4418
.11	.0438	.60	.2258	1.09	.3621	1.58	.4430
.12	.0478	.61	.2291	1.10	.3643	1.59	.4441
.13	.0517	.62	.2324	1.11	.3665	1.60	.4452
.14	.0557	.63	.2357	1.12	.3686	1.61	.4463
.15	.0596	.64	.2389	1.13	.3708	1.62	.4474
.16	.0636	.65	.2422	1.14	.3729	1.63	.4485
.17	.0675	.66	.2454	1.15	.3749	1.64	.4495
.18	.0714	.67	.2486	1.16	.3770	1.65	.4505
.19	.0754	.68	.2518	1.17	.3790	1.66	.4515
.20	.0793	.69	.2549	1.18	.3810	1.67	.4525
.21	.0832	.70	.2580	1.19	.3830	1.68	.4535
.22	.0871	.71	.2612	1.20	.3849	1.69	.4545
.23	.0910	.72	.2642	1.21	.3869	1.70	.4554
.24	.0948	.73	.2673	1.22	.3888	1.71	.4564
.25	.0987	.74	.2704	1.23	.3907	1.72	.4573
.26	.1026	.75	.2734	1.24	.3925	1.73	.4582
.27	.1064	.76	.2764	1.25	.3944	1.74	.4591
.28	.1103	.77	.2794	1.26	.3962	1.75	.4599
.29	.1141	.78	.2823	1.27	.3980	1.76	.4608
.30	.1179	.79	.2852	1.28	.3997	1.77	.4616
.31	.1217	.80	.2881	1.29	.4015	1.78	.4625
.32	.1255	.81	.2910	1.30	.4032	1.79	.4633
.33	.1293	.82	.2939	1.31	.4049	1.80	.4641
.34	.1331	.83	.2967	1.32	.4066	1.81	.4649
.35	.1368	.84	.2996	1.33	.4082	1.82	.4656
.36	.1406	.85	.3023	1.34	.4099	1.83	.4664
.37	.1443	.86	.3051	1.35	.4115	1.84	.4671
.38	.1480	.87	.3079	1.36	.4131	1.85	.4678
.39	.1517	.88	.3106	1.37	.4147	1.86	.4686
.40	.1554	.89	.3133	1.38	.4162	1.87	.4693
.41	.1591	.90	.3159	1.39	.4177	1.88	.4700
.42	.1628	.91	.3186	1.40	.4192	1.89	.4706
.43	.1664	.92	.3212	1.41	.4207	1.90	.4713
.44	.1700	.93	.3238	1.42	.4222	1.91	.4719
.45	.1736	.94	.3264	1.43	.4236	1.92	.4726
.46	.1772	.95	.3289	1.44	.4251	1.93	.4732
.47	.1808	.96	.3315	1.45	.4265	1.94	.4738
.48	.1844	.97	.3340	1.46	.4279	1.95	.4744

Table 2 (*continued*)

z	A	z	A	z	A	z	A
1.96	.4750	2.45	.4929	2.94	.4984	3.43	.4997
1.97	.4756	2.46	.4931	2.95	.4984	3.44	.4997
1.98	.4762	2.47	.4932	2.96	.4985	3.45	.4997
1.99	.4767	2.48	.4934	2.97	.4985	3.46	.4997
2.00	.4773	2.49	.4936	2.98	.4986	3.47	.4997
2.01	.4778	2.50	.4938	2.99	.4986	3.48	.4998
2.02	.4783	2.51	.4940	3.00	.4987	3.49	.4998
2.03	.4788	2.52	.4941	3.01	.4987	3.50	.4998
2.04	.4793	2.53	.4943	3.02	.4987	3.51	.4998
2.05	.4798	2.54	.4945	3.03	.4988	3.52	.4998
2.06	.4803	2.55	.4946	3.04	.4988	3.53	.4998
2.07	.4808	2.56	.4948	3.05	.4989	3.54	.4998
2.08	.4812	2.57	.4949	3.06	.4989	3.55	.4998
2.09	.4817	2.58	.4951	3.07	.4989	3.56	.4998
2.10	.4821	2.59	.4952	3.08	.4990	3.57	.4998
2.11	.4826	2.60	.4953	3.09	.4990	3.58	.4998
2.12	.4830	2.61	.4955	3.10	.4990	3.59	.4998
2.13	.4834	2.62	.4956	3.11	.4991	3.60	.4998
2.14	.4838	2.63	.4957	3.12	.4991	3.61	.4999
2.15	.4842	2.64	.4959	3.13	.4991	3.62	.4999
2.16	.4846	2.65	.4960	3.14	.4992	3.63	.4999
2.17	.4850	2.66	.4961	3.15	.4992	3.64	.4999
2.18	.4854	2.67	.4962	3.16	.4992	3.65	.4999
2.19	.4857	2.68	.4963	3.17	.4992	3.66	.4999
2.20	.4861	2.69	.4964	3.18	.4993	3.67	.4999
2.21	.4865	2.70	.4965	3.19	.4993	3.68	.4999
2.22	.4868	2.71	.4966	3.20	.4993	3.69	.4999
2.23	.4871	2.72	.4967	3.21	.4993	3.70	.4999
2.24	.4875	2.73	.4968	3.22	.4994	3.71	.4999
2.25	.4878	2.74	.4969	3.23	.4994	3.72	.4999
2.26	.4881	2.75	.4970	3.24	.4994	3.73	.4999
2.27	.4884	2.76	.4971	3.25	.4994	3.74	.4999
2.28	.4887	2.77	.4972	3.26	.4994	3.75	.4999
2.29	.4890	2.78	.4973	3.27	.4995	3.76	.4999
2.30	.4893	2.79	.4974	3.28	.4995	3.77	.4999
2.31	.4896	2.80	.4974	3.29	.4995	3.78	.4999
2.32	.4898	2.81	.4975	3.30	.4995	3.79	.4999
2.33	.4901	2.82	.4976	3.31	.4995	3.80	.4999
2.34	.4904	2.83	.4977	3.32	.4996	3.81	.4999
2.35	.4906	2.84	.4977	3.33	.4996	3.82	.4999
2.36	.4909	2.85	.4978	3.34	.4996	3.83	.4999
2.37	.4911	2.86	.4979	3.35	.4996	3.84	.4999
2.38	.4913	2.87	.4980	3.36	.4996	3.85	.4999
2.39	.4916	2.88	.4980	3.37	.4996	3.86	.4999
2.40	.4918	2.89	.4981	3.38	.4996	3.87	.5000
2.41	.4920	2.90	.4981	3.39	.4997	3.88	.5000
2.42	.4922	2.91	.4982	3.40	.4997	3.89	.5000
2.43	.4925	2.92	.4983	3.41	.4997		
2.44	.4927	2.93	.4983	3.42	.4997		

Table 3   Integrals

(C is an arbitrary constant.)

1. $\int x^n \, dx = \dfrac{1}{n+1} x^{n+1} + C \qquad (\text{if } n \neq -1)$

2. $\int e^{kx} \, dx = \dfrac{1}{k} e^{kx} + C$

3. $\int \dfrac{a}{x} \, dx = a \ln|x| + C \qquad (a \neq 0)$

4. $\int \ln|ax| \, dx = x(\ln|ax| - 1) + C$

5. $\int \dfrac{1}{\sqrt{x^2 + a^2}} \, dx = \ln\left|x + \sqrt{x^2 + a^2}\right| + C$

6. $\int \dfrac{1}{\sqrt{x^2 - a^2}} \, dx = \ln\left|x + \sqrt{x^2 - a^2}\right| + C$

7. $\int \dfrac{1}{a^2 - x^2} \, dx = \dfrac{1}{2a} \cdot \ln\left|\dfrac{a + x}{a - x}\right| + C \qquad (a \neq 0)$

8. $\int \dfrac{1}{x^2 - a^2} \, dx = \dfrac{1}{2a} \cdot \ln\left|\dfrac{x - a}{x + a}\right| + C \qquad (a \neq 0)$

9. $\int \dfrac{1}{x\sqrt{a^2 - x^2}} \, dx = -\dfrac{1}{a} \cdot \ln\left|\dfrac{a + \sqrt{a^2 - x^2}}{x}\right| + C \qquad (a \neq 0)$

10. $\int \dfrac{1}{x\sqrt{a^2 + x^2}} \, dx = -\dfrac{1}{a} \cdot \ln\left|\dfrac{a + \sqrt{a^2 + x^2}}{x}\right| + C \qquad (a \neq 0)$

11. $\int \dfrac{x}{ax + b} \, dx = \dfrac{x}{b} - \dfrac{b}{a^2} \cdot \ln|ax + b| + C \qquad (a \neq 0, b \neq 0)$

12. $\int \dfrac{x}{(ax + b)^2} \, dx = \dfrac{b}{a^2(ax + b)} + \dfrac{1}{a^2} \cdot \ln|ax + b| + C \qquad (a \neq 0)$

13. $\int \dfrac{1}{x(ax + b)} \, dx = \dfrac{1}{b} \cdot \ln\left|\dfrac{x}{ax + b}\right| + C \qquad (b \neq 0)$

14. $\int \dfrac{1}{x(ax + b)^2} \, dx = \dfrac{1}{b(ax + b)} + \dfrac{1}{b^2} \cdot \ln\left|\dfrac{x}{ax + b}\right| + C \qquad (b \neq 0)$

15. $\int \sqrt{x^2 + a^2} \, dx = \dfrac{x}{2}\sqrt{x^2 + a^2} + \dfrac{a^2}{2} \cdot \ln\left|x + \sqrt{x^2 + a^2}\right| + C$

16. $\int x^n \cdot \ln|x| \, dx = x^{n+1}\left[\dfrac{\ln|x|}{n + 1} - \dfrac{1}{(n + 1)^2}\right] + C \qquad (n \neq -1)$

17. $\int x^n e^{ax} \, dx = \dfrac{x^n e^{ax}}{a} - \dfrac{n}{a} \cdot \int x^{n-1} e^{ax} \, dx + C \qquad (a \neq 0)$

Answers to Selected Exercises

CHAPTER 1
SECTION 1.1 (page 9)

1. natural number, whole number, integer, rational number, real number   **3.** integer, rational number, real number
5. rational number, real number   **7.** irrational number, real number; -1.73 **9.** irrational number, real number; 9.42
11. true   **13.** true   **15.** commutative property (multiplication)   **17.** identity property (addition)
19. commutative property (addition)   **21.** associative property (multiplication)
23. distributive property and commutative property (multiplication)

27. 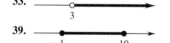   **29.**    **31.**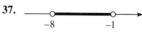

33. 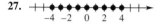   **35.** 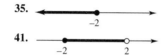   **37.**

39. 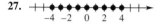   **41.**

43. 5   **45.** -58 **47.** -12 **49.** 15.88   **51.** -54 **53.** $-1/2$ **55.** 2   **57.** -4 **59.** 2   **61.** -4 **63.** 17   **65.** 4   **67.** -19
69. $=$ **71.** $<$ **73.** $=$ **75.** $=$ **77.** $=$ **79.** $=$ **85.** Let x represent the number of dollars contributed annually by foreign
students to the California economy. Then $x > 1,000,000,000$.   **87.** Let x represent the percent of foreign students now in the United
States from Middle Eastern countries. Then $x < 7.5$.   **89.** Let x represent the percent of foreign students in the United
States that are in California. Then $x > 13$.   **91.** 100; 0   **93.** 67; 11

SECTION 1.2 (page 21)

1. 5   **3.** 7   **5.** $-19/52$ **7.** $-9/8$ **9.** 2   **11.** $40/7$ **13.** $26/3$ **15.** $-12/5$ **17.** $-59/6$ **19.** $-9/4$ **21.** $x = (5a - b)/6$
23. $x = 3b/(5 + a)$ **25.** $x = (-3a + 3)/(1 - a^2 - a)$ or $x = (3a - 3)/(a^2 + a - 1)$ **27.** $v = k/P$ **29.** $g = (V - V_0)/t$
31. $B = (2A/h) - b$ or $B = (2A - bh)/h$ **33.** $R = (r_1 r_2)/(r_1 + r_2)$ **35.** .72   **37.** -13.26 **39.** 68°F   **41.** 15°C   **43.** 37.8°C
49. 13%   **51.** $247   **53.** $4000   **55.** $205.41   **57.** (a) Let x represent the height. (b) $16x + 288 + 36x = 496$ (c) 4 ft
59. 6 cm   **61.** 1629 mi   **63.** 15 min   **65.** $21,000   **67.** $70,000 for land that made a profit, $50,000 for land that produced a loss
69. $800   **71.** 400/3 liters

SECTION 1.3 (page 28)

3. $[-4, \infty)$ **5.** $(-\infty, 0)$ **7.** $(-\infty, 8/3]$

9. $(-\infty, -7]$ **11.** $(-\infty, 3)$ **13.** $(-1, \infty)$

15. $(-\infty, 1]$ **17.** $(1/5, \infty)$ **19.** $(-5, 6)$

21. $[7/3, 4]$

23. $[-11/2, 7/2]$

25. $(-\infty, 1]$ or $[6, \infty)$

27. $(-\infty, -2)$ or $[0, \infty)$

29. $(-\infty, 4)$ or $(8, \infty)$

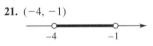

31. $[-17/7, \infty)$

33. $(.1745, \infty)$

35. $(-\infty, 1.50)$

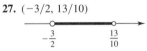

37. $(-\infty, .50]$

39. (a) Let x represent the number of mg per liter of lead in the water. **(b)** $.038 \le x \le .042$ **(c)** yes **41.** If x represents the scores, then $x < 88$ or $x > 112$; $x < 76$ or $x > 124$. **43.** 83 points **45.** $[500, \infty)$ **47.** $[45, \infty)$ **49.** impossible to at least break even **51.** at least 629 mi **53.** $[.68518, \infty)$ (Exact answer is $[37/54, \infty)$) **55.** $(14.4444, \infty)$ (Exact answer is $(130/9, \infty)$)

SECTION 1.4 (page 33)

1. $2, -3$ **3.** $-2, 8$ **5.** $5/2, 7/2$ **7.** $-4/3, 2/9$ **9.** $-7, -3/7$ **11.** $-7/8, 11/2$
13. $(-\infty, -5]$ or $[5, \infty)$ **15.** $(-1, 1)$ **17.** no solution

19. $[-10, 10]$ **21.** $(-4, -1)$ **23.** $(-\infty, -2/3)$ or $(2, \infty)$

25. $(-\infty, -8/3]$ or $[2, \infty)$ **27.** $(-3/2, 13/10)$

31. $-140 \le C \le -28$ **33.** $|F - 730| \le 50$ **35.** $|x - 2| \le 4$ **37.** $|z - 12| \ge 2$
39. If $|x - 2| \le .0004$, then $|y - 7| \le .00001$ **41.** $|x - 40| \le .05$ **43.** $(-\infty, .\overline{3})$ or $(3, \infty)$ **45.** no solution
47. $(-\infty, -.75)$ or $(8, \infty)$ (Answer is exact.)

SECTION 1.5 (page 39)

1. 81 **3.** $125/27$ **5.** -36 **7.** 81 **9.** 448 **11.** -2048 **15.** 2^7 **17.** $(-5)^7$ or -5^7 **19.** -3^9 **21.** $(2z)^{11}$ **23.** $-x^3 + x^2 + 3x$
25. $-6y^2 + 3y + 10$ **27.** $-10x^2 + 4x - 2$ **29.** $-18m^3 - 27m^2 + 9m$ **31.** $12z^3 + 14z^2 - 7z + 5$ **33.** $12k^2 - 20k + 3$
35. $6y^2 + 7y - 5$ **37.** $18k^2 - 7kq - q^2$ **39.** $4.34m^2 + 5.68m - 4.42$ **41.** $-k + 3$ **43.** $7x^2 - 4x - 12$ **45. (a)** 4 **(b)** 4 **(c)** 7
49. $5x^3 - 4x^2 - 8x + 1$ **51.** $x^2 - 7x + 40$

SECTION 1.6 (page 45)

1. $12x(x - 2)$ **3.** $r(r^2 - 5r + 1)$ **5.** cannot be factored **7.** $6z(z^2 - 2z + 3)$ **9.** $5p^2(5p^2 - 4pq + 20q^2)$
11. $2(2y - 1)^2 (5y - 1)$ **13.** $(x + 5)^4(x^2 + 10x + 28)$ **15.** $(2a + 5)(a - 1)$ **17.** $(x + 8)(x - 8)$ **19.** $(3p - 4)^2$
21. $(r + 2t)(r - 5t)$ **23.** $(m - 3n)^2$ **25.** $(2p + 3)(2p - 3)$ **27.** $3(x - 4z)^2$ **29.** cannot be factored
31. $(x - y)(3 - a)$ **33.** $(3a + 5)(a - 6)$ **35.** $(7m + 2n)(3m + n)$ **37.** $(4y - x)(5y + 11x)$ **39.** cannot be factored
41. $(y - 7z)(y + 3z)$ **43.** $(m^2 + 3)(n + 2)(n - 2)$ **45.** $(11x + 8)(11x - 8)$ **47.** $2a^2(4a - b)(3a + 2b)$
49. $3x^3(3x - 5z)(2x + 5z)$ **51.** $(2z + 1 + t)(2z + 1 - t)$ **53.** $(m + n - 1)(m - n + 1)$
55. $m^3(m - 1)^2(m^2 + m + 1)^2(5 + 3m^2 - 3m^5)$
59. $(a - 6)(a^2 + 6a + 36)$ **61.** $(2r - 3s)(4r^2 + 6rs + 9s^2)$
63. $(4m + 5)(16m^2 - 20m + 25)$ **65.** $(10y - z)(100y^2 + 10yz + z^2)$

SECTION 1.7 (page 51)

1. $x/5$ **3.** $4/(7p)$ **5.** $5/4$ **7.** $4/(w + 3)$ **9.** $(y - 4)/(3y^2)$ **11.** $2(x + 2)/x$ **13.** $(m - 2)/(m + 3)$ **15.** $(x + 4)/(x + 1)$
17. $2p/7$ **19.** $9a^3/4$ **21.** $15/(8c)$ **23.** $3/4$ **25.** $2/9$ **27.** $3/10$ **29.** $2(a + 4)/(a - 3)$ **31.** $(k + 2)/(k + 3)$ **35.** $-1/(15z)$
37. $4/3$ **39.** $(12 + x)/(3x)$ **41.** $(8 - y)/(4y)$ **43.** $137/(30m)$ **45.** $(3m - 2)/[m(m - 1)]$ **47.** $14/[3(a - 1)]$ **49.** $23/[20(k - 2)]$
51. $(7x + 9)/[(x - 3)(x + 1)(x + 2)]$ **53.** $y^2/[(y + 4)(y + 3)(y + 2)]$ **55.** $(k^2 - 13k)/[(2k - 1)(k + 2)(k - 3)]$
57. $(x + 1)/(x - 1)$ **59.** $-1/[x(x + h)]$ **61.** $(2 + b - b^2)/(b - b^2)$

SECTION 1.8 (page 56)

1. 1 **3.** $1/6$ **5.** $1/32$ **7.** $-1/64$ **9.** .00200 **11.** 9 **13.** $625/16$ **17.** $1/4^5$ **19.** $1/9$ **21.** 4^3 **23.** $1/7^7$ **25.** 8^5
27. $1/10^8$ **29.** 5 **31.** z^3 **33.** $p/4$ **35.** q^5/r^2 **37.** $2^3/(5^2p^7)$ **39.** $r/7^2$ **41.** $1/(12k^4)$ **43.** $x^3/(2^2y^3)$ **45.** m^7/p^2 **47.** 5 **49.** 6
51. 0 **53.** $73/8$ **55.** $1/6$ **57.** 9 **59.** $1/6$ **61.** $3/2$ **63.** $-5/9$ **65.** $b + a$

SECTION 1.9 (page 63)

1. 7 **3.** 1.7 **5.** 9 **7.** 15.5 **9.** -16 **11.** $3/5$ **13.** $1/27$ **15.** $27/64$ **17.** $4/3$ **19.** 5^2 **21.** 2 **23.** $1/16^{11/5}$ **25.** $1/9^{32/15}$
27. $2^{5/6}p^{3/2}$ **29.** $1/(m^{14/15}n^{6/5})$ **31.** $2p + 5p^{5/3}$ **33.** $-3m^2 - 12m - 8$ **37.** 4 **39.** -3 **41.** $4\sqrt{3}$ **43.** $-3\sqrt[4]{2}$ **45.** $-4\sqrt{3}/3$
47. $-\sqrt[3]{12}/2$ **49.** $2x^4z^2\sqrt{2z}$ **51.** $mn^3p^4\sqrt{n}$ **53.** $\sqrt{6x}/(3x)$ **55.** $(x^2y\sqrt{xy})/3$ **57.** $\sqrt[3]{2}$ **59.** $\sqrt[18]{x}$ **61.** $5\sqrt{2}$ **63.** $7\sqrt{6}$
65. $23\sqrt{2k}$ **67.** $5\sqrt[3]{2}$ **69.** 0 **71.** 3 **73.** 34 **75.** $58 + 5\sqrt{5}$ **77.** $(\sqrt{5} - 1)/2$ **79.** $-1 - \sqrt{3}$ **81.** $(3\sqrt{r} + r)/(9 - r)$
83. (a) 14 **(b)** 85 **(c)** 58 **(d)** 11 **85.** 50 cm

SECTION 1.10 (page 74)

1. $-3, 12$ **3.** $-7, -8$ **5.** $3/2, 1$ **7.** $-1/2, 1/3$ **9.** $5/2, 4$ **11.** 5, 2 **13.** $4/3, -4/3$ **15.** 0, 1 **17.** $2 \pm \sqrt{7}$
19. $(1 \pm 2\sqrt{5})/4$ **21.** $3, -4$ **23.** $(1 \pm \sqrt{2})/2$ **25.** $(-5 \pm \sqrt{17})/4$; $-.219, -2.281$ **27.** $(-1 \pm \sqrt{5})/4$; .309, $-.809$
29. $(-3 \pm \sqrt{19})/5$; .272; -1.472 **31.** no real number solutions **33.** $5/2, 1$ **35.** $4/3, 1/2$ **37.** no real number solutions
39. $0, -1$ **41.** $\pm\sqrt{5}$ **43.** $\pm\sqrt{6}/2$ **45.** $11/3, 3/2$ **47.** $-1, -15/2$ **49.** $-5, 3/2$ **51.** $-4 \pm 3\sqrt{2}$
53. (a) $(-r^2 + 3r - 12)/[r(r - 2)]$ **(b)** $4, -3$ **55. (a)** $150 - x$ **(b)** $x(150 - x) = 5000$ **(c)** length: 100 m; width: 50 m
57. 1 ft **59.** 176 mph **61.** 30 bicycles **63. (a)** 2 sec **(b)** 1/2 sec or 7/2 sec **(c)** It reaches the given height twice—once on the way
up and once on the way down. **65.** $t = \pm\sqrt{2Sg}/g$ **67.** $h = \pm(d^2\sqrt{kL})/L$ **69.** $R = (-2Pr + E^2 \pm E\sqrt{E^2 - 4Pr})/(2P)$
71. no real number solutions

SECTION 1.11 (page 80)

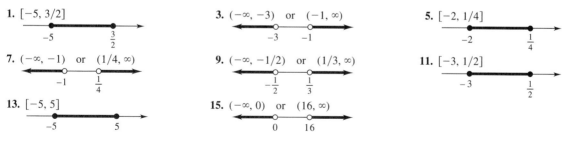

1. $[-5, 3/2]$

3. $(-\infty, -3)$ or $(-1, \infty)$

5. $[-2, 1/4]$

7. $(-\infty, -1)$ or $(1/4, \infty)$

9. $(-\infty, -1/2)$ or $(1/3, \infty)$

11. $[-3, 1/2]$

13. $[-5, 5]$

15. $(-\infty, 0)$ or $(16, \infty)$

17. $[-3, 0]$ or $[3, \infty)$ **19.** $(-\infty, -1/2)$ or $(0, 4/3)$ **23.** $(-\infty, 1)$ or $[3, \infty)$ **25.** $(7/2, 5)$ **27.** $(-\infty, 2)$ or $(5, \infty)$
29. $(-\infty, -3/2)$ or $[-13/9, \infty)$ **31.** $(-\infty, -1)$ **33.** excluded **35.** 5 is included, 2 is excluded **37.** $[200, \infty)$
39. $(0, 5/4)$ or $(6, \infty)$ **41.** $(-3/2, 4)$ **43.** $(-\infty, .56721)$ or $(1.81989, \infty)$ **45.** $(-\infty, .5)$ or $(5, 6)$

CHAPTER 1 REVIEW EXERCISES (page 84)

1. 0, 6   **3.** $-12, -6, -9/10, -\sqrt{4}, 0, 1/8, 6$ **5.** Whole numbers, integers, rational numbers, real numbers   **7.** integers, rational numbers, real numbers   **9.** irrational numbers, real numbers   **11.** $-7, -3, -2, 0, \pi, 8$ **13.** $-|3 - (-2)|, -|-2|, |6 - 4|, |8 + 1|$
15. -3 **17.** -1 **19.**

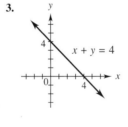

21.

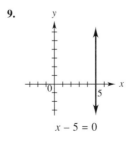

23. -8 **25.** $-3/4$ **27.** 3   **29.** -2 **31.** no solution   **33.** $x = 1/(5a - 1)$ **35.** $x = (c - 3)/(3a - ac - 2)$ **37.** $500
39. $60,000 at 8%, $40,000 at 5%   **41.** $(3/8, \infty)$ **43.** $(-\infty, 1/4]$ **45.** $[-1/2, 2]$ **47.** 12, -6 **49.** $-38, 42$ **51.** 2, 1/5
53. $[-8, 8]$ **55.** no solution   **57.** $(-\infty, 2]$ or $[5, \infty)$ **59.** $[-9/5, 1]$ **63.** $4x^4 - 4x^2 + 13x$ **65.** $2q^4 + 14q^3 - 4q^2$
67. $15z^2 - 4z - 4$ **69.** $16k^2 - 9h^2$ **71.** $36x^2 + 36xy + 9y^2$ **73.** $64x^3 - 144x^2 + 108x - 27$ **75.** $6k^3 - 11k^2 + 6k - 1$
77. $k(2h^2 - 4h + 5)$ **79.** $a^2(3a + 1)(a + 4)$ **81.** $(2y - 1)(5y - 3)$ **83.** $(2a - 5)^2$ **85.** $(12p + 13q)(12p - 13q)$
87. $(2y - 1)(4y^2 + 2y + 1)$ **89.** $(7x^2)/3$ **91.** 4   **93.** $[(y - 5)(4y - 5)]/(6y)$ **95.** $(m - 1)^2/[3(m + 1)]$ **97.** $1/6z$
99. $64/(35q)$ **105.** $1/10^2$ or $1/100$ **107.** $-1/3$ **109.** $-3^3/2^3$ or $-27/8$ **111.** $1/7^6$ **113.** $1/6^5$ **115.** $1/k^5$ **117.** $4/9$
119. $2^3 = 8$ **121.** $7/12$ **123.** 1   **125.** $2^{17/6}p^{31/6}$ **127.** 5   **129.** not a real number   **131.** $3\sqrt{7}$ **133.** $2a\sqrt[4]{4ab^3}$ **135.** $(x\sqrt{6xz})/(2z)$
137. $\sqrt[12]{10}$ **139.** $12\sqrt{7}$ **141.** $\sqrt[4]{mn^2}$ **143.** 4   **145.** $82 - \sqrt{14}$ **147.** $(16 + 4\sqrt{2} + 4\sqrt{5} + \sqrt{10})/11$ **149.** $(-1 \pm \sqrt{7})/2$
151. $3, -5/2$ **153.** $-2 \pm \sqrt{5}$ **155.** $(-1 \pm \sqrt{3})/3$ **157.** $1/2, 1/6$ **159.** $-8/3, 2$ **161.** $\pm\sqrt{3}/3$ **163.** $\pm\sqrt{3}/3$ **165.** $-8, -5$
167. $-3/4, 2$ **169.** $E = [\pm(r + R)\sqrt{pR}]/R$ **171.** $z = (h \pm \sqrt{h^2 + 4kt})/(2k)$ **173.** $(-\infty, -5]$ or $[1, \infty)$ **175.** $[-2, 7/3]$
177. $(-\infty, -3)$ or $(4, \infty)$ **179.** $[0, 2)$ **181.** $(-\infty, -1)$ or $(1, 3)$ **185.** North: 30 mph; West: 40 mph

CASE 1 (page 88)

1. $700 + 85x$ **3.** The $700 refrigerator costs $300 more over 10 years.

CHAPTER 2
SECTION 2.1 (page 98)

1.

3.

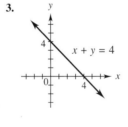

5.

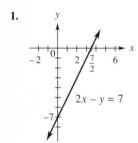

7.

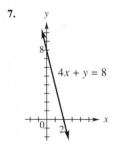

9.

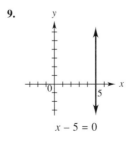

11.

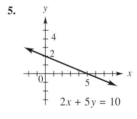

13.

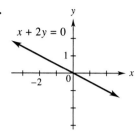

$x + 2y = 0$

15.

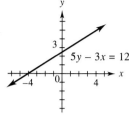

$5y - 3x = 12$

17.

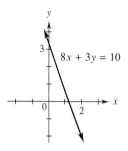

$8x + 3y = 10$

19.

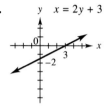

$x = 2y + 3$

23. $140   **25.** 10 items; 10 items

27. **(a)** $16 **(b)** $11 **(c)** $6 **(d)** 8 **(e)** 4 **(f)** 0 **(h)** 0 **(i)** 40/3 **(j)** 80/3 **(l)** 8 **(m)** $6

(g) (k)

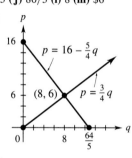

$p = 16 - \frac{5}{4}q$

$p = \frac{3}{4}q$

$(8, 6)$

29. **(b)** 125 **(c)** 50¢ **(d)** [0, 125)

(a)

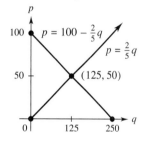

$p = 100 - \frac{2}{5}q$

$p = \frac{2}{5}q$

$(125, 50)$

31. **(a)** $135 **(b)** $205 **(c)** $275 **(d)** $345

(e)

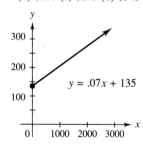

$y = .07x + 135$

33. **(a)** 200,000 policies $(x = 200)$ **(c)** Revenue: $12,500; Cost: $15,000

(b)

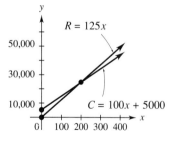

$R = 125x$

$C = 100x + 5000$

35. b   **37.** $3308.33

SECTION 2.2 (page 109)

1. 3/8 **3.** $-1/2$ **5.** $-3/2$ **7.** undefined **9.** 2 **11.** 5/9 **13.** 2 **15.** .5785 **17.** The product of their slopes is -1 if neither is vertical. Thus their slopes are negative reciprocals. **19.** parallel **21.** perpendicular **23.** neither **25.** 1/2, 5/2 **27.** 2/3, $-14/9$ **29.** 2/3, 0 **31.** undefined, no y-intercept

33.

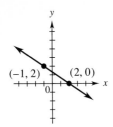

35.

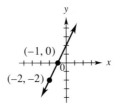

37.

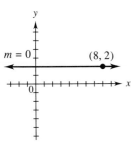

39.

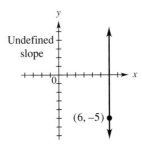

41. $4y = -3x + 16$ **43.** $2y = -x - 4$ **45.** $4y = 6x + 5$ **47.** $y = -x + 6$ **49.** $2y = 3x + 12$ **51.** $3y = -2x + 3$ **53.** $x = -8$
55. $y = 3$ **57.** $y = 11x + 256$; 1988 **59.** $y = 4x + 120$; 4 ft **61.** $y = 10x - 100$; 1950 **63.** $y = -.3125x + 14.25$; 42,500 sq mi

SECTION 2.3 (page 117)

1. function **3.** not a function **5.** function **7.** not a function **9.** function **13.** $(-\infty, \infty)$; $(-\infty, \infty)$ **15.** $(-\infty, \infty)$; $[-1, \infty)$
17. $(-\infty, 0]$; $[3, \infty)$ **19.** $(-\infty, 1)$ or $(1, \infty)$; $(-\infty, 0)$ or $(0, \infty)$ **21.** $(-\infty, \infty)$; $[0, \infty)$ **23.** (a) 6 (b) 6 (c) 6 (d) 6 **25.** (a) 48 (b) 6 (c) 0
(d) $2a^2 + 4a$ **27.** (a) $\sqrt{7}$ (b) 0 (c) $\sqrt{3}$ (d) $\sqrt{a + 3}$ ($a \geq -3$) **29.** (a) $5 - p$ (b) $5 + r$ (c) $2 - m$ **31.** (a) $\sqrt{4 - p}$ ($p \leq 4$)
(b) $\sqrt{4 + r}$ ($r \geq -4$) (c) $\sqrt{1 - m}$ ($m \leq 1$) **33.** (a) $p^3 + 1$ (b) $-r^3 + 1$ (c) $m^3 + 9m^2 + 27m + 28$ **35.** Assume all denominators
are nonzero. (a) $3/(p - 1)$ (b) $3/(-r - 1)$ (c) $3/(m + 2)$ **37.** 2 **39.** (a) about \$44 billion; about 56 billion (b) about \$37 billion;
about 70 billion (c) about \$14 billion more exports **41.** (a) \$11 (b) \$11 (c) \$18 (d) \$32 (e) \$39 (g) x (h) $S(x)$

SECTION 2.4 (page 126)

1. (a) 14.4° (b) 122° **3.** −40° **5.** (a) $y = 82,500x + 850,000$ (b) \$1,840,000 (c) 2003 **7.** (a) $y = -.15x + 31.2$ (b) −.15 million
per year **11.** (a) 2,000 (b) 2,900 (c) 3,200 (d) yes (e) 300 **13.** (a) 32.5 min (b) 145 min (c) 1.25; too slow **15.** (a) 3 (b) 3.5 (c) 1
(d) 0 (e) −2 (f) −2 (g) −3 (h) −1 **17.** Let $C(x)$ be the cost of renting a saw for x hours. Then $C(x) = 12 + x$. **19.** Let $P(x)$ be the
cost (in cents) of parking for x half-hours. Then $P(x) = 30x + 35$. **21.** $C(x) = 30x + 100$ **23.** $C(x) = 120x + 3800$ **25.** (a) \$5.25
(b) \$66,250 **29.** (a) $C(x) = 10x + 500$ (b) $R(x) = 35x$ (c) 20 **31.** (a) $C(x) = 18x + 250$ (b) $R(x) = 28x$ (c) 25
33. about 95 units in 1977 **35.** (a) Canada: $f(x) = 1.56x + 12.1$; Japan: $f(x) = 10.34x + 4.7$ (b)
(c) (.84, 13.4) The investments of the Canadians and Japanese were both about \$13.4 million
near the end of 1980.

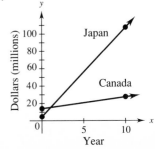

37. Break-even point is about 467 units; do not produce the item.   **39.** Break-even point is -200 units; this product will never make a profit, so don't produce it.   **41.** (a) $C(x) = .1x + .5$ (b) $1.2 million (c) $.1 million
43. (a)–(b)   (c) $y = 8x + 50$

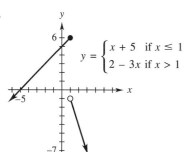

Year	Sales (actual)	Sales (Predicted from Equation of (c))	Difference, Actual Minus Predicted
0	48	50	-2
1	59	58	1
2	66	66	0
3	75	74	1
4	80	82	-2
5	90	90	0

(e) 106 (f) 122 **45.** c

SECTION 2.5 (page 138)

1. function   **3.** not a function   **5.** function

7.

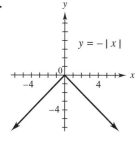

$y = |x - 4|$

9.

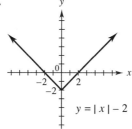

$y = |-4 - x|$

11.

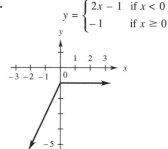

$y = |2x + 5|$

13.

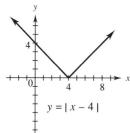

$y = -|x|$

15.

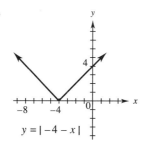

$y = |x| - 2$

19.

$y = \begin{cases} 2x - 1 & \text{if } x < 0 \\ -1 & \text{if } x \ge 0 \end{cases}$

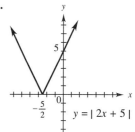

21.

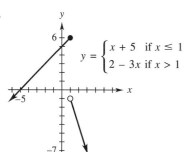

$y = \begin{cases} x + 5 & \text{if } x \le 1 \\ 2 - 3x & \text{if } x > 1 \end{cases}$

23.

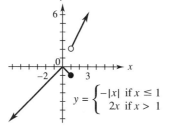

$y = \begin{cases} -|x| & \text{if } x \le 1 \\ 2x & \text{if } x > 1 \end{cases}$

25.

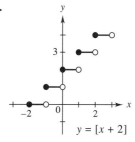

$y = [x + 2]$

27.

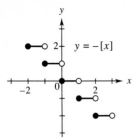

29.

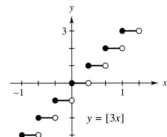

31.

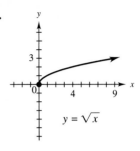

33.

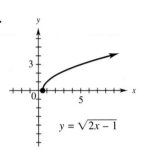

35.

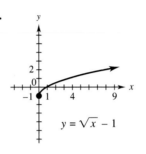

37. (a) $29 **(b)** $29 **(c)** $33 **(d)** $33
(e)

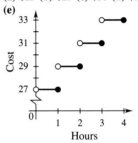

39.

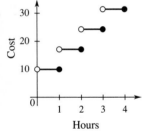

41.

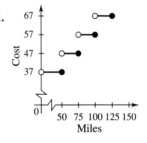

43. They rose faster from 1985 to 1990 than from 1960 to 1985.

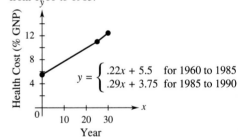

45. (a)

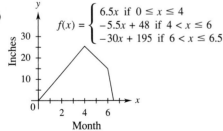

47. (a)

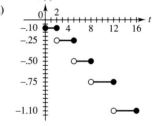

(b) at the beginning of February; 26 in.
(c) begins in early October; ends in mid-April

(b) Domain: $0 \le t \le 16$;
range: $\{-.10, -.25, -.50, -.75, -1.10\}$

49. (a) yes **(b)** years from 1984 to 1989 **(c)** approx. $[55, 1200]$

CHAPTER 2 REVIEW EXERCISES (page 142)

1.

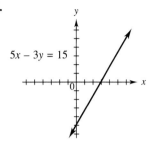

3.

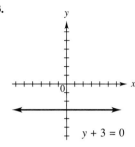

7. $-5/6$ **9.** $1/4$ **11.** 4   **13.** 0   **15.** -3 **17.**

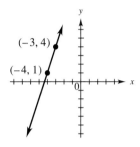

19. $3y = 2x - 13$ **21.** $4y = -5x + 17$ **23.** $x = -1$ **25.** $3y = 5x + 15$ **27. (a)** 1.17 units, 4.5 units **(b)** 2 units, 2 units **(c)** 2.5 units, .5 unit **(d)** \$15 **(e)** 2 units   **31.** $(-7, 10)$; -7 is the input, 10 is the output   **33.** function   **35.** not a function   **37.** function
39. (a) -21 **(b)** 11 **(c)** $3 - 4p$ **(d)** $-1 - 4r$ **41. (a)** -34 **(b)** 6 **(c)** $8 - p - p^2$ **(d)** $-r^2 - 3r + 6$ **43.** $f(x) = .88x + 4.2$; 35%; yes   **47. (a)** $C(x) = 180x + 2000$ **(b)** \$180 **(c)** \$200   **49. (a)** $C(x) = 46x + 120$ **(b)** \$46 **(c)** \$47.20   **51. (a)** 5 units **(b)** \$200 **(c)** $P(x) = 20x - 100$

53.

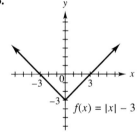

$f(x) = |x| - 3$

55.

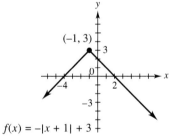

$(-1, 3)$

$f(x) = -|x + 1| + 3$

57.

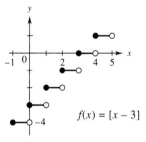

$f(x) = [x - 3]$

59.

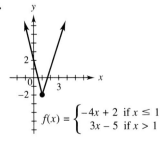

$f(x) = \begin{cases} -4x + 2 & \text{if } x \le 1 \\ 3x - 5 & \text{if } x > 1 \end{cases}$

61.

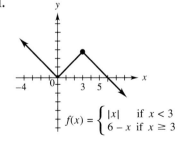

$f(x) = \begin{cases} |x| & \text{if } x < 3 \\ 6 - x & \text{if } x \ge 3 \end{cases}$

63. (a) yes **(b)**

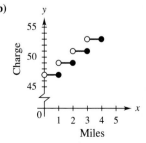

Charge

Miles

(c) domain: $(0, \infty)$; range: $\{47, 49, 51, 53, \ldots\}$

CASE 2 (page 146)

1. $2x - 7y = 93$ or $y = .286x - 13.3$ (approximately) Not much better as an approximation; for 1997; $y \approx 14.4$.

CASE 3 (page 148)

1. 4.8 million units **3.** In the interval under discussion (3.1 to 5.7 million units), the marginal cost always exceeds the selling price.

CHAPTER 3

SECTION 3.1 (page 156)

1. (5, 2); upward **3.** (1, 2); downward **5.** $(-6, -35)$; upward **7.** $(-1, -1)$; upward **9.** x-intercepts 1, 3; y-intercept 9
11. x-intercepts $-1, -3$; y-intercept 6
13. (0, 0), $x = 0$ **15.** $(-2, 0), x = -2$ **17.** $(1, -3), x = 1$

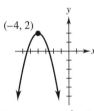

$f(x) = -x^2$

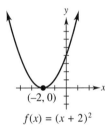

$f(x) = (x + 2)^2$

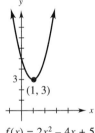

$f(x) = (x - 1)^2 - 3$

19. $(-4, 2), x = -4$ **21.** (2, 2), $x = 2$ **23.** (1, 3), $x = 1$

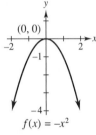

$f(x) = -(x + 4)^2 + 2$

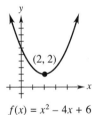

$f(x) = x^2 - 4x + 6$
$f(x) = (x - 2)^2 + 2$

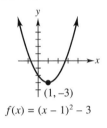

$f(x) = 2x^2 - 4x + 5$
$f(x) = 2(x - 1)^2 + 3$

25. (3, 3), $x = 3$ **27.** $x = 1, f(x) = -16$

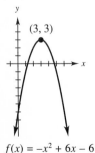

$f(x) = -x^2 + 6x - 6$
$f(x) = -(x - 3)^2 + 3$

29. (e) The graph of $f(x) = ax^2$ is the graph of $k(x) = x^2$ stretched away from the x-axis by a factor of a.
(a)-(d)

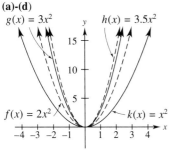

31. (e) Changing the sign of a reflects the graph in the x-axis.
(a)-(d)

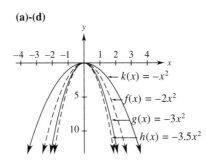

33. (e) The graph of $f(x) = x^2 - c$ is the graph of $k(x) = x^2$ shifted c units downward.
(a)-(d)

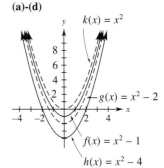

35. $f(x) = 3x^2$

SECTION 3.2 (page 160)

1. (a) \$16; \$8: \$32   **(c)** (5, 7) **(d)** 5; \$7 per box

(b)

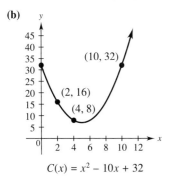

$$C(x) = x^2 - 10x + 32$$

3. (a) \$.60 **(b)** \$800   **5.** 16 ft; 2 sec   **7. (a)** $R(x) = 150x - (x^2/4)$ **(b)** 300 **(c)** \$22,500
9. (a) $R(x) = x(500 - x)$ **(c)** \$250 **(d)** \$62,500   **11.** 20 ft by 30 ft, with the short sides parallel to the river.
(b)   **13. (a)** \$640 **(b)** \$515 **(c)** \$140 **(e)** 800 **(f)** \$320

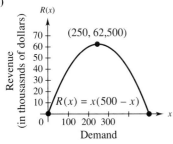

(d)

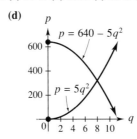

15. 10; \$20   **17. (a)** $S = 100 - x$ **(b)** $p = 50 + x$ **(c)** $R(x) = (100 - x)(50 + x)$ **(d)** 25 **(e)** \$5625   **19. (a)** 50 **(b)** 60 **(c)** 80 **(d)** 90
(e) 100 **(f)** 20   **21.** $6\sqrt{3}$ cm   **23. (a)** 2.3 or 17.7 **(b)** 15 **(c)** 10 **(d)** 120 **(e)** $0 < x < 2.3$ and $x > 17.7$ **(f)** $2.3 < x < 17.7$

SECTION 3.3 (page 168)

1.

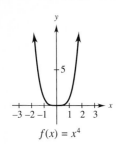

$f(x) = x^4$

3.

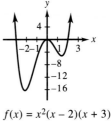

$f(x) = x^5$

7. (a) yes **(b)** no **(c)** no   **9. (a)** yes **(b)** yes **(c)** no

11.

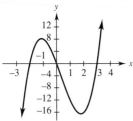

$f(x) = 2x(x - 3)(x + 2)$

13.

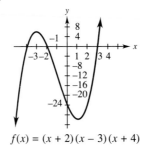

$f(x) = (x + 2)(x - 3)(x + 4)$

15.

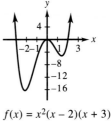

$f(x) = x^2(x - 2)(x + 3)$

17.

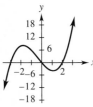

$f(x) = x^3 + x^2 - 6x$

19.

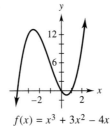

$f(x) = x^3 + 3x^2 - 4x$

21.

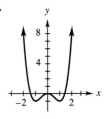

$f(x) = x^4 - 2x^2$

23. (a) 1.0 tenths of a percent, or .1%
(b) 2.0 tenths of a percent, or .2%
(c) 3.3 tenths of a percent, or .33%
(d) 3.1 tenths of a percent, or .31%
(e) 0.8 tenths of a percent, or .08%
(f)

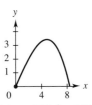

$A(x) = -.015x^3 + 1.058x$

(g) Between 4 and 5 hours, closer to 5 hours
(h) From about $1\frac{1}{2}$ hours to about $7\frac{1}{2}$ hours

25. (a) 0; 108; 28; 10   **(b)**

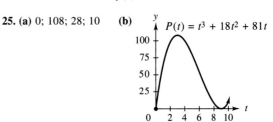

$P(t) = t^3 + 18t^2 + 81t$

(c) Increasing for years 0 to 3 and from the 9th year on; decreasing for years 3 to 9.

27–30 Many correct answers, including the following.   **27.** $-9 \le x \le 3$ and $-20 \le y \le 40$ **29.** $-8 \le x \le 4$ and $-100 \le y \le 150$

31.

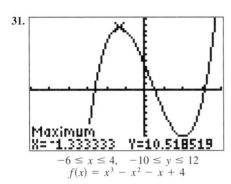

$-6 \le x \le 4,\quad -10 \le y \le 12$
$f(x) = x^3 - x^2 - x + 4$

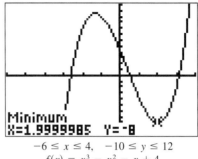

$-6 \le x \le 4,\quad -10 \le y \le 12$
$f(x) = x^3 - x^2 - x + 4$

SECTION 3.4 (page 175)

1. $x = -5, y = 0$

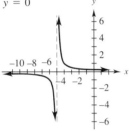

$$f(x) = \frac{1}{x + 5}$$

3. $x = -5/2, y = 0$

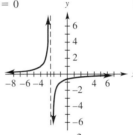

$$f(x) = \frac{-3}{2x + 5}$$

5. $x = 1, y = 3$

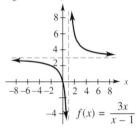

$$f(x) = \frac{3x}{x - 1}$$

7. $x = 4, y = 1$

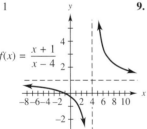

$$f(x) = \frac{x + 1}{x - 4}$$

9. $x = 3, y = -1$

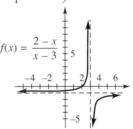

$$f(x) = \frac{2 - x}{x - 3}$$

11. $x = -1/2, y = 1/2$

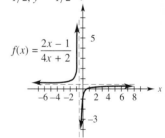

$$f(x) = \frac{2x - 1}{4x + 2}$$

13. $x = -4, y = -2/5$

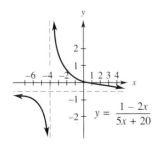

$$y = \frac{1 - 2x}{5x + 20}$$

15. $x = -2, y = -1/3$

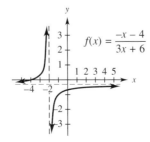

$$f(x) = \frac{-x - 4}{3x + 6}$$

17. $x = -2, x = 1$ **19.** $x = -1, x = 5$
21. (a) $4300 **(b)** $10,033.33
(c) $17,200 **(d)** $38,700 **(e)** $81,700
(f) $210,700 **(g)** $425,700 **(h)** No
(i)

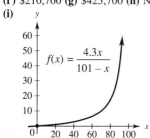

$$f(x) = \frac{4.3x}{101 - x}$$

23. (a) 125,000 **(b)** 187,500 **(c)** 200,000
(d)

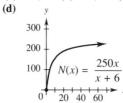

$$N(x) = \frac{250x}{x + 6}$$

(e) The part in the first quadrant, because x and $N(x)$ are both positive
(f) $y = 250$; 250,000

25. (a) 6 min **(b)** 1.5 min **(c)** .6 min
(d) $A = 0$
(e)

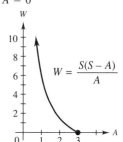

$$W = \frac{S(S - A)}{A}$$

(f) W becomes negative. The waiting time approaches 0 as A approaches 3. The formula does not apply for $A > 3$ because there will be no waiting if people arrive more than 3 min apart.

27. (a)

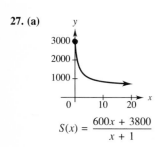

$$S(x) = \frac{600x + 3800}{x + 1}$$

(b) Yes; $600 **(c)** After 7 days

29. 500,000 gal of gasoline, 100,000 gal of oil

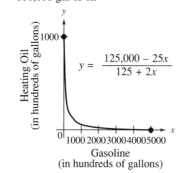

$$y = \frac{125,000 - 25x}{125 + 2x}$$

31. (a) yes; $t = 25$
(c)

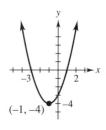

$$N(t) = \frac{600 - 7t}{4t - 100}$$

33. (a) $x = -2, x = 1$ **(b)** On most graphing calculators the window with $-4.7 \le x \le 4.7$ and $-6 \le y \le 8$ shows an accurate graph (with no erroneous vertical line segments). On other calculators, try $-4.7 \le x \le 4.8$ or $-6.3 \le x \le 6.3$.

CHAPTER 3 REVIEW EXERCISES (page 178)

1. upward; $(2, 6)$ **3.** downward; $(-1, 8)$ **5.**

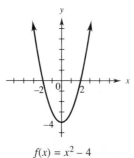

$$f(x) = x^2 - 4$$

7.

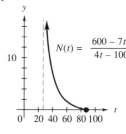

$$f(x) = x^2 + 2x - 3$$
$$f(x) = (x + 1)^2 - 4$$

9.

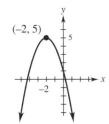

$(-2, 5)$

$f(x) = -x^2 - 4x + 1$
$f(x) = -(x + 2)^2 + 5$

11.

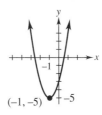

$(-1, -5)$

$f(x) = 2x^2 + 4x - 3$
$f(x) = 2(x + 1)^2 - 5$

13. minimum value -11 **15.** maximum value 7   **17.** $(0, 5/4), (6, \infty)$ **19.** Between 350 and 1850   **21.** 50 m $\times$ 50 m

23.

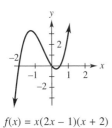

$f(x) = x^3 + 1$

25.

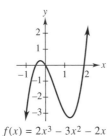

$g(x) = x^3 - x$

27.

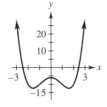

$f(x) = x(x - 2)(x + 3)$

29.

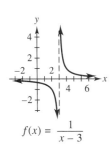

$f(x) = x(2x - 1)(x + 2)$

31.

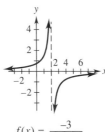

$f(x) = 2x^3 - 3x^2 - 2x$

33.

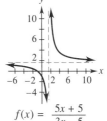

$f(x) = x^4 - 5x^2 - 6$

35.

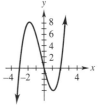

$f(x) = \dfrac{1}{x - 3}$

37.

$f(x) = \dfrac{-3}{2x - 4}$

39.

$f(x) = \dfrac{5x + 5}{3x - 5}$

41. (a) About $10.83 **(b)** About $4.64 **(c)** About $3.61 **(d)** About $2.71

43. (a) $(10, 50)$

(e)

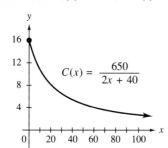

$$C(x) = \frac{650}{2x + 40}$$

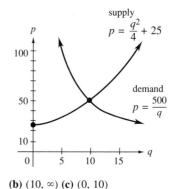

supply

$$p = \frac{q^2}{4} + 25$$

demand

$$p = \frac{500}{q}$$

(b) $(10, \infty)$ **(c)** $(0, 10)$

CHAPTER 4

SECTION 4.1 (page 186)

1. quadratic **3.** exponential **5.** Graph lies entirely above the x-axis and falls from left to right. It falls relatively steeply until it reaches the y-intercept 1 and then falls slowly, with the positive x-axis as a horizontal asymptote. **7.** Graph lies entirely above the x-axis and rises from left to right. The negative x-axis is a horizontal asymptote. The graph rises slowly until it reaches the y-intercept 1 and then rises quite steeply.

9.

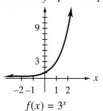

$f(x) = 3^x$

11.

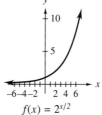

$f(x) = 2^{x/2}$

13.

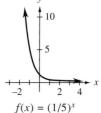

$f(x) = (1/5)^x$

15. (a)–(c)

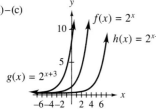

$f(x) = 2^x$

$h(x) = 2^{x-4}$

$g(x) = 2^{x+3}$

17.

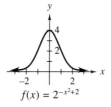

$f(x) = 2^{-x^2+2}$

19.

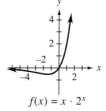

$f(x) = x \cdot 2^x$

21. 2 **23.** 3 **25.** 1.5 **27.** 1 **29.** 1/4 **31.** 1/9 **33.** 4, -4 **35.** 2 **37.** 0, 5/3 **39.** $-1, 5$

43. (a)

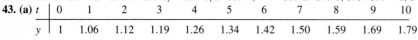

t	0	1	2	3	4	5	6	7	8	9	10
y	1	1.06	1.12	1.19	1.26	1.34	1.42	1.50	1.59	1.69	1.79

(b)

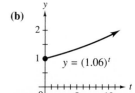

$y = (1.06)^t$

45. (a) About \$87,838 **(b)** About \$38.46
47. (a) 4.2 billion **(b)** 6.2 billion **(c)** 7.6 billion **(d)** 13.7 billion   **49. (a)** $f(t) = 2^t$ **(b)** First option pays \$9,000,000 for the month. Second option pays more, because on the last day alone the pay is 2^{30} cents = \$10,737, 418.24.
51. \$19,420.26 **53. (a)** \$100,000 **(b)** About \$225,000 **(c)** About \$1,818,000 **(d)** About \$2,488,000.   **55. (a)** About 207 **(b)** About 235 **(c)** About 249 **(d)**
57. (a) 7 **(b)** 9 **(c)** 33   **59. (a)**

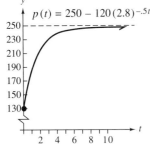

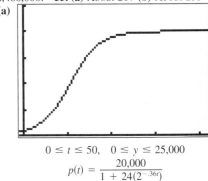

$0 \le t \le 50, \quad 0 \le y \le 25{,}000$

$$p(t) = \frac{20{,}000}{1 + 24(2^{-.36t})}$$

(b) 800 **(c)** from $t = 7$ to $t = 19$
(d) Never; limited food supply and oxygen might explain this.

SECTION 4.2 (page 194)

1. (a) 5000 **(b)** About 16,600 **(c)** About 607,552 **(d)** About 73,823,908   **3. (a)** 100 g **(b)** About 70.9 g **(c)** About 50.3 g **(d)** About 11.6 g
5. (a) 100 **(b)** About 111 **(c)** About 182 **(d)** About 332   **7. (a)** \$21,665.74 **(b)** \$29,836.49 **(c)** \$44,510.82 **(d)** About 13.7 years
9. (a) 1000 **(b)** About 1882 **11. (a)** 100,000 units **(b)** About 83,527 units and 58,275 units **(c)** No   **13. (a)** 446 (actual 410) **(b)** 1227 (actual 1250) **(c)** 4075 (actual 4077) **15. (a)** 77 **(b)** 72; 62 **(c)** Not much; the average score is about 4.   **17. (a)** 11% **(b)** 45.2% **(c)** 61.0% **(d)** 74.8% **(e)** 77.8% **(f)** 79.5% **(g)** No; $y = 80$ is a horizontal asymptote of the graph.   **19.** 6.25°C   **21.** 11.6 lb per square inch
23. (a)

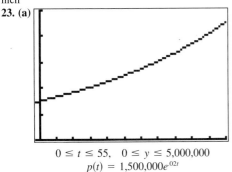

(b) during 2009 ($t = 14.4$)   **25.** In 1.98 years

SECTION 4.3 (page 204)

1. $10^5 = 100{,}000$ **3.** $e^{1.0986} = 3$ **5.** $3^4 = 81$ **7.** $\log .001 = -3$ **9.** $\ln 25 = 3.2189$ **11.** $\log_3(1/9) = -2$ **13.** 3   **15.** -2
17. 2   **19.** 3   **21.** -2 **23.** 1/2   **25.** 3.78   **27.** $\log 5$ **29.** $\ln 72$ **31.** $\log(x^3/y^2)$ **33.** $\ln(3x^2 + 14x + 8)$ **35.** $\log\left(\dfrac{x^3\sqrt{x+2}}{(x+1)^4}\right)$
37. .43   **39.** .26   **41.** .10   **43.** $2u + 5v$ **45.** $3u - 2v$ **47.** 3.5145   **49.** 2.4851   **51.** Many correct answers, including $b = 1$, $c = 2$ **53.** $a + 3c$ **55.** $2c + 3a + 1$ **57.** 1/5   **59.** 3/2   **61.** 16   **63.** 8/5   **65.** 1/3   **67.** 10   **69.** 5.2378   **71.** 5

73.

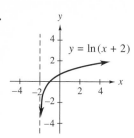

$y = \ln(x + 2)$

75.

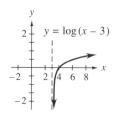

$y = \log(x - 3)$

77. (a) About 8 **(b)** About 12
79. (a) $125,000 **(b)** About $211,000 **(c)** About $235,000 **(d)** About $307,000 **(e)**

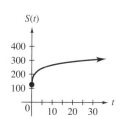

$S(t) = 125 + 83\log(5t + 1)$

81. About $60,000; about $450,000   **83.** $b^{\log_b x} = x$, so $\ln(b^{\log_b x}) = \ln x$. By property (c) for natural logarithms, $(\log_b x)(\ln b) = \ln x$. Dividing both sides by $\ln b$ shows that $\log_b x = (\ln x)/(\ln b)$, which completes the proof.

SECTION 4.4 (page 212)

1. 1.465   **3.** 2.710   **5.** -1.825 **7.** $-.128$ **9.** .805   **11.** $-.123$
13. ±2.141 **15.** 1.386   **17.** $\dfrac{\log d + 3}{4}$ **19.** $\dfrac{\ln b + 1}{2}$ **21.** 5
23. 4   **25.** No solution   **27.** 11   **29.** 3   **31.** No solution   **33.** $-2, 2$ **35.** 1, 10   **37.** $\dfrac{4 + b}{4}$ **39.** $\dfrac{10^{2-b}-5}{6}$
43. (a) 25 g **(b)** About 4.95 years   **45.** About 2476 years ago   **47. (a)** 3 **(b)** 5 **(c)** 7 **(d)** k **49. (a)** 21 **(b)** 70 **(c)** 120
(d) 140   **51.** .629   **53. (a)** About 527   **(b)** About 3466   **(c)** About 6020   **(d)** About 1783   **(e)** $r \approx .8187$ **55. (a)** 100 **(b)** About
521 **(c)** About 1564 **(d)** In about 27.2 days   **57. (a)** 871 lb; 912 lb **(b)** 126 lb

CHAPTER 4 REVIEW EXERCISES (page 216)

1. -1 **3.** 2
5.

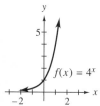

$f(x) = 4^x$

7.

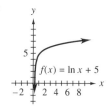

$f(x) = \ln x + 5$

9. (a) 0 **(b)** About 55 **(c)** About 98 **(d)** About 100   **11.** $\log 47 = 1.6721$ **13.** $\ln 39 = 3.6636$ **15.** $10^3 = 1000$ **17.** $e^{4.5581} = 95.4$
19. 3   **21.** 7.4   **23.** 4/3   **25.** 2.1531   **27.** $\log 20k^4$

29. $\log\left(\dfrac{b^2}{c^3}\right)$ **31.** 1.416   **33.** -2.807 **35.** -3.305 **37.** 2.115   **39.** .747   **41.** 28.463   **43.** ±1.097 **45.** 8   **47.** 6   **49.** 2

51. No solution   **53.** 3/2   **55.** 5/8   **57. (a)** .8 m **(b)** 1.2 m **(c)** 1.5 m **(d)** 1.8 m   **59. (a)** 10 g **(b)** About 140 days **(c)** About 243 days

CASE 4 (page 219)

1. 23.6;47.7;58.4. The estimates are a bit low.

CASE 5 (page 221)

1. .0315   **3.** About 157,000; forgery   **5.** About 58,000; forgery

CHAPTER 5

SECTION 5.1 (page 227)

3. $231.00   **5.** $286.75   **7.** $119.15   **9.** $241.56   **11.** $304.38   **13.** $3654.10   **17.** $46,265.06   **19.** $28,026.37   **21.** $8898.75
23. $48,140.65   **25.** 10.7%   **27.** 11.8%   **29.** $27,894.30   **31.** 6.8%
33. $34,438.29   **35.** $6550.92   **37.** 11.4%   **39.** $13,683.48; yes   **41.** (c)

SECTION 5.2 (page 236)

3. $1593.85   **5.** $1515.80   **7.** $13,213.16   **9.** $10,466.35   **11.** $2307.95   **13.** $1968.48   **15.** $26,545.91   **17.** $45,552.97
21. 4.04%   **23.** 8.16%   **25.** 12.36%   **27.** $8954.58   **29.** $3255.55   **31.** $11,572.58   **33.** $1000 now   **35.** $13,693.34
37. $7743.93   **39.** $142,886.40   **41.** $123,506.50   **43.** $8763.47   **45.** about $111,000   **47.** $30,611.30   **49.** about $1.946 million
51. $11,940.52   **53.** about 12 yr   **55.** about 18 yr   **57.** (a)

SECTION 5.3 (page 244)

1. 135   **3.** 6   **5.** 2315.25   **7.** 15   **9.** 6.24   **11.** 594.048   **13.** 15.91713   **15.** 21.82453   **17.** 22.01900   **21.** $4625.30
23. $327,711.81   **25.** $112,887.43   **27.** $1,391,304.39   **29.** $7118.38   **31.** $447,053.44   **33.** $36,752.01   **35.** $137,579.79
37. $3750.91   **39.** $236.61   **41.** $775.08   **43.** $4462.94   **45.** $148.02   **47.** $2290.90   **49.** $66,988.91   **51.** $130,159.72
53. $284,527.35   **55.** $1349.48   **57. (a)** 7 yr **(b)** 9 yr
59. (a) $120 **(b)** $681.83 except the last payment of $681.80 **(c)**

Payment Number	Amount of Deposit	Interest Earned	Total
1	$681.83	$ 0.00	$ 681.83
2	681.83	54.55	1418.21
3	681.83	113.46	2213.50
4	681.83	177.08	3072.41
5	681.80	245.79	4000.00

SECTION 5.4 (page 251)

1. (c)   **3.** 9.71225   **5.** 12.65930   **7.** 14.71787   **11.** $8415.93   **13.** $9420.83   **15.** $210,236.40   **17.** $103,796.58   **21.** $1603.01
23. $3109.71   **25.** $12.43   **27.** $681.02   **29.** $916.07   **31.** $86.24   **33.** $13.02   **35.** $48,677.34   **37. (a)** $158 **(b)** $1584
39. $573,496   **41.** about $77,790

We have used the computer program "Explorations in Finite Mathematics" for the following amortization schedules. Answers found using other computer programs or calculators may differ slightly.

43.

Payment Number	Amount of Payment	Interest for Period	Portion to Principal	Principal at End of Period
0	———	———	———	$72,000.00
1	$9247.24	$2160.00	$7087.24	64,912.76
2	9247.24	1947.38	7299.85	57,612.91
3	9247.24	1728.39	7518.85	50,094.06
4	9247.24	1502.82	7744.42	42,349.64

45.

Payment Number	Amount of Payment	Interest for Period	Portion to Principal	Principal at End of Period
0	———	———	———	$20,000.00
1	$2404.83	$700.00	$1704.83	18,295.17
2	2404.83	640.33	1764.50	16,530.68
3	2404.83	578.57	1826.25	14,704.42
4	2404.83	514.65	1890.17	12,814.25

47. (a) $4025.90 **(b)** $2981.93

49.

Payment Number	Amount of Payment	Interest for Period	Portion to Principal	Principal at End of Period
0	———	———	———	$37,947.50
1	$5783.49	$3225.54	$2557.95	35,389.55
2	5783.49	3008.11	2775.38	32,614.17
3	5783.49	2772.20	3011.29	29,602.88
4	5783.49	2516.24	3267.25	26,335.63
5	5783.49	2238.53	3544.96	22,790.67
6	5783.49	1937.21	3846.28	18,944.39
7	5783.49	1610.27	4173.22	14,771.17
8	5783.49	1255.55	4527.94	10,243.23
9	5783.49	870.67	4912.82	5,330.41
10	5783.49	453.08	5330.41	0

SECTION 5.5 (page 255)

1. $45,401.41 **3.** $350.26 **5.** $5958.68 **7.** $2036.46 **9.** $8258.80 **11.** $47,650.97; $14,900.97 **13.** $16,876.98 **15.** $23,243.61 **17.** $32,854.98 **19.** $8362.60 **21.** $1228.20 **23.** $9843.39 **25.** $6943.65 **27.** $26,850.21

29.

Payment Number	Amount of Payment	Interest for Period	Portion to Principal	Principal at End of Period
0	———	———	———	$8500.00
1	$1416.18	$340.00	$1076.18	7423.82
2	1416.18	296.95	1119.23	6304.59
3	1416.18	252.18	1164.00	5140.59
4	1416.18	205.62	1210.56	3930.03
5	1416.18	157.20	1258.98	2671.05
6	1416.18	106.84	1309.34	1361.71
7	1416.18	54.47	1361.71	0

31. 5.1% **33.** $8017.02 **35.** $2572.38 **37.** $89,659.63; $14,659.63 **39.** $6659.79 **41.** $4587.64 **43.** $212,026

CHAPTER 5 REVIEW EXERCISES (page 259)

1. $426.88   **3.** $62.91   **7.** $78,717.19   **9.** $739.09   **13.** $70,538.38; $12,729.04   **15.** $5282.19; $604.96   **17.** $12,857.07
19. $1923.09   **21.** 4, 2, 1, 1/2   **23.** -32 **25.** 5500   **29.** $137,925.91   **31.** $25,396.38
33. To receive funds deposited regularly for some goal   **35.** $2619.29   **37.** $916.12   **39.** $31,921.91   **41.** $14,222.42
43. $3628.00   **45.** $546.93   **47.** $896.06   **49.** $2696.12   **51.** $10,550.54   **53.** 8.21%   **55.** $5596.62   **57.** $24,818.76; $2418.76
59. $3560.61   **61.** (d)

CASE 6 (page 262)

1. $50(1 + i)^2 + 70(1 + i) - 127.40 = 0$; 4.3%   **3. (a)**

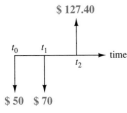

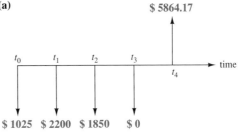

(b) 5.2%

5. (a) $50(1 + i)^2 + 50(1 + i) - 90 = 0$; $-.068, -2.932$ **(b)** $-.068$ is reasonable, -2.932 is not.

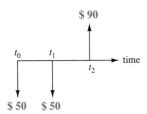

CHAPTER 6

SECTION 6.1 (page 274)

1. yes   **3.** $(11/5, -7/5)$ **5.** $(28, 22)$ **7.** $(2, -1)$ **9.** No solution   **11.** $(4y + 1, y)$ for any real number y **13.** No solution
15. $(12, 6)$ **17.** $(7, -2)$ **19.** dependent   **21.** inconsistent   **23.** independent   **25.** $(1, 2, -1)$ **27.** $(2, 0, 3)$ **29.** No solution
31. $(0, 2, 4)$ **33.** $((1 - z)/2, (11 - z)/2, z)$ for any real number z **35.** $(3 - z, 4 - z, z)$ for any real number z **37.** $(z, -2z + 1/2, z)$
for any real number z **39.** $(1, 2)$ **41.** $a = 3/4, b = 1/4, c = -1/2$ **45.** Boeing 300; GE 100   **47.** $27   **49.** 400 main floor, 200
balcony   **51.** $3000 at 8%, $6000 at 9%, $1000 at 5%   **53.** 5 bags of brand A, 2 bags of brand B, 3 bags of brand C   **55.** 15 X, 10
Y, 20 Z   **57.** either 10 buffets, 5 chairs, no tables, or 11 buffets, 1 chair, 1 table.   **59. (a)** $y = .01x^2 - .3x + 4.24$ **(b)** 15 platters;
$1.99

SECTION 6.2 (page 286)

1. $\begin{bmatrix} 2 & 1 & 1 & | & 3 \\ 3 & -4 & 2 & | & -5 \\ 1 & 1 & 1 & | & 2 \end{bmatrix}$ **3.** $\begin{aligned} 2x + 3y + 8z &= 20 \\ x + 4y + 6z &= 12 \\ 3y + 5z &= 10 \end{aligned}$ **5.** $\begin{bmatrix} 1 & 2 & 3 & | & -1 \\ 2 & 0 & 7 & | & -4 \\ 6 & 5 & 4 & | & 6 \end{bmatrix}$ **7.** $\begin{bmatrix} -4 & -3 & 1 & -1 & | & 2 \\ 0 & -4 & 7 & -2 & | & 10 \\ 0 & -2 & 9 & 4 & | & 5 \end{bmatrix}$

9. $(-3, 4)$ **11.** $(1, 0, -1)$ **13.** $((-9z + 5)/23, (10z - 3)/23, z)$ for any real number z **15.** $(-1, 23, 16)$ **17.** No solution
19. $(-7, 5)$ **21.** $(-3, z - 17, z)$ for any real number z **23.** $(-1, 1, -3, -2)$ **25.** No solution   **27.** Garcia 20 hr, Wong 15 hr
29. 5 trucks, 2 vans, 3 station wagons   **31. (a)** Let x_1 be the units purchased from the first supplier for Canoga Park, x_2 be the units from
the first supplier for Wooster, x_3 be the units from the second supplier for Canoga Park, and x_4 be the units from the second supplier for
Wooster.   **(b)** $x_1 + x_2 = 75$; $x_3 + x_4 = 40$; $x_1 + x_3 = 40$; $x_2 + x_4 = 75$; $70x_1 + 90x_2 + 80x_3 + 120x_4 = 10{,}750$ **(c)** $(40, 35, 0, 40)$

33. Three possibilities: 7 trucks, no vans, 2 station wagons; or 7 trucks, 1 van, 1 station wagon; or 7 trucks, 2 vans, no station wagons. **35.** 2340 of the first species, 10,128 of the second species, 224 of the third species (rounded)

SECTION 6.3 (page 294)

1. 2×3; $\begin{bmatrix} -7 & 8 & -4 \\ 0 & -13 & -9 \end{bmatrix}$ **3.** 3×3; square matrix; $\begin{bmatrix} 3 & 0 & -11 \\ -1 & -\frac{1}{4} & 7 \\ -5 & 3 & -9 \end{bmatrix}$ **5.** 2×1; column matrix; $\begin{bmatrix} -7 \\ -11 \end{bmatrix}$

7. B is a 5×3 zero matrix. **9.** False; corresponding elements are not all equal. **11.** True **13.** $x = 9, y = -4, z = -2$
15. $z = 18, r = 3, s = 3, p = 3, a = 3/4$
17. $\begin{bmatrix} 9 & 12 & 0 & 2 \\ 1 & -1 & 2 & -4 \end{bmatrix}$ **19.** $\begin{bmatrix} 5 & 13 & 0 \\ 3 & 1 & 8 \end{bmatrix}$ **21.** $\begin{bmatrix} 3 & -7 & 7 \\ 4 & 4 & -8 \end{bmatrix}$

23. $X + T = \begin{bmatrix} x & y \\ z & w \end{bmatrix} + \begin{bmatrix} r & s \\ t & u \end{bmatrix} = \begin{bmatrix} x + r & y + s \\ z + t & w + u \end{bmatrix}$ (a 2×2 matrix)

25. $X + (T + P) = \begin{bmatrix} x + (r + m) & y + (s + n) \\ z + (t + p) & w + (u + q) \end{bmatrix} = \begin{bmatrix} (x + r) + m & (y + s) + n \\ (z + t) + p & (w + u) + q \end{bmatrix} = (X + T) + P$

27. $P + O = \begin{bmatrix} m + 0 & n + 0 \\ p + 0 & q + 0 \end{bmatrix} = \begin{bmatrix} m & n \\ p & q \end{bmatrix} = P$ **29.** $\begin{bmatrix} 18 & 2.7 \\ 10 & 1.5 \\ 8 & 1.0 \\ 10 & 2.0 \\ 10 & 1.7 \end{bmatrix}$; $\begin{bmatrix} 18 & 10 & 8 & 10 & 10 \\ 2.7 & 1.5 & 1.0 & 2.0 & 1.7 \end{bmatrix}$

31. (a) $\begin{bmatrix} 5.6 & 6.4 & 6.9 & 7.6 & 6.1 \\ 144 & 138 & 149 & 152 & 146 \end{bmatrix}$ **(b)** $\begin{bmatrix} 10.2 & 11.4 & 11.4 & 12.7 & 10.8 \\ 196 & 196 & 225 & 250 & 230 \end{bmatrix}$ **(c)** $\begin{bmatrix} 4.6 & 5.0 & 4.5 & 5.1 & 4.7 \\ 52 & 58 & 76 & 98 & 84 \end{bmatrix}$
(d) $\begin{bmatrix} 12.0 & 12.9 & 13.7 & 14.5 & 12.8 \\ 221 & 218 & 254 & 283 & 250 \end{bmatrix}$ **33. (a)** Chicago: $\begin{bmatrix} 4.05 & 7.01 \\ 3.27 & 3.51 \end{bmatrix}$; Seattle: $\begin{bmatrix} 4.40 & 6.90 \\ 3.54 & 3.76 \end{bmatrix}$
(b) $\begin{bmatrix} 4.24 & 6.95 \\ 3.42 & 3.64 \end{bmatrix}$ **(c)** $\begin{bmatrix} 4.42 & 7.43 \\ 3.38 & 3.62 \end{bmatrix}$ **(d)** $\begin{bmatrix} 4.41 & 7.17 \\ 3.46 & 3.69 \end{bmatrix}$

SECTION 6.4 (page 304)

1. AB 2×4; BA not defined **3.** AB 3×3; BA 2×2 **5.** AB 3×2; BA not defined **7.** $\begin{bmatrix} -4 & 8 \\ 0 & 6 \end{bmatrix}$ **9.** $\begin{bmatrix} 24 & -8 \\ -16 & 0 \end{bmatrix}$

11. $\begin{bmatrix} -22 & -6 \\ 20 & -12 \end{bmatrix}$ **13.** $\begin{bmatrix} 13 \\ 25 \end{bmatrix}$ **15.** $\begin{bmatrix} -2 & 10 \\ 0 & 8 \end{bmatrix}$ **17.** $\begin{bmatrix} -4 & 1 \\ 2 & -3 \end{bmatrix}$ **19.** $\begin{bmatrix} 3 & -5 & 7 \\ -2 & 1 & 6 \\ 0 & -3 & 4 \end{bmatrix}$ **21.** $\begin{bmatrix} 16 & 11 \\ 37 & 32 \\ 58 & 53 \end{bmatrix}$

23. $\begin{bmatrix} 7 & 0 \\ -3 & -2 \end{bmatrix}$ **25.** $\begin{bmatrix} 0 & -8 & -2 \\ 4 & 16 & 18 \end{bmatrix}$ **27.** $\begin{bmatrix} 4 & -7/2 \\ 4 & 21/2 \end{bmatrix}$ **29.** $AB = \begin{bmatrix} -30 & -45 \\ 20 & 30 \end{bmatrix}$, but $BA = \begin{bmatrix} 0 & 0 \\ 0 & 0 \end{bmatrix}$

31. $(A + B)(A - B) = \begin{bmatrix} -7 & -24 \\ -28 & -33 \end{bmatrix}$, but $A^2 - B^2 = \begin{bmatrix} -37 & -69 \\ -8 & -3 \end{bmatrix}$

33. $(PX)T = \begin{bmatrix} (mx + nz)r + (my + nw)t & (mx + nz)s + (my + nw)u \\ (px + qz)r + (py + qw)t & (px + qz)s + (py + qw)u \end{bmatrix}$. $P(XT)$ is the same so $(PX)T = P(XT)$.

35. $k(X + T) = k\begin{bmatrix} x + r & y + s \\ z + t & w + u \end{bmatrix} = \begin{bmatrix} k(x + r) & k(y + s) \\ k(z + t) & k(w + u) \end{bmatrix} = \begin{bmatrix} kx + kr & ky + ks \\ kz + kt & kw + ku \end{bmatrix} = \begin{bmatrix} kx & ky \\ kz & kw \end{bmatrix} + \begin{bmatrix} kr & ks \\ kt & ku \end{bmatrix} = kX + kT$.

37. (a) $\begin{bmatrix} 900 & 1500 & 1150 \\ 600 & 950 & 800 \\ 750 & 900 & 825 \end{bmatrix}$ **(b)** $[1.50 \quad .90 \quad .60]$ **(c)** $[1.50 \quad .90 \quad .60]\begin{bmatrix} 900 & 1500 & 1150 \\ 600 & 950 & 800 \\ 750 & 900 & 825 \end{bmatrix} = [2340 \quad 3645 \quad 2940]$ **(d)** \$8925

39. (a)

$$\begin{array}{c} \\ S \\ C \end{array}\begin{array}{c} \begin{array}{ccc} CC & MM & AD \end{array} \\ \begin{bmatrix} .5 & .4 & .3 \\ .2 & .3 & .3 \end{bmatrix} \end{array}$$

(b)

$$\begin{array}{c} \\ SD \\ MC \\ M \end{array}\begin{array}{c} \begin{array}{cc} S & C \end{array} \\ \begin{bmatrix} 3 & 3 \\ 2 & 3 \\ 1 & 4 \end{bmatrix} \end{array}$$

(c)

$$\begin{array}{c} \\ SD \\ MC \\ M \end{array}\begin{array}{c} \begin{array}{ccc} CC & MM & AD \end{array} \\ \begin{bmatrix} 2.1 & 2.1 & 1.8 \\ 1.6 & 1.7 & 1.5 \\ 1.3 & 1.6 & 1.5 \end{bmatrix} \end{array}$$

(d) \$1.60 **(e)** \$1200 in Managua

41. (a) $\begin{bmatrix} 20 & 52 & 27 \\ 25 & 62 & 35 \\ 30 & 72 & 43 \end{bmatrix}$ The rows represent the amounts of fat, carbohydrates, and protein, respectively, in each of the daily meals.

(b) $\begin{bmatrix} 75 \\ 45 \\ 70 \\ 168 \end{bmatrix}$ The rows give the number of calories in one exchange of each of the food groups.

43. $\begin{bmatrix} 0 \\ \frac{3}{2} \end{bmatrix}$ **45.** $\begin{bmatrix} 44 & 75 & -60 & -33 & 11 \\ 20 & 169 & -164 & 18 & 105 \\ 113 & -82 & 239 & 218 & -55 \\ 119 & 83 & 7 & 82 & 106 \\ 162 & 20 & 175 & 143 & 74 \end{bmatrix}$ **47.** Cannot be found   **49.** No

SECTION 6.5 (page 313)

1. No   **3.** Yes   **5.** No   **9.** $\begin{bmatrix} 2 & -3 \\ -1 & 2 \end{bmatrix}$ **11.** No inverse   **13.** $\begin{bmatrix} 2 & -3 \\ -\frac{1}{2} & 1 \end{bmatrix}$ **15.** $\begin{bmatrix} 3 & 3 & -1 \\ -2 & -2 & 1 \\ -4 & -5 & 2 \end{bmatrix}$

17. $\begin{bmatrix} 2 & 1 & -1 \\ 8 & 2 & -5 \\ -11 & -3 & 7 \end{bmatrix}$ **19.** No inverse   **21.** $\begin{bmatrix} \frac{7}{4} & \frac{5}{2} & 3 \\ -\frac{1}{4} & -\frac{1}{2} & 0 \\ -\frac{1}{4} & -\frac{1}{2} & -1 \end{bmatrix}$ **23.** $\begin{bmatrix} \frac{1}{2} & \frac{1}{2} & -\frac{1}{4} & \frac{1}{2} \\ -1 & 4 & -\frac{1}{2} & -2 \\ -\frac{1}{2} & \frac{5}{2} & -\frac{1}{4} & -\frac{3}{2} \\ \frac{1}{2} & -\frac{1}{2} & \frac{1}{4} & \frac{1}{2} \end{bmatrix}$

25. $AB = \begin{bmatrix} 19 & -4 \\ 33 & -7 \end{bmatrix}$; $(AB)^{-1} = \begin{bmatrix} 7 & -4 \\ 33 & -19 \end{bmatrix}$ **27.** No   **29.** $A^{-1} = \begin{bmatrix} \frac{1}{a} & 0 \\ 0 & \frac{1}{d} \end{bmatrix}$

31. $A^{-1} = \dfrac{1}{ad - bc}\begin{bmatrix} d & -b \\ -c & a \end{bmatrix}$;

$$AA^{-1} = \begin{bmatrix} a & b \\ c & d \end{bmatrix} \cdot \frac{1}{ad - bc}\begin{bmatrix} d & -b \\ -c & a \end{bmatrix} = \frac{1}{ad - bc}\begin{bmatrix} a & b \\ c & d \end{bmatrix}\begin{bmatrix} d & -b \\ -c & a \end{bmatrix} = \frac{1}{ad - bc}\begin{bmatrix} ad - bc & 0 \\ 0 & ad - bc \end{bmatrix} = \begin{bmatrix} 1 & 0 \\ 0 & 1 \end{bmatrix}$$

33. Many correct answers, including $\begin{bmatrix} 1 & 0 \\ 0 & 1 \end{bmatrix}$, $\begin{bmatrix} -1 & 0 \\ 0 & -1 \end{bmatrix}$ and $\begin{bmatrix} -1 & 0 \\ c & 1 \end{bmatrix}$ for any real number c.

Answers to Exercises 35–39 may vary slightly depending on the calculator or computer software used.

35. $\begin{bmatrix} -.0447 & -.0230 & .0292 & .0895 & -.0402 \\ .0921 & .0150 & .0321 & .0209 & -.0276 \\ -.0678 & .0315 & -.0404 & .0326 & .0373 \\ .0171 & -.0248 & .0069 & -.0003 & .0246 \\ -.0208 & .0740 & .0096 & -.1018 & .0646 \end{bmatrix}$ **37.** $\begin{bmatrix} .0394 & .0880 & .0033 & .0530 & -.1499 \\ -.1492 & .0289 & .0187 & .1033 & .1668 \\ -.1330 & -.0543 & .0356 & .1768 & .1055 \\ .1407 & .0175 & -.0453 & -.1344 & .0655 \\ .0102 & -.0653 & .0993 & .0085 & -.0388 \end{bmatrix}$

39.

$$C^{-1}C = \begin{bmatrix} 1 & -2\times10^{-14} & -8\times10^{-15} & 5\times10^{-14} & -3\times10^{-14} \\ 3\times10^{-14} & 1 & -5\times10^{-14} & 0 & 1\times10^{-14} \\ -2\times10^{-14} & 3\times10^{-14} & 1 & -1\times10^{-14} & -2\times10^{-14} \\ 1\times10^{-14} & -1\times10^{-14} & -2\times10^{-15} & 1 & -2\times10^{-14} \\ 2\times10^{-14} & 4\times10^{-14} & 2.4\times10^{-14} & -4.2\times10^{-14} & 1 \end{bmatrix}$$

SECTION 6.6 (page 323)

1. $\begin{bmatrix} 8 \\ 6 \end{bmatrix}$ **3.** $\begin{bmatrix} \frac{1}{2} & 1 \\ \frac{3}{2} & 1 \end{bmatrix}$ **5.** $\begin{bmatrix} 11 \\ -3 \\ 5 \end{bmatrix}$ **7.** $\begin{bmatrix} -6 \\ -14 \end{bmatrix}$ **9.** $(-31, 24, -4)$ **11.** $(-31, -131, 181)$ **13.** $(15, -5, -1)$

15. $(-7, -34, -19, 7)$ **17.** $\begin{bmatrix} \frac{32}{3} \\ \frac{25}{3} \end{bmatrix}$ **19.** $\begin{bmatrix} 6.43 \\ 26.12 \end{bmatrix}$ **21.** $\begin{bmatrix} 6.67 \\ 20 \\ 10 \end{bmatrix}$ **23. (a)** $\begin{bmatrix} 72 \\ 48 \\ 60 \end{bmatrix}$ **(b)** $\begin{bmatrix} 2 & 4 & 2 \\ 2 & 1 & 2 \\ 2 & 1 & 3 \end{bmatrix}\begin{bmatrix} x_1 \\ x_2 \\ x_3 \end{bmatrix} = \begin{bmatrix} 72 \\ 48 \\ 60 \end{bmatrix}$

(c) 8 type I, 8 type II, and 12 type III   **25. (a)** $12,000 at 6%, $7000 at 7%, and $6000 at 10%   **(b)** $10,000 at 6%, $15,000 at 7%, and $5000 at 10%   **(c)** $20,000 at 6%, $10,000 at 7%, and $10,000 at 10%   **27.** About 893 tons of wheat, about 1179 tons of oil
29. Manufacturing 600, agriculture 300   **31.** Agriculture 5200/23, manufacturing 8260/23, transportation 6630/23

33. $\begin{bmatrix} 43 \\ 100 \end{bmatrix}, \begin{bmatrix} 29 \\ 63 \end{bmatrix}, \begin{bmatrix} 54 \\ 117 \end{bmatrix}, \begin{bmatrix} 100 \\ 227 \end{bmatrix}, \begin{bmatrix} 53 \\ 121 \end{bmatrix}, \begin{bmatrix} 28 \\ 61 \end{bmatrix}$ **35.** I like math.   **37. (a)** 21   **(b)** 25

39. (a)

	S	J	NO	H
S	0	1	2	1
J	1	0	1	0
NO	2	1	0	1
H	1	0	1	0

(b) 2   **(c)** 3   **(d)** 2

CHAPTER 6 REVIEW EXERCISES (page 328)

1. $(-5, 7)$ **3.** $(0, -2)$ **5.** No solution; inconsistent system.   **7.** $(-35, 140, 22)$ **9.** 8000 standard, 6000 extra large
11. $7000 in the first; $11,000 in the second   **13.** 50 from source I, 150 from source II, 100 from source III
15. $(0, 3, 3)$ **17.** No solution   **19.** $(-79, 99, -8)$ **21.** 2×2; $a = 2$, $b = 3$, $q = 9$, $c = 5$; square matrix
23. 1×4; $m = 12$, $z = -8$, $k = 4$, $r = -1$; row matrix   **25.** 2×2; $m = 7/2$, $r = 2$, $k = 11/2$, $x = -7$; square matrix

27. $\begin{bmatrix} 8 & 8 & 8 \\ 10 & 5 & 9 \\ 7 & 10 & 7 \\ 8 & 9 & 7 \end{bmatrix}$ **29.** $\begin{bmatrix} -2 & -3 & 2 \\ -2 & -4 & 0 \\ 0 & -1 & -2 \end{bmatrix}$ **31.** $\begin{bmatrix} 2 & 30 \\ -4 & -15 \\ 10 & 13 \end{bmatrix}$ **33.** Not defined

35. Next day $\begin{bmatrix} 2310 & -\frac{1}{4} \\ 1258 & -\frac{1}{4} \\ 5061 & \frac{1}{2} \\ 1812 & \frac{1}{2} \end{bmatrix}$; two-day total $\begin{bmatrix} 4842 & -\frac{1}{2} \\ 2722 & -\frac{1}{8} \\ 10,035 & -1 \\ 3566 & 1 \end{bmatrix}$ **37.** $\begin{bmatrix} 18 & 80 \\ -7 & -28 \\ 21 & 84 \end{bmatrix}$ **39.** $\begin{bmatrix} 13 & 43 \\ 17 & 46 \end{bmatrix}$ **41.** $\begin{bmatrix} 222 & 632 \\ -77 & -224 \\ 231 & 672 \end{bmatrix}$

43. (a) $\begin{bmatrix} \frac{1}{4} & \frac{1}{2} \\ \frac{1}{3} & \frac{1}{3} \end{bmatrix}$ **(b)** Cutting 34 hr; shaping 46 hr.   **45.** Many correct answers, including $\begin{bmatrix} 1 & 2 \\ 3 & 4 \end{bmatrix}$.   **47.** $\begin{bmatrix} -\frac{1}{4} & \frac{1}{6} \\ 0 & \frac{1}{3} \end{bmatrix}$ **49.** No inverse

51. $\begin{bmatrix} \frac{1}{4} & \frac{1}{2} & \frac{1}{2} \\ \frac{1}{4} & -\frac{1}{2} & \frac{1}{2} \\ \frac{1}{8} & -\frac{1}{4} & -\frac{1}{4} \end{bmatrix}$ **53.** No inverse   **55.** $\begin{bmatrix} -\frac{7}{19} & \frac{4}{19} \\ \frac{3}{19} & \frac{1}{19} \end{bmatrix}$ **57.** $\begin{bmatrix} 1 & 1 \\ -2 & -3 \end{bmatrix}$ **59.** No inverse   **61.** $\begin{bmatrix} 18 \\ -7 \end{bmatrix}$

63. $\begin{bmatrix} -22 \\ -18 \\ 15 \end{bmatrix}$ **65.** (2, 2) **67.** (2, 1) **69.** (−1, 0, 2) **71.** No inverse; no solution for the system

73. 18 liters of the 8%; 12 liters of the 18% **75.** 30 liters of 40% solution, 10 liters of 60% solution

77. 80 bowls, 120 plates **79.** $12,750 at 8%, $27,250 at 8 1/2%, $10,000 at 11%

81. $\begin{bmatrix} 218.09 \\ 318.27 \end{bmatrix}$ **83. (a)** $\begin{bmatrix} 1 & -\frac{1}{4} \\ -\frac{1}{2} & 1 \end{bmatrix}$ **(b)** $\begin{bmatrix} \frac{8}{7} & \frac{2}{7} \\ \frac{4}{7} & \frac{8}{7} \end{bmatrix}$ **(c)** $\begin{bmatrix} 2800 \\ 2800 \end{bmatrix}$

85. Agriculture $140,909, manufacturing $95,455 **87. (a)** $\begin{bmatrix} 54 \\ 32 \end{bmatrix}, \begin{bmatrix} 134 \\ 89 \end{bmatrix}, \begin{bmatrix} 172 \\ 113 \end{bmatrix}, \begin{bmatrix} 118 \\ 74 \end{bmatrix}, \begin{bmatrix} 208 \\ 131 \end{bmatrix}$ **(b)** $\begin{bmatrix} 2 & -3 \\ -\frac{1}{2} & 1 \end{bmatrix}$

CASE 7 (page 334)

1.
$$PQ = \begin{bmatrix} 1 & 2 & 0 & 2 & 1 & 1 \\ 0 & 1 & 0 & 1 & 0 & 0 \\ 1 & 1 & 0 & 1 & 2 & 1 \end{bmatrix}$$

3. Yes; the third person had no contacts with the first group.

CASE 8 (page 337)

1. (a) $A = \begin{bmatrix} .245 & .102 & .051 \\ .099 & .291 & .279 \\ .433 & .372 & .011 \end{bmatrix}$ $D = \begin{bmatrix} 2.88 \\ 31.45 \\ 30.91 \end{bmatrix}$ $X = \begin{bmatrix} x_1 \\ x_2 \\ x_3 \end{bmatrix}$ **(b)** $I - A = \begin{bmatrix} .755 & -.102 & -.051 \\ -.099 & .709 & -.279 \\ -.433 & -.372 & .989 \end{bmatrix}$ **(d)** $\begin{bmatrix} 18.2 \\ 73.2 \\ 66.8 \end{bmatrix}$

(e) $18.2 billion of agriculture, $73.2 billion of manufacturing, and $66.8 billion of household would be required (rounded to three significant digits).

CHAPTER 7

SECTION 7.1 (page 346)

1.

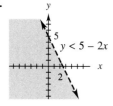

3.

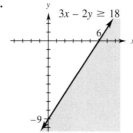

5.

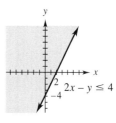

7.

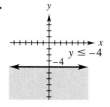

9.

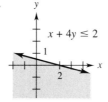

11.

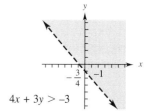

13.

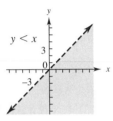

$2x - 4y < 3$

15.

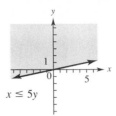

$x \le 5y$

17.

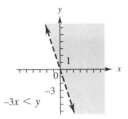

$-3x < y$

19.

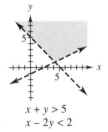

$y < x$

21.

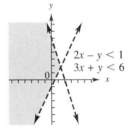

$x - y \ge 1$
$x \le 3$

23.

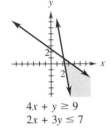

$4x + y \ge 9$
$2x + 3y \le 7$

25.

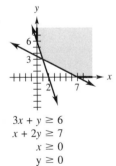

$x + y > 5$
$x - 2y < 2$

27.

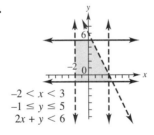

$2x - y < 1$
$3x + y < 6$

29.

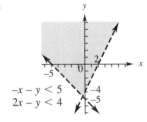

$-x - y < 5$
$2x - y < 4$

31.

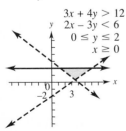

$3x + y \ge 6$
$x + 2y \ge 7$
$x \ge 0$
$y \ge 0$

33.

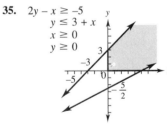

$-2 < x < 3$
$-1 \le y \le 5$
$2x + y < 6$

35. $2y - x \ge -5$
$y \le 3 + x$
$x \ge 0$
$y \ge 0$

37. $3x + 4y > 12$
$2x - 3y < 6$
$0 \le y \le 2$
$x \ge 0$

39. $2 < x < 7$
$-1 < y < 3$

41. (a)

	Number	Hours Spinning	Hours Dyeing	Hours Weaving
Shawls	x	1	1	1
Afghans	y	2	1	4
Maximum number of hours available		8	6	14

(b) $x + 2y \le 8;\ x + y \le 6;\ x + 4y \le 14;\ x \ge 0;\ y \ge 0$

41. (c)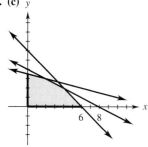

43. $x \geq 3000$; $y \geq 5000$;
$x + y \leq 10,000$

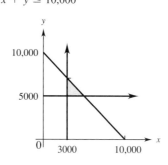

45. $x + y \geq 3.2$; $.5x + .3y \leq 1.8$;
$.16x + .20y \leq .8$; $x \geq 0$; $y \geq 0$

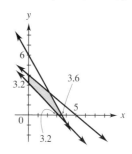

SECTION 7.2 (page 354)

1. Maximum of 65 at (5, 10); minimum of 8 at (1, 1)   **3.** Maximum of 9 at (0, 12); minimum of 0 at (0, 0)   **5. (a)** No maximum; minimum of 16 at (0, 8) **(b)** No maximum; minimum of 18 at (3, 4) **(c)** No maximum; minimum of 21 at (13/2, 2) **(d)** No maximum; minimum of 12 at (12, 0)   **7.** Maximum of 42/5 at $x = 6/5$, $y = 6/5$ or (6/5, 6/5)   **9.** Minimum of 13 at $x = 5$, $y = 3$ or (5, 3).
11. Maximum of 235/4 at $x = 105/8$, $y = 25/8$ or (105/8, 25/8)   **13.** No maximum; minimum 9   **15.** Maximum 22; no minimum
17. (a) (18, 2) **(b)** (12/5, 39/5) **(c)** No maximum

SECTION 7.3 (page 361)

1. $6x + 4y \leq 90$, where $x =$ number of canoes, $y =$ number of row boats.   **3.** $150x + 750y \leq 8500$, where $x =$ number of radio spots, $y =$ number of TV ads.   **5.** $x + y \geq 30$, where $x =$ number of pounds of cashews, $y =$ number of pounds of peanuts.
   7. Send 35 systems from Meadville to Superstore, 45 from Meadville to ValueHouse, 15 from Cambridge to ValueHouse, and none from Cambridge to Superstore for a minimum cost of $1015.   **9.** Ship 20 refrigerators to Warehouse A and 80 to Warehouse B for a minimum cost of $1040.   **11.** Produce 6.4 million gal of gasoline and 3.2 million gal of fuel oil for a maximum revenue of
$11,200,000.   **13.** 30 Maxivites and 120 Healthovites   **15.** 250,000 hectares to each crop for a maximum profit of $132,500,000
   17. 30 deluxe, no regular
19. 8/7 units of Species I and 10/7 units of Species II will meet the requirements with a minimum expenditure of 6.57 units of energy; however, a predator probably can catch and digest only whole numbers of prey.   **21.** 5 humanities, 12 science   **23.** (b)   **25.** (c)

SECTION 7.4 (page 375)

1. (a) 3 **(b)** x_3, x_4, x_5 **(c)** $4x_1 + 2x_z + x_3 = 20$
$5x_1 + x_2 + x_4 = 50$
$2x_1 + 3x_2 + x_5 = 25$

3. (a) 3 **(b)** x_4, x_5, x_6 **(c)** $3x_1 - x_2 + 4x_3 + x_4 = 95$
$7x_1 + 6x_2 + 8x_3 + x_5 = 118$
$4x_1 + 5x_2 + 10x_3 + x_6 = 220$

5.

x_1	x_2	x_3	x_4	x_5	z	
2	3	1	0	0	0	6
4	1	0	1	0	0	6
5	2	0	0	1	0	15
−5	−1	0	0	0	1	0

7.

x_1	x_2	x_3	x_4	x_5	x_6	z	
1	2	3	1	0	0	0	10
2	1	1	0	1	0	0	8
3	0	2	0	0	1	0	6
−1	−5	−10	0	0	0	1	0

9. 3 in row 2, column 1   **11.** 6 in row 1, column 1
13.

x_1	x_2	x_3	x_4	x_5	z	
−1	0	3	1	−1	0	16
1	1	$\frac{1}{2}$	0	$\frac{1}{2}$	0	20
2	0	$-\frac{1}{2}$	0	$\frac{3}{2}$	1	60

15.

x_1	x_2	x_3	x_4	x_5	x_6	z	
$-\frac{1}{2}$	$\frac{1}{2}$	0	1	$-\frac{1}{2}$	0	0	10
$\frac{3}{2}$	$\frac{1}{2}$	1	0	$\frac{1}{2}$	0	0	50
$-\frac{7}{2}$	$\frac{1}{2}$	0	0	$-\frac{3}{2}$	1	0	50
2	0	0	0	1	0	1	100

17. (a) Basic: x_3, x_5; nonbasic: x_1, x_2, x_4 **(b)** $x_1 = 0, x_2 = 0, x_3 = 16, x_4 = 0, x_5 = 29, z = 11$ **(c)** Not maximum
19. **(a)** Basic: x_1, x_2, x_5; nonbasic: x_3, x_4, x_6 **(b)** $x_1 = 6, x_2 = 13, x_3 = 0, x_4 = 0, x_5 = 21, x_6 = 0, z = 18$ **(c)** Maximum
21. Maximum is 30 when $x_1 = 0, x_2 = 10, x_3 = 0, x_4 = 0, x_5 = 16$.   **23.** Maximum is 8 when $x_1 = 4, x_2 = 0, x_3 = 8, x_4 = 2$,
$x_5 = 0$. **25.** No maximum   **27.** No maximum   **29.** Maximum is 34 when $x_1 = 17, x_2 = 0, x_3 = 0, x_4 = 0, x_5 = 14$ or when $x_1 = 0$,
$x_2 = 17, x_3 = 0, x_4 = 0, x_5 = 14$.   **31.** Maximum is 26,000 when $x_1 = 60, x_2 = 40, x_3 = 0, x_4 = 0, x_5 = 80, x_6 = 0$.
33. Maximum is 64 when $x_1 = 28, x_2 = 16, x_3 = 0, x_4 = 0, x_5 = 28, x_6 = 0$.   **35.** Maximum is 250 when $x_1 = 0, x_2 = 0, x_3 = 0$,
$x_4 = 50, x_5 = 0, x_6 = 50$.   **37. (a)** Maximum is 24 when $x_1 = 12, x_2 = 0, x_3 = 0, x_4 = 0, x_5 = 6$. **(b)** Maximum is 24 when $x_1 = 0$,
$x_2 = 12, x_3 = 0, x_4 = 0, x_5 = 18$. **(c)** The unique maximum value of z is 24, but this occurs at two different basic feasible solutions.

SECTION 7.5 (page 382)

1.

x_1	x_2	x_3	x_4	x_5	
2	1	1	0	0	90
1	2	0	1	0	80
1	1	0	0	1	50
-12	-10	0	0	0	0

where x_1 is the number of Siamese cats and x_2 is the number of Persian cats.

3.

x_1	x_2	x_3	x_4	x_5	x_6	
-2	3	1	0	0	0	0
1	0	0	1	0	0	6700
0	1	0	0	1	0	5500
1	1	0	0	0	1	12,000
-8.5	-12.10	0	0	0	0	0

where x_1 is the number of trucks and x_2 the number of fire engines.
5. Make no 1-speed or 3-speed bicycles; make 2295 10-speed bicycles; maximum profit is $55,080.   **7.** 48 loaves of bread and 6 cakes
for a maximum gross income of $168   **9.** 100 Basic Sets, no Regular Sets, and 200 Deluxe Sets for a maximum profit of $15,000
11. 4 radio ads, 6 TV ads, no newspaper ads, for a maximum exposure of 64,800   **13. (a)** (3) **(b)** (4) **(c)** (3)
15. 40 Siamese and 10 Persians for a maximum gross income of $580   **17.** 6700 trucks, 4466 fire engines
for a maximum profit of $110,989 (4467 fire engines is not a feasible solution.)

SECTION 7.6 (page 392)

1.
$$\begin{bmatrix} 3 & 1 & 0 \\ -4 & 10 & 3 \\ 5 & 7 & 6 \end{bmatrix}$$
3.
$$\begin{bmatrix} 3 & 4 \\ 0 & 17 \\ 14 & 8 \\ -5 & -6 \\ 3 & 1 \end{bmatrix}$$

5. Maximize $z = 4x_1 + 6x_2$
    subject to: $3x_1 - x_2 \le 3$
    $x_1 + 2x_2 \le 5$
    $x_1 \ge 0, x_2 \ge 0$

7. Maximize $z = 18x_1 + 15x_2 + 20x_3$
    subject to: $x_1 + 4x_2 + 5x_3 \le 2$
    $7x_1 + x_2 + 3x_3 \le 8$
    $x_1 \ge 0, x_2 \ge 0, x_3 \ge 0$

9. Maximize $z = -8x_1 + 12x_2$
    subject to: $3x_1 + x_2 \le 1$
    $4x_1 + 5x_2 \le 2$
    $6x_1 + 2x_2 \le 6$
    $x_1 \ge 0, x_2 \ge 0$

11. Maximize $z = 5x_1 + 4x_2 + 15x_3$
subject to: $x_1 + x_2 + 2x_3 \leq 8$
$x_1 + x_2 + x_3 \leq 9$
$x_1 \qquad + 3x_3 \leq 3$
$x_1 \geq 0, x_2 \geq 0, x_3 \geq 0$

13. $y_1 = 0, y_2 = 100, y_3 = 0$; minimum is 100   **15.** $y_1 = 0, y_2 = 12, y_3 = 0$; minimum is 12
17. $y_1 = 0, y_2 = 0$, and $y_3 = 2$; minimum is 4   **19.** $y_1 = 10, y_2 = 10$, and $y_3 = 0$; minimum is 320
21. $y_1 = 0, y_2 = 0$, and $y_3 = 4$; minimum is 4   **23.** 4 servings of A, 2 servings of B for a minimum cost of $1.76
25. Make 15 tables and 45 chairs for a minimum cost of $4080.
27. (a) 1 bag of Feed 1, 2 bags of Feed 2 **(b)** $6.60 per day for 1.4 bags of Feed 1 and 1.2 bags of Feed 2
29. (a) Minimize $w = 100y_1 + 20,000y_2$ **(b)** $6300 when $x_1 = 52.5, x_2 = 0, x_3 = 0$ **(c)** $5700, $x_1 = 47.5, x_2 = 0, x_3 = 0$
subject to: $y_1 + 400y_2 \geq 120$
$y_1 + 160y_2 \geq 40$
$y_1 + 280y_2 \geq 60$
$y_1 \geq 0, y_2 \geq 0$

SECTION 7.7 (page 405)

1. (a) Maximize $z = 5x_1 + 2x_2 - x_3$
subject to $2x_1 + 3x_2 + 5x_3 - x_4 = 8$
$4x_1 - x_2 + 3x_3 + x_5 = 7$
$x_1 \geq 0, x_2 \geq 0, x_3 \geq 0, x_4 \geq 0, x_5 \geq 0.$

(b)

x_1	x_2	x_3	x_4	x_5	
2	3	5	−1	0	8
4	−1	3	0	1	7
−5	−2	1	0	0	0

3. (a) Maximize $z = 2x_1 - 3x_2 + 4x_3$
subject to $x_1 + x_2 + x_3 + x_4 = 100$
$x_1 + x_2 + x_3 - x_5 = 75$
$x_1 + x_2 - x_6 = 27$
$x_1 \geq 0, x_2 \geq 0, x_3 \geq 0, x_4 \geq 0, x_5 \geq 0, x_6 \geq 0.$

(b)

x_1	x_2	x_3	x_4	x_5	x_6	
1	1	1	1	0	0	100
1	1	1	0	−1	0	75
1	1	0	0	0	−1	27
−2	3	−4	0	0	0	0

5. Maximize $z = -2y_1 - 5y_2 + 3y_3$
subject to: $y_1 + 2y_2 + 3y_3 \geq 115$
$2y_1 + y_2 + y_3 \leq 200$
$y_1 + y_3 \geq 50$
$y_1 \geq 0, y_2 \geq 0, y_3 \geq 0.$

y_1	y_2	y_3	y_4	y_5	y_6	
1	2	3	−1	0	0	115
2	1	1	0	1	0	200
1	0	1	0	0	−1	50
2	5	−3	0	0	0	0

7. Maximize $z = -y_1 + 4y_2 - 2y_3$
subject to: $7y_1 - 6y_2 + 8y_3 \geq 18$
$4y_1 + 5y_2 + 10y_3 \geq 20$
$y_1 \geq 0, y_2 \geq 0, y_3 \geq 0.$

y_1	y_2	y_3	y_4	y_5	
7	−6	8	−1	0	18
4	5	10	0	−1	20
1	−4	2	0	0	0

9. Maximum is 480 when $x_1 = 40, x_2 = 0$.   **11.** Maximum is 114 when $x_1 = 38, x_2 = 0, x_3 = 0$.   **13.** Maximum is 90 when $x_1 = 12$, $x_2 = 3$, or when $x_1 = 0, x_2 = 9$.   **15.** Minimum is 40 when $y_1 = 10, y_2 = 0$.   **17.** Minimum is 26 when $y_1 = 6, y_2 = 2$.

19.

y_1	y_2	y_3	y_4	y_5	y_6	y_7	y_8	
1	0	1	0	−1	0	0	0	32
0	1	0	1	0	−1	0	0	20
1	1	0	0	0	0	1	0	25
0	0	1	1	0	0	0	1	30
14	22	12	10	0	0	0	0	0

21.

y_1	y_2	y_3	y_4	y_5	y_6	y_7	
1	1	1	−1	0	0	0	10
1	1	1	0	1	0	0	15
1	$-\frac{1}{4}$	0	0	0	−1	0	0
−1	0	1	0	0	0	−1	0
.30	.09	.27	0	0	0	0	0

23. Ship 5000 barrels of oil from supplier 1 to distributor 2; ship 3000 barrels of oil from supplier 2 to distributor 1. Minimum cost is $175,000. **25.** Make no commercial loans and $25,000,000 in home loans for a maximum return of $3,000,000. **27.** Use 800,000 kg for whole tomatoes and 80,000 kg for sauce for a minimum cost of $3,460,000. **29.** Order 1000 small test tubes and 500 large ones for a minimum cost of $210.

CHAPTER 7 REVIEW EXERCISES (page 409)

1.

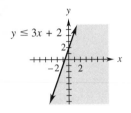

$y \leq 3x + 2$

3.

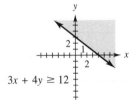

$3x + 4y \geq 12$

5.

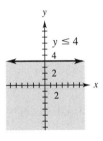

$y \leq 4$

7.

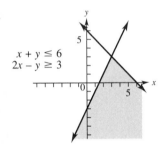

$x + y \leq 6$
$2x - y \geq 3$

9.

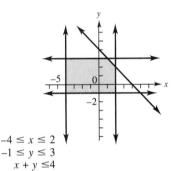

$-4 \leq x \leq 2$
$-1 \leq y \leq 3$
$x + y \leq 4$

11.

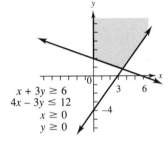

$x + 3y \geq 6$
$4x - 3y \leq 12$
$x \geq 0$
$y \geq 0$

13. Let x represent the number of batches of cakes, and let y represent the number of batches of cookies. Then

$$2x + (3/2)y \leq 15$$
$$3x + (2/3)y \leq 13$$
$$x \geq 0$$
$$y \geq 0.$$

15. Minimum of 8 at $(2, 1)$; maximum of 40 at $(6, 7)$ **17.** Maximum is 15 when $x = 5$, $y = 0$.

19. Minimum is 60 when $x = 0$, $y = 20$.   **21.** Make 3 batches of cakes and 6 of cookies for a maximum profit of $210.   **23. (a)** Let x_1 = number of item A, x_2 = number of item B, x_3 = number of item C.   **(b)** $z = 4x_1 + 3x_2 + 3x_3$

(c) $2x_1 + 3x_2 + 6x_3 \leq 1200$

$x_1 + 2x_2 + 2x_3 \leq 800$

$2x_1 + 2x_2 + 4x_3 \leq 500$

$x_1 \geq 0, x_2 \geq 0, x_3 \geq 0$

25. (a) Let x_1 = number of gallons of Fruity and x_2 = number of gallons of Crystal.   **(b)** $z = 12x_1 + 15x_2$

(c) $3x_1 + x_2 \leq 110$

$2x_1 + 2x_2 \leq 120$

$2x_1 + x_2 \leq 90$

$x_1 \geq 0, x_2 \geq 0$

27. (a) $3x_1 + 5x_2 + x_3 \qquad\qquad = 47$ **(b)**

$x_1 + x_2 \qquad + x_4 \qquad = 25$

$5x_1 + 2x_2 \qquad\qquad + x_5 \qquad = 35$

$2x_1 + x_2 \qquad\qquad\qquad + x_6 = 30.$

$$\begin{array}{cccccc} x_1 & x_2 & x_3 & x_4 & x_5 & x_6 \end{array}$$

$$\left[\begin{array}{cccccc|c} 3 & 5 & 1 & 0 & 0 & 0 & 47 \\ 1 & 1 & 0 & 1 & 0 & 0 & 25 \\ 5 & 2 & 0 & 0 & 1 & 0 & 35 \\ 2 & 1 & 0 & 0 & 0 & 1 & 30 \\ \hline -2 & -7 & 0 & 0 & 0 & 0 & 0 \end{array}\right]$$

29. (a) $x_1 + x_2 + x_3 + x_4 \qquad\qquad = 100$ **(b)**

$2x_1 + 3x_2 \qquad\qquad + x_5 \qquad = 500$

$x_1 \qquad + 2x_3 \qquad\qquad + x_6 = 350.$

$$\begin{array}{cccccc} x_1 & x_2 & x_3 & x_4 & x_5 & x_6 \end{array}$$

$$\left[\begin{array}{cccccc|c} 1 & 1 & 1 & 1 & 0 & 0 & 100 \\ 2 & 3 & 0 & 0 & 1 & 0 & 500 \\ 1 & 0 & 2 & 0 & 0 & 1 & 350 \\ \hline -4 & -6 & -3 & 0 & 0 & 0 & 0 \end{array}\right]$$

31. Maximum is 80 when $x_1 = 16$, $x_2 = 0$, $x_3 = 0$, $x_4 = 12$, $x_5 = 0$, $x_6 = 0$.   **33.** Maximum is 35 when $x_1 = 5$, $x_2 = 0$, $x_3 = 5$, $x_4 = 35$, $x_5 = 0$, $x_6 = 0$.   **35.** Maximize $z = -18y_1 - 10y_2$ with the same constraints.   **37.** Maximize $z = -6y_1 + 3y_2 - 4y_3$ with the same constraints.   **39.** Minimum of 40 at $(10, 0)$ **41.** Minimum of 1957 at $(47, 68, 0, 92, 35, 0, 0)$ **43.** $(9, 5, 8, 0, 0, 0)$; minimum is 62   **45.** Get 250 of A, none of B, none of C for a maximum profit of $1000.   **47.** Make 60 gallons of Crystal and no Fruity for a maximum profit of $900.   **49.** Build 3 atlantic and 3 pacific models for a minimum cost of $21,000.

CASE 9 (page 414)

1. 1: .95; 2: .83; 3: .75; 4: 1.00; 5: .87; 6: .94

CASE 10 (page 416)

1. (a) $x_2 = 58.6957$, $x_8 = 2.01031$, $x_9 = 22.4895$, $x_{10} = 1$, $x_{12} = .25$, $x_{13} = .15$ for a minimum cost of $12.55 **(b)** $x_2 = 71.7391$, $x_8 = 3.14433$, $x_9 = 23.1966$, $x_{10} = 1$, $x_{12} = .25$, $x_{13} = .15$ for a minimum cost of $15.05

CHAPTER 8

SECTION 8.1 (page 426)

1. False   **3.** True   **5.** True   **7.** True   **9.** False   **13.** $\subseteq$ **15.** $\nsubseteq$ **17.** $\subseteq$ **19.** $\subseteq$ **21.** 8   **23.** 64   **27.** $\cap$ **29.** $\cap$
31. $\cup$ **33.** $\cup$ **35.** $\{3, 5\}$ **37.** $\{7, 9\}$ **39.** $\emptyset$ **41.** U or $\{2, 3, 4, 5, 7, 9\}$ **43.** All students in this school not taking this course
45. All students in this school taking accounting and zoology   **47.** C and D, A and E, C and E, D and E **49.** B' is the set of all stocks on the list whose price to earnings ratio is less than 20; $B' = \{\text{ATT, Hershey}\}$.   **51.** $(A \cap B)'$ is the set of all stocks on the list with a dividend less than or equal to $2 or a price to earnings ratio of less than 20 (or both); $(A \cap B)' = \{\text{ATT, Gen Mill, Hershey, PepsiCo}\}$. (Note: IBM is not included because it has no price to earnings ratio.)   **53.** $M \cap E$ is the set of all male, employed applicants.

55. $M' \cup S'$ is the set of all female or married applicants.   **57.** {Showtime, Cinemax}   **59.** {HBO, Showtime, Cinemax}
61. {HBO, Disney Channel, Movie Channel}   **63.** {i, m, h}   **65.** U or {s, d, c, g, i, m, h}

SECTION 8.2 (page 434)

1.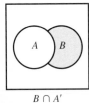

$B \cap A'$

3.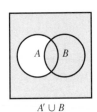

$A' \cup B$

5.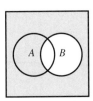

$B' \cup (A' \cap B')$

7. Ø

9.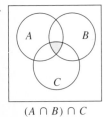

$(A \cap B) \cap C$

11.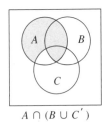

$A \cap (B \cup C')$

13.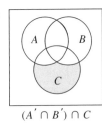

$(A' \cap B') \cap C$

15.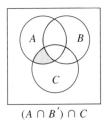

$(A \cap B') \cap C$

17. (a) 342 **(b)** 192 **(c)** 72 **(d)** 86   **19. (a)** 89% **(b)** 66% **(c)** 4% **(d)** 1% **(e)** 4%   **21. (a)** 50 **(b)** 2 **(c)** 14   **23. (a)** 54 **(b)** 17 **(c)** 10 **(d)** 7
(e) 15 **(f)** 3 **(g)** 12 **(h)** 1   **25. (a)** 40 **(b)** 30 **(c)** 95 **(d)** 110 **(e)** 160 **(f)** 65   **29.** 9   **31.** 27

33.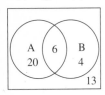

35.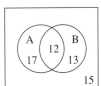

37.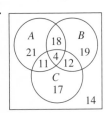

39.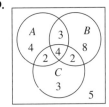

SECTION 8.3 (page 443)

3. $\{1, 2, 3, 4, \ldots, 30\}$ **5.** {buy, lease} **7.** {January, February, March, $\ldots$, December} **11.** $\{y_1, y_2, y_3, w_1, w_2, w_3, w_4, b_1, b_2, b_3, b_4,$ $b_5, b_6, b_7, b_8\}$ **(a)** $\{y_1, y_2, y_3\}$ **(b)** $\{b_1, b_2, b_3, b_4, b_5, b_6, b_7, b_8\}$ **(c)** $\{w_1, w_2, w_3, w_4\}$ **(d)** $\emptyset$ **13.** {ccc, ccw, cwc, cww, wcc, wcw, wwc, www} **(a)** {www} **(b)** {ccw, cwc, wcc} **(c)** {cww} **15.** $\{(1, 1), (1, 2), (1, 3), (1, 4), (1, 5), (2, 1), (2, 2), (2, 3), (2, 4), (2, 5), (3, 1),$ $(3, 2), (3, 3), (3, 4), (3, 5), (4, 1), (4, 2), (4, 3), (4, 4), (4, 5)\}$ **(a)** $\{(2, 1), (2, 2), (2, 3), (2, 4), (2, 5), (4, 1), (4, 2), (4, 3), (4, 4), (4, 5)\}$ **(b)** $\{(1, 2), (1, 4), (2, 2), (2, 4), (3, 2), (3, 4), (4, 2), (4, 4)\}$ **(c)** $\{(1, 4), (2, 3), (3, 2), (4, 1)\}$ **(d)** $\emptyset$ **17.** 1/6 **19.** 1/3 **21.** 1/13 **23.** 1/26 **25.** 1/52 **27.** 1/13 **29.** 1/5 **31.** 9/25 **33.** 21/25 **35.** 9/25 **37. (a)** 3/5 **(b)** 1/15 **(c)** .18

SECTION 8.4 (page 453)

1. no **3.** no **5.** yes **7.** .143 **9. (a)** 1/2 **(b)** 1/2 **(c)** 7/10 **(d)** 0 **11. (a)** 3/5 **(b)** 7/10 **(c)** 3/10 **13. (a)** 3/13 **(b)** 7/13 **(c)** 3/13 **15. (a)** .72 **(b)** .70 **(c)** .79 **17.** 0 **19.** .90 **21.** .84 **23. (a)** 1 to 5 **(b)** 1 to 2 **(c)** 1 to 1 **(d)** 1 to 5 **25.** 13 to 37 **27.** 8/25 **31.** possible **33.** not possible because the sum of the probabilities is less than 1 **35.** not possible because one probability is negative **37. (a)** .51 **(b)** .67 **(c)** .39 **(d)** .12 **39. (a)** .45 **(b)** .75 **(c)** .35 **(d)** .46 **41.** .527 **43.** .466 **45.** .515 **47.** 1/4 **49.** 1/2 **51.** .90 In Exercises 53–57, answers will vary. We give the theoretical probabilities. **53.** The theoretical probability is $5/32 = .15625$ **55.** The theoretical probability is $1/221 = .004525$. **57.** The theoretical probability is 7/8.

SECTION 8.5 (page 469)

1. 1/3 **3.** 1 **5.** 1/6 **7.** 4/17 **9.** 11/51 **11.** .012 **13.** .245 **19.** .815 **21.** .1698 **23.** .7456 **25.** 7/10 **27.** .0238 **29.** .43 **31.** The probability that a person in that group is white, given that the person is male is .956. **33.** The probability that a person in that group is black, given that the person is a woman is .083. **35.** .527 **37.** .042 **39.** 6/7 or .857 **41.** Yes **43.** .05 **45.** .25 **47.** .287 **49.** .248 **51.** .084 **53.** .4955 **55.** .0361 **57. (a)** $(.98)^3 \approx .94$ **(b)** Not very realistic **59. (a)** 3 backups **61.** No, they are dependent.

SECTION 8.6 (page 477)

1. 1/3 **3.** $2/41 \approx .0488$ **5.** $21/41 \approx .5122$ **7.** $8/17 \approx .4706$ **9.** .0135 **11.** .342 **13.** .006 **15.** .091 **17.** .556 **19.** .192 **21.** 82.4% **23. (c)** **25.** .481 **27. (a)** .161 **29.** .043

CHAPTER 8 REVIEW EXERCISES (page 481)

1. False **3.** False **5.** True **7.** True **9.** {New Year's Day, Martin Luther King's Birthday, Washington's Birthday (Presidents' Day), Memorial Day, Independence Day, Labor Day, Columbus Day, Veteran's Day, Thanksgiving, Christmas} **11.** $\{1, 2, 3, 4\}$ **13.** $\{B_1, B_2, B_3, B_6, B_{12}\}$ **15.** $\{A, C, E\}$ **17.** $\{A, B_3, B_6, B_{12}, C, D, E\}$ **19.** The set of all students in the class who are redheaded males. **21.** The set of all students in the class who are male or younger than 21. **23.** The set of all students in the class who are not A students and do not have red hair.
25.

27.

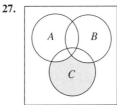

29. 5 **31.** $\{1, 2, 3, 4, 5, 6\}$ **33.** {(red, 10), (red, 20), (red, 30), (blue, 10), (blue, 20), (blue, 30), (green, 10), (green, 20), (green, 30)} **35.** {(2, blue), (4, blue), (6, blue), (8, blue), (10, blue)} **37.** no **39.** $E \cup F$ **45.** 3/13 **47.** 1/2 **49.** 1 **51.** 1 to 25 **53.** .86

55.

	N_2	T_2
N_1	$N_1 N_2$	$N_1 T_2$
T_1	$T_1 N_2$	$T_1 T_2$

57. 1/2   **59.** $5/36 \approx .139$ **61.** $5/18 \approx .278$ **63.** 1/3   **65.** .66   **67.** .71   **69. (a)** 3/11 **(b)** 7/39   **73.** 3/22   **75.** 2/3   **77.** 5/11
79. (a) row 1 : 400; row 2 : 150; row 3 : 750, 1000 **(b)** 1000 **(c)** 300 **(d)** 250 **(e)** 600 **(f)** 150 **(g)** those who purchased a used car, given that the buyer was not satisfied **(h)** 3/5 **(i)** 1/4

CASE 11 (page 485)

1. .076   **3.** .051

CHAPTER 9
SECTION 9.1 (page 496)

1. 12   **3.** 56   **5.** 8   **7.** 24   **9.** 792   **11.** 156   **13.** 6,375,600   **15.** 240,240   **17.** 352,716   **19.** 2,042,975
21. $10^9 = 1,000,000,000$; yes   **23.** $10^9 = 1,000,000,000$ **25.** 840   **27. (a)** 17,576,000 **(b)** 17,576,000 **(c)** 456,976,000   **29.** 800
31. 72,000,000   **33.** 60,949,324,800   **35.** 66   **37.** 210   **39.** 240   **41.** 2,118,760   **43.** 1,621,233,900   **45. (a)** 9 **(b)** 6 **(c)** 3; yes, from both   **49. (a)** 84 **(b)** 10 **(c)** 40 **(d)** 74   **51. (a)** 961 **(b)** 29,791   **53. (a)** 1,120,529,256 **(b)** 806,781,064,320   **55.** 558
57. 15,120   **59.** 1   **61.** 55,440   **63. (a)** 7,059,052 **(b)** $3,529,526 **(c)** No   **65. (a)** 840 **(b)** 180 **(c)** 420   **67. (a)** 362,880
(b) 6 **(c)** 1260

SECTION 9.2 (page 505)

1. .424   **3.** 7/9   **5.** 5/12   **7.** .147   **9.** .020   **11.** .314   **13.** 1326   **15.** .851   **17.** .765   **19.** .941   **23.** $\dfrac{\binom{4}{2}\binom{4}{3}\binom{44}{8}}{\binom{52}{13}}$

25. (a) 8.9×10^{-10} **(b)** 1.2×10^{-12} **27.** 3.3×10^{-16} **29. (a)** .218 **(b)** .272 **(c)** .728   **31. (a)** .003 **(b)** .005 **(c)** .057 **(d)** .252
33. $1 - {}_{365}P_{100}/(365)^{100}$ **35.** .350   **37.** There were 10 balls with 9 of them blue.   **39.** Answers will vary. We give the theoretical answers here. **(a)** .0399 **(b)** .5191 **(c)** .0226

SECTION 9.3 (page 512)

1. .067   **3.** .066   **5.** .934   **7.** .0000000005   **9.** .269   **11.** .875   **13.** 1/32   **15.** 13/16   **19.** .129   **21.** .065   **23.** .663   **25.** .349
27. .736   **29.** .218   **31.** .000729   **33.** .118   **35.** .121   **37.** .599   **39.** .955   **41.** .107   **43.** .678   **45. (a)** .000036 **(b)** .000052 **(c)** 0
47. (a) .148774 **(b)** .021344 **(c)** .978656

SECTION 9.4 (page 520)

1. No   **3.** Yes   **5.** No   **7.** No   **9.** Yes

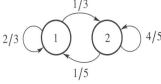

11. No   **13.** Not a transition diagram   **15.**

$$\begin{array}{c} \\ A \\ B \\ C \end{array} \begin{bmatrix} \begin{array}{ccc} A & B & C \\ .6 & .2 & .2 \\ .9 & .02 & .08 \\ .4 & 0 & .6 \end{array} \end{bmatrix}$$

17. No   **19.** Yes

21. No   **23.** $[1/3 \quad 2/3]$ **25.** $[3/11 \quad 8/11]$ **27.** $[.4633 \quad .1683 \quad .3684]$ **29.** $[1/4 \quad 1/4 \quad 1/2]$ **31.** 38%   **33.** 17/18

35.
$$\begin{array}{c} \\ R \\ P \\ W \end{array} \begin{array}{ccc} R & P & W \\ \left[\begin{array}{ccc} 1/2 & 1/2 & 0 \\ 1/4 & 1/2 & 1/4 \\ 0 & 1/2 & 1/2 \end{array}\right] \end{array} \quad [1/4 \quad 1/2 \quad 1/4]$$ **37.** Homeowners: .769; renters: .231; homeless: 0

39. (a) 42,500 in G_0; 5000 in G_1; 2500 in G_2 **(b)** 37,125 in G_0; 8750 in G_1; 4125 in G_2 **(c)** 33,281 in G_0; 11,513 in G_1; 5206 in G_2
(d) $[28/59 \quad 22/59 \quad 9/59]$ or $[.475 \quad .373 \quad .152]$; in the long run, the probabilities of no accidents, one accident, and more than one accident are .475, .373, and .152 respectively.

41. (a) $\begin{bmatrix} .348 & .379 & .273 \\ .320 & .378 & .302 \\ .316 & .360 & .323 \end{bmatrix}$ **(b)** .348   **(c)** .320   **43. (b)** $\begin{bmatrix} 5/12 & 13/36 & 2/9 \\ 5/12 & 13/36 & 2/9 \\ 1/4 & 5/12 & 1/3 \end{bmatrix}$ **(c)** 2/9

45. The first power is the given transition matrix; $\begin{bmatrix} .23 & .21 & .24 & .17 & .15 \\ .26 & .18 & .26 & .16 & .14 \\ .23 & .18 & .24 & .19 & .16 \\ .19 & .19 & .27 & .18 & .17 \\ .17 & .20 & .26 & .19 & .18 \end{bmatrix}$;

$\begin{bmatrix} .226 & .192 & .249 & .177 & .156 \\ .222 & .196 & .252 & .174 & .156 \\ .219 & .189 & .256 & .177 & .159 \\ .213 & .192 & .252 & .181 & .162 \\ .213 & .189 & .252 & .183 & .163 \end{bmatrix}$; $\begin{bmatrix} .2205 & .1916 & .2523 & .1774 & .1582 \\ .2206 & .1922 & .2512 & .1778 & .1582 \\ .2182 & .1920 & .2525 & .1781 & .1592 \\ .2183 & .1909 & .2526 & .1787 & .1595 \\ .2176 & .1906 & .2533 & .1787 & .1598 \end{bmatrix}$;

$\begin{bmatrix} .21932 & .19167 & .25227 & .17795 & .15879 \\ .21956 & .19152 & .25226 & .17794 & .15872 \\ .21905 & .19152 & .25227 & .17818 & .15898 \\ .21880 & .19144 & .25251 & .17817 & .15908 \\ .21857 & .19148 & .25253 & .17824 & .15918 \end{bmatrix}$; .17794

47. $[.171898 \quad .176519 \quad .167414 \quad .187526 \quad .296644]$ **49.** $[0 \quad 0 \quad .102273 \quad .897727]$

SECTION 9.5 (page 533)

1. (a)

Number of accidents	0	1	2	3	4
Probability	2/7	3/7	1/7	0	1/7

3. (a)

Winning Scores	1	2	3	4	5	6	7
Probability	1/7	3/7	1/7	0	0	0	2/7

(b)

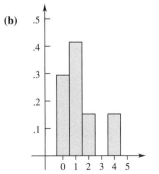

(b)

5.

Number	28	29	30	31	32	33	34	35
Probability	1/14	1/14	1/7	3/14	1/7	1/7	1/7	1/14

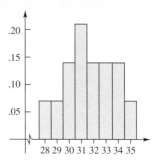

7.

Number of even numbers	0	1	2
Probability	3/10	3/5	1/10

9.

Number of Points	2	3	4	5	6	7	8	9	10	11	12
Probability	1/36	1/18	1/12	1/9	5/36	1/6	5/36	1/9	1/12	1/18	1/36

11.

Number of Black Balls	0	1	2
Probability	2/5	8/15	1/15

13. 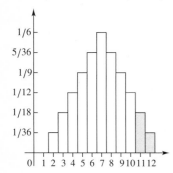   **15.** 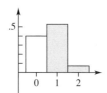   **17.**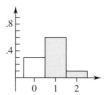

19. 6.5   **21.** 23.83   **23. (a)** Yes: the probability of a match is still 1/2. **(b)** 60¢ **(c)** − 60¢   **25.** 5.6   **27.** 30   **29.** −32¢
31. 4/7 ≈ .57   **33.** 1/2   **35.** 4.4435   **37.** −82¢   **39.** −5¢   **41.** −50¢   **45. (c)**   **47.** $30,000 at Age 55
49.

Account Number	Expected Value	Total	Class
3	2000	22,000	C
4	1000	51,000	B
5	25,000	30,000	C
6	60,000	60,000	A
7	16,000	46,000	B

51. (a) $94.0 million for seeding, $116.0 million for not seeding **(b)** Seed

SECTION 9.6 (page 540)

1. (a) Choose the coast. **(b)** Choose the highway. **(c)** Choose the highway; $38,000. **(d)** Choose the coast.
3. (a) Make no repairs. **(b)** Make repairs. **(c)** Make no repairs; $10,750.
5. (a)   **(b)** Do not overhaul before shipping. **7. (a)** No campaign **(b)** Campaign for all **(c)** Campaign for all

$$\begin{array}{c} \\ \text{Overhaul} \\ \text{Don't} \end{array} \begin{array}{cc} \text{Fails} & \text{Doesn't} \\ \left[\begin{array}{cc} -8600 & -2600 \\ -6000 & 0 \end{array} \right] \end{array}$$

9. Emphasize the environment; 14.25.

CHAPTER 9 REVIEW EXERCISES (page 544)

1. 120   **3.** 220   **5.** 120   **7. (a)** 72 **(b)** 72   **11.** 3/8   **13.** 11/16   **15.** .245   **17.** .362   **19.** .091   **21.** .455   **23.** yes   **25.** no
27. .000041   **29.** .182   **31. (a)** [.54   .46] **(b)** [.6464   .3536] **(c)** 2/3

33. (a)

x	1	2	3	4	5	6
$P(x)$	.05	0	.1	.25	.4	.2

(b)

35. (a)

Number	2	3	4	5	6	7	8	9	10	11	12
Probability	1/36	1/18	1/12	1/9	5/36	1/6	5/36	1/9	1/12	1/18	1/36

(b)

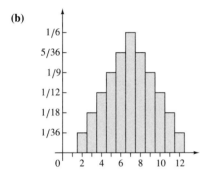

37. .6   **39.** $1.29   **41.** −28¢   **43. (a)** 3/13 or .23 **(b)** 3/4 or .75   **45. (a)** Oppose **(b)** Oppose **(c)** Oppose: 2700 **(d)** Oppose; 1400
47. (c)

CASE 12 (page 548)

1. 00100 indicates a cavity on the distal surface with all other surfaces healthy.
10100 indicates cavities on the occlusal and distal surfaces, with all other surfaces healthy.
11111 indicates cavities on all five surfaces.

CHAPTER 10
SECTION 10.1 (page 557)

1. (a)–(b)

Interval	*Frequency*
0–24	4
25–49	3
50–74	6
75–99	3
100–124	5
125–149	9

(c)–(d)

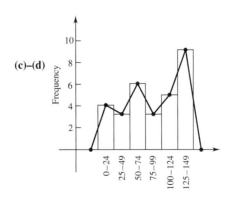

3. (a)–(b)

Interval	Frequency
70–74	2
75–79	1
80–84	3
85–89	2
90–94	6
95–99	5
100–104	6
105–109	4
110–114	2

(c)–(d)

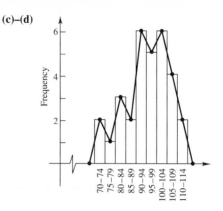

7. 16 **9.** 27,955 **11.** 7.7 **13.** 6.7 **15.** 17.2 **17.** 51 **19.** 130 **21.** 29.1 **23.** 9 **25.** 68 and 74 **27.** 6.3
31. 94.9; 90–94 and 100–104 **33. (a)** Mean = 1.9; median = 1; mode = 1 **35. (a)** Mean = 17.2; median = 17.5; no mode
37.   **39. (a)** About 17.5% **(b)** About 8% **(c)** 20–29 **41.** $3.29 **43. (a)** 23.98 **(b)** 21.47

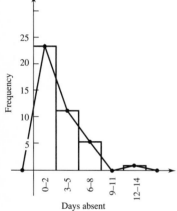

SECTION 10.2 (page 565)

3. 6; 2.2 **5.** 12; 4.7 **7.** 53; 21.8 **9.** 46; 16.1 **11.** 45.2 **13.** 10.9 **15.** at least 3/4 **17.** at least 24/25 **19.** At least 93.8%
21. No more than 11.1% **23.** Britelite: $\bar{x}$ = 26.2, s = 4.4; Brand X: $\bar{x}$ = 25.5, s = 7.2 **(a)** Britelite **(b)** Britelite
25. (a) 7.52; 1987 **(b)** .84 **(c)** 5 **27. (a)** 1.83; .76 **(b)** 1.09 **(c)** .88
29. (a) 299.4; Asia **(b)** 252.0; no; none; yes; Canada and Mexico
31. (a) 1/3; 2; −1/3; 0; 5/3; 7/3; 1; 4/3; 7/3; 2/3 **(b)** 2.1; 2.6; 1.5; 2.6; 2.5; .6; 1.0; 2.1; .6; 1.2 **(c)** 1.13 **(d)** 1.68 **(e)** 4.41; −2.15; The
process is out of control. **33.** Mean = 5.0876; standard deviation = .1087 **35.** Mean = 7.3571; standard deviation = .1326
37. Mean = 74.8; standard deviation = 13.4

SECTION 10.3 (page 577)

5. 9.10% **7.** 48.21% **9.** 7.7% **11.** 47.35% **13.** 92.42% **15.** 1.64 or 1.65 **17.** −1.04 **19.** .4052 **21.** .8185 **23.** .0475
25. .0021 **27.** 60.4 mph **29.** 6.68% **31.** 38.3% **33.** 99.38% **35.** 189 **37.** 15.87% **39.** 82 **41.** 70 **43.** .5000 **45.** .9050
47. .8106 **49.** .6736 **51.** .1820 **53.** .5398 **55.** .0796 **57.** .9933 **59.** .1527 **61.** .0051

SECTION 10.4 (page 585)

3. (a)

x	0	1	2	3	4	5	6
$P(x)$	.016	.094	.234	.313	.234	.094	.016

(b) 3 (c) 1.2

5. (a)

x	0	1	2	3	4	5
$P(x)$	.470	.383	.125	.020	.002	5×10^{-5}

(b) .7 (c) .78

7. 12.5; 3.49 9. 51.2; 3.2 13. .1747 15. .0240 17. .9032 19. .8665 21. .0956 23. .0760 25. .6443 27. .0146 29. .0222
31. .9945 33. (a) .0001 (b) .0002 (c) .0000 35. (a) .1686 (b) .0304 (c) .9575

CHAPTER 10 REVIEW EXERCISES (page 588)

3. (a)

Interval	Frequency
450–474	5
475–499	6
500–524	5
525–549	2
550–574	2

(b)–(c)

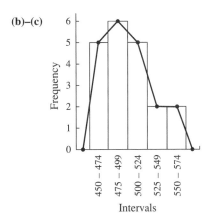

5. 73 7. 34.9 11. 38; 36 and 38 13. 55–59 17. 67; 23.9 19. 6.2
23. (a) 98.76% (b) Chebyshev's inequality would give "at least 84%." 25. .6591 27. .0606
31. (a) Diet A: (mean = 2.7) (b) Diet B: ($s = .95$) 33. 2.27% 35. 68.26% 37. .8315 39. .0305 41. 1.000 43. .1428
45. .0806 47. .8729

CASE 14 (page 594)

1. $\overline{p} = .73$ 3. $\overline{p} = .41$

CHAPTER 11
SECTION 11.1 (page 604)

1. (a) 3 (b) 0 3. (a) Does not exist (b) $-1/2$ 5. (a) 2 (b) Does not exist 7. (a) 1 (b) -1 11. 1 13. 2 15. 8 17. 2 19. 4
21. 3/2 23. (a) (b) Does not exist (c) 3 (d) Does not exist

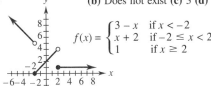

$$f(x) = \begin{cases} 3 - x & \text{if } x < -2 \\ x + 2 & \text{if } -2 \le x < 2 \\ 1 & \text{if } x \ge 2 \end{cases}$$

25. 41   **27.** 9/7   **29.** 6   **31.** −5   **33.** 1/2   **35.** Does not exist   **37.** $\sqrt{5}$ **39.** Does not exist   **41.** −1/9   **43.** 1/10   **45.** $1/(2\sqrt{5})$
47. (a) 25 **(b)** 5.2 **(c)** 7   **49. (a)** 3 **(b)** Does not exist **(c)** 2 **(d)** 16 months   **51. (a)** Approx .0444 **(b)** Approx. .0667 **(c)** Approx. .0667
53. 0   **55.** 8   **57.** Does not exist

SECTION 11.2 (page 615)

1. 7   **3.** 8   **5.** 1/3   **7.** −1/3   **9.** 20 ft per sec   **11.** 60 ft per sec   **13.** 7 ft per sec   **15. (a)** 2 **(b)** 3 **(c)** 2 **(d)** 1/2 **(e)** 1/2
19. (a) 1; from catalog distributions of 10,000 to 20,000, sales will have an average increase of $1000 for each additional 1000 catalogs
distributed. **(b)** 3/5; from catalog distributions of 20,000 to 30,000, sales will have an average increase of $600 for each additional 1000
catalogs distributed. **(c)** 2/5; from catalog distributions of 30,000 to 40,000, sales will have an average increase of $400 for each additional
1000 catalogs distributed. **(d)** As more catalogs are distributed, sales increase at a smaller and smaller rate.   **21. (a)** 3 **(b)** 0 **(c)** −9/5
(d) Sales increase in years 0–4, stay constant until year 7, and then decrease.   **23. (a)** −3/2 **(b)** −1 **(c)** 1 **(d)** −1/2 **(e)** 1988–1989
(f) stabilizing after a decline from 1985 to 1987.   **25. (a)** 7/3 ft per sec **(b)** 1 ft per sec **(c)** 2 ft per sec **(d)** 1 ft per sec **(e)** 11/6 ft per sec
(f) 5/4 ft per sec   **27. (a)** −25 boxes per dollar **(b)** −20 boxes per dollar **(c)** −30 boxes per dollar   **29. (a)** Approx. .133 billion per year
(b) Approx .333 billion per year **(c)** Approx .467 billion per year **(d)** 1.15 billion per year **(e)** The rate of increase in 1990–1992 was
approximately 8.6 times larger than the rate in 1981–1984.   **31. (a)** −.3 word per min **(b)** −.9 word per min

SECTION 11.3 (page 630)

1. 35   **3.** 1/8   **5.** 1/8   **7.** $y = 6x - 4$ **9.** $5x + 4y = 20$ **11.** $3y = 2x + 18$ **13.** 2   **15.** 1/5   **17.** $f'(x) = -8x + 11$; −5; 11;
35   **19.** $f'(x) = 8$; 8; 8; 8   **21.** $f'(x) = 2/x^2$; 1/2; does not exist; 2/9   **23.** $f'(x) = -4/(x - 1)^2$; −4; −4; −1/4   **25.** −6; 6
27. 0; 2; 3; 5   **29. (a)** x_5 **(b)** x_4 **(c)** x_3 **(d)** x_2 **31.**

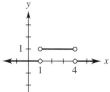

33. (a) Distance **(b)** Velocity   **35. (a)** $-4p + 4$ **(b)** −36; demand is decreasing at the rate of about 36 items for each increase in
price of $1.   **37. (a)** .48x, $0 \le x \le 30,000$ **(b)** 48; the cost of producing the 101st taco is approximately 48. **(c)** 48.24 **(d)** The exact
cost of producing the 101st taco is .24 greater than the approximate cost.   **39 (a)** 20 **(b)** 0 **(c)** −10 **(d)** At 5 hr   **41. (a)** −.15%; the
debt is decreasing at the rate of about .15% per month. .35%; the debt is increasing at the rate of about .35% per month. 0; the debt is not
changing at this point. **(b)** February to March, April to May, and September to October **(c)** It is positive, which means that the amount of
outstanding loans is increasing.   **43.** −.005; .006; −.00125; the derivative of −.005 indicates that the temperature is decreasing .005° for
each foot at 500 ft. At about 1500 ft, the temperature is increasing .006° per foot, and at 5000 ft, the temperature is decreasing .00125° per
foot.
45. (a)

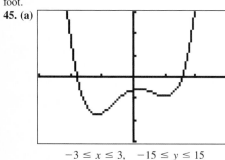

$-3 \le x \le 3, \quad -15 \le y \le 15$

(b)

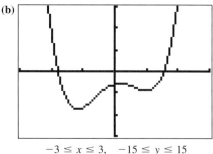

$-3 \le x \le 3, \quad -15 \le y \le 15$

(c) The graphs appear identical, which suggests that $f'(x) = g(x)$.   **47. (d)**

SECTION 11.4 (page 645)

1. $f'(x) = 8x - 6$ **3.** $y' = 6x^2 + 6x - 5$ **5.** $g'(x) = 4x^3 + 9x^2 - 6$ **7.** $f'(x) = 9x^5 - 2x^{-5}$ or $9x^5 - 2/x^5$
9. $y' = -48x^{2.2} + 3.8x^9$ **11.** $y' = 36t^{1/2} + 2t^{-1/2}$ or $36t^{1/2} + 2/t^{1/2}$ **13.** $y' = 4x^{-1/2} + (9/2)x^{-1/4}$ or $4/x^{1/2} + 9/(2x^{1/4})$
15. $g'(x) = -30x^{-6} + x^{-2}$ or $-30/x^6 + 1/x^2$ **17.** $y' = -20x^{-3} - 12x^{-5} - 6$ or $-20/x^3 - 12/x^5 - 6$
19. $f'(t) = -6t^{-2} + 16t^{-3}$ or $-6/t^2 + 16/t^3$ **21.** $y' = -36x^{-5} + 24x^{-4} - 2x^{-2}$ or $-36/x^5 + 24/x^4 - 2/x^2$
23. $g'(x) = 4x^{-3/2} - (5/2)x^{-1/2} + 1$ or $4/x^{3/2} - 5/(2x^{1/2}) + 1$ **25.** $y' = -6x^{-5/2} - 4x^{-3/2} + 2x$ or $-6/x^{5/2} - 4/x^{3/2} + 2x$
27. $y' = (-3/2)x^{-5/4}$ or $-3/(2x^{5/4})$ **29.** $y' = (-5/3)t^{-2/3}$ or $-5/(3t^{2/3})$ **31.** $dy/dx = -40x^{-6} + 36x^{-5}$
or $-40/x^6 + 36/x^5$ **33.** $(-9/2)x^{-3/2} - 3x^{-5/2}$ or $-9/(2x^{3/2}) - 3/x^{5/2}$ **35.** -28 **37.** $11/16$ **39.** (b)   **41.** $0; y = 0$
43. $7; y = 7x - 3$ **45.** (d)   **47.** (a) \$15 (b) \$119 (c) \$207 (d) \$319   **49.** (a) $C'(x) = 1 + 28x^{-1/3}$ (b) About \$9190 (c) About \$9150
(d) Yes   **51.** (a) $P(x) = 201\sqrt[3]{x} - .1x^2 - 3x - 40$ (b) About \$353; about \$406; about \$405; about \$300.
(c) $P'(x) = 67x^{-2/3} - .2x - 3$ (d) About \$9.43; about \$2.09; about $-$\$2.06; about$-$\$8.06   **53.** (a) 100 (b) 1 (c) $-.01$; the percent of
acid is decreasing at the rate of .01 per day after 100 days.   **55.** (a) $8t + 1000/\sqrt{t}$ people per year (c) About 456 people per year; about
396 people per year; about 384 people per year   **57.** (a) 30 (b) 0 (c) -10 **59.** (a) 264 (b) 510 (c) 97/4 or 24.25 (d) The number of
matings per hour is increasing by about 24.25 matings at a temperature of 16°C.   **61.** (a) $v(t) = 16t + 3$ (b) 3; 83; 163
63. (a) $v(t) = 6t^2 + 12t$ (b) 0; 210; 720   **65.** (a) -32 ft per sec; -64 ft per sec (b) In 3 sec (c) -96 ft per sec   **69.** (a) None (b) None
(c) $(-\infty, \infty)$ (d) Since the derivative is always negative, the graph of $g(x)$ is always decreasing.

SECTION 11.5 (page 654)

1. $y' = 9x^2 + 2x - 6$ **3.** $y' = 120x^3 - 54x^2 + 10$ **5.** $y' = 42x^5 - 112x^3 + 6x^2 - 8$
7. $y' = 50x^4 + 80x^3 - 39x^2 + 16x + 5$ **9.** $y' = 18x - 12$ **11.** $y' = 4x^3 - 4x$ **13.** $y' = 3x^{1/2} - 3x^{-1/2}/2 - 2$ or
$3\sqrt{x} - 3/(2\sqrt{x}) - 2$ **15.** $y' = -12 + 15x^{-1/2}$ or $-12 + 15/\sqrt{x}$ **17.** $y' = -7/(x - 4)^2$ **19.** $y' = -6/(3x - 5)^2$
21. $f'(t) = (t^2 - 2t - 1)/(t - 1)^2$ **23.** $y' = (-4x^2 - 22x - 12)/(x^2 - 3)^2$ **25.** $k'(x) = (4x^4 - 48x^3 - 2x + 6)/(4x^3 + 1)^2$
27. $y' = (x^2 - 4x - 12)/(x - 2)^2$ **29.** $y' = (5\sqrt{x}/2 - 3/\sqrt{x})/x$ or $(5x - 6)/(2x\sqrt{x})$
31. $f'(p) = (24p^2 + 32p + 29)/(3p + 2)^2$ **33.** $g'(x) = (10x^4 + 18x^3 + 6x^2 - 20x - 9)/[(2x + 1)^2(5x + 2)^2]$
35. In the first step, the numerator should be $(x^2 - 1)2 - (2x + 5)(2x)$.   **37.** $y = -2x + 9$
39. (a) \$22.86 per unit (b) \$12.92 per unit (c) $(3x + 2)/(x^2 + 4x)$ hundred dollars per unit (d) $\overline{C}'(x) = (-3x^2 - 4x - 8)/(x^2 + 4x)^2$
41. (a) $N'(t) = \dfrac{4200t}{(30 + t^2)^2}$ (b) 8.28 words per min.; 6.94 wpm; 4.71wpm; 2.49 wpm; .97 wpm   **43.** (a) $G'(20) = -1/200$; go faster.
(b) $G'(40) = 1/400$; go slower.   **45.** (a) $N'(t) = 6t^2 - 80t + 200$ (b) -56 million per hr (c) 46 million per hr (d) The population first
declines, and then increases.

SECTION 11.6 (page 664)

1. 275   **3.** 107   **5.** $24x + 4; 24x + 35$ **7.** $-64x^3 + 2; -4x^3 + 8$ **9.** $1/x^2; 1/x^2$ **11.** $\sqrt{8x^2 - 4}; 8x + 10$
13. If $f(x) = x^5$ and $g(x) = 4x + 3$, then $y = f[g(x)]$.   **15.** If $f(x) = \sqrt{x}$ and $g(x) = 6 + 3x$, then $y = f[g(x)]$.
17. If $f(x) = (x + 3)/(x - 3)$ and $g(x) = \sqrt{x}$, then $y = f[g(x)]$.   **19.** If $f(x) = x^2 + x + 5$ and $g(x) = x^{1/2} - 3$, then $y = f[g(x)]$.
21. $y' = 9(3x + 4)^2$ **23.** $y' = 72(3x + 2)^3$ **25.** $y' = -128x(8x^2 + 6)^3$ **27.** $y' = 36(2x + 5)^{1/2}$
29. $y' = (-21/2)(8x + 9)(4x^2 + 9x)^{1/2}$ **31.** $y' = 16(4x + 7)^{-1/2}$ or $16/\sqrt{4x + 7}$ **33.** $y' = -(2x + 4)(x^2 + 4x)^{-1/2}$ or
$-(2x + 4)/\sqrt{x^2 + 4x}$ **35.** $y' = 2(x + 1)(x - 3) + (x - 3)^2$ **37.** $y' = 70(x + 3)(2x - 1)^4(x + 2)$
39. $y' = [(3x + 1)^2(21x + 1)]/(2\sqrt{x})$ **41.** $y' = -2(x - 4)^{-3}$ or $-2/(x - 4)^3$ **43.** $y' = [2(4x + 3)(4x - 7)]/(2x - 1)^2$
45. $y' = (-5x^2 - 36x + 8)/(5x + 2)^4$ **47.** $y' = (3x^{1/2} - 1)/[4x^{1/2}(x^{1/2} - 1)^{1/2}]$ **49.** (a) -2 (b) $-24/7$ **51.** (d)
53. $D(c) = (-c^2 + 10c - 25)/25 + 500$ **55.** $A[r(t)] = A(t^2) = \pi t^4$; this function represents the area of the oil slick as a function of
time t after the beginning of the leak.   **57.** $C'(x) = \dfrac{15x}{\sqrt{50 + 15x^2}}$ **59.** \$77.46 per additional employee
61. (a) $R'(x) = \dfrac{5(300 - 4x)}{\sqrt{300x - 2x^2}}$ (b) \$10.61; \$2.89; $-$\$2.89; $-$\$10.61   **63.** (a) $F'(t) = \dfrac{150\,t}{(8 + t^2)^{3/2}}$ (b) About 3.96, 1.34, .36, .09
65. (a) $-.5$ (b) $-1/54 \approx -.02$ (c) $-.011$ (d) $-1/128 \approx -.008$ **67.** (a) $(-1, 1)$ (b) $x = -1$ (c) $(-\infty, -1), (1, \infty)$
(d) The derivative is 0 when the graph of $G(x)$ is at a low point. It is positive where $G(x)$ is increasing and negative where $G(x)$ is
decreasing.

SECTION 11.7 (page 671)

1. $y' = 3e^{3x}$ **3.** $f'(x) = 10e^{2x}$ **5.** $g'(x) = 20e^{-5x}$ **7.** $y' = 2xe^{x^2}$ **9.** $f'(x) = xe^{x^2/2}$ **11.** $y' = -18xe^{3x^2+5}$
13. $y' = 2x(1 - x)e^{-2x}$ **15.** $y' = (-9x^2 + 18x - 4)e^{-3x}$ **17.** $y' = 2x/(1 + x^2)$ **19.** $y' = (-8x + 3)/(-4x^2 + 3x)$
21. $y' = 1/(2x + 1)$ **23.** $f'(x) = \dfrac{6x^2 - 6x + 8}{(2x - 3)(x^2 + 4)}$ **25.** $y' = -23/[(6 - x)(3x + 5)]$ **27.** $y' = 3(15x^2 - 2)/[2(5x^3 - 2x)]$
29. $y' = -2x^2/(2 - x^2) + \ln(2 - x^2)$ **31.** $y' = (2x - 1)e^x/(2x + 1)^2$ **33.** $y' = (1 - 3\ln|x|)/x^4$
35. $y' = \dfrac{-20/x + 8 - 8\ln|x|}{(5 - 2x)^2}$ **37.** $y' = [3x^3 \ln|x| - (x^3 - 1)]/[2x(\ln|x|)^2]$ **39.** $y' = 1/[2(x - 3)\sqrt{\ln(x - 3)}]$
41. $y' = (xe^x \ln|x| - e^x + 1)/[x(\ln|x|)^2]$ **43.** $y' = (xe^x + xe^{-x} - e^x + e^{-x})/x^2$ or $[e^x(x - 1) + e^{-x}(x + 1)]/x^2$
45. $f'(x) = \dfrac{4e^{3x+2}}{4x - 5} + 3e^{3x+2} \ln(4x - 5)$ **47.** $y' = \dfrac{2000e^{.4x}}{(7 - 10e^{.4x})^2}$ **49.** $y' = 1250e^{-.5x}/(12 + 5e^{-.5x})^2$ **51.** (d) **53.** 1/8
55. (a) $dR/dx = 100 + 50(\ln x - 1)/(\ln x)^2$ (b) $112.48 **57.** (a) $C'(x) = 100$ (b) $P(x) = (50x/\ln x) - 100$ (c) $12.48
59. (a) $P'(t) = 50{,}000[(1 + .2t)(-.04)e^{-.04t} + .2e^{-.04t}]$ (b) gaining in 1985, 1995; losing in 2005, 2015.
61. (a) About $122,297 (b) About $177,139 (c) About $299,130 (d) About $10,939 per year (e) About $11,203 per year (f) About
$13,377 per year   **63.** (a) Increase at $54,065 per year (b) Increase at $13,116 per year (c) Decrease at $7935 per year (d) Decrease at
$19,517 per year (e) Between years 10 and 15   **65.** (a) 7.42 (b) 21.22 (c) .79   **67.** (a) .005 (b) .0007 (c) .000013 (d) $-.022$ (e) $-.0029$
(f) $-.000054$

SECTION 11.8 (page 680)

1. $x = 0, 2$ **3.** $x = 1$ **5.** $x = -3, 0, 3$ **7.** $x = 0, 6$ **9.** Yes, no **11.** No, no, yes **13.** Yes, no, yes **15.** Yes, no, yes
17. No, yes, yes **19.** Yes, no **21.** 1 **23.** Discontinuous at $t = m$; differentiable everywhere except $t = m$ and the endpoints.
25. (a) 6 cents (b) Does not exist (c) Any three of the years 1935, 1943, 1949, 1967, 1973, 1989, or 1990 **27.** (a) $15 (b) $25 (c) $50
(d) $92.50 (e) At $t = 20$ **29.** $(-4, 6)$ **31.** $(-\infty, -2]$ **33.** $[-25, 0]$ **35.** $(\pi, 10)$ **37.** $(-4, \infty)$
39. $(-\infty, 0)$ **41.** $(-3, 0), (0, 3)$

CHAPTER 11 REVIEW EXERCISES (page 685)

1. 4 **3.** -1 **5.** 13/2 **7.** 4 **9.** -13 **11.** 1/6 **13.** 1/4 **15.** $-3/2$ **17.** 21 **19.** About $-.3846$ **21.** $y' = 2$ **23.** $y' = 4x - 1$
25. -2; $y + 2x = -4$ **27.** 2/9; $2x - 9y = 2$ **29.** 3/4; $4y = 3x + 7$
31. (a) $150 (b) $187.50 (c) $189 (d)

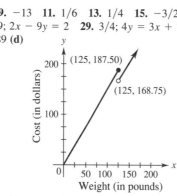

(e) At $x = 125$

35. $y' = 10x - 7$ **37.** $y' = 14x^{4/3}$ **39.** $f'(x) = -3x^{-4} + (1/2)x^{-1/2}$ or $-3/x^4 + 1/(2x^{1/2})$ **41.** $y' = 15t^4 + 12t^2 - 7$
43. $y' = 16x^{3/4} + 6x^{-1/4}(2x + 3)$ **45.** $f'(x) = \dfrac{-2x^2 + 4}{(x^2 + 2)^2}$ **47.** $y' = (2 - x + 2\sqrt{x})/[2\sqrt{x}(x + 2)^2]$
49. $y' = (x^2 - 2x)/(x - 1)^2$ **51.** $f'(x) = 12(3x - 2)^3$ **53.** $y' = 1/\sqrt{2t - 5}$ **55.** $y' = 18x(3x - 4)^2 + 2(3x - 4)^3$

57. $f'(u) = \dfrac{(2u + 3)(6u - 4) - 6(3u^2 - 4u)}{(2u + 3)^4}$ **59.** $y' = (x^2 - 4x + 4)/(x - 2)^2 = 1$ **61.** $y' = -12e^{2x}$ **63.** $y' = -6x^2e^{-2x^3}$

65. $y' = 10xe^{2x} + 5e^{2x}$ or $5e^{2x}(2x + 1)$ **67.** $y' = \dfrac{2x + 4}{x^2 + 4x + 1}$ **69.** $y' = \dfrac{x - 1/x - 2x \ln 4x}{(x^2 - 1)^2}$ **71.** $-1/[x^{1/2}(x^{1/2} - 1)^2]$

73. $dy/dt = (1 + 2t^{1/2})/[4t^{1/2}(t^{1/2} + t)^{1/2}]$ **75.** $-2/3$ **79.** None **81.** No, yes, yes **83.** No, yes, no **85.** Yes, no, yes
87. Yes, yes, yes **89.** (a) $25 (b) $0 (c) $-$25 **91.** (a) 55/3; sales will increase by 55 million dollars when 3 thousand more dollars are spent on research. (b) 65/4; sales will increase by 65 million dollars when 4 thousand more dollars are spent on research. (c) 15; sales will increase by 15 million dollars when 1 thousand more dollars are spent on research. (d) As more is spent on research, the increase in sales is decreasing. **93.** (a) -9.5; costs will decrease by $9500 for the next $100 spent on training. (b) -2.375; costs will decrease by $2375 for the next $100 spent on training. (c) Decreasing **95.** (a) About 19 per day (b) About 33 per day (c) About 99 per day

CASE 15 (page 690)

1. (a) $E = .5$; inelastic; A percentage change in price will result in a smaller percentage change in demand. (b) $E = 8$; elastic; A percentage change in price will cause a greater percentage change in demand. **3.** The elasticity is negative.

CHAPTER 12

SECTION 12.1 (page 703)

1. Relative minimum of -4 at $x = 1$ **3.** Relative maximum of 3 at $x = -2$
5. Relative maximum of 3 at $x = -4$; relative minimum of 1 at $x = -2$
7. Relative maximum of 3 at $x = -4$; relative minimum of -2 at $x = -7$ and $x = -2$
9. Relative maximum of 1 at $x = 0$; relative minimum of -3 at $x = 2$
11. Relative maximum of 2 at $x = -3$; relative minimum of -2 at $x = -1$
13. Relative maximum of 827/96 at $x = -1/4$; relative minimum of $-377/6$ at $x = -5$
15. Relative maximum of 50/3 at $x = -2$; relative minimum of -25 at $x = 3$
17. Relative maximum of 3 at $x = 0$; relative minimum of -45 at $x = 2$
19. Relative minimum of -2 at $x = 7/2$ **21.** No relative extrema **23.** Relative minimum of 0 at $x = 0$
25. Relative maximum of $1/e$ at $x = -1$ **27.** Relative minimum of 2 at $x = 0$
29. By symmetry and absolute value there is a relative maximum of $1/e$ at $x = -1/e$; relative minimum of $-1/e$ at $x = 1/e$
31. Absolute maximum at $x = -2$; absolute minimum at $x = -5$ and $x = 1$ **33.** Absolute maximum at $x = 2$; absolute minimum at $x = -2$ **35.** Absolute maximum at $x = -4$; absolute minimum at $x = -7$ and $x = -2$
37. Absolute maximum at $x = 6$; absolute minimum at $x = -4$ and $x = 4$ **39.** Absolute maximum at $x = 6$; absolute minimum at $x = 4$ **41.** Absolute maximum at $x = \sqrt{2}$; absolute minimum at $x = 0$ **43.** Absolute maximum at $x = -3$ and $x = 3$; absolute minimum at $x = 0$ **45.** Absolute maximum at $x = 0$; absolute minimum at $x = -1$ and $x = 1$ **49.** (a) 6 (b) $324
51. 100 units **53.** 1,000,000 tires; $700,000 **55.** 12°C **57.** 10 **59.** 10 min **61.** Relative maximum of -44.73 at $x = -.12$; relative minimum of -167.99 at $x = 4.23$ **63.** Relative maximum of 161.98 at $x = 3.35$; relative minima of -709.87 at $x = -6.77$ and -240.08 at $x = 10.92$

SECTION 12.2 (page 711)

1. $f''(x) = 6x - 10$; -10; 2; -28 **3.** $f''(x) = 12(x + 2)^2$; 48; 192; 12 **5.** $f''(x) = 2/(1 + x)^3$; 2; 2/27; $-1/4$
7. $f''(x) = -(x + 4)^{-3/2}/4$ or $-1/[4(x + 4)^{3/2}]$; $-1/32$; $-1/(4 \cdot 6^{3/2})$; $-1/4$
9. $f''(x) = (-6/5)x^{-7/5}$ or $-6/(5x^{7/5})$; $f''(0)$ does not exist; $-6/(5 \cdot 2^{7/5})$; $-6/[5(-3)^{7/5}]$
11. $f''(x) = 2e^x$; 2; $2e^2$; $2e^{-3}$ or $2/e^3$ **13.** $f'''(x) = 72x + 12$; $f^{(4)}(x) = 72$ **15.** $f'''(x) = 6/x^4$; $f^{(4)}(x) = -24/x^5$
17. $f'''(x) = 40e^{2x}$; $f^{(4)}(x) = 80e^{2x}$ **19.** $f'''(x) = 2x^{-3}$ or $2/x^3$; $f^{(4)}(x) = -6x^{-4}$ or $-6/x^4$
21. $f'''(x) = -x^{-2}$ or $-1/x^2$; $f^{(4)}(x) = 2x^{-3}$ or $2/x^3$ **23.** Relative minimum at $x = 1/3$.
25. Relative maximum at $x = -1/2$; relative minimum at $x = 2/3$ **27.** No critical numbers, no relative maxima or minima
29. Relative maximum at $x = -5$; relative minimum at $x = 4$

31. Relative maximum at $x = 0$; relative minima at $x = 2$ and $x = -2$ **33.** Relative maximum at $x = -\sqrt{3}$; relative minimum at $x = \sqrt{3}$
35. Relative maximum at $x = -3$; relative minimum at $x = 3$ **37.** No critical numbers; no relative maxima or minima
39. $v(t) = 12t + 2$; $a(t) = 12$; $v(0) = 2$ cm/sec; $v(4) = 50$ cm/sec; $a(0) = a(4) = 12$ cm/sec^2
41. $v(t) = 9t^2 - 8t + 8$; $a(t) = 18t - 8$; $v(0) = 8$ cm/sec; $v(4) = 120$ cm/sec; $a(0) = -8$ cm/sec^2; $a(4) = 64$ cm/sec^2
45. One of many possible functions is:

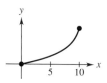

47. $40,000 in advertising produces maximum sales of 180,000 cameras.   **49. (a)** -96 ft/sec **(b)** -160 ft/sec **(c)** -256 ft/sec **(d)** -32 ft/sec^2
51. 8   **53.** 10,000   **55. (a)** At 4 hr **(b)** 1160 million

SECTION 12.3 (page 721)

1. 8 in by 8 in   **3. (a)** 54.8 mph **(b)** $73.03   **5. (a)** $200 + 20x$ **(b)** $80 - x$ **(c)** $R(x) = 16,000 + 1400x - 20x^2$ **(d)** 35 **(e)** $40,500
7. (a) 90 **(b)** $405   **9. (a)** $1200 - 2x$ **(b)** $A(x) = 1200x - 2x^2$ **(c)** 300 m **(d)** 180,000 sq m   **11.** 20,000 sq ft   **13.** $1000
15. (a) $R(x) = 100,000x - 100x^2$ **(b)** 500 **(c)** 25,000,000¢   **17.** 4 in. by 4 in. by 2 in.   **19.** 10 cm; 10 cm
21. Width $= 2\sqrt{6} + 2$ in ≈ 6.90 in; length $= 3\sqrt{6} + 3$ in ≈ 10.35 in   **23.** 1 mi from A   **25.** 10   **27.** 60   **29.** 5000
31. (c)   **33.** 12 times   **35.** $r \approx 2.10$ in; $h \approx 4.20$ in

SECTION 12.4 (page 745)

1. Increasing on $(1, \infty)$; decreasing on $(-\infty, 1)$ **3.** Increasing on $(-\infty, -4)$ and $(-2, \infty)$; decreasing on $(-4, -2)$
5. Increasing on $(-7, -4)$ and $(-2, \infty)$; decreasing on $(-\infty, -7)$ and $(-4, -2)$
7. Increasing on $(-4, \infty)$; decreasing on $(-\infty, -4)$ **9.** Increasing on $(-\infty, -1/3)$ and $(2, \infty)$; decreasing on $(-1/3, 2)$
11. Increasing on $(-\infty, -4)$ and $(-4, \infty)$ **13.** Decreasing on $(-\infty, 5)$ **15.** Increasing on $(-\infty, -1)$ and $(2, \infty)$; decreasing on $(-1, 2)$
17. Concave upward on $(2, \infty)$; concave downward on $(-\infty, 2)$; point of inflection at $(2, 3)$ **19.** Concave upward on $(-\infty, -1)$ and
$(8, \infty)$; concave downward on $(-1, 8)$; points of inflection at $(-1, 7)$ and $(8, 6)$ **21.** Concave upward on $(2, \infty)$; concave downward on
$(-\infty, 2)$; no points of inflection   **23.** Concave upward on $(-\infty, \infty)$; no points of inflection   **25.** Concave upward on $(-4/3, \infty)$; concave
downward on $(-\infty, -4/3)$; point of inflection at $(-4/3, 425/27)$ **27.** Concave upward on $(3, \infty)$; concave downward on $(-\infty, 3)$; no
points of inflection   **29.** Concave upward on $(-\infty, -5)$ and $(1, \infty)$; concave downward on $(-5, 1)$; points of inflection at $(-5, -1248)$
and $(1, 0)$ **31. (a)** Exercise 21 at $x = 2$; Exercise 22 at $x = 0$ **(b)** Exercise 21 has horizontal asymptote $y = 3$; Exercise 22 has slant
asymptote $y = -x$ **33.** 5/3   **35.** 4   **37.** 1/3   **39.** 0
There are many possibilities for Exercises 41–43, including the following.

41.

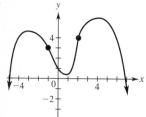

43.

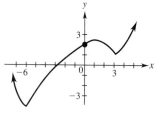

45. No points of inflection

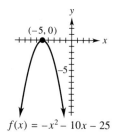

$f(x) = -x^2 - 10x - 25$

47. Point of inflection at $x = 1/3$

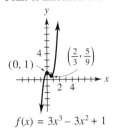

$(0, 1)$ $\left(\dfrac{2}{3}, \dfrac{5}{9}\right)$

$f(x) = 3x^3 - 3x^2 + 1$

49. Point of inflection at $x = -3/2$

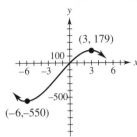

$(3, 179)$

$(-6, -550)$

$f(x) = -2x^3 - 9x^2 + 108x - 10$

51. Point of inflection at $x = -7/12$

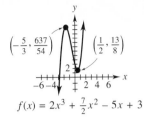

$\left(-\dfrac{5}{3}, \dfrac{637}{54}\right)$ $\left(\dfrac{1}{2}, \dfrac{13}{8}\right)$

$f(x) = 2x^3 + \dfrac{7}{2}x^2 - 5x + 3$

53. No points of inflection

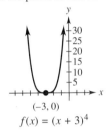

$(-3, 0)$

$f(x) = (x + 3)^4$

55. Points of inflection at $x = -\sqrt{3}$ and $x = \sqrt{3}$

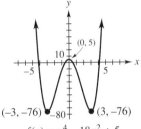

$(0, 5)$

$(-3, -76)$ $(3, -76)$

$f(x) = x^4 - 18x^2 + 5$

57. No points of inflection or relative extrema

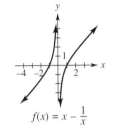

$f(x) = x - \dfrac{1}{x}$

59. No points of inflection

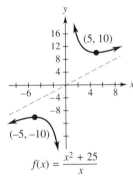

$(5, 10)$

$(-5, -10)$

$f(x) = \dfrac{x^2 + 25}{x}$

61. No points of inflection or relative extrema

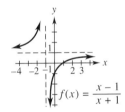

$f(x) = \dfrac{x - 1}{x + 1}$

63. (a) Nowhere **(b)** Everywhere   **65.** (0, 8000)   **67.** After 10 days   **69. (a)** (0, 3) **(b)** (3, ∞) (Note: x must be at least 0.)   **71.** 0; the concentration of the drug in the bloodstream approaches 0 as the number of hours after injection increases.   **73.** 75; the number of pieces of work a new employee can do gets closer and closer to 75 as the number of days of training increases.   **75.** (14, 4288)   **77.** Relative maximum at $(-.0241, 1.0000)$; relative minimum at $(.6908, .9818)$; point of inflection at $(1/3, 5351/5400)$ **79.** Relative maxima at $(-80.7064, 8671701.5635)$ and $(.3982, 5.0017)$; relative minima at $(-1.6166, -8.8191)$ and $(1.9249, -1.7461)$; points of inflection at $(-60.4796, 5486563.4612)$, $(-.7847, -2.7584)$, and $(1.2643, 1.2972)$

CHAPTER 12 REVIEW EXERCISES (page 750)

5. Increasing on $(-7/2, \infty)$; decreasing on $(-\infty, -7/2)$ **7.** Increasing on $(-\infty, -2/3)$ and $(1, \infty)$; decreasing on $(-2/3, 1)$ **9.** Decreasing on $(-\infty, 3)$ and $(3, \infty)$ **11.** Relative maximum of 101 at $x = -3$; relative minimum of -24 at $x = 2$

13. Relative maximum of 1 at $x = 0$; relative minima of -44 at $x = -3$ and $-4/3$ at $x = 1$
15. Relative maximum of $3/e$ at $x = 1$ **17.** Absolute maximum of $29/4$ at $x = 5/2$; absolute minimum of 5 at $x = 1$ and $x = 4$
19. Absolute maximum of 39 at $x = -3$; absolute minimum of $-319/27$ at $x = 5/3$ **21.** $f''(x) = 40x^3 - 24x$; 16; -272
23. $f''(x) = -80e^{4x}$; $-80e^4$; $-80e^{-8}$ **25.** Does not exist **27.** 1/5

29.

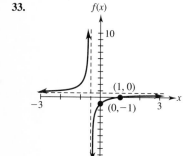

$f(x) = -2x^3 - \frac{1}{2}x^2 + x - 3$

31.

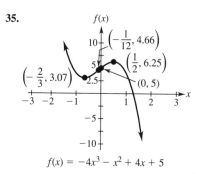

$f(x) = x^4 - \frac{4}{3}x^3 - 4x^2 + 1$

33.

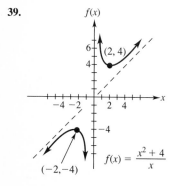

$f(x) = \dfrac{x - 1}{2x + 1}$

35.

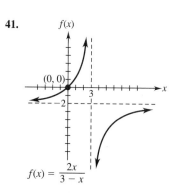

$f(x) = -4x^3 - x^2 + 4x + 5$

37.

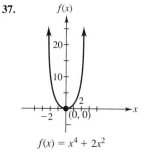

$f(x) = x^4 + 2x^2$

39.

$f(x) = \dfrac{x^2 + 4}{x}$

41.

$f(x) = \dfrac{2x}{3 - x}$

43. (a) $2000 **(b)** $400 **45.** 3 m by 3 m by 3 m **47.** 225 m by 450 m **49.** 126 **51.** 225 m by 225 m

CASE 16 (page 754)

1. 12 knots

CASE 17 (page 756)

1. $Z'(m) = -C_1/m^2 + DC_3/2$ **3.** About 3.33   **5.** $Z(m^+) = Z(4) = \$11,400; Z(m^-) = Z(3) = \$11,300$

CHAPTER 13
SECTION 13.1 (page 766)

1. They differ only by a constant.   **5.** $5x^2 + C$ **7.** $8p^3/3 + C$ **9.** $100\,x + C$ **11.** $5z^2/2 - z + C$ **13.** $z^3/3 - 2z^2 + 2z + C$
15. $x^4/4 - 14x^3/3 + 10x^2 + 3x + C$ **17.** $4y^{3/2} + C$ **19.** $12t^{5/2}/5 + 2t^{3/2} + C$ **21.** $112t^{3/2}/3 + 4t^{9/2} + C$
23. $-12/x^2 + C$ **25.** $-1/y - 4\sqrt{y} + C$ **27.** $-3x^{-2} + 4\ln|x| + C$ **29.** $4e^{3u}/3 + C$ **31.** $-15e^{-.2x} + C$
33. $3\ln|x| - 8e^{-.5x} + C$ **35.** $\ln|t| + 2t^3/3 + C$ **37.** $e^{2u}/2 + u^2/8 + C$ **39.** $x^3/3 + x^2 + x + C$
41. $6x^{7/6}/7 + 3x^{2/3}/2 + C$ **43.** $f(x) = 2x^3 - 2x^2 + 3x + 1$ **45.** $C(x) = .2x^3/3 + 5x^2/2 + 10$ **47.** $C(x) = 2x^{3/2}/3 - 8/3$
49. $C(x) = x^3/3 - x^2 + 3x + 6$ **51.** $C(x) = \ln|x| + x^2 + 7.45$ **53.** $C(x) = 3e^{.01x} + 5$ **55.** $p = 50 - 3x/2 - x^2/3$
57. (a) $f(t) = -e^{-.01t} + k$ **(b)** .095 unit

SECTION 13.2 (page 774)

3. $(12x - 1)^3/12 + C$ **5.** $-2/[3(3t + 1)] + C$ **7.** $-1/\sqrt{x^2 + 2x - 4} + C$ **9.** $2(r^3 + 3)^{3/2}/9 + C$ **11.** $-3e^{5k}/5 + C$
13. $2e^{2w^3}/3 + C$ **15.** $e^{4t-t^2}/2 + C$ **17.** $2e^{\sqrt{y}} + C$ **19.** $-4\ln|2 + 5x|/5 + C$ **21.** $(1/2)\ln|e^{2t} + 1| + C$
23. $-1/[8(2x^2 + 8x)^2] + C$ **25.** $\left(\dfrac{1}{r} + r\right)^2/2 + C$ **27.** $(x^3 + 3x)^{1/3} + C$ **29.** $2(u - 1)^{3/2}/3 + 2(u - 1)^{1/2} + C$
31. $(1/6)\ln|3x^2 + 12x + 8| + C$ **33.** $2(5t - 1)^{5/2}/125 + 2(5t - 1)^{3/2}/75 + C$ **35.** $(x^2 + 1)^4/4 + C$ **37.** $(x^2 + 12x)^{3/2}/3 + C$
39. $(1 + \ln x)^3/3 + C$ **41. (a)** $C(x) = 100/(x + 10)$ **(b)** No   **43. (a)** $S(t) = 27.5e^{.16t} - .2$ **(b)** 4.3 yr

SECTION 13.3 (page 784)

1. 60   **3.** 93   **5.** 13   **7.** 46   **9.** 2.083̄   **11.** $\displaystyle\int_0^8 [1/(x + 2)]\, dx$ **13.** 28; 32   **15.** 8; 10   **17.** 12.8; 27.2   **19.** 2.67; 2.08

23. (a) 3 **(b)** 3.5 **(c)** 4   **25.** About 53 billion barrels   **27.** About \$83,000   **29.** About 37 liters   **31.** About 1300 ft
33. (a) About 690 BTUs **(b)** About 180 BTUs   **35.** About 2

SECTION 13.4 (page 796)

1. 52   **3.** 13   **5.** 1/3   **7.** 76   **9.** 4/3   **11.** $5\ln 5 + (12/25) \approx 8.527$ **13.** $20e^{-.2} - 20e^{-.3} + 3\ln 3 - 3\ln 2 \approx 2.775$
15. $e^{10}/5 - e^5/5 - 1/2 \approx 4375.1$ **17.** $91/3 = 30.\overline{3}$ **19.** $447/7 \approx 63.857$ **21.** $(\ln 2)^2/2 \approx .24023$ **23.** 49

25. $1/4 - 1/(3 + e) \approx .075122$ **27.** $(6/7)(128 - 2^{7/3}) \approx 105.39$ **29.** A loss for the first two years; $\displaystyle\int_0^4 (x^2 - 4)\, dx$

31. $23/3 = 7.\overline{6}$ **33.** 9.5   **35.** $e^2 + e^{-1} - 3 \approx 4.757$ **37.** 1   **39.** $23/6 = 3.8\overline{3}$ **41.** $(e + e^3 - 2e^2)/2 \approx 4.01$
43. -12 **45. (a)** \$22,000 **(b)** \$62,000 **(c)** About 4.5 days   **47. (a)** $(9000/8)(17^{4/3} - 2^{4/3}) \approx \$46,341$
(b) $(9000/8)(26^{4/3} - 17^{4/3}) \approx \$37,477$ **(c)** It is slowly increasing without bound.   **49. (a)** \$3857 **(b)** When $t = 2$ (year 2)

51. (a) About 414 barrels **(b)** About 191 barrels **(c)** Decreasing to 0   **53. (a)** $\displaystyle\int_0^{10} 1.2e^{.04t}\, dt$ **(b)** $30e^4 - 30 \approx 14.75$ billion **(c)** About 12.8

years **(d)** About 14.4 years

SECTION 13.5 (page 807)

1. $280; $11.67 **3.** $2500; $30,000 **5.** No **7. (a)** 8 yr **(b)** About $148 **(c)** About $771 **9. (a)** 39 days **(b)** $3369.18 **(c)** $484.02
(d) $2885.16 **11. (a)** 5 **(b)** About 15.6 tons **13.** 13 years **15.** $1999.54 **17.** $40.50
19. (a)

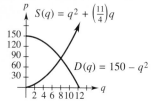

21. (a)

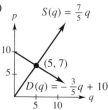

(b) (8, 86) **(c)** $341.33 **(d)** $429.33 **(b)** (5, 7) **(c)** $7.50 **(d)** $17.50 **23.** $5.76 million
25. (a) .019; the lower 10% of the income producers earn 1.9% of the total income of the population. **(b)** .275; the lower 50% of the
income producers earn 27.5% of the total income of the population. **(c)** .819; the lower 90% of the income producers earn 81.9% of the
total income of the population.
(d)

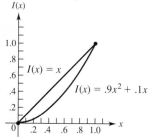

(e) .15; the amount of inequality of income distribution. **27. (a)** 10 items **(b)** 10 hr per item
The answers in Exercises 31–33 may vary slightly. **31.** 161.2 **33.** 5.516

SECTION 13.6 (page 811)

1. $-4 \ln|(x + \sqrt{x^2 + 36})/6| + C$ **3.** $\ln|(x - 3)/(x + 3)| + C \ (x^2 > 9)$ **5.** $(4/3)\ln[(3 + \sqrt{9 - x^2})/x] + C \ (0 < x < 3)$
7. $-2x/3 + 2\ln|3x + 1|/9 + C$ **9.** $(-2/15)\ln|x/(3x - 5)| + C$ **11.** $\ln|(2x - 1)/(2x + 1)| + C \ (x^2 > 1/4)$
13. $-3\ln|(1 + \sqrt{1 - 9x^2})/(3x)| + C \ (0 < x < 1/3)$ **15.** $2x - 3\ln|2x + 3| + C$ **17.** $1/[25(5x - 1)] - (\ln|5x - 1|)/25 + C$
19. $x^5[(\ln|x|)/5 - 1/25] + C$ **21.** $(1/x)(-\ln|x| - 1) + C$ **23.** $-xe^{-2x}/2 - e^{-2x}/4 + C$ **25.** $2094.71
27. About 15,100 microbes

SECTION 13.7 (page 818)

1. $y = -x^2 + x^3 + C$ **3.** $y = 3x^4/8 + C$ **5.** $y^2 = x^2 + C$ **7.** $y = Me^{x^2}$ **9.** $y = Me^{(x^3 - x^2)}$ **11.** $y = Mx$ **13.** $y = Me^x + 5$
15. $y = -1/(e^x + C)$ **17.** $y = x^3 - x^2 + 2$ **19.** $y = -2xe^{-x} - 2e^{-x} + 44$ **21.** $y^2 = 2x^3/3 + 9$ **23.** $y = e^{x^2 + 3x}$
25. $y^2/2 - 3y = x^2 + x - 4$ **29. (a)** 108.4 **(b)** 121.2 **(c)** No; if $x = 8$, the denominator becomes 0.
31. (a) $dy/dt = -.25y$ **(b)** $y = Me^{-.25t}$ **(c)** 4.8 yr **33.** About 1.35×10^5 **35. (a)** About 178.6 thousand **(b)** About 56.4 thousand
(c) About 50.0 thousand **37.** 54.6 grams **39.** About 10 **41.** $q = C/p^2$
43. (a)

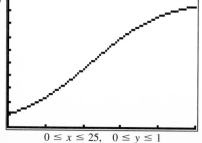

(b) at (11.0, .5)

CHAPTER 13 REVIEW EXERCISES (page 822)

1. $x^3/3 - 3x^2/2 + 2x + C$ **3.** $2x^{3/2} + C$ **5.** $2x^{3/2}/3 + 9x^{1/3} + C$ **7.** $2x^{-2} + C$ **9.** $-3e^{2x}/2 + C$
11. $2 \ln|x - 1| + C$ **13.** $e^{3x^2}/6 + C$ **15.** $(3 \ln|x^2 - 1|)/2 + C$ **17.** $-(x^3 + 5)^{-3}/9 + C$ **19.** $\ln|2x^2 - 5x| + C$
21. $-e^{-3x^4}/12 + C$ **23.** $2e^{-5x}/5 + C$ **27.** 20   **29.** 28   **33.** $72/25 \approx 2.88$ **35.** $2 \ln 3$ or $\ln 9 \approx 2.1972$
37. $2e^4 - 2 \approx 107.1963$ **39.** $(2/3)(22^{3/2} - 2^{3/2}) \approx 66.907$ **41.** $e^2 - 1 \approx 6.3891$ **43.** $C(x) = 10x - x^2 + 4$
45. $C(x) = (2x - 1)^{3/2} + 145$ **47.** 36,000   **49.** About 26.3 yr   **53.** About 142   **55.** $-\ln|(5 + \sqrt{25 + x^2})/x| + C$
57. $15[x/2 + 5 \ln|2x - 5|/4] + C$ **59.** $y = x^4/2 + 3x^2 + C$ **61.** $y^2 = 3x^2 + 2x + C$ **63.** $y = -5e^{-x} - 5x + 22$
65. $y = 2e^{5x-x^2}$ **67. (a)** $dA/dt = .05A - 1000$ **(b)** \$9487.29   **69. (a)** About 111 items **(b)** Not exactly, but for practical purposes, yes
71. $.2 \ln y - .5y = -.3 \ln x + .4x + C$; $x = 3/4$ unit, $y = 2/5$ unit

CASE 17 (page 827)

1. About 102 years   **3.** About 45.4 years

CASE 18 (page 829)

1. 10.65   **3.** 441.46

CHAPTER 14

SECTION 14.1 (page 838)

3. $4; -22; -20; 12$ **5.** $3; \sqrt{33}; \sqrt{17}; 8$

9.

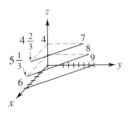

11.

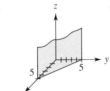

13.

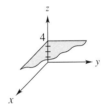

15.

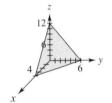

17.

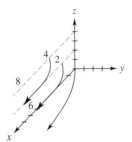

19. $y = 500^{10/3}x^{-7/3} \approx 10^9/x^{7/3}$

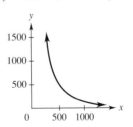

21. 1.12; the IRA account grows faster.   **23.** $f(L, W, H) = L + 2H + 2W$ **25. (a)** $18x + 9h$ (b) $-6y - 3h$ **27.** (c)   **29.** (e)   **31.** (b)

33.

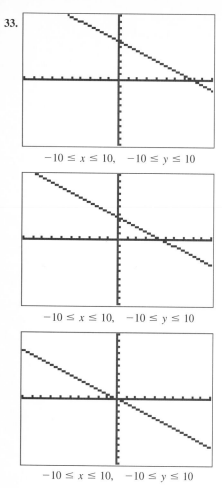

$-10 \le x \le 10, \quad -10 \le y \le 10$

$-10 \le x \le 10, \quad -10 \le y \le 10$

$-10 \le x \le 10, \quad -10 \le y \le 10$

35.

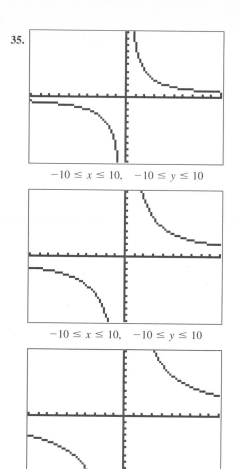

$-10 \le x \le 10, \quad -10 \le y \le 10$

$-10 \le x \le 10, \quad -10 \le y \le 10$

$-10 \le x \le 10, \quad -10 \le y \le 10$

SECTION 14.2 (page 847)

1. (a) $24x^2 - 8xy$ **(b)** $-4x^2 + 18y$ **(c)** 48 **(d)** -40 **3.** $f_x = -2xy + 12x^3; f_y = -x^2; 100; -16$ **5.** $f_x = 2e^{2x+y}; f_y = e^{2x+y}; 2e^3; e^{-5}$
7. $f_x = 2/(e^{x+2y}); f_y = 4/(e^{x+2y}); 2; 4/e^2$ **9.** $f_x = (y^3 - x^2 - 6xy^2)/(x^2 + y^3)^2; f_y = (6x^2y - 3xy^2 - 3y^4)/(x^2 + y^3)^2; -17/9;$
about .083   **11.** $f_x = \dfrac{2 - 2xy}{2x - x^2y}; f_y = \dfrac{-x}{2 - xy}; 3/4; 2/7$ **13.** $f_x = 2x^2ye^{2xy} + 2xe^{2xy}; f_y = 2x^3e^{2xy}; -4e^{-4}; -128e^{-24}$
15. $f_{xx} = 20y^3 - 30x; f_{xy} = f_{yx} = 60xy^2; f_{yy} = 60x^2y$ **17.** $h_{xx} = -8y^2; h_{xy} = h_{yx} = -16xy + 14y; h_{yy} = -6 - 8x^2 + 14x$
19. $R_{xx} = \dfrac{24y}{(2x + y)^3}; R_{xy} = R_{yx} = \dfrac{-12x + 6y}{(2x + y)^3}; R_{yy} = \dfrac{-12x}{(2x + y)^3}$ **21.** $z_{xx} = 0; z_{yy} = 4xe^y; z_{xy} = z_{yx} = 4e^y$ **23.** $r_{xx} = -1/(x + y)^2;$
$r_{yy} = -1/(x + y)^2; r_{xy} = r_{yx} = -1/(x + y)^2$ **25.** $z_{xx} = 1/x; z_{yy} = -x/y^2; z_{xy} = z_{yx} = 1/y$ **27.** 1; $-1/4$ **29.** $x = -4, y = 2$
31. $x = 0, y = 0;$ or $x = 3, y = 3$ **33.** $f_x = 2x; f_y = z; f_z = y + 4z^3; f_{yz} = 1; 3; 1$ **35.** $f_x = 6/(4z + 5); f_y = -5/(4z + 5);$
$f_z = -4(6x - 5y)/(4z + 5)^2; f_{yz} = 20/(4z + 5)^2$ **37.** $f_x = (2x - 5z^2)/(x^2 - 5xz^2 + y^4); f_y = 4y^3/(x^2 - 5xz^2 + y^4);$
$f_z = -10xz/(x^2 - 5xz^2 + y^4); f_{yz} = 40xy^3z/(x^2 - 5xz^2 + y^4)^2$ **41. (a)** 2.04% **(b)** 1.26% **(c)** .05%: the rate of change of the
probability for an additional semester of high school math; .003%: the rate of change of the probability per unit of change in the SAT
score.   **43. (a)** \$206,800 **(b)** $f_p = 132 - 2i - .02p; f_i = -2p;$ the rate at which sales revenue is changing per unit of change in price
(f_p) or interest rate (f_i) **(c)** A sales revenue drop of \$18,800   **45. (a)** 46.656 **(b)** $f_x(27, 64) = .6912$ and is the rate at which production is
changing when labor changes by 1 unit from 27 to 28 and capital remains constant; $f_y(27, 64) = .4374$ and is the rate at which production
is changing when capital changes by 1 unit from 64 to 65 and labor remains constant. **(c)** Production would be increased by about 69 units
when labor is increased by 1 unit.   **47.** $.7x^{-3}y^{.3}$ or $.7y^{.3}/x^{.3}$; $.3x^{.7}y^{-.7}$ or $.3x^{.7}/y^{.7}$ **49. (a)** About 5.71 **(b)** About 6.84 **(c)** $1/(a - v)$
(d) $b/(a - v)^2$ **51. (a)** $F_m = (gR^2)/r^2$; the rate of change in force per unit change in mass; $F_r = (-2mgR^2)/r^3$; the rate of change in
force per unit change in distance.

SECTION 14.3 (page 857)

3. Relative minimum of -22 at $(3, -1)$ **5.** Relative minimum of 0 at $(-2, -2)$ **7.** Saddle point at $(-3, 4)$ **9.** Relative maximum of 17 at $(2/3, 4/3)$ **11.** Saddle point at $(2, -2)$ **13.** Saddle point at $(1, 2)$ **15.** Relative minimum of -201 at $(6,18)$; saddle point at $(0, 0)$ **17.** Saddle point at $(0, 0)$; relative minimum of -5157 at $(72, 12)$ **19.** Saddle point at $(0, 0)$ **21.** Relative maximum of $9/8$ at $(1, 1)$; saddle point at $(0, 0)$ **23.** Relative minima of $1/16$ at $(1, 0)$ and $(-1, 0)$; saddle point at $(0, 0)$ **25.** Relative maxima of $17/16$ at $(0, 1)$ and $(0, -1)$; relative minima of $-15/16$ at $(1, 0)$ and $(-1, 0)$; saddle points at $(0, 0)$, $(1, 1)$, $(1, -1)$, $(-1, 1)$, and $(-1, -1)$ **27.** $P(12, 40) = 2744$ **29.** $P(12, 72) = \$2,528,000$ **31.** $C(12, 25) = 2237$ **33.** 3 m by 3 m by 3 m   **35.** 18 inches by 18 inches by 36 inches   **37.** 1000 tons of grade A ore and 2000 tons of grade B ore for maximum profit of \$1400

CHAPTER 14 REVIEW EXERCISES (page 860)

1. 32; 156   **3.** $-6/5$; $-8/3$
7.

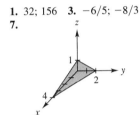

9.

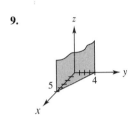

11. (a) $-4x + 5y$ **(b)** 3 **(c)** 5   **15.** $f_x = 12x^2y + 10y^4$; $f_y = 4x^3 + 40xy^3$ **17.** $f_x = (12y - 3x^2 + 4xy^2)/(x^2 + 4y)^2$; $f_y = (-4x^2y - 8y^2 - 12x)/(x^2 + 4y)^2$ **19.** $f_x = 2(y + 1)^2e^{2x+y}$; $f_y = (y + 1)(y + 3)e^{2x+y}$ **21.** $f_x = 3x^2y^2/(1 + x^3y^2)$; $f_y = 2x^3y/(1 + x^3y^2)$ **23.** $f_{xx} = 24xy^2$; $f_{xy} = 24x^2y - 8$ **25.** $f_{xx} = 8y/(x - 2y)^3$; $f_{xy} = (-4x - 8y)/(x - 2y)^3$ **27.** $f_{xx} = 2e^y$; $f_{xy} = 2xe^y$ **29.** $f_{xx} = (-2x^2y^2 - 4y)/(2 - x^2y)^2$; $f_{xy} = -4x/(2 - x^2y)^2$ **31.** Relative minimum at $(0, 1)$ **33.** Saddle point at $(0, 2)$
35. Relative minimum at $(1, 1/2)$; saddle point at $(-1/3, 11/6)$
37. $F(L, W, H) = 2LW + 2WH + 2LH$ **39. (a)** Relative minimum at $(11, 12)$ **(b)** \$431

Index

Students

**We would appreciate it if you would take a few minutes to answer these questions.
Then cut the page out, fold it, seal it, and mail it. No postage is required. Thank you!**

Which chapters were taught?

How much algebra did you have before this course?   How long ago?

Years in high school (circle)   0   1/2   1   1-1/2   2 or more   _____ last 2 years

Courses in college   0   1   2   3   _____ last 3-5 years

                                                                                                                                    _____ 5 years or more

We would appreciate knowing of any errors you found in the book. (Please supply page numbers.)

How would you characterize the quality of the explanations? If there were any which were unclear,
please give us the page numbers.

What did you think of the examples (were there enough, right amount of detail, etc.) and exercises?
If there is room for improvement, please let us know where.

How helpful were the marginal problems, chapter summaries, and cases in the text? Do you have any
creative ideas on what else might help you study still more effectively?

What is your major or career goal? _____ _____

What additional math courses, if any, do you plan on taking? _____

What kind of calculator do you own (model and type): _____ Did you use it in class? _____

What did you like most about the book?

What did you like the least about the book?

Name _____

College _____ State _____

NO POSTAGE
NECESSARY
IF MAILED
IN THE
UNITED STATES

BUSINESS REPLY MAIL

FIRST CLASS   PERMIT NO. 1537   NEW YORK, NY

Postage will be paid by

HarperCollins Publishers
College Division   Attn: MATH GROUP
1900 East Lake Avenue
Glenview, Illinois 60025

Teachers

WHAT DO YOU THINK OF MATHEMATICS W/Applications 6E

**We would appreciate it if you would take a few minutes to answer these questions.
Then cut the page out, fold it, seal it, and mail it. No postage required. Thank you!**

For what course (give title and brief description) and student population did you consider this book?
Annual enrollment _____ .

What chapters did you look at in considering this book?

How would you assess the organization and coverage?

How would you characterize the quality of the exposition? Where, if any, can improvements be made?

How would you assess the quality and quantity of the examples/applications? Where, if any,
can improvements be made?

How would you assess the pedagogical quality of the chapters (marginal problems, summaries, etc.)?
Any suggestions for improvement?

Your current book (author/title) is _____ .

How does this book compare?

Do you have any comments on the supplements? Ideas for additions?

If you adopted this book, what chapters would you cover? Are you willing to serve as a possible reviewer?
Do you have (or are you aware of someone who has) any plans to write a book?

What did you like most about the book?

- FOLD HERE -

What did you like the least about the book?

College _____   State _____

- FOLD HERE -

NO POSTAGE
NECESSARY
IF MAILED
IN THE
UNITED STATES

BUSINESS REPLY MAIL

FIRST CLASS   PERMIT NO. 1537   NEW YORK, NY

Postage will be paid by

HarperCollins Publishers
College Division   Attn: MATH GROUP
1900 East Lake Avenue
Glenview, Illinois 60025

The **derivative** of the function f is the function denoted f' whose value at the number x is defined to be the number

$$f'(x) = \lim_{h \to 0} \frac{f(x+h) - f(x)}{h},$$

provided this limit exists.

Rules for Derivatives Assume all indicated derivatives exist.

Constant Function If $f(x) = k$, where k is any real number, then

$$f'(x) = 0.$$

Power Rule If $f(x) = x^n$, for any real number n, then

$$f'(x) = n \cdot x^{n-1}.$$

Constant Times a Function Let k be a real number. Then the derivative of $y = k \cdot f(x)$ is

$$y' = k \cdot f'(x).$$

Sum or Difference Rule If $y = f(x) \pm g(x)$, then

$$y' = f'(x) \pm g'(x).$$

Product Rule If $f(x) = g(x) \cdot k(x)$, then

$$f'(x) = g(x) \cdot k'(x) + k(x) \cdot g'(x).$$

Quotient Rule If $f(x) = \dfrac{g(x)}{k(x)}$, and $k(x) \neq 0$, then

$$f'(x) = \frac{k(x) \cdot g'(x) - g(x) \cdot k'(x)}{[k(x)]^2}.$$

Chain Rule If y is a function of u, say $y = f(u)$, and if u is a function of x, say $u = g(x)$, then $y = f(u) = f[g(x)]$, and

$$\frac{dy}{dx} = \frac{dy}{du} \cdot \frac{dy}{dx}.$$

Chain Rule (alternate form) Let $y = f[g(x)]$. Then

$$y' = f'[g(x)] \cdot g'(x).$$

Generalized Power Rule Let u be a function of x, and let $y = u^n$ for any real number n. Then

$$y' = n \cdot u^{n-1} \cdot u'.$$

Natural Logarithmic Function If $y = \ln |g(x)|$, then

$$y' = \frac{g'(x)}{g(x)}.$$

Exponential Function If $y = e^{g(x)}$, then

$$y' = g'(x) \cdot e^{g(x)}.$$